Intraplate Volcanism in Eastern Australia and New Zealand

Compiled and edited by
R.W. Johnson
*Bureau of Mineral Resources,
Geology and Geophysics*

Associate Editors:
J. Knutson, Bureau of Mineral Resources,
Geology and Geophysics
S.R. Taylor, Australian National University

Published in association with the
Australian Academy of Science

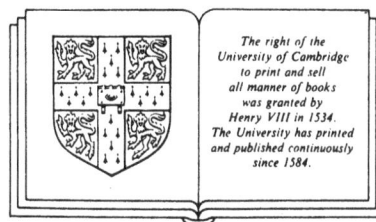

The right of the
University of Cambridge
to print and sell
all manner of books
was granted by
Henry VIII in 1534.
The University has printed
and published continuously
since 1584.

Cambridge University Press
Cambridge
New York   Port Chester   Melbourne   Sydney

Working Group on Intraplate Igneous Activity in Australasia, Sub-Committee for the International Lithosphere Program, Australian Academy of Science

R.A.F. Cas
M.B. Duggan
A. Ewart
R.W. Johnson
J. Knutson
R.W. Le Maitre
I. McDougall
I.A. Nicholls

S.Y. O'Reilly
R.C. Price
I.E.M. Smith
P.J. Stephenson
F.L. Sutherland
S.R. Taylor
P. Wellman
J.F.G. Wilkinson

CAMBRIDGE UNIVERSITY PRESS
Cambridge, New York, Melbourne, Madrid, Cape Town, Singapore,
São Paulo, Delhi, Dubai, Tokyo

Cambridge University Press
The Edinburgh Building, Cambridge CB2 8RU, UK

Published in the United States of America by Cambridge University Press, New York

www.cambridge.org
Information on this title: www.cambridge.org/9780521123228

First published 1989
This digitally printed version 2009

A catalogue record for this publication is available from the British Library

National Library of Australia Cataloguing in Publication data

Intraplate volcanism in eastern Australia and New Zealand

    Bibliography.
    Includes index.
    ISBN 0 521 38083 9.
    1. Volcanism — Australia, Eastern. 2. Volcanism — New
    Zealand. I. Johnson, R.W. (Robert Wallace). II. Knutson, J.
    III. Taylor, Stuart Ross, 1925–   . IV. Australian Academy of
    Science.
551.2'1'0994

Library of Congress Cataloguing in Publication data

Intraplate volcanism in eastern Australia and New Zealand /
    complied and edited by R.W. Johnson; associate editors,
    J. Knutson, S.R. Taylor.
        p.   cm.
    "Published in association with the Australian Academy of
    Science."
    Bibliography: p.
    Includes index.
    ISBN 0–521–38083–9
    1. Volcanism — Australia. 2. Volcanism — New Zealand.
    3. Geology. Stratigraphic–Cenozoic. I. Johnson, R.W.
    (R. Wally) II. Knutson, J. (Jan) III. Taylor,
    Stuart Ross. 1925–   . IV. Australian Academy of Science.
QE527.I58 1989
551.2'1'0994–dc20   89–32136
                                        CIP

ISBN 978-0-521-38083-6 Hardback
ISBN 978-0-521-12322-8 Paperback

Additional resources for this publication at www.cambridge.org/9780521123228

Contents

CONTENTS

CONTENTS

Plates, figures and tables

Chapter 5

Cover illustration

The upper parts of Cainozoic volcanoes at the border
between northern New South Wales and southern
Queensland are shown clearly on this colour-enhanced
LANDSAT satellite image. Tweed volcano is seen on the
right near the east-Australian coast, and Main Range
volcano is on the left (compare with Figs. 3.4.1,7,9). Image
supplied courtesy of the Australian Centre for Remote
Sensing, Canberra.

Common abbreviations used in this volume
(excluding the elements of the periodic table)

AFC	Assimilation/fractional crystallisation
F (or F-value)	Fraction of melt remaining after magma crystallisation
K_D (or D)	Partition coefficient
Mg-ratio	$100Mg/(Mg+Fe^{2+})$ (atomic proportions)
mg-ratio	$100Mg/(Mg+total\text{-}Fe)$ (atomic proportions)
MORB	Mid-ocean-ridge basalt
N-MORB	Normal mid-ocean-ridge basalt
N (subscript)	Normalised trace-element value
OIB	Ocean-island basalt
R (or R-ratio)	Ratio of the mass of assimilate to the mass of crystal cumulate (used in AFC modelling)
REE	Rare-earth elements
s.s.	Solid solution
ε_{Nd}	Deviation of the value of the isotopic ratio $^{143}Nd/^{144}Nd$ from that of a Bulk Earth value

Recommended method of citation

Reference to any contribution in this volume should be by the names of the appropriate author, or authors, identified in the above Contents listing. Contributions can be referred to at any of the hierarchal levels of title headings — from chapters, down through sections and sub-sections, to non-enumerated (italicised) headings. Co-ordinators of Chapters 1, 3, 4, and 7 should be identified as such, rather than as authors.

Contributors

P.W. Baillie, Tasmania Department of Mines, P.O. Box 56, Rosny Park, Tasmania 7018.

P. Bishop, Department of Geography, University of Sydney, Sydney, New South Wales 2006.

R.J. Blong, School of Earth Sciences, Macquarie University, North Ryde, New South Wales 2113.

D.F. Branagan, Department of Geology and Geophysics, University of Sydney, Sydney, New South Wales 2006.

M.C. Brown, School of Applied Science, Canberra College of Advanced Education, P.O. Box 1, Belconnen, Australian Capital Territory 2616.

R.A.F. Cas, Department of Earth Sciences, Monash University, Clayton, Victoria 3168.

B.W. Chappell, Department of Geology, Australian National University, G.P.O. Box 4, Canberra, Australian Capital Territory 2601.

J.B. Colwell, Division of Marine Geosciences and Petroleum Geology, Bureau of Mineral Resources, G.P.O. Box 378, Canberra, Australian Capital Territory 2601.

A. Cundari, Department of Geology, University of Melbourne, Parkville, Victoria 3052.

R.A. Day, School of Earth Sciences, Macquarie University, North Ryde, New South Wales 2113.

M.B. Duggan, Geoscience Planning and Information Branch, Bureau of Mineral Resources, G.P.O. Box 378, Canberra, Australian Capital Territory 2601.

R.A. Duncan, College of Oceanography, Oregon State University, Corvallis, Oregon 97331, U.S.A.

M.A. Etheridge, Division of Petrology and Geochemistry, Bureau of Mineral Resources, G.P.O. Box 378, Canberra, Australian Capital Territory 2601.

A. Ewart, Department of Geology and Mineralogy, University of Queensland, St Lucia, Queensland 4067.

J.A. Gamble, Department of Geology, Victoria University of Wellington, Private Bag, Wellington, New Zealand.

I.L. Gibson, Department of Earth Sciences, University of Waterloo, Waterloo, Ontario N2L 3G1, Canada.

C.M. Gray, Department of Geology, La Trobe University, Bundoora, Victoria 3083.

D.H. Green, Department of Geology, University of Tasmania, G.P.O. Box 252C, Hobart, Tasmania 7001.

A. Greig, Department of Earth Sciences, Monash University, Clayton, Victoria 3168.

A. Grenfell, Science Department, Brisbane College of Advanced Education, Kelvin Grove Campus, Victoria Park Road, Kelvin Grove, Queensland 4059.

W.L. Griffin, Division of Exploration Geoscience, Commonwealth Scientific and Industrial Research Organization, P.O. Box 136, North Ryde, New South Wales 2113.

J.D. Hollis, Division of Earth Sciences, Australian Museum, P.O. Box A285, Sydney South, New South Wales 2000.

R.W. Johnson, Division of Petrology and Geochemistry, Bureau of Mineral Resources, G.P.O. Box 378, Canberra, Australian Capital Territory 2601.

E.B. Joyce, Department of Geology, University of Melbourne, Parkville, Victoria 3052.

J. Knutson, Division of Petrology and Geochemistry, Bureau of Mineral Resources, G.P.O. Box 378, Canberra, Australian Capital Territory 2601.

R.J. Korsch, Division of Continental Geology, Bureau of Mineral Resources, G.P.O. Box 378, Canberra, Australian Capital Territory 2601.

J.H. Latter, Geophysics Division, Department of Scientific and Industrial Research, P.O. Box 1320, Wellington, New Zealand.

S.R. Lishmund, Geological Survey of New South Wales, Department of Mineral Resources, G.P.O. Box 5288, Sydney, New South Wales 2001.

G.S. Lister, Department of Earth Sciences, Monash University, Clayton, Victoria 3168.

D.J. Martin, School of Earth Sciences, Macquarie University, North Ryde, New South Wales 2113.

D.R. Mason, Australian Mineral Development Laboratories, P.O. Box 114, Eastwood, South Australia 5063.

M.A. Menzies, Department of Geology, Royal Holloway and Bedford New College, University of London, Egham Hill, Egham, Surrey TW20 0EX, United Kingdom.

W.F. McDonough, Abteilung Geochemie, Max-Planck-Institut für Chemie, Saarstrasse 23, Postfach 3060, D 6500 Mainz, Federal Republic of Germany.

I. McDougall, Research School of Earth Sciences, Australian National University, G.P.O. Box 4, Canberra, Australian Capital Territory 2601.

E.A.K. Middlemost, Department of Geology and Geophysics, University of Sydney, Sydney, New South Wales 2006.

P.A. Morris, Geological Survey of Western Australia, Department of Mines, Western Australia Government, Mineral House, 66 Adelaide Terrace, Perth, Western Australia 6000.

I.A. Nicholls, Department of Earth Sciences, Monash University, Clayton, Victoria 3168.

S.Y. O'Reilly, School of Earth Sciences, Macquarie University, North Ryde, New South Wales 2113.

G.M. Oakes, Geological Survey of New South Wales, Department of Mineral Resources, G.P.O. Box 5288, Sydney, New South Wales 2001.

N.A. Ortez (deceased), School of Earth Sciences, Macquarie University, North Ryde, New South Wales 2113.

R.J. Pankhurst, British Antarctic Survey, Natural Environment Research Council, Madingley Road, Cambridge CB3 0ET, United Kingdom.

N.J. Pearson, School of Earth Sciences, Macquarie University, North Ryde, New South Wales 2113.

R.C. Price, Department of Geology, La Trobe University, Bundoora, Victoria 3083.

A.D. Robertson, Queensland Department of Mines, P.O. Box 194, Brisbane, Queensland 4001.

J.A. Ross, c/- Department of Geology and Mineralogy, University of Queensland, St Lucia, Queensland 4067.

R.L. Rudnick, Abteilung Geochemie, Max-Planck-Institut für Chemie, Saarstrasse 23, Postfach 3060, D 6500 Mainz, Federal Republic of Germany

R.W. Schön, Department of Geology and Mineralogy, University of Queensland, St Lucia, Queensland 4067.

R.J. Sewell, Geological Survey of New Zealand, c/- Department of Geology, University of Canterbury, Christchurch 1, New Zealand.

M.J. Sheard, Geological Survey of South Australia, Department of Mines and Energy, P.O. Box 151, Eastwood, South Australia 5063.

I.E.M. Smith, Department of Geology, University of Auckland, Private Bag, Auckland, New Zealand.

P.J. Stephenson, Department of Geology, James Cook University of North Queensland, Townsville, Queensland 4811.

N.C. Stevens, Department of Geology and Mineralogy, University of Queensland, St Lucia, Queensland 4067.

A.J. Stolz, Department of Geology, University of Tasmania, G.P.O. Box 252C, Hobart, Tasmania.

S.-S. Sun, Division of Petrology and Geochemistry, Bureau of Mineral Resources, G.P.O. Box 378, Canberra, Australian Capital Territory 2601.

F.L. Sutherland, Division of Earth Sciences, Australian Museum, P.O. Box A285, Sydney South, New South Wales 2000.

S.R. Taylor, Research School of Earth Sciences, Australian National University, G.P.O. Box 4, Canberra, Australian Capital Territory 2601.

S.D. Weaver, Department of Geology, University of Canterbury, Christchurch 1, New Zealand.

P. Wellman, Division of Geophysics, Bureau of Mineral Resources, G.P.O. Box 378, Canberra, Australian Capital Territory 2601.

D. Wyborn, Division of Petrology and Geochemistry, Bureau of Mineral Resources, G.P.O. Box 378, Canberra, Australian Capital Territory 2601.

Foreword

Young volcanic rocks are widespread in eastern Australia, despite the popular misconception that 'there are no volcanoes in Australia'. Certainly, the extent of the Cainozoic volcanism in eastern Australia was recognised by geologists only late last century, but today the fact that it comprises one of the great basaltic provinces of the world (in extent if not in volume) is recognised only scarcely overseas — even among petrologists. New Zealand, on the other hand, has a volcanic 'image' on account of the active andesitic volcanoes in the Taupo Volcanic Zone of central North Island. However, less well known — except perhaps for Dunedin volcano and the Auckland area — is the intraplate volcanism in both North Island and South Island and in the seas to the east and southeast.

A broad appreciation of the diversity of eruptive styles, and of the wide range of volcanic rocks, within many of the intraplate provinces of eastern Australia and New Zealand existed by about 1950, but detailed studies — at least by modern standards — were few. Especially notable among the pioneering investigations were those by H.C. Richards (1916; volcanic rocks of southeastern Queensland), W.R. Browne (1933; Mesozoic and Cainozoic igneous activity in New South Wales), A.B. Edwards (1938; Newer Volcanics of Victoria), and W.N. Benson (1941–46; Dunedin volcano). These early studies were oriented chemically in different degrees, but were limited by a lack of detailed mineralogical, trace-element, isotopic, and other data.

Subsequent work coincided with the advent of modern chemical analytical techniques, the application of isotopic studies, and the integration of geophysical data with pressure-temperature estimates for upper-mantle and lower-crustal xenoliths. These techniques applied in conjunction with research in experimental petrology, collectively added new dimensions to understanding the more important controls of magma generation and the low- to high-pressure crystallisation behaviour of magma compositions.

Problems abound in the study of the intraplate volcanic areas of eastern Australia and New Zealand. What is the significance of their distribution? Why do some of the rocks form prominent constructional features such the scenic Warrumbungle volcano in northern New South Wales, whereas others make up thinly spread plains that go unnoticed by the casual observer. Why do the east-Australian volcanic areas run parallel to the coast? Why is there a well-defined southward younging of some of the volcanoes but not of others? Is the spacing of the centres significant? How does all of this relate to plate tectonics and, particularly, to the opening of the Tasman and Coral Seas by seafloor spreading which took place at about the same time as the older volcanism in eastern Australia? What relation do the New Zealand examples bear to the immediately adjacent Indo-Australian/Pacific plate boundary? Indeed, to what extent are they all truly 'intraplate' if some are so close to an active plate boundary? How do these rocks fit into modern petrological concepts, and how do they compare with occurrences elsewhere in the world? Finally, what do the xenoliths and megacrysts in the volcanic rocks tell us about the underlying lower crust and upper mantle?

The above questions and many more are addressed in this volume. Indeed, a primary purpose of the volume is to direct the attention of earth scientists to these problems, and to stimulate and focus directions of future research.

One aspect of petrogenesis merits particular comment. Many of the mafic volcanic associations are dominated by distinctive major- and trace-element and isotopic signatures. There is, in my opinion, no clear concensus whether such basaltic diversity has resulted from fractional crystallisation of a range of parental or primitive magmas, or is simply a consequence of the generation of primary mantle-derived magmas that range widely in composition. This problem should constitute an important aspect of future studies.

The volume provides a clear illustration of the importance of multidisciplinary collaboration in investigations of volcanic processes. It is based on contributions by volcanologists, petrologists, geochemists, geochronologists, geophysicists, and geodynamicists. It is, moreover, not merely a compendium of already-published data, although this, and the bibliography, form an indispensible starting point for future work. Rather, it also incorporates much recent, previously unpublished data and interpretations.

The volume represents a first attempt to bring together the vast amount of data now available on east-Australian and New Zealand intraplate volcanism. I believe it will be of particular interest to earth scientists overseas, if only because it is illustrative of the extent of the volcanic activity both in time and space, and of the diversity of rock types within the different provinces. The compilation, of course, represents only a summary of current knowledge, and some models and interpretations inevitably will be modified as more data become available. However, I suggest that this volume highlights the very significant contributions to the study of volcanic processes by Australian and New Zealand geoscientists over the past three decades.

J.F.G. Wilkinson
Armidale

Preface

Planning for this volume began shortly after a workshop meeting on the Cainozoic volcanism of eastern Australia, that was held at the Bureau of Mineral Resources, Geology and Geophysics (BMR), Canberra, on 9 December 1981. Twenty-seven invited participants from east-Australian universities, BMR, and state geological surveys and museums, attended the meeting. Ten key topics that form the basis of this volume were identified, and BMR subsequently established its East Australian Volcano Study, known informally as the 'EAVS project'.

1981 was also the year that the International Lithosphere Program (ILP) was launched by the Inter-Union Commission on the Lithosphere (established by the International Council of Scientific Unions). The Australian Academy of Science through its National Committee for the Solid-Earth Sciences set up a Sub-Committee that identified four areas of major emphasis in relation to Australian participation in the ILP. One of these areas is 'Intraplate igneous activity in Australasia'. Dr Taylor and I in 1982 were asked to establish, and be chairmen of, a Working Group on this subject.

Three important decisions were adopted at the first meeting of the Working Group that was held in early 1983. First, the main activity of the group would be preparation of this volume for publication on the occasion of Australia's Bicentenary in 1988. Second, the members of BMR's EAVS project would co-ordinate production of the volume. Third, the volume would include consideration of the Cainozoic intraplate volcanoes of the New Zealand region. The book subsequently became known informally as the 'ILP volume'.

Australasia encompasses too great a region for comprehensive consideration of all its intraplate volcanism in a single volume, so the areas of interest are restricted primarily to eastern Australia, New Zealand, and the volcanic islands east and southeast of South Island. Excluded from consideration are: the volcanic rocks of Norfolk and Lord Howe islands in the Tasman Sea; known 'intraplate'-type rocks in Papua New Guinea, the Solomon Islands, and Fiji; the kimberlitic and lamproitic rocks of the Kimberley region of northwestern Australia; intraplate volcanoes in the Antarctic and Sub-Antarctic territories of both Australia and New Zealand (Heard Island, Gaussberg, and Erebus) and the volcanoes of Marie Byrd Land and the Transantarctic Mountains; and seamounts in Australasian waters. Dredging of some of the seamounts in the Tasman Sea was undertaken during the time this volume was being prepared, and some results are incorporated here, but these seamounts are not considered in any detail.

Another limitation that had to be defined was the temporal scope of the volcanism. Cainozoic volcanism was to be the major subject of interest, but the intraplate style of volcanism in eastern Australia and New Zealand began in the late Mesozoic so the beginning of the Cainozoic is clearly an arbitrary starting point. Nevertheless, the base of the Cainozoic is a convenient one for eastern Australia because by this time the volcanism was taking place well away from the active plate margin in the Tasman Sea. Intraplate volcanism younger than 65 Ma therefore remains the dominant topic of discussion. However, Mesozoic intraplate igneous activity is not ignored entirely.

Principal subjects of interest to most Working Group members are the geology and geochronology of intraplate volcanic areas in eastern Australia and New Zealand, and the petrology, geochemistry, and petrogenesis of the volcanic rocks. But these topics cannot be considered in isolation and, ultimately, must be assessed with reference to plate-tectonic concepts and to data on the nature of the subcontinental mantle and crust obtained from geophysical and xenolith studies. This volume represents an attempt to span all of these diverse disciplines by providing a baseline summary of current knowledge on the Cainozoic intraplate volcanism and tectonics and on the nature of the crust and mantle of eastern Australia and New Zealand. The ILP volume can be regarded both as a sourcebook and as an interpretative account of the data in terms of current geochemical, petrological, and tectonic concepts.

The main contributors to this volume are Working Group members, several of whom are also chapter co-ordinators. The general approach taken in each of the seven chapters was determined largely by chapter co-ordinators who solicited the help of others where required. Many contributors (particularly to Chapter 3) were generous in supplying more material than could be incorporated in a single volume, and they are thanked for accepting that major surgery had to be undertaken on their submissions.

Contributions were reviewed at three levels. I reviewed all manuscripts in a first stage, soliciting help from others where second opinions were required. Completed chapters or parts of chapters then were reviewed by independent referees, mainly Working Group members, but also P. Bishop, D.S. Coombs, J.P. Cull, D. Denham, B.J. Drummond, B.F. Houghton, A.L. Jaques, K. Lambeck, J. McPhie, P.H. Nixon, N.C. Stevens, S.-S. Sun, E.M. Truswell, J.K. Weissel, and R.W. Young. Finally, four 'overview' referees,

R.J. Arculus, C.D. Ollier, A.J. Stolz, and J.F.G. Wilkinson, reviewed two or more related chapters, passing comment on both content and the degree to which chapters could be better co-ordinated. All of these reviewers are thanked most sincerely for putting aside their own work and tackling the reviewing of ILP-volume contributions to meet deadlines that were not always convenient to them. Professor Wilkinson was invited to prepare the Foreword on the occasion of his retirement after many years of research on the petrology of the Cainozoic volcanic rocks of eastern Australia.

Special thanks are extended to my Associate Editors, Drs Knutson and Taylor. Dr Taylor was involved in all of the important executive decision-making from 1982 onwards, and reviewed all of Chapters 1–6. My colleague at BMR, Jan Knutson, had the difficult task of compiling the diverse and numerous contributions to Chapter 3, but she also volunteered to compile the bibliography for the entire volume. Dr Knutson was aided through the use of two computer programs, FINDREF and AUTOREF, written by Dr R.J. Ryburn of BMR, but even so the bibliographic compilation was a formidable one. However, she tackled it with characteristic determination and tenacity. Dr Knutson was also an invaluable sounding board for numerous 'second-opinion' requests that I pestered her with, and which she responded to with unfailing commonsense and patience. She and I jointly compiled the Index.

The line diagrams for the volume were produced by draftspeople of the BMR Drawing Office, using partial financial support from the Australian Academy of Science and the Department of Geology and Mineralogy, University of Queensland. The drafting for the project was under the general supervision of Mr I.B. Hartig, and the principal draftspeople of the ILP-volume taskforce were G.A. Clarke, I.B. Hartig, A.R. Convine, and J.W. Convine, assisted by L. Holland, V. Ashby, K. Ambrose, A.D. Jaensch, C.R. Johnson, and D.I. Simper. These draftspeople maintained a high professional standard throughout the long period of production of the volume, responding to seemingly never-ending requests with patience and good humour. Their work was exemplary.

Debra Valk of Gowrie in Canberra undertook all of the contract wordprocessing, starting in 1984, and was an indispensible part of the production team, cheerfully handling all requests within the capabilities of the ageing machine and fractious printer available to her. Barbara Kit of Brown and Co. Typesetters, Canberra, typeset the entire volume. Her interest, professionalism, and suggestions improved the presentation enormously. Stella Sakkeus composed the page proofs with precision, care, and dedication.

Unexpected financial profits from a field workshop to Warrumbungle volcano in December 1986, led by Dr M.B. Duggan and Dr Knutson, were used to partly offset the cost of the colour photographs presented in the volume. This funding is gratefully acknowledged.

BMR contributors to the volume publish with the permission of the Director of the Bureau of Mineral Resources, Canberra. S.R. Lishmund and G.M. Oakes publish with the permission of the Secretary of the Department of Mineral Resources, New South Wales; A.D. Robertson with the permission of the Director General of the Department of Mines, Queensland; M.J. Sheard with the permission of the Director General of the Department of Mines and Energy, South Australia; and P.W. Baillie with that of the Director of the Department of Mines, Tasmania. J.H. Latter publishes with the permission of the Director of the Division of Geophysics, DSIR, New Zealand, and R.J. Sewell with that of the Director of the New Zealand Geological Survey (DSIR).

The support and assistance of the BMR, both direct and indirect, was vital to the completion of this project.

R.W. Johnson
Canberra

CHAPTER 1

Framework for Volcanism

1.1 Introduction to Intraplate Volcanism

1.1.1 Preview

Aboriginal people in eastern Australia probably were witness to volcanic activity before European colonisation of their lands in 1788, although no eruptions have taken place since the occupation. New Zealand Maoris also would have seen volcanic eruptions, not just in the subduction-related Taupo Volcanic Zone, but probably also in the Auckland area of North Island. However, the extent of the Cainozoic volcanism in eastern Australia and the important distinction between the explosive-type volcanoes of the Taupo Volcanic Zone and volcanic areas elsewhere in New Zealand, were not realised until well into the twentieth century. Furthermore, volcanic rocks in eastern Australia evidently were not identified by Europeans until well after the arrival of the First Fleet in Australia two hundred years ago.

Late Mesozoic to Quaternary volcanic rocks extend in a broken belt 4400 km from Torres Strait in the north along the 'highlands' on the eastern side of Australia, to eastern Victoria where the volcanic areas trend westwards into South Australia, and southwards into the Bass Basin and Tasmania (Fig. 1.1.1; see also Fig. 1.1.5). The belt represents one of the world's most extensive volcanic zones. Volcanic rocks of similar age and composition, but of much smaller total volume, are found also in the Northland-Auckland area of North Island, New Zealand, in several areas in South Island, and on the Sub-Antarctic and Chatham islands east and southeast of New Zealand (see Figs. 1.1.7–8). All of this volcanism may be referred to as 'intraplate'.

The term 'intraplate' in general carries the implication that volcanism is so remote from plate boundaries that magma generation cannot be attributed realistically to energy releases at mid-ocean ridges, subduction zones, and 'leaky' transform faults. This may appear reasonable from the spatial relationship between the margins of the Indo-Australian plate and the pattern of volcanic-rock distribution in eastern Australia shown in Figure 1.1.1, and the term certainly is appropriate for the Sub-Antarctic and Chatham islands which are well out on the Pacific plate. However, the term is less applicable in its strict tectonic sense for the volcanic

areas in the South Island of New Zealand, which are all within 200 km of the Indo-Australian/Pacific plate boundary, an active margin of seismic-energy release that changes from strike-slip motion at the Alpine Fault of South Island to convergent motion at the Hikurangi Trench east of North Island (see Fig. 1.3.7). In addition, the late Mesozoic and early Tertiary volcanism of eastern Australia and New Zealand was synchronous with the beginning of seafloor spreading in the Tasman and Coral sea marginal basins about 80–50 Ma ago, so it took place immediately adjacent to an extensional plate boundary. However, most of the Cainozoic volcanic rocks of eastern Australia and New Zealand post-date the end of spreading in the Tasman and Coral seas. This volcanism at least, therefore, can be considered at truly intraplate in the tectonic sense. Determining the relationship between Cainozoic volcanism and the formation of the marginal basins is one of the objectives of this study. Cainozoic volcanism is considered in this volume in much more detail than is the Mesozoic volcanism (see Section 3.9) because it is more obviously intraplate in character.

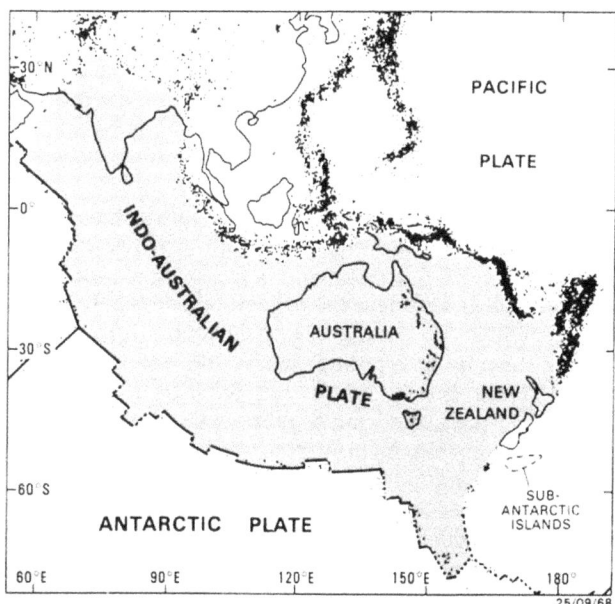

Figure 1.1.1 Indo-Australian plate showing distribution of continental intraplate volcanic areas of eastern Australia and New Zealand. Dots represent earthquakes defining margins of the plate. Intraplate and New Zealand earthquake epicentres are omitted.

'Intraplate' is used in this volume more in a petrological and geochemical sense than in a strictly tectonic one. The volcanic rocks to be described in following chapters have compositions distinctly different from those found at normal mid-ocean ridges and in arc-trench systems. Rather, they are similar to rocks present within continents, at rifts or similar structures, and at the trails of volcanoes in both oceans and continents — so-called 'hotspot' (or mantle-plume) traces and hotlines. The Tasmantid Guyots and other submarine volcanoes in the Tasman Sea are examples of traces of this type and are mentioned below, although the main subject of interest in this volume is the subaerial volcanism on the continents.

Evaluation of the nature of the continental crust and subcontinental mantle in eastern Australia and New Zealand is of critical importance to an understanding of the magmatism (Sections 1.4–5). The known surface geology is highly diverse, and in eastern Australia consists of a series of Palaeozoic fold belts and intervening basins beneath which Precambrian rocks are thought to exist (Section 1.3.1). The major watershed, known as the Great Divide or Great Dividing Range, runs down eastern Australia marking the crest of highlands that are mostly less than 1600 m high and which therefore hardly warrant the term 'range'. Nevertheless, the Great Divide and a prominent, east-facing, coastal escarpment which runs between the divide and the coast, were significant obstacles to westwards exploration by the early settlers of eastern Australia.

The timing of uplift of the highlands in eastern Australia has been a contentious subject, although most investigators now believe that uplift was in the late Mesozoic to early Tertiary (Section 1.3.2). Much of the volcanism is both contemporaneous with highlands uplift and spatially coincident with the highlands region. A similarly continuous linear zone of uplift is not recognised for New Zealand intraplate volcanic areas which, rather, may overlie lithospheric 'swells' (Section 1.5.2). The causes of uplift are discussed below (Sections 1.3.2, 1.7.3, 7.3), but here the process of underplating of the crust by magmas may be singled out in advance as a potentially important process of lithospheric evolution.

Extensive programmes of K-Ar dating have been undertaken of the late Mesozoic and Cainozoic volcanic rocks in eastern Australia and New Zealand, and there is now a clear appreciation of temporal trends for most of the volcanoes in the region (Section 1.7). These trends may be interpreted partly, but not entirely, in terms of the mainly northward passage of the Indo-Australian plate over one or more deep-mantle plumes.

Intraplate volcanism in eastern Australia and New Zealand is dominated by the effusion of basaltic lavas from numerous central vents (Chapter 2). These vents and their products are distributed over a wide area and give the impression of a vast aggregate volume. However, total thicknesses are not nearly so great as those in the major flood-basalt provinces of the world (Section 1.1.3). Most mafic volcanism in eastern Australia and New Zealand is of the 'plains-basalt' type, characterised by the eruption of relatively few, small-volume, lava flows from central vents or short fissures, and the building up of flat-lying or low-angle lava plains. Felsic volcanic activity at central vents produces domes and coulées which contribute to volcanic complexes of greater relief. Many of these complexes are eroded, have spectacular scenery, and for this reason have been declared state parks.

Studies of the physical volcanology of east-Australian and New Zealand intraplate volcanic areas are in their infancy. They lag behind the substantial amount of research in physical volcanological processes that has been undertaken in arc-trench systems, including the Taupo Volcanic Zone. Chapter 2 therefore is designed to review the world-wide literature relating to those volcanic processes most likely to apply to east-Australian and New Zealand intraplate areas, and to point out examples by cross-reference to Chapters 3 and 4. Some studies in physical volcanology have been undertaken in the youthful volcanic areas of western Victoria and northern Queensland and, accordingly, examples from these areas are given some prominence.

A great deal of geological data has been obtained during the last 30 years on the intraplate volcanic areas of eastern Australia and New Zealand, although some areas still remain poorly studied. These data are too numerous to present here, so an attempt is made to summarise information in a systematic fashion (Chapter 3, Sections 4.1–4). The geological literature on these areas is also too voluminous for inclusion in the bibliography of this volume. However, a computer-based bibliography has been compiled at BMR by J. Knutson and is available on request.

Intraplate volcanic rocks in eastern Australia and New Zealand have wide-ranging compositions, but are mainly alkali basalt, tholeiitic basalt, and hawaiite (Sections 4.6–8, Chapter 5; see Section 1.1.4 for a discussion on rock classification). Intermediate rocks such as mugearite and benmoreite are found also, and felsic rocks include rhyolite, trachyte, and phonolite, some of which are peralkaline. Distinctive geochemical features that must reflect source compositions include high abundances of Ti, Zr, Nb, Hf, and Ta (the high-field-strength elements) and large-ion-element abundances (K and so forth) which are higher than those found in mid-ocean-ridge rocks. There are also distinctive trace-element and isotopic differences.

Many hundreds of chemical analyses are available for major and trace elements and radiogenic and stable isotopes for whole-rock samples. Numerous electron-microprobe analyses of minerals from these rocks are available also, particularly for east-Australian rocks. A systematic account of this large amount of data has been compiled in Chapter 5 as a basis for petrogenetic discussion (Sections 7.5–7). Combined assimilation/crystal-fractionation (AFC) is regarded as the dominant process of magmatic differentiation.

New Zealand intraplate volcanoes contrast markedly in tectonic setting with those of eastern Australia, yet the ages and compositions of the rocks are similar in both regions. The volcanic areas of both regions can be regarded in one sense as a single volcanic 'province' in the same way that scattered volcanic areas such as Iceland, Greenland, and northwestern Scotland, are part of the North Atlantic or Thulean province, although the volumes of volcanic rock are much lower in eastern Australia and New Zealand (Section 1.1.3). Whether the intraplate volcanoes in both New Zealand and eastern Australia owe their origin to the same regional tectonic process is one of the questions addressed below (Section 7.8.7). The intraplate area in New Zealand best studied petrologically until quite recently was Dunedin volcano in South Island. However, a significant amount of new analytical work has been completed for Northland, Banks Peninsula, and the Sub-Antarctic and Chatham islands. Much of this new data is summarised in this volume (Sections 4.6–8).

The mafic magmas of eastern Australia and New Zealand have plucked rocks and megacrysts from the upper mantle and crust during their rapid rise through the lithosphere and brought them to the surface (Chapter 6). Numerous xenolith and megacryst localities are known in eastern Australia (together with minor examples in New Zealand) which may be regarded as one of the most extensive samplings of upper mantle and lower crust over such a wide area anywhere on Earth. These xenoliths and megacrysts are of importance because they are a direct sample of deep-lithosphere rocks that are otherwise unavailable to the petrologist, and because they carry a record of pressure and temperature conditions and volatile contents in deep-lithosphere sources. Interpretation of these data can be made in association with the results of seismic refraction, heat-flow, and other geophysical studies (Sections 1.4–5), and indications obtained of the 'stratigraphy' of rocks and temperature gradients in different parts of the lithosphere (Chapter 6, Section 7.2).

The first six chapters in this volume represent to a large extent the major descriptive parts of the study. The main problems and issues are identified in these chapters, and are drawn together and discussed, together with further data presentation, in a series of contributions in Chapter 7. These discussion sections are concerned primarily with relationships between magma genesis and tectonophysics.

The principal aim of the volume, as stated in the Preface, is to summarise what is known about intraplate volcanism in the region and to identify avenues for further research. Advances made before the early 1960s are summarised in the following section which serves as a comparison for the substantial amount of information on intraplate magmatism obtained in the last 30 years.

1.1.2. History of early work

First sightings

There seems little doubt that Australian Aborigines witnessed volcanic eruptions such as those that took place at Mount Gambier, South Australia, about 4600 years ago (Section 3.7.4). Mulvaney (1964) discussed the evidence for Aboriginal occupation of the Western Plains, Victoria, prior to volcanism, noting in particular the stone implements buried under tuff at Tower Hill (Keble, 1947; Gill, 1953). However, he cautioned against accepting Dawson's (1881) statement that such

events are enshrined in Aboriginal legend. Some legends that have survived, such as that told by Buller-Murphy (1958), come from areas essentially devoid of volcanism and deal with a sudden earth-moving event that is more easily interpreted as an earthquake and subsequent tsunami.

There appear to be no Maori legends relating to intraplate volcanism. The Maori name Rangitoto for the volcanic island in Auckland Harbour has one meaning of 'bloody sky', and Hochstetter added 'it could perhaps be concluded that in former centuries the natives had known the mountain in a state of full activity and were thereby led to this name, which at the present day they generally use also to designate black lava' (Fleming, 1959, page 187). Stone artifacts have been discovered beneath Rangitoto ash to the east of the volcano (Brothers & Golson, 1959). There are important legends dealing with eruptions at the larger volcanoes of the Taupo Volcanic Zone and Taranaki (Gregg, 1961), so perhaps the volcanic activity at Auckland was regarded as of little significance.

Lieutenant James Cook named several features that we now know to be of intraplate volcanic origin, during his circumnavigation of New Zealand and his voyage close inshore along the eastern coast of Australia. These included 'a pretty high mountain laying near the shore which on account of its figure I have named Mount Dromedary' (Reed, 1969, page 35). This is an eroded volcanic pile of Cretaceous age on the New South Wales coast (Fig. 1.1.2).

Eastern Australia's largest intraplate volcano is Tweed, in the centre of which is the prominent intrusive complex of Mount Warning (Fig. 1.1.3). This 'remarkable sharp peak'd mountain' and Point Danger, an outcrop of lava from Tweed volcano, were both named by Cook (Reed, 1969, page 55). John Oxley in 1823 recognised Small Island off Point Danger as of 'volcanic origin and the superincumbent rocks to be basaltic' (Uniacke, 1825, page 34), and compared the occurrence with the Giants Causeway in Northern Ireland. Oxley (1820) on an earlier journey had recognised the volcanic nature of the Warrumbungle and Liverpool ranges.

Cook also sighted and named the Glass Houses mountains (Fig. 1.1.3) — also referred to as the Glasshouse Mountains on more modern maps. Matthew Flinders observed these conspicuous peaks from the coast in 1799 and felt that 'as far as could be judged [they] had every appearance of being volcanic. That they were so, indeed, was in some measure corroborated by the quantity of pumice-stone which was lying at high-water mark upon the eastern shore of the river' (Mackaness, 1956, page 17–18). Flinders on another expedition, along the eastern coast of Tasmania, observed 'basaltic pillars or columns, several . . . much resembling tall chimnies, at the extremity of the point. I have called it Cape Basaltes. The cliffs are very high, steep and romantic' (Flinders, 1799, page 812). This is the occurrence of Jurassic dolerite at Cape Raoul. These sea-coast outcrops of dolerite had drawn the attention of French observers: Baudin (1802) referred to Cap des Pitons as 'remarkable for its jutting points, which look like bell-towers and give it a most distinctive appearance' (page 337).

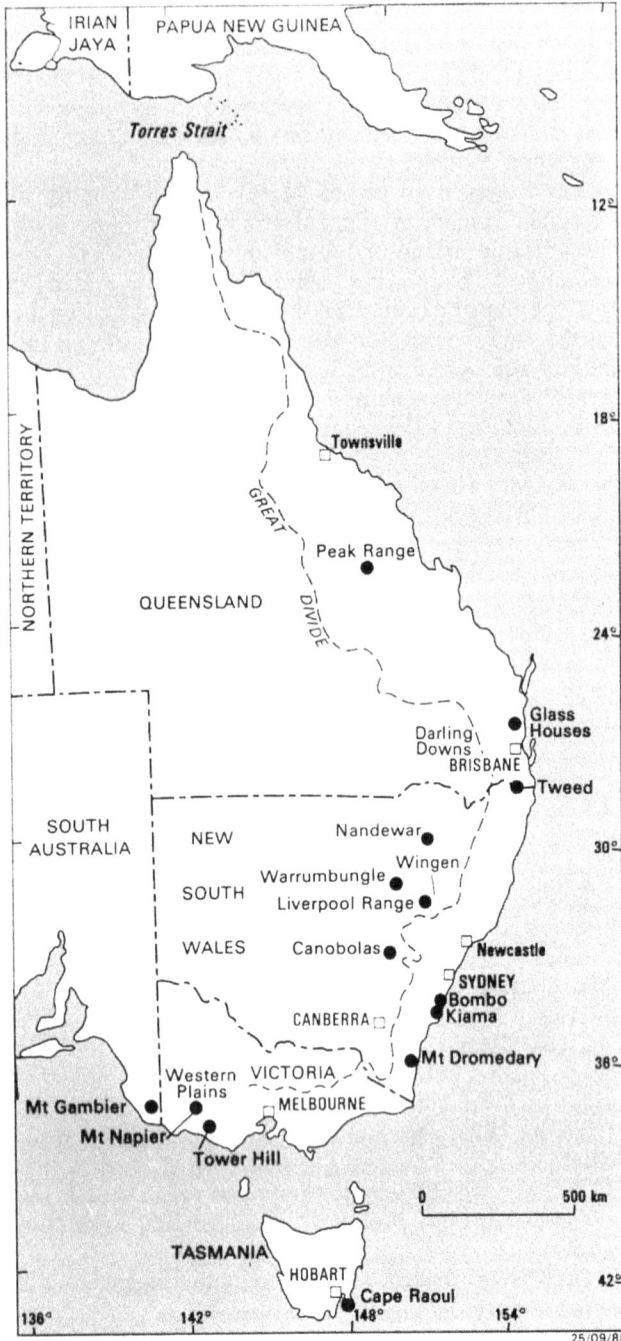

Figure 1.1.2 Localities referred to in early references to intraplate igneous activity in eastern Australia, shown in relation to present-day cities and state boundaries.

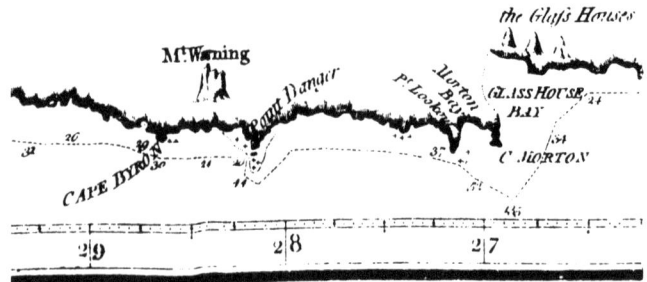

Figure 1.1.3 Facsimile of part of the chart produced by Lieutenant James Cook of his voyage up the eastern coast of Australia in 1770, showing Mount Warning (Tweed volcano) and the Glass Houses.

Active volcanism?

Bass (1797) observed Permian (non-intraplate) volcanic rocks at Kiama and there is a hint he felt the volcanic activity to be recent (indeed the present-day visitor might well feel the same). Flinders' conjunction of pumice pebbles and the Glass Houses peaks speaks more directly of present-day activity, but he was in error as the pumice found on east-Australian beaches derives from volcanic activity out in the southwest Pacific. Exploring parties in Australia in the next 25 years or so found no indications of modern volcanic activity.

The presence of sulphurous fumes on a 'burning mountain' at Wingen (Fig. 1.1.2) near the head of the Hunter Valley in New South Wales about 1829, excited Wilton (1830), Mitchell (1838), and other naturalists, and attracted a range of observers to study the phenomenon. The locality received some notoriety in Europe as a volcano or pseudo-volcano, but its discovery came too late to support the ideas of the last of the followers of Werner. It is, in reality, a coal seam on fire.

René P. Lesson in 1824 noted that certain parts of the western road from Sydney were 'laid with dolerite taken from an eminence 5 miles from Parramatta' (Mackaness, 1965, page 146). The Reverend Richard Taylor visited the 'quarry of whinstone at Prospect Hill' (now within the city limits of Sydney) twelve years later: 'I was much interested in observing the manner in which this volcanic formation protruded itself through the sand rock . . . the strata is nearly vertical, whilst that of the sandstone is almost horizontal' (Taylor, 1836). Today we recognise this igneous mass as an intrusion — albeit under a thin cover and intruded shortly after sedimentation had ceased in the Triassic.

Strzelecki in his book (1845), his report (1840), and more particularly on his unpublished map sections (Branagan, 1986), summarised the known extent of volcanism in southeastern Australia, commenting that 'numberless streams of lava, the trachitic [sic] rocks and others . . . give evident proofs of volcanic agency' (Strzelecki, 1840, page 14) which he recognised had been active over a long span of geological time.

The Australian landscape, before the 'burning mountain' discovery, had an established reputation for aridity, and perhaps longevity and stability, and no currently active volcanoes, but travellers to New Zealand brought rumours of a very different landscape there. Samuel Stutchbury early in 1826 was at the Bay of Islands, North Island (see Fig. 1.1.7), where he observed 'to the South westwd some exceeding black rocks, which I afterwards found was basaltic, externally bearing the appearance of *Scoria* . . . At the River Kedi Kedi [Kerikeri] the water passes over immense masses of Basalt containing chrysolite or Olivine . . . The Bays and Rivers generally I should suppose to have been formed by volcanic eruption, thus accounting for the jumbled incongruous state of the strata' (Stutchbury, 1826, page 83). Lesson two years earlier also described the rocks in the Bay of Islands. He noted that 'several recently extinct volcanoes have been reported in the interior, and there are many pumices and obsidians. Lake Roto-dorua and its hot springs are obviously an extinct crater' (Sharp, 1971, page 81).

Cook in 1769 had named White Island 'because as such it always appeared to us' (Reed, 1951, page 53). White Island was the first active volcano in New Zealand to become known to Europeans, although it is not of intraplate origin.

The volcanic character of the Auckland area was first recognised by a European in 1839 when Ernst Dieffenbach (1843) noted: 'as we approach Auckland several regular volcanic cones rise over the tablelands which stretches across the island to the habour of Manukao ... Rangitoto is [a] very remarkable island ... and has on its summit three cones ... in the middle cone is a very perfect crater' (pages 276 and 283).

Victoria

A better understanding of the Cainozoic volcanism of eastern Australia began with the expedition by Thomas Mitchell to western Victoria in 1836, where very young volcanism could be seen (*Fig. 1.1.4*). Mitchell was attracted by the perfect form of Mount Napier. He found on climbing to its top a circular vent containing cellular rocks — lava and scoria. 'The igneous character of these was so obvious that one of the men thrust his hand into a chasm to ascertain whether it was warm' (Mitchell, 1838, volume 2, page 249). The result was negative, but Mitchell was struck by the 'remarkable uniformity in size of the trees on the hill, there were very few dead or fallen'. Furthermore, the rocks had very sharp edges and Mitchell felt 'some might conclude that the volcano had been in activity at no very remote period' (volume 2, page 249). He also observed the 'Mammeloid hills' farther to the east, consisting completely of lava which was vesicular and, he felt, old. The westernmost part of this 'province' had been named 36 years earlier by Lieutenant Grant (Grant, 1803) when, from the deck of the *Lady Nelson*, he saw 'two high mountains a considerable way in shore. One of them was very like the Table Hill at the Cape of Good Hope' (page 68).

Knowledge of the physical extent of the west-Victorian lavas was well established by 1846 (Westgarth, 1846). Westgarth in 1853 wrote: 'the extinct volcanoes which exist in extraordinary numbers ... form a most interesting feature. In many instances the craters are perfectly defined, leaving not the slightest doubt as to their former character and doings' (pages 43–44). The related volcanoes in southeastern South Australia were described geologically first by Thomas Burr in 1846, and in more detail by J.E. (Tenison) Woods in 1862.

Dana (1849) was aware of the significance of the Victorian region. He designated it as 'the single volcanic region on the Australian Continent' (page 114), although he noted the presence of what he thought were minor occurrences of basalt elsewhere. Jukes (1850) made some personal observations of the Victorian basalts, but was less specific than Dana. He identified basalt occurrences on his map by 'colour(s) dabbed on roughly about the place the rock ... was observed' (page 3). However, Ludwig Leichhardt (1847) had described several volcanic occurrences in different parts of Queensland, which were by no means insignificant.

Leichhardt noted that Peak Range resembled 'very much the chain of extinct volcanoes of Auvergne' (Stephenson et al., 1980, page 349). His course along the Burdekin River was hindered as 'a wild field of broken basaltic lava rendered it impossible for us to follow its banks ... the lava was very cellular, the basalt of the tableland solid. The whole appearance of this interesting locality showed that the stream of lava was of much more recent date than the rock of the tableland' (Leichhardt, 1847, pages 239 and 244).

Formal geological mapping in Victoria was carried out by Alfred Selwyn and his staff from 1852 onwards. Selwyn's 1863 map of Victoria 'geologically coloured', at 8 miles to an inch, has four volcanic units identified as Upper Volcanic (Pliocene), Lower Volcanic (Pliocene), and Older Volcanic (Miocene and older), each containing (1) basalt and dolerite, (2) anamesite (a rock intermediate in texture between basalt and dolerite), (3) lava, and (4) ash, conglomerate, and breccia. Selwyn gave no age for his older 'trap or hypogene', in which he included greenstone, diorite, felstone, and feldspar porphyry.

Selwyn's work laid a firm foundation, but the ages of the Victorian basalts continued to cause discussion. Tate & Dennant in 1893 claimed that the 'older basalts of Southern Victoria and previously referred to the Miocene' (Tate, 1893, page 69) were pre-Eocene. Tate (1893) also drew attention to the New England (New South Wales) lavas and tuffs shown by David (1887) to be Eocene. He thought that the newer basalts and ash beds of Victoria and South Australia were 'post-pliocene' because they overlay 'deposits of Diprotodon Period' (Tate, 1893, page 69).

Gregory (1903) recognised 'Upper and Lower Kainozoic' volcanic rocks. He could put no age on the younger succession, but was sceptical of the idea that they post-dated Aboriginal occupation. Gregory pronounced the volcanoes as quite extinct because of the absence of fumaroles and hot springs.

Hochstetter and Haast

The Austrian Ferdinand Hochstetter made himself familiar with the results of Selwyn's surveys in Victoria. Hochstetter, aged 28, arrived in Auckland in 1859 as a member of the *Novara* Expedition (Fleming, 1959). The local authorities prevailed upon his superiors to allow Hochstetter to carry out a geological survey of the Auckland province, and the results were superb.

Previous observations on the geology of the North Island of New Zealand had been made by men such as Ernst Dieffenbach, the first 'professional' scientist to spend a lengthy period in New Zealand. Dieffenbach was influential because his writings were taken up by European scientists such as Leopold von Buch and Alexander Humboldt. There arose consequently 'an erroneous opinion ... that the whole North Island is predominantly volcanic in structure although it is true that the North Island owes its present form and surface shape chiefly to volcanism', wrote Hochstetter (Fleming, 1959, page 110).

Hochstetter recognised two or more epochs of volcanism — based on compositional differences — and four zones, those of Taupo, Taranaki, Auckland, and the Bay of Islands. He recognised that the Auckland and Bay of Islands basaltic rocks were significantly younger than the eruptions from Taranaki and the main body of the Taupo zone. He also delineated three older volcanic districts whose age relationships were uncertain. Hochstetter believed the 'basaltic' areas were not only petrologically similar to each other but also to the west-Victorian rocks and were of the same age.

The first important study of volcanism in the South Island of New Zealand was begun late in 1860 when Julius von Haast carried out mapping for the Christchurch-Lyttelton tunnel which was being cut through part of the Banks Peninsula volcanic pile (see Figs. 1.1.8, 4.3.1B). Haast showed the detailed stratigraphy of the volcanic sequence, the lenticularity of some flows, and the extent of particular tuffaceous layers. The later publication of his work caused considerable interest in European circles (Haast, 1948).

Expanding fields

Stutchbury between 1851 and 1855 examined a wide belt of country in eastern Australia from Sydney to Gladstone, outlining the central volcanoes of Canobolas, Warrumbungle, and Nandewar, and the more extensive lava fields of the Liverpool Range, New England, and the southern tablelands of Queensland (Branagan, 1975). The Reverend W.B. Clarke at the same time examined parts of the New England and southern tablelands of New South Wales (Branagan, 1975).

Expansion of geological mapping followed the establishment of the Geological Survey of New South Wales in 1879 under C.S. Wilkinson. Edgeworth David's New England work was particularly significant, as was later mapping of the numerous Sydney Basin intrusions by Morrison (1904), Carne (1908), and Willan (1925). Andrews (1910) also played a significant role by linking studies of volcanism with geomorphology. Mapping by the Survey was supported from the 1880s by excellent chemical work and petrological studies carried out by G.W. Card, G.A. Stonier, and J.H. Mingaye (Adrian, 1976, pages 36 and 42).

Queensland

Leichhardt's work in Queensland was followed shortly after by that of Jukes (1850). Jukes recognised in the Torres Strait area the volcanic nature of rocks of Murray Island, Darnley Island, and Bramble Kay, which he considered were '. . . of subsequent origin [to] the coral reefs here, [they] may yet date back into some tertiary period . . . along the North Coast of New Guinea runs a lofty volcanic chain, from which these look like offshoots' (Jukes, 1850, pages 31–32). Jukes also commented on the common occurrence of pumice high on the coastline beaches and attributed it to New Zealand or Tasman Sea volcanism.

R. Daintree, C.D. Aplin, R.L. Jack, A.G. Maitland, and W.H. Rands, during the early years of the Geological Survey of Queensland, outlined regions of volcanism and recorded details of specific areas (Jack & Etheridge, 1892). These data were incorporated on the geological maps published by Daintree (1872) and Jack (1886) and were collated in Jack & Etheridge (1892). They recognised an older and newer volcanic series akin to Leichhardt's original observations.

Australian petrology

Jack & Etheridge (1892) incorporated the first petrographic illustrations of Queensland rocks described by A.W. Clarke. However, Daintree (1872, 1875), Jack, and Rands had separately and earlier used the petrological microscope to further their study of the rocks. Interest in the petrological and chemical aspects of the Victorian volcanic rocks began with G.H. Ulrich (1875), C. Newbery (1878), and A.W. Howitt (1876), and was taken up again by Skeats (1909) and Skeats & James (1937). These studies were eclipsed by Edwards (1938a, 1939) who must be regarded as the major researcher up to that time of the petrology of Australian Cainozoic volcanism (see also Edwards, 1938b). His work on the Mesozoic dolerites of Tasmania (1942) was also a major contribution.

David moved to the University of Sydney in 1892 and petrological research became established there. David worked with W.A. Anderson and others. Volcanism received attention from several workers, particularly H.I. Jensen (1903, 1906a,b, 1907a,b) and W.M. Benson (1911). J.M. Curran (1891, 1899) and C.A. Süssmilch (1905, 1922, 1923) — both of the Sydney Technical College — also contributed with petrology and fieldwork, leading to general but erroneous agreement of a virtual absence of volcanism in New South Wales during the Mesozoic. Numerous, minor, extrusive and intrusive occurrences were placed in the Tertiary because of petrological similarities. Süssmilch in 1923 suggested there were three distinct epochs: newer basalts were found as plateau remnants, older basalts as residuals, and the third and oldest consisted of early Tertiary alkaline lavas and tuffs. The rocks of the third epoch were associated with extinct cones, where there was evidence of differentiation.

W.R. Browne as late as 1950 was able to write that little was known of the petrology of the Quaternary lavas of eastern Australia, except in Victoria. However, by this time the extent of Cainozoic volcanism in Queensland was established — perhaps half of the 'stupendous display of igneous activity' (Browne, 1950, page 571) extending from Torres Strait to Tasmania. New South Wales seemed to have only older basalts (Miocene or earlier, according to Browne) derived from fissures, together with plugs of dolerite and isolated alkaline occurrences.

H.C. Richards, who had cut his teeth on some Victorian intrusions in the Dandenong Ranges, moved to Queensland in 1910. Little had been done on southern volcanic rocks there since A.C. Gregory had outlined their extent on the Darling Downs in 1879. Richards described the volcanic rocks of the southeastern part of the state in a classic paper in 1916, and later expanded his study to cover all of Queensland (Richards, 1926).

New Zealand petrology

The different character of the volcanic rocks of the Auckland area (intraplate) and the Taupo Volcanic

Zone (subduction-related) was more clearly understood fifty years after Hochstetter's work, thanks to the use of the microscope and chemical analysis. Marshall (1912) recognised these differences, but he considered that there were all gradations between the basanites of Auckland and the rhyolites of Taupo. Marshall felt there were three possible explanations: different original magmas, an original heterogeneous composition of the crust, or differentiation (the interpretation he preferred). Marshall, like Hochstetter, thought that the Auckland province was similar to western Victoria, whereas the alkaline rocks of Dunedin were similar to those of East Africa and eastern Australia. Later important work on the Auckland and Northland basaltic rocks was carried out by Bartrum (1925, 1949).

Marshall made some important contributions to the study of the Banks Peninsula volcanism in New Zealand's South Island, but this area became almost the exclusive 'province' of R. Speight who published a series of papers from 1908 to the 1940s directed particularly to its petrography (for example, Speight, 1908, 1924, 1944). The volcanic rocks of many parts of the South Island were examined by Haast (1948), Hector (1865), and Hutton (1874) in the 1860–70s. Marshall (1894, 1906) studied Dunedin volcanism, writing a classic statement of their nature in 1914. These contributions formed the basis of work by Benson (1941a,b) and his students from the 1920s onwards. C.A. Cotton's geomorphological studies of New Zealand volcanoes were influential during the 1930–40s (for example, Cotton, 1944).

Speight contributed to the study of the Sub-Antarctic Antipodes and Auckland islands, collaborating with A.M. Finlayson (1909), and Dieseldorf (1901) made the first significant observations on the Chatham Islands. The volcanic sequence on the Campbell Islands was recognised early this century by Marshall (1912).

Tectonics and petrogenesis

The petrographic classification of east-Australian and New Zealand volcanic rocks depended initially on their field occurrence, simple mineral association, and visible texture. Ideas about classification and petrogenesis became firmer in the early 1900s following the publication of the work of Cross et al. (1903) whose methods were eagerly taken up by Australian and New Zealand workers.

The easily accessible Prospect intrusion (Sydney) exposed by quarrying, was studied for its petrographic differences which were thought to be the result of magmatic differentiation. Papers by Jevons et al. (1911, 1912) were followed by studies made by Browne (1925) and his students. Jensen, one of the workers at Prospect, already had drawn attention to the alkaline volcanic centres of eastern Australia, examining the Glass Houses (1903, 1906a), Warrumbungle volcano (1906b, 1907b), and Nandewar volcano (1907a), and defining an alkaline province (1908). Browne (1933) outlined evidence for at least two magmatic kindreds in the Mesozoic-Cainozoic igneous rocks of New South Wales: (1) tholeiitic types (including the quartz-dolerites and Browne's 'calcic

basalts'); (2) widespread alkali basalts and their derivatives.

Attention paid to the xenoliths and megacrysts in volcanic rocks and shallow intrusions (Chapter 6) has proved fruitful. Xenoliths from Tertiary dykes intruding Permian lavas at Bombo on the New South Wales south coast (Fig. 1.1.2) were first described by Süssmilch (1905). Occurrences in diatremes at Hornsby and Dundas (both in the Sydney area) were discussed by Benson (1911).

Possible relationships between volcanism and tectonism were discussed in print by only a few workers in the early years. A limiting factor was a lack of knowledge of the crust both on-shore and off-shore. W.H. Bryan as late as 1950 (quoted by Browne, 1950) was convinced the Tasman Sea floor was sialic, continuing the idea of Tasmantis (Süssmilch & David, 1919) foreshadowed by Clarke (1878). Browne (1950) felt there was evidence for geosynclinal deposition off eastern Australian during the Tertiary and that the foundering of this mass (to form the Tasman Sea floor) produced a compensatory uplift of the eastern highlands and, consequently, volcanism. Browne, on the other hand, recognised the attractions of some aspects of continental drift and collision to explain some of the geology (Du Toit, 1937), although he was not personally in favour of such mechanisms.

Andrews (1910) believed that from late in the Tertiary the whole of eastern Australia had been uplifted essentially as a single unit. He suggested the name 'Kosciusko Period' for this event, and from this idea the concept of the 'Kosciusko uplift' evolved. However, this concept now is not supported generally by geomorphologists (see Section 1.3.2). Andrews linked the volcanism to uplift but attempted a wider synthesis by equating the igneous and sedimentary successions and landscape of eastern Australia with a similar history for the western coastline of the United States. Jensen (1908) a little earlier suggested the alkaline volcanic rocks were associated with continental areas of great permanence, and that most were Cretaceous to early Tertiary in age. Süssmilch (1923) felt that the different episodes of Tertiary volcanism preceded uplift, and that compositional differences had no particular relation to tectonic conditions.

David (1932) believed in more dramatic and extensive activity, writing that the artesian-basin floor 'heaves and undulates in sympathy with the great contemporary orogeny in New Zealand' (page 171). He felt that in post-Cenomanian time there were immense pressures along the present Queensland coast and that 'mighty thrusts from the floor of the Pacific' (page 171) produced the Tasmantides — a great range of submarine volcanoes.

More modern perceptions of these relationships are discussed further in this chapter and in Chapter 7.

1.1.3   Volcano distribution and classification

Introduction

An appreciation of the extent and nature of intraplate volcanism in eastern Australia and New Zealand that had developed steadily up until about the 1950s (Section 1.1.2) began to grow at a greater rate in the

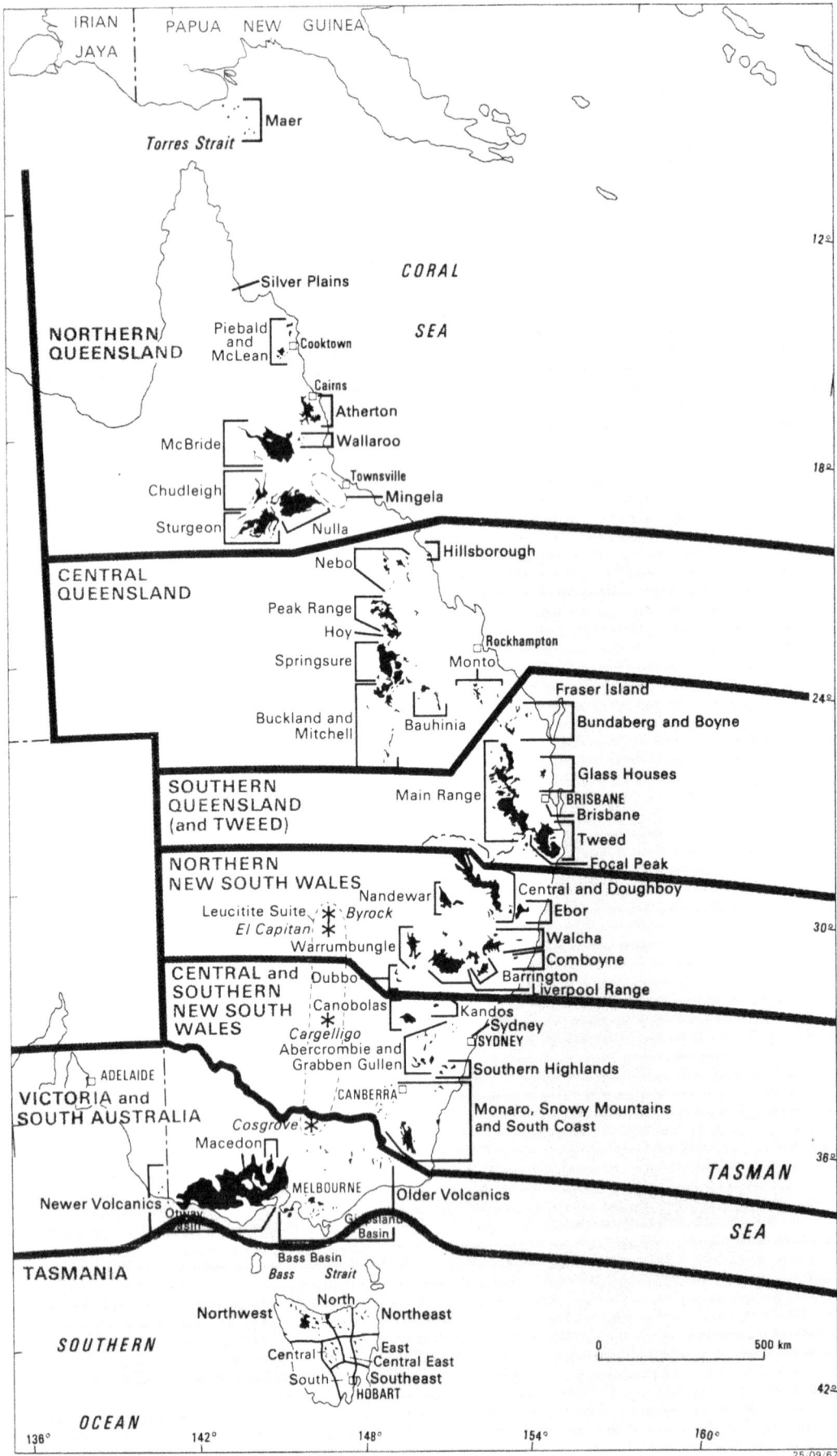

Figure 1.1.5   Intraplate volcanic areas of eastern Australia, divided into regions and provinces as described in Chapter 3 (adapted in part from O'Reilly & Griffin, 1987, fig.1).

1960s as new directions of research opened up — particularly the application of K-Ar dating methods to the volcanic rocks, and the development of rapid instrumental methods for chemical analysis of rocks and minerals (especially X-ray fluorescence spectrography and spectrometry, and electron-microprobe analyses). Geological mapping at about this time, particularly by BMR at 1:250 000 scale in eastern Australia, also led to a better understanding of the extent and distribution of intraplate volcanism in the region.

The first published east-Australian basalt date (33 Ma) was obtained for an analcime-rich rock from Spring Mount, New South Wales, by Cooper et al. (1963). Additional geochronological work culminated in the important paper by Wellman & McDougall (1974a).

Distribution in eastern Australia

East Australian intraplate volcanic rocks stretch 4400 km from the Maer Islands in Torres Strait, near the border with Papua New Guinea, along the highlands of eastern Australia to Victoria, and then through Bass Strait to Tasmania (Fig. 1.1.5). The volcanic belt splits into two in eastern Victoria: one part heads south to Tasmania; the other extends westwards into western Victoria and southeastern South Australia where it is defined by the youthful and extensive Newer Volcanics. This area includes the youngest known volcanic centre of eastern Australia — Mount Gambier, in South Australia (about 4600 yr B.P.; Section 3.7.4). The volcanic areas of northern Queensland are also youthful (as young as 13 000 yr B.P.) and, together with the Newer Volcanics, they are the most westerly of all the east-Australian volcanic areas (Fig. 1.1.5).

The belt of volcanism along eastern Australia is by no means continuous. There are many large gaps between individual areas, particularly the 500 km gap south of the Maer Islands. However, some gaps are filled by minor lava outcrops too small to show on a map of the scale used in Figure 1.1.5, and other minor outcrops may have eroded away completely. The east-Australian volcanic belt is less than 100 km wide north of Cairns. It is only about 150 km wide at the border between New South Wales and Queensland where it reaches its most easterly point near Cape Byron, the most easterly point in Australia.

The belt as a whole swings southwards from northern Queensland in a large arc that closely parallels the eastern coast of Australia, curving out and back through more than 12 degrees of longitude (Fig. 1.1.5). All areas are within 500 km of the coast, except for some leucitite centres (see below). This broad, curved trend is quite unlike the narrow, linear trends defined by oceanic hotspot traces such as the Hawaii/Emperor Seamount chain.

A relationship between volcanism and highlands uplift may be inferred for eastern Australia, but only in New South Wales do the volcanic areas (excluding the leucitite centres — see below) closely follow the trace of the axis of highlands uplift represented by the Great Divide (Fig. 1.1.6). In contrast, most of the volcanic areas in Queensland are to the east of the divide which, north of the New South Wales border, swings inland more than 400 km from the coast. The belt as a whole

therefore follows the eastern coastline more closely than it does the Great Divide.

The distribution in detail of the main volcanic areas in the belt is highly complex, and no systematic pattern is apparent along the entire belt (Fig. 1.1.5). North Queensland volcanic areas form an elongate cluster that peters out northwards around Cooktown. Central Queensland areas are dominated by a strong southward linearity from Nebo to Mitchell, but Bauhinia and Monto form an orthogonal trend towards the coast

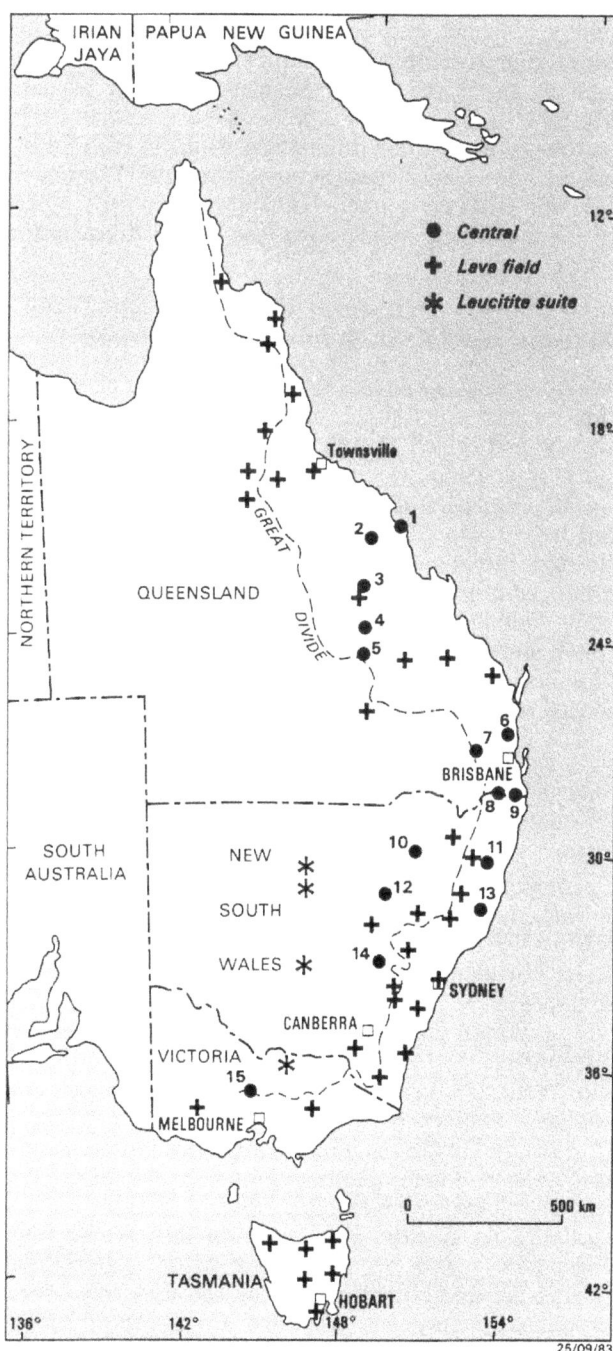

Figure 1.1.6 Central-volcano, lava-field, and leucitite-suite provinces of eastern Australia shown in relation to the Great Divide. Numbers 1–15 correspond to the following central volcanoes: 1, Hillsborough; 2, Nebo; 3, Peak Range; 4, Springsure; 5, Buckland; 6, Glass Houses; 7, Main Range; 8, Focal Peak; 9, Tweed; 10, Nandewar; 11, Ebor; 12, Warrumbungle; 13, Comboyne; 14, Canobolas; 15, Macedon-Trentham.

where, to the north, there are also volcanic areas at Rockhampton and Hillsborough. A quite well-defined eastward-facing arc of volcanic areas extends from the coast at Bundaberg, through Main Range, to Tweed volcano on the coast again, but volcanic areas are present east of the arc, particularly in the Glass Houses area.

Volcanic areas in northern New South Wales (excluding Tweed) have a crude, probably fortuitous, ring-like distribution (Fig. 1.1.5). Volcanic areas in southern New South Wales and eastern Victoria are relatively small and scattered (the largest is in the Monaro area south of Canberra). The most extensive area by far is the Newer Volcanics field of western Victoria. Intraplate volcanic rocks are known from petroleum-exploration drilling operations in Bass Strait, but their overall distribution remains unclear. Tasmanian volcanic areas are scattered mainly in the northern part of the island, extending down the centre towards the south.

Volcano classification in eastern Australia

Wellman & McDougall (1974a) proposed a classification for east-Australian volcanoes that has been used widely and which is adopted in this volume. Three types of volcano are recognised (Fig. 1.1.6):

(1) *Central volcanoes* are predominantly basaltic but have felsic (rhyolite, trachyte, or phonolite) lava flows or intrusions (or both). The volcanic rocks were produced from central vents (more precisely, clusters of vents) and in many cases they built up large volcanoes (more precisely, volcanic complexes) such as Nandewar and Warrumbungle. Tweed and Main Range are the largest central volcanoes in eastern Australia.

(2) *Lava fields* are basaltic, and were thought to have formed from 'a diffuse dyke and pipe swarm up to 100 km across' (Wellman & McDougall, 1974a, page 53). Lava fields generally are extensive and thin, but in some cases — for example, Liverpool Range and Barrington — lava piles up to 1000 m thick were formed. Lava fields such as the Newer Volcanics and north-Queensland areas are widespread and characterised by an abundance of small scoria and lava cones and maars.

(3) 'High potassium mafic' areas (Wellman & McDougall, 1974a) are dominated by leucitite minor intrusions and rare lavas that are petrologically and spatially distinct from all the other volcanic areas in eastern Australia. The high-potassium areas are discussed in Section 3.6.6 and together are referred to in this volume as the *leucitite suite*.

Some clarification is required for this terminology. First, the lava fields are thought to be examples of 'plains-basalt' volcanism, so they too may have been produced mainly by eruptions from central vents (Section 2.3), in some examples building substantial shield volcanoes not significantly different from the mafic shields of some of the 'central' volcanoes. Second, some volcanoes classified as 'central' have no overall shield or conical shape. The Glass Houses and Peak Range districts in Queensland, for example, are cases where felsic plugs and lava residuals are not clustered about the centre of a prominent mafic shield.

The volcanological distinction between the three volcano types therefore is not especially clear.

The most important distinctions between the three volcano types have been petrological: 'central' volcanoes are distinguished from 'lava field' volcanoes by the presence of significant amounts of felsic rock, and both are distinguished from 'high-potassium mafic' volcanoes by the absence of leucitite. Nevertheless, there are a few areas where the felsic rocks are so rare or are of uncertain age relative to the mafic rocks, that they cannot be classified readily — for example, Dubbo province and Fraser Island.

Striking differences in age trends (Wellman & McDougall, 1974a; Section 1.7), petrology and geochemistry (Chapter 5, Sections 7.6–7), and distribution are found between the three types of volcanic area. Most lava-field areas are on, or to the east of, the Great Divide (Fig. 1.1.6). The Central and Snowy Mountains (and Dubbo) areas are to the west of the divide, but even these are within 100 km of it. In contrast, the central and leucitite-suite volcanoes are not limited by the trace of the Great Divide (Fig. 1.1.6).

Central volcanoes in central Queensland define a southwards trend that curves inland from Hillsborough at the coast, intersecting the Great Divide in the Buckland area (the Mesozoic Rockhampton area is also at the coast). Main Range in southeastern Queensland is on the Great Divide, but Glass Houses, Tweed, Ebor, and Comboyne are all to the east of it. Nandewar, Warrumbungle, and Canobolas are well to the west of the divide. The leucitite-suite centres of Byrock, El Capitan, Cargelligo, and Cosgrove also are well inland (Byrock is 600 km from the coast), but Macedon central volcano — the most southerly of the central volcanoes — is on the divide. In general, therefore, central volcanoes in Queensland are on or to the east of the Great Divide whereas most of those in New South Wales and Victoria, including the leucitite centres, are on or to the west of it.

Volumes in eastern Australia

Estimating volumes of volcanic rock is a difficult task because of the erosion and removal of uncertain amounts of material and because of difficulties in estimating remaining thicknesses. However, Wellman & McDougall (1974a; after Wellman, 1971) made a bold effort at estimating original volumes for east-Australian areas.

The total-volume estimate is a little over 20 000 km³, consisting of about 9000 km³ for the central volcanoes and the remainder for lava fields. The volume of the leucitite suite in comparison is negligible (the original average thickness over the isolated areas is only about 20 m). Wellman & McDougall (1974a) concluded that the rate of total volcanism was roughly constant during the past 60 Ma, although the central and lava-field types began their activity at different times, and most lava-field volcanism cut out before central-type activity began (Section 1.7).

20 000 km³ is a small volume compared to values for other volcanic regions. For example, the Columbia River flood basalts of the western USA have a total volume of about 195 000 km³, the Deccan Traps of

northwestern India are more than 10^6 km³ in volume, and the largest volcano on Earth, Mauna Loa in Hawaii, has a reported volume of 40 000 km³, twice the total volume of the volcanic rocks in eastern Australia (comparative volumes taken from Cas & Wright, 1987). The total volume for eastern Australia may appear small, but it is by no means insignificant. The volume is spread out along more than 4000 km, and is clearly a major magmatic manifestation of large-scale tectonic processes. Furthermore, the volume of intraplate magmas is considerably greater than 20 000 km³ if magmatic underplating is a major cause of highlands uplift.

Distribution in New Zealand

The amount of intraplate volcanic rock in New Zealand is at least an order of magnitude less than that in eastern Australia, but as in Australia the volcanic areas are distributed widely throughout the region. Three areas of New Zealand intraplate volcanism are considered in this volume (see Sections 4.2–4):

(1) The Northland and Auckland areas, extending south of Auckland City and the Waikato River to Ngatutura, are in the extreme northwestern part of North Island and represent the only known intraplate volcanism in the island (Fig. 1.1.7). The Northland and Auckland areas are more than 200 km apart, and each is divisible into three sub-areas, one of which in the Auckland area is Auckland City itself. All of the volcanoes are on the Indo-Australian plate.

(2) Intraplate volcanic areas are widespread in the South Island of New Zealand and are mainly on the Pacific plate (Fig. 1.1.8). The largest volcanic centres are on the southeastern coast — on Banks Peninsula and in the Dunedin area — and smaller areas are found inland southwest of Marlborough, through Canterbury, to Geraldine, as well as at the coast at Timaru.

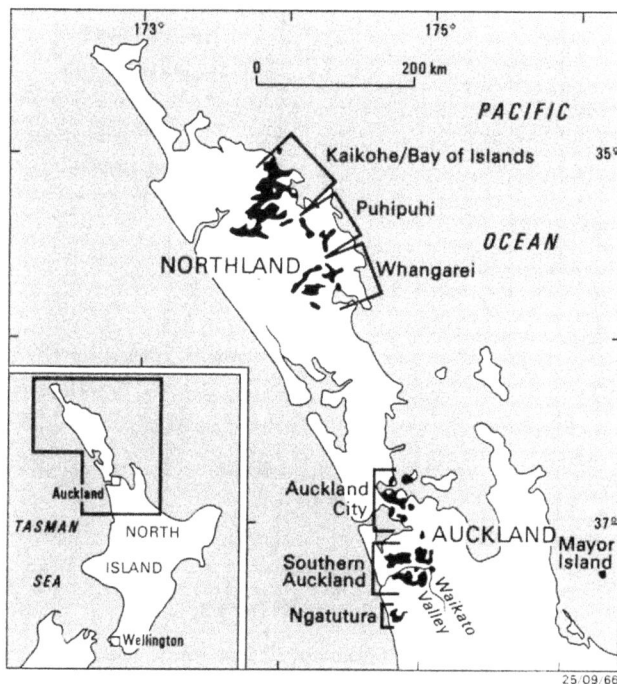

Figure 1.1.7 Intraplate volcanic areas of North Island, New Zealand, divided into provinces as described in Section 4.2.

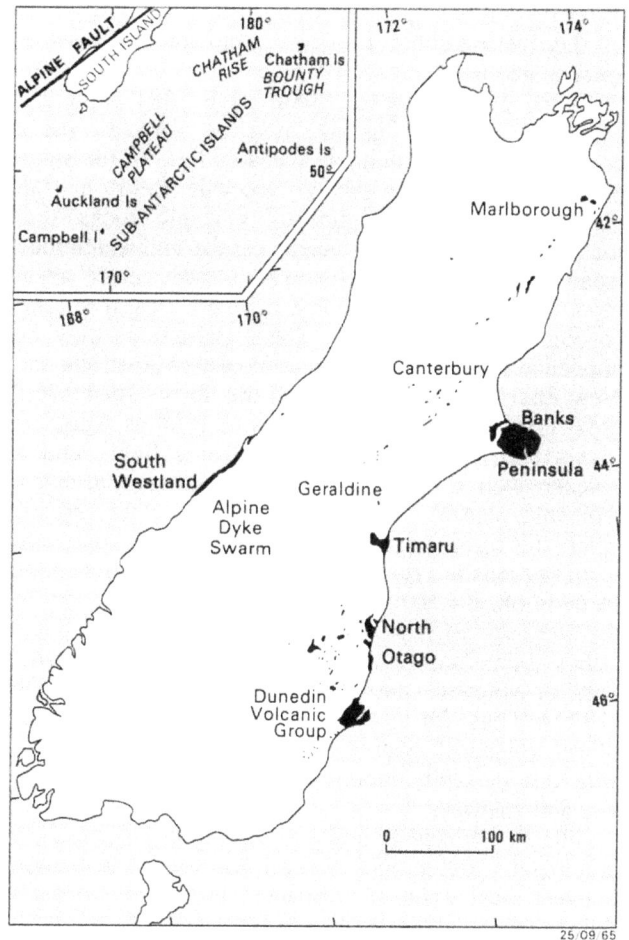

Figure 1.1.8 Intraplate volcanic areas of South Island, and the Sub-Antarctic and Chatham islands (see also Fig. 1.3.7), divided into provinces as described in Sections 4.3–4.

Intraplate volcanic rocks represented by the Alpine Dyke Swarm and in the South Westland area are within only a few kilometres of the active Alpine Fault, which represents the Indo-Australian/Pacific plate boundary in South Island.

(3) The Sub-Antarctic Islands to the southeast of South Island are also on the Pacific plate (Fig. 1.1.8). They consist of the Auckland, Campbell, and Antipodes islands on the Campbell Plateau. The Chatham Islands are to the north of the Sub-Antarctic Islands and rise from Chatham Rise. Campbell Plateau and the Chatham Rise are underlain by continental crust (Section 1.5).

Provinces

The term 'province' has been used widely in the literature on east-Australian and New Zealand intraplate volcanic regions. For example, Wellman & McDougall (1974a) referred to central-volcano provinces, lava-field provinces, and high-potassium-mafic provinces. However, a consistent definition of 'province' is difficult to find in the volcanological literature because it has been used and defined in different ways (see, for example, Rock, 1981). All the intraplate volcanic areas of eastern Australia and New Zealand can be regarded, at one extreme, as a single province, even though different, widely spaced, volcanic areas may have distinctly different ages and rock types. This corresponds to the 'petrographic province' concept of classical igneous geology.

Fisher & Schmincke (1984, page 368) defined a 'volcanic province' as an area of volcanic rocks consisting of more than one volcanic field of the same or different time spans; a 'volcanic field' as 'an association of consanguineous volcanic rocks that includes more than one volcanic centre' and which may 'span long periods of time'; and a 'volcanic centre' as an 'area that includes one source or many closely spaced sources of volcanic rock'. These definitions may be applied to several volcanic areas in eastern Australia and New Zealand. For example, all the north-Queensland areas comprise a 'province', an area such as Atherton in northern Queensland is a 'volcanic field', and a volcano such as Lake Eacham in the Atherton area is a 'volcanic centre'. However, there are other regions where this three-stage hierarchy is difficult to apply and is an unrealistic representation of actual relationships. In addition, the term 'province' is used extensively in the literature on east-Australian intraplate areas, but it tallies more closely with the definitions of 'volcanic field' or 'volcanic centre' given by Fisher & Schmincke (1984). Thus, for example, Warrumbungle and Nandewar volcanoes each have been termed 'provinces', which is certainly inappropriate as both are clearly volcanic centres.

The Fisher-and-Schmincke definition has much to commend it, but in order that no confusion is generated between past usage of 'province' in relation to east-Australian volcanic areas, we have adopted the term loosely in this volume simply as a convenient grouping of rocks in a single area that is sufficiently different in petrology or age (or both) from neighbouring groups to be regarded as geologically distinctive. This is a pragmatic compromise rather than a desirable definition. The 'convenient groupings' or provinces are listed by name as Sections in the Contents for Chapter 3 (for example: 3.2.2 Maer; 3.5.2 Nandewar), although in some cases two or more provinces are described for convenience in a single Section (for example, 3.3.7. Buckland and Mitchell).

1.1.4   Rock classification and analytical data bases

Rock names in the foregoing sections are used in a general sense, and there is a need to define more precisely the rock names used in the remainder of this volume. The matter of rock nomenclature was discussed at length by the petrologists and geochemists of the ILP Working Group, but no unanimous agreement could be found on the best scheme to be used. However, most considered that the classification of alkaline rocks proposed by Coombs & Wilkinson (1969) should be adopted as a starting point because of its wide usage in Australia and New Zealand. The value of adopting the newly developed TAS system (for example, Le Maitre, 1984) was discussed, but most preferred not to use it mainly because it was a largely untried system when the volume was being planned in 1982–3.

The Coombs-and-Wilkinson classification is based on a plot of Differentiation Index (DI; Thornton &

Tuttle, 1960), which is the sum of the CIPW normative minerals quartz (*qz*), orthoclase (*or*), albite (*ab*), nepheline (*ne*), leucite (*lc*), and kalsilite (*ks*), against whole-rock normative-plagioclase content, $100An/(An+Ab)$. A simple grid is constructed on the plot and used to assign rock names (Figs. 1.1.9–10). The main difficulty with this system is the well-known one of sensitivity to the oxidation state of iron used in the CIPW classification. Another is that the use of normative-plagioclase content is not universally popular

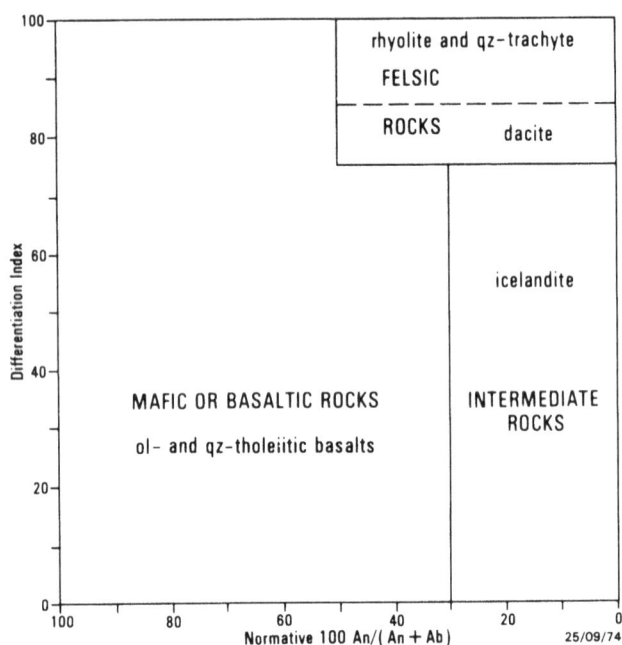

Figure 1.1.9 Classification of subalkaline rocks, where *qz*-tholeiitic basalt is quartz normative, *ol*-tholeiitic basalt has no normative quartz, *qz*-trachyte contains up to 10-percent normative quartz, and rhyolite contains more than 10-percent normative quartz.

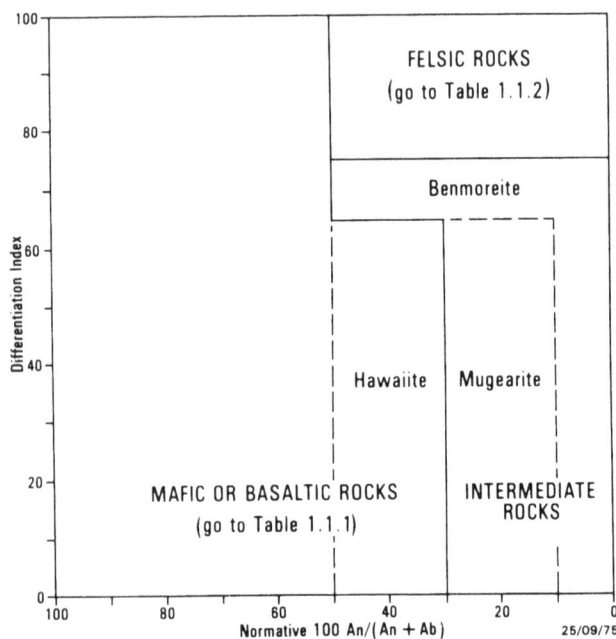

Figure 1.1.10 Classification of alkaline rocks. The prefix '*ne-*' should be used where hawaiite, mugearite, and benmoreite contain 5-percent or more normative nepheline, and the prefix 'K-' may be used where K_2O/Na_2O is 0.5 or greater.

as a fundamental parameter for the classification of igneous rocks, as it can have confusing implications. For example, the term hawaiite was proposed for certain alkaline intermediate rocks (Macdonald, 1960), but many hawaiites in the Coombs-and-Wilkinson scheme, though appearing intermediate because of modest $100An/(An+Ab)$ values, can be quite strongly mafic on the basis of MgO contents and are essentially basaltic (see Fig. 1.1.10).

A further and perhaps critical problem is that the Coombs-and-Wilkinson classification does not accommodate subalkaline rocks — that is, those containing significant amounts of hypersthene in the norm. Naming these rocks caused particular problems. However, a grid for subalkaline rock names was devised that is similar to the one used for alkaline compositions (Figs. 1.1.9). Names such as 'tholeiitic andesite' have been used in the past for quartz-normative intermediate rocks in intraplate environments, but are not favoured here because of the preference for using 'andesite' for rocks from arc-trench systems rather than those from intraplate areas. The term 'icelandite' (Carmichael, 1964), although not entirely satisfactory, is used therefore as an alternative (see below). A cut-off at a $100An/(An+Ab)$ value of 30 is chosen for the boundary between qz-tholeiitic basalt (mafic) and icelandite (intermediate) in order to conform with the hawaiite/mugearite boundary adopted for the alkaline rocks. However, some rocks that are clearly intermediate in terms of their SiO_2 contents classify as qz-tholeiitic basalts (see, for example, Section 4.6.3). The term 'basaltic icelandite' would be appropriate for such rocks (R.C. Price, personal communication, 1984), but was not adopted by the Working Group because it is not used widely in the literature.

The following procedure was adopted on the basis of majority opinion. However, we stress that some Working Group members disagreed with some of the definitions and naming of certain rock compositions, and even with the adoption of the DI-versus-$100An/(An+Ab)$ plot as the basis of the classification scheme. The procedure has four steps.

(1) Standardise the oxidation state of Fe and recalculate totals: Fe_2O_3/FeO values are set equal to 0.20 for all rocks; and the sum of the following eleven oxides (only) is recalculated to 100 percent — SiO_2, TiO_2, Al_2O_3, Fe_2O_3, FeO, MnO, MgO, CaO, Na_2O, K_2O, and P_2O_5.

(2) Calculate the CIPW norm.

(3) Plot normative compositions in the grid shown in Figure 1.1.9, but only if the rocks have *both* of the following characteristics: (1) normative quartz *or* more than 10-percent normative hypersthene; (2) molecular Na_2O+K_2O is less than molecular Al_2O_3 — that is, the rocks are non-peralkaline. Then use the names shown in lower-case lettering on Figure 1.1.9 and in the caption.

(4) Plot normative compositions in the grid shown in Figure 1.1.10 if the conditions outlined in step 3 do not apply. Then go to the appropriate Tables 1.1.1–2, where necessary. Some strongly undersaturated rocks such as nephelinite may have neither albite nor anorthite, in which case $100An/(An+Ab)$ is zero. However, these rocks are clearly basaltic and are

Table 1.1.1 Names of mafic (or basaltic) rocks

Transitional basalt	Up to 10-percent hypersthene.
Alkali basalt	Up to 5-percent nepheline.
Basanite	More than 5-percent nepheline.
Hawaiite	As in Figure 1.1.10.
Nephelinite	More than 5-percent normative nepheline and no, or less than, 5-percent normative albite, and has normative nepheline in excess of normative leucite.
Leucitite, melilitite, and analcimite	These strongly undersaturated rocks are uncommon and are sufficiently distinctive both petrographically and chemically that field boundaries need not be defined for them for the purposes of this volume.

Table 1.1.2 Names of felsic rocks

Peralkaline rhyolite	More than 10-percent normative quartz. The terms comendite and pantellerite are not used.
Peralkaline qz-trachyte	Up to 10-percent normative quartz.
Ne-trachyte	Up to 10-percent normative nepheline.
Phonolite	More than 10-percent normative nepheline. 'Peralkaline' should be used as a prefix for ne-trachyte and phonolite where molecular Na_2O+K_2O is equal to, or greater than, molecular Al_2O_3.

plotted on the left-hand vertical axis of the grid rather than in the benmoreite field.

Rocks in steps 3 and 4 are classified initially as mafic (or basaltic), intermediate, or felsic.

Two collections of chemical analyses of east-Australian volcanic rocks were used during the course of producing this volume. One was developed by A. Ewart at the University of Queensland and forms the basis of Chapter 5. The second collection is a computer database of analyses at BMR, which was used to produce the chemical variation diagrams presented in Chapter 3. Both collections of data are essentially the same. Copies of the BMR data set may be obtained by writing to the Director of BMR.

The expressions 'Mg-ratio' and 'mg-ratio' are used throughout the volume. Mg-ratio refers to $100Mg/(Mg+Fe^{2+})$ where Fe^{2+} (atomic) is adjusted on the basis of Fe_2O_3/FeO equal to 0.2. The term 'mg-ratio', on the other hand, refers to $100Mg/(Mg+total\text{-}Fe)$ where 'total-Fe' is *all* the Fe in the rock or mineral expressed as Fe^{2+} (atomic).

1.2 Plate Tectonic Setting

1.2.1 Introduction

The tectonic events that changed the boundaries of the eastern part of the Indo-Australian plate (Fig. 1.2.1) from late Cretaceous time to the present are reviewed in this section. Forces exerted at plate boundaries such as ridge push and trench pull are likely to affect the stress regime within plates (Cloetingh & Wortel, 1985; Richardson et al., 1979), and intraplate volcanic activity may take place where sufficiently high tensional stresses are developed within the plate (Pilger, 1982; see also Section 1.6.1). The timing of changes at the boundaries of the Indo-Australian plate are compared with the age and distribution of Cainozoic intraplate volcanism in Section 1.7.

1.2.2   Regional tectonics

Australia, New Zealand, and Antarctica were grouped together in eastern Gondwanaland in late Mesozoic time, and occupied a high southern latitudinal position. South America, Africa, Madagascar, and India, to the west had begun to separate and drift apart between about 160 and 120 Ma ago (Norton & Sclater, 1979; Veevers, 1984). The eastern and northern margins of eastern Gondwanaland at this time were dominated by collision with oceanic plates spawned in the Pacific basin. In fact, the fold belts of eastern Australia and New Zealand may record a series of subduction episodes from Palaeozoic through Mesozoic time (Crook & Feary, 1982; Veevers, 1984; Section 1.3.1). Portions of the Mesozoic convergent margins are preserved today in Papua New Guinea, New Caledonia, and New Zealand. This period of predominantly compressional tectonism was followed in late Cretaceous time by a series of extensional events that left rifts and small marginal basins along the northern and eastern margin of the Indo-Australian plate (Falvey & Mutter, 1981; Fig. 1.2.1). Extension also produced large marginal basins such as the Tasman Sea and the South Fiji Basin. A convergent boundary remained between the Australian and Pacific-basin plates during this period.

Australia and Antarctica began to separate about 95 Ma ago (Cande & Mutter, 1982; Veevers, 1984). The initial separation, from 95 to 44 Ma, was extremely slow (about 6 mm/yr), but then accelerated to average about 58 mm/yr from 44 Ma ago to the present. Spreading between southern New Zealand, including the Campbell Plateau, and Antarctica began about the same time as the Australia-Antarctica separation. The oldest identified marine magnetic anomaly (A34, about 84 Ma) is close to the southeastern edge of the Campbell Plateau (Molnar et al., 1975), and this spreading continues today at the Pacific-Antarctic ridge.

The Tasman Sea also began opening at about this same time (about 95 Ma) as the Lord Howe Rise separated from eastern Australia (Weissel & Hayes, 1977; Veevers, 1984; Fig. 1.2.1). This is marked by some igneous activity adjacent to the southeastern coast of Australia (96–90 Ma; McDougall & Roksandic, 1974; Embleton et al., 1985) and on the Lord Howe Rise (96 Ma; McDougall & Van der Lingen, 1974). Fission-track cooling ages on apatite (Morley et al., 1981), vitrinite reflectance (Shibaoka & Bennett, 1976), and palaeomagnetic overprinting (Schmidt & Embleton, 1981) have been taken as evidence for uplift between about 100 and 70 Ma. Again, spreading began only slowly at about 95 Ma ago. The major period of opening and generation of new oceanic crust took place

between about 84 and 53 Ma (Weissel & Hayes, 1977).

Other marginal basins, such as the New Caledonia and D'Entrecasteaux basins (Kroenke & Rodda, 1984; Lapouille, 1982) probably formed at the same time. The Coral Sea opened between about 63 and 53 Ma ago (Weissel & Watts, 1979), possibly as a northward extension of the Tasman Sea spreading. The Loyalty Basin developed between about 53 and 42 Ma (Maillet et al., 1982), and spreading in the South Fiji Basin took place between 36 and 26 Ma (Malahoff et al., 1982). Finally, the northern- and eastern-most marginal basins — Woodlark, North Fiji, and Lau — have been forming since the Pliocene (Kroenke & Rodda, 1984). Thus, the eastern margin of the Indo-Australian plate has moved into the Pacific basin as an ever-expanding series of arcs and marginal basins whose ages decrease away from the continent, as proposed by Carey (1938) and Glaessner (1950).

The northern boundary of the Indo-Australian plate is complex, but essentially dominated by convergence and sinistral shear relative to the Pacific plate, involving several microplates and lithospheric subduction. Convergence to the east with the Pacific plate is marked by more straightforward westward subduction beneath the Tonga-Kermadec volcanic arc and the North Island of New Zealand. Convergence and dextral shear between the two plates characterise the boundary in southern New Zealand and south along the Macquarie Ridge to the spreading ridge which now is further separating Australia from Antarctica. Most of southern New Zealand, together with the Campbell Plateau and Chatham Rise are part of the Pacific plate.

The timing of these events may be compared (Fig. 1.2.2) with the timing of intraplate volcanism on the Indo-Australian plate. Lava-field volcanic activity took place throughout the east-Australian highlands beginning about 70 Ma ago, and the main eruptive pulse was between about 55 and 30 Ma ago (Wellman & McDougall, 1974a; Sections 1.1.3 and 1.7). There was relatively little lava-field volcanism between about 30 and 5 Ma ago when a resurgence of activity took place. This volcanism does not appear to correlate readily with changes in plate-boundary forces during the same period. The earliest central-volcano activity in eastern Australia began at about 34 Ma and progressed southwards up to the present day. A similar trend is recorded in the parallel seamount chains of the Tasman Sea (Section 1.7.2). These linear, age-progressive, volcanic provinces most likely correlate with the onset of rapid spreading between Australia and Antarctica, as the timing is similar.

1.2.3   Plate motions relative to the mantle

Some of the models proposed for the origin of the east-Australian Cainozoic intraplate volcanism involve motion of the Indo-Australian plate relative to the mantle (see Section 1.7.3). The volcanism in these models is considered to be associated with hotspots or other thermal perturbations, regarded as fixed within the mantle, beneath the moving plate. Calculations of the hypothetical motion of Australia over the mantle during the last 100 Ma now can be made. The only assumption in this analysis is that hotspots — identified by age-progressive, linear volcanism — do not move relative to one another, and therefore provide a reference frame for plate motions. The calculated motion (both direction and velocity) for the Indo-

Australian plate then can be compared directly with the distribution of intraplate volcanism.

The single assumption that the hotspots are fixed applies to the world-wide distribution of hotspots, so that if the motion of one plate relative to this reference

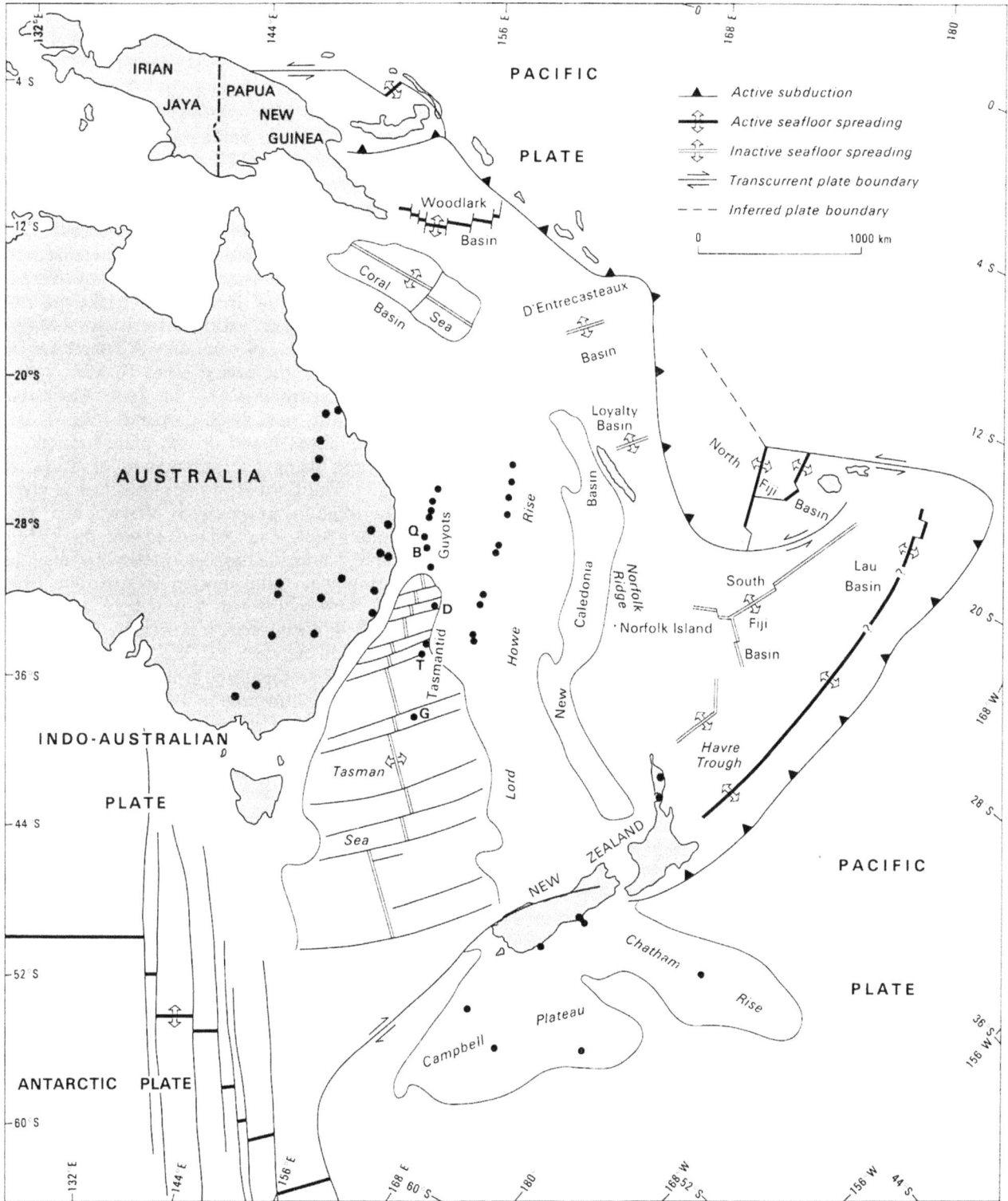

Figure 1.2.1 Generalised plate-tectonic features of the eastern part of the Indo-Australian plate. Subduction and transcurrent motions dominate the plate margin in the north and east, whereas spreading-ridge segments form the southern plate margin. Filled circles represent main intraplate volcanoes in the Tasman Sea and New Zealand region but central-type and leucitite-suite volcanoes only in eastern Australia (lava-field provinces are not shown; see Fig. 1.1.6). Abbreviations for Tasmantid Guyots are: G, Gascoyne; T, Taupo; D, Derwent Hunter; B, Britannia; Q, Queensland.

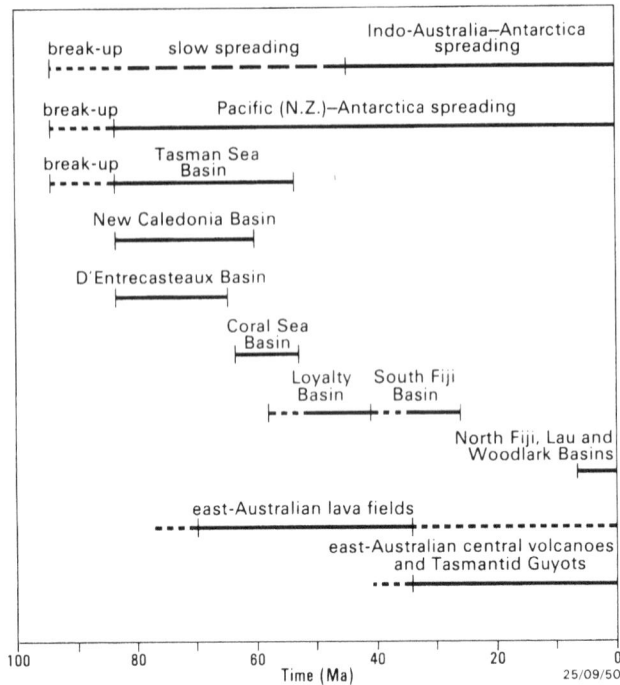

Figure 1.2.2 Plate-tectonic events on the eastern part of the Indo-Australian plate (compare with Fig. 1.2.1).

The motion of the Indo-Australian plate relative to the fixed hotspots was calculated for the last 100 Ma (Table 1.2.1) using two published sets of relative motion between Australia and Antarctica (Weissel et al., 1977; Stock & Molnar, 1982). The Antarctic plate motion is from Duncan (1981). Stage poles and cumulative rotations are listed for 5 Ma intervals.

Differences in velocity and direction of motion of the Indo-Australian plate during this period were calculated for a point (Canberra) in eastern Australia (Fig. 1.2.3). The two published relative motions give similar results. The Indo-Australian plate moved slowly northwards from 100 Ma to 80 Ma as spreading began away from Antarctica. Then, between 80 and 55 Ma, the plate moved slowly *southwards*. Significantly, this also corresponds to the time of spreading in the Tasman Sea. Australia moved slowly to the northeast from 55 Ma to 40 Ma, and then gradually increased its northward velocity. The plate maintained an average velocity of about 60 mm/yr from 40 Ma to the present, switching from a slightly northwesterly to a slight northeasterly direction of motion. A decrease in velocity took place in the period 20 to 10 Ma.

Graphical reconstructions of the Indo-Australian and Antarctic plates have been prepared (Fig. 1.2.4) for 20 Ma time slices based on the plate motions in Table 1.2.1, and using the parameters of Stock & Molnar (1982). The Lord Howe Rise position is given by adding the rotations proposed by Weissel & Hayes (1977). Southern New Zealand moved with the Pacific plate. The same features of the plate-motion calculations are seen: virtually no total motion of Australia from 100 to 80 Ma; slow southwards motion between 80 and 60 Ma; then slow northwards followed by more rapid northwards motion between 60 Ma and the present. The Lord Howe Rise separated from Australia during the period of slow southwards motion and then moved northwards as part of the Indo-Australian plate. Southern New Zealand, including the Campbell Plateau, moved north-northwestwards out of the Ross Sea area

frame can be described, then all other plate motions can be calculated by adding the appropriate relative motions to the starting plate. The African plate was chosen as the starting plate because it has a large number of well-defined hotspot tracks and a clear pattern of seafloor spreading around it. Duncan (1981) and Morgan (1981) each proposed similar motions for the African plate over hotspots beneath it. The motion of the Antarctic plate is found by adding Africa-Antarctica relative motion to the African plate motion. The motion of the Indo-Australian plate then follows by adding Indo-Australia/Antarctica relative motion to the motion of the Antarctic plate.

Table 1.2.1   Absolute motion of the Indo-Australian plate, 100 Ma to present[1]

Period (Ma)	Stage pole			Cumulative rotation		
	Lat (°N)	Long (°E)	Angle (°CCW)[2]	Lat (°N)	Long (°E)	Angle (°CCW)[2]
0–5	23.30	37.40	3.52			
5–10	23.30	37.40	3.52	23.30	37.40	7.04
10–15	27.08	27.90	2.86	24.61	34.78	9.87
15–20	27.05	27.74	2.86	25.29	33.28	12.71
20–25	17.55	21.49	3.76	23.93	30.26	16.39
25–30	17.62	21.31	3.76	23.04	28.37	20.10
30–35	17.83	21.25	3.77	22.41	27.09	23.83
35–40	17.99	35.77	2.72	21.80	27.87	26.52
40–45	8.75	66.41	1.50	20.90	29.83	27.70
45–50	4.02	−79.12	−1.31	19.78	32.26	28.09
50–55	0.61	−77.63	−1.11	19.01	34.40	28.46
55–60	7.06	−87.98	1.40	20.36	32.05	27.84
60–65	5.32	−87.33	1.46	21.81	29.52	27.25
65–70	11.92	−84.62	1.57	23.55	26.64	26.83
70–75	24.18	−73.78	2.00	26.30	22.90	26.93
75–80	25.01	−74.07	2.07	28.98	19.01	27.19
80–100	13.81	41.28	7.51	24.98	23.27	34.11

[1] Calculated from the motion of the Antarctic plate relative to the mantle (Duncan, 1981) and Indo-Australia/Antarctica relative motions (Stock & Molnar, 1982; Cande & Mutter, 1982).
[2] Counter-clockwise rotation where viewed from outside the Earth.

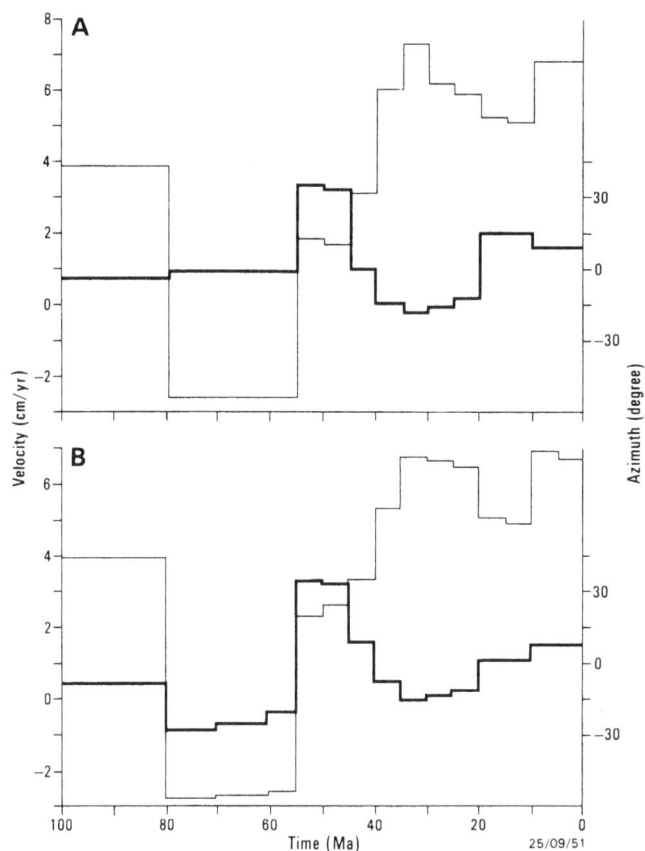

Figure 1.2.3 Changes in azimuth and velocity of the Indo-Australian plate during the last 100 Ma relative to the hotspot reference frame (Table 1.2.1). Azimuth (thick line) and velocity (thin line) were calculated for a point in eastern Australia (Canberra) from rotation poles (5 Ma intervals) for two models of Indo-Australia/Antarctic relative motion (Weissel et al., 1977; Stock & Molnar, 1982). Similar features are shown for both models.

Figure 1.2.4 Plate reconstructions for Australia, Antarctica, and southern New Zealand in the hotspot reference frame for 20 Ma intervals. Rotations used are given in Table 1.2.1.

between 100 and 43 Ma (western Antarctica was farther south at this time; see Duncan, 1981) and then west-northwestwards between 43 Ma and the present, as prescribed by Pacific plate motion relative to the mantle (Duncan & Clague, 1985).

A further advantage of the mantle-fixed reference frame in examining the plate-tectonic evolution of the Indo-Australian plate is in its application to calculating convergence vectors at plate boundaries. In other words, the direction and velocity of convergence between the Indo-Australian plate and oceanic plates in the Pacific basin during the last 100 Ma can be quantified. The only assumption, again, is that hotspots beneath the Indo-Australian and Pacific plates remained stationary. A formal analysis of this convergence history is beyond the scope of this volume, but would follow similar studies of plate convergence in the eastern Pacific (Wells et al., 1984; Duncan & Hargraves, 1984). An accurate model would depend upon the orientation of the plate margin and on identification of any small oceanic plates in the southwestern Pacific that may have been subducted beneath the Indo-Australian plate. These are possible complications, but the general elements of Indo-Australia/Pacific plate relative motion in the mantle-fixed reference frame are as follows:

(1) The Indo-Australian plate moved northwards from 100 to 80 Ma at about 40 mm/yr. An oceanic plate (called the Phoenix plate) existed to the north and east of Australia at this time (Hilde et al., 1976), and its motion was south-southwestwards away from the northward-moving Pacific plate. Thus, depending on the orientation of the margin of the Indo-Australian plate, the boundary was convergent to obliquely convergent and included a large component of dextral shear (southwards transport along the margin).

(2) Australia moved slowly southwards (about 30 mm/yr) from 80 to 55 Ma as the Tasman and Coral seas opened. The Pacific plate lay to the north and east, and moved north-northwestwards at about 60 mm/yr (Duncan & Clague, 1985). Oblique subduction in the east therefore took place with sinistral shear and northward transport along the plate margin while the opening of the Coral Sea and formation of the D'Entrecasteaux Basin were taking place along the northern margin.

(3) Australia moved slowly northeastwards (20–30 mm/yr) between 55 and 43 Ma as the Pacific plate continued moving northwestward (about 50 mm/yr). Nearly perpendicular convergence along the eastern margin therefore took place at a moderate rate (about 40 mm/yr), while along the northern margin there was a significant component of sinistral shear.

(4) There was a major change in Pacific plate motion at 43 Ma. The velocity of the plate increased and its direction became more east-west. Australia at the same time began to move more rapidly northward (about 60 mm/yr) with the result that the eastern margin underwent rapid east-west convergence (about 70 mm/yr). Northeast-southwest convergence along the northern margin may have allowed some sinistral shear and westward transport.

Palaeomagnetic measurements are also a record of the northward motion of the Australian plate which is

independent of the estimates based on the hotspot
reference frame. Idnurm (1985a) recently revised the
Australian palaeomagnetic polar wander path for late
Mesozoic and Cainozoic time and the new results (Fig.
1.2.5) are now compatible with the Indian polar-
wander path for the same period.

The new palaeomagnetic data correspond to a more
southerly position for the Indo-Australian plate throughout the Cainozoic compared to that derived from
hotspot tracks (Idnurm, 1985b). Palaeomagnetic data,
of course, correspond to motion relative to the geomagnetic pole, which is thought to coincide (on
average) with the Earth's spin axis, whereas the
hotspot tracks are a record of motion relative to the
mantle. The discrepancy between the palaeomagnetic
and hotspot reference frames can be resolved in three
ways: (1) departures of the geomagnetic field from
the geocentric axial dipole model; (2) random but
significant motions of the hotspots; and (3) true polar
wander.

The first of these alternatives would violate the
central assumption of the palaeomagnetic method and
there is no evidence that this is the case. The second is

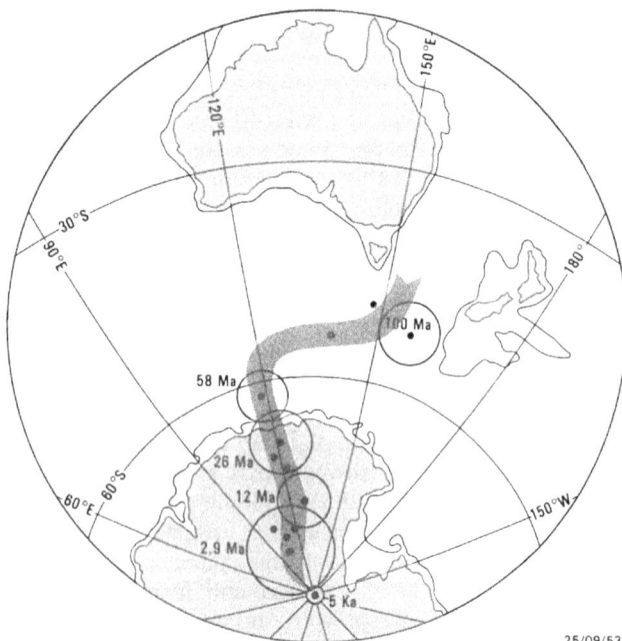

Figure 1.2.5 Late Mesozoic and Cainozoic apparent polar-
wander path for Australia (after Idnurm, 1985a). Circles
represent 95-percent uncertainty for dated poles shown as dots at
the centres of the circles. Other dots represent undated poles
whose uncertainty circles are less than 4 degrees. Poles shown
were measured on dated marine sediments (Idnurm, 1985a) and
pre-Cainozoic rocks remagnetised by weathering or heating at an
undated time during the late Cretaceous and Cainozoic. Poles
derived from dated volcanic rocks (McElhinny et al., 1974) in
continuous sections are not used here to determine polar wander
because they are considered (Idnurm, 1985a) to have had larger
errors than those published (owing to either a quiescent
geomagnetic field or small time intervals between some flows).
The palaeomagnetic poles from dated marine sediments yield
palaeolatitude changes for Australia that differ from those
derived from hotspots or seafloor-spreading data. Idnurm (1985b)
showed that part of this discrepancy is caused by significant non-
dipole components in the geomagnetic field throughout most of
the Cainozoic, and part because of a small net true polar wander
caused by either a net drift of the lithosphere plates or differential
movement between the hotspot and mantle reference frames.

unacceptable because of the lack of significant motion
between hotspots (Duncan, 1981; Morgan, 1981).
True polar wander, or motion of the Earth relative to its
spin axis, has been concluded from similar comparisons
on other plates, and is considered to be the likely
explanation for the discrepancy noted by Idnurm
(1985b). The direction and magnitude of Cainozoic
true polar wander determined from the Indo-Australian
plate are consistent with results from other parts of the
world. This complexity does not detract from the
usefulness of the hotspot tracks as records of plate
motion over the mantle.

The plate motions discussed in this section are used
in Section 1.7.3 to evaluate the different origins
proposed previously for intraplate volcanism. First,
however, summary descriptions are given of the east-
Australian and New Zealand lithosphere.

1.3   Geological Framework

1.3.1   Geology of eastern Australia

Introduction

Cainozoic intraplate volcanism in eastern Australia
developed on the composite Tasman Fold Belt system
and on Proterozoic blocks in northern Queensland
(Georgetown and Coen inliers) and Tasmania. The
Tasman Fold Belt system occupied part of the eastern
margin of Gondwanaland and is thought to have formed
by interaction with the Ur (Proterozoic) and Palaeo-
Pacific plates. Its age span is late Proterozoic to
Cretaceous.

The main elements of the Tasman Fold Belt system
(Fig. 1.3.1) are: (1) the Kanmantoo Fold Belt in South
Australia, western Victoria, and New South Wales; (2)
the Lachlan Fold Belt in Victoria, New South Wales,
and Tasmania; (3) the Thompson Fold Belt in Queens-
land, which is separated from the Lachlan Fold Belt by
a curvilinear junction near the New South Wales and
Queensland border (Murray, 1986), and shown on the
gravity-trend map of Wellman (1976a); (4) the
Hodgkinson/Broken River Fold Belt in northern
Queensland; and (5) the New England Fold Belt in
Queensland and New South Wales, which is separated
from the Lachlan and Thompson Fold Belts by the
Sydney-Bowen Basin.

Principal geological features of the Precambrian and
of the fold belts of eastern Australia are summarised
here as a basis for understanding the nature of the
continental crust through which the largely Cainozoic
intraplate magmas were erupted.

Precambrian

The oldest elements in east-Australian geology are the
Georgetown and Coen inliers in northern Queensland.
They are separated from the Hodgkinson/Broken River
Fold Belt by the Palmerville and Burdekin River
faults.

The oldest rocks in the Georgetown Inlier are a
westward-younging sequence of early Proterozoic
metasedimentary and minor metabasaltic rocks, the
Etheridge Group (Withnall et al., 1980). The Etheridge

Group was deformed and metamorphosed at about 1570 Ma and again at 1470 Ma (Black et al., 1979), and ranges in grade from lower-greenschist facies in the west to granulite facies in the east. Extensive S-type syntectonic granites intruded the amphibolite-facies rocks during the 1470 Ma event, and large I-type

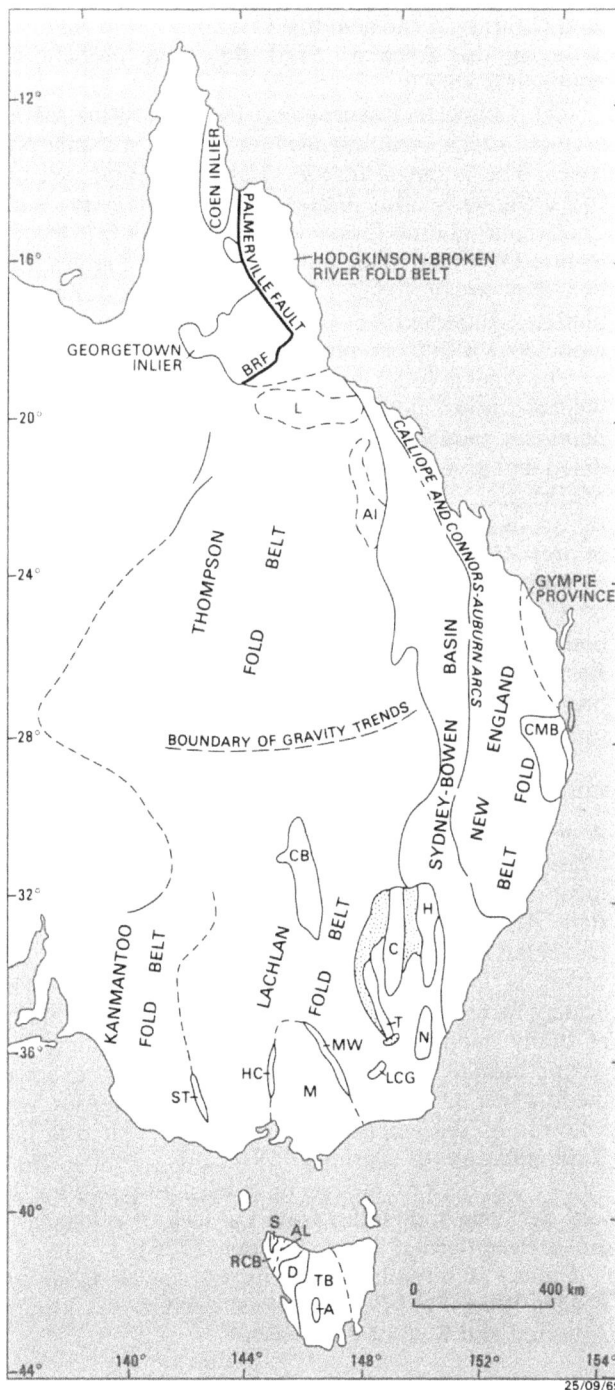

Figure 1.3.1 Pre-Mesozoic elements of east-Australian geology. BRF: Burdekin River Fault. L: Lolworth-Ravenswood Block. AI: Anakie Inlier. CMB: Clarence-Moreton Basin. CB: Cobar Basin. Stippling represents the Molong Volcanic Belt. C: Cowra Trough. H: Hill End Trough. T: Tumut Trough. N: Ngunawal Basin. LCG: Limestone Creek Graben. MW: Mount Wellington Greenstone Belt. M: Melbourne Trough. HC: Heathcote Greenstone Belt. ST: Stavely Greenstone Belt. S: Smithton Trough. AL: Arthur Lineament. RCB: Rocky Cape Block. D: Dundas Trough. TB: Tyennan Block. A: Adamsfield Trough.

granite batholiths were emplaced in the eastern half of the region about 400 Ma ago. They and the Proterozoic rocks are in places overlain by fluviatile sedimentary rocks of late Devonian to Carboniferous age, and Carboniferous to Permian cauldron-subsidence related felsic volcanic rocks (Branch, 1966; Oversby et al., 1980). The Coen Inlier contains a similar stratigraphic and metamorphic sequence to the Etheridge Group in the Georgetown Inlier (Willmott et al., 1973). It also contains granitic rocks similar to the Precambrian S-types of the Georgetown Inlier (Withnall et al., 1980). Cainozoic basaltic volcanism is abundant in the eastern part of the Georgetown Inlier close to its boundary with the Hodgkinson/Broken River Fold Belt (Section 3.2). Crustal weaknesses along this boundary may have controlled the positions of centres of eruption.

Two blocks of Precambrian sedimentary rocks dominate the geology of western Tasmania (Collins & Williams, 1986). The Tyennan Block occupies most of the central highlands of Tasmania and contains mainly quartzites and mica schists with deformation dated at about 800 Ma (Raheim & Compston, 1977). The Rocky Cape Block in northwestern Tasmania is divided into two parts by a northeast-trending metamorphic belt, the Arthur Lineament (Banks, 1965). The Rocky Cape Group west of the lineament consists of gently folded supermature orthoquartzites and laminated mudstones. East of the lineament are more complexly deformed greywackes and rare mafic rocks.

Eocambrian and Cambrian sediments and volcanic rocks unconformably overlie the Precambrian rocks of western Tasmania and have collected in a number of supposed distinct troughs (Dundas, Smithton, and Adamsfield troughs). Late Cambrian movements along faults produced local unconformities, but the Cambrian sequences are mostly structurally conformable with younger Ordovician to early Devonian shelf deposits. The rocks of western Tasmania were deformed during the middle Devonian and intruded by S- and A-type granites. The presence of late Proterozoic basement for the Lachlan Fold Belt is inferred from isotopic data for granitic rocks in the belt (see below). However, no such evidence exists for the New England Fold Belt. Tertiary intraplate volcanism was developed extensively in western Tasmania, particularly in the north, where the Dundas Trough appears to be a focus for magmatism (Section 3.8).

Kanmantoo Fold Belt

The Kanmantoo Fold Belt consists of turbiditic strata, some volcanic rocks, and shallow-water sediments. These were deformed, regionally metamorphosed, and intruded by granites near the end of the Cambrian during the Delamerian Orogeny. The western third of the Newer Volcanics (Section 3.7.4) were extruded onto the Kanmantoo Fold Belt, but there is no evidence of basement structures controlling volcanism.

Lachlan Fold Belt

The Lachlan Fold Belt is a complex orogenic belt that has a dominantly meridional structural grain and in Victoria is over 700 km wide. Sediments within the Lachlan Fold Belt range in age from Cambrian to

Carboniferous. The most recent models for the tectonic development of the fold belt (Crawford et al., 1984; Scheibner, 1985; Degeling et al., 1986) involve the rifting of Proterozoic continental crust, formation of backarc basins, subsequent closure of those basins, and collision of the rifted blocks giving rise to a number of orogenies or fold episodes.

The Cambrian rocks of the Lachlan Fold Belt are mafic to intermediate in composition (Crawford et al., 1984). However, gneisses and migmatites of the Gundowring terrane southeast of Albury, which recently have been called basement to the Ordovician rocks of the Lachlan Fold Belt (Fleming et al., 1985) may represent Proterozoic crystalline basement.

The Cambrian volcanic rocks are overlain by the widespread Ordovician quartz-rich greywacke-slate association (Packham, 1969) in both Victoria and New South Wales. Sedimentation continued throughout the Ordovician, and there was little difference in lithology because of the constant composition of the detrital input to the fold belt from the south and west. Interbedded within the greywackes are submarine volcanic rocks in a broad belt extending from the Upper Murray area on the New South Wales and Victorian border through central New South Wales to Dubbo (Molong Volcanic Arc). Volcanism generally is thought to be associated with contemporaneous westward-dipping subduction (Scheibner, 1985), and rocks on the southern coast of New South Wales possibly represent the trench complex (Packham, 1973; Prendergast, 1987).

The crustal development of the Lachlan Fold Belt changed dramatically at about the end of the Ordovician. High-temperature and low-pressure metamorphism (Vallance, 1967) and crustal anatexis began initially in the belt immediately west of the Molong Volcanic Arc, where uplift associated with the anatexis is known as the Benambran Orogeny. Several middle to late Silurian sedimentary troughs developed in association with the rising of great masses of granitic magma to form batholiths on the highs between the troughs. There is isotopic (Compston & Chappell, 1979) and geochemical (Wyborn & Chappell, 1979) evidence that almost all the granitic magma was derived by melting of continental crustal basement rocks of the Lachlan Fold Belt. Evolved isotopic systems for the granites (McCulloch & Chappell, 1982), and an abundance of inherited zircons of late Precambrian age in I-type granites (Williams et al., 1983) are consistent with the development of much, if not all, of the fold belt on a thin Precambrian crystalline basement.

Cainozoic intraplate volcanism is widespread in the Lachlan Fold Belt, but not as abundant as in some of the other regions. Major faults are known to control some of the volcanism — for example, the Wedderburn Line (Section 1.6.2) and the Shoalhaven Fault (Wyborn & Owen, 1986). Other faults possibly associated with intraplate volcanism include the Long Plain, Gilmore, and Kiewa faults.

Thompson Fold Belt

The Thompson Fold Belt (Kirkegaard, 1974; Murray & Kirkegaard, 1978) is a poorly known, largely concealed assemblage of rocks of probable early Palaeozoic age exposed in central Queensland. Two areas contain surface exposures — the Lolworth-Ravenswood Block and the Anakie Inlier (Fig. 1.3.1). Isotopic ages in the 500–400 Ma range have been obtained for rocks from drillholes in basement granites beneath the Eromanga Basin in southwestern Queensland (Murray, 1986), so the Thompson Fold Belt is as wide as the Lachlan Fold Belt and probably is continuous with it.

The Lolworth Ravenswood Block contains mainly granitic rocks and includes inliers of metamorphic rocks. The granites intrude sedimentary and volcanic rocks (Seventy Mile Range Group) containing early Ordovician marine fossils (Henderson, 1983). Middle to late Devonian shelf sediments overlie the granites, and they are intruded by Carboniferous to Permian plutons. Intermediate to felsic volcanic rocks are associated with these plutons.

The Anakie Inlier contains greenschist-facies, mafic, volcanic rocks including pillow lavas overlain by a dominant sequence of albite-muscovite-quartz schists dated by a single K-Ar date at 466 Ma (Webb & McDougall, 1968). Massive granodiorite intruding schists in an outlier 50 km west of Springsure has dates of 460–450 Ma (Webb, 1969).

Intraplate volcanism produced extensive areas of Cainozoic basalts along the meridional-trending boundary between the Anakie Inlier and the Bowen Basin. This boundary may be the fundamental structural break between the New England Fold Belt and the rest of the Australian continent.

Hodgkinson/Broken River Fold Belt

A deformed and faulted sequence of sedimentary rocks consisting mainly of Silurian-Devonian turbidite sediments makes up the Hodgkinson/Broken River Fold Belt. Allochthonous blocks of early Silurian to early Devonian limestones are present in the western facies, which is known as the Chillagoe Formation. The eastern facies — the Hodgkinson Formation — consists of thinly bedded turbidites and minor conglomerate, spilitic basalt, and chert. The major orogenic event in the fold belt, at the end of the early Carboniferous, was followed by the emplacement of granites dated at late Carboniferous to Permian (Richards, 1980). Both forearc and backarc models have been proposed for the fold belt, but both suffer from the lack of evidence of arc-derived detrital input (Murray, 1986).

Cainozoic intraplate volcanism in the Hodgkinson/Broken River Fold Belt is concentrated mostly around Atherton and west of Cooktown.

New England Fold Belt

The New England Fold Belt consists of fault-bounded blocks each thought to represent an assemblage caused by several episodes of plate convergence in the middle to late Palaeozoic and early Mesozoic. Its precise relationship with the Lachlan and Thompson fold belts is not clear because the contact mostly is concealed beneath the Sydney-Bowen Basin.

The oldest rocks in the fold belt are Cambro-Ordovician, mafic, volcaniclastic rocks and ocean-floor sediments (Cawood, 1976) which probably represent ocean-floor deposits east of the Lachlan Fold Belt. Late Silurian to middle Devonian volcanic rocks and volcaniclastic sediments and limestones in the Rockhampton area are older than the main arc development in the New England Fold Belt, which took place from middle Devonian to late Carboniferous, and there is general agreement that this arc developed as a continental-margin arc. The arc in the New South Wales section of the fold belt is thought to be buried beneath the Sydney-Bowen Basin. Uplift and deformation of late Carboniferous to early Permian age and followed by intrusion of granitic rocks in the period 280 to 260 Ma, constitute the main orogenic event of the New England Fold Belt.

New England granites all have high ε_{Nd} values (-1 to $+6$) and low initial Sr ratios (0.703–0.706; Hensel et al., 1985) corresponding to young sources. There is no evidence of a Precambrian component to any rocks in the New England Fold Belt including the sediments, so it is a true, continental-margin, accretion domain.

The Gympie province or terrane (Harrington, 1983) to the east of the main part of the New England Fold Belt, contains shallow-water, late Carboniferous to early Triassic sedimentary and volcanic rocks. The belt is thought to have collided with the main part of the New England Fold Belt at the end of the early Triassic (Flood, 1983), producing folding and greenschist-facies metamorphism. Post-tectonic granites intrude the sequence. This province is thought to correlate with similar rocks in New Zealand (Harrington, 1983; Cawood, 1984).

Post-collision rifting at the end of the middle Triassic produced continental basins, including the Clarence-Moreton Basin. The area then acted as a stable cratonic region in the Jurassic, but renewed early Cretaceous granitic and felsic volcanic activity broke out along what is now the Queensland coastal belt between Brisbane and Townsville. Some of this activity produced alkaline complexes (Paine et al., 1970; Harrington & Korsch, 1985a).

Cainozoic intraplate volcanism is widespread in the New England Fold Belt, and its greatest development is southwest of Brisbane and in the Clarence-Moreton Basin. There is an abundance of faulting in the fold belt, including strike-slip displacements proposed along some faults (Harrington & Korsch, 1985b; Murray, 1986), so the presence of the volcanism may be related to the faulting.

Summary

East Australian pre-Cainozoic geology is represented by a complex of tectonic elements dating back to the middle Proterozoic. Late Proterozoic geology is obscured almost completely by the apparent development of early Palaeozoic fold belts on top of the Proterozoic crust. Adequate tectonic models for this early Palaeozoic geology are lacking still, because magmatism consisted mainly of melting and reworking of older basement with little addition of new material from the mantle. In contrast, the late Palaeozoic to Mesozoic development of the region (before the opening

of the Tasman and Coral seas) conforms to a more conventional, continental-margin, plate-tectonic model: arc-trench type magmatism and crustal growth result from abundant additions of newly-formed mantle-derived magma.

1.3.2. Geomorphology and evolution of the eastern highlands

Introduction

The geomorphology of the eastern highlands of Australia is described briefly in this section, and current models for their formation are reviewed. The review is by no means exhaustive. It concentrates more fully on the southeastern highlands in New South Wales and Victoria, because this is the area to which the bulk of published studies relates. Additional discussion on eastern-highlands evolution is included in Section 7.3.

Geomorphology

The eastern highlands are in the far east of the Australian continent (Fig. 1.3.2). North of about latitude 18°S and south of about 26°S the highlands consist of a broad asymmetrical arch that rises gradually from beneath Mesozoic and Cainozoic intracratonic basins west of the highlands (Figs. 1.3.3C–D). The highlands in intervening areas are more symmetrical and their crest is rather more distant from the continental margin (Figs. 1.3.3A–B). This intervening, more symmetrical stretch of highlands coincides offshore with a much wider continental shelf (Fig. 1.3.2; see Section 7.3.3). The continental shelf elsewhere along the eastern margin is very narrow, and is among the narrowest shelves in the world.

Nowhere are elevations of the highlands surface great. They are generally less than 1000 m in Queensland, and less than 2000 m elsewhere, except in the New South Wales and Victorian Alps in the southeastern corner of the continent. Moreover, the surface relief of the highlands tends to be rather low and characterised by extensive upland 'surfaces'. There is no general agreement on the origin of these surfaces (Young, 1977, 1981; Ollier, 1982a; see Section 7.3), although locally they obviously are associated with extensive Tertiary lava fields.

The highlands are not of uniform elevation along their length, but rather are characterised by alternating higher regions separated by lower saddles. These higher areas in the southeast, for example, are associated with the northern, central, and southern tablelands, and are separated by saddles in the headwaters of the Hunter River and around Goulburn (New South Wales). The northern, central, and southern tablelands are formed on New England, Sydney Basin/Lachlan, and Lachlan fold-belt rocks, respectively, and the greatest elevations are on Palaeozoic areas. Indeed, there are areas in the central tablelands where Palaeozoic rocks rise above adjacent Sydney Basin rocks, much as would have been the case in late Sydney Basin (middle Triassic) times (Dulhunty, 1964).

The major relief of the highland belt is associated with a more or less steep fall from the upland plateau surface to a coastal plain of different widths. This coastal escarpment (Figs. 1.3.2-6) appears to be dominantly of erosional origin and presumably is forming by scarp retreat from the original Tasman Sea

breakup fault (Ollier, 1982a). Its age can be established locally by its relationship with basalts found above, below, or draping (Fig. 1.3.6) the escarpment. The escarpment in northern New South Wales and south-eastern Queensland has eroded back into middle Tertiary lava flows on the plateau top (Fig. 1.3.5). Young & McDougall (1982) described Oligocene basalts on the coastal plain and on the plateau surface behind the escarpment in southern New South Wales. They concluded that escarpment erosion at this locality must have pre-dated lava extrusion (see, however, M.C. Brown, 1983).

Westerly drainage of the highlands is by the long, low-gradient rivers of the Murray-Darling system and the Lake Eyre basin, both of which have catchment areas of about 10^6 km². These rivers have been in existence throughout much of the Tertiary at least, delivering sediment to the sedimentary basins of the interior (Macumber, 1978; Woolley, 1978). The easterly drainage consists generally of low-gradient upper and lower reaches, on the plateau top and coastal plain, respectively. These are separated by a steeper

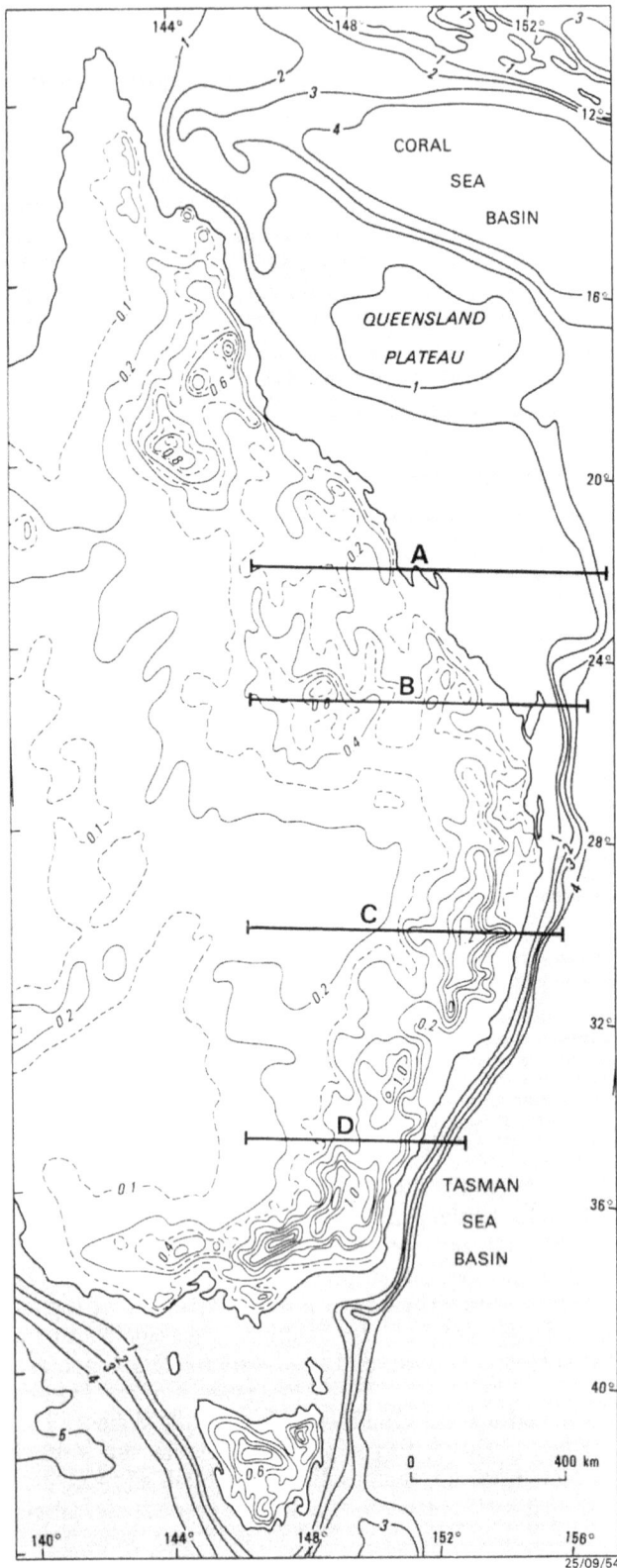

Figure 1.3.2   0.2 km contour intervals (0.1 km dashed) for the highlands of eastern Australia and 1 km isobaths for the Tasman Sea and Southern Ocean, adapted from 11 × 11 km grid derived from gravity-station altitudes (contours smoothed; after Wellman, 1987, fig.1). East-west lines represent cross sections shown in Figure 1.3.3.

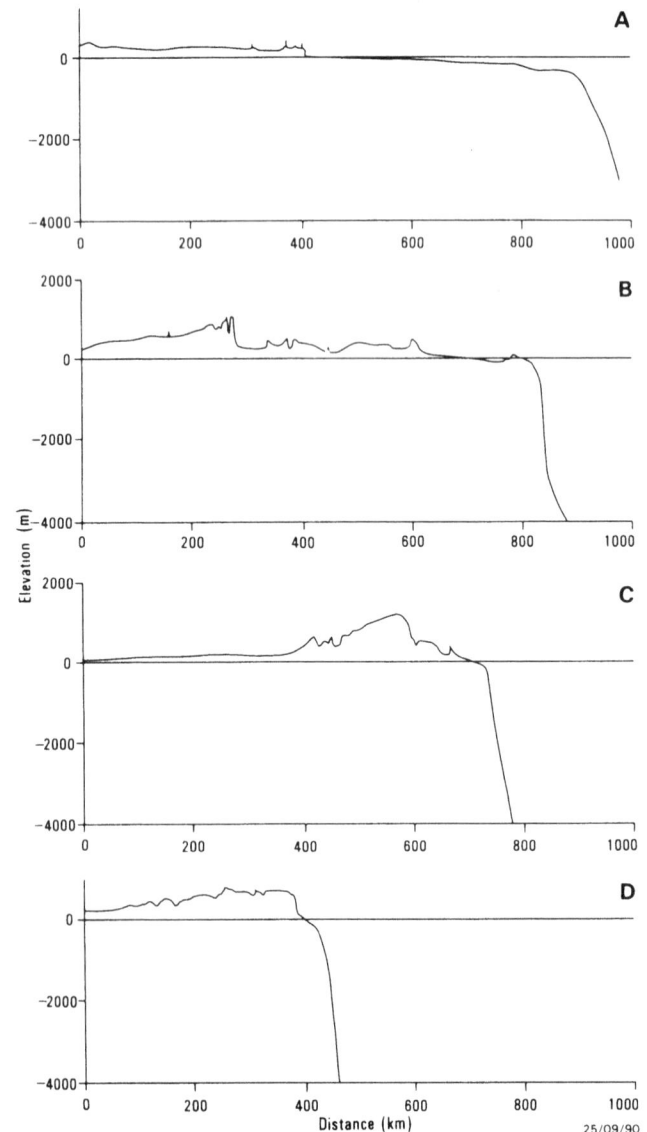

Figure 1.3.3   Cross sections through the eastern highlands along the lines shown in Figure 1.3.2.

fall across the coastal escarpment (Ollier, 1982a; Figs. 1.3.5–6). River systems in central and southern New South Wales probably date from quite early in the Tertiary (Bishop, 1986).

Evolution

Published interpretations up to the early 1970s of the geomorphology and evolution of the eastern highlands were based on the framework of W.M. Davis's Cycle of

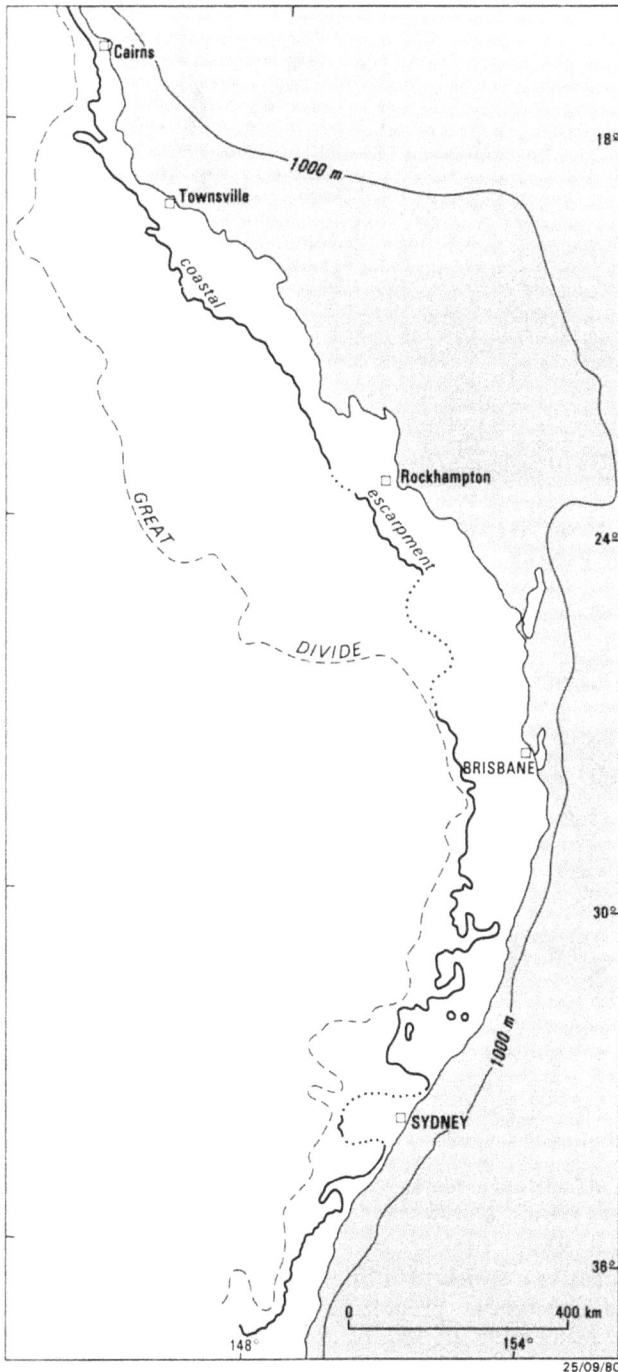

Figure 1.3.4 Part of the coastal escarpment or Great Escarpment (Ollier, 1982a), shown in relation to the Great Divide of eastern Australia (adapted from Ollier, 1982a, fig. 5). Dotted line represents places where the escarpment is absent or obscure. The thin continuous line represents the 1000 m isobath that approximates the edge of the continental shelf.

Erosion (Browne, 1969; see also the summary by P. Wellman, 1979b, fig. 2). These interpretations emphasised very recent (late Pliocene or Pleistocene) uplift of multiple erosion surfaces ('peneplains'), some of which were thought to be duricrusted. Dissatisfaction with the cyclical interpretations of highlands evolution can be related to several factors, including (1) the early results of potassium-argon (K-Ar) dating of Tertiary lavas in the highlands (the dating programmes of I. McDougall, P. Wellman, and others), and (2) a general, increasing scepticism about cyclical interpretations of landscape history. Geochronologists clearly demonstrated an antiquity of the highland surfaces that was considerably greater than appreciated hitherto, and the scepticism led to a re-thinking of the origin of the surfaces themselves. If the surfaces did not form as peneplains near sea-level, the corollary was that their current elevation may not necessarily relate to uplift of the highlands.

Renewed interest in highlands history during the last decade has resulted in a spate of studies of their evolution. These recent studies have been undertaken from and using a wide range of viewpoints and approaches, and a clear consensus on highland evolution has yet to emerge. Lambeck & Stephenson (1986) recently noted: 'That the south[east]ern highlands of Australia exist is perhaps the only statement ... with which all readers will agree' (page 253).

The more or less well-established data relevant to the age of the eastern highlands, relate to known highlands relief at different times during the Tertiary (the age of the coastal escarpment relative to the Tertiary basalts of the plateau top is noted above). Relief at other localities in the highlands has been shown to be high early in the Tertiary (for example, at least 400–500 m in the New South Wales southern tablelands: Snowy River valley — Wellman & McDougall, 1974b; Cathcart — Veitch, 1987; 850 m in the New South Wales northern tablelands: Mount Royal Range (east of Liverpool Range) — Martin et al., 1987). A minimum relief of about 250 m in the early Tertiary has been established on the plateau surface in northern, central, and southern New South Wales (Francis & Walker, 1978; Young & McDougall, 1985; Taylor et al., 1985).

At least three broad groups of recent models for highlands uplift can be identified, as discussed below.

Passive erosion of older highlands

Lambeck & Stephenson (1986) argued for a stable, passive highlands persisting from the last of the Palaeozoic or early Mesozoic orogenic deformations of southeastern Australia, 300 to 200 Ma ago. They proposed, in effect, that erosion, over about 200 Ma, of a sufficiently substantial highland belt, at meaningful Mesozoic and Cainozoic erosion rates, would have resulted in a highland belt very similar to the present eastern highlands. The predictions of the extent of the former highlands are based on the present highlands, and the model is 'run backwards' to predict the former extent of the eastern highlands. Thus, the ancestral highlands must perforce be similar in extent and shape to the present eastern highlands.

Figure 1.3.5 East-facing coastal escarpment looking north from Mount Mitchell near Cunninghams Gap. Lava flows of Main Range volcano are visible in the southern flank of Mount Cordeaux on the left.

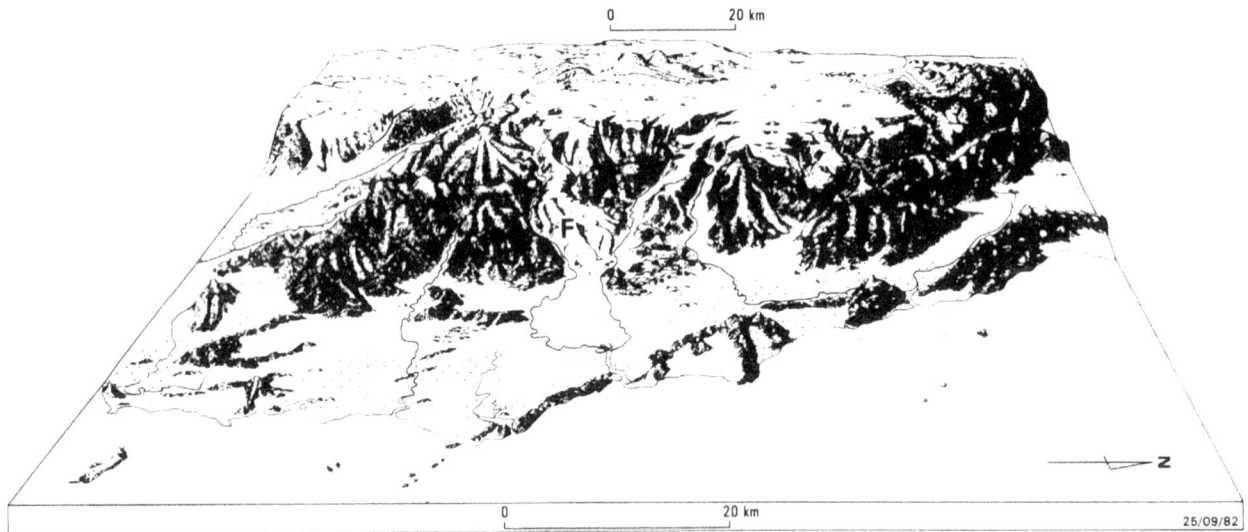

Figure 1.3.6 Sketch of the coastal escarpment east of the Atherton Tableland in northern Queensland (adapted from De Keyser, 1964, fig. 6). Lavas about 2 Ma old (Ollier, 1982a) from the Atherton volcanic province have flowed down the escarpment (see F).

Whether there was ever a substantial highland belt in late Palaeozoic to early Mesozoic times on the site of the present eastern highlands, and whether the erosion model used by Lambeck & Stephenson (1986) is appropriate, remain matters of debate (see, for example: Summerfield, 1986; Wellman, 1987). However, an important aspect of the Lambeck and Stephenson model is the emphasis on passive, isostatic uplift of the eastern highlands in response to erosional unloading. Such isostatic rebound is a form of uplift (albeit passive) which will maintain highland elevation at a substantial proportion of initial elevation for a considerable period of time. Isostatic rebound will figure also in river incision (see below). Moreover, such isostatic uplift must account for some at least of the stripping demanded by reported coal-maturation ranks, palaeomagnetic overprints, and apatite fission-track ages (Schmidt & Embleton, 1981; Middleton & Schmidt, 1982; Moore et al., 1986; see also Branagan, 1983).

Dynamic uplift

P. Wellman (1979b, 1987) proposed a model for eastern-highlands history on the basis of (1) the height and age relations between the Cainozoic and the present channels of major rivers that drain the highlands, and (2) the distribution of Mesozoic and Cainozoic marine sediments in the highlands. The distribution and age of these sediments, largely in Queensland but including the Sydney Basin and a Tasmanian basin, are evidence that uplift at these localities must have been after the early Cretaceous in Queensland and essentially after the Triassic in the other localities.

The Cainozoic channels were those infilled by dated basaltic lavas, and post-basalt incision by the rivers has left these channels at different elevations above the present river beds. P. Wellman (1979b) argued that this incision was solely the result of dynamic uplift, the major rivers being able 'quickly [to] reach an equilibrium with the base level outside the highland' (page 2). Bishop & Young (1980) argued that incision could not be assumed simply to be the result of uplift, and P. Bishop et al. (1985) interpreted the details of post-basalt incision in the upper Lachlan River, New South Wales, as the result of normal erosional development, without any necessary component of dynamic uplift.

Wellman (1987) acknowledged that river incision may not be the result of dynamic uplift, but it has remained his preferred model, especially for the river reaches near the highland margins. He pointed to localities at the margins where his expectation would be that Cainozoic channels would not be so elevated above the present channels (if uplift were not taking place) and interpreted the observed, greater-than-expected elevation as indicative of dynamic uplift. However, these localities are precisely those that would be expected to be most sensitive to isostatic rebound, because they are on the transition between the isostatically rising highlands and the more stable or subsiding extra-highlands areas, and indeed the present rivers have oversteepened profiles in this zone (Bishop, 1987). Moreover, the incision at several of these localities is consistent with the low amounts of isostatic rebound expected by Wellman (1987). Nevertheless, Wellman drew attention to a critical issue in an understanding of highlands evolution in identifying the need to separate the amounts of dynamic and passive uplift.

Passive subsidence punctuated by dynamic uplift

Jones & Veevers (1982, 1983) proposed a very different model on the basis of the patterns of east-Australian highlands volcanism, and of fluctuations in sedimentation and sea-level changes in the basins flanking the highlands. They noted a fundamental change in the tectonic regime of eastern Australia 95–90 Ma ago: the former regime was characterised by Pacific-rim type orogenic tectonism, whereas the latter was a 'relatively placid' aftermath.

They identified a highland belt coincident with, but broader than and as high as, the present highlands probably by 80 Ma ago (Jones & Veevers, 1983), and certainly by the start of the Cainozoic (Jones & Veevers, 1982). They interpreted the subsequent history of this highland belt as being one of overall passive subsidence (highland shrinking) and low rates of sediment supply to the flanking basins, punctuated by relatively short periods of dynamic uplift. These short periods of uplift are signalled by pulses of highlands basaltic volcanism and accompanied by flanking-basin subsidence and increased rates of sediment supply to the flanking basins.

C.M. Brown (1983) disagreed with Jones & Veevers (1982), arguing that basin sedimentation was related to global sea-level fluctuations, rather than to tectonic fluctuations in the highlands and basins.

Summary and evaluation

A wide range of conclusions about eastern-highlands history and evolution is demonstrated from the foregoing review. Highlands of considerable antiquity (at least late Mesozoic) are considered in two theories. The possibility of such an age is acknowledged in a third theory but, on the basis of the history of sedimentation and fluvial incision in the area of the current highlands, more continuous uplift through to the present is preferred. The overall Cainozoic highland decline proposed by Jones & Veevers (1982, 1983) rests on the perceived association between low rates of highland volcanism and low rates of sediment supply to the flanking basins. The simple, passive erosion and rebound proposed by Lambeck & Stephenson (1986), on the other hand, relies on the ability of their model to produce the present eastern highlands using an ancient highland eroded at reasonable rates.

On balance, therefore, the eastern highlands probably have considerable antiquity, but an adequate demonstration of the age is not yet available. All three models are critically dependent on initial assumptions. However, Wellman's interpretation offers considerable promise, because it is based, in part, on clearly demonstrable uplift of marine sediments. These data constrain the final uplift in the Lambeck-and-Stephenson type interpretation to be after the middle Triassic. The history of incision of the highland river systems is fundamental to an understanding of eastern-highlands evolution, and an essential task is the identification of the different forces 'driving' eastern-highlands river incision, including both passive and dynamic tectonics, as well as simple erosion as part of landscape evolution.

Tasman Sea formation

The relationship between Tasman Sea formation and eastern-highlands evolution has not been clarified fully, but an attempt to establish such a relationship is given in Section 7.3. There is no fission-track evidence in support of thermal or tectonic/erosional effects in the highlands belt associated with rifting prior to breakup (Moore et al., 1986). Thus, models incorporating simple Atlantic-type or East African rifting models for Tasman Sea formation as an essential element of highlands evolution do not appear to be applicable (see: Ollier, 1982a; Bishop, 1986). Other models for Tasman Sea rifting may be applicable (Lister et al., 1986; Wellman, 1987; see Section 7.3) but the pre-breakup heating of only a very narrow strip of coastal south-eastern Australia must be borne in mind.

Mechanisms of highlands uplift

Uplift mechanisms in the models of Lambeck & Stephenson (1986) and Jones & Veevers (1982, 1983) relate to early orogenies, and Jones & Veevers (1982) did not propose a mechanism for the minor uplift pulses associated with Cainozoic volcanism. Wellman (1987) reviewed the principal mechanisms proposed to account for the dynamic model of eastern-highlands uplift, and concluded that different mechanisms may have acted in concert to produce the uplift. These can be summarised simply as (1) asymmetric lithospheric stretching and concomitant heating associated with Tasman Sea formation, (2) Cainozoic isostatic rebound due to erosional unloading, and (3) underplating associated with Cainozoic volcanism. The pattern of basement apatite fission-track ages (Moore et al., 1986) is consistent with the view that any heating associated with Tasman Sea formation was restricted to a very narrow coastal strip — that is, underplating must be the principal uplift mechanism if Wellman's model of Cainozoic uplift is preferred.

Highlands unity

There are at least three major issues concerning eastern-highlands evolution that await resolution. Two of these — the timing and mechanism of the uplift (including the nature of the links with Tasman Sea formation), and the roles of passive and dynamic uplift in shaping the highlands — are discussed above. A third important issue is Cainozoic unity of the eastern highlands.

The three models of eastern-highlands evolution reviewed here are based on the assumption that the highlands have acted as one unit throughout their history — that is, evidence of uplift at one locality is evidence of eastern-highlands uplift as a whole. There is no clear reason to believe that this assumption is necessarily valid (Jennings, 1972). The assumption has most impact where a link is sought between highlands volcanism and uplift. Jones & Veevers (1982) and Wellman (1987) interpreted eastern-highlands volcanism as indicative of dynamic uplift of the highlands. If this relationship is real, and the highlands indeed have acted as a unit throughout the Cainozoic, then a link between time-transgressive volcanism throughout the eastern highlands and overall highlands uplift must be established.

1.3.3 Geological setting of New Zealand

Regional setting

New Zealand is at the zone of convergence and strike-slip motion between the Pacific and Indo-Australian plates (Fig. 1.1.1). The present-day boundary in North Island is marked by the Hikurangi Trench (Brodie & Hatherton, 1958) which passes northeastwards into the Kermadec Trench (Fig. 1.3.7). The principal volcanoes related to this subduction system are of basaltic, andesitic, and dacitic composition and are found to the

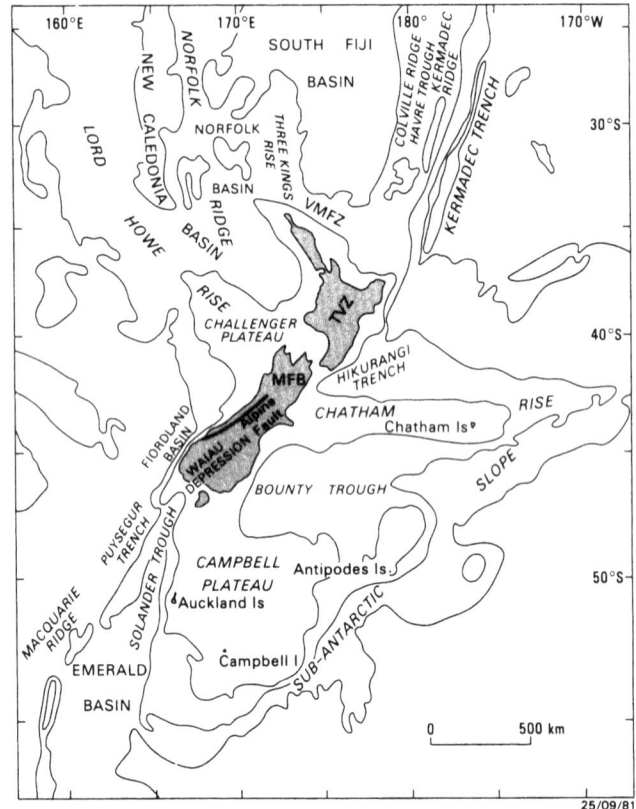

Figure 1.3.7 Bathmetry of region around New Zealand showing main tectonic features. VMFZ: Vening Meinesz Fracture Zone. TVZ: Taupo Volcanic Zone. MFB: Marlborough Fault Belt.

north (Edgecumbe, Whale Island, White Island) and south (Tongariro Volcanic Complex) of the currently extensional Taupo Volcanic Zone where magmas are overwhelmingly rhyolitic. Subduction-related volcanoes stand on the Kermadec Ridge, to the west of which is the Havre Trough, interpreted as a backarc zone of crustal thinning, rifting, and subsidence, analogous to the Taupo Volcanic Zone.

Older structures are found west of the Havre Trough. The Colville Ridge on the eastern side of the South Fiji Basin, an area of active spreading about 32 to 25 Ma ago (Watts et al., 1977), terminates to the south at about the 1500 m isobath on the New Zealand continental shelf. Colville Ridge is intersected by the west-northwest-striking Vening Meinesz Fracture Zone (Van der Linden, 1967) which separates Northland Peninsula to the south from the Three Kings Rise to the north. Northland and the Three Kings Rise, as well as the Norfolk Basin, Norfolk Ridge, and New Caledonia Basin, successively to the west, are not related directly to the present-day subduction zone. West and southwest of the New Caledonia Basin is the broad structure of the Lord Howe Rise, which continues to the southeast as the Challenger Plateau adjoining South Island.

The Hikurangi Trench on the Pacific/Indo-Australian plate boundary passes southwestwards into the right-lateral strike-slip Hope Fault, the most southerly member of the complex Marlborough Fault Belt that crosses the northern part of the South Island. The Marlborough Fault Belt merges southwestwards into the Alpine Fault, a major right-lateral transform structure linking the westward-dipping subduction zone in the North Island with the eastward-dipping subduction zone of the Fiordland/Macquarie Ridge system (Smith, 1971; Scholz et al., 1973; Davey & Smith, 1983). This system represents underthrusting of the Pacific plate by the Indo-Australian plate. It is marked by the Fiordland Basin and Puysegur Trench, east of which and towards the south is Macquarie Ridge, and farther east the Emerald Basin and Solander Trough. The Solander Trough continues on land in the south of the South Island as the Waiau Depression, a possible zone of rifting and subsidence similar to the Taupo Volcanic Zone in the north (Norris & Carter, 1980, 1982; Anderson, 1980). A single Quaternary volcano, Solander Island, is about 140 km southeast of the Fiordland Basin axis.

East of the depression defined by the Emerald Basin, Solander Trough, and Waiau Depression is the broad, quasi-continental, intraplate area that includes the Campbell Plateau. North of the plateau is the Bounty Trough and Chatham Rise (Brodie, 1957; Austin et al., 1973a,b) as well as parts of the South Island east and southeast of the Southern Alps. This wide region is bounded towards the southeast by the Sub-Antarctic Slope, leading down into the deep water of the southwestern Pacific Basin.

The tectonic and structural evolution of the region was discussed by Cullen (1967, 1970), Christoffel (1971), Weissel et al. (1977), Lewis (1980), Sporli (1980), Katz (1982), Crook & Feary (1982), Grindley & Davey (1982), Ballance et al. (1982), Cole (1984), and Walcott (1984a,b). Convergence rates of about 60 mm/yr for the Pacific and Indo-Australian plates north of the North Island, 50 mm/yr east of Hawkes Bay, 40 mm/yr in the central South Island, and 30 mm/yr at the Puysegur Trench, were reported by Walcott (1978). Crustal extension across the Taupo Volcanic Zone at the Bay of Plenty coast was measured at 7 ± 4 mm/yr by Sissons (1979) on the basis of geodetic data that span the last 70 years (see also Walcott, 1984c). H.W. Wellman (1979) presented a map of uplift rates for the whole of South Island, and Pillans (1986) presented an uplift map for North Island.

Geology

The 'terrane' approach taken in this account of the geology of New Zealand is based on the work of Korsch & Wellman (1988). Other workers who have used the terrane approach include D.G. Bishop et al. (1985), Coombs (1985), Norris & Craw (1987), and Landis & Blake (1987). New terrane syntheses are being undertaken by R.A. Cooper and J.D. Bradshaw (personal communications, 1988).

New Zealand rocks older than middle Cretaceous can be divided into two provinces, separated by the Median Tectonic Line (Landis & Coombs, 1967; Fig. 1.3.8). Precambrian to Devonian rocks are present only in the western province, whereas no rocks are older than Carboniferous in the eastern province.

The western province is now in two parts about 500 km apart because of dextral faulting on the Alpine

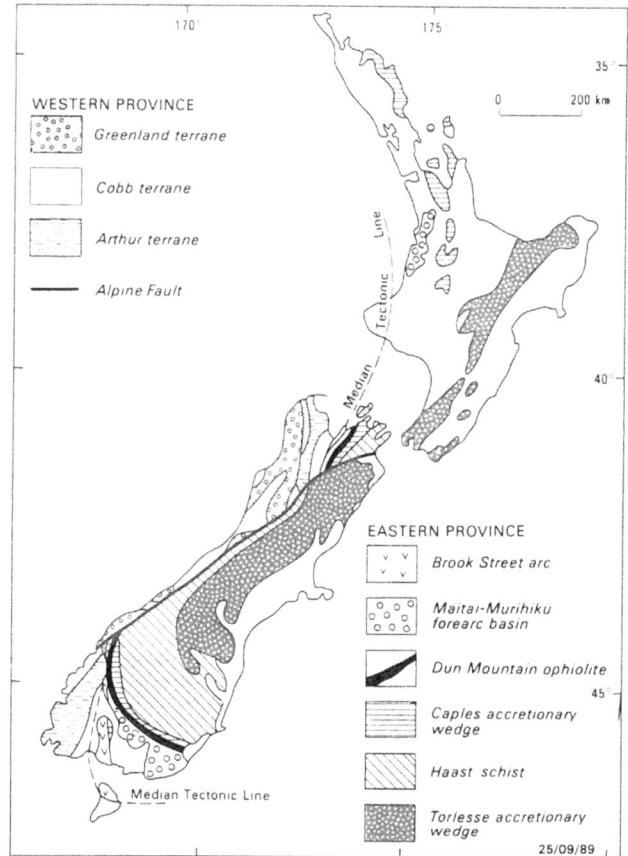

Figure 1.3.8 Major Palaeozoic and Mesozoic geological units of New Zealand.

Fault. Two terranes were postulated for this province by D.G. Bishop et al. (1985) and three by Coombs (1985). Three slightly different terranes were proposed by Korsch & Wellman (1988) who termed them the Greenland, Cobb, and Arthur terranes. These match the three sedimentary belts of R.A. Cooper (1979) but also contain crystalline rocks.

The Greenland terrane consists of: (1) the Constant Gneiss which is the only known Precambrian unit in New Zealand (680 ± 21 Ma; Adams, 1975), except for a possible Vendian volcanic unit in northwestern Nelson; (2) the Greenland Group which is a thick, monotonous, quartzose turbidite sequence of early Ordovician age (Cooper, 1974; Adams, 1975); and (3) the Karamea Batholith which consists of late Devonian and Cretaceous biotite granodiorites and biotite-muscovite granites (Tulloch, 1983).

The Cobb terrane in northwestern Nelson consists of complexly deformed Vendian or early Cambrian to late Ordovician mafic to intermediate volcanic rocks and volcanogenic sediments associated with ultramafic to mafic intrusive rocks (R.A. Cooper, 1979; Grindley, 1980). Several metamorphic units in Fiordland were suggested as higher-grade correlatives of the Cobb terrane by Korsch & Wellman (1988). A complex Cretaceous history for the region is inferred from recent evidence of granulites and arc-type igneous rocks of early Cretaceous age from Fiordland (Mattinson et al., 1986).

The Arthur terrane consists of quartz-rich sandstone, siltstone, and thick limestone, ranging in age from early

Ordovician to early Devonian (R.A. Cooper, 1979) and intruded by three belts of igneous rocks: the Riwaka Complex (a late Devonian, mafic to ultramafic intrusive suite), the Rotoroa Complex (intermediate to mafic intrusion of uncertain age), and the Cretaceous Separation Point Granite.

The initially independent histories of the three terranes, the mutual obliquity of trends in the Cobb and Arthur terranes, the formation of nappes in the Cobb terrane, and the several deformational, metamorphic, and plutonic events that have been recognised, correspond to a complex history. The three terranes may represent separate allochthonous terranes that were amalgamated during the Silurian or Devonian.

Six tectonostratigraphic units, described below from west to east, make up the Permian to middle Cretaceous eastern province (Fig. 1.3.8). There is, in general, reasonable agreement on the recognition of terranes within this province (Coombs et al., 1976; Howell, 1980; Bradshaw et al., 1981; D.G. Bishop et al., 1985; Korsch & Wellman, 1988).

The Brook Street magmatic-arc complex consists of island-arc mafic rocks, more felsic volcanic rocks, and related intrusions of Permian and Jurassic age (Williams & Smith, 1979; Johnston et al, 1987). The Maitai-Murihiku forearc basin lies to the east. Sediments of the middle to late Permian Maitai sequence rest unconformably on the Dun Mountain Ophiolite and consist of about 4 km of mainly volcanic-derived siltstone. The early Triassic to late Jurassic Murihiku sequence is up to 10 km thick, is also volcanic-derived, and is generally coarser-grained than the Maitai sequence.

The Dun Mountain Ophiolite consists of pillow lavas and rare mafic dykes that grade down through gabbros to ultramafic rocks that are now mostly serpentinites (Blake & Landis, 1973; Coombs et al., 1976; Davis et al., 1980). U-Pb isotopic dates on zircon from plagiogranite in the ophiolite are early Permian (Kimbrough & Coombs, 1983; Dickins et al., 1986). The ophiolite has been strongly tilted to the west and its basal (eastern) side is faulted against the Patuki Melange (Johnston, 1981), a unit here included in the Caples terrane.

The Caples terrane consists of mainly unfossiliferous volcanic-derived turbidites, and minor metabasalt, chert, and limestone. It includes the Caples Group in western Otago, the Pelorus Group in Nelson, and the Waipapa Group in Northland, as well as distinctive melange zones, and is interpreted as an accretionary wedge (for example: Carter et al., 1978; Korsch & Wellman, 1988). This terrane in Nelson has been subdivided into four discrete terranes by Landis & Blake (1987). The Caples terrane is sparsely fossiliferous, but accretion was probably mostly during the Permian. However, the turbidites of the Waipapa Group are Jurassic whereas the limestones associated with sea-floor basalt are Permian (Sporli, 1978).

The Haast Schist is between the Caples terrane and the Torlesse Complex and is a strongly deformed unit that has been regionally metamorphosed up to amphibolite facies. The Caples/schist and schist/Torlesse boundaries are gradational, and the boundary between the Caples and Torlesse was thought to be within the schist belt. However, Norris & Craw (1987)

proposed that a new unit — the Aspiring terrane, consisting of metavolcanics and pelagic sediments — is present between the two. Boundaries between the three lithological units are complex and consist of large nappes and shear zones. The Haast Schist therefore is not a terrane but consists of the Aspiring terrane plus the higher-grade metamorphic zones of the Torlesse and Caples terranes.

The Torlesse Complex (Korsch, 1984) ranges in age from Permian to early Cretaceous, and is the most extensive unit in New Zealand. It is an intensely deformed, sparsely fossiliferous, sedimentary sequence that is considered generally to be an ancient accretionary wedge (for example: Sporli, 1978; MacKinnon, 1983; Korsch & Wellman, 1988). Sandstones are quartzo-feldspathic in composition and derived from a crystalline plutonic source with a minor volcanic contribution, in marked contrast to the Caples accretionary wedge which is volcaniclastic in composition.

The eastern province therefore consists of several, subparallel, geological units that are consistent with formation in association with a westward-dipping subduction zone that was a margin of Gondwanaland. Under debate at present is the original spatial relationships of these units, particularly the relationship of the partly coeval but compositionally different Caples and Torlesse terranes. An allochthonous origin for the Torlesse has been suggested (for example: Howell, 1980; Bradshaw et al., 1981; Norris & Craw, 1987). The continental crust (basement to the younger sedimentary rocks and intraplate volcanoes) had attained its present extent by the middle Cretaceous.

The final phase in the development of the geology of New Zealand took place from the middle Cretaceous to the present day and involved the separation of the New Zealand region from Gondwanaland followed by a series of complex events that led to the development of microplates along the present day Indo-Australian/Pacific plate boundary (Section 1.2).

Development of continental rifts (failed arms) in the late Cretaceous were associated with the opening of the Tasman Sea and led to extensive rift-basin development (some basins filled with thick Tertiary sequences) particularly along the present western margin of the North Island and South Island (Kamp, 1986a,b; Korsch & Wellman, 1988).

Subduction at the Hikurangi Trench along the eastern margin of New Zealand, particularly the North Island, led to the development of an extensive accretionary wedge which is still actively accreting (Davey et al., 1986). This has produced complex deformation and melange formation in late Cretaceous to Miocene successions (for example, Pettinga, 1982) and to the deposition of trench-slope basins on top of the accretionary wedge (for example, Van der Lingen & Pettinga, 1980). The subduction-related plate boundary changes to a complex transform, the Alpine Fault system, in the South Island.

The intraplate volcanoes in Northland and Auckland are on geological basement of the Waipapa Group (Caples terrane) although some of the thick late Cretaceous and Tertiary sediments on top of basement have been shown to be allochthonous (Ballance & Sporli, 1979). The large intraplate volcanic centres of

Banks Peninsula and Dunedin in the South Island are on a basement of Torlesse Complex and Haast Schist, respectively. The Chatham Islands volcanoes also are on the eastern extension of the Haast Schist. In contrast, the intraplate volcanoes of the Sub-Antarctic Islands are found on crystalline basement of the western province (Gamble & Adams, 1985).

1.4 Upper Mantle, Crust, and Geophysical Volcanology of Eastern Australia

1.4.1 Introduction

The nature of the lithosphere in eastern Australia beneath the surface of exposed geology (Section 1.3.1) may be inferred from a wide range of geophysical data. Results of this type are summarised here, together with accounts of *in-situ* stress measurements, climate, and palaeomagnetism. Topography and uplift in the eastern highlands are discussed in Section 1.3.2.

1.4.2 Gravity and magnetic anomalies

Gravity and magnetic anomalies of wavelength 50–100 km are a reflection of differences in upper-crustal density and apparent susceptibility. The dominant trend in the gravity and magnetic anomalies is north-south in the volcanic areas of eastern Australia (BMR, 1976a,b), reflecting the north-south trends in the underlying Palaeozoic basement (Wellman, 1976a) and later sedimentary basins.

Areas of intraplate volcanism in both eastern Australia and New Zealand (Section 1.5) have topographic and gravity highs of the order of 200 km across. The topographic high in eastern Australia is shown in Figure 1.3.2, and the gravity high is given by Wellman & Murray (1979) and Wellman (1979a, fig. 5). Stephenson (1986) related areas of broad uplift, 50–100 km across and up to 600 m high, to areas of late Cainozoic volcanism in northern Queensland. However, there is some uncertainty here, as for elsewhere in eastern Australia, concerning what part of the topographic and gravity high is directly related to the volcanism, or the cause of volcanism, and what part is related to other causes such as effects at the margin of the continent.

There is a good correlation between gravity anomaly and altitude in eastern Australia: a positive free-air anomaly exists over the highlands (Wellman, 1976b). Lithospheric rigidity is relatively low judging by the exactness of the correlation. Stephenson & Lambeck (1985) inferred a viscoelastic lithosphere having flexural rigidity of 10^{23} Nm and viscosity time constant of 25 Ma. Their results also can be represented by an elastic crust of flexural rigidity of 10^{18}–10^{19} Nm. The low rigidity means that $0.5° \times 0.5°$ areas are essentially in isostatic equilibrium (P. Wellman, 1979a). The amplitude of the gravity high over the highlands is consistent with isostatic compensation of the topography being complete at 40 to 60 km depth (P. Wellman, 1979a) which is the depth of the base of the crust defined by seismic-refraction work.

A long-wavelength magnetic anomaly high follows the eastern highlands both in maps derived from satellite (Langel et al., 1982) and from near-surface observations (Wellman et al., 1985). The cause of this magnetic high is unknown.

1.4.3 Heat-flow

Cull (1982) and Cull & Conley (1983) analysed Australian heat-flow information from short holes in basement, water bores in sedimentary basins, and oil exploration wells. The regional differences in heat-flow found by averaging estimates over a $3° \times 3°$ grid are shown in Figure 1.4.1A. The eastern highlands have a smoothed heat-flow value of about 50–80 mW/m². However, areas of extensive Quaternary volcanic activity in northeastern Queensland and western

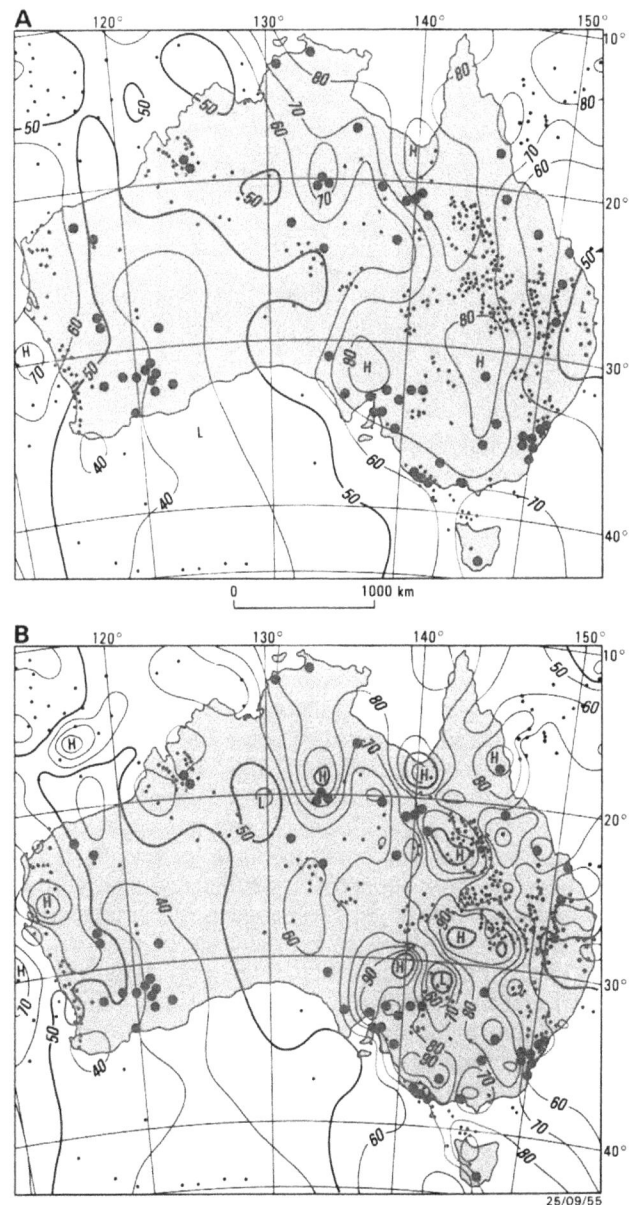

Figure 1.4.1 Heat-flow values in mW/m² contoured using 3° (A) and 1° (B) grids (adapted from Cull, 1982, figs. 1 and 2, and Cull & Conley, 1983, figs. 1,8, and 9). Measurement sites shown as small filled circles. Larger filled circles represent sites where the heat-flow error is less than 10 percent.

Victoria have mean heat-flow values in a 1° × 1° grid (Fig. 1.4.1B) of about 90 mW/m². High values of heat-flow in southeastern Australia previously reported by Sass & Lachenbruch (1979) now are known to be unrepresentative of the area, so there is less of an observational basis for their model of large-scale crustal intrusion. The world average heat-flow is about 60 mW/m², so heat-flow in parts of the eastern highlands is near world average, and in the areas of extensive Quaternary igneous activity is about 1.5 times as great.

1.4.4   Crustal seismic velocities

Seismic-refraction data have been used to determine seismic-velocity profiles in the crust (Fig. 1.4.2) along the sections shown in Figure 1.4.3. The crust/mantle boundary in this section is defined seismically as where the seismic velocity approaches 8 km/s. Velocities in the upper crust where Palaeozoic rocks crop out range from 5.6 km/s near the surface to about 6.3 km/s at depths of 10–15 km, and in the lower crust from 6.3–6.5 km/s to greater than 7 km/s at the base of the crust at about 35–55 km. Adjacent profiles do not have the same velocities or depth of velocity change, so the velocities differ laterally. Velocity changes with depth in the highland area are generally gradual, not abrupt, and there are zones of velocity decrease with depth. There are sharp velocity discontinuities away from the highlands in the Eromanga Basin area. There is a general correlation of higher topography, and regional outcrops of basement, with deeper crust. The area of volcanism and highlands may correspond to parts of the crust where there are thick layers at the base of the crust that have a velocity of about 7.3 km/s. Drummond & Collins (1986) showed that the velocity of the lower crust increases with increasing crustal thickness and age, and they attributed the increase in crustal thickness and average velocity in Proterozoic and Palaeozoic areas to underplating of post-Archaean crustal rocks to the base of the crust through time.

The following interpretation of results from deep-seismic reflection surveys (Figs. 1.4.2,4) is summarised from Finlayson & Mathur (1984). Murray and Eromanga basin profiles have reflections from sedimentary rocks, few reflections from the upper part of the non-sedimentary crust and from below the crust, but have numerous coherent reflections over horizontal distances of about 3 km between the mid-crustal velocity increase at 8 s and the base of the crust at 12–13 s. Profiles in the eastern highlands (Bowen Basin and Lachlan Fold Belt) differ in having reflections throughout the crust. There are reflections in the Denison Trough (part of the Bowen Basin) from 1 to 3 s in the sedimentary section, from 4 to 5 s in the upper crust, and particularly strong and continuous reflections from 8 to 12 s from near the base of the crust. Scattered reflections beneath the Gundary Plains site in the Lachlan Fold Belt are found throughout the top two-thirds of the crust, and in this depth range there are more continuous reflections at 7–9 and 10–11 s This corresponds to a layered structure at about 20–33 km at depths where a velocity gradient is inferred from seismic-refraction results.

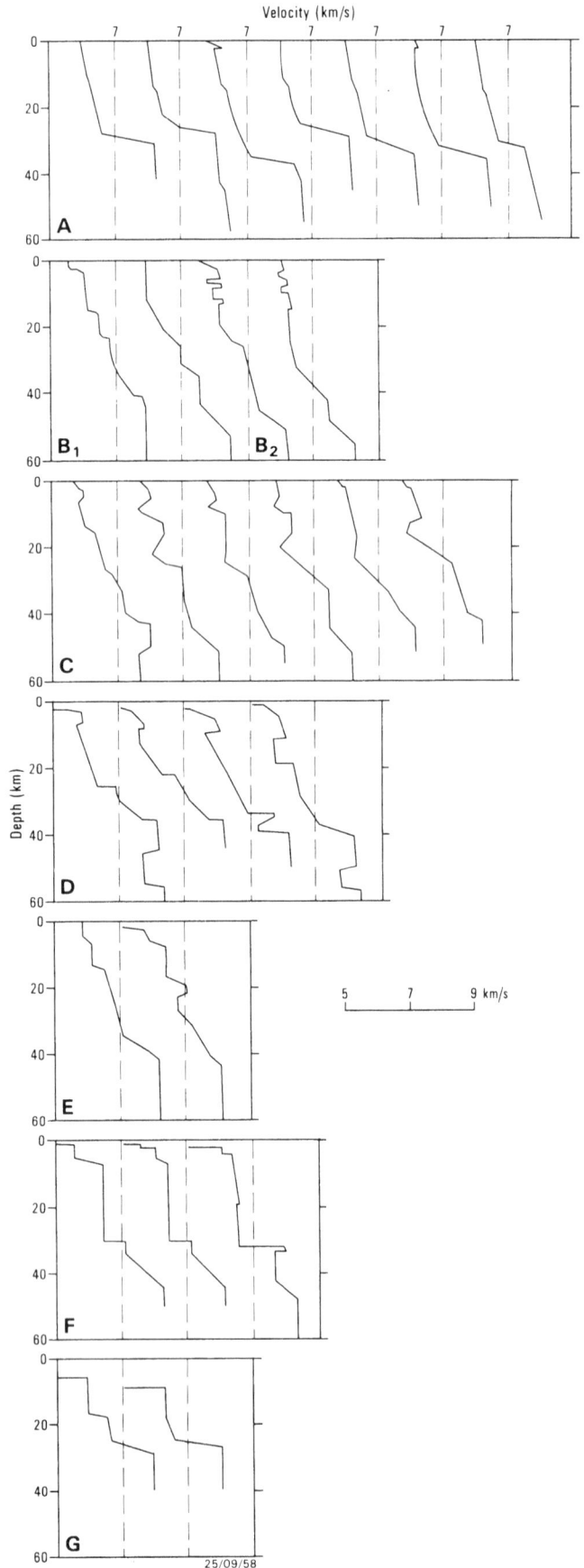

Figure 1.4.2   Seismic-velocity depth models for profiles whose positions are given in Figure 1.4.3 (after Drummond & Collins, 1986, fig.2).

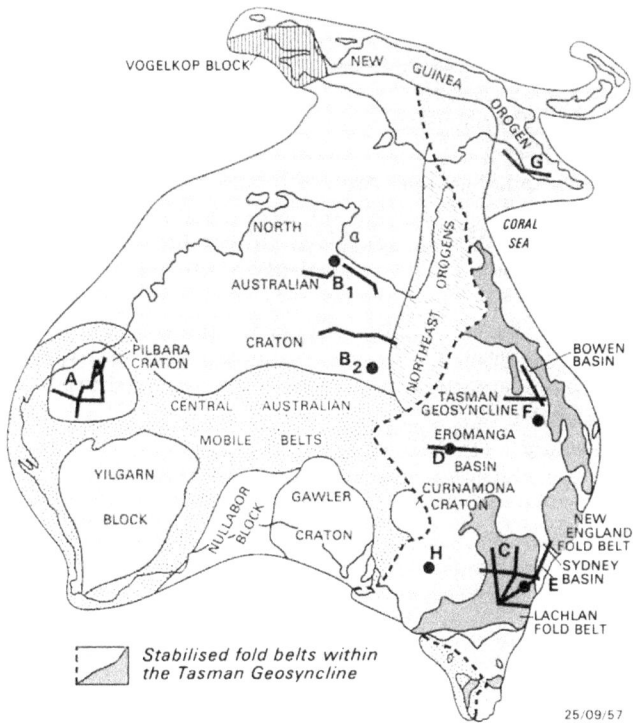

Figure 1.4.3 Positions of seismic-velocity profiles (short thick lines) used in Figure 1.4.2 and seismic-reflection profiles (filled circles) in Figure 1.4.4, shown in relation to the major geological provinces of Australia (after: Drummond & Collins, 1986, fig.1; Finlayson & Mathur, 1984, fig.2).

Finlayson & Mathur (1984) made the following generalisations on the reflection data from the few sites occupied. There is a smaller depth to the base of the crust in the major sedimentary areas, as well as sharper velocity changes and a smaller thickness of lower crust having velocities over 7 km/s, compared to regions where non-sedimentary rock crops out. Reflection characteristics range widely in the upper 20 km of the crust which consists largely of deformed rocks similar to those at the surface. The lower crust generally has strong and coherent reflecting horizons. The lower-crustal reflections, seismic velocities, and rock compositions are consistent with underplating and lower-crustal intrusion causing significant crustal thickening, particularly under regions of outcropping basement. The subcrustal lithosphere has very few seismic reflecting horizons.

1.4.5 Seismicity

The main band of seismicity (Fig. 1.4.5) follows the eastern highlands, and there is only minor earthquake activity on the margin of the lowland to the west and on the continental shelf to the east. Lambeck et al. (1984) showed that the earthquakes in the southeastern part of the highlands in New South Wales have a depth of less than 32 km, and generally less than 20 km. Nearly horizontal compressive axes that have azimuths ranging from northeast to southeast are shown from focal-plane solutions (Fig. 1.4.6). Events deeper than 15 km have thrust-faulting mechanisms, and events shallower than 5 km have predominantly strike-slip mechanisms. One 21 km-deep earthquake 50 km west of Melbourne in

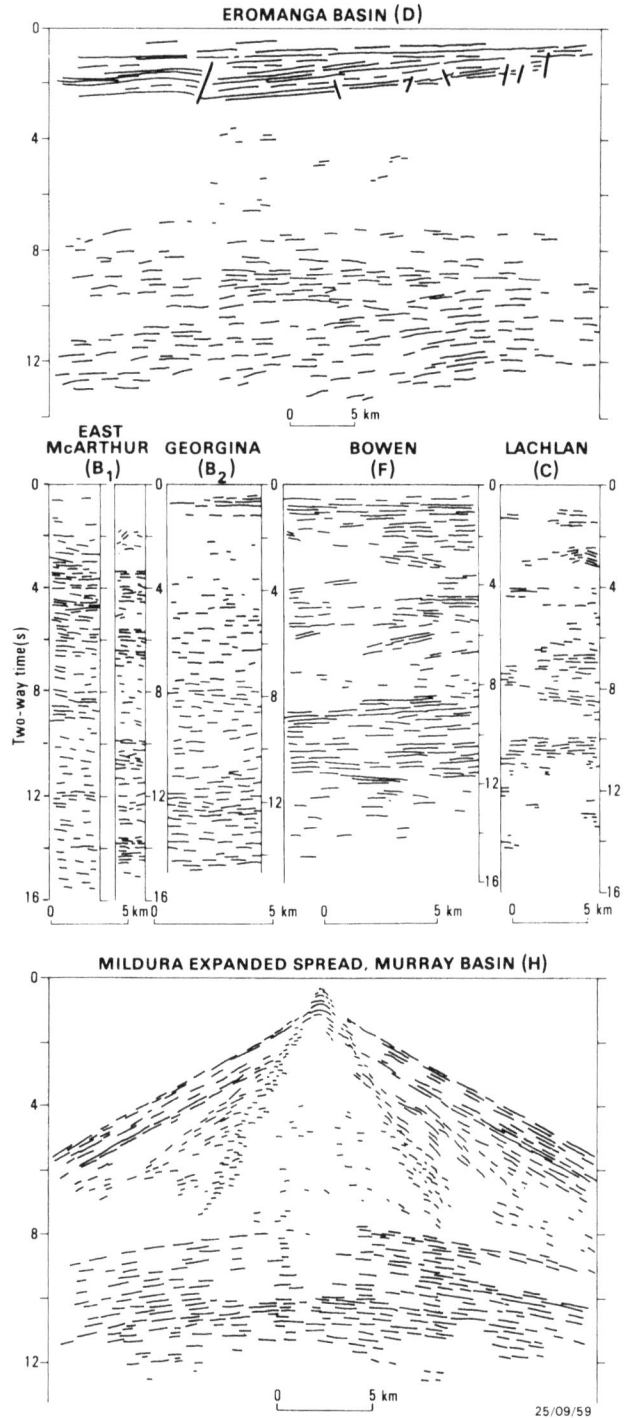

Figure 1.4.4 Vertical reflecting horizons on seismic-reflection profiles whose positions are shown in Figure 1.4.3 (after Finlayson & Mathur, 1984, figs. 4,6,7).

the area of Quaternary volcanism has a thrust-fault solution and a southeasterly compression axis.

Line A-B of Figure 1.4.5 is the position of any present-day central-volcano provinces predicted by Wellman (1983) from a regression of position and age of east-Australian central-volcano provinces. Earthquake epicentres plot near this line both immediately east of Bass Strait and farther east out in the Tasman Sea. The two areas of earthquake activity are thought to represent two centres of present-day intraplate magmatism, one northeast of Tasmania, the other on the southward extension of the Tasmantid Guyots (see Section 1.7.2). Denham (1985) showed that a 1983 earthquake in the Tasman Sea earthquake cluster at a depth of 25 km had a thrust-fault solution and an east-southeast pressure axis.

1.4.6   *In-situ* stress measurements

The results of stress measurements for eastern Australia are summarised in Figure 1.4.6. Maximum compressive stresses for nine sites in southeastern New South Wales range from 3 to 18 MPa (Denham et al., 1979). Axes of maximum compression at inland measurement sites are easterly, and at the few sites on the coastline are north-northeast to northeast. Maximum compressive stress in the west at Cobar, Broken Hill, and Mount Isa, is 22–30 Mpa and axes of maximum compression are east to east-southeast. Borehole information, over-coring, hydrofracturing, and earthquake focal-mechanism solutions are all consistent with the interpretation that east-Australian crust is in strong compression, although there are local differences in direction.

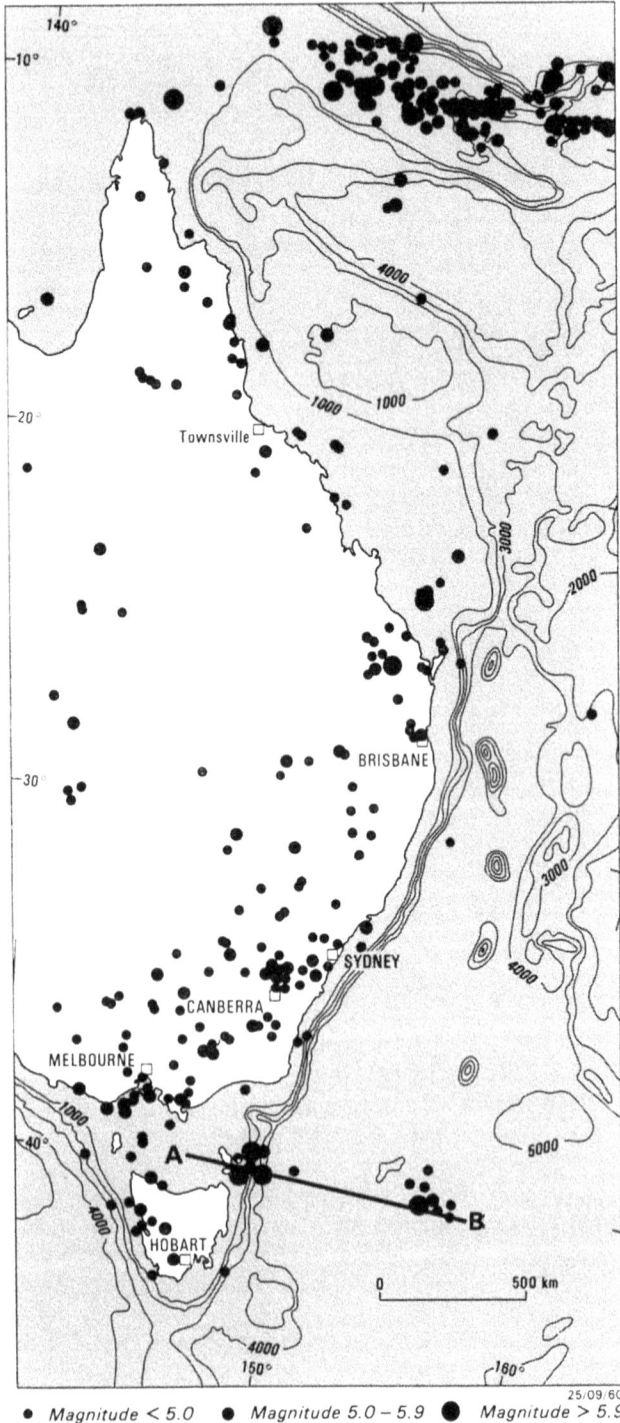

● *Magnitude < 5.0*   ● *Magnitude 5.0 – 5.9*   ● *Magnitude > 5.9*

Figure 1.4.5 Earthquake epicentres of magnitude 4.0 or greater for 1873–1987 (BMR Earthquake Data File). Line A-B is the position of the 0 Ma hotspot volcanoes predicted from a least-squares fit to the ages of the east-Australian central volcanoes (Wellman, 1983).

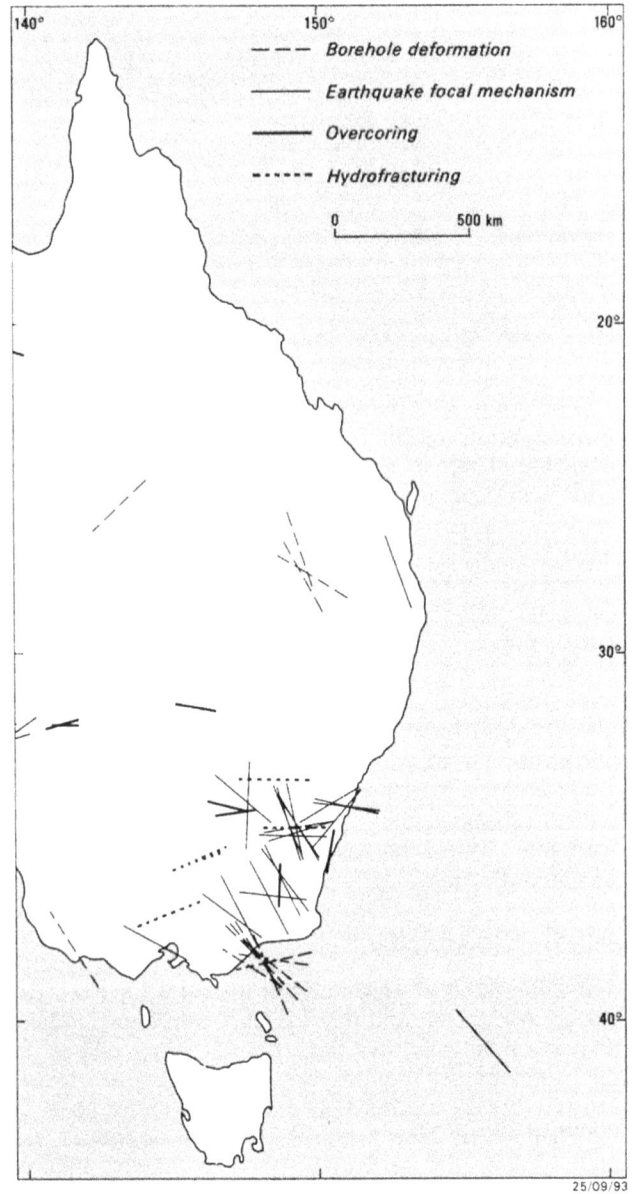

Figure 1.4.6 Directions of principal compressional stress in the horizontal plane, from borehole deformation, earthquake focal mechanisms, overcoring, and hydrofracturing (after Denham, in press).

1.4.7 Lithosphere seismic studies

Lithosphere seismic-velocity models have been established for both eastern and central Australia from seismic refraction and from seismic surface-wave dispersion (Fig. 1.4.7). Some of the discrepancies between models in the same area were discussed by Finlayson (1982). Vp velocities are higher in the uppermost mantle beneath Proterozoic crust compared

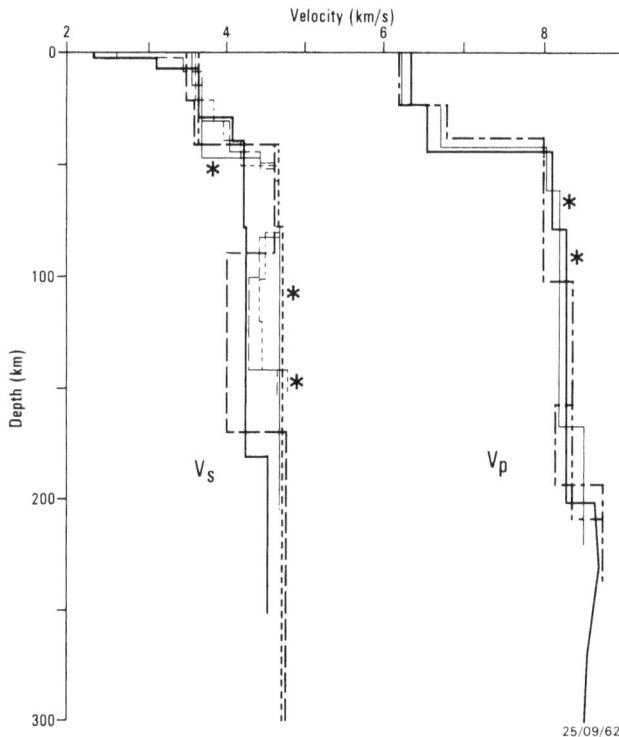

Figure 1.4.7 Upper-mantle velocity models for Australia (after: Finlayson, 1982, fig. 7; Ellis & Denham, 1985, fig. 3). Asterisks refer to profiles in the Precambrian of central and western Australia. Other profiles are in the Phanerozoic of eastern Australia.

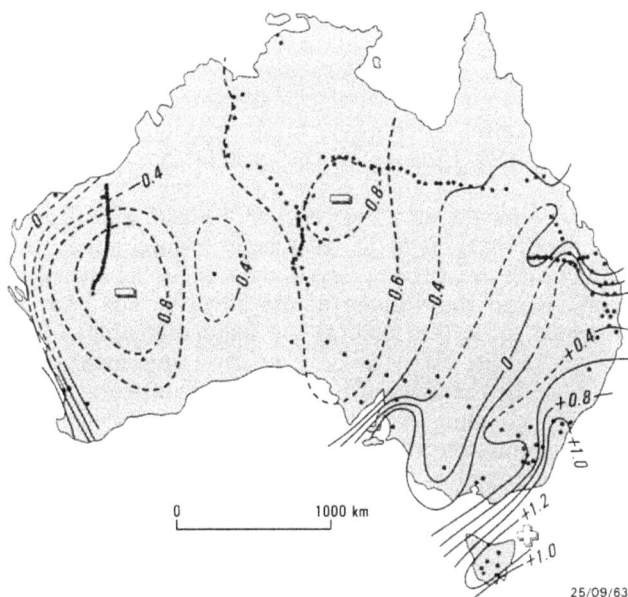

Figure 1.4.8 P-wave seismic station residuals (seconds) for Australia (after Drummond et al., in press, fig. 1). Filled circles represent measurement stations.

to Phanerozoic crust. There is some velocity increase in all of the models between 160 and 210 km, and in three models of Phanerozoic crust there is evidence for a low-velocity zone in the range 90–190 km.

Anomalies in the travel time of teleseismic P waves were mapped by Drummond et al. (in press; Fig. 1.4.8). Residuals more negative than –0.4 s generally are found in Archaean and Proterozoic crust whereas more positive residuals appear over Phanerozoic crust. Residuals over Phanerozoic crust are progressively more positive towards Sydney and Tasmania. The changes in residuals cannot be explained by velocity changes in the crust, so part of the cause must be in the mantle, including velocity differences as deep as 300 km. Higher residuals in the Phanerozoic areas may be caused by the low-velocity zone found by seismic-refraction and surface-wave dispersion. The large positive residuals found in Tasmania may be caused by high temperatures associated with a mantle plume producing the Cainozoic hot-spot volcanism in eastern Australia.

1.4.8 Electrical conductivity

Deep electrical conductivity studies have been carried out in Victoria and southern New South Wales. Bennett & Lilley (1973) showed that the main horizontal differences are caused by differences in conductivity as depth increases under the ocean and continent, and by higher conductivities in a small area of central western Victoria. The west-Victorian conductivity anomaly is inferred to be due to a 'zone with conductivity of order $1 (ohm.m)^{-1}$, at a depth possibly much less than 50 km' centred about 38.2°S, 143.5°E (Bennett & Lilley, 1973, page 204). This 'conductive zone is considered to represent a crustal or upper mantle zone of elevated temperature and partial melt' (page 204) associated with the most recently active part of the area of voluminous Quaternary volcanic rocks.

Tammemagi & Lilley (1971) reported on a traverse of magnetotelluric soundings in southern New South Wales. They inferred resistivities at two sites in the eastern highlands of 45–50 ohm.m in the top 10–20 km and a resistivity of about 225 ohm.m below. In addition, a thin surface layer of low resistivity and a layer 400–550 km thick of thousands of ohm.m, underlain by a layer of tens of ohm.m, is inferred for two sites in the Murray Basin. Hence, electrical conductivity in the depth range 100–300 km is greater under the highlands than under the lowland to the west. Conductivity depth models for the lowlands were inferred also by Lilley et al. (1981). They found conductivity of less than 0.1 S/m to 300 km, and conductivity of more than 0.1 S/m below. Lilley et al. (1981) showed that the relatively high conductivity at depths of 300 km is consistent with partial melting and with a low-velocity zone in this part of eastern Australia, in contrast to the evidence for no melting in central Australia (Fig. 1.4.9).

1.4.9 Sub-volcanic intrusions

The following major volcanoes in eastern Australia and New Zealand are dissected and have exposures of a

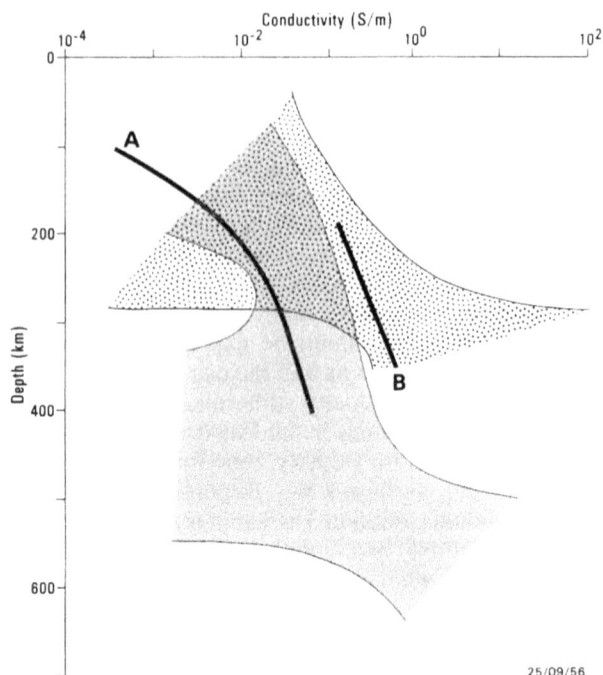

Figure 1.4.9   Limits to conductivity models that fit magnetic variometer observations (after Lilley et al., 1981, figs. 4,7,9). Stippling represents southern Australia. Screen represents central Australia. Line A is the conductivity profile predicted for a normal continental lithosphere. Line B represents the conductivity expected where partial melt is present in the lower lithosphere.

central complex of ring dykes, plugs, and cone sheets: Main Range (see Section 3.4.4), Focal Peak (3.4.6), Tweed (3.4.7), Ebor (3.5.4), Akaroa (4.3.3), Carnley volcano, Auckland Islands (4.4.2), and Campbell Island volcano (4.4.3). The following summary of the nature and emplacement of major intrusive complexes beneath some intraplate volcanoes in the region is based on the study reported by Wellman (1986) who assessed topographic, magnetic, gravity, and geological data. Minor high-level intrusions within lava piles and near the top of the basement are not considered.

Basement at some east-Australian volcanoes is observed to have been uplifted and tilted radially outwards rather than moved downwards. Many of these volcanoes have sub-volcanic land surfaces that can be mapped where streams have cut valleys through to the underlying basement. Present-day stream levels are more or less at the level of the pre-volcanic surface. Pre-volcanic land surfaces had a relatively low relief, and post-volcanic deformation was either minor, or was involved in epeirogenic uplift, so the relief of the sub-basaltic surface, after the regional slope has been removed, roughly represents the deformation of the pre-volcanic surface related to the igneous activity. Some large volcanoes have a conical uplift of the sub-volcanic surface centred under the volcano. The uplift diameter at one-half amplitude is 8–20 km and the maximum amplitude is over 0.6 km. This uplift is attributable to underlying intrusive complexes.

The volcanic/basement contacts at New Zealand and Sub-Antarctic Islands volcanoes are mainly below sea-level and so are not mapped easily. Uplift of the basement is demonstrated by the presence of pre-volcanic rocks at Lyttelton, Dunedin, Carnley, and Campbell Island, but only in the area of the volcanoes themselves.

Magnetic anomalies in the vicinity of most east-Australian volcanoes have been measured and are shown on national 1:250 000–scale aeromagnetic maps. Aeromagnetic profiles over Lyttelton volcano were given by Gerard & Lawrie (1955). Large dipole anomalies at the centres of the volcanoes have diameters at one-half amplitude of 6–12 km, amplitudes of 200–1500 nT, and are caused by either reverse (only one example, Canobolas volcano) or normal magnetisation. These anomalies too are attributable to the sub-volcanic intrusive complexes.

The centres of the large east-Australian volcanoes, and Dunedin, Akaroa, and Lyttelton volcanoes in New Zealand, are also the sites of circular, positive, anomalies on Bouguer gravity-anomaly maps (Reilly, 1972; Wellman, 1973; Ross, 1974). These anomalies were not caused by the volcanic rocks because the effects of their mass above sea-level are largely removed by Bouguer and topographic corrections. They are attributed, rather, to the intrusive complexes.

Contour maps have been constructed for many volcanoes of (1) the upper surface of the volcanic rocks, (2) the surface between the volcanic rocks and the sub-volcanic surface, and (3) residual Bouguer gravity anomalies (Fig. 1.4.10). The volume of volcanic rock is given by the differences shown on the maps between the upper and lower surfaces of the volcanic rocks, and the uplift is found by removing a flat regional surface from the sub-volcanic surface. The residual gravity anomaly is obtained by removing a smooth 'regional' from the Bouguer gravity anomaly, and the magnetic anomaly is found by inspection of the aero-magnetic-anomaly map. Estimates of these derived quantities are listed for each volcano in Table 1.4.1.

Volumes for the rocks of central volcanoes increase as both uplift volume and excess mass increase (Wellman, 1986). In addition, the ratios of the respective values for these parameters are constant to within experimental error, and the volume of uplift is found to be about one-third of the lava volume (Wellman, 1986, fig. 2). Uplift data for Barrington volcano (a lava-field type centre) is consistent with the same relationships for the central volcanoes, but the gravity anomaly is low, possibly because of the granite that crops out there.

Space for the intrusive complex beneath the studied volcanoes is thought to have been formed solely by uplift of the crust above, and to the side of, the intrusive complex, so the volume of the intrusive complex is assumed to be the same as the volume of uplift. The complexes are likely to have nearly radial symmetry because the contours of uplift and residual gravity are roughly circular. An approximately equidimensional shape seems most likely.

Each complex can be modelled as a sphere where the depth of the centre of the mass is 1.3 times the radius at one-half the maximum amplitude of contours of the residual gravity anomaly. The centres of mass are found to be 6–8 km below the present land surfaces which is about the same as the depths to the pre-volcanic surfaces. The diameters of the complexes are 4–10 km, if the complexes contain only minor country

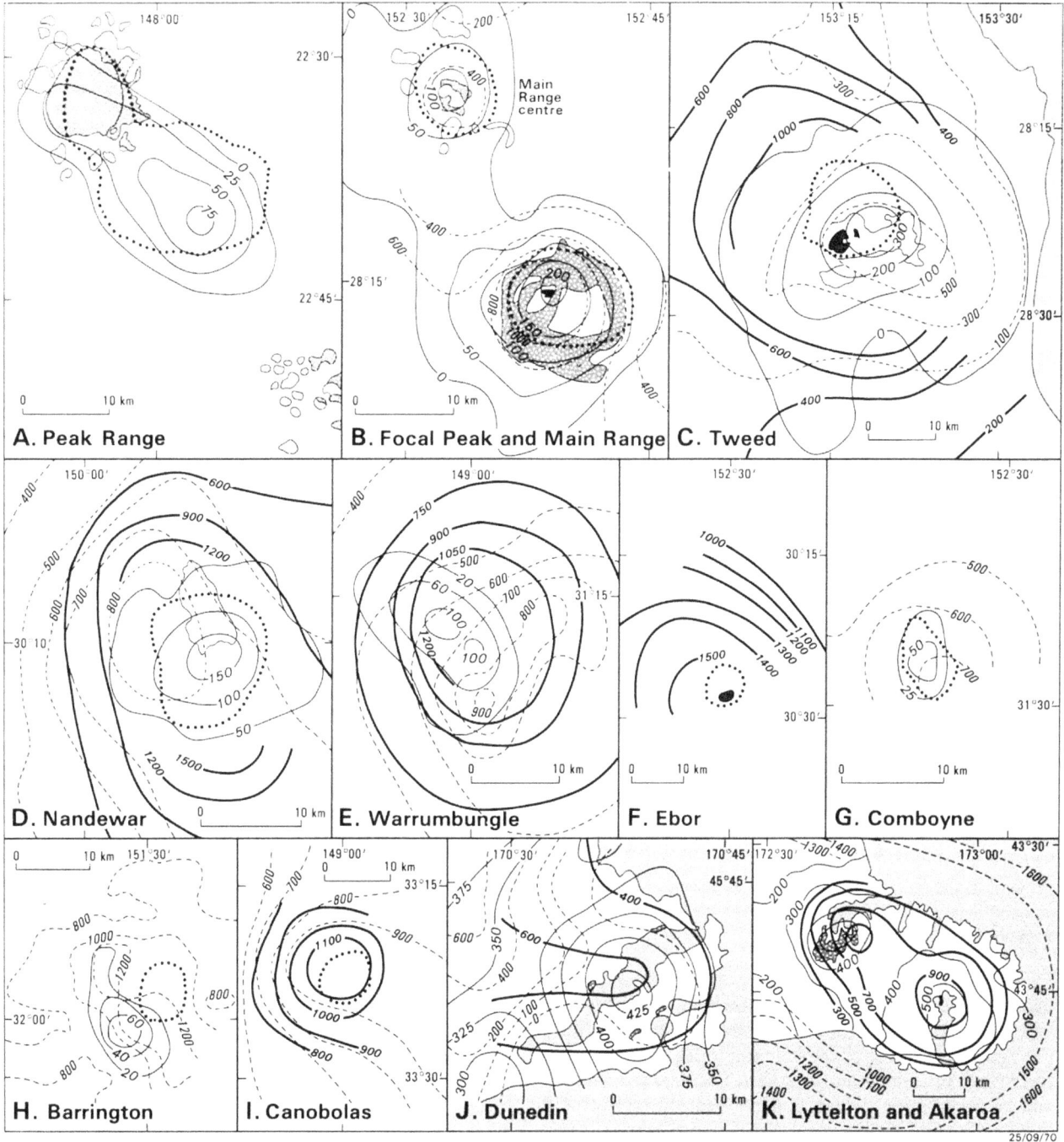

Figure 1.4.10   Relationships at eleven east-Australian and New Zealand volcanoes, between the volcanic piles and their respective intrusive complexes. Upper level of volcanic rocks (in metres) represented by thick continuous lines. Bases of the volcanic pile (in metres) represented by thin dashed lines. Residual gravity anomaly in A-I (in mGal) and isostatic anomaly in J-K represented by thin continuous lines. Outer margin of magnetic anomaly caused by the intrusive complex shown by dotted line. Stippled pattern represents strongly uplifted basement. Large mafic intrusions are shown solid, and other intrusions (mainly rhyolite plugs) are shown by a thin outline.

rock (and assuming the uplift volume is the volume of the intrusive complex). The tops of the complexes therefore would be up to 4 km below the volcanic and original land surfaces, and the bases at depths of less than 13 km. Complexes therefore would be wholly within the upper crust. The uplift caused by the complex can be modelled as the expansion of a sphere in an elastic half-space.

Geophysical and uplift information is available for volcanoes in western Scotland (Rhum: McQuillin &

Tuson, 1963; Skye and other centres: Bott & Tuson, 1973), Uganda (Trendall, 1965), Kenya (Le Bas, 1977), and Hawaii (Koolau; Strange et al., 1965). The intrusive complexes listed in Table 1.4.1 have centres of mass (at about 6–8 km) which are similar to values of 8 km for Skye, 7.5 km for Rhum, 8 km for Koolau, and 6 km inferred for Ugandan volcanoes (from the 10 km-diameter diameter at one-half maximum uplift amplitude). East Australian and New Zealand volcanoes have an uplift over the intrusive complex that is

Table 1.4.1 Volume, uplift, and residual gravity and magnetic anomalies for nine volcanoes

	Main Range	Tweed	Nandewar	Warrum-bungle	Comboyne	Barrington	Canobolas	Dunedin	Lyttelton
Volcanic rock									
volume (km³)	?1000	3000	400	210	?50	?1500	70	800	?240
diameter (km)[1]		30	19	22			8	?22	?17
maximum thickness (km)	0.9	1.2	1.0	0.5	?0.2	0.7	0.3	1.0	0.6
Uplift									
volume (km³)		450+	180	210	43	210	110		
diameter (km)[1]		20	20	16	11	13	15		
maximum uplift (km)		0.6+	0.4	0.4	0.3	0.5	0.5	0.6	
Residual gravity anomaly									
excess mass (mGal.km²)	4200	10600	3600	3100				4700	5600−
excess mass (kg.10¹⁰)	100	250	86	74				110	130−
diameter (km)[1]	15	12	11	15				9	13
depth to centre (km)	10	8	7	10				6	8
maximum anomaly (mGal)	13	40	16	13	3–5	5–10	?0	31	13
Residual magnetic anomaly									
diameter (km)[1]		8	6	none	9	8	6		12
amplitude (nT)		+150[2]	+1150		+1450	+750	−240		+40[3]
Outcrop of intrusion or greatly uplifted country rock — diameter (km)	6	9			15	2+		12	10

[1] Diameter at one-half maximum amplitude.
[2,3] Aeromagnetic surveys at 0.15 m above terrain, except for (2) 2.9 km and (3) 3.05 km.

found also at east-African volcanoes and which may be present in the Scottish volcanoes (Watson, 1985, fig. 5).

1.4.10 Climate

Cainozoic climate is relevant to the following three aspects of volcanology: (1) weathering of the tops of volcanic units before they are covered by the next unit; (2) deep weathering of the volcanic piles, which is well developed in central and northern Queensland; (3) the rate of denudation, which is dependent on rainfall and vegetation and which controls the proportion of volcanic material remaining at the present day (see also Section 2.8).

Temperatures were shown by Kemp (1978) using palynological results to have been higher in the early Cainozoic compared to the late Cainozoic. Sluiter (in press) as a result of later, more detailed studies, gave estimated temperatures and rainfall figures for central Australia of 18–19°C and 1800 mm in the late Palaeocene, 20°C and 2000–2100 mm in the early Eocene, and 17–18°C and 1800–2000 mm in the middle Eocene. Truswell et al. (1985) estimated a mean temperature of 18–19°C and rainfall of 1500 mm in the late Oligocene to middle Miocene for the western Murray basin. Much of the coastal land was covered by a closed rain forest, and precipitation was reasonably high up until the late Miocene, judging by the types of land flora. This equable and wet climate may have caused fast soil development, resulting in the weathering found on the tops of some flows within volcanic sequences. Rainfall was higher than at the present, but the closed rain forest may have resulted in deep soil and less run off, and so a lower middle Cainozoic erosion rate than in the late Cainozoic.

Deep weathering profiles of two ages are developed on some types of decomposable rocks in southwestern Queensland and, to a lesser extent, the eastern highlands. The profiles are up to 95 m thick. Older profiles are thicker and have a palaeomagnetic age of 60 ± 10 Ma, whereas younger profiles have a palaeomagnetic age of 30 ± 15 Ma (Idnurm & Senior, 1978). These periods of early Cainozoic weathering, together with more recent weathering, are the cause of the deep weathering of volcanic piles that is a serious obstacle to sampling fresh rock from middle Cainozoic or earlier volcanism in central and northern Queensland.

1.4.11 Palaeomagnetic volcanology

Palaeomagnetic studies on Cainozoic volcanic rocks were carried out initially on samples from isolated flows (for example: Coombs & Hatherton, 1959; Robertson, 1966), but relative and isotopic dating generally were poorly controlled. Studies on isolated flows dated as post Miocene (McDougall et al., 1966; Aziz-ur-Rahman, 1971) are useful in providing an average palaeomagnetic pole for the last 6 Ma.

The most useful results have come from later studies of steep stream sections in the central parts of the major

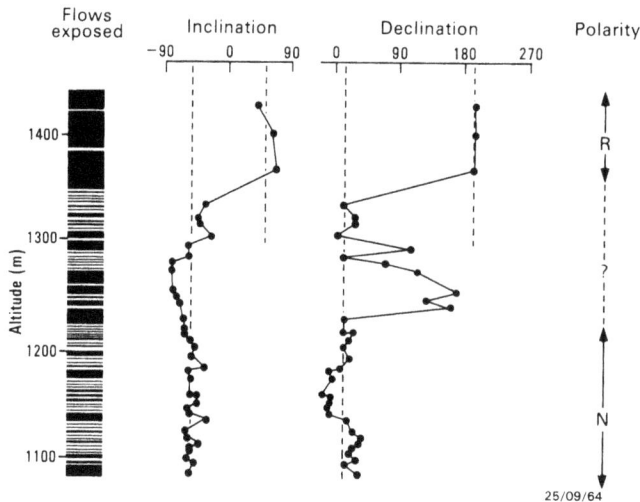

Figure 1.4.11 Palaeomagnetic directions for lava flows of Nandewar volcano (after Wellman et al., 1969, fig. 3).

volcanoes (Wellman et al., 1969 — see Fig. 1.4.11; Evans, 1970; McElhinny et al., 1974; Wellman, 1975; Hoffman, in press). Parts of these sections are well exposed, rocks are relatively unaffected by later weathering and protected from remagnetisation by lightning strikes, and stratigraphy and dating are unambiguous. Sections are 300–800 m thick and overall exposure is about 50 percent. Each flow or flow unit typically consists of a massive, grey, flow base overlain by reddish clinker 0.5–1 m thick. The magnetisation of the clinker top of the flow is the same as the flow centre, so magnetisation was acquired after the whole flow solidified.

Palaeomagnetic directions can be divided into normal polarity, transitional, and reverse polarity. The transitional directions are caused by intermediate magnetism during a geomagnetic field reversal or a geomagnetic excursion. There are 1–4 reversals of polarity within the lava sections of the major volcanoes. Heirtzler et al. (1968) found an average of three reversals per Ma between 0 and 45 Ma, and one reversal per Ma between 45 and 80 Ma. The 1–4 reversals in the major volcanoes (for example, Fig. 1.4.11) are roughly consistent with the K-Ar geochronological results of Wellman & McDougall (1974a) who showed that the volcanoes were active for 1–3 Ma.

Accurate time horizons are required in mapping volcanoes, but the determination of accurate time horizons by K-Ar dating is expensive, and the method is not sufficiently accurate for the older volcanoes. Use should be made of the accurate time horizons provided by the magnetic reversals and magnetic excursions. Remanent magnetic directions can be determined in the field using a Brunton compass and the more oxidised upper part of the flows.

Palaeomagnetic results provide some information on the time interval between flows. Major changes in composition up sections, such as the base of the three, upper, thick flows in Nandewar volcano and the base of the Binna Burra Rhyolite in Tweed volcano, are generally horizons of polarity change. These horizons of lava-composition change therefore evidently represent relatively long periods without volcanism.

Flow tops generally are not weathered, and many flows are of similar composition to those below, leading to the suspicion that some groups of superimposed thin flows of similar composition may be flows units of a single extrusive event, or that there was rapid extrusion of several flows of similar composition. The similarity in palaeomagnetic pole directions of successive flows has been used to infer that successive flows were extruded over a short interval of time. Little or no pole movement was inferred for the twelve flows forming a valley fill near Nerriga (McElhinny et al., 1974), for groups of 3–8 flows in Nandewar, Liverpool, and Barrington volcanoes (Idnurm, 1985a), and for the Stoddart succession on Banks Peninsula (Evans, 1970).

Another measure of flow rate can be derived using the time interval of a magnetic polarity transition, which is 1000–4000 yr according to McElhinny (1973). About 50 percent of the reversals studied at sections near the centres of large volcanoes have transitional directions. There is therefore a 50-percent chance of a flow having been erupted during any 1000–4000 yr interval.

Some information on the composition and oxidation state of the iron-titanium oxides is provided by palaeomagnetic studies. Massive, grey, flow centres in general have relatively low remanent magnetic intensity and coercivity, whereas the red blocky tops of flows have relatively high intensity and high coercivity. However, no extensive studies relating magnetic minerals and magnetic properties have been undertaken. The range of magnetic properties observed during demagnetisation is discussed by Wellman et al. (1969) and Wellman (1971).

Hoffman (1984) described a case of self-reversal found in some samples of a flow from Liverpool volcano. The initial mineral in these samples was a titanomagnetite rich in titanium. This was not oxidised during cooling, so it had too low a Curie temperature for a stable thermoremanent magnetisation. A stable, chemical, remanent magnetisation was acquired through low-temperature oxidation of the titanomagnetite to titanomaghemite, at some much later time, after the geomagnetic field direction had changed.

1.4.12   Summary

(1) Areas of Cainozoic volcanic rocks are generally coincident with areas of higher topography. There is circumstantial evidence that the highland uplift and volcanism have the same cause — namely, effects related to the asymmetrical breaking of the lithosphere at the time of formation of the Tasman Sea and Coral Sea basins. Topographic doming in northern Queensland apparently is directly related to the late Cainozoic volcanism.

(2) Present-day heat-flow in east-Australian areas of Quaternary volcanism is about 1.5 times the world average, whereas the area of Cainozoic volcanism as a whole has an average, present-day heat-flow near world average.

(3) The lithosphere is abnormally weak, according to the results of gravity and topography studies. Earthquakes take place in a belt along the highlands, but are confined largely to the upper 20 km of the crust.

An elastic/brittle upper crust above a more plastic lower lithosphere is inferred from these observations.

(4) Earthquake focal-mechanism solutions for southeastern Australia, and *in-situ* stress measurements, are indicative that areas of late Cainozoic volcanism currently are under compression (reverse and transcurrent faulting).

(5) The Phanerozoic lithosphere in Australia seems to differ from the Precambrian lithosphere. It has a low-velocity layer in the range 100 to 200 km, and strong layering (determined by seismic-reflection studies) forms the lower part of the crust, as though it was underplated.

(6) There is seismic evidence for underplating of the Phanerozoic crust after cratonisation, and this underplating appears to be strongly developed under the eastern highlands.

(7) Seismic P-wave residuals have high positive values in Tasmania, consistent with the existence of hot lithosphere and a mantle plume in that area.

(8) Central volcanoes are underlain by mafic intrusive complexes about one third the volume of the overlying volcanoes. The intrusive complex has a roughly spherical shape, and a centre of mass about 6–8 km deep.

1.5   Upper Mantle and Crust of New Zealand

1.5.1   Topography and bathymetry

Continental crust of the New Zealand region is crossed by the seismically active lithospheric boundary that separates the Indo-Australian and Pacific plates (Fig. 1.2.1). This plate boundary and its processes have a band of major geophysical anomalies that extends 100–200 km from the boundary. The Cainozoic intraplate volcanoes of New Zealand are mainly outside this zone of plate-boundary processes, in areas where geophysical anomalies are relatively small. Geophysical anomalies in the New Zealand region associated with the intraplate volcanoes are not well defined, because of their small size, and because much of the continental crust is below sea-level, making some geophysical measurements difficult. This section is a summary of geophysical results for the whole of the New Zealand region, together with interpretations only on the significance of results for the intraplate areas.

Known intraplate volcanoes on the Indo-Australian plate are on the topographic high of the Northland to Auckland area (Fig. 1.1.7). They are present on the Pacific plate, along the margin of the topographic high of South Island, and in the Chatham Rise and Campbell Plateau area, on relative bathymetric highs that are generally near the edge of the continental crust (Fig. 1.1.8). The original volume of volcanic material is small compared to the volume of the topographic features. The topographic highs on the Pacific plate represent regional uplift of basement, and thinning of the sediment layers towards the volcanoes is seen on seismic-reflection profiles (Davey, 1977; R.H. Herzer, personal communication, 1986). The areas of these intraplate volcanoes therefore have been relative topographic highs throughout the Cainozoic.

1.5.2   Gravity and magnetic anomalies

Gravity anomalies on land were mapped by Reilly et al. (1977) and Whiteford (1979), and at sea by Bowin et al. (1981) and Davey & Watts (1983). Most of the areas of exposed and dredged Cainozoic intraplate volcanism are relative highs of free-air gravity anomalies. These highs in part are caused by upper-crustal structures: the association of volcanism with regional basement highs; thickening of sediments away from the basement highs; and mafic intrusions beneath the centres of the major central volcanoes locally having domed the basement (see Section 1.4.9). However, the gravity highs may be caused in part by a deeper (lower-crustal or mantle) regional structure related both to the formation of the basement uplift and the volcanism.

Most of the large volcanoes in the South Island of New Zealand and Campbell Plateau are well separated from other major geological structures. The volcanoes are about 300 km apart, and there are separate topographic and gravity highs associated with each volcanic centre. Major volcanoes on the Campbell Plateau (Auckland and Campbell islands) do not rise directly from the plateau, but are on broad, flat-topped topographic highs about 200 km across and 0.4 km high, which have free-air gravity anomalies of a similar or shorter wavelength. There is no broad topographic high at Banks Peninsula, because the area is one of Pleistocene sedimentation, but the peninsula is seen from seismic-reflection surveys of the sedimentary section to rest on a broad structural high (R.H. Herzer, personal communication, 1986) that was formed at a similar time to the volcanoes. There is also a corresponding gravity-anomaly high (Reilly, 1966). The highs are best defined at Campbell Island where the ratio of gravity anomaly to topographic anomaly is about 65 mGal/km. A depth of isostatic compensation of about 90 km (that is, within the lithosphere) is obtained using the equations of Crough (1978) and a swell diameter of 200 km at one-half maximum amplitude. Long-wavelength topographic and gravity highs can be attributed either to lithospheric heating or to underplating at the base of the crust. The one-half wavelength observed (200 km) is much shorter than that found elsewhere (Crough, 1978, 1981, 1983), if this is caused by lithospheric heating by hotspots.

Small-scale magnetic maps for most of the area are given by Geophysics Division (1969–79) and Davey & Robinson (1978, 1981). More detailed magnetic maps and profiles for the main intraplate volcanic areas are as follows: Banks Peninsula and Mernoo Bank (Gerard & Lawrie, 1955); Otago Peninsula (Woodward, 1976). Magnetic anomalies in New Zealand have been used only in general to map the extent of concealed volcanic rocks and the positions of the related intrusions.

1.5.3   Seismic velocities and crustal structure

Crustal thicknesses of 30–40 km for New Zealand (Thomson & Evison, 1962), 20 km for the Campbell Plateau (Adams, 1962), 20 km for the Lord Howe Rise, and 15–20 km for the New Caledonia Basin and Norfolk Ridge (Officer, 1955) were interpreted from surface-wave dispersion studies. These thicknesses are roughly consistent with the following seismic-refraction crustal thicknesses (Stern et al., 1986): (1) 32 km west of Lake Pukaki in the South Island (Calhaem et al., 1977); (2) 35 km near Wellington (Garrick, 1968); (3) 25 km in Northland, and underlying material from 25 to 45 km increasing in velocity from 7.6 to 7.9 km/s (T.A. Stern, personal communication, 1986); (4) 15 km in the central volcanic area of crustal extension in the North Island, underlain by anomalously low-velocity mantle that has a velocity of 7.4 km/s (Stern & Davey, 1985).

P-wave residuals in the New Zealand region have large differences of up to 3 s (Randall, 1971; Robinson, 1976). These are interpreted as being caused mainly by the high velocity of underthrust oceanic lithosphere.

1.5.4   Seismicity

Earthquakes in the New Zealand region are located routinely by means of the New Zealand National Seismograph Network, if they are within 10 degrees of

Wellington, and if either their magnitudes exceed about 4.0 (ML) or if they are felt (Adams, 1968). Most of these earthquakes are concentrated in a band along the plate boundary. Earthquake distribution in Figure 1.5.1 is for the selected intraplate regions (as defined in Section 1.1.3), so it excludes most earthquakes that took place in New Zealand.

Earthquakes in the Northland-Auckland region (Eiby, 1964) are (with one exception) restricted to, or are near, the land area. There are three minor concentrations of earthquakes. The cause of these Northland earthquakes is not certain, but they are clearly a feature of this relative basement and topographic high. The major concentration of earthquakes on the Pacific plate away from the plate boundary is along the axis and the northern margin of the Chatham Rise. The Chatham Rise earthquakes may be related to stresses associated with the abrupt change in crustal thickness.

1.5.5   Heat-flow

Pandey (1981) presented heat-flow data for much of New Zealand, including the Northland, Auckland and Huntly areas (10 values), the Canterbury region (five values), and the South Canterbury Bight and Great South Basin (seven values; Fig. 1.5.2).

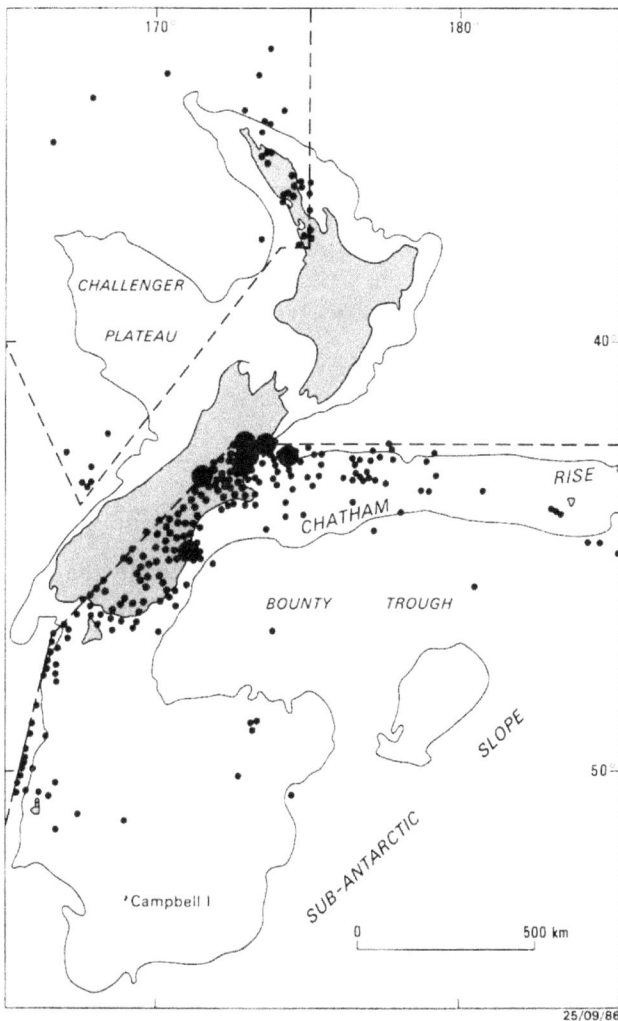

Figure 1.5.1   Earthquake epicentres for 1835–1984 for the intraplate areas of the New Zealand region (positions from the New Zealand Department of Scientific and Industrial Research, Geophysics Division, 1986). Small filled circles represent earthquake magnitudes of less than 6.0. Larger filled circles represent magnitude 6 or greater.

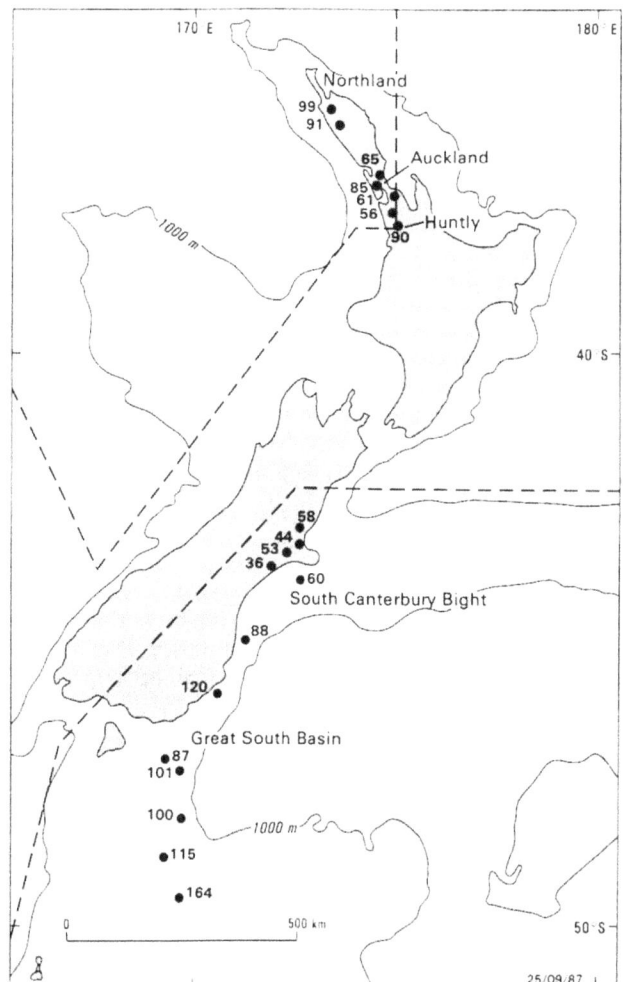

Figure 1.5.2   Average heat-flow values in mW/m² of the same intraplate areas shown in Figure 1.5.1 (after Pandey, 1981).

The Northland-Auckland-Huntly region has a mean value of 81.4 ± 15.7 mW/m². Two determinations for Northland average about 95 ± 6 mW/m² (about 52 percent higher than the world mean continental heat-flow of 62.3 mW/m²; Jessop et al., 1976). Pandey's four determinations for the Auckland area (Orewa to Pukekohe) have a wider spread than elsewhere in the region — from about 56 to 85 mW/m², averaging about 67 ± 13 mW/m², which is slightly higher than the mean continental value. Four closely-spaced values for the Huntly area average about 90 ± 8 mW/m² (about 44 percent higher than the mean value). The geothermal gradient throughout this region averages about 35 ± 7°C/km. Pandey (1981) derived a high reduced heat-flow of 61.8 mW/m² for this region (including five values not far from Huntly, but outside the zone of Cainozoic intraplate volcanism as here defined), and inferred that the major source of heat is in the upper mantle. He calculated that melting conditions exist at a depth of about 55 km. Average radioactive heat generation is 1.21 ± 0.53 μW/m³ for the Northland to Huntly area yielding a still higher value for reduced heat-flow of 67.8 mW/m².

Heat-flow is low in the Canterbury region (Pandey, 1981). Five heat-flow determinations average 50.3 ± 10.2 mW/m² (about 20 percent less than the mean continental value). Average radioactive heat generation is 2.42 ± 0.82 μW/m³, and reduced heat-flow is therefore 26.1 mW/m², which is lower than elsewhere in New Zealand. The geothermal gradient averages about 24 ± 4°C/km. Pandey inferred normal stable lithosphere beneath this region. The Great South Basin and part of the South Canterbury Bight — shown in Figure 1.5.2 extending from about 100 km north of Dunedin to nearly 50°S — has notably high heat-flow, averaging 110.7 ± 26.5 mW/m² which is 78 percent higher than the mean continental value. The reduced heat-flow is very high at 94.8 mW/m², using an average radioactive heat generation of 1.59 ± 0.45 μW/m³, and the geothermal gradient averages about 39 ± 11°C/km. Pandey inferred melting conditions at a depth of only about 35 km, which he took as the base of the lithosphere in this region. This assumes that there is in fact a partially molten layer, and that no significant heat transfer has taken place by mass movement.

1.5.6 Crustal strain

Crustal strain in the New Zealand region is dominated by the effects of oblique interaction of the Pacific and Australian lithospheric plates. Late Cainozoic poles of rotation for this movement are only 7–15° southeast of New Zealand, so the oblique interaction is different along different parts of the plate boundary, and there are strike-slip and either divergent or convergent components of movements. Discussions of Cainozoic strain were given by Walcott (1978, 1984c) and Sporli (1980).

1.5.7 Summary

(1) The average crustal thickness beneath North Island and South Island is 30–40 km, whereas surrounding ridges and plateaux (including the Northland Peninsula, Chatham Rise, and Campbell Plateau) have a thinner crust of 20–30 km.

(2) Areas of late Cainozoic intraplate volcanism have been basement highs throughout the late Cainozoic, and at present are regional basement highs and positive regional gravity highs that have a diameter (at one-half maximum amplitude) of about 200 km. These highs could be caused by lithosphere heating or by underplating at the base of the crust.

(3) Northland and Great South Basin areas have heat-flow values sufficiently high to be indicative of partial melting in the lower lithosphere. Heat-flow near Banks Peninsula is relatively low, and is not consistent with partial melting in the lower lithosphere.

1.6 Regional Influences on Magma Emplacement

1.6.1 Initiation and tectonic setting of volcanic provinces

Substantial volumes of magma may be erupted away from plate boundaries, but volcanism in many intraplate settings receives less attention than does the more 'dynamic' plate-margin type of volcanism. However, conditions that control the rise of large volumes of magma to the surface are similar in all tectonic settings (for example, Cas & Wright 1987). The rise of magma through the crust is facilitated most readily by the existence of an extensional stress field in the lithosphere (Shaw, 1980). These stress fields clearly exist in divergent plate-margin settings, but they are found also in both the arc and backarc regions of convergent plate boundaries and in strike-slip settings (where conditions are transtensional). Extensional stress fields also exist in intraplate settings, as considered in this section (see also Section 1.7.3).

Voluminous outpourings of magma have two requirements. First, and obviously, there must be an ample supply of magma, corresponding to a significant subsurface heat source that is most likely to be of mantle origin. Second, the lithosphere and especially the crust must have an appropriate stress-field configuration in order that large volumes of magma can rise to the surface. Large volumes of basalt apparently can reach the Earth's surface in areas of continental crust without significant contamination, or without being accompanied by significant proportions of crust-derived magmas, so mafic magmas evidently can rise rapidly through the lithosphere, preventing significant interaction with the crust. This situation may be feasible only where the crust and lithosphere are in an extensional, transtensional, or 'atectonic' stress state (Cas & Wright, 1987). Implied, rapid, magma-ascent rates are dealt with further in Section 6.1.5 where xenolith entrainment and transport are considered.

Conditions in the lithosphere are most amenable to the eruption of large volumes of magma where the minimum principal stress component is horizontal, and where the maximum principal stress component is vertical — that is, a state of extension (maximum stress component is equivalent to the lithostatic load) — or the maximum principal stress component is horizontal (a state of strike-slip; Shaw 1980). Subvertical fractures can be propagated and magma can dilate the fractures and rise to the surface where the magma fluid pressure exceeds both the minimum principal stress component and the tensile strength of the country rock.

Extensive Cainozoic basaltic provinces such as the Newer Volcanics of southeastern Australia (Section 3.7.4), the north-Queensland provinces (Section 3.2), and the Northland-Auckland area of New Zealand (Section 4.2), are therefore evidence that the lithosphere may have been relatively 'relaxed' at the time of formation of these volcanic provinces. However, note that measurements of *present-day* stress in southeastern Australia are consistent with a compressional stress field (Section 1.4.6). Basalts in northern Tasmania are associated mostly with Tertiary rifts and faults (Section 3.8.3), and the importance of extensional tectonics during development of the intraplate volcanic provinces of New Zealand has been recognised (Sections 4.9.2–3).

Intraplate continental volcanism is found in three main tectonic settings:

(1) *Rifting or newly rifted continents.* Examples are the East African Rift system (Mohr & Wood, 1976; Mohr, 1983), the Deccan Traps (Choubey, 1973; Subbarao & Sukheswala, 1981), Karoo volcanism (Bristow & Saggerson, 1983), and the late Mesozoic and early Cainozoic volcanism discussed in this volume (see point 3 below). Volcanism in this setting clearly takes place under the influence of an extensional lithospheric stress field, as discussed above. Two distinct types of volcanism are associated with this tectonic setting (see also Section 7.4.3). First, extensive flood or plateau-basalt volcanism affects areas marginal to and beyond the rift. Second, where a narrow rift is well developed such as in East Africa, compositionally diverse volcanism is notable within the rift valley. The volcanism is mafic to felsic in composition, but is characteristically bimodal. Alkaline to peralkaline magmas are volumetrically significant. The volcanoes themselves are also diverse, and include stratovolcanoes, calderas, and maars (for example, Di Paola, 1972) as well as fissure-vent systems.

(2) *Basin-and-range provinces.* The type example in the western United States is inland from, and closely associated with, an active transpressional plate margin — the major, transcurrent-transform San Andreas fault system. Normal, convergent-type, plate-margin continental arcs exist to the north and south of this transcurrent system (Cascades arc, Middle America/Mexican arc, respectively). Basin-and-range type magmatic activity in the type example therefore is in close spatial association with both transcurrent-transform tectonic activity and with more linearly focused, plate-margin, arc-type magmatic and tectonic activity. There are different hypotheses for the origin of the extension and volcanism associated with the Basin and Range province, including the transtensional effects of the San Andreas fault system, the effects of a subducted segment of the East Pacific spreading ridge underneath the North American continent, and continental backarc spreading (for example, Eaton, 1982, 1984). The volcanism is diverse in character, and includes plateau (or flood) and plains-basaltic volcanism

(for example: Columbia River Plateau, Swanson et al., 1975; Snake River Plain, Greeley, 1982a,b), as well as central-volcano type volcanism within the basin-and-range terrain. The volcanism typically is bimodal but includes alkaline, calcalkaline, and tholeiitic mafic to felsic rocks (for example, Christiansen & Lipman, 1972). Intermediate and felsic products in fact are abundant and commonly are associated with prominent stratovolcanoes and caldera volcanic centres.

(3) *Intraplate continental volcanism well away from the influence of active plate margins.* Two types of volcanism can be recognised in this category (see also Sections 7.4.3–4). First are the lines of volcanoes believed to be related to the passage of lithosphere over some form of mantle hotspot or hotline. The volcanoes of the Cameroon area in western Africa represent a possible example (for example, Fitton, 1987). Second is the volcanism that develops on the uplifted flanks of some rifted continental margins after periods of rifting and sea-floor spreading — for example, the volcanism in Marie Byrd Land, Antarctica (see Section 7.4.3). Both types of volcanism are thought to be represented in eastern Australia (Sections 1.1.1, 1.7.2). The central-type felsic volcanoes of eastern Australia define a southwards younging hotspot trail (or trails), and much of the dominantly mafic, lava-field type volcanism appears to be related to post-rifting uplift that led to the formation of the Great Divide paralleling the east-Australian margin (Sections 7.3–4).

'Intraplate' volcanism in eastern Australia has been active since the late Mesozoic, and therefore originally may have been related to, or was a by-product of, the separation of the continental blocks of Antarctica, and the Norfolk Ridge, Lord Howe Rise, and New Zealand assemblage, from Australia (point 1 above; Section 1.2). Continental volcanism in eastern Australia therefore can be treated in both the context of intraplate magmatism well away from the active plate margin (middle Tertiary onwards and of hotspot and uplifted-flanks types) and originating in or near a tectonically active rift setting (late Mesozoic to early Tertiary). Extensional stress-field conditions may have existed in both situations (Section 1.7.3).

The Northland and Auckland volcanic provinces in New Zealand are currently well behind the active plate-margin zone, but are superimposed on Tertiary arc systems (Sections 1.3.3, 4.9.2). They are in an area of possible extension around and behind the present arc system of the North Island (Sporli, 1980; Cole, 1984; Walcott, 1984b). There may be some similarities with the setting of basin-and-range related volcanism, including the fact that the arc system to the east of the Northland-Auckland province, is within the influence of the transcurrent-transform plate margin of the Alpine Fault. However, the scale of volcanism is greatly different, and basin-and-range style, horst-and-graben topography is absent.

The localised Tertiary volcanic rocks of the South Island appear to have developed in relation to an extensional tectonic setting closely related to the developing transcurrent-transpressional Alpine Fault system (Section 4.9.3). The relatively small volume of magma associated with the South Island Tertiary volcanism may well be a reflection of Alpine Fault transpressional influences. Nevertheless, there are local extensional tectonic effects associated with South Island and Campbell Plateau volcanism away from the immediate vicinity of the Alpine Fault system (Section 4.9.3). Note also the existence of a Miocene dyke swarm along 110 km of the Alpine Fault (Section 4.3.7). The oblique orientation (east-west) of the dykes to the trend of the fault zone (northeast) is consistent with dyke intrusion under the influence of a local, pull-apart, strike-slip regime within the Alpine Fault system.

1.6.2   Location of vents

Intraplate volcanism is, in part at least, a response to intraplate extensional forces, so volcanism should to some degree be controlled by crustal fractures. This is observed in the alignment of eruption centres and points at all scales and may also be represented by fissure-vent systems. Wellman & McDougall (1974a) and Sutherland (1978, 1983) noted at the regional scale linear traces of volcanic centres in eastern Australia that have definite age trends, but whether the volcanic centres are developed on single, crustal-fracture systems is doubtful.

Alignment of eruption points and vents at a more local scale almost certainly reflects basement-fracture control. There are numerous examples of this in the Newer Volcanics province of Victoria and South Australia (see also Section 1.3.1). The scoria-cone complex in the middle of the major Tower Hill volcanic complex appears to define two linear trends interpreted by Edney (1984a) to represent basement fractures along which basaltic magmas rose. Aligned scoria cones, spatter cones, and a crater lava lake at the Mount Eccles volcanic centre also appear to represent aligned eruption points along a crustal fracture. The Wedderburn Line in Victoria is a major, north-south trending lineament that marks structural discordance between the rock systems on either side. It is marked by the Avoca River valley that has been filled in large part by an extensive lava flow, and in the south is marked by an alignment of several basalt eruption points. The Wedderburn Line therefore appears to be a major fault along which basaltic volcanism has taken place. The Red Rock volcanic complex, the Alvie scoria-cone complex, and the Warrion Hill lava complex, all north of Colac, are closely spaced along a northeasterly trend, indicative of development above a basement fracture-fissure zone.

P. Wellman (personal communication, 1987) noted that about half of the central volcanoes of eastern Australia, as well as Barrington volcano and the north-Queensland lava-field provinces, are on, or close to, a major fault. Other eruptive centres may lie close to unrecognised major faults. The north-south position of any central volcano is determined by the position of the apparently migrating hotspot source, whereas the east-west position of many provinces is localised on the mainly north-south faults. The Nulla, Sturgeon, and Chudleigh provinces of northern Queensland are associated with a topographic dome on the southeastern boundary of the Precambrian Georgetown Inlier. The Cape Hillsborough province is near the western

boundary fault of the Cainozoic Hillsborough Basin, and the Springsure and Buckland provinces immediately overlie the western boundary fault of the Denison Trough. Tweed volcano is on a dislocation of magnetic anomalies caused by an east-west fault. Ebor volcano is only 5–10 km from the mainly transcurrent Demon Fault, Nandewar volcano is on Mooki-Hunter Thrust, and Barrington volcano is 10 km from the Peel Fault.

Many vent systems in continental basaltic provinces are fissures that in some instances are known to extend many tens and perhaps even more than 100 km (for example, the Columbia River Plateau; Swanson et al., 1975; Greeley, 1982a; Hooper, 1982a,b). The fissure vents may be fed by complex dyke systems whose trend may be an indicator of the orientation of the palaeo-stress field at the time of propagation. This approach has been used in several volcanic provinces with some apparent success (for example: Nakamura, 1977; Zoback et al., 1981; Eaton, 1982, 1984). Piiger (1982) proposed that the principal state of tension in northern Queensland was northwest-southeast, in southern Queensland was east-west, and in central New South Wales was southwest-northeast. This change in orientation of the tension axis with latitude is the same as that in the present theoretical stress field calculated by Cloetingh & Wortel (1985), as discussed in Section 1.7.3.

Most vent fissure systems are not active along their whole length for prolonged periods, except possibly for flood-basalt provinces, and activity is soon restricted to several points along the fissure system which then becomes a system of aligned central-volcano eruption points such as spatter cones and scoria cones (for example, the Mount Eccles system described above). Fissure systems may be difficult to identify because they may be buried by the lavas produced from them, and no major fissure systems of regional significance have been identified yet in eastern Australia and New Zealand.

1.6.3   Influence of water

The distribution of both subsurface and surface water exerts a major influence on the formation of phreatomagmatic eruption centres such as maars, tuff rings, and tuff cones, and the style of eruption (Section 2.5). Extensive subsurface groundwater reservoirs will interact with rising magmas and lead to the formation of a large number of phreatomagmatic centres at the expense of magmatic explosive centres such as scoria cones and lava volcanoes. Joyce (1975) showed for the Newer Volcanics basalt province of Victoria and South Australia that phreatomagmatic centres such as Tower Hill, Purrumbete, Leura, and Red Rock, are concentrated in the southern half of the field where rising basaltic magmas passed through major aquifers in the Tertiary sedimentary successions of the Otway Basin (Fig. 1.6.1). Phreatomagmatic eruption centres are rare to the north where the basement consists only of relatively impermeable Palaeozoic metasedimentary basement. Maars in northern Queensland are developed only in the Atherton province possibly because of the high rainfall that ensures a nearly constant high groundwater table.

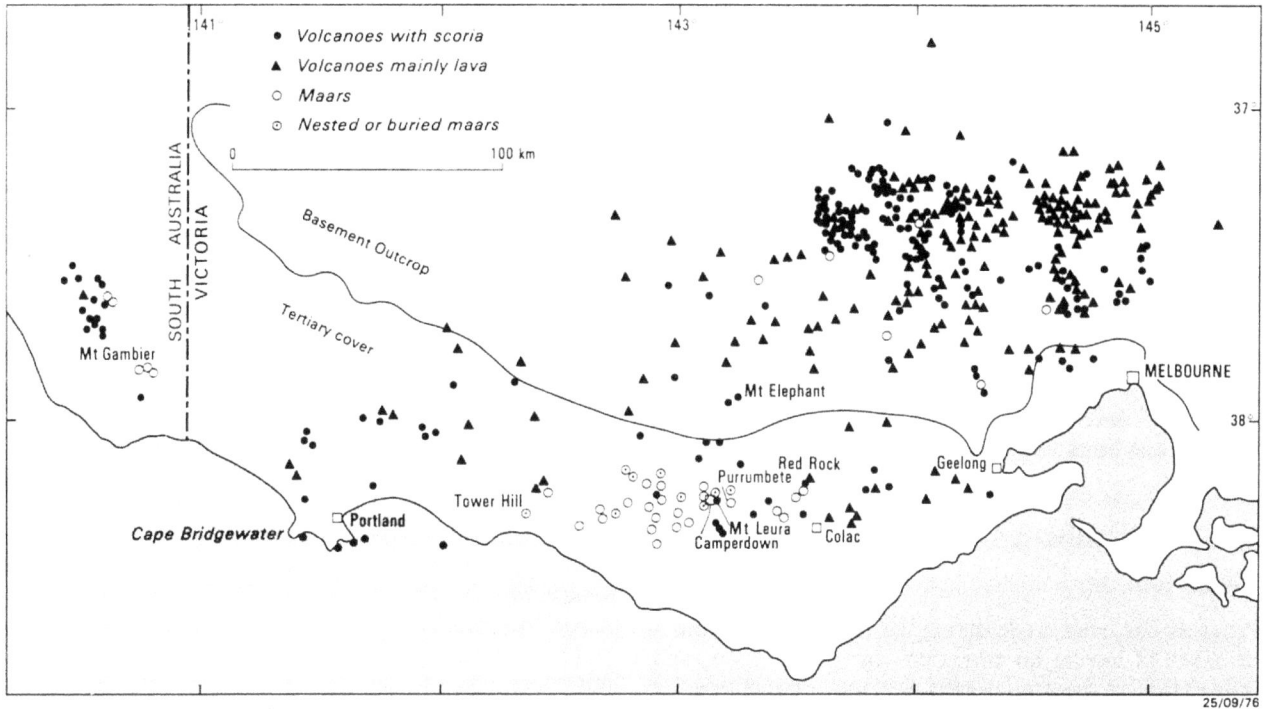

Figure 1.6.1 Distribution of four volcano types in the Newer Volcanics of western Victoria and southeastern South Australia, showing particularly the correlation between phreatomagmatic eruption centres (maars) and the aquifers of the Tertiary sedimentary cover of the Otway Basin (adapted from Joyce, 1975, fig. 2).

Volcanic eruptions and the types of volcanic centres that develop will be significantly affected by the influence of seawater where volcanism takes place in epicontinental seas such as Torres Strait or the Campbell Plateau off New Zealand where large areas of continental crust are submerged. Seawater may interact explosively with erupting submarine magma leading to phreatomagmatic explosive activity and phreatomagmatic eruption centres such as Surtseyan volcanic centres (Thorarinsson, 1967; Cas et al., 1986). This has been noted as an influence on phreatomagmatic volcanism in the formation of volcanic islands in Torres Strait, northern Queensland (Section 3.2.2). In addition, basaltic volcanism on the early Tertiary continental shelf seas of central Otago in New Zealand produced numerous Surtseyan tuff cones (Coombs et al., 1986; Cas et al., 1986). Surtseyan tuff cones are present also in the Antipodes and Chatham islands on the submerged Campbell Plateau (Section 4.4).

1.7 Volcanic Time-Space Relationships

1.7.1 Introduction

Intraplate volcanism in general takes place far removed from plate boundaries and therefore is not explained readily in terms of current plate-boundary interactions. Ultimately, however, intraplate volcanism may reflect a tensional regime which results either directly from plate motions or indirectly in response to structures or conditions imposed by plate-tectonic processes (Section 1.6.1). The Cainozoic intraplate volcanism of the

eastern Indo-Australian plate and of southern New Zealand on the Pacific plate, is found in a range of geological environments, including Palaeozoic fold belts, late Cretaceous to early Tertiary oceanic lithosphere, and rifted continental fragments. The timing and spatial distribution of this volcanism are related in this section to specific geological events and regional plate-tectonic patterns.

The volcanism can be divided conveniently into four regions which have had significantly different plate-tectonic histories — the east-Australian highlands from northern Queensland to Tasmania, the Tasman Sea, Norfolk Island, and New Zealand including the Campbell Plateau and Chatham Rise. Age relationships of the volcanism within each region are reviewed first in this section (giving particular attention to any systematic time-related progression in the location of the volcanism), and the consistency of published ages and the duration of volcanism in individual provinces are evaluated. The plate-tectonic evolution of the eastern part of the Indo-Australian plate was discussed in Section 1.2 with reference to changes in plate motion that may have produced the tensional conditions apparently necessary for intraplate volcanism. An analysis of the Cainozoic motion of the Indo-Australian plate in the hotspot, or mantle-fixed, reference frame was presented in Section 1.2. The proposed origins for the Cainozoic intraplate volcanism are examined in this section in relation to time-space relationships and plate-tectonic processes, and the strengths and weaknesses of each are assessed.

The conclusion is reached that part of the intraplate volcanism in the east-Australian region was a response to a regional tensional stress field, perhaps imposed by plate-boundary forces. Another, age-progressive part reflects motion of the Indo-Australian plate relative to the sub-lithospheric mantle. Other less significant factors contributing to the intraplate volcanism may be a response to erosional rebound of the eastern highlands and a thermal pulse resulting from opening of the Tasman Sea (see also Section 7.3). Two parallel seamount chains in the Tasman Sea almost certainly record the motion of the Indo-Australian plate in relation to stationary hotspots over the last 30 to 40 Ma. The intraplate volcanism of southern New Zealand is more enigmatic as it is found over a broad area, but it does appear to record the motion of the Pacific plate relative to the mantle. However, this activity may have resulted from a tensional regime caused by dextral shear along the Indo-Australian/Pacific plate boundary, rather than reflecting passage of a hotspot.

1.7.2 Distribution and age of volcanism

Geochronology

Progress has been made during the last 25 years from a situation of having no numerical age information, and little definitive biostratigraphic age data, concerning the Cainozoic volcanism of the region, to where there is now a large number of isotopic ages. Results are available from virtually all the volumetrically significant volcanic provinces, as well as from many of the smaller provinces. The great majority of these data were obtained through application of the K-Ar method, mainly on whole-rock samples of lavas. This dating technique — based upon the accumulation of radiogenic argon ($^{40}Ar^*$) produced from the decay of the naturally occurring radioactive isotope of potassium, ^{40}K — is particularly useful for dating relatively young igneous rocks because of its great sensitivity. A K-Ar age on an igneous rock is a good estimate for the time since crystallisation and cooling, provided underlying basic assumptions are met. The two main assumptions are that all pre-existing $^{40}Ar^*$ was lost from the rock at the time of eruption and that $^{40}Ar^*$ generated by decay of ^{40}K since cooling has been trapped in the rock. The first assumption generally is met for subaerially erupted lavas that are free of xenolithic material, but departures from the second assumption are probably the main cause of incorrect ages.

The physical measurements required to formally calculate a K-Ar age can be made with high precision (about 1-percent standard deviation), but the geological validity of the measured age needs to be established. Much reliance is placed upon consistency or otherwise of results on rocks from a given area in assessing the reliability of K-Ar ages. If the measured ages for a particular volcanic province fall within a narrow range, and especially if there is consistency with the stratigraphy and between ages on separated minerals, then a good estimate for the timing of the eruptions is likely to have been made. If the range of measured ages in a province is large, there may be difficulty in determining whether volcanic activity has extended over a long period, or whether the spread in apparent age is the result of variable loss of radiogenic argon from the samples, leading to low apparent ages. Careful sample selection helps minimise such problems.

High-temperature, potassium-bearing minerals generally have good retention properties for $^{40}Ar^*$, and normally will begin to lose argon by diffusion only above temperatures of about 200°C. However, many of the Cainozoic igneous rocks are lavas that are too fine-grained for mineral separation, so most age measurements have been made on whole-rock samples. The ideal kind of whole-rock sample for K-Ar dating is well crystallised and free of alteration, belonging to category A of Wellman & McDougall (1974b). However, many samples used for dating do not fall in the ideal category as they contain a proportion of glass or poorly crystallised mesostasis in which potassium is concentrated. Some leakage of $^{40}Ar^*$ might have taken place in such cases, especially if glass is devitrified or altered, as diffusional loss of $^{40}Ar^*$ readily takes place from such materials, possibly even at ambient temperature. Unfortunately, judgements on the suitability of samples for K-Ar dating purposes remain somewhat subjective. Thus, in assessing published data, the degree of consistency of the results from a given area is an important criterion.

The location, age range, and mean age for the individual volcanic provinces are presented in Tables 1.7.1–3. Ages have been recalculated where necessary to conform with the currently recommended decay constants (Steiger & Jager, 1977). The critical table of Dalrymple (1979) is useful for this purpose. The numerical time scale of Harland et al. (1982) is used here, as shown in the following where the estimated ages for the boundaries are given in parentheses: Cretaceous-Tertiary (65 Ma), Palaeocene-Eocene (54.9 Ma), Eocene-Oligocene (38.0 Ma), Oligocene-Miocene (24.6 Ma), Miocene-Pliocene (5.1 Ma) and Pliocene-Pleistocene (2 Ma).

The amount of published data available in 1985 when this review was completed, ranges widely from quite detailed geochronological studies to just one or two reconnaissance measurements, and this should be borne in mind. Here, the broader aspects of the distribution of volcanism in space and time are examined. More comprehensive discussions of additional geochronological data from many of the provinces are given in Chapters 3 and 4. These additional data include previously unpublished dates as well as new data obtained since 1985.

Eastern Australia

Wellman & McDougall (1974a) recognised three types of volcanic province (Section 1.1.3). Location and age data for the central-volcano and leucitite-suite provinces are summarised in Table 1.7.1, and those for lava-field provinces are given in Table 1.7.2.

Relatively large volumes of basaltic lava were erupted from a vent area to produce a substantial volcano in many of the central-volcano provinces. An essential distinguishing characteristic of this type of province is the presence of some felsic flows, but felsic and mafic intrusions also may be present. The composition of the volcanism is dominantly mildly alkaline, although some subalkaline rocks have been recognised. Minor differences exist between the compilation of data for central-volcano provinces in Table 1.7.1 compared

Table 1.7.1 Summary of location, age, and volume data for central-volcano and leucitite provinces of eastern Australia

Province	Latitude °S	Longitude °E	Age range Ma	Average age Ma (± 1 s.d.)	Number of ages	Material dated[1]	Volume (km³)[2]	Reference
Central-volcano provinces								
Hillsborough	21.0	149.0	31.3–34.1	33.2 ± 0.8	10	WR, Pl, AF, Bi	25	McDougall & Slessar (1972)
Nebo	21.4	148.2	28.4–35.1	31.0 ± 2.1	13	WR	100	Sutherland et al. (1977)
Peak Range	22.8	148.0	27.4–35.2	31.0 ± 2.4	7	WR, AF	800	Wellman (1978)
Springsure	24.0	148.1	23.9–33.6	27.3 ± 2.1	20	WR, AF	360	Webb & McDougall (1967); Ewart (1982a).
Glass Houses	26.7	152.9	25.2–27.3	25.8 ± 0.9	5	WR, Pl, AF	25	Webb et al. (1967); Ewart (1982a).
Main Range	27.9	152.4	22.6–27.2	24.2 ± 1.2	12	WR, AF	840	Webb et al. (1967); Ewart (1982a).
Tweed	28.4	153.3	20.5–24.2	22.7 ± 1.2	12	WR, AF	4000	Webb et al. (1967); McDougall & Wilkinson (1967); Ewart (1982a).
Nandewar	30.2	150.1	17.4–21.0	18.3 ± 0.9	13	WR, AF	600	Stipp & McDougall (1968a); Wellman et al. (1969).
Ebor	30.4	152.4	18.2, 19.2	18.7 ± 0.7	2	WR, Pl	270	McDougall & Wilkinson (1967); Wellman & McDougall (1974b).
Warrumbungle	31.3	149.0	13.6–17.1	15.0 ± 1.1	11	WR	400	Dulhunty & McDougall (1966); McDougall & Wilkinson (1967); Dury et al. (1969); Wellman & McDougall (1974b).
Comboyne	31.6	152.5	16.1–16.4	16.3 ± 0.2	3	WR	40	McDougall & Wilkinson (1967); Wellman & McDougall (1974b).
Canobolas	33.4	149.0	11.2–13.0	12.0 ± 0.6	12	WR	50	Wellman & McDougall (1974b)
Western Victoria[3]	37.7	143.3	<0.59–6.99	3.4 ± 1.7	41	WR, AF	1300	McDougall et al. (1966); Aziz-ur-Rahman & McDougall (1972); Wellman (1974b); Singleton et al. (1976).
Leucitite-suite provinces[4]								
Byrock	30.7	146.3	16.8	16.8	1	Bi		Sutherland (1985)
El Capitan	31.2	146.2	12.2, 15.7	15.7?	2	WR		Wellman et al. (1970); Sutherland (1985).
Cargelligo	33.4	146.5	10.4–14.8	12.5 ± 1.4	8	WR, Lc, Bi		Wellman et al. (1970)

[1] WR — whole rock; Pl — plagioclase; AF — alkali feldspar; Bi — biotite; Lc — leucite.
[2] Volume estimates from Wellman (1971).
[3] Consists of the felsic Macedon-Trentham area, the non-felsic Newer Volcanics, and the Cosgrove occurrence (Wellman, 1974b, 1983) where the rocks now are known to be leucitites (Section 3.6.6).
[4] Total volume of leucitite-suite provinces is about 10 km³ (Wellman, 1971).

Table 1.7.2 Summary of location, age, and volume data for lava-field provinces in eastern Australia

Province	Latitude °S	Longitude °E	Age range Ma	Average age Ma (± 1 s.d.)	Number of ages	Volume (km³)[1]	Reference
Piebald	15.1	145.1	<3			3[2]	Stephenson et al. (1980)
McLean	15.8	144.8	<4			9	Stephenson et al. (1980)
Atherton	17.5	145.5	<0.1–3			5	Stephenson et al. (1980)
Wallaroo	18.0	145.4	<5			20	Stephenson et al. (1980)
McBride	18.3	144.6	<0.1–2.8	0.7 ± 0.7	28	160	Griffin & McDougall (1975)
Chudleigh	19.5	144.3	<5			60	Stephenson et al. (1980)
Nulla	19.7	145.3	<0.1–4.7	1.8 ± 1.3	23	225	Wyatt & Webb (1970)
Sturgeon	20.3	144.2	<5			150	Stephenson et al. (1980)
Rockhampton[3]	23.3	150.4	67.3–71.7	69.9 ± 2.3	3	180	Harding (1969); Wellman (1978).
Bundaberg	24.8	152.4	0.6–1.1	0.9 ± 0.2	3	12	Wellman (1978)
Bauhinia	24.8	149.5	21.3–24.8	23.0 ± 1.6	5	85	Harding (1969)
Mitchell	26.0	148.2	20.6–23.8	22.6 ± 1.2	9	50	Exon et al. (1970)
Bunya Mountains[4]	26.9	151.8	22.7–23.7	23.3 ± 0.4	5	400	Webb et al. (1967); Ewart (1982a).
Brisbane	27.7	152.6	16.2–62.7	?	16	30	Webb et al. (1967); Green & Stevens (1975).
Central (older)	29.8	151.6	32.0–34.3	33.4 ± 1.2	3	1440	McDougall & Wilkinson (1967);
Central (younger)	30.0	151.3	19.0–22.6	21.0 ± 1.8	3		Wellman & McDougall (1974b).
Doughboy	30.3	152.2	37.5, 44.6	41.1	2	20	Wellman & McDougall (1974b)
Bunda Bunda	31.1	152.4	70.4, 72.2	71.3	2	120	Wellman & McDougall (1974b)
Walcha	31.4	151.8	44.9–57.7	50.3 ± 6.6	3	700	Wellman & McDougall (1974b)
Barrington	31.8	151.2	44.1–54.6	50.0 ± 3.8	14	1500	Wellman et al. (1969); Wellman & McDougall (1974b).
Liverpool west	31.8	150.1	31.8–35.5	34.2 ± 1.2	7	1500	Wellman et al. (1969)
Liverpool east	31.9	150.7	38.9–42.7	40.8 ± 1.8	4	1500	Wellman & McDougall (1974b)
Dubbo[5]	32.2	149.2	12.2–15.2	13.3 ± 1.3	5	8	Dulhunty (1971); Wellman & McDougall (1974b).
Airly[6]	32.9	150.1	34.6–42.7	39.0 ± 3.8	6	25	Dulhunty (1971); Wellman & McDougall (1974b).
Sydney	33.0	150.6	19.6–57.6+	?	22		Embleton et al. (1985)
Abercrombie (younger)	34.2	149.3	14.4–26.0	18.8 ± 2.7	27	200	Wellman & McDougall (1974b); Young & Bishop (1980); P. Bishop et al. (1985).
Abercrombie (older)					4		
Southern Highlands	34.5	149.5	36.4–50.8	44.0 ± 6.9	12	80	Wellman & McDougall (1974b)
Nerriga[7]	35.2	149.8	40.8–50.6	45.2 ± 2.8	14	2	Wellman & McDougall (1974b); Ruxton & Taylor (1982); Young & McDougall (1985).
Snowy Mountains	35.5	148.3	17.7–22.8	20.6 ± 1.8	10	24	Wellman & McDougall (1974b)
South Coast	35.7	150.3	26.6–31.8	29.5 ± 1.9	5	2	Wellman & McDougall (1974b); Young & McDougall (1982).
Monaro	36.5	149.2	36.9–54.9	45.6 ± 6.7	9	480	Wellman & McDougall (1974b)
Uplands	36.8	147.6	2.32, 2.33	2.3	2	2	Wellman (1974)
Toombullup[8]	36.9	146.3	37.1–44.3	39.6 ± 3.2	4	10	Wellman (1974)
Bogong	37.1	147.2	25.8–37.2	32.0 ± 4.2	5	140	Wellman (1974)
Bonang	37.2	148.7	38.7, 42.5	40.6	2	2	Wellman (1974)
Gelantipy	37.2	148.3	33.6–42.8	37.3 ± 3.2	9	20	Wellman (1974)
Howitt	37.2	146.7	32.7–35.7	34.2 ± 1.5	3	7	Wellman (1974)
Bacchus Marsh	37.6	144.4	54.8–64.3	60.3 ± 4.9	3	10	Wellman (1974)

continued

Table 1.7.2   Summary of location, age, and volume data for lava-field provinces in eastern Australia *(continued)*

Province	Latitude °S	Longitude °E	Age range Ma	Average age Ma (± 1 s.d.)	Number of ages	Volume (km³)[1]	Reference
Aberfeldy	37.8	146.4	27.0, 28.6	27.8	2	2	Wellman (1974)
Neerim	38.0	146.0	19.8–25.1	22.4 ± 2.2	4	10	Wellman (1974)
La Trobe	38.5	146.3	50.6–58.7	55.1 ± 4.1	3	700	Wellman (1974)
Flinders	38.5	145.3	39.9–48.3	43.8 ± 3.5	4	700	Wellman (1974)
Northwest[9]	41.0	145.6	13.3, 26.3	?	2		Sutherland & Wellman (1986)
North	41.1	147.2	30.7		1		Sutherland & Wellman (1986)
Northeast (younger)	41.1	147.8	16.0–16.4	16.2 ± 0.2	3		McClenaghan et al. (1982)
Northeast (older)	41.2	147.9	46.2–47.3	46.9 ± 0.6	3		Sutherland & Wellman (1986)
Northwest	41.2	146.5	38.1		1		Cromer (1980)
Central (younger)	41.9	146.7	22.4–24.2	23.1 ± 0.8	5		Sutherland et al. (1973)
Central (older)	42.2	146.6	30.1, 35.4	32.8	2		Sutherland & Wellman (1986)
Central-east	42.3	147.3	24.3–27.6, 36.3	25.6 ± 1.4	4, 1		Sutherland & Wellman (1986)
South	42.8	147.4	23.0–30.2	26.0 ± 3.2	4		Sutherland & Wellman (1986)

[1] Volume estimates mainly from Wellman (1971).

[2] Volumes for provinces north of 21°S are from Stephenson et al. (1980).

[3] Includes felsic rocks and strictly speaking should be classified as a central volcano.

[4] Includes minor felsic rocks and can be considered either as a lava field separate from Main Range or as part of the Main Range central-volcano province which has a similar age (Section 3.4.4).

[5] Considered as a central volcano by Wellman & McDougall (1974a) but is classified here as lava-field type because of the uncertain age relationship of the felsic to the more mafic rocks at Dubbo.

[6] Part of the Sydney province described in Section 3.6.4.

[7] Part of the South Coast province described in Section 3.6.6.

[8] Toombullup and the following nine areas are part of the Older Volcanics of Victoria (Section 3.7.2). Some of these areas (identified by Wellman, 1974a) are different to those distinguished as a result of more recent detailed work reported in part in Section 3.7.2. In particular, Bonang is considered now to be part of the Gelantipy field. Also, the La Trobe province includes the area previously called Thorpdale by Wellman (1974a).

[9] Cainozoic volcanism in Tasmania totals about 400 km³ (Sutherland, 1969a).

to those made previously by Wellman & McDougall (1974a) and Wellman (1983). In particular, age results have been pooled from each province to obtain an overall mean age, rather than attempting to distinguish subprovinces.

The central-volcano provinces are distributed over a relatively broad zone, extending southward from about 21°S in central Queensland to Victoria (Fig. 1.1.6). There is also alignment of individual groups of volcanic centres in narrower longitudinal zones. Many of the provinces have been well dated. There is commonly quite good agreement between K-Ar ages on high-temperature, alkali-feldspar separates and whole-rock samples (Table 1.7.1). Plagioclase and biotite also have been used for dating in a few cases. The mean age for most provinces provides a good estimate for the timing of igneous activity, as the duration of volcanism in individual provinces generally is restricted to less than 5 Ma.

Central-volcano volcanism has a remarkably systematic southward younging (Wellman & McDougall, 1974a). Regression of the mean age of each of the 13 provinces against latitude (Table 1.7.1, Fig. 1.7.1) yields the following relationship:

$$\text{Age (Ma)} = 69.66 - 1.708\ (\pm 0.086)\ x,$$

where x is the latitude. The correlation coefficient for this regression is 0.987, and the predicted latitude for zero age is 40.8°S. The rate of migration of the volcanism calculated from the slope of the least-squares-fit line is 65 ± 3 mm/yr, which is indistinguishable from the value previously derived by Wellman & McDougall (1974a). The significance of this age-progressive volcanism is discussed in more detail below and is interpreted most readily in terms of passage of the Australian continent across a broadly defined magma source or hotspot within the upper mantle

beneath the crustal plate. Age data from the small leucitite provinces of New South Wales are on essentially the same regression line (Fig. 1.7.1). Indeed, incorporation of these results in the regression changes the slope of the line by less than 0.5 percent. No central-volcano province older than 35 Ma has been recognised. Note also that from 35 Ma ago to the present, volcanic activity in this region was dominated by eruptions from central volcanoes.

Lava-field provinces consist mainly of mafic lavas, and have a wide diversity of form and size. Areally large lava fields are exemplified by the young provinces in northern Queensland and by the old (Eocene) Monaro province. Thick lava piles are found in some cases, possibly representing former shield volcanoes — for example, in the Liverpool Range. Some provinces consist of remnants of lava flows scattered over a wide area, or of valley-filling flows. These may have been more extensive in the past. The composition of the lavas is dominantly alkaline, but tholeiitic basalt is known in some provinces.

Volcanic activity in some lava-field provinces was restricted to an interval of a few million years. In others at least two widely separated periods of activity are recognised, and in a minority a large range of ages has been found, the meaning of which remains unclear (Table 1.7.2). The volume of magma in individual provinces ranges from a few cubic kilometres to several hundred cubic kilometres, even where allowance is made for erosional losses. The lava fields do not have any simple or obvious age-distribution pattern, although most of the activity took place before the central-volcano provinces were formed for any given latitude (Fig. 1.7.1). Note, however, there are some younger lava fields, especially in northern Queensland. This absence of age-progressive volcanism for the lava-field provinces was recognised by Wellman & McDougall (1974a). Sutherland (1981, 1983), however, interpreted virtually all the activity in terms of a hotspot model (see below).

Finally, Wellman & McDougall (1974a) showed that the overall rate of eruption was more or less constant throughout the Cainozoic, noting that lava-field volcanism dominated the activity until about 35 Ma ago, and that subsequent volcanism was produced mainly from central volcanoes. The total volume of Cainozoic volcanism was estimated to be about 20 000 km³ (Wellman & McDougall, 1974a).

Tasman Sea

The Tasmantid Seamounts or Guyots (Standard, 1961; Conolly, 1969) extend over more than 1300 km in a north-trending line of submarine mountains. They rise from depths of more than 4500 m in the middle of the Tasman Basin, between the eastern margin of the Australian continent and the Lord Howe Rise (Fig. 1.2.1). The seamounts are built upon oceanic crust formed between about 80 and 50 Ma ago by seafloor-spreading processes (Weissel & Hayes, 1977; Shaw, 1978). These seamounts are known to be of volcanic origin because of their general shape and the recovery of basaltic material from some of them (for example: David, 1932; Slater & Goodwin, 1973). Most of the

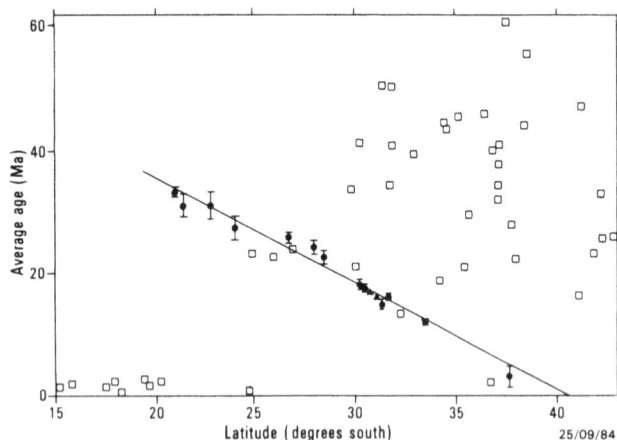

Figure 1.7.1 Age:distance relationships for the intraplate volcanism of eastern Australia. Average ages for the central volcanoes (filled circles; bar represents one standard deviation) are progressively younger to the south, corresponding to a rate of motion of 65 mm/yr of the Indo-Australian plate over the mantle. 'Zero-age' volcanism is predicted for the vicinity of Bass Strait (see also Fig. 1.4.5.). Ages for the leucitite-suite province (filled triangles) were not used in the regression but follow the same trend. No spatial correlation is evident for the lava-field provinces (open squares).

volcanoes were built above sea-level, and subsequently bevelled by erosion, followed by subsidence to depths ranging from 90 m for Gascoyne in the south to more than 400 m for Recorder Guyot in the north. Vogt & Conolly (1971) suggested that because of the progressively greater submergence to the north along the chain, the volcanoes become older in that direction, and proposed that the volcanoes are hotspot or plume trails. McDougall & Duncan (1988) recently reported results of a test of the hotspot-trail hypothesis based on K-Ar and ^{40}Ar/^{39}Ar age measurements on basaltic samples dredged from Gascoyne, Taupo, Derwent Hunter, Britannia, and Queensland seamounts in 1985 and 1986. Their results provide strong support for the hypothesis, as discussed below.

Individual seamounts of the Tasmantid chain are as much as 60 km in diameter at their bases, rising from the sea floor with slopes commonly in the range 10 to 20°, and locally much steeper. Some seamounts are nearly circular in plan, and others are elongate sub-parallel to the trend of the volcanic chain. Volcanic rocks were recovered in 1985–6 by dredging of the slopes of a number of the seamounts. The samples are basaltic and most contain olivine microphenocrysts. Plagioclase and less commonly clinopyroxene are additional phenocryst minerals in some cases. A minority of the samples has a well crystallised basaltic groundmass. Most samples contain glass or poorly crystallised mesostasis making up as much of 80 percent by volume. Eggins (1988) showed that these basalts range in composition from ol-tholeiitic basalt to alkali basalt and hawaiite. Dating was undertaken on the petrographically least altered and best crystallised samples. However, most samples contain some glass, so that the measured ages generally are regarded as minimum values because of the propensity for glass to alter and to leak radiogenic argon.

A fairly regular increase of measured age northwards along the chain is evident in Figure 1.7.2. The two samples dated from a single dredge haul on Gascoyne Seamount, the southernmost volcano, yielded measured ages in the range 6.4 to 7.2 Ma. A single sample from Queensland Guyot, about 1000 km north of Gascoyne, gave a conventional K-Ar age of 21 Ma and a ^{40}Ar/^{39}Ar total-fusion age of 24 Ma. Ages on samples dredged from Taupo, Derwent Hunter, and Britannia have measured ages intermediate between those for Gascoyne and Queensland seamounts (Fig. 1.7.2). Least-squares linear regression of the age-versus-distance data yield the best fit lines shown in Fig. 1.7.2 for the two sets of results. The inverse of the slope of each line provides an estimate of the average rate of migration of the volcanism along the chain. A mean of the two estimates is 67 ± 5 mm/yr. The predicted location of the volcanic focus at the present time is at 40.4°S latitude.

These results represent clear and unambiguous evidence for systematic migration of volcanism south-wards along the Tasmantid Guyots chain over a period in excess of 15 Ma. There is no direct evidence for volcanic activity at the predicted current position of the volcanic focus, and no significant feature is evident on the bathymetric charts that might be ascribed to a submarine volcano. However, a magnitude–6 earth-

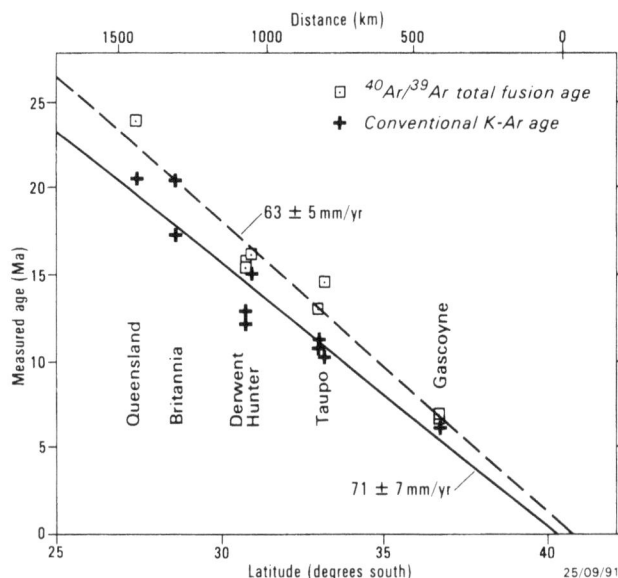

Figure 1.7.2 Measured K-Ar ages and ^{40}Ar/^{39}Ar total-fusion ages for volcanic rocks from the Tasmantid Guyots, versus latitude (after McDougall & Duncan, 1988, fig. 3). Best-fit straight lines are shown through the two sets of data.

quake took place on 25 November 1983 at an estimated depth of 25 km at a location given as 40.45°S, 155.51°E (Denham, 1985) which essentially is at the predicted volcanic focus (see also Fig. 1.4.5). The age-progressive nature of the volcanism in a systematic manner along the chain, together with the seismic data, provide strong evidence that the Tasmantid Guyots represent a hotspot track. Note that the rate of migration of the volcanism and its direction are indistinguishable from those observed for the central-volcano provinces in continental eastern Australia.

The Lord Howe seamounts form a line parallel with and east of the Tasmantid Guyots, on the flanks of the Lord Howe Rise (Fig. 1.2.1). Lord Howe Island at the southern end of the chain is an eroded basaltic shield volcano that was built above sea-level between about 6.9 and 6.4 Ma ago in the late Miocene (McDougall et al., 1901). Reefs and banks delineate the chain to the north, although the bevelled summit of Gifford Guyot is about 300 m below sea-level. The Lord Howe seamount chain also is likely to be a hotspot trail.

Norfolk Island

Norfolk Island is in the southwest Pacific Ocean in latitude 29°S, longitude 168°E, on the narrow Norfolk Ridge that extends from New Zealand to New Caledonia (Fig. 1.2.1). The island is only about 8 by 5 km in plan and rises to an altitude of 315 m. It consists almost entirely of basaltic lavas and some interbedded tuffs, all locally erupted over a time interval from about 3.2 to 2.4 Ma ago in the late Pliocene (Jones & McDougall, 1973; McDougall & Aziz-ur-Rahman, 1972). The volcanism is regarded as intraplate in character because it is far from plate boundaries. Norfolk Island clearly is younger than the ridge on which it is built, although the origin of the ridge is still uncertain.

New Zealand

Cainozoic intraplate volcanism is distributed sporadically throughout the region of continental crust of the South Island of New Zealand, Campbell Plateau, and Chatham Rise on the Pacific plate, and in the Northland-Auckland region of North Island on the Indo-Australian plate (Figs. 1.1.7–8; Table 1.7.3). The most voluminous volcanism has produced the large eroded volcanoes of Banks Peninsula (Lyttelton and Akaroa), Dunedin, Auckland Islands, and Campbell Island. These shield volcanoes were built mainly of alkali-basalt lavas, but more fractionated lavas — including hawaiite, mugearite, and trachyte, or their intrusive equivalents — are found in each. These volcanoes are therefore quite similar to those in eastern Australia which are included in the central-volcano group, as noted previously by Wellman (1983).

All the provinces on the Pacific plate listed in Table 1.7.3 are alkaline apart from the Timaru Basalt which is subalkaline (Sections 4.3–4). There is evidence for volcanism at two widely separated times in the Chatham Islands, and volcanism in the Auckland Islands continued over a relatively long interval. Adams (1981) recognised a broad pattern of decreasing age to the east for the New Zealand intraplate volcanism on the Pacific plate. He interpreted the pattern in terms of passage of the continental block across a linear magma source in the mantle (see below).

The intraplate volcanism of North Island is restricted to the Northland-Auckland region and may be divided into two distinct provinces: Northland, covering about 2500 km², and Auckland, covering about 400 km² (Section 4.2). The total volume of volcanic rocks, mainly basalts, is quite small — perhaps in the order of 70 km³. Activity is confined to an interval from late

Pliocene essentially to the present day (Table 1.7.3). Northland province consists mainly of transitional basalts and minor, more felsic lavas, whereas Auckland province is characterised by basalts of a more alkaline character.

1.7.3   Evaluation of proposed origins for intraplate volcanism

Introduction

The foregoing reviews of the age distribution of intraplate volcanism and discussion of relative and absolute plate motions (Section 1.2) can be used now to evaluate the many models proposed for the origin of Cainozoic intraplate volcanism in the east-Australian region. No single model can account for all of the volcanism, and some models are quite specific in their applicability. A wide range of proposed mechanisms has emerged for the volcanism in eastern Australia which is underlain by geologically complex lithosphere (Section 1.3.1). The proposed origins and their strengths and weaknesses, are examined in this section. A new model involving detachment faulting is presented in Section 7.3.

Hotspots and hotlines

The hotspot model has been successful in interpreting age-progressive linear volcanic provinces elsewhere (for example, McDougall & Duncan, 1980). Vogt & Conolly (1971) first proposed that the Tasmantid Guyots seamount chain of the Tasman Sea represents a record of the northward motion of the Indo-Australian plate over sub-lithospheric, mantle-fixed, melting anomalies. They predicted a northward progression in the age of the volcanoes based upon the increasing depth to the volcano summits — a result of thermal

Table 1.7.3 Summary of location and age data for basaltic (mainly alkaline) intraplate volcanic areas of New Zealand and Campbell Plateau

Province	Latitude °S	Longitude	Age range Ma	Average age Ma (± 1 s.d.)	Number of ages	Reference
Indo-Australian plate						
Northland	35.0–35.8	173.9E–174.6E	0.02–2.3		5	Stipp (1968)
Auckland	36.7–37.5	174.7E–175.1E	0.01–ca. 1.7		numerous	Stipp (1968); McDougall et al. (1969).
Pacific plate						
Lyttelton volcano	43.6	172.7E	10.6–12.2	11.0 ± 0.6	8	Stipp & McDougall (1968b)
Diamond Harbour	43.7	172.7E	6.0–8.4	6.8 ± 0.9	20	Stipp & McDougall (1968b); Evans (1970).
Akaroa volcano	43.8	172.9E	8.2–9.4	8.7 ± 0.4	38	Stipp & McDougall (1968b); Evans (1970).
Timaru	44.4	171.2E	2.6		1	Mathews & Curtis (1966)
Waipiata	45.2	170.4E	13.1, 16.5		2	McDougall & Coombs (1973)
Dunedin volcano	45.8	170.6E	10.4–13.4	11.9 ± 1.0	9	McDougall & Coombs (1973)
Chatham Islands						
northern	43.7	176.6W	35.6–41.1	37.7 ± 3.0	3	Grindley et al. (1977)
northern	43.7	176.6W	5.1–5.3	5.2 ± 0.1	3	Grindley et al. (1977)
Pitt Island	44.3	176.3W	2.7, 6.3		2	Grindley et al. (1977)
Antipodes Island	49.7	178.8E	0.3, 0.5		2	Cullen (1969)
Auckland Islands						
Ross volcano	50.6	166.2E	12.3–19.2	15.3 ± 1.9	12	Adams (1983)
Carnley volcano	50.8	166.1E	16.5–24.7	21.2 ± 3.2	12	Adams (1983)
Campbell Island	52.5	169.1E	6.5–7.8	7.1 ± 0.3	21	Adams et al. (1979)

subsidence increasing with time. The orientation of the seamount chains fits well with the seafloor-spreading and palaeomagnetic evidence for the motion of the Indo-Australian plate away from Antarctica.

Mafic lava-field provinces were produced sporadically throughout the eastern highlands from 70 Ma to the present, and mafic central volcanoes and associated felsic volcanism and some leucitite-suite lavas, progressively increase in age from south to north (0 to 34 Ma). Wellman & McDougall (1974a) related the first group to unspecified tensional stresses in the lithosphere, and the second to two hotspots now underlying Bass Strait. They calculated the southward migration of the activity to be 66 ± 5 mm/yr from the ages and distribution of the age-progressive portion of the volcanism, and this matches the rate of separation of Australia from Antarctica. Wellman & McDougall (1974a) concluded that this volcanism was related to movement of the Indo-Australian plate over the mantle. Two hotspots were proposed to account for the width of the band of central-volcano activity (about 600 km) even though there is no range in age for volcanic centres longitudinally.

Sutherland (1981) embraced the hotspot model fully in proposing that all the east-Australian volcanism younger than 55 Ma is caused by seven hotspots. However, his suggested rates of migration of activity are faster than the rate of movement of Australia over the mantle or relative to Antarctica. Several of the proposed hotspots now should be south of Australia, but there is no evidence for them. In addition, his proposed east-Australian hotspot tracks do not match the orientation of the Tasman seamount chains.

Wellman (1983) and Sutherland (1985) most recently advanced a hotline model in which volcanic activity swept southwards over a wide longitudinal swath as the Indo-Australian plate moved northwards across a stationary east-west thermal anomaly in the upper mantle. The origin of the lava-field volcanism is not addressed, but the model is used to explain the rather wide band of age-progressive central volcanism in eastern Australia and the supposed age-progressive seamount chains of the Tasman Sea. One problem noted by Wellman (1983) is that the Lord Howe seamount chain stops several hundred kilometres north of the youngest activity in the Tasmantid chain, and the dated volcanic activity at Lord Howe Island (McDougall et al., 1981) is nearly 10 Ma younger than central volcanoes at the same latitude in eastern Australia. The hotline therefore must have a highly irregular shape if this model is correct.

Sutherland (1985) argued that the stationary melting anomaly has the shape of the locus of spreading in the Coral Sea and D'Entrecasteaux basins, and that it originated as a mantle upwelling when these marginal basins developed (about 80–53 Ma). The upwelling persisted as the Indo-Australian plate moved northwards over the region, forming the age-progressive trails seen in eastern Australia and the Tasman Sea. No indication is given as to why the seamount chains should be narrow, well-defined chains like other oceanic lineaments associated with hotspots, rather than a wide band of volcanic activity similar to that found in eastern Australia.

Pilger (1982) re-examined the Wellman & McDougall (1974a) age data and proposed that all the volcanism originated from an extensional stress regime that began about 70 Ma ago. A compressional stress field developed as Australia moved northwards, and this moved progressively southwards and terminated the volcanism in the observed age sequence. Thus, no hotspots or hotlines are required according to Pilger (1982). Instead, the volcanism originated from changing intraplate stress conditions which in turn reflect motion relative to the mantle. Pilger also argued that if the Tasman seamount chains are hotspot tracks on the same plate then their orientation does not parallel the distribution of central volcanoes in eastern Australia. However, a significant number of lava-field provinces were active well after the age-progressive central-volcano activity swept past, as discussed above (see also Wellman, 1983). In fact, lava-field volcanism terminated mainly about 15 Ma ago along much of the highlands, including Tasmania. Sutherland (1983) pointed out that by adding the leucitite-suite centres of central New South Wales, the distribution of age-progressive volcanism in eastern Australia was sufficiently broad to match the orientation of the Tasman seamount chains. Thus, the main criticism made by Pilger (1982) of the hotspot model for the age-progressive portion of the east-Australian intraplate volcanism was removed.

Finally, Adams (1981, 1983) showed that Cainozoic volcanic centres on the continental lithosphere of southern New Zealand and the Campbell Plateau decrease in age from northwest to southeast (25 to 0 Ma) in concert with Pacific plate motion over stationary hotspots. This age progression applies only to major alkali-basalt shield volcanoes and not to other tholeiitic-basalt lava-field occurrences, as is the case with the east-Australian volcanism (Adams, 1981; Wellman, 1983). The volcanic centres are scattered over a broad region (Fig. 1.1.8) and at least three hotspots would be required to account for their distribution. Alternatively, Adams (1981) contended that the age distribution of this intraplate volcanism is more consistent with a linear, sub-lithospheric thermal anomaly which may have developed from an episode of early Oligocene rifting to the west of the Campbell Plateau. The southeastwards sweep of intraplate volcanism from southern New Zealand across the Campbell Plateau would have resulted, then, from motion of this region of the Pacific plate across a stationary mantle upwelling. Farrar & Dixon (1984) suggested that the site of mantle upwelling earlier was associated with part of the Indian-Antarctic ridge, now overridden by the Pacific plate.

The volcanism for either the hotspot or hotline models must have an age progression and spatial distribution that matches those predicted from calculations of Indo-Australian plate motion over the mantle, as discussed above. Thus, hotspot tracks derived from the present-day positions and calculated motions of the Indo-Australian (Table 1.2.1) and Pacific plates (Duncan & Clague, 1985) are superimposed on the volcano distribution map in Figure 1.7.3.

The hotspot underlying Bass Strait is defined by the presence of anomalous highly conductive mantle

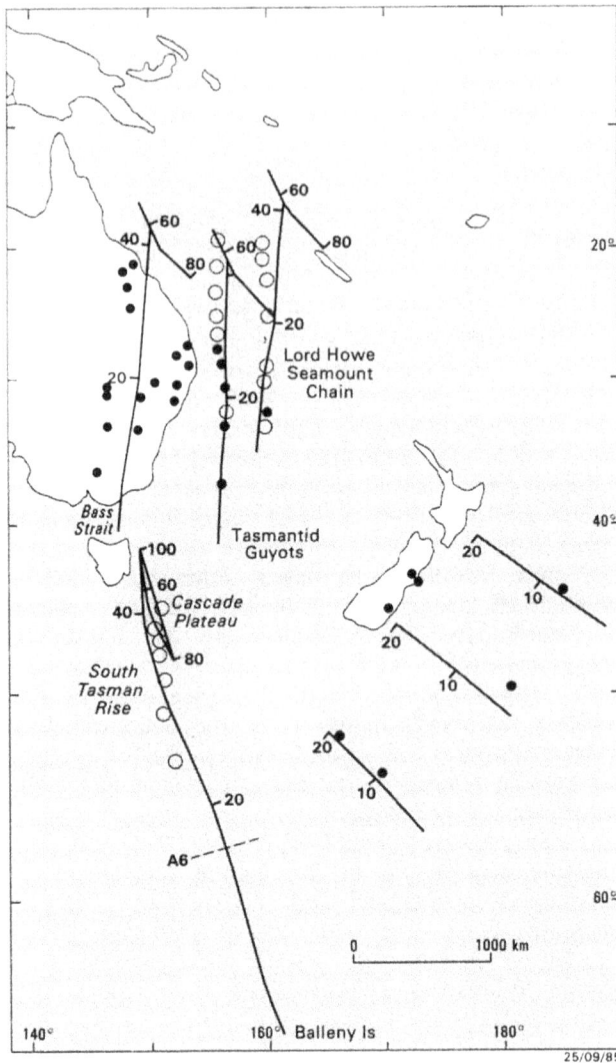

Figure 1.7.3 Hotspot tracks on the eastern part of the Indo-Australian plate and the New Zealand part of the Pacific plate match the age and distribution of lines of intraplate central (and leucitite-suite) volcanoes. Filled circles represent dated volcanoes (Tables 1.7.1,3). Open circles represent undated seamounts. Predicted paths and ages of volcanism were calculated from the plate-rotation parameters in Table 1.2.1 and given by Duncan & Clague (1985). Present-day hotspot positions are beneath the Balleny Islands, Bass Strait, and the southern end of the Tasmantid and Lord Howe seamount chains. Southern New Zealand intraplate volcanism appears to track plate motion over the mantle, but is not reconciled easily with a hotspot origin.

(Lilley, 1976), by Quaternary uplift of northern Tasmania (Colhoun, 1978), by seismicity (Section 1.4.5), and by positive seismic P-wave anomalies (Section 1.4.7). The hotspot south of Gascoyne Seamount is defined by a prominent zone of seismicity to the east of Tasmania (Section 1.4.5), whereas the hotspot south of the Lord Howe seamount chain is defined arbitrarily.

Volcanism on the continental shelf south and southeast of Tasmania that was unaccounted previously may be explained by the inclusion of the Balleny hotspot in the modelling. The Balleny Islands are Quaternary alkali-basalt volcanoes (Mawson, 1950) that rise from abyssal oceanic floor about 300 km northwest of the Antarctic continent (Fig. 1.7.3). There is limited age

control, but an age migration to the northwest along the 160 km lineament is suggested (Embleton, 1984). The hotspots shown southeast of New Zealand in Figure 1.7.3 are hypothetical and are intended to represent only the expected geometry and age progression of tracks on the Campbell Plateau.

The orientations of seamount chains in the Tasman Sea are matched remarkably well by the modelled tracks (McDougall & Duncan, 1988; see also Section 1.7.2). The volcanic pedestal that includes Norfolk and Philip islands rises from the otherwise submerged Norfolk Ridge (Fig. 1.2.1). Volcanism was active here in the late Pliocene (Jones & McDougall, 1973) but apparently is not manifest elsewhere. Future dredging and marine geophysical work may result in the identification of related volcanism along the ridge, but for the present this volcanism remains an isolated occurrence.

Only one hotspot track was modelled for comparison with the east-Australian age-progressive centres. This is not to say that two or more separate melting anomalies could not have been involved, but they are not required by the data. The width of the volcanic region is as much as 600 km and volcanism appears to have shifted from a western line to an eastern line and back to the west again as it migrated from central Queensland to southern Victoria (Fig. 1.1.6). The ages of the volcanoes decrease in a perfectly regular pattern along this track, and a broad zone of magma generation (600 km diameter) rather than many discrete point sources, is suggested. The predicted track for this broad hotspot follows the pattern of central volcanoes, and predicted ages match the measured ages to an extraordinary degree.

The predicted path of the Balleny hotspot could account for volcanic centres on the Cascade Plateau and South Tasman Rise (Fig. 1.7.3) which are evident on SEASAT images of the region (Haxby et al., 1983). Dredged volcanic conglomerate from the Cascade Plateau, southeast of Tasmania, contains foraminifera of Middle to Upper Eocene age (P.G. Quilty, personal communication, 1985), which is in good agreement with the predicted age. Seamounts south of the South Tasman Rise would be expected to range in age from 36 to 20 Ma, and to terminate when the Australia-Antarctic spreading ridge crossed over the hotspot, roughly at magnetic-anomaly 6 time about 20 Ma ago (see A6 in Fig. 1.7.3). The position of the Balleny hotspot along the east coast of Tasmania at 100 Ma possibly may explain the minor alkalic hypabyssal rocks at Cygnet and Cape Portland of this age (McDougall & Leggo, 1965; McClenaghan et al., 1982).

Volcanism in response to intraplate stress

Pilger (1982) offered an alternative to the hotspot model for age-progressive volcanism in eastern Australia, in which volcanism took place throughout the highlands region from about 70 to 35 Ma ago in response to a regional tensional stress field. Volcanic activity from 35 Ma to the present ceased in an age-progressive fashion from north to south as the plate entered a compressional regime. However, the additional lava-field age data included in the summary by Wellman (1983) do not appear to support Pilger's model of age-progressive cessation of volcanism (see also Fig. 1.7.1). Also, the causes of the changing intraplate stress pattern were not specified by Pilger, although they could include plate-margin forces, plate drag over the mantle, and changes in the radius of curvature of the earth with latitude. Why the volcanism is focused along the eastern margin of the continent also was not addressed.

The present-day intraplate stress field of the Indo-Australian plate was calculated by Cloetingh & Wortel (1985). They considered the effects of ridge push and slab pull as a function of the age of the plate interior and the age of subducting oceanic lithosphere along the plate margin, and found that most of the plate should be under compression caused by resistance along the Himalayan and the Banda Arc collision zones. However, the calculated stress field along the east-Australian continental margin changes from east-west tension in northern Queensland, to lesser northeast-southwest tension in eastern New South Wales, to slight compression in the Bass Strait area (compare with Section 1.4.6). The region of greatest expected tension on the continent coincides with Miocene and younger volcanic rocks in northern Queensland, whereas the young end of the main age-progressive volcanic trend (Bass Strait) is expected to be under compression.

The previous state of intraplate stress is much more difficult to estimate. Pilger (1982) provided a summary of palaeostress indicators, including dyke and normal-fault azimuths, and aligned volcanic centres, which he interpreted to indicate that the principal axis of tension was northwest-southeast in northern Queensland, east-west in southern Queensland, and southwest-northeast in central New South Wales from the Upper Eocene to the present. These data are not independent of the volcanism, so the argument is rather circular. However, the change in orientation of the axis of tension with latitude is identical to that of the present theoretical stress field (Cloetingh & Wortel, 1985) so this intraplate stress regime may have persisted for much of the Cainozoic. The period 80 to 40 Ma ago was one of slow southwards, then northwards, movement of the Indo-Australian plate. This relative immobility may have promoted a tensional regime and an increased vulnerability of the plate to volcanism. The beginning of rapid northwards plate motion coincides rather well with the start of the southward age progression of the central volcanism. A tensional intraplate stress field in eastern Australia could have resulted from increased rates of plate convergence to the north and east, from about 40 Ma onwards.

Lithosphere flexure during erosional rebound

Stephenson & Lambeck (1985) and Lambeck & Stephenson (1986) proposed that the landscape of the southern highlands of eastern Australia has been dominated by erosional unloading and simultaneous rebound since the end of orogeny about 250 to 200 Ma ago (see Section 1.3.2). They dismissed the possibility that volcanism in Cainozoic time has rejuvenated uplift through thermal expansion of the lithosphere (P. Wellman, 1979b), and argued that the volcanism was volumetrically small compared with typical hotspot

lineaments which have reset the thermal decay of lithosphere passing over them (Crough, 1983). In addition, heat-flow in the eastern highlands is not anomalously high (Sass & Lachenbruch, 1979). Instead, Stephenson and Lambeck argued that the volcanism was an expression of a tensional regime that resulted from the flexure of the lithosphere, produced during erosional rebound of the highlands. They calculated a maximum tensional stress oriented northwest-southeast in late Cainozoic time, which might account for some of the widespread lava-field activity, but it does not explain the age-progressive volcanism.

The Stephenson-and-Lambeck model has been successful in accounting for the old erosional surfaces in the highlands (Young & McDougall, 1985) and for the extremely low erosion rates (Bishop, 1985b). However, the timing of the volcanism is not predicted precisely. Nor is post-orogenic volcanism a general feature of old, rebounding mountain chains such as the Appalachians or Urals. In addition, the predicted tensional stress field for southeastern Australia is at odds with the present northwest-southeast compressional stress field observed from seismological and *in-situ* stress measurements (Denham et al., 1981; Section 1.4.5; see also Fig. 1.4.6).

Thermal pulse from opening of the Tasman Sea

Karner & Weissel (1984) argued that the thermal and flexural properties of southeastern Australia have been reset relative to older portions of the continent. They proposed that rifting during opening of the Tasman Sea led to rapid initial isostatic uplift of the eastern highlands caused by thermal expansion of the lithosphere. Details of their model are yet to appear, but Karner & Weissel (1984) contended that a thermal pulse moved westwards from the emplacement of hot oceanic lithosphere next to the continent in the late Cretaceous to early Cainozoic. This thermal event initially reduced the flexural rigidity and may have promoted melting and volcanism from the coast into the highlands.

The absence of rift-induced topography and volcanism along other Australian passive margins is related to the time interval of rift development and the nature of the transition between continental and oceanic lithosphere. However, there is some evidence that Tasman Sea rifting began slowly, like that along the southern margin (Veevers, 1984). Cainozoic volcanism is found throughout the eastern highlands, yet the Tasman Sea opening extended only as far north as southern Queensland (Fig. 1.2.1). Finally, there is no explanation for the age-progressive part of the volcanism.

1.7.4 Summary and conclusions

Cainozoic intraplate volcanism in eastern Australia, the Tasman Sea, and southern New Zealand can be related to a range of plate-tectonic mechanisms. No single model can account for all the volcanism. The consistency of several models has been examined in the foregoing by calculating the motion of the Indo-Australian plate relative to the mantle-fixed (hotspot) reference frame (Section 1.2.3). In addition, insights have been obtained on changing patterns of relative motions and plate-boundary forces through time, and future areas of research have been clarified.

The seamount chains of the Tasman Sea are most likely the result of two stationary hotspots. The geometry and age distribution along the volcanic trails matches exceedingly well that predicted from the calculations of Indo-Australian plate motion in the hotspot reference frame (McDougall & Duncan, 1988). Tectonic activity may be revealed by bathymetric and seismological studies of the suspected current hotspot positions, as seen in the French Polynesian region (Talandier & Okal, 1984). The volcanic centres of the Cascade Plateau and South Tasman Rise may be related to early Cainozoic activity above the Balleny hotspot. Existing age control is poor, but is consistent with this correlation. Further sampling and age determinations are crucial to establish a connection with hotspot volcanism.

The Cainozoic volcanism in the eastern highlands of Australia is more problematical. The onset of the widespread lava-field volcanism at 80 to 70 Ma ago coincides with slow motion of the Indo-Australian plate over the mantle and the initiation of opening of the Tasman Sea. This period of slow motion probably contributed to a tensional pattern of intraplate stress which was relieved mainly by rifting and seafloor spreading in the Tasman Sea and in other marginal basins bounding the eastern border of the plate from about 80 to 50 Ma ago. This volcanism continued to the present day, but it diminished considerably in volume from about 35 Ma ago compared with the earlier part of the Cainozoic (Wellman & McDougall, 1974a). Intraplate tension in eastern Australia possibly was maintained from about 40 Ma by plate-boundary forces exerted at zones of rapid subduction along the northern and eastern margins of the plate. Tensional stress from erosional rebound of the east-Australian highlands (Stephenson & Lambeck, 1985) and lateral heating from the opening of the Tasman Sea (Karner & Weissel, 1984) may have contributed to lava-field volcanic activity, but neither mechanism adequately accounts for the timing or regional extent of this intraplate volcanism.

The age-progressive, central-volcano portion of east-Australian volcanism appears to be the track of a fixed melting anomaly in the sub-lithospheric mantle. The geometry, although diffuse, and the well-defined age sequence of volcanic centres, are consistent with the hotspot model. This volcanism may be related to a single hotspot, but if so its region of influence is much broader than that seen in the seamount chains just to the east, and the lithosphere must exert a considerable control on the surface expression of volcanism over hotspots. Important future studies should be the geophysical elucidation of the dimensions and character of the anomalous mantle beneath Bass Strait.

Sutherland (1981) contended that the Pliocene to Pleistocene volcanism in northern Queensland had a southward age progression, consistent with Indo-Australian plate motion over an additional hotspot. An intraplate stress mechanism may be equally likely because this province coincides with the region of maximum tension in the continent, as calculated by Cloetingh & Wortel (1985). Migration of the Indo-Australian plate over a stationary mantle upwelling initiated by spreading in the Coral Sea (Sutherland, 1981, 1983, 1985) is not a viable mechanism for eastern Australia and the Tasman seamount age-progressive centres because the calculated ('back-tracked') volcanic paths never cross this region. The regions of zero-age volcanism at the southern ends of the age progressive lineaments are not colinear, so in this sense the hotline model (Wellman, 1983) is no different from the multi-hotspot model.

A stationary, linear-melting anomaly may be appropriate for intraplate volcanism in southern New Zealand and the Campbell Plateau (Adams, 1981; Wellman, 1983; Farrar & Dixon, 1984). Alternatively, dextral shear at the Pacific/Indo-Australian plate boundary is producing tensional stress across the Campbell Plateau which allows the eruption of sub-lithospheric magmas. The migration of volcanism should be roughly consistent with Pacific-plate motion. Additional sampling and dating of submarine volcanic centres are necessary, as well as an assessment of the state of intraplate stress in this region.

CHAPTER 2

Physical Volcanology

2.1   Introduction

Physical processes operate at all scales in influencing the location, development, and nature of intraplate volcanism. They dictate where a volcanic province develops and why (Section 1.6). They also control the location of eruption points, the types of eruption centres that develop, the style of eruption, and the nature of the products.

There are few detailed studies published on the physical volcanology of Cainozoic intraplate volcanic centres and deposits in eastern Australia and New Zealand, so this review is a general one. Comparisons are made here between the features of east-Australian and New Zealand areas and those found elsewhere. Many of the examples come from the Newer Volcanics province of southeastern Australia, because it is a young, well preserved, volcanic area, and its features have been relatively well documented. Significant areas of submerged continental crust have been host to intraplate volcanism in eastern Australia and New Zealand, and therefore volcanism in aqueous environments (lakes and the sea) also is dealt with here.

2.2   Chemical and Physical Diversity of Magmas

2.2.1   Chemical diversity

Magma compositions may be either relatively uniform or diverse in intraplate continental volcanic terrains. Some provinces such as the Columbia River Plateau province have magma compositions that are remarkably uniform and basaltic in character, despite the large area occupied by the province and, especially, the huge volumes of magma erupted (see below). Magmatism lasted for only a few million years during the middle Miocene (Swanson et al., 1975), so magma discharge rates were probably high, comparable with those calculated for the eruption of large-volume ignimbrite eruptions (for example, Cas & Wright, 1987).

Diversity in magma composition in intraplate volcanic provinces is recognised, for example, in the Basin and Range province of the western United States (Christiansen & Lipman, 1972). Mafic to felsic rocks encompassing all of alkaline, subalkaline, and calcalkaline suites are known, but the Basin and Range province is characterised most commonly as bimodal. Similarly, the Snake River Plain which is best known as the type example of the plains basaltic province (Greeley, 1982a,b; see below) is in fact distinctly bimodal in character (Leeman, 1982). The Snake River Plain cross cuts the northern end of the Basin and Range province in Idaho, postdates the middle Miocene Columbia River Plateau basalts, and has a history of both voluminous rhyolitic and basaltic volcanism from the late Miocene to the present day. Rhyolitic products include large-volume ignimbrites erupted from caldera centres (Christiansen, 1982, Hildreth et al., 1984), but have been subordinate to basaltic volcanism during the Quaternary. Magma discharge or eruption rates in the Snake River Plain province appear to have been less, and the province has had a longer life span, than those for the Columbia River Plateau. Consequently there appears to have been significant interaction between the mantle-derived basaltic magmas and the crust, leading to crustal melting and the generation of rhyolitic magmas (Leeman, 1982). The Columbia River basalts, in contrast, appear to have ascended rapidly and to have undergone little, if any, crustal interaction.

Mafic magmas in east-Australian and New Zealand Cainozoic volcanic terrains are generally more voluminous than intermediate and felsic ones. This is true at the scale of provinces (such as the Newer Volcanics province of Victoria and South Australia, and the north-Queensland provinces), but is not true necessarily for small volcanic areas, such as the Glass Houses area of southeastern Queensland which is dominated by rhyolite and trachyte (Section 3.4.3). Some of the most marked compositional differences are found in some of the large shield volcanoes of eastern Australia, where trends from early mafic volcanism to later peralkaline felsic volcanism are common (for example: Focal Peak volcano, Section 3.4.6; Canobolas, Section 3.6.2), and where complex associations of magma compositions are represented (for example, Nandewar volcano, Section 3.5.2). New Zealand intraplate volcanism also is overwhelmingly dominated by mafic rocks. However, considerable compositional differences are found in individual volcanic centres such as Dunedin volcano (Section 4.3.6).

2.2.2 Physical properties

Comprehensive discussions of the physical properties
and rheological behaviour of magmas were presented
by Williams & McBirney (1979), Fisher & Schmincke
(1984), McBirney & Murase (1984), and Cas &
Wright (1987). Only those factors that are most
relevant to understanding eruption and lava-flow
behaviour are discussed here — namely, magma
viscosity (which is dependent on many other magma
properties) and magma-discharge or eruption rate.

Mafic lavas represent high-temperature, low-viscosity, mostly volatile-poor magmas characterised by
low degrees of melt polymerisation. These properties
affect the style of eruption, and the mobility and form of
mafic lavas. Measured eruption temperatures are in the
range 1000–1200°C (Macdonald, 1972) and typical
viscosities are 6.5–9.4×10^2 Pa s (Shaw et al., 1968;
Pinkerton & Sparks, 1978; Cas & Wright, 1987).
Mafic magmas therefore generally have low viscosities
and are mobile as lavas. Flows are capable of travelling
great distances — greater than 100 km for some north-
Queensland flows (Section 3.2) — provided magma-
discharge rates are high enough. Walker (1973a)
suggested that magma-discharge rate was probably the
most important factor affecting the distances that lava
flows travel.

Felsic magmas, in contrast, are generally lower in
temperature, more highly polymerised, higher in
viscosity, and higher in volatiles, and are therefore
much less mobile and flow shorter distances. For
example, rhyolitic magmas have estimated eruption
temperatures in the range 700–900°C and have
experimentally determined viscosities of 10^8–10^{11} Pa s
at these temperatures (Murase & McBirney, 1973).
Rhyolite lavas have higher viscosities and yield
strengths, and so are erupted at low magma-discharge
rates and rarely flow more than a few kilometres from
vent. High-aspect-ratio (ratio of average thickness of
flow to diameter of circle covering the same area as the
flow) lavas and domes such as those at the Focal Peak,
Tweed, and Glass Houses volcanic centres, are
produced. Intermediate magmas are in general more
viscous than mafic ones, and may produce high-aspect-
ratio lavas and domes where the available magma
volume and the magma discharge rate are low. Some
relatively low-aspect-ratio rhyolite flows are known on
the southern flank of Tweed volcano (M.B. Duggan,

personal communication, 1987). The flows are mostly
about 100 m thick and may be several kilometres in
length.

The peralkaline rhyolites found in several east-
Australian Cainozoic volcanic centres may have had
viscosities up to two orders of magnitude less than non-
peralkaline rhyolites (Cas & Wright, 1987). These are
less common in the New Zealand intraplate volcanic
provinces and fields (for example, Smith et al., 1986).
Peralkaline rhyolites therefore may flow more easily
than other rhyolites, and may produce relatively low-
aspect-ratio lavas.

2.3 Plains and Flood Basalt Provinces

Two major types of intraplate, continental, basaltic
provinces have been recognised: (1) plains-basalt
provinces, typified by the Snake River Plain of the
western United States (Greeley, 1977, 1982a,b); (2)
flood- or plateau-basalt provinces, typified by the
Columbia River Plateau basalts, also in the western
United States (for example: Swanson et al., 1975;
Basaltic Volcanism Study Project, 1981; Camp et al.,
1982; Hooper, 1982b).

Plains-basalt provinces are dominated by smaller-
volume lavas. Fissure systems are known, but most
eruptions are from central vents, especially lava-shield
volcanoes, where discharge rates are low (Fig. 2.3.1).
Discharge rates are not known specifically, but the
rates for volcanoes such as Mount Etna of 5×10^3 m³/s
(Pinkerton & Sparks, 1976) also may be typical in
plains-basalt provinces (Cas & Wright, 1987). Lavas
in the Snake River Plain are up to 10 m thick, and
compound and elongate valley-fill flows are common.
Lava tubes and canals (see below) are important means
of flow propagation in these lavas. Plains lavas may
form low-profile, coalescing shields (Fig. 2.3.1) which
may be aligned along distinctive rift zones or fissures
(Greeley, 1977, 1982a,b). Pyroclastic deposits are
minor, and soils and sediments may be intercalated
with flows.

Flood-basalt provinces are characterised by thicker
(15–35 m), regionally extensive, sheet-like lavas that
have low aspect ratios. Simple lavas are most common.
The thick lavas spread across topography, drowning it,
so forming a flat, plateau-like morphology. Flood
basalts almost certainly are fed by fissure vents, and
magma-discharge rates can be in the order of 10^6 m³/s
(Swanson et al., 1975; Cas & Wright, 1987). Interflow
soils, lacustrine sediments (including basaltic peperites),
and fluvial sediments are also common (for example,
Schmincke, 1967).

East Australian and New Zealand Cainozoic basaltic
provinces are clearly not of the flood-basalt type. Some
lavas such as those in northern Queensland and
western Victoria are far reaching, but they are small to
moderate volume, elongate, valley-fill lavas (Figs.
2.3.2–3) rather than vast sheet-like flood lavas. East
Australian Cainozoic basalt areas are more like plains-
basalt provinces. The Newer Volcanics appear to have
larger numbers of scoria cones, spatter cones, tuff rings
and maars (Section 3.7.4; Fig. 1.6.1.) than apparently

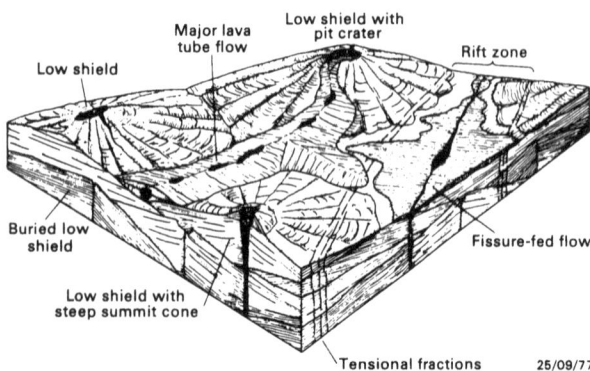

Figure 2.3.1 Volcanic relationships between shield volcanoes
in plains-basalt provinces (after Greeley, 1982a, fig. 4b).

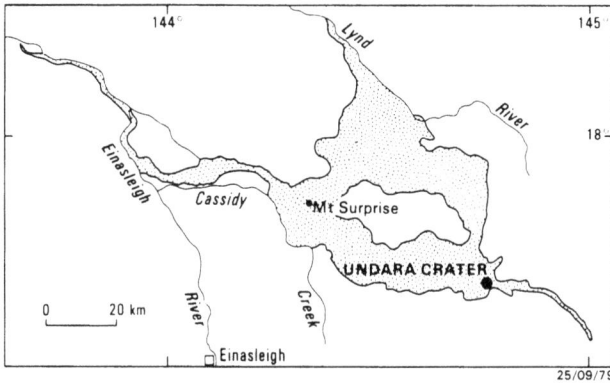

Figure 2.3.2 Far-flowing north-Queensland lava flows (stippling; after Stephenson & Griffin, 1976b, fig. 3).

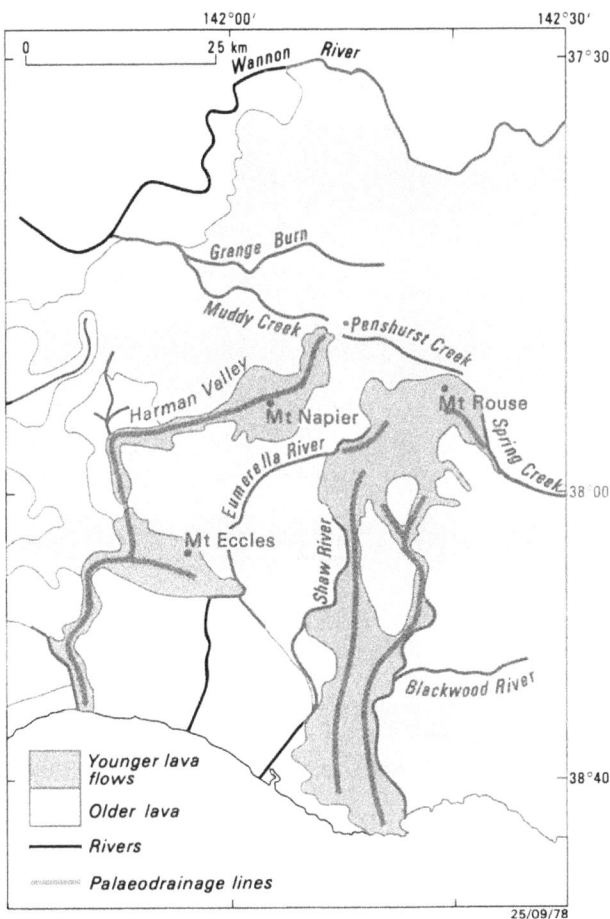

Figure 2.3.3 Far-flowing Victorian lavas, Newer Volcanics province (after Ollier, 1985a).

are found in the Snake River Plain which, in contrast, is dominated by fissure vents and lava shields. Low-angle lava-shield volcanoes also are abundant in the Newer Volcanics province.

The primary control distinguishing flood-basalt from plains-basalt provinces is the short-term supply of mafic magma and the eruption or discharge rate. Availability of large volumes of magma and high discharge rates are likely to lead to flood-basalt eruptions of the fissure type. Lower rates of magma supply and lower discharge rates are likely to lead to central or point-source eruptions of the plains-basalt type, many of which may be aligned along basement fractures (for example, the Mount Eccles centre, Victoria). Discharge rate will be controlled to a large degree by the local availability of magma, but may be enhanced significantly where both a high availability of magma and appreciable extension rates exist (Section 1.6).

The flood-basalt province of the Columbia River Plateau occupies an area of at least 200 000 km², the Ethiopian flood basalt province covers an area of about 480 000 km², and the Cretaceous Deccan flood-basalt province of northwestern India exceeds 500 000 km², whereas the Snake River Plain — the type plains-basalt province — covers about 100 000 km², although the Quaternary part is much less. In contrast, the Miocene-Quaternary Newer Volcanics province of Victoria and South Australia covers only 15,000 km², containing over 400 eruption points (Joyce, 1975). The Pliocene-Quaternary basaltic provinces of northern Queensland occupy 20,000 km², and over 300 vents have been identified (Stephenson & Griffin, 1976a). East Australian provinces clearly are much smaller than are the flood-basalt provinces cited above, and much closer in size to the Quaternary part of the Snake River Plain province. However, as documented in Chapter 3, there are also many widespread smaller basaltic fields in eastern Australia which in total can be viewed as a very large, long-lived, volcanic terrane of dispersed provinces, fields, and centres influencing an area of about 1.6×10^6 km². The Northland-Auckland volcanic areas in New Zealand occupy only about 2500 km².

Durations of volcanism in intraplate provinces range widely. The Columbia River Plateau province is thought to have been active for only several million years in the middle Miocene (Williams & McBirney, 1979) as noted above. On the other hand, the Snake River plain has been active since the Miocene (Greeley, 1982a,b; Leeman, 1982), the Newer Volcanics province has been active since the late Miocene (Aziz-Ur-Rahman & McDougall, 1972), and the north-Queensland area since the Pliocene (Stephenson & Griffin, 1976a).

2.4 Lava-Forming Eruptions and Flow Features

2.4.1 Eruption styles

Subaerial eruptions

Eruptions can be subdivided into two essential types — lava-forming and explosive. The salient aspects of lava-forming eruptions and lava-flow processes, and the principal features of lavas, will be discussed in the following sections.

Subaerial lava-forming eruptions have two different styles. The first is a passive, effusive eruption, and the second is a more spectacular, Hawaiian-style, fire- or lava-fountaining (Macdonald, 1972). Fountaining is characteristic of the eruption of low-viscosity magmas, especially basalts, under conditions of high magma-discharge rates. Basaltic lava eruptions seldom are solely of one or the other style, and they can fluctuate between effusive and fountain-style eruptions. Passive,

Figure 2.4.1 Succession of thin clastogenic lavas separated by agglutinate horizons, representing the tops and bottoms of lava layers, at Bridgewater volcano, Newer Volcanics province, Victoria.

Figure 2.4.2 Agglutinate deposits of the Mount Leura volcanic complex, Newer Volcanics province, near Colac, Victoria.

effusive eruption of lava results from a low, steady, discharge rate relative to a given vent size, together with discharge of magma characterised by low volatile contents, relatively high viscosity, or both.

Fire fountaining cannot be identified as a process from an examination of old lavas flows because spatter fragments absorb back into a lava flow after deposition. However, relict outlines of spatter fragments may be visible near vent in some lavas of fire-fountain origin. These lavas have been termed 'clastogenic' (Cas & Wright, 1987). Excellent examples are found at the Red Rock volcanic complex, the Mount Leura volcanic centre, and at the Bridgewater volcano (Fig. 2.4.1) in the Newer Volcanics province of Victoria. Lava clots discharged at low rates can cool sufficiently to form agglutinated spatter deposits (Fig. 2.4.2) which may build up a spatter cone or rampart around the vent (for example, the Mount Eccles spatter cones, the summit rim of Mount Napier, and agglutinates of the Red Rock Complex, all in the Newer Volcanics province, Victoria). Spatter deposits are too viscous to flow as lavas, apart from small-scale, downslope creep and 'dribbling'.

Eruption styles of felsic and intermediate lavas and domes may be less spectacular than those of basaltic eruptions, mostly because the eruption rate is significantly less and the magmas are more viscous.

Eruptions may last for months to years, and the growth of the lava flow or dome may be almost imperceptible. However, felsic-to-intermediate magmas may be significantly more volatile rich than basaltic magmas, so their eruptions may be punctuated or terminated by violent explosive activity. The association of domes and pyroclastic rocks in the summit regions of Focal Peak (Section 3.4.6), Nandewar (3.5.2), and Canobolas (3.6.2), are good examples of explosive activity associated with dome-forming eruptive events.

Lavas predominate in the intraplate volcanic terrains of eastern Australia and New Zealand. Pyroclastic products are minor and, excluding phreatomagmatic pyroclastic deposits (those produced through explosive interaction of magma with external water), by far the greater number of eruptions were of effusive, lava-forming type. However, both explosive (including phreatomagmatic) and lava-forming eruptions may take place in the relatively shallow epicontinental seas of continental volcanic settings, including eastern Australia and New Zealand, forming Surtseyan tuff cones, as well as lava-shield volcanoes.

Subaqueous eruptions

Subaqueous basaltic lava eruptions produce either pillow lavas where magma-discharge rates are low, or massive sheet flows where discharge rates are high.

Figure 2.4.3   Cross-section through the Eocene-Oligocene pillow lava exposed at Oamaru, South Island, New Zealand.

Pillows develop through the slow propagation of pods of lava. The outer skin of each pillow chills during propagation, and further injection of magma into the pillow from the complex, interior, lava-feeder tube system, causes the pillow to inflate further, and the skin of the pillow to split, so allowing a new pillow to bud along either transverse or longitudinal spreading cracks (Yamagishi, 1985). Pillows form where eruption or discharge rates are low, and massive lavas where magma or discharge rates are high, all other factors being equal. Furthermore, a stable boundary layer, including in most cases a thin layer of steam, must prevent dynamic mixing between water and the erupting magma in order that non-explosive eruptions take place in shallow seas.

Excellent exposures of marine basaltic pillow lavas and associated quench-fragmented hyaloclastite deposits are found, for example, in cliff exposures of the Oamaru Surtseyan volcano in the South Island of New Zealand (Fig. 2.4.3). This Eocene-Oligocene tuff cone is dominated by phreatomagmatic pyroclastic deposits, but a more quiescent pillow-lava forming event also must have taken place (Coombs et al., 1986; Cas et al., 1986). Nearly contemporaneous marine pillow lavas are reported also from Canterbury and Marlborough provinces in the South Island of New Zealand (Section 4.3.2).

Pillow-like basalts also can form intrusively where fluidal basaltic magma invades water-saturated, unconsolidated sediments. Oligocene basalts in central Otago formed pillows where they intruded unconsolidated

carbonate clastic deposits of the Totara Limestone at Totara Terraces. Similar intrusive pillow basalts are exposed at Aireys Inlet in the Lower Tertiary marine succession of the Otway Basin of Victoria (R.A. Day and R.E. Fordyce, personal communications, 1986) where, again, the host is unconsolidated carbonate sediment. Intrusive pillow masses are found also in the Tamar valley of Tasmania (Section 3.8.3). Massive flows also can intrude unconsolidated sediments.

Pillow lavas and associated quench fragmented debris may form where basaltic lava flows, or is erupted, into lakes or dammed river valleys — for example, dammed by an earlier lava flow, causing ponding of river water against the natural dam. New South Wales examples were described by Bishop (1985a), and numerous valley-fill lava successions are known in Tasmania (Section 3.8.3). Non-marine pillows and associated debris have been recorded also in similar situations in the Columbia River flood-basalt province (for example, Camp et al., 1982).

2.4.2   Eruption-discharge rates

Magma-discharge rates differ greatly depending on the eruption type and vent geometry. Rates for large flood-basalt eruptions, such as those for the Columbia River Plateau flows, are high. For example, rates calculated by Swanson et al. (1975) for the Roza Member are up to 10^6 m^3/s. The flows of this member are considered to have originated from a vent fissure system extending at least tens of kilometres. Rates for point-source eruption centres range from as low as 0.3 m³/s (for example, Mount Etna, 1975; Pinkerton & Sparks, 1976) to as high as 5 × 10³ m³/s for the 1783 Laki eruption in Iceland (Thorarinsson, 1968; Cas & Wright, 1987).

Walker (1973a) suggested that of the three principal factors contributing to lava-flow mobility and morphology — magma-discharge rate, physical properties, and slope — magma-discharge rate was the most important, and that the distance travelled by a lava was proportional to the discharge rate. Stephenson & Griffin (1976b) used this relationship and suggested that discharge rates in the order of 1000 m³/s were involved in maintaining the 0.19 Ma Undara flow of northern Queensland which has a length of 160 km and a volume of 10–23 km³ (Figs. 2.3.2, 3.2.6).

Little is known about eruption rates for rhyolitic lavas. However, eruption rates for dacitic lavas range between 0.5 and 3.5 m³/s (Newhall & Melson, 1983; Cas & Wright, 1987). This slow rate is in large part a reflection of the higher viscosity of felsic magmas. Exceptions to these low magma eruption rates are implied for the extraordinarily extensive rhyolites described as lavas by Bonnichsen (1982) from the Bruneau-Jarbridge eruptive centre in the Snake River Plain (Idaho) volcanic province.

2.4.3 Eruption durations

The durations of individual lava-forming eruptions are also wide-ranging, depending on the continued availability of magma and the eruption rate. Large flows, such as those of flood-basalt provinces, despite their great volumes, may have been erupted in short periods of days to a few weeks at the most, because of their high eruption rates (for example, Swanson et al., 1975). However, single basaltic lava-forming eruptions lasting several weeks to months or longer are not unusual, as exemplified by the 1973 Heimaey eruption in Iceland and by recent eruptions on Mount Etna. Stephenson & Griffin (1976b) estimated that the total eruptive phase that produced the long Undara flow of northern Queensland, may have lasted more than three months, assuming an average magma eruption rate of 1000 m³/s. The main flow may have formed in about a week from sustained discharge, and the smaller associated flows may have formed during the remaining time as a result of more feeble eruptions and intervening pauses.

Other basaltic eruptive phases can be longer lasting, resulting in the formation of multiple, smaller-volume lavas whose individual eruptions were short-lived and separated by periods of repose. Eruptive activity on the small shield volcanoes of Hawaii may be typical of such prolonged eruption phases. Hawaiian volcanoes such as Mauna Loa consist of large shield volcanoes up to 100 km in diameter whose active life is in the order of a million years or more. Smaller parasitic shield volcanoes in Hawaii, such as Mauna Ulu, have active lives perhaps of about ten years.

Shield volcanoes in continental basaltic provinces are smaller than the large Hawaiian shield volcanoes. Small shields in continental provinces, by analogy with their small Hawaiian counterparts (Mauna Ulu), typically may be produced by short, punctuated, eruptive phases lasting several years. The small shield volcanoes of the Newer Volcanics province of south-eastern Australia and of the Northland province of New Zealand (Fig. 2.4.4) were probably produced in no more than a few years. Some shield-forming eruptions may have been as short as several months, if discharge rates were similar to those of scoria cones. However, the eruptive intervals that produced the large, compositionally diverse, central volcanoes of eastern Australia during the Cainozoic may have lasted for thousands to hundreds of thousands of years. Some, such as the Nandewar volcano of New South Wales, had an active life of at least a million years which is similar to the life span of the largest Hawaiian shield volcano (see also Section 1.4.11).

Eruption durations for intermediate to felsic lavas also may be short- or long-lived. Individual domes, for example, may develop for weeks to years before final emplacement ceases. The 1982 dacite dome in the crater of Mount St. Helens is still growing (1986), and periods of dome growth on the Usu volcano in Japan in historical times have lasted for several years. Relatively long durations almost certainly were involved in the emplacement of many of the felsic lavas and domes in east-Australian volcanic centres such as Canobolas and Focal Peak during the Cainozoic. Lava-forming eruption periods may last months to years, but repose periods between eruptive periods almost invariably are much longer and could last weeks to thousands of years.

2.4.4 Eruption centres

Fissure-vent systems

Flood-basalt lavas originate from linear fissure-vent systems, because of their required high magma-discharge rates (see above). Other, less extensive lavas such as plains basalts also may have fissures as sources, but the eruption rates are much lower. For example, the 1973 lava flow at Heimaey, Iceland, during most of its movement originated from a fissure-vent system several hundred metres long that had numerous simultaneously active points along its length (Gunnarsson, 1973). Lavas in the Snake River Plain also originated from fault-controlled fissure systems (Greeley, 1982a,b; Kuntz et al., 1982).

The Tyrendarra flow of the Newer Volcanics province in southeastern Australia issued from the Mount Eccles centre, a complex of lava canals, spatter cones, scoria cones, and a lava lake (Boutakoff, 1963; Ollier & Joyce, 1973). The principal source of the lava flows appears to have been the elongate, north-northwest trending 'crater' lake of Lake Surprise, a former lava lake that was 700 m long and up to 150 m wide. Spatter and scoria cones are on the same north-northwest trend, corresponding to alignment along a basement fracture-fissure system (Section 1.6.2). Lake Surprise could be described best as a fissure crater lake. The Tyrendarra flow (possibly 5000 to 7000 yr B.P.; Gill, 1978) is thought to have flowed at least 50 km during a low sea-level stand. There is also an older flow in the Tyrendarra valley about 19 000 years old (Ollier, 1981). Much of it is now submerged below sea-level. Similarly, lava flows and cones have formed along

Figure 2.4.4 Whatatiri shield volcano (right middle distance) in front of a scoria cone, Northland volcanic province, New Zealand.

fault-fissure systems in the Whangarei volcanic field of the Northland province in New Zealand (Smith et al., 1986). Most north-Tasmania basalts are associated with major Tertiary faults and rifts (Section 3.8.3).

Lava-shield volcanoes

Shield volcanoes are also major sources of lavas in intraplate provinces. Eruptions on large Hawaiian-type basaltic shield volcanoes may involve fissure-vent systems. However, eruptions from small shield volcanoes such as those characteristic of plains-basalt provinces, represent eruptions from central vents. Some of these shields may develop along basement fissure-fracture systems (Greeley, 1982a,b). Lava shields are made up of multiple, overlapping, thin, small-volume lavas, corresponding to fluctuating, low-eruption rates. About half of the 400 recognised eruption points in the Newer Volcanics province are low-angle shield or lava-cone volcanoes (Joyce, 1975). Some shields produce more extensive lavas that flow well beyond the margins of the shield itself. The fundamental architecture of the younger part of the Snake River Plains basaltic province in Idaho consists of coalescing low-angle shields and lava mounds (Greeley, 1982a,b; Fig. 2.3.1).

Little work has been done to document the physical characteristics of shield lava centres in eastern Australia, but Greeley (1982a,b) described those from the Snake River Plain as having slopes less than 0.5°, except at the summit where slopes may rise to 5°. The shields are up to 15 km in diameter and have volumes of less than 7 km³. Small shields such as those in east-Australian

and New Zealand basaltic fields are probably monogenetic.

Large shield volcanoes of the Hawaiian type (Macdonald, 1972) are more complex, longer-lived, polygenetic centres whose life spans may be measurable in terms of hundreds of thousands of years to a million years. The shield volcanoes of eastern Australia without question were long-lived polygenetic centres. This is reflected by the diversity of products, including low-aspect-ratio, shield-forming lavas, high-aspect-ratio domes, and pyroclastic deposits in some cases, as well as by the diversity of compositions in all the centres (Chapter 3). However, the shield volcanoes of eastern Australia are significantly smaller than the large Hawaiian shields. Mauna Loa has a diameter of 100 km and a volume of 40 000 km³, but the largest Australian shield volcano, Tweed, has a diameter of 100 km (Section 3.4.7) and a volume of only 4000 km³ (Middlemost, 1985). The other shield volcanoes in eastern Australia do not exceed 1000 km³ (Middlemost, 1985).

Middlemost (1985) likened some east-Australian shield volcanoes to the Turkana-type shield volcanoes of the northern part of the Kenya rift. These were first distinguished as a type of shield volcano by Webb & Weaver (1975). Turkana-type shield volcanoes are trachytic and have stratiform, low-angle (5°) slopes. They are up to 50 km in diameter and are distinguished by a complex summit region marked by plugs, dykes, and cones. Calderas and craters are minor features. The Focal Peak, Nandewar, and Canobolas volcanic centres conform with this general description.

Details of the internal structure of the east-Australian shield volcanoes are not fully documented, but the description of the Nandewar volcano in Section 3.5.2 provides a general idea of the character of these volcanoes, as does the general model for the east-Australian shields proposed by Middlemost (1985). Nandewar consists mainly of hawaiite, trachyte, and peralkaline rhyolite. Lava and pyroclastic horizons dip 5–10° away from the centre. The central point is marked by a high-level monzonitic intrusion and abundant dykes, corresponding to the probable source area of the extrusive rocks. The mafic flows are 2–10 m thick, whereas the trachytic flows are between 10 and 20 m thick. Some flows extend for 6–8 km.

Small lava shield volcanoes are common in the Northland-Auckland region of New Zealand (Fig. 2.4.5). Rangitoto Island in Auckland (Waitemata) Harbour, the youngest volcanic centre in the province (activity is known to have taken place about 1200 A.D.; Section 4.2.3) is largely a shield volcano (Fig. 2.4.5). It has an early phreatomagmatic eruptive history, and later Hawaiian and Strombolian satellitic centres (Ballance & Smith, 1982). Morphologically well-preserved shield volcanoes with slopes of 8-17° are preserved in the Alexandra Volcanic Group south of Auckland at Pirongia and Kaioi (Briggs, 1986).

Larger shield volcanoes are found around Dunedin in the South Island (Coombs et al., 1986; Section 4.3.6) and in the Auckland Islands on the Campbell Plateau (Section 4.4.2). Dunedin volcano has been studied extensively. It is a Miocene volcano that has a present-day preserved diameter of 25 km and relief of 700 m. Lavas predominate, but pyroclastic deposits — including base-surge deposits — have been reported. The lavas are diverse in composition, ranging from basalt to trachyte and phonolite. Mafic aa lavas have been recorded, whereas many domes are trachytic. Dated samples range widely from 13 to 10 Ma, indicative of a polygenetic history, and even older basalts (up to 21 Ma) are known also (Section 4.3.6). Successions of tens of thin basalt, hawaiite, mugearite, and trachyte lavas and rare tuffs at Akaroa, and a similar, multiple-flow succession of basalt in the western Purau valley, are examples of shield volcanism on Banks Peninsula, South Island (B.F. Houghton, personal communication, 1987).

Lava cones

Lava cones are intermediate in character between spatter-scoria cones (see below) and lava shields. Stephenson & Griffin (1976a) recorded the presence of minor pyroclastic deposits in lava cones of northern Queensland. Hawaiian-style lava fountaining was presumably the most important eruption style, and thin clastogenic lavas are probably prominent among the thin lava successions.

North Queensland cones are low-profile features that rise up to 80 m above surrounding lava plains. The craters are up to 700 m across and have flat floors bounded by rims up to 40 m high. These lava cones are the sources of the long lava flows of northern

Figure 2.4.5 Auckland City in North Island, New Zealand, has numerous, youthful, pyroclastic cones. The youngest volcano is Rangitoto, in Auckland Harbour, shown here in the distance. It has a basal phreatomagmatic succession, overlain by a prominent lava shield and, finally, by the craters and scoria cones of the summit region (see also Fig. 4.2.2).

Queensland (Stephenson & Griffin, 1976b). Large-volume lavas may have issued from central lava lakes, at times through breaches, and the low-relief rims may have built up by spill-over from the lava lake and by Hawaiian-style, lava-fountain eruptions. The Warrion Hill lava pile in Victoria, north of Colac, also may be a lava cone, as well as Mount Hamilton in western Victoria.

Scoria and spatter cones

Cones of scoria (see also Section 2.5.2) and spatter may mark the sources of lavas, involving eruptions from two main types of site. First, lavas may be fed by Hawaiian-style lava fountaining from the summit vent and crater and may spill through breaches in the crater walls. The extensive Harman Valley flow of the Newer Volcanics province in Victoria may have originated in this way. The summit of Mount Napier (Fig. 2.4.6) consists of a scoria-spatter cone containing a marked crater-wall breach and a crater rim of spectacular driblet, spatter deposits. This scoria-spatter cone developed above a basalt shield volcano, corresponding to a change from an initial, effusive, shield-building phase to more dynamic explosive activity.

Second, lavas may be erupted from base-of-cone vents or fissures. For example, very small, local lavas derived from scoria cones in the central, scoria-cone complex of the Tower Hill maar centre of the Newer Volcanics province were erupted from base-of-cone vents (Edney, 1984a). Another example is the lava flow from Mount Fox, a pyroclastic cone 45 km southwest of Ingham, in Queensland (Section 3.2.11).

Both sources of lava flows from scoria cones have been noted by Smith et al. (1986) in the Kaikohe volcanic field, in the Northland province of New Zealand. One Tree Hill cone in the Auckland province has had both summit and flank eruptions of lava (B. F. Houghton, personal communication, 1987).

Even maar centres (see below) may produce phases of Hawaiian- and Strombolian-style eruptive activity from one or more eruption points, leading to the formation of agglutinates and clastogenic lavas. Examples are the Red Rock and Tower Hill volcanic complexes in the Newer Volcanics province of Victoria, and Crater Hill in the Auckland province. C.D. Ollier (personal communication, 1988) noted that complex centres may have definable sequences or cycles of eruption consisting of an initial maar-forming explosive phase, followed by a lava eruption that may partly bury the maar, followed by a final explosive scoria-cone forming phase. Examples are found at Mounts Leura and Warrnambool, Victoria.

2.4.5 Lava types and geometry

Pahoehoe and aa lavas

The two principal lava types in subaerial basaltic fields are pahoehoe lava and aa lava (Macdonald, 1972). Pahoehoe lava is fluidal, mobile, relatively fast moving, and the more far-reaching. Typical ropey pahoehoe surface features are relatively fragile, are eroded readily, and are preserved rarely in ancient lavas. Pahoehoe ropey lava surfaces are uncommon even in the late Tertiary and Quaternary provinces of northern

Figure 2.4.6 Mount Napier, Newer Volcanics province, Victoria, is a composite lava shield and superimposed scoria cone. Mount Napier, shown in the distance, was the source of the extensive Harman Valley lava flow filling the valley in the foreground.

Queensland and Victoria. Examples have been noted in the Toomba flow, northern Queensland (Stephenson & Griffin, 1976b), and on isolated surface exposures of the Harman Valley lava and at Cape Duquesne in the Bridgewater volcanic centre, both in Victoria. The best examples are preserved on the young island volcano Rangitoto in Auckland Harbour, New Zealand (Ballance & Smith, 1982; Section 4.2.3).

The mobility of pahoehoe flows is demonstrated by the great extent of flood-basalt lavas (for example, Swanson et al., 1975). Plains basalts also are commonly far-flowing pahoehoe lavas, but they do not spread laterally as much as flood basalts because of their smaller volumes. Examples include the previously mentioned north-Queensland flows which flowed 80–160 km from source over average slopes of less than 0.5° (Stephenson & Griffin, 1976b; Section 3.2; Fig. 2.3.2). Similar, shorter flows are found in Victoria (the Tyrendarra flow — 50 km; the Harman Valley flow — 20 km, Boutakoff 1963; the Mount Rouse lava flows — 60 km, Ollier, 1985a; Fig. 2.3.3).

The extent of some pahoehoe flows is illustrated well by the Roza Member of the Columbia River Plateau flood-basalt province, which covers 40 000 km² and has a volume greater than 1500 km³ (Swanson et al., 1975; Greeley, 1982a). The Roza Member is a composite of several flows, but the areas covered and volumes of individual flows are still large. Even the 'small'-volume lavas of the Ice Harbour member, were up to 10 km³. Lavas in plains-basalt provinces are moderate to small in volume by comparison. The Wapi flow in the Snake River Plain covers 300 km² and has a volume of 23 km³ (Greeley 1982a). Most of the other flows are smaller, covering several hundred square kilometres and having volumes generally less than 10 km³. The Tyrendarra flow of Victoria is similar to these: it covered about 300 km² originally (Boutakoff, 1963).

Aa lava is a more viscous, slower-moving, mafic to intermediate lava type. This higher viscosity, in part at least, is a function of lower flow temperatures, because pahoehoe lavas have been observed to change downstream to aa lavas (Peterson & Tilling, 1980). The pahoehoe-to-aa transition is also a function of shear rate: a high rate of shear is required to begin the transition to aa if viscosity is low, and vice versa. Aa lavas are characterised by highly irregular, hackly, spinose tops, which also are susceptible to erosion. Aa lavas as such have not been recognised from east-Australian fields, but have been documented from Rangitoto volcano in Auckland Harbour (Ballance & Smith, 1982) and noted on Dunedin volcano (Section 4.3.6).

Block and clastogenic lavas

Two other subaerial lava types are block lavas and clastogenic lavas. Block lavas result from the autobrecciation of viscous lava, commonly producing a marginal aggregate of smooth-faced to spinose, slaggy blocks surrounding a cohesive interior (Macdonald, 1972). None have been documented yet in eastern Australia or New Zealand. Clastogenic lavas are the products of Hawaiian-style lava fountaining, and are intermediate between true, reconstituted, coherent lavas, which flow freely away from the vent, and

Figure 2.4.7   Mantle-form geometry of a vesicular basalt layer, Mount Eden, Auckland volcanic province. The layer has the lithological aspect of a clastogenic lava. Its geometry distinguishes it as a densely welded, at-vent, agglutinate, air-fall deposit.

agglutinates or spatter deposits which are *in situ* accumulations of spatter. Some densely welded agglutinates also may resemble lavas, but may mantle irregular near-vent topography, like welded air-fall deposits (Cas & Wright, 1987; Fig. 2.4.7). Clastogenic lavas are preserved in excellent coastal exposures at Cape Bridgewater in Victoria as a succession of thin flows up to 0.5 m thick separated by thinner spatter horizons (Fig. 2.4.1), and are found as localised lavas, mostly associated with agglutinates, in the multiple vent Red Rock volcanic complex near Colac in Victoria.

Intermediate-felsic lavas

The characteristics of intermediate and felsic lavas were reviewed recently by Cas & Wright (1987). These lavas do not flow great distances from the vent because of high viscosity and low magma-discharge rates. Many in fact sit astride the vent, forming high-aspect-ratio domes, coulées, or spires, such as the Glass Houses of Queensland (Section 3.4.3) and the Macedon-Trentham domes in Victoria (Section 3.7.3). The sides of domes and coulées are steep-sided, whereas the tops — particularly of coulées, although generally flat — are locally irregular and hummocky. Both the margins and tops may be extensively autobrecciated (for example, the domes of the Focal Peak, Nandewar, and Canobolas volcanic centres), and some may develop marginal debris fans, such as the Miwaka phonolite domes of Dunedin volcano, New Zealand (B.F. Houghton, personal communication, 1987).

The interiors of intermediate to felsic lavas generally have well-defined flow banding, corresponding to relatively passive laminar flow during emplacement. However, extensive zones of autobrecciation may be present internally, and well-developed flow banding may be truncated, passing abruptly and laterally into zones of autobrecciation (for example, the rhyolites at Cape Hillsborough in central Queensland; Section 3.3.2). The banding may be flow folded and can dip steeply inward back into the vent, defining ramp structure.

Exceptions to high-aspect-ratio intermediate-to-felsic lavas may be caused either by a lower-than-normal viscosity or an exceptionally high magma-discharge rate (or both). Bonnichsen (1982) described thick, extraordinarily extensive rhyolitic lavas from the Snake River Plain, some of which seem to extend for up to 40 km — distances more similar to those of ignimbrites than to rhyolite lavas. Bonnichsen suggested that the rhyolites were formed by anatexis of the crust induced by large volumes of mantle-derived basaltic magma which raised the temperature of the rhyolitic magma well above liquidus temperature. High magma-discharge rates also probably contributed to the unusual mobility of these rhyolite lavas.

2.4.6 Propagation of lava flows

Basaltic magmas are so mobile that they are capable of flowing over long distances on slopes of less than 0.5°. Relatively narrow flows appear to propagate forward, largely by the development of an interior magma feed system of lava tubes and channels. The margins cool first and, in accord with its Bingham characteristics, may freeze, so forming natural *levées* (or 'initial' levées in the terminology of Sparks et al., 1976) which act to confine and to some degree insulate the interior of the flow — for example, the Tyrendarra lava flow, Mount Eccles and the Harman Valley lava flow, Newer Volcanics province, Victoria). The initial levées will extend downstream as the flow advances so defining an interior channel system through which lava is fed to the front of the flow. These form in both pahoehoe and aa lavas. *Channels* or *canals* are marked on many lava flows because, where magma supply ceases, channels drain and the interior lava surface level falls or deflates, leaving well-defined lateral, confining levée ridges.

Marginal cooling of the lava flow also may affect the top surface of the flow to such a degree that a solid roof layer forms over the channel, so totally enclosing and insulating the interior of the flow. These *lava tubes* or *tunnels* may be drained where the magma supply wanes, and extensive *lava caves* are produced. The well documented lava caves of some north-Queensland lava flows (Atkinson et al., 1975; Figs. 2.4.8–9) and in the Harman Valley lava flow in the Newer Volcanics of Victoria (Ollier & Brown, 1965) are good east-Australian examples of this mode of propagation. The tubes and caves are considered to have been essential to the long distances of propagation of the flows because of the insulation against heat loss provided by sealed lava tubes. Collapse of parts of such roof systems can create entrances or 'skylights' to the caves (Fig. 2.4.10). More extensive collapses can expose the whole tube, giving the appearance of a canal (for example, some of the Mount Eccles and Byaduk caves, Newer Volcanics province; Fig. 2.4.11). Some skylights are created during flow and can become the vents for lava extrusion during times when the tube is full of lava. Lava tubes and caves can be 20 m or more in height and width, especially where lava flows are confined within valleys.

Intermediate lavas also may develop marginal levées. Felsic lavas and domes appear to spread laterally as more lava is erupted successively from the vent in the interior of the flow. New lava rises along subvertical flow paths, and is succeeded by more lava, that ramps up against it as the flow slowly spreads laterally (for example, Cas & Wright, 1987).

2.4.7 Other features of lava

Other notable surface features of basaltic lavas include transverse pressure ridges, tumuli, hornitos, and columnar jointing. *Pressure ridges* develop perpendicular to the direction of flow and form by updoming of parts of the plastic surface crust, much like pahoehoe flow folds, but on a larger scale. They form across the relatively fast-moving interior of flows, and many become convex downstream because of the velocity gradient between the slow-moving sides and the fast-moving centre of the lava. Pressure ridges are well preserved on the surface of the Harman Valley lava flow on the road between MacArthur and Byaduk, western Victoria. The so-called 'stoney rises' lava flows of the Newer Volcanics province in Victoria, best

Figure 2.4.8   Barkers Cave in the Undara lava-tube system, northern Queensland.

Figure 2.4.9   Barkers Cave in the Undara lava-tube system, northern Queensland, showing lava-level lines and a lateral gutter adjacent to final lava fill.

developed west of Colac, represent a combination of pressure ridges, overlapping lava lobes, autobrecciated lava surfaces, and collapse structures resulting from the withdrawal of magma from the interior lava-tube systems. Lava tubes also may become inflated and their surfaces elevated by the late-stage injection of lava (G.P.L. Walker, personal communication, 1986).

Tumuli are circular to elliptical domes or blisters that also form on the surface of the lava and which may be 20 m or more in diameter. They result from local fluid-pressure points inside the lava causing the plastic crust to dome up. The fluid pressure could be caused by magma, or locally high concentrations or pockets of gas, which could be either magmatic gas or vapourised groundwater over which the lava has flowed. The tumuli that are so well preserved on the surface of the Harman Valley lava flow at Wallacedale, western Victoria (Fig. 2.4.12) are notable for the exceptionally vesicular nature of the lava in the tumuli, suggesting that the tumuli developed in response to local gas-pressure points. Ollier (1967a) suggested that they developed where the Harman Valley lava flow began to move over water-saturated ground at Condah swamp. Tumuli domes can be hollow where the magma supply wanes and magma withdrawal takes place within the lava tube. Outward-radiating columnar joints are developed in the roofs of tumuli on the Harman Valley lava flow. The shrinkage caused by cooling in some cases has been sufficient to produce collapse of the roof of the dome, accompanied in many cases by withdrawal of magma from the interior of the flow.

Hornitos are small, rootless, spatter mounds that can grow to metres in height. They commonly develop

LAVA-FORMING ERUPTIONS AND FLOW FEATURES

Figure 2.4.10 Entrance to lava cave formed by collapse of the roof of a lava tube at Byaduk, Newer Volcanics province, Victoria.

Figure 2.4.11 Lava canal formed by collapse of the roof of a lava tube at Mount Eccles, Newer Volcanics province, Victoria.

above small openings in the roof of a lava tube (skylight), caused by the pressure of surging magma in the interior lava tube, or perhaps by volatilisation of trapped groundwater at the base of the flow. The preservation potential of hornitos is very low, and none have been reported from east-Australian fields.

However, Rangitoto Island in Auckland Harbour has examples of hornitos (Ballance & Smith, 1982).

Columnar jointing is a product of the progressive cooling of lavas and intrusions. Cooling at the margins produces polygonal contraction cracks which then propagate perpendicular to the cooling front as the front

Figure 2.4.12   Tumulus in the Harman Valley lava flow, Wallacedale, Victoria.

Figure 2.4.13   Columnar jointing with transverse cracks in basaltic lavas at Diggers Rest, Organ Pipes National Park, Newer Volcanics province, Victoria. Height of outcrop is about 8 m.

advances, producing polygonal columns (Fig. 2.4.13). The inwards growth of the columns appears to be achieved in some examples by a step-by-step advance, represented by transverse joints or segmentation perpendicular to the column axes (Fig. 2.4.13). Columnar jointing in thin sheet-like bodies is simple and upright. A two- or multiple-tiered layering may develop in thicker flows. The lower layer consists of simple, upright, columnar jointing and is known as the *colonnade*. The upper tier consists of irregular, generally partially or wholly radiating clusters of columns and is known as *entablature* (Fig. 2.4.14). This may form in response to irregularities in the top surface of lavas, so producing an irregular upper cooling surface and front, or where external water (for example, river water from dammed valleys) permeates the top of the flow and creates local cooling points. Radiating or rosette-like clusters of columns may form from the cooling of enclosed feeder lava tubes. Lava flows in the Organ Pipes National Park (Fig. 2.4.13) and in the Campaspe River (Fig. 2.4.14) in the Newer Volcanics province in Victoria have excellent columnar jointing.

2.4.8   Hyaloclastite deposits

Hyaloclastite deposits result from the quench fragmentation of magma where it comes into contact with water. They can form where lava flows are erupted into water or flow from land to water, or where magma intrudes wet unconsolidated sediments. Lava flowing from land and fragmenting into a standing body of water may produce a lava delta consisting of crudely defined foresets of hyaloclastite material and topsets of the lava. The base of the lava consists of a highly irregular zone of lava breccia and irregular pods of lava ('flow-foot breccia'; Jones & Nelson, 1970) and is important in recording the ancient water level at the time the lava flowed into water.

Figure 2.4.14   Entablature above colonnade columnar jointing, Campaspe River, Newer Volcanics province, Victoria (note person for scale).

Figure 2.4.15   Hyaloclastite associated with pillow-lava breccia, Oamaru, South Island, New Zealand.

Hyaloclastite deposits and flow-foot breccias have been recognised in several places and settings in eastern Australia and New Zealand. Australian examples are associated most commonly with the passage of lava into lakes dammed by earlier lava flows. Bishop (1985a) recorded them from highland valleys in New South Wales, and Sutherland (1980b) documented occurrences in Tasmania.

Minor hyaloclastite material is found associated also with basalt that intruded unconsolidated carbonate clastic sediments in marine Oligocene strata at Totara Terraces, near Oamaru, in the South Island of New Zealand, and in Eocene marine strata of the Otway Basin at Aireys Inlet in Victoria. Hyaloclastite material at Oamaru is also prominent in thick, massive to diffusely stratified beds at the top of a Surtseyan volcano preserved in coastal exposures (Coombs et al., 1986; Cas et al., 1986). The hyaloclastite contains numerous pillow-lava fragments (Fig. 2.4.15). The hyaloclastite is interpreted to represent mass-flow resedimented debris that resulted from the passage of a subaerial lava (possibly the terminal eruptive event in the history of the volcano) into the surrounding sea. This produced large volumes of hyaloclastite which was then resedimented by mass-flow processes onto the deeper-water apron of the volcano.

2.5 Pyroclastic Eruptions

2.5.1 Introduction

Explosive volcanic eruptions involve different modes of fragmentation, different styles of eruption, and a diversity of products. These will be reviewed briefly in this section (for more details see, for example, Fisher & Schmincke, 1984). A useful review of the terminology applied to pyroclastic-eruption styles, transport processes, and deposits was presented by Wright et al. (1980).

Fragmentation processes in volcanic terrains may be subdivided into three general categories: pyroclastic, autoclastic, and epiclastic. Pyroclastic processes involve explosive fragmentation. Specific types of explosion include magmatic, phreatomagmatic, and phreatic. Autoclastic fragmentation involves non-explosive fragmentation and includes autobrecciation or flow brecciation, and quench fragmentation. Epiclastic fragmentation includes normal surface processes such as erosion and gravitational collapse.

2.5.2 Magmatic eruptions and their deposits

Eruptions

Magmatic explosions are driven by the exsolution and explosive expansion of magmatic volatiles, especially water and carbon dioxide. The controls on magmatic explosive activity were reviewed by Sparks (1978).

Magmatic explosions take place in two distinct situations. The vent in the first situation is open, and vesiculating magma interfaces directly with the atmosphere. Fragmentation may be caused by high-level, gas-pressure induced fragmentation or by shearing of the vesiculated magma caused by high exit velocities during Strombolian, Sub-plinian and Plinian eruptions (Bennett, 1974; Wilson et al., 1980). The vent in the second situation either is blocked, perhaps by a thick, cold cap of lava, or else there is a deeper-seated magma chamber without a magma conduit to the surface. Failure and fracturing of the chamber roof or lava cap will take place if the fluid pressure exceeds the minimum principal stress component and the tensile strength of the country rock or lava cap. Release of confining pressure — that is, the decompression resulting from roof failure — may be rapid enough to initiate rapid exsolution and bubble growth and explosive eruption. Vulcanian eruptions involve the build-up of gas pressure immediately beneath a lava cap that develops in a vent, and explosive activity is marked by short, violent explosions which represent short-lived, vent-clearing eruptions.

Products

The nature and origin of the products of magmatic explosions have been discussed elsewhere (Heiken & Wohletz, 1985; Macdonald, 1972; Cas & Wright, 1987, and references therein) so only the principal aspects will be summarised here.

The diagnostic products of basaltic magmatic explosions are scoria fall deposits. These generally are massive and structureless, and can be tens to hundreds of metres thick near the vent. They result from continuous, maintained, Strombolian-type eruptions and produce scoria cones at the vent and sheet-like stratified scoria deposits farther away (Fig. 2.5.1). Faint stratification may represent slight differences in eruption rate or instabilities in the tephra-charged eruption column (Fig. 2.5.2). Most scoria deposits have a limited dispersal.

Scoria cones are common in basaltic fields in intraplate settings in eastern Australia and New Zealand (Chapters 3 and 4). They are particularly well preserved in northern Queensland, in the Newer Volcanics province in Victoria (*Fig. 2.5.3*) and South Australia, and in the Northland-Auckland region of New Zealand (Fig. 2.4.4). Quarrying in the Auckland area (for example, Mount Eden) and in the Newer Volcanics province has exposed tens of metres of scoria deposits in numerous centres (for example, Fig. 2.5.4).

More distal, sheet-like, scoria deposits are also common, and are well-exposed intercalated with phreatomagmatic deposits at Tower Hill, Mount Leura (Fig. 2.5.1), and the Red Rock volcanic complex. Spindle-shaped bombs are found also in scoria deposits near the vent. Normal or reverse (or both) grading is found in scoria-fall deposits, reflecting changes in the height of the eruption column and dispersal of tephra. Normal grading reflects declining column height, intensity of explosive eruption, and dispersal, whereas reverse grading reflects the opposite.

Spatter deposits predominate where the eruptions are Hawaiian rather than Strombolian in style (for example, Bridgewater volcano, Red Rock volcanic complex, and Mount Napier, all in Victoria; Fig. 2.5.5). Other distinctive grain types include smooth-surfaced glass fragments called acneliths (Red Rock), and tear-

Figure 2.5.1 Stratified Strombolian scoria deposits (dark) mantling the tuff ring of the Leura tuff-ring/scoria-cone complex, Camperdown, Victoria. The underlying deposits (light) are of base-surge origin.

like to delicate needles and threads of volcanic glass (Pele's hair) which result from the rapid cooling of sprays of highly fluidal basaltic lava. Deposits from Vulcanian eruptions may contain vesiculated juvenile clasts, but most are dominated by dense blocks of lava resulting from the violent explosive disintegration of the lava cap that has sealed a vent.

Intermediate-to-felsic magmatic explosions produce vesiculated pumice and glass shards. Pumice may be found in massive to stratified air-fall deposits, in pyroclastic-flow deposits, and in surge deposits, including not only base-surge deposits, but also those of ground-surges and ash-cloud surges (see below). Pumiceous deposits commonly are associated with more violently explosive eruptions such as Sub-plinian, Plinian, and Ultraplinian eruptions (see below). Normal and reverse grading may be related also to column height, eruption intensity, and dispersal, as discussed for scoria deposits. Pumiceous deposits from Cainozoic intraplate settings in eastern Australia and New Zealand are not well known, but have been reported from Focal Peak and Canobolas volcanoes and at Cape Hillsborough in Queensland.

2.5.3 Phreatic and phreatomagmatic eruptions and their deposits

Phreatic explosions are steam-generated and do not involve the ejection of fresh magma. Carbon dioxide also may be an important explosive agent in some cases (for example, Chivas et al., 1987). The expanding steam in many examples is generated from groundwater,

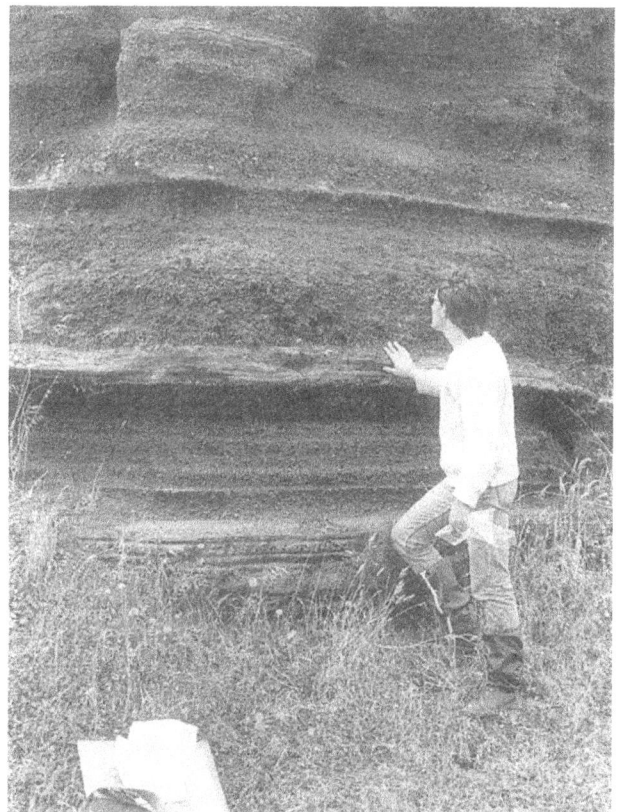

Figure 2.5.2 Maar-rim succession at Coragulac maar, Red Rock volcanic complex, near Colac, Victoria, showing crudely stratified Strombolian scoria deposits. Coragulac maar contains anomalously large proportions of scoria in the maar rim.

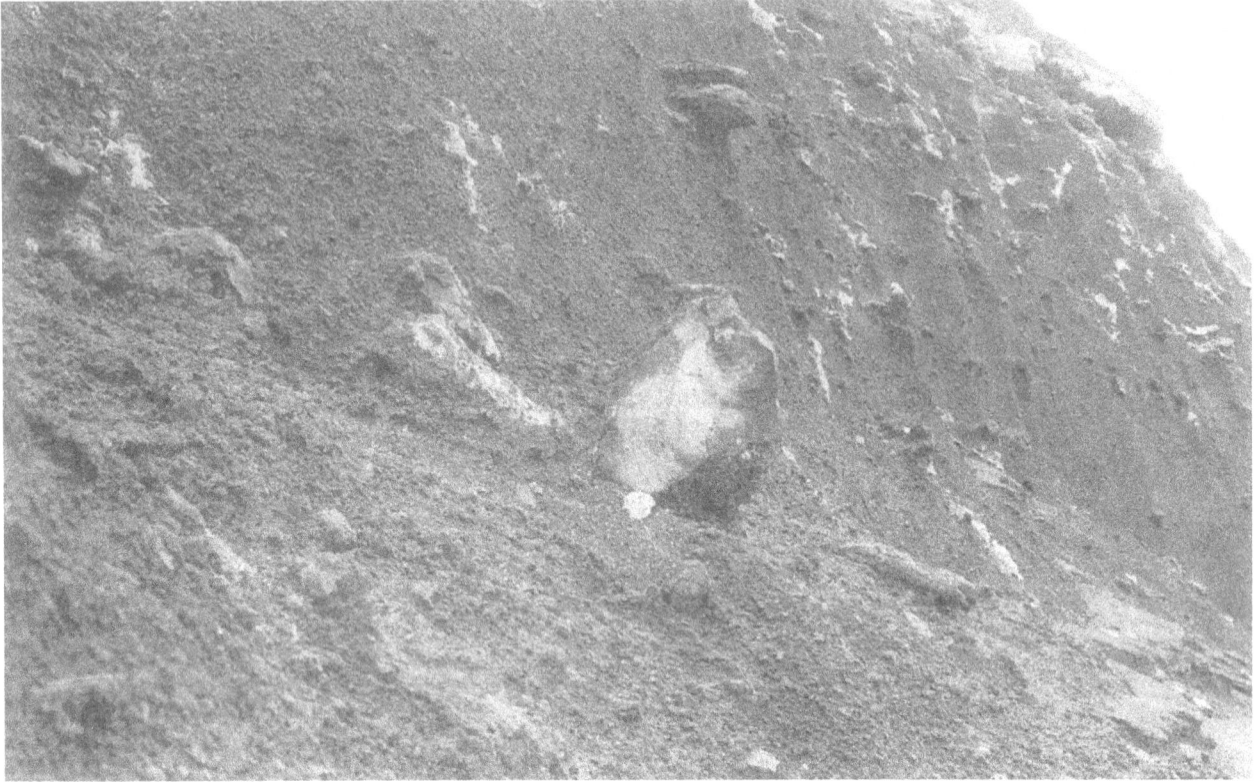

Figure 2.5.4   Massive to crudely stratified scoria and bombs at Mount Elephant.

Figure 2.5.5   Fluidal-bomb accumulation in agglutinate deposits, Red Rock volcano complex, near Colac, Victoria.

either from a water table or probably most commonly from hydrothermal systems (for example, Cas & Wright, 1987). Ejected solids are country-rock fragments, which may be either volcanic or non-volcanic, depending on the earlier history of the area. Hydrothermal alteration products such as clays also will be a significant component where the subsurface rock has been significantly hydrothermally altered. Deposits are thick near-vent sheets of a poorly sorted, clay-supported breccia containing a range of rock types that have been explosively excavated from the subsurface succession. Purely phreatic explosive deposits have not been documented from eastern Australia, and the only known New Zealand examples are from a hydrothermally active area in the Northland province (Section 4.2.2).

Phreatomagmatic explosions involve the dynamic explosive interaction between magma and an external water source such as groundwater or a surface body of water such as a lake or the sea, and the ejection of a significant juvenile magmatic component. The efficiency and intensity of explosive activity depends critically on the water/magma mass ratio and are highest at values of 0.3 according to experimental results (Wohletz & Sheridan, 1983; Heiken & Wohletz, 1985; Wohletz, 1986). There is probably in practice a complete spectrum between purely phreatic explosions and purely magmatic ones. Magma also contributes significant mechanical energy directly through the explosive expansion of magmatic volatiles, as discussed above. External water in this situation simply may act as a catalyst to enhance what would otherwise be incipient magmatic explosions (Cas & Wright, 1987). Houghton & Schmincke (1986) also pointed out that simultaneous Strombolian magmatic explosive eruptions and phreatomagmatic explosive eruptions can take place from closely spaced vents in the same volcanic centre.

Pyroclasts produced by phreatomagmatic explosive activity are diverse in their textural characteristics, corresponding to the intensity of the explosive activity, the nature of the interaction with the external water, and the degree of vesiculation of the magma prior to interaction with the external water (Heiken & Wohletz, 1985). Phreatomagmatic fall deposits almost invariably are finer grained than are magmatic scoria deposits at a given distance from vent (corresponding to more thorough explosive fragmentation). Pyroclasts from phreatomagmatic explosions can be blocky and non-vesicular or highly vesiculated, presumably reflecting an advanced stage of vesiculation in the magma prior to explosive interaction with external water, as discussed above.

Concentrically structured, spherical, accretionary lapilli and larger cored lapilli (Moore & Peck, 1962; Fig. 2.5.6) also are present in many phreatomagmatic centres (for example, Tower Hill, Red Rock, and Bridgewater). Phreatomagmatic deposits tend to alter because of the interaction with external water, as mentioned above, and to consolidate and cement rapidly even within only several years of eruption (for example, Jakobsson, 1978). Basaltic glass (sidero-melane) alters to palagonite, even in continental settings, and typical cements include zeolites, clays, and calcite.

Basaltic phreatomagmatic deposits typically are better and more thinly bedded than are scoria-fall deposits, especially near the vent, and are deposited by both pyroclastic-fall and surge mechanisms (Fig. 2.5.7). Better stratification results from the punctuated, pulsatory nature of the explosive interaction between water and magma.

Phreatomagmatic deposits are widespread in Caino-zoic volcanic fields in eastern Australia and New Zealand where, in both instances, only basaltic deposits are known. However, existing descriptions are not detailed. Palagonitised phreatomagmatic deposits including base-surge deposits are known in the Maer province in Torres Strait in northern-most Queensland. Some of the eruptions appear to have begun sub-aqueously and to have involved eruptions through reefs (Section 3.2.2). The phreatomagmatic eruptions therefore appear to have been triggered by explosive interaction between subaqueously erupted basalt and seawater, suggesting that they were Surtseyan in origin. A likely phreatomagmatic cone including base-surge deposits is recorded near Stevens Island, and maars and a diatreme in the Atherton area are known in northern Queensland (Section 3.2). These maars would have originated from explosive interaction with groundwater.

Diatremes are pipe-like bodies filled with variably bedded phreatomagmatic pyroclastic debris and bedded rafts. They are especially well exposed in the Sydney area although are dispersed more extensively throughout the Sydney Basin area, as well as beyond (Crawford et al., 1980; Sections 3.6.3, 3.9.3). They are seen in several quarries and consist of phreatomagmatic fall

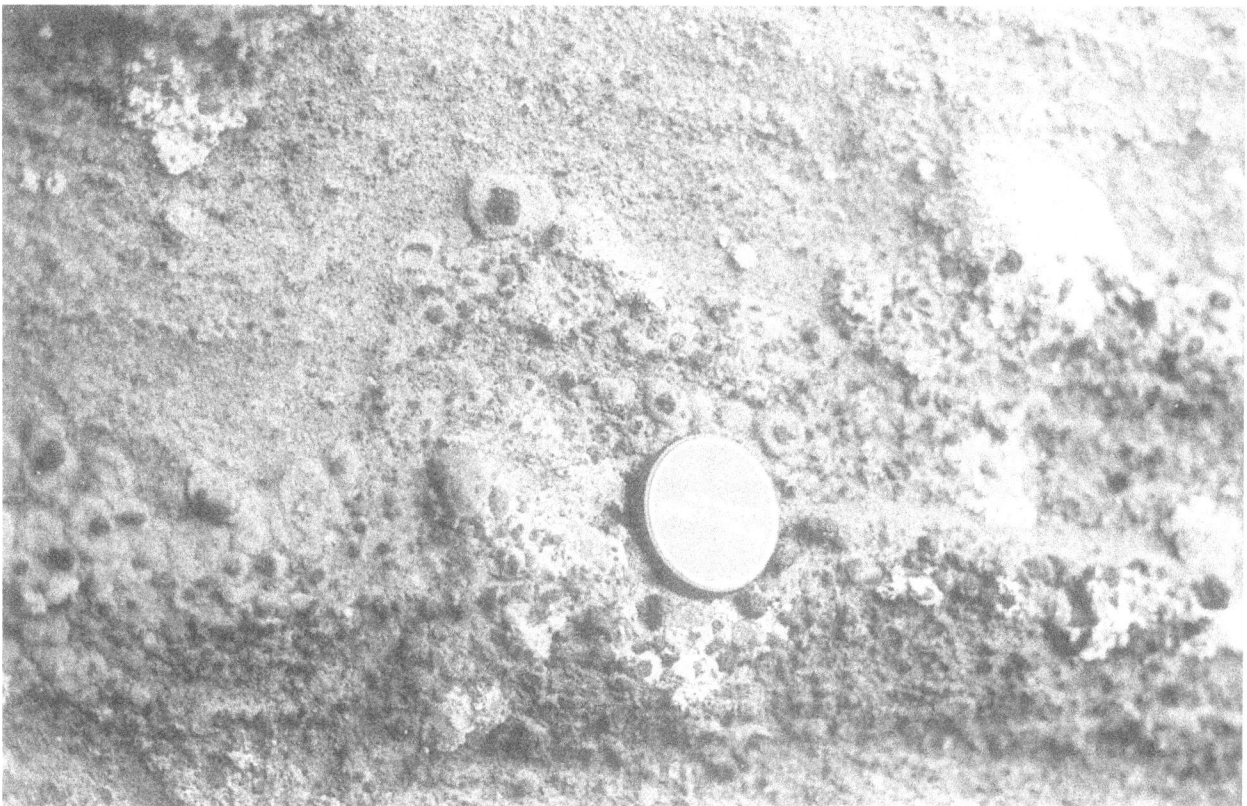

Figure 2.5.6   Large, cored, accretionary lapilli, including cores of highly vesicular scoria encrusted by finer phreatomagmatic ash, at Bridgewater volcano, near Portland, Victoria.

Figure 2.5.7 Bedded, phreatomagmatic, fine tuffs and lapilli tuffs of surge and fall origin at the Red Rock volcanic complex, Purdigulac maar, Russells Pit, near Colac, Victoria. Note the large ballistic blocks.

Figure 2.5.8 Base-surge deposits of the Purrumbete maar, Newer Volcanics province, near Camperdown, Victoria. Classical dune-form structures have marked climbing cross-stratification, corresponding to rapid rates of tephra accumulation.

Figure 2.5.9 Dune-form base-surge deposits, Oamaru volcano, South Island, New Zealand.

and surge deposits that, locally, are significantly affected by block subsidence and sliding into the diatreme from original maar or tuff-ring deposits at the surface.

Centres such as Tower Hill, Purrumbete, Red Rock, and Leura, above the aquifers in the Tertiary sector of the Otway Basin in Victoria (Fig. 1.6.1), represent the best preserved phreatomagmatic volcanic centres and landforms in eastern Australia or New Zealand. Base-surge and fall deposits are exposed in quarries in these maar and tuff-ring volcanic centres. The base-surge deposits include classic dune-form structures, including typical, low-angle, cross-stratification and common, climbing dune-forms (Fig. 2.5.8).

Several examples of phreatomagmatic basalt deposits are known in Tasmania and the Bass Basin (Section 3.8). Pyroclastic rocks intersected in drill core within the marine Lower Miocene of the Bass Basin, may be Surtseyan phreatomagmatic deposits. Phreatomagmatic deposits on the Tasmanian mainland also appear to be associated with ancient river valleys (for example, in the Mersey/Forth River system) and with ancient lakes (the Great Lake area).

Phreatomagmatic deposits and centres in New Zealand are found in both the Quaternary Auckland province (Ballance & Smith, 1982) and in the Tertiary mafic-intermediate successions of Otago (Coombs et al., 1986). The Auckland province faces directly onto, and has developed within, the seas of the Hauraki Graben. Phreatomagmatic centres exist in their own right, such as the maars and tuff rings of Panmure Basin, Lake Pupuke, Onepoto Basin, Tank Farm, and

Orakei Basin (Ballance & Smith, 1982; Section 4.2.3). However, some centres such as Rangitoto Island in Auckland Harbour are more complex. Rangitoto appears to have had an initial phreatomagmatic history, as lava was erupted on the seafloor of Auckland Harbour before emerging above sea-level as a shield volcano, but details of the deposits are not documented.

The Tertiary intraplate volcanic rocks of the South Island of New Zealand include numerous examples of phreatomagmatic deposits and volcanic centres. The most spectacular of these are exposed along the coastline of central Otago, especially south of the harbour of Oamaru, and farther south at Kakanui and Moeraki (Coombs et al., 1986). These volcanic rocks are part of a Cretaceous to early Tertiary transgressive marine succession, and represent Eocene-Oligocene, Surtseyan, continental-shelf volcanoes. The volcanoes developed in shallow seas and some were ephemerally emergent structures (for example: Cas et al., 1986; Coombs et al., 1986). Base-surge deposits, as well as both subaerial and submarine (water-settled) fall deposits, are well represented (Fig. 2.5.9). The younger Miocene Dunedin volcano (Coombs et al., 1986; Section 4.3.6) also contains base-surge deposits which are well exposed along the coast north of Dunedin at Warrington (C.A. Landis, personal communication, 1986). Phreatomagmatic pyroclastic deposits are present also on the volcanic islands of the Campbell Plateau (Section 4.4). Felsic phreatomagmatic pyroclastic deposits have not been recognised in east-Australian or New Zealand intraplate volcanic provinces.

2.6   Pyroclastic Transport Processes

2.6.1   Pyroclastic-fall processes and products

Ballistic ejecta

Ballistic blocks and, in the case of basaltic eruptions, spindle-shaped bombs are ejected from all explosive eruptions. Their trajectories are determined by their exit or muzzle velocity from the vent, and by their terminal-fall velocity. The dynamics of ballistic transport were reviewed comprehensively by Wilson (1972) who included tabulations of the ranges of ballistics based on density, muzzle velocity, and angle of launch. Ballistic blocks are identified most readily by their generally large size (centimetres to metres) and, in many instances, by impact or sag structures, especially in wet phreatomagmatic tephra deposits. Excellent examples are especially well exposed in some of the basaltic phreatomagmatic centres of the Newer Volcanics province in Victoria, such as at Tower Hill and the Red Rock complex (Fig. 2.5.7).

Airfall deposits

An eruption plume or column of tephra, gas, vapour, and entrained air rises above the vent during sustained explosive eruptions. The dynamics of plume development were discussed by Settle (1978), Wilson et al. (1978), Sparks & Wilson (1982), and Sparks (1986), and have been summarised by Cas & Wright (1987).

Both magmatic and phreatomagmatic explosive eruptions produce eruption columns and airfall deposits. Both deposit types are well represented in east-Australian and New Zealand volcanic areas. Basaltic magmatic airfall deposits are represented in scoria-cone accumulations such as Mounts Elephant (Fig. 2.5.4), Leura, and Eccles, and Red Rock in the Newer Volcanics province, and at Mount Eden, Auckland. They are found also interstratified in the tuff-ring successions of phreatomagmatic centres such as the Tower Hill maar, the Red Rock maars (Fig. 2.5.2), and the Leura tuff ring (Fig. 2.5.1). In addition, they are dispersed more widely over the landscape beyond the limits of volcanic centres, but are not well exposed.

There is some difficulty in distinguishing thin, massive, and planar-stratified fall deposits from similar facies of base-surge deposits in the maars and tuff rings of the Newer Volcanics province (for example, Tower Hill, Leura, and Red Rock; Fig. 2.5.7). However, grain-size and sorting characteristics (for example, Walker, 1984) can help distinguish between the two. So too can evidence of low-angle truncations. Airfall and surge deposition in near-vent depositional environments take place simultaneously, producing unusual, wavy, stratified and cross-stratified surge-modified fall deposits (for example, Bridgewater volcano, Victoria).

Cones in the Northland and Auckland provinces of New Zealand (Section 4.2.2–3; Smith et al., 1986) contain numerous basaltic scoria-fall deposits. Phreatomagmatic basaltic fall deposits are best exposed in the coastal sections of central Otago, around the townships of Oamaru and Kakanui, as parts of Tertiary

Figure 2.6.1   Distal, fine-grained, water-lain basaltic tuff (dark layer) within a shelf bioclastic succession in the Eocene-Oligocene Waiareka Volcanics, Totara Terraces, near Oamaru, South Island, New Zealand.

continental-shelf volcanoes. Both true airfall and water-settled fall deposits are represented in the successions of the volcanic centres (Fig. 2.5.9). Distal fine-grained ashes are also preserved, intercalated with highly fossiliferous calcarenites (Fig. 2.6.1).

More felsic pyroclastic-fall deposits have not been documented in detail from east-Australian intraplate centres, although their presence has been noted or implied in several centres (for example, Focal Peak, Nandewar, and Canobolas).

Classification of fall deposits

Differences in the style of explosive eruptions affect not only the nature of the products but also the dispersal of the tephra. Walker (1973b) proposed a quantitative scheme for comparing the general dispersal and grain-size characteristics of pyroclastic-fall deposits as a basis for standardising comparisons and nomenclature. The scheme (see Walker, 1973b, for details) is based on the dispersal (D, called the dispersal index) of the fall deposits and on their degree of fragmentation (F, called the fragmentation index).

The following D-F fields and air-fall deposits have been defined: Hawaiian, Strombolian, Sub-plinian, Plinian, and Ultraplinian, for magmatic explosive eruptions, and Surtseyan and Phreatoplinian for phreatomagmatic or hydrovolcanic explosive eruptions. An additional Vulcanian field is defined, but there is current debate on whether these eruptions are magmatic or phreatomagmatic, or whether they can be both. Houghton et al. (1986a) pointed out that there is, in real terms, some overlap between the fields of different deposit types and suggested that a third parameter V, vesicularity, be used to help distinguish, in particular, magmatic from phreatomagmatic pyroclastic fall deposits. Walker (1980, 1981b) also introduced the use of 'area plots' which depict the area enclosed by isopach or grain-size isopleth contours.

The D-F approach in classifying deposits is a useful one where exposure is good, deposits are unconsolidated, and vegetation is sparse. However, exposure in eastern Australia and New Zealand is limited, and the D-F concept has not been applied so far. Rather, most deposits are described on the basis of their lithologies and structures in limited quarry and cliff exposures. Following are brief descriptions of the general eruption dynamics and deposits produced by the eruption types listed above.

Hawaiian-style eruptions involve fire- or lava-fountaining (Macdonald, 1972) and together with their deposits were described briefly in Section 2.4.1.

Strombolian eruption styles and activity were considered by Blackburn et al. (1976), Wilson (1980), Wilson & Head (1981), and Cas & Wright (1987). Eruptions consist of a series of discrete explosions separated by periods of less than 0.1 seconds to several hours. A rapid succession of explosions (say, every 1 to 2 seconds) leads to a maintained eruption column, driven by convection, that may reach heights of 5–10 km, as observed during the 1973 Heimaey eruption (see Blackburn et al., 1976). Thick massive scoria deposits are typical products which, around the vent, can reach several hundred metres in thickness.

Vulcanian eruptions typically consist of discrete cannon-like explosions separated by tens of minutes to hours (Schmincke, 1977; Self et al., 1979; Wilson, 1980; Cas & Wright, 1987). A series of small eruption columns up to 10 km in height may form, producing plumes strung out downwind. There is some debate on the role of external water in Vulcanian eruptions (see references cited above). Vulcanian pyroclastic-fall deposits from individual eruptions are thin, small volume (greater than 1 km^3), commonly stratified ash deposits that, near-vent, contain large ballistic bombs and blocks, some having breadcrusted and jointed surfaces. The deposits are mostly intermediate in composition (andesite and dacite) but basaltic ones are known also — for example, Tower Hill in western Victoria (Edney, 1984a). Thick, near-vent breccias may form locally, but deposits are generally so thin and fine-grained that they are soon eroded by wind and water. Eruptions may continue for a few years, and thin, bedded sequences accumulate near-vent but generally are not thick.

Plinian pyroclastic-fall deposits are a common product of highly explosive eruptions of high-viscosity magmas. These are generally andesitic to rhyolitic, or phonolitic and trachytic compositions, although rare basaltic scoria fall deposits which have Plinian dispersal patterns are known (Williams, 1983; Walker et al., 1984). The deposits of individual eruptions may attain thicknesses of 10 to 25 m within 10 km of the vent, but maximum thicknesses can be much smaller.

Sub-plinian fall deposits resemble Plinian deposits but they have a smaller dispersal and volume. The term *Ultraplinian* was defined by Walker (1980) for the most widely dispersed type of Plinian fall deposits. Plinian, Sub-plinian, and Ultraplinian pyroclastic-fall deposits have not been recognised in east-Australian and New Zealand intraplate settings.

The term *Surtseyan* was used by Walker (1973b) to describe the type of airfall deposits which result from similar activity to that observed during the explosive eruption of the island volcano of Surtsey in 1963–64. The Surtseyan D-F field, as defined by Walker (1973b), has been used since in a general way to group basaltic fall deposits that have similar dispersal and fragmentation resulting from different types of hydro-volcanic explosion. Kokelaar (1983, 1986) pointed out there may be significant differences between true Surtseyan activity, where water floods into the top of an open vent, and other phreatomagmatic activity involving trapped groundwater (for example, Newer Volcanics phreatomagmatic centres).

Maars, tuff rings, and tuff cones formed by phreato-magmatic activity are built up largely from the deposits of base surges (Figs. 2.5.1,7–9) and airfall deposits, as discussed above. Air-fall deposits near the vent are found interbedded with base-surge deposits and the distinction between airfall layers and planar-bedded base-surge deposits is problematic (Cas & Wright, 1987). Both modes of deposition may have taken place simultaneously as ash from a previous explosion or maintained column, falls around the vent into newly generated base surges, producing near-vent, surge-modified fall deposits.

Phreatoplinian deposits are extremely widely dispersed phreatomagmatic deposits of felsic composition (Self & Sparks, 1978; Walker, 1981a). No basaltic examples are known, and no deposits of Phreatoplinian character have been recognised in the intraplate provinces of eastern Australian and New Zealand.

2.6.2   Pyroclastic flows

Eruption columns in some instances become so overloaded with pyroclastic debris that the bulk density of the column exceeds that of the atmosphere. The column collapses gravitationally and, under the influence of the potential energy and of the fluidising effects of entrained volatiles and ingested air, the mass of pyroclastic debris, gas, and vapour moves laterally as a dynamic, gas-supported, high-particle-concentration, plug-flow system. Pyroclastic flows are commonly associated with intermediate to felsic Plinian eruptions and, on a small scale, with dense-column, intermediate, Vulcanian-Strombolian eruptions. They are not recognised as significant transporting agents in basaltic terrains. Fisher & Schmincke (1984) and Cas & Wright (1987) gave comprehensive treatments of pyroclastic flows and their deposits.

There are few pyroclastic-flow deposits in east-Australian and New Zealand intraplate terrains, and their characteristics have not been described in detail. They have been reported from the Hillsborough rhyolitic succession in Queensland (Section 3.3.2), and at Canobolas (Section 3.6.2) where trachytic pyroclastic-flow deposits, including welded types with eutaxitic textures, have been noted.

2.6.3   Pyroclastic surges

Pyroclastic surge is a type of pyroclastic flow characterised by a low-particle concentration, and is recognised commonly as a discrete pyroclastic-transport process. Pyroclastic fragments in surges are supported by turbulence of the interstitial gas and vapour. Three types of surge are recognised: base surge, ground surge, and ash-cloud surge. The last two types are spawned by pyroclastic flows and have not been recognised in intraplate terrains in eastern Australia and New Zealand. They are discussed fully by Cas & Wright (1987).

Base surges are so named because they appear to develop from the bases of phreatomagmatic and phreatic eruption columns (Moore, 1967). They may in some instances originate from partial collapse of an eruption column, but more commonly are initiated as lateral blasts from the vent, spreading radially as an expanding, turbulent, collar-like cloud, much like analogues from nuclear explosions. Some of the best descriptions of the processes and products are in early publications such as those of Moore (1967), Fisher & Waters (1970), Crowe & Fisher (1973), Schmincke et al. (1973), and Sheridan & Updike (1975).

Typical structures in base-surge deposits include low-profile dune forms (Fig. 2.5.8) and low-angle cross-stratification (for example, Walker, 1984). Horizontal stratification and both massive ashes and massive lapilli-ashes are other common facies types (Cas & Wright, 1987). Cross-stratification and horizontal stratification are common products of particulate, tractional transport in epiclastic sedimentary systems, and the same transport process appears to operate in base surges. However, cross-stratification in surge deposits is almost always at an angle lower than that of repose (Fig. 2.5.8), so a significant lateral shear stress may have operated at the bed surface. Conditions were perhaps akin to, although certainly not the same as, high-flow regime antidune stages in aqueous sedimentary settings.

2.7   Explosive Eruption Centres

2.7.1   Introduction

Basaltic explosive-eruption centres superficially can be subdivided into magmatic and phreatomagmatic explosive centres. Basaltic magmatic explosive centres characteristically build scoria cones, whereas phreatomagmatic centres can be more diverse, including maars, tuff rings, and tuff cones.

2.7.2   Scoria cones

Small, steep-sided cones of scoria typically result from a single, Strombolian, eruptive phase. C.A. Wood (1980a) reviewed the characteristics of scoria cones and evaluated the nature of the eruptions producing them. He found that basal diameters (W_{co}) for 910 cinder cones range from 0.25 to 2.5 km (median value 0.8 km, mean 0.9 km). Mount Elephant is the largest scoria cone in the Newer Volcanics province in Victoria (*Fig. 2.5.3*) and has a maximum basal diameter of 1.3 km. Heights (H_{co}) for 83 fresh cones reported by C.A. Wood (1980a) equal 0.18 W_{co}, and the crater diameters (W_{cr}) equal 0.40 W_{co}. Cone heights therefore can be up to several hundred metres. Mount Elephant has a height of 240 m, equal to 0.18 W_{co}.

The sides of fresh, relatively undegraded cones slope at an angle that is close to the angle of repose (about $32-33°$), so contemporaneous downslope tumbling and grain-flow redeposition of the falling tephra are important ancillary depositional processes on scoria cones. Most cones are circular in plan, unless developed over a fissure in which case they may be elongate. The craters are bowl-shaped. They are rapidly infilled by downslope redeposition of tephra, both contemporaneously and post-eruptively. Some cones may develop numerous vents, such as the Mangere Mountain cone in the Auckland field.

Strombolian eruptions typically produce scoria cones, but Hawaiian-style lava fountaining is not uncommon and agglutinate, and even clastogenic, lavas may form. Vulcanian explosions also may take place at scoria cones, perhaps towards the end of the explosive phase. Sub-plinian to Plinian eruptions characterised by high maintained columns relate more to high-discharge fissure eruptions (for example, Walker et al., 1984). C.A. Wood (1980a) noted that scoria cones commonly are the source of lava flows as well (Section 2.4.4). Phreatomagmatic explosions also may take place under appropriate conditions.

Eruption rates can be so high that scoria-cone heights may reach 100 m in one day. Eruption durations range up to fifteen years and have a median value of 30 days based on observed eruptions from 42 centres (C.A. Wood, 1980a). 93 percent of the observed eruptions stopped within a year. Little is known about the duration of scoria-cone-forming eruptions in eastern Australia and New Zealand, but Mount Elephant has no evidence of time breaks.

The internal structure of scoria cones is dominated by massive beds (Fig. 2.5.4). Stratification may coincide with changes in the eruption style, eruption rates, or wind direction. Some cones such as Mount Elephant contain little or no stratification (Fig. 2.5.4), or only subtle, widely spaced layering, corresponding to an essentially continuous, short-lived eruption, and therefore high discharge rates. Scoria cones are widespread in both the Newer Volcanics and north-Queensland provinces, punctuating the otherwise very flat landscape (*Fig. 2.5.3*), and are also common in the Northland and Auckland provinces in New Zealand.

2.7.3   Maars

There has been considerable confusion on the distinction between maars, tuff rings, and tuff cones, but the usage adopted here is that of Wood (in press). Other important reviews include those of Ollier (1967b), Lorenz (1973, 1986), Wohletz & Sheridan (1983), and Cas & Wright (1987).

Maars are low-profile volcanic centres characterised by a high $W_{cr}:W_{co}$ value, crater depth-to-width ratios of 1:5 for the freshest centres, and crater widths of up to 3–4 km (Wood, in press). They have steep to vertical inner rims caused by collapses into the vent, and gently sloping outer flanks (Figs. *2.7.1–2*). Craters commonly lie below the general ground surface. Maars result from phreatomagmatic and phreatic explosive activity generated below ground level. Breccia pipes and diatremes probably represent the deeper root zones of maars (for example, Lorenz, 1986). Many modern maars contain lakes.

Maar-rim successions may contain considerable bedrock clasts as a result of the subsurface explosive activity, as well as juvenile material. Some are dominated by bedrock clasts. The deposits typically are well-bedded, fine to coarse tuffs and lapilli-tuffs, deposited by a combination of air-fall and base-surge processes. Maar-rim successions may be in excess of 100 m in thickness. Accretionary lapilli (Fig. 2.5.6) and bomb-impact sag structures in wet cohesive ashes are characteristic (Fig. 2.5.7). The inner walls are steeply inward-dipping owing to collapse of parts of the maar rim back into the crater and leading to crater infilling. Normal surface processes assist this infilling. Preserved inward-dipping beds are rare to minor because of this inward collapse, but are known at Tower Hill and Purrumbete in Victoria.

Maars are widespread in the southern half of the Newer Volcanics province. Some of the largest, such as Tower Hill (*Fig. 2.7.1*) and Purrumbete in Victoria, and Mount Gambier in South Australia (Fig. 3.7.4), are amongst the largest in the world (for comparison, see the statistics of Wood, in press). Maars are found

both as discrete individual centres, and as nested complexes — for example, the Red Rock complex (Fig. 2.7.2) and the area surrounding Mount Leura, both in Victoria. In addition, as pointed out above, Strombolian eruptions also may take place periodically from maar centres. These produce not only sheets of scoria that are interbedded in the maar rim and dispersed over the surrounding landscape, but also may build up significant scoria cones and, in some centres, multiple-cone complexes — for example, Tower Hill and Red Rock, Victoria (see Section 2.7.6). The building of these scoria cones represents periods when the magma was able to reach the surface without interaction with groundwater, corresponding to either lining of the conduit by congealed lava which acted as a seal, or temporary exhaustion or withdrawal of the groundwater supply, perhaps used up during immediately preceding phreatomagmatic eruptions (for example, Edney, 1984a), or because of increased eruption rate.

2.7.4   Tuff rings

Tuff rings have higher profiles than do maars, and are thought to originate from explosions closer to, or at, the ground surface. They result, like maars, from phreatic and phreatomagmatic explosive activity and have high $W_{cr}:W_{co}$ ratios (Wood, in press). Inner rims also have steep slopes, but inward-dipping beds are common. The outer slopes, unlike those of maars, are steep. Tuff-ring successions in general contain a higher proportion of magmatic material than do maars. Centres such as Mount Leura near Camperdown in Victoria are probably tuff rings rather than maars. Mount Leura also has a scoria cone complex.

2.7.5   Tuff cones

Tuff cones are also phreatic-phreatomagmatic centres but, as the name implies, they are steep-sided and have high profiles. Steep inward- and outward-dipping beds are found which have dips of up to 25°. They commonly form where surface water — a lake or sea — overlies the vent when eruptions begin. They are dominated by juvenile magmatic tephra. Country-rock material is accessory. The deposits are the products of both pyroclastic-fall and surge processes (see Section 2.6), but because of the inhibiting effects of the water body and the wetness of the tephra, the material is deposited closer to vent. $W_{cr}:W_{co}$ values are notably small compared to those for maars and tuff rings. Magmatic explosive eruption phases also may take place, so tuff cones may be composite in character. Composite tuff-scoria cones appear to be present in the multiple-eruption-point Red Rock complex (Fig. 2.7.2), but exposures are limited.

The best-known tuff cone is probably the marine island volcano of Surtsey which formed in shallow marine waters (depths of 130 m) off the coast of Iceland between 1963 and 1967 (Thorarinsson, 1967). Equivalent Surtseyan tuff cones are preserved in the Eocene-Oligocene Waiareka-Deborah Volcanics of Otago in the South Island of New Zealand (Coombs et al., 1986; Cas et al., 1986). The internal features of a range of Surtseyan volcanoes are preserved in coastal

Figure 2.7.2   Multiple maars of the Red Rock volcanic complex which also has a complex of scoria and tuff cones.

exposures, including simple monogenetic cones — for example, Bridge Point, south of Kakanui — and more complex polygenetic ones — for example Oamaru — which have evidence of multiple eruptions. The volcanic succession at Bridge Point volcano consists of a basal succession of bedded lapilli tuffs, truncated by the discordant base of a volcaniclastic debris-flow deposit, and followed by fossiliferous, micritic, and glauconitic volcaniclastic deposits. These represent a degradation stage, as a small (1 km wide) marine tuff cone was wave planated, much of it presumably during storm activity.

The Oamaru volcano succession contains numerous erosional surfaces bounding discrete packages of water-laid tuffs and lapilli-tuffs, surge deposits (suggesting emergence of the cone; Fig. 2.5.9), pillow lava (Fig. 2.4.3), and pillow breccias interpreted as redeposited flow-foot breccias (Fig. 2.4.15). The erosion surfaces

Figure 2.7.3   Coastally eroded Bridgewater volcano near Portland, Victoria.

Figure 2.7.4 Phreatomagmatic surge-modified fall deposits at Bridgewater volcano, Victoria, showing ripple-like accumulations of scoria in finer phreatomagmatic ash.

commonly are mantled by bioclastic sediment. Other sedimentary processes including wave-current reworking, storm-surge resedimentation, and subaqueous mass flow resedimentation, were important influences during the history of the volcano. In addition, a significant change from subalkaline to alkaline volcanism is recorded by Coombs et al. (1986). Oamaru volcano therefore appears to have had a long, complex history marked by multiple eruptive phases and long sedimentary intervals.

The remnant Bridgewater volcano near Portland in Victoria (Fig. 2.7.3) was also a tuff cone, but appears to have evolved into a lava-shield volcano. The tuff-cone succession consists of near-vent phreatomagmatic surge, airfall, and surge-modified airfall deposits (Fig. 2.7.4), as well as a stack of multiple clastogenic lavas (Fig. 2.4.1) and cross-cutting dykes (Fig. 2.7.5). The main lava succession consists of thin (up to 2–3 m thick), laterally continuous sheet-like pahoehoe lavas. Coulson (1941) and Boutakoff (1963) speculated that the Bridgewater volcano was a nested complex of multiple, closely stacked, tuff cones. Boutakoff (1963) also suggested that the tuff-cone complex grew offshore (like Surtsey) prior to the building of a dune-spit complex linking Bridgewater to the mainland during the late Tertiary and Pleistocene during lower sea-level stands. However, intercalated marine horizons in the volcanic succession are absent, so this interpretation is difficult to evaluate. Nevertheless, the existence of the Bridgewater succession as a major coastal promontory amidst late Tertiary to Holocene dune fields (including the Bridgewater Formation) is a strong possibility.

Figure 2.7.5 Basalt dyke cross-cutting bedded phreatomagmatic tuffs and lapilli tuffs at Bridgewater volcano, Victoria.

Likely marine Surtseyan tuff-cone successions are found also in Torres Strait in northern Queensland (Section 3.2.2) and have been intersected in drill holes in the Miocene of the Bass Basin (Section 3.8.2). Possible, non-marine, tuff-cone successions that formed in lava-ponded valley lakes are known also in Tasmania (Section 3.8.3). Tuff cones are also present in the Antipodes and Chatham islands off New Zealand (Section 4.4.4–5).

2.7.6   Complex volcanoes

Some eruption centres are best described as volcanic complexes, in the sense that they consist of multiple-eruption points and substantial combinations of both phreatomagmatic and magmatic products. A range in the proportions of magmatic and phreatomagmatic deposits has been identified in selected centres from the southern half of the Newer Volcanics province in Victoria, where rising basaltic magmas passed through regionally extensive subsurface aquifers in Tertiary sedimentary deposits of the Otway Basin.

The Tower Hill maar (phreatomagmatic) of Victoria (*Fig. 2.7.1*) contains a major scoria cone complex (magmatic) in its centre (Edney, 1984a). Similarly, the maar/tuff-ring of Mount Leura contains a major, central, scoria-cone complex. The Red Rock complex near Colac in Victoria is more complicated in consisting of multiple maars and a large scoria- and tuff-cone complex (Fig. 2.7.2). It has at least 30 identifiable eruption points. Both Mount Gambier and Mount Schank (Fig. 2.7.6) in South Australia are also complex centres, made up from maar-forming, scoria-cone forming, and lava-forming eruptions. Crater Hill in the Auckland province evolved from a maar/tuff-ring to a scoria-cone centre, then to a shield volcano, and finally to a scoria-cone centre again (B.F. Houghton, personal communication, 1987). Similarly, Motakorea Island in the Auckland province began with phreato-magmatic activity, followed by scoria-cone building, and lava eruptions once the lava pile had emerged above sea-level (Fig. 2.7.7).

Lava-forming eruptions also may take place from explosion centres and, conversely, lava-shield volcanoes may produce explosive phases of activity. These may involve simply Hawaiian-style lava fountaining, or Strombolian-style, scoria-cone-forming eruptions — for example, on small basaltic shields such as Mount Napier, Victoria, and Rangitoto, Auckland. Explosive activity at the large polygenetic shields has involved eruption of magmas of intermediate and felsic compositions. Focal Peak volcano, for example, is a rhyolite-lava complex that includes pyroclastic deposits (Section 3.4.6). There are also pyroclastic deposits of hawaiitic and trachytic composition in the Nandewar shield volcano of New South Wales (Section 3.5.2) and trachytic pyroclastic deposits, including pyroclastic-flow deposits, are associated with the central region of trachytic domes and plugs of Canobolas volcano

Figure 2.7.6   Mount Schank in southeastern South Australia, viewed from the east, and showing a central crater straddling two earlier structures — a maar in the south and a scoria cone in the north. An early lava flow west of the craters has been blanketed by later tephra falls and provides one of the very few hard-rock outcrops in the area — hence its active quarrying.

Figure 2.7.7 Motokorea (or Browns) Island is typical of many of the monogenetic eruption centres of the Auckland volcanic province. Early phreatomagmatic deposits form the low ridge behind the scoria cones in the centre of the island. Lava flows underlie the flat areas in the foreground and to the right. These flows emanated from vents at the base of the cone or are rootless flows resulting from Hawaiian-type fire fountaining. Photograph courtesy of Whites Aviation, Auckland.

(Section 3.6.2). No detailed work has been undertaken on any of these pyroclastic deposits, but Strombolian, Vulcanian, Sub-plinian and even Peléan eruption styles may have been involved. All these would produce deposits of relatively local, near-vent extent. There is no indication that widely dispersed deposits associated with highly explosive Plinian eruptions were produced.

Rhyolitic calderas are uncommon in intraplate settings. However, one of the better known examples, the Yellowstone Plateau volcanic field, is found along the eastern margin of the Snake River Plain (Leeman, 1982; Christiansen, 1982, Hildreth et al., 1984) and is the youngest rhyolitic volcanic centre in the bimodal Snake River Plain volcanic province. No equivalent caldera centres are known in intraplate settings in eastern Australia and New Zealand (the major calderas in New Zealand are part of the subduction-related Taupo Volcano Zone). However, the probable remnants of a small summit caldera have been recognised in the Focal Peak shield volcano (Section 3.4.6). Summit calderas are not unusual in basaltic shield volcanoes, but their origins are more likely to be associated with long-term, lava-eruption events than with catastrophic ignimbrite-forming ones.

2.8 Erosion and Sedimentation

Epiclastic processes are important in all volcanic terrains (Cas & Wright, 1987). They operate contemporaneously with and after volcanism. These surface processes are dynamic on high-profile volcanoes such as stratovolcanoes. They are less dynamic in intraplate basaltic terrains characterised by generally low-profile volcanic centres, by relatively low proportions of clastic deposits, and by the topographic smoothing effects of fluidal basaltic lavas.

Tephras deposited on steep-sided cones are re-deposited immediately by individual-particle free-fall and tumbling, by creep, and by grain-flow. The angles of repose of scoria- and tuff-cone slopes are a response to this. The infilling of a valley by a lava flow causes, at a larger scale, reorganisation and relocation of the drainage system of that valley (for example, Ollier, 1969, 1988), producing immediate erosional rejuvenation and acceleration of the rate of erosional downcutting. Succeeding lava-flow geometries will be controlled in some measure by this continual downcutting, although the volume of the lava flow, its rate of supply and propagation, and the way in which the volume of the flow is accommodated by the topography, will be principal controls (Cas & Wright, 1987).

The main effects of surface processes post-dating volcanism in subaerial settings will be degradation of individual centres and regional downcutting of the terrain as a whole. Regional downcutting to some degree is controlled by absolute relief within the margins of the terrain, but more so — at least initially — by the relief of the terrain relative to the regional base-level. The relief in well-developed continental intraplate volcanic provinces is low because of the smoothing effects of basaltic lavas and the isolated and low-relief nature of the volcanic landforms, even where low-angle shield volcanoes are involved. The elevation of the province above regional base-level may be significant, however, depending on pre-existing relief and tectonics. Downcutting rates in provinces such as the Columbia River Plateau flood-basalt province were appreciable because of this, and have produced a significant incised topography. Walker (1984) calculated downcutting rates of 58 m/Ma in the elevated basaltic highlands of Iceland in a region where the mean elevation is 400 m. Rates in east-Australian and New Zealand provinces are likely to be significantly less than this, despite the milder climate, because of the relatively low elevations above base-level, except perhaps in the older Tertiary successions of the highlands. However, post-volcanic denudation rates in the highlands of southeastern Australia may be only 8 m/Ma (Bishop, 1985b; Wellman, 1979b).

Denudation at the more local scale of individual centres, can take place in two ways. Significant sectors of scoria cones may collapse penecontemporaneously with volcanic activity, producing thick, local, debris flows (for example, within the scoria cones of the Tower Hill volcanic complex, Victoria; Edney, 1984a). Post-eruptive degradation, however, is surprisingly slow, even for loosely packed scoria-cone piles, given that they are the volcanic centres with the steepest slopes and that spinose scoria is itself susceptible to erosion. Kieffer (1971) suggested that, even after 1 Ma, cones may have some relief and slopes in the order of 10°, but after 4 Ma only vent plugs remain. Scoria cones seldom reach heights of more than 200 m, so down-levelling of the order of 100 m/Ma, or less, is still exceptionally low relative to that on stratovolcanoes (see Cas & Wright, 1987). However, rates of degradation differ from region to region depending on climate and regional relief.

C.A. Wood (1980b) suggested that scoria cones are degraded largely through the weathering of tephra to clay, and then the development of radial gullies leading to incision, lateral transport, slope lowering, and concomitant growth of the width of the cone and apron. The implication of this is that cone degradation does not begin significantly until the scoria pile becomes impermeable. Rainfall initially soaks into the porous pile. It does not run off over its surface, and therefore is not capable of transporting the surface tephra downslope. Once the cone becomes impermeable through cementation, clay and soil formation, and development of a vegetation mat, surface run-off, erosional gullying, and lateral transport are likely to take place. The small surface and catchment area of the cone also contributes

Figure 2.8.1 Erosional surface draped by bioclastic limestone and wave-reworked lag pebbles at Oamaru volcano, South Island, New Zealand.

to low degradation rates. High torrential downpours in tropical climates, however, may lead to rapid initial degradation of young unconsolidated cones.

Little or no reference to erosional processes and products from subaerial intraplate settings in eastern-Australia and New Zealand exists in the literature. However, possible resedimented pyroclastic deposits are known on the flanks of the Warrumbungle volcano (Section 3.5.5) and interbedded conglomerates and cross-bedded sandstone are reported from the crater of the Lyttelton composite volcano of Banks Peninsula, New Zealand (Section 4.3.3). The scant record of erosional products may be an artificial result of non-recognition, but could be a real reflection of the lack of any significant transported particulate volcanic sediment in continental intraplate volcanic settings where basaltic volcanism predominates. This is especially true where the climate is temperate to humid (for example, Hawaii). Basaltic terrains do not shed large volumes of clastic sediments. Neither streams nor shorelines are choked with abundant volcanic sediment. Only in cold climates (for example, Iceland) are there significant volumes of sediment. The implications are that in temperate to humid climates, mafic rocks are more susceptible to chemical weathering than they are to erosion, and that large volumes of rock are removed in solution, in surface and groundwaters.

Epiclastic processes play an important part in the history of marine intraplate volcanoes. Downslope surface creep and grain-flow, contemporaneous with volcanism and post-dating it, are important where there are steep subaerial and subaqueous slopes, such as at Oamaru and Bridge Point volcanoes in the South Island of New Zealand. Degradation of emergent to shallow-water cones is effected by storms, surface currents, and gravitational collapse, both during periods of volcanic inactivity and after volcanism has ceased altogether. The Surtseyan volcanic centres of Oamaru and Bridge Point contain excellent examples of the effects of these processes (Cas et al., 1986; Cas & Landis, 1987; Coombs et al., 1986). The Oamaru volcanic succession contains numerous erosional surfaces, some draped by bioclastic layers and wave-reworked lag pebbles (Fig. 2.8.1), that correspond to a time gap of at least several months in eruptive activity. Storm-generated, hummocky and swaley, cross-stratified volcaniclastic sediments, some containing glauconite, are also present (Fig. 2.8.2), and there are more distal, deeper-water,

volcano-apron turbidites. The Bridge Point volcanic succession consists of a lower, bedded, pyroclastic section of probably water-lain tuffs and lapilli tuffs, overlain by a succession of resedimented volcanic sediments, including a major debris flow deposit (Cas & Landis, 1987) that probably represents an initial sector-collapse event of the small cone. The debris flow is overlain by turbidites and a succession of bioclastic, glauconitic volcanic sandstones and mudstones, which represent a submarine, platform-margin talus succession resulting from storm planation, faunal colonisation, and reworking.

2.9 Volcanic Hazards

2.9.1. Introduction

No eruptions have taken place from the intraplate volcanoes of eastern Australia and New Zealand since European exploration and colonisation, and there is an understandably low perception of the hazards and risks posed by intraplate volcanoes in the region. However, the volcanism that began in the late Mesozoic more than 65 Ma ago has continued intermittently into the late Quaternary, and there are six areas where the volcanism is so youthful that the possibility of future intraplate eruptions in the region cannot be ruled out completely. The geology of the six areas is described in Chapters 3 and 4. The areas are, in approximately decreasing order of importance:

(1) Auckland province, particularly the Auckland City field which includes Rangitoto, a volcano thought to have been active since at least 1200 A.D. (Section 4.2.3.; see below);

(2) Newer Volcanics province of western Victoria and southeastern South Australia, including Mount Gambier, the youngest known volcanic centre in eastern Australia (4600–4300 yr B.P.; Section 3.7.4);

(3) Kerikeri Volcanics of Northland province, New Zealand, which include flows dated as young as 1800–1300 yr B.P., as well as Ngawha, an active geothermal system (Section 4.2.2);

(4) Bundaberg and Boyne areas of southeastern Queensland which include, for example, the exceptionally well-preserved volcanic cones at Coalstoun Lakes from which the Barambah Basalt lava flow (K-Ar date of 0.6 Ma) was extruded (Section 3.4.2);

(5) Some of the volcanic provinces of northern Queensland, particularly Nulla province where the Toomba lava flow has been dated at 13 000 yr B.P. (Section 3.2.9);

(6) Antipodes Islands (Sub-Antarctic Islands) where young ages are inferred from many juvenile volcanological features (Section 4.4.4).

Basalt is the dominant eruptive product in these areas and the style of eruption ranges from Hawaiian to Strombolian to phreatomagmatic (Surtseyan) and phreatic, as described and discussed in Sections 2.4–7. This section is concerned with the hazards that these styles of eruptive activity are likely to pose and the consequences that may be expected.

Very few hazard assessments have been made of the intraplate volcanoes of eastern Australia and New

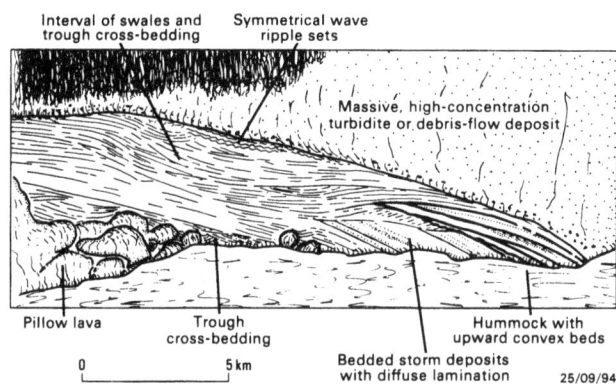

Figure 2.8.2 Storm-generated, hummocky and swaley cross-stratification at Oamaru volcano, South Island, New Zealand.

Zealand. Indeed, the only significant ones for eastern Australia appear to be the unpublished studies by Hilmansyah (1985) on the Newer Volcanics and Sheard (1986b) on southeastern South Australia.

The volcanological features that can be used in assessing volcanic hazards are similar for each of the six areas listed above. The greatest risk is in the Auckland province where youthful volcanic cones are conspicuous features in and around the city of Auckland, so this province can be regarded as a type example for hazard assessment in the region.

2.9.2   Auckland

Forty-eight eruptive centres have been identified in the Auckland City area (Figs. 2.4.5, 4.2.2), and more than 70 are known in the Southern Auckland field (Section 4.2.3). Most eruptions have been mainly effusive, but also producing scoria cones and commonly lava flows from breached craters or vents low on the cone. The largest cones, Mounts Eden and One Tree Hill, are nearly 200 m high and have basal diameters of about 0.7 km. Most lava flows are relatively short but at least two reach more than 10 km from source. The flows are up to about 20 m thick, and accumulations of lava reach more than 60 m in places. Lava tubes are found in some flows. Most flows are of the pahoehoe type.

Possibly one third of the eruptive centres in the Auckland area has evidence of phreatomagmatic activity, in particular those that formed either at low elevations or were erupted through wet Pleistocene sediments (or both). Typical maars in the area have diameters of 0.3–0.8 km, and associated tuff rings reach 1 to 2.2 km in diameter (Searle & Mayhill, 1981; Dibble et al., 1985; Cassidy et al., 1986).

Authors of published accounts have emphasised the lack of a general tephra cover in the Auckland City field, but Searle & Mayhill (1981) noted extensive tephra cover in at least some areas distant from vents. Furthermore, modern eruptions of similar style have produced extensive areas of tephra. For example, Paricutin (Mexico) tephra deposits cover an area of about 350 km² to a depth of greater than 150 mm (Segerstrom, 1950), and the 1973 Heimaey (Iceland) eruption covered an area of more than 10 km² to a depth of greater than 400 mm (Clapperton, 1973a,b; Preusser, 1973).

There is some evidence that the younger eruptions at Auckland may have produced greater volumes of material than earlier ones. Cassidy et al. (1986) tentatively concluded that the next eruption might be larger than average, and stressed that an eruption 'would have a major effect upon daily life in New Zealand's largest city' (page 60). They remarked that there is no basis on which to predict the time or even the approximate location of a future eruption.

2.9.3   Hazard identification

The range of eruption styles observed in the late Quaternary intraplate centres of eastern Australia and New Zealand is quite limited, but several different hazards are possible, even likely (see also Blong, 1984).

Earthquakes would probably be limited to Modified Mercalli intensities of VIII or less, and may cause damage as magma rises toward the surface. The area affected is unlikely to be symmetrical around the vent, and ground accelerations (and therefore damage) are largely dependent on soil and subgrade conditions. Earthquakes also may trigger slope failures, leading to further damage. However, earthquakes associated with intraplate volcanism are likely to be initiated at relatively shallow depths, and any damage is likely to be confined to an area within a few kilometres radius of the vent.

Ground deformation also may be produced by a rising magma body. Tilt is commonly only a few microradians, but surface deformation may take place, producing deflections and damage to buildings and other structures. The area at risk is likely to be limited to about 1 km or less from the vent, and could be asymmetrical depending on local conditions. This area in any case will be subject to far greater peril during subsequent phases of the eruption.

Ballistic projectiles (see Section 2.6.1) are unlikely to affect areas more than about 3 km from the vent, and in most cases smaller areas are likely to be at risk. The area affected is likely to be circular and centered on the vent. Few building materials can resist penetration by projectiles more than 100 mm in diameter, travelling at terminal velocity. Larger missiles will not achieve thermal equilibrium with the atmosphere so that ignition of combustible materials is also possible.

Tephra falls are likely to affect a wide area, particularly in the downwind direction from a vent. Most of the tephra, particularly the dominant fraction smaller than 64 mm diameter (lapilli-sized) would reach thermal equilibrium with the atmosphere so that ignition of combustible materials on contact is unlikely to be a problem. An area of some thousands of square kilometres may be assumed to receive at least a dusting of tephra. The hazard produced by tephra fall as a general rule decreases exponentially with distance from the vent, but the rate of decrease may be significantly lower in the downwind direction.

Pyroclastic flows formed as a result of eruption-column collapse are rare in eruptions of intraplate volcanoes (Section 2.6.2). Pyroclastic flows are likely to be limited to only a segment of the area around the vent and maximum travel distances would rarely be more than a few kilometres, but flow velocities may reach more than 60 m/s and emplacement temperatures more than 100°C. Pyroclastic flows are therefore extreme hazards to life and property.

Pyroclastic surges (Section 2.6.3) may be either hot or cold (for example, Cas & Wright, 1987). The area affected is likely to be roughly circular and deposits will thin rapidly away from the vent. Surge deposits in extreme cases, such as the 1965 eruption of Taal volcano (non-intraplate), may cover an area of more than 15 km². Deposits from individual surges may reach more than 1 m in thickness, but maximum grain-size rarely exceeds 100 mm. Emplacement velocities probably reach tens of metres per second, so pyroclastic surges are likely to be the most serious (life-threatening) hazard associated with intraplate volcanism.

Lava flows, from both vents and fissures, are one of the most common hazards produced by intraplate eruptions. Flows are relatively mobile, and lava tubes are common, so flow distances are increased. Flow-front velocities are rarely more than a few kilometres per hour, but temperatures may exceed $1000°C$, thus exceeding ignition temperatures for many man-made materials.

Lightning strikes also may be severe during phreatomagmatic and phreatic eruptions. Ignition or other damage is likely at numerous points within a few kilometres of the vent and in portions of the area where tephra falls.

Lahars (mudflows) may be generated in areas where fine-grained tephra accumulates in substantial thicknesses. They are likely to be found particularly where tephra-fall and surge deposits have accumulated on steep slopes to thicknesses of more than 150 mm. Lahars may be generated after the eruption where rainfall intensities exceed 10 mm per hour, or during phreatic and phreatomagmatic eruptions where the eruption cloud contains large volumes of water. Lahars in most cases associated with intraplate volcanism are likely to be relatively small, limited in area, and confined to drainage lines.

Volcanic gases, particularly CO_2, may be produced from some vents. Areas at particular risk are those of low ground and limited air drainage. Gases can be a significant problem in some volcanic areas, but the hazard is difficult to evaluate in the case of eastern Australia and New Zealand. However, there is some evidence of CO_2 accumulation in the Gambier area (Wopfner & Thornton, 1971; Chivas et al., 1987).

Volcano-related tsunamis are likely to take place at eruption centres in coastal areas where eruption rates are rather high or where there is partial collapse of the volcanic edifice. These hazards therefore are likely to be rare and confined to limited areas.

2.9.4 Assessing hazard impact

Each of the hazards listed above has a different frequency of occurrence in intraplate eruptions and affects areas of different sizes and distributions in relation to the vent. Some hazards, such as earthquakes, are likely to take place early in an eruption cycle, whereas others — for example, lahars — may be produced only after significant eruptive activity has ceased. Some areas may be affected severely by a range of hazards. Others, particularly those more distant from the vent, may suffer only slightly from a single hazard.

Clearly, hazards differ by several orders of magnitude in severity, and there is a general decrease in significance with increasing distance form the vent. This may not be an inviolate rule as finer-grained tephra, for example, can cause greater disturbance to flowering plants (or pollinating insects) than coarser tephra closer to the vent. Consequences also may differ dramatically depending upon the items at risk. For example, pyroclastic surges and flows are likely to be fatal to humans and other animals, except perhaps near the margins, but vegetation may regenerate within a matter of weeks. Similarly, lava flows commonly destroy

buildings in their paths, but human deaths are relatively rare because flow fronts generally move slowly.

Hazards that are life-threatening or which severely damage man-made structures are considered commonly to be of the most significance. However, other hazards cannot be ignored. The types of consequences that deserve consideration include:

(1) human (deaths, injuries, psychological trauma);
(2) social (familial, organisational, institutional);
(3) the built environment (buildings, communication and transport links, utilities);
(4) economic (production and distribution);
(5) environmental (physical, biosystemic, and heritage effects).

An attempt to rank the potential impact of each of the above listed hazards as 'high', 'medium', or 'low' in relation to each of the above is given in Table 2.9.1. An assumption is made that items *can* be damaged, and consequently the table is weighted towards the consequences of an eruption in Auckland rather than, say, in northern Queensland. A qualitative estimate is also made in the table of the frequency of occurrence of each hazard.

The values in Table 2.9.1 are in large part 'informed estimates'. Few data are available in most cases to reach definite conclusions about potential effects. Furthermore, each of the five consequences considered is a generalisation of a number of attributes.

Table 2.9.1 Frequency of occurrence and potential consequences from hazards associated with volcanic eruptions in relation to (1) humans, (2) social systems, (3) built environment, (4) economic conditions, and (5) environmental phenomena

Hazard	Frequency	(1)	(2)	(3)	(4)	(5)
Earthquakes	M	M	L	M	M	L
Ground deformation	L	L	L	M	M	L
Ballistic projectiles	H	H	M	H	H	H
Tephra falls	H	M	H	L/M	H	L/M
Pyroclastic flows	L	H	H	H	H	M
Pyroclastic surges	M	H	H	H	H	M
Lava flows	H	L	M	H	H	M
Lightning strikes	H	M	M	M	M	L
Lahars	L	L	L	M	L	L
Volcanic gases	M?	H	H	L	M	L
Tsunamis	L	L/M	L	M	L/M	L

Abbreviations: H, High; M, Medium; L, Low.

Exceptional circumstances may be imagined which surpass the rankings given in Table 2.9.1. Examples may relate to the extinction of a rare species by even a small lava flow, or the ignition of a large number of significant dwellings (perhaps some on the heritage list) by a single lightning strike, ballistic projectile, or lava flow.

Sensible generalisations of the significance of the matrix in Table 2.9.1 are difficult to make because individual perceptions of 'value' are different. Can the loss of a dozen human lives be compared with the disappearance of a single insect species or with the destruction of the grandest community building? Nevertheless, ballistic projectiles, pyroclastic flows, and pyroclastic surges each rank 'high' in four categories and would appear to be the most significant hazards in terms of potential consequence. No other hazard is perceived to have more than two 'high' rankings.

Pyroclastic flows probably have a very low incidence of occurrence in intraplate volcanism, and therefore the other two hazards would seem to be the more significant. Ballistic projectiles are produced almost certainly much more frequently than pyroclastic surges, but the areas affected are likely to be significantly less. Tephra falls do not rate highly in this scheme, but their impact may be of long duration and the effects may be widespread, covering areas one or two orders of magnitude larger than either of the other two hazards. Furthermore, even a light dusting of airfall tephra can produce enormously costly effects in areas of intensive single-crop agriculture and in urban areas totally dependent on modern electronic communication systems. Similarly, even a small lava flow down a topographic depression in which high-density housing, a major arterial road, or important public buildings are located, can produce economic consequences out of all proportion to the area affected. Methods that can be used to mitigate at least some of the adverse effects of intraplate and other volcanic eruptions were outlined by Blong (1984).

CHAPTER 3

East Australian Volcanic Geology

3.1 Introduction

Contributors to this chapter describe the geology of the Cainozoic volcanic areas of eastern Australia, giving emphasis to distribution, ages, and rock types. The chapter is designed as a systematic catalogue of geological information and therefore contains a minimum of interpretation. The aim is to present uniform, summary descriptions, including references that can be used as a guide for obtaining additional information. Much more geological data are available than can be presented here, and some areas are known in much greater detail than others. The chapter also includes a summary of the Mesozoic intraplate igneous activity that preceded the much more extensive Cainozoic volcanism. Gemstone occurrences in eastern Australia are dealt with in the final section. Geological information on New Zealand provinces is provided in Chapter 4.

Eastern Australia is divided for the purposes of description into seven geographical regions whose boundaries have no particular geological significance. Each of these broad regions is further divided into individual volcanic 'provinces' (see Section 1.1.3) that are generally of restricted age and small area. Not all provinces are readily distinguishable, so small, ill-defined, potential provinces in the same general area are grouped in some cases under a single heading. Also considered together are those provinces (for example, Southern Highlands and Kandos) that have two (or more) separate basaltic episodes of widely different ages in which both the older and younger rocks have similar geochemical signatures.

The petrology and geochemistry of the Cainozoic volcanism are considered fully in Chapter 5 and these topics are discussed only briefly in this chapter. However, some representative whole-rock chemical analyses are given for each region to provide an indication of the range of rock compositions (Tables 3.2.1–3.8.1). The range of compositions within individual provinces is wide, particularly in the central-type provinces where rocks range from basaltic through to highly evolved felsic compositions. In contrast, lava-field provinces mostly contain only mafic volcanic rocks. Rock names used are those defined in Section 1.1.4.

The ratio $100Mg/(Mg+Fe^{2+})$ is used as a measure of how close mafic rock compositions approach those of primary magmas (Fe_2O_3/FeO weight percent normalised to 0.20 before calculation of Mg-ratio). Generalised plots of Differentiation Index (DI) versus $100An/(An+Ab)$ and of the system Ne-Ol-Di-Hy-Qz (CIPW normative) are provided for all provinces, although many sets of data points are summarised simply as fields. Only mafic and some intermediate compositions (DI less than 60) are plotted in the Ne-Ol-Di-Hy-Qz diagrams.

Numerous K-Ar age determinations are quoted throughout the chapter and shown as ranges in the accompanying diagrams. These include many dates from published sources that were used in the compilation discussed in Section 1.7. However, many other dates are unpublished and from the data files of individual contributors. These were not used in Section 1.7 and are treated informally in this chapter. Ages older than 10 Ma are rounded off to the nearest whole number. Several contributors intend publishing these geochronological data elsewhere.

3.2 Northern Queensland

3.2.1 Introduction

Nine discontinuous areas of Cainozoic volcanic rocks are found in the northern part of eastern Australia (Fig. 3.2.1). The region extends nearly 1200 km north from the Sturgeon province west of Townsville to the Maer province in Torres Strait, and the total area of volcanic rocks is just over 23 000 km². These provinces are all mafic, and most are dominated by numerous relatively small volcanoes and extensive lava fields. The range of radiometric-age determinations is Eocene to late Pleistocene, but most individual provinces span less than 10 Ma. This range in ages is reflected by wide contrasts in erosion — from remnant plugs without remnant flows, to volcanoes that have craters and associated lava fields in excellent preservation. Six of the provinces are near the crest of the Great Divide, whereas most of the others are to the east. The average thickness of mafic volcanic rocks across provinces is thought to be about 30 m, and total volume of lava is estimated to be less than 700 km³. Eruption centres in most provinces are within restricted areas 60–80 km long and 35–45 km in diameter. The outlines of several of the younger vent areas (McBride, Chudleigh, Sturgeon, Nulla) trend northeastwards, whereas the

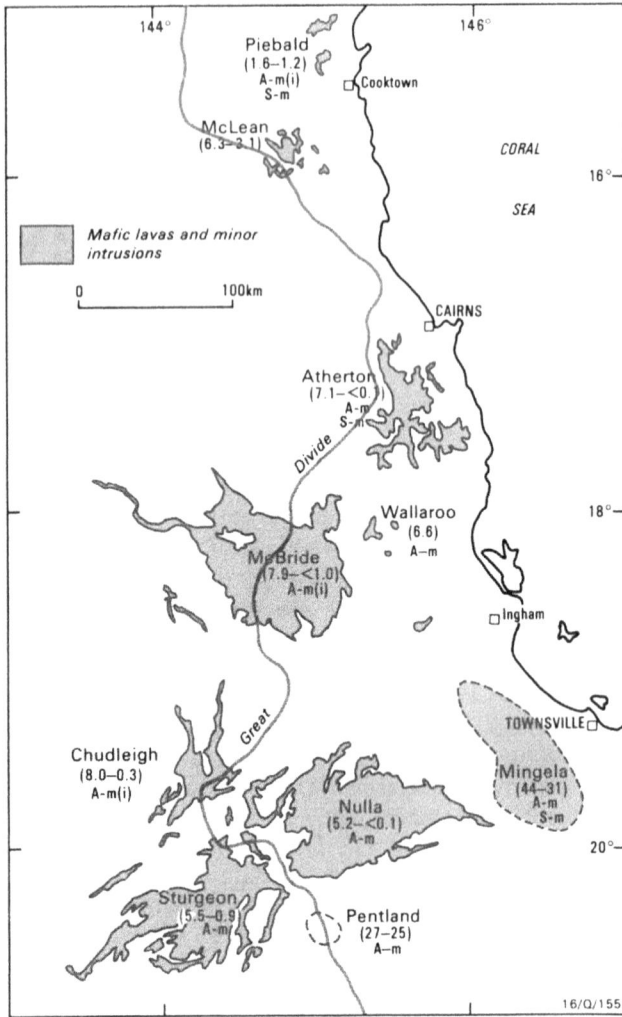

Figure 3.2.1 Distribution, rock types, and ages of volcanic provinces in northern Queensland. Rock associations: A, alkaline; S, subalkaline. Rock types: m, mafic; (i), minor intermediate. Ages (in brackets) in Ma.

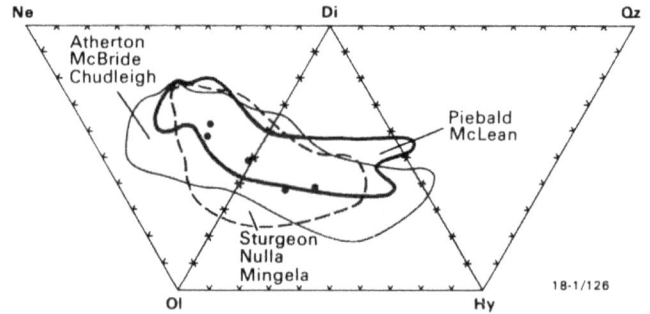

Figure 3.2.2 Compositional fields for north-Queensland provinces using the plotting procedure outlined in Section 1.1.4. Filled circles refer to Pentland and Burdekin samples.

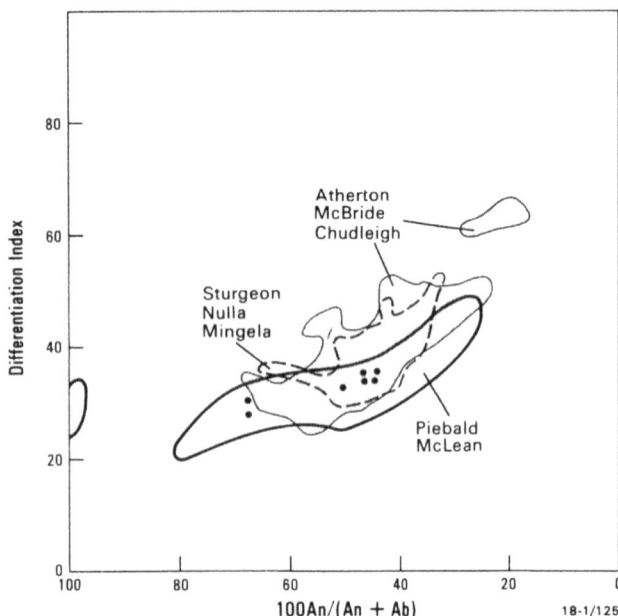

Figure 3.2.3 Compositional fields for north-Queensland rocks projected in the system Ne-Ol-Di-Hy-Qz. Filled circles refer to Pentland and Burdekin samples.

older Mingela province is elongate northwestwards. The structural controls localising the provinces are obscure and presumably are deep in the lithosphere. Aspects of the volcanism were reviewed by Stephenson et al. (1980). Generalised compositional fields for north-Queensland rocks are shown in Figures 3.2.2 and 3.2.3, and selected chemical analyses are given in Table 3.2.1. The provinces contrast in the relative abundances of different basaltic types and percentages of potassic rocks.

3.2.2 Maer

Maer province consists of eight small islands in the northeastern part of Torres Strait, across an area of up to 140 km² (Fig. 1.1.5). The largest volcano is Maer, one of the Murray Islands, which is 3 km long, 2 km wide, and rises 200 m above sea-level. Three volcanic centres are recognised, dated at 3–1 Ma, and whose total volume is about 1 km³.

Most of the Maer-province islands are Quaternary pyroclastic cones, some of which have restricted mafic flows. They are known as the Maer Volcanics (Whitaker & Willmott, 1969; Willmott et al., 1973). A few smaller islands consist only of mafic flows that support dense tropical forest, whereas the pyroclastic cones are naturally grassed. The tuffs are palagonitic and some limestone fragments are present in most of the pyroclastic cones that erupted evidently through reef material. The eruptions at each centre appear to have begun with subaqueous emission of volcanic ash, followed by mafic flows where the vents were closed off from the sea (Willmott et al., 1973). Cross-bedding and scouring in some of the cone materials may be the result of base-surge type eruptions.

Maer is the northernmost province of the east-Australian Cainozoic volcanic belt. Its alkaline mafic character is similar to that of many of the other provinces in eastern Australia. The range of mafic rock types in the Maer province is limited, consisting mostly of mildly nepheline- to hypersthene-normative alkali, transitional, and tholeiitic basalts and hawaiite. Mg-ratios range from 65 to 60. No xenoliths or megacrysts have been reported.

3.2.3 Silver Plains, Piebald, and McLean

The Piebald and McLean provinces (De Keyser & Lucas, 1968) are inland north and southwest of

Table 3.2.1   Selected chemical analyses for north-Queensland provinces

	1	2	3	4	5	6	7	8	9	10
SiO_2	47.19	48.65	54.34	47.62	50.89	50.48	48.17	47.61	44.48	47.61
TiO_2	1.84	1.81	0.16	2.12	2.02	2.09	2.16	1.85	2.08	1.72
Al_2O_3	15.41	15.97	19.46	16.24	15.29	15.23	14.94	15.56	14.81	14.95
Fe_2O_3	3.45	2.38	4.15	3.18	6.40	2.65	1.95	2.25	1.82	2.68
FeO	6.78	7.94	1.55	6.82	5.55	7.42	8.39	7.78	9.33	7.42
MnO	0.18	0.15	0.19	0.15	0.17	0.13	0.14	0.22	0.16	0.15
MgO	8.59	8.77	0.98	6.85	5.70	6.41	7.70	7.54	9.56	7.95
CaO	9.41	8.04	1.89	9.82	7.86	8.85	8.64	7.85	10.26	8.69
Na_2O	3.77	4.05	8.18	3.74	3.68	3.65	3.83	4.35	3.37	2.86
K_2O	1.54	2.06	4.46	2.19	1.27	1.58	1.96	1.74	1.65	1.15
P_2O_5	0.63	0.67	0.29	0.57	0.37	0.46	0.60	0.65	0.75	0.45
S	<0.02			<0.02	<0.02	<0.02	<0.02		0.03	0.02
H_2O+	0.53	0.19	1.34	0.45	0.58	0.47	0.41	1.84[1]	1.20	2.34
H_2O-	0.20	0.05	1.60	0.21	0.48	0.17	0.22		0.38	1.22
CO_2	0.30	0.07	0.14	0.27	0.12	0.09	0.42		0.14	0.44
Total	99.82	100.80	98.73	100.23	100.38	99.68	99.53	99.24	100.02	99.65
Ba	395	240	128	270	210	325	265	462	340	305
Rb	31.5	19	36	14	16.5	19.5	16	31	20	17.5
Sr	620	708	848	765	487	640	775	927	850	905
Pb	4	4	12	3	3	3	3	4	2	3
Th	3.0	5	21	1	3	5	4	4	5	4
U	<1		2	<1	<1	<1	<1	1	<1	<1
Zr	162	208	965	163	129	167	196	245	171	149
Nb	38.5	37	236	34.5	24	31	38	51	46	27
Y	18	19	20	15	21	17	18	23	17	18
La	24	21	66	19	18	23	22	35	37	22
Ce	55	47	116	42	34	52	53	72	75	48
V	207	112		182	178	136	137	137	163	139
Cr	232	267	34	176	266	193	172	217	221	198
Ni	144	184	8	94	207	88	134	144	150	132
Cu	63	62	13	69	52	42	50	37	55	40

1. Basanite from Atherton province (P.J. Stephenson & B.W. Chappell, unpublished data, sample EAV234).
2. Ne-hawaiite from McBride province (Stephenson et al., 1980, sample 3).
3. Phonolite from McBride province (Stephenson et al., 1980, sample 5).
4. Basanite from Chudleigh province (P.J. Stephenson & B.W. Chappell, unpublished data, sample EAV250).
5. Ol-tholeiitic basalt from Chudleigh province (P.J. Stephenson & B.W. Chappell, unpublished data, sample EAV243).
6. Hawaiite from Sturgeon province (P.J. Stephenson & B.W. Chappell, unpublished data, sample EAV255).
7. Ne-hawaiite from Nulla province (P.J. Stephenson & B.W. Chappell, unpublished data, sample EAV240).
8. Ne-hawaiite from Mingela province (J. Knutson & M.B. Duggan, unpublished data, sample MGL1).
9. Basanite from Pentland-Burdekin province (P.J. Stephenson & B.W. Chappell, unpublished data, sample EAV226).
10. Transitional basalt from Pentland-Burdekin province (P.J. Stephenson & B.W. Chappell, unpublished data, sample EAV223).

[1] Loss on ignition.

Cooktown, respectively, and there are more restricted volcanic areas near Silver Plains, 250 km northeast of Cooktown (Fig. 1.1.5).

The Piebald province consists of relatively restricted basaltic areas at the head of Starke River, in the Morgan-McIvor valley, and west of Hopevale. The lavas overlie alluvium and nearly horizontal Mesozoic sandstone. The volcanoes are mildly eroded, and two K-Ar determinations confirm young ages (Bald Hills, 1.6 Ma; Mount Piebald, 1.2 Ma). The most prominent volcano is Mount Piebald, 3 km southwest of Hopevale (Morgan, 1968). A total of 14 vents has been recognised, some of which are preserved as well dissected, pyroclastic cones. Others are residual basaltic hills that may represent small shield-volcano remnants.

The McLean province is larger than the Piebald, and also includes a scattering of relatively small, isolated lava fields and remnants. The lavas overlie alluvium and the steeply folded Upper Palaeozoic Hodgkinson

Formation. The volcanic activity has a wider time span than does the Piebald province, and the preservation of volcanic forms ranges from residual to well-preserved cones, and valley-floor lava fields. Seven age determinations for the older volcanoes range from 6.3 to 3.1 Ma, but the undated younger cones and lavas are probably less than 1 Ma.

Morgan (1968) described the main characteristics of the McLean province and recognised 18 vents. The majority of the younger volcanoes are scoria cones, but several broader volcanoes are probably composite. The much older basaltic rocks east and southeast of Lakeland Downs are denuded lava cones. Several dissected volcanoes and their lavas, 10 km south of Lakeland Downs, overlie the plateau above the Byerstown Range escarpment. 4.6 Ma nephelinite flows at Toms Hole are overlain in a limited area by diatomite and well-indurated mottled sandstone, and the rocks are downfaulted against the adjacent Palaeozoic sedimentary rocks with over 10 m displacement. This is the only known locality in north Queensland where faulting of Cainozoic volcanic rocks can be demonstrated.

Many of these older centres have nephelinites. A possible maar structure was recognised by De Keyser & Lucas (1968) adjacent and to the east of the highway at the edge of the Byerstown Range escarpment. It now contains no basaltic remnants, but a former volcano may have existed as megacrysts of corundum and zircon are found there (M. Forbes and D. Jones, personal communication, 1986). A hill of nephelinite dated at 3.7 Ma near Balclutha forms a lava field several kilometres across in the Silver Plains area 250 km northeast of Cooktown. Willmott et al. (1973) referred to another nephelinite occurrence near Silver Plains.

Variably undersaturated nephelinite (seven analyses have normative leucite), and basanite, alkali basalt, and hawaiite are dominant in the Cooktown area, and there are minor occurrences of mugearite and feldspathoidal mugearite. However, several rocks including those of Mount Piebald have a tholeiitic character (greater than 10-percent normative hypersthene). Mg-ratios of the volcanic rocks range from 70 to 47.

3.2.4 Atherton

The areas of basalt east of Mount Garnet in the Cairns-Innisfail hinterland comprise the Atherton province (De Keyser & Lucas, 1968). The province has an area of 1800 km² and contains 52 known centres of eruption (Fig. 3.2.4). It covers extensive areas of the Atherton Tableland and its eastern parts have relatively high rainfall. These wetter parts of the province are characterised by deep red soils (Isbell et al., 1976), tropical rain forest, and somewhat limited outcrop. Flows west of Ravenshoe are less deeply weathered and support open forest.

The province has a range of volcano types, including lava shields, cinder cones, maars, and one diatreme.

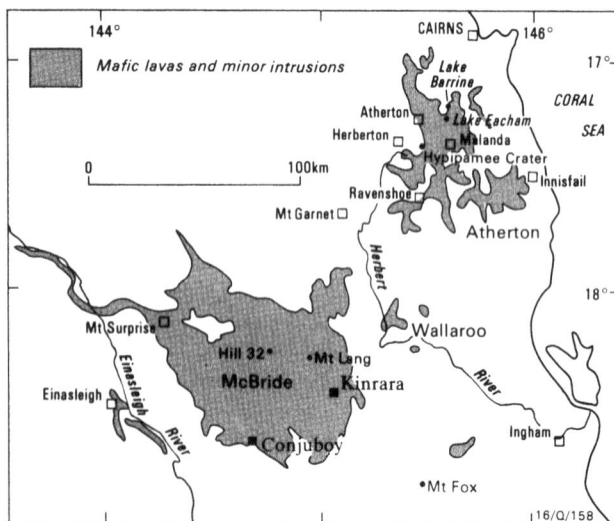

Figure 3.2.4 Distribution of volcanic rocks in the Atherton, Wallaroo, and McBride provinces of northern Queensland (see Fig. 3.2.1).

The southern flows in the Mount Garnet area follow the Herbert valley south and pass into the smaller Wallaroo volcanic province straddling the Herbert River 40 km south of Mount Garnet (Best, 1960; White, 1965). The Malanda cone is the oldest one dated in the province (3.1 and 2.9 Ma). There are numerous cinder cones, a prominent example of which is Mount Quincan, 3 km south of Yungaburra.

Atherton is the only area in northern Queensland in which maars have developed. The higher rainfall and groundwater conditions seem likely to have been key factors in determining this eruptive style, rather than explosive magmas. Lakes Eacham (Fig. 3.2.5) and Barrine occupy young maars, up to 1 km in diameter and 65 m deep (Timms, 1976). The flanks of Barrine have bedding features characteristic of base-surge eruption, but are seldom visible except after road work. The maars range in age from 200 000 to 10 000 yr, and several contain sediments recording vegetation changes (for example, Kershaw, 1970, in press).

The Hypipamee Crater diatreme 12 km south of Herberton has exceptional features and must have formed during a violent explosive event. There is no cone and the 60 m-wide crater has impressive walls of granite rising 55 m sheer from a crater lake 87 m deep. Basalt is not visible in the walls, but granite blocks and basalt bombs containing granite surround the crater for some distance, and inconspicuous scoriaceous lapilli are found in the rain-forest soil.

Stevens Island in the South Barnard Islands includes an offshore volcano (Jones, 1979). It is an eroded pyroclastic cone having some bedding features suggestive of deposition by surges. The pyroclastic rocks are cut by mafic dykes, and mafic flows are present on adjacent islands and reefs farther south.

The Atherton-province lavas rest on a range of Palaeozoic metamorphic and volcanic rocks and granites. They overlie Tertiary sediments and younger alluvium in many places south and east of Malanda. Warping of drainage is inferred, possibly in connection with the diversion of Herbert drainage from a westerly course into its present path down the coastal gorge, caused by uplift in the McBride province.

The Atherton province is characterised topographically by broad lava plains on the tableland, modified by erosion in the south. The younger cones, west of Tolga, produced flows that extended to Mareeba, 30 km to the north. One volcano has been dated at 1.7 Ma, and the older volcano that partly surrounds it is 1.8 Ma old. The Ravenshoe volcano produced lavas that flowed down the Millstream for 20 km and formed the upper level at the Millstream Falls (dated at 1.2 Ma). Indications of at least two complete gorge fillings of Johnstone River drainage by mafic flows totalling up to 300 m thick were described by Stephenson & Griffin (1976b). Former courses of the Wild River south of Herberton feature the Herberton Deep Lead (Cuttler, 1972) which contains stanniferous gravels beneath basalt cappings. A northern tributary of the Wild River has a 7.1 Ma flow preserved well above the 100 m-deep Wild River gorge, which is the oldest known lava in the province.

There are several small volcanoes and young lava fields in the coastal region, including Green Hill, near

Gordonvale (0.99 Ma), and four dated volcanoes in the Innisfail district (0.99–0.64 Ma).

Mafic plugs are present well outside the Atherton province in the Petford-Almaden district, 65 km west of Atherton (Lawrence, 1973; Reddicliffe, 1974). The extent of this plug field is not known, but an early Cainozoic age is indicated for it by one age determination (39 Ma).

Basaltic rocks in the Atherton area range from basanite (Table 3.2.1, column 1) and alkali basalt to tholeiitic basalt. Most rocks are moderately under-saturated, and Mg-ratios range from 76 to 48.

Localities of xenolith-bearing basalts include Lakes Eacham and Barrine, Mount Quincan, and Gillies Crater. Host rocks are mostly stratified scoria — for example, the Mount Quincan basanite. Mantle peridotite xenoliths up to 20 cm across are dominant. Black and green pyroxenites as well as amphibole-pyroxenites are also present, but lower-crustal granulites and minor schists are less common. Megacryst species include pyroxene, amphibole, spinel, and anorthoclase.

3.2.5 Wallaroo

Wallaroo is a small province 50 km south of the Atherton province and apparently separate from it. The Wallaroo lavas are up to 50 m thick and flowed down a former Herbert valley. They were erupted from four known volcanoes, one of which has been dated at 6.6 Ma. Rock types based only on petrographic identification are alkali basalt and possibly hawaiite.

3.2.6 McBride

Volcanic rocks of the McBride province about 150 km southwest of Atherton cover a roughly circular area close to 5500 km² and about 80 km in diameter (Fig. 3.2.4). This province is characterised by broad lava plains and numerous cones, especially in the central area. The most recent comprehensive study was made by Griffin (1977) who identified 164 volcanic centres.

The McBride province as a whole is an impressive volcanic landform in that it forms a broad topographic dome. The highest point is 1020 m at Undara Crater close to the centre of the province, and the margins of the lava plains have an average height above sea-level of nearly 400 m. The thickest parts therefore could be as much as 600 m, although granite inliers are known in the central part where the volcanic province is considerably thinner. The dome landform may be an uplift feature similar to another recognised in the region of the Chudleigh and Sturgeon provinces 250 km farther south, rather than a volcanically constructed feature (see also Section 1.4.2). The central part of the dome in the McBride province contains most of the known vents which define a zone extending northeastwards. The mapped lavas flowed radially away from the central region. Some cover large areas and have travelled considerable distances.

McBride and the other three principal north-Queensland inland provinces (Chudleigh, Sturgeon, and Nulla) are noteworthy for the length of some of the basaltic flows. A 160 km flow from Undara is the longest one known, but a flow from Barkers Crater in the Chudleigh area may be of similar length. Stephenson & Griffin (1976b) concluded the great length of flows must be attributed to high and persistent rates of extrusion and to favourable topography.

Vents in the McBride province have a range in size and type, from shield volcanoes to small pyroclastic or scoria cones. Many prominent cones in the province are steep, large, predominantly pyroclastic cones. Kinrara is the youngest volcano in the province (an apparent K-Ar age of 70 000 to 50 000 years is regarded as a maximum by Griffin & McDougall, 1975).

Undara volcano has lava fields that cover 1550 km². The cone is somewhat unusual compared to other major centres in this province because it has only a minor pyroclastic component. Its crater is a steep-sided depression 340 m across and 40 m deep (Fig. 3.2.6). Undara formed several long flows: one, 90 km long, entered the Lynd River in the north; another — the 160 km flow — entered and moved down the Einasleigh River to the northwest. Several branches of the Undara lavas contain evidence of major lava-tube systems (Atkinson et al., 1975; Figs. 2.4.10–11).

Geochronological data define a pattern of spasmodic but essentially continuous volcanic activity in the McBride province over the last 2.7 Ma. However, there are flows in parts of the province that appear to be significantly older, including two whose age determinations are close to 8 Ma. The older rocks are represented by plugs in the southwest and by mesas and plateaux around the province margins in the southwest and north. A ridge mesa near Carpentaria Downs has been dated at 7.9 Ma.

The underlying rocks in the McBride province mostly belong to the Georgetown Inlier, which includes Proterozoic granites and a range of multiply deformed metamorphic rocks and some Palaeozoic granites. Damming of drainage in some places by the older lavas influenced diatomite development, especially near Conjuboy (White & Crespin, 1959). There are rich Cainozoic vertebrate faunas in sub-basaltic fluvio-lacustrine sediments 20 km southeast of Conjuboy (McNamara, in press; Gaffney & McNamara, in press).

Basaltic rocks in the McBride province are pre-dominantly nepheline-normative and have Mg-ratios ranging from 72 to 49. Rock types include nephelinite, basanite, hawaiite (Table 3.2.1, column 2), and mugearite. There is one occurrence of phonolite — at Phonolite Hill, 8 km south of Undara Crater (Table 3.2.1, column 3). The basaltic rocks are mostly glassy to fine-grained, containing abundant microphenocrysts of euhedral and skeletal olivine. Slightly less mafic hawaiite and mugearite tend to be vuggy, coarser in grain size, and have prominent, radial, acicular plagioclase and titaniferous pyroxene phenocrysts. Mantle and lower-crustal xenoliths are present at several localities and have considerable compositional diversity (Rudnick et al., 1986). In particular, the volcanic rocks at localities such as Hill 32 and Mount Lang contain a wide range of lower-crustal and upper-mantle xenolith types. These include felsic granulite, amphibole and mica-bearing granulite,

Figure 3.2.6   Aerial view of McBride province showing Undara Crater in foreground, Racecourse Knob shield volcano on left-hand horizon, and older volcanoes in middle distance. A lava tube starting northwest of the crater, is seen as a narrow zone of dark vegetation running towards the upper right edge (see also Figs. 2.4.10–11).

peridotite, and pyroxenite (some garnet-bearing). Megacrysts of pyroxene, amphibole, mica, and spinel are also common.

3.2.7   Chudleigh

The large, nearly continuous basaltic area west of Charters Towers was subdivided by Twidale (1956) into the Chudleigh, Sturgeon, and Nulla provinces (Fig. 3.2.7). The geographic distinction between these provinces is partly arbitrary, and discontinuous plateau remnants between the three in the Clarke Hills and Wando Vale area are evidence of former continuity. Original topographic control of lava-flow directions and later erosion have led to semi-isolation of the three named provinces. A broader area of volcanism may be inferred also from the distribution of eruptive centres, but there is justification for distinguishing the three named provinces on the basis of density of centres and their likely ages. Plateau remnants in the Clarke Hills and Wando Vale area are significantly older than the adjacent provinces (older than 7 Ma, compared to the oldest adjacent activity of up to 5.5 Ma).

Chudleigh province straddles the Great Divide and its flow systems extend down former valleys to the north, east, and southwest (Fig. 3.2.7). The total area of the province is about 2000 km² and 46 volcanic centres have been recognised. One flow extends over 100 km down the Einasleigh River from Barkers Crater (Stephenson & Griffin, 1976a,b), as did a much older flow from an unknown source. Two rocks from the

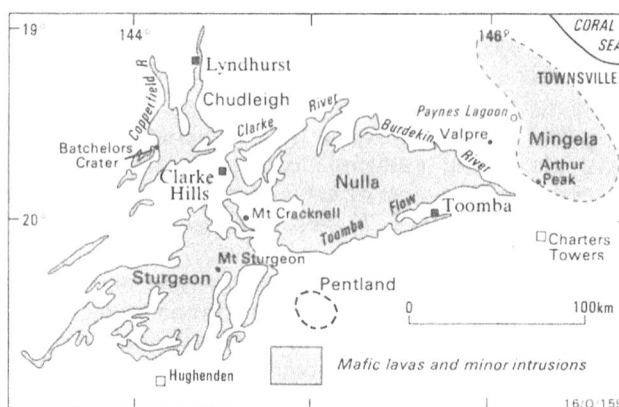

Figure 3.2.7   Distribution of volcanic rocks in the Chudleigh, Sturgeon, Nulla, and Mingela provinces of northern Queensland (see Fig. 3.2.1).

volcano (and nearby) gave an age of 0.25 Ma, and a similar result (0.26 Ma) was obtained on the basalt in the Copperfield River near Einasleigh. The province as a whole is characterised by broad, partly dissected lava plains between numerous pyroclastic cones, some composite cones, and several lava shields. Many of the volcanoes have relatively young features, and age determinations range from 8.0 to 0.36 Ma. The basement includes Proterozoic metamorphic rocks of the Georgetown Inlier, tightly folded Palaeozoic sediments, granites, and Jurassic sediments.

Moderately to strongly nepheline-normative basaltic rocks are dominant in the Chudleigh province. Rock types include nephelinite, basanite, and mildly under-saturated alkali basalt, hawaiite, and minor mugearite (Table 3.2.1, columns 4–5). Sodic and potassic types are represented equally which contrasts with the adjacent McBride province where most rocks are sodic. Mg-ratios are evenly distributed between 70 and 50. Basanite in the Chudleigh province is typically fine-grained, containing abundant euhedral and skeletal olivine microphenocrysts. Coarser-grained, more plagioclase-rich hawaiite contains titaniferous pyroxene and is commonly vuggy.

Upper-mantle and lower-crustal xenoliths and mega-crysts are common at several localitites and, in contrast to the McBride province, upper-mantle rather than lower-crustal xenoliths are dominant (Irving, 1980; Kay & Kay, 1983; Rudnick et al., 1986). Felsic granulites are rare. Host rocks range from *ne*-hawaiite, as at Batchelors Crater where locally up to 70 percent of the rock is composed of xenoliths, to pyroclastic cones such as Airstrip Crater and Sapphire Hill. Mantle xenoliths include abundant peridotite and pyroxenite up to 20 cm across, some composite. Lower-crustal xenoliths include corundum-bearing and garnet-bearing granulites. Megacrysts include pyroxene, amphibole, mica, spinel, anorthoclase and corundum.

Numerous mesa-basalt remnants are present in the Clarke River headwaters between the continuous basalt plateaux and plains of the Chudleigh and Nulla provinces. These mesa basalts once must have formed a continuous sheet bridging the provinces. Age deter-minations for some of the mesa basalts range from 8.9 to 7.8 Ma, exceeding the ages obtained from the neighbouring provinces (6.3 Ma or less). At least four eruptive centres can be recognised topographically among the mesas, and this area around Clarke Hills appears to circumscribe a sub-province of somewhat older activity, although younger volcanoes and valley lavas (4.5–0.7 Ma) are known within it.

3.2.8 Sturgeon

Sturgeon is the farthest inland of the north-Queensland provinces (Figs. 3.2.1,7). It features wide lava plains adjacent to the Great Divide, that form plateaux above the heads of the Flinders River and its tributaries draining southwards and the Stawell and Dutton rivers draining westwards. Many of these streams are incised considerably towards the edges of the province, and are filled progressively with successive flows. The southern edge of the province forms an inclined scarp, up to 100 m high on Lower Cretaceous sedimentary rocks of the Great Artesian Basin. The basement elsewhere also includes Proterozoic and Lower Palaeozoic meta-morphic and granite rocks and Permian to Jurassic sedimentary rocks.

Sturgeon province forms part of a broad topographic dome, interpreted as a Miocene uplift (Stephenson & Coventry, 1986). Volcanism was initiated after this uplift, following incision of the marginal drainage. Flows followed the drainage lines, and erosion estab-lished new valleys after each eruption, subsequently producing inverted relief. The lava units — especially

in the southern and western parts of the province — now form a series of narrow and elongate plateaux, the heights of which decrease with age (Coventry et al., 1985).

The area of volcanic rocks in the Sturgeon province is about 75 km² and 46 volcanic centres have been recognised. Age determinations correspond to intensive volcanic activity around 5.5 Ma which continued sporadically up until 0.9 Ma ago.

The youngest Sturgeon flow (0.92 Ma) terminated near Hughenden and was erupted 120 km farther north (Stephenson & Griffin, 1976b). The only volcanic source so far confirmed for extensive 5.5 Ma flows forming high mesas northwest of Hughenden is Mount Sturgeon (5.4 Ma). Mount Cracknell, an isolated mesa 20 km northeast of Sturgeon, may conceal or be adjacent to an unrecognised centre. Mount Cracknell (5.4 Ma) now stands 140 m above its surroundings. The 110 km Beckford flow is a well-defined line of narrow mesas for which five successive ages down the flow average 3.3 Ma (standard deviation 0.08 Ma).

The oldest activity for this region is difficult to establish, because evidence is likely to have been removed by erosion or covered by later eruptions. However, clear evidence for much older basaltic activity is provided by the occurrence of basalt clasts of unknown source in lateritised sediments exposed at the tableland edge northwest of Eaglehawk Tank. The age of lateritisation is uncertain, but could be around 30 Ma.

Most basaltic rocks in Sturgeon province are slightly to moderately undersaturated in silica and have Mg-ratios in a range similar to those from McBride and Chudleigh provinces (about 70-50). Hawaiite, including nepheline-normative types, is the dominant rock type (Table 3.2.1, column 6). Rocks more evolved than hawaiite have not been found.

3.2.9 Nulla

Nulla province is 150 km inland from Townsville, close to the crest on the eastern flank of the Great Divide (Fig. 3.2.1). It has an area close to 7500 km², and most of its 46 recognised vents produced lavas that flowed east to northeast towards the Burdekin valley. Age determinations correspond to effectively continuous, spasmodic activity in the Nulla province over the period of 5.2 Ma up to the youngest volcano, Toomba, believed to be only 13 000 yr B.P. (Stephenson et al., 1978b). Toomba has well preserved flow morphology and a vent complex of northeast-trending fissures and minor pyroclastic cones.

Inliers of rocks beneath the lavas are exposed along some of the incised streams. Fluvial and lacustrine sediments in limited exposures (Archer & Wade, 1976) are interpreted to have been deposited in a basalt-dammed lake. The sediments contain rich vertebrate faunas considered to be early Pliocene.

Basaltic rocks in the Nulla province are predomin-antly nepheline-normative and have Mg-ratios greater than 60 (Table 3.2.1, column 7). The 120 km-long Toomba flow is made up of hawaiite and *ne*-hawaiite containing 3–7 percent normative nepheline. Very few xenolith localities have been identified in the Nulla province, and these have only small peridotite inclusions. The younger flows of Nulla are systematically richer in K_2O (and related trace elements) compared to older ones.

3.2.10 Mingela

Mingela province (Stephenson & Griffin, 1976a) includes 22 known plugs and several dykes that are concentrated in the Arthur Peak and Mingela areas within a 50 km southeast-trending belt, inland from Townsville (Fig. 3.2.7; Table 3.2.1, column 8). Flows are absent. The plugs are up to 400 m in diameter and mostly form small hills. The most prominent hill, Arthur Peak, has the youngest age (35, 31 Ma) whereas the remaining plugs are 44 to 41 Ma old. They intrude a range of country rocks, including metamorphic rocks of possible Proterozoic age, Lower Palaeozoic granites, Devonian limestone, and associated sedimentary rocks. Several of the Mingela plugs are composite, such as FR2 (Stephenson et al., 1978a). This plug is 400 m across and has a fine-grained hawaiite margin containing megacrysts and ultramafic xenoliths, and a 200 m-wide tholeiitic basalt core that lacks inclusions. Virtually all the alkaline plugs and dykes contain ultramafic xenoliths and megacrysts. Several similar plugs in a belt extending over 100 km northwest of Mingela may be affiliated with the Mingela province.

The main xenolith locality in Mingela province is the prominent plug of Arthur Peak where a range of rocks (Mg-ratios of 67–57) contain abundant peridotite and pyroxenite xenoliths, as well as pyroxene, amphibole, spinel, and anorthoclase megacrysts. The Arthur Peak rocks are massive, fine-grained, and (excluding megacrysts and xenoliths) generally aphyric. They range in composition from basanite to alkali basalt.

3.2.11 Other areas

Small developments of volcanic rocks are present outside the major basaltic areas described above. These are: inland from Ingham in the Mount Fox district; west of Townsville in the eastern Burdekin watershed near Paynes Lagoon and Valpre; and in the Torrens Creek and Pentland district.

The Mount Fox area is on a dissected plateau 45 km southwest of Ingham. Sutherland (1977a) described its volcanic features, including plugs (two dated as 22 and 21 Ma) and undated flow remnants in the same area. Mount Fox volcano is a much younger pyroclastic cone, over 120 m high, containing a shallow infilled crater and having a flow that extends southwestwards from the southern base of the cone. Older flow remnants up to 45 m thick overlie unconsolidated Tertiary sediments, including stanniferous leads, and mostly form ridge cappings along former drainage lines. Young (1.5 Ma) lherzolite-bearing basanite flows descend the coastal scarp into the Stone River valley.

There are other basaltic flows 50 km south at The Knobs (Sutherland, 1977a) and near Ruxton close to Blue Range. The Ruxton basalts overlie the major Ruxton tin deep lead that trends westwards, virtually opposite in direction to the present Burdekin River

drainage. The youngest basalt flow has an age of 26 Ma. Another isolated field of lavas extends south of Lake Lucy (two flows dated at 27 and 19 Ma), but their sources are not known. An isolated hawaiite plug (29 Ma) is found much farther west, beyond the McBride province at Stockmans Hill, Einasleigh.

Basalts in the Paynes Lagoon and Valpre area are limited in extent (less than 10 km²). Two possible volcanic centres are inferred on topographic evidence, but more topographic data are required to account for the total lava distribution. The flows are weathered and form low plateaux overlying Tertiary sediments at Blueberry Hill and silcrete north of Valpre. Two age determinations (27 Ma and 26 Ma) are similar to results for the Ruxton basalts referred to above and to plug ages in the Torrens Creek and Pentland district (see below). These dates are somewhat older than those obtained for the plugs and flow well to the west of Mount Fox. There are deeply weathered basalts in the Taravale area, 30 km north of Blueberry. These cap the southern dissected margin of the Paluma Tablelands and could be of similar age.

Six residual plugs are known in the Torrens Creek and Pentland district and have been called the Pentland

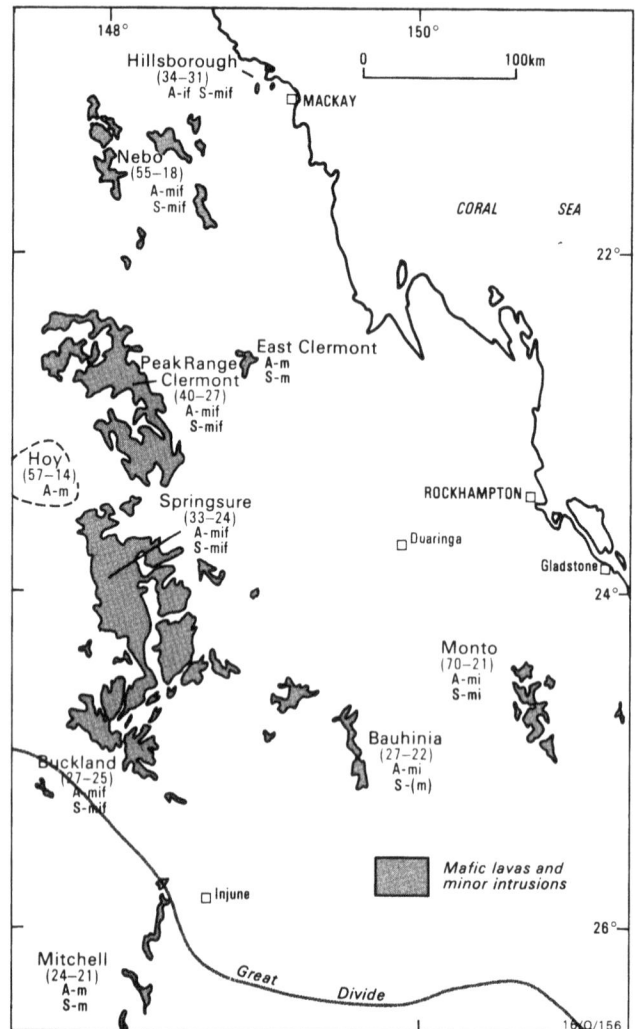

Figure 3.3.1 Distribution, rock types, and ages of volcanic provinces in central Queensland. Rock associations: A, alkaline; S, subalkaline. Rock types: m, mafic; i, intermediate; f, felsic; (m), minor mafic. Ages (in brackets) in Ma.

province (Fig. 3.2.7; Table 3.2.1; columns 9–10). They intrude either Mesozoic sediments near the exposed margin of the Great Artesian Basin or the underlying Palaeozoic Cape River Beds. The plugs are up to 1.5 km across and two are nephelinite containing small peridotite inclusions. Age determinations are similar for four of the plugs and range from 27 to 25 Ma.

3.3 Central Queensland

3.3.1 Introduction

Late Cretaceous to Cainozoic volcanic rocks in central Queensland are mostly within the north-northwest trending Bowen Basin. This elongate enclosure of Permian and Triassic beds separates a Devonian-Carboniferous orogen to the west from the Silurian-Triassic New England Fold Belt to the east (Dickens & Malone, 1973; Day et al., 1983; Section 1.3.1).

The volcanic belt starts in the northeast, outside the Bowen Basin, at the Hillsborough province which straddles the early Tertiary Hillsborough Basin and its western margin of Devonian-Carboniferous beds (Fig. 3.3.1). A chain of central-volcano provinces (Nebo, Peak Range, Springsure, and Buckland), extends to the southern end of the Bowen Basin. These larger provinces are flanked by several other basaltic provinces. Hoy province is outside the Bowen Basin and extends across early Palaeozoic metamorphic rocks and granites into the late Palaeozoic Drummond Basin. The isolated Monto province occupies a similar position on the southeastern margin of the Bowen Basin, and is on a structurally complex, highly faulted region of late Palaeozoic to early Mesozoic rocks. Generalised geochemical trends for the central Queensland provinces are shown in Figures 3.3.2 and 3.3.3, and selected chemical analyses are given in Table 3.3.1.

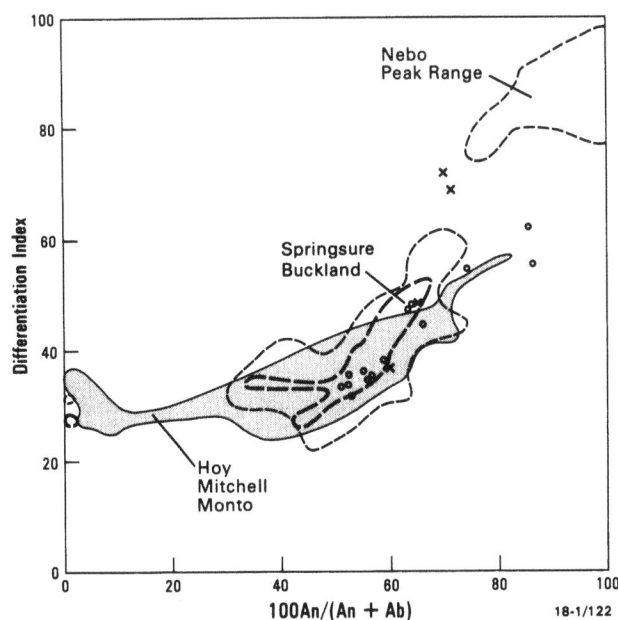

Figure 3.3.2 Compositional fields for central-Queensland provinces using the plotting procedure outlined in Section 1.1.4. Crosses refer to Hillsborough samples. Open circles refer to Bauhinia samples.

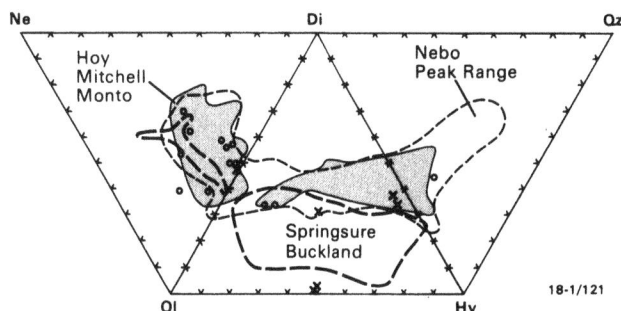

Figure 3.3.3 Compositional fields for central-Queensland rocks projected in the system Ne-Ol-Di-Hy-Qz. Crosses refer to Hillsborough samples. Open circles refer to Bauhinia samples.

Volcanism in some provinces has a restricted time span of up to 8 Ma, but in others multiple episodes cover periods of up to 50 Ma. The provinces range from essentially alkaline volcanism (Hoy) through mixed alkaline and subalkaline episodes (East Clermont, Duaringa, Monto, Bauhinia, and Mitchell) to those of central-volcano type (Hillsborough, Nebo, Peak Range, Springsure, and Buckland). The central volcanoes of the region belong to the south-migrating hotspot chain (Wellman & McDougall, 1974a).

3.3.2 Hillsborough

The Hillsborough province (Fig. 3.3.1) contains both intrusions and volcanic sequences dominated by felsic rocks (Slessar, 1970; Stephenson et al., 1980; Champion, 1984; Stephenson, 1985). Two main centres are exposed: the Cape Hillsborough volcanic sequence on the coast and the Mount Jukes intrusive complex inland. Both are early Oligocene (34–31 Ma; McDougall & Slessar, 1972). Similar rocks to those at Cape Hillsborough are known around Mount Mandurana, 10 km east of the Jukes complex.

The Cape Hillsborough volcanic sequence is near the western margin of the Hillsborough Basin, an asymmetric graben filled with up to 3000 m of Palaeocene to Middle Oligocene freshwater mudstone, shale and sandstone, oil shale, and ostracod limestone (Gray, 1976). The volcanic rocks are well exposed along the precipitous coast of the 5 km-long cape and rise to 270 m above sea-level. They overlie Hillsborough Basin sediments and dip southwards at 5–15°. The succession is complicated by faulting and is downfaulted against Palaeozoic and Mesozoic rocks immediately west of the Hillsborough promontory. The volcanic rocks appear to represent the residual flank of a central volcano whose centre lies to the north or northeast at no great distance. Several small pyroclastic and lava plugs are preserved at Hillsborough but represent only minor vents.

Porphyritic rhyolite and peralkaline rhyolite alternate with felsic pyroclastic deposits and reworked volcaniclastic deposits in a sequence slightly thicker than 300 m. Individual flows are up to 30 m thick. Many are spherulitic and some perlitic. Well-bedded felsic pyroclastic deposits include massive tuffs which are probably pyroclastic flows and a remobilised welded tuff near the top of the sequence. Basalt lavas and pyroclastic rocks are found in the lower part of the

Table 3.3.1   Selected chemical analyses for central-Queensland provinces

	1	2	3	4	5	6	7	8	9	10
SiO_2	44.85	56.48	45.17	74.08	47.87	48.49	68.00	45.26	53.87	38.74
TiO_2	2.23	1.60	2.57	0.11	1.91	2.94	0.11	1.89	1.64	2.43
Al_2O_3	14.93	15.29	14.97	10.80	14.24	16.42	14.67	14.63	14.89	10.01
Fe_2O_3	5.24	2.59	2.17	3.27	1.69	3.88	1.14	2.19	0.79	3.59
FeO	6.25	5.28	8.56	0.20	9.04	8.77	1.30	8.29	8.21	8.85
MnO	0.14	0.11	0.17	0.01	0.17	0.12	0.07	0.15	0.13	0.74
MgO	5.45	5.08	8.94	0.02	8.60	4.14	0.05	7.59	6.68	12.50
CaO	7.12	6.20	8.90	0.05	8.85	6.17	0.79	7.67	8.71	12.72
Na_2O	4.19	3.82	3.53	5.43	3.95	3.99	4.39	4.79	3.03	4.14
K_2O	3.53	2.12	1.89	4.05	1.02	1.89	5.84	2.45	0.74	1.73
P_2O_5	1.68	0.41	0.88	0.02	0.49	0.86	0.01	0.98	0.24	1.56
S				0.02	0.02					
H_2O+	5.19[2]	1.54[2]	1.29	0.34	1.33	2.76[1]	3.71[1]	4.05[2]	1.45[2]	2.74[2]
H_2O-			0.80	0.40	0.73					
CO_2			0.14	0.11						
Total	100.80	100.52	100.00	98.91	99.89	100.43	100.08	99.94	100.38	99.75
Ba	317	299	395	6	295	361	9	232	145	640
Rb	35	43	30	510	20.1	34	166	18	14	30
Sr	4789	440	885	1	576	579	4	1061	321	1560
Pb	2	11	3	26		3.7	18	20	13	24
Th	8	18	6	96	3.2	4.5	22	7	<2	15
U	1	2	1	13	0.5			<2	<2	<2
Zr	125	162	287	4280	130	359	465	349	128	350
Nb			55	418	36	57	219	77	18	153
Y	22	17	25	83	21.6	41	119	18	22	42
La		26	39	135	26.0			49	17	109
Ce		55	84	276	49.5	78	173	106	42	184
V	87	80	150	1	140	105		75	139	160
Cr	46	143	232	2	208	2		204	264	514
Ni	40	99	165	<1	103	11	4	180	166	342
Cu	41	39	49	<1	82			52	73	68

1. *Ne*-hawaiite from Nebo province (F.L. Sutherland, unpublished data, sample FLSIC).
2. *Qz*-tholeiitic basalt from Nebo province (F.L. Sutherland, unpublished data, sample FLS104A).
3. Basanite from Peak Range province (J. Knutson & B.W. Chappell, unpublished data, sample EAV87).
4. Peralkaline rhyolite from Peak Range province (J. Knutson & B.W. Chappell, unpublished data, sample EAV84).
5. Hawaiite from Hoy province (A. Ewart, unpublished data, sample AK71).
6. Hawaiite from Springsure province (Ewart, 1982a, sample 38747).
7. Rhyolite from Springsure province (Ewart, 1982a, sample 38739b).
8. *Ne*-hawaiite from Buckland province (F.L. Sutherland, unpublished data, sample BCCG).
9. *Qz*-tholeiitic basalt from Bauhinia province (F.L. Sutherland, unpublished data, sample CQ67D)
10. Melilite nephelinite from Monto province (F.L. Sutherland, unpublished data, sample COP).

[1] Total H_2O.
[2] Loss on ignition.

sequence and also in several of the minor vents. Coarse conglomerates locally contain fragments of pyroclastic rocks and may represent debris flows. The whole sequence has been affected by hydrothermal alteration. Local areas are characterised by close fracturing, occupied by chalcedonic veins, and some tuffs are extensively opalised.

The basalts at Hillsborough range from alkali basalt to more common transitional to *qz*-tholeiitic basalt, and have low Mg-ratios (50–30). Some lavas are icelandite (59–65 percent SiO_2). Felsic lavas containing Na-rich amphibole or pyroxene are uncommon, but six out of 10 analysed rhyolites contain normative acmite.

The Mount Jukes intrusive complex surrounds the precipitous Mounts Jukes (547 m) and Blackwood (590 m) and consists of the following four main units that intrude the Permian Carmila Beds and Calen Coal Measures and which are aligned north-northeast (Fig. 3.3.4):

(1) The Neilson Leucogabbro is 3 km across and is the oldest unit. It is deeply weathered and occupies a broad annular valley zone around the central Jukes granite which intrudes it. The gabbro produced a relatively narrow hornfels aureole which stands up as a prominent circular rim. The presence of undisturbed country-rock bedding trends and large xenoliths of country rock are evidence that the leucogabbro was emplaced as a boss-shaped intrusion by stoping.

(2) The quartz syenite at Blackwood covers 12 km² in the south of the complex, and has partly displaced the Calen Coal Measures. The syenite is medium-grained and granophyric. It contains in places xenoliths up to 40 cm across and having features consistent with synchronous intrusion of mafic and felsic melts. Ring dykes and dykes form minor phases of the syenite intruding the leucogabbro.

(3) The Jukes Granite is nearly circular and over 1 km across. It occupies the central part of the Nielsen

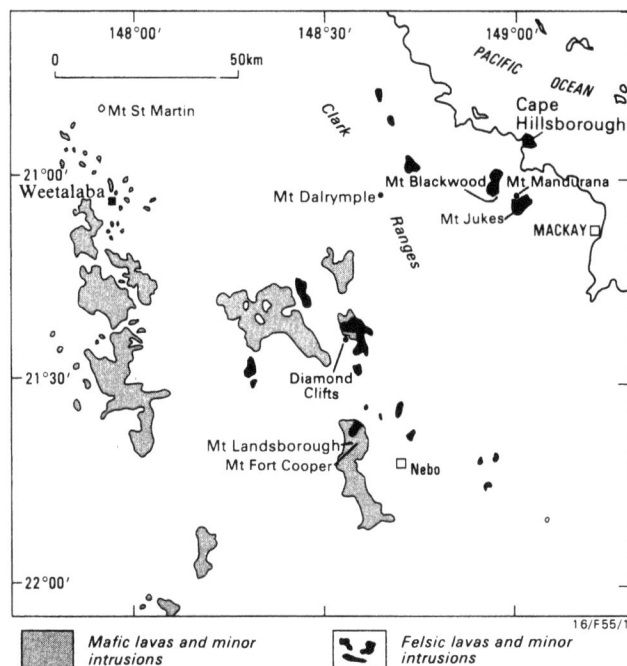

Figure 3.3.4 Distribution of volcanic rocks in the Hillsborough and Nebo provinces of central Queensland (see Fig. 3.3.1).

Leucogabbro. The spectacular peak of Mount Jukes itself is flat-topped with very steep flanks and its present shape may approximate that of the original intrusion. Radial to concentric microgranite dykes (spherulitic and different porphyritic types) intrude the surrounding rocks.

(4) The Seaforth Microgranite may be the youngest phase of the complex. It forms a small intrusion, 500 m across, and cuts the hornfelsed rim of the Nielsen Leucogabbro at the northern end of the complex.

The gabbros are transitional to tholeiitic in composition. They contain low normative quartz or olivine and are characterised by high alumina (greater than 20 percent). Champion (1984) modelled fractionation trends from the leucogabbro composition to the microgranite and found trace-element differences to be consistent with fractionation control. The more felsic rocks have characteristics typical of A-type granites (Whalan et al., 1987). The Hillsborough and Mount Jukes areas each represent a relatively coherent rock suite and although not chemically identical they are generally similar in major- and minor-element chemistry. The two appear to be volcanic and intrusive equivalents.

3.3.3 Nebo

Volcanic rocks in the Nebo area mostly crop out within the northern closure of the Bowen Basin, extending over 200 km in a northwest-trending distribution pattern (Fig. 3.3.4). They are found along the eastern side of the Basin and at higher levels they cap the granite highlands of the Clark Range up to heights of about 1000 m. Mafic volcanic rocks predominate to the west, whereas felsic rocks are prominent to the east (Malone et al., 1964; Jensen et al., 1966; Sutherland et al., 1977; Sutherland, 1980a; Stephenson et al., 1980).

The volcanic rocks have a range of ages. The oldest ones identified are found northeast of the main basaltic belts as isolated remnants of hawaiite overlying Tertiary sediments and silcrete around Mount Dalrymple. They may represent a single flow of early Eocene age around 55 Ma.

The main basaltic areas are infillings of an ancestral drainage system. Most areas on the western side of the province are preserved as extended ridges and mesas, but large tracts are deeply weathered and in places have lateritic soil profiles. *Ol*-tholeiitic basalts are common, but alkali and transitional basalts are found at the base and interbedded in the sequence, and rare *qz*-tholeiitic basalts are present.

Lava plains form the central area of the province, and are little dissected by headward erosion of the main drainage. The surface is erosional and some higher peaks are plugs protruding through the flows. Alkali basalt and hawaiite are common, but there are also rare *ol*-tholeiitic basalts. A nephelinite plug (a later intrusion, dated at 18 Ma) has a marginal breccia containing diverse fragments of the underlying flow sequence. Radiometric ages for lavas near the base of the flows are greater than 31 Ma.

Alkali basalt, hawaiite, and porphyritic transitional basalt form the Clark Range and Diamond Cliffs region, where they have been dated as early Oligocene — around 35–31 Ma. Similar basalts are shed off the highest parts of the plateau around Mount Dalrymple. Alluvial gem zircons have a fission-track age of 36 Ma, and may be from a localised pyroclastic source at Mount Bruce, a centre for some of the later flows.

Felsic rocks intrude and overlie the basalt sequences and surrounding areas in an elongate zone through Clark Range, Exevale, Diamond Cliffs, Mount Britton, and Mount Landsborough. They include plugs, dykes, possible domes (rare), and some lava caps of trachyte and rhyolite. Intrusions are dominant between Clark Range and Exevale, and any breccias are confined mainly to their margins. These rocks are mostly trachyte. In contrast, largely rhyolitic pyroclastic rocks are prominent around two main centres at Diamond Cliffs and Mount Britton, and are found together with minor porphyritic trachytic-pitchstone dykes. The breccias at Diamond Cliffs are exposed to 200 m thickness. They are composed of fragments of massive and flow-banded rhyolite, pitchstone, and some ignimbrite. Landsborough volcano has been reduced to low-lying dykes and plugs of rhyolite. Many have steep flow-banding parallel to intrusion margins. Outlying flow remnants are found on peaks to the south at Mount Donaldson and The Peak, and at Mount Landsborough where the remnant overlies *ol*-tholeiitic basalt and strongly porphyritic, transitional basalt.

Mount Fort Cooper south of Mount Landsborough forms a prominent mesa of mafic rock, 2 km across and 200 m thick, overlying the eroded tholeiitic flow sequence. A chilled base of basanite passes into coarse rock, including some of olivine-rich hawaiite composition, and then into rock that has veins and patches of pegmatoidal material which farther up constitute the central top part of the body. Mount Fort Cooper may represent an old lava lake and belong to a volcanic rock group younger than the felsic rocks. It has a minimum

K-Ar age of about 16 Ma. These younger basalts are largely in the northern parts of the province. Alkali basalt (29 Ma) overlies oil shale at Plevna, and plugs and flow remnants form local peaks on Redcliffe Tableland and at lower elevations around Weetalaba, where they are dated between 25 and 21 Ma. The Redcliffe-Weetalaba rocks include alkali basalt, hawaiite, and rare *ne*-mugearite. Many carry upper-mantle and lower-crustal xenoliths and megacrysts, and are commonly K-rich.

The youngest exposed volcanic rocks form the isolated flank of a centre at Mount St Martin. Its rocks are dated at 3 Ma and consist of basal volcanic breccia and agglomerates overlain by a flow remnant. The rock has a K-rich *ne*-hawaiite composition, but consists of two phases — a highly undersaturated lava and trachyte. It contains an exceptionally abundant suite of xenoliths which include large pieces of underlying Permian country rock, but also diverse granulites, pyroxenites, and peridotites of a transitional lower-crust/upper-mantle column (Griffin et al., 1987).

The Nebo province is dominated by an Oligocene, central-volcano sequence in which alkali basalt and hawaiite are subordinate to fractionated sodic and K-rich alkaline, transitional, and both *ol*- and *qz*-tholeiitic basalts (Table 3.3.1, columns 1–2). Some of the most nepheline-normative rich rocks form a largely mantle-derived suite younger than (early Miocene) and unrelated to the central-volcano suite. There was also minor, late, alkaline volcanism which formed the Pliocene Mount St Martin.

3.3.4 Peak Range

Peak Range extends as a linear feature for about 100 km and covers an area of about 2500 km² (Fig. 3.3.5). The region extends in elevation from about 280 m to a maximum of 807 m at Browns Peak in the

central part of the range. Basement rocks range in age from Lower Palaeozoic to Cretaceous, and represent several structural units (Mollan, 1965). These include the Anakie Inlier, a structural high consisting of: Lower Palaeozoic metamorphic rocks and Devonian granite; the Bowen Basin, a Lower Permian to Triassic downwarp; and a block of Upper Devonian to Lower Carboniferous volcanic rocks that form a basement high in the northern part of the range. Most of the peaks in the area are sited around this basement high (see also Section 1.4.9 and Fig. 1.4.10A). The axis of the Peak Range is thought to be on a major pre-Permian fault that is exposed in faulted Permian-Triassic rocks to the southeast (Mollan, 1965).

Peak Range (Veevers et al., 1964b; Mollan, 1965; Wellman, 1978) consists of a chain of prominent hills of trachyte, rhyolite, and basalt that are surrounded by extensive flat-lying basaltic lava plains. Pyroclastic deposits are rare, but this may be a result of erosion and blanketing by deeply weathered soil profiles. Residual peaks and mesas in the central part of Peak Range consist of thick sequences of basaltic flows standing high above the basaltic lava plains. Removal of a large volume of basaltic material is implied. The ridged peaks and mesas in the central part of the range (Fig. 3.3.6) consist of up to 50 flows in a 300 m sequence of mafic and intermediate rock types — mainly hawaiite, *ol*-tholeiitic and *qz*-tholeiitic basalts, and lesser amounts of icelandite and dacite (Mollan, 1965). Flow tops are commonly vesicular, and amygdaloidal minerals include zeolites (mostly chabazite), calcite, and chalcedony. Deeply weathered flows within the volcanic pile are indicative of at least localised breaks in volcanic activity. The age of the province is essentially older than the period of deep weathering and lateritisation in central Queensland (dated around 27–25 Ma; Grimes, 1980), so secondary alteration and weathering is common in the basalts.

Water bores drilled through the plains basalts adjacent to the Peak Range penetrated up to 100 m of basalt, but more commonly the lava sequences are less than 30 m thick. These low-lying flows consist mainly of hawaiite and tholeiitic basalt.

Basaltic dykes up to 3 m across are most common in the central part of the Peak Range. They mostly trend northwestwards and cut through the plains basalts. Some dykes are composite, consisting of a core of *ol*-tholeiitic basalt flanked and cut by hawaiite. There is a close spatial relationship between the dyke swarms and thick remnants of flat-lying basalts in the central part of the Range, so this area may be the main focus of eruption for the basaltic lavas. However, numerous vents were probably scattered over a wide area, because the thin lava plain basalts and the associated plugs and dykes are widely distributed.

Upper-mantle and lower-crustal xenoliths and megacrysts are abundant in some of the outlying, undersaturated, basaltic bodies (for example, Luxor Quarry and Campbell Peak), but are less common in the central part of the region.

Larger outlying intrusions are dominantly teschenite that has some coarse- and medium-grained banding. Fayalite-bearing trachyte is found in the southern part of the Peak Range. It forms the crater of Mount

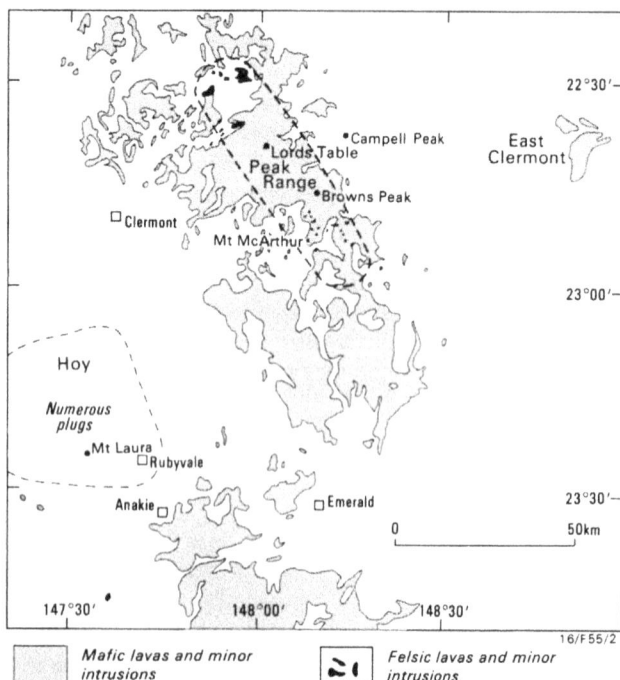

Figure 3.3.5 Distribution of volcanic rocks in the Peak Range and Hoy provinces of central Queensland (see Fig. 3.3.1).

Figure 3.3.6   Anvil Peak (left) and Lords Table Mountain, Peak Range province, showing stacked lava flows of *ol-* and *qz*-tholeiitic basalt, icelandite, and dacite, forming prominent mesa topography.

McArthur where the main body consists of peralkaline trachyte and rhyolite.

Felsic rocks, in contrast to the mafic ones, are mostly well exposed and preserved, and they form the numerous peaks in the northern and southern parts of the range (Fig. 3.3.7). The peaks are domes, plugs, pinnacles, dykes, and restricted flows that consist mostly of trachyte and rhyolite. A 2 m-thick, volcanic-breccia flow overlies bedded sediments and dips away from Scotts Peak at about 30° in the southern part of the range. Minor tuffs appear to be associated with subalkaline rhyolitic domes only in the north. Rhyolite dykes, some glassy, are also confined mainly to the northern part of Peak Range.

Tholeiitic-basalt extrusions evidently were common during the earliest volcanism at Peak Range. However, both tholeiitic and alkali-basalt flows are found inter-layered in the thick sequences of the central part of the range, and there is an apparent change in the magmatic affinities of lavas in the northern and southern parts of the range. Lavas to the south are dominantly alkaline, whereas those to the north are dominantly tholeiitic. This contrasts with the apparent dominance of tholeiitic basalts forming the lava plains in the southern area and alkali basalts in the north. Similarly, peralkaline rhyolite is found almost exclusively in the south, whereas subalkaline rhyolite is dominant in the north.

Mafic magmatism mostly preceded felsic, judging by field relationships and mafic inclusions in comendites. However, a dyke of xenolith-bearing *ne*-hawaiite at

Campbell Peak intrudes phonolite, and fayalite trachyte intrudes peralkaline rhyolite at Mount McArthur.

The Peak Range volcanic province contains both nepheline- and quartz-normative rocks, ranging in composition from primary and near-primary mantle-derived magmas to highly evolved types (Table 3.3.1, columns 3–4). Mg-ratios of analysed rocks range up to 71. The mafic to intermediate alkaline and tholeiitic rocks have a wide range in texture — from fine- to coarse-grained, aphyric to porphyritic. Fayalite, aegirine, riebeckite, aenigmatite, sanidine, anorthoclase, and quartz are present in the more felsic rocks.

Wellman (1978) concluded that tholeiitic-basalt lavas in the Peak Range and Clermont area are 40 to 34 Ma old, and that alkali basalts and felsic intrusions are 30 Ma old in the north and less than 27 Ma old in the south. This apparent southward shift of volcanism in the province forms part of the larger-scale, southward migration of volcanic activity described by Wellman & McDougall (1974a) for the whole of eastern Australia. The *ne*-hawaiite dyke cutting the Campbell Peak phonolite is dated at 33 Ma and the Beacon Hill basanite at 35 Ma (F.L. Sutherland, personal communication, 1986).

3.3.5   Hoy

Hoy province is to the west of the larger Peak Range and Springsure provinces (Fig. 3.3.1). It differs from other central-Queensland provinces in the abundance

Figure 3.3.7   Scotts Peak, Ropers Peak (peralkaline rhyolite), and Malvern Hill (fayalite trachyte), from left to right form prominent hills rising to 804 m in the southern part of the Peak Range.

of volcanic necks of undersaturated basalt (Veevers et al., 1964a). Over 70 plugs are found within a nearly circular area 50 km in diameter, and there are isolated, outlying intrusions, volcanoes, and flows. The province lies across the faulted eastern margin of the Devonian-Carboniferous Drummond Basin against the older Devonian Retreat Granite. The majority of the plugs intrude the granite, but some in the south cut early Palaeozoic metamorphic rocks of the Anakie Inlier and some western plugs cut sandstone in the Drummond Basin. Nearly two-thirds of the plugs form a north-northeast trending zone 10 km wide near the centre of the field.

The plugs typically form hills that range from sharp cones to low rises. A few breccia bodies are found near Rubyvale and are known at some plug margins, but the majority of the plugs are massive, fine-grained basalt. The plugs range in size up to 400 m in diameter and rise up to 250 m above their surroundings. The most prominent plug in the province, Mount Leura, is nearly circular and 400 m across. It forms the summit cone on the mountain, and is surrounded by a raised collar of granitic country rock that is most prominent around the southern flank. Some elongate plugs such as Mount Dumbell may be composite, composed of two or more intrusions. A few plugs are zoned (contrasting central and marginal rocks), but others — including Mounts Leura and Pleasant — are homogeneous.

Hoy plugs appear to be petrographically similar. Surprisingly, age determinations on Hoy plugs are consistent with four periods of activity. The oldest measured basalt is Policemans Knob (56 Ma), west of Rubyvale. Other plugs are early Oligocene (31 Ma), late Oligocene (three plugs, 28–26 Ma), and early

Miocene (three plugs, 19–18 Ma). The younger plugs have more prominent topography. Policemans Knob is more subdued.

The basaltic rocks are nearly all fine grained, regardless of location within a plug, but some plugs contain fine doleritic facies and all contain phenocrysts of olivine. They commonly contain a mixed assemblage of plagioclase, olivine, clinopyroxene, spinel, and other oxide xenocrysts. Basanite constitutes nearly half of 52 analysed rocks from 44 centres. Alkali basalt (19 percent), ne-hawaiite (15 percent), and hawaiite (12 percent) are all represented (Table 3.3.1, column 5), but only single examples of nephelinite and transitional basalt have been found. Potassic rocks make up 46 percent of the total and 70 percent of these are basanites.

The province is notable for its association with regional alluvial gem deposits which yield abundant sapphires and zircons (Section 3.10.2). Sapphire and zircon have been found also as rare inclusions in several plugs including Mounts Leura and Pleasant. The question of the genesis of these sapphires and zircons — cognate or xenocrystal — has not been fully resolved, but they do have igneous-like crystal forms.

There are restricted basalt flow areas adjacent to the Hoy province, to the south (Spring Creek headwaters) and 20 km east of the province (Fork Lagoons). Age determinations of the Spring Creek hawaiite mesas are 33 and 27 Ma, corresponding to possible extrusion from Hoy plugs of this age. The Fork Lagoons flow forms an elongate, meandering mesa system which represents a former valley-filling flow. It has a K-Ar age of 28 Ma and its ol-tholeiitic basalt composition is distinct from those of the Hoy basalts. Extensive flows surround Anakie Hill, 15 km southeast of the Hoy province. The hill is composed of nephelinite which has a K-Ar age of 15 Ma. It encloses a crater-like basin open to the south and has weathered, coarse nephelinite on its floor that possibly represents a lava lake. Mount Scholfield (19 Ma), 30 km south of the Hoy province and 1 by 1.5 km across, contains hawaiite, *ne*-hawaiite, dolerite, and gabbro with pegmatoid veins.

Hoy rocks have a relatively restricted nepheline-normative character over a considerable time span (57–14 Ma). The subalkaline Fork Lagoons field is similar in age (29–26 Ma) to the voluminous, more felsic activity in the adjacent southern Peak Range and Springsure central volcanoes.

3.3.6   Springsure

A large apron of basaltic lavas (7000 km²) surrounds a restricted central area (350 km²) of felsic intrusions and volcanic rocks termed the Minerva Hills Volcanics (Veevers et al., 1964a; Mollan et al., 1969) in the Springsure area (Fig. 3.3.8). The basalts reach 330 m thickness around Springsure and exceed 600 m in elevation. They thin out over the Springsure Anticline, and are interrupted by a median strip of gently folded Permian beds. The basalts merge with those of the Buckland province to the south and, with them, spill over eastwards into the Bauhinia province.

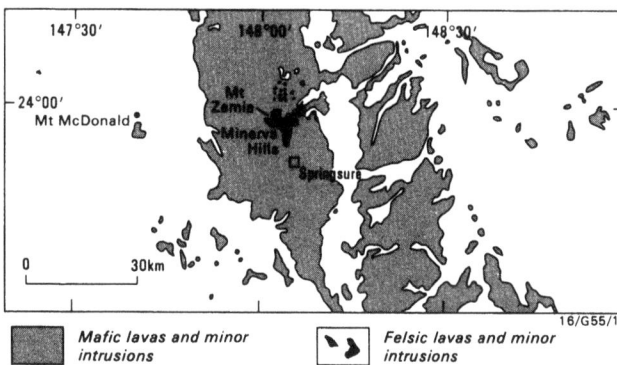

Figure 3.3.8   Distribution of volcanic rocks in the Springsure province of central Queensland (see Fig. 3.3.1).

The main part of the Springsure province forms a divide that is undergoing dissection by westerly and easterly flowing tributaries draining into flanking north-flowing rivers. The topography is characterised by mesa-like plateaux and remnants standing above rolling plains reduced to basaltic soils. Many basalt hills have stepped relief from differential erosion of the flow stacks. The felsic outcrops around Minerva Hills form more rugged relief, including spectacular peaks and precipitous flow caps north of Springsure.

The oldest volcanic rock identified is a *ne*-mugearite dyke 5.4 km east of Mount McDonald. Zircons derived from the dyke have an early Oligocene fission-track age (35 Ma). Related rocks may exist under the thick flood basalts closer to Springsure, but their lower levels are generally strongly weathered and poorly exposed.

The main basalt sequence interdigitates with and overlaps extensive Tertiary sediments and lateritic

surfaces northeast and east of Springsure. Outlying flows on Mount McDonald (522 m) are above silcrete developed on Carboniferous beds at around 400 m elevation. A central intrusive felsic complex has domed up the Permian beds, and the basaltic lavas generally dip away from this focus. However, the western-plateau basalts appear to have a slight regional dip eastwards, possibly caused by tilting.

Three stages of extrusion are represented around Springsure: older mafic rocks, interbedded mafic and felsic rocks, and younger mafic ones. The older basalts are predominantly hawaiites, but include tholeiitic basalts. These pass up into more evolved K-rich hawaiites and mugearites among the basalts associated with subsequent felsic centres. Basalt ages range from early to late Oligocene (33–26 Ma) and felsic rocks are late Oligocene (29–27 Ma; Webb & McDougall, 1967; Harding, 1969; Mollan et al., 1969; Ewart, 1982a). Agglomeratic tuffs, up to 60 m thick as well as scoriaceous ash, lapilli, bombs and, in some cases, dense basalt blocks, are exposed in places, corresponding to multiple explosive centres.

The rhyolitic and trachytic Minerva Hills Volcanics consist of a wide range of volcaniclastic rocks and igneous intrusions. These include pyroclastic deposits, autobreccias, radiating, multiple dyke complexes with autobrecciated margins, and arcuate, domal, and crescentic intrusions, some of which have caused arching of Permian beds. A major intrusion forms Saint Peter (0.8 km long, 300 m high). The Minerva Hills Volcanics are on a major fault system, and a gravity anomaly in the area is indicative that finger-like feeders may have risen from an underlying magma chamber.

East of Minerva Hills is a north-trending zone of dome-like intrusions, major plugs, and multiple dykes similar to those of Minerva Hills. Their associated agglomerates contain pink to light-grey, flow-banded rhyolites and are interbedded with basalts. A large southern flow extends over 10 km, and major hummocks in the flow may mark underlying feeders. Its southern extremities overlie trachytic tuffs and agglomerates, forming nosean-trachyte, and perlitic and spherulitic basal sections that are veined by precious opal (Richards, 1918).

Younger basalt flows form bold cappings east of Springsure and on Mount Boorambool where they overlie the trachyte making up Mount Zamia. These rocks are predominately hawaiites, but include tholeiitic basalts, and are dated between 26 and 24 Ma (Mollan et al., 1969; Ewart, 1982a). Over twelve flows, mostly hawaiites, form Mount Catherine, and its prominent profile is capped by two massive valley-filling flows. Some strongly oxidised basalts contain hyalite opal (Richards, 1918). Some of the basalts are feldsparphyric and yield large, gem-quality labradorite crystals.

Mount Sterculia is a massive nephelinite peak that contains abundant upper-mantle xenoliths. It has a K-Ar age of 26 Ma. A hawaiite dyke near Tanderra Homestead is notable for abundant mantle peridotites and mafic to felsic gabbroic cumulates, but its precise affinities to the Springsure basalts are unknown.

The Springsure rock suites are dominated by hypersthene-normative basalts that mainly fall into two groups — less mafic more sodic rocks, and more felsic hawaiites which include K-rich types (Table 3.3.1, column 6). Rare qz-tholeiitic basalt forms the youngest lava flow dated in the province, capping Mount Boorambool. Strongly nepheline-normative rocks form only isolated peripheral occurrences and contain mantle-derived peridotite xenoliths. The strongly evolved felsic rocks appear to fall into two slightly different series — less siliceous qz-trachyte transitional to rhyolite, and more SiO_2-rich rhyolites (Table 3.3.1, column 7).

3.3.7 Buckland and Mitchell

Volcanic rocks in the Buckland area are perched on tablelands formed by relatively resistant Jurassic horizons (Mollan et al., 1969, 1972). They are at higher elevations (500–1230 m) than the Springsure province to the north, except where lavas have descended valleys cut into underlying Triassic and Permian beds at the conjunction of the provinces. Buckland province covers about 15 000 km² in two distinctive parts — a large eroded shield (6000 km²) to the east, and a sparse scattering of intrusive and flow remnants to the west (Fig. 3.3.9). The shield sits on the Great Divide and is dissected by radial drainage. Drainage has incised deep gorges where basalts overlie massive Jurassic sandstones.

The Buckland shield consists of radial extensions of lavas that probably represent valley fills of the original drainage before topographic inversion by subsequent erosion. Numerous flows form sequences up to 330 m thick. The lowest flows in the northeast are thick, coarse-grained, alkali basalts and hawaiites, one of which is dated at 27 Ma (Elliot, 1973). They are overlain by fine-grained, olivine-rich, alkali basalts,

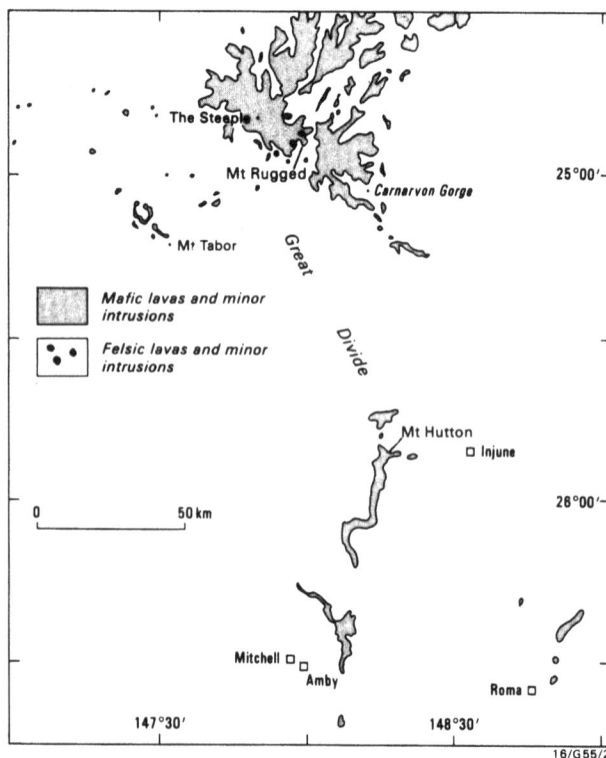

Figure 3.3.9 Distribution of volcanic rocks in the Buckland and Mitchell provinces of central Queensland (see Fig. 3.3.1).

intervals of flow-foot breccias, and a range of massive to vesicular hawaiites and tholeiitic basalts which include feldspathic and porphyritic types. Some lower basalts carry upper-mantle or cumulate peridotites, pyroxenites, and gabbro xenoliths, as well as olivine, clinopyroxene, orthopyroxene, spinel, and plagioclase megacrysts. The sequence to the west is coarse hawaiite, and alkali basalt and hawaiite interspersed with feldsparphyric hawaiite.

The southern sequence is exposed in a 300 m-thick section in Carnarvon Gorge. Basal, weathered, transitional or ol-tholeiitic basalt is overlain by a massive flow up to 30 m thick that has a chilled basal zone of K-rich ne-hawaiite (Table 3.3.1, column 8). This contains abundant upper-mantle peridotites and pyroxenites, lower-crustal granulite xenoliths, and anorthoclase inclusions that have a K-Ar age of 27 Ma (Griffin et al., 1987; unpublished data). The flow is overlain by coarse ankaramite, flow-foot breccias (ol-tholeiitic basalt), and over ten flows of hawaiite and transitional and ol-tholeiitic basalt. Narrow dykes exposed in Carnarvon Gorge and massive dyke-like plugs are commonly feldsparphyric hawaiite and ol-tholeiitic basalt. A small ring and radial dyke structure of nephelinite contains upper-mantle peridotite inclusions and has unknown relationships to the shield lavas. The central part of the shield has hawaiite and younger ol-tholeiitic basalt and also several trachyte and rhyolite intrusions (Bryan, 1983).

The felsic intrusions include a microsyenite ring dyke that formed around a parental gabbro plug late in the shield construction (Bryan, 1969a, 1983). Other felsic bodies are related to, or probably hidden under, small domal uplifts in the Jurassic beds. An eroded trachyte body known as 'The Steeple' includes an anorthoclase-rich phase that has a K-Ar age of 25 Ma.

The western parts of Buckland province contain about 60 intrusive bodies and a similar number of dissected flow remnants. The largest intrusions include stocks and the central, basin-like sill or cone sheet of Tabor Gabbro (Mollan et al., 1972). The Tabor Gabbro may represent an early mafic episode. Most other intrusions are plugs, plus some dykes, and some form northwest-trending alignments of up to six or seven nephelinite bodies or close clusters of alkali basalt. Feeders for the larger flow remnants are known northwest of Mount Tabor where a northwest-trending dyke passes into a massive columnar hawaiite flow.

The west-Buckland lava flows are dominated by compositions ranging from nephelinite to alkali basalt and hawaiite, some containing interstitial analcime. Rare, more strongly fractionated rocks approach ne-mugearite and ne-benmoreite compositions. Many contain upper-mantle peridotites and associated pyroxenites and lower-crustal xenoliths.

The Buckland province has some evidence of volcanic activity that pre- and post-date the major shield-forming episode. The highest point is outside the central area of felsic activity, and another anomalously 'high' residual is present farther north. These may represent additional centres of basaltic activity a few million years after the main felsic activity, as in the Springsure province. Scattered, much less voluminous western activity resembles that of the multi-episode

Hoy province 150 km to the north. A common northwest trend in dykes and aligned centres corresponds to a prominent direction of structural weakness in the basement below the western bodies. Basalts at Mount Rugged appear to form a separate sequence to the main shield lavas. The sequence of vesicular and porphyritic flows and agglomerate is capped by partly eroded pyroclastic beds, whose preservation and form are indicative of a relatively young vent.

Basaltic lavas in the Mitchell area (Exon et al., 1970; Exon, 1971a,b; Mollan et al., 1972) fill palaeo-drainages in two regions where they overlie Jurassic and Cretaceous beds of the Eromanga and Surat basins. The most extensive basalts follow old river courses over a distance of 120 km. The sequences reach over 30–80 m in thickness and have massive to vesicular flows 6–30 m thick. The lavas are alkali basalt and *ol*-tholeiitic basalt.

Small basalt caps between 350–450 m elevation mark an old river course through Cretaceous beds northeast of Roma. They include similar basalts to the western sequence. A dolerite intrusion in Jurassic beds 34 km north of Roma is probably a feeder for a transitional basalt. Small intrusions are found also between the two basalt sequences. A dyke-like intrusion 16–17 km northeast of Injune is an alkali basalt and the only rock found to carry peridotite inclusions.

Basalts dated in both the western and eastern regions of the Mitchell province have early Miocene ages of between 24 to 21 Ma, but the younger ages may have resulted from argon loss (Langford-Smith et al., 1966; Exon et al., 1970).

The Buckland province petrographically extends from undersaturated to saturated rocks that fall into three separate groups. The bulk of the rocks are dominated by transitional to slightly quartz-normative tholeiitic basalts which form the main part of the Buckland shield. The second group consists of nephelinite, basanite, alkali basalt, and hawaiite containing upper-mantle and other inclusions. The third group includes strongly K-rich felsic rocks forming late-stage isolated intrusions within the Buckland shield. Mitchell province is dominated by tholeiitic basalts, distinct from the main Buckland basalts, that were extruded within a short time-span (24–21 Ma). There appears to be a trend of increasing silica-saturation towards the south.

3.3.8 Bauhinia

The Bauhinia basalts form scattered outcrops over an area of about 500 km² (Fig. 3.3.10). The flows fill Tertiary drainages cut into Triassic-Jurassic sediments on the western limb of the Mimosa Syncline (Olgers et al., 1966; Mollan et al., 1972).

The thickest basalt caps the Expedition Range in the western part of the province where at least seven flows and some interbedded basal tuffs form a 310 m-thick sequence at Mount Nicholson. The lavas pass from pyroxene-phyric, hypersthene-normative hawaiite through alkali basalts and hawaiite and rare *ol*-tholeiitic basalt, into capping, strongly feldsparphyric hawaiite.

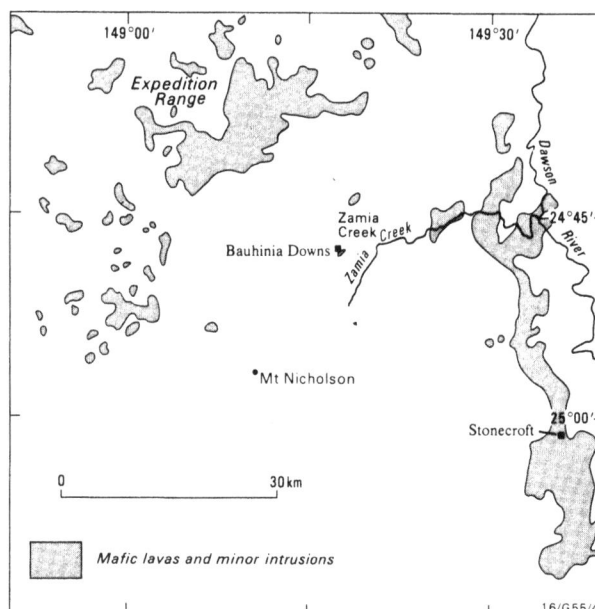

Figure 3.3.10 Distribution of volcanic rocks in the Bauhinia province of central Queensland (see Fig. 3.3.1).

The northern extremities of the Mount Nicholson flows intermingle with flows descending east from the Buckland and Springsure provinces that connect by a narrow pass through Expedition Range. A coarse-grained mafic lava 3 km west-southwest of Bauhinia Downs was dated at 23 Ma (Harding, 1969), so that late Oligocene to early Miocene ages probably are typical of these northern flows. More alkaline lavas and related plugs (R.E. Pogson, personal communication, 1986) crop out in Zamia Creek. They include K-rich mugearites and characteristically contain upper-mantle and lower-crustal inclusions and megacrysts. One flow remnant directly overlies the hawaiite dyke dated at 27 Ma, so considerable erosion and an early Miocene or younger age are likely for the later alkaline activity.

The eastern part of the Bauhinia province consists of elongate valley fills extending north and south of a volcanic centre at Stonecroft Homestead. Agglomerate exposed in the eroded core of the vent is 50 m thick and contains conspicuous volcanic bombs as well as abundant upper-mantle and lower-crustal inclusions. Dykes fed a flanking flow of hawaiite 70 m thick. The lower Stonecroft flows include tholeiitic basalts (Table 3.3.1, column 9), and have been weathered, lateritised, and eroded prior to eruption of younger flows of *ol*-tholeiitic and alkali basalt. Ages of 25, 24, and 22 Ma are recorded in the upper flows north of Stonecroft (Harding, 1969), and the whole sequence is probably late Oligocene to early Miocene in age. The Stonecroft volcano and minor nearby plugs appear to be emplaced on a structural line of weakness along the Mimosa Syncline. Mount Slopea, 40 km southeast of Stonecroft, is an isolated laccolith that passes from a chilled margin of basanite carrying upper-mantle xenoliths into differentiated coarser phases of hawaiite that has a K-Ar age of 23 Ma.

Bauhinia-province basalts were erupted over a fairly limited time span (27–22 Ma) and range from nepheline- to quartz-normative sodic compositions.

Tholeiitic basalts comprise the least fractionated lavas. The Mount Slopea laccolith of related age is nepheline-normative, chemically distinct from the extrusive rocks of the province, and contains evolved dyke and pegmatoidal differentiates.

3.3.9 Monto

Basalts crop out over 1000 km^2 in the Monto and Mundubbera areas (Fig. 3.3.11; see also Fig. 3.4.4; Dear et al., 1971; Whitaker et al., 1975). Minor, late Cretaceous and some early Tertiary basalts are present, but the greatest volume of basalt was erupted in the middle Tertiary. These later basalts flowed into separate watersheds controlled by a divide through the Dawes Range which was in substantially the same position as that of the present divide. The basalts are in a structurally complex and strongly faulted region of Palaeozoic and early Mesozoic rocks. Scattered basalt remnants also extend east of the fault system towards the Boyne-Yarrol fault system, and flow residuals are found at Binjour Plateau and Mount Redhead to the south.

The oldest dated basalt is a plug of *ne*-hawaiite containing upper-mantle inclusions northwest of Mundubbera (70 Ma; Green, 1974). Early Tertiary weathered basalt was drilled at the base of the Bileola Basin (Noon, 1982) and may correspond in age with dolerite dated south of Bileola (47 Ma; Green, 1974). Extensively weathered and in places lateritised flows form residual plateaux and knobs. These are predominantly alkali basalt and hawaiite, and they descend southwards from Dawes Range. Similar basalts are found on Binjour Plateau and at Mount Redhead where a hawaiite interbedded with laterites has a late Oligocene age (27 Ma). A dolerite intruding the oil-shale sequence northeast of the Monto basalts has a similar age (27 Ma; Henstridge & Missen, 1981). Oligocene basalts also intrude the oil-shale sequence to the east of the province (28 Ma; Henstridge & Hutton, 1986).

Unlateritised flows dominate the northern part of the Monto province. They overlie deeply weathered Tertiary sediments in the Callide valley, and the thickest section has at least six flows totalling 270 m in thickness. *Ol*- and *qz*-tholeiitic basalts are typical amongst the basal flows, whereas alkali basalt and hawaiite prevail in the upper flows. An *ol*-tholeiitic basalt flow overlying lateritised Tertiary sediments is dated at the Oligocene-Miocene boundary (25 Ma; Queensland Department of Mines Annual Report, 1980; C.G. Murray, personal communication, 1985).

Tholeiitic-basalt flows are also common in the unlateritised sequences to the south around Monto, but the volume and thickness of lava are less. *Ol*-tholeiitic basalts farther east form old valley fills, and a dyke may be one of the feeders. These tholeiitic basalts are interspersed with valley fills of K-rich alkali basalt and mugearite. The mugearite contains abundant upper-mantle and lower-crustal inclusions and is dated at 25 Ma. It forms dissected remnants of an extensive flow system, apparently erupted from the prominent plug of similar rock at Mount Weary.

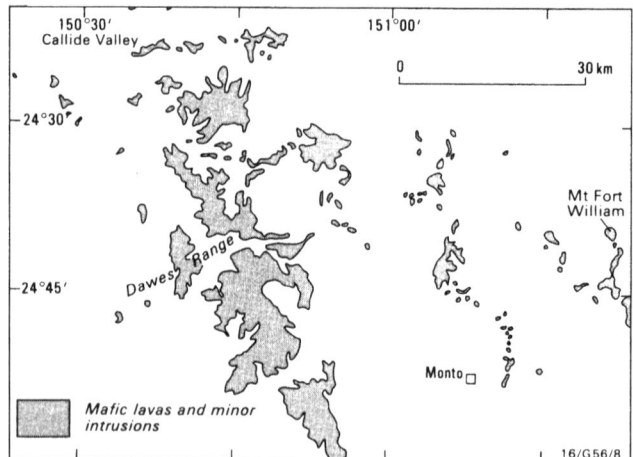

Figure 3.3.11 Distribution of volcanic rocks in the Monto province of central Queensland (see Fig. 3.3.1).

Small plugs, dykes, and flows of alkaline rocks intrude or overlie the main Monto basalts or form isolated bodies to the south and southeast of Monto. Most contain abundant upper-mantle inclusions. Nephelinite forms flow caps dropping southwards in altitude from 650 m to 400 m. A cap at Mount Fort William is dated at 21 Ma. Olivine melilitite, basanite, alkali basalt, and hawaiite plugs intrude the flow sequence in several areas.

Restricted exposures of basalt flows to the northwest of Monto and south and southwest of Duaringa form minor scattered bodies in an area of 600 km^2. The basalt flows are lateritised and interbedded with middle to late Eocene sediments in the Duaringa Basin which

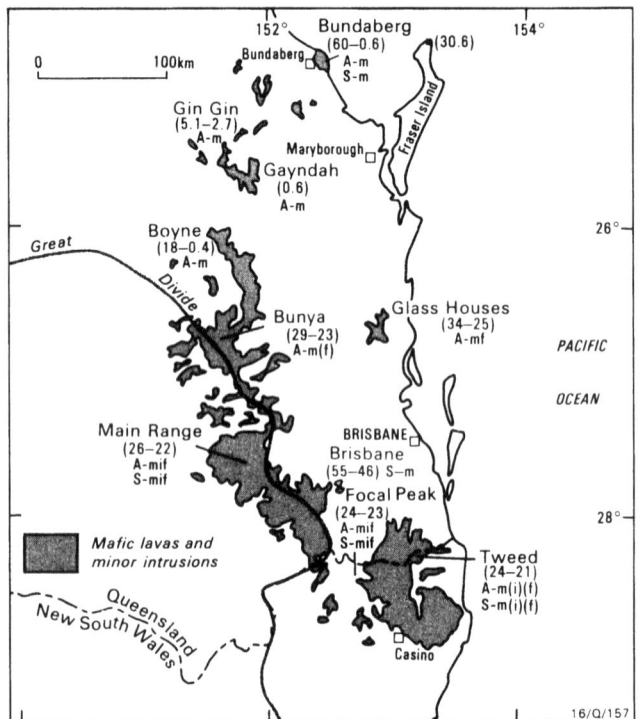

Figure 3.4.1 Distribution, rock types, and ages of volcanic provinces in southern Queensland and of Tweed volcano. Rock associations: A, alkaline; S, subalkaline. Rock types: m, mafic; i, intermediate; f, felsic; (i), minor intermediate; (f), minor felsic. Ages (in brackets) in Ma.

are up to 1300 m thick, and some basalts form small high-level residuals (Malone et al., 1969; Day et al., 1983). Flows drilled in the Duaringa Basin include *ol*-tholeiitic basalt. Undersaturated rocks are found amongst plugs intruding a fault bounding the Gogango Range 35 km north of Duaringa (Dunstan, 1901), and K-rich hawaiite containing upper-mantle inclusions forms flow remnants on Blackdown Tableland 45 km west-southwest of Duaringa.

The basaltic rocks of Monto Province have a range of compositions, falling into a more saturated group of transitional and tholeiitic lavas and a nepheline-normative group of lavas and small intrusions (Table 3.3.1, column 10). The former dominate lavas of the small shield volcanoes that erupted in the main period of volcanism around 25 Ma, and are characterised by sodic compositions. The undersaturated group has a wider range of differentiates and includes K-rich types, many containing upper-mantle inclusions. The undersaturated rocks are found in the basal and uppermost lavas of the main flow sequences. Duaringa province has a similar spread in hypersthene- and quartz-normative lavas to the Monto province but without much development of nepheline-normative rocks.

3.4 Southern Queensland and Tweed Volcano

3.4.1 Introduction

Volcanic provinces in southern Queensland, including all of Tweed volcano, are dominated by large central-type volcanoes made up of tholeiitic, transitional, and mildly alkaline basalts, ranging through to peraluminous and peralkaline rhyolites (Fig. 3.4.1).

There was intermittent, short-lived, dominantly mafic, lava-field type volcanism in the vicinity of Bundaberg and to the west in the area of Gin Gin, Childers, and Coalstoun Lakes, from 60 Ma to 0.6 Ma.

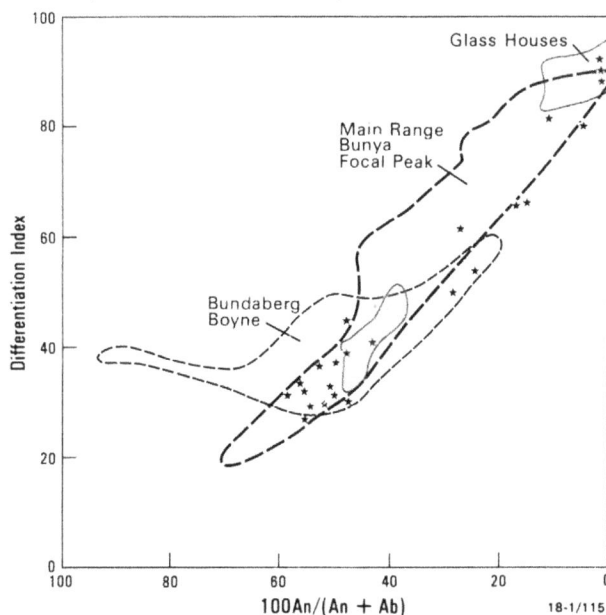

Figure 3.4.2 Compositional fields for south-Queensland and Tweed provinces using the plotting procedure outlined in Section 1.1.4. Stars refer to Tweed samples.

These lava-field provinces contrast with most other east-Australian Cainozoic provinces, having formed to the east of both the coastal escarpment and the Great Divide. In addition, they include, together with the north-Queensland and Newer Volcanics provinces in western Victoria and South Australia, some of the youngest volcanic activity in eastern Australia. Generalised compositional fields for the southeastern Queensland and Tweed volcano provinces are shown on Figures 3.4.2–3, and selected chemical analyses are given in Table 3.4.1.

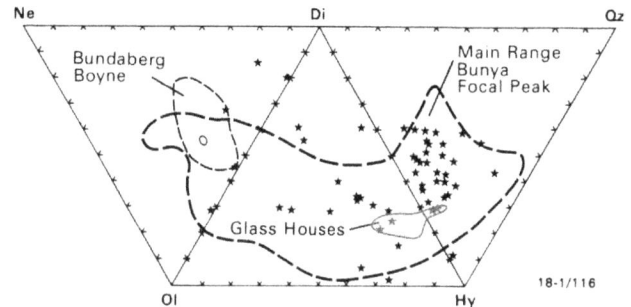

Figure 3.4.3 Compositional fields for south-Queensland and Tweed rocks projected in the system Ne-Ol-Di-Hy-Qz. Stars refer to Tweed samples.

3.4.2 Bundaberg and Boyne

Six periods of short-lived mafic volcanism from about 60 Ma through to the late Pleistocene are recorded in the southern part of the Bundaberg and the northern

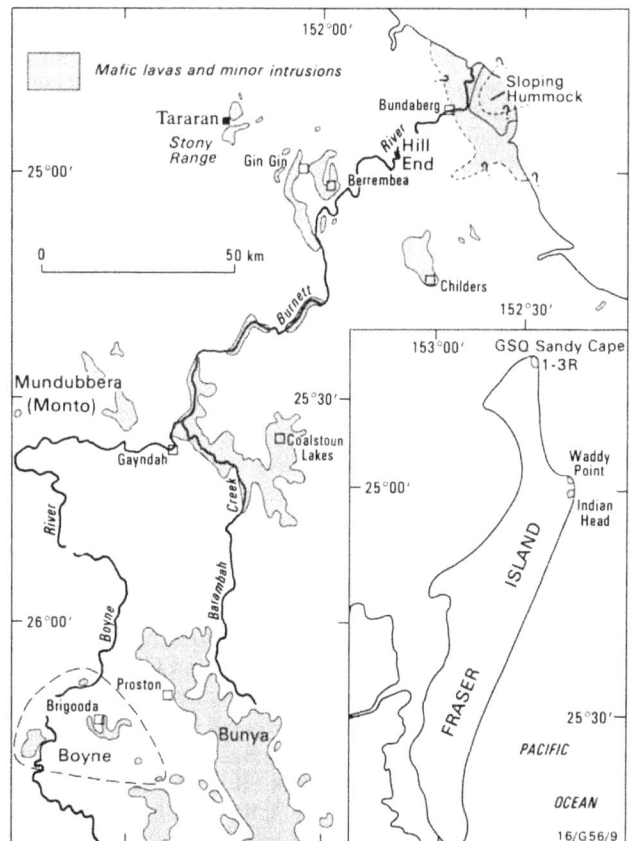

Figure 3.4.4 Distribution of volcanic rocks in the Bundaberg and Boyne provinces of southern Queensland (see Fig. 3.4.1). Dashed line in Bundaberg area represents subsurface extent of Pemberton Grange Basalt.

Table 3.4.1 Selected chemical analyses for south-Queensland and Tweed-volcano provinces

	1	2	3	4	5	6	7	8	9	10
SiO_2	40.67	48.41	72.46	52.01	42.83	66.28	47.01	48.94	48.85	74.10
TiO_2	2.80	1.82	0.15	1.67	2.53	0.45	3.05	2.17	2.69	0.06
Al_2O_3	13.33	15.24	11.46	14.75	13.71	13.83	13.82	16.41	15.01	12.11
Fe_2O_3	5.46	2.60	2.02	1.84	4.07	3.24	4.03	6.69	8.32	0.31
FeO	8.16	7.97	2.03	7.91	9.43	1.87	7.43	4.29	3.30	0.64
MnO	0.19	0.14	0.13	0.13	0.18	0.13	0.16	0.13	0.15	0.01
MgO	7.51	6.91		7.99	8.35	0.24	8.35	5.34	5.83	0.03
CaO	9.84	7.51	0.12	7.53	8.68	1.68	8.66	7.15	6.67	0.60
Na_2O	5.29	5.45	5.65	3.33	3.60	3.74	3.12	4.21	3.78	2.80
K_2O	2.08	1.68	4.71	0.70	2.21	5.30	1.30	1.44	1.52	5.49
P_2O_5	1.48	0.89	0.37	0.33	0.92	0.03	0.54	0.55	0.67	
S	0.06	<0.02								
H_2O+	1.59	0.46	0.01[1]	2.28[1]	3.14[2]	2.67[2]	2.04[1]	2.01[1]	2.32	3.34
H_2O-	0.84	0.30							0.88	0.34
CO_2	0.74	0.15								
Total	100.04	99.53	99.11	100.47	99.65	99.46	99.51	99.33	99.99	99.83
Ba	440	445	25		267	971		323	370	3
Rb	6.5	26.5	365	16	32	106	22	23	21	475.7
Sr	1500	1100	1	402	1120	76	538	430	504	1.7
Pb	5	3	47		4	17		3.2		
Th	11	6	38		5.0	10.6		2	1.6	44
U	2	1							0.62	10.8
Zr	308	148	2245	148	293	1044	189	213	233	124
Nb	94	52	315	11	63	30	33	23	24	32
Y	23	17	220	21	19	49	33	31	29	104
La	72	34							21	21
Ce	152	69	380		90	103		41	49	56
V	165	108			151	6		161	141	1.2
Cr	66	132			207	1		24	67	
Ni	79	123	4					61	67	1
Cu	42	91							34	11

1. Nephelinite from Bundaberg-Boyne province (J. Knutson & B.W. Chappell, unpublished data, sample EAV301).
2. Ne-hawaiite from Bundaberg-Boyne province (J. Knutson & B.W. Chappell, unpublished data, sample EAV271).
3. Peralkaline rhyolite from the Glass Houses province (Ewart et al., 1980, sample 38672).
4. Qz-tholeiitic basalt from the Glass Houses province (Ewart et al., 1980, sample 38687).
5. Basanite from Main Range province (Grenfell, 1984, sample TBC02).
6. Trachyte from Main Range province (Grenfell, 1984, sample RCS01).
7. Hawaiite from Bunya Mountains (Ewart et al., 1980, sample 38784).
8. Hawaiite from Focal Peak province (Ewart, 1982a, sample 33037).
9. Ol-tholeiitic basalt from Tweed volcano (Ewart et al., 1977, sample OR21).
10. Rhyolite from Tweed volcano (Ewart et al., 1977, sample Q71).

[1] Total H_2O.
[2] Loss on ignition.

part of the Maryborough 1:250 000 sheet areas (Fig. 3.4.4). This volcanism is represented now by outcrops of restricted extent and thickness. The composition of the basalts changed with successive periods of eruption from tholeiitic in the Palaeocene to early Eocene, through moderately to strongly undersaturated compositions in the late Miocene to Pliocene, to a range of magma compositions during the Pleistocene.

Lavas in the Bundaberg, Gin Gin, and Childers area were extruded from a series of vents or fissures along and adjacent to the contact between the late Triassic to middle Cretaceous Maryborough Basin and the Permian to early Triassic Gympie Block (Robertson, 1985).

The oldest volcanic activity in the area is represented by the subsurface Pemberton Grange Basalt at the base of the Bundaberg Trough and is overlain by early Eocene sediments. Lateritised Gin Gin Basalt exposed mainly to the east and south of Gin Gin in a shallow basin between the Bullyard and Electra faults, may be of the same age (A.D. Robertson, 1979, 1985). Interbedded basalt in sediments considered by Ellis & Whitaker (1976) to be equivalent to sediments in the Lowmead Graben, crops out to the north of Gin Gin.

Volcanic rocks have been recorded also in drill core at a depth between 421 and 592 m in GSQ Sandy Cape 1–3R from the northern end of Fraser Island. Basaltic dykes similar in composition to the volcanic rocks in GSQ Sandy Cape 1–3R cut trachytes of middle Oligocene age (31 Ma) at Waddy Point. Grimes (1982) assigned a tentative age of Oligocene to early Miocene to the volcanic rocks in GSQ Sandy Cape 1–3R, on the basis of the Waddy Point age and ages determined from fossils in sediments above and below the basalts. Lateritised volcanic rocks in the vicinity of Childers and Goodwood have been assigned a similar age (Day et al., 1983; Robertson, 1985).

Extrusion of nepheline-rich lavas (Table 3.4.1, column 1) during the late Miocene to late Pliocene was controlled structurally by both the Electra and Bullyard faults. The nephelinite at Hill End (Robertson & Murray, 1978) to the east of the Bullyard Fault appears to be an exception. Phreatomagmatic activity during the early stages of eruption of the Tararan melanephelinite produced a well-developed pyroclastic cone. Pyroclastic deposits are associated also with the Maroondan melanephelinite that intrudes the Gin Gin Basalt west of the Bullyard Fault and the nephelinite at Hill End (recorded only from drill holes; Robertson & Murray, 1978).

The Tararan melanephelinite and the nepheline-rich lavas of Stony Range carry abundant upper-mantle and lower-crustal xenoliths. The Tararan lavas contain xenoliths of pyroxenite, hornblendite, and granulite, along with megacrysts of anorthoclase, kaersutite, garnet, spinel, clinopyroxene, and mica. The Stony Range lavas contain xenoliths of lherzolite, pyroxenite, and granulite as well as megacrysts of clinopyroxene, ilmenite, and spinel. Anorthoclase from Tararan gave a K-Ar age of 3.0–2.7 Ma, and an age of 5.1 Ma was determined on a whole-rock sample from the Hill End nephelinite.

Three Pleistocene volcanic eruptions have been recorded from near Bundaberg, from Berrembea southeast of Gin Gin, and from Coalstoun Lakes. The Hummock Basalt (1.1–0.9 Ma; Wellman, 1978) was extruded from a vent at Sloping Hummock east of Bundaberg in the now inactive Bundaberg Trough. The basalt crops out over an area of about 215 km² (Robertson, 1982) between the Burnett and Elliott Rivers on the mainland, and extends for an unknown distance offshore. The maximum thickness recorded for the Hummock Basalt is 55 m in a water bore about 1 km south of the vent. The Berrembea Basalt to the southeast of Gin Gin was erupted from an ill-defined fissure vent. The different flows of restricted areal extent terminate in well-defined flow edges overlying Gin Gin Basalt. An age of late Pleistocene was assigned tentatively to this basalt by Robertson (1985).

The youngest isotopically dated volcanic event recorded in the region is centred on Coalstoun Lakes where the Barambah Basalt was extruded from at least three centres (Table 3.4.1, column 2). Basalt flowing from these centres spread out over a valley 3–5 km wide to the south of the vents, following the valley down to the confluence with Barambah Creek and hence into the Burnett River as far as the old Paradise Goldfield, a distance of about 100 km. At least one lava tunnel is known, 1.6 km southwest of Dundurrah (Willmott, 1976). The Mount Le Brun cone rises to a height of 330 m, contains two unbreached crater lakes, and is largely made up of basaltic agglomerate, spheroidal volcanic bombs (some containing crustal xenoliths), and vesicular and pahoehoe basalts. Bores in the vicinity of Mount Le Brun penetrated basalts to a depth of 75 m. A K-Ar age for the Barambah Basalt is 0.6 Ma (Wellman, 1978).

A small region (1000 km²) of scattered vents and distinctive undersaturated basalts is found near the Boyne River, south of its junction with the Burnett River and west of Proston. These volcanic rocks are younger than the more extensive and widely lateritised ones of the northern Main Range and Bunya Mountains

that form a more saturated series to the east and southeast (30–22 Ma; Ewart et al., 1980; Ewart & Grenfell, 1985; Griffin et al., 1987). The Boyne volcanic rocks largely intrude and overlie Permian-Triassic intrusions of the granitic Boondooma Igneous Complex (Day et al., 1983).

Most of the volcanic rocks were extruded during two separate episodes from a single vent, and flowed for limited distances along drainage systems incised in a deeply weathered landsurface. The rocks range from K-rich basalts through K-rich ne-hawaiites to K-rich ne-mugearite. The less fractionated ones are porphyritic in olivine and pyroxene (including a coarse ankaramitic basalt) and the more fractionated ones carry megacrysts of amphibole, and pyroxene or anorthoclase (or both). The suite is termed the Brigooda Basalt (Robertson et al., 1985), most of which is probably early Miocene in age (amphibole K-Ar age of 18 Ma). Explosive outbursts in the late Pleistocene produced several breccia deposits (anorthoclase K-Ar age 0.45 ± 0.04 to 0.381 ± 0.025 Ma). The breccias are apparently sources for gem garnet, derived as xenocrysts from garnet pyroxenites (Griffin et al., 1987).

A composite ne-hawaiite diatreme at Ballogie of early Miocene age (anorthoclase megacryst K-Ar age 16.0 Ma) also intruded granitic basement and has been interpreted as outgassing above a rising diapir (Hollis et al., 1983). The extruded lava includes ne-hawaiite, and typically is less fractionated and more sodic than the majority of the Brigooda Basalt. The breccia phase is rich in megacrysts and upper-mantle xenoliths, and a fissure vent of basanite/ne-hawaiite southeast of Ballogie contains upper-mantle xenoliths and pyroxene and anorthoclase megacrysts. A similar rock also forms low isolated knobs adjacent to Boyne River near Corrunovan Homestead. Small plugs and flow remnants near Nanango contain garnet and anorthoclase megacrysts.

3.4.3 Glass Houses

The Glass Houses (or Glasshouse Mountains) are a group of dome-shaped hills and conical and spine-like peaks of fine-grained igneous rocks that rise abruptly from the coastal plain of southeastern Queensland 50 to 66 km north of Brisbane (Figs. 3.4.1, 5-6). They are intrusions — and in part extrusions — of late Oligocene peralkaline rhyolites and trachytes (Table 3.4.1, column 3). Other hills of peralkaline rhyolite similar in appearance and rock type to the Glass Houses are present to the north in the Noosa Heads area. The rhyolite and trachyte intrude Mesozoic sandstones and volcanic rocks in the Glass Houses and Mount Coolum areas, and early Triassic phyllite northwest of Cooroy.

Basaltic flows in the Maleny-Mapleton district (Table 3.4.1, column 4) between the Glass Houses and Cooroy are the same age, or a little older, than the intrusions, and preserve a late Oligocene surface that has a gentle easterly or northeasterly slope at elevations between 190 and 240 m above sea-level. Most of the Glass Houses rise above this level, implying that the upper parts of the more prominent peaks were probably extrusive and represent the upper parts of cumulodomes.

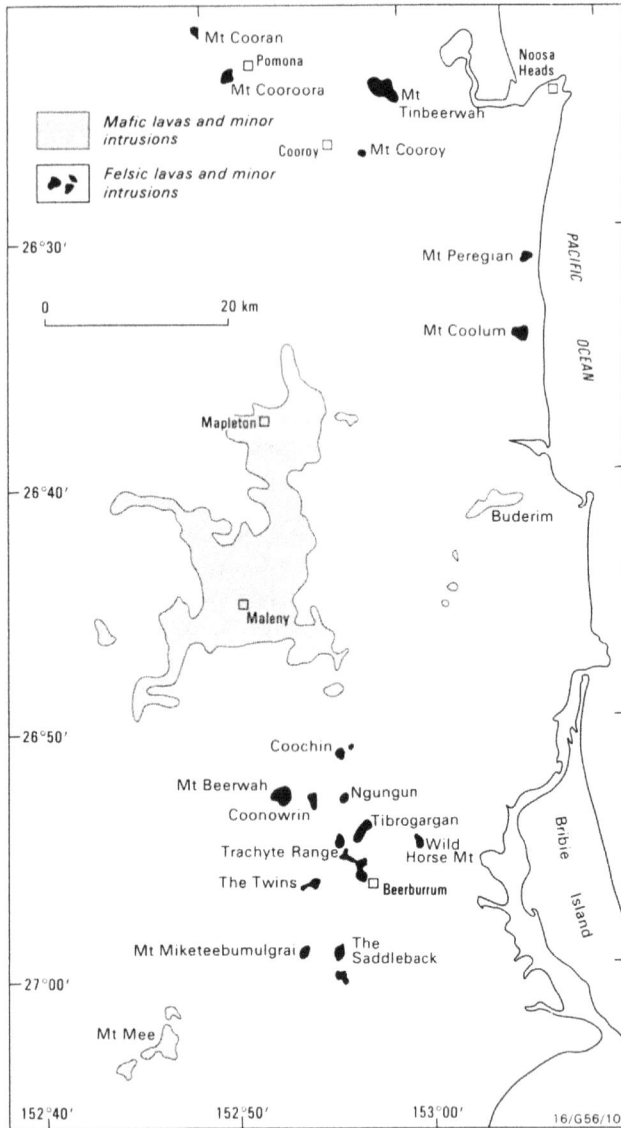

Figure 3.4.5 Distribution of volcanic and intrusive rocks in the Glass Houses province of southern Queensland (see Fig. 3.4.1).

Lower masses of peralkaline rhyolite and trachyte are interpreted as sills, dykes, and laccoliths or other dome-shaped intrusions. No associated pyroclastic rocks have been recognised. They would have been above the late Oligocene surface, if any were present originally.

The main rock types are peralkaline rhyolite, porphyritic anorthoclase trachyte, and peralkaline qz-trachyte. Most of the high peaks and several low hills and ridges are composed of peralkaline rhyolite. The rocks typically are pale grey, fine-grained, and are either aphyric or have only a few, scattered, small phenocrysts of anorthoclase-sanidine. Sodic amphibole is visible in hand specimens of the coarser, spotted type of rhyolite from Mount Ngungun (Bryan & Stevens, 1973), from the Trachyte Range, and from Mount Tibrogargan. Groundmass sodic amphibole belongs to the fluor-bearing riebeckite-arfvedsonite series, aegirine is subordinate to amphibole, and aenigmatite is distributed erratically (Ewart & Grenfell, 1985). The

rock of Mount Beerwah, the highest peak (556 m), is a peralkaline qz-trachyte.

Dome-shaped Mounts Beerburrum and Miketee-bumulgrai are composed of almost identical, porphyritic, anorthoclase trachyte which is dark blue-grey where fresh, but is readily oxidised to brown. Additional phenocrysts and microphenocrysts include ferroaugite, fayalite pseudomorphed by iddingsite, and titano-magnetite. Other trachytes, presumably dykes or the upper parts of sills or laccoliths, are found at the southern margin of the Glass Houses. The only mafic intrusions of similar age are alkali basalts in the Trachyte Range area.

Cliff exposures of near-vertical columnar jointing are obvious in most of the large peralkaline-rhyolite and trachyte masses, — for example, at Coonowrin, Beerwah, and Ngungun, and there may be only limited areas of gently dipping columns, mostly in the smaller dykes.

Several of the Glass Houses rocks and Mount Coolum have been dated by the K-Ar method at 27 Ma (peralkaline trachyte of Mount Beerwah), 26, and 25 Ma (peralkaline rhyolite and trachyte of Mount Beerburrum). The basalt of Trachyte Range also falls within this age range.

The northern plugs of peralkaline rhyolite in the Coolum-Cooroy district near Noosa Heads are similar in hand specimen. Mount Cooroy, a conical mountain between Mount Coolum and Cooroy, is composed of relatively coarse-grained rocks reported by different authors as monzonite, mangerite, and syenite. A K-Ar age from its base is similar to that of the Glass Houses and Mount Coolum (Noon, 1972). The existence nearby of small masses of syenite and arfvedsonite granite at lower levels is evidence that the Mount Cooroy intrusion is related to the Glass Houses alkali rhyolite and trachyte. Biotite-bearing rhyolite in the same district (Mounts Tinbeerwah and Peregian) also may be Tertiary (Ewart & Grenfell, 1985) but no age dating has been carried out.

Basaltic lava flows to the north of the Glass Houses form a dissected tableland in the Maleny-Mapleton district (Fig. 3.4.5), unconformably overlying Mesozoic sedimentary and volcanic rocks and Palaeozoic meta-morphic rocks. The maximum thickness of the basalts is 180 m where they filled a pre-basalt valley, and elsewhere the thickness is up to 60 m (Evans, 1976; Ewart & Grenfell, 1985). Individual flows are generally less than 10 m thick and those exceeding 15 m have columnar jointing. Red palaeosols between flows are prominent, corresponding to long time intervals. Whole-rock ages range from 34 to 28 Ma (Evans, 1976). An additional date of 25 Ma was obtained on anorthoclase megacrysts in one of the stratigraphically highest flows (Ewart, 1982a). Flows are mostly massive, and contain scarce phenocrysts, dominantly of plagioclase. The lavas are oversaturated or near-saturated, and have been classed as hawaiites (Ewart et al., 1980). The basalts may have been erupted from a north-south fissure system and flowed eastwards to Buderim and beyond (Ferguson, 1969). Another poss-ible source, suggested by Willmott (1983), is in the Glass Houses where two small basaltic intrusions of similar age have been found.

3.4.4 Main Range

The Tertiary volcanic rocks of the Main Range, including the Bunya Mountains, form a continuous belt extending north-northwest from the State border for 220 km, parallel to and close to the Great Divide (Figs. 3.4.1,7). The belt extends north beyond the Bunya Mountains, and south to near Tooloom in northeastern New South Wales. The volcanic rocks cover an area of about 4900 km², and their present volume is estimated at about 1000 km³. Wellman (1986) showed that Main Range coincides with a gravity high thought to have been caused by emplacement of an intrusive complex (Section 1.4.9; Fig. 1.4.10B).

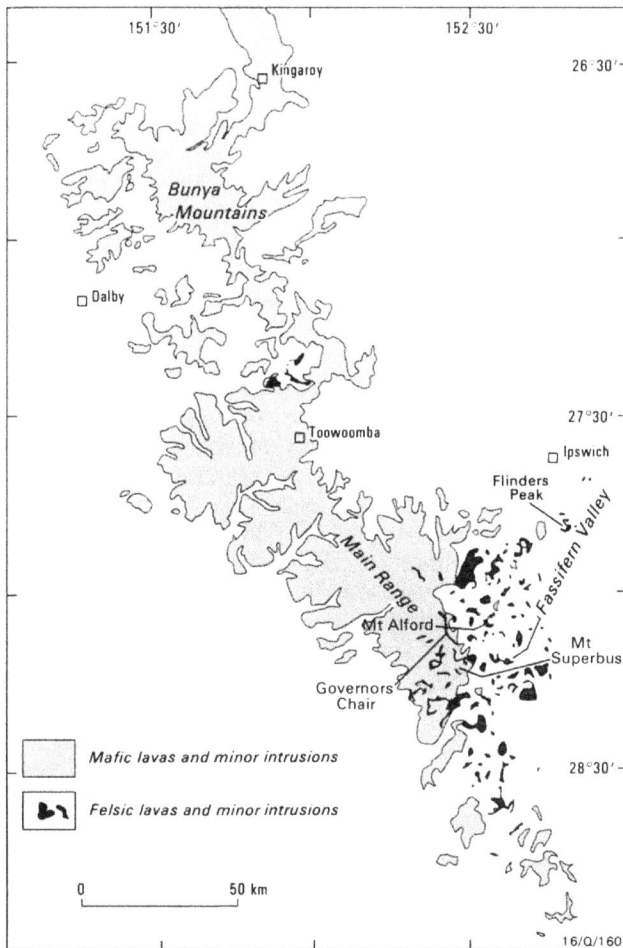

Figure 3.4.7 Distribution of volcanic and intrusive rocks in the Main Range province of southern Queensland (see Fig. 3.4.1).

The belt may be divided physiographically into several parts: (1) the southern Main Range which has high peaks, is strongly dissected on the western slopes, and is fronted by the coastal escarpment; (2) the northern Main Range, plateau country with mesas around Toowoomba, also bounded on the east by the coastal escarpment; (3) the Bunya Mountains, an elliptical, dissected, dome-like mass; and (4) lower, undulating, lateritised basalt country to the north extending through Kingaroy to Proston. The basement rocks are mainly near-horizontal Mesozoic sedimentary rocks south of Toowoomba, and granites and Palaeozoic metamorphic rocks farther north. Investigations of

Main Range volcanic geology were carried out by Stevens (1965, 1969), Russell (1965), and Grenfell (1984), and of the Bunya Mountains by Bryan (1983). A summary of the geology and geochemistry of both areas was presented by Ewart & Grenfell (1985)

Two volcanic formations are recognised in the southern Main Range — namely, the Governors Chair Volcanics (lower) and the Superbus Basalt (upper). These account for a maximum total thickness of nearly 900 m of lavas with some pyroclastic rocks and minor sediments. The formations are of nearly equal thickness, and both are distributed extensively in the southern Main Range. Correlative volcanic formations in the northern Main Range are the Meringandan Volcanics and the Toowoomba Basalt, respectively.

K-Ar dates for the lower formations are in the range 26–24 Ma (Grenfell, 1984; Webb et al., 1967). The ages of upper-formation rocks correspond to a more protracted history of at least 5 Ma, although most of these were erupted between 24 and 22 Ma. Relatively young ages of 22 Ma correspond to sporadic, inclusion-bearing lavas, in accordance with their stratigraphic position at or near the top of the succession.

The Governors Chair Volcanics have a complete gradation across a compositional spectrum extending from mildly alkaline basalt through hawaiite, mugearite, benmoreite, metaluminous trachyte, and peralkaline trachyte, to peralkaline rhyolite. Lavas dominate, but basaltic agglomerate, trachytic tuff, and rare ignimbrite are present. Mafic, benmoreite plus melatrachyte, and leucotrachyte plus peralkaline-rhyolite lava groups are estimated to be present in the volumetric proportions 97:2:1, respectively (Grenfell, 1984). The formation contains conspicuous metaluminous and peralkaline trachyte horizons. The Tregony Conglomerate Member is a predominantly fluviatile and lacustrine deposit, and represents an intercalated, discontinuous unit that crops out over about 60 km² in the southern Main Range.

Lava successions within the Governors Chair Volcanics are thickest in the eastern escarpment of the southern Main Range (*Fig. 3.4.8*). The formation thins westwards from the escarpment. Associated intrusions of predominantly felsic-intermediate composition are scattered through the Fassifern Valley in an elliptical 90 by 30 km belt to the east of, and contiguous with, the Main Range pile (Stevens, 1960; Fig. 3.4.7). The volcanic edifice implied for the Governors Chair Volcanics is a dominantly basaltic shield volcano having an elliptical base at least 50 km across, built over rhyolitic and trachytic plug-domes. Subsequent erosion removed the eastern flank of this constructional landform and revealed numerous plugs, dykes, and domes of the complex volcanic plumbing system.

The Meringandan Volcanics in the northern Main Range are correlated with the Governors Chair Volcanics. This formation consists of hawaiite, of which some flows are transitional to mugearite, and melanocratic trachyte. The total thickness recorded for the Meringandan Volcanics is 75 m (Stevens, 1969).

The Superbus Basalt conformably overlies the Governors Chair Volcanics in the southern Main Range, cropping out over 1600 km². The formation is exclusively mafic. Sparse phenocrysts include labradorite-andesine, olivine, augite-salite, and titano-magnetite. Hawaiite is the predominant rock type. Alkali basalt and mugearite are also present, together with sporadic normatively-undersaturated lavas including *ne*-benmoreite and the Inglewood leucite basanite. These nepheline-normative rocks typically contain high-pressure inclusions and are found at or near the top of the volcanic succession. The leucite basanite has affinities with other leucite-bearing lavas in eastern Australia.

The Toowoomba Basalt conformably overlies the Meringandan Volcanics in the Cooby Creek area, and it passes laterally into the Superbus Basalt in the southern Main Range. The formation is exposed over more than 2700 km² in the northern sector of the Main Range province. Hawaiite is the main lava type, but there are sporadic occurrences of inclusion-bearing *ne*-hawaiite at or near the top of the succession. Intercalated tuff and lapilli tuff in the Toowoomba area contain basaltic material together with quartz fragments derived from underlying Mesozoic sandstones.

Scattered necks and dykes in the Mesozoic sandstone to the east, and to the north and northeast of Toowoomba (Protheroe, 1981), apparently represent the feeders to the Toowoomba Basalt. The correlative Superbus Basalt, on the other hand, was fed chiefly by fissures that are identified as a series of dykes localised in a zone extending for about 30 km close to the present escarpment.

Analysed Main Range mafic lavas range from strongly alkaline to weakly alkaline (Ewart & Grenfell, 1985), and a spectrum of normative compositions exists from silica-undersaturated to oversaturated (Table 3.4.1, columns 5–6). Felsic types are invariably quartz normative. Both of the upper formations (Superbus and Toowoomba) in the Main Range group are characterised by the presence of two series, distinct both chemically and geochronologically. The earlier series (24–22 Ma) is composed predominantly of hawaiitic lavas, and there is subordinate transitional basalt, alkali basalt, and mugearite. Later volcanic rocks (21–18 Ma) have a moderately to strongly alkaline character (normative nepheline reaches 14.5 percent), and Mg-ratios in the range 74–51. They are nephelinite, basanite, leucite basanite, hawaiite (predominant), *ne*-hawaiite, and *ne*-benmoreite. There is an overall trend in Main Range volcanism of progressively more alkaline character in the younger rocks.

Nepheline-normative rocks containing high-pressure inclusions are sporadic throughout the Main Range province (Ewart & Grenfell, 1985). The host lavas range in composition from nephelinite to *ne*-benmoreite, and the inclusions locally may be very abundant. Megacrysts include clinopyroxene, orthopyroxene, anorthoclase, kaersutite, biotite, and Mg-ilmenite. The xenolith suites are dominated by spinel lherzolites, but include pyroxenites (some amphibole-bearing) and lower-crustal granulites.

The Mount Alford Ring Complex (Stevens, 1959, 1962) forms a group of hills in the Fassifern Valley. It is 3–4 km in diameter and consists of a central plug of microdiorite and subordinate granophyre cut by dacite and mafic, rhyolitic and trachytic ring dykes. The rhyolite dykes gave K-Ar dates of 25 to 24 Ma (Webb et al., 1967; Ewart, 1982a). The topographically prominent ring dyke of Minto Crags, south of Mount Alford, intrudes horizontal or gently folded sedimentary rocks and is peralkaline rhyolite, like some of the dykes cutting the ring complex. The Flinders Peak area at the northern end of the Fassifern intrusive belt is a prominent volcanic centre of sills, dykes, and plugs consisting mainly of trachyte with subordinate rhyolite and mafic dykes (Ewart & Grenfell, 1985). K-Ar age determinations for three trachytes range from 27 to 25 Ma (Webb et al., 1967; Ewart, 1982a).

The Bunya Mountains appear as a northern extension of the Main Range, but represent the eroded remnants of a shield volcano (Bryan, 1969b, 1983; Ewart et al., 1980; Ewart & Grenfell, 1985). A maximum diameter of 60 km for the shield has been estimated and the maximum total thickness for the lava pile probably exceeded 600 m. Numerous lava flows have generally outward radial dips away from the summit region. These rocks are broadly contemporaneous with rocks of the Toowoomba Basalt (K-Ar ages mostly in the range 25–23 Ma). A lava below a laterite profile northeast of the Bunya Mountains dated at 29 Ma may be from an older volcanic episode.

The Bunya Mountains are characterised by a scarcity of rock types more evolved than hawaiite. The lavas are mainly *ne*-hawaiites and hawaiites (Table 3.4.1, column 7) although two quartz-normative, potassic, tholeiitic basalts are present near the base of the volcanic sequence on the southern flank (Ewart et al., 1980). The most felsic lava found is a glassy trachyte (a minor plug or flow or both) on the southwestern flank of the volcano. Megacrysts of plagioclase, olivine, aluminous augite (including sub-calcic types) and spinel are present in the lowest and highest parts of the volcanic succession (Ewart & Grenfell, 1985).

3.4.5   Brisbane

East of the Main Range are basaltic lava flows of Palaeocene and Eocene age that are interbedded with, or rest on, lacustrine and fluvial sediments in small structural basins near Ipswich and in the southern and northern suburbs of Brisbane. The basalts are highly weathered and do not form prominent outcrops.

Palaeocene basalts southwest of Ipswich (Webb et al., 1967; Green & Stevens, 1975) appear to overlie Tertiary sediments. Two analysed rocks are *qz*-tholeiitic basalts. Basalts of Eocene age to the east of Ipswich (48–46 Ma) are interbedded with limestones and other sediments of the Silkstone Formation in two basins. The major development of basalt (up to 150 m) is in the lower part of the formation (Staines, 1960). Basalts interbedded with dolomitic rocks south of Ipswich may be of the same age.

Basalts in the southern suburbs of Brisbane are found near the base and top of the Corinda Formation (Houston, 1967). Ages reported for the two basalt members are 55 Ma (Palaeocene) and 47 Ma (Eocene), but low K₂O in one casts some doubt on the accuracy of the determination.

Other basalts, interbedded with calcareous and other sediments in the Petrie Basin, north of Brisbane, and basalts along the coastline of Moreton Bay, are probably within the same age range as those mentioned above. Again, two horizons of basalt, separated by sediments, have been intersected in several drill holes. A thick sequence of basalts beneath Wellington Point (up to 300 m) has been correlated with the whole of the Corinda Formation, and may be due to downwarping along an adjacent fault during extrusion (Houston, 1967). Volcanic rocks in the Brisbane area seem to be largely of *ol*- and *qz*-tholeiitic basalt, but there are few analyses and alkaline types may be present.

3.4.6 Focal Peak

McElroy (1962) postulated the presence of a volcanic shield in the Mount Barney area (Fig. 3.4.9), but not until 1974 was supporting evidence reported for the existence of a shield volcano and for a relationship with the earlier eruptive rocks of the Lamington Group, Albert Basalt, and the Mount Gillies Volcanics (Ross, 1974). This volcano in the Mount Barney area was named the Focal Peak shield volcano (Ross, 1974). Its age is about 23 Ma (Webb et al., 1967; Wellman & McDougall, 1974a; Ross, 1977) based on ages of the Mount Gillies Volcanics and the Albert Basalt.

The centre of the shield volcano is about 60 km west-northwest of Mount Warning, the centre of Tweed volcano which is slightly younger than the Focal Peak shield. The Mount Barney central complex is the

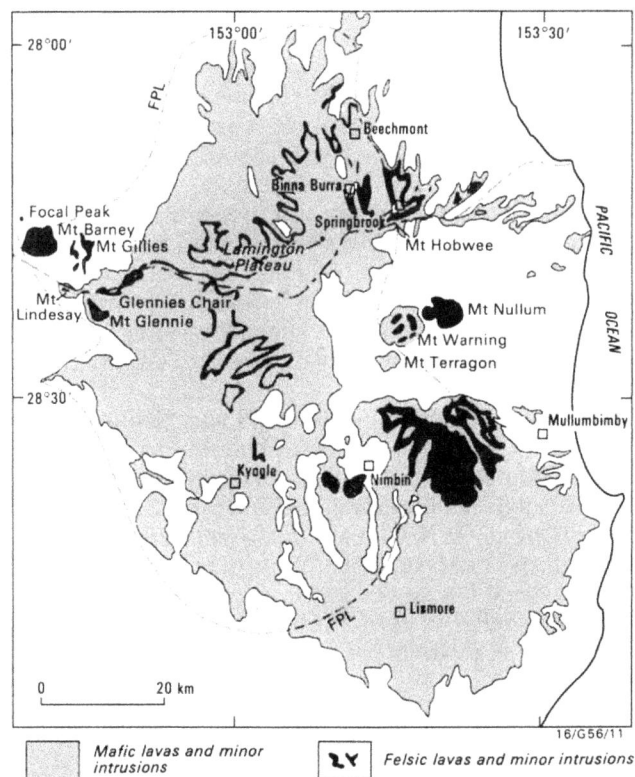

Figure 3.4.9 Distribution of volcanic and intrusive rocks in the Focal Peak and Tweed provinces of southern Queensland and northern New South Wales (see Fig. 3.4.1). Dashed line FPL represents possible former limit of basaltic lavas from Focal Peak volcano.

eruptive centre of the Focal Peak shield volcano, within which a central caldera is thought to exist at Focal Peak (Stephenson 1956, 1959; Ross, 1974, 1977). The complex corresponds with a zone of basement uplift (Section 1.4.9; Fig. 1.4.10B).

The eruptive sequence on the eastern flank of Focal Peak volcano consists of the Albert Basalt and the overlying Mount Gillies Volcanics, which are equivalent to the Hillview Rhyolite and the Mount Lindesay Rhyolite (McTaggart, 1962; Ross 1974, 1977; Ewart et al., 1987). The Chinghee Conglomerate in part overlies the Mount Gillies Volcanics, or is interbedded with it. Formations in the area between Mount Lindesay and Richmond Gap have low dips (less than 5°) to the east, reflecting the sloping flank of the shield.

The Albert Basalt on the eastern flank of the volcano is thickest (440 m) in the Mount Lindesay area, and it thins to the east where the rocks underlie the Beechmont Basalt of Tweed volcano. It has been correlated with the Kyogle Basalt (Duggan & Mason, 1978) of the Kyogle-Nimbin area, and it probably forms most of the basalt country between Kyogle and Mount Lindesay, and underlies the Lismore Basalt of Tweed volcano east of Nimbin.

The Albert Basalt consists of numerous flows averaging about 10 m in thickness, and associated fine-grained, agglomeratic, and doleritic intrusions. It is generally uniform in hand specimen and few flows are noticeably porphyritic, although microphenocryst olivine or plagioclase (or both) are common.

Peridotite nodules and megacrysts (clinopyroxene and orthopyroxene) are rare, but upper-mantle lherzolite xenoliths have been described from hawaiite in the Kyogle Basalt (Wilkinson & Binns, 1969). The lavas and fine-grained intrusions are mostly alkaline, and hawaiite (Table 3.4.1, column 8) predominates over alkali basalt, mugearite, basanite, *ol*-tholeiitic basalt, and icelandite. There are in addition to the postulated caldera at Focal Peak about nine other possible mafic plugs for the Albert Basalt in the Mount Gillies area (Ross, 1977). These feeders are commonly scoriaceous agglomerate, less commonly dolerite and basalt.

Rhyolitic lavas in the Hillview and Mount Lindesay areas are commonly autobrecciated and associated with pitchstone. Similar pitchstones are present in the Homeleigh Agglomerate northeast of Kyogle. The Homeleigh Agglomerate consists of blocks up to 1 m in diameter in a pyroclastic matrix of rounded pumice, felsic and altered mafic volcanic rocks, and crystal fragments (Duggan & Mason, 1978). It is classed as a member unit of the Kyogle Basalt and is overlain by the Georgica Rhyolite member and alkaline basaltic flows of the Kyogle Basalt. Basalt flows have been recorded also near the top of the Mount Gillies Volcanics near Chinghee Creek.

Mount Gillies and Campbells Folly on the margin of the Mount Barney central complex are composed largely of intrusive and extrusive flow-banded rhyolite, rhyolitic tuff, and agglomerate (Ross, 1985a). Stephenson (1956, 1959) regarded the rhyolites of Mount Gillies as part of the Mount Barney central complex, although he thought they might be somewhat younger.

Rhyolitic volcanoes in the Mount Gillies, Mount Glennie, and Levers Plateau area are commonly fissure vents and less commonly plugs (Ross, 1977). Plugs are found at Glennies Chair, on the west of Levers Plateau and near Mount Gillies. These are believed to have been only minor feeders for the rhyolites of the Mount Gillies Volcanics, and the fissure vents were the main sources. Fissure vents are prevalent in the Mount Gillies and Campbells Folly area where there are between 700 and 2800 rhyolitic dykes, together with related extrusions.

There is a wide range of rhyolite types, the majority of which are devitrified and silicified (Ross, 1985b). Pitchstones are less common. All of the rhyolites are metaluminous and noticeably porphyritic.

The Chinghee Creek Conglomerate overlies the Mount Gillies Volcanics in the type area 2.5 km south of Chinghee Creek village, but overlaps it to the southeast, and is found in the Tweed Range without the underlying rhyolites, according to mapping by McTaggart (1962). Ross (1977) reported a thickness of up to 130 m in the Chinghee Creek area where a gorge — filled with Chinghee Conglomerate — cuts through the underlying rhyolite of the Mount Gillies Volcanics into the Albert Basalt. The formation consists mainly of conglomerate and coarse, feldspathic, cross-bedded sandstones which are either uncemented or poorly cemented. Boulders and cobbles in the conglomerates are up to 60 cm in diameter, and are of rhyolite (probably from the Mount Gillies Volcanics), granophyre from Mount Barney, Mesozoic sedimentary rocks, some basalt, and rare riebeckite-bearing peralkaline rhyolite. The well-rounded boulders and cobbles in the conglomerates and the cross bedding of the sandstones may be evidence that it is a fluvial deposit, at least in the type area.

The Mount Barney central complex (Stephenson, 1956, 1959) consists of a range of igneous rocks, mainly intrusive, forming a mountainous region 20 by 15 km. Significant doming and upfaulting along ring faults have taken place in the complex, forming the Barney dome.

An important part of the complex is the Mount Barney Granophyre which is intrusive on its eastern side into steeply dipping Carboniferous strata. The granophyre and Carboniferous and Mesozoic strata to the east are surrounded by the oval-shaped Barney ring fault which is offset to the east of the centre of the dome. Stephenson (1956) estimated upfaulting of at least 2400 m on a basis of the offsetting of the Barrier Granophyre ring dyke by the ring fault. The only radiometric age for the Mount Barney Granophyre is 24 Ma. Elevation and erosion of the Mount Barney Granophyre must have been rapid for the rock to be preserved in the Chinghee Conglomerate underlying lavas from Tweed volcano of age 22 Ma (Webb et al., 1967; Stevens, 1970). The Barrier Granophyre, an incomplete and faulted ring dyke, has intruded Mesozoic strata east of Mount Barney, and has been displaced by the Barney ring fault.

Stephenson (1956) described a complex of different types of dioritic rocks surrounding Focal Peak. Ross (1977) called these gabbro, monzonite, and minor syenite and noted that they graded from one type into the other. The intrusion is here referred to as the Focal Peak pluton.

Trachytic agglomerate on the summit of Focal Peak and intermediate and basaltic volcanic rocks to the northeast belong to the 'central volcanics' or 'vent volcanics' of Stephenson (1956). The basaltic rocks have evidence of thermal metamorphism at their margin, presumably caused by the Focal Peak pluton below.

A zone of inward-dipping doleritic cone sheets (Ballow cone sheets) surrounds the central volcanic rocks and the Focal Peak pluton in a zone about 6.5 km in diameter. The Montserrat Granophyre (or microsyenite according to Ross, 1977) within this zone forms a discontinuous ring of intrusions. The central area may have been a caldera, similar to the summit calderas of Hawaiian shield volcanoes, if the ring of intrusions marks the position of a ring fault along which subsidence took place.

The outer zone of the Mount Barney central complex includes several, large sill-like or laccolithic intrusions of rhyolite and microsyenite. Minor doleritic, trachytic, rhyolitic, and granophyric dykes are also present. The rhyolites of the Mount Gillies and Campbells Folly area may be included here.

3.4.7 Tweed volcano

Introduction

Tweed volcano is a major, 100 km-wide, shield volcano containing the central, high-level, intrusive complex of Mount Warning. The general shape of the shield volcano is recognisable, despite the 20 Ma age of the youngest lavas, except for part of the eastern side of the shield and for the erosion caldera, 30 km in diameter, which separates the central intrusive complex from the lavas on the flanks of the shield (Fig. 3.4.9). The present area covered by the lavas is nearly 4000 km² and their maximum thickness about 900 m (excluding lavas from the Focal Peak volcano — Section 3.4.6). The original height of the shield is estimated to have been 1900 ± 150 m above present sea-level (Solomon, 1964) and the original volume of lava 3200 km³. Wellman (1986) showed that Tweed volcano rests on a basement high considered to have formed by emplacement of the central intrusive complex (Section 1.4.9; Fig. 1.4.10C).

Previous literature, field relations, petrology, and geochemistry were summarised by Ewart et al. (1987). Information on the volcanic formations on the southern (New South Wales) side of the shield volcano is mostly from Duggan & Mason (1978). Volcanism is dated at 24–21 Ma and consists of a generalised volcanic succession of a lower mafic sequence, followed by rhyolitic units, in turn overlain by a younger mafic sequence. A separate peralkaline rhyolite dyke phase is also present.

Mount Warning intrusive complex

The 8 × 5 km central intrusive complex is on an unconformable junction between folded and metamorphosed Carboniferous sedimentary and volcanic rocks and west-dipping Mesozoic strata, beginning with Triassic volcanic rocks, followed by Clarence-Moreton Basin sedimentary rocks.

Figure 3.4.10 Mount Warning showing the central 'trachyandesite' spine surrounded by syenite (upper slopes) and gabbro (lower slopes). An encircling ring-dyke complex of syenite and monzonite forms the prominent ridges on the left and right.

The gabbroic rocks surrounding the 1125 m peak of Mount Warning appear to be slightly older than most of the volcanic sequence. They form a belt around a mass of syenite, which in turn encloses a core of trachyandesite. The intrusive sequence is: early gabbro, laminated gabbro, monzonite, central syenite, trachyandesite (K-mugearite), icelandite dyke-sill phase, multistage ring dyke, and peralkaline rhyolite dykes.

The monzonite was intruded forcibly between the gabbro and hornfelsed country rocks. It extends irregularly around the eastern and southern rims of the complex and outcrops of it form the high peaks of Mount Uki. The central syenite consists of both non-peralkaline and peralkaline rock types intruding and completely surrounded by gabbro. Peralkaline types contain sodic pyroxene, sodic amphibole, and aenigmatite. The strongly jointed porphyritic trachyandesite (K-mugearite), which forms the summit of Mount Warning, represents the core of the volcano (Fig. 3.4.10).

The ring dyke forms an almost complete elliptical, narrow, steep-sided ridge encircling the summit region. Numerous basaltic, doleritic, and icelanditic dykes and sills pervade much of the southern and eastern parts of the complex, except for the central syenite and trachytic ring dyke. The basaltic dykes are cut by peralkaline rhyolite and, to a lesser extent, by trachyte dykes and veins. These represent the youngest recognised intrusive phase and intrude all rock types in the complex, including the youngest ring-dyke phase.

Rhyolitic rocks and the fine-grained Mount Nullum granite to the east of Mount Warning are thought to be members of the intrusive complex. The Mount Nullum intrusion changes from an outer fayalite-augite-hornblende microgranite to an inner zone of hornblende-biotite granite. It encloses a lenticular body of monzonite and contains abundant monzonite xenoliths.

Volcanic sequence

The volcanic succession of Tweed volcano and surrounding areas has been collectively designated the Lamington Volcanics or Lamington Group (Table 3.4.1, columns 9–10). It consists of a sequence of basaltic, icelanditic, and rhyolitic eruptives, the lower formations of which (Albert Basalt, Kyogle Basalt, and Mount Gillies Volcanics; Table 3.4.2) are believed to be associated with the slightly older Focal Peak volcano (Section 3.4.6; Ross, 1974).

The Lismore Basalt (southern area) and Beechmont Basalt (northern area) are considered to be the first lavas erupted from Tweed volcano. The age determined for the base of the Beechmont Basalt is 22 Ma, and a similar basalt from Burleigh Heads, on the eastern flank of the shield, was dated at 24 Ma. Slightly younger and older dates were determined from basaltic rocks on the shield flanks (McDougall & Wilkinson, 1967). The Lismore Basalt is the underlying rock of wide areas of low relief around Lismore and to the south and east. Much of the upper part of the succession in this area

Table 3.4.2 Correlation and age of Tertiary volcanic and intrusive rocks, Tweed volcano (maximum thicknesses in metres).

Queensland	Age (Ma)		New South Wales
Hobwee Basalt (590 m)	20.6		Blue Knob Basalt (350 m)
Binna Burra Rhyolite (300 m)	20.8 20.9 21.3		Nimbin Rhyolite (500 m)
Beechmont Basalt (270 m)	21.8 22.3		Lismore Basalt (250 m)
Basalts on eastern flank of shield	23.2 23.5	22.9 23.4 23.7	Mount Warning gabbro and K-mugearite
Rhyolite intrusion, Surprise Rock	23.6		

has been removed by erosion, and remains where protected by the overlying flows. Both the Lismore and Beechmont basalts are predominantly tholeiitic, and there are only minor occurrences of alkaline types (Duggan & Mason, 1978). Thin sedimentary beds, largely shales and diatomite, are intercalated amongst the lowest lavas of the Beechmont Basalt (Willmott, 1983).

The Nimbin Rhyolite to the south, and its northern counterparts (the Binna Burra and Springbrook Rhyolites) crop out in the Lamington Plateau area and to the northeast, and conformably overlie the Lismore and Beechmont basalts, respectively. They consist of thick, aphyric and porphyritic, rhyolite flows. Tuffs are found at the base and, in some instances, at the top of the formation. These rhyolite sequences (including only very minor dacite) are up to 500 m thick and individual flows are up to 150 m thick. The rhyolites are devitrified, apart from a vitreous zone up to 10 m thick at the base and also at the top of some flows. The Nimbin Rhyolite has some intercalated basaltic flows. The pyroclastic rocks in the Nimbin Rhyolite retain evidence of reworking and include blocks of rhyolite and tholeiitic volcanic rocks and sedimentary material. The tuffs in the Binna Burra Rhyolite are mostly of air-fall origin, but one pyroclastic flow is recognised in this area.

The rhyolites are relatively potassic. K_2O typically is in the range 5–6 weight percent and K_2O/Na_2O values are about 2, as reflected in the modal abundance of sanidine. The separate Binna Burra and Springbrook types may be distinguished on trace-element and mineralogical criteria, but no clear distinction is possible on major-element data (Ewart et al., 1977).

The uppermost basalt formation on the southern side of Mount Warning is the Blue Knob Basalt which forms residual cappings on many of the higher areas near Nimbin. The Blue Knob Basalt was erupted onto a highly irregular land surface shaped by the Nimbin Rhyolite and probably has a maximum thickness of about 300 m (Duggan & Mason, 1978). No vents have been identified, but its petrological similarity to the underlying Lismore Basalt may be indicative of a common source. The Blue Knob Basalt correlates with the Hobwee Basalt on the northern side of Mount

Warning which has a maximum thickness of about 600 m. There are at least 20 flows in the cliffs near Mount Hobwee each averaging about 30 m thick.

The Lismore/Beechmont and Blue Knob/Hobwee basalts consist predominantly of both olivine- and quartz-normative lavas, including some containing pyroxenes and feldspars of high-pressure origin (Duggan & Wilkinson, 1973). They are characterised by the absence of hypersthene and pigeonite (except in most felsic compositions) and have a wide range of textural types. The transitional lavas have relatively evolved chemical characteristics. Basalts *sensu stricto* are very rare and Mg-ratios rarely exceed 65. They are typically andesine-normative and close to silica saturation.

Scattered sporadically throughout the Lismore/ Beechmont basalts, and rarely in the Blue Knob/ Hobwee basalts, are some flows of more typical tholeiitic chemistry (higher normative quartz and lower normative diopside/hypersthene values) including tholeiitic basalt and icelandite. The more evolved icelandites have more Fe-rich olivine phenocrysts and a larger proportion of residual glass.

Alkaline volcanism in the southern Tweed area subsequent to the Kyogle/Albert Basalt was confined mainly to small sills, dykes, and flows of alkali basalt and hawaiite (Duggan & Mason, 1978). The intrusions cut the overlying units. High-pressure xenoliths are rare in the Tweed lavas and restricted to these minor products of alkaline magmatism. Small pyroxenite xenoliths are present in a *ne*-hawaiite dyke cutting the Nimbin Rhyolite 10 km west-northwest of Mullumbimby, and pyroxene megacrysts are found in a hawaiite plug 3.5 km west of Byron Bay. A hawaiite flow within the Lismore Basalt 6 km north-northeast of Clunes contains small but abundant pyroxene megacrysts. Peralkaline-rhyolite dykes, in the northern Tweed area, south of Binna Burra, similar to those in the central complex, have intruded Beechmont Basalt and Binna Burra Rhyolite.

3.5   Northern New South Wales

3.5.1   Introduction

Nine volcanic provinces are recognised in northern New South Wales. The region extends from near the Queensland border south to the Hunter-Goulburn valley (Fig. 3.5.1). The volcanism ranges in age from 70 to 12 Ma and has a great diversity in volcanic style and petrological character — from widespread lava fields to large, basalt-dominated volcanoes, to central-type volcanoes containing felsic rocks.

The Nandewar, Warrumbungle, and Comboyne provinces all contain lavas and intrusions of similar composition — namely, near-saturated to transitional basalts ranging to peralkaline rhyolite — and all are of similar and restricted age (21 to 13 Ma). In contrast, saturated peralkaline rocks are rare or absent in the Dubbo province and phonolite and undersaturated trachyte are prominent felsic rock types. Lava-field provinces in which alkali basalt is the dominant lava type, include the Central, Walcha, Liverpool Ranges, and Barrington Tops areas. These are generally older,

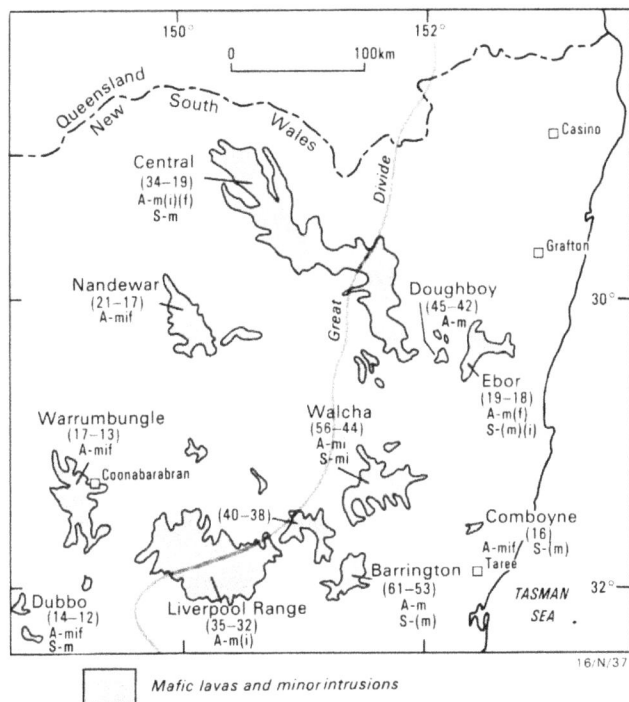

Figure 3.5.1 Distribution, rock types, and ages of volcanic provinces in northern New South Wales. Rock associations: A, alkaline; S, subalkaline. Rock types: m, mafic; i, intermediate; f, felsic; (m), minor mafic; (i), minor intermediate; (f), minor felsic. Ages (in brackets) in Ma.

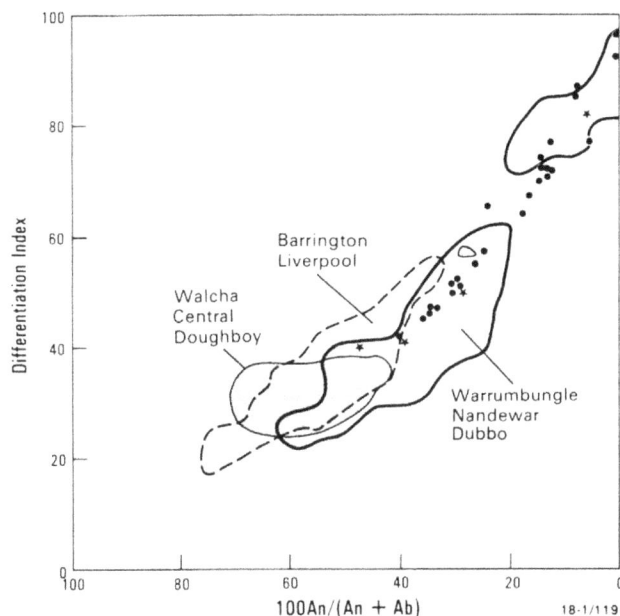

Figure 3.5.2 Compositional fields for northern New South Wales provinces using the plotting procedure outlined in Section 1.1.4. Filled circles refer to Comboyne samples. Stars refer to Ebor samples.

and have a greater range in age (mostly between 70 and 30 Ma). Four provinces — Central, Ebor, Walcha, and Barrington — are underlain directly by Palaeozoic rocks of the New England Fold Belt. The other provinces overlie non-marine sedimentary rocks of Mesozoic age. Generalised compositional fields for the northeastern New South Wales provinces are shown in Figures 3.5.2 and 3.5.3, and selected chemical analyses are given in Table 3.5.1.

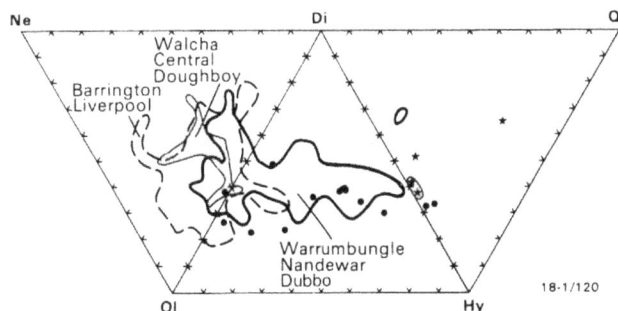

Figure 3.5.3 Compositional fields for northern New South Wales rocks projected in the system Ne-Ol-Di-Hy-Qz. Filled circles refer to Comboyne samples. Stars refer to Ebor samples.

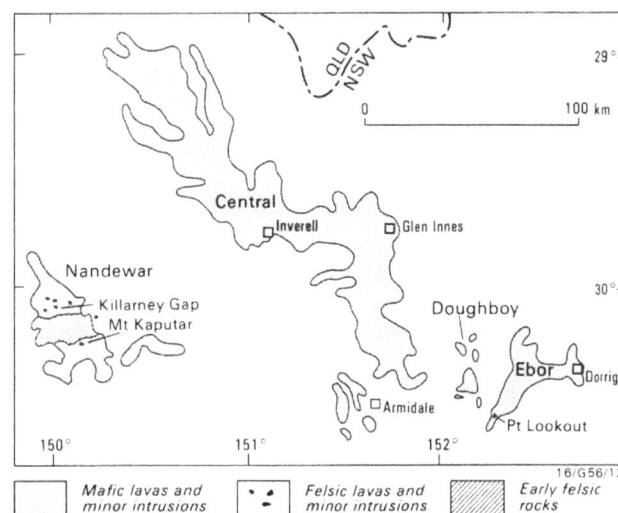

Figure 3.5.4 Distribution of volcanic and intrusive rocks in the Nandewar, Central, Doughboy, and Ebor provinces of northern New South Wales (see Fig. 3.5.1).

3.5.2 Nandewar

The products of Nandewar volcano crop out in an elongate belt about 50 km long and up to 30 km wide covering an area of about 800 km^2 (Fig. 3.5.4). Their present-day maximum thickness is estimated to be about 800 m based on a section from Bullawa Creek towards Mount Kaputar. However, the original thickness must have been considerably greater than this, because the intrusive plug-like body cropping out at Mount Kaputar lookout is one of the highest points in the complex. Wellman (1986) demonstrated that Nandewar volcano rests on a basement uplift caused by the emplacement of the central intrusive complex (Section 1.4.9; Fig. 1.4.10D). Approximate relative proportions of the main lava types in Nandewar volcano (total volume about 150 km^3) are: hawaiite (Table 3.5.1, column 1), K-hawaiite, and K-mugearite (70 volume percent); mafic and peralkaline trachyte (4); rhyolite and peralkaline rhyolite (2); and older rhyolites and trachytes (24).

The oldest rocks in Nandewar volcano, dated at 21 Ma (Stipp & McDougall, 1968a), include a trachyte flow overlying a teschenite sill north of Killarney Gap. This flow forms a prominent bluff and is capped by K-hawaiite and intruded by a small peralkaline-rhyolite plug.

EAST AUSTRALIAN VOLCANIC GEOLOGY

Table 3.5.1 Selected chemical analyses for northern New South Wales provinces

	1	2	3	4	5	6	7	8	9	10
SiO_2	46.99	46.98	52.68	49.22	57.90	61.10	45.53	48.22	44.19	74.24
TiO_2	2.65	1.92	2.19	2.65	0.55	0.07	1.94	1.62	2.66	0.15
Al_2O_3	14.55	12.85	14.92	14.36	17.16	17.14	14.91	15.14	15.11	10.93
Fe_2O_3	1.77	2.29	3.54	2.87	3.45	3.35	3.96	2.28	4.69	3.34
FeO	8.90	8.83	6.87	6.96	3.93	1.83	8.49	8.71	6.08	1.43
MnO	0.13	0.16	0.15	0.13	0.33	0.10	0.16	0.17	0.22	0.05
MgO	8.18	11.66	4.52	6.37	0.87	0.04	9.62	8.72	8.07	0.05
CaO	8.97	8.22	8.01	8.00	2.24	0.84	9.36	8.40	9.03	0.20
Na_2O	2.80	2.50	3.73	3.33	5.04	7.94	2.99	3.28	3.76	4.98
K_2O	1.43	1.51	1.18	2.49	5.57	5.18	1.36	0.99	2.07	4.63
P_2O_5	0.47	0.37	0.41	0.72	0.27	0.02	0.61	0.44	1.10	0.02
S		<0.02	<0.02	<0.02	<0.02	<0.02		0.02	<0.02	
H_2O+	2.03	1.57	0.66	0.63	1.12	1.63	1.99[1]	1.70	1.58	0.36[2]
H_2O-	0.83	0.72	0.81	0.97	1.12	0.32		0.53	1.07	
CO_2		0.09	0.10	0.78	0.40	0.11		0.16	0.17	
Total	99.70	99.67	99.77	99.48	99.95	99.67	100.92	100.38	99.80	100.38
Ba	297	245	250	695	1020	<2	493	255	645	2
Rb	26	34	17.5	41	129	520	21	19.5	30.5	416
Sr	659	468	428	805	110	4	1020	590	1280	5
Pb	4	1	3	5	11	22	2	3	5	30
Th	2	3.0	3	6	12	34	3	3	9	44
U		<1	<1	<1	3	6	2	<1	<1	10
Zr	178	139	192	291	790	1900	233	144	273	2848
Nb	44	27	20	48	109	291		31	123	
Y	20	17	22	26	49	213	24	20	25	95
La	33	18	20	36	71	374	32	20	56	
Ce	63	40	45	85	145	630	66	40	109	
V	205	138	167	143	12	<1	199	158	170	<1
Cr	176	246	86	144	2	2	328	235	125	<1
Ni	120	1230	14	127	<1	<1	173	159	110	16
Cu	50	67	10	34	15	7	74	63	43	<1

1. Hawaiite from Nandewar province (Stolz, 1985, sample 49000).
2. Hawaiite from Doughboy province (M.B. Duggan & B.W. Chappell, unpublished data, sample EAV212).
3. Qz-tholeiitic basalt from Ebor province (M.B. Duggan & B.W. Chappell, unpublished data, sample EAV214).
4. Hawaiite from Warrumbungle province (M.B. Duggan & B.W. Chappell, unpublished data, sample EAV30).
5. Trachyte from Warrumbungle province (M.B. Duggan & B.W. Chappell, unpublished data, sample EAV147).
6. Peralkaline ne-trachyte from Warrumbungle province (M.B. Duggan & B.W. Chappell, unpublished data, sample EAV23).
7. Alkali basalt from Liverpool Range province (Schön, 1985, sample 64984).
8. Hawaiite from Walcha province (M.B. Duggan & B.W. Chappell, unpublished data, sample EAV217).
9. Basanite from Barrington province (D.R. Mason & B.W. Chappell, unpublished data, sample EAV48).
10. Peralkaline rhyolite from Comboyne province (Knutson, 1975, sample 6640).

[1] Total H_2O.
[2] Loss on ignition.

The slightly porphyritic and, in some examples, vesicular, older rhyolites are found as domes and flows and appear to have been extruded along a north-northwest trending fracture, possibly an extension of the Hunter-Mooki thrust which strikes from the southeast towards the volcano. They were extruded onto an irregular basement of Palaeozoic sedimentary and volcanic rocks on the eastern side, and onto Mesozoic sedimentary rocks on the western side of the complex. The rhyolites are up to 600 m thick in the region to the west of Mount Lindesay, and occupy much of the central region of the volcano where they have been exposed by erosion of the main shield-forming lavas.

Abbott (1965, 1969) argued that many of the rhyolitic domes in the central region were endogenous, and had intruded the main shield-forming lavas. However, several of these rhyolites are vesicular and have associated breccias, probably meaning that they were extruded — possibly as large exogenous domes — during the early stages of volcanism. Furthermore, the absence of xenoliths of shield-forming lavas in the rhyolites and the lack of evidence for buckling and deformation in the Mesozoic sandstones adjacent to the domes, does not support a model of forceful rhyolite emplacement.

The main shield-forming lavas range from hawaiite, K-hawaiite and K-mugearite, to trachyte and peralkaline rhyolite. Their eruption appears to have been largely from a central vent thought to have been in the vicinity of Mount Lindesay because traceable flow boundaries and pyroclastic horizons dip radially away from this general area at about 5 to 10°. The presence in the region surrounding Mount Lindesay of a small, high-level, monzonitic intrusion and abundant small dykes ranging in composition from K-rich hawaiite to rhyolite, represents a concentration of intrusive magmatic activity in this area and supports the interpretation that this is close to the feeder zone for the volcanic rocks.

Extrusion of the main shield-forming lavas took place over the restricted interval 18 to 17 Ma (Stipp & McDougall, 1968a; Wellman et al., 1969; Wellman & McDougall, 1974b). The earliest of these were hawaiite, followed by K-hawaiite, K-mugearite, trachyte, and peralkaline rhyolite, then more K-mugearite and trachyte. The flows range in thickness from about 2 to 20 m. The mafic flows commonly are relatively thin (2–10 m), whereas the trachytic flow units typically are the thickest (10–20 m). Several of the more massive flow units persist laterally for considerable distances, cropping out continuously for 6–8 km as prominent bluffs in youthful valleys on the southern flank of the volcano.

Flows of hawaiite, K-hawaiite, and K-mugearite are most common in the 500 m of volcanic rocks exposed from the base to the top of Mount Grattai in the northern sector, whereas trachyte is relatively scarce compared with the southern part of the volcano.

Relatively abundant pyroclastic rocks are found intercalated with flows in the southern part of Nandewar volcano. Most tuffaceous rocks associated with hawaiite, K-hawaiite, and K-mugearite, are extensively weathered and typically composed of highly vesicular, lapilli-sized fragments of nearby lavas, crystal debris, and ash. On the other hand, the pyroclastic rocks associated with trachyte characteristically are agglomeratic. They consist of ejected blocks up to 1 m in diameter, ranging in composition from K-hawaiite to rhyolite and set in a matrix of smaller clasts, lapilli-sized fragments, and ash.

Following, and possibly even during, the waning stages of eruption of the main shield-forming lavas, alkaline and peralkaline trachyte dykes, ring-dykes, and plugs intruded the volcanic pile and surrounding basement rocks. An age of 17 Ma (Stipp & McDougall, 1968a) was determined on peralkaline trachytes that intrude basement rocks 10 km west and 12 km southwest of Mount Kaputar. Rocks of similar composition found as ring-dyke (Mount Yulladunida) and plug-like (Mount Kaputar) intrusions in the main shield must be similar in age or slightly younger, and clearly have cooled and crystallised under a shallow cover. These trachytic intrusions are confined largely to the southern part of the volcano. This contrasts with the major, younger group of rhyolite and peralkaline-rhyolite extrusions and intrusions within the northern sector.

There may have been two distinct episodes of rhyolitic magmatism. A peralkaline-rhyolite plug, dated at 18 Ma, intrudes the older rhyolites 5 km northeast of Killarney Gap. This age, together with another date of 17 Ma determined on an alkali rhyolite from the northern sector (Stipp & McDougall, 1968a) means that these magmas were emplaced during the main shield-forming episode.

Peralkaline rhyolite is found as high-level ring dykes, dykes, and plugs, that have intruded both the volcanic pile and basement rocks. Alignment of several small intrusions along north-northwesterly trends in the eastern part of the complex is evidence for strong structural control on their emplacement. The younger rhyolites are typically vesicular and generally more leucocratic than are the peralkaline rhyolites which are rarely vesicular. Contact relationships between the relatively erosion-resistant younger rhyolites and adjacent rocks are commonly unclear, but the rhyolites were extruded probably as exogenous domes as they are commonly vesicular.

The mafic rocks of Nandewar volcano are mostly transitional hawaiite or K-rich hawaiite in composition (Stolz, 1985). There is a continuum of compositions, from hawaiite through to rhyolite (Fig. 3.5.2). Intermediate compositions are represented by K-mugearites. Mafic rock types include both sodic and moderately potassic (K_2O/Na_2O less than 0.5) types. However, the more felsic members of the suite are all moderately potassic. Mg-ratios range up to 65, but are mostly less than 60. The dominant hawaiites and K-hawaiites are generally aphyric or slightly porphyritic containing phenocrysts of olivine, plagioclase, and clinopyroxene. Megacrysts of aluminous augite and bronzite are found in several K-hawaiites. Alkali feldspar and apatite are additional phenocrysts in the more evolved K-mugearites and mafic trachytes. Arfvedsonite, acmitic pyroxene and, more rarely, aenigmatite are present in the peralkaline rocks (Stolz, 1986).

3.5.3   Central and Doughboy

The Central and Doughboy area is an elongate belt of predominantly mafic volcanic rock about 240 km in length and up to 70 km wide, capping the New England Tablelands in northern New South Wales (Figs. 3.5.1,4). It includes the Central province of McDougall & Wilkinson (1967), which occupies by far the greater proportion of the outcrop area, and the Doughboy province (Binns, 1966) that forms the Doughboy Range west of Ebor. The two provinces are combined here on the basis of their petrological and geological similarity and geographic proximity. No detailed field or petrological survey of the Central and Doughboy provinces has been undertaken, but a considerable amount of data is available on specific areas and localities in individual publications and, especially, in theses from the University of New England.

Lava flows throughout the province are flat-lying and form a relatively thin veneer typically less than 100 m thick but reaching a maximum of about 300 m. Up to 150 m of basalt flows are present in the Doughboy Range (Binns, 1969).

The volcanic rocks over most of the outcrop area rest unconformably on highly irregular erosion surfaces developed on Permian granites of the New England Batholith and associated highly deformed sediments and volcanics. They overlap flat-lying to shallow-dipping quartzose sedimentary rocks along the eastern margin of the Mesozoic Surat Basin northwest of Inverell. Voisey (1942) demonstrated that the pre-basalt topography had a relief of at least 420 m. The basalts locally are separated from Permian basement by lacustrine sediments. An Upper Eocene age has been assigned to the sediments in the Armidale area on the basis of palaeobotanical evidence (Slade, 1964). These sediments in some cases form deep leads which in the past have been important sources of alluvial gold and cassiterite.

The alluvial deposits of sapphire (gem corundum) in the Inverell and Glen Innes area also are derived from erosion of the Cainozoic volcanic rocks. Sparse corundum megacrysts have been observed in basaltic lava flows, but much of the sapphire may be derived from associated pyroclastic units (Lishmund, 1987; Pecover, 1987; Section 3.10.2).

The province is dominated by mildly to strongly undersaturated volcanic rocks, including alkali basalt and basanite, and minor hawaiite, mugearite, nephelinite, analcimite, and some pyroxene-rich types approaching ankaramite (Wilkinson, 1966, 1969). Associated high-level intrusive rocks are teschenite and alkali dolerite. The greatest development of strongly undersaturated rocks (mainly nephelinite) is probably in the Inverell area (Duggan, 1972) where they form the uppermost lavas in the sequence and are underlain by alkali basalt. Low-pressure fractionation of the nephelinites has produced schlieren of malignite compositions (Wilkinson, 1977b). Analcimite is found west of Glen Innes (Wilkinson, 1962; Wilkinson & Binns, 1977).

Evolved alkaline rocks including hawaiite and mugearite are comparatively rare. One small body of phonolite, probably at least partly intrusive, forms Sugarloaf Mountain, but evolved felsic rocks are otherwise absent. Tholeiitic basalts form a minor but important component of the succession in the Inverell area where they appear to form the lowermost flows (Duggan, 1972; Wilkinson & Duggan, 1973). A petrographically similar basalt residual 30 km east of Guyra (Binns, 1966) also may be tholeiitic.

Peridotite xenoliths are relatively common in alkaline rocks throughout the province, in places accompanied by spinel pyroxenite (Duggan, 1968; Wilkinson, 1973). A range of megacryst species at different localities includes Al-rich augite and bronzite, spinel, kaersutite, titanbiotite, andesine, anorthoclase, and rare zircon (Binns, 1969; Binns et al., 1970).

The basalts of the New England Tableland have been subdivided into two age groups on the basis of their relationship to associated lacustrine sediments and on the presence or absence of an overlying zone of deep lateritisation (David, 1887; Browne, 1933; Owen, 1954). These were termed the Older and Newer series by Browne (1933). K-Ar age data (Cooper et al., 1963; McDougall & Wilkinson, 1967; Wellman & McDougall, 1974b) define a bimodal distribution of ages for the province (excluding basalts of the Doughboy Range which form a separate, older group), but field control on the dated samples is not sufficient to determine whether the two radiometric age groups correspond to the Older and Newer series. Samples giving ages in the older group (34–32 Ma) are all from the central portion of the province south and west of Glen Innes. Ages in the younger group (22–19 Ma) are for samples restricted to the southern (Armidale area) and northwestern (Inverell area) parts of the province. Laterite surfaces are widespread in the Inverell area, and the absence of dated samples from the older group may mean that lateritisation is not restricted to basalts of the older group. The basalts of the Doughboy Range were dated at 45–42 Ma by Wellman & McDougall (1974b).

The predominant alkaline basaltic rocks in the Central and Doughboy provinces are closely similar to those in lava-field provinces elsewhere in eastern Australia (Table 3.5.1, column 2). Mg-ratios range from 69 to 46 consistent with the restricted compositional range of these lavas. Evolved rocks are rare, but include the Spring Mount analcimite and the phonolite at Sugarloaf Mountain. The tholeiitic rocks of the Inverell area are K-poor and have sub-calcic pyroxenes in the groundmass.

3.5.4   Ebor

The volcanic rocks of the province around Ebor and Dorrigo represent the eroded remnants of a central-type volcano (Fig. 3.5.4). The province extends from just south of Point Lookout in an arcuate belt up to 20 km wide, north through the village of Ebor and east to the township of Dorrigo. Few field and laboratory data are available on the province. No detailed mapping has been undertaken and only two chemical analyses have been published previously. The magmatic affinities of the rocks were discussed briefly by McDougall & Wilkinson (1967) and Wilkinson (1969), and K-Ar ages (19–18 Ma) were given by McDougall & Wilkinson (1967) and Wellman & McDougall (1974b).

The lava flows of the province are perched on the edge of an east- to south-facing portion of the coastal escarpment (Ollier, 1982b). The sequence is well exposed along the length of the scarp as a succession of relatively thick flows (estimated flow thickness about 50 m on average, but no sections have been measured), up to 300 m thick, dipping at low angles into the scarp face. The outcrop pattern corresponds to that of an original 50 km-diameter shield that had a height of about 500–600 m above basement and a volume of about 300 km^3. Extensive erosion has removed an estimated 60 percent of the original surface area and 90 percent of the original volume, and and scarp retreat has cut deep gorges into the volcano and basement from the south and east. Volcanic rocks are confined now almost exclusively to the Ebor-Dorrigo plateau, except for a few isolated outliers on higher peaks to the south.

Development of the present-day topography appears to have been controlled at least in part by differences in the underlying basement rock types. Thus, the southern and central parts of the original shield, now almost totally removed by scarp retreat, are underlain by the Permian Nambucca beds predominantly of mica-rich slate and phyllite, whereas most of the northern part is underlain by the Devonian Moombil Beds dominated by siliceous argillite.

A prominent radial drainage pattern is preserved over the whole of the presumed original area of the volcano (Ollier, 1982b). A small gabbroic intrusion near the centre of this drainage system is about 2 km in diameter and surrounded by a prominent hornfels ridge. This body has not been studied in detail because of the extreme difficulty of access and poor outcrop in dense rainforest. However, dykes of trachyte and feldsparphyric basalt are known to be present, in addition to the gabbro. K-Ar dating of the body has proved difficult, but a recent fission-track determination of 18 Ma is consistent with the gabbro representing a major feeder system for the volcanic sequence (Gleadow & Ollier, 1987). A magnetic anomaly exists around the intrusion (Fig. 1.4.10F).

The full range of magma compositions for the Ebor province is unknown. Two published analyses are of trachyte and qz-tholeiitic basalt (Wilkinson, 1969), and compositions in the range qz-tholeiitic basalt (Table 3.5.1, column 3) through icelandite to rare qz-trachyte approaching rhyolite, are indicated by recent analyses. Mafic lavas are dominated by porphyritic types containing plagioclase phenocrysts up to 3 cm in length. These rocks closely resemble the 'high-Al tholeiitic andesite' from Brunswick Heads in the Tweed volcano (Duggan & Wilkinson, 1973) and elsewhere. Microphenocrysts of olivine are present in most of the rocks, but clinopyroxene is restricted to the groundmass. Pyroxene as a low-pressure phenocryst is absent, so the lavas may have equilibrated at elevated pressures near the crust/mantle boundary (Duggan, 1974; Ewart et al., 1980).

A flow of trachyte thought to be part of the Ebor-Dorrigo sequence, is exposed west of Point Lookout. It consists of microphenocrysts of olivine, magnetite, anorthoclase, and minor apatite in a groundmass of anorthoclase laths, clinopyroxene, and opaque minerals. The thickness and extent of the trachyte are unknown.

3.5.5 Warrumbungle

Warrumbungle volcano occupies an approximately circular outcrop area about 50 km in diameter immediately west of the township of Coonabarabran (Fig. 3.5.5). Outliers of basaltic rocks of the same age as those of the main volcano extend in a broad belt to the southeast, towards the western limit of the Liverpool Range (Dulhunty, 1973b).

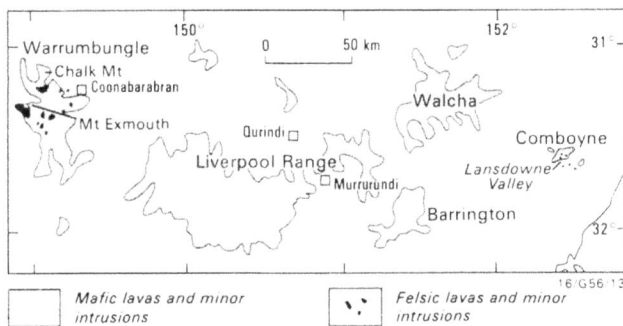

Figure 3.5.5 Distribution of volcanic and intrusive rocks in the Warrumbungle, Liverpool, Walcha, Barrington, and Comboyne provinces of northern New South Wales (see Fig. 3.5.1).

The first documentation of Warrumbungle Range was by the explorer John Oxley (1820) and the first geological study of the volcano was by Jensen (1906b, 1907b), but no other detailed studies have been published. However, Faulks (1962) and Hubble (1983) described in unpublished theses portions of the southern part of the shield. Hockley (1972, 1973, 1974) suggested that more than one magmatic lineage may be present, but no supporting analytical data are given.

Warrumbungle volcano is deeply dissected, and has a pronounced radial drainage pattern and some spectacular volcanic scenery (Figs. 3.5.6–7). The volcano was originally a fairly symmetrical low shield reaching a height of about 1000 m above surrounding basement, judging by its constructional remnants.

However, the apparent height probably has been exaggerated by significant updoming of the basement beneath the central part of the volcano (Wellman, 1986; Section 1.4.9; Fig. 1.4.10E). Major constructional remnants of the original shield are preserved at several points around the shield, including Mount Woorut, Mount Exmouth, and Blackheath Mountain, and stratified lava sequences are exposed at most of these localities.

The central part of the volcano is characterised by an abundance of dykes, plugs, and domes of feldspathoid or quartz-bearing peralkaline trachyte or of highly leucocratic trachyte (Fig. 3.5.7). The plugs and dykes intrude a thick and relatively poorly bedded and poorly sorted sequence of tuffs and breccias composed of angular trachyte fragments mostly less than 10 cm but up to 50 cm in diameter. The pyroclastic rocks are well exposed at several points along the main Grand High Tops walking track. Low to moderate dips may represent initial dips or may have resulted from emplacement of nearby plugs and domes. Well-bedded pyroclastic rocks appear to be confined to the outer flanks of the volcano and may represent a redeposited facies. Basement rocks of Warrumbungle volcano are predominantly quartzose sandstone of the Jurassic Pilliga Sandstone which forms part of the sequence in the southeastern part of the Coonamble Basin, and which is exposed extensively in the eroded, central part of the volcano.

Localised deposits of lacustrine sediments are developed at several localities in addition to the widespread occurrence of pyroclastic rocks, especially on the outer flanks of the volcano. For example, diatomite deposits presumably deposited in ephemeral lakes resulting from damming by lava flows, are found at Chalk Mountain and elsewhere (Griffin, 1961; Herbert, 1968). They crop out at Chalk Mountain at several localities around the mountain in flat-lying sequences up to 15 m thick, including mudstone and well-bedded pyroclastic rocks (the Chalk Mountain Formation; Holmes et al., 1983) that are underlain and capped by hawaiite flows. The diatomite beds have yielded a range of fossils, including insects, fish, birds, and the leaves and fruit of eucalyptus species.

K-Ar age data (Dulhunty & McDougall, 1966; McDougall & Wilkinson, 1967; Dulhunty, 1973b; Wellman & McDougall, 1974b) correspond to a total range of 17–13 Ma for the Warrumbungle volcano. Dated flows of 17 and 15 Ma in the valley of the Castlereagh River may mean that volcanism took place over most of the total time span in the southern part of the volcano. Dated trachyte samples are confined to a narrow age range (16–15 Ma; McDougall & Wilkinson, 1967; Wellman & McDougall, 1974b).

The range of rock types in Warrumbungle volcano is strongly bimodal, with maxima in the hawaiite-mugearite and trachyte ranges. Alkali basalt and benmoreite are extremely rare. Hawaiite (Table 3.5.1, column 4) and mugearite are the dominant rock types in the outer flanks of the shield, especially in the southern and northern areas. They form a thin plateau-covering sequence in the southern part and extending southwards, but thicker sequences are exposed on Blackheath Mountain and nearby peaks. Principal outcrops in the north are on a series of north-trending ridges including Chalk Mountain, and a series of isolated residuals toward the northwest. Hawaiite flows at a few localities contain a range of xenolith and megacryst types.

Trachyte (Table 3.5.1, column 5) is the dominant rock type in the central part of the volcano, and its different compositions are to some extent reflected in the mode of outcrop. More mafic trachyte, characterised by Fe-rich olivine and ferroaugite, is found in shield-building sequences of fairly thick (up to 50 m) flows in eastern and western areas. Similar trachyte elsewhere forms small conical hills which probably represent small lava domes.

A second and equally important type of trachyte is peralkaline and relatively leucocratic (Table 3.5.1, column 6). This makes up most of the important plugs, dykes, and domes in the central part of the volcano and on the southern and eastern flanks of the volcano. Peralkaline trachyte is also by far the dominant rock type making up the widespread and abundant pyroclastic rocks. Some of the trachytes have unusual mineralogical features including Zr-rich sodic pyroxene (Duggan, 1988) and Ti-depleted aenigmatite.

The mildly alkaline lavas of Warrumbungle volcano are dominated by nepheline- and hypersthene-normative hawaiites, mugearites, and trachytes. No significant geochemical differences are evident between the nepheline- and hypersthene-normative mafic or felsic rocks, other than differences in silica saturation. Both the mafic and felsic rocks are moderately potassic (K_2O/Na_2O values mostly greater than 0.5). Pyroxene megacrysts and Al-Ti-augite pyroxenite xenoliths are found at some localities. Both nepheline- and quartz-normative felsic rocks are approximately equally developed. The abundance of undersaturated trachytes appears to be unique among Cainozoic central-type volcanoes in eastern Australia. Mg-ratios range up to about 64 and, except for the absence of rocks in the benmoreite field, there is a continuum of compositions from basalt (rare) through to rhyolite (Fig. 3.5.2).

3.5.6   Liverpool Range

The Liverpool Range is the largest volcanic province in New South Wales (Fig. 3.5.5). Its volcanic rocks cover the northern and northeastern boundary between the Sydney Basin and the Oxley and Gunnedah basins, about 200 km northwest of Sydney. Present-day exposure is 6000 km² so, assuming a typical thickness of 300–400 m, the original volume of volcanic material must have been in the order of 4000–6000 km³. There is a surprisingly small range in rock compositions. All rocks are mafic-ultramafic containing 44–51 percent SiO_2. The majority of analysed samples are nepheline normative, and all are petrographically alkaline. Alkali basalt predominates (Table 3.5.1, column 7), basanite is common, and hawaiite and mugearite are rare.

Extensive K-Ar dating has been undertaken (Wellman et al., 1969; Wellman & McDougall, 1974b) and two age groups have been identified: one late Eocene (to the east, 40–38 Ma) and the other early Oligocene (35–32 Ma, to the west). There are also topographic and minor geochemical differences (Schön, 1985), and two distinct volcanic sources may have existed for these groups.

Liverpool Range volcanic rocks rest on top of the Upper Jurassic Pilliga Sandstone which in turn unconformably overlies the Narrabeen Group Triassic sequence. Basalt near the contact commonly is weathered, and tends to be amygdaloidal, containing concentrations of alkali-rich zeolites that are considered to have been produced by the interaction of meteoric fluids with the basalt immediately after eruption (Schön, 1986).

The creeks responsible for erosion of the western part of the range are cutting back into what appears to be a plateau surface that has a present height of about 800 m above sea-level. In contrast, the eastern part of the range has the form of a curved 'spine' of adjacent peaks, as do the extremities of the western part of the range. The plateau is that part of the Liverpool Ranges, near the junction of the eastern Warrumbungle Range and the Liverpool Range, where the summits form a broad surface whose local relief is less than 150 m, and whose mean height is 300 m above the surrounding plain. The maximum height reached by the hills that rise above the general plateau surface is 1197 m.

Most of the Liverpool Range, in stark contrast to the broad plateau, is composed of a narrow (1 km) 'spine' or ridge that does not preserve a flat summit. The spine curves gradually northeastward from the eastern side of the plateau. Individual peaks that make up the ridges and spines have an accordant summit level at about 950–1000 m. The crestal spine of the Liverpool Range is interpreted to be the erosional remnant of a once more-extensive lava plateau.

A second type of spine also can be recognised, making up about 20 percent of the total, and only developed in the southwestern part of the Liverpool Range. These spines all run south-southwestwards (towards Coolah), and are essentially interfluves consisting of sediment capped by basalt. Flows exposed in the interfluves thin progressively towards the south.

Three types of extrusive events, based on the character of the erupted materials, are recognised in the Liverpool Range:

(1) Massive lava flows, each 2–4 m thick and distinguishable by differences in colour and degree of erosion, form the top of the sequence. These flows, although light in colour, are basaltic, and constitute about half of the exposed sequence. Some amygdaloidal flows are present, but generally the magma seems to have had a low volatile content. Slab jointing and possible pahoehoe flows have been identified.

(2) Columnar layers, totalling 204 m thick, of the same material as (1), but jointed forming columns up to 1 m in diameter, are interbedded with the more massive flows. No intrusive boundaries are apparent, and the layers appear to be merely slowly cooled lava flows. Columnar flows are more restricted than the other flow types, and some are compositionally more evolved. Ardglen quarry, at the eastern edge of the range, has excellent sections through columnar mugearite flows.

(3) Massive units, up to 15 m thick, of highly altered, light-coloured amygdaloidal to agglomeratic flows, constitute the base of the sequence in many areas. Individual flows are not distinguishable, and the material now consists of a compact clay and zeolite (thomsonite, natrolite) mixture. Caves up to 80 m long

have formed at the top of the units with their roofs in the overlying unaltered basalt (Osborne, 1979).

Numerous igneous bodies crop out around the margins of the main Liverpool Range. Some of these bodies demonstrably are erosional remnants of Liverpool Range flows, and are found at the end of the ridges that extend from the main mass (see above). More commonly, these bodies appear unrelated geochemically and petrographically to the alkaline rocks of the Liverpool Range, having features more characteristic of tholeiitic basalts. Several periods of igneous activity in the area are suggested from K-Ar dating (for example: Wellman & McDougall, 1974b; Dulhunty, 1976; Embleton et al., 1985), particularly during the Jurassic (198–132 Ma) and during several parts of the Tertiary — especially in the Oligocene-Eocene, contemporaneous with the Liverpool Range volcanism.

The rocks of the Liverpool Range are predominantly sodic basanite and alkali basalt porphyritic in olivine. Development of more evolved hawaiites and mugearites is limited. Two broad compositional groups can be identified. The older (39–32 Ma) eastern group consists of rocks containing more than 2-percent normative nepheline, whereas the younger (36–30 Ma) western group contains a wide range of compositions from undersaturated through to hypersthene-normative. Mg-ratios for the saturated/near-saturated and undersaturated rocks average 60 and 62, respectively. There is a general absence of lavas having 'primary' Mg-ratios (about 70), but one basanite has an Mg-ratio of 58 and contains xenoliths of spinel lherzolite and xenocrysts of diopsidic clinopyroxene.

3.5.7 Walcha

The Walcha province is an area of lava-field type volcanic rocks extending 60 km south from Walcha and southeast about 60 km beyond Yarrowitch (Fig. 3.5.5). The lava flows around Walcha form only relatively thin residuals (less than 100 m), but in the south and southeast along and near the crest of the Great Divide and Tia Range, the lava pile is up to 600 m thick and estimated to have been originally in the range 100–200 km² in area. The present volume of lavas is about 20 km³.

K-Ar ages for three samples from the southeastern part of the province are 56, 47, and 44 Ma (Wellman & McDougall, 1974b). More data and better field control are required to determine whether two periods of volcanism are represented. The dates are closely similar to ages obtained for Barrington volcano to the southwest (55–44 Ma; McDougall & Wilkinson, 1967; Wellman et al., 1969; Wellman & McDougall, 1974b). Flows rest on a basement composed predominantly of poorly differentiated Ordovician to Permian sediments of the New England Fold Belt that have a range in degree of metamorphism, including one high-grade regional metamorphic terrain (the Tia Complex; Gunthorpe, 1970).

No detailed geological or petrological studies of Walcha volcanic rocks have been undertaken, except in connection with occurrences of megacrysts or ultramafic xenoliths in flows near Walcha and Tia (Binns et al., 1970; Stolz, 1984). The rocks are predominantly alkaline, ranging from alkali basalt through hawaiite (Table 3.5.1, column 8) to mugearite. Analysed lavas include K-basanite, alkali basalt, hawaiite (for example, Binns et al., 1970), and ne-mugearite (Stolz, 1984).

The basalts have a wide range of silica saturation from moderately undersaturated (more than 10-percent normative nepheline) basanite, ne-hawaiite, and ne-mugearite, to hypersthene-normative hawaiite and ol-tholeiitic basalt. The rocks are predominantly sodic. Analysed rocks have K_2O/Na_2O values in the range 0.24–0.42, except for one sample. The rocks appear to be chemically similar to the alkaline rocks of the Central province. High-pressure megacrysts are known from three localities (Binns et al., 1970) and may be relatively common in the province. Garnet-pyroxenite xenoliths in a ne-mugearite dyke southwest of Walcha were described by Stolz (1984).

3.5.8 Barrington

Cainozoic igneous rocks of the Barrington province crop out mostly on the Barrington Tops Plateau and its escarpment, although outliers are found to the west and southwest (Fig. 3.5.5). The original total volume of this lava-field type volcanism may have been about 700 km³, but the current total volume is only about 120 km³. K-Ar ages range from 55 to 44 Ma (Wellman et al., 1969; Wellman & McDougall, 1974b). Much of the sequence appears to have been erupted rapidly, although a few intrabasalt palaeosols correspond to periods of quiescence.

Basement for the Cainozoic rocks consists of deformed Carboniferous and Devonian sedimentary and volcanogenic rocks (Baker, 1983; see also references therein) and Permian granites of the Barrington Tops Granodiorite (Mason & Kavalieris, 1984; Eggins, 1984). Significant relief on the basement was centred on the granites (Galloway, 1967; Pain, 1983) that formed rounded hills around which the lavas accumulated and, in part, covered. Wellman (1986) showed that Barrington volcano coincides with a basement uplift and with gravity and magnetic anomalies (Section 1.4.9; Fig. 1.4.10H).

The Cainozoic volcanic rocks originally may have formed a low-angle shield which subsequently was extensively modified by erosion (Pain, 1983). Sub-horizontal lava flows 2–10 m thick dominate the pile. A volcanic plug of alkali gabbro is found at Careys Peak and there is a sill-like body of theralite at Hunters Springs. Flow directions in lavas from the southern part of the field (recorded from elongate olivine phenocrysts and felted groundmass plagioclase laths) are consistent with the Careys Peak plug having acted as a vent.

The great majority of the rocks is basalt, and there are only a few samples having slightly evolved hawaiitic compositions. Ol-tholeiitic basalt lavas were erupted first, and accumulated in topographic lows in the northern part of the province. A basal sequence of at least four flows up to 50 m thick is present in the Sempills Creek region, but tholeiitic basalt is absent from higher in the pile.

Alkaline rock types constitute the bulk of the volcanic pile. They include transitional basalt, alkali basalt, basanite (Table 3.5.1, column 9), and nephelinite.

Many are porphyritic in olivine or clinopyroxene or both. Some ankaramites contain up to 40 volume percent titanaugite phenocrysts up to 1 cm across. Nephelinite is rare, but good exposures are found on the western slopes of Mount Royal. Small (less than 3 cm) peridotite inclusions, mainly of spinel lherzolite, are present in some of the finer-grained alkaline lavas. Most of the analysed rocks have Mg-ratios in the 70–60 range and so approach primary or near-primary magma compositions.

The differentiated Square Top intrusion (59 Ma; McDougall & Wilkinson, 1967), near Nundle, consists of theralite-tinguaite (Wilkinson, 1965) and may belong to the Barrington province, or possibly to the Walcha province.

3.5.9   Comboyne

The Comboyne Plateau lavas and high-level intrusions had an estimated original diameter of about 20 km and a total volume about 50 km³ (Fig. 3.5.5). Most of the Cainozoic igneous rocks are on the Comboyne Plateau, although there are important outcrops on the eastern side of the adjoining Bulga Plateau, in the Lansdowne valley, Lorne Basin, and to the north of Wingham. Cainozoic igneous rocks overlie the Triassic Camden Haven Group which in turn unconformably overlies Palaeozoic strata that were deformed during the Hunter-Bowen Orogeny. Post-Triassic faulting is found in parts of the Lorne Basin (Stewart, 1953). The tendency for alkaline plugs to be grouped in the Lansdowne valley and on the western rim of the Lorne Basin may be an indication that there was some tectonic control during emplacement. Wellman (1986; Section 1.4.9, Fig. 1.4.10G) suggested that a cogenetic intrusive complex is responsible for basement uplift and for a magnetic anomaly centred near Mount Gibraltar on the southern scarp of the plateau.

Lavas and high-level intrusions of the Comboyne central-type province are typically transitional in character, ranging from mildly alkaline to subalkaline compositions. Rock types are hawaiite, *ol-* and *qz-*tholeiitic basalt, mugearite, icelandite, benmoreite, dacite, *qz*-trachyte, and peralkaline rhyolite (Table 3.5.1, column 10).

Hawaiite and *ol-* and *qz*-tholeiitic basalt are the most mafic rocks in the area, and generally form boulder slopes low on the periphery of the Comboyne Plateau, particularly on the western and northwestern escarpments. Boulder slopes of these rocks are present also on the northwestern rim of the Lorne Basin and on the Bulga Plateau. The most spectacular occurrence of hawaiite is the Lansdowne valley Mount Baldy plug which rises steeply from the valley floor. Mugearite and icelandite have been identified from near the southeastern escarpment, and from a boulder slope relatively low on the northwestern escarpment. The absence of field relationships precludes positive identification of these as either high-level intrusions or lava flows, although the fine grain-size of the groundmass is consistent with the latter. Mugearite flows are found also to the west of Hannon Vale in the Lorne Basin where they form terraces overlying possibly Triassic calcalkaline rhyolite (Knutson, 1975).

Benmoreite and dacite crop out extensively at a relatively high elevation on the Comboyne Plateau, and their wide range in grain-size is probably an indication that these lavas formed both high-level intrusions and lava flows. A small area of benmoreite and dacite low on the southern escarpment of the Comboyne Plateau is isolated from the major occurrences of similar rocks. Here porphyritic and vesicular benmoreite and dacite appear to intrude the Carboniferous Giro Beds and are in close proximity to a volcanic-breccia pipe that contains fragments of vesicular benmoreite and dacite, peraikaline rhyolite, and sedimentary material, ranging up to 1.5 m in diameter, but most commonly about 1.0 cm. A second breccia pipe on the eastern escarpment of the Comboyne Plateau is surrounded on three sides at least by peralkaline rhyolite.

Qz-trachyte forms the prominent hills on the Comboyne Plateau and the eastern escarpment of the plateau. These trachytes are probably intrusive, but the fine grain-size of the rock matrix is an indication that final crystallisation took place at very shallow levels.

Comboyne Plateau peralkaline rhyolites tend to be concentrated towards the southern margin of the plateau as stratigraphic highs, and in some places they overlie benmoreite and dacite. The peralkaline rhyolites are mainly fine-grained and they have well-developed flow banding reflecting the alignment of both mafic and felsic minerals. Peralkaline-rhyolite plugs in the Lansdowne Valley, and the western rim of the Lorne Basin, have a range in composition and texture, but the majority are relatively coarse-grained, pointing to their intrusive nature. One dome-like body consists of flow-banded, non-porphyritic peralkaline rhyolite, and was probably formed from lava flows emanating from the summit. Brecciated feldspar and quartz with undulose extinction are present in some of the plugs, indicating emplacement as a crystal mush.

The Lansdowne valley Mount Oliver peralkaline rhyolite plug has been dated at 16 Ma (McDougall & Wilkinson, 1967), and K-Ar dating of a fine-grained Comboyne Plateau trachytic rock by Wellman (1971) also gave an age of 16 Ma.

The actual order of emplacement of the Comboyne lavas and high-level intrusions can be inferred only from field relationships and geochemistry. The low stratigraphic position of the Comboyne hawaiites and *ol-* and *qz*-tholeiitic basalts is consistent with their being erupted prior to the more evolved lavas. The increase in degree of differentiation with stratigraphic level is indicative that volcanism proceeded from hawaiite and *ol-* and *qz*-tholeiitic basalt towards more evolved lava types.

Volcanic rocks in the Lorne Basin and Bulga Plateau initially may have been more extensive as remnants are found now only as isolated topographic highs. Erosion, including major coastal escarpment retreat (Pain & Ollier, 1986), evidently has removed much of the volcanic material. Comboyne lavas have large areal extent and limited total thickness, and therefore flows of restricted volume were probably emitted from several volcanic centres.

The transitional lavas and high-level intrusions of the Comboyne province consist mainly of intermediate and felsic rock types, and the most primitive rock analysed has an Mg-ratio of only 51. Most of the rocks are hypersthene-normative, and a continuum of compositions exists from hawaiite to peralkaline rhyolite (Fig. 3.5.2). The series is sodic although the more evolved rock types have strong enrichment in K_2O relative to Na_2O. Hawaiites are mostly fine- to medium-grained containing phenocrysts of olivine, clinopyroxene, and plagioclase. Megacrysts and cumulates consisting of orthopyroxene, olivine, clinopyroxene, plagioclase, ilmenite, titanomagnetite, and apatite are present in some hawaiites and, more rarely, in the mainly aphyric mugearites. There is experimental evidence that these megacrysts and cumulates crystallised at a depth of about 20–24 km prior to rapid eruption (Knutson & Green, 1975). Anorthoclase, sanidine, riebeckite/arfvedsonite, hedenbergite/aegirine, and less commonly aenigmatite are present in the more evolved trachytes and rhyolites.

3.5.10 Dubbo

Alkaline and associated rocks of the Dubbo province (Fig. 3.5.1) overlie or intrude early to middle Mesozoic sedimentary units of the Great Artesian Basin (Dulhunty, 1973a). Felsic rocks to the south of the area overlying or intruding Palaeozoic units (Offenberg et al., 1968) have similar morphology, petrography, and geochemistry, and probably represent related igneous activity.

The only radiometric dates available are for potassic hawaiites cropping out just north of Dubbo. These are 14–12 Ma (Dulhunty, 1973b; Wellman & McDougall, 1974b). However, there is strong geomorphological, petrological, and geochemical evidence that there were at least two magmatic episodes — the first in the Eocene to Lower Oligocene, and the second in the Miocene (Dulhunty, 1973b). Some felsic domes appear analogous to Jurassic felsic rocks from the Mullaley area in northeastern New South Wales (Bean, 1974) and may be of a similar age. Some of the topographically high felsic and mafic rocks are inferred to be older erosional remnants representing lavas erupted over an early Tertiary erosion surface of relatively mature topography and uplifted to different degrees later in the Tertiary. The younger Miocene flows occupy topographic lows and have differences in elevations caused by the undulating Miocene erosion surface.

Felsic rocks dominate the southwestern parts of the Dubbo area, whereas basaltic rocks dominate in the northeastern parts. Older felsic masses are found as elevated domes of porphyritic rocks containing fine-grained, trachytic to subtrachytic groundmasses. Undersaturated syenite plugs are closely associated with phonolite flows and may represent the high-level intrusive parts of the eruptive centres for these magmas. One exposed sill (or possibly a thick flow) of analcime trachyte forms a columnar-jointed, thick (more than 30 m) capping at Geurie Hill (Ortez, 1976).

The basaltic flows can be divided into elevated and low topographic types corresponding to the age differences described above. The older igneous activity is dominated by both saturated and undersaturated felsic rocks, rare analcime-mugearite intrusions, and analcime-hawaiite sills. Some of these may be con-

temporaneous with the inferred Eocene to Lower Oligocene basaltic flow remnants on topographic highs. The Miocene basaltic activity consists of ol-tholeiitic basalt, basanite, alkali basalt, and highly potassic hawaiite. Many of the basaltic rocks are primary or primitive. Sodic and potassic groups can be distinguished on the basis of K_2O/Na_2O values. Compositional data for primary, nepheline-normative basaltic rocks of this region (at the western margin of the main belt of Tertiary volcanism) where plotted with available data for rocks from farther east, define a trend of increasing K_2O/Na_2O as distances increase westward.

Xenoliths and megacrysts are widespread and locally abundant in the basaltic rocks (Wass & Irving, 1976). Xenoliths consist of Cr-diopside lherzolite, wehrlite, pyroxenite, lower-pressure gabbroic rocks, and very rare granulite. Abundant megacryst species include clinopyroxene, orthopyroxene, olivine, anorthoclase, and plagioclase. One ol-tholeiitic basalt contains orthopyroxene and olivine megacrysts which may represent disaggregated Cr-diopside lherzolite xenoliths.

3.6 Central and Southern New South Wales

3.6.1 Introduction

Cainozoic provinces in central and southern New South Wales (Fig. 3.6.1) range in age from about 57 Ma in the Sydney province to as young as 10 Ma for the Cargelligo leucitite. All but one of these provinces contain primary or near-primary alkali basalt, basanite, nephelinite, or leucitite as the dominant lava types.

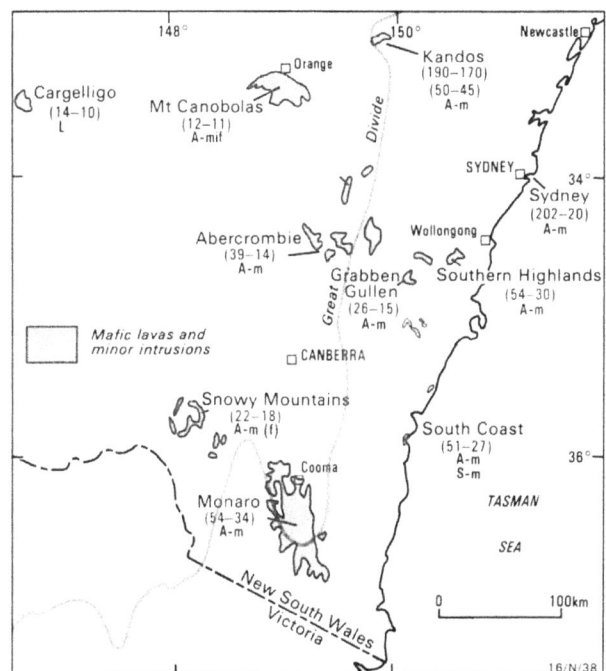

Figure 3.6.1 Distribution, rock types, and ages of volcanic provinces in central and southern New South Wales. Rock associations: A, alkaline; S, subalkaline. Rock types: m, mafic; i, intermediate; f, felsic; (f), minor felsic; L, leucitite. Ages (in brackets) in Ma.

These typically form thin flows that make up widespread lava-field type sequences. In contrast, extrusive and high-level intrusive rocks in the Canobolas area consist mainly of more evolved mildly alkaline to transitional rocks that form a large, central-type volcanic edifice. Generalised compositional fields are given in Figures 3.6.2 and 3.6.3, and selected chemical analyses are given in Table 3.6.1.

3.6.2 Canobolas

The Middle to Upper Miocene rocks of the Canobolas volcanic complex crop out to the south and west of the town of Orange in central New South Wales, and cover an area of about 825 km² (Fig. 3.6.4). The rocks are mainly hawaiite, qz-tholeiitic basalt, mugearite, benmoreite, trachyte, peralkaline trachyte, rhyolite, and peralkaline rhyolite. They are part of the moderately well-preserved remains of a shield volcano (Süssmilch & Jensen, 1909; Middlemost, 1981) which has low-angle, stratiform flanks, surrounding a large central

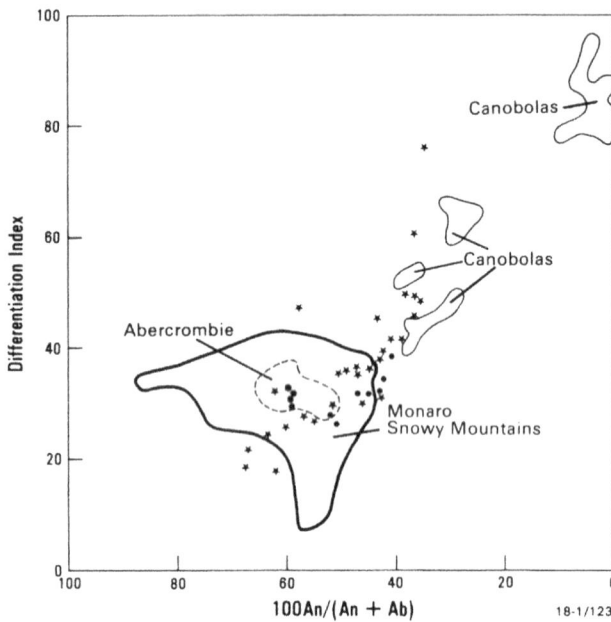

Figure 3.6.2 Compositional fields for central and southern New South Wales provinces using the plotting procedure outlined in Section 1.1.4. Stars refer to Southern Highlands samples. Filled circles refer to South Coast samples.

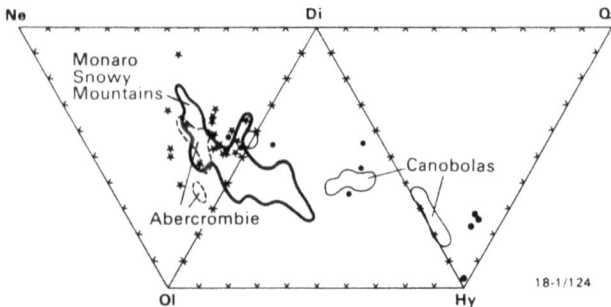

Figure 3.6.3 Compositional fields for central and southern New South Wales rocks projected in the system Ne-Ol-Di-Hy-Qz. Stars refer to Southern Highlands samples. Filled circles refer to South Coast samples.

zone containing a range of volcanic domes, dykes, and small cones. The estimated volume of Canobolas volcano is about 50 km³ and, at the present level of erosion, trachyte is the most abundant rock type in the core of the complex and hawaiite is dominant on the flanks. Canobolas rests on a basement high (Wellman, 1986; Section 1.4.9; Fig. 1.4.10I).

The rocks that formed the flanks of the volcano are best preserved to the northwest and southeast of the volcanic complex. Flows in the northwest normally radiate out from the central zone, and the dips are shallow. Most flows are only 2–3 m thick, but a few are greater. Pyroclastic beds, palaeosols, and layers of diatomaceous earth are found interlayered within some flows. The main rock types in the northwestern part of the complex are hawaiite and mugearite. Extensive flows of lava in the southeastern area extend radially outwards from the central zone for over 20 km. These lavas are also mainly hawaiite and mugearite, and in some areas they are over 150 m thick. The hawaiites mostly have vitrophyric textures and their essential minerals are olivine, augite/salite, and plagioclase (An₅₂). Most rocks also contain Fe-Ti oxides.

Extensive erosion has removed much of the southern part of the volcanic edifice. The rocks that remain are mainly hawaiite, mugearite, and trachyte. More trachyte crops out to the north of this area, where it is overlain by hawaiite.

An 8 km-long tabular body of trachyte in the Panuara area attains a maximum width of 3 km and a thickness of 30 m. It was first described by Stevens (1954). The base of this flow decreases in elevation from north to south, so the lava probably erupted from a vent in the central zone of the volcanic complex. Landscape in the southwestern part of the complex is dominated by north-trending ridges of Palaeozoic supracrustal rocks which impeded the free movement of hawaiite to trachyte flows. Trachyte and, more particularly, trachytic pyroclastic-flow deposits, are the predominant rock types in this area.

The structurally and petrographically complex central zone of the Mount Canobolas volcano once contained an extensive system of eruptive vents. These are now represented by an intricate pattern of domes,

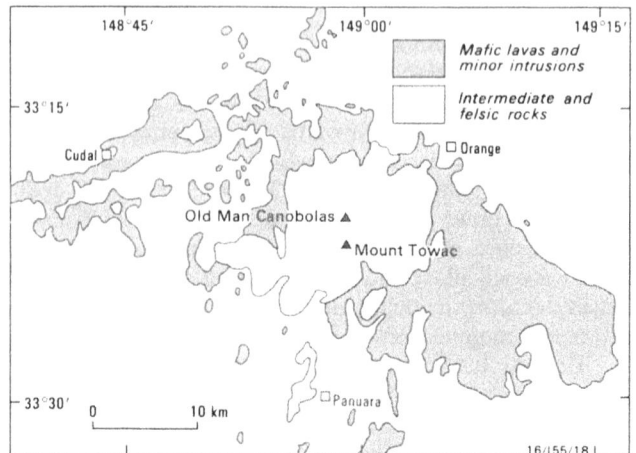

Figure 3.6.4 Distribution of volcanic rocks in the Canobolas province of central New South Wales (see Fig. 3.6.1).

Table 3.6.1   Selected chemical analyses for central and southern New South Wales provinces

	1	2	3	4	5	6	7	8	9	10
SiO_2	44.62	47.92	44.80	45.36	46.47	44.53	45.12	50.03	43.08	44.97
TiO_2	2.50	2.11	2.43	1.46	1.88	2.12	2.35	1.70	4.11	4.68
Al_2O_3	14.73	15.64	14.20	13.39	14.33	14.14	14.89	14.71	9.97	8.60
Fe_2O_3	1.87	1.95	2.61	2.35	2.30	2.42	1.87	1.52	3.88	3.02
FeO	9.35	9.74	8.81	9.30	9.16	7.92	9.35	8.29	7.25	6.67
MnO	0.19	0.21	0.17	0.19	0.16	0.17	0.17	0.14	0.17	0.12
MgO	9.69	6.74	9.67	12.85	8.81	9.90	8.72	7.29	12.78	13.64
CaO	10.05	6.81	8.94	7.93	9.54	10.10	9.00	9.77	9.40	7.89
Na_2O	4.01	5.00	3.28	2.75	3.01	3.86	3.37	3.14	2.14	0.88
K_2O	1.85	2.55	1.93	0.99	1.19	1.58	1.57	0.52	4.08	7.09
P_2O_5	0.94	1.29	0.55	0.36	0.52	1.27	1.05	0.29	1.12	0.87
S						<0.02	<0.02	0.03		
H_2O+	1.49[1]	1.77[1]	2.06[1]	2.56[1]	1.49	0.53	1.68	0.80	1.14[2]	0.86[2]
H_2O-					0.89	0.52	0.73	0.67		
CO_2					0.20	0.44	0.20	0.91		
Total	101.29	101.73	99.45	99.49	99.95	99.50	100.07	99.81	99.12	99.29
Ba	479	460	546	250	319	980	480	145	1170	1100
Rb	28	20	30	16	22	25	32	7.5	114	218
Sr	1094	1456	701	628	607	1420	910	415	1210	940
Pb	15	10			6	6	4	2		
Th	11	9			6	12	4	1.5		
U	5	2				3	1.5	0.5		
Zr	286	462	218	179	172	239	191	106	550	515
Nb			63	40	70	116	58	17.5		
Y	33	27	28	26	22	21	23	17		
La			32	22	36	82	33	14		
Ce			62	43	70	149	74	29		
V	167	110	200	156	141	161	143	145		
Cr	464	147	262	358	189	292	203	228	419	330
Ni	836	85	155	277	141	225	169	138	369	437
Cu	100	27			54	56	75	52		

1. Basanite from Southern Highlands province (Wass, 1980, sample 45065).
2. *Ne*-hawaiite from Southern Highlands province (Wass, 1980, sample 45094).
3. Basanite from Abercrombie province (Morris, 1986, sample 64386).
4. Alkali basalt from Abercrombie province (Morris, 1986, sample 64400).
5. Alkali basalt from Monaro province (Kesson, 1972, sample K65).
6. Basanite from Snowy Mountains province (J. Knutson & B.W. Chappell, unpublished data, sample EAV203).
7. Basanite from South Coast province (J. Knutson & B.W. Chappell, unpublished data, sample EAV312).
8. *Ol*-tholeiitic basalt from South Coast province (J. Knutson & B.W. Chappell, unpublished data, sample EAV315).
9. Leucitite from Lake Cargelligo, New South Wales (Cundari, 1973, sample LC–9).
10. Leucitite from El Capitan, New South Wales (Cundari, 1973, sample CPT–11-II).

[1] Loss on ignition.
[2] Total H_2O.

plugs, and dykes that occupy a roughly circular-shaped area covering about 210 km². Most of this area is higher than 800 m, and it contains a complicated drainage pattern that is dominated by a range of resistant constructional landforms, such as volcanic domes, plugs, and small cones. Trachytic domes, many of which are devoid of vegetation, dominate the central zone, and give the false impression that trachytic lava is the predominant type of rock found in this area. However, the central zone exposed in road cuts and areas of recent gullying is seen to contain a large volume of felsic pyroclastic materials. Mugearite and benmoreite are also found, and mostly overlie trachyte. Relationships between the different rock units in many parts of the central zone are difficult to unravel because rocks of different ages have been juxtaposed by volcanic subsidence. Mugearite and benmoreite are generally porphyritic to cumulophyric rocks in which plagioclase is the most common phenocryst. Some benmoreite has plagioclase mantled by sanidine, and sanidine or anorthoclase (or both) are commonly found in the groundmass. Augite and Fe-Ti oxides are common mafic minerals.

The landscape in the northwestern sector of the central zone is dominated by the presence of more than twenty, relatively small trachytic volcanic domes that protrude through a sequence of shallow-dipping mafic to intermediate rocks. There is more felsic pyroclastic material in the eastern and southern sectors of the central zone, but the landscape is dominated by solitary volcanic domes, or complex clusters of coalescing domes. These domes mostly consist of trachytic or, less commonly, rhyolitic rocks, and they tend to rest on or intrude felsic pyroclastic materials. The mean diameter of the exposed parts of these domes is about 400 m (Fig. 3.6.5).

Figure 3.6.5   Columnar jointing in the peralkaline rhyolite of Bald Dome in the central zone of the Mount Canobolas volcanic complex.

Small outcrops of obsidian are found at several localities in the southern part of the central zone. These rocks are particularly well exposed on the side of a ridge about 700 m southeast of Mount Towac, where they are seen to consist of narrow alternating bands of light and dark brownish-grey glass. This glass contains about 70-percent SiO_2, and over 8-percent total alkalis. Many specimens are slightly peralkaline. Some of the obsidians grade into tuffs mainly composed of deformed and flattened glass shards. These vitroclastic tuffs are ignimbrites, and some of the obsidians are regarded as strongly welded tuffs.

Pyroclastic materials that range from unconsolidated tuffs to coherent welded rocks are abundant and widely distributed throughout the central zone. Most of this material is of trachytic composition, and most of the widely distributed crystal tuffs contain a high proportion of alkali feldspar, mostly anorthoclase. The different types of pyroclastic materials are observed readily on the western flanks of Old Man Canobolas where both pyroclastic-fall and pyroclastic-flow deposits are found. Some of these bedded units locally are disrupted by bomb-sags that generally contain irregular fusiform volcanic bombs. The pyroclastic-flow rocks commonly have eutaxitic textures, and carry fragments of pumice and glass that have been distorted and stretched. Small discontinuous outcrops of volcanic breccia are found also within the central zone. These breccias mostly contain angular fragments of trachyte set in a felsic glassy or fine-grained trachytic groundmass. However, some of them — such as those found in the Mount Towac area — contain fragments of peralkaline rhyolite and coarser-grained syenitic rocks.

The Mount Canobolas volcanic complex was once part of a large shield volcano, about 50 km in diameter, and more-or-less similar in form to the Plio-Pleistocene trachytic volcanoes of the South Turkana region of Kenya (Middlemost, 1985). The flanks of the volcano were composed mainly of flows of hawaiite and mugearite, whereas the large central zone contained over 50 volcanic vents that produced a range of differentiated rocks — from mugearite, to benmoreite and trachyte, to alkali rhyolite. The highest known Mg-ratio is 52, so there is an apparent absence of volcanic rocks having primary or near-primary compositions. The mildly alkaline to transitional magma evidently evolved within a large, high-level, thermally and compositionally zoned magma chamber.

3.6.3   Sydney

The Sydney province is taken to include all the basaltic volcanic rocks and intrusions (including diatremes) within the Sydney Basin, excluding those in provinces otherwise specifically defined (Kandos, Barrington, and so on) that also lie within the Sydney Basin (Fig. 3.6.6).

Diatremes are particularly abundant and are distributed over an area of about 20 000 km^2 (Crawford et al., 1980). Many of these were considered originally to be of mainly Cainozoic age but, on the basis of more recent spore and pollen studies (Helby & Morgan, 1979), many are as old as early Jurassic (see Section 3.9.3). Some diatremes have associated plugs and small flows, and most are less than 500 m in diameter. They contain abundant fragments of country rocks and

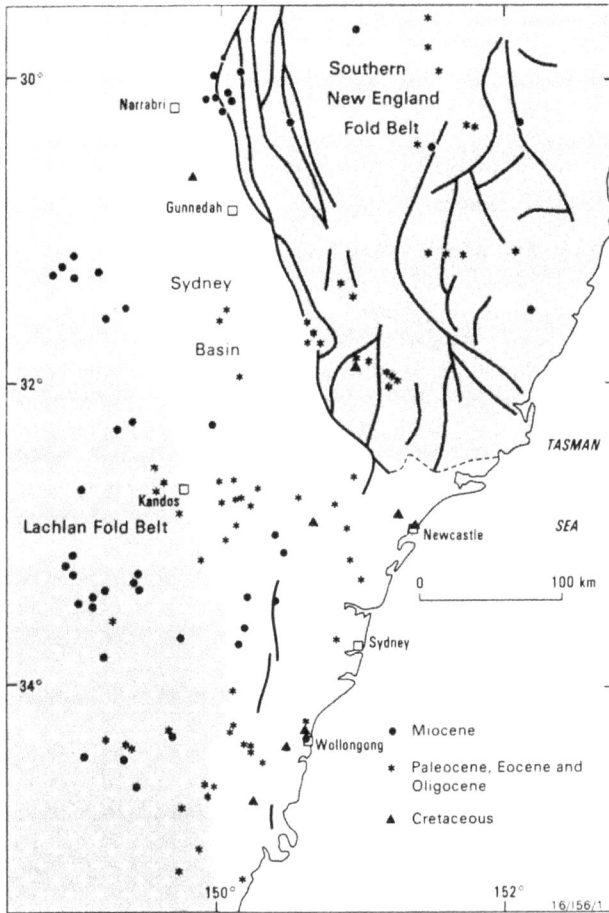

Figure 3.6.6 Distribution of volcanic centres in the Sydney Basin, eastern Lachlan, and southern New England fold belts.

sporadic upper-mantle and lower-crustal ultramafic and mafic xenoliths, including peridotite and lherzolite (W.A. Robertson, 1979; Emerson & Wass, 1980; Crawford et al., 1980; Embleton et al., 1985). In addition to these diatremes, a series of dykes and plugs east of the coastal lineament (Scheibner, 1976) extends from north of Broken Bay along the coast to its intersection with the structural hinge line (Bembrick et al., 1973) near Wollongong. Palaeocene, Eocene, and Oligocene volcanic rocks, such as the plugs and flows east of the Kandos province, are related mainly to northwesterly or east-northeasterly lineaments (Scheibner, 1973).

The basaltic rocks of the Sydney province form a geochemically coherent suite of basaltic, basanitic, and hawaiitic types. Primary magmas are relatively abundant.

3.6.4 Southern Highlands, Grabben Gullen, Abercrombie, and Kandos

Kandos province is within the Permian Sydney Basin, the Southern Highlands province partly within and partly beyond the western margin of the Sydney Basin, and the Grabben Gullen and Abercrombie provinces are on Palaeozoic sequences. However, all of these provinces are close to one another and are associated with the same broad region of tectonic uplift, related to the Great Divide, in eastern New South Wales (Fig. 3.6.1).

Basaltic rocks from each of the Southern Highlands, Grabben Gullen, and Kandos provinces have been well characterised geochemically, geomorphologically, and by radiometric dating, and they conform well with the criteria for discrete provinces used by Wass (1980). Fewer data are available for the Sydney and Abercrombie provinces, so interpretations for these must be regarded as preliminary.

Both the Southern Highlands and Kandos provinces have a bimodal distribution of ages. Southern Highlands Tertiary volcanic activity took place from 54 to 30 Ma (Wellman & McDougall, 1974b). Wellman & McDougall suggested that this volcanism may have centred around two main events at 45 and 35 Ma, but this may be because of the relatively small number of dated samples. The Kandos basalts have Tertiary ages ranging from 50 to 45 Ma, as well as Jurassic ages of 190 to 170 Ma (O'Reilly & Griffin, 1984; Embleton et al., 1985). These Tertiary and Jurassic products are geochemically similar, as in the Southern Highlands province.

Younger ages of 26 to 15 Ma (Young & Bishop, 1980) distinguish the small Grabben Gullen province from the adjacent Abercrombie and Southern Highlands provinces. However, western outliers of the Southern Highlands (or possibly southern outliers of the Abercrombie) have ages of 50 to 49 Ma at the northern margin (Bishop, 1984) so, again, the repetitive nature of thermal events in at least some regions is highlighted. Abercrombie province includes the Blue Mountains basalts as a northeastern extension. The age range for this province given by Wellman & McDougall (1974b) is about 39 to 14 Ma. However, two ages exceeding 21 Ma may be western extensions of the Southern Highlands province.

The Southern Highlands basaltic rocks are found generally as isolated flows, necks, dykes, and rare sills. Individual flows rarely exceed a few metres in thickness and most are less than 1.5 km² in extent. Rare flows up to 2 km long are preserved as hill cappings. Most isolated outcrops are not erosional remnants of large-scale flows, as they are geochemically and petrographically distinct from neighbouring outcrops. Localised flow sequences have been built up at rare eruption centres such as Rileys Peak, Mount Wayo, and Mount Wanganderry.

The Kandos basaltic rocks are found most commonly as columnar-jointed, high-level vent fillings that form topographic highs. Small flows, dykes, sills, diatremes, and rare flow sequences or large flows are also present. The geomorphology of these is well described in the meticulous and accurate account of the volcanic geology of the western coalfields by Carne (1908). A more recent work is that of Bradley et al. (1984).

The Grabben Gullen basalts form both lava flows of moderate volume that represent residuals of valley fills, and lava flows of small volumes together with associated local volcanic necks and diatremes (Bishop, 1984; O'Reilly & Griffin, 1984). The lavas rest unconformably on Palaeozoic rocks, or conformably over thin Cainozoic sediments, or are found as cappings over river gravels.

The Abercrombie Province basalts generally appear to form isolated flows and intrusions, but only the Oberon area (Morris, 1986) and the Blue Mountains basalts (Carne, 1908; O'Reilly & Griffin, 1984; Rickwood et al., 1983) have been studied in detail. The Oberon basaltic rocks (Morris, 1986) are mainly individual, primitive flows 3–10 m thick, unconformably resting on Ordovician-Silurian volcanic rocks and sediments and Carboniferous granites of the Lachlan Fold Belt (Packham, 1969). The Blue Mountains basalts were considered once to be remnants of a single flow (David, 1896), but on the basis of age and geochemistry they represent separate volcanic episodes (for example, Rickwood et al., 1983). Many of the Blue Mountains flows have features consistent with successive separate eruptions from a fractionating mid-crustal magma chamber.

All of these provinces contain flows, dykes, necks, or diatremes that carry high-pressure xenoliths together with megacrysts at some localities. These are mainly mantle-derived Cr-diopside spinel lherzolite, subordinate Al-augite type xenoliths (including the amphibole/apatite series), and rare crustal granulites. There are two significant features of the xenoliths in these provinces: (1) all the Cr-diopside lherzolites have thoroughly recrystallised mosaic microstructures ('coarse granular' using the nomenclature of Harte, 1977), except for some of the Mogo Hill xenoliths (Emerson & Wass, 1980) which have evidence of strong deformation and incipient recrystallisation; (2) mantle-derived xenoliths are dominated by volatile-free assemblages. Amphibole-bearing rocks (± mica ± apatite) are recorded only from a separate sub-province within the Southern Highlands (O'Reilly & Griffin, 1984) and from Umbiella Creek pipe at the southeastern margin of the Kandos province (Allen, 1972; Lovering & White, 1964).

Published geochemical data are available only for the Southern Highlands, Kandos, Grabben Gullen (O'Reilly & Griffin, 1984; Wass, 1980), and the Oberon and Blue Mountains basalts of the Abercrombie province (Carne, 1908; Rickwood et al., 1983; Morris, 1986). The Southern Highlands, Kandos, and Grabben Gullen provinces are dominated by primary and primitive melts ranging from rare nephelinite through abundant basanite, rare alkali basalt, and abundant hawaiite and ne-hawaiite (Table 3.6.1, columns 1–2). The primary or near-primary rocks are all geochemically distinctive and are interpreted to represent discrete batches of magma formed as separate (but related within provinces) volumes of partial melt.

The Oberon basanites and alkali basalts (Table 3.6.1, columns 3–4) are also commonly near-primary and considered to be discrete partial-melt batches (Morris, 1986). The Blue Mountains basalts have relatively low Mg-ratios and Ni values and contain no mantle-derived xenoliths. These observations are consistent with some fractionation at middle- to upper-crustal levels, and are supported by decreasing Ni and Cr contents concomitant with decreasing Mg-ratios.

The dominantly primitive, K-rich, basalts from the Southern Highlands area are all alkaline to strongly alkaline, except for one. There appears to be a continuous chemical range among the basaltic rocks. More felsic compositions are lacking except for one

flow of K-rich hawaiite. Mg-ratios range from 73 to 29, and most are greater than 60. The province overall has extreme ranges and high abundances in minor and trace elements.

3.6.5 Monaro, Snowy Mountains, and South Coast

The southern New South Wales area contains numerous, mainly small and isolated, occurrences of lava-field type basalts. The area extends from Canberra to the New South Wales and Victorian border, and includes basalts in the Burrunjuck area to the west of Canberra, the coastal area to the east of Canberra, the Monaro area, and the Tumut, Kiandra, and Tumbarumba areas in the Snowy Mountains (Fig. 3.6.1). Monaro is the largest and the best studied province (Kesson, 1972; Brown et al., 1988). The other areas are of limited extent and have not been studied extensively.

The Monaro province has multiple flows of basalt that cover hundreds of square kilometres and that have aggregate thicknesses of up to 400 m (*Fig. 3.6.7*). The cover is almost continuous between Cooma and Ando to the south, and around this area there are elongate areas of basalt outcrop that fill deep palaeovalleys. Taylor et al. (1985) mapped near Cooma, and Veitch (1987) and Brown et al. (1988) studied the south-Monaro area. They showed that relief at the time of the eruptions was up to at least 500 m, and that erosion rates since then have been low. Young & McDougall (1985) obtained similar results from the Shoalhaven and Ulladulla regions.

Alluvial and lacustrine sediments are found at the base of, and interbedded with, the flows of the Monaro province. These are thickest and most extensive around the southern and western margins of the province where pre-basalt valleys contain lacustrine sequences of clay, silt, coal, and sand, up to 70 m thick. At least three weathering profiles including red bauxite caps are present within the volcanic sequence between Cooma and the Maclaughlin River. Hyaloclastite deposits consisting of fragments of basaltic glass and lacustrine clay, and enclosing small pillows, are found between Bombala and Cathcart, and well-bedded basaltic pyroclastic deposits are present near Red Cliff, northwest of Bombala.

The volcanic rocks have a limited compositional range and are mainly flows of alkali basalt (Table 3.6.1, column 5), basanite, and minor nephelinite. Several plugs and dykes of nephelinite, basanite, and alkali gabbro have been located on the southern Monaro. The most alkaline of these, near Nimmitabel, contain kaersutite and biotite megacrysts. Upper-mantle and lower-crustal xenoliths are common in the nephelinite and basanite plugs. Mg-ratios for the basalts are mostly between 70 and 60. The lowest recorded value of 35 is for a phaneritic intrusive nephelinite from between Bombala and Cathcart.

The Kiandra basalts to the west are less voluminous than those of Monaro, and are restricted mainly to valley flows 30 to 100 m in total thickness. These are now reduced by erosion to a series of disconnected elongate outcrops that form small plateaux and narrow-topped ridges (Mackenzie & White, 1970). The basalts appear to have originated from several centres — including Round and Tabletop mountains which are probably volcanic plugs — as well as dykes. These basaltic rocks are alkaline lavas, mainly basanite, alkali basalt, and nephelinite, and are similar petrographically and geochemically to the Monaro basalts. All contain mainly olivine and, in some, clinopyroxene phenocrysts (or xenocrysts, or both) in a groundmass of titanaugite, plagioclase, magnetite, and different amounts of interstitial alkali feldspar, nepheline, and analcime.

Other basalt localities in the Snowy Mountains include the many small areas of lava flows that rest on plateau surfaces near the Bago Range, Cabramurra, and Tumbarumba. The flows descend almost to the Tooma River, and range in altitude from 450 to 1200 m. These appear to be predominantly alkali basalt and basanite (Table 3.6.1, column 6) whose Mg-ratios are mostly 70–60. Similar, widely dispersed basalts in the Wee Jasper, Burrunjuck, and Honeysuckle Range area also top plateaus and partly fill valleys.

Scattered outcrops of basanite, alkali basalt, hawaiite, and ol-tholeiitic basalt flows mostly up to 30 m thick are also common in the southern coastal area of New South Wales (Table 3.6.1, columns 7–8). Outcrops are known at Moruya, Pambula, Central Tilba, Batemans Bay, Nerriga, Ulladulla, Milton, and Tuross Lakes (Purvis, 1965; Etheridge, 1967; Kesson, 1968; Raine, 1968; Burton, 1973; Govey, 1974; Kidd, 1975; Elphinstone, 1983). These basalts are commonly vesicular, porphyritic, and iddingsitised, and most appear to have flowed along valley floors as extensive sheets. Remnants of basaltic lava flows to the west of Ulladulla can be traced halfway across the lowland between the present coastline and the foot of the escarpment (Young & McDougall, 1985). Thin sequences of basaltic flows overlie sediments near Nerriga and between Bungonia and Lake Bathurst. A 100 m-thick sequence of tholeiitic basalt near Nerriga fills a steep-sided valley. Alkali and transitional basalts form flow sequences up to 90 m thick in the Araluen area southeast of Canberra. These crop out along the Shoalhaven Fault, along which 30 m of vertical movement has taken place since emplacement of the basalts (D. Wyborn, personal communication, 1986).

The southern New South Wales basaltic lavas have a wide range of K-Ar ages (Wellman & McDougall, 1974b). Monaro basalts are the oldest. K-Ar ages range from 54 to 34 Ma, and pollen from sediments underlying and interbedded with basalts low in the sequence correspond to a somewhat older late Palaeocene age (E. Truswell, personal communication, 1987). The closely similar Snowy Mountains basalts to the west are significantly younger at 22 to 18 Ma. Basalts in the Nerriga area are dated at 51 to 40 Ma, in the Moruya area 35 to 27 Ma, and in the Ulladulla area 30 to 29 Ma (Wellman & McDougall, 1974b; Young & McDougall, 1982). The basalts in the Araluen area southeast of Canberra are dated at 19 Ma (D. Wyborn, personal communication, 1986).

3.6.6 East Australian leucitite suite

The known outcrops of the east-Australian leucitite suite are within a 90 km-wide strip extending 640 km south from Byrock, 75 km southeast of Bourke in New South Wales, to Cosgrove in Victoria (Fig. 1.1.5). However, about 90 percent of the total outcrop area — conservatively estimated to be about 114 km² — is concentrated within 1300 km² in the Lake Cargelligo and Tullibigeal district of New South Wales (Cundari, 1973). The total volume of lava is difficult to estimate because a substantial proportion of rock may have been removed by erosion, but probably it would not have exceeded 3 km³.

The lavas generally rest unconformably on a peneplained basement complex consisting mainly of Palaeozoic geosynclinal sediments and granites, but they also have been reported to overlie horizontally bedded Cainozoic sediments (for example, Branagan, 1969). North- to northwest-trending faults and lineaments are prominent in the basement formations. These were inherited from intense folding during the Palaeozoic orogenies and subsequent post-orogenic stabilisation associated with block faulting and uplift (Browne, 1969).

K-Ar ages of the leucitite lavas range between 16 and 10 Ma in New South Wales (Wellman et al., 1970; Wellman, 1983), and 6 Ma for the Cosgrove leucitite (Wellman, 1974). Apatite fission-track ages are 14–13 Ma for Begargo Hill (Cundari et al., 1978) and 7 Ma for Cosgrove (A.J.W. Gleadow, personal communication, in Birch, 1976).

Lava flows are the dominant volcanic form. These are dissected to different degrees by gully erosion, forming hummocky ridges that have low-angle flanks (for example, the Tullibigeal field). Differential removal of softer rocks from the flanks of valley flows by lateral streams resulted in relief inversion at El Capitan (Cundari, 1973; Cundari & Ollier, 1970). Lava mounds with lobate boundaries, seldom rising more than 50 m above the surrounding plains, are common — for example, in the Lake Cargelligo field. Arcuate domal structures form prominent landmarks at Mount Bygalore (Cundari, 1973) and 17 km southwest of Condobolin. Pyroclastic material is found at Flagstaff Hill where a well-preserved scoria cone is nested in a leucitite flow. The volcanic activity of the east-Australian leucitite suite was predominantly from small vents that were sealed and buried after one major lava eruption.

The lava, where massive and fresh, is remarkably uniform in lithology, both within outcrops and throughout the suite. It is dark blue to dark grey, aphyric to microphyric, and contains golden mica flakes and iddingsitised olivine that commonly are distinguishable in hand specimen. Coarser-grained pegmatoid types containing well-developed pyroxene crystals up to 2 cm in length are found at Begargo Hill (Cundari, 1973) and at Cosgrove (Birch, 1979).

Minerals identified in polished thin section are clinopyroxene (33–45 volume percent), leucite (19–36 percent), olivine (3–22 percent), and Fe-Ti oxides (6–22 percent). Lower amounts of modal leucite (15 percent) and higher clinopyroxene (55 percent) are reported for Cosgrove by Birch (1979). Mica and amphibole are found in accessory amounts (3–11 percent), and alkali feldspar or nepheline (or both) are common (14–17 percent) at some localities (for example, Condobolin, Begargo Hill; Cundari, 1973), but are generally less than 5 volume percent. Perovskite is notable (4–8 percent) in some specimens from El Capitan (Cundari 1973), and aenigmatite and sodalite are present in trace amounts at Cosgrove (Birch, 1979). These lavas are classified as melanocratic leucitites, in view of the high proportion of mafic minerals (average 73 volume percent for 28 samples).

The leucitite lavas are notable for their extreme compositions and narrow compositional differences (Cundari, 1973; Table 3.6.1, columns 9–10). Bulk-rock compositions have high Ti, Mg, K, P, H_2O+, Cr, Ni, Zr, Rb, Sr and Ba. High Mg-ratios, averaging 79, and high Ni and Cr values correspond to those of near-primary magmas.

3.7 Victoria and South Australia

3.7.1 Introduction

Young volcanic activity was first recognised in the central highlands and western plains of Victoria by the explorer Major Mitchell in 1836 (see Section 1.1.2), and has been the subject of intermittent study ever since. The most recent summaries are those of Ollier (1967a), Joyce (1975), and Ollier & Joyce (1976). Bibliographic listings were given by Joyce (1974, 1982). The Cainozoic volcanic rocks of Victoria are divided into three main groups: the Older Volcanics, the Newer Volcanics, and Macedon-Trentham (Fig. 3.7.1). The early to middle Tertiary Older Volcanics are confined mainly to the eastern part of the state (Singleton & Joyce 1969), and even earlier activity of Lower to Upper Jurassic age has been described.

Activity in the late Tertiary to Holocene Newer Volcanics is of lava-field type and analogies have been drawn with Strombolian-type activity in which both explosive and effusive eruptions are important. A major feature of the younger activity has been the development of maar volcanoes. The Newer Volcanics province is estimated to cover 15 000 km^2 from nearly 400 points of eruption (Joyce, 1975). The Older Volcanics style of activity may have been similar to that identified in the Newer Volcanics province, but few centres of eruption have been found.

Macedon-Trentham volcanic rocks are distinguished from both the Older and Newer Volcanics by their

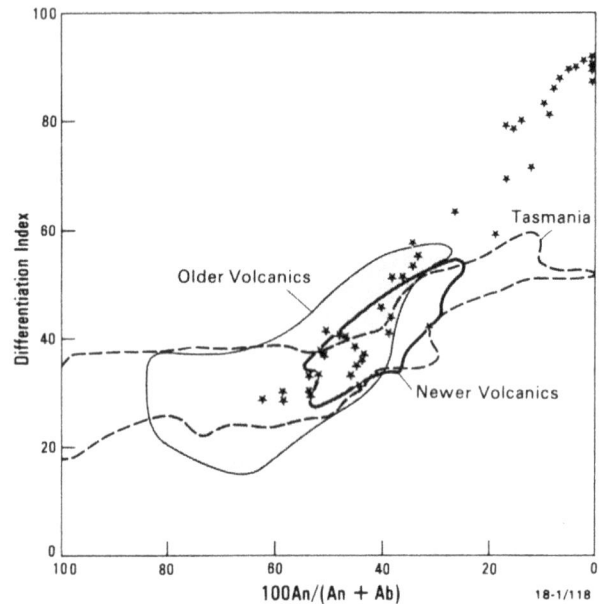

Figure 3.7.2 Compositional fields for Victorian, South Australian, and Tasmanian provinces using the plotting procedure outlined in Section 1.1.4. Stars refer to Macedon-Trentham samples.

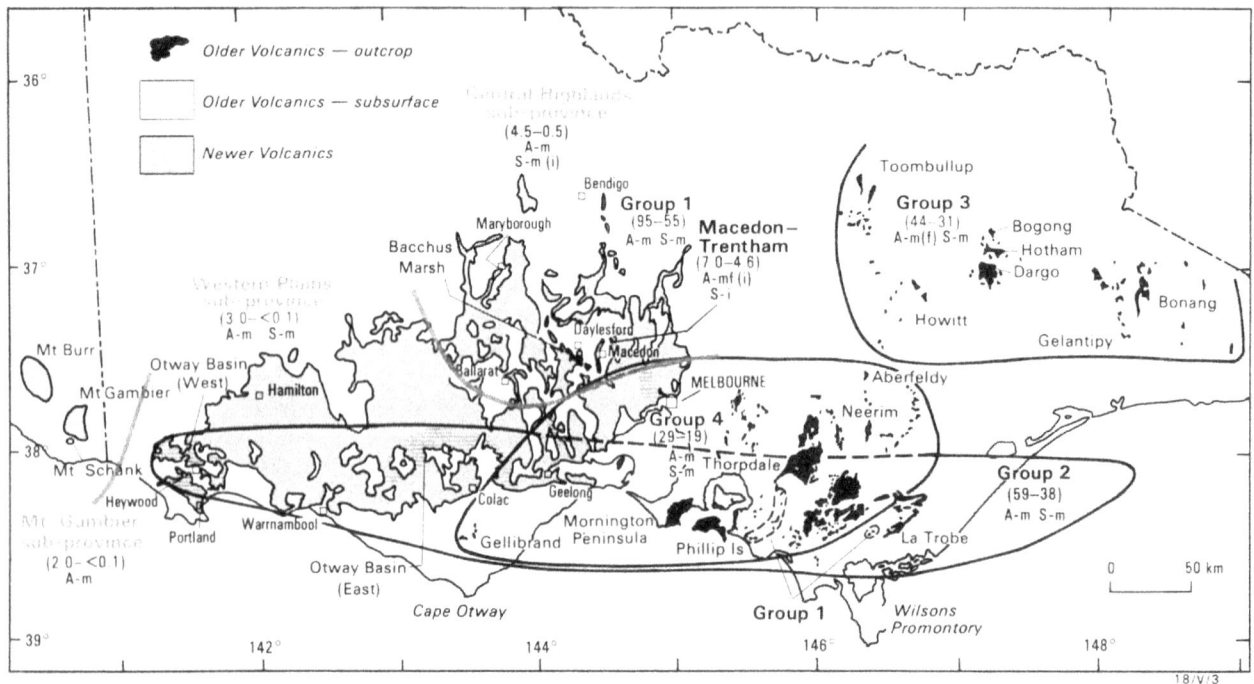

Figure 3.7.1 Distribution of Cainozoic volcanic rocks in Victoria and South Australia. Groups 1–4 refer to the Older Volcanics. Sub-provinces refer to the Newer Volcanics. Rock associations: A, alkaline; S, subalkaline. Rock types: m, mafic; i, intermediate; f, felsic; (i), minor intermediate; (f), minor felsic. Ages (in brackets) in Ma. Volcanic centres in the Newer Volcanics are shown in Figure 1.6.1.

Table 3.7.1 Selected chemical analyses for Victorian and South Australian provinces

	1	2	3	4	5	6	7	8	9	10
SiO$_2$	45.75	52.31	47.57	61.43	66.81	51.88	49.63	45.59	48.34	50.39
TiO$_2$	1.87	1.65	2.06	0.62	0.11	1.69	2.18	2.52	2.35	1.81
Al$_2$O$_3$	14.05	15.05	14.47	16.83	16.09	14.45	15.45	12.28	13.41	13.87
Fe$_2$O$_3$	2.98	1.53	2.34	2.11	2.25	2.71	3.47	1.85	1.89	3.32
FeO	7.81	7.67	9.48	3.26	0.97	7.75	6.94	10.00	9.44	7.01
MnO	0.18	0.02	0.19	0.13	0.07	0.16	0.16	0.18	0.17	0.16
MgO	12.74	7.84	9.49	0.97	0.08	7.49	6.41	12.65	8.88	8.91
CaO	8.89	8.85	8.69	2.07	0.27	8.71	6.78	8.14	9.58	8.97
Na$_2$O	3.44	3.59	3.29	5.93	6.87	3.25	5.43	3.86	3.69	3.34
K$_2$O	0.65	0.57	1.74	5.06	5.09	0.78	2.91	2.01	1.68	1.33
P$_2$O$_5$	0.41	0.23	0.80	0.27	0.06	0.30	0.64	0.91	0.57	0.41
S										
H$_2$O+	1.52		1.82	0.47	0.41	0.62	0.39	0.11	0.17[1]	0.43
H$_2$O−	0.29		0.27	0.33	0.21	0.44	0.28	0.04		0.23
CO$_2$			0.04	0.13	0.02	0.05	0.54	0.06	0.18	0.09
Total	100.58	99.31	102.25	99.61	99.31	100.28	101.21	100.20	100.35	100.27
Ba	247	121	426			246	860	450	455	401
Rb	13	11	27	176	441	26	67	40	34	31
Sr	610	410	650	236	4	377	995	835	642	529
Pb	2.2	1.3	4.2			4	5	4	3	3
Th	2.0	1.5	4.4			2.4				3.6
U	0.76	0.42	1.1			1.0				
Zr	225	122	205	991	1009	144	480	331	202	177
Nb	39.1	17.5	58.3	152	408	18	89	66	44	33
Y	25.2	22.0	23.8	33	58	23	21	24	21	23
La	22.9	10.9	36.1			14.9	56	49	34	22.2
Ce	51.8	25.4	80.6			32.4	117	105	70	46.1
V						167	101	133	173	175
Cr	360	320	360			374	144	412	249	403
Ni	360	190	200			135	131	388	185	154
Cu						43	28	43	56	55

1. Basanite from Older Volcanics province (R.A. Day, unpublished data, sample 1459).
2. *Ol*-tholeiitic basalt from Older Volcanics province (R.A. Day, unpublished data, sample G48).
3. Hawaiite from Older Volcanics province (R.A. Day, unpublished data, sample N29A).
4. Dacite from Macedon-Trentham province (O'Hanlon, 1975, sample 2980).
5. Peralkaline rhyolite from Macedon-Trentham province (O'Hanlon, 1975, sample 2982).
6. *Ol*-tholeiitic basalt from Newer Volcanics province (Price et al., 1988, sample F1/18).
7. *Ne*-mugearite from Newer Volcanics province (McDonough et al., 1985, sample 2101).
8. *Ne*-hawaiite from Newer Volcanics province (McDonough et al., 1985, sample 2156).
9. *Ne*-hawaiite from Newer Volcanics province (McDonough et al., 1985, sample 2183).
10. Hawaiite from Newer Volcanics province (R.C. Price, unpublished data, sample F1/49).

[1] Total H$_2$O.

closer geochemical affinity with the mildly alkaline to transitional central-type provinces, including the presence of a major component of high-level felsic intrusions. Generalised compositional fields for the Victorian and South Australian provinces are shown in Figures 3.7.2 and 3.7.3, and selected chemical analyses are given in Table 3.7.1.

3.7.2 Older Volcanics

Introduction

Fifteen separate fields of Older Volcanics can be identified on the basis of outcrop distribution, borehole data, and petrological affinities (Fig. 3.7.1). Most are lava fields 10 to 130 km across. Dykes and small plugs are important only locally. All are within 200 km of the coast, and there is a total east-west extent of 700 km. There is a continuum of compositions from nephelinite to *qz*-tholeiitic basalt (Table 3.7.1, columns 1–3), but

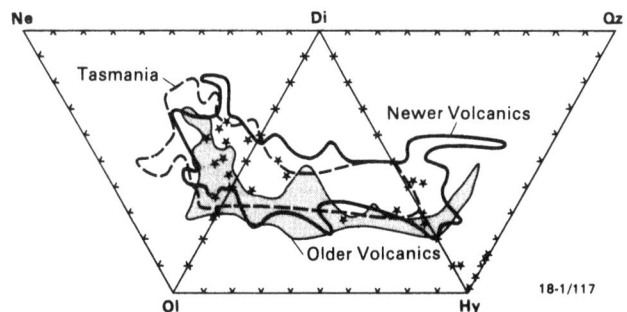

Figure 3.7.3 Compositional fields for Victorian, South Australian, and Tasmanian rocks projected in the system Ne-Ol-Di-Hy-Qz. Stars refer to Macedon-Trentham samples.

ranges and relative proportions differ widely between volcanic fields and, regionally, there is no correlation between basalt type and either age or location. However, six fields became more alkaline or increasingly fractionated during their lifespan.

Geological settings include: (1) dissected lava fields covering high-altitude plains in uplifted areas such as the eastern highlands; (2) old valley-filling flows now forming ridge tops mainly on the margins of the highlands; (3) flows preserved in downfaulted areas, especially in the Ballan Graben area west of Melbourne, the Sorrento Graben south of Melbourne, and in the Central Deep of the Gippsland Basin; and (4) flows associated with Tertiary sedimentary deposits in the Gippsland and Otway basins.

The best exposures are in the eastern highlands where good drainage has minimised the effects of alteration by groundwater and lavas remain resistant to erosion. Good exposures of a thick lava pile are found also along the coast of the Mornington Peninsula south of Melbourne. Poor drainage or burial beneath a cover of sediments (some of them marine) in most other areas and especially in the Tertiary basins, has resulted in extensive alteration, and present-day exposures in downfaulted areas and basins are commonly poor and deeply weathered. Pyroclastic rocks are invariably deeply weathered.

Age and distribution

Seventy-three K-Ar age determinations available for the Older Volcanics were compiled by Day (1983). The dates correspond to a long history of almost continuous volcanism from possibly 95 Ma up to 19 Ma. Individual volcanic fields are of more limited duration — in the range 10 to 2 Ma (average of 6 Ma). Four major groups can be distinguished on the basis of age and location (Day, 1983; Fig. 3.7.1):

Group 1 (95–55 Ma): Poowong, Bacchus Marsh
Group 2 (59–38 Ma): La Trobe, Flinders, Otway Basin (two fields)
Group 3 (44–31 Ma): Toombullup, Howitt, Bogong, Gelantipy
Group 4 (29–19 Ma): Gellibrand, Aberfeldy, Neerim, Thorpdale, Melbourne

These fall into distinct regions and have similar geological settings, except for Group 1 where activity is the most limited and two small fields spanning the Mesozoic-Cainozoic boundary are 140 km apart. Group 2 contains the greatest volume of products. It consists of a 70 km-wide, east-west belt of four separate fields extending for 620 km as deposits in the Gippsland and Otway basins and intervening structural depressions and ridges. The Flinders lava field south of Melbourne is the most voluminous. Over 75 percent of the deposits are in the subsurface.

Group 3 is generally younger and less voluminous than Group 2. The four separate fields are confined entirely to the eastern highlands in a region 270 km east-west by a maximum of 115 km north-south. Group 4 is the second-most extensive. It forms a broad zone of five fields within south-central Victoria, extending 315 km east-west and 95 km north-south. More than 50 percent of the deposits are subsurface. They occupy a higher stratigraphic position in the basins than those of Group 2, and K-Ar ages are 30 to 11 Ma younger. Group 4 fields are also younger than any of those nearby in Group 3.

Group 1

The Mesozoic *Poowong* intrusions and flows (Section 3.9.4) represent the earliest episode of, and make up one of the smallest fields in, the Older Volcanics and, together with the Bacchus Marsh lavas, constitute Group 1.

The Cainozoic lavas of *Bacchus Marsh* are preserved in a small northwest-trending graben, cropping out in an area 20 km east-west by 5 km north-south. They overlie Ordovician and Permian basement and are covered in the southern half of the graben by Tertiary sands and lignites or by the Newer Volcanics.

Measured thicknesses of at least 160 m are of lava flows (Jacobson & Scott, 1937) — *ol*-tholeiitic and transitional basalts at the base, then alkali basalt and basanite, followed by basanite and nephelinite at the top (Day, 1983). The flows were probably fed by numerous, closely related dykes whose trends are parallel to the axis of the graben, intruding both the nearby basement rocks and the flows.

Group 2

Group 2 volcanism coincides with, or slightly predates, uplift and downwarping of Palaeozoic basement and Cretaceous sediments of the Strzelecki Basin during early development of the Gippsland Basin.

The *La Trobe* field extends 150 km east-west, passing onto the continental shelf in the downfaulted Central Deep of the Gippsland Basin, and 95 percent of its area is subsurface. Only the western extremity crops out. The La Trobe and Thorpdale fields (Group 4) overlap in the subsurface of the La Trobe valley north of the Balook Block (Hocking, 1976). Rock types in the western part of the field range from basanite to *ol*-tholeiitic basalt.

Flinders is the most voluminous field containing at least 700 km³ of volcanic rock (Wellman, 1974). It is dominated by a main lava field in its central portion, composed of basanite (Table 3.7.1, column 1) and alkali-basalt lavas over 600 m thick and intruded by K-rich hawaiite dykes (Day, 1983). Less extensive K-rich hawaiite flows are found farther west on the Bellarine Peninsula, and numerous alkaline intrusions are found in a large area of uplifted Cretaceous rocks to the east (eastern dykes and plugs). Bellarine Peninsula flows and minor pyroclastic rocks are the youngest preserved in the Flinders field. A flow close to the basement has a K-Ar age of 39 Ma.

Outcrops of the main lava field are along the eastern side of the Mornington Peninsula and on Phillip and French islands, covering at least 1000 km². Flows are entirely basanite and minor alkali basalt. Most of the volcanic rocks on Phillip Island are obscured by alluvium, but there are good coastal exposures where flows are similar to those on the Mornington Peninsula. However, weathered, massive, or crudely layered scoriaceous pyroclastic rocks are also important on Phillip Island. Beds are up to 45 m thick, and Edwards (1945) estimated that their volume equals that of the lava flows.

A total of 93 intrusions in Cretaceous sediments of the Narracan Block were mapped in early surveys (Kitson, 1903). Basanite dykes up to 50 m wide can be traced for over 1 km. Dykes trend 155–170° (the same direction as the hawaiite dykes of the main lava field). These dykes have identical petrography and chemistry to the main lava-field rocks, including equivalents to the late-stage K-rich hawaiitic dykes.

Volcanism in the two *Otway Basin* fields was limited compared to other Group 2 fields. Two areas 90 km apart contain subsurface flows and intrusions (Abele et al., 1976). Rock types in both fields are mainly transitional basalt to alkali basalt, and there are single occurrences of basanite and *ol*-tholeiitic basalt in the western field (Day, 1983).

Group 3

The eastern highlands already had at least 1000 m of relief at the time of volcanic activity (Hills, 1940; Wellman, 1974) and erosion since has been extensive. P. Wellman (1979b) favoured continuous uplift throughout the period of volcanism. Lava fields were not erupted continuously across the highlands, and their location coincides with areas traversed by large (and probably deep) fractures of Palaeozoic age.

The *Toombullup* flows are distributed around a localised source at the southern end of the field where their estimated present-day thickness reaches 250 m (Day, 1983). Rock types range continuously through basanite, alkali basalt, and minor hawaiite, to transitional basalt and *ol*-tholeiitic basalt. Basanite is present in about the same proportion as the other rock types combined.

The main body of *Howitt* lava flows is found in two adjacent areas on the Howitt High Plains at an elevation of about 1400 m. They extend over a north-south distance of 17 km and have a maximum, present-day thickness of about 100 m. The overall proportions of hawaiite, alkali basalt, and transitional basalt are about equal, and nephelinite is known at only two localities. Scattered residuals of transitional basalt like that of the main body are present up to 40 km to the southwest.

Bogong contains the most extensive highland area of volcanic rocks. Three adjacent areas covered by flows remain, along with numerous intervening residuals. They cover a deeply dissected region at elevations between 1500 m in the north and 1200 m in the south. The northernmost outcrops contain the only known evidence for the source of the Bogong lavas where there are several basanite plugs and dykes associated with nearby flow residuals and intercalated scoriaceous beds. Elsewhere only minor lacustrine and fluvial sediments are associated with flows.

The Bogong High Plains outcrop is about 5 km across and 150 m thick (Beavis, 1962) and is the northernmost large lava residual. Its northern margin consists of at least seven identical basanite flows. Only a third of the Hotham Heights residual contains basanite. One section in the centre of the field farther south consists of transitional basalt overlying basement, followed by alkali basalt and hawaiite, and then basanite and hawaiite, corresponding to a temporal trend towards greater alkalinity. Basanite also overlies alkali basalt elsewhere.

The Dargo High Plains consist of another large area of Bogong flows, extending a further 15 km south of Hotham. Only half the Dargo flows examined are basanite, and the rest are of alkali basalt, hawaiite, and rare mugearite.

Basement rocks in the vicinity of the Bogong field have been intruded by numerous lamprophyric dykes and small plugs of alkaline affinity, including tinguaite and phonolite. These were regarded as part of the Older Volcanics by Edwards (1938a), Crohn (1950), Beavis (1962), and Tattam (1976), but may have the same Jurassic age as a nearby phonolitic plug dated by McDougall & Wellman (1976).

Gelantipy flow residuals are within a large region that covers an east-west distance of 120 km, and which includes five separate fields. All are noteworthy because of the predominance of *qz*- and *ol*-tholeiitic basalt (Day, 1983; Table 3.7.1, column 2). They occupy large valleys ancestral to the present-day rivers in an area that had a relief of at least 1000 m.

The Nunniong High Plain is the westernmost field. It has an elevation of 1200 m, and is covered by a 20 km north-south sheet of flows beginning with *qz*-tholeiitic basalt, then *ol*-tholeiitic basalt, and finally basanite and hawaiite (Day, 1983). Farther east the Gillingal flows of *qz*- and *ol*-tholeiitic basalt cover the divide between the Timbarra and Buchan rivers for a north-south distance of 55 km, decreasing in altitude from 800 to 200 m.

The Gelantipy-Wulgulmerang field is the largest of the five Gelantipy fields, extending 35 km north-south and decreasing gradually in altitude from 800 to 400 m. All flows examined consist of *qz*- and *ol*-tholeiitic basalt.

Bonang outcrops represent two residuals 27–40 km east of the main Gelantipy field, and together with another at Club Terrace a further 20 km to the east, are also small remnants of valley flows. They consist of *qz*- and *ol*-tholeiitic basalts, and near Bonang are associated with nephelinite and transitional basalt.

Group 4

The main Group 4 volcanic rocks are associated with sediments in the western Gippsland Basin (Neerim and Thorpdale fields), but flows also filled former south-flowing valleys along the southern margin of the adjacent eastern highlands (Aberfeldy and Neerim fields), and some extend beneath Tertiary sediments. Areas in the eastern margin of the Otway Basin (Gellibrand, Melbourne) are limited to isolated single flows, related multiples of flows, and small intrusions and associated pyroclastic rocks. These small occurrences are important because of the information they provide about Tertiary stratigraphy.

Gellibrand field is represented by two small areas in the eastern Otway Basin. The main occurrence includes a small area containing six plugs 400–1000 m across of virtually identical transitional basalt intruding Tertiary sediments of the Eastern View Formation. A coastal exposure 50 km farther east at Aireys Inlet contains *ol*-tholeiitic basalt and pyroclastic rocks included in the Demons Bluff Formation (Abele & Page, 1974).

Aberfeldy field is a small group of at least three former valley flows capping a prominent north-south ridge between the Aberfeldy and Thompson rivers. The source of the flows may be nearest the largest northern outcrop where the lowest flow is transitional basalt extending 5 km southwards. It is overlain by alkali basalt (having distinctive glomeroporphyritic plagioclase) that extends southwards for at least 20 km. This flow is overlain by hawaiite confined to the highest point on the northernmost outcrop.

The *Neerim* and adjacent *Thorpdale* fields together represent by far the largest proportion of the Group 4 volcanic rocks. They have similar ages, and lavas of both fields pass beneath Tertiary sediments in the La Trobe valley, but they are not a continuous body (Gloe, 1976) and there are also noticeable petrographic and chemical differences between them (Day, 1983).

Flows in the Neerim field overlie basement along the northern margin of the La Trobe valley and Westernport Sunkland. Warragul volcanic field is the main area of outcrop. Outcrops extend for an east-west distance of 140 km.

Half of the Neerim samples analysed are transitional basalt, a quarter are hawaiite (Table 3.7.1, column 3), and the remainder range through *ol*-tholeiitic basalt, alkali basalt, and basanite. Half the lavas range into relatively K-rich types — a distinctive feature of the Neerim field. A single, prominent, nephelinite plug about 200 m across intrudes the main Warragul lavas and may be evidence for the development of limited, later, more alkaline activity.

Thorpdale consists of the main Thorpdale volcanic field and nearby flows between the Narracan and Balook blocks in an area 50 by 25 km. More limited outcrops are along the southern margin of the La Trobe valley. The main Thorpdale field is less than 60 m thick and includes minor pyroclastic rocks and interbedded sediments (Douglas, 1979). Rock types are similar to the Neerim field: transitional basalt makes up about half the outcrops, and the rest consists of alkali basalt, *ol*-tholeiitic basalt, and minor hawaiite and basanite.

Outcrops in the vicinity of *Melbourne* consist of four isolated areas of virtually the same age. The flows of the Tullamarine and Melbourne-South areas are found over a 25 km north-south distance in the valley of the Maribyrnong River and Moonee Ponds Creek, and continue beneath Melbourne into the Yarra River delta. The Greensborough basalts contain the youngest of all the Older Volcanics, and are distinct from the Tullamarine and Melbourne-South flows. They consist of transitional basalt and alkali basalt. The Maude Basalt may consist of a single extensive sheet of *ol*-tholeiitic basalt 15–30 m thick and extending north-south for at least 15 km, enclosed for most of this distance by the Maude Limestone (Bowler, 1963; Abele & Page, 1974).

The Curlewis hawaiite plugs and associated pyroclastic rocks are exposed on the northern coast of the Bellarine Peninsula (Coulson, 1933). One K-Ar dated hawaiite plug is 18 Ma younger than flows nearby that form a western extension of the Flinders field.

3.7.3 Macedon-Trentham

The only recognised felsic volcanic rocks of Cainozoic age in Victoria are represented by small trachytic flows, domes, plugs, and spines, that have a maximum relief of about 100 m in the Lancefield, Mount Macedon, Gisborne, and Trentham-Blackwood areas northwest of Melbourne (Fig. 3.7.1). These, and associated occurrences of more mafic rocks, were referred to as the Woodend province by Ewart et al. (1985). The province straddles the Great Divide in the central highlands of Victoria where the early Tertiary uplift of Ordovician to Devonian sequences was followed by erosion prior to Cainozoic volcanism (Singleton, 1973).

The rocks of the Macedon-Trentham province are closely associated spatially with the extensive products of the Newer Volcanics Central Highlands sub-province (Section 3.7.4), but their age range of 7.0–4.6 Ma (Wellman, 1974; Dasch & Millar, 1977; Ewart et al., 1985) is slightly older than that of the oldest known Newer Volcanic rocks (about 4.5 Ma), and clearly older than the earliest, nearby, Newer Volcanics rocks (about 3.0 Ma). Initial $^{87}Sr/^{86}Sr$ values of 0.7052–0.7127 obtained for domes and flows of the Lancefield and Mount Macedon area by Dasch & Millar (1977) also appeared to clearly distinguish these from the Newer Volcanics (initial $^{87}Sr/^{86}Sr$ 0.7038–0.7054; Section 3.7.4). However, more recent values reported by Ewart (1985) for trachytic rocks from both the Mount Macedon and Trentham areas (initial $^{87}Sr/^{86}Sr$ 0.7042–0.7051) are within the Newer Volcanics range.

The Macedon-Trentham province rocks represent a mildly alkaline to transitional series similar both petrographically and chemically to those of the large central-type volcanoes of New South Wales and Queensland (Ewart, 1985; Ewart et al., 1985). They, rather than the much more extensive Newer Volcanics lava-field occurrences in Victoria and South Australia, are probably the latest manifestation of the hotspot trace represented by these central volcanoes (Section 1.7). Their transitional character is illustrated in Figure 3.7.3 where analysed samples span the range from mildly nepheline- to mildly quartz-normative compositions (Table 3.7.1, columns 4–5). The *ne*-trachytes of the series represent one of the rare east-Australian occurrences of this rock type (those of Warrumbungle volcano represent another; Section 3.5.5).

O'Hanlon (1975), Ewart et al. (1985), and D.A. Wallace (personal communication, 1987) identified rock types from this province ranging from K-rich basanite through alkali and transitional basalts, K-*ne*-hawaiite, hawaiite, mugearite, and benmoreite to *ne*- and *qz*-trachytes. The 'limburgites' recognised by earlier authors (Edwards, 1938a; Edwards & Crawford, 1940) are glassy K-rich basanites, alkali to transitional basalts, or more mafic hawaiites with glassy ground-masses. Other equivalent rock types are anorthoclase basalt and olivine basalt (alkali to transitional basalt), woodendite (K-*ne*-hawaiite), andesine basalt (hawaiite), macedonite, oligoclase basalt, and oligoclase trachy-basalt (mugearite to benmoreite), anorthoclase-olivine trachyte and anorthoclase trachyte (benmoreite to mafic *ne*-trachyte), and soda trachyte and sölvsbergite (*ne*-trachyte to peralkaline *qz*-trachyte).

Most of the volcanic hills of the Macedon-Trentham province are small, consist of a single rock type, and were formed during short-lived volcanic events. The form of the bodies reflects the viscosity of associated lavas. The most felsic bodies (the peralkaline qz-trachytes, containing 65–68 percent SiO_2) form steep-sided plugs and spines, such as Camels Hump, Hanging Rock, and Brocks Monument in the Lancefield and Mount Macedon area (Singleton, 1973; Stewart, 1985). The ne-trachytes, containing 58–62 percent SiO_2, form prominent domes having steep to gentle slopes, including Babbington Hill, Blue Mountain, and Mount Wilson in the Trentham-Blackwood area (these are probably lava domes or coulées). Benmoreites, mugearites, hawaiites, and alkali to transitional basalts form relatively extensive lava flows. There are at least two centres of voluminous mugearite eruption, and a benmoreite in the Jim Jim area north of Woodend has flowed around topographic highs of basalt and overlies hawaiite. A flow of hawaiite originating in the twin craters of Rocky Hill, near Lancefield, is overlain by both mugearite and alkali to transitional basalt (O'Hanlon, 1975). The vents of 'olivine basalt' flows are represented by numerous small hills in the south-eastern and northwestern parts of the Mount Macedon and Gisborne district. The mafic, glassy rocks of the former 'limburgite' and 'woodendite' range have less than 50-percent SiO_2, form only small domes and restricted lava flows (Melbourne Hill, Western Hill, the Kings Quarry basanite near Hesket, Racecourse Hill at Woodend, and a 'limburgite' south of Mount Gisborne). Pyroclastic deposits are rare, but originally they may have been more extensive.

The relative age sequence for the Mount Macedon and Gisborne area determined by O'Hanlon (1975) is alkali to transitional basalt, mugearite, benmoreite, hawaiite, to alkali-transitional basalt. The stratigraphic positions of basanites and trachytes are unclear from field relationships, but the peralkaline qz-trachyte plugs in the Mount Macedon area and the ne-trachyte domes and flows of both the Hesket and Trentham-Blackwood areas have ages (6.3–5.8 Ma; Ewart et al., 1985) amongst the oldest for the province, whereas the Kings Quarry basanite has one of the youngest (4.6 Ma; Wellman, 1974).

The basalt-to-benmoreite range has essentially similar mineralogy, consisting of the phenocryst assemblage olivine + augite + plagioclase ± anorthoclase ± Fe-Ti oxides, set in groundmasses of the same minerals, biotite, and apatite (O'Hanlon, 1975; Stewart, 1985). The trachytic rocks have a range of structures, textures, and mineral assemblages (Ferguson, 1978c). More mafic ne- to qz-trachytes have flow textures and contain phenocrysts of anorthoclase rimmed by soda-sanidine, microphenocrysts of fayalitic olivine, augite, and Ti-magnetite, in groundmasses of alkali feldspar, soda-augite, and rare quartz (for example, at Camels Hump and Turritable Falls). More felsic peralkaline qz-trachytes lack flow textures and contain phenocrysts of anorthoclase, fayalitic olivine, ferroaugite, and titanomagnetite, in groundmasses of alkali feldspar, arfvedsonitic amphibole, aegirine-augite, aenigmatite, and quartz (Hanging Rock and Brocks Monument). The mineral chemistry of the more felsic rocks (including details of aenigmatites, amphiboles, and unusual Na-Ti pyroxenes) was dealt with by Ferguson (1978c) and Ewart (1985). The clinopyroxenes have an initial rapid Fe-enrichment (ferroaugite-hedenbergite), followed by rapid Na-enrichment (hedenbergite-aegirine), characteristic of the felsic rocks of other east-Australian central-volcano suites.

The most mafic basanites and K-ne-hawaiites of the Macedon-Trentham province have Mg-ratios greater than 65, corresponding to near-primary, mantle-derived magmas (analyses of Edwards, 1938a). Lherzolite xenoliths are rare but are present in the Racecourse Hill K-ne-hawaiite (woodendite). Other basalts have lower Mg-ratios, corresponding to an origin from more mafic parent magmas by fractional crystallisation or crustal assimilation (or both). The influence of minor crustal assimilation in the origin of more felsic rocks is indicated by $\delta^{18}O$ data in the range +6.3 to +7.1 per mil, which correlate positively with $^{87}Sr/^{86}Sr$ (Ewart, 1985).

3.7.4 Newer Volcanics

Age and distribution

The Newer Volcanics province covers about 15 000 km^2, and contains nearly 400 known eruption points (Fig. 1.6.1). Its major geomorphical feature, the Western Plains of Victoria, existed prior to volcanism that produced only a thin veneer (normally less than 60 m) on a late Tertiary erosion surface. The small Uplands lava field, the Central Highlands, and the northern margin of the Western Plains (Wellman, 1974) are underlain by folded Ordovician-Devonian rocks and Silurian-Devonian granites, whereas the remainder of the Western Plains and the South Australian portion of the province are underlain by shelf carbonate rocks of Miocene-Pliocene age.

The earliest exposed lavas of the province have been dated at around 4.5 Ma (McDougall et al., 1966). These include a major subalkaline (ol- and qz-tholeiitic basalt to icelandite) component in addition to the moderately alkaline types (hawaiite-basanite) characteristic of the province. A Western Plains subdivision into 'plains basalts' and 'trickle basalts' (which flowed down valleys cut into plains sequences) has been suggested (E.D. Gill, personal communication, 1986). The best preserved volcanic complexes of the plains activity are exposed in coastal sections near Portland, and they include thick lava piles dated at 3.0–2.4 Ma (Aziz-ur-Rahman & McDougall, 1972). Similar ages (3.0–2.2 Ma) were obtained for plains basalts of the Macedon-Trentham district (Section 3.7.3) and the Melbourne district (VandenBerg, 1973). Younger plains basalts near Warrnambool have been dated at 2.0–1.9 Ma by McDougall & Gill (1975).

The most prominent volcanic features of the Newer Volcanics province are scoria cones, maars, tuff rings, and major valley lava flows that belong to much younger phases of activity (about 500 000–4000 yr B.P.) involving mainly alkaline magmas (ne-hawaiite to basanite). Eruptions were generally of small volume and low explosivity. Even the largest constructional features, the Mount Elephant and Smeaton Hill scoria cones, stand only about 250 m above the surrounding lava plains and are little more than 1 km in diameter. The overall proportion of pyroclastic material estimated for the province is only 1 percent (Ollier & Joyce, 1973).

Relative ages of the youthful volcanic features have been determined using the degrees of weathering and soil development observed on lava flows (Ollier & Joyce, 1964; Gill, 1978). Absolute ages have been determined by K-Ar dating of lavas and ^{14}C dating of buried charcoal or organic carbon, or carbonate in soils or sedimentary deposits, especially lake sediments (for example: McDougall et al., 1966; Aziz-ur-Rahman & McDougall, 1972; Gill, 1978; Barton & Polach, 1980; McKenzie et al., 1984; Edney et al., 1985). The oldest K-Ar date so far obtained for material of a well-preserved volcanic complex is 1.8 Ma for a lava flow within the main scoria cone of Mount Rouse (Ollier, 1985a), although this appears inconsistent with K-Ar dates of 0.45–0.31 Ma obtained by McDougall & Gill (1975) for large lava flows apparently from the Mount Rouse complex. The youngest date (by radiocarbon dating on charcoal in soils beneath tuff layers) was obtained for Mount Gambier (4600–4300 yr B.P.; see below). This is inconsistent with a date on organic material in lake sediments of 6000–5000 years obtained by Barton & McElhinny (1980), but is confirmation that Mount Gambier is probably the youngest centre of the Newer Volcanics province (*Fig. 3.7.4*). Additional radiocarbon dates and palaeomagnetic results from sediment successions in maar lakes in the Camperdown-Warrnambool area are evidence that some maars were in existence before 10 000 yr B.P. (Barton & Polach, 1980; Edney et al., 1985), and that some may be older than 100 000 yr B.P. (P. De Deckker, personal communication, 1986).

The Newer Volcanics province may be subdivided into at least three major sub-provinces: Central Highlands, Western Plains, and Mount Gambier in southeastern South Australia (see below). Each of these may be further subdivided, and areas of the order of 50–100 km across may have an obvious homogeneity in terms of age of volcanism, eruption type, geomorphology, and petrology (see, for example, Ollier & Joyce, 1964).

The Central Highlands sub-province includes the majority of Newer Volcanics lava-cone/dome and scoria-cone/dome occurrences, associated with an extensive network of valley lava flows and lava fields that are smaller and less accessible than counterparts in the Western Plains province. The total number of recognised eruption points in this sub-province is about 250. The Geelong-Melbourne region contains most of the lava cones and domes of the Western Plains sub-province, associated with an extensive lava field (the Werribee Plains). Regions within the Victorian Western Plains District east and west of the Mount Shadwell scoria-cone complex (at the town of Mortlake) have marked differences. The eastern region includes almost all the Newer Volcanics maars and tuff rings, as well as most of the largest scoria cones (most recent activity probably around 25 000 to 20 000 yr B.P.) which typically are extremely rich in xenoliths and megacrysts (for example, Mounts Elephant, Leura, Noorat, Shadwell, Lake Bullenmerri; Irving, 1974b,c). The western region contains several very young complexes (most recent activity 7000–5000 yr B.P.) whose products totally lack ultramafic xenoliths and megacrysts (Mount Eccles, Mount Napier, Tower Hill).

Recent grid sampling of the Melbourne-Kilmore area of the Central Highlands sub-province and parts of the Western Plains sub-province and has shown that 'geochemical sub-provinces', whose lavas differ in key trace-element and isotopic ratios, also may be recognised (see below). The eastern and western regions of the Western Plains sub-province, with their distinctive ages and styles of volcanism, abundance of xenoliths, and so on, also have distinctive trace-element and Sr-isotope signatures. Further geochemical sub-provinces have been recognised in the areas immediately north and west of Melbourne.

The Western Plains sub-province has some of the richest and most diverse of known continental xenolith and megacryst occurrences. Most are associated with large scoria cones in the Colac-Camperdown-Mortlake area. Other rich localities are the Anakies (Werribee Plains) and Mount Franklin (Central Highlands) scoria-cone complexes.

Xenoliths of upper-mantle wall-rock origin at all major localities are mainly members of the Cr-diopside suite, and lherzolites are much more abundant than harzburgites and dunites (Section 6.2). Xenoliths of the less abundant Al-Ti-augite suite are also present at many localities, especially Mounts Leura, Porndon, and Shadwell, and the Anakies, and xenoliths of obvious shallow crustal origin are present at most Newer Volcanics centres. These include sedimentary (limestones, mudstones, sandstones), igneous (basalts, dolerites, and common granites), and rare metamorphic rock types. The best known occurrences are of granitic blocks in which there is evidence of extensive breakdown of biotite and quartz (Mount Elephant, the Anakies, and Mount Noorat; Le Maitre, 1974). Megacrysts of a wide range of minerals have been reported from at least 18 Newer Volcanics localities (Irving, 1974b; Wass & Irving, 1976).

Volcanic processes and products in intraplate continental settings are discussed in detail in Chapter 2, where many examples from the Newer Volcanics province are used.

Central Highlands sub-province

The Central Highlands sub-province is distinctive on account of its high proportion of lava flows and the scarcity of products of phreatomagmatic eruption. The activity of about 250 volcanoes (about two-thirds of Newer Volcanic centres) produced an extensive system of valley flows and small lava plains in a dissected upland region consisting largely of Palaeozoic meta-sediments and granites. Most of this activity appears to date from the period 4.5–2.0 Ma, contemporary with emplacement of 'plains' sequences and the Portland-Heywood group of centres in the Western Plains sub-province. About 75 percent of nearly 200 Newer Volcanics centres composed mainly of scoria are found in the Central Highlands sub-province, and there is a concentration towards its western margin — particularly in the Daylesford-Ballarat district where Coulson (1954) identified 123 centres. More than 80 percent of more than 140 Newer Volcanics lava volcanoes are in the central and eastern parts of the sub-province. There are only six poorly known examples of maars or tuff rings.

The *Melbourne-Kilmore* area includes about fifteen Newer Volcanics centres built on, or surrounded by, an extensive lava plain formed by flows that moved southwards in pre-volcanic stream valleys (Hanks, 1955; Croker, 1984). Most centres are small (20-50 m), but four scoria volcanoes are similar in size (50–120 m) to those of the Western Plains. The largest, Mount Fraser, is a complex of four large, and up to six small, scoria cones surrounding a crater-like depression.

The centres produced a wide range of tholeiitic to alkaline lavas, including *ol*-tholeiitic basalts (Mount Ridley), hawaiites (Mount Fraser), and rare strongly alkaline types such as basanites (Bald Hill), classified as olivine nephelinites by Irving & Green (1976). Some centres such as Mount Fraser are not directly related geochemically to the underlying lava flows which belong to five isotopic domains corresponding roughly to distinct flow sequences confined by pre-volcanic interfluves. These domains are dominated by *qz*- and *ol*-tholeiitic basalts and hawaiites (see below). K-Ar dates of 4.6 and 2.2 Ma for flows in the north and east of the area (McKenzie et al., 1984) are an indication of the age range of most Newer Volcanics activity, but younger dates (for example, 0.8 Ma) were obtained for Yarra valley flows in the Melbourne area (VandenBerg, 1973).

The *Sunbury-Macedon-Malmsbury* area northwest of Melbourne contains a very wide range of Pliocene to Pleistocene volcanic rocks. It includes the distinctive Gisborne-Macedon-Trentham felsic alkaline suite (mugearite, benmoreite, *ne*-trachyte, and *qz*-trachyte) which, with more mafic types (basanite, alkali basalt, and hawaiite), forms a separate older association (Section 3.7.3). The main phase of Newer Volcanics activity in the area is represented by basalts in the southeast (Sunbury-Gisborne-Lancefield) and northwest (Trentham-Malmsbury-Redesdale area).

'Typical' Newer Volcanics products of the area initially were classified petrographically by earlier investigators (Edwards, 1938a; Edwards & Crawford, 1940; Singleton, 1973) as labradorite or andesine basalts, subdivided on the presence of relatively fresh or iddingsitised olivine, and referred to type localities (Footscray, Gisborne, Malmsbury, and Trentham types; Edwards, 1938a). These rocks and associated more felsic types ('trachybasalts', 'oligoclase basalts') are found mainly as lava flows confined to deeper valleys and making up small lava fields that have strongly influenced the regional stream pattern. Older valley flows in many cases are represented by narrow erosion residuals, bounded by lateral streams, which overlie 'deep lead' gravels or sands (many gold-bearing and extensively mined in the 19th century) of the pre-volcanic drainage. The best examples of this type involve *ol*- and *qz*-tholeiitic basalts and icelandites in the area of the present-day Loddon-Campaspe-Coliban river system (D.A. Wallace, personal communication, 1987). Other examples of well-preserved large valley-filling flows include hawaiites or mugearites originating from Spring Hill (near Trentham), Mount Bullengarook (near Gisborne), and Melbourne Hill (near Lancefield). The Spring Hill flow was dated at 3.3 Ma (Wellman, 1974) and the Mount Bullengarook flow at 3.6–3.3 Ma (McKenzie et al., 1984).

The most extensive lava fields are in the Sunbury-Lancefield area (Clarke, 1984; O'Neill, 1984). Associated lava or lava-scoria cones are relatively rare and small (for example, Mounts Aitken, Holden, and Sunbury Hill). Relatively few geochemical data are available for this area, but are indicative that the most abundant lava types of the plains are *ol*- and *qz*-tholeiitic basalts ranging to hypersthene-normative hawaiites, whereas hypersthene- and nepheline-normative hawaiites are more common in the associated cones (Edwards, 1938a; Edwards & Crawford, 1940; Irving & Green, 1976). Available K-Ar dates from lavas in the south-eastern part of this area, close to Melbourne, are in the range 4.5–2.5 Ma (Page, 1968; VandenBerg, 1973).

The *Daylesford-Ballarat-Maryborough* area contains the majority of Central Highlands volcanoes. The features of valley lava flows, lava fields, and associated lava or scoria cones were discussed by Coulson (1954) and Yates (1954), and summarised more recently by King (1985). Edwards (1938a) outlined the general geology and physiography of the area, and provided most of the petrographic and geochemical information available, until a recent study by D.A. Wallace (personal communication, 1987).

Coulson (1954) noted the topographic contrast between small, confined lava fields in steep country to the east of Daylesford and more extensive fields in the more open Creswick-Ballarat area to the west. He recognised '100-odd' eruption points in the Daylesford-Creswick area, mostly small craterless scoria cones, domes or mounds, rising 60–150 m (for example, Bullarook Hill and Mount Moorookyle). Others have well-developed craters, some of them breached. The most prominent is Mount Franklin, an important locality for megacrysts, particularly anorthoclase, and upper-mantle xenoliths.

The Ballarat area (Yates, 1954; King, 1985) has lava fields with diameters of tens of kilometres, that have produced relatively level plains related to the pre-volcanic drainage pattern and hence covering 'deep leads'. Younger volcanoes on this field, and others north and south of Ballarat, are mainly scoria cones (Mounts Buninyong, Mercer, and Warrenheip), but there are also large lava-cones/domes (Mounts Blowhard, Hollowback, and Pisgah) and well-defined lava discs (Lawaluk and Mondilibi). Probable maars or tuff rings include structures at Callender Bay (on the margin of the volcanically-dammed Lake Burrumbeet), Stockyard Hill (Lake Goldsmith), Lake Learmonth, and Hardie Hill. The well-preserved forms of flows and cones may be indicative of ages of less than 100 000 yr, but the few available K-Ar dates on plains lavas near Ballarat and flows apparently from Mount Rowan and Smeaton Hill are within the range 2.9–2.1 Ma (King, 1985).

Rocks from the volcanic centres and lava flows in the Daylesford-Ballarat-Maryborough area are dominantly tholeiitic basalts and hawaiites that are often more potassic than equivalent types in the Western Plains sub-province (D.A. Wallace, personal communication, 1987). The more extensive lava fields are mainly subalkaline (abundant *ol*- or K-*ol*-tholeiitic basalts, and related *qz*-tholeiitic basalts, and icelandites), and these types are also present in lava cones and domes. The younger scoria and scoria/lava-cones commonly contain more alkaline types (abundant hawaiite and K-hawaiite to rare *ne*-hawaiite and K-*ne*-hawaiite — for example, Mounts Franklin, Moolort, and Warrenheip). There are no obvious spatial differences amongst the different rock types in the region.

Western Plains sub-province

The Western Plains sub-province extends more than 300 km from the plains near Melbourne to a group of 3–2 Ma old volcanoes west of Heywood (Mounts Vandyke, Deception, and Red Hill). It includes the widest range of volcanological and petrological features of the Newer Volcanic province. The eastern half includes most of the best examples of large scoria cones, maars, tuff rings, and xenolith/megacryst localities (Warrnambool-Derrinallum-Colac area), as well as an extensive lava field containing most of the lava volcanoes of the sub-province (Geelong, Bacchus Marsh, and Melbourne area). The western half of the sub-province contains both the oldest well-exposed products (coastal complexes near Portland) and the youngest (complexes within the Portland-Hamilton-Warrnambool area).

Plains of the *Geelong, Bacchus Marsh, and Melbourne* area are composed largely of lava flows, the youngest of which are related to recognised lava cones and domes. Large scoria volcanoes like those of the Warrnambool-Derrinallum-Colac region farther west are rare, but the most prominent examples are those of the Anakies group, north of Geelong.

Plains and valley lava flows in the Geelong area originated both from vents nearby (Mounts Duneed, Moriac, and Pollock) and farther north (Green Hill and the Anakies; Spencer-Jones, 1970). Relative age relationships of these flows were studied by Coulson (1933, 1941). Only Mount Duneed, whose peak is formed by a small scoria cone, and the Anakies, include a major pyroclastic component. The Anakies are three scoria cones or mounds up to 100 m in height whose ejecta include a wide range of Cr-diopside lherzolite xenoliths, mafic granulite xenoliths, megacrysts (especially anorthoclase), and granite blocks derived from a pluton exposed nearby. Cr-diopside suite xenoliths are found also at Green Hill and Mount Moriac.

The 2000 km² Werribee Plains between Geelong and Melbourne and extending northward almost to Bacchus Marsh, are formed of flow sequences up to 80 m thick (Bolger, 1977) resting on an erosion surface developed on Miocene-Pliocene sands and clays. These sequences consist of numerous flows and interbedded tuffs, sand sheets, and soil horizons (VandenBerg, 1973). The northern margins of the plains between Geelong and Bacchus Marsh may include lavas from younger volcanoes within the topographic boundaries of the Central Highlands sub-province (Green Hill, and Mounts Bullengarook, Darriwil, Gorong, and Wallace; Roberts, 1984). Lava cones and domes and rarer composite scoria-lava domes of the central Werribee Plains were described by Condon (1951).

The Werribee Plains sequences include both tholeiitic basalts (Table 3.7.1, column 6) and differentiates (*ol-* and *qz*-tholeiitic basalts and icelandites) and alkaline basalts (alkali basalts or mafic hawaiites). At least two 'domains' on the eastern edge of the plains have been recognised on the basis of isotopic mapping (see below), one dominated by hawaiites in the Laverton area, the second extending from Diggers Rest to Williamstown and dominated by tholeiitic basalts and icelandites. Lavas from late-eruption points are generally more alkaline. Most lava or scoria cones on or near the northern margin of the plains are composed of hawaiite, K-hawaiite, or *ne*-hawaiite (Mounts Bullengarook, Cotteril, Gorong, Moriac, Wallace, and Green Hill; see Irving & Green, 1976). Two of the three Anakies cones also contain hawaiite or *ne*-hawaiite, but the eastern cone is composed of *ne*-mugearite (Table 3.7.1, column 7).

The *Warrnambool-Derrinallum-Colac* region includes 15 of the largest scoria volcanoes of the Newer Volcanics province, most of which are amongst the best-known upper-mantle xenolith and megacryst localities in eastern Australia — for example, Mounts Elephant, Leura, Noorat, Porndon, and Shadwell (Irving, 1974b,c). This restricted area also contains many of the best examples of maars and tuff rings of the Newer Volcanic province. Joyce (1975) noted these are closely associated with the thickest parts of the Cainozoic sedimentary sequences which underlie much of the Western Plains. Aquifers within limestone and sandstone units probably supplied water for phreatomagmatic eruptions (Section 1.6.3).

Two typical scoria volcanoes of the area are Mounts Elephant and Shadwell. Mount Elephant is a near-circular dome of coarse Strombolian deposits containing large *ne*-hawaiite lava blocks rich in megacrysts of Al-augite and Al-bronzite, abundant small bombs with Cr-spinel lherzolite cores, and rarer bombs cored by partially melted granite (Le Maitre, 1974). Mount Shadwell scoria deposits contain basanite lava blocks with abundant megacrysts (augite, anorthoclase) and small peridotitic xenoliths, and abundant bombs cored by a wide range of peridotitic and pyroxenitic fragments.

The largest simple maar of the Newer Volcanics province is Lake Purrumbete, almost 3 km in diameter and with a crater now occupied by a 45 m-deep lake (Edney, 1984b). Several other maars near Camperdown have been studied intensively. The Lake Bullenmerri and Lake Gnotuk complex is best known for its wide range of peridotitic, pyroxenitic, and gabbroic xenoliths (Hollis, 1981; Griffin et al., 1984), including abundant garnet-bearing pyroxenites of deep-crustal or shallow-mantle origin (see Chapter 6). Sediment sequences on the floors of Lakes Bullenmerri and Gnotuk, and in the nearby Lake Keilambete maar, have been radiocarbon dated (Bowler & Hamada, 1971; Dodson, 1974; Barton & Polach, 1980), and the dates corrected using the respective magnetic stratigraphies (Barton & McElhinny, 1981). The oldest dates obtained are within the range 11 500 to 9500 yr B.P., similar to those for lake and swamp sediments at the Tower Hill maar farther west.

Many of the volcanoes in the Warrnambool-Derrinallum-Colac region have involved multiple eruption points and types of products. The extreme example is the Red Rock complex, with up to 14 eruption points associated with maar-forming activity and 28 eruption points within its scoria-cone complex (see: Leach, 1977; Cas et al., 1984a). Phreatomagmatic and magmatic eruptive phases are well defined in other volcanic complexes of the region. Staughtons Hill consists of a maar (Keayang Swamp), several broad scoria mounds (for example, Mount Cunnies Hill) and a small spatter-rimmed crater (Lake Mumblin). A scoria-cone complex at Mount Leura with a restricted lava flow is within and partly buries a 2.5 km-diameter tuff ring (Edney & Nicholls, 1984). Mount Porndon was the source of lava flows making up nearly 200 km² of stoney rises topography that has a maximum relief of about 10 m. These flows, which are hawaiites having small amounts of either normative hypersthene or nepheline, are probably younger than 20 000 yr. Their extent and flow directions were outlined by Ellis (1971). Cr-diopside suite xenoliths and megacrysts are abundant in the cone sequences.

Information on the ages of scoria cones and volcanic complexes of the region is limited to relative maghemite dates and some uncertain radiocarbon dates on soil carbonate and freshwater fossils (Gill, 1978) associated with lava flows and pyroclastic layers. The Mount Leura complex is about 22 000 yr B.P., whereas the Red Rock complex may be as young as 7800 yr B.P. and therefore of similar age to the complexes of the Portland-Hamilton-Warrnambool area farther west.

The main lava types of the region are hawaiite and *ne*-hawaiite (Table 3.7.1, column 8). Many of the *ne*-hawaiites were classified as nepheline basanites by Irving & Green (1976). Strongly alkaline lavas having An-rich normative plagioclase in a few cases, are classified as basanites. *Ol*-tholeiitic basalts are rare. A high proportion of all these rock types is olivine-rich (greater than 20-percent normative olivine), and therefore geochemically primitive.

Magmatic activity at three major complexes in the *Portland-Hamilton-Warrnambool* area — Mounts Eccles, Napier, and Rouse — produced major early lava flows up to 60 km in length, and later scoria cones. A fourth, Tower Hill, is an unusually large maar containing nested scoria cones. The form and eruptive history of Mount Eccles and Tower Hill were studied most recently by Edney (1984a) and Edney et al. (1985), and age relations of lava flows from Mounts Eccles, Napier, and Rouse were discussed by Ollier (1981, 1985a). The oldest of these complexes is Mount Rouse which consists of a scoria cone of minor interbedded hawaiite to *ne*-hawaiite lavas and a small lava-rimmed crater (Ollier & Joyce, 1964). Apparently relatively old flows from the complex having degraded 'stoney rises' surfaces gave K-Ar dates of 0.45–0.31 Ma (McDougall & Gill, 1975), whereas a lava from the main cone gave 1.82 ± 0.04 Ma (Ollier, 1985a). Whitehead (1986) recognised four groups of lavas on the basis of distinctive ⁸⁷Sr/⁸⁶Sr ranges (overall range 0.70381–0.70446).

The Harman valley lava from Mount Napier is a mafic hawaiite (Irving & Green, 1976). The younger lavas of the associated shield include plagioclase-rich types. Stream damming by the shield apron produced Buckleys Swamp, whose basal peat has been dated at 7240 ± 140 yr B.P. (Gill & Elmore, 1973). The shield is capped by several scoria cones, the largest containing a breached summit crater and well-formed spatter

rampart. The narrow range of Sr-isotope ratios of Mount Napier lavas may be an indication they are related by fractionation, and only minor contamination by included granitic material (Whitehead, 1986).

There is some evidence from boreholes in the major hawaiite lava flow from Mount Eccles (the Tyrendarra flow) for the presence of two units separated by a weathering surface (Gill, 1979). Part of the flow was emplaced at a time of low sea-level, and now extends beneath Portland Bay. Radiocarbon ages range from around 29 000 to 6235 ± 120 yr B.P., but apparent inconsistencies between dates remain to be resolved.

The 2.4–3.2 km-diameter maar crater at Tower Hill contains a scoria cone complex and moat lake. Plant material beneath its ash apron was dated at 8700–6500 yr B.P. (for example, Gill, 1978), but sediments within the lake and a scoria-cone crater have yielded ages of 11 400 and 9980 yr B.P. (Edney et al., 1985).

The old volcanic centres in the *Portland-Heywood* region are mostly shield volcanoes, up to 2 km across and less than 200 m high, and may contain a high proportion of lava flows (Sukhyar, 1985; Stewart, 1985). Cape Bridgewater volcano is a multi-vent complex resting upon a thick lava sequence (Cape Duquesne), and surge and air-fall tuffs and breccias of mainly phreatomagmatic origin form several overlapping cone structures (Cas et al., 1984b). Cliff and quarry exposures at Cape Grant include a sill of mafic hawaiite and a small cone containing air-fall, mass-flow, and spatter deposits, buried by *ol*-tholeiitic basalt lava flows. Cape Nelson includes bedded tuffs and breccias, layers of accretionary lapilli, sequences of thin *qz*- and *ol*-tholeiitic basalt to hawaiite lava flows, and associated sills (Sukhyar, 1985).

The three coastal complexes contain closely related rock types. Initial ⁸⁷Sr/⁸⁶Sr values fall within a narrow range (0.70426-0.70448) and major- and trace-element trends correspond to a simple fractionation series from hawaiite or *ol*-tholeiitic basalt parent magmas. The more evolved members are *qz*-tholeiitic basalt and icelandites which are dominant at Capes Bridgewater and Duquesne. Some of these have ⁸⁷Sr/⁸⁶Sr values reaching 0.7050 (Stewart, 1985).

There are fewer data for the inland complexes Mounts Clay, Deception, Eckersley, Kincaid, and Vandyke, but these centres appear to be more petrologically diverse than those of the coast, and they do not appear to be closely related genetically. Lavas range from *ne*-hawaiite (Mount Kincaid) through *ol*- and *qz*-tholeiitic basalt to icelandite. Mount Vandyke has distinctive felsic icelandites containing high incompatible-element abundances. Initial ⁸⁷Sr/⁸⁶Sr values for Mount Eckersley lavas are unusually high (0.70489–0.70492; Stewart, 1985).

Mount Gambier sub-province

An extensive area lacking volcanic deposits separates the Mount Gambier and Western Plains sub-provinces. Both include two distinct groups of volcanoes, one older than 2 Ma, and the younger active within the last 10 000 years. However, the older centres of the Mount Gambier sub-province include rock types that are more strongly alkaline than those of the contemporary

Western Plains group (Irving & Green, 1976), and the very young Mount Gambier complex has ultramafic xenoliths that are notably rare or absent from the youngest Western Plains volcanoes.

The volcanoes of the Mount Gambier sub-province rest on a karst limestone surface. Fifteen, older, northwestern centres known as the Mount Burr group (Sheard, 1983) are associated with three northwest-trending lineaments and a horst-like basement high known informally as the Mount Burr range. The group is partly to completely buried by Pleistocene dune sands of the Bridgewater Formation, and includes cones and domes composed mainly of pyroclastic fall deposits, lava flows, and maars or tuff rings of phreatomagmatic base surge and air-fall origin. The total area of pyroclastic deposits is about 100 km². The estimated age range of the Mount Burr group is 2 Ma to 20 000 yr B.P. (Sheard, 1983). Lavas at three centres of the group (The Bluff and Mounts McIntyre and Watch) lack feldspar and contain abundant groundmass analcime, and were classified as olivine analcimites by Irving & Green (1976).

Mounts Gambier (*Fig. 3.7.4*) and Schank are two young volcanic complexes in the southeast of the province (the Southern Volcanic group of Sheard, 1983), and consist mainly of *ne*-hawaiite (Table 3.7.1, column 9) and K-*ne*-hawaiite respectively (Irving & Green, 1976). Dominant phreatomagmatic activity at Mount Gambier produced a number of overlapping maars with characteristic bedded surge deposits, of which the most spectacular is the Blue Lake crater (Sheard, 1978). Minor magmatic activity produced lava flows and scoria cones, the largest on Mount Gambier itself. A single series of volcanic events within the period 4600–4300 yr B.P. is indicated by recent dating (Blackburn et al., 1982, 1984; F.W.J. Leaney, personal communication, 1987).

The Mount Schank complex is dominated by a large steep-sided cone. Two craters extend almost to the original ground surface and are formed of bedded ash-lapilli deposits largely of phreatomagmatic origin (Sheard, 1986a). Charcoal beneath tuffs has been dated at 18 100 yr B.P. (Polach et al., 1978), and palaeomagnetic data (Barbetti & Sheard, 1981) are indicative of an age either greater than 7000 yr B.P. or between 5000 and 1000 yr B.P. In addition, Smith & Prescott (1987) obtained an age of 4930 ± 540 yr B.P. by thermoluminescence dating of former beach sands overlain by lava flows. This leaves open the possibility that Mounts Gambier and Schank were active almost simultaneously.

Petrology

Older 'plains' lava sequences in the Western Plains sub-province are predominantly subalkaline. More than 60 percent of about 300 analyses are *ol*- or *qz*-tholeiitic basalts and icelandites. The remainder are mostly hypersthene- to weakly nepheline-normative hawaiites. About 2 percent are more strongly alkaline and silica-undersaturated (*ne*-hawaiite and *ne*-basanite) and are probably related to young centres. The older centres of the Portland-Heywood area that are coeval with plains activity have similar proportions: nearly

60 percent of 74 samples analysed by Sukhyar (1985) and Stewart (1985) are *ol*- and *qz*-tholeiitic basalts and icelandites, and the remainder hawaiites. In contrast, about 120 samples from young Western Plains eruptive centres analysed by Irving & Green (1976) and Price et al. (1988) include less than 10-percent subalkaline types. Over 60 percent are strongly alkaline types (mainly *ne*-hawaiite) and the remainder mostly hawaiites. Seventy percent of the 55 analyses presented by Irving & Green (1976) have more than 5-percent normative nepheline, and under 5 percent have more than 10-percent normative hypersthene.

The Central Highlands sub-province appears to be intermediate in age and eruptive style to the 'plains' and 'cone' stages of the Western Plains, and also in its proportions of subalkaline and alkaline rocks. The plains lava sequence beneath Ballarat consists almost exclusively of *ol*- and *qz*-tholeiitic basalts (Yates, 1954). About 55 percent of more than 60 rocks analysed by Yates, and by Irving & Green (1976) from lava and scoria cones and related flows, are hawaiites. *Qz*- and *ol*-tholeiitic basalts (about 25 percent) and *ne*-hawaiites and *ne*-basanites (about 15 percent) are the only other common types. The 80 occurrences studied by D.A. Wallace (personal communication) are mostly lava or scoria cones and related flows. These include almost equal proportions of *ol*-tholeiitic basalt and hawaiite (each nearly 40 percent). Icelandites and *ne*-hawaiites (each about 10 percent) are other significant types. Both the plains and cone stages in the Central Highlands therefore seem to contain higher proportions of subalkaline types than do the Western Plains equivalents.

A major characteristic of the Newer Volcanics province as a whole is an overwhelming dominance of sodic rocks (Irving & Green, 1976; Sukhyar, 1985; Stewart, 1985; Yates, 1954). This is reflected in the very high proportion of hawaiites and *ne*-hawaiites (Table 3.7.1, columns 8–10), to the near exclusion of transitional basalts and alkali basalts. Most rocks of the Mount Gambier and Western Plains sub-provinces are also sodic, whereas more potassic types are a feature of the central and eastern parts of the Central Highlands sub-province, centred on the Macedon-Trentham area.

Sr-isotope domains

Sr isotopes have been used in recent studies of the Newer Volcanics to identify 'domains' of similar lava composition. The method involves sampling from outcrops and cuttings on a 5 km grid. The range of initial $^{87}Sr/^{86}Sr$ values is sufficiently large that different lava units can be distinguished, and the domains therefore may aid field mapping.

Resolution is clear where a domain has a Sr-isotope ratio range that does not overlap with those of its neighbours (allowing for experimental uncertainties). Additional information such as petrographic data or trace-element ratios may be required if there is overlap. Domains that are uniform in both petrography and isotopic ratios probably represent single flows. Others with some petrographic differences may consist of related flows from a single centre. Fifty domains have been recognised to date within an area of 17 000 km², and up to 100 may be resolvable.

Readily distinguishable, young lava flows define two types of domain:

(1) Roughly circular aprons. The 12 km-diameter hawaiite apron of Mount Porndon is isotopically uniform ($^{87}Sr/^{86}Sr$ 0.70443–0.70456), and distinct from both the underlying plains sequence (0.7041–0.7043) and the overlying scoria-cone complex (0.70387; McDonough et al., 1985). Hawaiites of the 15 km-diameter apron at the nearby Warrion-Hill/ Red-Rock complex have Sr-isotope ratios (0.70425–0.70445) which contrast with those of plains tholeiitic basalts to the east (0.7047, 0.7050). Similar aprons are found at Mounts Eccles, Elephant, Rouse, and Napier. Late scoria cones at Mounts Rouse and Napier are isotopically indistinguishable from underlying apron flows (Whitehead, 1986).

(2) Elongate valley flows originating from centres such as Mounts Eccles, Napier, and Rouse. Anastomosing hawaiite flows from Mount Rouse are relatively unradiogenic (0.7037–0.7039) and distinct from underlying older flows (0.7043–0.7048).

A simple model for the older plains sequences, using these young occurrences as a guide, involves numerous, elongate, overlapping flows controlled initially by ancient, largely northward-trending, stream patterns. Circular aprons about eruptive centres should lie at the head of some of these flows, or be randomly superimposed upon them. The modifying effects of erosion and emplacement of younger cover sequences must be taken into account in applying this model to domains recognised only by isotopic characteristics.

An approach of this type has been used for the lava plains north of Melbourne where five, approximately northward-trending, isotopic domains corresponding to palaeovalleys are separated primarily by basement ridges. These domains are developed mainly in ol- and qz-tholeiitic basalts and rarer hawaiites. The first four from east to west have $^{87}Sr/^{86}Sr$ ranges of 0.7041–2 (12 samples), 0.7046–7 (9), 0.7048–51 (7), and 0.70465 (2). The range for the plains sequence of the remaining domain (Woodstock) at the northern limit of the area, is 0.7044–5 (7 samples). This domain includes two prominent cones, Mount Fraser (hawaiite, $^{87}Sr/^{86}Sr$ 0.7038; McDonough et al., 1985) and Bald Hill (*ne*-basanite) which, on isotopic or petrographic grounds, may not be related directly to the lava sequences on which they rest.

Three clear-cut domains have been recognised farther west. The most important domain of the Melbourne area extends from Diggers Rest southeast to Williamstown, and is dominated by a series of massive ol- and qz-tholeiitic basalt and icelandite flows that have $^{87}Sr/^{86}Sr$ values of 0.7050–0.7054. Two domains in the Laverton area consisting dominantly of hawaiite ($^{87}Sr/^{86}Sr$ 0.7043 and 0.7044) probably represent two large flows from a single vent.

Domain mapping within the Western Plains subprovince has revealed regional correlations. Values for about 300 samples from the region east of Mortlake have a range of 0.7040–0.7058. The greatest range is found from the area north and northeast of Geelong, where values for eight samples cannot be grouped into domains. The range of Sr-isotope ratios for 115 samples from the region west of Mortlake is narrower,

0.7037–0.7047, and only 13 samples have ratios above 0.7045. Many of the low values are those of young lavas from Mounts Eccles, Napier, and Rouse, but 60 samples from older flows have similar values. Regional differences are thought to reflect fundamental distinctions in magma sources or conditions of magma genesis, possibly connected with the nature of the lithosphere.

Domain mapping at the local scale is also possible within the Western Plains. The Mortlake area has seven domains at the boundary between the major regions outlined above, and Sr-isotopic ratios range from 0.7040 to 0.7050. A value of 0.70392 for the Mount Shadwell scoria-cone complex is compatible with only one neighbouring terrain, at the limit of analytical uncertainty.

3.8 Tasmania and Bass Strait

3.8.1 Introduction

Intraplate volcanic rocks continue southwards from the southeastern coast of Australia onto Tasmania. Volcanism in this region is not restricted to land areas, as intraplate volcanic rocks are known from offshore drilling to be represented in at least three of the basins in the Bass Strait region (Fig. 1.1.5).

3.8.2 Bass Strait basins

The intracratonic Bass Basin trends northwestwards between northern Tasmania and Victoria. It is separated from the Otway and Torquay basins to the west and northwest by the King Island and Mornington Rise and from the Gippsland Basin to the east by the Bassian Rise. It once was considered a huge volcanic crater that fed the basaltic lavas on the adjacent Tasmanian coast (Noetling, 1911), but as a result of oil-exploration surveys it is now known to contain up to 10.5 km of possibly Jurassic to Holocene sediments and subordinate volcanic rocks (Robinson, 1974; Brown, 1976; Smith, 1986; Williamson et al., 1987a).

The basin was initiated in the early Cretaceous by extension or rifting which produced a set of west-northwest trending, rotational, normal faults and an orthogonal set of near-vertical transfer (transform) faults (Etheridge et al., 1985a,b). The resultant half grabens were modified by later faulting, folding, and uplift. The northern margin of the basin was controlled, and at times reactivated, by the Gambier-Gabo Lineament (Harrington et al., 1973; Baillie, 1985), and the geological development of the southern margin was a complex interplay of basement control, tectonism, volcanism, and sea-level changes (Sutherland, 1973; Williams, 1978; Moore et al., 1984). The early thermal history of the basin, related to sagging and mantle upwelling, is detailed by Middleton (1982).

Drilling has been confined largely to the Cainozoic section and knowledge of volcanism in the Mesozoic section is poorly known. However, weathered mafic volcanic rocks are known from the Konkon-1 and Durroon-1 wells. These may be related to the Cretaceous volcaniclastic Otway Group sedimentary rocks in the Otway Basin and the Strzelecki Group in the Gippsland Basin. The Boobyalla Sub-basin (Moore et al., 1984) is known to contain early Cretaceous volcaniclastic sediments and middle Cretaceous basaltic rocks. Basalts at the top of the Lower Cretaceous section in Durroon-1 are unconformably overlain by Upper Cretaceous sediments. Corresponding volcanic rocks are found onshore in northeastern Tasmania, in the Boobyalla area, and elsewhere, and consist of 'trachyandesite' flows, lamprophyric dykes, and dioritic complexes (Jennings & Sutherland, 1969; Sutherland & Corbett, 1974; Moore et al., 1984; Baillie, 1984). Extensive, possibly latest Cretaceous to Palaeocene volcanic rocks are known from several wells and this phase of volcanism may be confined to the central northeastern sector of the basin. Strongly altered mafic rocks were recovered in the Aroo-1, Yolla-1, Tilana-1, Bass-2, and Chat-1 wells.

An intrusion in early Eocene beds in Cormorant-1 well contains alkali picrite that has considerable alteration of olivine and pyroxene as well as late-stage crystallisation of biotite and amphibole. The picrite is associated with coarse alkali gabbro containing plagioclase, analcime, and some biotite. The suite was probably intruded, fractionated, and altered within wet sediments. A K-Ar age for the biotite is 22 Ma, relating the intrusion to the Lower Miocene volcanism (Sutherland & Wellman, 1986). Biotite from a similar body in the Yolla–1 well gave an age of 24 Ma (Wheeler & Kjellgren, 1986), and apparently similar rocks from Tilana–1 gave ages of 28 and 21 Ma (Amoco, 1986). Similar intrusive bodies were found in the Koorkah, Seal, Toolka, and Squid wells. Many of the volcanic rocks in Bass Basin seen on seismic profiles appear to be fault controlled, and some may reach the sedimentary floor of Bass Strait.

The Gippsland Basin underlies the continental shelf and slope between Victoria and Tasmania and extends onshore into the La Trobe valley of eastern Victoria (James & Evans, 1971; Threlfall et al., 1976; Thompson, 1986). South Platform is largely within Tasmanian waters, separated from the Central Deep by the Foster Fault, and constrained to the south by the Bassian Rise (Baillie, 1987). This platform forms the continental shelf of northeastern Tasmania. Mullet-1 and Bluebone-1 penetrated thin shelf sequences into granitic basement of Carboniferous age (about 280 Ma) which therefore is apparently younger than the typical Devonian granite basement of northeastern Tasmania.

A pattern of magnetic anomalies obtained from aerial magnetic surveys of Bass Basin may relate to strips of volcanic rocks, and some small circular anomalies east of the Furneaux Islands and off northeastern Tasmania may mark basaltic plugs or centres (Beattie, 1978; personal communication, 1983). Volcanic horizons are seen in seismic lines throughout the South Platform sequence within sediments or over basement. Sailfish-1 penetrated a region where volcanic rocks were spread over an area of 50 km² associated with at least six possible cone-like centres. The well intersected over 200 m of mafic volcanic breccias or pillow lavas. Deposition of the late Cretaceous to late Eocene La Trobe Group throughout much of the remainder of the Gippsland Basin, appears to have been accompanied initially by volcanism (Robertson et al., 1978).

The offshore Otway Basin is west of the Bass Basin, and the Tasmanian sector (west of King Island) is part of the Mussell Platform (Williamson et al., 1987b). Little is known of the volcanic history of the region, although the presence of basalt flows has been postulated on the continental slope west of Tasmania (Hinz et al., 1986).

3.8.3 Tasmania

Introduction

Tasmania has about 400 km³ of Cainozoic basalts, in lava fields, eroded valley fills, and other scattered remnants. The basalts mainly extend diagonally across the island. They are absent in the southwest and form only minor outcrops on the larger Bass Strait islands (Fig. 3.8.1). Basalt is found overlying sediments in Upper Cretaceous to Lower Tertiary rift basins or along fault zones, but considerable amounts either obscure or are not related obviously to these structures.

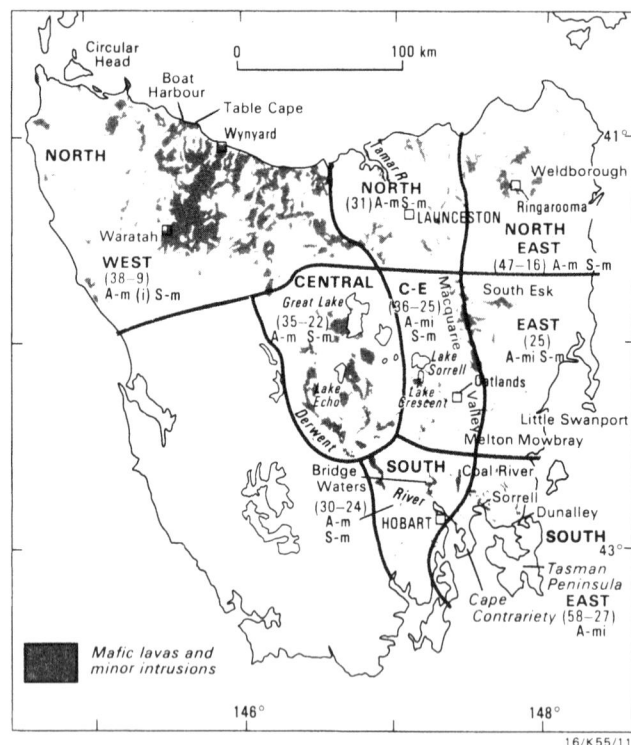

Figure 3.8.1   Distribution, rock types, and ages of volcanic provinces in Tasmania. C-E refers to Central-East province. Rock associations: A, alkaline; S, subalkaline. Rock types: m, mafic; i, intermediate; (i), minor intermediate. Ages (in brackets) in Ma.

All the dated basalts (Lower Eocene to Upper Miocene, 58–8 Ma; Sutherland & Wellman, 1986; Baillie, 1986a,b, 1987) are younger than the major epeirogenic faulting and uplift. The distribution, nature, and petrology of the basalts were reviewed by Edwards (1950), Spry & Banks (1962), and Sutherland (1969a). Generalised geochemical trends for Tasmania are shown in Figures 3.7.2 and 3.7.3, and selected chemical analyses are given in Table 3.8.1.

Table 3.8.1 Selected chemical analyses for Tasmanian provinces

	1	2	3	4	5	6	7	8	9	10
SiO_2	44.46	39.27	51.73	44.55	36.69	42.93	44.96	46.81	48.31	39.41
TiO_2	2.00	3.03	1.57	2.21	2.71	2.73	2.22	2.72	1.23	2.79
Al_2O_3	13.25	8.60	13.14	14.61	9.32	12.95	13.27	14.15	15.18	11.60
Fe_2O_3	3.63	5.07	1.73	4.70	5.44	3.65	4.41	3.33	5.67	3.57
FeO	9.41	8.34	9.51	7.26	8.44	9.92	9.34	8.85	5.21	9.64
MnO	0.18	0.23	0.16	0.17	0.21	0.21	0.18	0.17	0.21	0.20
MgO	9.63	13.21	8.79	8.68	14.91	9.01	6.65	7.03	5.26	11.14
CaO	8.57	14.69	8.86	11.03	13.23	9.49	7.87	7.65	5.95	13.58
Na_2O	3.85	3.36	2.60	2.55	3.82	4.69	4.45	3.22	6.89	3.09
K_2O	1.53	1.25	0.31	0.79	1.44	1.44	2.36	1.81	2.58	1.63
P_2O_5	0.70	1.51	0.21	0.42	1.32	0.93	1.27	0.95	0.62	1.03
S										
H_2O+[1]	2.14	0.93	1.07	2.39	1.76	1.59	2.03	2.70	2.17	2.09
CO_2						0.04			0.11	0.12
Total	99.35	99.49	99.68	99.36	99.29	99.58	99.01	99.39	99.39	99.89
Ba	135	514	102	218	566	510	660	395	165	720
Rb	9	23	8	6	32	31	62	36	24	46
Sr	794	1297	249	580	1258	1150	1295	1000	1000	830
Pb		20				8		7	6	11
Th	3	11	1	3	11	6	11	6	6	4
U	<1	2	<1	<1	3	<5	2	2	<5	<5
Zr	201	442	92	152	268	410	507	425	730	250
Nb	44	111	13	42	116	110	116	69	130	115
Y	19	35	18	25	28	39	39	34	22	25
La								64		
Ce								125		
V	165	254	159	235	233	190	57	105	125	340
Cr	325	521	392	299	442	260	234	153	160	240
Ni		246				260		118	115	170
Cu		85				72		37	41	100

1. *Ne*-hawaiite from Northwest Tasmania (F.L. Sutherland unpublished data, sample DR1180).
2. Melilite nephelinite from Northwest Tasmania (F.L. Sutherland unpublished data, sample BHT).
3. *Qz*-tholeiitic basalt from North Tasmania (F.L. Sutherland unpublished data, sample DR11808).
4. Alkali basalt from Northeast Tasmania (F.L. Sutherland unpublished data, sample DR11804).
5. Melilitite from Central Tasmania (F.L. Sutherland, unpublished data, sample LJMT).
6. Basanite from Central-East Tasmania (Tasmanian Mines Department, sample HT).
7. *Ne*-hawaiite from Central-East Tasmania (F.L. Sutherland unpublished data, sample DR10478).
8. Hawaiite from Central-East Tasmania (F.L. Sutherland, unpublished data, sample T6).
9. *Ne*-mugearite from Central-East Tasmania (Tasmanian Mines Department, sample DR12488).
10. Nephelinite from South Tasmania (Tasmanian Mines Department, sample IR131).

[1] Total H_2O.

Northwest

The greatest concentration of basalt in Tasmania is in the northwest, west of the Tamar Tertiary rift (see below) and east of the Arthur Lineament (Figs. 1.3.1, 3.8.1). Lavas infill old valleys cut by major drainage systems that flowed north into Bass Basin. The basalts in places rest on Permo-Triassic beds intruded by Jurassic dolerite, or on the Tertiary sediments of rift basins. Basalt lava plains west of the dissected valley fills, extend over 250 km² from about 750 m above sea-level and pass into more dissected profiles eastwards to about 100 m above sea-level.

The northwest contains the known range of Tasmanian basalt types, including basanite, alkali basalt, *ne*-hawaiite (Table 3.8.1, column 1), hawaiite, and *ol*- and *qz*-tholeiitic basalts. There is rare *ne*-mugearite and olivine-melilite nephelinite (Table 3.8.1, column 2) near Boat Harbour, and a nephelinite flow extends over 15 km from Forth to Devonport. The older sequences are best dated east of Devonport, where the Thirlstone

Basalt is 38 Ma (Cromer, 1980) and the Moriarty Basalt in the overlying lead is 26 Ma (Baillie, 1986b). Basalts west of this are stratified with Lower Miocene marine beds that are possibly facies equivalents to non-marine deposits within leads in coastal sections (Burns, 1965). No basalts below the marine beds are dated, but the olivine-melilite nephelinite in the lateritised Boat Harbour sequence has an age of 26 Ma and an unlateritised massive basalt valley fill at Table Cape is 13 Ma (Sutherland & Wellman, 1986). Zircon fission-track ages from the Boat Harbour centre are indicative of reset eruption ages of 13 and 9 Ma.

Subaerial flows predominate and many have classic cooling columns (Spry, 1962). Aquagene volcanic rocks appear where lava flows blocked major rivers during build-up of large eruptive centres. Transitional basalt and *ol*-tholeiitic basalt are typical at Mersey-Forth (Spry, 1955) where a thick lower flow is overlain by hyaloclastite tuffs and breccias, flow-foot breccias, and subaerial-flow caps.

Many feeders for lavas are probably buried under thick and heavily lateritised sections, but a few dykes and plugs are exposed in faults or folded basement rocks (Burns, 1965; Jennings, 1979). Basanitic pyroclastic rocks and lava flows are exposed below Miocene marine beds between Somerset and Wynyard, and coarse basalts form prominent landmarks at the Mount Hicks and Table Cape centres (Gee, 1971, 1977). The Table Cape centre has partly 'welded' nephelinite tuffs containing conspicuous megacrysts and lower-crustal and upper-mantle xenoliths. The overlying lava is a 165 m-thick analcime basanite in which lherzolitic upper-mantle xenoliths are concentrated towards the base and pegmatoidal veins and patches of fractionated rocks increase towards the top (Gee, 1971). Lavas in far northwestern Tasmania are restricted to a few valley courses that drained Precambrian-Cambrian sections, but centres containing aquagene volcanic rocks are prominent along the coast (Sutherland, 1980b).

Hyaloclastite tuffs below pillow lavas on the western flank of Flat Topped Bluff volcano may be equivalent to the tuffs below breccias of the Cape Grim centre. The tuffs have soft-sediment deformation, fold and fault structures, and climbing ripple lamination consistent with deposition in shallow water. The Cape Grim tuffs and breccias underlie a channel fill of Lower Miocene marine beds, and younger massive basalt at around present sea-level.

The northwest aquagene centres have a general north-south trend and probably are controlled structurally by faults bounding the coast and the major Precambrian-Cambrian blocks south of Marrawah. Circular Head (Fig. 3.8.2) is a prominent neck of fractionated teschenite (Edwards, 1941; Cromer, 1972) that intrudes pyroclastic rocks (Gill & Banks, 1956). Its 12 Ma age date is older than the 8.5 Ma date for adjoining lava flows of the peninsula (Baillie, 1986a). These are the youngest lavas known in Tasmania.

Basalts on western Bass Strait islands include alkali basalt or hawaiite on Three Hummock Island, and a basanite plug on central King Island containing inclusions of wehrlite and pyroxenite cumulates.

North

Basalts in the north in the Tamar area are found mostly along Tertiary rifts and faults (Tamar Trough). Many follow courses cut in extensive Jurassic dolerite sheets in the Permian-Triassic sedimentary cover or excavated in Tertiary rift beds. Eruptive centres appear to cluster at changes in rift structure or form sporadic alignments along bounding faults.

The Tamar lavas are largely alkali basalt, hawaiite, and transitional basalt. Coarse-grained basalts in flows over 70 m thick developed a range of textures and compositions during cooling and differentiation. Minor flows and plugs of nephelinite, basanite, *ne*-hawaiite, and *ol*-tholeiitic basalt are also present (Sutherland, 1969b, 1971). Alkali basalt and hawaiite are common in the Longford Sub-basin, and transitional basalt underlies laterites.

Figure 3.8.2 The Nut, a teschenite neck intruding pyroclastic rocks (exposed on shore in foreground), at Circular Head, Stanley, northwestern Tasmania, in the region of the youngest dated volcanic rocks in Tasmania (13 to 8.5 Ma).

Ol- and *qz*-tholeiitic basalts (Table 3.8.1, column 3) are more abundant east of the Tamar Trough although only a few eruptive centres are known (Marshall, 1969; Sutherland, 1973). Fractionated alkaline lavas and nephelinites commonly overlie Tertiary sediment or form higher caps in the upper parts of drainages.

Basalt ages in northern Tasmania are based largely on palynological data. The older flows are alkali basalts underlain by Middle Eocene, or interbedded with Middle to Upper Eocene, beds (Matthews, 1983). The lower Tamar flows overlie unweathered Upper Oligocene sediments, and coarse basalts may have Miocene ages (Sutherland, 1971). A *qz*-tholeiitic basalt in Pipers Brook is dated as Lower Oligocene (31 Ma), and overlying Tertiary sediments and alkaline lavas may be Upper Oligocene to Middle Miocene (Sutherland & Wellman, 1986).

Northeast

Basalts in northeastern Tasmania occupy drainages and basins cut in Devonian granites and surrounding Ordovician-Devonian beds, except in the far northeast where there are downfaulted cover rocks of largely Jurassic dolerite (Brown et al., 1977; Jennings & Sutherland, 1969).

High-level basalts and pyroclastic rocks up to 380 m thick around Weldborough are 500–800 m above sea-level. Agglomerates and tuffs up to 150 m thick mark several eruptive centres. Flows about 230 m thick overlie the pyroclastic rocks and descend northwards as eroded remnants. A small swarm of outlying dykes to the west may be related to a similar basalt capping. The lavas are alkali basalt (Table 3.8.1, column 4), hawaiite, and transitional basalt (Brown, in McClenaghan et al., 1982).

Lower-level basalts in the Ringarooma River valley extend over 40 km as flows towards the north and northeast. The flows include nephelinite, glomeroporphyritic basanite, and alkali basalt.

Flows at similar levels in the Scottsdale Sub-basin are highly weathered to lateritic soils and may be older than the Ringarooma flows. An upper nephelinite flow contains abundant upper-mantle and lower-crustal inclusions — including sanidine (Or_{46}), an unusual megacryst phase for Tasmania.

Small exposures of basalt at Cape Portland include basanite and *ol-* and *qz*-tholeiitic basalt flows and pyroclastic rocks. Remnants in the Furneaux group of islands are alkali basalt, hawaiite, and nephelinite which contains upper-mantle xenoliths and megacrysts (Sutherland & Kershaw, 1971; Frey et al., 1978).

The northeastern basalts range from Middle Miocene valley flows, dating around 16 Ma, to high-level basalts which are Eocene (47 Ma; K-Ar and zircon fission-track methods; Sutherland & Wellman, 1986; Yim et al., 1985).

Central

Extensive basalt remnants fill old depressions and tributary courses of the ancestral Derwent River which cut into the resistant dolerite sheet that caps the Central Plateau to heights of 1000–1400 m above sea-level (Banks, 1972; Sutherland, 1972).

Aquagene volcanic rocks are abundant at Great Lake around centres that developed after massive lava flows blocked the Ouse outlet to over 90 m depth (Sutherland & Hale, 1970; Sutherland et al., 1973; Sutherland, 1980b). The aquagene units exposed during exceptionally low drought levels of Great Lake include basal hyaloclastite tuffs. These have many soft-sediment deformation structures, and evidently formed during shallow-water phreatic eruptions. Flow-foot breccias containing pillow-like lenses have delta-like, overlapping bed forms as if lava was erupted into lake water from several emerged vents. Aquagene sections are capped by subaerial flows or cut by dykes (or both). Some lavas descended into the basin from higher levels to the west around Lake Augusta.

Lava sequences away from Great Lake are not related easily to their eruptive sources. Feeders clearly existed northwest of Bronte and northeast of Lake Echo, and a few basalt dykes are found in the Nive River. Restricted exposures of flow-foot breccias (Sutherland, 1980b) probably represent long tributary flows entering the main Derwent River course.

Ol-tholeiitic and transitional basalts dominate several sequences and thick *qz*-tholeiitic basalt flows are present. Alkali basalt, basanite, and nephelinite underlie or cap some sequences (Sutherland & Hale, 1970; Sutherland, 1980b). Rare isolated plugs of olivine (monticellite) melilitite (Table 3.8.1, column 5) and melilite nephelinite are found at Shannon Tier (Edwards, 1950) and Laughing Jack Marsh. The nephelinite is one of the few rocks in the region containing abundant upper-mantle xenoliths (Sutherland, 1972).

Radiometrically dated rocks around Great Lake (Sutherland et al., 1973; Sutherland & Wellman, 1986) have Upper Oligocene to Lower Miocene ages (25–22 Ma). The melilitite plugs at Shannon Tier and Laughing Jack Marsh are older at 30 and 35 Ma, respectively.

The other basalt sequences may have similar ages on the basis of degree of dissection. Thus, the main Upper Derwent infilling caused by lavas spilling down its northern tributaries is probably an Upper Oligocene to Lower Miocene event. This diverted the Derwent out of the early Tertiary sedimentary rift structure to a more southerly course cut in dolerite and Triassic sediments.

Central-East

Basalts on the plateau around Lake Sorell and Lake Crescent and on the slopes descending east and south to Jordan drainage, have a lower volume and a much greater proportion of exposed pyroclastic vents, plugs, and dykes than those in the Central Plateau and Upper Derwent region. They are typified by a range of alkaline rocks (Table 3.8.1, columns 6–9), whereas tholeiitic lavas dominate the latter region. In contrast, again, the lower exposures in Lake Sorell and Lake Crescent have no aquagene rocks, except for a partly glassy flow that descends to Lake Sorell.

Flow remnants around the twin lakes consist of hawaiite, K-rich hawaiite, *ne*-hawaiite, mugearite, olivine-nepheline melilitite, nephelinite, and mafic *ne-*

benmoreite. A plug north of Lake Sorell is a K-rich nephelinite containing abundant upper-mantle xenoliths and unusual crustal xenoliths of sodalite-malignite, ijolite, and nepheline syenite.

Scattered flows and centres of similar petrology to those around Lake Crescent extend south down the Jordan valley to Melton Mowbray. Here a distinctive, porphyritic, alkali basalt 90 m thick lies 210 m above the bed of Jordan River. Several flows at Bow Hill are capped by K-rich *ne*-hawaiite and basanite which contains xenoliths of garnet lherzolite (Sutherland et al., 1984a; J. Adam, personal communication, 1987). Centres are commonly sited on fault lines, or on dolerite margins, and a few have pyroclastic rocks. K-rich *ne*-hawaiite, *ne*-mugearite, hawaiite, and mugearite are common, and many of them contain abundant upper-mantle and lower-crustal xenoliths.

Sutherland & Wellman (1986) dated the alkali-basalt residual at Melton Mowbray as 36 Ma (Lower Oligocene), a basanite plug near Jericho as 28 Ma (Upper Oligocene), and flow caps on pyroclastic rocks west of Oatlands as 25–24 Ma (Lower Miocene). The nephelinite plug north of Lake Sorell and a complex lava cap on Wild Pig Tier also have K-Ar ages of around 25 Ma.

East

Substantial valley fills occupy former courses of the Macquarie and South Esk rivers which drain much of the area before flowing northwards into the Tamar and Longford rift basins. *Ol*-tholeiitic basalt is the most common rock type, but *qz*-tholeiitic basalt is found also, together with small developments of alkali basalt, hawaiite, and nephelinite. Centres are rarely exposed, but a flow feeder northeast of Campbell Town contains many fragments of dolerite.

Smaller volumes of similar basalts are found south of the Macquarie drainage in the headwaters of Little Swanport River. One pyroclastic centre including an *ol*-tholeiitic basalt flow is notable for its upper-mantle xenoliths — a very rare occurrence of Cr-diopside peridotite xenolith in a tholeiitic host (Sutherland, 1974). This flow is dated at 25 Ma (Upper Oligocene; Sutherland & Wellman, 1986).

The lavas centred around Campbell Town mark a change in trend of the Tamar rift structure where the main downthrows are taken up on other fault trends. *Ol*-tholeiitic basalt extends 30 km upstream along the South Esk River. This line of tholeiitic rocks continues southwards 80 km to a small vesicular pipe of glassy *qz*-tholeiitic basalt. This trend may mark a deep fracture, possibly the extension of the major Tamar basement fault.

South

Voluminous and multiple flows related to the Derwent and Coal River grabens are found in southern Tasmania (Alwar, 1960; Sutherland, 1976; Sutherland, 1977b). The association resembles the eastern region in containing abundant tholeiitic basalts, but it lacks *qz*-tholeiitic compositions. Flanking flows and plugs are alkali basalt, basanite, and nephelinite (Table 3.8.1,

column 10), and there is little development of fractionated alkaline rocks.

Some centres along the Derwent graben developed aquagene rocks where flows entered impounded sections of old Derwent courses. Pillow lavas, flow-foot breccias, and hyaloclastite tuffs, interbedded with lavas erupted from centres on subsidiary faults of the main graben structures between Bridgewater and Hobart, caused minor diversions of the Derwent channel.

Infills in the Coal River graben near Campania include *ol*-tholeiitic basalt of Miocene age overlying a narrow valley fill of 24 Ma alkali basalt erupted from a dyke in a small pyroclastic vent.

Small centres and minor flows of alkaline rocks fringe the tholeiitic-basalt fills. The most substantial flow at New Norfolk is a basanite that overlies middle Tertiary sediments and a lower *ol*-tholeiitic basalt. An alkali-basalt centre nearby is underlain by pyroclastic rocks and contains upper-mantle peridotite xenoliths and clinopyroxene and spinel megacrysts. Small nephelinite plugs are present near the large Dromedary-Cascades downthrow. This structure deflects the easterly course of the Derwent southwards, and scattered plugs, dykes, and necks extend down on its eastern side to Hobart. Two centres have kaersutite megacrysts (rare in Tasmania) in tuffs and a *ne*-hawaiite plug that also contains lherzolite xenoliths. Crystals from the tuffs are dated at 30 Ma, so this is the oldest known southern centre. A nephelinite containing abundant upper-mantle and lower-crustal xenoliths near Rekuna may be one of the youngest on physiographic grounds. Isolated remnants of alkali basalt and hawaiite south of Hobart infill Mountain River down to its junction with Huon River.

Southeast

The southern Derwent and Coal River tholeiitic basalts infill parallel grabens, separated by a complex horst scattered with alkaline rocks. In contrast, the volcanic rocks to the south and east pass into a more fractionated alkaline association that has numerous centres and a greater abundance of pyroclastic rocks.

A composite centre located on a fault bend of the Derwent graben south of Hobart has folded flows and pyroclastic rocks that collapsed back into the vent while still plastic (Spry, 1955). The upper flows are mafic *ne*-benmoreite, considered to represent extreme end-members of fractionated nephelinite magma (Sutherland, 1974). Flow remnants of similar petrology are found at a higher level on the opposite east Derwent shore. They fall on the perimeter of a circular fracture identified from satellite imagery, which encloses the collapsed vent (Burns, in Sutherland, 1976).

Other centres south of Hobart extend to D'Entre-casteaux Channel on the western side of the Derwent and towards South Arm on the eastern side. Most of these centres are restricted to hawaiite, mugearite, *ne*-hawaiite, or *ne*-mugearite compositions and some have upper-mantle peridotite xenoliths. The Kaoota centre is composed of *ne*-mugearite containing abundant upper-mantle peridotite and late-state pegmatoid (Sutherland, 1985). Cape Contrariety vent contains coarse, basal, pyroclastic rocks and blocks torn from fault-bounded dolerite walls, and is intruded by a small hawaiite pipe made 'picritic' by abundant upper-mantle peridotite debris. Cape Contrariety lavas are less evolved than the mugearitic lavas in the nearby collapsed Rokeby centre (Green, 1960; Sutherland, 1985).

Alkaline centres east of Coal River graben cluster within and around a block of faulted Permian beds uplifted against Triassic beds and Jurassic dolerite intrusions. An inner complex of pyroclastic vents and reworked tuffs is partly buried by lavas descending into the drainage to below sea-level east of Sorell. Initial lavas produced from the vents are upper-mantle xenolith bearing, K-rich *ne*-hawaiite and mugearite, and later flows are hawaiite and mugearite without xenoliths. A glomeroporphyritic lava erupted from a high-level plug overlies weathered *ol*-tholeiitic basalt in Coal River graben, and precedes more fractionated hawaiite and mugearite in the Sorell sections.

The alkaline volcanic rocks between Sorell and Tasman Peninsula pass into less-fractionated rocks towards the coast. Small plugs and limited flows are found along northwest-trending faults (Blake, 1958). They include alkali basalt, basanite, nephelinite and minor K-rich hawaiite and *ne*-hawaiite, and some contain upper-mantle peridotite and lower-crustal granulite, or cognate composite inclusions.

More abundant basalts fill valleys largely dissected in Triassic beds north of Tasman Peninsula. Older Tertiary sediments and basalt flows descend eastwards and younger flows cap Ragged Tier giving local sequences up to 150 m thick. They form a mildly fractionated series and probably erupted from centres under Ragged Tier. Weathered, zeolitised, and faulted alkali basalts or hawaiites are exposed 5 km west of Dunalley. Younger alkaline mafic rocks crop out in pyroclastic centres and as dykes and flows closer to Dunalley, and contain upper-mantle and lower-crustal xenoliths and clinopyroxene, oligoclase-labradorite, and spinel megacrysts.

Basalts extend offshore in southeast Tasmania as seamounts on the East Tasman Rise. Volcaniclastic fragments in sandstones on Soela seamount at depths of 823–990 m at 43°56'S, 150°21'E have foraminiferal faunas, and the underlying basalt has a minimum Upper Eocene age (P.J. Quilty, personal communication, 1986).

A K-Ar age of 27 Ma was obtained for the mafic *ne*-benmoreite south of Hobart (Sutherland & Wellman, 1986) and one of 58 Ma for basalt capping the Bream Creek sequence (Baillie, 1987), and a range of ages seems likely for the southeastern volcanic region. Considerable differences in elevation (0–1300 m above sea-level) between some adjacent centres, and exhumed courses in many of the valley fills, are evidence that the fault-related topography and drainage was well established before volcanism.

Petrology

Tasmanian basalt suites range widely in composition from highly undersaturated to saturated types. Primary-basalt compositions (Mg-ratio greater than or equal to 68) are represented amongst melilitites, nephelinites,

basanites and probably alkali basalts. Most basalts (about 90 percent) represent fractionated types (Mg-ratios 68–45). Fractionation lineages are well developed amongst the undersaturated rocks, apart from melilite-bearing types. Basanites pass through hawaiite to *ne*-mugearite, alkali basalts through hawaiites to mugearites, and *ol*-tholeiitic basalts to hypersthene-normative hawaiites. K-rich types are found in all the undersaturated lineages, but generally are absent from saturated basalts.

Both unfractionated and fractionated alkaline lineages have members carrying upper-mantle peridotite (some with pyroxenite) xenoliths. Upper-mantle peridotite is also found in one fractionated *ol*-tholeiitic basalt host. Some lavas carry high-pressure megacrysts, commonly in addition to upper-mantle and lower-crustal xenoliths. Clinopyroxene, opaque oxides, and olivine are the most common megacryst species, but isolated occurrences contain kaersutite, alkali feldspar (anorthoclase, sanidine, albite), or orthopyroxene as the dominant megacrysts. Cumulates are sporadic within the range of alkaline rocks. Unusual sodalite-malignite xenoliths in rare highly undersaturated hosts are related to extreme high-level plutonic fractionation of nephelinite magma.

Undersaturated basalts erupted over the greater part of Tasmania's dated volcanic activity (58–8 Ma), but tholeiitic types were most common in the late Eocene to early Miocene, and the more fractionated alkaline types in the late Oligocene to early Miocene. The oldest pre–58 Ma valley flows are transitional to *ol*-tholeiitic basalt.

3.9 Mesozoic Intraplate Volcanism and Related Intrusions

3.9.1 Introduction

Intraplate volcanism in eastern Australia became widespread in the Cainozoic, but actually began much earlier — especially in Tasmania with the emplacement of extensive dolerite bodies (Section 3.9.4), and in New South Wales where it was confined mostly to intracratonic Mesozoic sedimentary basins (Sutherland, 1975, 1987; McDougall & Wellman, 1976; Veevers, 1984; Jaques et al., 1985; Section 1.3.1). Reliably dated rocks older than 200 Ma (Upper Triassic) are predominantly of subduction-related type and pre-date the complete stabilisation of the Tasman orogen — for example, in southeastern Queensland (Evernden & Richards, 1962), the Lorne Basin in New South Wales, and the Benambra Complex in Victoria (McDougall & Wellman, 1976; Tattam, 1976). However, subsequent volcanism was mainly of intraplate character (Wellman, 1983; Sutherland, 1985) which may be traced back to late Palaeozoic to early Mesozoic thermal events (Sutherland, 1987).

3.9.2 Queensland

A well-defined province of late Mesozoic volcanic rocks extending to the base of the Tertiary is found around Rockhampton in eastern Queensland (Fig. 3.9.1). The Rockhampton province covers an ovoid

Figure 3.9.1 Distribution of Mesozoic intraplate volcanic areas and related intrusions in eastern Australia.

area of about 3000 km² northeast and west of Rockhampton, and consists of a mixed suite of felsic and mafic rocks (Murray, 1975a). The rocks mostly intrude and overlie Carboniferous to Permian sediments, volcanic rocks and plutons of the Rockhampton block, and a Permian serpentinite body to the west. They also overlie Mesozoic rocks — including Triassic volcanic rocks, Jurassic sediments, and Lower Cretaceous Coal Measures — near the Stanwell Fault, a cross-cutting fault in the Rockhampton Block. The Fitzroy River system became diverted between the volcanic areas, has dissected them substantially, and is now below the basal flows.

The volcanic rocks at Rockhampton mainly form an elongate plateau extending 35 km as a northwest-

southeast divide from Mount Salmon to Native Cat Range. They rise from near river level to 475 m. Some lower flows of alkali basalt, hawaiite, and transitional basalt infill the old easterly flowing drainage. These basalts are overlain by, intercalated with, and over-lapped by trachytes that reach 90 m in thickness near Mount Salmon, where bedded and lapilli tuffs are capped by porphyritic sanidine-trachyte flows fed from Mount Salmon and other local peaks. A small area of volcanic rocks south of the Stanwell Fault consists of alkali basalt intruded by small plugs and vent breccias of porphyritic trachyte which were feeders for flows now preserved as remnant aprons. Spherulitic rhyolites associated with centres around Mount Hay are notable for unusual chalcedonic and spectacular agate infillings (Kay, 1981). The Mount Hay basalt-felsic remnants may be early Cretaceous on palaeobotanical evidence, and therefore older than most of the volcanic rocks in the area, except possibly for the trachyte-rhyolite flow remnants at Mount Lion (Willmott et al., 1986).

Nineteen trachyte plugs dominate the volcanic area northwest of Rockhampton extending north to Yeppoon, and intrude either the Devonian-Permian basement rocks or basalt flows in the central part of the complex. The basalts include feldsparphyric transitional types and the plugs are typically riebeckite-sanidine trachytes. The plugs range from small rounded spires to large elongate masses such as Mount Wheeler and the triple-peaked Pine Mountain. Riebeckite trachytes within basalt sequences noted in drillholes north of Rock-hampton are presumably flows (Willmott et al., 1986).

K-Ar data on the rocks are sparse. The trachyte plug at Jim Crow Mountain has been dated at 71 Ma, and an altered basalt west of Rockhampton has a minimum age of 67 Ma (Harding, 1969; Wellman, 1978). Scattered intrusions in surrounding regions having related ages or petrological features include a nepheline-syenite intrusion and associated dykes near Baralaba (72 Ma; Olgers et al., 1966; Harding, 1969; A.D. Robertson, personal communication, 1986) and plugs of possible hawaiite (72 Ma) and *ol*-tholeiitic basalt near Mundubbera (Whitaker et al., 1975). Some small syenite, trachyte, and rhyolite intrusions in the St Lawrence, Port Clinton, and Gladstone areas are thought also to be late Cretaceous (Malone et al., 1969; Kirkgaard et al., 1970; Murray, 1975a,b).

Mesozoic volcanism also took place in the Clarence-Moreton Basin in southeastern Queensland and north-eastern New South Wales, but apparently was very restricted. The Towallum Basalt, a thin unit of tholeiitic basalt underlying the Walloon Coal Measures is early Jurassic in age (183 Ma; Flint et al., 1976).

3.9.3 New South Wales

Mesozoic intraplate volcanism in New South Wales was centred in particular on the Sydney and Coonamble basins (Figs. 3.6.6, 3.9.1; Martin, 1986). A large central-type volcano in the Mullaley area between Gunnedah and Coonabarabran, termed the Garrawilla Volcanics, consists of pyroclastic rocks, mafic flows and plugs, and intrusive and extrusive domes of phonolite and phonolitic trachyte (Bean, 1974, 1975). The rocks of the complex are predominantly Upper

Triassic and Lower Jurassic (Dulhunty & McDougall, 1966; Dulhunty, 1973a), and they rest on and, in some cases, intrude Permian and Triassic rocks of the Coonamble Basin. They are in turn overlain by non-marine Jurassic sediments at least on the outer flanks of the volcanic succession. Bean (1974) suggested that the complex remained exposed as an island or west-jutting peninsula during sedimentation in a Jurassic lake.

The Mullaley area was clearly the main focus of Jurassic volcanism in the Coonamble Basin, but rocks of a similar age are present over a much wider area (Dulhunty et al., 1987), including the Coonabarabran-Binnaway area to the west (Dulhunty, 1973b) and the lower reaches of the Nandewar Range to the north. A series of teschenite sills intruding the Permian coal measures in the area, including the differentiated Black Jack sill (Wilkinson, 1958), are probably contemporaneous with the mafic intrusions of the Mullaley area (the Glenrowan Intrusives; Bean, 1974). Similar intrusions extend southward to the Quirindi, Scone, and Muswellbrook districts (Raggatt & Whitworth, 1930, 1932; Gamble, 1984; Martin, 1985; Dulhunty et al., 1987) and their ages may extend back to the late Permian. Minor mica-bearing lamprophyres and alnöites crop out in the Scone-Gloucester region and range in age from Jurassic to Cretaceous (Jaques & Perkin, 1984; Sutherland, 1985). Late Cretaceous (70 Ma) basalts cap hills in the Bunda Bunda area west of Kempsey.

Mesozoic volcanism in the Sydney Basin includes some Jurassic basalt flows, numerous dykes along the coastline near Sydney, the Prospect alkaline-dolerite intrusion west of Sydney, and a series of flows and high-level intrusions of phonolite and nepheline microsyenite in the northwestern part of the basin. The central part of the Sydney Basin has been intruded also by numerous diatremes which span a wide age range but fall mainly in the Jurassic or Tertiary (Crawford et al., 1980). However, older and younger groups are difficult to distinguish on the basis of geomorphology, basalt or diatreme type, or geochemistry. Numerous dykes, plugs and sills in the southern part of the Sydney Basin including syenite, trachyte, and basaltic rocks (including both alkaline and tholeiitic types), range in age from 202 to 177 Ma. Basanitic and nephelinitic dykes intruding Permian volcanic rocks near Kiama (Wass, 1979; Wass et al., 1980) and the Mount Dromedary Complex near Narooma range from Triassic (191 Ma; Wass & Shaw, 1984) to Cretaceous (94 Ma; McDougall & Wellman, 1976).

Other dated Mesozoic rocks in southern New South Wales include the Myalla Road Syenite near Cooma (Joplin, 1971), mafic dykes west of Cooma (Lovering & White, 1964) and at Murrumburrah (Harvey & Joplin, 1941), and the breccia pipes at Delegate (Lovering & Richards, 1964).

The Prospect intrusion (Wilshire, 1967) is an elongate, dish-shaped mass about 2500 m long, 1000 m wide, and up to 130 m thick intruding Triassic sediments of the central Sydney Basin and has been dated at 168 Ma by Evernden & Richards (1962). The strongly differentiated body is dominated by picrite in the lower half and alkaline dolerite in the upper part

where veins and schlieren of pegmatitic syenite and aplite are also abundant.

Permian and Triassic sediments in the northwestern part of the Sydney Basin are intruded by a series of high-level sills and laccoliths of nepheline microsyenite or tinguaite (Day, 1961; Dulhunty, 1976) including the Mount Stormy, Bald Mountain, Pinnacle Mountain and Murrumbo Laccoliths and the Wollar Sill. Dated intrusions range in age from 179 to 163 Ma, but spatially associated flows of trachyte and sills of teschenite are significantly older (216–185 Ma).

3.9.4 Other areas

The Tasmanian dolerite sills and dykes of Middle Jurassic age (175 Ma; McDougall, 1961; Schmidt & McDougall, 1977) cover an area of at least 15 000 km^2 and in places exceed 300 m in thickness (Fig. 3.9.1). They transgressively intrude pre-Carboniferous folded basement and Upper Palaeozoic and Triassic sediments (Carey, 1958; Sutherland 1966; Leaman, 1975). The dolerites, as typified by the Mount Wellington sill (Edwards, 1942) and the Red Hill intrusion (McDougall, 1963), are silica saturated and many are differentiated towards granophyric compositions. They have close affinities with the Ferrar dolerites in Antarctica and other voluminous tholeiitic intrusive and volcanic areas such as the Karoo, Paraná, and Deccan basalts, and their occurrence is clearly related to large-scale rifting and to the breakup of Gondwanaland.

Mesozoic intraplate igneous activity elsewhere in eastern Australia appears to have been scattered. Jurassic ages have been recorded for volcanic rocks from a few localities (McDougall & Wellman, 1976; McKenzie et al., 1984), including a lamprophyre dyke on King Island (140 Ma), subsurface hawaiite and a trachyte from the Dundas Tablelands of western Victoria (Evernden & Richards, 1962; Bowen, 1975), monchiquites, hawaiite, and phonolites from central Victoria (Tattam, 1976), and tholeiitic basalts on Kangaroo Island (Tilley, 1921). Cretaceous alkaline intrusions and minor flows are found in Tasmania (Sutherland & Corbett, 1974) and extensive volcanic rocks in the Bass, Otway, and Gippsland basins include similar rocks. Some Cretaceous dates for Victorian basalts are considered to be anomalously old (McKenzie et al., 1984).

The main Bass Basin rift contains over 14 km of sediments and subordinate volcanic rocks. Only the upper 4.5 km is drilled so that the volcanic record in the Jurassic to Lower Cretaceous section is poorly known, but includes weathered basalt (Konkon-1 and Duroon-1 wells). These are related to rocks in the Otway Group (Otway Basin) and Strzelecki Group (Gippsland Basin). The Duroon Sub-basin contains early Cretaceous volcaniclastic sediments of the lower Bass Basin section, but formed as a further rift in the middle Cretaceous, associated with shoshonitic and basaltic activity. Basalts (102 m thick) are present between the topmost Lower Cretaceous and Upper Cretaceous beds (Duroon-1). This rift is exposed onshore in northeastern Tasmania at Boobyalla and is marked by 'trachyandesite' flows, lamprophyric dykes, and dioritic complexes on the margins (103–91 Ma; Jennings & Sutherland,

1969; Sutherland & Corbett, 1974; Moore et al., 1984; Baillie, 1984).

The earliest volcanic activity assigned to the predominantly Cainozoic Older Volcanics in eastern Victoria in fact was initiated in the Mesozoic between 95 and 85 Ma. Ten coarse-grained doleritic plugs intrude Lower Cretaceous sediments in the Poowong area, and two flows have been intersected in a borehole in the central La Trobe valley. All intrusions and flows are petrographically and chemically similar hawaiite and *ne*-hawaiite. The intrusions are in the central regions of faulted, dome-like structures of the Gippsland Basin and the largest is 1400 m across (Kitson, 1917). Little is known about the flows, but they have been correlated with the intrusions because of their similar age and chemistry. Both flows underlie the Morwell coal seam in the La Trobe valley (Hocking, 1976).

Mesozoic intraplate igneous activity in New Zealand was apparently limited. Documented occurrences are restricted to the Marlborough province in the northeast of South Island where two mafic to ultramafic complexes in the northern part of the inland Kaikoura Range intrude steeply dipping Jurassic greywackes. The larger body, centred on Mount Tapuaenuku, has not been studied in detail, but the smaller Blue Mountain Complex (Challis, 1965; Grapes, 1975) is an elongate ultramafic to gabbroic ring complex about 1.5 km long and 1.0 km wide exposed in vertical section over about 800 m. The rocks have close affinities with nearby Upper Cretaceous alkali basalts in the Awatere and Clarence valleys and an alkaline dyke swarm cutting both intrusions (Challis, 1961, 1963).

3.10   Gemstones

3.10.1   Introduction

Cainozoic volcanic terrains in eastern Australia previously have attracted little attention from mineral explorers, despite an association with significant commercial deposits of sapphire. In contrast, companies involved in the less glamorous, but nevertheless highly profitable, construction-materials industry have maintained close interest in Tertiary and other volcanic rocks. Construction materials such as aggregate and road materials are the major mineral resource won from volcanic material in eastern Australia. However, gemstones may become more important in the near future, particularly as a result of recent advances in understanding of the origin of sapphires.

3.10.2   Sapphires

Sapphire, corundum, and associated minerals — notably zircon and pleonaste — are present in alluvial and weathered deposits virtually throughout the eastern highlands (Fig. 3.10.1). These deposits have been mined in several areas, notably the Anakie-Sapphire area in Queensland, and the Inverell and Glen Innes area and, to a much lesser extent, Crookwell-Oberon area in New South Wales. Sapphire also is reported commonly in Tertiary or older alluvium, but these

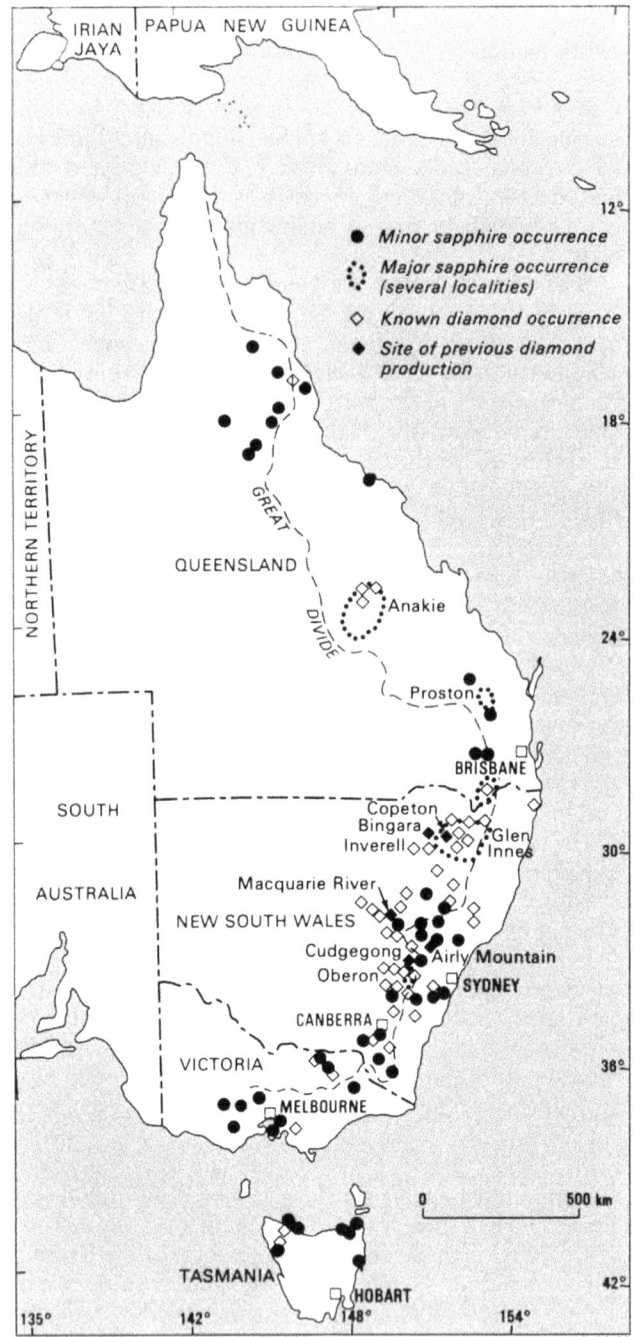

Figure 3.10.1. Sapphire and diamond occurrences in eastern Australia.

deposits to date have not been exploited commercially.

The strong spatial association of sapphire with alkali basalt in Queensland has been noted by several authors, including Dunstan (1902), Stephenson (1976), and Broughton (1979), and in New South Wales by numerous authors from the time of Curran (1897) to MacNevin (1972). This led previous investigators to suspect the basalts as the source of sapphire, a conclusion supported by the occasional discovery of sapphire (or corundum) in basalt fragments, such as in New South Wales at Frazers Creek and Waterloo near Inverell (MacNevin, 1972), in basalt-derived soil at Argyle, Inverell (MacNevin, 1972), and with zircon in basaltic soils in the Chudleigh, McLean, Atherton, and McBride provinces, northern Queensland (Stephenson et al., 1980). Stephenson (1976) also reported sapphire megacrysts and a sapphire-anorthoclase xenolith from basalt at Mount Leura in the Hoy province (Section 3.3.5). Further apparently supportive evidence is provided by the association of pleonaste with sapphire in alluvial deposits, and by the common occurrence of pleonaste in basalt in eastern Australia. The sapphire on this basis was thought to be of lower-crustal or upper-mantle origin and to be xenocrystic or xenolithic in the basalt.

Sapphire has been reported also from diatremes of different ages and compositions in eastern Australia — for example, associated with magnetite and 'basic plagioclase' in basalt from the Green Hills neck, Blue Mountains (Browne, 1933), from other diatremes in the Sydney Basin (CRA Exploration Pty Ltd, 1983), and from olivine-rich bombs from the Pleistocene Bullenmerri diatreme near Camperdown, Victoria (E. Scheibner, personal communication, 1985). Furthermore, the New South Wales Department of Mineral Resources (1987) recently has established an association between previously unrecognised volcaniclastic rocks formed at the onset of Tertiary volcanic activity in the New England area (and elsewhere in the eastern highlands of Australia) and sapphires and associated gemstones (Lishmund & Oakes, 1983). These volcaniclastic rocks were recognised first in 1982, and previously had escaped attention because they are altered to clay and only rarely crop out, or because their features have been obscured by ferruginisation.

Volcaniclastic rocks emplaced (and reworked at least in part) during middle Tertiary explosive volcanic activity in the New England region in places contain sapphire, some having rich concentrations. Deposits of this type are being mined currently near Elsmore ('Braemar'; Fig. 3.10.2) and in the Kings Plains and Reddestone Creek areas. S.R. Lishmund and G.M. Oakes subsequently identified similar rocks in all of the sapphire-bearing regions of New South Wales, and in the Proston, Springsure, and Anakie areas of Queensland (only small remnants are preserved near Anakie). Volcaniclastic rocks have been identified also at many localities where sapphires had been recorded previously in 'basaltic soil' or 'laterite', and are found near Duncans Creek, near Nundle, where sapphire had been reported from weathered basalt (Sutherland et al., 1984b).

Sapphires found in alluvial settings have different forms — from unabraded euhedral crystals to heavily abraded rounded fragments. In contrast, sapphires from the volcaniclastic rocks commonly have little signs of abrasion, but may be embayed or pitted (or both), or highly irregular in outline. There is a wide range of colours, but particular colours tend to dominate in certain areas, notably in volcaniclastic settings. This may be indicative of different sources. Zircons associated with sapphires in volcaniclastic rocks commonly have very high 'polish' and well-preserved crystal faces.

Volcaniclastic rocks in the Inverell and Glen Innes area are found in a well-defined interval (up to 45 m thick, but mostly much thinner, averaging 4 to 5 m) at,

Figure 3.10.2 'Grey-white facies' volcaniclastic rocks in pit at 'Braemar', north of Elsmore, New South Wales.

or near the base of, the volcanic pile. These rocks generally are entirely altered to clay, but original textures commonly are well preserved. The rocks are predominantly alkali basaltic in character, but include a lesser, more felsic component (Barron, 1987).

Two facies of volcaniclastic rocks have been identified in the Inverell region (New South Wales Department of Mineral Resources, 1987) and can be distinguished readily by texture, chemistry, colour, and the degree of sedimentary processing. The most widespread facies is termed the 'red-brown facies' and consists of red to reddish brown to khaki, matrix-supported, volcanic breccias, tuffs, volcanic mudstones, and other rocks which are mostly only poorly bedded. They have few features that could be attributed to fluvial reworking, they tend to mantle basement topography, and in places appear to be intrusive into basement on a small scale. These rocks contain relatively low concentrations of sapphire. They are considered to be a series of ash-fall or ash-flow (or both) deposits formed during explosive volcanic episodes centred on maar volcanoes. Numerous volcanic centres have been identified in the Inverell and Glen Innes area, ranging from entirely teschenitic or basaltic plugs and laccoliths (for example, Ben Lomond, Spring Mount) to breccia-filled diatremes with or without associated basaltic intrusion, and subdued, maar-like surface features (New South Wales Department of Mineral Resources, 1987).

The other major volcaniclastic unit is a 'grey-white facies' consisting of whitish to grey or dark-grey volcanic breccias, mudstones, clays, and other rocks that are commonly well bedded and layered. These rocks may be sorted and graded (mainly upward fining), contain concentrations of heavy minerals in defined layers, and have relatively abundant quartz and other detritus.

The grey-white facies rocks are interpreted as having a depositional history similar to that of the red-brown facies but involving a degree of reworking — probably in a low-energy, fluvial, or lacustrine environment within or near to the source vents. This reworking commonly resulted in a marked enrichment in the heavy-mineral fraction, notably sapphire, which in places is found in extraordinarily rich concentrations (for example, the 'Braemar' and Kings Plains mines).

The two facies commonly are in close proximity to each other, and in places appear intergradational, but the exact relationship is still uncertain. Barron (1987) suggested on the basis of petrological, chemical, and field data that mild sedimentary processing accompanied by an exchange of SiO_2 for Fe_2O_3 would account for many of the observed differences.

The grey-white facies is considered a major exploration target, as are the diatremes, suspected maars, controlling basement structures, and those parts of the present and palaeodrainage systems that may have allowed secondary concentration of the valuable heavy minerals.

3.10.3 Diamonds

Diamonds have been reported in alluvial settings, mainly Holocene, at numerous localities in eastern Australia (Fig. 3.10.1), commonly in the vicinity of, or in catchments that contain, Tertiary alkali basalts. These deposits have been prospected in many areas, but have been exploited commercially at only four localities, all of which are in New South Wales: the Macquarie River (about 4000 carats produced as a by-product of extensive gold dredging), Bingara (about 34 000 carats from Tertiary deep leads), Cudgegong (about 2000 carats from deep leads), and Copeton (an estimated 300 000 carats from deep leads).

Diamonds, as well as having a strong association with Tertiary alkali-basalt occurrences, are associated commonly with sapphire, although many authors (for example, Stellar Mining NL, 1971) noted an apparent inverse relationship in the abundances of diamond and sapphire in Tertiary and Holocene alluvial deposits: diamonds are more abundant than sapphire in the former and *vice versa* in the latter. Some investigators (for example, MacNevin, 1977) suggested that many of the Holocene occurrences may be derived from redistribution of the deep-lead deposits.

Diamonds have been reported occasionally from the Queensland sapphire fields as small 'waterworn stones' (Chalmers, 1967), from alluvial gold deposits at Gilberton (northern Queensland), and from alluvial cassiterite deposits at Stanthorpe. Local fossickers also report occasional diamonds near Proston (Kingaroy area) where volcaniclastic rocks similar in nature and occurrence to those in the Inverell and Glen Innes area are to be found. Sapphire is reported commonly in the Proston area.

Only rarely have diamonds been found in the other eastern states. They have been found in Victoria in auriferous alluvial deposits in the Beechworth, Mansfield, Chiltern, and Rutherglen districts (Chalmers, 1967; Baragwanath, 1948), and in Tasmania in alluvial deposits in the Donaldson and Savage River areas (as clear to straw-coloured octahedra to ⅓ carat). Diamond was reported also to be present in a peridotite at Bald Hill in the Corinna area (Tasmanian Department of Mines, 1970).

A significant proportion of the diamonds in New South Wales have crystal faces, but the majority are rounded, thought to be caused by abrasion (MacNevin, 1977). However, the rounded habit of many diamonds may have been caused by resorption during ascent from the mantle (Cull & Meyer, 1986). Most diamonds in New South Wales are small (in the range of 3 to 5 to the carat), and straw yellow appears to be the most common colour. Only a small proportion (probably less than 5 percent) are of gem quality, and the largest stone reported was a 28.3-carat, distorted crystal from Mount Werong (MacNevin, 1977). New South Wales diamonds have a reputation for superior hardness, particularly stones from Copeton, Bingara, and Airly Mountain. The hardness is thought to be caused by repeated imperfections or knots in the crystal structure.

The origin of east-Australian diamonds remains unknown, and no source has yet been identified that could account for the significant occurrences (such as Copeton and Bingara), despite many years of exploration. This lack of success in part may be owing to inappropriate exploration concepts and techniques — for example, the assumption that an east-Australian source would be kimberlitic, the ambiguity of indicator minerals in alkali-basalt terrain, and inadequate allowance for the episodic, multi-phase nature of most diatremes.

Geothermal gradients determined from studies of xenolithic and xenocrystic material from eastern Australia appear to have been too high since the late Mesozoic for diamonds to survive at the depths inferred (Section 7.2.8). However, the anomalous accumulations of diamond in areas such as Copeton are indicative that 'windows' of low geothermal gradient may have existed. Sobolev (1984) reported Copeton diamonds to have inclusions of eclogitic paragenesis, which may be relevant to the apparent east-Australian association with sapphire and alkali basalt. Diamond has been reported from a dolerite dyke (Copeton), as well as from peridotite in Tasmania, and persistently from alkali basaltic and other diatremes elsewhere — for example, Ruby Hill near Bingara (Pittman, 1901), a basaltic diatreme at Prince Charlie Creek near Gloucester (Stockdale Exploration, 1971), the AUK 1 intrusion near Gloucester (A. Chubb, personal communication, 1977), alluvium adjacent to the Umbiella Creek diatreme near Glen Alice (CRA Exploration Pty Ltd, 1983), from clay overlying the diatreme at Diggers Creek, Mittagong (Wilkinson, 1891), and from stream sediments or weathered material derived wholly or in part from a number of other diatremes in the Sydney Basin.

Differences in the size and character of diamonds from one field to the next, and within one field, may be evidence for multiple sources. Persistent reports of an association with alkali basaltic and other diatremes may be an indication that sources other than kimberlites or lamproites may be involved. The reported occurrences at Proston associated with volcaniclastic rocks may be relevant. Note also that diamonds have been recovered from the tailings of a sapphire plant at Kings Plains, Inverell (T. Nunan, personal communication, 1986) which treats volcaniclastic material that has evidence of some fluvial re-working, and from a previously unrecorded lithic tuff overlying Tertiary alluvium at Copeton (A. Gaddes, personal communication, 1985). Similar volcaniclastic rocks are found in all of the diamond provinces of New South Wales. These rocks and their source vents may be the source of the diamonds, as well as the sapphires, of eastern Australia.

3.10.4 Other gemstones

Opal, garnet, a range of silica gems, peridot (gem olivine), and gem labradorite, have been reported in association with Tertiary volcanic rocks (particularly alkali basalts) in many parts of eastern Australia. Only the opal deposits have attracted commercial interest, although the other gems are sought avidly by fossickers.

Precious opal has been mined from Tertiary volcanic rocks at Rocky Bridge Creek (near Blayney, New South Wales) and at Tintenbar (northern coast of New South Wales), and an unsuccessful recent attempt was made to exploit volcanic opal from near Mullumbimby. The opal is found as amygdales in basalts — highly vesicular basalts in the case of the Rocky Bridge Creek occurrences. Unfortunately the cut gems generally proved unstable, as their usually high water contents caused 'crazing' when the stones dehydrated after exposure to air for long periods.

CHAPTER 4

New Zealand Intraplate Volcanism

4.1 Introduction

New Zealand straddles a tectonically active boundary between the Pacific and Indo-Australian plates (Section 1.3.3). A westward-dipping Wadati-Benioff zone together with characteristically arc-type volcanoes define the active convergent system that trends northeastwards from central North Island. The plate boundary south of New Zealand is marked by an eastward-dipping Wadati-Benioff zone and the isolated arc-type volcano, Solander Island (Reay, 1986). Plate interaction between these opposed sections of the boundary is transform in character, but including a strong compressive component, along the Alpine and related fault systems in South Island and southern North Island.

Volcanism attributable directly or indirectly to interaction between the Pacific and Indo-Australian plates is widespread in the late Cainozoic successions of northern North Island. However, a distinct, dominantly basaltic volcanic association on the Northland-Auckland peninsula in the northwest of the island (Fig. 1.1.7) has intraplate petrographic and geochemical characteristics (Section 1.1.3). This association is Pliocene to Holocene in age and has no obvious tectonic relationship to the active plate boundary. It is considered here as an intraplate volcanic association that has developed on the Indo-Australian plate well behind the plate margin.

Widespread Cainozoic volcanism is found in New Zealand's South Island and in an area extending south and east to the Sub-Antarctic and Chatham islands (Figs. 1.1.8, 1.3.7). This is the second and more significant area of intraplate volcanic activity in New Zealand. It is almost entirely on the Pacific plate.

This chapter is a review of intraplate volcanism in the New Zealand region. Many of the data presented are previously unpublished, form the basis of continuing research, and are given here in summary form.

4.2 North Island

4.2.1 Introduction

Volcanic activity has been a major factor in the late Cainozoic development of North Island. Much of the activity can be related directly to convergence

between the Pacific and Indo-Australian plates, and is represented by the Miocene basalt-andesite-rhyolite association of the Northland peninsula and by similar suites in the Pliocene and Quaternary volcanic associations of the Coromandel Peninsula and central North Island.

Other North Island volcanic successions appear to be related indirectly to plate convergence or represent a magmatic response to spreading behind the currently active arc. The Alexandra Volcanics consist of several, large, late Tertiary volcanic centres in the Waikato region of North Island (Fig. 4.2.1) and are an association of alkali basalt together with tholeiitic and calcalkaline basalt. Briggs & Goles (1984) and Briggs

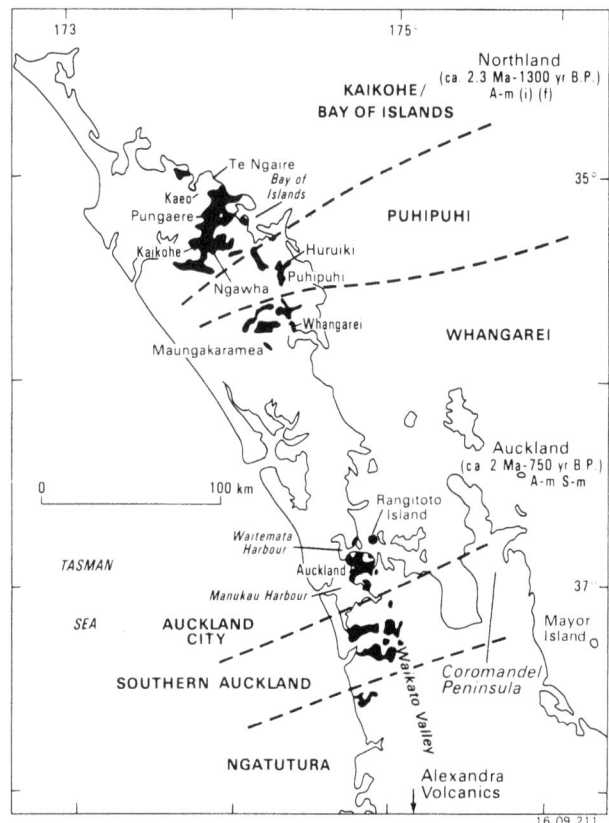

Figure 4.2.1 Distribution of volcanic rocks in the Northland and Auckland provinces of North Island. Rock associations: A, alkaline; S, subalkaline. Rock types: m, mafic; (i), minor intermediate; (f), minor felsic. Ages in brackets.

(1986) interpreted these rocks to be the result of partial melting of a mantle source in a region of perturbation behind the currently active Taupo Volcanic Zone of central North Island. The predominantly basaltic rocks forming the islands just off the eastern coast of Coromandel Peninsula (Skinner 1986; Fig. 4.2.1) are not well known but appear to be distinct from the essentially arc-type rocks of the peninsula itself. They, and the peralkaline rhyolites of nearby Mayor Island (Ewart et al., 1968), may represent magmatism associated with backarc spreading (Smith et al., 1977; Cole, 1978).

In contrast, late Tertiary and Quaternary volcanic associations in two areas in the northwestern part of North Island do not appear to be related to plate-boundary processes (Figs. 1.1.7, 4.2.1). These are the Northland and Auckland intraplate volcanic provinces. The Northland province is the youngest in a complex sequence of volcanic associations (Smith et al., 1986) but is temporally and geochemically distinct from the underlying volcanic formations. It is 500–700 km northwest of the present-day convergent boundary in an area of relative tectonic stability. The Auckland intraplate province is 350–400 km behind the present convergent boundary and is spatially and geochemically distinct from the Alexandra Volcanics of the Waikato region immediately to the south.

The late Cainozoic intraplate basaltic groups in North Island fall naturally into two provinces whose ages overlap, but which are geographically separate and have distinct petrographic characteristics. The Northland province contains basalts transitional between alkali basalt and tholeiitic basalt. It includes minor intermediate rock types and rare rhyolite. The Auckland province contains an exclusively basaltic spectrum of rock types ranging from nephelinite and basanite through alkali to tholeiitic basalt.

4.2.2 Northland

General geology

The Northland peninsula extends for 350 km from Auckland to Cape Reinga at the northern tip of North Island (Fig. 4.2.1). The peninsula, although narrow, contains a greater range of rock units than seen elsewhere in North Island, reflecting a complex and diverse geological history (see also Section 1.1.3). The exposed basement of the peninsula is a sequence of sandstone, siltstone, and argillite containing rare spilitised basaltic rocks, manganiferous cherts, marbles, and conglomerates of late Palaeozoic to early Mesozoic age. These basement rocks are exposed mainly along the eastern side of the peninsula and dip westwards. The basement sediments are overlain in the northern part of the peninsula by a diverse and complex sequence of mainly clastic sediments ranging in age from late Cretaceous to Miocene. Some of them rest unconformably on the basement. Others have been interpreted as allochthonous thrust sheets (Ballance & Sporli, 1979). Dominantly basaltic rocks and associated intrusions, together with rare ultramafic bodies, form massifs in the northern part of the peninsula. They are interpreted as Cretaceous and lower Tertiary oceanic

crust that was obducted onto Northland in the late Oligocene (Brothers & Delaloye, 1982).

Arc-type volcanism began in the Miocene at several separate centres along the length of Northland. This type of volcanism migrated southeastward during the late Miocene and Pliocene and was replaced by intraplate basaltic volcanism during late Pliocene to Quaternary times (Brothers, 1986). The complex geological sequence observed in the Northland peninsula reflects processes along a major plate boundary which alternated between convergent and divergent through the Cainozoic.

Intraplate volcanic rocks

Predominantly basaltic Pliocene and Quaternary volcanic rocks are a feature of the Northland peninsula (Fig. 4.2.1). They crop out over an area of about 2500 km^2, extending 74 km northwards from Maungakaramea near Whangarei to Te Ngaire near Kaeo in the north. The basaltic rocks are referred to collectively as the Kerikeri Volcanics (Kear & Hay, 1961; Thompson, 1961), and associated rhyolites have been mapped as part of the Parahaki Volcanics (Bowen, 1974). The Kerikeri Volcanics were erupted over a period of several million years. Volcanic activity is thought to have begun about 2.3 Ma ago on the basis of radiometric ages given by Stipp (1968), and the most recent eruption was only about 1800–1300 years ago on the basis of radiocarbon dating of carbonised wood from beneath young basalt flows at Te Puke in the Bay of Islands (Kear & Hay, 1961).

The basalts are flow remnants forming outliers that cap older sedimentary formations, multiple flow sequences forming small plateaux, small lava cones, scoria cones, and lava flows filling recent valley systems. Weathering profiles distinguish older from younger basaltic volcanics in Northland, but there is no significant temporal change in the nature of the activity or petrographic character of the rocks.

Intermediate volcanic rocks are a rare component of the Northland province. They are interspersed with the basalts as lava flows and in pyroclastic sequences, and are not distinguishable readily from them using field criteria.

Rhyolite has long been recognised as a part of the volcanic associations in Northland (for example: Bell & Clarke, 1909; Ferrar, 1925), but magmatic affinities and age are not well known. Two groups of rhyolite appear to be a part of the Pliocene-Recent basaltic association. These are peralkaline rhyolite in the Kaikohe and Pungaere areas, and non-peralkaline rhyolite found in the Te Pene and Hurikuri areas (Ashcroft, 1986a).

Volcanic fields

Three distinct but in part overlapping fields are recognised in the Northland province (Fig. 4.2.1). These are the Kaikohe/Bay of Islands, Puhipuhi, and Whangarei fields (Ashcroft, 1986a). The Kaikohe/Bay of Islands field is the largest contiguous area of Pliocene-Holocene volcanic rocks in Northland. It includes an extensive area of older lava flows to the north which form a plateau of overlapping flows. This area presumably contains some of the oldest lavas in the Northland area. The Kaikohe/Bay of Islands field contains tholeiitic basalt, transitional basalt, and hawaiite, together with minor intermediate rocks and rare rhyolite.

The Puhipuhi field to the north of Whangarei consists of lava plateaux and flow remnants overlying basement rocks. Basalt on Puhipuhi plateau overlies (without an erosional break) lacustrine sediments that contain early Pliocene pollen. There are no recognisable volcanic landforms in the field, and all of the rocks are assumed to be Pliocene or possibly Lower Pleistocene. The basaltic rocks of the Puhipuhi field are alkali and transitional basalt and subordinate hawaiite. Rhyolite is found beneath basalts at Huruiki in the east of the field.

The Whangarei field consists of a group of eruptive centres southeast of Whangarei, a northeast-aligned group of scoria cones to the north of the city, and a group of lava flows that apparently emanated from vents aligned along a fault to the east of the city. The basaltic rocks in the Whangarei area are hawaiite and subordinate tholeiitic and rare transitional basalt. Intermediate rock types are rare.

Hydrothermal activity and mineralisation are associated with the Pliocene-Quaternary volcanic rocks in Northland. Siliceous sinter, silicified lacustrine sediments, hydrothermal breccias, chert, and mercury-bearing sulphide deposits at Puhipuhi are the manifestations of a now extinct geothermal system that was active during the late Pliocene or early Quaternary (White, 1983). The Ngawha area (Browne et al., 1981) is an active geothermal system in the Kaikohe volcanic field. The heat source in both of these geothermal fields is thought to have been rhyolitic rather than basaltic intrusive bodies (Heming, 1980a).

Rock types

The Pliocene-Quaternary volcanic rocks of Northland are typically medium-to-dark grey, sparsely to moderately porphyritic rocks of basaltic composition (Table 4.2.1, columns 1–4). Petrographic diversity within the suite is the result of different proportions of phenocrysts and minor textural differences. The rocks are mainly

Table 4.2.1 Selected chemical analyses for North Island provinces, New Zealand

	1	2	3	4	5	6	7	8	9	10
SiO_2	50.03	50.62	50.46	56.63	73.83	41.69	43.39	46.67	45.05	45.42
TiO_2	0.94	0.96	1.57	1.45	0.12	2.66	2.96	2.55	3.29	2.56
Al_2O_3	17.62	17.43	17.39	14.57	10.10	11.85	12.54	13.30	14.36	13.35
Fe_2O_3	8.83[1]	8.80[1]	8.56[1]	9.95[1]	4.60[1]	4.79	3.96	5.94	2.28	1.08
FeO						8.70	8.50	6.70	10.01	11.13
MnO	0.14	0.14	0.15	0.26	0.08	0.23	0.19	0.19	0.21	0.19
MgO	8.22	7.69	6.90	3.81	0.06	11.06	12.30	10.51	7.82	9.41
CaO	9.37	9.13	10.42	5.15	0.14	10.49	10.67	9.70	8.76	8.63
Na_2O	3.50	3.46	4.21	3.80	5.41	4.40	2.72	2.20	3.29	4.13
K_2O	0.24	0.34	0.20	2.33	4.16	1.48	1.08	0.89	1.56	1.69
P_2O_5	0.25	0.25	0.29	0.37	0.02	1.08	0.53	0.51	0.85	0.69
LOI[2]	0.43	0.29		1.39	0.50	0.88	0.92	0.46	1.98	0.83
H_2O-	0.55	0.57	0.22	0.36	0.08					
Total	100.12	99.68	100.37	100.07	99.10	99.31	99.76	99.62	99.46	99.11
Ba	83	106	<5	439	<5	448	382	287	288	365
Rb	<1	<1	7	65	597	24	16	14	18	24
Sr	312	324	299	248	<1	876	519	502	894	890
Pb					55				5	4
Th					122					
Zr	113	126	142	377	2016	292	184	187	327	276
Nb	3	4	<1	22	260				61	50
Y	20	22	27	38	217				27	26
La	<5	18	7	35	171					
V	152	148	167	109	<5	253	288	235	243	199
Cr	276	268	217	147	<5	257	447	351	200	256
Ni	120	112	58	95	9	284	282	231	120	197
Cu	46	43	56	39	<5	79	89	84	97	59

1. Transitional basalt, Northland province (I.E.M. Smith, unpublished data; AU37702).
2. Ol-tholeiitic basalt, Northland province (I.E.M. Smith, unpublished data; AU37700).
3. Hawaiite, Northland province (I.E.M. Smith, unpublished data; AU37677).
4. Qz-tholeiitic basalt, Northland province (I.E.M. Smith, unpublished data; AU37714).
5. Peralkaline rhyolite, Northland province (I.E.M. Smith, unpublished data; AU37602).
6. Nephelinite, Auckland province (Heming & Barnet, 1986; AU30654).
7. Basanite, Auckland province (Heming & Barnet, 1986; AU30692).
8. Transitional basalt, Auckland province (Heming & Barnet, 1986; AU30691).
9. Ne-hawaiite, Auckland province (Utting, 1986; WU25401).
10. Ne-hawaiite, Auckland province (Utting, 1986; WU25525).

[1] Total Fe as Fe_2O_3.
[2] Loss on ignition.

fresh, and they range chemically from alkali basalt to tholeiitic basalt. The most abundant rock type overall is hawaiite (50 percent), followed by tholeiitic basalt (22 percent), transitional basalt (12 percent), and alkali basalt (10 percent). Gabbroic inclusions and megacrysts are found in many of the Kerikeri Volcanics and appear to be especially common in hawaiite (Section 4.5.2).

Intermediate rocks represent about 10 percent of the volcanic rocks in the Northland province. They resemble petrographically the basaltic rocks, and cannot be recognised in the field on the basis of a distinctive eruptive style. The chemical compositions of the intermediate rocks form a continuous range with those of the tholeiitic and transitional basalts. They may be referred to as basaltic icelandite and icelandite, although they are qz-tholeiitic basalts in terms of the classification used in this volume (Section 1.1.4).

Peralkaline and other rhyolites are a minor component of the volcanic association in Northland. They are separated from the basaltic and intermediate rocks of the association by an appreciable SiO_2 gap. Peralkaline rhyolite is either crystalline or aphyric obsidian consisting of colourless glass containing up to 5-percent microlites of anorthoclase and arfvedsonite (Table 4.2.1, column 5). Crystalline specimens consist of a fine-grained aggregate of anorthoclase, quartz, riebeckite, and minor Fe-Ti oxides. Non-peralkaline rhyolites are found as sparsely porphyritic obsidian containing 3–4 percent phenocrysts of alkali feldspar (up to 2 mm across) in a colourless glass together with microlites of alkali feldspar and brown amphibole.

There are some differences in the relative proportions of rock types in the different fields of the Northland province. The Kaikohe/Bay of Islands field contains tholeiitic and transitional basalt and hawaiite, together with minor intermediate rocks and rare rhyolite. The basaltic rocks of the Puhipuhi field are alkali and transitional basalt and subordinate hawaiite, and rhyolite is present beneath basalts at Huruiki in the eastern part of the field. The basaltic rocks in the Whangarei area are hawaiite, subordinate tholeiitic basalt, and rare transitional basalt. Intermediate rock types are rare and rhyolites are unknown.

Eruptive style

The nature of the eruptions that produced the volcanic rocks of Northland province has not been studied in detail. Outcrops of the older rocks are almost entirely of extensive, moderately thick, massive lava and there is apparently little accompanying pyroclastic material. Earlier activity therefore was dominantly effusive. The lack of evidence for well defined eruptive centres may mean that the lava flows issued from fissure-like vents, building up the extensive plateaux of the northern Kaikohe/Bay of Islands field and the Puhipuhi field.

More recent activity in Northland produced prominent cones and accompanying valley-filling lava fields. The cones are either low-angle lava shields (height:base ratio 1:20) or steep-sided scoria mounds (height:base ratio 1:5) typically 50 to 150 m in height (see Fig. 2.4.4). The shields are characterised by low-angle, smooth slopes rising to rounded summits that contain no discernible craters. The scoria cones are steep-sided

and typically have well-developed summit craters. They were built by Hawaiian to Strombolian pyroclastic eruptions. Many of these young cones have breached craters from which lava flowed to fill local valley systems. Some of these flows extend up to 10 km from their source and reach thicknesses of 10–15 m. There is no evidence of phreatomagmatic explosive activity in any of the deposits from eruptive centres in the Northland fields

4.2.3 Auckland

Auckland City

The Auckland volcanic province extends from Auckland City south to the lower Waikato Valley (Fig. 4.2.1). There are three spatially distinct fields within this area — Auckland City, Southern Auckland, and Ngatutura — although as a group they have common compositional characteristics and overlap in age.

The Auckland City (Fig. 4.2.2) field is small (140 km^2; 7 km^3 erupted volume). The volcanic centres were mapped and described by Hochstetter (1864), Searle (1964) and, most recently, by Kermode (1986). The field contains 48 recognised eruptive centres within an area more or less centred on Auckland City. The earliest activity was at least 60 000 years ago and may have been as much as 100 000 years ago. Rangitoto Island, the youngest centre, began activity about 1200 A.D. according to Brothers & Golson (1959), and may have continued for several hundred years (Law, 1975; Robertson, 1983; Fig. 4.2.2).

The Auckland City volcanoes were produced by a range of eruptive styles — from lava effusion to phreatomagmatic and Strombolian-Hawaiian eruptive activity. The resulting land forms include lava fields, tuff rings, and scoria cones. Thirty-four of the centres are principally phreatomagmatic in origin, ranging from water-dominated, containing little or no juvenile magmatic material, to magma-dominated. Subsequent weak lava fountaining at 23 of these centres built cones within the tuff rings. Many of these small cones are complex 'castle and moat' structures. Twenty-six of the Auckland City centres are small, steep-sided, scoria cones ranging in size from a few hundred square metres to more than 8×10^6 m^2. Fifteen lava fields and ten minor lava flows extend beyond their original eruptive centres, and lava was extruded at seven other centres forming tholoids or lava ponds. The longest lava flow extends nearly 10 km from source, and the thickest reached 60 m. The total area of lava is about 75 km^2, and the largest individual area is the lava fields of Rangitoto Island (23 km^2).

Southern Auckland

The Southern Auckland field consists of extensive areas of Upper Pliocene to middle Quaternary basaltic rocks in the Pukekohe-Tuakau-Onewhero region south of Auckland City. These have been mapped as the Bombay and Franklin basalts (Kear, 1960; Schofield, 1967) and have been referred to as the Pukekohe field (Schofield, 1967) or the South Auckland field (Rafferty & Heming, 1979). The field mainly occupies the

Figure 4.2.2 The Auckland City volcanic field, showing in the foreground the summit scoria cones of Rangitoto Island surmounting a broad lava apron (see also Fig. 2.4.5). Auckland City in the middle distance is built on and around numerous, small, monogenetic eruptive centres. Miocene arc-type volcanic rocks form the Waitakere Ranges in the background. Photograph supplied courtesy of Whites Aviation, Auckland.

lowlands of the lower Waikato Valley and the Manukau lowlands. Volcanic activity ranges in age from 1.56 Ma to 0.51 Ma (Stipp, 1968; Robertson, 1976).

The Southern Auckland field contains over 70 distinct eruptive centres within an area of about 190 km², including about 44 cones and 30 tuff rings. The tuff rings range in size from 0.5 to 2.5 km and are commonly in clusters or are nested. Their distribution is presumably related to the presence of abundant subsurface water at the eruption site. Some contain basaltic material resulting from Strombolian activity late in the eruption sequence.

There are two types of cone. The most common cones are small steep-sided scoria mounds each associated only with a small lava flow. The second type is a much larger cone that has a lower profile. Pyroclastic material where present in the second type is minor, whereas flows are more voluminous, commonly forming an apron around the cone. Kear (1960) and Schofield (1967) used the terms Bombay and Franklin basalts to refer to flows having recognisable volcanic landforms (Bombay) and those with a deeper weathering profile and no recognisable land form (Franklin).

Ngatutura

Ngatutura is the southernmost of the three fields in the Auckland province. It consists of 16 mapped eruptive centres (Dow, 1954; Spratt & Rodgers, 1975; Utting, 1986) in the Ohuka Valley near the western coast, south of the Waikato River mouth. These rocks are considered to have been produced during late Pliocene to early Pleistocene times, on the basis of K-Ar dating (Stipp, 1968) and stratigraphic relationships (Rodgers et al., 1975).

The centres are represented by either eroded scoria mounds surrounded by an apron of lava flows, or by isolated lava flows without an associated scoria mound. The volume of material erupted from each centre is small, and most have produced only a single lava flow. Eruptive activity is inferred to have been Hawaiian to Strombolian. Heming (1980c) interpreted a tuffaceous breccia exposed in a coastal section at Ngatutura Point as a diatreme formed by phreatomagmatic activity from this centre.

Rock types

Auckland City rocks are predominantly alkali basalt and basanite, together with subordinate nephelinite, transitional basalt, and tholeiitic basalt (Table 4.2.1, columns 6-10). They range in age from Pliocene in the south to Holocene in the north. Searle (1961) distinguished several types of basaltic rocks on the basis of petrography, and identified the most common types as olivine basalt and an augite-rich picrite basalt.

Olivine basalt (more than 5-percent modal olivine, and more than 35-percent modal plagioclase) dominates the lava of the youngest cones. Picrite basalt is found as flows in the smaller centres and among the earliest flows of the largest centres. Minor rock types include nepheline-bearing blocks in the tuff of the Domain centre, an amphibole-bearing basalt from Three Kings centre, a few picrite basalts containing phlogopite, and some rutile-bearing basalts (Heming & Barnet, 1986). Ultramafic xenoliths ranging from dunite to harzburgite are present in pyroclastic deposits of some older centres but are notably absent from the youngest ones.

The most common rock type in the Auckland field on the basis of chemical composition is basanite. Alkali basalt is less common, and transitional and tholeiitic basalts are found in the youngest centres. Nephelinite has been reported from some of the oldest centres. Heming & Barnet (1986) suggested there is a systematic temporal distribution of rock types in which the most silica-undersaturated rocks characterise the oldest centres and the transitional and tholeiitic rocks make up the youngest ones. However, Houghton et al. (1986b) studied one centre in detail and showed a range from early basanite to later alkali basalt within one eruption sequence, so within-centre differences may be as important as differences between centres.

Two petrographic types are recognised in the Southern Auckland field (Rafferty, 1977; Rafferty & Heming, 1979). The first contains common microphenocrysts and rare phenocrysts of olivine and clinopyroxene. The second type is more coarsely porphyritic, contains abundant phenocrysts of olivine and clinopyroxene and, in some, ultramafic xenoliths and megacrysts (Section 4.5.2). Southern Auckland rocks are mainly alkali basalt and basanite. Tholeiitic compositions are rare.

Rock types in the Ngatutura field are typically fine grained and contain phenocrysts of olivine and titaniferous clinopyroxene. Lavas from individual centres are similar, but there are some differences in the proportions of phenocrysts between centres. The volcanic rocks are hawaiite and *ne*-hawaiite in terms of their normative mineralogy.

Eruptive style

Studies of volcanic processes in the Auckland volcanic province in the main have been descriptive (Searle, 1964). Only at Crater Hill in the Auckland City field has there been any detailed volcanological work on the nature of the deposits and the eruption mechanisms (Houghton et al., 1986b).

The volcanic deposits of the Auckland province are found mainly in clearly defined centres as lava shields and pyroclastic cones, together with valley-filling lava flows in different stages of preservation. The eruptive style is Hawaiian to Strombolian. It ranges from lava effusion in high-standing areas to purely phreatomagmatic activity in some centres — typically those in low-lying areas where magmas have interacted with groundwater. A mixture of purely magmatic and phreatomagmatic activity is common in many centres. A consistent pattern in the eruptive activity of an individual centre is from explosive phreatomagmatic or

Strombolian activity (or both) early in an eruption sequence, to less violent Hawaiian activity (or the emission of lava flows) later in the eruption sequence.

4.3 South Island

4.3.1 Introduction

The South Island of New Zealand is marked by widely scattered areas of Cainozoic volcanism (Figs. 1.1.8, 4.3.1) of predominantly mafic, alkaline to subalkaline composition. All of this volcanism appears to be of intraplate continental character. Major episodes of activity were in the Paleocene to Lower Eocene, Upper Eocene to Lower Oligocene, and in the Miocene, and minor episodes in the Pliocene. The best-known volcanic areas are the large Miocene shields of Banks Peninsula and Dunedin volcano. Other occurrences include the mafic lava fields of Canterbury, Marlborough, and Otago, and small volumes of lava and minor intrusions in South Westland and the Southern Alps. Seven areas of volcanism are considered below.

4.3.2 Canterbury and Marlborough

The oldest Cainozoic volcanic rocks in Marlborough (Fig. 1.1.8) consist of minor tuffs and pillow lavas interbedded with Lower Eocene flysch deposits and marls (Morris, 1987) in Woodside Creek in the southwest. Little is known geochemically of these rocks, although Lensen (1963) described them as teschenite and albite dolerite. Pillowed tholeiitic basalt and agglomerate of Palaeocene to Lower Eocene age (View Hill Volcanics; Gregg, 1964) represent the earliest period of activity in central Canterbury. Outcrops of these rocks are confined to small areas around the Oxford district, but tuffs of similar age are found in several onshore and offshore drillholes to the southeast. The volcanism, therefore, may have been widespread.

The only record of Upper Eocene volcanism is found in Marlborough where pillowed basalt and palagonitised tuff are present within micritic limestone (Suggate, 1958; Reay, 1980; Morris, 1987). These are termed the Grasseed Volcanics and crop out in Grasseed and Bluff streams and in the Kekerengu and Waima Rivers.

Submarine, alkaline to transitional and tholeiitic basalts characterise the major episode of Oligocene volcanism. The most voluminous volcanic deposits are in northern Canterbury and are termed the Cookson Volcanics (Fig. 4.3.1A; Gregg, 1964; Coote, 1987). Minor occurrences in Seymour Stream, Marlborough, are probable correlatives (Morris, 1987). The Cookson Volcanics consist mainly of tholeiitic to alkaline, pillowed basalt flows, calcareous, massive and stratified coarse to fine tuff, and hyaloclastite breccia. Rhyolite plugs were thought previously to be of Cretaceous age, but are now included within the group which has an estimated minimum volume of 100 km³ (Coote, 1987). Transitional basalts of Upper Oligocene age are found in the Oxford district and Castle Hill areas of inland central Canterbury (Fig. 4.3.1A). These are mostly

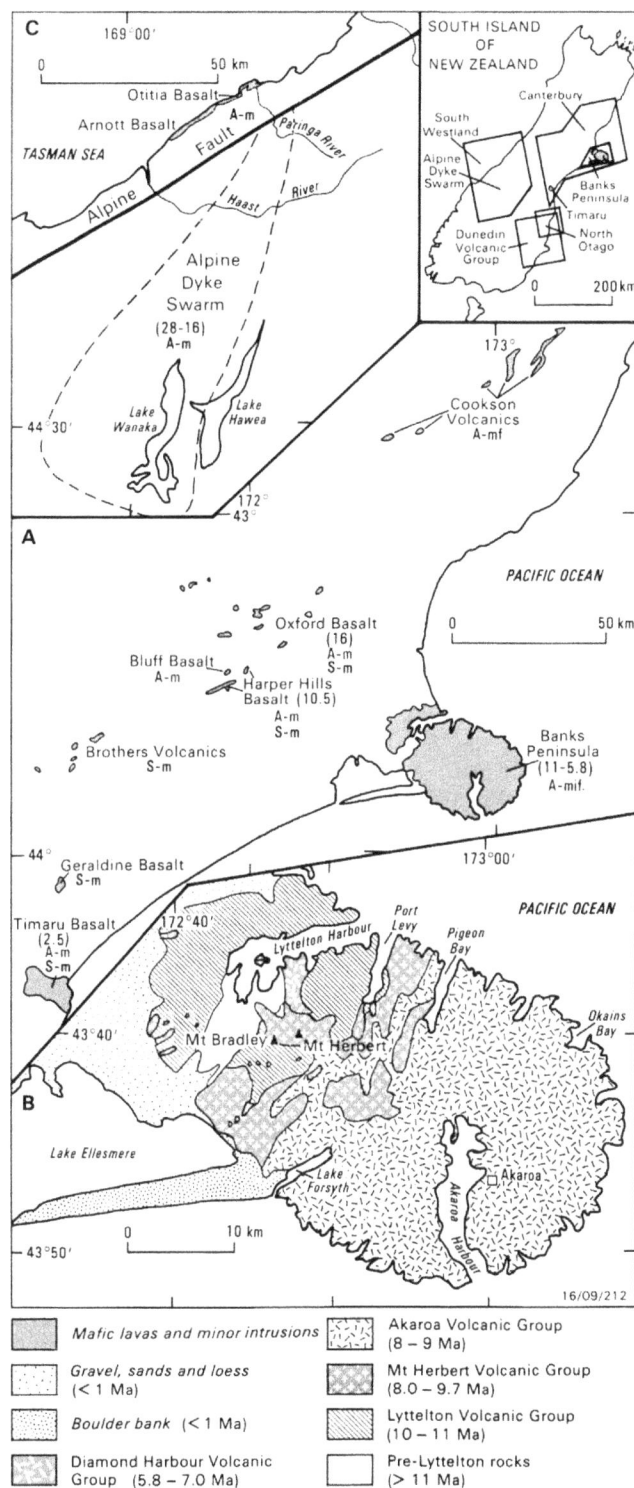

Figure 4.3.1 Distribution of volcanic rocks in (A) Canterbury, (B) Banks Peninsula, and (C) South Westland including the Alpine Dyke Swarm. Rock associations: A, alkaline; S, sub-alkaline. Rock types: m, mafic; i, intermediate; f, felsic. Ages (in brackets) in Ma.

palagonitised tuffs and associated feeder dykes interbedded with Oligocene limestone. A nephelinite plug near Oxford has a K-Ar age of 30 Ma (McLennan & Weaver, 1984). Several small areas of pillowed basalt and palagonitised tuff interbedded with Oligocene limestone and greensand are known in the Mount Somers

district of southern Canterbury. These constitute the Brothers Volcanics (Field & Browne, 1986) and are dominantly of transitional tholeiitic chemistry.

Products of the third episode (Miocene) of volcanic activity are abundant in the Canterbury province but absent from Marlborough. These are alkaline to sub-alkaline rocks and include the large composite volcanoes of Banks Peninsula (see below). The first eruptions were largely submarine and are recorded by mildly alkaline basaltic ash and hyaloclastite breccia (Wairiri Volcaniclastite, Sandpit Tuff, Bluff Basalt) interbedded with Middle Miocene greensand and marine sandstone in the Harper Hills district of inland Canterbury. These deposits are overlain by tholeiitic-basalt flows of the Harper Hills Basalt (Carlson et al., 1980) which have a K-Ar age of 10.5 Ma (P. Whitla, personal communication, 1986; Fig. 4.3.1A). Tholeiitic dykes are exposed nearby in the Glentunnel area, and are thought to have fed these lava flows. Similar tholeiitic-basalt flows containing abundant segregation veins are found in the Oxford district and have a K-Ar age of 16 Ma (C.J. Adams, personal communication, 1986). These are known as the Oxford Basalt (Fig. 4.3.1A). A nephelinite sill in the Little Lottery River, northern Canterbury, has a K-Ar age of 15 Ma and carries a range of xenoliths (see Section 4.5.3).

4.3.3 Banks Peninsula

The largest accumulation of Miocene volcanic rocks along the east coast of the South Island is on Banks Peninsula, southeast of Christchurch City (Figs. 1.1.8, 4.3.1B). This volcanic promontory has been recognised since the pioneering work of Haast (1879; see Section 1.1.2) as the eroded remnants of two large composite volcanoes — the Lyttelton volcano in the northwest (Fig. 4.3.2) and the younger Akaroa volcano in the southeast (Fig. 4.3.1B). The present diameters of the volcanoes are 25 and 35 km, respectively, but on the basis of evidence from boreholes both extend for some distance beneath the adjacent plains, and they probably had original diameters close to 35 and 50 km. Banks Peninsula covers an area of 1200 km² and is founded on continental crust at the western end of the Chatham Rise (Fig. 1.3.7). The volcanic evolution of Banks Peninsula was summarised by Weaver et al. (1985), Weaver & Sewell (1986), and Sewell (1988).

Cainozoic volcanic activity began with the eruption of icelandite, dacite, and peraluminous rhyolite (Table 4.3.1, columns 1–2) on a basement of Triassic greywacke and Miocene sediments. These rocks are known as the Governors Bay Volcanics, and they have a Rb-Sr age of 11 Ma (Barley et al., 1988) the same as the overlying, oldest, Lyttelton lavas. They crop out mostly within the eroded crater of Lyttelton volcano as lava flows and exogenous domes beneath the main Lyttelton pile (Fig. 4.3.1B). Pillow lavas in these and the overlying succession are absent, and the eruptions were probably subaerial. The icelandites are strongly porphyritic containing phenocrysts of calcic plagioclase, salite, hypersthene, and rare olivine. Rhyolites are holocrystalline or partially devitrified obsidian, and alkali feldspar commonly has been hydrothermally altered to clay minerals. Bipyramidal black quartz

Figure 4.3.2   A view of Lyttelton volcano looking east-northeastwards down Lyttelton Harbour which occupies the eroded centre of the volcano. Dip slopes in the foreground are dissected by radial drainage channels. The prominent dip slope in the right middle distance is underlain by basaltic lavas of the Diamond Harbour Group which have flowed into the crater. Photograph by D.L. Homer, New Zealand Geological Survey.

phenocrysts in rhyolites are particularly distinctive, and sanidine and biotite phenocrysts are also present.

The main period of volcanic activity that formed Banks Peninsula was thought until recently to consist of three phases: Lyttelton, Akaroa, and Diamond Harbour (Liggett & Gregg, 1965). However, a fourth phase of activity — Mount Herbert — has been recognised in the region of overlap between Lyttelton and Akaroa volcanoes (Sewell, 1985). This is considered to represent an intermediate stage in the southeastward migration of activity from the Lyttelton to the Akaroa centre (Fig. 4.3.1B). The four main groups of rocks have alkaline to transitional chemistry (Weaver & Sewell, 1986).

The Lyttelton Volcanic Group (Sewell, 1988) consists dominantly of hawaiite flows, minor basalt, mugearite, benmoreite, trachyte and dacite lava flows, and interbedded pyroclastic deposits of different thicknesses. The volcanic products reflect a Hawaiian style of eruptive activity, although locally thick sections of stratified airfall pyroclastic material represent parasitic cones of Strombolian type. The basalts, hawaiites, mugearites, benmoreites, and trachytes (Table 4.3.1,

columns 3–5) define an alkaline trend, whereas the dacites are of potassic subalkaline composition (Weaver & Sewell, 1986). Most rocks were produced between 11 and 10 Ma ago (Stipp & McDougall, 1968b). Radial dykes appear to have been emplaced throughout the period of activity (Shelley, 1987, 1988), and the main cone is thought to have been constructed to a height of about 1500 m above sea-level.

Most Lyttelton mafic lavas are porphyritic and contain olivine, clinopyroxene, and plagioclase phenocrysts, and Fe-Ti oxide microphenocrysts. Plagioclase is subordinate in the basalts but is the most abundant phenocryst mineral in hawaiites and mugearites, reaching 30 percent by volume of the total rock. Partially resorbed kaersutite megacrysts are present in the youngest Lyttelton mafic flows and in some hawaiites of the Mount Herbert Volcanic Group. Lyttelton trachytes have feldspar phenocrysts ranging from sodic oligoclase to anorthoclase in composition which generally are accompanied by ferroaugite. Hedenbergite and rarely fayalite are present in peralkaline trachytes as microphenocrysts, and aenigmatite and arfvedsonite are ubiquitous groundmass minerals.

Table 4.3.1   Selected chemical analyses for South Island provinces, New Zealand

	1	2	3	4	5	6	7	8	9	10
SiO_2	62.79	77.29	48.46	56.00	61.18	44.72	53.10	42.12	47.77	55.81
TiO_2	1.39	0.09	3.18	1.89	0.28	2.61	1.50	3.38	2.15	0.21
Al_2O_3	15.65	12.49	16.54	17.99	18.06	13.93	13.48	14.38	18.07	19.11
Fe_2O_3	5.35[1]	0.83[1]	12.41[1]	10.04[1]	5.04[1]	12.28[1]	1.87	4.44	4.34	4.28
FeO							9.36	8.71	6.20	1.90
MnO	0.04	0.01	0.19	0.15	0.13	0.18	0.15	0.22	0.21	0.25
MgO	1.80	0.15	4.16	1.26	0.35	8.49	7.50	6.71	2.90	0.35
CaO	4.11	0.31	8.94	4.94	1.13	10.98	8.94	11.10	7.12	1.18
Na_2O	4.01	3.54	3.71	4.47	7.41	3.48	2.69	3.86	6.26	9.43
K_2O	3.73	4.80	1.21	2.20	5.29	1.53	0.28	1.75	2.27	4.80
P_2O_5	0.22	0.01	0.59	0.47	0.06	0.67	0.14	0.92	0.74	0.04
S								0.11	0.04	0.04
LOI^2	1.16	0.80	1.03	0.73	0.66	1.10	0.44	0.54[3]	0.66[3]	1.13[3]
H_2O-								0.52	1.15	0.90
CO_2								0.22	0.24	0.11
Total	99.80	100.32	100.42	100.14	99.59	99.97	99.45	99.00	100.10	99.54
Ba	584	98	203	569	121	521	39	512	699	40
Rb	127	500	28	56	157	40	7	39	54	274
Sr	307	19	522	444	36	778	183	922	959	169
Pb	22	16	7	16	18		6	4	6	37
Th	17	70	5	15	30	7	1	5	7	71
U								2	3	18
Zr	359	176	222	446	1032	238	77	308	323	1554
Nb	42	65	51	85	197	77	9	83	93	295
Y	40	144	34	56	51	27	22	29	30	56
La	61	28	31	77	128	55	7	57	73	
Ce	112	73	71	142	212	99	12		142	
V	90	8	165	41	2		167	214	93	5
Cr	67	8	21	12	9	368	307	114		7
Ni	25	13	20	9	10	198	177	85	6	3
Cu								47	18	6

1. K-dacite from Governors Bay Volcanics, Banks Peninsula (B. Thiele and S.D. Weaver, unpublished data; sample 1619).
2. K-rhyolite from Governors Bay Volcanics, Banks Peninsula (B. Thiele and S.D. Weaver; unpublished data, sample 1681).
3. Hawaiite from Lyttelton volcano, Banks Peninsula (Weaver & Sewell, 1986; sample 29).
4. Mugearite from Lyttelton volcano, Banks Peninsula (S.D. Weaver, unpublished data; sample 46c).
5. Trachyte from Lyttelton volcano, Banks Peninsula (Weaver & Sewell, 1986; sample 6a).
6. Basanite from Diamond Harbour Group, Banks Peninsula (Weaver & Sewell, 1986; sample 2434).
7. Qz-tholeiitic basalt from Timaru (Sewell & Gibson, 1988; sample P45801a).
8. Basanite from Dunedin Volcanic Group (Price & Chappell, 1975; Price & Taylor, 1973; sample OU30442).
9. Ne-hawaiite from Dunedin Volcanic Group (Coombs & Reay, 1986; Price & Taylor, 1973; sample OU22487).
10. Phonolite from Dunedin Volcanic Group (Coombs & Reay, 1986; sample OU30453).

[1] Total Fe as Fe_2O_3.
[2] Loss on ignition.
[3] H_2O^+.

The Mount Herbert Volcanic Group (Sewell, 1988) is a volcanic complex of mildly alkaline basalt and hawaiite plugs, lava flows, and intercalated clastic deposits within central Banks Peninsula. These rocks are dated at between 9.7 and 8.0 Ma. There is evidence for a lake that occupied the crater of Lyttelton volcano. Eruptions took place initially from vents within the crater, but later shifted southeastwards, culminating in the accumulation of a thick pile of lava in the vicinity of Mount Herbert, now the highest point on the Peninsula. Base-surge deposits in the upper parts of the succession are evidence for episodic phreatomagmatic eruptions. Hawaiite is the dominant rock type in the Mount Herbert lavas. Most rocks are aphyric, but others contain olivine and clinopyroxene phenocrysts. Some lavas contain kaersutite megacrysts and distinctive red-brown dichroic apatite microphenocrysts.

Eruption of dominantly alkaline lavas of the Akaroa volcano began about 9.0 Ma ago and continued for

1 Ma during which time a large composite cone was constructed (Stipp & McDougall, 1968b). The original height of the volcano is estimated to have been more than 1800 m above sea-level. The rocks consist mostly of basalt and hawaiite, although a complete lineage from basalt to trachyte is recorded. The Akaroa Volcanic Group includes pyroclastic deposits of Strombolian type and, rarely, base-surge deposits. A syenite-gabbro plug is present on Onawe Peninsula near the geometric centre of the volcano, within Akaroa Harbour (see also, Section 1.4.9, Fig. 1.4.10K), and a dyke swarm similar to that of Lyttelton volcano radiates out from it. Akaroa volcanic rocks are petrographically and mineralogically similar to those of Lyttelton volcano.

Eruption of undersaturated lavas took place from several monogenetic cones on the flanks and within the crater of Lyttelton volcano during the transitional period between the end of Akaroa volcanism and beginning of Diamond Harbour volcanism. These include basanite, alkali basalt, and hawaiite, some of which contain crust-derived xenoliths, and have been dated at between 8.1 and 7.3 Ma (Stipp & McDougall, 1968b; Sewell, 1988). Interbedded conglomerate and cross-bedded sandstone on the floor of the crater are evidence that drainage was effected through a breach in the crater wall at the site of the present entrance to Lyttelton Harbour.

The Diamond Harbour Volcanic Group (Sewell, 1988) consists mainly of basanite (Table 4.3.1, column 6) to transitional-basalt lavas that represent the final period of volcanism on the peninsula. Eruption took place from several monogenetic cones on the flanks and within the crater of Lyttelton volcano between 7.0 and 5.8 Ma (Stipp & McDougall, 1968b; Sewell, 1988). The Diamond Harbour flows are much more voluminous compared with mafic lavas erupted during the transitional period, and typically have the features of Hawaiian-type effusion. They were erupted mostly as valley infillings and broad sheets, the best known forming the low-angle, northward-dipping slope into Lyttelton Harbour. Most Diamond Harbour mafic rocks contain olivine and clinopyroxene but no plagioclase phenocrysts.

The craters of both Lyttelton and Akaroa volcanoes were deeply eroded and breached by the sea at the end of the volcanism. A minimum estimate for the volume of the Banks Peninsula volcanoes is 1800 km³ (Sewell, 1988). Akaroa represents about 1200 km³, and Lyttelton about 350 km³.

4.3.4   Timaru and Geraldine

Basalts of Pliocene age are represented by two relatively thin sheets in the vicinity of Timaru and Geraldine in southern Canterbury (Fig. 4.3.1A). Both sheets consist of several individual flows having a range of compositions from tholeiitic to transitional in terms of their normative and modal mineralogy (Duggan & Reay, 1986). The volume of the Timaru Basalt is estimated to be 1.3 km³, and that of the Geraldine Basalt 0.05 km³ (Gair & Rickwood, 1965). Mathews & Curtis (1966) obtained a K-Ar age of 2.47 ± 0.37 Ma for the Timaru Basalt. The basalt outcrops at Timaru and Geraldine are noteworthy because they include the only known occurrences of hypersthene-rich, olivine-poor tholeiitic basalts along the eastern coast of South Island.

Duggan & Reay (1986) recently described the petrology of the Timaru Basalt which crops out over an area of 130 km² immediately to the west of Timaru. The basalt consists of a sheet, typically less than 5 m in thickness, but up to 25 m thick near its source, extending eastward for about 16 km from Mount Horrible. Earliest eruptions produced flows of transitional basalt which are restricted in distribution to the northeastern and eastern portion of the area. These were followed sequentially by eruptions of *ol-* and *qz*-tholeiitic basalts (Table 4.3.1, column 7) extending eastward from Mount Horrible.

The Geraldine flows crop out over an area of 12 km² immediately west of the township of Geraldine and are of tholeiitic-basalt composition. Timaru and Geraldine basalts contain hypersthene microphenocrysts some of which are mantled by pigeonite. Olivine microphenocrysts are present in some rocks but augite and plagioclase are restricted to the groundmass.

4.3.5   North Otago

Three periods of Cainozoic volcanism are recognised in North Otago (Fig. 4.3.3). One of these is known only from the presence of tuffs of Paleocene age (probably 65–54 Ma) in exploration wells drilled off the Otago and Canterbury coasts. A 100 m thickness of glassy basaltic tuff was intersected in the Endeavour–1 well (Fig. 4.3.3), and Coombs et al. (1986) suggested that this may represent the flanks of a large submarine volcano. No contemporaneous volcanic products have been recognised onshore in Otago, but *ol*-tholeiitic basalts of Palaeocene age in Canterbury — termed the View Hill Volcanics (Section 4.3.2) — are likely correlatives. If this is so, the Palaeocene volcanic field extended from offshore Otago to central Canterbury, a distance of at least 250 km.

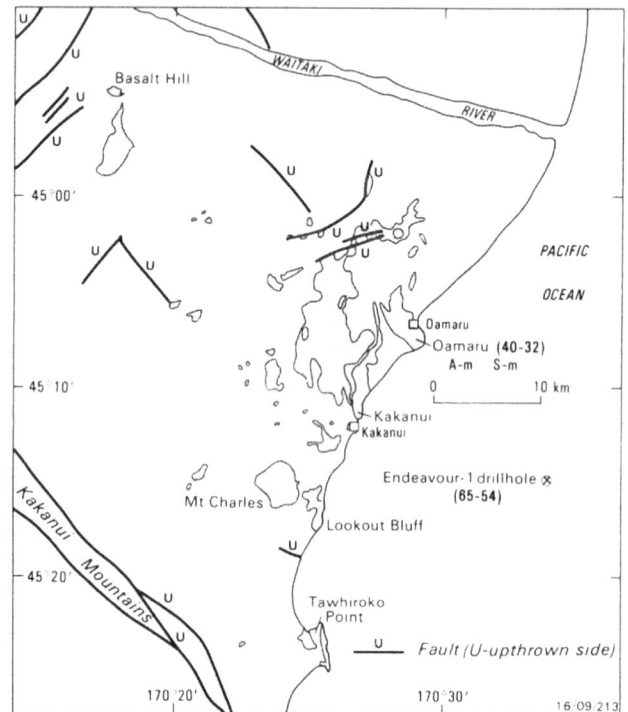

Figure 4.3.3   Distribution of volcanic rocks in the North Otago province. Rock associations: A, alkaline. S, subalkaline. Rock types: m, mafic. Ages (in brackets) in Ma.

The major period of volcanism in North Otago is represented by extensive basaltic tephra, pillow lavas, and shallow intrusions of Upper Eocene to Lower Oligocene age (Fig. 4.3.3). The volcanic rocks have been divided previously into Waiareka (older) and Deborah (younger) volcanic formations (Gage, 1957), and the two are separated by the Totara Limestone. However, the formations are not clearly distinguishable and the pattern is probably one of episodic volcanism, beginning at about 40 Ma and continuing until about 32 Ma, according to Coombs et al. (1986), who propose the informal collective term 'Waiareka-Deborah volcanics'.

Volcanism was largely submarine on a shallow continental shelf. Products include pillow lavas, hyaloclastite breccias and tuffs, pyroclastic surge deposits, and marine volcaniclastic sediments. The magnificent occurrence of pillow lavas at Boatmans Harbour, Oamaru, is of particular interest (see, for example, Fig. 2.4.3) and was the first to be recognised in New Zealand (Park, 1905). Intrusions include dykes and bulbous bodies formed where basaltic magma has invaded soft marine sediment. Coombs et al. (1986) interpreted the tuffs as the products of Surtseyan eruptions from shallow submarine vents that ejected pyroclastic material into the atmosphere. Some materials were deposited directly around vents by fall-out or as surges, but many deposits were reworked by wave and current action. The total volume of magma erupted or emplaced as shallow intrusions during the Waiareka-Deborah episode of activity is estimated to be of the order of 100 km^3.

Waiareka-Deborah rocks have undergone pervasive alteration because of their subaqueous emplacement or intrusion into water-saturated sediment, and are heavily charged with secondary minerals. Most of the early basaltic volcanism was tholeiitic (Coombs et al., 1986). Whole-rock compositions are olivine-normative whereas glasses are quartz-normative. Basanitic ash was reported by Coombs & Dickey (1965) from just above the tholeiitic pillow lavas at Boatmans Harbour.

The Kakanui Mineral Breccia (Dickey, 1968a,b) is a shallow-marine volcaniclastic deposit that includes coarse debris flows notable for their abundance of mantle xenoliths. The magma from which the Kakanui Mineral Breccia was formed is represented by clasts of melanephelinite containing microphenocrysts of titanaugite and olivine. The predominant xenolith type is a coarse, four-phase, spinel lherzolite (Section 4.5.3).

Tholeiitic and strongly alkaline basaltic compositions are represented in tephra of the Waiareka-Deborah volcanics, but the relative proportions are unknown. Tholeiitic basalts appear to be of normal intraplate composition, and the alkaline rocks range from basanitic to nephelinitic compositions.

The third period of Cainozoic volcanism in North Otago is represented by outlying alkali basalts and nephelinites of the Dunedin Volcanic Group which reach to an area about 25 km northwest of Oamaru (Coombs et al., 1986).

4.3.6 Dunedin Volcanic Group

Coombs et al. (1986) proposed the term Dunedin Volcanic Group for all late Tertiary volcanic rocks of eastern and central Otago. The group in essence is the Dunedin volcano and all peripheral intrusive bodies, flows, and tephra which are present at distances of up to 95 km from the centre of the volcano at Port Chalmers (Fig. 4.3.4). Rocks of the Dunedin Volcanic Group rest on early to middle Tertiary marine strata and are overlain by terrestrial sediments deposited during the current Kaikoura orogenic phase.

Dunedin volcano (Fig. 4.3.5) is a major shield that has a diameter of about 25 km and a present-day relief of 700 m. It was constructed during the period 13–10 Ma, although activity at outlying vents began at

about 21 Ma in central Otago (see above). Dunedin volcano for most of its history was subaerial, but the presence of tuff in underlying marine beds may mean that initial eruptions were offshore. The stratigraphy and evolution of the volcano were summarised by Benson (1959, 1968). Quartz-normative trachytes of Benson's Initial Eruptive Phase were shown subsequently to be underlain by basalt (Coombs et al., 1986). Each of the three Main Eruptive Phases consists of a spectrum of compositions from basanite and alkali basalt to phonolite and trachyte (Table 4.3.1, columns 8–10). Mafic compositions predominate, but evolved rocks — such as benmoreite and phonolite — appeared throughout the history of the volcano. Many flows — particularly of basalt and hawaiite — were produced as thin aa units. Most benmoreite, phonolite, and trachyte is found as lava domes and many have a strong fissility that defines their internal structure (Price & Coombs, 1975).

A wide range of pyroclastic deposits is found in Dunedin volcano. The Port Chalmers breccia consists of carbonate-cemented volcanic clasts, schist fragments, and plutonic blocks, and occupies a vent area of about 3 km^2. Elsewhere, Strombolian deposits and thick cross-stratified surge beds, of Surtseyan type, have been recognised. A substantial positive gravity anomaly is centred on the Port Chalmers and Portobello area (Reilly, 1972), probably corresponding to an intrusive complex beneath the volcanic superstructure (see Section 1.4.9, Fig. 1.4.10J).

Most of the outlying vents of the Dunedin Volcanic Group appear to represent short-lived monogenetic

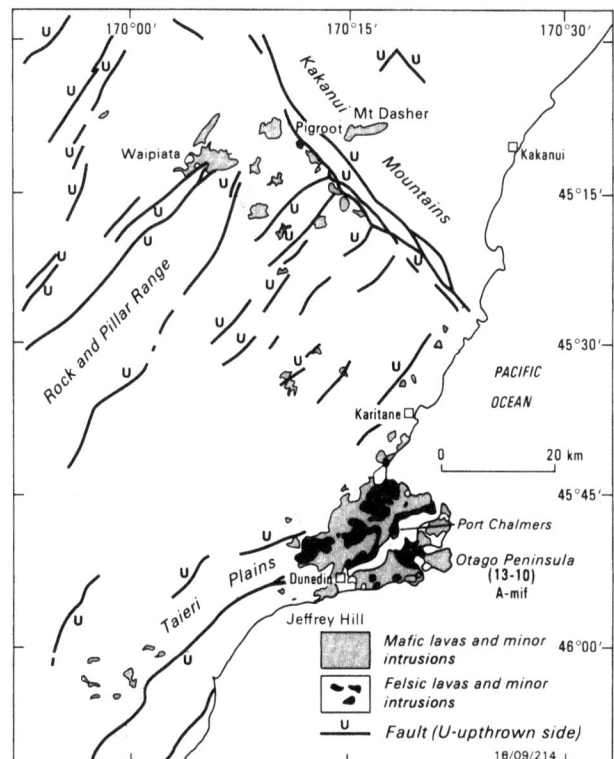

Figure 4.3.4 Distribution of rocks in the Dunedin Volcanic Group. Rock associations: A, alkaline. Rock types: m, mafic; i, intermediate; f, felsic. Ages (in brackets) in Ma.

Figure 4.3.5 A view of Dunedin volcano across Otago Harbour looking northwestwards to Dunedin City. Otago Peninsula in the foreground consists largely of lavas of the second Main Eruptive Phase. Phonolite flows of the third main phase cap Flagstaff (668 m) in the background. Photograph by D.L. Homer, New Zealand Geological Survey.

eruptions. A sheet of basalt near Waipiata (Fig. 4.3.4) extends over an area 13 by 10 km, but most occurrences are much smaller remnants of flows or shallow intrusive bodies. Initial activity at several vents was phreatomagmatic, probably because of low relief and poor drainage existing before development of the present-day fault-block topography. K-Ar ages of volcanic rocks from the peripheral vents range from over 20 to slightly less than 10 Ma (C.J. Adams and D.S. Coombs, personal communications, 1988).

Most rocks are of mafic composition. Basanite is the dominant lithology, and alkali basalt and nephelinite are subordinate. Phonolite is found at Karitane, 'trachyandesite' at Brinns Point, and mugearite at Scroggs Hill and Jeffrey Hill southwest of Dunedin (Fig. 4.3.4). Mafic rocks at many localities contain spinel-lherzolite and harzburgite xenoliths (Section 4.5.3). The common presence of mantle xenoliths in the peripheral vents of the Dunedin Volcanic Group, attests to the rapid rise of a range of magma compositions from mantle depths. In contrast, mantle xenoliths are largely absent from Dunedin volcano, beneath which mantle-derived melts evidently were trapped and differentiated in crustal magma chambers.

Coombs & Wilkinson (1969) postulated a spectrum of lineages for the Dunedin volcano. The main ones are a moderately undersaturated alkali-basalt to trachyte lineage and a strongly undersaturated basanite to phonolite lineage, both including sodic and moderately potassic rock types. Mafic rocks have phenocrysts of olivine, clinopyroxene, and calcic plagioclase in different proportions. Intermediate rocks have clinopyroxene and sodic plagioclase phenocrysts, and in many — especially the moderately potassic types — partially resorbed kaersutite megacrysts are common. Sodalite, nepheline, and alkali-feldspar phenocrysts are present in phonolites, together with clinopyroxene, kaersutite, and rare fayalitic olivine. A summary of the petrography of the principal rock types of the Dunedin volcano was given by Price & Chappell (1975).

Dunedin volcanism took place in a continental intraplate setting that appears to have been mildly extensional (see also Sections 1.6.1, 4.9.3). Faults defining the fault-block topography of the region (Fig. 4.3.4) are of reverse type. These postdate the volcanism and are related to the compressive tectonics of the Kaikoura Orogeny which, in central South Island, began about 10 Ma ago.

Figure 1.1.4 One of the crater lakes of geologically youthful volcano Mount Gambier, South Australia, as portrayed in a picture by G.F. Angas painted in 1844–5. The illustration is from the artist's *South Australia Illustrated* (London, 1847). Reproduced with the permission of the National Library of Australia, Canberra.

Figure 2.5.3 Mount Elephant, the largest scoria cone in the Newer Volcanics province, looms up behind a homestead in western Victoria in this painting by Eugene von Guérard. The painting, called 'Larra' was completed in 1857, and is now in a private collection. It is reproduced here with the permission of the owner.

These plates are available in colour as a download from www.cambridge.org/9780521123228

Figure 2.7.1 Tower Hill volcanic complex, Newer Volcanics province, Victoria, showing scoria-cone complex within a maar.

Figure 3.2.5 Lake Eacham, Atherton province, occupies a young maar about 1 km in diameter and up to 65 m deep. Stratified scoria forming the maar rim hosts upper-mantle and lower-crustal xenoliths. Lava fields and scoria cones are visible in background. Photograph courtesy of Murray Views, Gympie, Queensland.

Figure 3.4.6 Peralkaline rhyolite and trachyte plugs of the Glass Houses province showing, from left to right, Mount Coonowrin (Crookneck), Mount Tibberoowuccum, Mount Ngungun, Mount Coochin (in distance), and Mount Tibrogargan.

Figure 3.4.8 Basalt-capped coastal escarpment vegetated by burnt Xanthorrhoea sp. in the foreground, at Cunninghams Gap, Main Range province. The base of the Gap marks the boundary between underlying Governors Chair Volcanics and the overlying Superbus Basalt.

Figure 3.5.6 Central area of Warrumbungle volcano showing the Breadknife peralkaline-trachyte dyke in the foreground. The cleared area in the middle distance is erosion-exposed Jurassic basement.

Figure 3.5.7 Grand High Tops, Warrumbungle volcano, showing Belougery Spire (trachyte) on the left skyline and Crater Bluff (peralkaline trachyte) on the right. The Breadknife dyke (peralkaline trachyte) is below and to the left of Crater Bluff.

Figure 3.6.7 Near flat-lying flows forming terraced hills in Monaro province. This view is southwest of Mount Cooper towards Wangollic Hill.

Figure 3.7.4 Aerial view of Mount Gambier volcanic province, South Australia, looking southeastwards. The town of Mount Gambier is on the left (north). From rear forward are the Blue Lake, Valley Lake, and Brownes Lake maar centres. The playing field in bottom left-hand corner is sited on an older cluster of maars, now partially buried by later tephra. The maar complex has developed on one of many northwest-trending crustal faults controlling the Otway Basin geometry.

Figure 6.2.7 Porphyroclastic microstructure in a spinel harzburgite from Mount Shadwell (62734). Intensely strained olivine porphyroblasts having closely spaced, commonly curved kink bands are set in a matrix of equant polygonal neoblasts. Enstatite (pale grey) is strongly deformed and cleavages have opened to allow growth of fine olivine neoblasts. Field of view (with crossed polars) is 35 × 22 mm.

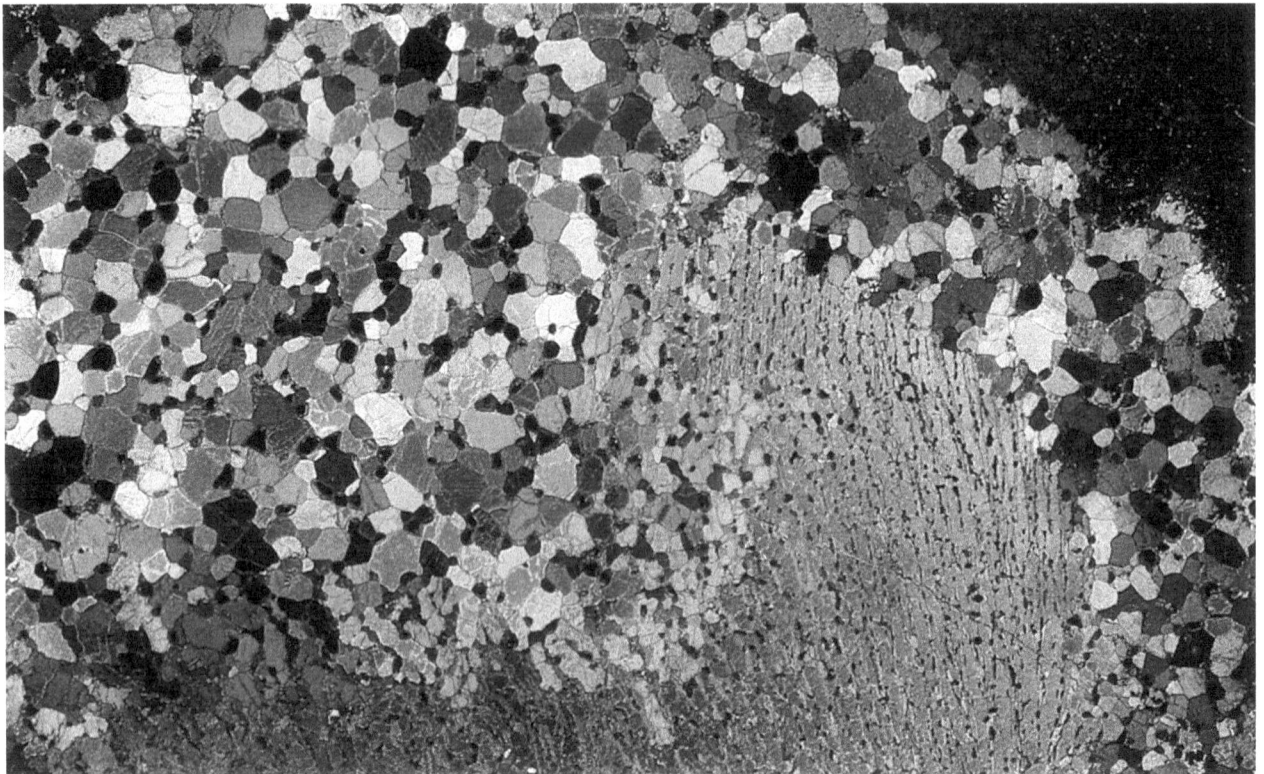

Figure 6.2.12 Porphyroclastic microstructure in garnet websterite from Lake Bullenmerri (81482). A lamellar augite porphyroclast of 5 cm maximum diameter has recrystallised to a mosaic of polygonal augite, bronzite, and garnet neoblasts. The porphyroclast has exsolved lamellae of garnet (black) and orthopyroxene, and has strong undulose extinction. Note the transition zone at the porphyroclast margin consisting of semi-aligned fine-grained neoblasts. Field of view (with crossed polars) is 35 × 22 mm.

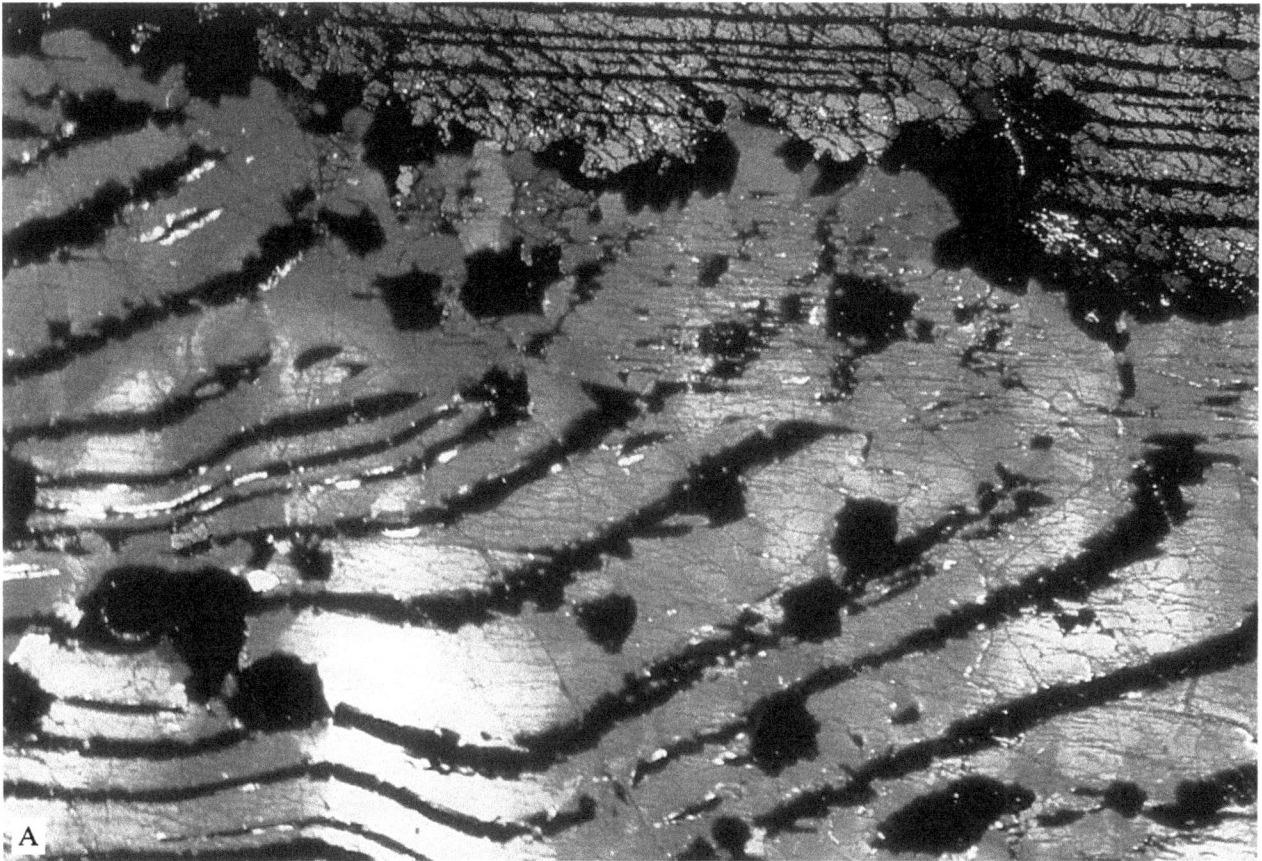

Figure 6.2.14   (A) Relict clinopyroxenes in metapyroxenite showing deformed exsolution lamellae of garnet, orthopyroxene, and spinel. Width photographed (with crossed polars) is 2.5 cm. (B) Same sample in plane polarised light. Spinel was present originally as inclusions in the primary, unexsolved clinopyroxene. Garnet rims developed on these original spinel grains during re-equilibration.

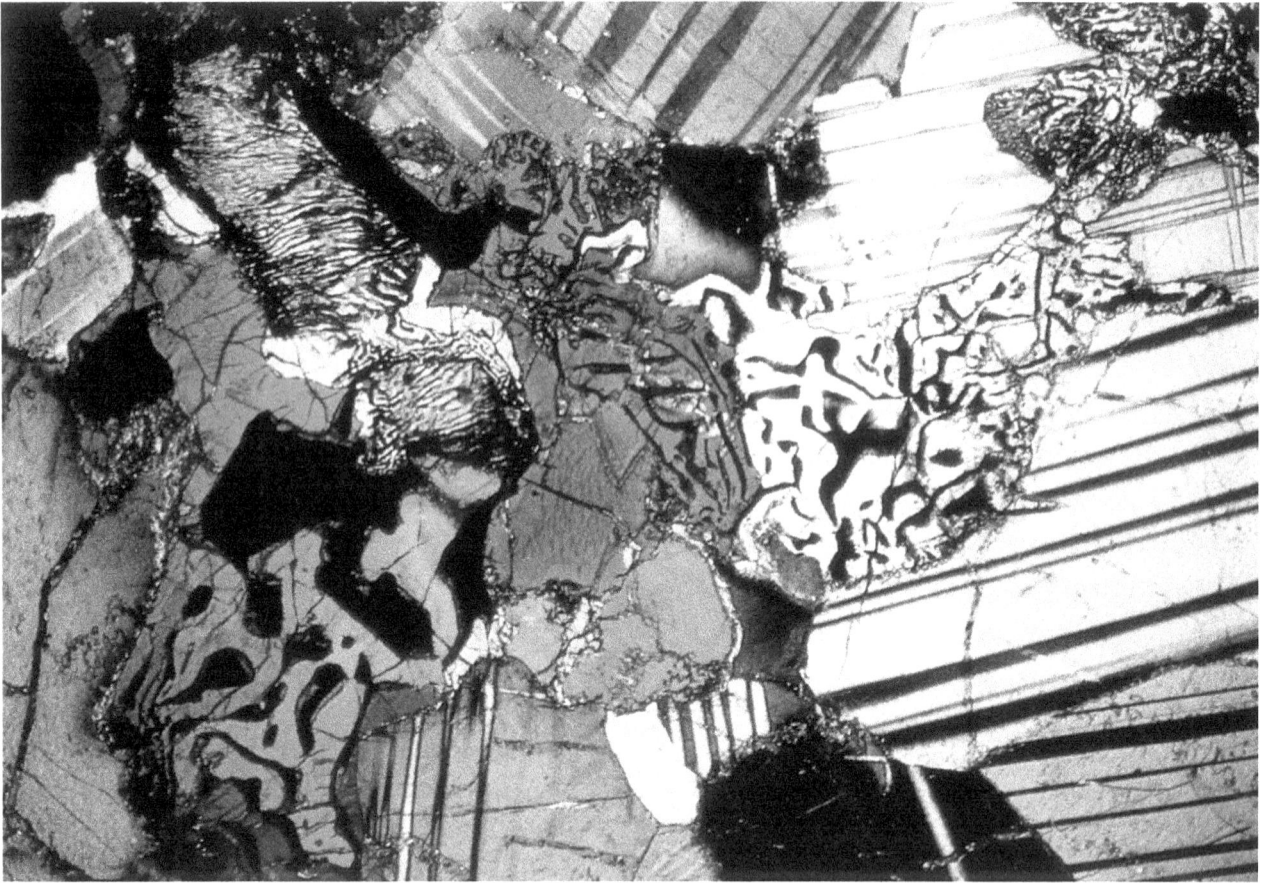

Figure 6.3.1 Coarse-grained symplectitic intergrowth of spinel with clinopyroxene and orthopyroxene in foliated granulite from Dundas (Sydney). Width photographed (with crossed polars) is 3.5 mm.

4.3.7   Alpine Dyke Swarm

Cooper (1986) documented a lamprophyre dyke swarm extending over 110 km from the Alpine Fault in southern Westland to northwestern Otago (Fig. 4.3.1C). Dykes, sills, and diatremes cut Haast Schist and probably were emplaced in the period 28–16 Ma. The dykes have a predominant east-west orientation and are typically less than 2 m wide. A range of strongly undersaturated alkaline lithologies is present. Lamprophyre is the most abundant, and is associated with tinguaite, trachyte, and carbonatite. The lamprophyres represent hydrated and carbonated nephelinitic magma that fractionated in a crustal magma chamber producing phonolitic derivatives. Carbonatite is considered by Cooper (1986) to be the product of late-stage liquid immiscibility.

4.3.8   South Westland

Two episodes of mildly alkaline, mafic, Cainozoic volcanism are recognised in South Westland — the late Cretaceous to Palaeocene Arnott Basalt and late Eocene Otitia Basalt (Sewell & Nathan, 1987). Rocks of both groups are typically fine grained, and because of submarine eruption and subsequent deep burial, most are at least incipiently altered. Onshore outcrops (Fig. 4.3.1C) are limited largely to the area between the Paringa and Haast Rivers, but a line of magnetic anomalies revealed by geophysical surveys extends as far north as the mouth of the Karangarua River. Seismic profiles farther offshore have several steep-sided, roughly circular features that are discordant with the surrounding sedimentary rocks, and which may represent buried volcanic centres (Esso, 1968).

The Arnott Basalt was erupted close to the axis of rifting during separation of the Australia and New Zealand continental blocks, on the basis of plate-tectonic reconstructions for the southwestern Pacific (Weissel et al., 1977; Grindley & Davey, 1982). Such a setting is in accord with the mildly alkaline composition of the flows. The Otitia Basalt was erupted during a period of crustal extension, starting about 45 Ma and indicated onshore by the formation of small, fault-bounded basins as well as regional subsidence leading to marine transgression (Nathan et al., 1986). These lavas also have intraplate characteristics. The outcrop areas of both units (and their inferred extension along magnetic anomalies) have a northeasterly trend parallel to the general structural grain in South Westland, including the Alpine Fault and offshore South Westland fault zone (Nathan et al., 1986).

4.4   Sub-Antarctic and Chatham Islands

4.4.1   Introduction

Campbell Plateau and Chatham Rise to the south and southeast of New Zealand are extensive areas beneath a shallow epicontinental sea (Figs. 1.1.8, 1.3.7). Lithological and structural links to continental New Zealand were well documented by Adams (1983), Adams & Cullen (1978), Adams et al. (1979), Beggs (1978), Denison & Coombs (1977), Gamble & Adams (1985), Gamble et al. (1986), Grindley (1961), Grindley et al. (1977), and Watters & Fleming (1975). The late Cainozoic volcanic edifices forming much or part of island groups such as the Auckland Islands, Campbell Island, Antipodes Islands, and the Chatham Islands therefore can be related temporally and provincially to the volcanic shields of Lyttelton, Akaroa, and Dunedin on the eastern coast of the South Island (Section 4.3). The volcanoes grew on high points of the pre-Cainozoic basement of metasediments and granites, so a deep structural control is inferred for their setting.

4.4.2   Auckland Islands

The Auckland Islands (Fig. 1.1.8) are about 400 km south of New Zealand near the western edge of the Campbell Plateau (Gamble et al., 1986). There are four main islands: Auckland (the largest), Enderby Island, Adams Island, and Disappointment Island, together with numerous smaller islands and sea stacks. Fiord-like inlets indent the southern and eastern coastlines of the islands, and precipitous cliffs over 300 m high line the western coast making access difficult to the high 'felsenmeer' plateaux (more than 500 m above sea-level) which form the central part of the islands. There appear to be two coalesced volcanic shields — Carnley volcano in the south, and Ross volcano in the north (Wright, 1966b; Gamble & Adams, 1985).

The islands consist predominantly of thin laterally discontinuous lava flows that accumulated to a maximum thickness of 680 m on Adams Island, south of Auckland Island. A few more extensive flows, up to 30 m thick, form prominent plateaux — such as Fleming Plateau (Gamble & Adams, 1985).

Mafic rocks form the greater proportion (about 80 percent) of the lava flows (Table 4.4.1, columns 1–2). Intermediate and felsic rocks (Table 4.4.1, columns 3–4) account for the remainder, and tend to be more abundant towards the base of the succession as stumpy flows in localised centres. Lava flows in Carnley Harbour are cut by an intense swarm of dykes and inclined sheets and sills, petrographically identical to the extrusive rocks. These dykes and sheets decrease in abundance upwards through the volcanic pile and are interpreted as feeders to the lava flows. Gabbroic rocks cropping out in the heart of Carnley Harbour intersect and have metamorphosed volcanic rocks at the base of the sequence, and are interpreted as the roof zone of a subvolcanic magma chamber (Gamble & Adams, 1985).

Adams (1983) reported whole-rock K-Ar age determinations on rocks from the Ross and Carnley shields. Ages range from around 25 to 12 Ma, but the older ages are suspect because of possible contamination from Mesozoic basement granite that crops out on Musgrave Peninsula (Gamble & Adams, 1985). Ages for the two centres may well overlap.

The rocks are porphyritic to sparsely porphyritic. Olivine and clinopyroxene are the principal phenocryst minerals in the mafic rocks, together with minor plagioclase and Fe-Ti oxides. Plagioclase and Fe-Ti

Table 4.4.1   Selected chemical analyses for Sub-Antarctic and Chatham islands, New Zealand

	1	2	3	4	5	6	7	8	9	10
SiO_2	45.85	48.93	51.24	72.06	46.96	62.66	41.89	41.71	40.46	40.30
TiO_2	3.12	2.44	1.86	0.04	3.01	0.60	4.15	3.88	3.34	3.99
Al_2O_3	14.59	12.01	15.94	14.46	12.68	16.64	11.32	12.85	13.32	9.81
Fe_2O_3	3.43	3.72	4.07	1.38	2.34	2.96	17.65[1]	15.47[1]	6.44	5.33
FeO	9.40	7.99	7.18	0.28	10.59	3.24			6.72	7.76
MnO	0.19	0.17	0.22	0.01	0.19	0.13	0.19	0.20	0.16	0.18
MgO	7.95	9.58	2.56	0.08	10.97	0.83	9.48	7.88	7.70	10.75
CaO	9.60	9.99	5.57	1.24	9.74	2.48	11.45	10.84	11.03	12.10
Na_2O	2.74	2.40	4.59	3.87	2.14	5.63	2.32	4.80	2.14	3.18
K_2O	1.04	1.21	2.58	5.09	0.89	4.94	0.75	1.56	1.11	1.29
P_2O_5	0.59	0.42	1.22	0.05	0.48	0.16	0.52	1.18	0.68	1.01
S										
LOI[2]	1.39	0.93	2.59	0.70		0.95	−0.27	−0.41	5.62	4.27
Total	99.89	99.79	99.62	99.26	99.99	101.22	99.45	99.96	98.72	99.97
Ba	322	276	702	754	253	1048	209	494	549	570
Rb	21	34	58	196	23	73	17	45	24	24
Sr	670	457	614	227	542	114	525	1182	784	766
Pb	<2	2	8	25			<2	8	2	
Th	3.9	5.3	13	20	6	10	4	11	12	5
U	0.8	1.3		5.0						
Zr	229	232	543	590	192	998	228	393	211	367
Nb	61	48	101	106	50	114	45	100		99
Y	33	32	61	64	28	60	30	40	26	34
La	36	35	76	114	38	126	25	83	58	61
Ce	73	72	167	237	73	239	70	156	118	125
V	268	241	54	<2	238	21	366	244	318	214
Cr	222	490	5	<2	350	6	301	124	189	349
Ni	149	150	<5	2	265	12	138	117	122	287
Cu	65	41	<5	2			58	64	59	51

1. Hawaiite from Carnley volcano, Auckland Islands (Gamble & Adams, 1985; Gamble et al., 1986; sample 82262).
2. Transitional basalt from Carnley volcano, Auckland Islands (Gamble & Adams, 1985; sample 82272).
3. Mugearite from Carnley volcano, Auckland Islands (Gamble & Adams, 1985; Gamble et al., 1986; sample 82210).
4. Rhyolite from Carnley volcano, Auckland Islands (Gamble & Adams, 1985; Gamble et al., 1986; sample 82215).
5. Alkali basalt, Campbell Island (Morris, 1984; Gamble et al., 1986; sample 39796).
6. Trachyte, Campbell Island (Morris, 1984; Gamble et al., 1986; sample 39772).
7. Basanite, Antipodes Island, Campbell Plateau (Gamble et al., 1986; sample 81/7).
8. Basanite, Antipodes Island, Campbell Plateau (Gamble et al., 1986; sample 81/15).
9. Basanite, Momoe-a-toa, Chatham Islands (Gamble et al., 1986; sample 14467).
10. Basanite, Momoe-a-toa, Chatham Islands (Gamble et al., 1986; sample 14465).

[1] Total as Fe_2O_3.
[2] Loss on ignition.

oxides are more common and olivine and clinopyroxene less common in more evolved, intermediate rocks. Sodic plagioclase, Fe-Ti oxides, and rare quartz and sanidine are phenocrysts in the felsic lavas. Rare obsidian has been recorded from the saddle between the Ross and Carnley shields (C.J. Adams, personal communication, 1986).

The Carnley and Ross rocks are chemically identical. Mafic lavas and intrusions range from mildly alkaline basalt (less than 5-percent normative nepheline) to *ol*-tholeiitic basalt, hypersthene-normative hawaiite, and icelandite. Most of the evolved rocks are quartz-normative trachyte and rhyolite, although rare nepheline-normative trachytes are present.

4.4.3   Campbell Island

Campbell Island is the most southerly of New Zealand's Sub-Antarctic islands (Figs. 1.1.8, 1.3.7). Roughly two-thirds of the island are covered by Cainozoic lavas

that rest on volcaniclastic breccias of the Shoal Point Formation (Oliver et al., 1950; Beggs, 1978; Morris, 1984). These underlie limestone of the Tucker Cove Formation, and sandstones, conglomerates, and mudstones of the Garden Cove Formation, and range from late Cretaceous to Miocene. They rest unconformably on basement schist that may be as old as Ordovician (Adams et al., 1979).

The lavas of Campbell Island are predominantly mafic, and both mildly alkaline (nepheline-normative) and subalkaline (hypersthene-normative) compositions are present (Table 4.4.1, column 5). Intermediate and felsic compositions range from mugearite to benmoreite, trachyte (Table 4.4.1, column 6), and rhyolite. Morris (1984) suggested that low-pressure fractional crystallisation controlled the mafic-intermediate lineage of Campbell Island volcano, and that the rhyolites evolved along a separate petrogenetic path. A coarse-grained biotite gabbro petrographically similar to the gabbro in Carnley volcano (Auckland Islands) crops out at the head of Perseverance Harbour.

Radiometric (K-Ar) ages for whole-rock samples from Campbell Island cluster around 7.4–7.0 Ma, but range from 11 to 6.5 Ma, so the Campbell Island centre is appreciably younger than the Auckland Islands.

4.4.4 Antipodes Islands

The Antipodes Islands are the smallest and geologically least well known of the Sub-Antarctic Islands. They consist of Antipodes Island (about 4 km across), Bollons Island (about 1 km) 1.5 km to the north, and several small islets and sea-stacks. No pre-Cainozoic basement is exposed on the islands. A young age (less than 1 Ma) is inferred from many juvenile volcanological features, and confirmed by available radiometric dates (Cullen, 1969; C.J. Adams, personal communication, 1986).

The volcanic rocks are dominantly pyroclastic in origin, together with lava flows and associated minor intrusions. Major scoria cones have been recognised in the vicinity of Mount Galloway and Mount Waterhouse in the centre of Antipodes Island. Other tuff cones are partially eroded by the sea, and steep cliff sections of pyroclastic rocks are exposed in Perpendicular Head (in the northwest) and Albatross Point (southeast).

The lavas of Antipodes Island range from strongly porphyritic ankaramite to near-aphyric glassy rocks. Olivine, clinopyroxene, and plagioclase are the principal phenocryst minerals, and kaersutitic amphibole is sparse. The lavas are all nepheline-normative and range from basanite (Table 4.4.1, columns 7–8) to *ne*-hawaiite and *ne*-mugearite. There are no records of ultramafic xenoliths.

4.4.5 Chatham Islands

The Chatham Islands, about 900 km east of Christchurch near the eastern end of the Chatham Rise, are New Zealand's largest group of off-shore islands (Figs. 1.1.8, 1.3.7). They contain appreciable areas of late Mesozoic and Cainozoic volcanic rocks that were studied by Grindley et al. (1977) and Morris (1982). Late Cretaceous volcanic rocks dominate the coastline of southern Chatham Island and are the most voluminous volcanic rocks on the islands (Morris, 1985). Periods of Eocene-Oligocene (40–35 Ma) and Miocene-Pliocene (6–2.6 Ma) volcanism have been recognised from palaeostratigraphic and K-Ar dating studies (Grindley et al., 1977). The centres of activity are generally small, localised, and related to fracture patterns in the underlying basement of Chatham Island schist. Volcanic activity was generally explosive, building up tuff and scoria cones. Lava flows are somewhat rare but are particularly well developed near Cape Young and Momoe-a-toa in northern Chatham Island.

The lavas are generally porphyritic and partly glassy, containing abundant olivine, clinopyroxene, and rare plagioclase phenocrysts. The Eocene-Oligocene and Miocene-Pliocene rocks are basanites and alkali basalts (Table 4.4.1, columns 9–10). Evolved compositions (*ne*-mugearite and phonolite) are rare, accounting for only 5 percent of the erupted volume.

4.5 Xenoliths and Megacrysts

4.5.1 Introduction

Upper-mantle and lower-crustal xenoliths reported from the intraplate volcanic rocks of both North Island and South Island and from the outlying Chatham and Sub-Antarctic Auckland Islands, are described briefly in this section. Examples of cognate inclusions and upper-crustal materials are considered also, but emphasis is given to inclusions from deeper levels. A summary account of megacrysts and mantle xenoliths in New Zealand was presented recently by Reay & Sipiera (1987). A more detailed account of samples from the upper mantle and lower crust of eastern Australia is given in Chapter 6 (see also Section 7.2).

4.5.2 North Island

Xenoliths and megacrysts are found in the rocks of both the Northland and Auckland provinces, although ultramafic xenoliths of mantle origin are rare in Northland.

Lherzolite xenoliths are found in a Miocene (11.9 Ma) nephelinite dyke near Arapohue at the southern end of the Northland province (Black & Brothers, 1965). Other examples of large (up to 25 cm) inclusions are known only from two localities — the Waitangi and Waimimiti flows — where the rocks are coarse-grained gabbros and leucogabbros. Mineral compositions are similar to those of the host-rock phenocrysts and these inclusions therefore are assumed to be cognate.

Many of the Northland lavas contain pyroxene and plagioclase crystals significantly larger than the associated phenocryst phases. These crystals, except for the case of the 'big feldspar' basalts from Puhipuhi, are interpreted as megacrysts. Megacrysts are most abundant in hawaiites. They are rare in the rocks of the Whangarei field. Quartzo-feldspathic and sedimentary material from the underlying formations is found as inclusions in some centres but is not common.

Ultramafic and mafic inclusions have been reported from six localities in the Auckland province (Rodgers et al., 1975). These are found at Shoal Bay/Lake Pupuke in the Auckland City field as fragments, small lava-encrusted masses, and scattered grains in tuff and lapilli beds surrounding explosion craters. They are not found generally in the lava flows, although mafic xenocrysts are present in flows from a nearby centre. Plagioclase, diopside, and amphibole megacrysts are found in the South Auckland field where inclusions too have been reported from Stevensons quarry at Ramarama and Rooses quarry at Bombay.

Ultramafic xenoliths are an important constituent of many lava flows in the Ngatatura field (Rodgers & Brothers, 1969; Rodgers et al., 1975; Utting, 1986). The suite is diverse and includes dunite, harzburgite, lherzolite, saxonite, websterite, clinopyroxenite, plagioclase peridotite, and gabbro. The xenoliths are mostly 5–20 mm in size and have hypidiomorphic to allotriomorphic granular textures.

Xenoliths of the underlying sedimentary formations are moderately common in some Auckland centres, as well as metamorphic inclusions found in the pyroclastic deposits of the St Heliers centre and which presumably represent the deep basement beneath Auckland.

4.5.3   South Island

Xenoliths and megacrysts are known in southern New Zealand from the Kakanui Mineral Breccia in North Otago, in plugs and flows of Dunedin volcano and in associated outlying parts of the Waipiata Volcanics, and in the volcanic rocks of Banks Peninsula and northern Canterbury.

The inclusion-bearing breccia at Kakanui forms the upper part of the Waiareka and Deborah volcanic formations of Upper Eocene to Lower Oligocene age. This is a shallow-water volcaniclastic unit in which alkali basaltic lapilli tuffs were followed by an eruption of melanephelinite (the Mineral Breccia Member) carrying a range of mafic to ultramafic xenoliths and megacrysts (Dickey, 1968a,b). Xenolith types recognised at Kakanui include spinel lherzolite, garnet and spinel pyroxenites, hornblendite, gabbro, and granulite (Dickey, 1968b, Mason, 1968; White et al., 1972; Reay & Sipiera, 1987). Reported megacryst species (including some polycrystalline aggregates) are anorthoclase, augite, kaersutite, ilmenite, titanbiotite, pyrope, and apatite (Mason, 1966; Dickey, 1968b; Reay & Sipiera, 1987). Further data on individual species were reported for ilmenite (Reay & Wood, 1974), kaersutite (Merrill & Wyllie, 1975; Wallace, 1977; Saito et al., 1978), and augite (McCallister et al., 1976), and on minerals from xenoliths by Philpotts et al. (1972), White et al. (1972), and McCallister et al. (1976). Schist-derived quartzose xenoliths are extraordinarily abundant in some of the intrusive bodies of Moeraki Peninsula which are part of the Waiareka Volcanics (Coombs & Reay, 1986).

Inclusions of demonstrably high-pressure origin are rare in the central part of Dunedin volcano but are much more common and widespread in the products of the peripheral vents of the Dunedin Volcanic Group of northern and central Otago. Spinel lherzolites have been recorded in basanitic rocks from Mount Dasher and surrounding areas (Brown, 1955), Mount Stoker (Brown, 1964), the Kokonga area (Turner, 1942), and in the Hindon area where there are examples also of lherzolite veined by pyroxenite. Lherzolite xenoliths are present in more felsic rocks (mafic *ne*-benmoreite/ phonolite) at Flat Hill (Reay & Sipiera, 1987) and at Pigroot or Trig L (Wright, 1966a; Price & Green, 1972) where they are accompanied by corona-textured, two-pyroxene metagabbros (Price & Wallace, 1976).

Few data are available on inclusions in the central part of Dunedin volcano. Highly oxidised Cr-diopside bearing xenoliths (possibly lherzolites) are known at Pilot Point east of Purakanui in an altered trachyandesite host, and peridotite xenoliths are found also (but are extremely rare) in the felsic Port Chalmers Breccia in the core of the complex. The Port Chalmers Breccia also contains xenoliths of gabbro, syenite, and nepheline syenite (Allen, 1974). A basanite at Saddle Hill on the southern flanks of the volcano contains an assemblage of pyroxenite xenoliths whose compositional characteristics span the range between typical Group I

(Cr-diopside) and Group II (Al-augite) xenoliths (see Section 6.1.3). Megacrysts of clinopyroxene characterised by a range of Cr, Al, and Ti contents and Mg-ratios are present also. Kaersutite megacrysts are found in many rocks from Dunedin volcano (Kesson & Price, 1972).

Coote (1987) described a range of xenoliths from a nephelinite sill (15 Ma old) in the Little Lottery River, northern Canterbury. These include lherzolites and harzburgites of mantle origin and orthopyroxenites and websterites that are believed to be cognate cumulates.

No mantle-derived xenoliths have been found in the volcanic rocks of Banks Peninsula, although cognate xenoliths are known from several localities. Basaltic lava flows of Akaroa volcano exposed near sea-level in Haycocks Bay contain inclusions of olivine gabbro, clinopyroxenite, wehrlite, and diorite (C.J. Dorsey, personal communication, 1987). Sewell (1985) described inclusions of olivine gabbro, clinopyroxenite, wehrlite, and peralkaline syenite in hawaiites of the Mount Herbert Volcanic Group. Some gabbroic inclusions contain kaersutite and red-brown apatite which are present also as megacrysts and microphenocrysts, respectively, in mafic lavas of the Mount Herbert and Lyttelton volcanic groups. Gabbro and clinopyroxenite inclusions are found also in lavas of the Diamond Harbour Volcanic Group on Quail Island. Quartzose xenoliths are ubiquitous in the Governors Bay Volcanics and represent recrystallised Tertiary sediments which immediately underlie the Miocene volcanic sequence.

An alkaline dyke swarm cutting the Haast Schist in southern Westland (Cooper, 1986) carries xenoliths at some localities. These include spinel lherzolites and amphibole-apatite xenoliths similar to those described by Wass et al. (1980) from Kiama in New South Wales.

4.5.4   Sub-Antarctic and Chatham islands

Data are sparse on xenoliths and megacrysts in volcanic rocks from the islands offshore from New Zealand. However, a diverse suite of lower-crustal and upper-mantle xenoliths consisting of granulites, spinel lherzolite, and pyroxenite are known on the Chatham Islands (Gamble et al., 1986). In addition, kaersutite megacrysts up to 15 cm across are numerous at Momeo-a-toa in northern Chatham Island. Rare, partially resorbed kaersutite phenocrysts have been reported from a benmoreite from Campbell Island (Morris, 1984), and rare kaersutite is known also in rocks from the Antipodes Islands. Lherzolite xenoliths are present in the Mount Eden plug in the Auckland Islands (Wright, 1968) where they are accompanied by megacrysts of augite and bronzite (Green & Hibberson, 1970; see Section 7.5.2).

4.6   Petrology and Major Element Chemistry

4.6.1   Introduction

This section is a summary of the petrology and geochemistry of volcanic rocks from the New Zealand intraplate provinces. Much of the information is unpublished material from student theses and current research projects. The scope of the available data is different between the different provinces and, inevitably, coverage is not uniform. Mineralogical studies of intraplate volcanic rocks using the electron microprobe are still in progress in New Zealand and the currently available database is much smaller than that for east-Australian provinces (Section 5.3). Known compositions for the common rock-forming minerals are similar to those described for the same rock types in eastern Australia.

The intraplate provinces are considered as two geographically separate groups: the Northland and Auckland provinces on the Indo-Australian plate; and the South Island and Sub-Antarctic and Chatham islands provinces which are mainly on the Pacific plate. The provinces also may be subdivided into two further categories in a manner similar to that adopted for east-Australian provinces (Section 1.1.3):

(1) Basaltic *lava fields* are relatively thin volcanic successions fed from monogenetic centres and are found commonly as scattered erosional remnants. These consist predominantly of mafic lavas and associated tuff rings, scoria cones, and phreatomagmatic deposits. Mafic compositions range mainly from nephelinite to *qz*-tholeiitic basalt. Two or more geographically separate fields may be recognised in some provinces.

(2) *Central-volcano* provinces are large composite shields characterised by mafic lavas of mildly undersaturated to transitional composition, and by associated felsic flows and intrusions including phonolite, trachyte, and rhyolite.

Division into these two categories is unclear in some provinces, or for other purposes inappropriate. For example, the Dunedin Volcanic Group, as defined by Coombs et al. (1986), includes both the Dunedin central volcano and all late Tertiary volcanic rocks of eastern and central Otago all of which are considered to represent peripheral vents.

The Northland and Auckland provinces are geographically separate regions, each consisting of several identifiable concentrations of eruptive centres (Section 4.2). The style of volcanism and the age range in both provinces are similar, and both provinces may be defined as lava-field type. The different fields in each province have differences in the distribution of rock types and the age of eruptive activity (Section 4.2), but these are not of importance in the context of this volume. However, there are significant differences between the petrology of the two North Island provinces (see below).

The petrological data base for the Northland province is founded on a systematic and representative collection of the exposed rocks. Some of the material has been published (Ferrar, 1925; Heming, 1980a,b; Ashcroft, 1986a) but most is in unpublished theses (Mansergh, 1965; Muhlheim, 1974; Ashcroft, 1986b; Yimir, 1986). Petrological data for the Auckland province include both published (Rodgers et al., 1973; Rafferty & Heming, 1979; Heming & Barnet, 1986) and unpublished (for example, Utting, 1986) material. The data provide an indication of the general petrological

nature of the province but are not considered to be truly representative. Selected chemical analyses of North Island volcanic rocks are presented in Table 4.2.1.

Volcanic provinces in the South Island and Sub-Antarctic and Chatham islands range in age from Palaeocene to Quaternary. Large central volcanoes are present on Banks Peninsula and Otago Peninsula and on Auckland and Campbell islands (Sections 4.3–4). The North Otago, Canterbury (including Timaru and Geraldine), and Marlborough provinces are of lava-field type, and for the purposes of the following sections are grouped together as a single province, despite a range in age from Palaeocene to Pliocene. The volcanic rocks of the Chatham Islands, South Westland, and the Alpine Dyke Swarm probably are classified best as lava-field as there is no evidence that large central edifices were constructed.

Reasonably comprehensive whole-rock geochemical databases exist for Banks Peninsula, the Dunedin Volcanic Group, Canterbury, and North Otago provinces, and the Chatham Islands, although many data as yet are unpublished. Data for the other provinces presented here cannot be taken as fully representative. Published sources of geochemical data for the South Island and Sub-Antarctic and Chatham islands provinces are as follows: Canterbury and North Otago (McLennan & Weaver, 1984; Coombs et al., 1986; Duggan & Reay, 1986; Coote, 1987; Sewell & Gibson, 1988), Banks Peninsula (Price & Taylor, 1980; Falloon, 1982; Thiele, 1983; Sewell, 1985; Weaver & Sewell, 1986), Dunedin Volcanic Group (Coombs & Wilkinson, 1969; Price & Taylor, 1973; Price & Chappell, 1975; Coombs & Reay, 1986), Alpine Dyke Swarm (Cooper, 1979, 1986; Barreiro & Cooper, 1987), South Westland (Sewell & Nathan, 1987), Sub-Antarctic Islands (Cullen, 1969; Morris, 1984; Gamble et al., 1986), and Chatham Islands (Morris, 1979, 1985). Selected chemical analyses of South Island and Sub-Antarctic islands volcanic rocks are presented in Tables 4.3.1 and 4.4.1.

Petrological differences between the New Zealand volcanic provinces are summarised in histograms together with plots of normative mineralogy (Figs. 4.6.1–7).

4.6.2 Mafic rocks

Mafic lavas (hawaiite and rocks containing less than 54-percent SiO_2) predominate in all New Zealand Cainozoic provinces. They range from nephelinite through basanite, alkali basalt, *ne*-hawaiite, and hawaiite, to *ol*- and *qz*-tholeiitic basalt. No strongly potassic rocks or melilitites such as are found in eastern Australia, are known in New Zealand. Carbonatite is found as a volumetrically minor component (about 1 percent) of the lamprophyre dyke swarm of the Southern Alps (Cooper, 1986). Felsic rocks are volumetrically significant only in the large central volcanoes, but in these too mafic compositions predominate.

The dominant rock type in the Northland province of North Island is hawaiite. Other mafic rocks range from basanite to *qz*-tholeiitic basalt, but undersaturated compositions are a minor component of the suite.

Intermediate rocks (greater than 54-percent SiO_2) are comparatively rare (9 samples or 4.5 percent of the analysed suite). These rocks range up to 60 weight-percent SiO_2 but because $100An/(An+Ab)$ values are greater than 30 they are qz-tholeiitic basalts using the classification scheme adopted in this volume and are plotted as such in Figure 4.6.1A. A perhaps more appropriate name would be basaltic icelandite (see Section 1.1.4). High normative-An contents in these essentially intermediate rocks appear to be caused by high CaO/Na_2O values (1.1 to 2.3) in samples that have moderately high Al_2O_3 contents (15 to 18 percent).

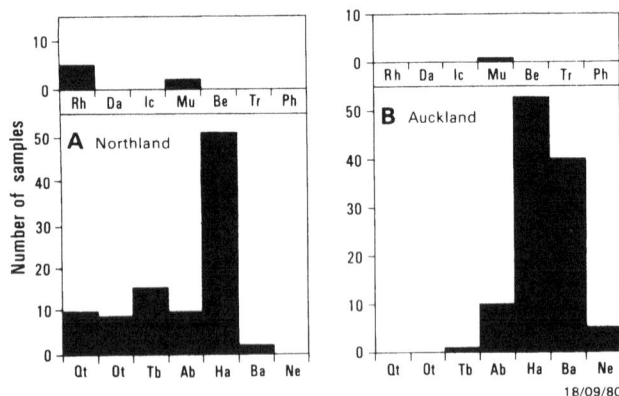

Figure 4.6.1 Histograms of analysed rock types from North Island provinces as a percentage of all the analysed rock types in each province. Abbreviations: Qt, qz-tholeiitic basalt; Ot, ol-tholeiitic basalt; Tb, transitional basalt; Ab, alkali basalt; Ha, hawaiite; Ba, basanite; Ne, nephelinite; Rh, rhyolite; Da, dacite; Ic, icelandite; Mu, mugearite; Be, benmoreite; Tr, trachyte; Ph, phonolite.

The Auckland province is dominated by ne-hawaiite and hawaiite. The relative proportions of rock types differ in each field, but ne-hawaiites predominate overall (Fig. 4.6.1B). Basaltic rock types range from nephelinite to transitional basalt, and basanite is the second-most abundant rock type. There is a lower SiO_2 mode for Auckland compared with Northland province (Fig. 4.6.3). Northland mafic rocks straddle the critical plane of silica undersaturation in the system Ne-Ol-Di-Hy-Qz (Fig. 4.6.5A) whereas Auckland-province samples cluster around the composition $Di_{40}Ol_{48}Ne_{12}$ (Fig. 4.6.5B).

Mafic rocks also predominate in each of the nine provinces of South Island and the Sub-Antarctic (and Chatham) islands, but there are significant differences in SiO_2 mode and in the distribution of rock types (Figs. 4.6.2,4,6). Hawaiite is the dominant rock type in the Canterbury/North Otago, Dunedin, Banks Peninsula, and Campbell Island provinces (Fig. 4.6.2). The hawaiites are hypersthene-normative, or else contain very little normative nepheline. The exception is the Dunedin Volcanic Group in which true ne-hawaiites are abundant and associated with other strongly undersaturated mafic rocks. Nephelinites and basanites are the dominant rock types on the Chatham Islands and in the Alpine Dyke Swarm, and are also volumetrically significant in the Dunedin province. The SiO_2 modes for the mafic rocks of these three provinces

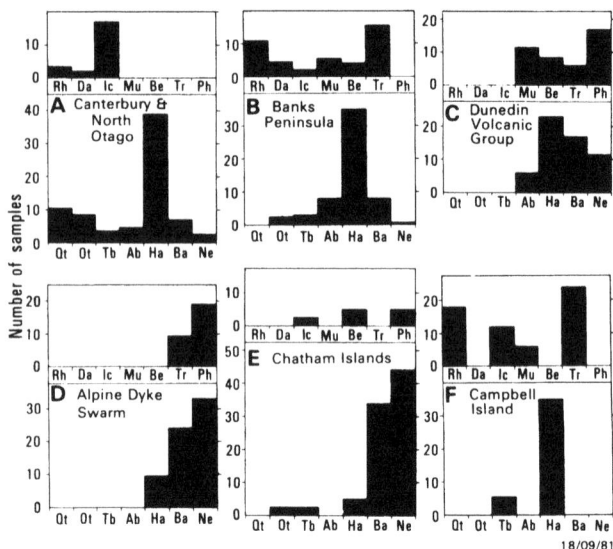

Figure 4.6.2 Histograms of analysed rock types from South Island and the Sub-Antarctic and Chatham islands as a percentage of all the analysed rocks in each province. Abbreviations as in Figure 4.6.1.

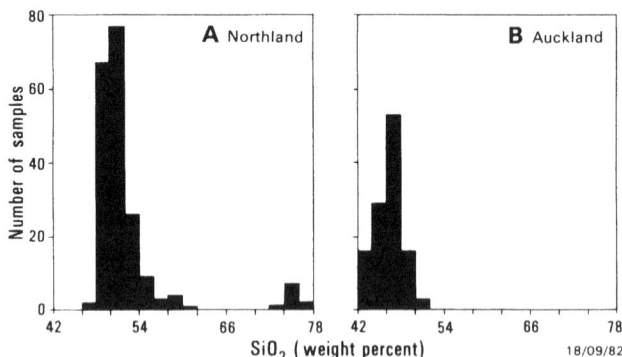

Figure 4.6.3 SiO_2 histograms for rocks from North Island provinces. Analyses recalculated anhydrous and summed to 100 percent.

therefore are distinctly lower than those of Banks Peninsula and Campbell Island (Fig. 4.6.4).

The Canterbury/North Otago province has the greatest range of mafic rock types — from qz-tholeiitic basalt to nephelinite (Fig. 4.6.2). This is not surprising as several different volcanic fields of different ages are included here. Tholeiitic basalts are found mainly in inland Canterbury and in the Timaru-Geraldine lava sheets (Duggan & Reay, 1986; Sewell & Gibson, 1988). The Waiareka-Deborah volcanic rocks of North Otago (Coombs et al., 1986) are a bimodal tholeiitic-nephelinitic suite, and so too are the Cookson Volcanics of northern Canterbury (Coote, 1987). The bimodal nature is illustrated clearly in the combined SiO_2 histogram of the Canterbury and North Otago rocks (Fig. 4.6.4A).

Canterbury/North Otago volcanic rocks define a discontinuous trend from nepheline-rich to quartz-normative compositions (Fig. 4.6.6A). Nephelinites have particularly high normative diopside/olivine ratios compared with Dunedin and Chatham Islands samples. Alkali and transitional basalts are similar to those of

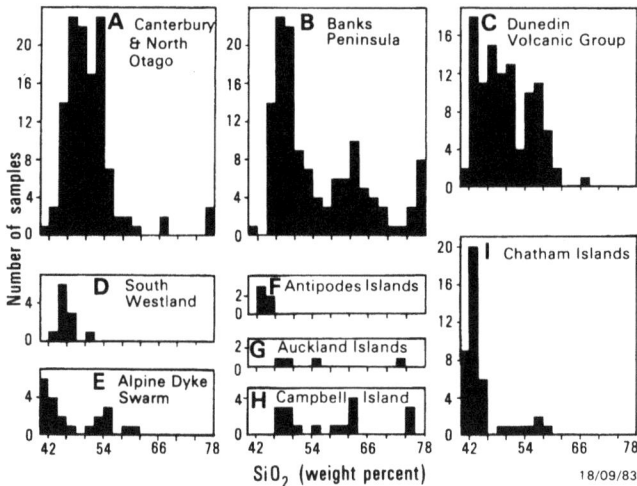

Figure 4.6.4 SiO_2 histograms for rocks from South Island and Sub-Antarctic islands. Analyses recalculated anhydrous and summed to 100 percent.

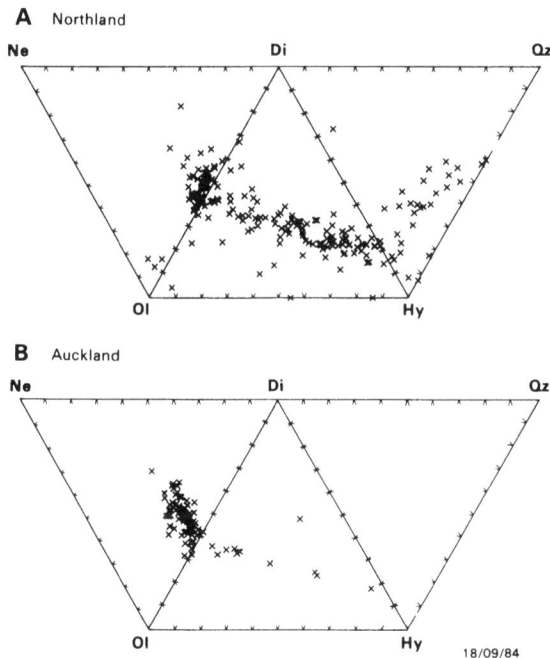

Figure 4.6.5 Normative compositions for rocks from North Island provinces projected in the system Ne-Ol-Di-Hy-Qz.

Banks Peninsula. Particularly noteworthy are the *ol*- and *qz*-tholeiitic basalts straddling the Di-Hy join (Fig. 4.6.6A). *Qz*-tholeiitic basalts appear to be absent from the other South Island and Sub-Antarctic Islands provinces.

Dunedin province mafic rocks are strongly undersaturated and widely dispersed around a composition $Di_{40}Ol_{35}Ne_{25}$ (Fig. 4.6.6C). In contrast, Banks Peninsula data points straddle the critical plane of silica undersaturation (Fig. 4.6.6B). Lyttelton samples plot close to the Di-Ol join and range from alkali basalt to *ol*-tholeiitic basalt. Mount Herbert and Akaroa samples are dominantly alkali basalt to hawaiite, and are scattered around a composition $Di_{40}Ol_{50}Ne_{10}$. Diamond Harbour rocks are mostly basanites, have higher diopside/olivine ratios than do other Banks Peninsula rocks, and are dispersed around a composition

$Di_{50}Ol_{35}Ne_{15}$. There is an overall evolutionary trend on Banks Peninsula towards more undersaturated and more mafic compositions with time.

Campbell Island mafic rocks are mostly hypersthene- and nepheline-normative hawaiites, similar in composition to Banks Peninsula hawaiites. Auckland Islands mafic rocks are probably similar but there is only a small number of analysed samples. Chatham Islands samples are strongly undersaturated and dispersed about a composition $Di_{50}Ol_{30}Ne_{20}$. Antipodes Islands rocks are similar in composition, but few samples have been analysed.

4.6.3 Intermediate and felsic rocks

The Northland province of North Island is a bimodal suite dominated by basaltic compositions but including minor peralkaline and non-peralkaline rhyolites (Fig. 4.6.1A; Ashcroft, 1986a,b; Smith, unpublished data). There is a distinct gap between intermediate rocks associated with the basaltic rocks and ranging up to 61-percent SiO_2, and the rhyolites (72-percent SiO_2 or more; Fig. 4.6.3A). Intermediate and felsic rocks are absent from the Auckland province, except for one *ne*-mugearite.

Northland rhyolites are found in the Bay of Islands field. Fresh samples are Group B pantellerites (Macdonald & Bailey, 1973), but some slightly altered rocks have lost alkalies and are not strictly peralkaline, although they retain general peralkaline petrographic features. The non-peralkaline rhyolites are found in both the Bay of Islands and Puhipuhi fields. These rhyolites are geochemically distinct from older calcalkaline rhyolites in Northland which form part of the Miocene arc-type volcanic suite. They have a close spatial relationship with the intraplate basaltic fields and are thought to be linked to them by crustal fractionation processes, although in the absence of intermediate members of the series this has not been demonstrated.

Intermediate and felsic rocks are volumetrically significant in the central volcanoes of Dunedin, Banks Peninsula, and Campbell Island. There is overall a wide range of felsic compositions, from phonolite and *ne*- and *qz*-trachytes to K-rhyolite and peralkaline rhyolite (Fig. 4.6.7). Compositions of felsic rocks in any one centre reflect those of the associated mafic rocks, and differentiation in most cases can be traced back through intermediate rocks.

Several lineages and fractionation trends have been recognised in Dunedin volcano (Coombs & Wilkinson, 1969; Price & Chappell, 1975; Coombs et al., 1986; Coombs & Reay, 1986). A sodic, moderately nepheline-normative, alkali basalt, hawaiite, mugearite, benmoreite, to trachyte/phonolite lineage is similar to the Hawaiian trend (Macdonald & Katsura, 1964), but a sodic, strongly undersaturated lineage — basanite, *ne*-hawaiite, *ne*-mugearite, *ne*-benmoreite, to phonolite — is dominant. Moderately potassic types in both of these lineages are recognised. One is similar to the Tristan da Cunha series (Baker et al., 1964) and gives rise to 'tristanites' and K-trachytes. The other is a sanidine-basanite to K-phonolite lineage. Dunedin phonolites and *ne*-trachytes are metaluminous to

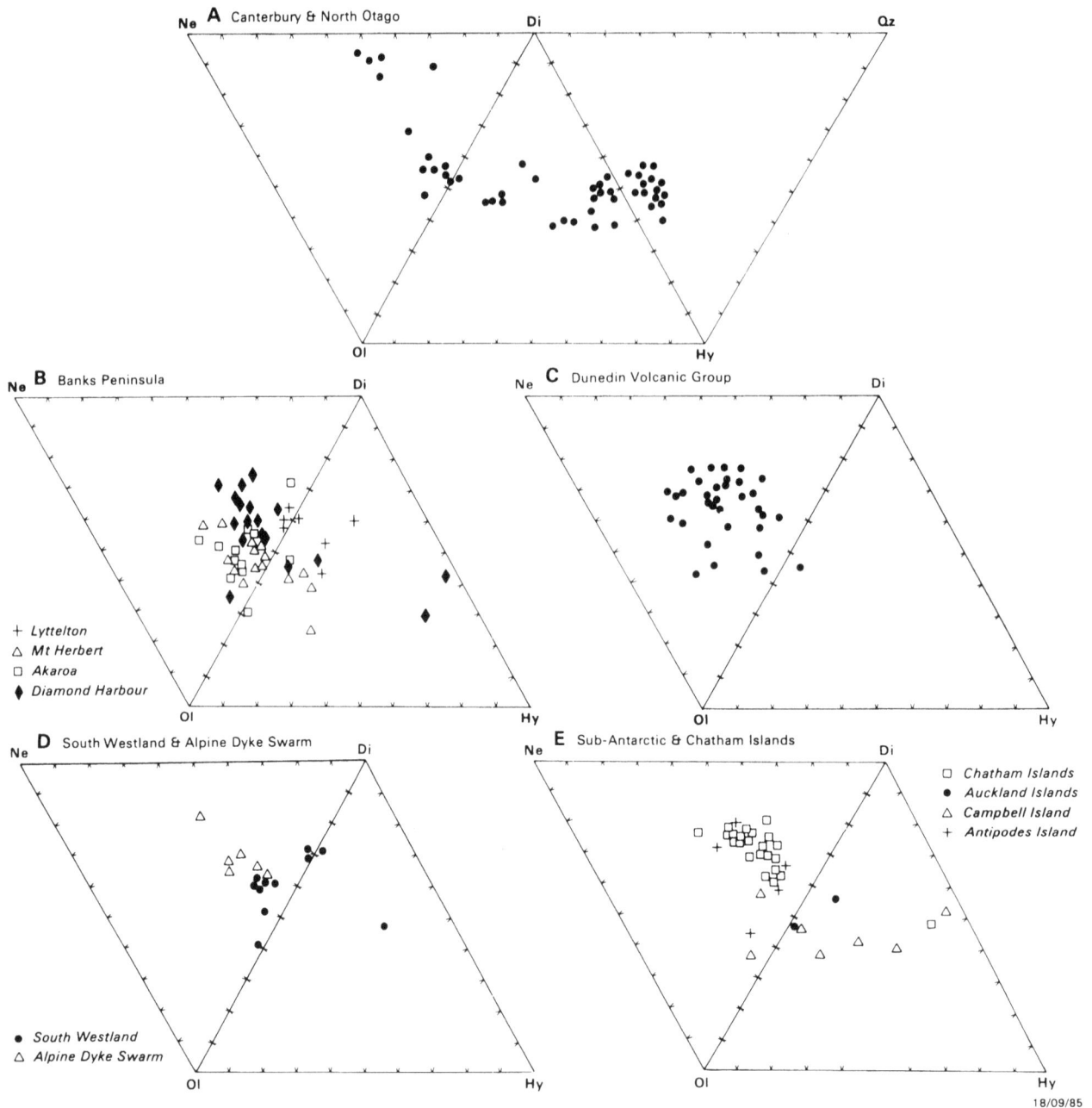

Figure 4.6.6 Normative compositions for mafic rocks (less than 54-percent SiO₂) from South Island and the Sub-Antarctic and Chatham islands projected in the system Ne-Ol-Di-Hy-Qz.

peralkaline in composition. Phonolites of the Alpine Dyke Swarm and rare phonolites on the Chatham Islands are geochemically similar to those of Dunedin volcano.

The main lineage of Lyttelton and Akaroa volcanoes on Banks Peninsula is a sodic series from alkali and transitional basalt to trachyte (Weaver & Sewell, 1986). Intermediate rocks are both hypersthene- and weakly nepheline-normative reflecting the same features in the mafic rocks. Trachytes are both nepheline- and quartz-normative, and metaluminous to weakly per-alkaline. The Governors Bay Volcanics are a sub-alkaline potassic series — icelandite, K-dacite, to peraluminous K-rhyolite. Some Lyttelton rocks appear to be members of this series. Strongly undersaturated intermediate to felsic rocks such as *ne*-benmoreites and

phonolites are not found on Banks Peninsula despite the presence of basanites and rare nephelinites in the Diamond Harbour Group.

Campbell Island volcanic rocks make up a series from transitional basalt, through hypersthene-hawaiite, to metaluminous *qz*-trachyte and peralkaline rhyolite (Morris, 1984). The same series (transitional basalt to trachyte) is present in the Antipodes Islands, using as a basis the sub-standard analyses given by Wright (1970, 1971). However, the Auckland Islands rhyolite is potassic and peraluminous, and is similar to the Governors Bay rhyolites of Banks Peninsula. Similar, high-silica, peraluminous, K-rhyolites are found in the Cookson Volcanics of northern Canterbury (Coote, 1987) together with K-dacites and icelandites, and associated with hypersthene-normative mafic rocks.

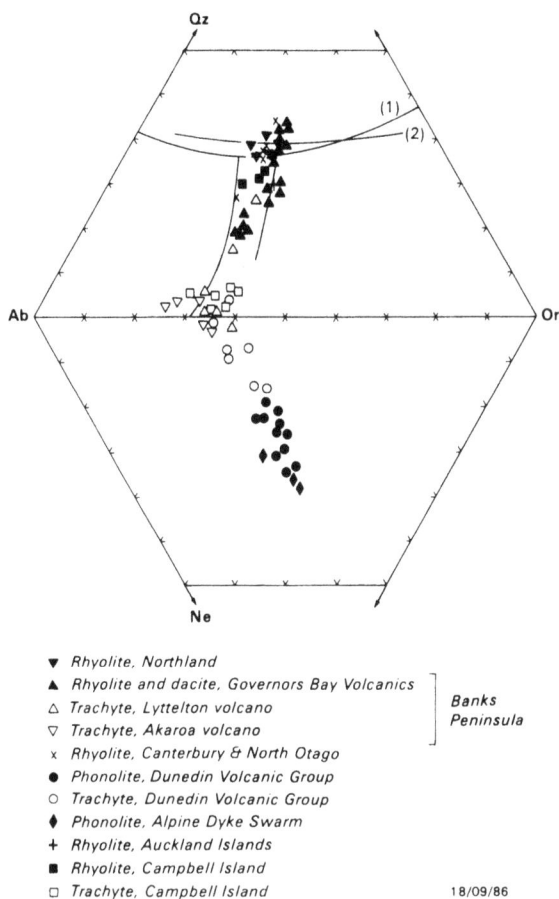

Figure 4.6.7 Normative compositions for New Zealand felsic rocks projected in the system Qz-Ab-Or-Ne. Curve (1) represents quartz-feldspar and two-feldspar boundary curves at 1 kb water-vapour pressure (Tuttle & Bowen, 1958). Curve (2) represents the same curve as in (1) but for the composition $(Ab-Or-Qz)_{97}An_3$ (James & Hamilton, 1969b).

Legend for Figure 4.6.7:
▼ Rhyolite, Northland
▲ Rhyolite and dacite, Governors Bay Volcanics
△ Trachyte, Lyttelton volcano   ⎤ Banks
▽ Trachyte, Akaroa volcano   ⎦ Peninsula
× Rhyolite, Canterbury & North Otago
● Phonolite, Dunedin Volcanic Group
○ Trachyte, Dunedin Volcanic Group
◆ Phonolite, Alpine Dyke Swarm
+ Rhyolite, Auckland Islands
■ Rhyolite, Campbell Island
□ Trachyte, Campbell Island
18/09/86

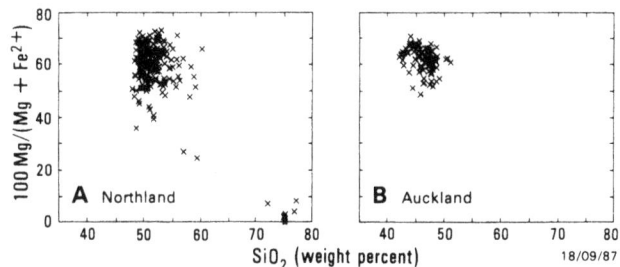

Figure 4.6.8 Mg-ratio:SiO_2 relationships for rocks from North Island provinces.

4.6.4 Major-element chemistry and normative mineralogy

Mg-ratio:SiO_2 relationships

Mg-ratios for Northland rocks correlate negatively with SiO_2 contents, although the plotted data points are scattered (Fig. 4.6.8A). The highest Mg-ratios (greater than 65) are found for transitional basalts, tholeiitic basalts, and hawaiites, which represent potential parental magmas for several petrological lineages. Ashcroft (1986a) evaluated these lineages qualitatively and proposed the existence of hawaiitic and tholeiitic fractionation trends. Mg-ratios for Auckland rocks have a limited range (71–50). The highest Mg-ratios (greater than 65) are found in alkali basalts and basanites (Heming & Barnet, 1986). The Auckland province data plot on the low-SiO_2 side of, and almost completely separate from, the Northland data because of the more undersaturated character of the Auckland rocks (Fig. 4.6.8B).

The familiar negative correlation between Mg-ratio and SiO_2 is found in all the provinces of South Island and the Sub-Antarctic and Chatham islands, but the trends are very scattered (Fig. 4.6.9). Rocks that have

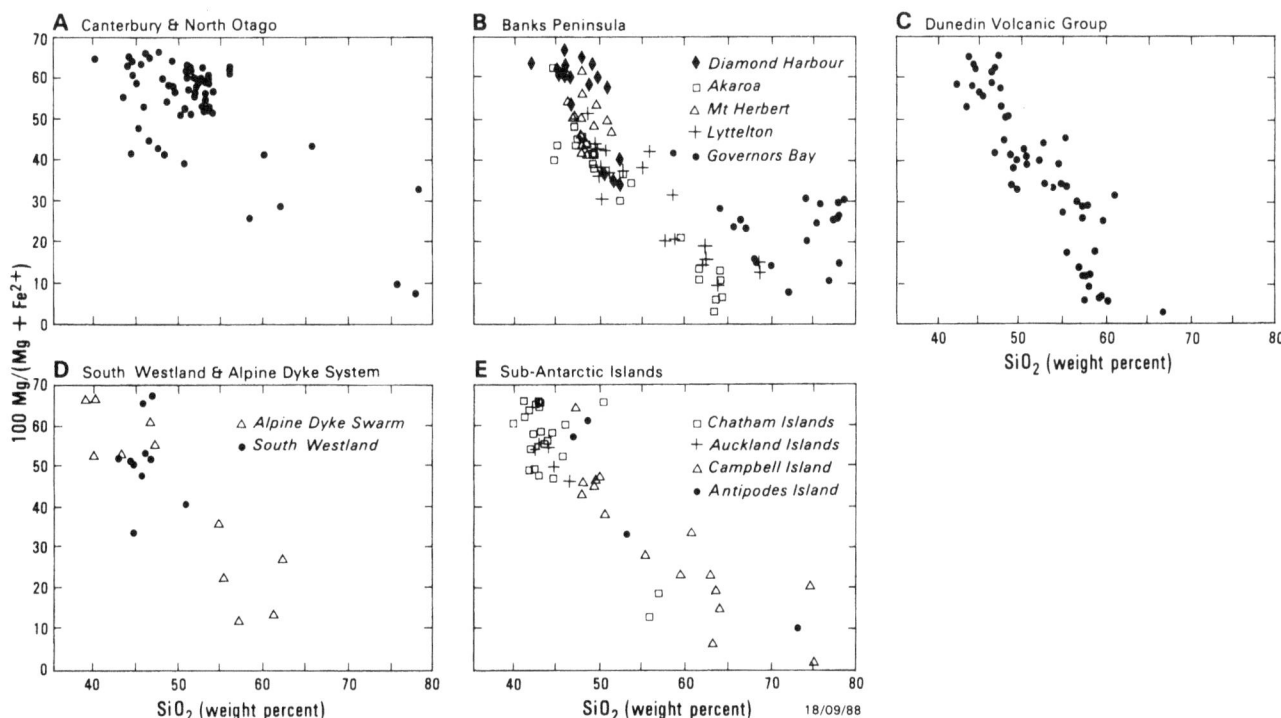

Figure 4.6.9 Mg-ratio:SiO_2 relationships for rocks from South Island and the Sub-Antarctic and Chatham islands.

the highest Mg-ratios (greater than 65) in each province are nephelinites and basanites. The highest values (greater than 60) for the rocks of Banks Peninsula are those of Diamond Harbour basanites. Mount Herbert and Akaroa nepheline-normative mafic lavas have intermediate values, and hypersthene-normative mafic lavas of Lyttelton and Akaroa have Mg-ratios of about 40 (Fig. 4.6.9B). These differences are found at more or less constant SiO_2 content, so Mg-ratios in mafic rocks may correlate with degree of silica undersaturation as well as with SiO_2, at least on Banks Peninsula. However, tholeiitic basalts of Canterbury and North Otago (Fig. 4.6.9A) have high Mg-ratios decreasing to 64 in *ol*-tholeiitic basalts and to 62–55 in *qz*-tholeiitic basalts. The high values (for felsic compositions) and the wide range of Mg-ratios for Governors Bay rhyolites (Fig. 4.6.9B) reflect the exceedingly low levels of Fe in these rocks.

$Na_2O+K_2O:SiO_2$ relationships

Northland and Auckland province rocks have different total-alkali/silica relationships (Fig. 4.6.10). Alkalies in Northland rocks correlate positively with SiO_2, but there is a scatter in the trends corresponding to relatively alkaline and more subalkaline lineages. In contrast, total-alkali contents in Auckland rocks do not correlate with SiO_2 (Fig. 4.6.10B). However, Houghton

et al (1986b) documented for the Crater Hill centre a negative correlation between total-alkali content and SiO_2. This reflects a fractionation trend from alkali-basalt parent magma having a high Mg-ratio, towards basanitic differentiates with lower Mg-ratios.

The strongly alkaline and silica-poor nature of the Dunedin volcanic rocks compared with those of Banks Peninsula is illustrated in Figure 4.6.11. The wide scatter defined by the Dunedin samples accommodates the several lineages referred to above. The subalkaline potassic trend of the Governors Bay Volcanics and the alkaline sodic trend of the other Banks Peninsula groups are separated clearly (Fig. 4.6.11B). Campbell Island samples overlap significantly with those of the Banks Peninsula sodic lineage, whereas rocks of the Alpine Dyke Swarm and the Chatham Islands match those of the Dunedin province. A few samples from Canterbury and North Otago are alkaline, but Figure 4.6.11A is dominated by a subalkaline trend that contrasts sharply with the Dunedin and main Banks Peninsula lineages.

DI:normative-An relationships

The predominance of hawaiites in the Northland province is also illustrated by a plot of Differentiation

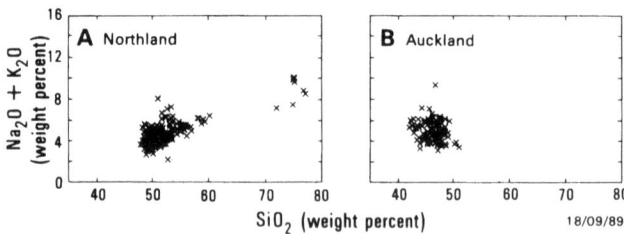

Figure 4.6.10 $Na_2O+K_2O:SiO_2$ relationships for rocks from North Island provinces.

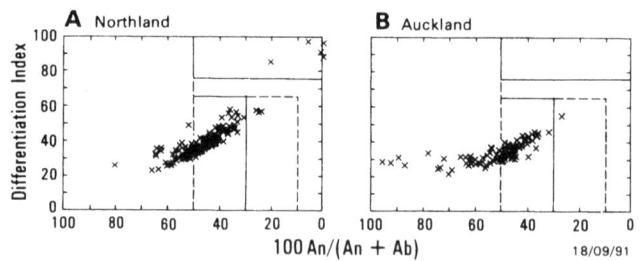

Figure 4.6.12   Differentiation Index versus normative-plagioclase relationships for rocks from North Island provinces. Grid is from the rock-classification scheme given in Figures 1.1.9–10.

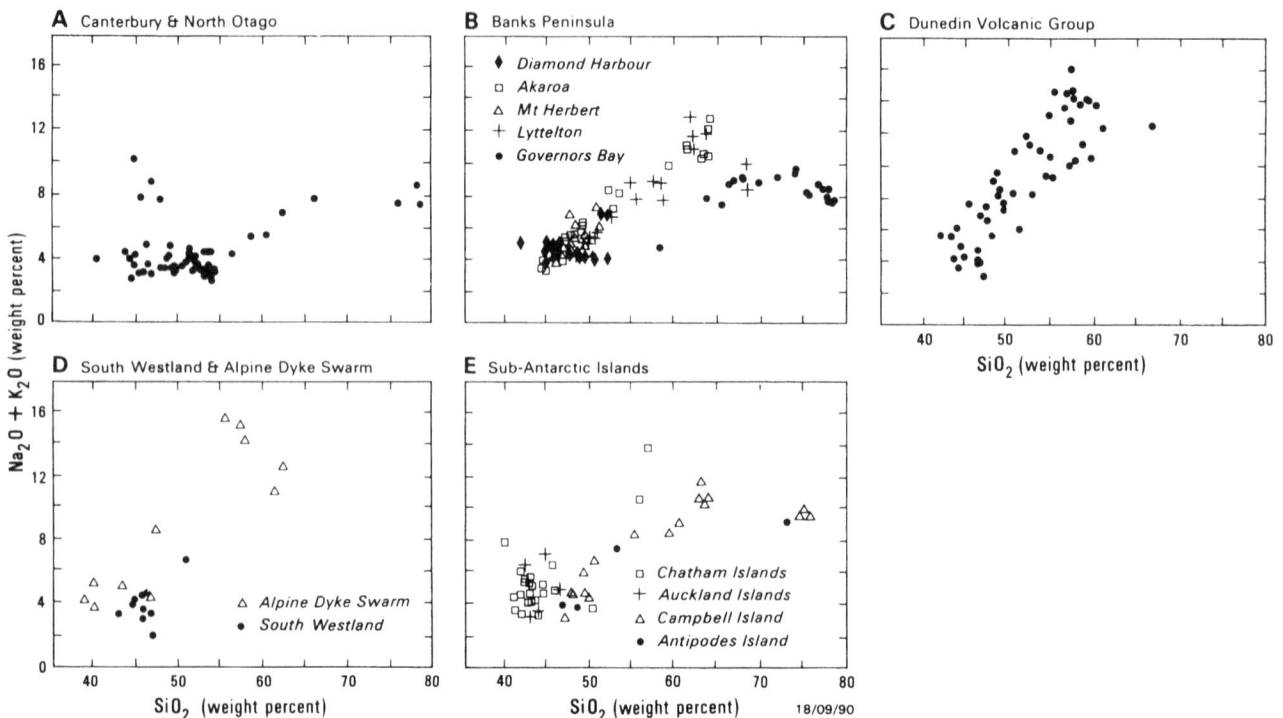

Figure 4.6.11 $Na_2O+K_2O:SiO_2$ relationships for rocks from South Island and the Sub-Antarctic and Chatham islands.

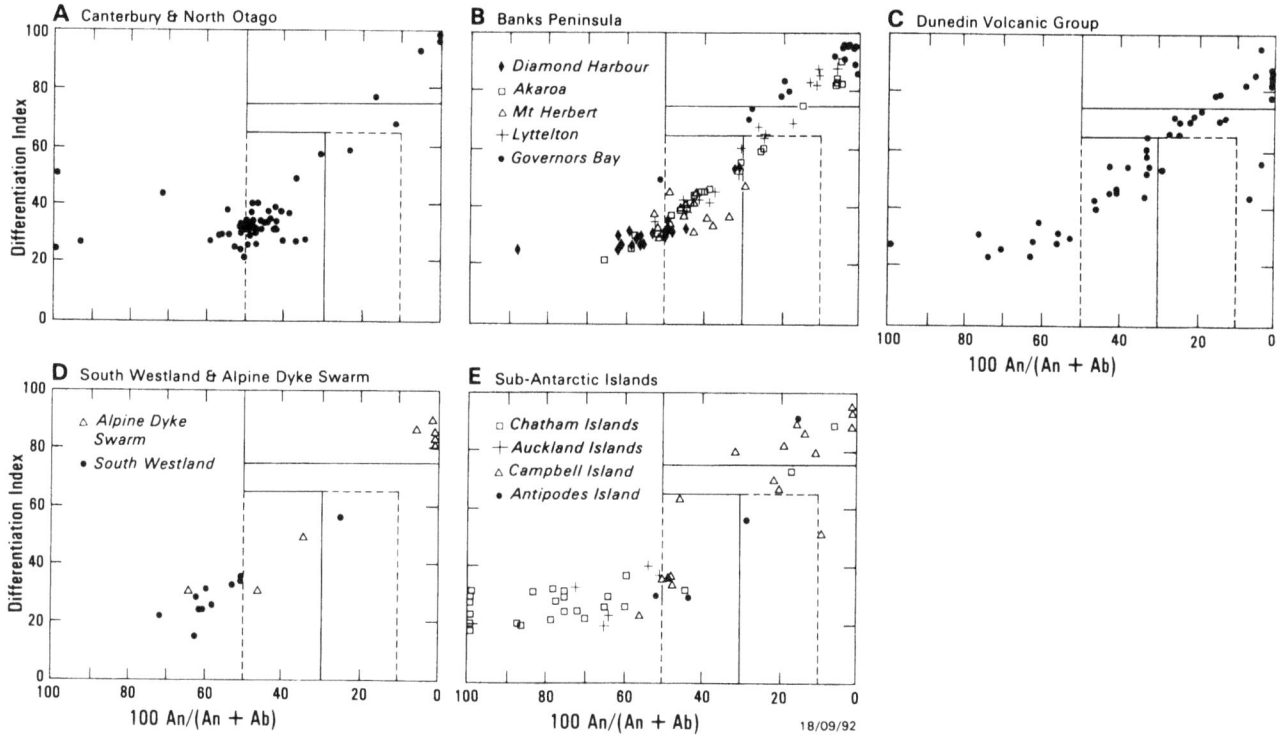

Figure 4.6.13 Differentiation Index versus normative-plagioclase relationships for rocks from South Island and the Sub-Antarctic and Chatham islands. Grid is from the rock-classification scheme given in Figures 1.1.9–10.

Index versus 100An/(An+Ab) values (Fig. 4.6.12A), as well as the continuity of compositions through the mafic range. Felsic rocks plot in a separate field. In contrast, the Auckland province suite lacks the felsic group, but has a greater compositional spread towards nephelinites that have high proportions of normative anorthite (Fig. 4.6.12B).

There is also an abundance of hawaiite in the central volcanoes of the Dunedin and Banks Peninsula provinces (Fig. 4.6.13B,C). Many rocks in the

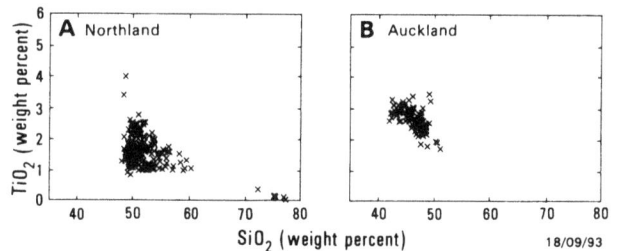

Figure 4.6.14 TiO$_2$:SiO$_2$ relationships for rocks from North Island provinces.

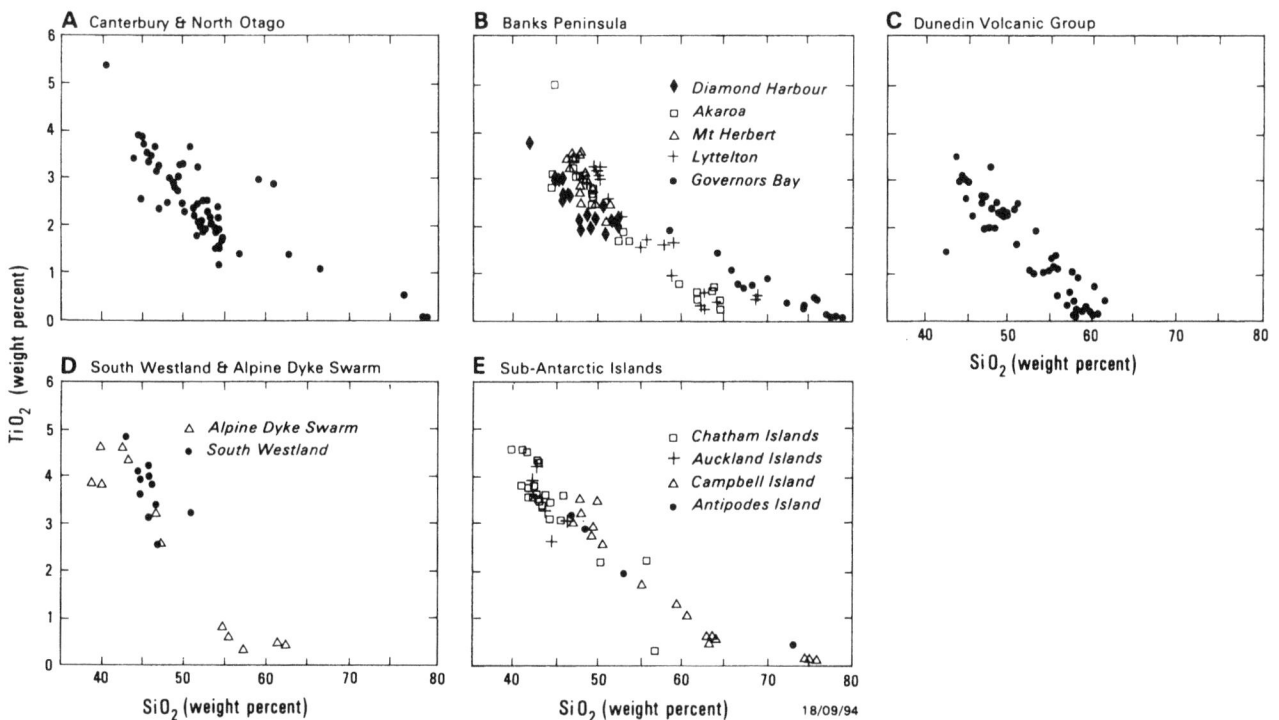

Figure 4.6.15 TiO$_2$:SiO$_2$ relationships for rocks from South Island and the Sub-Antarctic and Chatham islands.

Canterbury/North Otago area are quartz-normative, and include icelandites that plot in the same field as the true hawaiites (Fig. 4.6.13A). Few samples for all the provinces plot in the mugearite field. Benmoreites are well represented at Dunedin, but this is not the case on Banks Peninsula where intermediate compositions are scarce, although there is no well-defined 'Daly Gap' (Chayes, 1977). Two samples fall well off the trend defined by Dunedin rocks (Fig. 4.6.13C). These are mafic 'phonolites' (strictly speaking they are *ne*-benmoreites) of the Pigroot area which contain abundant lherzolite fragments (Price & Green, 1972).

TiO_2:SiO_2 relationships

There are significant differences in the absolute abundances of TiO_2 and TiO_2:SiO_2 relationships for rocks from different areas (Figs. 4.6.14–15). Low-SiO_2 Northland rocks range mainly from 1 to 4 weight-percent TiO_2 (Fig. 4.6.14A). TiO_2 is lower in alkali and transitional basalts and higher in tholeiitic basalts and mafic hawaiites. TiO_2 abundances decrease with fractionation. In contrast, Auckland rocks have higher TiO_2 (typically 2.2–3.3 percent) than Northland rocks, and there is an inverse correlation between TiO_2 and SiO_2 (Fig. 4.6.14B).

There is a strong inverse correlation between TiO_2 and SiO_2 also in all the provinces of South Island and the Sub-Antarctic and Chatham islands (Fig. 4.6.15). TiO_2 values differ in mafic rocks from the different centres of Banks Peninsula, comparing samples that have the same silica content. TiO_2 values in Lyttelton rocks are greater than those in Mount Herbert and Akaroa rocks which, in turn, are greater than those in Diamond Harbour rocks. TiO_2 therefore increases as degree of silica saturation increases, the opposite of the most common trend. These differences appear to have been established during mantle partial melting and may imply heterogeneous mantle sources, whereas the normal TiO_2:SiO_2 inverse relationship originated through the crystal fractionation of titanomagnetite and titaniferous clinopyroxene.

4.7   Trace Element Geochemistry

4.7.1   Mafic rocks

Normalised trace-element diagrams may be used to illustrate the main compositional differences within each of the main provinces (Figs. 4.7.1–2). Some unpublished trace-element data obtained recently for Northland are considered to be representative of the province. Data for Auckland province are from published papers (Rafferty & Heming, 1979; Heming & Barnet, 1986) and an unpublished thesis (Utting, 1986). The only REE data (unpublished) are from the Crater Hill centre. Sources of trace-element data for rocks from South Island and the Sub-Antarctic Islands are as follows: Canterbury — Sewell & Gibson (1988); Banks Peninsula — Price & Taylor (1980), Weaver, Sewell, and Gibson (unpublished data); Dunedin Volcanic Group — Price & Chappell (1975), Price & Taylor (1973); South Westland — Sewell & Nathan (1987); Alpine Dyke Swarm — Cooper (1986); and

Sub-Antarctic Islands — Gamble et al. (1986). Trace-element data for selected New Zealand volcanic rocks are given in Tables 4.2.1, 4.3.1, and 4.4.1.

Trace-element patterns for Northland province (Fig. 4.7.1) are generally similar, but there are some important differences in detail. Three of the patterns have a distinct, negative, Nb anomaly, and are also Ti depleted. Sr does not behave consistently. However, all patterns have light-REE enrichment and Hf depletion.

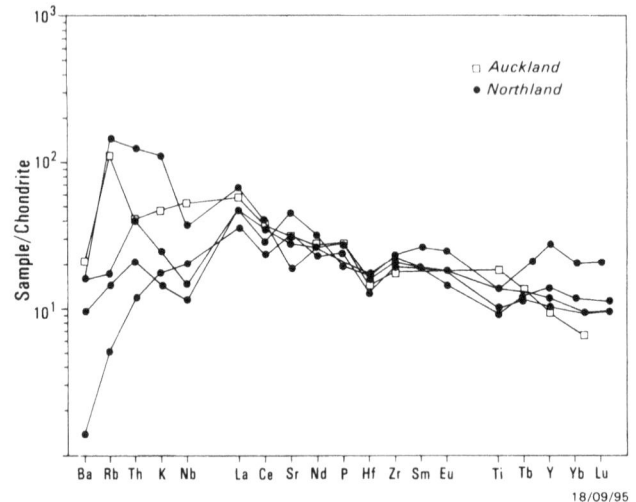

Figure 4.7.1 Normalised trace-element diagrams for selected mafic rocks from North Island provinces. Normalising factors are those of Thompson et al. (1983) for chondrites, excepting Rb, K, and P.

The one available Auckland pattern lacks Nb depletion, but otherwise is generally similar to those from Northland. Negative Nb anomalies characterise subduction-related basalts (for example: Gill, 1981; Thompson et al., 1983) and have been ascribed to the refractory behaviour of a Ti-rich phase, stabilised under hydrous conditions in the mantle wedge, and which retains both Nb and Ta (Saunders et al., 1980) during partial melting. Green & Pearson (1986, 1987) argued that Ti-rich phases are unlikely to be refractory and suggested that island-arc magma source regions must have been depleted in Nb and other high-field-strength elements during a previous melting episode. Whatever their origin, the negative Nb anomalies of the Northland basalts could be interpreted as a subduction-related signature. One Northland mafic rock is remarkably depleted in Ba and Rb (Fig. 4.7.1) and is similar in this respect to N-type MORB compositions. The very large range in concentrations of large-ion-lithophile elements in Northland mafic volcanic rocks is indicative of significant mantle source heterogeneity.

The overall shape of the element-abundance patterns for alkaline rocks from South Island and the Sub-Antarctic Islands are similar (Fig. 4.7.2), and they contrast sharply with the much flatter patterns and lower incompatible-element abundances for the tholeiitic basalts from Canterbury (Fig. 4.7.2A). Most patterns for both alkaline and subalkaline rocks have marked, negative K anomalies. Patterns for the Canterbury, Auckland Islands, Campbell Island, Antipodes Island, South Westland, and the Alpine Dyke Swarm generally have increasing enrichments (relative to mantle) from

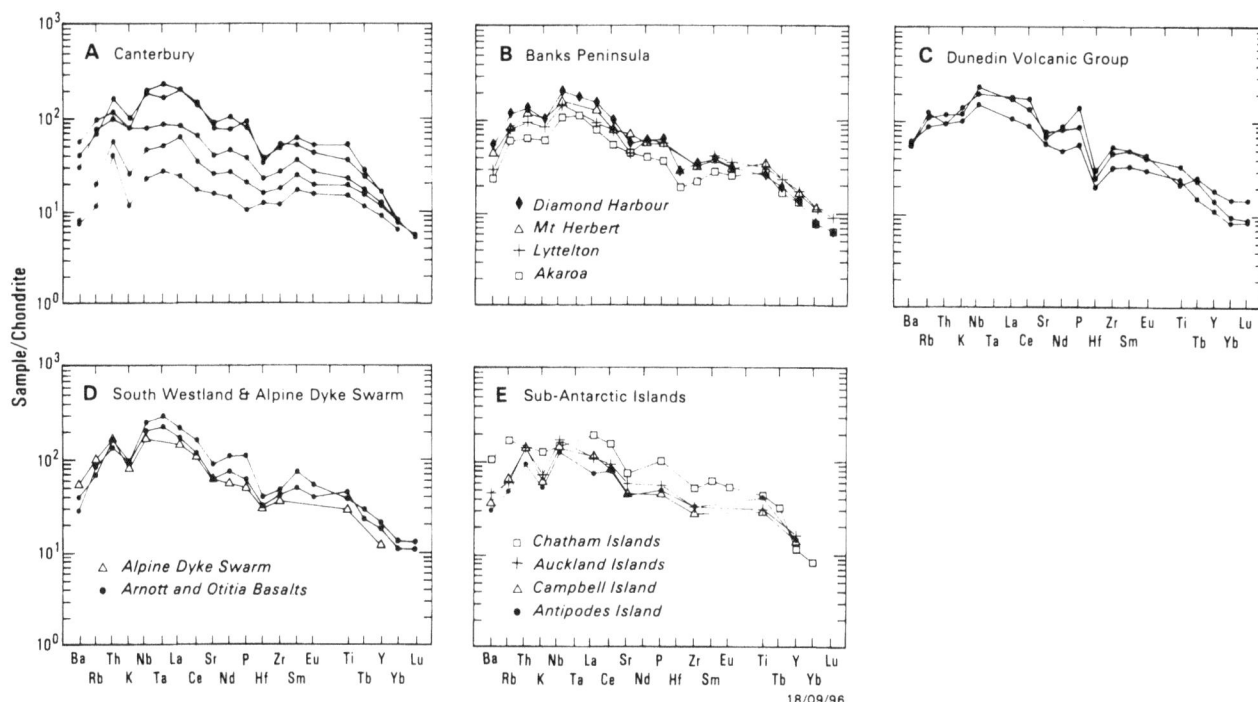

Figure 4.7.2 Normalised trace-element diagrams for selected mafic rocks from South Island and the Sub-Antarctic and Chatham islands. Normalising factors are those of Thompson et al. (1983) for chondrites, excepting Rb, K, and P.

Ba to Nb or Ta, and strong negative K anomalies, whereas moderate, negative K anomalies characterise Banks Peninsula rocks and only weak ones characterise the Dunedin Volcanic Group. All alkaline rocks have decreasing enrichment (relative to mantle) from La to Yb, reflecting control of REE abundances by refractory garnet in their mantle sources.

Strongly undersaturated rocks from Dunedin, Banks Peninsula, Canterbury, and South Westland also have marked negative anomalies in Hf and to a lesser extent Zr. This may be indicative of the refractory behaviour of a Zr-Hf mineral during small degrees of partial melting. However, small, negative Hf and Zr anomalies also can be identified in the subalkaline rocks of Canterbury (Fig. 4.7.2A) which probably formed by significantly larger degrees of melting. This feature, as well as negative K anomalies, therefore may be a regional source characteristic.

Northland and Auckland province rocks have different Zr/Nb patterns (Fig. 4.7.3). Northland rocks have a wide range of ratios (4.4–19.0) and no clear trends, whereas Auckland samples are characterised by Zr/Nb values between 4.0 and 6.2 that are typical of those of ocean-island basalts. However, there are few reliable Nb-abundance data for Auckland-province rocks, and those illustrated in Figure 4.7.3B are exclusively from the Ngatutura field (Utting, 1986).

There is a strong similarity in Zr/Nb versus Differentiation Index relationships between the rocks of Banks Peninsula, Dunedin, Campbell Island, South Westland, and the Alpine Dyke Swarm (Fig. 4.7.4). Zr/Nb values are relatively constant (about 4.0) as degree of differentiation increases, although only limited data are available. This feature may be reflected also in the rocks of the Auckland and Antipodes islands. There is a wide range in Zr/Nb values in samples from the Canterbury lava fields (Fig. 4.7.4A), reflecting the

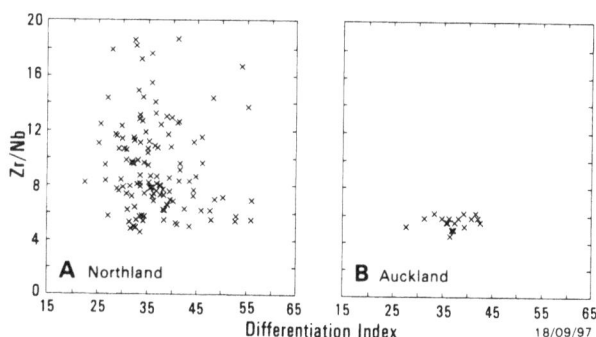

Figure 4.7.3 Zr/Nb versus Differentiation Index relationships for mafic rocks from North Island provinces.

bimodal alkaline/subalkaline nature of the province. Most Chatham Islands samples have Zr/Nb values of about 4.0, typical of alkaline rocks, but several samples have higher ratios (Fig. 4.7.4E) similar to those of the subalkaline rocks of Canterbury. However, these high-Zr/Nb Chatham Islands rocks are basanites, and therefore their high Zr/Nb values are unusual. Significant source heterogeneity is implied (Morris, 1985). There appears to be a subtle but systematic decrease in Banks Peninsula rocks in average Zr/Nb from Lyttelton rocks, through Mount Herbert and Akaroa lavas, to Diamond Harbour samples, which reflects the evolutionary trend towards more undersaturated compositions.

There is a weak, positive correlation between Zr/Y and Differentiation Index for rocks from the Northland and Auckland provinces (Fig. 4.7.5). However, most Auckland analyses plot in a field above that for most Northland data. This is probably a reflection of the relatively high Ti and Zr abundances in the Auckland suite.

Wide ranges are shown in plots of Zr/Y against Differentiation Index for South Island and Sub-Antarctic

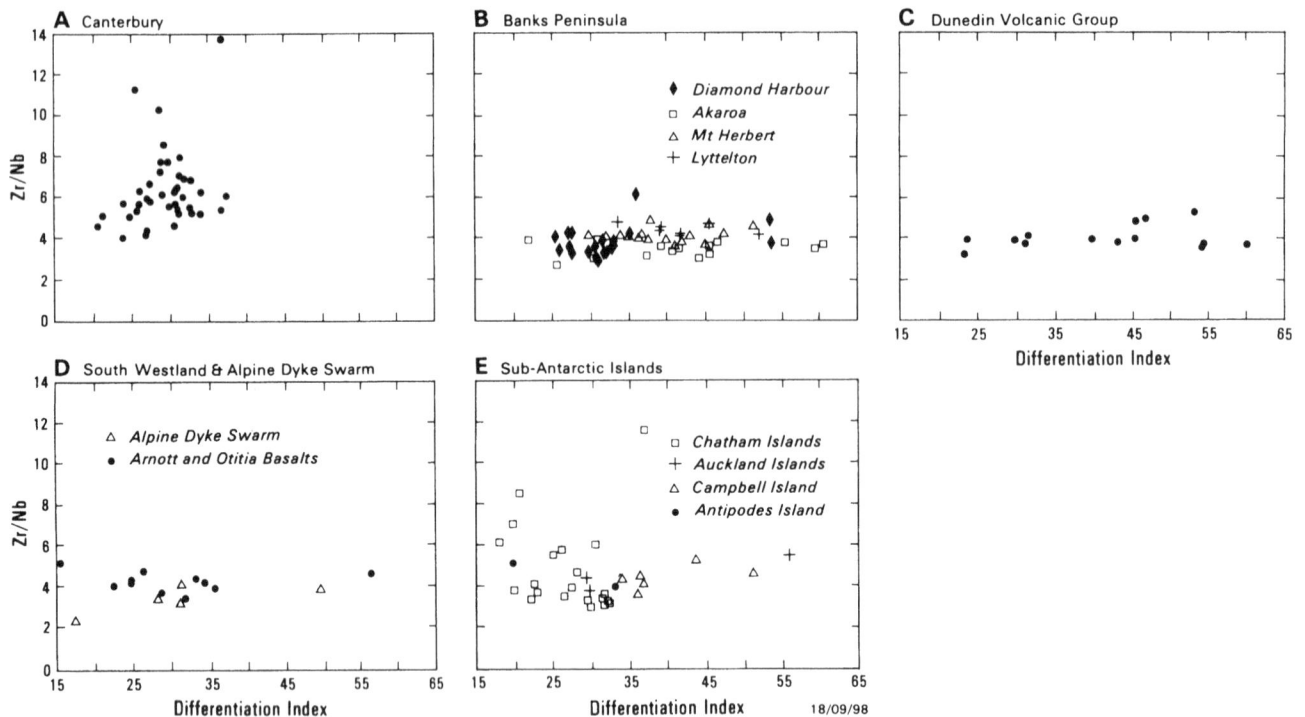

Figure 4.7.4 Zr/Nb versus Differentiation Index relationships for mafic rocks from South Island and the Sub-Antarctic and Chatham islands.

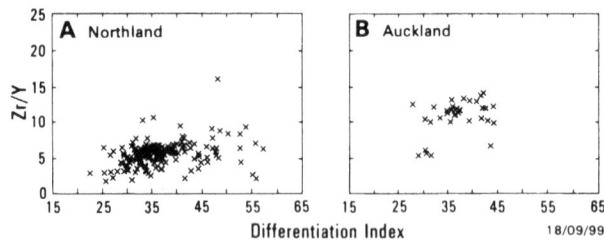

Figure 4.7.5 Zr/Y versus Differentiation Index relationships for mafic rocks from North Island provinces.

Islands rocks (Fig. 4.7.6). Dunedin, Banks Peninsula, and Campbell Island rocks define trends of slightly increasing Zr/Y as degree of differentiation increases. Zr/Y values of Akaroa samples (Banks Peninsula) increase slightly and smoothly with DI, whereas the Mount Herbert lavas define a more rapid increase (Fig. 4.7.6B). The undersaturated lavas of the Diamond Harbour rocks have distinctly higher Zr/Y values than do the transitional basalts of the same group. Strongly undersaturated mafic rocks of Dunedin,

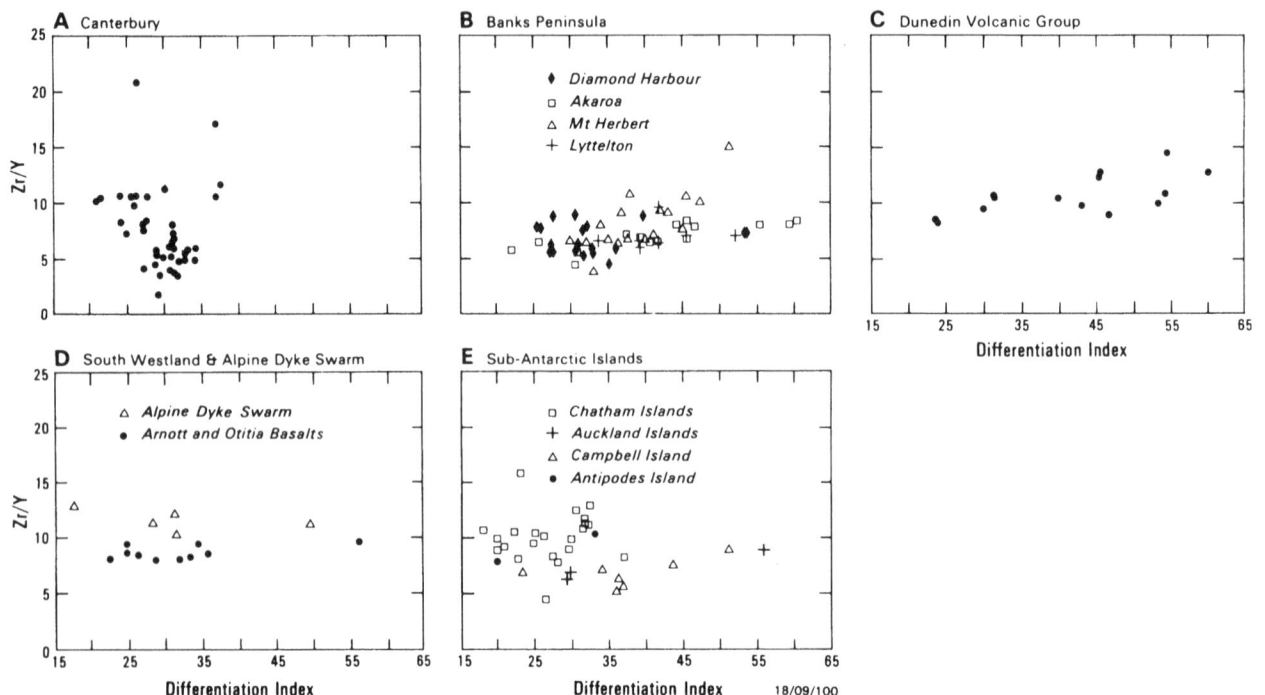

Figure 4.7.6. Zr/Y versus Differentiation Index relationships for mafic rocks from South Island and the Sub-Antarctic and Chatham islands.

the Alpine Dyke Swarm, and the Chatham Islands, have significantly higher ratios than do Banks Peninsula rocks. The bimodal Canterbury province has an extreme range in Zr/Y from 2 to 20, and subalkaline rocks have values of less than 5 (Fig. 4.7.6A). Two controls are apparent. Zr/Y values for the most mafic composition of each province clearly increase as degree of silica undersaturation increases, and evidently are governed by the percentage melting and amount of garnet in the mantle source. The slight but systematic increases of Zr/Y as Differentiation Index increases in mafic rocks of the central-volcano provinces continues through into intermediate and felsic compositions, and reflect lower bulk partition coefficients for Zr compared to Y during high-level fractionation.

4.7.2 Intermediate and felsic rocks

Relationships between felsic and intermediate compositions and the associated mafic rocks may be illustrated by REE patterns and supplementary data for Cr, Rb, Sr, and Zr (Figs. 4.7.7–8). Representative REE patterns for Northland mafic rocks have only moderate light-REE enrichment, typical of many continental subalkaline suites. $[Ce/Yb]_N$ values increase from 2.4 in an alkali basalt to 3.2 in a tholeiitic basalt, to 3.8 in a hawaiite (Fig. 4.7.7). The rhyolite analysis is of a peralkaline sample and has characteristic, strong light-REE enrichment, a flat heavy-REE pattern, and a strong, negative, Eu anomaly.

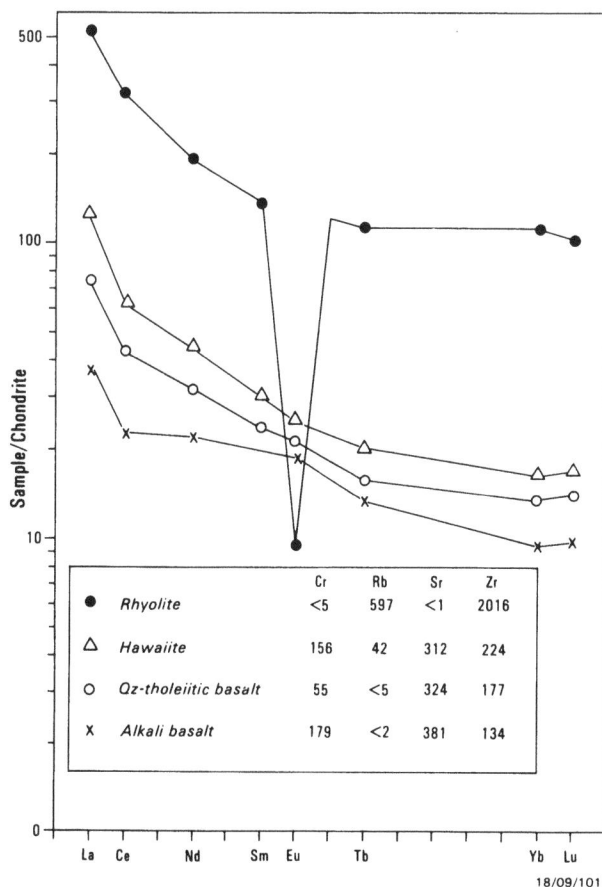

Figure 4.7.7 Chondrite-normalised REE patterns for Northland-province rocks (Smith, Weaver, and Gibson, unpublished data). Normalising factors are those of Nakamura (1974) with some values interpolated.

Banks Peninsula alkaline rocks have moderate to strong light-REE enrichment, as exemplified by Lyttelton lavas (Fig. 4.7.8C). $[Ce/Yb]_N$ values for the sodic, basalt-trachyte, main trend of Lyttelton volcano increase from 7.0 in the basalt to 10.8 in the trachyte which has a marked Eu anomaly (Eu/Eu* 0.30; Fig. 4.7.8C) indicative of prolonged plagioclase fractionation. The preferential removal of the middle trivalent REE apparent in the trachyte pattern is attributed to the crystallisation of kaersutite. Cr is strongly depleted, Sr remains constant before being drastically depleted in the trachyte, and Rb and Zr are concentrated through the series. The trace-element differences are compatible with crystallisation of olivine, clinopyroxene, kaersutite, plagioclase, magnetite, and apatite. Rocks of Akaroa volcano have similar trends.

The series icelandite-dacite-rhyolite of the subalkaline high-K Governors Bay Volcanics appear to be a fractionation sequence in which $[Ce/Yb]_N$ increases from 6.0 to 8.6 and Eu/Eu* decreases from 0.85 to 0.29 (Fig. 4.7.8B). Fractionating minerals include clinopyroxene, orthopyroxene, and plagioclase, and minor olivine, apatite, and magnetite. The high-SiO_2 rhyolites (Fig. 4.7.8C) are characterised by a range of generally low $[Ce/Yb]_N$ values (1.3–6.1) and extreme Eu/Eu* anomalies (0.027–0.07). These rhyolites strongly resemble the Binna Burra rhyolites of Tweed volcano, southeastern Queensland (Ewart et al., 1985; see also, for example, Section 5.4.3). The general patterns of trace-element behaviour in the Governors Bay rhyolites may be ascribed to protracted fractionation of assemblages of sodic plagioclase, sanidine, biotite, quartz, and accessory phases (see also Ewart, 1985). These models are supported by Sr- and Nd-isotopic data (see Section 4.8.2) which preclude derivation of the rhyolites by melting of old continental crustal materials.

Rocks of Dunedin volcano have the strong light-REE enrichment typical of alkaline compositions (Fig. 4.7.8D). The lineage alkali basalt, hawaiite, benmoreite, to phonolite has a progression of REE patterns compatible with fractional crystallisation of olivine, clinopyroxene, kaersutite, plagioclase, alkali feldspar, titanomagnetite, and apatite (Price & Taylor, 1973). $[Ce/Yb]_N$ values decrease from 11.1 in the basalt to 8.6 in the phonolites which have flat heavy-REE patterns. The convergence of the patterns in the middle-REE section is attributed to the crystallisation of kaersutite. Phonolites have large Eu anomalies (Eu/Eu* down to 0.24) and are strongly depleted in Sr, indicative of prolonged plagioclase fractionation. Cr is strongly depleted throughout the series, and Rb and Zr behave incompatibly and become significantly enriched (Fig. 4.7.8D). REE patterns for the basanite-to-benmoreite series are more fractionated, and $[Ce/Yb]_N$ values decrease from 14.6 to 10.1. Incompatible elements (REE, Rb, Zr, and so on) are enriched 1.5 times in basanites compared to alkali basalts. Evolution from basanite to ne-benmoreite evidently was controlled by the crystallisation of olivine, clinopyroxene, kaersutite, apatite, and titanomagnetite (Price & Taylor, 1973). There is a significant range in the composition of phonolites, and derivation from both alkali basalt and basanite magmas is considered probable.

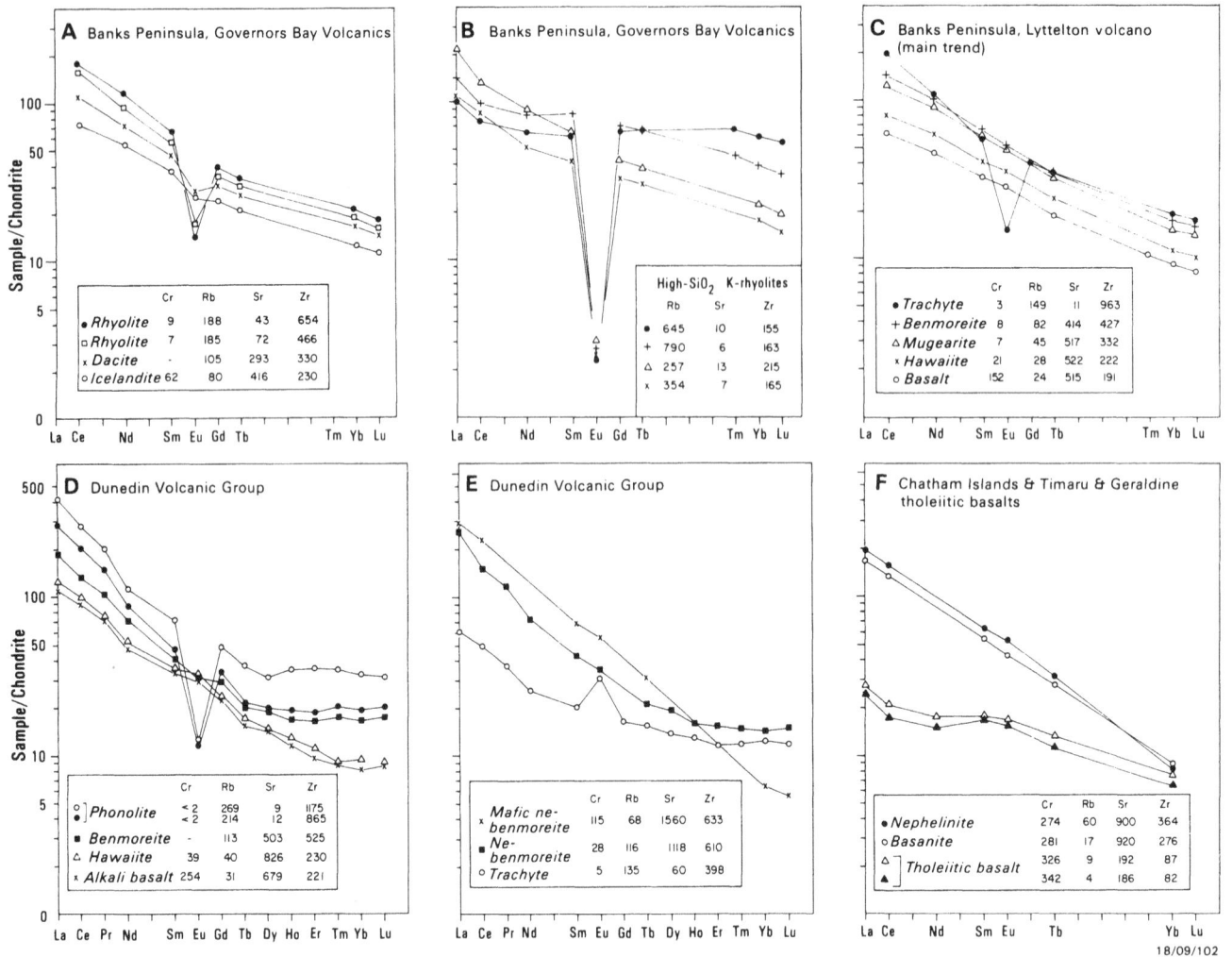

Figure 4.7.8 Chondrite-normalised REE patterns and other trace-element abundance data for rocks from South Island provinces. Data sources: (A-C) Weaver and Gibson (unpublished data); (D) Price & Taylor (1973); (E) Price & Taylor (1973), Irving & Price (1981); (F) Morris (1985), Sewell & Gibson (1988). Nephelinite and basanite in F are from the Chatham Islands, open triangle represents a Geraldine basalt, and filled triangle represents a Timaru basalt. Normalising factors are those of Nakamura (1974).

The earliest felsic rocks of Dunedin volcano are qz-trachytes (Koputai Trachyte) and their different geochemistry may be an indication that they had a different petrogenesis from the other felsic compositions. REE concentrations in a qz-trachyte are significantly lower than in Dunedin mafic rocks. The $[Ce/Yb]_N$ value is very low (4.0) and the sample has a distinct, positive, Eu anomaly (Eu/Eu* 1.86; Fig. 4.7.8E). Price & Compston (1973) stated that the trachytes have higher initial $^{87}Sr/^{86}Sr$ values than do other Dunedin rocks, but initial ratios recalculated by Coombs & Reay (1986) for phonolites and qz-trachytes are the same at about 0.7040. A possible explanation for the origin of the qz-trachytes is partial melting of cumulate alkali-feldspar crystallised from early Dunedin magma that was not erupted (Price & Taylor, 1973). Alternatively, trivalent REE may have been lost from the qz-trachytes during hydrothermal alteration.

The REE pattern for the Pigroot mafic 'phonolite' contrasts sharply with that of a typical Dunedin-volcano ne-benmoreite (Fig. 4.7.8E). The Pigroot rock has a distinctly linear pattern and a $[Ce/Yb]_N$ value of

46, compared to the dished pattern of the Dunedin sample which has a $[Ce/Yb]_N$ value of 10.4 and unfractionated heavy REE at significantly higher concentrations. The Pigroot lava is notable for its numerous lherzolite xenoliths and may be the product of direct melting of a mantle source followed by minor fractionation of olivine and kaersutite (Irving & Price, 1981).

$[Ce/Yb]_N$ values for a nephelinite and basanite from the Chatham Islands are 19.0 and 15.2, respectively (Morris, 1985; Fig. 4.7.8F). These figures and the general concentrations of incompatible elements are higher than for equivalent rock types from Dunedin volcano.

The weak light-REE enrichment ($[Ce/Yb]_N$ 2.7) and low incompatible-element abundances of Canterbury tholeiitic basalts (Fig. 4.7.8F) are typical of continental subalkaline magmas and contrast sharply with the highly fractionated patterns of the alkaline volcanic rocks described above. The high Cr (and high Mg and Ni) concentrations in the Geraldine and Timaru tholeiitic basalts testify to their primitive unfractionated nature.

4.8 Isotope Geochemistry

4.8.1 North Island

Sr- and limited Nd-isotopic compositions for mafic Northland samples are available (for example, Stipp, 1968). Limited Sr-isotope data only are available for the Auckland province (Stipp, 1968). The rocks from both provinces have similar ranges in initial $^{87}Sr/^{86}Sr$ (0.7027–0.7037) and there is a tendency for these to be concentrated at the lower end of the range (Fig. 4.8.1). There is no correlation observed in the available data between $^{87}Sr/^{86}Sr$ value and rock type. These ratios overall are low for intraplate basaltic rocks but this appears to be a feature of all of the New Zealand intraplate provinces (see below). The available Nd-isotope data for Northland province are limited, but ε_{Nd} appears higher (+8.0 to +6.6) compared to values for South Island provinces (see Fig. 4.8.4).

4.8.2 South Island and Sub-Antarctic and Chatham islands

Sr- and Nd-isotopic compositions are available for Banks Peninsula (Weaver and Pankhurst, unpublished data), Dunedin volcano (for example: Price & Compston, 1973; Coombs et al., 1986; Coombs &

Reay, 1986), and the Alpine Dyke Swarm (Barreiro & Cooper, 1987; Cooper, 1986). In addition, Coote (1987) reported Sr-isotopic results for the Cookson Volcanics, Canterbury. Some ratios were given also for Sub-Antarctic Islands samples by Gamble et al. (1986) and McDonough et al. (1986), and additional data are available for the Auckland Islands.

Mafic rocks of Dunedin volcano have very low initial $^{87}Sr/^{86}Sr$ values, mostly in the range 0.7027–0.7029, and all are less than 0.7032 (Fig. 4.8.2). Intermediate rocks in general have higher ratios of 0.7029–0.7034, and felsic rocks still higher ones of 0.7035–0.7043. This relationship is repeated with similar values for the Alpine Dyke Swarm. Cooper (1986) quoted values for mafic rocks in the range 0.7027–0.7035 (the value for one lamprophyre is 0.7046), and two trachytes have values of 0.7052–0.7064.

$^{87}Sr/^{86}Sr$ values for the bimodal alkaline-subalkaline Cookson Volcanics of northern Canterbury (Coote, 1987) are 0.7029–0.7033. Most samples, including both nephelinites and ol-tholeiitic basalts, have ratios of less than 0.7030. Mafic rocks of the Auckland Islands have $^{87}Sr/^{86}Sr$ values in the range 0.7029 to 0.7043 (Fig. 4.8.2). Felsic rocks have significantly higher values of 0.7059 in a trachyte and 0.7082 and 0.7093 in two rhyolites. Few data are available for the Antipodes Islands, Campbell Island, and Chatham Islands.

Initial Sr-isotopic ratios for Banks Peninsula rocks are slightly higher than those of Dunedin and the Alpine Dyke Swarm. All Banks Peninsula samples have $^{87}Sr/^{86}Sr$ values greater than 0.7029 (Fig. 4.8.2). Mafic rocks of Lyttelton and Akaroa volcanoes have ratios in the range 0.7029–0.7032, whereas Diamond Harbour and Mount Herbert samples range up to 0.7038. Average ratios for the different rock groups of Banks Peninsula are higher in felsic rocks compared to

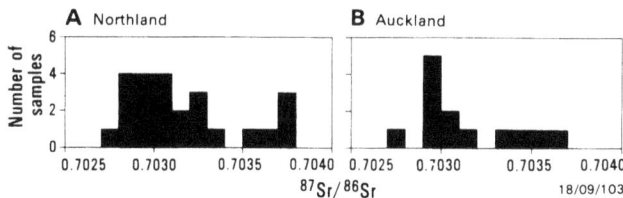

Figure 4.8.1. Histograms of initial $^{87}Sr/^{86}Sr$ values for North Island provinces. Data sources are Stipp (1968), and Smith, Weaver, and Pankhurst (unpublished data).

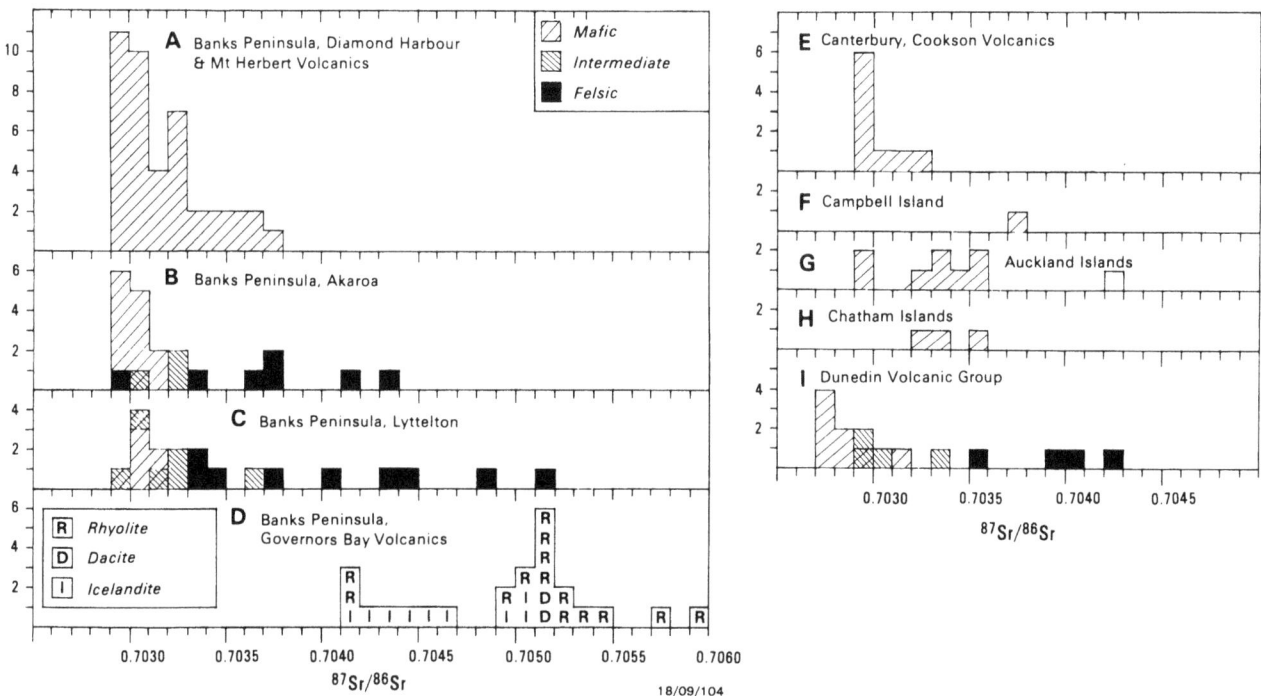

Figure 4.8.2 Histograms of initial $^{87}Sr/^{86}Sr$ values for South Island and the Sub-Antarctic and Chatham islands provinces. Data sources are Price & Compston (1973), Sewell (1985), Gamble et al. (1986), Coombs & Reay (1986), Coote (1987), and Weaver, Gamble, and Pankhurst (unpublished data).

intermediate ones which, in turn, are higher than in the mafic rocks. Akaroa and Lyttelton trachytes have $^{87}Sr/^{86}Sr$ values up to 0.7052 but, significantly, one Akaroa trachyte has a ratio of less than 0.7030 which is the same as for the associated mafic rocks. Sr-isotope ratios for the Governors Bay Volcanics of Banks Peninsula are significantly more radiogenic than are the alkaline suites, ranging from 0.7041 to 0.7060 (Fig. 4.8.2). In general, ratios in the rhyolites are higher than in the dacites which have values greater than those of the icelandites. However, two rhyolite samples have Sr-isotope ratios that match those of the least radiogenic icelandites.

The mafic rocks in the region as a whole — at least the alkaline and transitional suites — are characterised by relatively low initial $^{87}Sr/^{86}Sr$ values (generally less than 0.7032). Most mantle-derived magmas are not contaminated significantly by continental crust. In contrast, most felsic rocks have significantly higher initial $^{87}Sr/^{86}Sr$ values, reflecting the susceptibility of these low-Sr magmas to upper-crustal contamination. The relationship between the subalkaline Governors Bay icelandites and the overlying transitional Lyttelton mafic rocks is unclear, but possibly the former represent initial, mantle-derived, Lyttelton-type magmas that have undergone contamination as they reamed their way through continental crust. Cretaceous high-SiO_2 rhyolites on Banks Peninsula and in inland Canterbury are considered to have been generated by partial melting of Torlesse metasedimentary protolith (Barley, 1987; Barley et al., 1988), but the isotopic ratios of the Governors Bay rhyolites are less radiogenic than those of the metasediments. The isotopic data are indicative that these rhyolite magmas were derived by fractionation of associated intermediate compositions, even though there is isotopic evidence of significant crustal contamination.

Combined assimilation/fractional-crystallisation (AFC) is considered to be a likely petrogenetic process for all the mafic-felsic volcanic suites. This conclusion is supported by the $\varepsilon_{Nd}-^{87}Sr/^{86}Sr$ covariation shown in Figure 4.8.3. Felsic rocks, including the Lyttelton and Akaroa trachytes, and Governors Bay rhyolites, trend away from the mantle fan in a direction consistent with crustal contamination. Auckland Islands rocks define a similar trend from mafic through intermediate compositions to a high-SiO_2 K-rhyolite which has an initial $^{87}Sr/^{86}Sr$ value and ε_{Nd} value of 0.7093 and −1.4, respectively. These data are consistent with the interpretation that the rhyolite represents a minimum melt of a continental crustal source.

Banks Peninsula and Dunedin mafic magmas in Figure 4.8.3 plot within the mantle fan in the quadrant characterised by positive ε_{Nd} values and initial $^{87}Sr/^{86}Sr$ values less than that of Bulk Earth. This corresponds to a mantle source that has undergone a time-integrated depletion in large-ion-lithophile elements and light REE. The range for the Diamond Harbour basalts is consistent with a limited degree of mantle-source isotopic heterogeneity (Sewell, 1985), although other Banks Peninsula mafic rocks are isotopically rather similar. Dunedin basanites and the Pigroot ne-benmoreite plot just outside the Banks Peninsula envelope, but more data are required to confirm that the two

source regions are isotopically distinct. Sr- and Nd-isotopic analyses of mafic rocks quoted by McDonough et al. (1986) for the Chatham Islands (0.7033 and +4.0) and Antipodes Island (0.7029 and +4.5) all plot within the Banks Peninsula $\varepsilon_{Nd}-^{87}Sr/^{86}Sr$ envelope for mafic rocks.

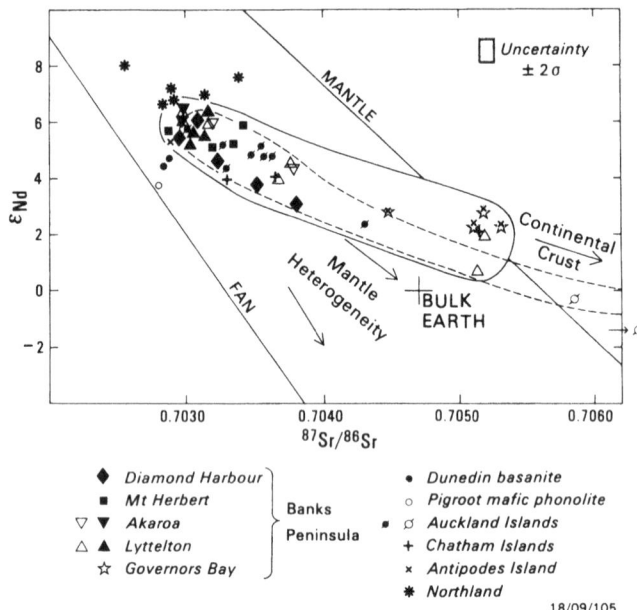

Figure 4.8.3 Nd- (ε_{Nd} units) and initial Sr-isotopic relationships for New Zealand intraplate volcanic rocks. Closed symbols represent mafic rocks and open symbols intermediate to felsic ones. The boundaries of the mantle fan are taken from Hawkesworth et al. (1987). Data sources are Coombs et al. (1986), McDonough et al. (1986), and Weaver, Pankhurst, Gamble, and Smith (unpublished data). Felsic rocks of the Governors Bay and Lyttelton groups extend outside the array, and are indicative of significant interaction with continental crust (solid field boundary). Samples of the Auckland Islands extend to even more extreme compositions (dashed field boundary).

4.8.3   Discussion

All Cainozoic, intraplate, mafic volcanic rocks in New Zealand have time-integrated, light-REE depleted and low Rb/Sr mantle source characteristics (Fig. 4.8.4). The Dunedin and Alpine Dyke Swarm fields significantly overlap the field for St Helena ocean-island basalts. Banks Peninsula data have greater overlap with the continental Marie Byrd Land field and the oceanic Hawaiian field. The mantle source for Northland basalts has among the highest ε_{Nd} and lowest $^{87}Sr/^{86}Sr$ values for continental alkaline magmas, and its composition overlaps with that for N-type MORB sources. Sources for most east-Australian basalts are characterised by less time-integrated depletion in large-ion-lithophile elements and light REE than are New Zealand sources.

The similarities in Sr- and Nd-isotopic composition between New Zealand mafic magmas and those of oceanic intraplate volcanic suites (St Helena, Hawaii) are evidence for the involvement of asthenospheric mantle, but do not support the proposed existence of the Dupal isotopic anomaly (Hart, 1984) in the mantle beneath the region. The light-REE enriched composition of New Zealand mafic magmas and their depleted isotopic signatures correspond to mantle source regions

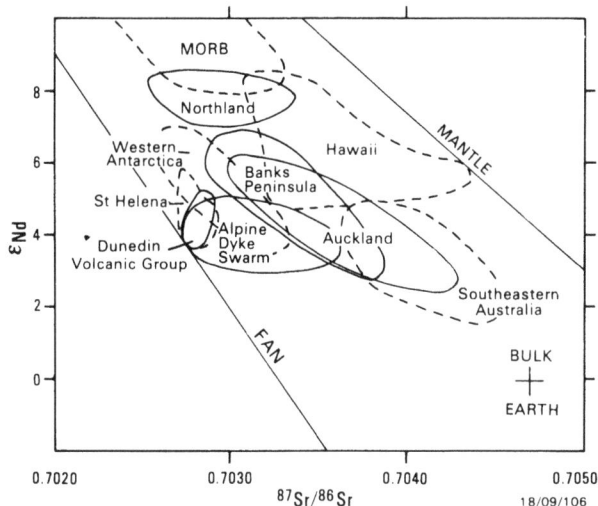

Figure 4.8.4 Generalised Nd- and Sr-isotopic compositional fields for New Zealand intraplate and other volcanic areas. New Zealand data sources are Coombs et al. (1986), Cooper (1986), Barreiro & Cooper (1987), and Smith and Weaver (unpublished data). Other data sources are: Western Antarctica — Futa & LeMasurier (1983); St Helena — White & Hofmann (1982), Cohen & O'Nions (1982a); southeastern Australia — McDonough et al. (1985); Hawaii and MORB — Hawkesworth et al. (1987).

that have undergone relatively recent episodes of metasomatism and incompatible-element enrichment.

4.9   Petrological Overview and Tectonic Relationships

4.9.1   Crust and mantle influences

There is a predominance of volcanic rocks of mafic composition in all New Zealand Cainozoic intraplate provinces, but hawaiites rather than basalts, are volumetrically most abundant in most, especially in the central volcanoes. This is probably indicative of fractionation of more primitive basaltic magmas during their rise through continental crust and during residence in magma chambers within the crust. The absence of mantle-derived xenoliths from the central-volcano provinces and their restriction to small monogenetic vents of strongly alkaline composition, are evidence in support of this conclusion.

The isotopic and trace-element compositions of New Zealand Cainozoic alkaline mantle-derived magmas are indicative that crustal contamination has been minimal. However, this is not generally the case for intermediate and felsic compositions which have elevated initial $^{87}Sr/^{86}Sr$ values and for which combined assimilation-fractional crystallisation (AFC) models of evolution are appropriate. The Auckland Islands, Campbell Island, and South Westland province are west of the Median Tectonic Line within the western tectonic province of New Zealand (see Section 1.3.3), whereas the petrological provinces of the North Island, South Island east of the Alpine Fault, Chatham Islands, and Antipodes Islands, are east of the line. However, no geochemical signatures indicative of different crustal compositions have been identified yet in the Cainozoic volcanic rocks.

New Zealand intraplate mafic rocks are characterised by low initial $^{87}Sr/^{86}Sr$ values and cannot have been derived from mantle with Dupal-type chemical characteristics (see Section 7.7.6). Differences in isotopic composition, exemplified by Figure 4.8.4, and incompatible-element characteristics, such as the magnitude of K anomalies (Fig. 4.7.1–2), are evidence for mantle heterogeneity of limited scale beneath the New Zealand region.

4.9.2   Tectonic framework of North Island

Northland basalts have Sr- and Nd-isotopic ratios similar to those of N-type MORB (Le Roex, 1987). High Zr/Nb values, low Zr/Y values, and low concentrations of large-ion-lithophile elements in some Northland samples are consistent with this observation. However, the negative Nb anomalies of Northland basalts and the different enrichments in large-ion-lithophile elements (Fig. 4.7.1) may represent a subduction-related signature. This would be consistent with the position of the Northland region adjacent to the active plate margin during the Miocene (Cole, 1986). Auckland province mafic rocks have typical ocean-island basalt chemical characteristics and are more than 400 km from the present plate boundary. Both Northland and Auckland basalts were erupted during late Pliocene to Holocene times in stable or mildly extensional tectonic settings, outside the zone of compressional deformation associated with the active plate boundary (Walcott, 1984b).

4.9.3   Tectonic framework of South Island and Sub-Antarctic and Chatham islands.

The Cainozoic intraplate mafic rocks of the South Island and Sub-Antarctic Islands are of ocean-island basalt composition which, especially in continental settings, are associated typically with extensional tectonic regimes. The early to middle Cretaceous compressional deformation of the late phase of the Rangitata Orogeny (Bradshaw et al., 1981) was followed in middle Cretaceous to late Oligocene times by extensional tectonism associated with the opening of the Tasman Sea and the separation of New Zealand from the Gondwana supercontinent (about 80–50 Ma). The Arnott Basalt, South Westland, of late Cretaceous to Palaeocene age was erupted close to the rifted margin of the New Zealand continental block during the final stages of Tasman Sea opening. Palaeocene to early Eocene volcanic rocks of both alkaline and subalkaline composition in Canterbury, Marlborough, and offshore northern Otago are poorly documented, but were erupted during tectonically stable or mildly extensional conditions. Extensional tectonics giving rise to graben formation and basin development prevailed during middle Tertiary times, and late Eocene to early Oligocene mafic volcanism of strongly alkaline to subalkaline composition was widespread and voluminous in Canterbury, North Otago, Marlborough, South Westland, and the Chatham Islands.

The change during the middle Tertiary from extensional to compressional tectonics corresponds to the propagation of the present-day plate boundary through southern continental New Zealand. Cooper et al. (1987) showed that magma intrusion during the late Oligocene to early Miocene of the Alpine Dyke Swarm was into tensional fractures and Riedel shears formed during initiation of the dextral transcurrent movement of the Alpine Fault.

Major volcanic centres of alkaline to subalkaline composition were active in middle Miocene times at Banks Peninsula, Dunedin, Auckland Islands, and Campbell Island. The central volcanoes of Dunedin and Banks Peninsula were on the margin of the zone of compressional deformation. Compressional tectonics had become well established by late Miocene times in Canterbury and Otago. Reverse faults in eastern and central Otago postdate the volcanism and represent reactivated normal faults of late Cretaceous age (Coombs et al., 1986). Eruptions continued at the Banks Peninsula volcanic complex (which is structurally undeformed) until about 6 Ma ago. In a general sense, therefore, development of the compressional tectonic regime may have 'turned off' the alkaline intraplate volcanism of southern New Zealand. Exceptions to this are the Pliocene volcanism of the Chatham Islands and the Quaternary activity of the Antipodes Islands. These centres are near the eastern margin of the Campbell Plateau, well outside the zone of compression, and where extensional tectonism may have persisted.

Both Adams (1981) and Farrar & Dixon (1984) noted the eastward younging of late Cainozoic alkaline volcanism across South Island, Campbell Plateau, and Chatham Rise. They suggested that this pattern was generated by the overriding by continental lithosphere of a spreading ridge (see also Section 1.7.3). A general eastward younging of Miocene to Holocene volcanism is apparent, but not for early Tertiary rocks, and there

is little evidence in favour of the recognition of long hotspot trails such as the one recording the northward drift of eastern Australia from Antarctica (Wellman & McDougall, 1974a; Section 1.7). There has been a migration of the focus of activity in Otago towards the southeast from the Alpine Dyke Swarm (28–16 Ma), to the peripheral vents of the Dunedin volcanic complex (20–10 Ma), to Dunedin volcano (13–10 Ma), at a rate of about 15–20 mm per year. There is also a migration of volcanism in Banks Peninsula towards the southeast from Lyttelton (11–10 Ma) to Mount Herbert (9.7–8.0 Ma), to Akaroa (9.0–8.0 ma) — a rate of about 12 mm per year, but Diamond Harbour activity (7.0–5.8 Ma) shifted northwestwards, back to the Lyttelton and Mount Herbert centres. The above rates are low compared with the 30–50 mm per year rates implied for the Otago and Canterbury regions using the model of Farrar & Dixon (1984).

The youngest intraplate volcanism of the South Island is the late Miocene subalkaline activity of the Timaru and Geraldine areas. These lavas are similar in composition to some early Tertiary subalkaline lavas of inland Canterbury and have been likened to P-type MORB (Sewell & Gibson, 1988). The Canterbury lavas also have the negative K anomalies characteristic of most alkaline mafic rocks of southern New Zealand and presumed to be a regional source characteristic (Section 4.7.1). The location of Timaru and Geraldine in the South Canterbury Bight is in line with the southern margin of the Chatham Rise, near the western edge of the Bounty Trough. The continental crust may be as thin as 15 km beneath the Bounty Trough, and the volcanism may be associated with continued crustal thinning in this region (Sewell & Gibson, 1988), although the Bounty Trough originated in the Cretaceous.

CHAPTER 5

East Australian Petrology and Geochemistry

5.1 Introduction

An attempt is made in this chapter to present an overview of the petrology, mineralogy, and trace and isotope geochemistry of the Cainozoic volcanic provinces of eastern Australia. The account is based on data from published papers, unpublished theses, and other unpublished sources. Much of the data is subdivided for comparative purposes according to state boundaries, except for the following: (1) northern Queensland, which by virtue of its different volcanic style is distinguishable from the volcanic provinces of southern and central Queensland; (2) the northeastern corner of New South Wales which is included in the southern- and central-Queensland division, as this region contains the large Tweed volcano that sits astride the state border. A further subdivision of volcanic provinces into the lava-field, central-volcano, and leucitite-suite categories of Wellman & McDougall (1974a) is used also (see Section 1.1.3).

Distinctions within many areas between the lava-field and central-volcano provinces are well defined. Good examples are the lava fields of western Victoria (Newer Volcanics) and northern Queensland and the large central volcanic complexes of Warrumbungle, Canobolas, Nandewar, Tweed, and Focal Peak, extending from southeastern Queensland to central New South Wales. On the other hand, some regions are not so clearly distinguished — as found, for example, throughout much of central and southern Queensland.

The broad approach taken in this chapter permits a regional overview in which large-scale trends and the characteristics of the east-Australian volcanic region as a whole can be determined. However, it has the obvious disadvantage that within-province differences largely are ignored.

A summary of the overall chemistry of the east-Australian Cainozoic volcanic rocks is provided in silica histograms (Fig. 5.1.1) and in a summary table of volcanic rock types (Table 5.1.1). The following three points are illustrated:

(1) The east-Australian region is dominated by mafic lavas, ranging from leucitite, melilitite, nephelinite, and analcimite, through basanite, alkali basalt, ne-hawaiite, and hawaiite, to ol- and qz-tholeiitic basalt. This extreme diversity is not uniformly distributed throughout the region. Melilitite and nephelinite are

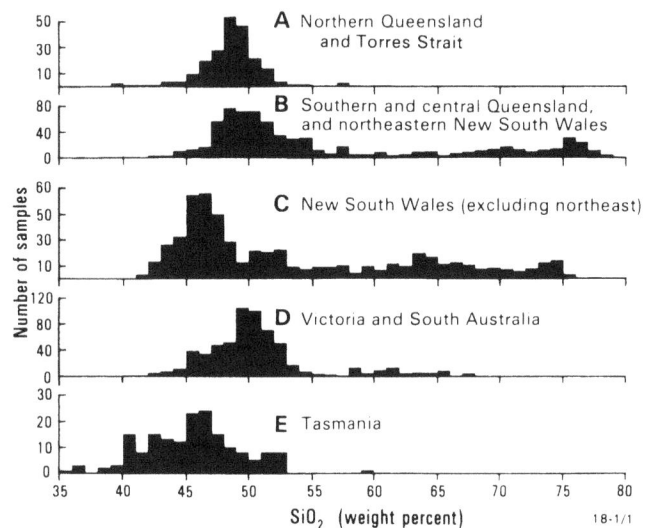

Figure 5.1.1 SiO_2 histograms for analysed Cainozoic volcanic rocks from eastern Australia. All data calculated water-free to 100 percent using the same Fe_2O_3/FeO value of 0.2 adopted for all of the major-element and normative diagrams presented in Chapter 5.

most common in Tasmania, whereas the dominant leucitite occurrences are in central New South Wales. Ne-hawaiite and hawaiite are the dominant mafic rock compositions, amounting to 33 percent of all analysed samples.

(2) Intermediate and felsic compositions are extremely rare in northern Queensland, in the Newer Volcanics of western Victoria, and in Tasmania. These compositions are best developed in central Victoria, New South Wales, and southern and central Queensland. Even here, however, there are marked regional differences. The most extensive development of rhyolite (including an associated large granophyre intrusion) is in southern Queensland, whereas the New South Wales and central-Victorian region is characterised by more abundant trachytic and peralkaline felsic volcanic rocks.

(3) Many of the regional differences noted in 1 and 2 can be attributed to the development of central-volcano provinces as opposed to lava-field provinces. Not only do the central complexes contain more evolved magma compositions, but their mafic lavas are characterised by higher relative proportions of hawaiite and tholeiitic

Table 5.1.1   Percentage distribution of analysed Cainozoic volcanic rocks from eastern Australia

	Northern Queensland	Southern and central Queensland including Tweed	New South Wales excluding Tweed	Victoria and South Australia	Tasmania	Total	Central volcano complexes	
							Warrumbungle, Nandewar, and Canobolas	Tweed and Focal Peak
Leucitite			6.2 (10.2)	0.5 (0.6)		1.6 (2.1)		
Melilitite, nephelinite, and analcimite	6.5 (6.9)[1]	1.2 (1.8)	3.5 (5.7)	0.6 (0.7)	26.4 (27.6)	4.1 (5.4)		
Basanite	11.6 (12.3)	4.7 (7.1)	18.6 (30.5)	4.2 (4.8)	18.7 (19.6)	9.6 (12.7)		1.0
Alkali basalt	2.3 (2.4)	2.3 (3.5)	10.8 (17.7)	3.3 (3.8)	10.4 (10.9)	5.3 (7.0)	0.5	3.0
Ne-hawaiite	35.4 (37.7)	6.3 (9.5)	4.3 (7.0)	13.5 (15.5)	14.8 (15.5)	11.1 (14.7)	0.5	2.5
Hawaiite	36.3 (38.6)	20.1 (30.2)	10.0 (16.4)	34.9 (40.0)	9.3 (9.7)	22.4 (29.7)	19.7	18.8
Transitional basalt and ol-tholeiitic basalt	1.9 (2.0)	17.3 (26.0)	4.5 (7.4)	21.5 (24.6)	8.2 (8.6)	13.0 (17.2)	3.0	18.3
Qz-tholeiitic basalt		14.6 (22.0)	3.1 (5.1)	8.8 (10.1)	7.7 (8.1)	8.4 (11.1)	1.5	15.2
Ne-mugearite and mugearite	4.2	1.4	3.9	2.0	2.2	2.5	5.9	1.0
Ne-benmoreite and benmoreite	0.9	0.8	3.0	0.2	2.2	1.3	4.4	
Icelandite		5.5	2.4	2.4		3.0	3.0	6.6
Ne-trachyte and phonolite	0.9		1.9	3.3		1.5	4.4	
Trachyte		6.2	9.6	4.6		5.6	22.2	1.5
Rhyolite		16.7	3.5			6.0	5.4	31.0
Peralkaline trachyte		0.3	8.9	0.2		2.3	22.2	
Peralkaline rhyolite		2.6	5.7			2.2	7.4	1.0
Number of analyses	215	727	565	634	182	2324	203	197

[1] Numbers in brackets are the frequency percentages calculated for mafic lavas only (leucitites through to tholeiitic basalts).

basalt (Table 5.1.1). The two major rhyolite-containing complexes of southeastern Queensland are notable for their greater development of near-saturated and over-saturated tholeiitic basalt compared to the large central complexes of New South Wales in which the evolved lavas are dominated by trachyte and peralkaline felsic compositions. Central-volcano provinces are absent from the Tasmanian, west-Victorian, and north-Queensland regions. These different characteristics are discussed in detail in the following sections.

5.2   Major Element Chemistry and Chemical Affinities

5.2.1   Silica saturation in mafic rocks

Silica saturation in mafic rocks (defined for the purpose of this section as those containing less than 56-percent SiO_2) is considered most appropriately in terms of the CIPW normative components plotted in a conventional

basalt saturation diagram (Figs. 5.2.1A-F). The most conspicuous feature of these plots is the continuity of compositions from nepheline-normative through to quartz-normative. A bimodal distribution seems to exist for the Tasmanian data, but the north-Queensland data are dominated by nepheline-normative compositions.

The total east-Australian data set has been contoured (Fig. 5.2.2) using the procedure suggested by Chayes (1972). A strong dominating peak is revealed within the nepheline-normative region, centred on a composition $Di_{40}Ol_{46}Ne_{14}$. A second, but less prominent concentration is present within the quartz-normative field, and the two peaks are linked by a narrow band of data points spanning the transitional olivine- and hypersthene-normative region. The general pattern of compositions illustrated in Figure 5.2.2 is very similar to that reported by Chayes (1972) for a continental Cainozoic basalt data set that includes relatively few east-Australian data. However, the peaks found by Chayes are displaced slightly towards the diopside apex compared to those in Figure 5.2.2, although this

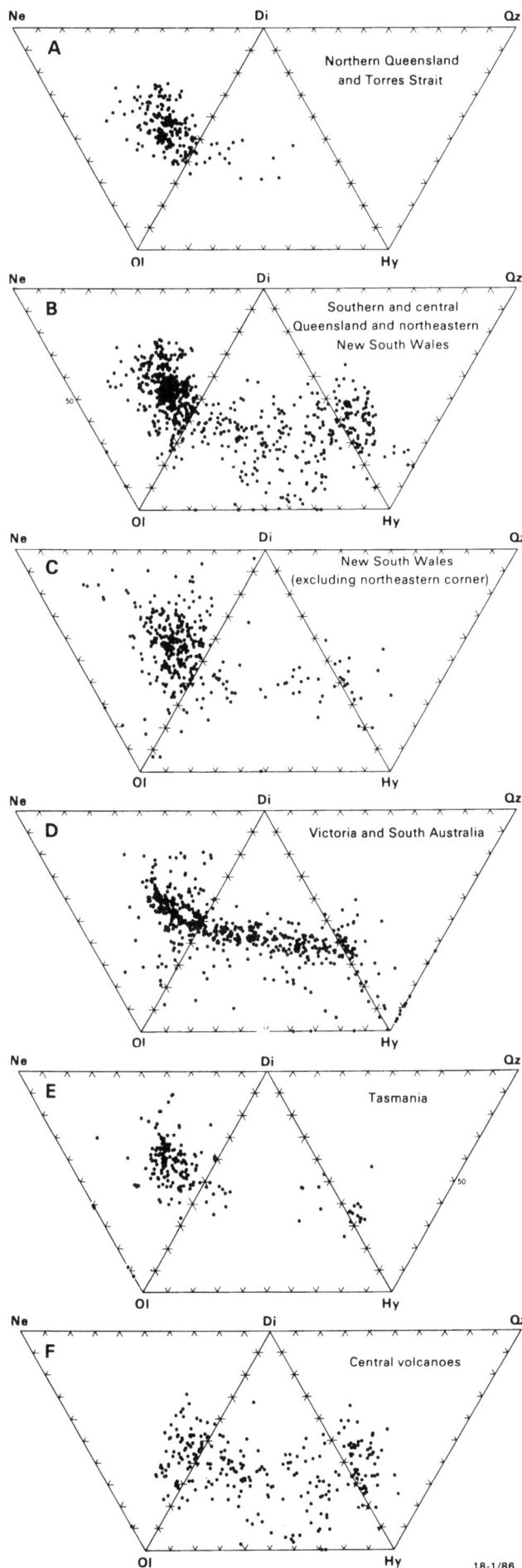

may be a function of differences in Fe_2O_3/FeO normalisation procedure.

Differences in mafic-rock chemistry between well-defined lava-field and central-volcano provinces can be evaluated with reference to data from northern Queensland and Tasmania (Figs. 5.2.1A,E), and from Warrumbungle, Nandewar, Canobolas, Tweed, and Focal Peak (central volcanoes; Fig. 5.2.1F). The main conclusions are: (1) transitional rocks are developed more characteristically in the central-volcano provinces; (2) the central volcanoes contain a much smaller proportion of undersaturated mafic magma compositions, as previously noted for Table 5.1.1.

5.2.2 Normative An, mg-ratio, and DI relationships

The compositional characteristics of the central-volcano and lava-field provinces have been investigated using other petrochemical criteria. $100Mg/(Mg+total-Fe)$ values (mg-ratios, calculated from atomic ratios) have been plotted as histograms (Fig. 5.2.3) and against Differentiation Index (DI; Thornton & Tuttle, 1960) in a series of diagrams, one of which is shown in Figure 5.2.4. DI values are plotted against normative $100An/(An+Ab)$ in Figure 5.2.5. The following points are significant:

(1) Broad negative correlations between mg-ratios and DI are found in all regions, although there is an extensive scatter of data points.

(2) There is an extensive overlap between nepheline- and quartz-normative lava types in terms of mg-ratio:DI covariance. However, the majority of analyses are nepheline-normative for rocks whose mg-ratios are high. A small number of quartz-normative compositions extend to higher mg-ratios than are observed for the olivine-normative types, although most are less magnesian.

(3) Tasmanian and north-Queensland rocks do not extend to low mg-ratios compared to values for the other regions. This is consistent with their narrow ranges of DI and SiO_2. The mafic magmas of Tasmania and northern Queensland apparently have been less modified by fractional-crystallisation processes, although the data fields and trends for these two provinces are distinctly different. Tasmanian magmas appear to be controlled more strongly by removal of Mg-olivine whereas north-Queensland magmas have trends more consistent with augite fractionation (Fig. 5.2.4).

(4) The volcanic rocks of the selected central complexes tend to be distinctly less magnesian (Fig. 5.2.4) than those of the lava-field provinces (northern Queensland and Tasmania). In addition, there is more extensive development of olivine- and quartz-normative types within the higher mg-ratio range for the central complexes. The inference, discussed more fully below, is that the magmas of the central complexes have undergone more extensive fractional crystallisation, extending their compositional ranges to extremely evolved compositions. Fractionation of some combina-

Figure 5.2.1 Normative compositions for east-Australian mafic (less than 56-percent SiO_2) volcanic rocks projected in the system Ne-Ol-Di-Hy-Qz.

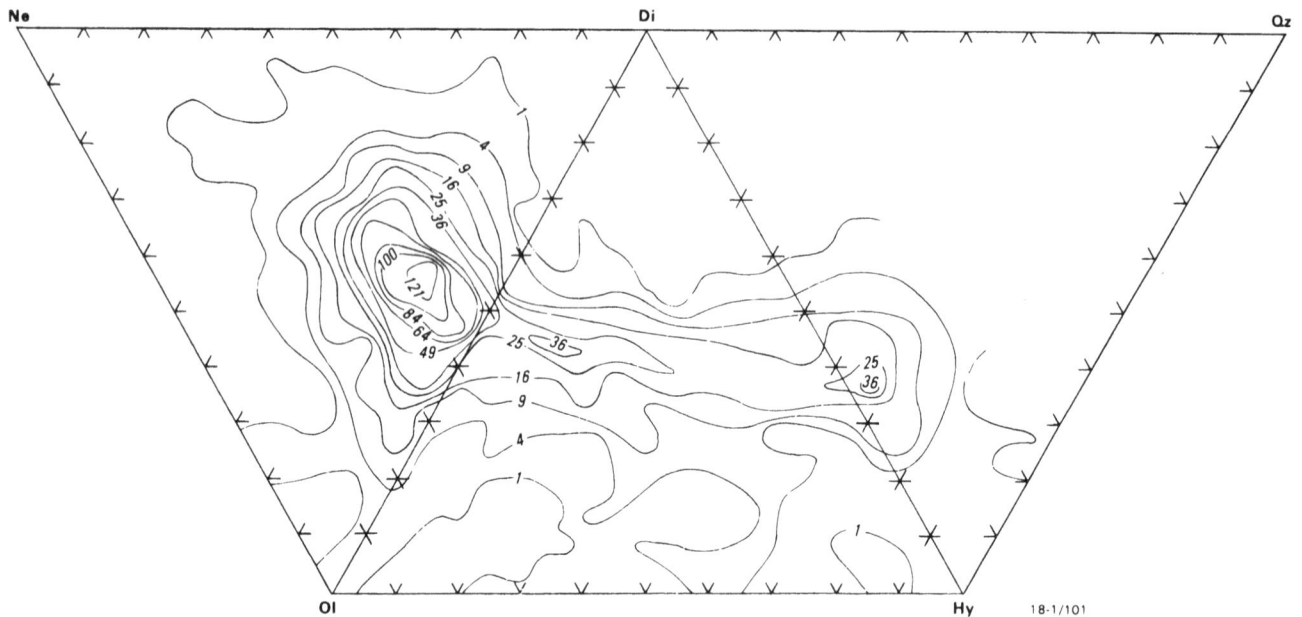

Figure 5.2.2 Contoured normative compositions for east-Australian rocks used in Figure 5.2.1 (1770 analyses). Contouring method is that used by Chayes (1972).

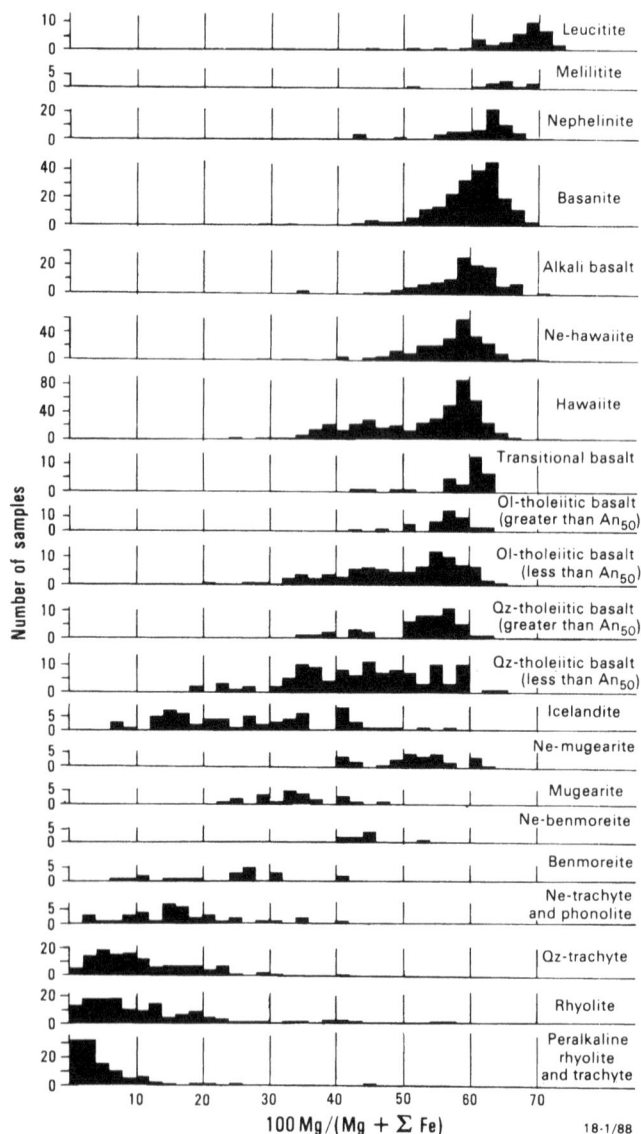

Figure 5.2.3 Histograms of mg-ratios for the main lava types of eastern Australia.

tion of olivine, clinopyroxene, and plagioclase is envisaged on the basis of the calculated fractionation trends for two olivine and augite compositions shown in Figure 5.2.4.

(5) The most highly evolved compositions are dominantly rhyolite, qz-trachyte, and their peralkaline equivalents. Undersaturated evolved compositions, however, are found within Warrumbungle and Nandewar volcanoes and in central Victoria. Some of the high-silica rhyolite compositions have what appear to be anomalously high mg-ratios (30–20), even though they are strongly depleted in Fe and Mg. This is believed to result from the presence of relatively magnesian xenocrysts within the rhyolites (see below).

(6) The progressive general decrease of mg-ratios as levels of silica undersaturation decrease is seen in the histograms (Fig. 5.2.3). This decrease also correlates with increasing SiO_2 in the felsic rocks. Hawaiite, ne-hawaiite, and alkali basalt have mg-ratio maxima in the range 60–58, whereas mg-ratios for most basanites, nephelinites, and melilitites are in the range 66–60. Most of the leucitites have mg-ratios greater than 66. Frey et al. (1978) suggested that basaltic magmas in equilibrium with mantle peridotite (Mg-ratios 89–88) should have Mg-ratios of about 75–68 (for up to 30-percent partial melting). These ratios are estimated to be equivalent to about 71–64 for the mg-ratios used here. The ranges, if appropriate, mean that most of the hawaiite, alkali-basalt, and basanite magmas have undergone olivine-dominated fractionation (note also their negatively skewed mg-ratio distributions in Fig. 5.2.3).

The ol- and qz-tholeiitic basalts have broad ranges of mg-ratio that extend to the different intermediate lava compositions. Extensive fractional crystallisation again is implied as an important control on the evolution of these magma types. Other features of note in Figure 5.2.3 include the strong concentration of low mg-ratios for peralkaline rocks, and the relatively broad spread of mg-ratios in the rhyolites, as noted above.

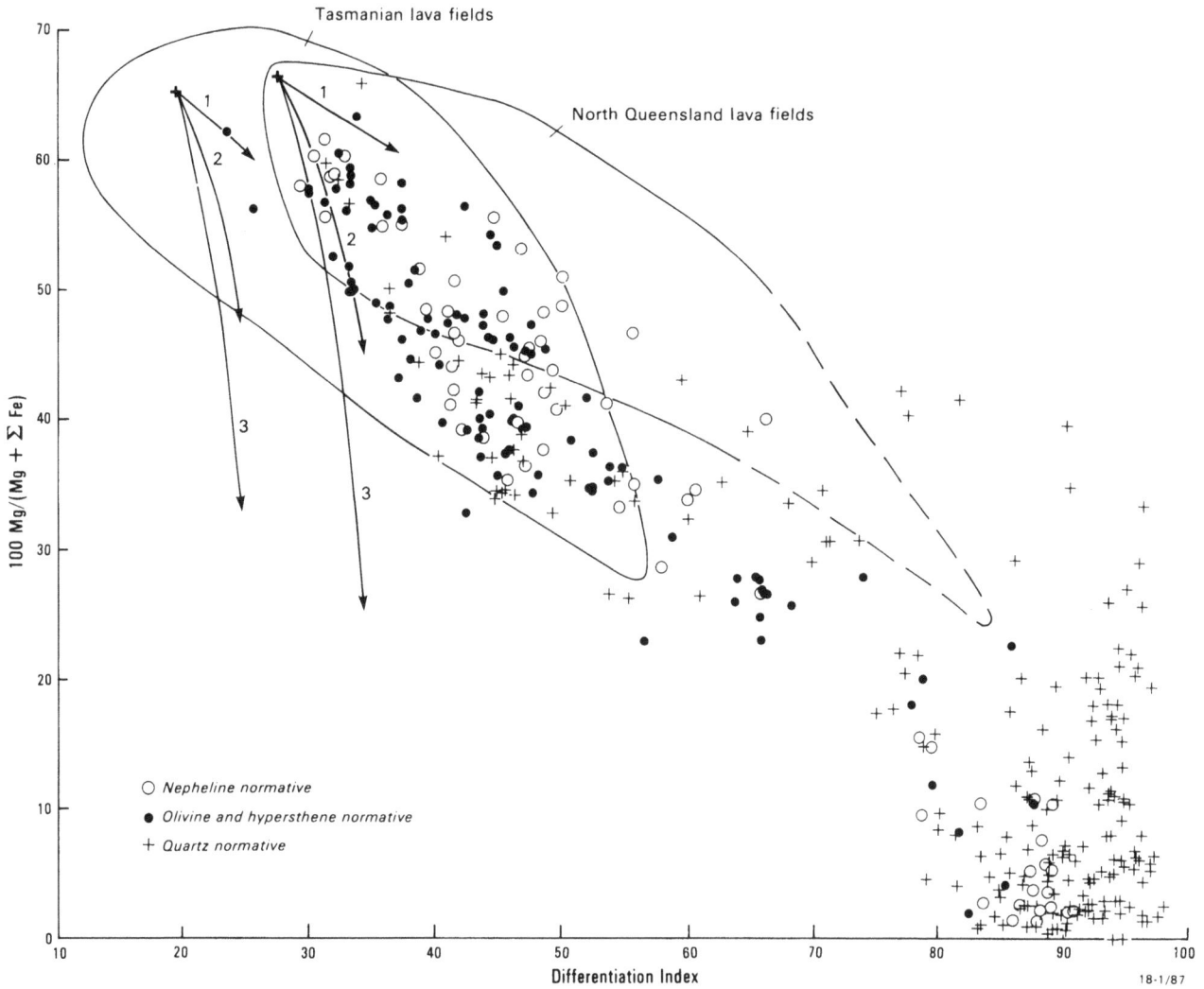

Figure 5.2.4 mg-ratio versus Differentiation Index relationships for rocks from central volcanoes in eastern Australia, shown in relation to generalised compositional fields for lava-field provinces in northern Queensland and Tasmania. Theoretical fractionation curves from two near-primary magma starting compositions are for the following mineral assemblages and degrees of fractionation (F): (1) clinopyroxene to F 0.7; (2) olivine (Fo$_{80}$) to F 0.8; (3) olivine (Fo$_{90}$) to F 0.8.

Di versus normative-An relationships (Fig. 5.2.5) are similar to those described above for mg-ratios. The negative correlations between these parameters are well defined and, again, there is extensive overlap between the nepheline-, olivine-, and quartz-normative compositions at 100An/(An+Ab) levels less than about 60 percent. Note, however, that at higher normative-plagioclase levels only the more extremely under-saturated types are present because of the procedures followed in the CIPW-norm calculation. The most significant aspect, however, is the almost complete continuity of normative-plagioclase compositions and DI from high to very low values. Moreover, the dominance of compositions within the normative andesine range is clear at the mafic end of the spectrum, which is reflected in part by the high proportion of hawaiites (Table 5.1.1). The Tasmanian and north-Queensland data again generally do not extend to strongly evolved (low normative-An) values. Nevertheless, normative andesine compositions are still conspicuous in both of these regions.

5.2.3 K$_2$O, TiO$_2$, and SiO$_2$ relationships

The range of K$_2$O values in the mafic rocks is also wide, extending from about 0.35 to 4.0 weight percent (leucitites excluded). A group of relatively low-K (less than about 0.75 percent K$_2$O) qz-tholeiitic basalts and, to a lesser extent, ol-tholeiitic basalts within the Tasmanian, New South Wales, and southern- and central-Queensland regions also can be distinguished. Intermediate and felsic rocks have relatively potassic compositions, such as potassic trachyte and rhyolite in southern and central Queensland, potassic trachyte and phonolite in the Victorian and New South Wales regions, and somewhat less potassic rhyolites in the New South Wales provinces (see also Ewart et al., 1988). The high-K$_2$O (and TiO$_2$) contents of the New South Wales leucitites are unique characteristics of this rock suite in eastern Australia (Fig. 5.2.6). K$_2$O/Na$_2$O values in the mafic rocks are most commonly less than 0.5 and only rarely greater than about 0.7 (excluding the leucitites).

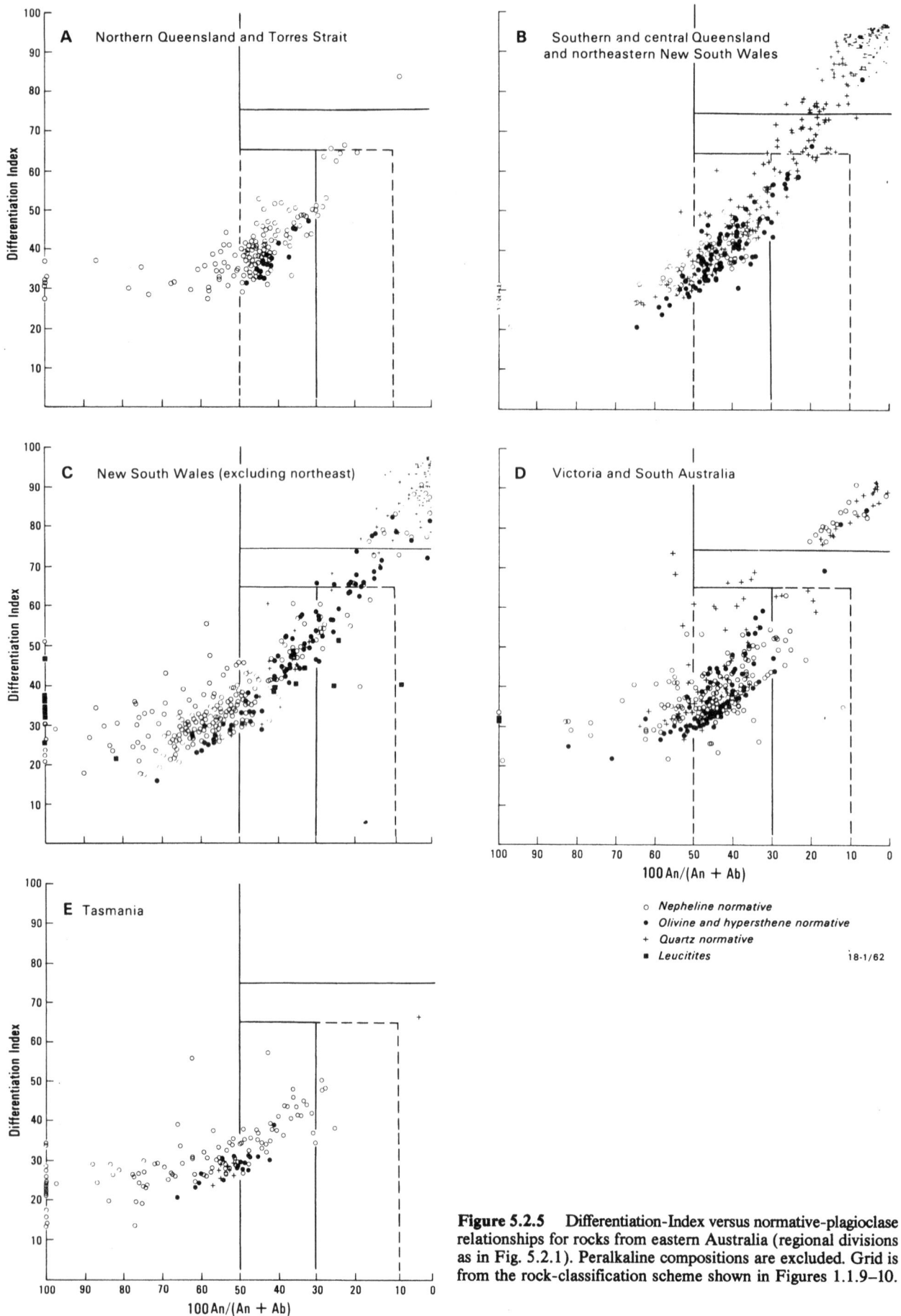

Figure 5.2.5 Differentiation-Index versus normative-plagioclase relationships for rocks from eastern Australia (regional divisions as in Fig. 5.2.1). Peralkaline compositions are excluded. Grid is from the rock-classification scheme shown in Figures 1.1.9–10.

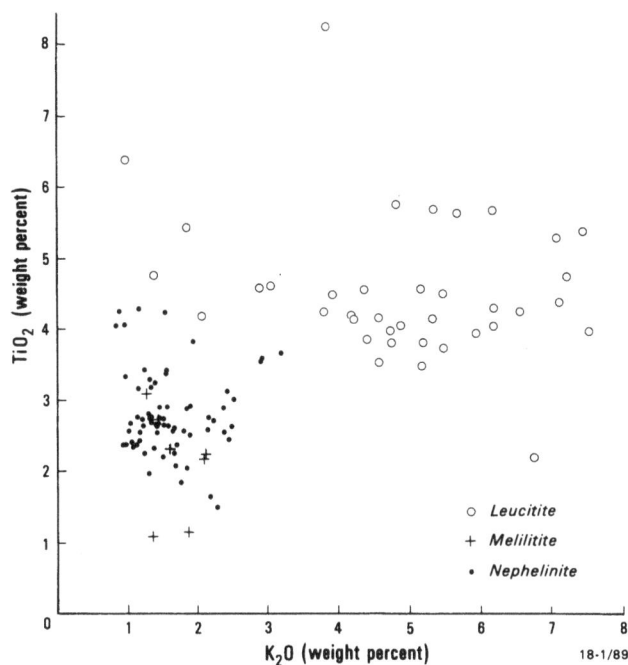

Figure 5.2.6 TiO$_2$:K$_2$O relationships for strongly undersaturated east-Australian volcanic rocks.

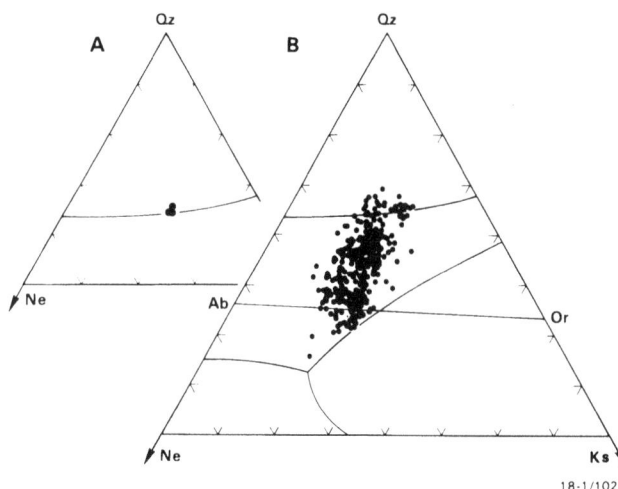

Figure 5.2.7 Normative compositions for east-Australian non-peralkaline intermediate and felsic rocks projected in the system Qz-Ne-Ks. (A) Average compositions for high-silica rhyolites from southeast Queensland. (B) Intermediate and felsic rocks (greater than 56-percent SiO$_2$), excluding the high-silica rhyolites of southeastern Queensland. Phase boundaries (1 bar) are those defined by Schairer (1950).

TiO$_2$ values decrease as SiO$_2$ increases, and this is well defined for the more evolved intermediate and felsic compositions. TiO$_2$ abundances in the mafic rocks range from about 1.0 to 4.0 weight percent (excluding the leucitites) and from about 1.0 to 1.8 percent in the low-K tholeiitic basalts. A feature of note, however, is the absence of a clear correlation between TiO$_2$ and K$_2$O within the different mafic-rock groups, except for the leucitites and *ol*-tholeiitic basalts.

5.2.4 Intermediate and felsic rocks

The majority of the evolved rocks (DI greater than about 60) of eastern Australia are silica oversaturated. Non-peralkaline compositions define a broad array in Figure 5.2.7 more or less coinciding with the thermal valley within the alkali-feldspar field, and extending to near the ternary minimum. However, a notable feature of the southern- and central-Queensland and New South Wales data field is the rather broad scatter. This is interpreted to be a primary magmatic feature rather than one caused by secondary processes, although such effects can be identified as having modified a small number of samples. Part of this compositional range may be attributable to assimilation/fractional-crystallisation (AFC) processes operating during the evolution of especially the trachytic and related intermediate magmas, as discussed below.

The major development of non-peralkaline rhyolite in eastern Australia is found in the northeastern corner of New South Wales and in the southern- and central-Queensland regions. Five main high-silica rhyolite types are recognised within the central-volcano complexes in the southeastern corner of Queensland and the northeastern corner of New South Wales: (1) Binna Burra type, found on the northern flanks of the Tweed volcano; (2) Nimbin type, found on the southern flanks of the Tweed volcano; (3) fayalite-bearing rhyolite type, most extensively developed on Focal Peak volcano, but also found in the Mount Alford complex; (4) ferrohypersthene-bearing rhyolite type, found within the northern flanks of Tweed volcano (Springbrook Rhyolite) and the Mount Alford complex; (5) biotite-bearing rhyolite of the Noosa area (Glass Houses province, Queensland).

These rhyolite types are distinguishable on the basis of both mineralogical and trace-element characteristics, as discussed below (see also Ewart et al., 1985). They are best classified as high-silica potassic rhyolites (normative quartz greater than 30 percent), and have close similarities to rhyolites from bimodal basalt-rhyolite occurrences in such regions as northwestern Scotland and northwestern USA (Ewart, 1979). Average compositions, or compositional fields, for these rhyolites are shown in Figures 5.2.7–8. The compositional trend projects close to the piercing point in the (Q-Ab-Or)$_{97}$An$_3$ system for 1 kb (James & Hamilton, 1969a,b), consistent with the interpretation that they are minimum-melt compositions.

The Springsure volcanic area of south-central Queensland embraces an extensive complex of rhyolite flows, dykes, plugs, and sills, known as Minerva Hills. These rocks are rhyolite according to the classification adopted in this volume, but are chemically somewhat intermediate between trachyte and the high-silica rhyolites discussed above. They are here referred to as low-silica rhyolite. Minerva Hills rhyolites have lower normative quartz (about 15 to 28 percent) and a wider range of Ab-Or values compared to the high-silica rhyolites (Fig. 5.2.8). Similar rhyolites are found in Nandewar volcano (Stolz, 1985).

Peralkaline chemistry characterises a limited number of strongly undersaturated lavas, but the majority of peralkaline rocks from the east-Australian region are trachytic (quartz-normative) and rhyolitic. They define an array close to the thermal valley within the alkali-feldspar field in Figure 5.2.9, extending close to the

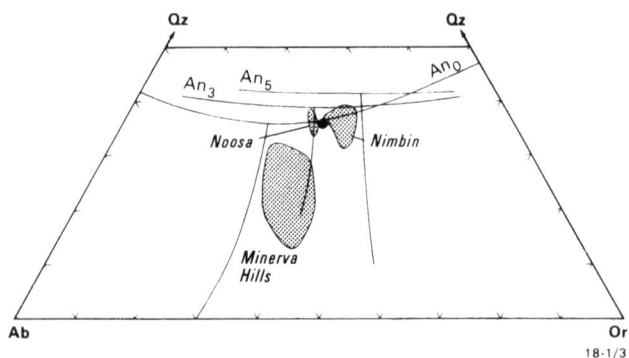

Figure 5.2.8 Average Qz-Ab-Or compositions and compositional fields for the main high-silica rhyolite types of southeastern Queensland and northeastern New South Wales (Nimbin Rhyolite of Tweed volcano, and the Noosa area) and the low-silica rhyolite of Minerva Hills, central Queensland. Filled circle represents the average composition of the ferrohypersthene-bearing, fayalite-bearing, and Binna Burra rhyolites of southeastern Queensland. Phase boundaries are after Tuttle & Bowen (1958) and James & Hamilton (1969a) for 1 kb and water-saturation.

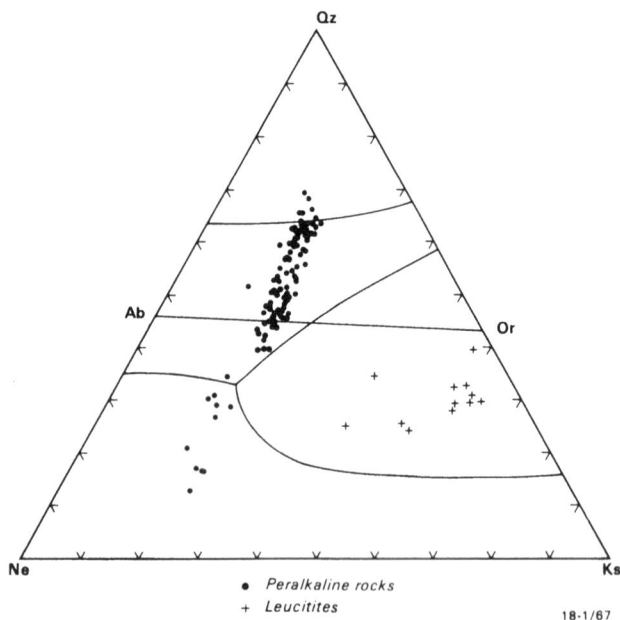

Figure 5.2.9 Normative compositions for east-Australian peralkaline rocks and leucitites projected in the system Qz-Ne-Ks. Phase boundaries (1 bar) are those defined by Schairer (1950).

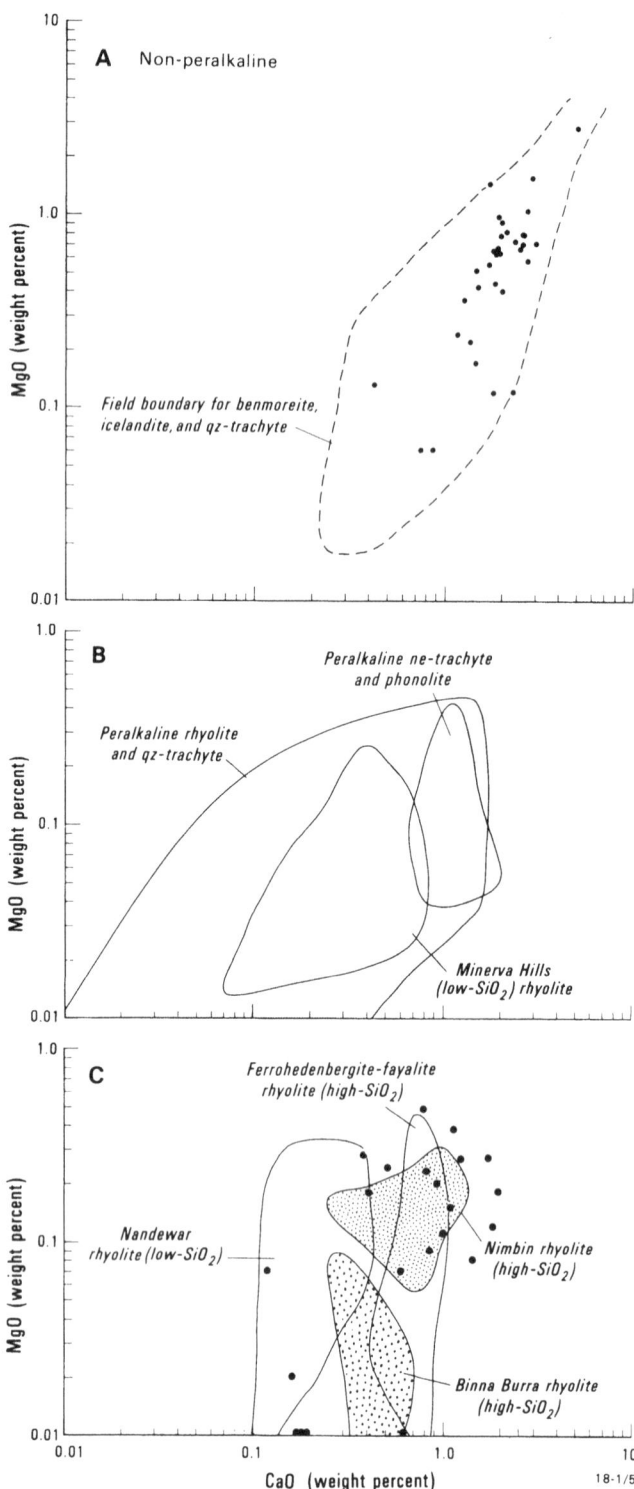

Figure 5.2.10 Generalised MgO:CaO relationships for east-Australian intermediate and felsic rocks. Filled circles in A represent *ne*-benmoreites, *ne*-trachytes, and phonolites, and in C represent individual data points for other southeast-Queensland high-silica rhyolites.

ternary minimum on the quartz-feldspar boundary curve. A particularly significant feature is the more constrained compositional grouping of the data points, which is best appreciated by comparison with the corresponding non-peralkaline fields in Figure 5.2.7.

The intermediate and felsic lavas are characterised by well-defined, progressive depletions of Mg and Ca as DI increases, and depletions are extremely pronounced in some of the rhyolites and peralkaline rhyolites (Fig. 5.2.10). Intermediate rocks — benmoreite, icelandite, and trachyte — define a regular, correlated depletion of Mg and Ca. Mg depletion is more pronounced than is Ca. The undersaturated rocks have a similar Mg-Ca depletion, but not reaching such low levels, and also tending to be even less depleted in Ca.

The behaviour of these two elements is indicative of crystal-liquid fractionation processes, as discussed below in more detail.

Mg-Ca behaviour is different between the peralkaline rocks and the different rhyolite types (Figs. 5.2.10B-C). Peralkaline rhyolites and trachytes develop extreme Mg and Ca depletion which may be a continuation of the trends observed in Figure 5.2.10A. However, peralkaline *ne*-trachyte and phonolite have not developed the same levels of depletion, especially of CaO, which is consistent with the non-peralkaline undersaturated trachyte data plotted in Figure 5.2.10A. In contrast, there is no clear covariance between MgO and CaO for the high-SiO$_2$ rhyolites which have a range of MgO values but nearly constant (possibly buffered) CaO contents. In fact, the low-SiO$_2$ rhyolites have somewhat more depleted CaO abundances. The rather different behaviour of MgO and CaO within the high-silica rhyolites is of some significance as the two rhyolite types also have differences in trace-element, isotopic, and mineralogical characteristics (see below).

5.2.5 Summary

East Australian Cainozoic volcanic rocks represent a highly diverse, complex, and heterogeneous group of magma compositions, ranging from extremely silica-undersaturated types, including melilitite and leucitite, through to highly fractionated peralkaline lavas and high-silica rhyolites. There is a continuity of chemical features through the series, such as in mg-ratios, DI, and normative 100An/(An+Ab). The importance of fractional crystallisation in controlling the evolution of the magmas is clear from the major-element chemistry. Nevertheless, an extreme range of mafic compositions exists, and these need to be evaluated in terms of different potential parental magmas and different magma sources. AFC processes operating for at least some of the non-peralkaline intermediate magmas are inferred also from the isotopic data (see below).

One aspect of particular significance is the differences in chemical composition between the central-volcano and lava-field provinces. The mafic lavas of the central-volcano complexes are, overall, significantly more fractionated (for example, higher DI), and there are very few high-Mg 'parental-type' representatives.

5.3 Mineralogy and Mineral Chemistry

5.3.1 Summary of mineral assemblages

Introduction

The following descriptions are based on extensive microprobe studies of samples representing the complete range of magma compositions found throughout the east-Australian region. Major data sources are Birch (1979, 1980), Cundari (1973), Duggan (1974), Ewart (1981, 1985), Ewart et al. (1977, 1980, 1985, 1987), Ferguson (1977b, 1978b,c), Grenfell (1984), Ross (1977), Stolz (1986), and Wass (1973), together with extensive previously unpublished data obtained by the author. Mineral assemblages are classified on the basis of whole-rock type. An extensive series of mineralogical chemical-variation diagrams has been generated, not all of which, however, are able to be presented here.

Mafic rocks

Phenocrysts in the mafic rocks are mainly olivine, and less commonly plagioclase, augite, and titanomagnetite.

Orthopyroxene is particularly uncommon, and is found as phenocrysts only sporadically in *qz*-tholeiitic basalt. Orthopyroxene megacrysts are found rarely also in tholeiitic and hawaiitic rocks. Groundmass minerals are dominated by plagioclase (extending through zoning and as discrete grains to alkali feldspars of anorthoclase-sanidine compositions), augite, olivine, titanomagnetite, and apatite. A wide range of subcalcic-augite and pigeonite compositions, and rare orthopyroxene, is found in many *qz*-tholeiitic and, less commonly, *ol*-tholeiitic basalts. Ilmenite typically is present in the groundmass of tholeiitic and hawaiitic rocks. It is less common in more alkaline rocks and is normally absent in basanite, nephelinite, and melilitite. Ilmenite is present in some leucitites. Nepheline is identifiable in the groundmass of melilitite, nephelinite, leucitite and, less commonly, basanite and *ne*-hawaiite. Melilite, as microphenocrysts or as groundmass grains, is characteristic of the melilitites, whereas leucitite contains phlogopite, richterite, and sanidine in addition to leucite.

Intermediate rocks, trachyte, and phonolite

The dominant Fe-Mg silicate phenocryst assemblage is Ca-rich pyroxene plus olivine. Titanomagnetite phenocrysts also may be present. Mugearite and benmoreite contain augite-ferroaugite and an intermediate olivine, and as SiO$_2$ increases (passing into icelandite and trachyte) the pyroxene compositions extend through ferroaugite to ferrohedenbergite and the olivine ranges through to fayalite. Phenocryst feldspars include sodic plagioclase and sanidine, and calcic anorthoclase in the more evolved rocks. Groundmass mineralogy is similar to that of the phenocryst assemblages. Some groundmass pyroxenes are enriched in Na, anorthoclase (to sodic sanidine) compositions dominate the feldspars, and apatite, zircon, titanomagnetite, and in some rocks ilmenite, may also be present. Biotite is extremely rare as a phenocryst, but is notable in a Canobolas trachyte where it coexists with ferroaugite and titanomagnetite. Sodalite and nepheline have been recorded in only one phonolite (Mount Wilson, Victoria).

Rhyolite

Four major phenocryst assemblages are recognised in the rhyolites:

(1) The assemblage ferrohedenbergite + fayalite + ilmenite ± titanomagnetite, plus some combination of plagioclase, sanidine or calcic anorthoclase, ± quartz, is found in the high-silica rhyolite of three southeast-Queensland areas (Focal Peak, Mount Alford, and Noosa). Chevkinite is present as a minor accessory mineral.

(2) The assemblage Fe-hastingsite ± fayalite ± ferrohedenbergite + allanite + titanomagnetite + ilmenite + anorthoclase is present in about one third of the low-silica rhyolites of the Minerva Hills volcanic centre, central Queensland, but is well preserved only in rare glassy rocks.

(3) The assemblage biotite ± allanite + plagioclase + sanidine + quartz + ilmenite ± titanomagnetite, is particularly characteristic of two groups of high-silica rhyolites in southeastern Queensland (the Binna Burra

rhyolites of Tweed volcano, and those of the Noosa area (Glass Houses province).

(4) Ferrohypersthene-eulite + plagioclase + sanidine + quartz + ilmenite ± titanomagnetite, is an assemblage characteristic of some high-silica rhyolites of Tweed volcano (Springbrook and Nimbin rhyolites) and the Mount Alford volcanic centre (southeastern Queensland).

Apatite and zircon are minor minerals in all of the above rhyolite phenocryst assemblages.

Peralkaline trachyte and rhyolite

Peralkaline rocks contain phenocrysts of hedenbergite, together with anorthoclase and quartz, but are characterised particularly by a groundmass mineralogy of Ca-poor anorthoclase to sodic sanidine, ± quartz, fluor-arfvedsonite/riebeckite, aegirine and aegirine-augite, ± aenigmatite. Some rocks also contain fayalite, and ilmenite or magnetite or both.

Hybrid rocks

A series of hybrid lavas is recognised from Tweed volcano that technically fall in the icelandite, dacite, and rhyolite rock groups. These lavas have a wide range of Ca-rich and Ca-poor pyroxene-phenocryst compositions, together with both Mg- and Fe-rich olivine, Ca and Na plagioclase, and sanidine.

5.3.2 Olivine and pyroxene

Olivine/host-rock relationships

The following conclusions can be drawn from a comparison of olivine compositions for the different volcanic-rock types (Fig. 5.3.1).

(1) Extensive, normal compositional zoning is characteristic of olivine phenocrysts from all rock types, ranging from 10–15 percent in leucitite and melilitite, to ranges in excess of 20 percent in hawaiitic and tholeiitic rocks. These ranges are found even within single rock specimens.

(2) Mg-rich olivines (Fo_{90-88}) are common in leucitite, melilitite, and nephelinite (in most of which phenocrystal olivine is present), but decreasing in amount through to hawaiite. Tholeiitic basalts contain less magnesian olivine compositions (Fo_{84-82}).

(3) Olivine in intermediate and felsic rocks is more Fe-enriched, reaching Fa_{100} in some rhyolites, but more magnesian xenocrystal olivine is found sporadically, even in the rhyolites.

Olivine is the liquidus phase in the majority of the mafic lavas (see below), and therefore whole-rock $100Mg/(Mg+Fe^{2+})$ values (Mg-ratios) should closely correlate with the most magnesian olivine-core composition, reflecting Mg-Fe olivine-liquid partitioning (K_D). A range of proposed Ol-liquid K_D values is plotted in Figure 5.3.1 (Roeder & Emslie, 1970; Ford et al., 1983; Takahashi & Kushiro, 1983). There is considerable discrepancy between equilibrium olivine compositions (represented by the most Mg-rich compositions) and whole-rock Mg-ratios. A relatively high proportion of data points for the nepheline-normative rocks of the lava-field provinces fall within the K_D limits shown, so plausibly these rocks represent

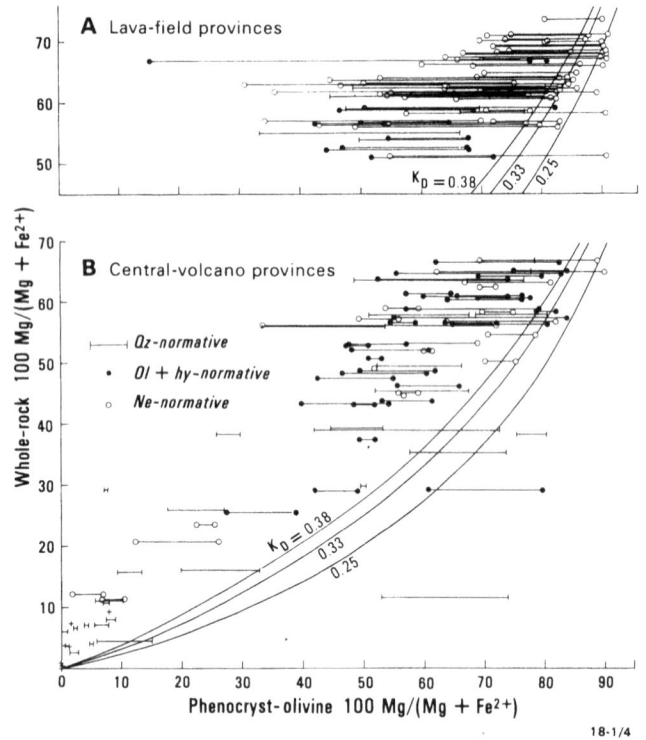

Figure 5.3.1 Whole-rock Mg-ratios for east-Australian samples (Fe_2O_3/FeO values adjusted to 0.2) plotted against the Mg-ratio ranges of their respective olivine phenocrysts. Three sets of Mg-Fe partition coefficients are based on data given by Roeder & Emslie (1970; data for about 1 kb), Ford et al. (1983), and Takahashi & Kushiro (1983; 15 kb data).

equilibrium liquid compositions. However, few of the data for the olivine- and quartz-normative compositions plot within the accepted K_D limits. These conflicting data fall into the following two groups:

(1) *Rocks containing olivine cores too magnesian for their Mg-ratios.* These olivine compositions are considered to be xenocrystic, near Fo_{90} in the more Mg-enriched, nepheline-normative, mafic lavas, and may represent relict olivines initially in equilibrium with their lherzolitic (or harzburgitic) source in the region of magma segregation, or disaggregated lherzolite debris. This is supported also by the low CaO values (less than 0.1 percent) of these olivines, consistent with those of analysed olivine in lherzolite nodules. The presence of xenocryst olivine in the less-magnesian rocks is thought to represent mixing or partial assimilation processes.

(2) *Rocks containing olivine not sufficiently magnesian for the whole-rock Mg-ratios (K_D values too high).* Most of the olivine-and quartz-normative rocks fall into this category. Possible explanations include: homogenisation (resorption) of more forsteritic olivine that formed during early equilibrium crystallisation; olivine accumulation; and supercooled crystallisation (see Maaløe & Hansen, 1982).

Most whole-rock compositions at the low-Mg-ratio end of the spectrum are noticeably more Mg-rich than the coexisting olivine. Ghiorso et al. (1983) discussed this relationship and suggested that olivine-liquid phase relationships in the fayalite region are more complex than recognised previously.

Perhaps one of the more significant aspects of the olivine data is the implication that relatively few of the olivine- and quartz-normative mafic rocks (Fig. 5.3.2) can be regarded as representing primary liquid compositions.

Pyroxene in mafic rocks

Data for phenocryst and groundmass pyroxenes are discussed here in relation to the main basaltic groups. Only three pyroxene-quadrilateral diagrams are shown (Figs. 5.3.2–4) for basaltic rocks, although this section is based on a more extensive set of pyroxene analyses.

Qz-tholeiitic basalts have the most complex groundmass and microphenocryst pyroxene compositions, and a wide range of calcic and subcalcic augite and pigeonitic compositions is present (Fig. 5.3.2). Subcalcic augite and pigeonite groundmass grains are most common in rocks that have the highest 100An/(An+Ab) and mg-ratio values, presumably representing quenching from highest eruptive temperatures among the *qz*-tholeiitic basalts, including especially the least potassic types. Groundmass augite in the *qz*-tholeiitic basalts is consistently of subcalcic composition, and groundmass orthopyroxene tends to appear mostly in the least-magnesian lavas. Augite phenocrysts are uncommon in the *qz*-tholeiitic basalts, except in the most fractionated compositions. Phenocrysts of orthopyroxene are present only sporadically.

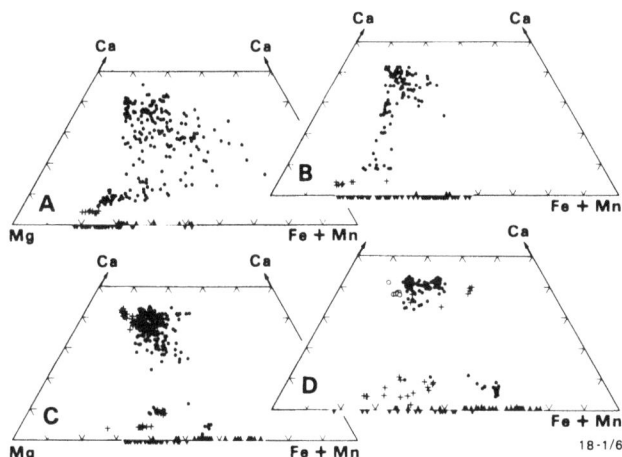

Figure 5.3.2 Groundmass and phenocryst pyroxene (filled circles and crosses, respectively) and olivine (upward- and downward-pointing triangles, respectively) compositions (atomic percent) for east-Australian *qz*-tholeiitic basalts. (A) Whole-rock mg-ratios greater than 55. (B) mg-ratios 50–55. (C) mg-ratios 40–50. (D) mg-ratios 30–40. Open circles represent megacrysts.

Ol-tholeiitic basalts contain groundmass subcalcic augite and pigeonite (Fig. 5.3.3), but again only in rocks that have relatively high mg-ratio and high 100An/(An+Ab) values. These commonly include low-potassium types that, again, are inferred to have relatively high quenching temperatures. Groundmass augite is generally more calcic, has a smaller range of compositions, and is less Fe enriched than found in the *qz*-tholeiitic basalts. Phenocrysts of augite are rare, although relatively high-pressure megacrysts of coexisting subcalcic augite and bronzite may be present.

Hawaiites are dominated by groundmass augite and contain no subcalcic or low-Ca pyroxene. The augite is typically calcic (salitic) in composition, and compositional ranges reflect host-rock mg-ratios. Exceptionally Fe-enriched, interstitial, groundmass-pyroxene and olivine grains are found in a few samples. Phenocryst augite (extending to subcalcic augite) is present sporadically in the more magnesian hawaiites.

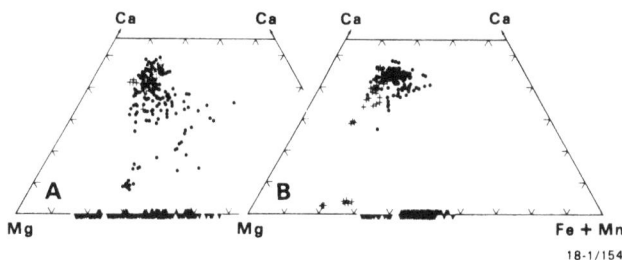

Figure 5.3.3 Groundmass and phenocryst pyroxene (filled circles and crosses, respectively) and olivine (upward- and downward-pointing triangles, respectively) compositions (atomic percent) for east-Australian *ol*-tholeiitic basalts. (A) Whole-rock mg-ratios 50 or more. (B) mg-ratios 40–50.

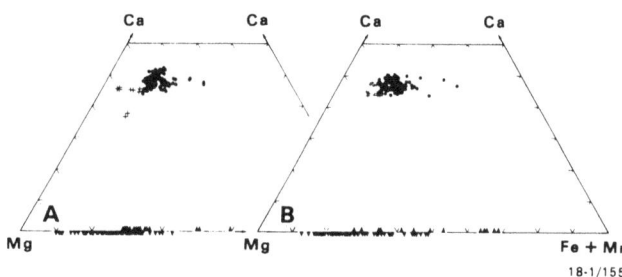

Figure 5.3.4 Groundmass and phenocryst pyroxene (filled circles and crosses, respectively) and olivine (upward- and downward-pointing triangles, respectively) compositions (atomic percent) for east-Australian (A) basanites (mg-ratios 52–66) and (B) alkali and transitional basalts (mg-ratios 45–68).

Ne-hawaiites also are dominated by groundmass calcic augite (salitic) that are relatively magnesian (extending from compositions near $Wo_{48}Fs_{10}En_{42}$). Phenocryst/megacryst diopside, diopsidic augite, and subcalcic augite coexisting with bronzite, are found in the most magnesian *ne*-hawaiites. The subcalcic augite and bronzite are believed to be higher-pressure megacryst minerals.

Transitional basalt, alkali basalt, and basanite contain calcic-augite/salite which, in the more magnesian basanites, extends to compositions above the diopside-hedenbergite join (Fig. 5.3.4). Limited Fe enrichment is found in a few of the more evolved basanites and alkali basalts. Phenocryst (and megacryst) diopside, diopsidic augite, augite, and subcalcic augite are found in transitional and alkali basalt and in the more magnesian basanites.

Leucitite, melilitite, and nephelinite are characterised by closely similar magnesian-pyroxene compositions, consisting of highly calcic types — namely, diopside, endiopside, salite — that commonly extend to compositions above the diopside-hedenbergite join. Phenocryst and microphenocryst diopside and augite are common in leucitite and melilitite.

Pyroxene in intermediate and felsic rocks

Mugearite, benmoreite, icelandite, and dacite. Mugearite, benmoreite, and non-hybrid icelandite are characterised by a single, Ca-rich, groundmass pyroxene which, in mugearite and benmoreite, is of intermediate Fe-Mg composition, similar to co-existing phenocryst augite (Fig. 5.3.5A). Icelandite represents a more advanced stage of fractionation and has marked Fe enrichment of both groundmass and phenocryst pyroxene (and olivine; Fig. 5.3.5C).

Icelandite and dacite lavas (extending to low-silica rhyolite) found in Tweed volcano, contain an extremely wide range of phenocryst and groundmass pyroxene and olivine compositions, including augite, ferroaugite, bronzite-hypersthene-ferrohypersthene, and pigeonite-ferropigeonite (Fig. 5.3.5B). These are clearly hybrid assemblages, and in one area on the northern flanks of the volcano, rhyolite lava has been modified by the incorporation and partial assimilation of mafic xenoliths

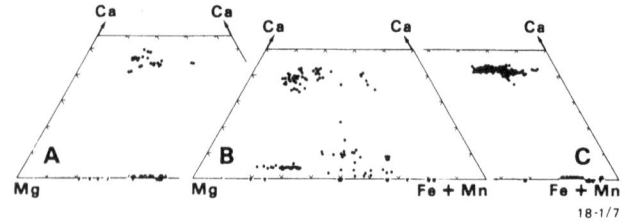

Figure 5.3.5 Groundmass and phenocryst pyroxene (filled circles) and olivine (upward- and downward-pointing triangles) compositions (atomic percent) in (A) east-Australian mugearites and *ne*-benmoreites, (B) hybrid icelandites and dacites from Tweed volcano, and (C) icelandites from Main Range and Mount Alford (southern Queensland).

(Ewart et al., 1977). Additional magma mixing also may have taken place. The resulting hybrid chemistry has allowed development of phenocryst pigeonite and orthopyroxene.

Table 5.3.1 Chemical analyses (electron microprobe) of selected minor and accessory minerals in east-Australian volcanic rocks

	1 Aegirine augite	2 Sodic amphibole	3 Rich-terite	4 Fe-rich hasting-site	5 Fe-rich biotite	6 Fluor-biotite	7 Sodalite	8 Melilite	9 Melilite	10 Aenig-matite
SiO₂	49.77	48.34	50.32	38.43	34.13	34.96	39.7	42.94	43.24	39.85
TiO₂	0.30	0.90	4.59	2.65	4.77	4.34	0.08	0.18	0.07	8.35
ThO₂										
ZrO₂	3.04	0.77								n.d.[1]
HfO₂										
Al₂O₃	0.53	0.47	2.47	8.39	12.68	15.13	30.8	6.81	6.28	1.22
La₂O₃										
Ce₂O₃		n.d.[1]								
Nd₂O₃										
Y₂O₃										
Nb₂O₅										
Fe₂O₃	11.65[2]						1.16	3.07[2]	2.84[2]	
FeO	14.66	33.28	8.47	33.33	32.69	26.09		1.43	1.57	42.15
MnO	1.05	1.68	0.14	0.68	0.17	0.81		0.07	0.02	0.67
MgO	0.83	0.15	16.83	n.d.	2.01	5.61		7.16	7.51	0.02
NiO		n.d.						0.11	0.05	
CaO	12.41	1.81	5.83	9.69			0.12	33.45	33.82	0.84
BaO								0.43	n.d.	
Na₂O	6.09	7.97	5.81	2.16	0.46	0.61	20.9	4.33	4.21	6.71
K₂O		1.17	1.95	1.48	8.55	8.70	0.19	0.06	0.02	
P₂O₅										
F		2.78	1.65	0.49	0.77	3.24				n.d.
Cl		n.d.				0.34	7.82			n.d.
Total	100.33	98.15[3]	97.37[3]	97.09[3]	95.91[3]	98.39[3]	99.01[3]	100.04	99.62	99.81

1. Phonolite, Mount Wilson, central Victoria (sample 7809/3). Groundmass.
2. Peralkaline trachyte, Grand High Tops, Warrumbungle volcano (sample 38994). Groundmass.
3. Leucitite, 4 km west of Lake Cargelligo (sample LC). Groundmass.
4. Low-silica rhyolite, Minerva Hills, central Queensland (sample 38739a). Phenocryst.
5. High-silica rhyolite, Binna Burra, Tweed volcano (sample 38675). Phenocryst.
6. High-silica rhyolite, Mount Tinbeerwan, Noosa area, Glass Houses province (sample 42982). Phenocryst.
7. Melilitite, Shekleton Creek, Boat Harbour, Tasmania (sample BHT). Groundmass.
8. Melilitite, Shekleton Creek, Boat Harbour, Tasmania (sample BHT). Groundmass.
9. Melilitite, Mount Coppin, Monto, central Queensland (sample CQ25). Groundmass.
10. Peralkaline rhyolite, Mount Ngungun, Glass Houses province (sample 38672). Groundmass.

[1] n.d.: not detected.
[2] FeO and Fe₂O₃ calculated assuming stoichiometry.
[3] Less O for F and Cl.

continued

Trachyte, phonolite, and low-silica rhyolite. Pyroxene and olivine in *ne*-trachyte and phonolite are distinguished from those in *qz*-trachyte by slightly different Mg-fractionation behaviour (Fig. 5.3.6), although progressive Fe-enrichment is observed in the trachyte pyroxenes and is correlated with decreasing whole-rock mg-ratios. Pyroxene in the silica-undersaturated trachytes has a smaller range of Mg/Fe values, and is more calcic than those in *qz*-trachyte. The main exception is in the most fractionated phonolite where pyroxene is enriched in Na (Table 5.3.1, column 1). Pyroxene (and olivine) in the most fractionated *qz*-trachytes and low-silica rhyolites extend to pure ferrohedenbergite and fayalite.

High-silica rhyolite. Two distinct ferromagnesian-phenocryst assemblages are found: ferrohedenbergite-fayalite bearing rhyolite, well developed in the Focal Peak volcanic centre (the Mount Gillies Rhyolite; Ross, 1977), but also found in the Mount Alford centre, and in dykes of the Noosa area, all in southeastern Queensland. Pyroxene and olivine compositions in

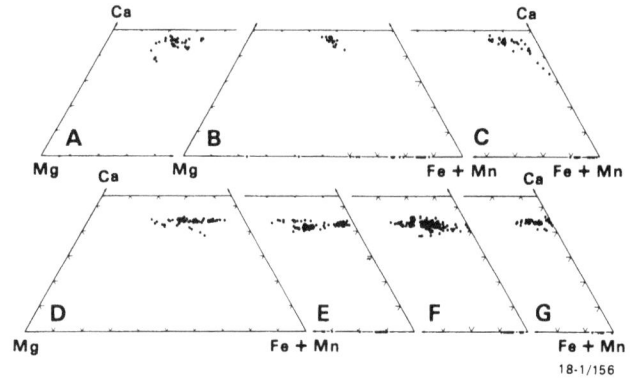

Figure 5.3.6 Microphenocryst and phenocryst pyroxene (filled circles) and olivine (triangles) compositions (atomic percent) for east-Australian felsic rocks. (A) *Ne*-trachytes (whole-rock mg-ratios 21 or more). (B) *Ne*-trachytes (mg-ratios 18–20.6). (C) *Ne*-trachytes and phonolites (mg-ratios 9.8–10.6). (D) *Qz*-trachytes (mg-ratios 18–24). (E) *Qz*-trachytes (mg-ratios 10–18). (F) *Qz*-trachytes (mg-ratios 5–10). (G) *Qz*-trachytes and low-silica rhyolites (mg-ratios less than 5).

Table 5.3.1 Chemical analyses (electron microprobe) of selected minor and accessory minerals in east-Australian volcanic rocks *(continued)*

	11 Ti-free aenig-matite	12 Apatite	13 Apatite	14 Zircon	15 Zircon	16 Chev-kinite	17 Allanite	18 Allanite	19 Monazite	20 Ram-sayite
SiO_2	40.55	0.76	0.68	32.38	32.3	19.28	30.40	29.60	0.35	34.6
TiO_2	n.d.[1]	n.d.	0.13	0.12	0.07	19.58	2.95	2.66	0.04	46.3
ThO_2				0.25	0.11	1.04	0.36	0.38	4.19	
ZrO_2	0.14			64.2	65.0	0.80	n.d.	n.d.	n.d.	
HfO_2				2.19	1.55	n.d.	n.d.	n.d.		
Al_2O_3	0.40	n.d.	n.d.			0.68	11.47	11.40	n.d.	n.d.
La_2O_3						12.6	8.0	8.6	18.2	
Ce_2O_3		0.25	0.32			20.8	12.2	13.1	30.2	
Nd_2O_3						8.2	4.0	3.7	11.0	
Y_2O_3				0.40	0.21	0.56	0.19	0.13	1.17	
Nb_2O_5	n.d.					0.73	n.d.	n.d.	n.d.	
Fe_2O_3	19.93[2]	0.35	0.75							0.89
FeO	29.41			0.45	0.35	9.87	17.35	17.13	n.d.	
MnO	1.17					0.10	0.42	0.23	n.d.	0.12
MgO	0.02	n.d.	n.d.					0.45	0.12	n.d.
NiO										n.d.
CaO	0.11	54.4	53.8			3.50	9.37	9.38	0.63	0.15
BaO										n.d.
Na_2O	7.20	n.d.	0.16					0.09	n.d.	n.d.	17.8
K_2O	0.07	n.d.	n.d.							n.d.
P_2O_5		39.5	40.3						28.81	
F		3.6	4.6			0.34	0.25			
Cl		n.d.	n.d.							n.d.
Total	99.00	97.34[3]	98.80[3]	99.99	99.59	97.94[3]	96.94[3]	96.76	94.71	99.96

11. Sodalite trachyte, Bingie Grumble Mountain, Warrumbungle volcano (sample WMB100; M.B. Duggan, personal communication, 1986).
12. Melilitite, Shekleton Creek, Boat Harbour, Tasmania (sample BHT). Groundmass.
13. Leucitite, 4 km west of Lake Cargelligo (sample LC). Groundmass.
14. High-silica rhyolite, Binna Burra, Tweed volcano (sample 38675). Phenocryst. High-Hf type.
15. High-silica rhyolite, Binna Burra, Tweed volcano (sample 38675). Phenocryst. Lower-Hf type.
16. High-silica rhyolite, Mount Gillies, southeastern Queensland (sample 33033). Microphenocryst.
17. Low-silica rhyolite, Minerva Hills, central Queensland (sample 38739c). Phenocryst.
18. High-silica rhyolite, Noosa area, Glass Houses province (sample AE79). Microphenocryst.
19. High-silica rhyolite, Noosa area, Glass Houses province (sample AE81). Microphenocryst.
20. Mela-nephelinite pegmatoid, near Bacchus Marsh, Victoria. After Ferguson (1977a).

[1] n.d.: not detected.
[2] FeO and Fe_2O_3 calculated assuming stoichiometry.
[3] Less O for F and Cl.

these rhyolites are strongly Fe-enriched, commonly pure hedenbergite and fayalite (Figs. 5.3.7B). Relatively few samples are known to contain more magnesian compositions. The second pyroxene assemblage is characterised by ferrohypersthene-eulite (Fs_{50-74}), normally not co-existing with either ferroaugite-ferro-hedenbergite or fayalite (only a rare sample from the Mount Gillies Rhyolite contains all three minerals; Ewart et al., 1976; Ewart, 1985; Fig. 5.3.7A). These orthopyroxene-bearing rhyolites are found in Tweed volcano (the Nimbin and Springbrook rhyolites) and the Mount Alford centre. Samples commonly also contain small numbers of augite and hypersthene to bronzite xenocrysts.

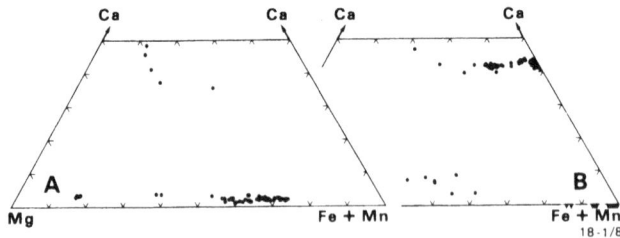

Figure 5.3.7 Phenocryst pyroxene (filled circles) and olivine (triangles) compositions (atomic percent) for two high-silica rhyolite types from southeastern Queensland. (A) Ferro-hypersthene-bearing rhyolites. (B) Ferrohedenbergite fayalite-bearing rhyolites.

Na enrichment in pyroxene

Peralkaline trachyte and rhyolite contain aegirine-augite and aegirine as groundmass minerals. Rare phenocrysts of ferroaugite-ferrohedenbergite commonly are rimmed by sodic pyroxene. In addition, sodic pyroxene is found in some phonolites, qz-trachytes, and microsyenites (whose bulk compositions are not peralkaline), typically as fine-grained groundmass or interstitial grains, reflecting late-stage development of peralkalinity during crystallisation. Another important occurrence of sodic pyroxene is in leucitite and 'mela-nephelinite' pegmatoids from Cosgrove and Bacchus Marsh, Victoria (Ferguson, 1977a; Birch, 1979). A summary of individual analyses of the different calcic and sodic-pyroxene compositions observed in five major volcanic and subvolcanic groups of lavas and intrusions is illustrated in Figure 5.3.8. Complete solid-solution exists between ferrohedenbergite and aegirine in the intermediate and felsic (peralkaline and non-peralkaline) rocks. However, Na enrichment in these rocks is present only where the calcic pyroxene is already strongly Fe enriched. Na enrichment in the pyroxene of the leucitite and nephelinite pegmatoids (and, to a lesser extent, the ne-trachyte and phonolite) takes place at an earlier and more magnesian stage of pyroxene crystallisation and, again, includes a complete range of compositions to aegirine.

Ewart et al. (1976) and Ewart (1981) explained Na enrichment in pyroxene in terms of the interplay between three major variables — namely, oxygen fugacity (fO_2) and the activities of Al_2O_3 and SiO_2 in the liquid. Increasing fO_2, for a given Al_2O_3 activity and temperature, favours increasing Na enrichment, whereas if crystallisation follows a constant fO_2 buffer, then decreasing Al_2O_3 activity favours increasing Na enrich-

ment. Decreasing SiO_2 activity tends to favour the earlier onset of Na enrichment, as seen in the leucitite and nephelinite pegmatoids. Crystallisation of the non-peralkaline, quartz-normative, felsic magmas of the region is inferred to have taken place at an fO_2 below the FMQ buffer (see below), shown by Ewart (1981) to be consistent with the onset of Na enrichment only after extreme Fe enrichment has taken place. Na enrichment in the pyroxene of the single Cainozoic phonolite studied (from Mount Wilson, Victoria; Fig. 5.3.8B) is at a slightly less extreme stage of Fe enrichment, again consistent with lower SiO_2 activity. Similar behaviour is found in Jurassic phonolites from the Mullaley District of New South Wales (Ewart, 1985). Ferguson (1977a,b) and Birch (1979) showed that the aegirine in the Victorian leucitite and 'mela-nephelinite' pegmatoids, and in a pegmatoid within a New South Wales leucitite, is a titanian variety.

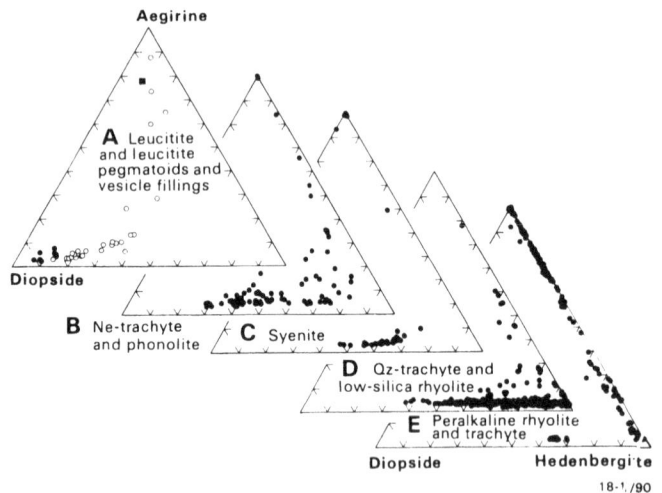

Figure 5.3.8 Na-enrichment trends for groundmass and phenocryst pyroxenes in east-Australian leucitites and 'mela-nephelinite' pegmatoids, peralkaline and non-peralkaline trachytes, micro-syenites, and rhyolites. Data for rocks from southeast-Queensland, Nandewar, Warrumbungle, and Canobolas volcanoes, and from central Victoria. (A) Filled circles represent leucitites. Open circles represent leucitite pegmatoids and vesicle fillings. Filled square represents a phonolitic segregation in a New South Wales leucitite.

Zr enrichment is a characteristic feature accompanying increasing Na from hedenbergite to aegirine, and abundances of 2–3 weight-percent ZrO_2 are found commonly (Ewart, 1981; Stolz, 1986). However, up to 14-percent ZrO_2 has been found in aegirines in peralkaline qz-trachytes from Warrumbungle volcano (M.B. Duggan, personal communication, 1986), supporting the existence of the Zr-pyroxene end-member component (Jones & Peckett, 1980).

Ti-Al relationships in pyroxene of mafic rocks

Pyroxene compositions plotted in Figure 5.3.9 are based on average groundmass and phenocryst data for individual rocks, although in many samples there are several distinct populations of, for example, ground-mass pyroxenes that have relatively small but distinct compositional differences. These correspond generally to coarser groundmass cores, groundmass-grain rims,

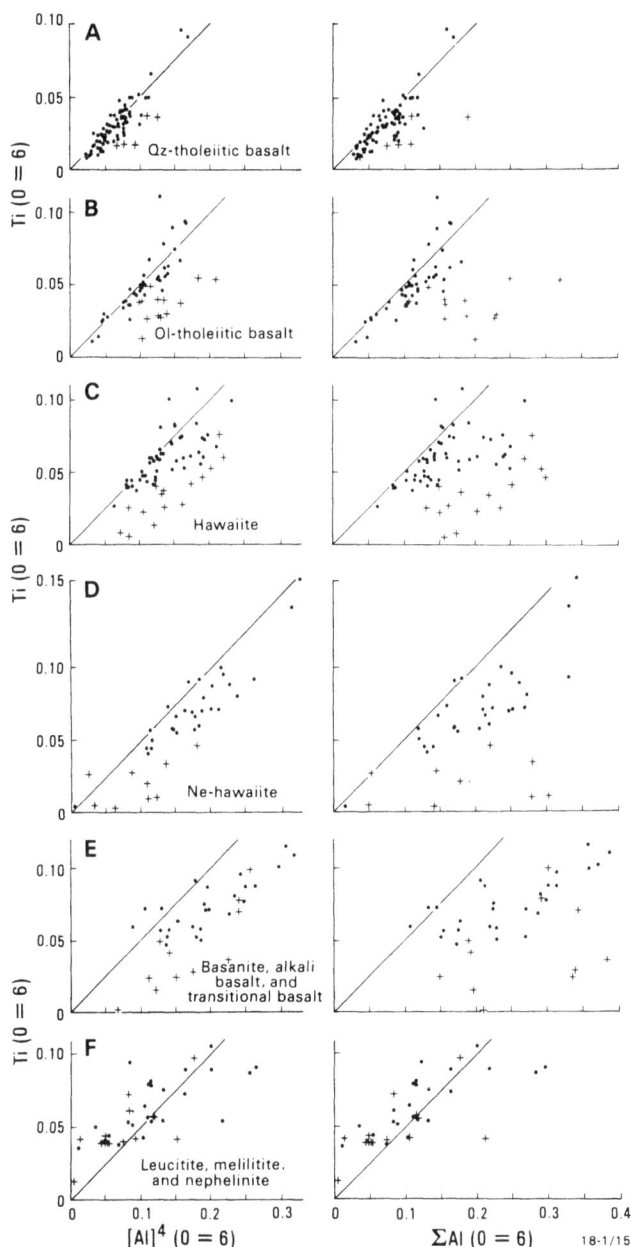

Figure 5.3.9 Ti:[Al]^4 and Ti:total-Al(O=6) relationships for pyroxene in east-Australian mafic rocks. Crosses represent phenocryst compositions. Filled circles represent groundmass compositions.

and very fine-grained interstitial groundmass grains. All such average data for the different textural and compositional pyroxene types are included in Figure 5.3.9. The data are subdivided on the basis of whole-rock compositions, and in each graph a line is presented representing the pyroxene component $R_2TiAl_2O_6$.

There is a systematic change in Ti-Al relationships for the groundmass pyroxenes through the range of basaltic compositions, both in terms of absolute abundances of Ti and Al and, more significantly, in terms of Ti/Al values. Pyroxenes in the tholeiitic basalts (Fig. 5.3.9A-B) conform closely to a Ti:[Al]^4 ratio of 1:2, which is consistent with these elements being controlled by the $R^{2+}Al_2O_6$ component. A progressive decrease in Ti:[Al]^4 takes place from hawaiitic to basanitic and alkali-basaltic pyroxene

compositions. This feature is illustrated even more strongly by the corresponding Ti and total-Al plots (Fig. 5.3.9).

The increase in total Al correlates, therefore, with increasing silica undersaturation of the host rocks, and is attributed to increasing solid-solution of the Ca-tschermak's component ($CaAlAlSiO_6$) following the reaction (see, for example, Carmichael et al., 1974):

$$CaAl_2SiO_6 + SiO_2 = CaAl_2Si_2O_8$$
pyroxene s.s.　　liquid　　anorthite

This, however, does not account fully for the decreasing Ti:[Al]^4 ratios as SiO_2 activity (in the liquid) decreases. This decrease is thought to be caused by increasing solid-solution of the $CaCrAlSiO_6$ component or, alternatively, to incorporation of some Ti into a $NaTiSi_2O_6$ or $Na(TiFe)Si_2O_6$ component (see, for example, Larsen 1976).

There is a tendency for the Ti and Al to decrease in the groundmass pyroxenes of leucitite, melilitite, and nephelinite, relative to values for basanite and alkali basalt (Fig. 5.3.9F). Carmichael et al. (1974) suggested that for melilite-bearing lavas, decreasing Al solid-solution may be represented by the reaction:

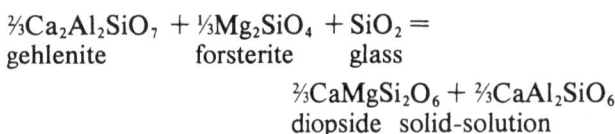

$$\tfrac{2}{3}Ca_2Al_2SiO_7 + \tfrac{1}{3}Mg_2SiO_4 + SiO_2 =$$
gehlenite　　　　forsterite　　　glass

$$\tfrac{2}{3}CaMgSi_2O_6 + \tfrac{2}{3}CaAl_2SiO_6$$
diopside solid-solution

However, igneous melilites are gehlenite-deficient (see below), so a more realistic reaction may be of the type:

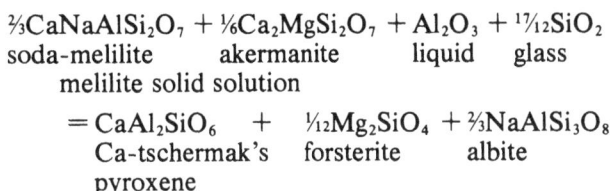

$$\tfrac{2}{3}CaNaAlSi_2O_7 + \tfrac{1}{6}Ca_2MgSi_2O_7 + Al_2O_3 + \tfrac{17}{12}SiO_2$$
soda-melilite　　akermanite　　　liquid　　glass
melilite solid solution

$$= CaAl_2SiO_6 + \tfrac{1}{12}Mg_2SiO_4 + \tfrac{2}{3}NaAlSi_3O_8$$
Ca-tschermak's　forsterite　　albite
pyroxene

A more complex relationship therefore is indicated, involving silica and alumina activities of the liquids. There is insufficient Si+Al in leucititic pyroxenes to be able to fill the T site to the stoichiometric 2.00 atoms per formula unit, indicative that Ti^{4+} occupies the T site, which is consistent with the results of recent crystallographic studies (Cundari & Saliulo, 1987). A similar Si+Al deficiency has been observed within the melilitite pyroxenes and, less systematically, the pyroxenes of nephelinite during the course of this study. These observations, together with the leucitite pyroxene data, can be related to the common increase of Ti:Al ratios to values greater than 0.5. A possible explanation for the decreasing Al solid-solution for the leucitite pyroxenes may relate to the coexistence of amphibole and phlogopite, both of which are extremely rare in other east-Australian rocks (see below).

One additional aspect shown by Figure 5.3.9 is the consistently higher [Al]^4 and total-Al solid-solution levels in the phenocrysts, compared to groundmass pyroxenes, in all mafic rock compositions (except for leucitite). This is interpreted to result from increased total-pressure effects during phenocryst crystallisation.

Cr in clinopyroxene of mafic rocks

Data based on routine microprobe analyses of individual crystals (plotted in Fig. 5.3.10) are grouped into four whole-rock compositional types, in each of which Cr behaviour in the pyroxenes is similar. Increasing Cr clearly correlates with increasing MgO (that is, decreasing $Fe/(Fe+Mn+Mg+Ca)$ values) in both phenocryst and groundmass pyroxenes, and phenocryst-core pyroxenes normally are strongly enriched in Cr relative to groundmass and phenocryst-rim compositions.

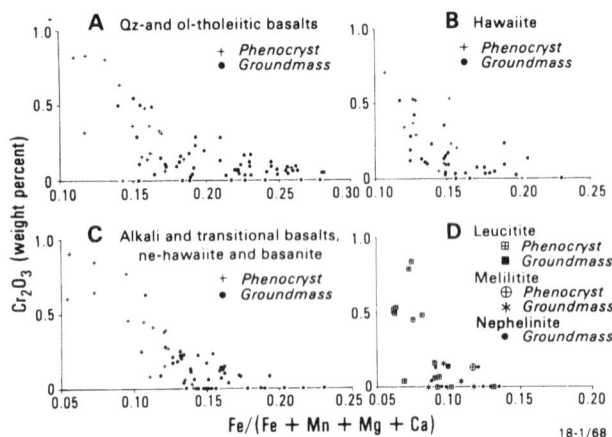

Figure 5.3.10 Relationships between Cr_2O_3 and $Fe/(Fe+Mn+Mg+Ca)$ (atomic percent) in calcic pyroxenes from east-Australian mafic rocks. Points based on average microprobe analyses for individual rocks.

Maximum Cr levels are similar in phenocryst pyroxene found throughout the whole-rock compositional range, from tholeiitic basalts to the strongly undersaturated magmas, and presumably represents initial precipitation from primitive magmas containing similarly high Cr abundances. However, whether these phenocrysts are strictly in equilibrium with the rock compositions in which they are found now may be questioned in view of the olivine data presented above. This is especially so for the tholeiitic basalts. Pyroxene in the tholeiitic rocks has higher abundance levels of Cr at higher Fe ratios than is observed in the pyroxenes of the more alkaline rocks, and tends to have less rapid Cr depletion as Fe increases. In contrast, the pyroxenes for melilitite, nephelinite, and leucitite (Fig. 5.3.10D) have a sharp decrease in Cr over relatively small changes of Fe. These differences may correlate with the progressively lower Ca levels in the pyroxenes as the degree of silica saturation of the rock increases, as discussed above.

Mn substitution in calcic and sodic pyroxenes

The behaviour of Mn in relation to Fe enrichment in the augite-ferrohedenbergite-aegirine series of pyroxenes is shown in Figure 5.3.11, based mainly on average data from the different rocks analysed. Mn in calcic pyroxene predictably correlates positively with total Fe (which is dominantly Fe^{2+}), although the slope of the data trend

is by no means constant. Mn values decrease abruptly but quite regularly in aegirine-augite and aegirine as Fe enrichment increases. This must reflect Mn^{2+}/Fe^{2+} substitution where progressively lower Fe^{2+} (and higher Fe^{3+}) is found in those pyroxenes more closely approaching pure aegirine. Attention is drawn also to the higher Mn levels observed in several of the ferroaugite and aegirine-augite crystals in *ne*-trachyte and phonolite (Fig. 5.3.11).

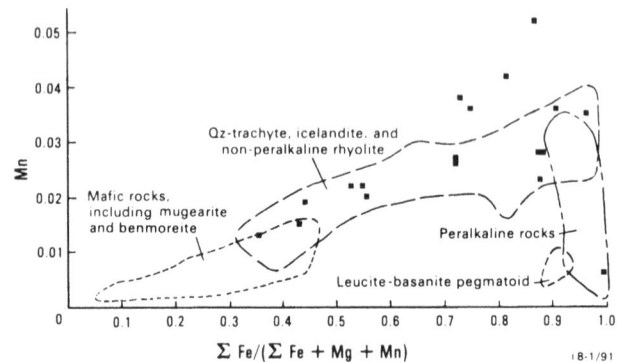

Figure 5.3.11 Mn (O=6) versus $Fe/(Fe+Mn+Mg)$ relationships (atomic percent) in calcic and sodic pyroxenes from east-Australian rocks. Based on average microprobe data for individual rock samples, except for aegirine and aegirine-augite compositions in peralkaline rocks. Filled squares represent compositions for pyroxenes in *ne*-trachytes and phonolites.

Ca in olivine of mafic rocks

A plot of CaO as a function of olivine mg-ratios for phenocryst cores, phenocryst rims, and groundmass grains (Fig. 5.3.12) is based on microprobe analyses of individual grains. The data again are grouped according to host-rock chemistry.

Stormer (1973) demonstrated that Ca solid-solution in olivine is a function of silica activity, temperature, and pressure. He further showed that olivine phenocrysts from nephelinites and basanites are strongly zoned towards more calcic compositions, whereas olivine phenocrysts in tholeiitic basalts have little Ca enrichment. This general relationship is confirmed in more detail in Figure 5.3.12 for a wide range of olivine compositions for east-Australian mafic rocks, where a progressive increase in the slope of Ca enrichment as Fe enrichment increases, is seen from tholeiitic, through hawaiitic, to strongly silica-undersaturated rocks. Note that in all the plots in Figure 5.3.12 the extension towards more Fe-rich compositions is based on phenocryst-rim and groundmass-olivine analyses — that is, the trends are not likely to be controlled by pressure effects (see: Stormer, 1973; Ferguson, 1978b). Rather, they are interpreted to result primarily from differences in SiO_2 activity in the liquid. In-situ crystallisation, even within the same chemical grouping of lavas, will produce changes in SiO_2 activity for individual samples. This presumably is responsible for some of the scatter of data points, as such variability is expected to increase in the more silica-undersaturated lava compositions.

A second feature is the notably low CaO contents of most, but not all, of the most magnesian olivine grains

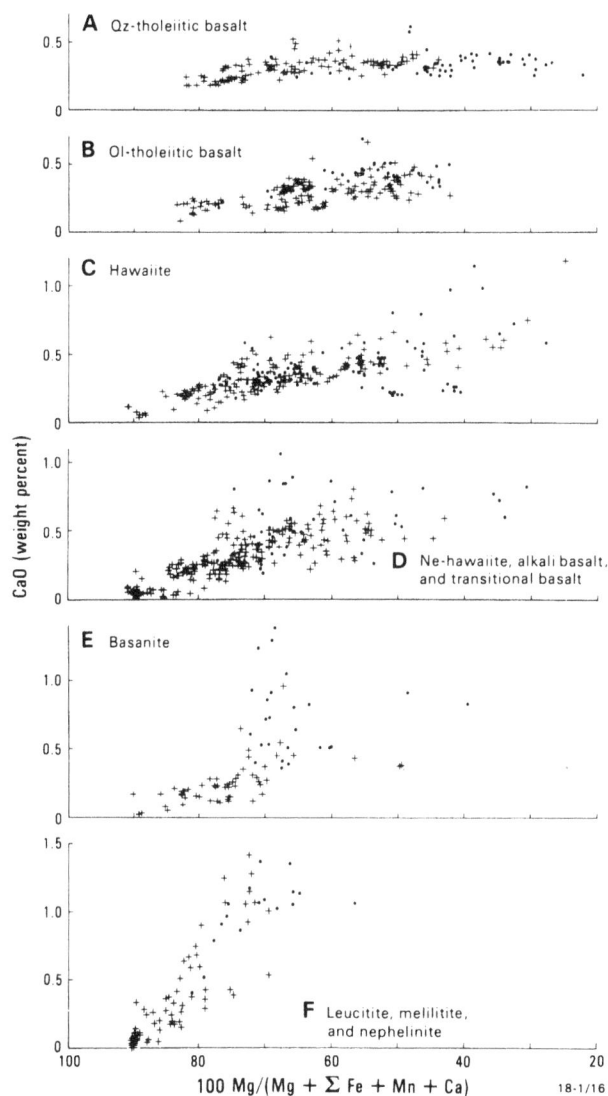

Figure 5.3.12   CaO content in olivine versus olivine composition (atomic percent) for east-Australian mafic rocks. Crosses represent phenocryst compositions. Filled circles represent groundmass and phenocryst-rim compositions.

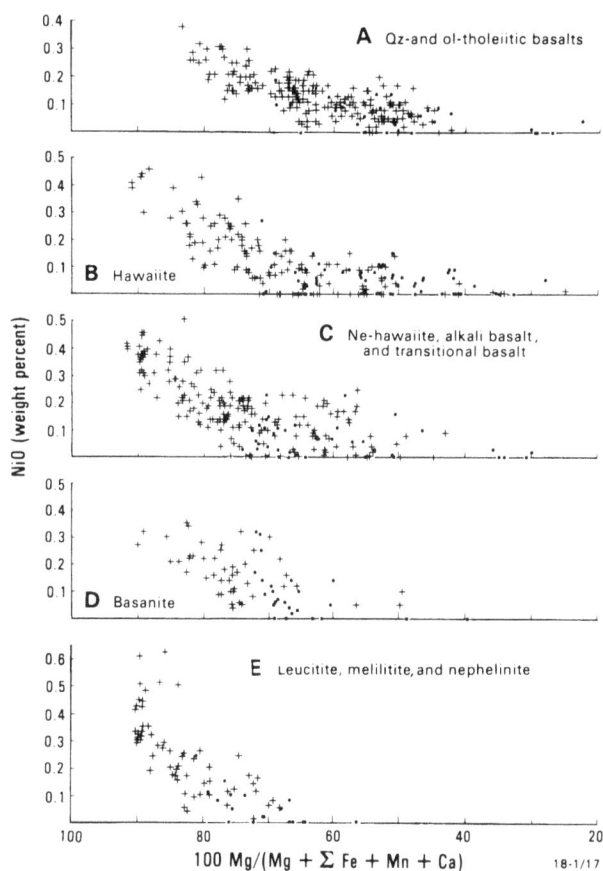

Figure 5.3.13   NiO content in olivine versus olivine composition (atomic percent) for east-Australian mafic rocks. Crosses represent phenocryst compositions. Filled circles represent groundmass and phenocryst-rim compositions.

found especially in the more alkaline lavas (Figs. 5.3.12C–F). These low values are typical of olivines from lherzolite nodules included in alkaline lavas throughout the east-Australian region (see also Simkin & Smith, 1970). The highly magnesian cores of these olivine phenocrysts therefore may be xenocrystal, which is in accord with the systematic presence of highly magnesian olivine in the alkaline lavas — too magnesian in many instances to be in equilibrium with the whole rock (see Fig. 5.3.1).

Ni in olivine of mafic rocks

Relevant data (Fig. 5.3.13) are based on routine microprobe analyses of individual grains in which phenocryst cores and rims, and groundmass grains, were analysed. The data are grouped into fields on the basis of mafic-rock compositions. The predictable, progressive decrease in Ni as Fe increases in the olivines is evident, but the degree of Ni depletion is not necessarily similar within the different rock groups.

There is a relatively low-angle Ni-depletion curve for olivine in the tholeiitic basalt and hawaiitic rocks, whereas Ni in the most undersaturated lavas (Fig. 5.3.13E) decreases strongly over quite restricted compositional intervals. These types of pattern are similar to those observed for Cr in pyroxene, as described above. However, this relationship appears to be somewhat anomalous in relation to olivine-liquid distribution coefficients for Ni, which have been shown to decrease as MgO increases in the melt (Hart & Davis, 1978). Note that because the most under-saturated lavas are significantly more magnesian than are the tholeiitic basalts, Ni in olivine in tholeiitic rocks should be depleted more sharply than Ni in the olivine in nephelinite, melilitite, and leucitite, assuming that only closed-system fractional crystallisation was involved. An additional factor is temperature (Leeman & Lindstrom, 1978), possibly implying different cooling histories for the tholeiitic and more alkaline rocks. Higher eruption temperatures have been obtained from geothermometric estimates for the tholeiitic rocks (see Section 5.3.11).

5.3.3   Feldspar

Mafic rocks

Phenocryst plagioclase in the tholeiitic basalts is typically labradorite to calcic andesine, uncommonly bytownite, and in most samples has relatively narrow rims zoned to sodic andesine (for example, Fig. 5.3.14). Phenocrysts in hawaiite and *ne*-hawaiite are also mainly labradorite to calcic andesine, commonly including slightly enhanced Or solid-solution related to the more potassic whole-rock compositions. Phenocryst plagioclase in these east-Australian mafic rocks generally does not have well developed oscillatory zoning, except adjacent to the rims. Calcic-anorthoclase/sanidine (Fig. 5.3.14D), anorthoclase (Fig. 5.3.14A), and potassic oligoclase (Fig. 5.3.14I) are found as megacryst/xenocryst minerals in some hawaiites and potassic *qz*-tholeiitic basalts, and are considered indicative of magma mixing.

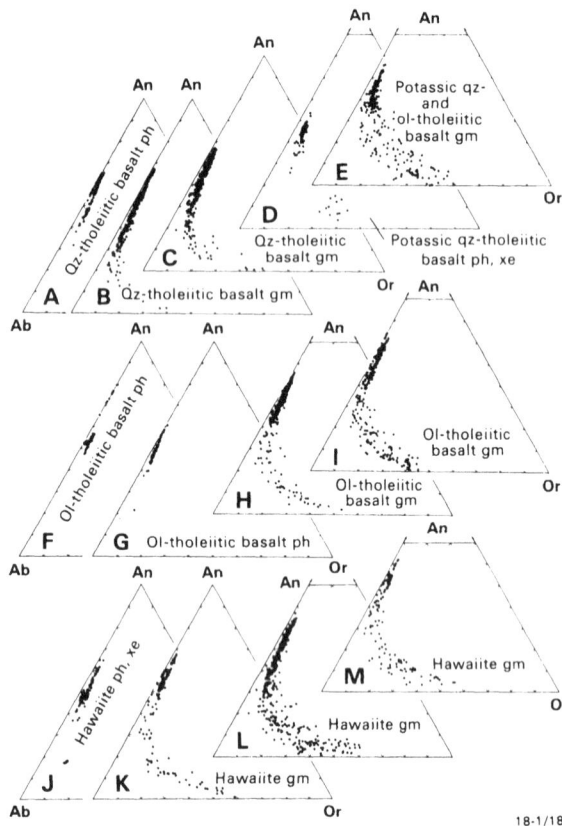

Figure 5.3.14 Phenocryst (ph), xenocryst (xe), and ground-mass (gm) feldspar compositions (molecular percent) for east-Australian tholeiitic basalts and hawaiites.

Groundmass feldspar has a continuous range of compositions from labradorite through to calcic anorthoclase, together with rare grains of sanidine, in the *qz*-tholeiitic basalts (excluding potassic types). *Ol*-tholeiitic basalt and hawaiite contain a greater proportion of groundmass feldspar extending through to calcic anorthoclase and sanidine compositions (to about Or_{60}). These trends are accentuated further in *ne*-hawaiite to basanite, although relatively fewer alkali feldspars are found in basanite. The presence of more potassic groundmass-feldspar compositions is attributed to the generally more potassic chemistry of the silica-undersaturated lavas.

The importance of whole-rock potassium is illustrated further by a comparison of *qz*-tholeiitic-basalt ground-mass feldspars (Figs. 5.3.14B–C) with those in the potassic *qz*-tholeiitic basalts (Fig. 5.3.14E). The

different groundmass-feldspar compositional ranges correlate with their different textural types. Groundmass feldspar in the majority of the lavas is present most abundantly as microlites or laths, the core areas of which are typically labradorite. Each crystal is normally zoned, and the change of composition becomes more marked towards the margins where compositions are commonly oligoclase or anorthoclase. The finest-grain interstitial feldspar is typically sanidine.

Two additional points are noteworthy in relation to the groundmass feldspars. First, there is a positive correlation between whole-rock normative-plagioclase composition and the most calcic groundmass plagioclase composition. Second, there is an increase in the maximum levels of ternary solid-solution in the sequence *qz*-tholeiitic, to *ol*-tholeiitic basalt, to hawaiite, *ne*-hawaiite, and basanite (Fig. 5.3.15).

Feldspar is absent in nephelinite and melilitite. Leucitite contains interstitial sanidine (Or_{79-88}, containing low An, mostly less than 1.0 percent).

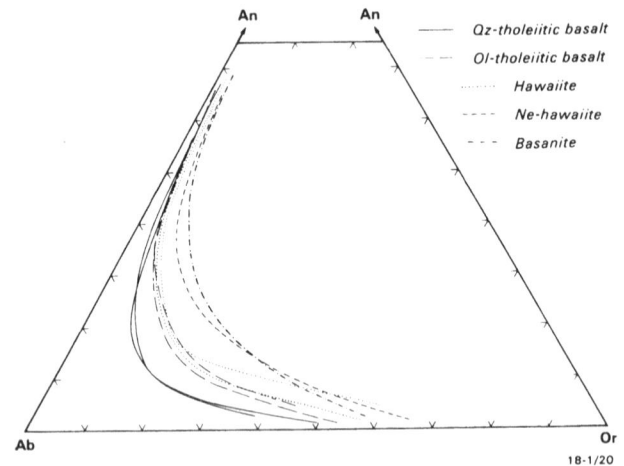

Figure 5.3.15 Limits of ternary solid-solution (molecular percent) for groundmass feldspars in the main mafic volcanic-rock types of eastern Australia.

Intermediate rocks

Feldspar phenocrysts have a wide range of compositions in the intermediate rocks, again reflecting the different K_2O abundances of the whole rocks. Benmoreite and icelandite characteristically have plagioclase ranging from andesine to oligoclase, whereas compositions in two analysed potassic mugearites range from labradorite and andesine through to calcic anorthoclase and anorthoclase rims. The phenocrysts have complex zoning patterns, including the distinct alternation of sodic plagioclase and anorthoclase zones in some crystals (Ewart, 1985), together with other abrupt compositional changes. Such lavas may have resulted from magma-mixing processes, involving trachyte and a second magma type more mafic than mugearite. A second example of a disequilibrium phenocryst-feldspar assemblage is provided by icelandite from Tweed volcano to which reference was made in Section 5.3.2. Feldspar phenocrysts in these lavas range from labradorite to sanidine.

Groundmass feldspar in the intermediate rocks typically extends from andesine through calcic anorthoclase to sodic sanidine. The plagioclase is less calcic in the icelandites than in the benmoreites.

Trachyte

Phenocryst feldspar is a characteristic mineral in trachyte. Assemblages consist of discrete plagioclase plus sanidine, but are more commonly potassic oligoclase, calcic anorthoclase, and sodic sanidine (Fig. 5.3.16). This complete compositional range may be found in a single phenocryst. Ewart (1985) showed that individual trachyte feldspars also may have discrete compositional breaks within the anorthoclase compositional range, corresponding to two (rarely three) distinct alkali-feldspar compositional groupings. These compositions evolved towards the alkali-feldspar minima during crystallisation.

Overall feldspar-compositional changes broadly conform to the phase relationships expected for the ternary-feldspar system (Tuttle & Bowen, 1958), but the feldspar-zoning patterns in the trachytes are complex, even within a single rock sample (Ewart, 1985). These changes include: anorthoclase cores and potassic envelopes that may have gradational, sharp, or oscillatory contacts; relatively large-scale reversals or oscillations between sodic and potassic compositions; plagioclase cores and sharply defined outer zones of sanidine; sanidine cores zoned towards more sodic rims; alternations between sodic plagioclase and anorthoclase-sanidine compositions, typically involving abrupt changes. The complexity of the crystallisation histories recorded in these feldspars is explained (Ewart, 1985) in terms of dynamic-crystallisation models such as those proposed, for example, by McBirney (1980), Turner & Gustafson (1981), and Sparks et al. (1984).

Groundmass-feldspar compositions in the trachytes are anorthoclase-sanidine, characterised by lower levels of anorthite solid-solution relative to the co-existing phenocrysts.

Rhyolite

Phenocryst-feldspar compositions for four types of high-silica rhyolite from southeastern Queensland, and low-silica rhyolite of the Minerva Hills complex, south-central Queensland, are shown in Figure 5.3.17. The following phenocryst assemblages are recognised in the high-silica rhyolites.

(1) *Sodic-plagioclase and sanidine* (Or_{55-65}), characteristic of the biotite-bearing, the ferrohypersthene-bearing, and the Binna Burra rhyolites. Plagioclase compositions range from sodic oligoclase (Fig. 5.3.17D; Binna Burra Rhyolite) to median oligoclase (Fig. 5.3.17C; biotite-bearing rhyolite) to calcic oligoclase-andesine (Fig. 5.3.17A; ferrohypersthene rhyolite). The progressive change in these plagioclase compositions is interpreted as fractional-crystallisation controlled and, as discussed below, correlates closely with the relative trace-element behaviour in these rhyolite types.

(2) *A single sanidine* (Or_{40-55}), commonly zoned to sodic sanidine or anorthoclase, and rarely having cores

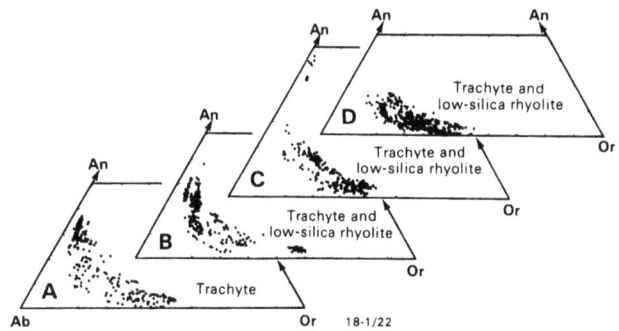

Figure 5.3.16 Phenocryst-feldspar compositions (molecular percent) for east-Australian trachytes and low-silica rhyolites. (A) Whole-rock normative-An contents greater than 14 percent. (B) Normative An greater than 9 and up to 14 percent. (C) Normative An greater than 6 and up to 9 percent. (D) Normative An greater than 0.1 and up to 6 percent.

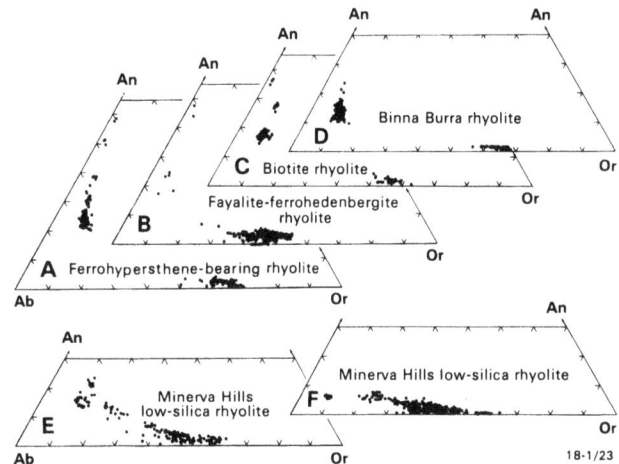

Figure 5.3.17 Phenocryst-feldspar compositions (molecular percent) for east-Australian high-silica rhyolites from southeastern Queensland (A–D) and low-silica rhyolites of the Minerva Hills, central Queensland (E–F). Rhyolites in E have normative-An contents of 4–8 percent and in F 1–3.5 percent.

of partially resorbed plagioclase (Fig. 5.3.17B). This type of feldspar assemblage has been recognised only in fayalite-ferrohedenbergite rhyolite. The sanidine has only relatively minor normal zoning. It is considered most likely that a primary plagioclase initially crystallised from these rhyolitic magmas, but was subsequently resorbed (Ewart, 1985).

The feldspar compositions in Minerva Hills low-silica rhyolite (Figs. 5.3.17E-F) are dominated by calcic-anorthoclase and sodic sanidine, but one sample contains phenocrysts of primary albite (Fig. 5.3.17F). Again, there is a decrease in anorthite and a general contraction in overall compositions towards the potassic-anorthoclase/sodic-sanidine region as whole-rock normative-An decreases. The feldspar compositional trends in these low-silica rhyolites are closely similar to those in trachyte but are clearly distinct from the feldspars in high-silica rhyolite. High-silica rhyolite, therefore, cannot be interpreted simply as the end product of fractional crystallisation of trachyte or low-silica rhyolite (Ewart, 1985). This is in accord with other mineralogical data, and with the trace-element and isotopic data presented below (see also Ewart, 1982a).

Peralkaline rocks

Peralkaline whole-rock compositions project close to the thermal valley extending from the binary minimum in the alkali-feldspar system into the ternary Qz-Ab-Or system (Tuttle & Bowen, 1958), as discussed in Section 5.3.1. East Australian rocks are typically only mildly peralkaline, and therefore phenocryst and microphenocryst feldspars precipitating from these magmas should concentrate within the anorthoclase/sodic-sanidine compositional range. This indeed is observed for most phenocryst and microphenocryst feldspars (Fig. 5.3.18) although, in detail, compositions clearly reflect the $K_2O/(K_2O+Na_2O)$ values of the host rocks. They are also low in Ca (see also Ewart, 1985). The data, therefore, are broadly consistent with the conclusion of Thompson & MacKenzie (1967) that synthetic liquids fractionating in the feldspar volume in the SiO_2-Al_2O_3-Na_2O-K_2O system will tend to move into a low-temperature zone and be held in it by crystallisation of a feldspar of restricted composition (Or_{30-40}). The natural feldspars, however, have wider compositional ranges (Fig. 5.3.18), even within individual samples, though the ranges decrease in more potassic lavas (Figs. 5.3.18A-B). Nicholls & Carmichael (1969) also noted greater compositional ranges in feldspars from comendite compared to those from more peralkaline pantellerite, and this is predictable from the fractionation curves deduced by Thompson & MacKenzie (1967) for liquids moving into the peralkaline low-temperature trough.

The compositions of the more potassic feldspars (Figs. 5.3.18A-B) overlap with those for fayalite-bearing, high-silica rhyolites described above (see Fig.

5.3.17). Ewart (1985) suggested that these more potassic peralkaline rhyolites may have developed from high-silica rhyolite by extended 'ternary minimum' fractional crystallisation. There is additional trace-element and isotopic data in support of this petrogenesis for at least one of the potassic peralkaline rhyolites from Tweed volcano (Ewart et al., 1977; Ewart, 1982a). The sodic and potassic peralkaline rocks, therefore, may have developed through separate evolutionary paths.

5.3.4 Amphibole

Nine types of amphibole occurrence are identifiable in relation to the host-magma types in which they are found (megacryst and xenocryst occurrences are excluded):

(1) Amphiboles are present as a ubiquitous ground-mass mineral — and, less commonly, as microphenocrysts — in peralkaline rhyolite and trachyte. These are mainly fluor-arfvedsonite to fluor-riebeckite (Fig. 5.3.19; Table 5.3.1, column 2) and are nearly all almost pure-Fe end-members (Fig. 5.3.20). Their fluorine contents are mostly near maximum limits, corresponding to little or no OH replacement. Cl is below detection limits. ZrO_2 contents reach levels of 0.7–1.3 percent, although are more commonly between 0.2 to 0.6 percent.

(2) Accessory groundmass amphibole found in leucitite — and, more commonly, in associated pegmatoidal assemblages — from New South Wales

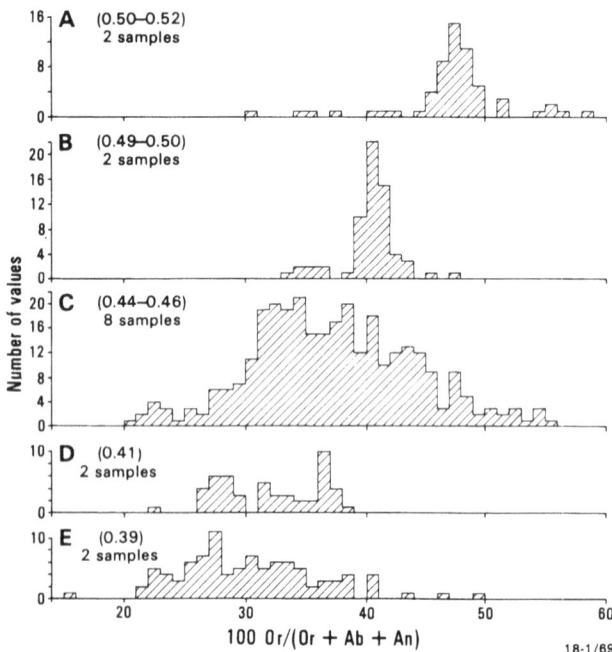

Figure 5.3.18 Histograms of phenocryst- and microphenocryst-feldspar compositions (molecular percent) for peralkaline trachyte and rhyolite from eastern Australia. Values in parentheses represent $K_2O/(Na_2O+K_2O)$ weight ratios for whole rocks.

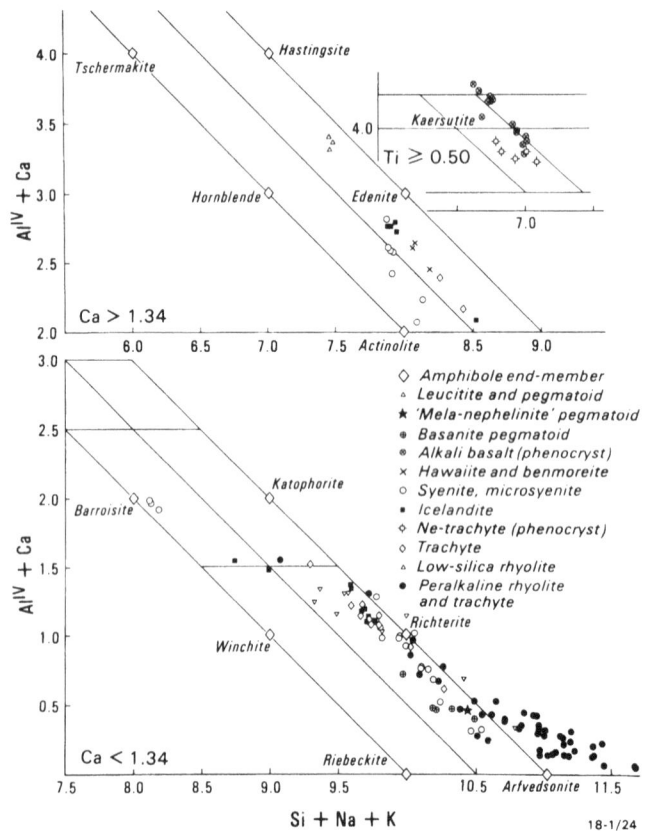

Figure 5.3.19 Groundmass and phenocryst amphibole compositions (atomic percent) for east-Australian volcanic rocks. Si+Na+K and Al^{IV}+Ca values are based on (O, OH, F) equal to 23, except for arfvedsonite where (O, OH, F) is assumed to be 24. Plotting procedure after the method of Giret et al. (1980).

x Leucitite and pegmatoids
* 'Mela-nephelinite' pegmatoid
▲ Basanite pegmatoid
▼ Alkali basalt (kaersutites)
○ Microsyenite and syenite
▽ Hawaiite and benmoreite
■ Icelandite
□ Ne-trachyte (phenocryst)
◇ Trachyte pegmatoid
⊞ Low-silica rhyolite
● Peralkaline rocks

18-1/70

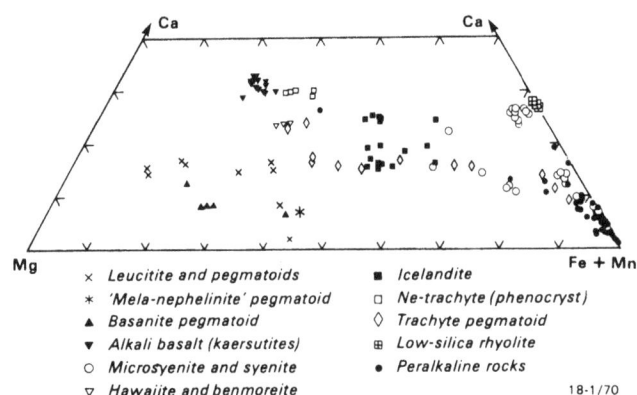

Figure 5.3.20   Ca:Mg:Fe+Mn relationships (atomic percent) in amphiboles from east-Australian volcanic rocks.

and Cosgrove, Victoria (Cundari, 1973; Ferguson, 1978a; Birch, 1979) are titaniferous richterites (Table 5.3.1, column 3). The Cosgrove examples are less titaniferous. They are magnesian, although there is limited Fe enrichment in the amphibole of some pegmatoidal facies (Fig. 5.3.20). The groundmass amphibole in some pegmatoids grades into magnesio-arfvedsonite. Fluorine abundances measured on a small number of grains in leucitite and the pegmatoids range between 1.3–2.4 percent (Ferguson, 1978d; Birch, 1979).

(3) Titaniferous magnesio-arfvedsonite — almost certainly a fluor variety — is found in a pegmatoidal phase of a single leucite basanite from the Main Range, southeastern Queensland (Grenfell, 1984). A similar amphibole is reported also from a 'mela-nephelinite' pegmatoid (quartz-bearing) from Bacchus Marsh, Victoria (Ferguson, 1977a).

(4) Kaersutitic phenocryst and groundmass amphibole are found in intrusive alkaline rocks from the Main Range (Grenfell, 1984). Kaersutite phenocrysts (low fluorine) are present also in a Jurassic *ne*-trachyte from the Mullaley District, New South Wales (Ewart, 1985). Their compositions are plotted also in Figures 5.3.19–20.

(5) Icelandite from several southeast-Queensland centres and from the Mount Warning intrusive complex (Tweed volcano) contains minor amphibole which ranges from groundmass grains of ferrorichterite and ferroedenite to microphenocrysts ranging in composition through winchite and barroisite to edenite — again, all of them relatively Fe rich (Figs. 5.3.19–20).

(6) Richteritic amphiboles are found in pegmatoidal phases of two central Victorian *ne*-trachytes (Ferguson, 1978c).

(7) Miocene syenite and microsyenite intrusions in southeastern Queensland, and the Mount Warning intrusive complex, commonly contain interstitial ferrorichterite that grades compositionally in some rocks to Fe-rich arfvedsonite. In addition, one micro-syenite contains late-stage ferroactinolite/ferroedenite. The crystallisation sequence ferrohedenbergite–sodic-ferrohedenbergite–ferrorichterite–ferroactinolite, has been determined for this rock using textural relationships (Grenfell, 1984). Interstitial ferrorichterite in these syenitic rocks is related to late-stage Na enrichment in the coexisting interstitial pyroxene. Grenfell (1984)

noted that a late-stage stabilisation of ferrorichterite at the expense of acmitic pyroxene is suggested by textural relationships:

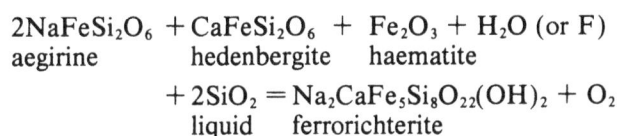

$$2NaFeSi_2O_6 + CaFeSi_2O_6 + Fe_2O_3 + H_2O \text{ (or F)}$$
aegirine   hedenbergite   haematite

$$+ 2SiO_2 = Na_2CaFe_5Si_8O_{22}(OH)_2 + O_2$$
liquid   ferrorichterite

(8) Phenocrysts of a pure Fe-amphibole are found in about one third of the low-silica rhyolite intrusions and lavas of the Minerva Hills complex of south-central Queensland. This amphibole is preserved freshly only in three glassy samples, and is classed as a pure Fe-hornblende-hastingsite amphibole (Ewart, 1982a, 1985; Table 5.3.1, column 4). It coexists with phenocrysts of allanite, rare fayalite, and ferrohedenbergite, as well as Fe-Ti oxides that have crystallisation temperatures of between 950 and 1015°C. Fluorine values are low (0.49–0.56 percent).

(9) Stolz (1986) recorded a range of groundmass amphiboles in rocks from Nandewar volcano. These include edenite (hawaiite and benmoreite), richterite and ferro-richterite grading to katophorite (trachyte), alumino-katophorite (peralkaline trachyte), and arfvedsonite grading to richterite (peralkaline rhyolite and trachyte).

Phenocryst amphibole is, overall, rare throughout the region, despite the preceding list of amphibole occurrences. Even occurrences of interstitial amphibole are sporadic, minor, and commonly associated with only pegmatoidal or other late-stage parageneses. This general paucity of amphibole evidently corresponds to low volatile levels in the different magmas, particularly in the intermediate to felsic types. The major exception — which is of considerable petrogenetic interest — is the fluor-arfvedsonite that is so characteristic of the peralkaline felsic rocks, and which is illustrative of the importance of fluorine in these magmas.

5.3.5   Mica

Mica, like amphibole, is uncommon in east-Australian Cainozoic rocks. A wide range of compositions extending from phlogopite to annite is illustrated by the plot of available microprobe data (again, excluding megacryst and xenocryst occurrences) presented in Figures 5.3.21A–B. The more important occurrences are as follows:

(1) A mostly groundmass mica, some poikilitic, is found in the leucitites and related pegmatoids of New South Wales and Cosgrove, Victoria (Cundari, 1973; Ferguson, 1978a; Birch, 1979). It is also present in late-stage vesicles and as rare phenocrysts. Compositions in leucitite are phlogopite, but grade into biotite, especially in some pegmatoids. Both Ti and Al increase as Fe enrichment increases. Phenocrysts, where present, are zoned to Fe-rich rims, and are more magnesian than coexisting groundmass grains. High BaO levels (3.7–5.8 percent) are recorded in phlogopite from the Cosgrove leucitite and pegmatoid, and fluorine abundances of 1.6–3.1 percent are recorded from the Cosgrove and New South Wales micas (Ferguson, 1978a; Birch, 1979).

(2) Groundmass phlogopite is present in a leucite basanite and related pegmatoid from Main Range, southeastern Queensland. Rare groundmass phlogopite and, less commonly, biotite have been found also in hawaiite and *ol*-tholeiitic basalt from the Main Range, Bunya Mountains, Springsure, and Anakie volcanic centres of southern and central Queensland, but development is only sporadic. The most Fe rich of these micas coexists with correspondingly Fe-enriched pyroxene and olivine.

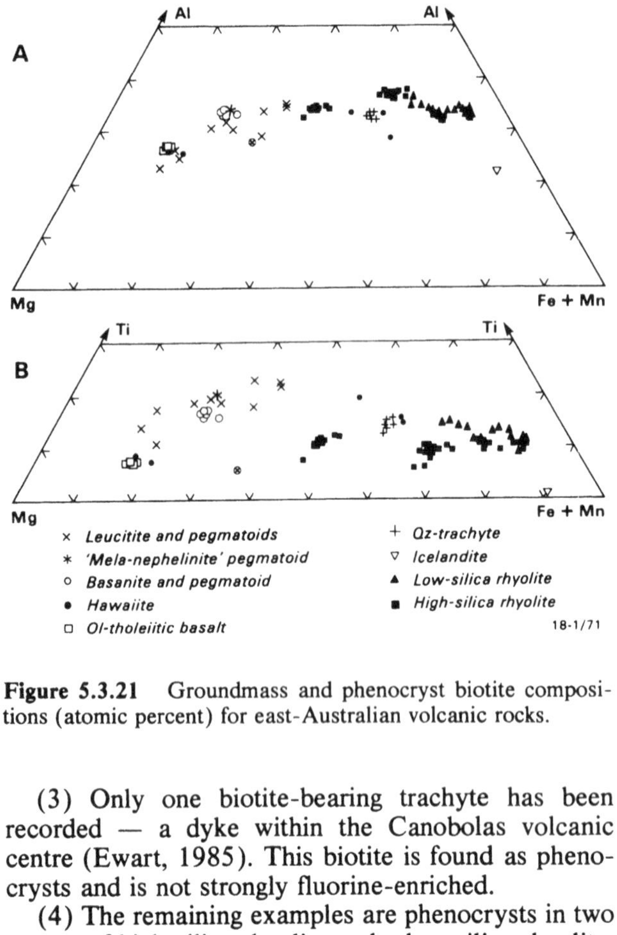

- ×   *Leucitite and pegmatoids*
- *   *'Mela-nephelinite' pegmatoid*
- ○   *Basanite and pegmatoid*
- ●   *Hawaiite*
- □   *Ol-tholeiitic basalt*
- +   *Qz-trachyte*
- ▽   *Icelandite*
- ▲   *Low-silica rhyolite*
- ■   *High-silica rhyolite*

18-1/71

Figure 5.3.21 Groundmass and phenocryst biotite compositions (atomic percent) for east-Australian volcanic rocks.

(3) Only one biotite-bearing trachyte has been recorded — a dyke within the Canobolas volcanic centre (Ewart, 1985). This biotite is found as phenocrysts and is not strongly fluorine-enriched.

(4) The remaining examples are phenocrysts in two groups of high-silica rhyolite and a low-silica rhyolite, all from southeastern Queensland. Two of these localities are characterised by strongly Fe-enriched biotites, extending to annite, and containing low fluorine (less than 1 percent). One occurrence is a Binna Burra rhyolite flow (Tweed volcano), and a second is a low-silica rhyolite at Flinders Peak centre (Ewart et al., 1977; Ewart 1981; Table 5.3.1, column 5). The other is two high-silica rhyolites in the Noosa area of the Glass Houses province — Mounts Tinbeerwah and Peregian — in which phenocrysts of more magnesian fluor-biotites are characteristic (Ewart, 1985; Table 5.3.1, column 6).

In only one of these localities — the Mount Tinbeerwah rhyolite — does biotite coexist with magnetite, ilmenite, sanidine, and quartz. Estimates of water fugacities (based on the method of Wones, 1972) range up to a maximum of 140 bars (Ewart, 1985). Low water pressures during crystallisation are inferred, a feature that seems to be common to most of the felsic rocks throughout eastern Australia (see below).

5.3.6   Feldspathoids

Nepheline

Nepheline is a ubiquitous, although generally minor groundmass mineral in leucitite and associated pegmatoids, in melilitite, and is also readily identifiable in the groundmass of most nephelinites. It is found less commonly as a groundmass mineral in *ne*-hawaiite and basanite. Cundari (1973) and Birch (1979) described the nepheline in leucitite as small anhedral grains and patches, commonly interstitial to leucite, and also poikilitic. Nepheline forms euhedral crystals in the pegmatoids, and small anhedral crystals — some of them interstitial — in nephelinite, melilitite, and the other alkaline lavas. Wilkinson (1965, 1977b) discussed feldspar-nepheline-analcime crystallisation trends in the theralite-tinguaite Square Top intrusion near Nundle, New South Wales, and nephelines in nephelinites and associated pegmatoids from Inverell (New South Wales).

Tie-lines joining nepheline compositions and co-existing whole-rock compositions — and, for the leucitites, coexisting alkali feldspar — are plotted in Figure 5.3.22. The nepheline from the New South Wales leucitite plots close to the 700°C isotherm for 1 kb water pressure (Hamilton, 1961), whereas Cosgrove nepheline is slightly less stoichiometric and falls above the 775°C curve. Nepheline coexisting with sanidine in both sets of data for the pegmatoids has lower Ne/Ks than does nepheline in the feldspar-free rocks. Cundari (1973) noted that the excess SiO_2 in the nepheline tends to correlate with excess SiO_2 in the coexisting leucite.

Marked compositional differences exist between the nepheline of melilitite, nephelinite, *ne*-hawaiite, and basanite, although compositions overlap with the range observed for leucitite. Nepheline in melilitite and nephelinite (both feldspar-free) is more potassic than in the *ne*-hawaiite and basanite, and K-enrichment is most pronounced in the nephelinites. It is also more Fe-rich and contains significantly lower amounts of excess silica, extending from near, to well below, the 775°C isotherm. These trends correlate with whole-rock compositional differences. Nepheline in *ne*-hawaiite and basanite not only contains higher excess silica, but has a wider compositional range in individual rocks. This is particularly so in the most silica-rich nepheline from a *ne*-hawaiite in which incipient alkali-feldspar unmixing presumably has taken place (Brown, 1970).

Compositions of nepheline in leucitite are somewhat different from those in the less potassic lavas in which there is evidence of stronger liquid compositional control on nepheline chemistry, both in terms of excess SiO_2 and Ne/Ks values. The problem may be complicated by subsolidus re-equilibration processes, especially as fluorine was a significant volatile phase within the leucitite magmas (as demonstrated by the groundmass mica, amphibole, and apatite compositions).

Sodalite

Sodalite is rare, but it is found in several specific rock types:

(1) Rare, small, euhedral crystals of sodalite are

enclosed by nepheline in Cosgrove leucitite pegmatoid (Birch, 1979), and in leucitite pegmatoid from Begargo Hill, New South Wales, where it is found in cavities mantling alkali feldspar and corroded nepheline (Ferguson, 1978a).

(2) Sodalite is a minor interstitial phase in the phonolite from Mount Wilson, central Victoria.

(3) It appears as an interstitial groundmass mineral in melilitite (Boat Harbour, Tasmania; Table 5.3.1, column 7), nephelinite (Mount Fort William, central Queensland), and *ne*-hawaiite (Oatlands, Tasmania). All of these occurrences represent late-stage parageneses, possibly involving reaction between nepheline and late NaCl-rich solutions (the sodalite-bearing rocks each contain groundmass nepheline, and are holocrystalline; Ferguson, 1978a).

The sodalites have no detectable SO_3 and very little K or Ca replacement of Na, and are therefore nearly pure sodalite. Sodalite in the melilitite is notable for its compositional range between grains in terms of the relative Si, Na, and Cl abundances, and therefore their stoichiometry.

Figure 5.3.22 Nepheline, sanidine, leucite, and coexisting whole-rock relationships (weight percent) in east-Australian volcanic rocks. Whole-rock compositions represented by symbol-free ends of lines in A and B. Data in B is from Cundari (1973) and in C from Birch (1979, 1980). Boundaries 775°C and 1068°C represent the estimated limits of nepheline solid-solution at these temperatures and at 1 kb (after Hamilton, 1961).

Leucite

All analysed leucites are chemically similar, and approximate ideal $KAlSi_2O_6$ stoichiometry. Two occurrences of leucite are recognised:

(1) Subhedral to anhedral crystals, typically less than 1 mm in diameter, are present in the New South Wales and Cosgrove leucitites (Cundari, 1973; Birch, 1979). Inclusions are common, both as disseminations (New South Wales) and in regular symmetrical patterns (Cosgrove). Weak birefringence and multiple twinning are well developed. Alteration is found locally.

(2) Leucite is present as a groundmass mineral in a leucite basanite from the Main Range, southeastern Queensland, and in linings on associated pegmatoidal patches. This leucite is optically similar to those noted above (Grenfell, 1984). Fe-Ti-oxide quenching temperatures of 995°C (basanite) and 810°C (pegmatoid) have been determined for this occurrence.

Analcime

Wilkinson (1963) described analcime from the Square Top intrusion and showed extensive replacement of Na+Al for Si. Cundari (1973) described primary analcime in analcimites (nephelinites) from Griffith and Harden, New South Wales. Birch (1979) noted analcime in a leucitite pegmatoid from Cosgrove where it is found as subhedral, turbid crystals, resembling leucite in general form. Analcime is reported to replace nepheline in the Cosgrove leucitite (Birch, 1980). Wilkinson (1977a) described analcime 'phenocrysts' from a vitrophyric analcimite from southeastern Queensland. These are interpreted as ion-exchanged leucite phenocrysts. Other occurrences were reported by Irving (1971) from Victoria and South Australia.

5.3.7 Fe-Ti-Cr oxides

Titanomagnetite and ilmenite

Titanomagnetite and ilmenite are found mainly as groundmass to microphenocryst grains, but some reach phenocryst sizes. The main occurrences are as follows:

(1) *Qz*- and *ol*-tholeiitic basalts normally contain coexisting magnetite and ilmenite, but there are notable exceptions in tholeiitic basalts from Tweed volcano and from some *qz*-tholeiitic basalts from central Queensland (spinel phase absent). The absence of a spinel in the Tweed rocks is attributed to equilibration at an fO_2 below magnetite stability and close to the WM buffer (Ewart et al., 1977).

(2) Coexisting oxides are found in virtually all hawaiites and in the majority of *ne*-hawaiites. However, a significant number of the *ne*-hawaiites, transitional, and alkali basalts contain only a spinel.

(3) Ilmenite is observed only rarely in basanite, and not at all in nephelinite and melilitite. These alkaline rocks are characterised by titanomagnetite alone. Carmichael et al. (1974) suggested that this may be explained by a reaction of the type:

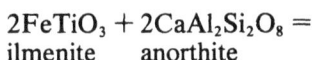

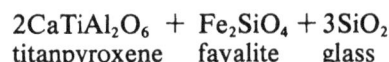

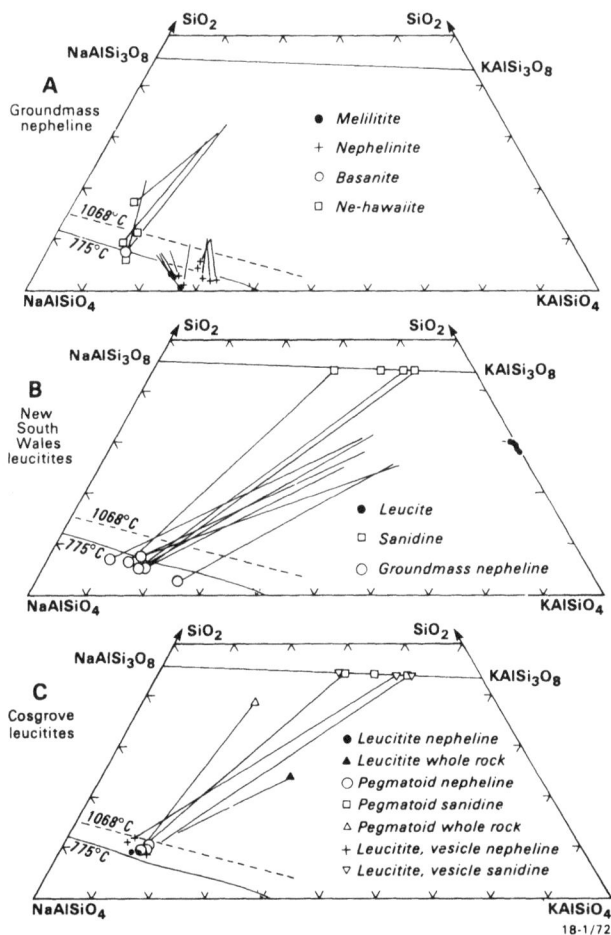

$$2FeTiO_3 + 2CaAl_2Si_2O_8 =$$
ilmenite   anorthite

$$2CaTiAl_2O_6 + Fe_2SiO_4 + 3SiO_2$$
titanpyroxene   fayalite   glass

Lowered silica activity in the liquid, therefore, should favour increasing titanpyroxene at the expense of ilmenite. However, Ti tends to decrease in the groundmass pyroxene of nephelinite and melilitite (relative to basanite and so forth) as noted above, and therefore a slightly different reaction of the following type may apply:

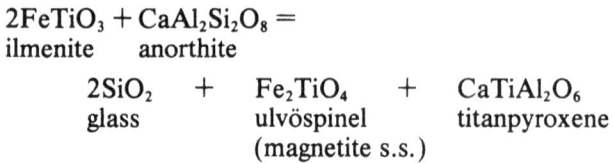

$$2FeTiO_3 + CaAl_2Si_2O_8 =$$
ilmenite   anorthite

$$2SiO_2 + Fe_2TiO_4 + CaTiAl_2O_6$$
glass   ulvöspinel   titanpyroxene
(magnetite s.s.)

A more specific partitioning of Ti between groundmass pyroxene and spinel in the absence of ilmenite is implied by this reaction.

(4) Both groundmass titanomagnetite and ilmenite are reported in leucitite and associated pegmatoids, but, based on the data in Cundari (1973) and Birch (1979), they evidently do not coexist in the same samples. One exception is a New South Wales leucitite, but the compositions fail to give a T-fO$_2$ solution, suggesting non-equilibrium compositions (or subsequent oxidation). Ferguson (1978a) suggested that titanomagnetite is more characteristic of leucitite. Birch (1980) recorded titanomagnetite in late-stage vesicles in the Cosgrove leucitite.

(5) Mugearite, benmoreite, and icelandite mostly contain coexisting oxides — again, except for the icelandite from the Tweed volcano in which ilmenite is found alone.

(6) Trachyte of both silica-oversaturated and under-saturated types normally contains only a phenocryst and groundmass titanomagnetite, although coexisting Fe-Ti oxides are found in a few of the most silica-rich qz-trachytes and in most of the low-silica rhyolites. Most dacites for which mineralogical data are available are from Tweed volcano, and they contain hybrid mineral assemblages (see above) that normally contain ilmenite alone.

(7) Biotite-bearing high-silica rhyolite of the Noosa area, Glass Houses province, contains co-existing Fe-Ti oxides. However, the majority of the remaining high-silica rhyolites from northern New South Wales and southeastern Queensland are characterised by ilmenite (R$_2$O$_3$ less than 3 percent) alone, except for a few samples of Binna Burra Rhyolite in which rare grains of extremely ulvöspinel-rich spinel are present — in one case, almost pure ulvöspinel (Ewart et al., 1977). These ilmenite-bearing rhyolites were interpreted by Ewart et al. (1977) to have crystallised under low fO$_2$, close to the WM buffer.

(8) Fe-Ti oxides are absent from many of the peralkaline rocks, but where present are mostly ilmenite in peralkaline rhyolite or, less commonly, titano-magnetite in peralkaline trachyte. Two peralkaline rhyolite dykes from Mount Warning (Tweed volcano) contain coexisting magnetite and ilmenite. However, no T-fO$_2$ solution can be obtained, which is indicative of non-equilibrium compositions.

Ulvöspinel solid-solution (calculated by the method of Carmichael, 1967) differs in titanomagnetite from different rock types (Fig. 5.3.23). Spinels in tholeiitic

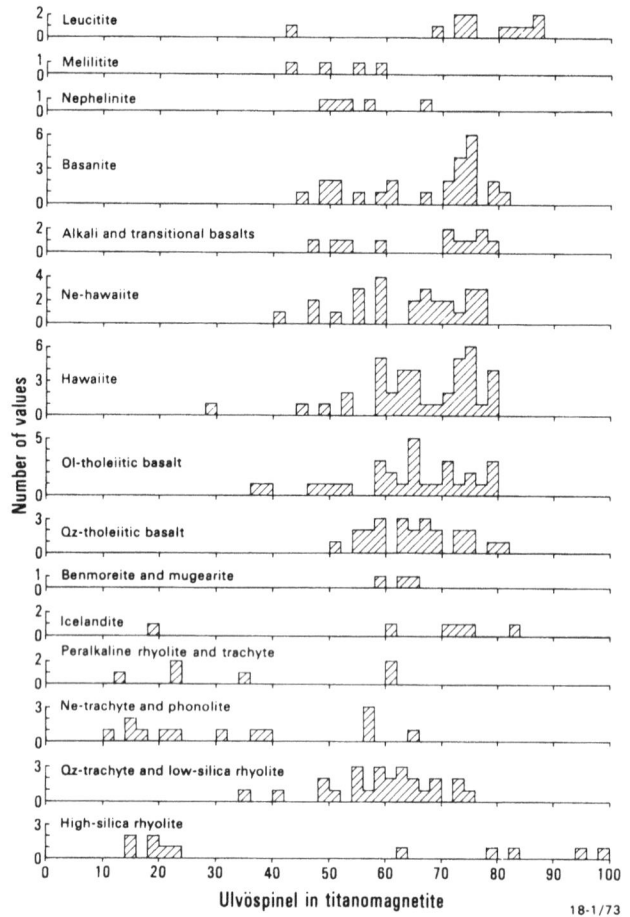

Figure 5.3.23 Histogram of calculated ulvöspinel solid-solution abundances (molecular percent; based on the method of Carmichael, 1967) in titanomagnetite from different east-Australian volcanic-rock types.

basalt and hawaiite are similar, and ulvöspinel solid-solution is mostly 54–83 percent. The spinels become somewhat less titaniferous as the degree of silica undersaturation of the rock increases. Ulvöspinel ranges are 49–67 and 43–59 percent in nephelinite and melilitite, respectively. This is possibly related to the reaction (noted above) involving Ti partitioning between ulvöspinel and titanpyroxene. However, the following reaction (El Goresy & Yoder, 1974) could be relevant for extremely silica-undersaturated magmas in determining titanomagnetite compositions:

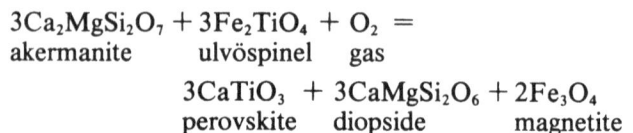

$$3Ca_2MgSi_2O_7 + 3Fe_2TiO_4 + O_2 =$$
akermanite   ulvöspinel   gas

$$3CaTiO_3 + 3CaMgSi_2O_6 + 2Fe_3O_4$$
perovskite   diopside   magnetite

Perovskite is present in New South Wales leucitites and in one melilitite (see below). Alternatively, the titanomagnetite compositions may simply reflect magnetite-olivine-liquid equilibrium:

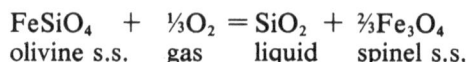

$$FeSiO_4 + \tfrac{1}{3}O_2 = SiO_2 + \tfrac{1}{3}Fe_3O_4$$
olivine s.s.   gas   liquid   spinel s.s.

A similar relationship of ulvöspinel contents in spinel is observed also where comparing ne- and qz-trachyte data (Fig. 5.3.23), and is suggestive of silica-activity control. Magnetite compositions in high-silica rhyolite are extreme (as noted above) and include xenocrystal compositions (see below).

Minor-element behaviour in both ilmenite and magnetite, illustrated in Figures 5.3.24–25, are correlated broadly with host-rock composition. This type of plot has the advantage of providing an indication of equilibrium versus disequilibrium coexisting compositions, by virtue of the tie-line slopes (Carmichael, 1967). However, information is not provided on such plots regarding the absolute abundances of these minor elements which are found to range widely in each of the spinel groups.

Most of the coexisting Fe-Ti oxide pairs in the mafic rocks have reasonably consistent tie-line slopes,

although there are a few discordant compositions, particularly in the tholeiitic rocks. There is evidence for a small relative shift of ilmenite composition towards MgO enrichment in the more undersaturated rocks. However, the major change is a general shift in spinel composition towards MgO enrichment and Al_2O_3 depletion, as host-rock undersaturation increases. This is not matched uniformly by absolute increases in MgO in these spinels compared to those in tholeiitic rocks. Spinel compositional fields are generalised in Figure 5.3.24H (ignoring discordant compositions), where the compositional shifts are more clearly defined, and spinel in the more undersaturated magmas is seen to be progressively more Mg-enriched. This chemical change may reflect a solid-solution spinel exchange of the following type:

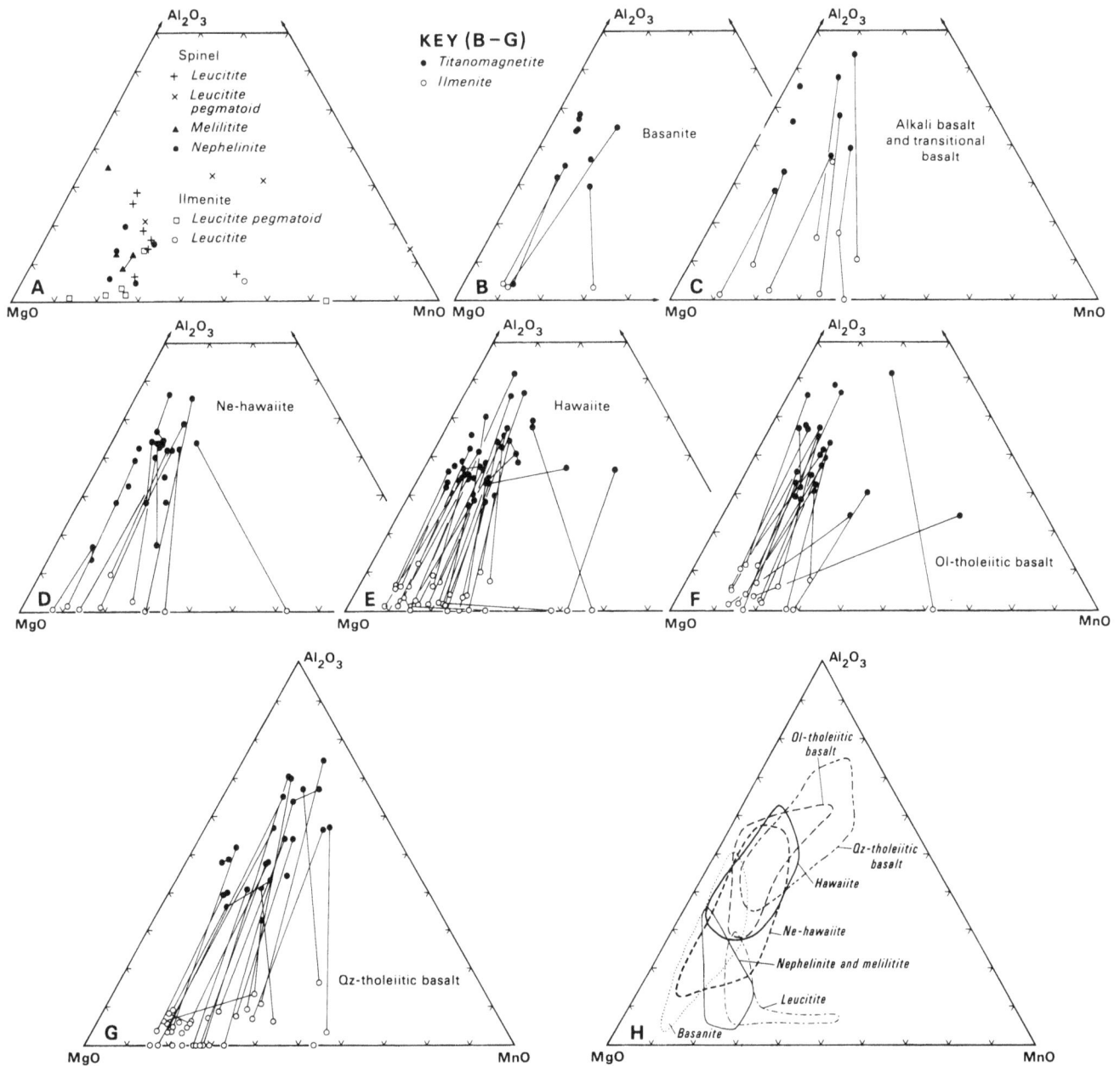

Figure 5.3.24 Minor-element compositions in titanomagnetite (filled symbols) and ilmenite (open symbols) in east-Australian mafic rocks. All data plotted are for optically homogeneous grains. Tie-lines join co-existing minerals. Generalised compositional fields shown for titanomagnetite in H are from data plotted in A-G, but excluding clearly discordant compositions.

$2FeAl_2O_4$ + $CaMgSi_2O_6$ + $\tfrac{1}{3}O_2$ =
hercynite   diopside   gas
(magnetite s.s.)

$MgAl_2O_4$ + $CaAl_2SiO_6$ + $\tfrac{2}{3}Fe_3O_4$ + SiO_2
spinel   Ca-tschermak's   magnetite liquid
(magnetite s.s.) pyroxene

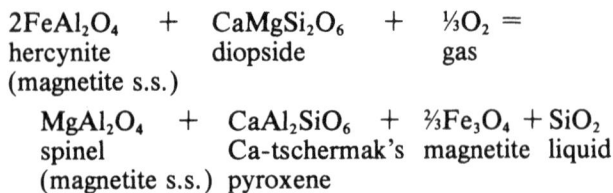

Silica activity of the liquid therefore should control sensitively the activities of the $MgAl_2O_4$ and Fe_3O_4 components (in titanomagnetite) relative to the $FeAl_2O_4$ component. These are evidently the types of chemical change observed in magnetite from oversaturated to undersaturated mafic rock types.

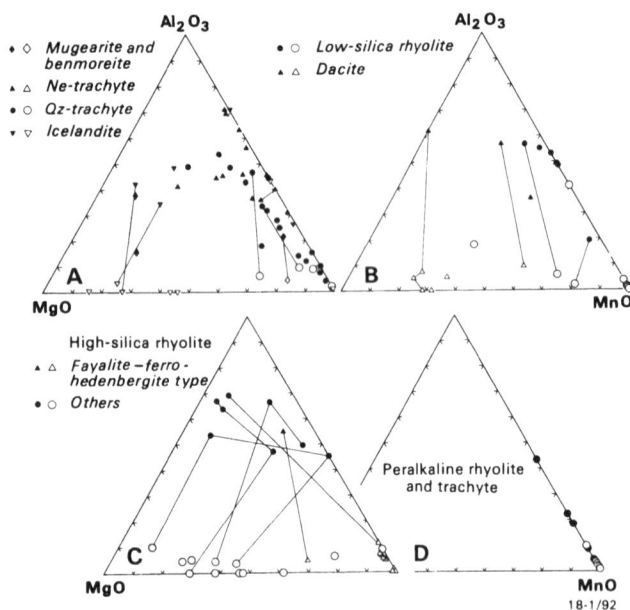

Figure 5.3.25   Minor-element compositions of titanomagnetite (filled symbols) and ilmenite (open symbols) in east-Australian intermediate and felsic rocks. Tie-lines join co-existing minerals.

Minor-element chemistry for the Fe-Ti oxides in the intermediate and felsic rocks is illustrated in Figure 5.3.25. The major changes in mugearite, benmoreite, icelandite, and trachyte are the progressive MnO enrichment, and MgO depletion, in both spinel and ilmenite, which clearly reflect MgO depletion in the host magmas, and which become most extreme in low-silica rhyolite and peralkaline rocks. Spinel in the peralkaline rocks is also depleted in Al_2O_3. These changes are attributable to fractional-crystallisation controls on magma chemistry. Attention is also drawn to relatively high Zn abundances in some of the oxides of the peralkaline rocks.

High-silica rhyolite contains Fe-Ti oxides that have much more complex minor-element patterns, and these are complicated further in many samples by the presence of several compositionally distinct phases (there are therefore discordant tie-lines between co-existing minerals). These are interpreted to include xenocryst Fe-Ti oxides. A notable exception is the ferrohedenbergite-fayalite rhyolite which normally contains only an ilmenite that is strongly enriched in MnO (and therefore similar to ilmenite in trachyte and

low-silica rhyolite) but does not have the xenocrystal Fe-Ti oxide compositions.

Cr-spinel and Cr-bearing titanomagnetite

True Cr-spinel (magnesiochromite-chromite) is found characteristically only as small inclusions (mostly less than 10 microns) in phenocryst olivine, and has been identified in all of the mafic rock types excepting leucitite and melilitite. Cr-spinel is found sporadically in rocks from both lava-field and central-volcano complexes, but tends to be more common in lavas from particular provinces. Wass (1973) described their presence in the Southern Highlands province (New South Wales). Cr-spinel is found most commonly in Queensland in lavas of the Maleny area (Glass Houses province), the Mitchell, Anakie, and central-coastal centres, and Cooktown.

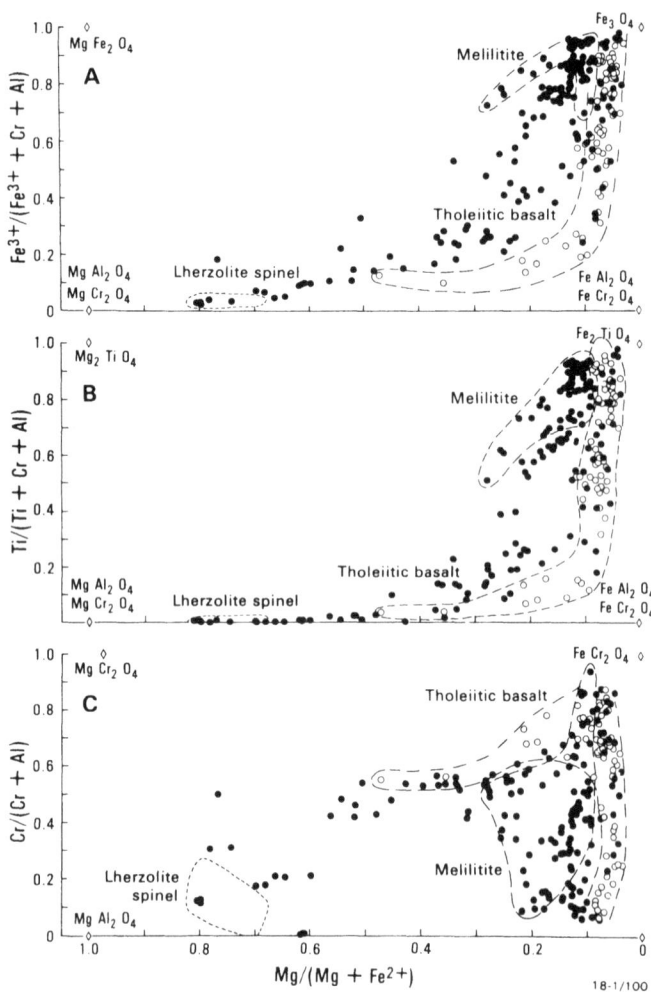

Figure 5.3.26   Cr-spinel and Cr-bearing titanomagnetite compositions (atomic percent, individual-grain analyses) for different basalt types. Fe^{2+} and Fe^{3+} calculated on the basis of spinel stoichiometry. Open symbols represent tholeiitic rocks. Filled circles represent other rock types.

Cr-bearing titanomagnetite (defined as containing more than 0.5 percent Cr_2O_3) is found commonly as inclusions in olivine and, more often, as discrete groundmass grains. There is complete solid-solution between the different spinel types, as shown in Figure 5.3.26 where several different compositional trends are identifiable. All plotted data are for optically homogeneous grains, and compositional trends represent inter-grain differences.

A close similarity in the dominant compositional trends involving Ti and Fe^{3+} is illustrated in Figures 5.3.26A–B. There is a broad trend between ulvöspinel-magnetite and $Fe(Al,Cr)_2O_4$ solid-solution components, and an important distinction in the relative degree of Mg enrichment as Al+Cr substitution increases (the ulvöspinel-magnetite enriched region of these compositions is defined by groundmass Cr-bearing titanomagnetite). The difference in the spinel Mg-enrichment trends correlates clearly with whole-rock compositional characteristics. Thus, spinel in the qz- and ol-tholeiitic basalts have Al+Cr enrichment accompanied by little change in $Mg/(Mg+Fe^{2+})$ values until extreme Al+Cr enrichment is reached. This 'tholeiitic trend' is emphasised in Figure 5.3.26A–B, and is similar to the 'basaltic trend 1' (Haggerty, 1976) found in Hawaiian tholeiitic lavas (Evans & Moore, 1968; Beeson, 1976). Spinel in the more alkaline lavas is enriched in Mg at an earlier stage as Al+Cr substitution increases. This is most pronounced in some melilitite groundmass spinels. Magma composition is evidently a major controlling factor, but other factors must also be involved. A possible explanation for the different spinel compositional trends is provided by the spinel exchange reaction given in the preceding Fe-Ti-oxide section. Decreasing silica activity (in the melt) should favour increasing activity of the $MgAl_2O_4$ component (in titanomagnetite), as well as decreasing $FeAl_2O_4$ activity (titanomagnetite).

A complete compositional continuity exists between the Cr-spinel in olivine and those within discrete lherzolite xenoliths included in alkali basalt (Fig. 5.3.26C). Two distinct trends are observed: one linking the $FeAl_2O_4$-$FeCr_2O_4$ solid-solution end-members (corresponding predominantly to groundmass titanomagnetite spinel), and a second trend linking the $Mg(Al,Cr)_2O_4$ and $Fe(Cr,Al)_2O_4$ solid-solution spinel end-members. Two interpretations of these trends include the following:

(1) The Cr-spinel trends represent a sequence of trapped, high-temperature, liquidus and sub-liquidus phases that differ in composition in response to changes in magma chemistry during fractional crystallisation (see, for example, Arculus, 1974). The most magnesian Cr-spinel therefore would represent either xenocrysts, or Cr-spinel and olivine precipitated initially in equilibrium within residual upper-mantle spinel lherzolite. In contrast, the titanomagnetite-dominated spinel represents a lower-temperature groundmass phase, precipitated at lower fO_2 (see, for example, Haggerty, 1976), and the Cr solid-solution is determined by changing residual-liquid Cr abundances.

(2) The primary spinels were modified by reaction with residual magmas during the later stages of crystallisation and quenching, extending to post-eruption subsolidus alteration of original chromites (see, for example: Evans & Moore, 1968; Arculus, 1974). This possibility is supported, at least in part, by application of the olivine-spinel geothermometer (Roeder et al., 1979) to Cr-spinel, Cr-rich titanomagnetite, and their coexisting (adjacent) olivine compositions. Only a few olivine-

spinel pairs (Fig. 5.3.27) give reasonable liquidus temperatures.

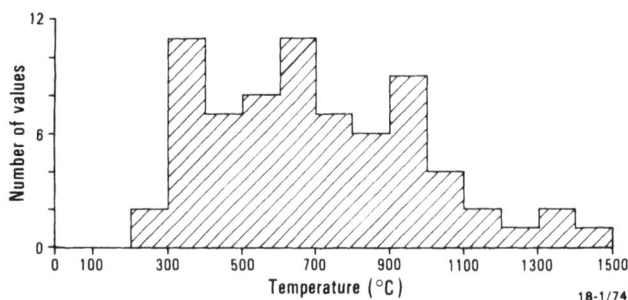

Figure 5.3.27 Histogram of equilibration temperatures calculated from coexisting olivine/Cr-spinel and Cr-bearing titanomagnetite pairs (method of Roeder et al., 1979).

Perovskite

Perovskite is present as very small, sporadically distributed grains in a melilitite from Boat Harbour, Tasmania. Cundari (1973) also noted its presence in New South Wales leucitites as a fine-grained groundmass mineral, and he gave a partial analysis of perovskite from a leucitite pegmatoid. The most significant feature of analysed perovskite (Table 5.3.2) is the extensive minor-element substitutions in the pegmatoidal perovskite, most notably the high, coupled substitution of Nb and Ta (for Ti) and the Na and Sr (for Ca), together with extreme light-REE enrichment.

Table 5.3.2 Averaged analyses of perovskite[1]

	1	2		1	2
SiO_2	0.47	0.06	ZrO_2	0.12	0.06
TiO_2	39.19	53.11	ThO_2	0.83	0.13
Al_2O_3	0.02	0.07	UO_2	0.03	0.02
CrO_3	<0.02	0.06	Nb_2O_3	7.29	0.34
La_2O_3	9.73	1.13	Ta_2O_3	0.54	0.10
Ce_2O_3	17.21	1.88	V_2O_3	0.05	0.11
Pr_2O_3	1.26	0.23	FeO	0.29	0.97
Nd_2O_3	4.06	0.58	MnO	0.01	0.03
Sm_2O_3	1.50	0.19	MgO	0.02	0.06
Eu_2O_3	0.49	0.03	CaO	5.17	37.23
Gd_2O_3	0.83	0.17	SrO	3.60	3.01
Tb_2O_3	0.04		BaO	0.14	0.13
Dy_2O_3	0.02		PbO	0.08	0.02
Ho_2O_3	0.03	0.03	Na_2O	7.23	0.80
Er_2O_3	0.02	0.04	K_2O	0.10	0.09
Yb_2O_3	0.05	0.04	Total	100.42	100.72

1. Leucitite pegmatoid (sample 70/1038). Three grains.
2. Leucitite (sample CPT6). Two grains.

[1] Analyses by electron microprobe (A. Cundari, personal communication, 1987). All values in weight percent. Y_2O_3, Sc_2O_3, and HFO_2 were not detectable (less than 0.02 weight percent). Specimen numbers are those used by Cundari (1973).

5.3.8 Melilite

Melilite in melilitite is present either as fine groundmass laths or as anhedral to subhedral tabular groundmass crystals, some of nearly microphenocryst size (Table 5.3.1, columns 8–9). Individual grain analyses from three melilitites are compared in Figure 5.3.28 with those of igneous melilite from other world occurrences. East Australian melilite is closely similar in composition

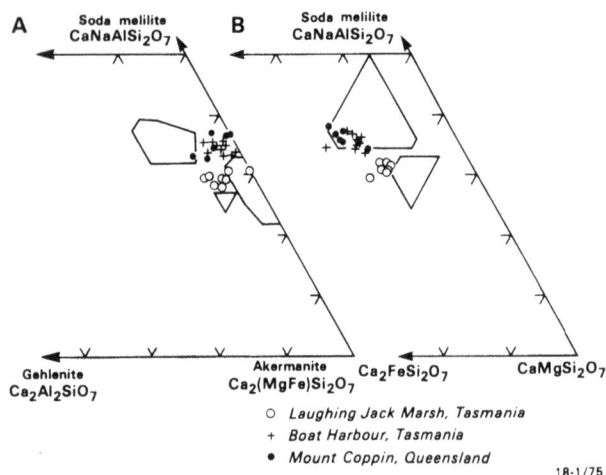

Figure 5.3.28 Groundmass melilite compositions (atomic percent) for three east-Australian localities, shown in relation to the compositional fields for igneous melilites reported by El Goresy & Yoder (1974).

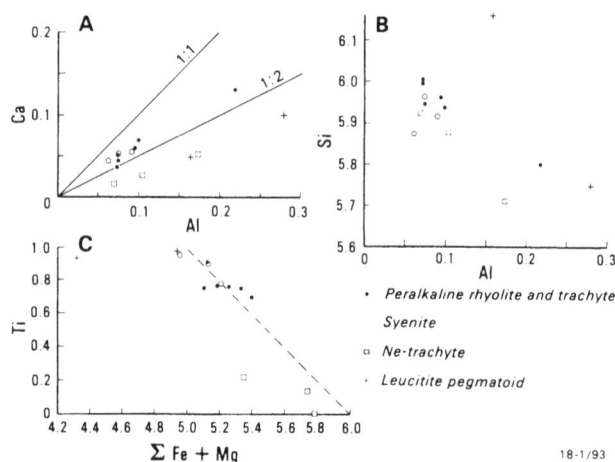

Figure 5.3.29 Aenigmatite compositions (based on O=20) in peralkaline trachytes and rhyolites, interstitially in syenites and microsyenites from southeastern Queensland, in a Warrumbungle *ne*-trachyte, and in the Cosgrove (Victoria) leucitite pegmatoid. Dashed line in C represents ideal $Na_2Fe^{2+}_4Fe^{3+}_2Si_6O_{20}$ substitution.

to these other melilites, and is characteristically dominated by soda-melilite and akermanite solid-solution components. No significant compositional zoning is found.

5.3.9   Accessory and minor minerals

Aenigmatite

Aenigmatite has three distinct occurrences:

(1) It is found as an accessory groundmass mineral in some peralkaline trachytes and rhyolites, coexisting with sodic pyroxenes and amphiboles. Nicholls & Carmichael (1969) noted that aenigmatite normally does not coexist with Fe-Ti oxides (particularly ilmenite + fayalite), and the only east-Australian samples in which a coexisting oxide phase has been found is in Nandewar peralkaline trachytes that contain either titanomagnetite (Abbott, 1967; see also Marsh, 1975) or ilmenite. Stolz (1986) recorded aenigmatite rimming both titanomagnetite and ilmenite in these Nandewar lavas.

(2) Aenigmatite is also present as a fine-grained, interstitial, late-stage mineral in Miocene microsyenite and syenite from southeastern Queensland and northeastern New South Wales (Tweed volcano). It coexists with Na-enriched pyroxene and richteritic-arfvedsonitic amphibole, corresponding to the late-stage development of peralkaline residual liquids, as described above.

(3) Birch (1979) recorded aenigmatite as sparse, small, deep reddish-brown grains on the margins of skeletal titanomagnetite in the leucitite pegmatoid from Cosgrove.

Available analyses of aenigmatite are plotted in Figure 5.3.29 (see also Table 5.3.1, columns 10–11) in terms of the coupled substitution $Na_xSi_z \rightleftharpoons Ca_xAl_z$ (Larsen 1977), but $Fe^{2+}Ti^{4+} \rightleftharpoons 2Fe^{3+}$ is also possible, with the formation of ferri-aenigmatite. The aenigmatites found in oversaturated peralkaline trachytes, rhyolites, and syenites have only small compositional ranges. In contrast, extensive solid-solution is observed in zoned aenigmatite in the undersaturated rocks of the Ilimaussaq

intrusion (Larsen, 1977). In addition, M.B. Duggan (personal communication, 1986) has shown strong antipathetic Ti-Fe solid-solution effects in aenigmatites in trachytes from Warrumbungle volcano in which the nepheline-normative rocks contain the low-Ti aenigmatites (enriched in the ferri-aenigmatite end-member; Table 5.3.1, column 11). Examples of three such compositions are illustrated in Figure 5.3.29.

Two analyses of aenigmatite in leucitite given by Birch (1979) also have a wider compositional range than those from the oversaturated peralkaline lavas and, in particular, have significant Mg substitutions. All other aenigmatites from eastern Australia have negligible Mg, in conformity with the highly Mg-depleted host-rock chemistry.

Rhönite

Rhönite has been recognised as a decomposition product of kaersutite and pargasite in alkaline lavas at five, widely separate localities in the New England region of northeastern New South Wales (R.W. Schön, personal communication, 1986). These rhönites have significant compositional differences amounting to between 9.8 to 28.0 percent solid-solution of the aenigmatite component.

Apatite

All of the main rock types — except rhyolite, peralkaline rocks, and some silica-rich trachytes (in which P_2O_5 is extremely depleted by fractional-crystallisation processes) — contain apatite as a groundmass accessory mineral and, in some rocks, as inclusions in phenocrysts. Analysed apatite from a range of mafic and intermediate rocks (Table 5.3.1, columns 12–13) is fluor-apatite containing only minor Cl and small but significant light-REE abundances (Ce_2O_3 data). Y_2O_3 and S were not detected. The phonolite from Mount Wilson (central Victoria) contains unusual, anhedral, interstitial apatite that is inferred to be a carbonate apatite, also notably high in SiO_2.

Zircon

Small, euhedral, prismatic crystals of zircon are found either in the groundmass or included in phenocrysts, especially in intermediate and felsic rocks, but not in peralkaline samples (see also Watson, 1979). Microprobe analyses of zircon were provided by Ewart (1981, 1985; Table 5.3.1, columns 14–15). Ewart (1985) showed a correlation between Zr/Hf values for zircon and the degree of trace-element fractionation (K/Rb values) for the whole-rock host rhyolites. All analysed zircon has significant Fe and Y contents. Heavy-REE enrichment is implied by the Y abundances.

Chevkinite

The rare-earth titanosilicate, chevkinite, is recorded only for fayalite-ferrohedenbergite high-silica rhyolite where it is found as non-metamict, euhedral, dark-brown prismatic microphenocrysts, in some samples included in pyroxene phenocrysts. Modal abundances are extremely low (less than about 0.001 volume percent). Identification was confirmed by X-ray diffraction, and partial analyses by microprobe were presented by Ewart (1981; see also Table 5.3.1, column 16). The chemical composition of this mineral is notable for its extreme light-REE enrichment, together with significant concentrations of Th, Zr, Nb, and Zn.

Allanite

Allanite — a rare-earth calcium-aluminosilicate — is found as non-metamict, euhedral, stubby prismatic to equant microphenocrysts that are either separate or included in sanidine phenocrysts. Analysed allanite (Ewart, 1981; Table 5.3.1, columns 17–18) has extreme light-REE enrichment and significant Th contents. Allanite has a restricted occurrence:

(1) In ferrohypersthene-bearing, vitrophyric rhyolite of the southern part of the Tweed volcano (Georgica and Nimbin rhyolites; Duggan, 1976);

(2) In biotite (annite)-bearing, vitrophyric rhyolite from northern Tweed volcano (Binna Burra Rhyolite);

(3) In a biotite-bearing, devitrified low-silica rhyolite from the Flinders Peak centre of Main Range province;

(4) In amphibole (Fe-hornblende-hastingsite)-bearing, vitrophyric, low-silica rhyolite of the Minerva Hills volcanic centre, south-central Queensland. Three of these rhyolite samples contain homogeneous Fe-Ti oxides that yield equilibration temperatures of 950–1015°C.

(5) In a fayalite-ferrohedenbergite-bearing, vitrophyric, high-silica rhyolite dyke in the Noosa area, Glass Houses province. Co-existing Fe-Ti oxides yield a temperature of 780°C, but these have been affected by late-stage hydration processes.

Monazite

The rare-earth phosphate, monazite, is recorded so far only from a fayalite-bearing, vitrophyric, high-silica rhyolite dyke in the Noosa area (Table 5.3.1, column 19). The monazite is found as euhedral microphenocrysts at abundances of less than 0.01 volume percent.

Ramsayite

The rare accessory mineral, ramsayite ($Na_2Ti_2Si_2O_9$), is recorded in both silica-oversaturated and undersaturated peralkaline rocks (Ferguson, 1977a, 1978b,c). Ferguson (1978a) noted its presence in leucite pegmatoid from Begargo Hill, New South Wales, and showed that equivalent leucitite pegmatoid from Cosgrove (Victoria) contains aenigmatite (Birch, 1979), but not ramsayite. Ferguson (1977a) also described ramsayite in a late-stage pegmatoidal clot in a 'mela-nephelinite' near Bacchus Marsh, Victoria, together with an analysis of this mineral phase (Table 5.3.1, column 20). Other coexisting minerals in the pegmatite are sanidine, salite, titanomagnetite, titanian-aegirine, quartz, magnesio-arfvedsonite, phlogopite, and rutile.

Sulphide

Sulphide minerals are conspicuously rare. Only three primary sulphide occurrences have been recorded during routine microprobe studies of about 300 rock samples, representing all compositional lava types from throughout eastern Australia. All three are found as minute (less than 5 microns) groundmass blebs of pyrrhotite included in titanomagnetite or ilmenite. They are rare in each sample. The rocks include two *qz*-tholeiitic basalts and a hawaiite, all from different central-Queensland areas. Pyrrhotite, nickeliferous pyrrhotite, and chalcopyrite have been recorded in mafic and ultramafic xenoliths found in a range of mafic volcanic rocks (for example, Ewart & Grenfell, 1985).

5.3.10 Geothermometry and oxygen geobarometry

The following methods have been used routinely during this study for the estimation of crystallisation temperatures and oxygen fugacity (Figs. 5.3.30–31):

(1) *Coexisting Fe-Ti oxides*. The preferred method of data reduction involves the recalculation procedures of Stormer (1983) and the fitting procedures of Ghiorso & Carmichael (1981) who used the original Buddington & Lindsley (1964) temperature/oxygen-fugacity curves.

(2) *Olivine/whole-rock compositions* for the mafic rocks, based on the calculation procedures and computer programs of Ghiorso et al. (1983). Use of magnesian-olivine core compositions is thought to provide an estimate of liquidus temperatures, as olivine seems to be the liquidus phase in virtually all the lavas to which this method has been applied. A second application of this geothermometric procedure is by combining groundmass calcic-plagioclase and whole-rock compositions. The most calcic plagioclase compositions (in the cores of groundmass laths) are chosen, from which estimates may be obtained of the initial quenching temperatures of the magmas. Leucite has been used in the calculations for leucitite.

(3) *Coexisting alkali- and plagioclase-feldspar phenocrysts* (Ghiorso, 1984). This method has been applied only to some trachyte and rhyolite samples.

Fe-Ti oxide (mainly groundmass) equilibration temperatures for the mafic lavas are between about 800 and 1190°C, and there is no systematic relationship

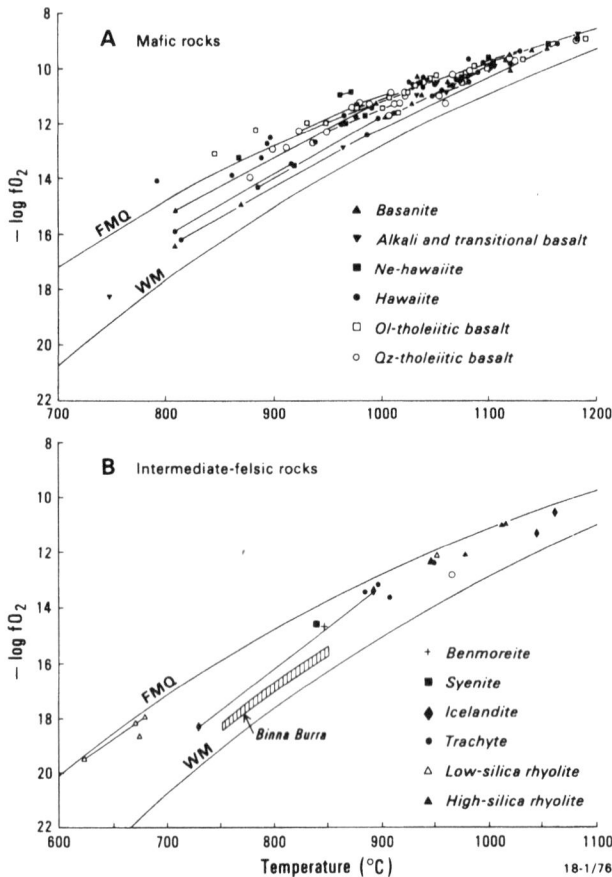

Figure 5.3.30 Equilibration temperatures and oxygen fugacities (fO_2) obtained from coexisting Fe-Ti oxides. FMQ and WM buffer curves are after O'Neill (1987, and personal communication, 1987). Tie-lines join pairs of data obtained from a single sample (different oxide compositions). The ruled area in B is an estimate inferred for a single high-silica rhyolite (Binna Burra) from Tweed volcano (modified after Ewart et al., 1977).

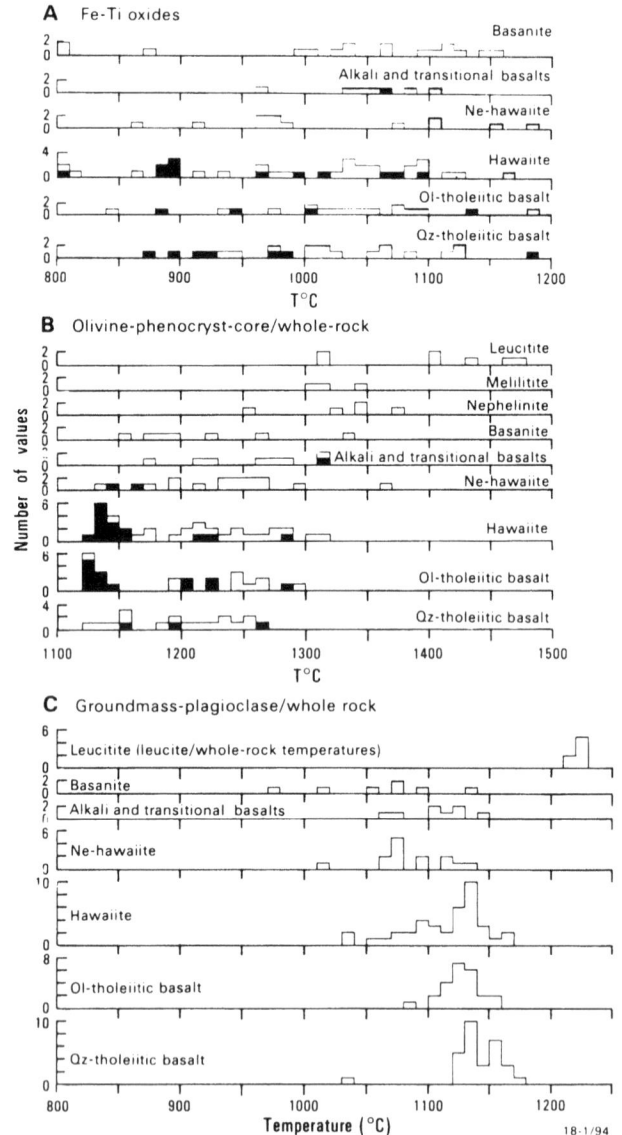

Figure 5.3.31 Comparative histograms of temperatures calculated by three different methods. (A) Coexisting Fe-Ti-oxide equilibration temperatures (see also Fig. 5.3.30). (B) Phenocryst-olivine-cores (most Mg-rich)/whole-rock compositions, inferred to give liquidus or near-liquidus temperatures. Solid areas in A and B represent central-volcano samples, and open areas represent lava-field provinces. (C) Groundmass-plagioclase (most calcic)/whole-rock compositions, inferred to give initial plagioclase-quenching temperatures. Values in B and C calculated using the method of Ghiorso et al. (1983).

with whole-rock compositions (Figs. 5.3.30–31). The wide range of equilibration temperatures therefore presumably represents different cooling histories for different samples. Perhaps the most significant feature is that Fe-Ti oxide data correspond to equilibration between the FMQ and WM fO_2 buffer curves. This appears to represent more reducing conditions than is apparently normal for other mafic volcanic suites (for example, Haggerty, 1976).

Relatively high temperatures (based on coexisting phenocryst Fe-Ti oxides) of 840 to 1060°C are calculated for most intermediate and felsic rocks (Fig. 5.3.30B). Some high-silica rhyolites have low equilibration temperatures (less than 700°C), but are considered to have been affected by subsolidus requilibration during post-eruption devitrification and hydration.

An estimated T-fO_2 region is shown in Figure 5.3.30B for the Binna Burra Rhyolite (high-silica type) in which rare, extremely ulvöspinel-rich spinel is found together with low-R_2O_3 ilmenite. Again, the coexisting oxides correspond to fO_2 equilibration below the FMQ buffer, approaching the WM curve. Most oversaturated trachytes in the region do not contain coexisting oxides, as discussed above, but they do contain phenocryst

titanomagnetite and fayalitic olivine. Calculations using the FMQ data (O'Neill, 1987) have been made, and these trachytes are estimated to have equilibrated between about 0.5 and 1 log unit below the FMQ buffer. This is entirely consistent with the fO_2 estimated from the coexisting oxide data plotted in Figure 5.3.30B. Attention also is drawn to the fact that most high-silica rhyolite in southeastern Queensland contains only an ilmenite that typically has low R_2O_3 (less than 3 percent, mostly less than 2 percent). Ewart et al. (1976) interpreted these rhyolites as having equilibrated at an fO_2 close to the WM buffer.

Table 5.3.3 Calculated feldspar-crystallisation temperatures for east-Australian trachytes and rhyolites

	1 kb		5 kb		n[1]
Ne-trachyte	956	(159)[2]	1024	(167)	4
Qz-trachyte	914	(147)	986	(157)	3
Low-silica rhyolite	878	(79)	936	(86)	7
High-silica rhyolite:					
1. Binna Burra type	768	(76)	828	(80)	3
2. biotite-bearing type (Noosa)	836	(17)	895	(15)	2
3. ferrohypersthene type	946	(77)	1006	(79)	5

[1] Number of samples.
[2] Values in brackets represent one standard deviation.

A list of temperatures calculated from coexisting plagioclase and alkali-feldspar phenocryst data, specifically for trachyte and rhyolite feldspars, is given in Table 5.3.3. A comparison of these temperatures with Fe-Ti-oxide equilibration temperatures is not possible, unfortunately, because only one sample for which feldspar temperatures can be calculated contains coexisting oxides. Nevertheless, the general temperature ranges obtained from the two-feldspar method agree with the ranges plotted in Figure 5.3.30. In particular, the lower crystallisation temperatures for the low-silica rhyolite compared with those for the trachyte are indicated by the method, as well as significant differences in temperatures for the different high-silica rhyolites containing different Fe-Mg-silicate minerals. The fayalite-ferrohedenbergite rhyolites do not contain two feldspars in equilibrium, and therefore temperature estimates for these rhyolites are available only for three samples containing coexisting Fe-Ti oxides. Values obtained are 945, 967, and 952°C, which represent concordant temperatures that are similar to those estimated for ferrohypersthene rhyolite. Estimates obtained from the olivine-clinopyroxene geothermometer (Powell & Powell, 1974) for coexisting phenocryst pairs in trachyte and rhyolite range between 890–1100°C (1 bar). Average values for high-silica rhyolite, low-silica rhyolite, and qz-trachyte are, respectively: 942°C (9 samples), 982°C (6 samples), and 962°C (6 samples).

Results obtained by application of the biotite geothermometer of Luhr et al. (1984) to the rare biotite-bearing rhyolite and trachyte rocks are as follows: Binna Burra rhyolite, 688°C (785°C where the Fe^{3+} determined for this mica is taken into account); Noosa (Glass Houses) rhyolite, 693 and 805°C (two separate occurrences); low-silica rhyolite, Flinders Peak, 760°C; Canobolas trachyte, 828°C. These values are lower than those obtained by other methods, and two of the rhyolite estimates appear unreasonably low. This is undoubtedly the result of difficulties in estimating Fe^{3+} and Fe^{2+}.

Ewart (1981, 1985) noted: the inferred, relatively high crystallisation temperatures of east-Australian felsic rocks in general; their relatively low fO_2 equilibration pressures; the extreme Fe-enriched nature of the Fe-Mg-silicate assemblages; and the general scarcity of hydrous minerals. These felsic magmas therefore are interpreted to have evolved under anhydrous or strongly water-undersaturated conditions.

Calculated olivine-crystallisation temperatures (Fig. 5.3.31B) have wide ranges in each mafic compositional group (reflecting different mg-ratios; see also Fig. 5.2.3), and higher liquidus temperatures for the progressively more undersaturated lavas. In addition, mafic lavas from the central-volcano complexes have lower calculated crystallisation temperatures which parallel their generally lower mg-ratios and their more fractionated or evolved chemistry.

Groundmass-plagioclase quench temperatures (Fig. 5.3.31C) progressively decrease as undersaturation of the mafic host rocks increases, and are remarkably consistent for the tholeiitic basalts. Perhaps therefore the tholeiitic magmas were erupted not only over a relatively narrow temperature range, but at slightly higher temperatures (on average) than the more undersaturated mafic lavas. Alternatively, these groundmass-plagioclase temperatures are dependent simply on lava chemistry: the more undersaturated magmas initially precipitate a less calcic plagioclase at slightly lower temperatures during quenching. Leucite temperatures calculated for the leucitites are concordant at between 1214 and 1230°C.

5.4 Trace Element Geochemistry

5.4.1 Introduction

The following section represents an attempt to summarise available trace- and minor-element geochemical data in order to identify the characteristic geochemical features and geochemical behaviour observed through the wide range of lava types from the east-Australian region. The main data sources used in this section are those of Cundari (1973), Duggan (1974), Ewart (1982a), Ewart et al. (1977, 1980, 1985), Ewart & Grenfell (1985), Farmer (1985), Forsyth (1984), Frey et al. (1978), Grenfell (1984), Griffin (1977), Kesson (1973), Knutson (1975), D.R. Mason (unpublished data), McClenaghan et al. (1982), Menzies & Wass (1983), P.A. Morris (unpublished data), Ross (1977), Stolz (1985; unpublished data), Sutherland (1980a), Wass (1980), Wass & Rogers (1980), together with a previously unpublished basaltic data set consisting of 76 samples on which integrated major-element, trace- and minor-element (by X-ray fluorescence and neutron-activation techniques), isotopic, and microprobe analyses have been undertaken.

Emphasis is given to the following aspects of trace-element geochemistry: (1) general 'tectonomagmatic' affinities; (2) the role of fractional crystallisation in determining the observed geochemical patterns for a large proportion of the lavas studied; (3) crustal and lithospheric contamination, or interaction, with the magmas, which is evaluated more extensively in the following isotopic Section 5.5. Most of the data plots have ancillary graphs illustrating the idealised effects of fractionation of specified minerals, or mineral assemblages, and for specific sets of partition coefficients (K_D), using the Rayleigh-fractionation equation. Partition coefficients are known to range widely, but can be

Table 5.4.1 Averaged partition coefficients (K_D) for trace elements for selected minerals

Olivine							Plagioclase				
	Basalt	Trachy-andesite	Trachyte	Low-silica rhyolite	High-silica rhyolite			Basalt	Trachy-andesite	Trachyte	Low-silica rhyolite	High-silica rhyolite
Rb	0.0002	0.01	0.01	0.015	0.015	Rb	0.026	0.045	0.045	0.05	0.05
Sr	0.0002	0.01	0.01	0.04	0.04	Sr	1.63	2.36	5.6	11.2	10.0
Ba	0.009	0.007	0.007	0.10	0.10	Ba	0.40	0.43	0.94	1.6	1.2
Zr	0.007	0.02	0.12	0.12	0.12	Zr	0.03	0.02	0.19	0.23	0.29
Nb	0.006	0.05	0.05	0.05	0.05	Nb	0.09	0.09	0.09	0.01	0.009
Ce	0.0047	0.037	0.068	0.068	7.70	Ce	0.048	0.10	0.20	0.19	0.28
Yb	0.0091	0.18	0.34	0.34	1.29	Yb	0.009	0.11	0.044	0.038	0.058
Ni	12.6	13.7	13.7	1.2	1.2	Ni	0.03	0.08	0.08	0.08	0.08
Cr	0.99	1.2	1.2	1.7	2.1	Cr	0.02	0.06	0.06	0.15	0.15
V	0.06	0.11	0.11	0.11	2.1	V	0.01	0.02	0.02	0.02	0.02
Sc	0.18	0.27	0.27	4.8	6.4	Sc	0.008	0.08	0.08	0.03	0.05

Clinopyroxene							Magnetite				
	Basalt	Trachy-andesite	Trachyte	Low-silica rhyolite	High-silica rhyolite			Basalt	Trachy-andesite	Trachyte	Low-silica rhyolite	High-silica rhyolite
Rb	0.027	0.027	0.026	0.026	0.047	Rb	0.004	0.004	0.004	0.004	0.004
Sr	0.10	0.11	0.20	0.15	0.58	Sr	0.04	0.04	0.04	0.04	0.04
Ba	0.017	0.026	0.13	0.05	0.50	Ba	0.07	0.07	0.12	0.10	0.85
Zr	0.14	0.31	0.31	0.75	0.75	Zr	0.33	0.33	0.40	0.65	0.55
Nb	0.04	0.08	0.08	0.08	0.04	Nb	13.1	13.1	10.7	10.7	10.7
Ce	0.18	0.57	1.43	5.1	10.8	Ce	0.016	0.29	0.57	0.75	12.6
Yb	0.56	1.6	3.52	6.3	4.55	Yb	0.018	0.23	0.44	0.60	1.60
Ni	2.5	4.5	4.5	6.3	3.1	Ni	1.4	5.2	9.6	19.0	1.2
Cr	6.4	7.5	2.4	7.5	2.9	Cr	10.0	6.6	14.7	10.0	32.0
V	1.5	1.7	3.0	4.7	10.3	V	17.5	27.6	24.7	52.0	249.0
Sc	2.4	3.6	13.2	23.4	90.0	Sc	12.4	3.9	6.6	21.0	21.0

correlated with the composition of the coexisting liquids. The sets of K_D values used here (Table 5.4.1) are from a data base set up by A.R. Duncan (University of Cape Town) and A. Ewart, in which the coefficients have been subdivided and averaged according to whole-rock chemistry (or to residual-liquid chemistry where the relevant data exist). The values were extracted from the literature and are based primarily on natural mineral data. The coefficients are subdivided into the following compositional groups: basalt, trachyandesite, trachyte, low-silica rhyolite (less than 74 percent SiO_2), and high-silica rhyolite. The criteria used for these rock subdivisions are based on those of the TAS classification system (Le Maitre, 1984).

5.4.2 General trace-element characteristics and tectonomagmatic affinities

Mafic rocks

A summary of the overall trace-element abundance patterns is illustrated in MORB (mid-ocean-ridge basalt)-normalised diagrams in Figure 5.4.1 where the data are grouped on the basis of their major-element compositions. Critical element ratios, distinguished according to basaltic compositional type, are given in Table 5.4.2, but only for the more magnesian sets of samples in which 'primary' geochemical parameters are most likely to have been preserved (that is, least modified by fractionation processes). The following

points can be highlighted with reference to Figure 5.4.1:

(1) The basaltic rocks have enrichments, relative to MORB values, of the incompatible elements from Sr through to Hf in the tholeiitic basalts and hawaiites, extending to TiO_2 in the more undersaturated types. These enrichment patterns are similar to those found in other intraplate environments (see, for example, Pearce et al., 1981, and the Karoo data presented by Cox, 1983). They have maximum relative enrichments between Ba (tholeiitic basalt and hawaiite) and Th (basanite, nephelinite, and melilitite). Relative depletions are observed in all of the patterns between Y, Yb, and Sc. This is believed to be most likely an inherited source characteristic in which garnet was left as a residual phase after partial melting.

(2) The patterns illustrated in Figure 5.4.1 are similar in terms of their general enrichments and depletions, except for the leucitites for which the patterns have a shift of maximum enrichment to Rb-Ba. This is interpreted to represent the original presence, and partial melting, of phlogopite in the mantle source.

(3) A general progressive increase in the degree of relative enrichment is found in the sequence tholeiitic basalt through to hawaiite, basanite, nephelinite, and melilitite, to leucitite. The least enriched lavas are the low-K_2O (less than 0.75 percent) tholeiitic basalts. The question of the relative roles of heterogeneous, incompatible-element enriched mantle sources, different degrees of partial melting (Frey et al., 1978), together

with crustal-interaction processes in producing these geochemical features, is still a major unresolved problem, and one which is considered further in this section and in Section 5.5.

Systematic differences and similarities in element ratios exist between the different basaltic magma types (Table 5.4.2). K/Rb values are generally lowest, and Rb/Sr, Rb/Zr, Ba/Th, Nb/U values are mainly higher, in the leucitites compared to the other basaltic types. However, there is no indication of significant differences between the basalt compositions in K/Rb, Rb/Sr, Zr/Nb, Ba/La, Sr/Nd, Ti/Nd, Ti/V, Nb/Ta, Rb/Cs, or Zr/Hf values. Zr/Hf values range between 45.6 and 71.0 (mostly 48–56). Systematic decreases in La and La/Yb are evident towards the tholeiitic-basalt compositions, and there is also an indication of decreasing Zr/Y. Ce/Pb and Nb/U values are lowest in the *qz*-tholeiitic basalts. The overall patterns of trace-element abundances clearly are similar to those of other intraplate volcanic provinces worldwide.

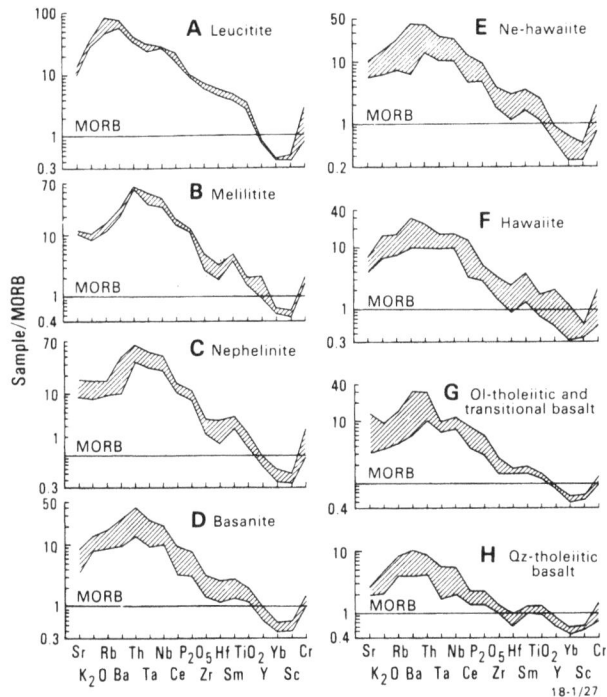

Figure 5.4.2   Hf:Th:Ta relationships for east-Australian mafic (A) and intermediate and felsic (B) rocks. The 'tectonomagmatic' boundaries are those defined by D.A. Wood (1980). Field *b* represents E-type MORB and tholeiitic within-plate basalts and differentiates. Field *c* represents alkaline within-plate basalts and differentiates. Field *d* represents destructive-plate-margin basalts and differentiates.

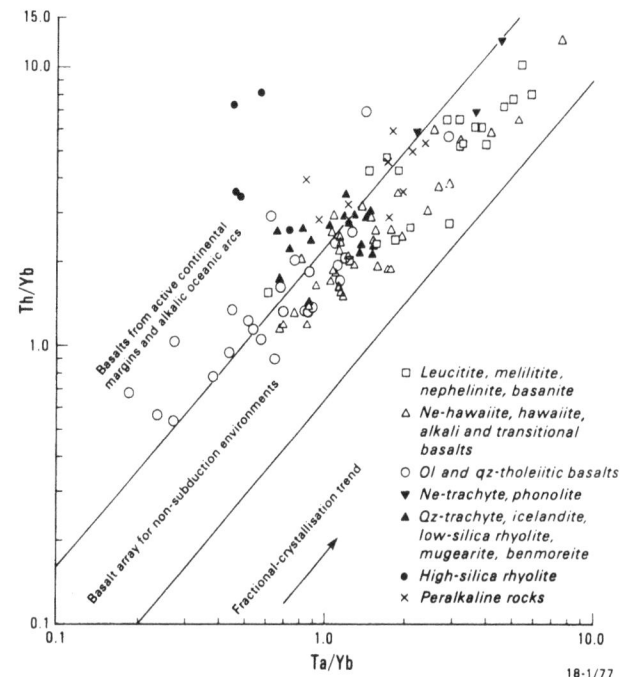

Figure 5.4.1   MORB-normalised element abundances for the main east-Australian mafic rock types (Ewart and Chappell, unpublished data; Frey et al., 1978). (A) mg-ratios greater than 66. (B) mg-ratios greater than 64.5. (C) mg-ratios greater than 61. (D) mg-ratios greater than 57. (E–G) mg-ratios greater than 55. (H) Low-K$_2$O (less than 0.75 weight percent) type and mg-ratios greater than 54. Normalising factors are from Pearce et al. (1981).

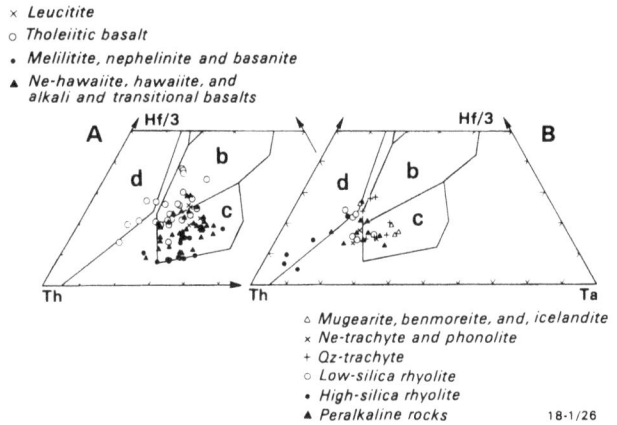

Figure 5.4.3   Th/Yb:Ta/Yb relationships for east-Australian mafic, intermediate, and felsic rocks. Basalt array and fields of active continental margins and alkalic oceanic arcs are those defined by Pearce (1983). Fractional-crystallisation trend based on oliv+cpx and oliv+plag assemblages (both in 0.5:0.5 proportions). *Abbreviations for mineral names used here and in following figures are as follows: oliv, olivine; cpx, clinopyroxene; opx, orthopyroxene; plag, plagioclase; oxide, Fe-Ti oxides; san, sanidine; qtz, quartz.*

The majority of data points representing east-Australian basaltic analyses plot in the appropriate fields of the tectonomagmatic discrimination diagrams (Figs. 5.4.2–3). Points representing alkaline rocks in Figure 5.4.2 are mostly within field *c* (alkaline within-plate basalts and differentiates), whereas the tholeiitic basalts mostly plot within field *b* (tholeiitic within-plate basalts and differentiates). The hawaiite data points fall in both fields. However, a small number of points, mostly for *qz*-tholeiitic basalts, extend outside these

intraplate fields into, or close to, the compositional fields of the arc-trench-type basalts (Figs. 5.4.2–3). These 'anomalous' points are characterised also by fairly distinctive isotopic compositions (see below). They are taken as evidence for a subduction-modified lithospheric origin, or contamination by Palaeozoic-Mesozoic subduction-related volcanic rocks and associated volcanogenic sediments.

Table 5.4.2   Selected element ratios and abundances of main basaltic compositional types from eastern Australia[1]

	Leucitite	Melilitite	Nephelinite	Basanite	Alkali basalt	Transitional basalt	Ne-hawaiite	Hawaiite	Ol-tholeiitic basalt	Qz-tholeiitic basalt	N-MORB[2]	Southwest Pacific basalt[3]
mg-ratios greater than 64												
K/Rb	138–487 (341)	380–455	381–669	374–579	493–644		(58)–390	408–442			716	354
Rb/Sr	0.074–0.36 (0.113)	0.018–0.025	0.013–0.026	0.029–0.049	0.026		0.036–0.052	0.033–0.044			0.017	0.046
La/Yb	49.7–71.3	18.2–42.7	32.4–60.0	22.7–30.3	8.9–10.9		31.0–41.9	13.0–23.6			0.97	7.5
La (ppm)	74–102	41–90	52–77	32–52	19–20		44–45	23–30				11.6
Rb/Zr	0.16–0.47 (0.25)	0.052–0.12	0.083–0.12	0.13–0.14	0.09–0.13		0.14–0.61	0.16–0.13			0.02	0.42
Rb/Ba	0.043–0.2 (0.094)	0.052–0.077	0.026–0.09	0.053–0.078	0.068–0.070		0.081–0.16	0.064–0.069			0.10	0.08
Rb/Cs	91–127	53–70	41–107	134	36–80		63	123			99[4]	Ca. 42[5]
Ce/Pb	53–96	87–210	53–189	82	23–38		41–53	25			27.5[4]	Ca. 5.1[5]
Nb/U	68–140	37–45	36–44	33–39	48–57		52–55	40			Ca. 47	Ca. 7.3
Nb/Ta	15.5–17.2	16.2–20.4	16.9–19.0	24.0	19.1–21.1		16.3–17.3	20.0				
Zr/Y	21–28	7.8–12.5	8.5–13.4	8.9–11.7	4.7–6.9		10–13	5.6–6.1			2.7	3.5
Ti/Zr	32–60 (43.3)	41–67	44–119	59–69	51–63		43–69	47–84			100	73.1
Zr/Hf	43.1–45.1	49.1–56.0	52.1–60.0	58.2	57.2–61.7		49.5–50.0	65.0			37.5	53.6
Zr/Nb	5.6–6.8	2.3–4.0	2.2–2.5	3.2–4.5	3.7–4.5		2.7–4.9	3.6			26	13.2
Ba/Sr	0.78–3.11 (1.19)	0.32–0.43	0.29–0.74	0.49–0.91	0.36–0.39		0.64–0.68	0.47–0.65			0.17	0.58
Ba/La	15.2–15.5	4.8–10.2	5.2–14.2	8.8–10.1	12.1–13.0		10.4–18.9	9.4–15.2			5.4	31.4
Ba/Nb	12–16	3.0–5.4	3.2–9.3	7.4–7.6	5.7–6.9		8.1–9.9	7.2			5.7	68.7
Ba/Th	151–184	33–49	35–78	48–81	71–88		64–133	61–121			60	383
P_2O_5/Ce	52–74	84–90	48–101	87–100	94–120		84–118	69–104			120	108
Sr/Nd	16–35	16–23	20–32	17–25	33–37		17–32	16–24			15	50.2
Th/U	6.2–13	4.2–4.4	4.1–5.1	3.6–4.0	3.9–5.8		4.5–7.3	2.6–5.2				17
Ti/V	117–142 (392)	67–97	66–78	85–94	53–61		69–76	42–76			32	
Ni (ppm)	267–520 (424)	236–458	195–355	308–447	299–356		409–445	346			130	104
Cr (ppm)	212–736 (431)	419–510	261–464	310–391	440–478		479–537	388–543			250	273
n[6]	3	4	6	3	2		2	2				

continued

Table 5.4.2 Selected element ratios and abundances of main basaltic compositional types from eastern Australia (continued)

	Leucitite	Melilitite	Nephelinite	Basanite	Alkali basalt	Transitional basalt	Ne-hawaiite	Hawaiite	Ol-tholeiitic basalt	Qz-tholeiitic basalt	Low-K qz-tholeiitic basalts
mg-ratios 55–64											
K/Rb	166–267		484–1072	333–1062	423–974	406–413	400–1396	398–898	360–498	314–581	314–581
Rb/Sr	0.074–0.36		0.010–0.026	0.017–0.071	0.017–0.048	0.018–0.041	0.012–0.047	0.021–0.045	0.024–0.049	0.033–0.098	0.033–0.052
La/Yb			35.6–43.8	12.7–23.8	10.6–15.4	14.6–21.3	14.2–47.0	11.3–24.7	8.8–13.7	3.7–19.8	3.7–6.1
La (ppm)			62–73	17–37	19–23	26–45	21–52	16–51	16–23	6.3–23	6.3–9.3
Rb/Zr	0.20–0.43		0.038–0.11	0.069–0.24	0.071–0.16	0.10–0.23	0.045–0.14	0.099–0.17	0.062–0.13	0.089–0.21	0.089–0.14
Rb/Ba	0.001–0.22		0.048–0.15	0.057–0.12	0.020–0.098	0.047–0.070	0.053–0.87	0.018–0.076	0.072–0.097	0.055–0.12	0.055–0.12
Rb/Cs			41–107	60–143	62–210	67	61–170	12–115	53	41–130	39–89
Ce/Pb			53–115	18–45	13–40	20	28–65	26–40		9.3–15	9.3–15
Nb/U			40–44	28–89	36–102	70	31–65	35–72	39	17–40	17–44
Nb/Ta			18.3	15.1–20.0	18.3–24.2	22.8	16.5–23.6	18.9–21.4	20.8	16.7–24.0	17.5–24.0
Zr/Y			11–12	7.1–11.6	5.1–6.3	4.8–9.4	7.1–22	2.3–18.6	5.1–7.5	1.4–12.7	4.5–5.5
Ti/Zr	20–51		46–64	45–87	67–87	55–97	29–76	43–104	80–93	54–102	79–102
Zr/Hf			58.2	49.2–55.6	54.4	55.5	45.4–57.7	44.8–55.0	60.0	44.7–52.6	46.0–52.6
Zr/Nb			2.4–3.3	3.6–4.9	3.6–5.1	5.7	3.2–5.3	3.4–5.2	6.7	6.3–9.9	6.3–9.9
Ba/Sr	0.75–3.2		0.18–0.22	0.20–0.70	0.43–0.83	0.38–0.59	0.17–0.62	0.28–2.6	0.34–0.62	0.32–0.85	0.32–0.62
Ba/La			2.8–4.5	6.0–18	11.1–22	13–14	4.2–13	7.9–24	7.1–11	6.7–23	11–23
Ba/Nb			1.7–2.6	3.9–8.3	5.9–18	8.2	2.9–8.7	4.8–10.2	9.9	6.2–16	6.2–16
Ba/Th			19–38	55–108	91–247	104–113	42–92	66–444	57–94	52–201	68–131
P₂O₅/Ce			79–95	98–123	86–116	83–85	88–124	97–121	81–100	82–156	116–156
Sr/Nd			19–22	22–30	22–27	20–47	20–32	9.2–31	17–23	8.7–28	21–26
Th/U			2.9–4.4	2.2–6.1	2.5–6.5	5.0	3.0–6.0	2.2–6.4	3.0–11.0	2.8–6.5	2.1–3.0
Ti/V			61–114	52–101	56–75	72–78	65–113	61–96	61–82	52–106	52–64
Ni (ppm)	102–351		187–405	161–188	119–149	183–328	135–375	90–280	142–190	106–174	142–174
Cr (ppm)	189–401		238–520	255–341	226–359	276–331	178–400	139–300	240–312	132–375	211–375
n⁶	5		3	4	3	2	8	9	3	8	4

[1] These ratios are based mainly on a specific set of samples for which an integrated set of major- and trace-element analyses has been obtained (Ewart and Chappell, previously unpublished data). Additional data are from Frey et al. (1978; excepting Zr/Hf), Menzies & Wass (1983), and Stolz (1985). Additional data for some ratios in the leucitites are from Cundari (1973) which are shown by the additional mean values quoted in brackets (28 samples).
[2] In part after Ewart (1982b).   [3] In part after Pearce et al. (1981).   [4] Newson et al. (1986).   [5] Basaltic Volcanism Study Project (1981).   [6] n refers to number of samples.

Felsic rocks

Four groups of evolved lavas — qz-trachyte, low-silica rhyolite, high-silica rhyolite, and peralkaline rocks — have differences in MORB-normalised abundance patterns in which there are systematic and spectacular enrichments and depletions of elements (Figs. 5.4.4A–D). Qz-trachytes generally have enrichments in the incompatible elements (Rb to Ta), but there are also strong superimposed Sr, P_2O_5, TiO_2, Sc, and Cr depletions, and some indication of Ba depletion. These same trends are accentuated further in the low-silica rhyolites, and to an even more extreme degree in the peralkaline lavas. The high-silica rhyolites also have these same depletion patterns, as well as notably lower relative enrichment in the Ta, Nb, Zr, and Hf group of elements. These extreme enrichment-depletion patterns are indicative of strong crystal-liquid fractionation control in the evolution of the felsic magmas. The strong depletions of Sr, Ba, P_2O_5, TiO_2, Sc, and Cr are qualitative evidence for feldspar, apatite, Fe-Ti oxide, and pyroxene fractionation (see below).

There are also some notable deviations in trace-element behaviour in the evolved-lava compositions compared to that for the mafic rocks. For example, data points in Figure 5.4.2B are scattered throughout fields c and d, and a few qz-trachytes fall within field b. The high-silica rhyolites have the most extreme deviations into the 'destructive plate margin' field, but other peralkaline and non-peralkaline lavas have similar general trends. The general conclusion to be drawn from these data arrays is that input of lithospheric and crustal components has been important in the evolution of a relatively high proportion of the evolved lavas, as well as for some of the qz-tholeiitic basalts.

Critical element ratios are listed in Table 5.4.3 for the major groups of the more evolved lavas. Those

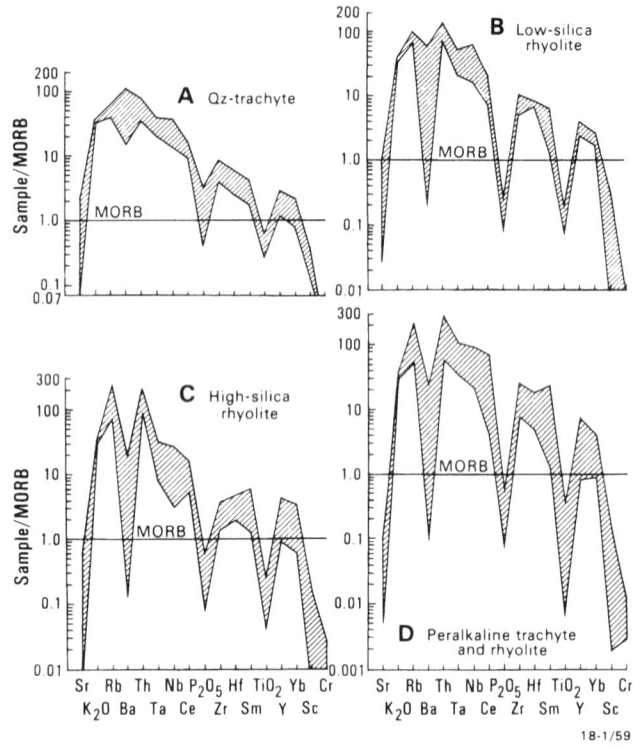

Figure 5.4.4 MORB-normalised element abundances for selected felsic rocks from eastern Australia (Ewart et al., 1977; Ewart et al., 1985; Stolz, 1985). Normalising factors are from Pearce et al. (1981).

elements that are depleted (Sr, Ba, Ti) predictably have wide ratio ranges. Note, particularly, the consistent Eu depletions, which become extreme in some of the peralkaline rocks and high-silica rhyolites. Zr/Hf values also decrease accompanying strong fractionation.

Table 5.4.3   Selected element ratios and abundances for felsic volcanic-rock types from eastern Australia[1]

| | Qz-trachyte | Ne-trachyte and phonolite | Peralkaline rhyolite and trachyte[2] | Potassic peralkaline rhyolite | Low-silica rhyolite | High-silica rhyolite | | |
						Ferro-hypersthene type	Fayalite-ferrohedenbergite type	Binna Burra type
K/Rb	345–687	174–410	88–389	247–251	223–320	171–202	105–284	81–99
Rb/Sr	0.37–16	0.27–6.4	10–600	38–97	1.7–190	2.9–10.8	10.5–32	277–2000
La/Yb	12.4–22.1	29.6–34.4	3.7–29	1.9–2.0	4.3–12.3	12.6–17.3	9.2–15.2	2.3–3.1
La (ppm)	49–94	75–105	28–285[3]	10–17	32–86	28–52	45–78	21–36
Eu/Eu*	0.36–1.4	0.33–0.84	0.032–0.44	0.19–0.22	0.17–0.53	0.078–0.32	0.11–0.14	0.0023–0.03
Rb/Zr	0.13–0.22	0.15–0.18	0.15–0.40	0.12–0.20	0.18–0.36	0.76–1.6	0.31–0.44	3.2–3.8
Rb/Cs	126–263	31–84	81–885	127	40–207	17–19	16–19	14–20
Ce/Pb	8.5–13.6	9.0–18	4.0–25	2.5–5.4	2.5–10.5	2.6–5.4	4.2–7.4	1.3–2.1
Nb/U	26.0–61.5	32–58	16–69	10–23	22–63	2.5–3.4	8.6–27	4.7–7.0
Nb/Ta	13–22	14–16	10.7–22		15–24	7.9–8.2	11–17	11.5
Zr/Y	6.4–18	22–40	4.1–28	11–19	3.9–12	3.9–10.0	4.3–8.9	1.13–1.19
Zr/Hf	43.9–66.0	46.7–51.8	29.8–58.5	49.0–49.1	27.5–52.8	21.9–44.2	28.3–39.9	21.0–24.3
Zr/Nb	5.6–15	4.8–7.6	2.9–9.5	14–20	2.1–16	6.4–25	9.5–11.0	2.3–2.5
Ba/La	3.2–24	0.29–9.5	0.013–5.5	1.9–2.1	0.098–17	1.4–14	0.079–2.2	0.14–0.50
Ba/Th	19–209	0.69–49	0.064–27	1.7–2.1	0.21–80	3.1–32	0.29–10.4	0.068–0.43
Sr/Nd	0.11–6.3	0.71–7.0	0.006–0.18	0.13–0.21	0.037–1.7	0.48–2.9	0.095–0.21	0.003–0.054
Th/U	3.3–7.1	6.5–7.4	3.8–9.9	4.0–4.7	5.2–7.9	4.1–6.0	3.4–5.9	4.1–4.7
Ti/V	280–1800	78–470	7–1400	1250	216–2700	100–430	440–1300	300–1100
n[4]	8	3	12	2	6	3	4	2

[1] Data from Stolz (1985), Ewart et al. (1977, 1985), and Ewart and Chappell (unpublished).   [2] Excluding 'potassic' types (see next column).
[3] One sample has La value of 754.   [4] n refers to number of samples.

Attention is drawn also to some markedly different ratios for the high-silica rhyolites, compared to other rhyolitic and trachytic types, notably Rb/Zr, Rb/Cs, and Nb/U.

A further point concerns two possible types of peralkaline lava. Ewart (1985) described a subgroup consisting of potassic and 'transitional' peralkaline rhyolites (as distinct from the more common sodic types), and showed that these contained more potassium-rich feldspar (correlating with higher total-rock K/Na values) than is found in the majority of peralkaline lavas. These feldspar compositions are similar to those observed in high-silica rhyolite. 'Potassic' peralkaline rhyolite is interpreted to have an independent origin involving ternary-minimum fractionation from non-peralkaline rhyolitic parents (Ewart, 1982a, 1985). Available data are summarised in Table 5.4.3 for the two peralkaline groups in which the 'potassic' type is seen to have some distinctive trace-element characteristics — notably, lower light-REE and La/Yb, and higher Zr/Nb values.

Ewart (1982a) identified the following three broad patterns of trace-element behaviour in the more evolved east-Australian lavas:

(1) Those behaving as incompatible elements which, typically, have relative enrichment patterns. Rb, Th and Pb are specific examples.

(2) Elements that are strongly depleted, especially in the rhyolites. Most notable amongst these are Ba, Sr, V, Ni, Cr, Mg, Ca, Mn, P, and Eu.

(3) The more highly charged cations, notably Zr, light REE, Y, and Nb. These normally become enriched in the melt during fractional crystallisation, but they also may have mutually 'decoupled' behaviour in some groups of lavas. This decoupling is attributed to the crystallisation and fractionation of minor microphenocryst minerals such as zircon, chevkinite, allanite, and apatite in the non-peralkaline lavas, and to sodic pyroxene and sodic amphibole in the peralkaline lavas.

5.4.3   Rare-earth elements

Mafic rocks

There is a general increase in light- to heavy-REE fractionation, and in light-REE abundances, in the sequence from tholeiitic basalt through to the more alkaline basaltic types (Fig. 5.4.5; Table 5.4.2). The REE are regular in their relative enrichment patterns. The most noticeable differences are in some of the tholeiitic basalts — especially the qz-tholeiitic basalts — in which there is a relative middle-REE enrichment. This is unexplained, but is most conspicuous in those lavas having the lowest total-REE abundances.

The effects of olivine, clinopyroxene, and plagioclase fractionation on specific nephelinite and qz-tholeiitic basalt REE patterns (one enriched, the other relatively depleted) have been calculated using basaltic K_D values (Fig. 5.4.5J–L). The major effect of fractionation clearly is to produce a uniform increase in REE abundances, and evidently only minor relative REE fractionation. This is emphasised further in Figure 5.4.5L where the results of least-squares modelling

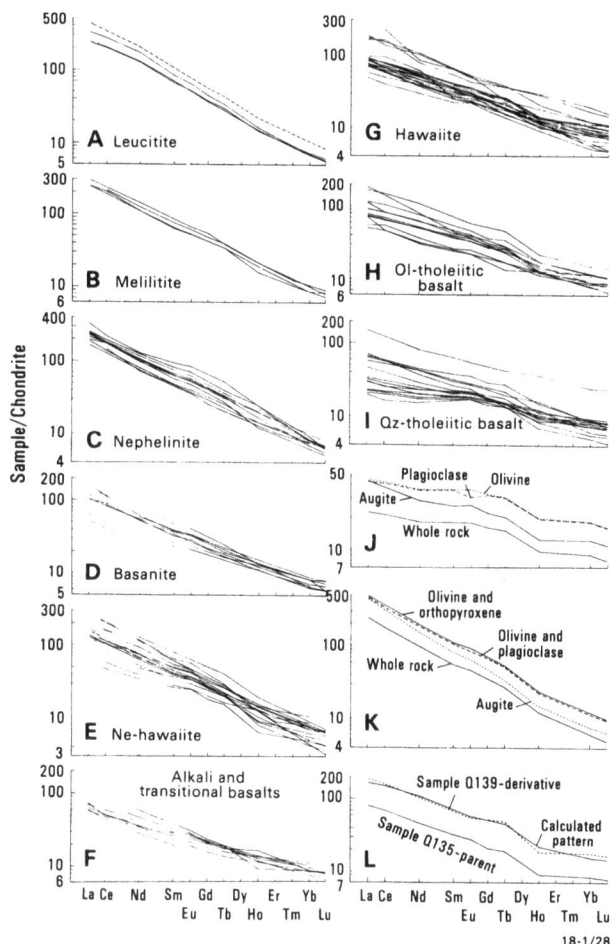

Figure 5.4.5   (A-I) Chondrite-normalised REE abundances in east-Australian mafic rocks (Ewart & Chappell, unpublished data; Frey et al., 1978; Menzies & Wass, 1983). (J) Oliv, plag, and cpx fractionation (F 0.5) of qz-tholeiitic basalt (sample GA3126) using basaltic partition coefficients. (K) Oliv+opx, and oliv+plag (both in 0.5:0.5 proportions) fractionation (F 0.5) of nephelinite sample (sample CQ88) using basalt partition coefficients. (L) REE abundances calculated from least-squares-fitted transitional basalt (sample Q135; mg-ratio 69.1) to ol-tholeiitic basalt (sample Q139; mg-ratio 35.0; F 0.408) using basalt partition coefficients and the mineral proportions shown in Table 5.4.4.

Table 5.4.4   Least-squares-modelling results for transitional-basalt (sample Q135) to ol-tholeiitic-basalt (sample Q139) fractionation[1]

Mineral	Weight fraction	
Olivine (Fo$_{76}$)	0.248	F = 0.408
Augite	0.316	R^2 = 1.11
Plagioclase (An$_{60}$)	0.392	(including
Ilmenite	0.019	independent FeO
Titanomagnetite	0.026	and Fe_2O_3
		estimates)

[1] See text and Figure 5.4.5L.

(Table 5.4.4) are shown for two closely associated basalts from the Springsure centre in central Queensland — a transitional basalt (Q135; mg-ratio 62.9) and an ol-tholeiitic basalt (Q139; mg-ratio 35.0). This particular example is also used in subsequent plots.

The calculated REE pattern obtained using this solution (basaltic K_D values, and Rayleigh fractionation) is reproduced reasonably well by the fractionation process. The major effect is total-REE enrichment, as noted above, and therefore the overall differences observed in the REE-abundance and fractionation patterns between the different basaltic types, are primarily source-inherited differences. This conclusion also was reached by Frey et al. (1978) from their detailed study of a selected set of southeast-Australian basaltic samples.

Intermediate and felsic rocks

Differences between the REE patterns observed in the major groups of evolved lavas (Figs 5.4.6) are described most appropriately in terms of La/Yb, Eu/Eu*, and La values (Table 5.4.3). The major characteristics of the observed patterns are likely to have been determined principally by crystal-liquid fractionation processes, although in the less evolved intermediate lavas, the patterns still may reflect partially the parental-magma compositions.

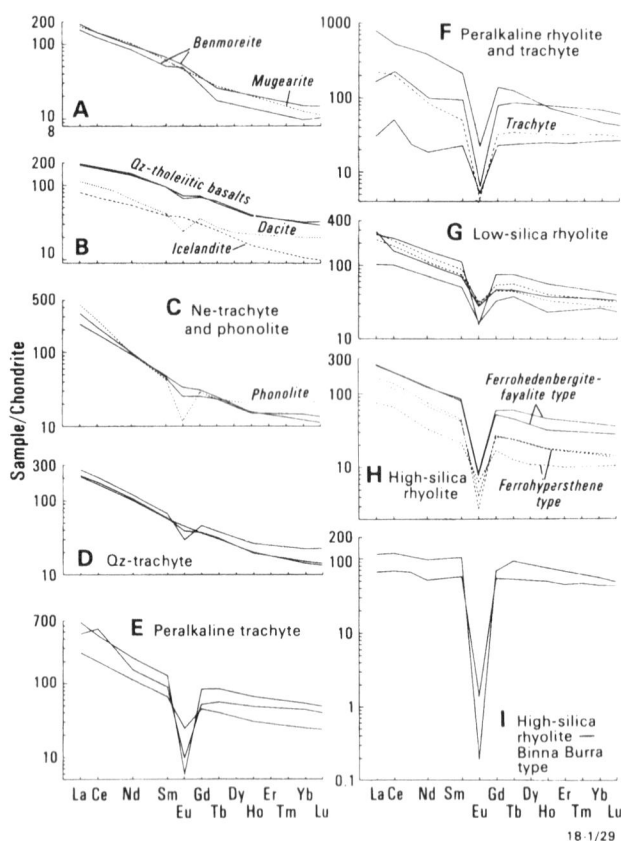

Eu depletion is the most conspicuous general characteristic of the REE in these rocks, although not in mugearite and benmoreite, and only slightly in some qz-trachytes. Furthermore, some trachytes from Nandewar volcano have marked positive Eu anomalies, indicative of feldspar accumulation (Stolz, 1985). Eu depletion is strongly developed in all the rhyolites and peralkaline rocks, and is most extreme in the Binna Burra high-silica rhyolite (Tweed volcano) where Eu/Eu* in one sample is estimated to be 0.0023 (Fig. 5.4.6I). However, equally significant is the fact that extreme Eu depletion commonly is accompanied by decreasing light-REE fractionation (Table 5.4.3; see also Ewart et al., 1985) and, therefore, decreasing total-REE abundances, although there are exceptions in the Nandewar peralkaline rhyolites (Stolz, 1985). A further significant point is that although the peralkaline lavas are consistently highly fractionated (for example, highly depleted Eu), they are not necessarily more strongly geochemically fractionated than are non-peralkaline rhyolites. This also is found consistently in relation to other trace elements (see below).

Eu depletion, especially in the most felsic rocks, was modelled using Rayleigh-fractionation calculations by Ewart et al. (1985), and a summary of the results is given in Table 5.4.5.

Conclusions are as follows:

(1) The observed Eu anomalies can be reproduced by feldspar-dominated, crystal-liquid fractionation.

(2) A positive correlation exists between La/Yb and Eu/Eu* values, as is observed in natural rocks.

(3) The relative enrichment of the light REE is strongly controlled by fractionation of non-feldspar silicate minerals, but obviously is sensitive also to differences in the light-REE partition coefficients. Moreover, the greater the relative proportion of feldspar in the fractionating mineral assemblage, the smaller the relative change in La/Yb during fractionation.

(4) Production of the extremely low La/Yb and Eu/Eu* values observed in Binna Burra high-silica rhyolite requires a significant proportion of Fe-silicate minerals in the fractionating assemblage.

(5) Development of the more strongly fractionated REE patterns observed in east-Australian felsic rocks requires high values for partition coefficients during crystal-liquid fractionation. Such high values are observed in highly felsic magmas (see above). This is indicative that the observed extreme fractionation effects must have developed in magmas that already had evolved to a felsic stage, either by prior fractionation, or through partial melting.

Feldspar fractionation clearly is important in producing Eu anomalies, and therefore correlations should exist between Eu/Eu* and other feldspar-controlled elements, notably Sr and Ba. Ewart et al. (1985) showed that these correlations exist, although they are not simple 'single-stage' correlations.

5.4.4 Transition elements

Introduction

Ni, Cr, V, and Sc abundances progressively decrease as silica increases. In detail, however, the behaviour patterns of each element are different.

Figure 5.4.6 Chondrite-normalised REE abundances in east-Australian evolved rocks (Ewart et al., 1977, 1985; Stolz, 1985). (A) Mugearite and benmoreite from Nandewar volcano. (B) Qz-tholeiitic basalt (mg-ratio 18.9 and 23.8), dacite from southeastern Queensland, and icelandite from Canobolas volcano. (C) Ne-trachyte and phonolite from central Victoria. (D) Qz-trachyte from southeastern Queensland. (E) Peralkaline trachyte and rhyolite from southeastern Queensland. (G) Low-silica rhyolite from Flinders Peak and Minerva Hills, southeastern Queensland. (H) High-silica rhyolite of fayalite- and ferro-hypersthene-bearing types from southeastern Queensland. (I) High-silica rhyolite of Binna Burra type from Tweed volcano.

Table 5.4.5 Modelling of Eu anomalies in east-Australian felsic rocks

Weight fractions of minerals[1]				Model 1	Model 2	Model 3
Sodic plagioclase ($An_{11.4}$ $Ab_{79.7}$ $Or_{8.9}$)				0.27	0.289	0.311
Sanidine ($Or_{62.0}$ $Ab_{36.7}$ $An_{1.3}$)				0.60	0.639	0.689
Ferrohedenbergite				0.018	0.019	
Fayalite				0.049	0.052	
Titanomagnetite				0.058		
Apatite				0.005		

Mineral assemblage		Maximum K_D			Minumum K_D	
	F = 0.7	F = 0.3	F = 0.1	F = 0.7	F = 0.3	F = 0.1
Calculated[2]						
Model 1						
La/Yb	8.2	0.77	0.04	19.9	15.5	11.1
Eu/Eu*	0.30	0.016	0.0003	0.50	0.09	0.01
Model 2						
La/Yb	12.7	3.37	0.61	21.2	18.9	16.4
Eu/Eu*	0.26	0.011	0.0002	0.44	0.065	0.005
Model 3						
La/Yb	21.0	18.5	15.6	20.8	17.9	14.7
Eu/Eu*	0.23	0.008	0.00008	0.41	0.05	0.003

Observed		La/Yb		Eu/Eu*		
Peralkaline rocks (sodic type)		3.7–29		0.03–0.44		
Peralkaline rocks (potassic type)		1.9–2.0		0.19–0.22		
Low-silica rhyolites		4.3–12.3		0.17–0.53		
High-silica rhyolites						
(1) ferrohypersthene type		12.6–17.3		0.08–0.32		
(2) fayalite type		9.2–15.2		0.11–0.14		
(3) Binna-Burra type		2.3–3.1		0.002–0.03		

[1] The weight fractions and mineral types are based on least-squares linear mixing calculations using different *qz*-trachyte parental compositions, together with observed phenocryst minerals (Ewart, 1982a).
[2] Partition coefficients for these calculations are maximum and minimum values from data presented by Mahood & Hildreth (1983), Hildreth (1979), and Irving (1978; apatite data). Parental REE data are for a mafic trachyte in which La/Yb is 22.1, Eu/Eu* is 0.97, and La is 66 ppm (sample 38989; see Ewart et al., 1985).

Ni-MgO

A compilation and summary diagram have been made of the available data for the basaltic and intermediate (mugearite, benmoreite, and icelandite) rocks from the east-Australian Cainozoic provinces (Fig. 5.4.7A). The more felsic rocks are not included because of their strong depletion in both Ni and MgO. The overall data trend follows a logarithmic curve indicative of olivine-controlled fractionation, and this is illustrated further in Figure 5.4.7A by the partial-melting and fractionation curves calculated by Hart & Allègre (1980). Two sets of curves are presented. One represents batch partial melting (5 and 20 percent) of a peridotite source containing 2300 ppm Ni, and an assumption is made that melts within a wide range of MgO contents can be produced by the melting process. The other set of curves represents fractional crystallisation. Ni partition coefficients are based on the equation: $K_D = (124/MgO) - 0.9$ (Hart & Davis, 1978). The generalised basaltic compositional fields shown in Figure 5.4.7A specifically distinguish the tholeiitic-basalt fields.

The observed Ni-MgO distributions correspond closely to the calculated melting and fractionation curves, and support the interpretation that the high-Ni bearing magmas represent near-primary melts. In addition, extensive olivine-dominated fractionation clearly has modified many of the basaltic lavas. An important feature of Figure 5.4.7 is the overall lower Ni and MgO abundances in the tholeiitic basalts. Part of the tholeiitic-basalt data field overlaps the low-MgO end of the melting curves (shown in Fig. 5.4.7A), but the overall trace-element data are indicative that the tholeiitic lavas have probably undergone extensive crystal fractionation, mainly olivine and clinopyroxene, and therefore do not represent primary melts.

More specific fractional-crystallisation curves presented in Figure 5.4.7B are for the following three lava compositions and conditions: (1) a nephelinite, fractionating assemblages olivine (Fo_{80}) + augite, olivine + orthopyroxene, and also olivine + plagioclase for reference (all minerals in equal proportions), and calculated to an F-value (fraction of liquid remaining) of 0.7 using two sets of olivine partition coefficients; (2) a *qz*-tholeiitic basalt, fractionating assemblages of augite, augite + plagioclase, orthopyroxene, and olivine (Fo_{70}); (3) the transitional-basalt to *ol*-tholeiitic-basalt model referred to above, involving fractionation of olivine, augite, plagioclase, and Fe-Ti oxides. The different sets of calculated curves correspond to Ni-MgO depletion along the same observed general trends, but are consistent with pyroxene or plagioclase (or both) as important additional minerals controlling the Ni-MgO behaviour. Fractionation involving specifically olivine and pyroxene, rather than plagioclase + olivine, is most consistent with the nephelinite data, in accordance with the observed mineralogy of this lava type.

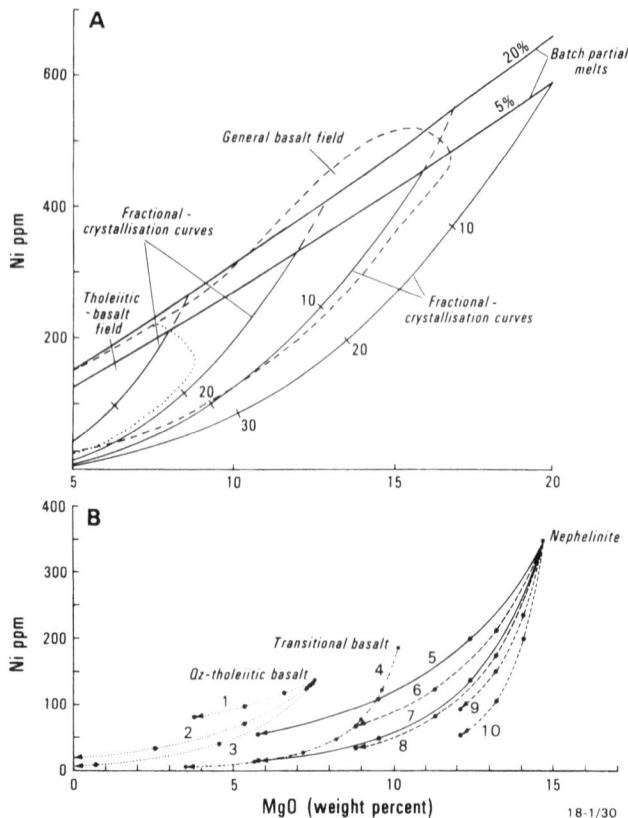

Figure 5.4.7 (A) Generalised Ni:MgO relationships in basalts from eastern Australia, superimposed on the partial-melting and olivine-fractionation curves given by Hart & Allègre (1980). The two batch partial-melting curves (5 and 20 percent) are for an assumed peridotite source containing 2300 ppm Ni. The four fractionation curves start from the 5-percent melting curve at initial liquids containing 8, 12, 16, and 20 percent MgO. Ni partition coefficients used are from Hart & Davis (1978). Numbers on the curves represent percentages of olivine crystallised. (B) Calculated Ni-MgO fractionation curves for a qz-tholeiitic basalt, transitional basalt, and nephelinite using two sets of olivine and clinopyroxene basalt partition coefficients. Fractionation trends are for (1) cpx (K_D 2.5), (2) opx (K_D 7.3), and (3) oliv (Fo_{70} and K_D 12.6) fractionation. Fractionation trend 4 represents the least-squares-fitted model for samples Q135 to Q139. Fractionation trends from nephelinite are for the following mineral assemblages: (5) oliv (Fo_{80}, K_D 9.5) + opx (K_D 3.0); (6) oliv (Fo_{80}, K_D 9.5) + cpx (K_D 1.9); (7) oliv (Fo_{80}, K_D 12.6) + opx (K_D 7.3); (8) oliv (Fo_{80}, K_D 12.6) + cpx (K_D 2.5); (9) oliv (Fo_{80}, K_D 9.5) + plag (K_D 0.03); and (10) oliv (Fo_{80}, K_D 12.6) + plag (K_D 0.03). Filled circles represent 0.1 fractionation intervals decreasing from 1.0 at the three points representing parental compositions.

Cr-MgO

The compilation of available Cr and MgO data shown in Figure 5.4.8A excludes data for rocks more evolved than mugearite and benmoreite. The correlated depletion of Cr and MgO is both expected and observed, although the trend is more nearly linear than logarithmic. There is also significantly more scatter of the data than is seen in the corresponding Ni-MgO plot, and the tholeiitic-basalt compositions again are restricted to lower maximum Cr levels than are the more alkaline basalts.

Generalised modelling of Cr behaviour (Fig. 5.4.8B) is based on the same nephelinitic and qz-tholeiitic

basalt compositions used for Ni (Fig. 5.4.7B), as well as similar mineral compositions, except for the notable addition of magnetite. Two sets of pyroxene partition coefficients are used also. The general form of the calculated fractionation curves for the mixed-phase assemblages of olivine + pyroxene ± magnetite ± plagioclase, corresponds broadly to the observed data trends, and these assemblages are also consistent with the Ni-MgO relationships described previously. However, the important point shown again is that the development of the majority of these east-Australian basaltic lavas, especially the tholeiitic basalts, has involved extensive fractional crystallisation.

V-MgO

V behaviour is more complex than that of either Cr or Ni and, as shown clearly in Figure 5.4.9 for mafic and intermediate rocks, there is an extensive range of V abundances. However, the net effect is V depletion as MgO decreases, and the depletion becomes most pronounced below about 3-percent MgO. The data in Figure 5.4.9 are grouped into central and lava-field provinces. The central volcanoes include the Tweed, Focal Peak, Glass Houses, Main Range (in part), and Springsure complexes of southeastern Queensland, as well as Comboyne, Nandewar, and Canobolas (New South Wales), and Macedon-Trentham. Rocks from both central volcanoes and lava fields commonly have an increase in V, between about 3 and 8-percent MgO, and this is most pronounced in the central-volcano tholeiitic basalts.

The relative importance of the major minerals in determining V behaviour during crystallisation is illustrated by the calculated curves shown in Figure 5.4.9C where three different model compositions are shown. The general conclusion to be drawn from these calculated curves is that magnetite + olivine ± augite must be in the fractionating assemblage in order that MgO and V are depleted concurrently, although pure augite fractionation may also cause limited V depletion. In addition, however, relatively Fe-enriched pyroxene ± olivine are required, as well as magnetite, in the fractionating assemblage in order that the relatively extreme V depletion in the lower-MgO lavas is produced, as shown by the qz-tholeiitic basalt curves in Figure 5.4.9C. This is consistent with the observed mineralogy (see above).

The scatter in V values for the basaltic lavas therefore is attributed to different mineral assemblages, particularly involving the presence of magnetite, during progressive fractional crystallisation of these lavas. The apparent V enrichment at intermediate MgO levels in the tholeiitic lavas from, especially, the central volcanoes (Fig. 5.4.9A) implies relatively delayed magnetite precipitation, but there is no clear evidence yet for systematic differences in fO_2 with which this behaviour can be correlated (see Fig. 5.3.30).

Sc versus mg-ratio

Sc decreases only slightly as mg-ratios decrease in the basaltic lavas, and this extends to some of the more evolved intermediate lavas (Fig. 5.4.10A; the data are neutron-activation analyses which are sensitive for this

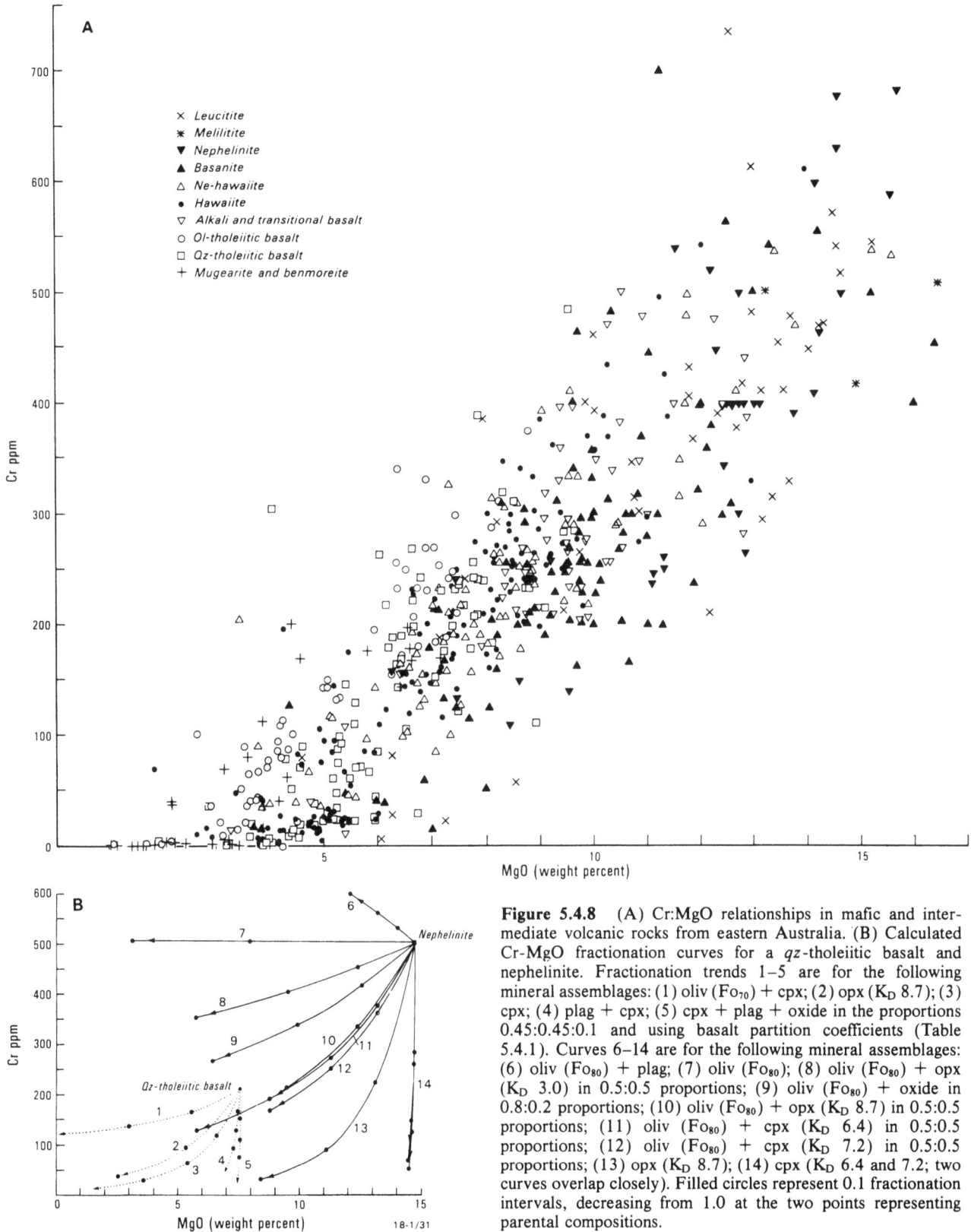

Figure 5.4.8 (A) Cr:MgO relationships in mafic and intermediate volcanic rocks from eastern Australia. (B) Calculated Cr-MgO fractionation curves for a qz-tholeiitic basalt and nephelinite. Fractionation trends 1–5 are for the following mineral assemblages: (1) oliv (Fo_{70}) + cpx; (2) opx (K_D 8.7); (3) cpx; (4) plag + cpx; (5) cpx + plag + oxide in the proportions 0.45:0.45:0.1 and using basalt partition coefficients (Table 5.4.1). Curves 6–14 are for the following mineral assemblages: (6) oliv (Fo_{80}) + plag; (7) oliv (Fo_{80}); (8) oliv (Fo_{80}) + opx (K_D 3.0) in 0.5:0.5 proportions; (9) oliv (Fo_{80}) + oxide in 0.8:0.2 proportions; (10) oliv (Fo_{80}) + opx (K_D 8.7) in 0.5:0.5 proportions; (11) oliv (Fo_{80}) + cpx (K_D 6.4) in 0.5:0.5 proportions; (12) oliv (Fo_{80}) + cpx (K_D 7.2) in 0.5:0.5 proportions; (13) opx (K_D 8.7); (14) cpx (K_D 6.4 and 7.2; two curves overlap closely). Filled circles represent 0.1 fractionation intervals, decreasing from 1.0 at the two points representing parental compositions.

element). However, Sc in rhyolite and peralkaline lavas becomes extremely depleted, even where only small changes take place in mg-ratio. This behaviour can be modelled by crystal-liquid fractionation, as shown in Figure 5.4.10B. The model compositions and mineral assemblages used are the same as those shown in Figure 5.4.9C.

There are only comparatively small changes in Sc abundance as mg-ratios decrease where the range of relatively magnesian mineral assemblages is used, starting from the nephelinite and ol-tholeiitic-basalt model parental compositions, together with basaltic and trachyandesitic partition coefficients. This is consistent with the observations. In contrast, the

Figure 5.4.10 (A) Sc:mg-ratio relationships in volcanic rocks from eastern Australia. (B) Calculated Sc:mg-ratio fractionation curves for a qz-tholeiitic basalt, transitional basalt, and nephelinite. Fractionation trends 1–3 from qz-tholeiitic basalt (Q174) containing an Fe-enriched mineral assemblage (plag, oliv, oxide, ferroaugite), using trachyandesite (1), trachyte (2), and low-silica-rhyolite (3) partition coefficients (to F 0.6; Table 5.4.1). Fractionation trends 4 and 5 represent the least-squares-fitted model for samples Q135 to Q139 (see text), extended to F 0.4, for basalt (4) and trachyandesite (5) partition coefficients. Fractionation trends 6–9 are for the following mineral assemblages and basalt partition coefficients: (6) oliv (Fo_{80}) + plag in 0.5:0.5 proportions (to F 0.5); (7) oliv (Fo_{80}) + opx (K_D 1.2) in 0.5:0.5 proportions (to F 0.6); (8) oliv (Fo_{80}) + cpx in 0.5:0.5 proportions (to F 0.5); (9) oliv (Fo_{80}) + oxide in 0.9:0.1 proportions. Filled circles represent 0.1 fractionation intervals decreasing from 1.0 at the three points representing parental compositions.

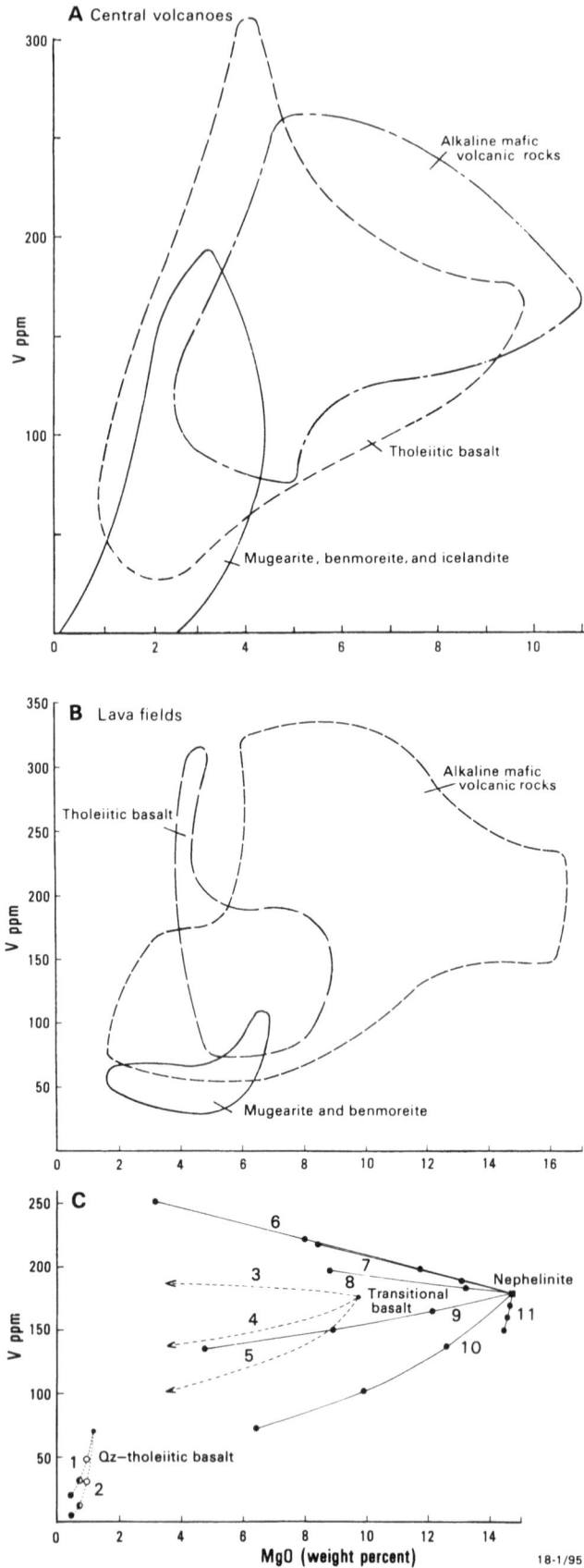

Figure 5.4.9 Generalised V:MgO relationships in mafic and intermediate rocks of central-volcano (A) and lava-field (B) provinces. (C) Calculated V-MgO fractionation curves for a qz-tholeiitic basalt, transitional basalt, and nephelinite. Fractionation trends are for plag (An_{50}) + oliv (Fo_{50}) + oxide + ferroaugite in the proportions 0.6:0.05:0.15:0.20 and using trachyandesite (1) and low-silica rhyolite (2) partition coefficients (Table 5.4.1).

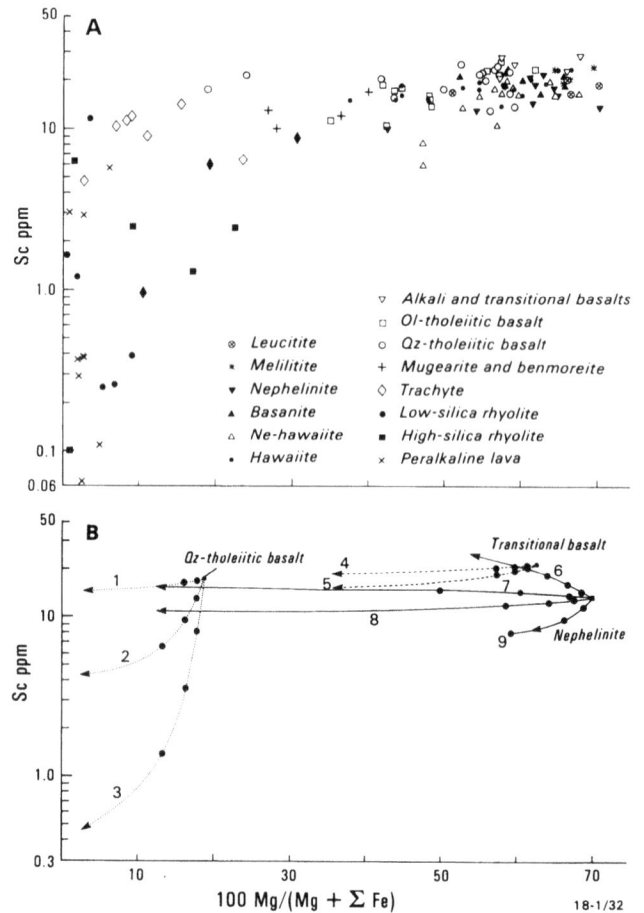

(Fig. 5.4.9 continued)

Fractionation trends 3–5 are for the least-squares-fitted model for samples Q135 to Q139 (see text) for basalt (3), trachyandesite (4), and trachyte (5) partition coefficients (to F 0.4). Fractionation trends 6–11 are for the following mineral assemblages and using basalt partition coefficients (Table 5.4.1): (6) oliv (Fo_{80}); (7) opx (K_D 0.05); (8) cpx + oliv (Fo_{80}) in the proportions 0.5:0.5; (9) oliv (Fo_{80}) + oxide in the proportions 0.9:0.1; (10) oliv (Fo_{80}) + oxide in the proportions 0.8:0.2; (11) cpx. Filled circles represent 0.1 fractionation intervals decreasing from 1.0 at the points representing parental qz-tholeiitic basalt and nephelinite.

calculated curves for the *qz*-tholeiitic-basalt model composition use a more Fe-enriched olivine (Fo_{30}) and ferroaugite in the fractionating assemblage (see also V-MgO discussion), resulting in much smaller relative mg-ratio changes during fractionation. However, the

critical importance of the increasing Sc partition coefficients, as found in the more felsic lavas, is illustrated by these curves. The rhyolite partition coefficients, in particular, have the capacity to produce rapid and strong Sc depletion, similar to the observed pattern of behaviour, by such crystal-liquid fractionation processes, involving only moderate amounts of crystal extract.

5.4.5   Rb-Sr, Rb-Ba, and K-Rb

Rb-Sr and Rb-Ba relationships (Figs. 5.4.11–12) are characterised by (1) a marked difference in abundance patterns between the mafic and the more evolved lava types, and (2) pronounced and progressive depletion of both Ba and Sr in the evolved lavas as Rb increases. This behaviour is most extreme for Sr. Rb and Ba abundances for icelandite, mugearite, and benmoreite

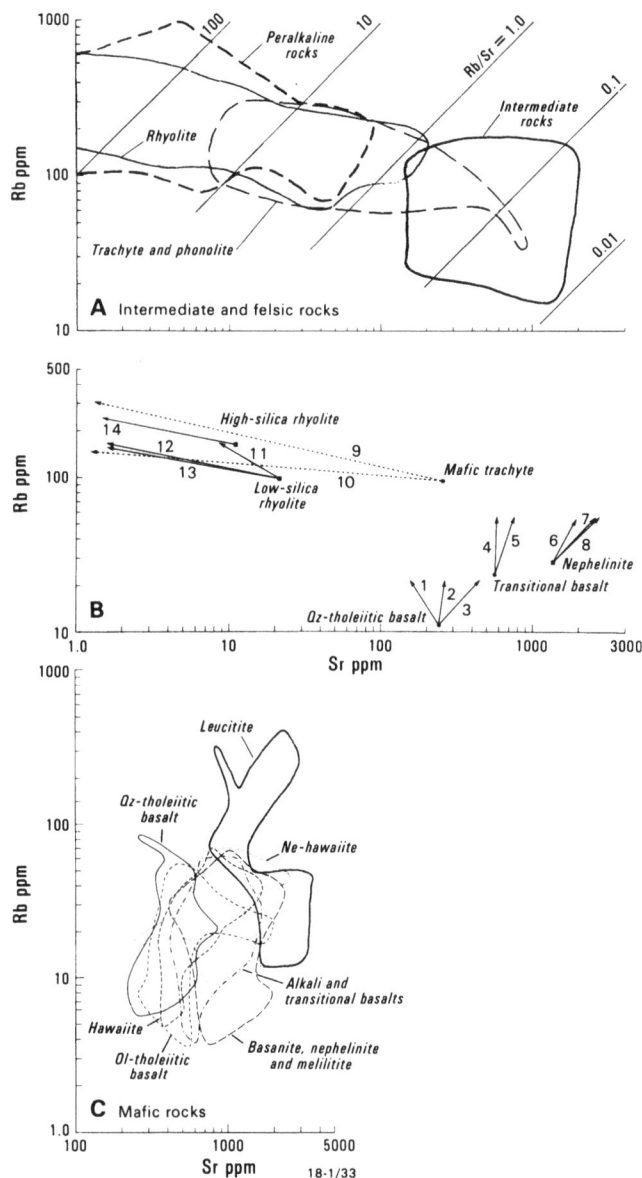

Figure 5.4.11   (A, C). Generalised Rb:Sr relationships in volcanic rocks from eastern Australia. (B) Calculated Rb:Sr fractionation curves for a *qz*-tholeiitic basalt, transitional basalt, nephelinite, mafic trachyte, low-silica rhyolite, and high-silica rhyolite. Fractionation trends 1–3 (F 0.5) are for (1) plag, (2) plag + cpx in 0.5:0.5 proportions, and (3) cpx, using basalt partition coefficients for Rb and Sr (Table 5.4.1). Fractionation trends 4–5 (F 0.41) represent the least-squares-fitted model for samples Q135 to Q139 (see text) using (4) trachyandesite and (5) basalt partition coefficients. Fractionation trends 6–8 (F 0.5) are for (6) plag + oliv in 0.5:0.5 proportions, (7) cpx, and (8) oliv. Fractionation trends 9–10 are for (9) san + plag (0.5:0.5 proportions, F 0.2) using trachyte partition coefficients, and (10) san + plag in (0.5:0.5 proportions, F 0.5) using low-silica rhyolite partition coefficients (Table 5.4.1). Fractionation trends 11–13 (F 0.5) are for (11) san + plag + qtz (in 0.56:0.12:0.32 proportions) using (11) trachyte, (12) high-silica rhyolite, and (13) low-silica-rhyolite partition coefficients. Trend 14 (F 0.6) is for san + plag + qtz (in 0.56:0.12:0.32 proportions) using high-silica-rhyolite partition coefficients.

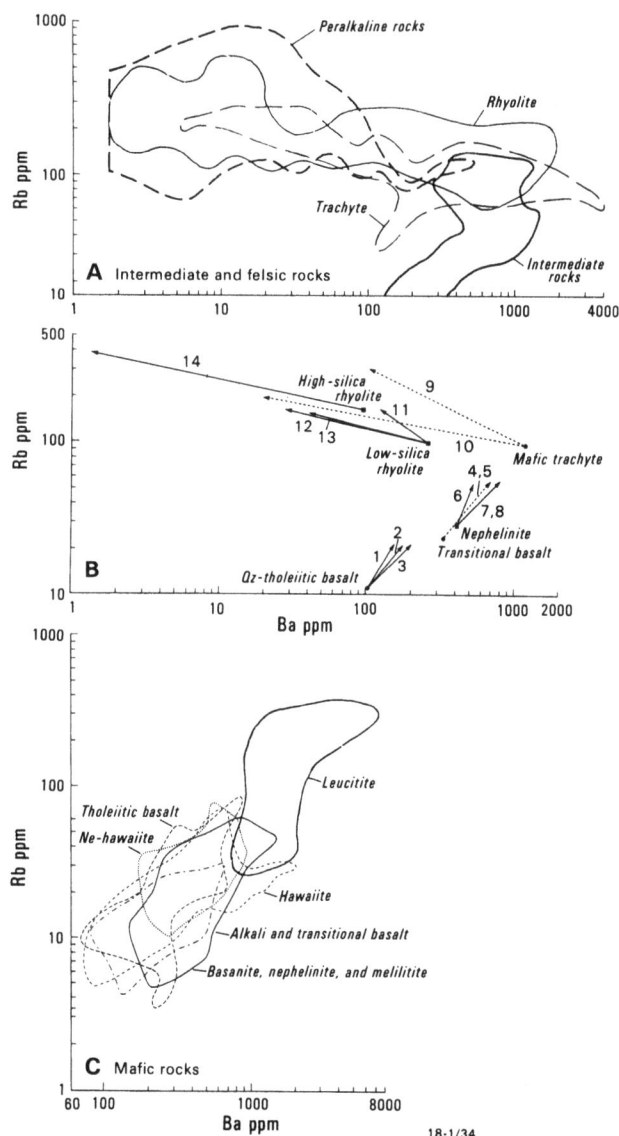

Figure 5.4.12   (A,C) Generalised Rb:Ba relationships in east-Australian volcanic rocks. (B) Calculated Rb:Ba fractionation curves are in the same notation and for the same rocks, mineral assemblages, and degrees of fractionation, as used in Figure 5.4.11B, except for trend 14 which is for fractionation to F 0.3. Partition coefficients for Rb and Ba also are for the same rock groups used in Figure 5.4.11.

correlate positively, as they do in the basalts. In contrast, Rb and Sr in both the intermediate and felsic lavas are negatively correlated. The resulting range of Sr and Rb abundances throughout the whole range of the east-Australian lava compositions is so great that Rb/Sr values extend over about five orders of magnitude. Rb enrichment, and Sr and Ba depletion, in the low and high-silica rhyolites and peralkaline lavas are extreme. Sr and Ba abundances of less than 5 ppm are common, and Sr values can be lower than 1 ppm. Rb abundances range widely within the same groups of rhyolite, from 110 to nearly 600 ppm.

A relatively well-defined positive Rb-Ba correlation exists for each of the major basalt groups shown in Figure 5.4.12C, although the different basalt compositional fields almost completely overlap in terms of their absolute Rb and Ba abundances. The notable exceptions are the leucitites which are strongly enriched in both Rb and Ba. Rb-Sr relationships through the major basaltic compositional groups (Fig. 5.4.11C) range widely, and clear correlations are not defined. Again, there is an extensive overlap of Sr abundances through the different basaltic types, except for the leucitites, but nevertheless a clear tendency exists for the more alkaline rocks to be characterised by higher Sr abundances than the tholeiitic basalts. The range of Ba values in the leucitites is greater than for Sr.

Results of modelling of the contrasting behaviour shown by the Rb-Sr and Rb-Ba data between the basaltic and the more evolved lava types, are given in Figures 5.4.11B and 5.4.12B. The contrasting behaviour may be explained in terms of (1) the increasing relative importance of feldspar fractionation in the more evolved magmas, and (2) the increasing magnitude of the effective feldspar Ba and Sr partition coefficients in the more silica-rich magmas. The three basaltic model compositions shown are those used previously — that is, a *qz*-tholeiitic basalt, a nephelinite, and the least-squares-fitted transitional basalt to *ol*-tholeiitic-basalt model (specimens Q135 and Q139). The correlated enrichment of Rb and Ba is shown clearly and the basaltic and trachyandesite partition coefficients for the proposed crystal extracts are sufficiently low that both Rb and Ba behave as incompatible elements. Plagioclase/melt coefficients for Sr are significantly larger and differ between basaltic and trachyandesite magma compositions (Table 5.4.1), so the net result is a wider range of derivative Sr abundances that depend on the relative proportion of plagioclase in the fractionating mineral assemblage.

Modelling of the data for the more evolved lavas has been undertaken for mafic trachyte, low-silica rhyolite, and high-silica rhyolite model parental compositions. The effects of sanidine+plagioclase (1:1) fractionation are illustrated for the first composition, using trachyte and low-silica-rhyolite partition coefficients. The rhyolites shown have the calculated fractionation assemblage near that of the quartz-feldspar ternary minimum composition, consisting of quartz, sanidine $(Or_{51.6}Ab_{45.4}An_{3.0})$, and plagioclase $(An_{15.0}Ab_{76.6}Or_{8.4})$ in the proportion 32:56:12 (based on observed assemblages and compositions and least-squares-fitted proportions). The enrichment of Rb and the depletion of Sr and Ba are duplicated in all cases illustrated, using

trachytic and rhyolitic partition coefficients. Also modelled is the more rapid depletion of Sr, for given fractionation parameters, compared to that for Ba. In fact, only the high Ba partition coefficients, comparable in magnitude to those found in high-silica rhyolites, are capable of producing the observed extreme Ba depletion found in some of the rhyolites.

Progressively decreasing K/Rb values accompany increasing Rb in the evolved lavas and, significantly, tend to remain distinct from the basaltic field (Fig. 5.4.13A). The observed relationship between K/Rb and Rb is presumed to result from the progressive buffering of K by feldspar crystallisation controlled at the quartz-feldspar ternary minima (or piercing points), together with continued Rb enrichment. K/Rb values are therefore a useful fractionation index for the most evolved rocks. Note also that the different ranges of fractionation, as defined by K/Rb values, equally

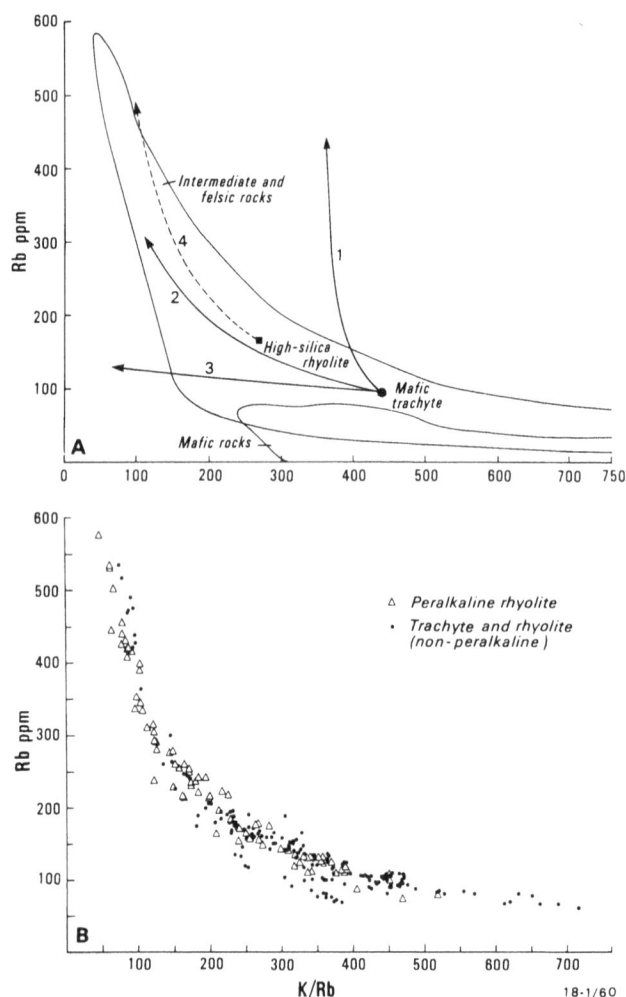

Figure 5.4.13 (A) Generalised Rb:K/Rb relationships in volcanic rocks from eastern Australia, showing calculated fractionation curves for a mafic trachyte and high-silica rhyolite. Fractionation trends 1–3 are for (1) plag (K_D 0.5) using rhyolite partition coefficients for Rb (fractionation to F 0.2), (2) san (K_D 0.51) + plag in 0.5:0.5 proportions using trachyte Rb partition coefficients (to F 0.2), (3) san (K_D 0.73) using rhyolite Rb partition coefficients (to F 0.5). Fractionation trend (4) is for san + plag + qtz in 0.56:0.12:0.32 proportions using high-silica-rhyolite Rb partition coefficients. (B) Peralkaline and non-peralkaline rhyolites and trachytes from eastern Australia, illustrating their overlapping compositional fields.

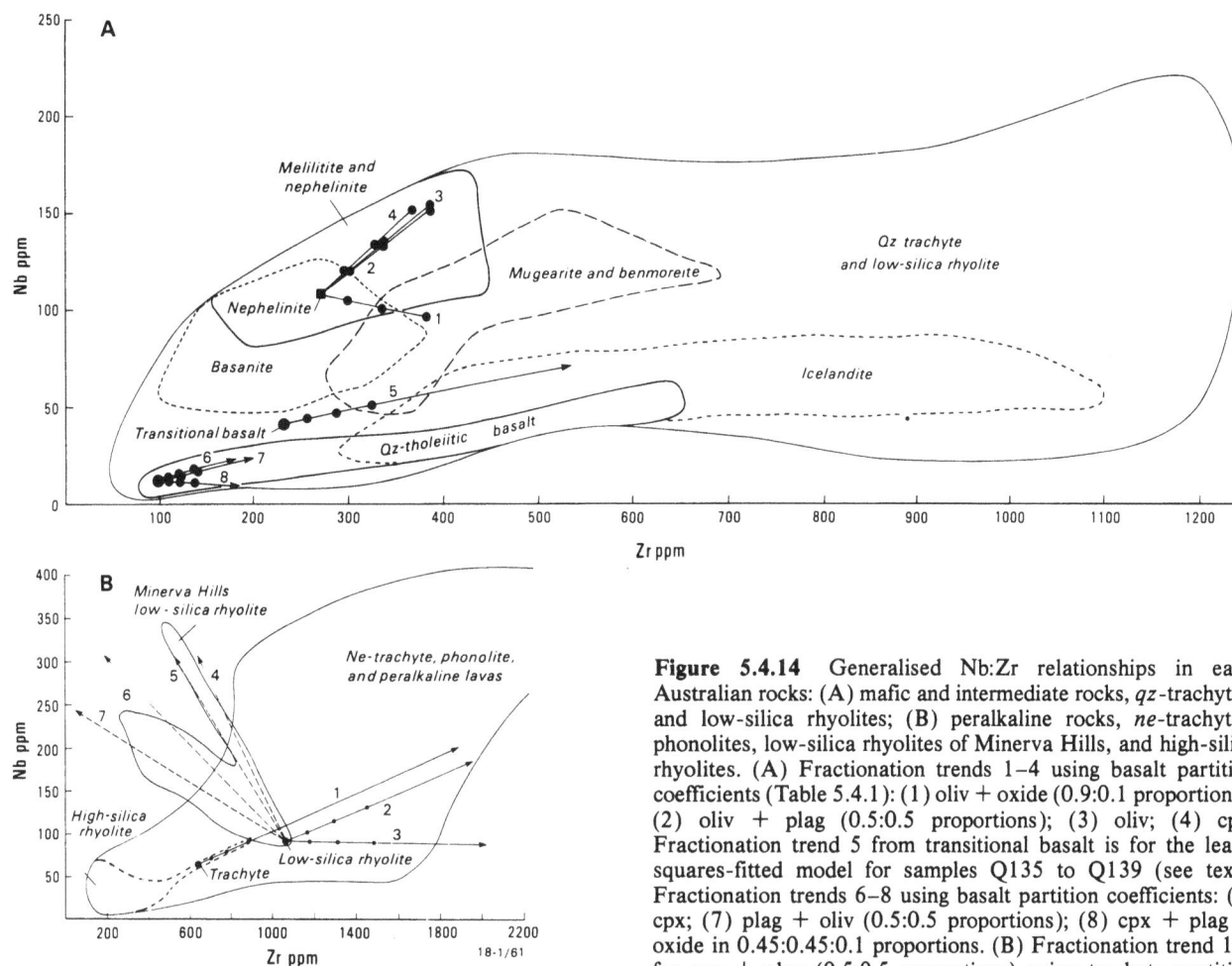

Figure 5.4.14 Generalised Nb:Zr relationships in east-Australian rocks: (A) mafic and intermediate rocks, qz-trachytes, and low-silica rhyolites; (B) peralkaline rocks, ne-trachytes, phonolites, low-silica rhyolites of Minerva Hills, and high-silica rhyolites. (A) Fractionation trends 1–4 using basalt partition coefficients (Table 5.4.1): (1) oliv + oxide (0.9:0.1 proportions); (2) oliv + plag (0.5:0.5 proportions); (3) oliv; (4) cpx. Fractionation trend 5 from transitional basalt is for the least-squares-fitted model for samples Q135 to Q139 (see text). Fractionation trends 6–8 using basalt partition coefficients: (6) cpx; (7) plag + oliv (0.5:0.5 proportions); (8) cpx + plag + oxide in 0.45:0.45:0.1 proportions. (B) Fractionation trend 1 is for san + plag (0.5:0.5 proportions) using trachyte partition coefficients. Fractionation trends 2–3 for Minerva Hills sample containing lowest Nb: (2) plag + san + qtz (0.16:0.47:0.37 proportions) using rhyolite partition coefficients: (3) san + plag + oxide (0.7:0.2:0.1 proportions; similar trends using either trachyte or low-silica-rhyolite partition coefficients). Fractionation trends 4–7 are for the same assemblage as in trend 3, except for containing four separate weight fractions of zircon: (4) 0.0026 (F 0.3); (5) 0.0027 (F 0.3); (6) 0.0030 (F 0.3); (7) 0.0035 (F 0.4). Filled circles on fractionation trends in both A and B represent fractionation intervals decreasing by 0.1 from 1.0 at the points representing parental compositions.

include peralkaline and non-peralkaline rhyolites and trachytes (Fig. 5.4.13B).

The effects of fractionation of plagioclase, sanidine, plagioclase + sanidine (1:1), and a 'ternary minimum' quartz-oligoclase-sanidine assemblage (see data used above) from a trachyte and a high-silica rhyolite, using trachytic and rhyolitic partition coefficients, are illustrated in Figure 5.4.13A. The importance of a mixed-feldspar assemblage in producing the general form of the observed trend is shown, as well as the feasibility of a quartz-feldspar, ternary-minimum fractionation process in extending the trend to relatively low K/Rb values. However, the highest Rb (and lowest K/Rb) values seem to require extremely protracted fractionation (F-values of less than 0.2).

5.4.6 Zr-Nb

The following points can be made with reference to the generalised Zr and Nb data (Fig. 5.4.14A) for basaltic and intermediate rocks, qz-trachytes, low-silica rhyolites (one group excepted), and in Figure 5.4.14B for peralkaline lavas, ne-trachyte and phonolite, high-silica rhyolite, and the rather unique low-silica rhyolites of the Minerva Hills complex, central Queensland.

(1) Broad correlations between Zr and Nb are observed for the different basalt groups, although different trends are defined. For example, the data for most alkaline lavas (nephelinite, basanite, melilitite) define relatively steep trends compared to those for the tholeiitic basalts, resulting from the inherently high Nb abundances in the more alkaline basalts. In contrast, Zr abundances for the different basalt groups completely overlap. The concurrent enrichment of Nb and Zr is illustrated in Figure 4.4.14A by the results of three sets of model calculations. Nb and Zr behave as incompatible elements, except for those assemblages containing major magnetite, and the slopes correlate clearly with initial Nb concentrations. Basaltic partition coefficients are used in these calculated curves.

(2) Data for intermediate rocks, trachyte, and low-silica rhyolite (Fig. 5.4.14A) extend the basaltic trends, but in common with the basalts there is a range of enrichment trends. Thus, icelandite, saturated and oversaturated mugearite and benmoreite, and many of the qz-trachytes, have relatively low levels of Nb

enrichment relative to Zr, as the degree of fractionation increases. This is consistent with trends for the tholeiitic basalts and icelandites. In contrast, nepheline-normative mugearite, benmoreite, and trachyte tend to be much more strongly enriched in Nb. However, low-silica rhyolites have a wide range of Nb and Zr abundances. Nb differences in these rhyolites may result from differences in the involvement of magnetite during fractionation.

(3) Peralkaline lavas tend to be consistently enriched in Zr but, as shown clearly by the available data (Fig. 5.4.14B), they are not necessarily more enriched in Zr than are some non-peralkaline low-silica rhyolites and trachytes. Increasing Zr correlates generally with increasing Nb, but no simple pattern is defined. However, some problems arise in accounting for the extremely high Nb enrichments (greater than 250 ppm) found in some of the peralkaline rhyolites and ne-trachytes in terms of crystal-liquid fractionation. The observed slopes of their Nb-Zr enrichment trends are slightly steeper than those calculated for fractionation from two model compositions — a trachyte and a low-silica rhyolite (Figs. 5.4.14B). These models involve mixed feldspar, feldspar+quartz, and feldspar+magnetite fractionating assemblages, using rhyolite and trachyte partition coefficients. The observed, strong Nb enrichment certainly seems to be inconsistent with the involvement of magnetite (and zircon) during fractionation, and a parental magma containing high Nb (greater then, say, 100 ppm) would seem to be an essential prerequisite. A further possibility is fluorine complexing within the peralkaline magmas (fluor-arfvedsonite is characteristic of these lavas), but this explanation is not so applicable to the trachytes. Several stages of Nb concentration during fractionation may be required.

(4) The characteristically low Nb and Zr abundances in high-silica rhyolite, compared to the other evolved magma types found through the region, is clearly illustrated in Figure 5.4.14B and is suggestive of a different petrogenesis. This is supported by isotopic results (Ewart, 1982a; see also Section 5.5).

(5) The low-silica rhyolites of the Minerva Hills complex (central Queensland) have unique Nb-Zr patterns in which Zr is progressively depleted as Nb increases. Two distinct subtrends are apparent, and the trajectory of increasing fractionation is taken to be that of increasing Nb. The simplest explanation for these patterns is the presence of zircon in a fractionating mineral assemblage, and this is supported by the existence of accessory zircon in these Minerva Hills rhyolites (Ewart, 1982a). The results of modelling zircon fractionation are shown in Figure 5.4.14B for four different zircon weight fractions (0.0026, 0.0027, 0.003, and 0.0035) in a sanidine+plagioclase+quartz (0.471:0.162:0.367) assemblage. The effect on Zr is dramatic, and the two Minerva Hills subtrends can be modelled in terms of two sets of zircon abundances in the proposed fractionating assemblage involving 0.0026–0.0027 and 0.003–0.0035 weight fractions. The starting composition is the least Nb-enriched rhyolite analysed, and production of the observed maximum Nb concentrations from this parent composition requires fractionation (F) to values of about 0.3–0.4. However, a problem raised by the Minerva

Hills data is why zircon fractionation has not affected other, apparently chemically similar, trachytes and low-silica rhyolites from other volcanic centres throughout eastern Australia, some of which are also zircon-bearing.

5.4.7 P_2O_5-Ce

Frey et al. (1978) drew attention to the strong correlation between P_2O_5 and the light REE, to the incompatible behaviour of P_2O_5 during partial melting, and used P_2O_5 as an index of degree of partial melting. Frey et al. (1978) considered a P_2O_5/Ce value of 75 as

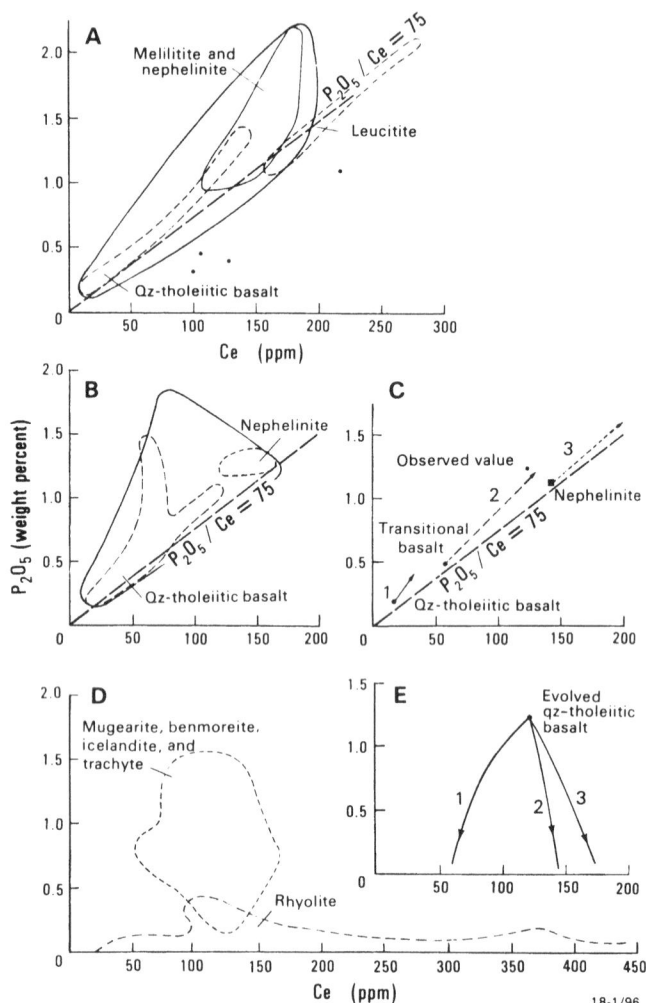

Figure 5.4.15   Generalised P_2O_5:Ce relationships in volcanic rocks from eastern Australia. (A) Instrumental neutron-activation and mass-spectrometric analyses of Ce (Frey et al., 1978; Menzies & Wass, 1983; Ewart and Chappell, unpublished data). Filled circles represent samples (Frey et al., 1978) that fall outside the generalised compositional field. (B) X-ray-fluorescence analyses of Ce. (C) Calculated fractionation curves for a qz-tholeiitic basalt, transitional basalt, and nephelinite using basalt partition coefficients (Table 5.4.1). Fractionation trend 1 (F 0.5) is for cpx + plag (0.5:0.5 proportions) ± oliv. Curve 2 represents the least-squares fitted model for samples Q135 to Q139 (see text). Fractionation trend 3 (F 0.7) is for oliv + cpx (0.5:0.5 proportions). (D) Generalised P_2O_5:Ce relationships in intermediate and felsic volcanic rocks from eastern Australia. (E) Fractionation curves (F 0.43) for oliv + plag + oxide + apatite (0.151:0.636:0.164:0.048 proportions) using (1) low-silica rhyolite, (2) trachyte, and (3) trachyandesite partition coefficients.

typical for primary alkali basalts (after Sun & Hanson, 1975). Two sets of P_2O_5–Ce data for east-Australian basalts are shown in Figures 5.4.15A-B.

Correlations clearly exist between P_2O_5 and Ce, and the relative differences in P_2O_5 and Ce abundances between the more undersaturated and the tholeiitic basalts are also well defined. However, data for the majority of lavas tend to plot above the line for P_2O_5/Ce equal to 75, and many samples have much higher values. The only exceptions are a small number of samples that have anomalously low ratios, possibly resulting from unusually early apatite fractionation (see below). Calculated fractionation paths are shown in Figure 5.4.15C for plagioclase+augite±olivine fractionation in a qz-tholeiitic basalt, and the previously described transitional basalt to ol-tholeiitic-basalt model fractionation sequence. Fractionation of feldspar, pyroxene, olivine, and Fe-Ti oxides (but not apatite) from basaltic magmas (basaltic partition coefficients) is predicted to result in increasing P_2O_5/Ce. Fractional crystallisation, therefore, is likely to be responsible for the higher P_2O_5/Ce values observed for many of the east-Australian basaltic rocks, relative to ratios expected from strictly 'primary' alkali basalts.

Large differences in the behaviour of both P_2O_5 and Ce in the more evolved rocks is revealed by a comparison of the data in Figure 5.4.15D — particularly the progressive depletion of P_2O_5, which becomes extreme in the rhyolitic and peralkaline rocks. This pattern of depletion correlates with the presence of apatite in the fractionating mineral assemblages. This is modelled in Figure 5.4.15E by means of a specific assemblage, calculated from least-squares procedures, linking the evolved (but parental) ol-tholeiitic basalt (mg-ratio 35.0) to a low-silica rhyolite derivative magma. The importance of apatite fractionation in controlling P_2O_5 abundances is confirmed, and the sensitive nature of the Ce partition coefficients is emphasised.

The absence of apatite in the most strongly fractionated rhyolites (including peralkaline types), caused by extreme P_2O_5 depletion, accounts for the trend towards Ce enrichment shown in Figure 5.4.15E. This trend must represent a very late fractionation process.

5.4.8 Summary

East Australian volcanic provinces contain a highly diverse, complex, and heterogeneous group of magma types. Also evident from the results of trace- and minor-element geochemistry is that this extreme diversity and heterogeneity is a prominent feature of the whole region. One end of the basalt spectrum is represented by relatively incompatible-element depleted qz-tholeiitic basalt (most especially from central Queensland), and at the other end highly enriched leucitite (central New South Wales). There seems to be little doubt, as emphasised by Frey et al. (1978) that gross trace-element heterogeneities must exist in the east-Australian upper mantle. Another feature, however, is that the basaltic magmas in general have been modified by fractional crystallisation, and that many of the observed trace-element differences can be attributed to olivine + augite ± plagioclase ± Fe-Ti oxide

fractionation. Only some of the most Mg-rich alkaline basalts can be interpreted as possible primary melts. 'Primary' tholeiitic basalts (in the sense of strictly unmodified mantle melts) seem to be absent from the analysed sample suite. This question is considered further in the following section on isotope geochemistry, together with a discussion on the role of the lithosphere, and the possible involvement of crustal contamination.

The role of fractional crystallisation is demonstrated even more clearly in the petrogenesis of the more evolved intermediate and rhyolitic lavas. Rhyolite and peralkaline trachyte and rhyolite in particular have patterns of extreme trace-element depletion and enrichment. The peralkaline lavas are consistently highly fractionated geochemically, but they are not necessarily more intensely fractionated than are some of the non-peralkaline rhyolites. Significant differences exist between the trace-element chemistry of low-silica and high-silica rhyolites, and these are interpreted to result from different origins (see below). An important aspect of the geochemistry of the felsic rocks is that it can be modelled in terms of crystal-liquid equilibria, using the Rayleigh-fractionation equation. However, examples of extreme fractionation appear to require that magmas have undergone what is best described as 'multistage' crystal fractionation. Some of the convective fractionation models proposed (see, for example: McBirney, 1980; Rice, 1981; Turner & Gustafson, 1981; Sparks et al., 1984) seem to provide possible mechanisms whereby such complex fractionation processes may take place.

5.5 Isotope Geochemistry

5.5.1 Introduction

Currently available isotopic data for east-Australian rocks are reviewed in this section and, where possible, correlated with trace-element geochemistry. Following are the major data sources used: Cooper & Green (1969), Ewart (1982a, 1985), Ewart et al. (1977, 1988), McDonough et al. (1986), W.F. McDonough (unpublished data), Menzies & Wass (1983), Nelson et al. (1986), O'Reilly & Griffin (1984), and Rudnick et al. (1986). Data are recalculated where necessary to a common basis: Nd-isotopic compositions are based on a normalisation value for $^{146}Nd/^{142}Nd$ of 0.63223; Sr-isotopic data are normalised to the E and A standard value of 0.70800.

General emphasis is given to: (1) overall isotopic characteristics of the different compositional groups of rocks; (2) regional isotopic differences; (3) relationships between trace-element and isotopic compositions and their significance in evaluating the extent of mixing, assimilation, and combined assimilation/fractional-crystallisation (AFC) processes.

Generalised mixing and AFC model curves are presented to illustrate gross trends. Mineral compositions and partition coefficients are those used in Section 5.4. The following generalised 'crustal' end-member compositions are used for this purpose (Table 5.5.1):

(1) Palaeozoic greywacke-shale sediments characteristic of the Lachlan and New England orogenic fold belts of eastern Australia and representing predominantly

felsic, volcanic/plutonic-derived sediments that have clear upper-crustal and active continental-margin geochemical signatures. Chemical and trace-element data are from Lohe (1980; south-Queensland sediments), and isotopic data from Ewart (1982a), Hensel et al. (1985), and previously unpublished data. Hensel et al. (1985) distinguished two general groups of sediment: (1) a predominantly mafic volcanogenic provenance from an early to middle Palaeozoic western source (zone A); (2) more felsic, volcanogenic-plutonic greywacke and siltstone, associated with mudstone interbedded locally with cherts and basalts (zone B). Only zone-B type sediment data are used below.

(2) Archaean, depleted, granulite-type lower-crustal rocks. Data are mainly from the extensive Lewisian granulite-facies results presented by Weaver & Tarney

Table 5.5.1 Crustal end-member compositions used in AFC and mixing-model calculations

	Average Palaeozoic greywacke/shale sediment[1]	Average Archaean granulite gneiss[2]	Andesite 'lower crust' model[2]
SiO$_2$	70.10	61.20	54.0
TiO$_2$	0.60	0.54	0.90
Al$_2$O$_3$	15.74	15.60	19.0
FeO[4]	3.86	5.31	9.0
MnO	0.05	0.08	0.1
MgO	1.26	3.40	4.1
CaO	1.52	5.60	9.5
Na$_2$O	2.71	4.40	3.4
K$_2$O	4.00	1.00	0.60
P$_2$O$_5$	0.09	0.18	0.20
Total	99.93	97.31	100.8
Rb	183	12	20
Ba	661	757	175
Sr	248	569	425
Zr	147	202	30
Y	31	9	20
Ni	13	58	35
V	37	88	230
Cr	37	3.6	65
Hf	4.5	3.6	1.6
Nb	5.3	5.0	4.0
Th	15	0.42	1
Pb	16	13	3
La	25	22	9.5
Ce	50	44	17
Nd	21	18.5	8.0
Sm	4.2	3.3	2.75
Yb	4.0	1.2	1.9
^{87}Sr/^{86}Sr	0.7131	0.7037	0.70341
^{206}Pb/^{204}Pb	19.10	15.38	18.897
^{207}Pb/^{204}Pb	15.65	15.15	15.568
^{208}Pb/^{204}Pb	39.29	35.81	38.527
^{143}Nd/^{144}Nd	0.51244	0.5120	0.51292

[1] Abundance data from Lohe (1980) for upper-crustal model. Isotopic data after Hensel et al. (1985), Ewart (1982a), and Ewart et al. (1988).
[2] Abundance data from Weaver & Tarney (1980a, 1981) for lower-crustal model. Isotopic data from Leeman et al. (1985), Stosch & Lugmair (1984), Harmon et al. (1984), Hamilton et al. (1979), and Moorbath et al. (1975).
[3] Abundance data from Weaver & Tarney (1980b) and Taylor & McLennan (1981). Isotopic compositions based on Palaeozoic mafic volcanic rocks from southeastern Queensland (Ewart, 1982a; Ewart et al., 1988).
[4] Total Fe as FeO.

(1980a, 1981). The isotopic data used are for Snake River Plain xenoliths, Eifel granulite xenoliths, and the Lewisian (Moorbath et al., 1975; Hamilton et al., 1979; Harmon et al., 1984; Stosch & Lugmair, 1984; Leeman et al., 1985).

(3) 'Lower crust' compositions based on an andesite crustal model (Weaver & Tarney, 1980b; Taylor & McLennan, 1981). Isotopic compositions are for Palaeozoic mafic volcanic rocks from southeastern Queensland (Ewart, 1982a; unpublished data).

5.5.2 Sr and Nd isotopes

Initial ^{87}Sr/^{86}Sr and ^{143}Nd/^{144}Nd values broadly have the expected covariance, except in the most evolved felsic rocks (Fig. 5.5.1). The most alkaline mafic lavas, excluding leucitite, tend to have the least radiogenic Sr- and most radiogenic Nd-isotopic compositions, although there are exceptions, most particularly from the Southern Highlands province (Menzies & Wass, 1983). Hawaiite and tholeiitic basalt have wide isotopic ranges, particularly for ^{143}Nd/^{144}Nd, and several samples deviate from the main mantle trend by having relatively unradiogenic Nd and Sr compositions. The leucitite data define a distinct compositional array characterised by relatively radiogenic Sr and unradiogenic Nd compositions (see also Nelson et al., 1986). Trachytic and rhyolitic samples define different trends, towards unradiogenic Nd- and strongly radiogenic initial Sr-isotopic compositions. The most marked ^{87}Sr/^{86}Sr enrichments are suggestive of upper-crustal signatures, and are especially characteristic of the high-silica rhyolites. Data are limited for peralkaline rocks, but both Sr- and Nd-isotopic ratios have wide ranges.

Simple mixing curves have been calculated (Fig. 5.5.1B) between the Palaeozoic greywacke-shale sediment composition, a nephelinite (enriched total-Sr and Nd abundances: 1384 and 59.1 ppm respectively), and a low-K$_2$O qz-tholeiitic basalt (Sr 241 ppm, Nd 10.9 ppm). Clearly, the Nd-Sr isotopic array for east-Australian rocks cannot represent simple-mixing relationships.

Basaltic Sr-Nd isotopic data may be compared in terms of the gross regional distribution of rocks (Fig. 5.5.2A) from which the following points are evident:

(1) Tasmanian alkaline lavas in general have the most radiogenic Nd and least radiogenic Sr compositions compared to those found elsewhere on the Australian mainland. However, Tasmanian qz-tholeiitic basalts are slightly more evolved isotopically (less radiogenic Nd, more radiogenic Sr). Tasmanian rocks also have distinctive Pb-isotope compositions (see below).

(2) Considerable isotopic differences are found within the different geographic regions of the Australian mainland, and generally follow the main mantle trend defined by ocean-island basalt data (Fig. 5.5.2B). Data for north-Queensland mafic rocks do not extend to markedly unradiogenic Nd-isotopic compositions, but do tend to be characterised by relatively radiogenic Sr compositions.

(3) New South Wales leucitites, as noted above, are distinctive amongst the mafic rocks of eastern Australia, but clearly are somewhat similar isotopically to rocks

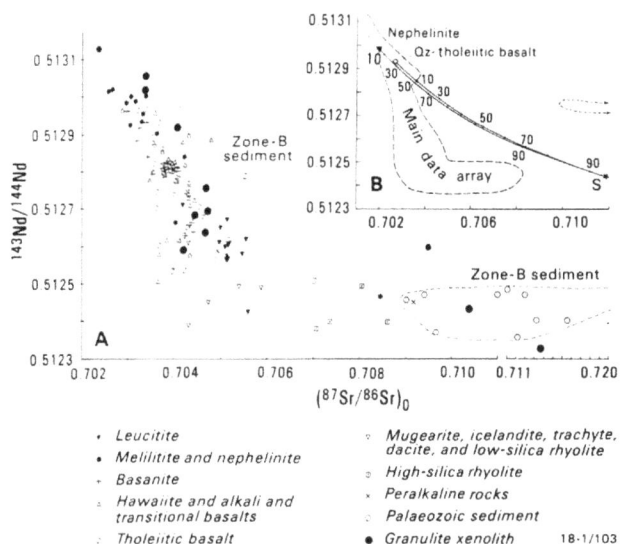

Figure 5.5.1 (A) Nd- and initial-Sr-isotope relationships for east-Australian volcanic rocks. Also shown are isotopic data for granulite xenoliths (Ewart, 1982a; Rudnick et al., 1986) and Palaeozoic greywacke-type sediments, including the zone-B sediments of Hensel et al. (1985). (B) Mixing curves between average greywacke sediments (point S), a primitive (high Mg-ratio) nephelinite (filled triangle), and a low-K_2O qz-tholeiitic basalt (open circle). Numbers along curves represent percentages of sediment in mixtures.

from Kerguelen, Tristan da Cunha, and Gough islands.

(4) The hawaiites and tholeiitic basalts that trend towards unradiogenic Nd- and Sr-isotopic compositions are found in the central-volcano complexes of southern Queensland and New South Wales. Central-volcano rocks are more fractionated in their overall chemistry, as discussed in Sections 5.2–4. The data plotted in Figure 5.5.2A are further evidence for AFC processes (see below), in part involving an unradiogenic-Nd and relatively unradiogenic-Sr end-member such as might be expected for some lower-crustal compositions (see, for example, Harmon et al., 1984).

5.5.3 Sr- and Nd-isotope and Rb:Sr and Sm:Nd relationships

Isotopic and daughter-parent element ratios discussed in this section are plotted, respectively, in terms of epsilon notation as well as the δ' notation proposed by O'Nions et al. (1979a). The ε fields in Figure 5.5.3A are indicative that the majority of east-Australian mafic rocks were derived from mantle sources that have time-integrated depletions of Rb and the light REE relative to estimates for Bulk Earth. Major exceptions are the leucitites which plot close to the Bulk Earth values. In contrast, the δ'Nd data fall mainly between -25 and -50 (notable exceptions are those for the qz-tholeiitic basalts), and the δ'Sr data range from -60 to more than $+300$ (the leucitites plot within the higher part of the range). These differences obviously are much greater than those for the isotopic data, and they imply strong decoupling between the isotopic compositions and the daughter-parent elements. Possible explanations include (1) relatively young (less than about 200 Ma) source metasomatism, (2) lithosphere or crustal (or

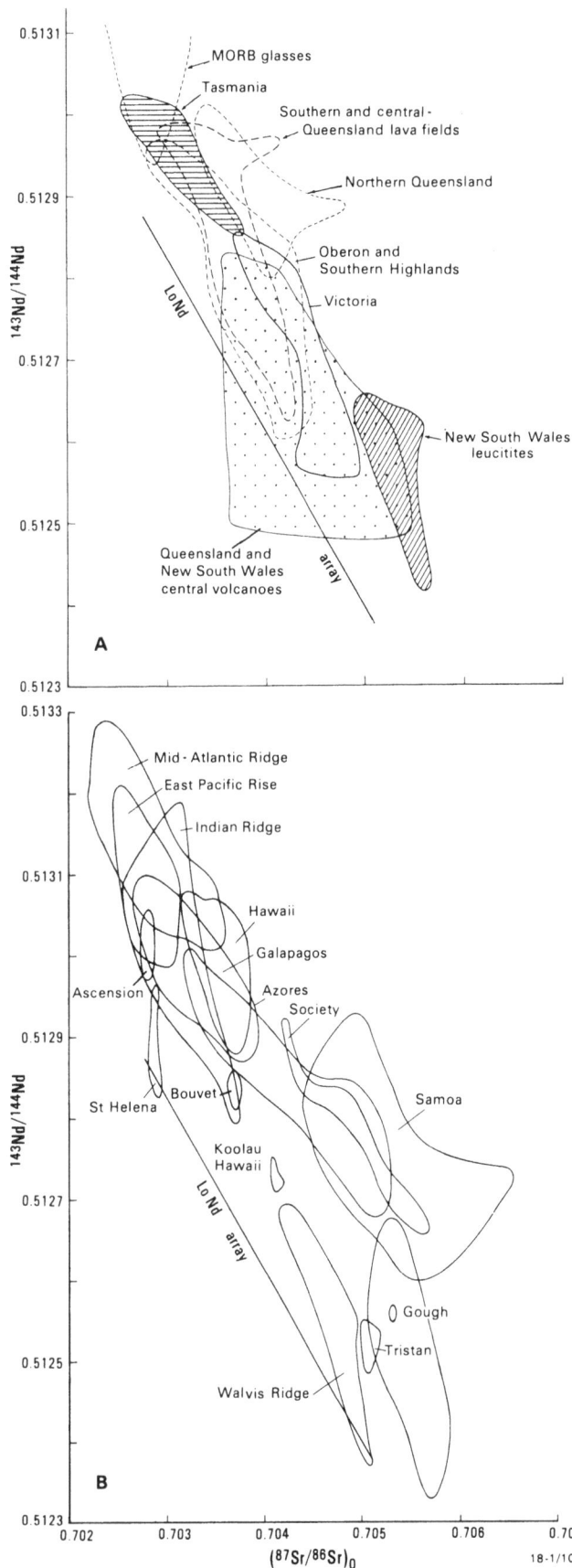

Figure 5.5.2 (A) Generalised Nd- and Sr-isotopic compositional fields divided into regional groups. MORB-glass data from Cohen et al. (1980), Dupré & Allègre (1980), and Cohen & O'Nions (1982b). (B) Generalised fields for mid-ocean-ridge and ocean-island basalts (after Zindler & Hart, 1986). LoNd array after Hart et al. (1986).

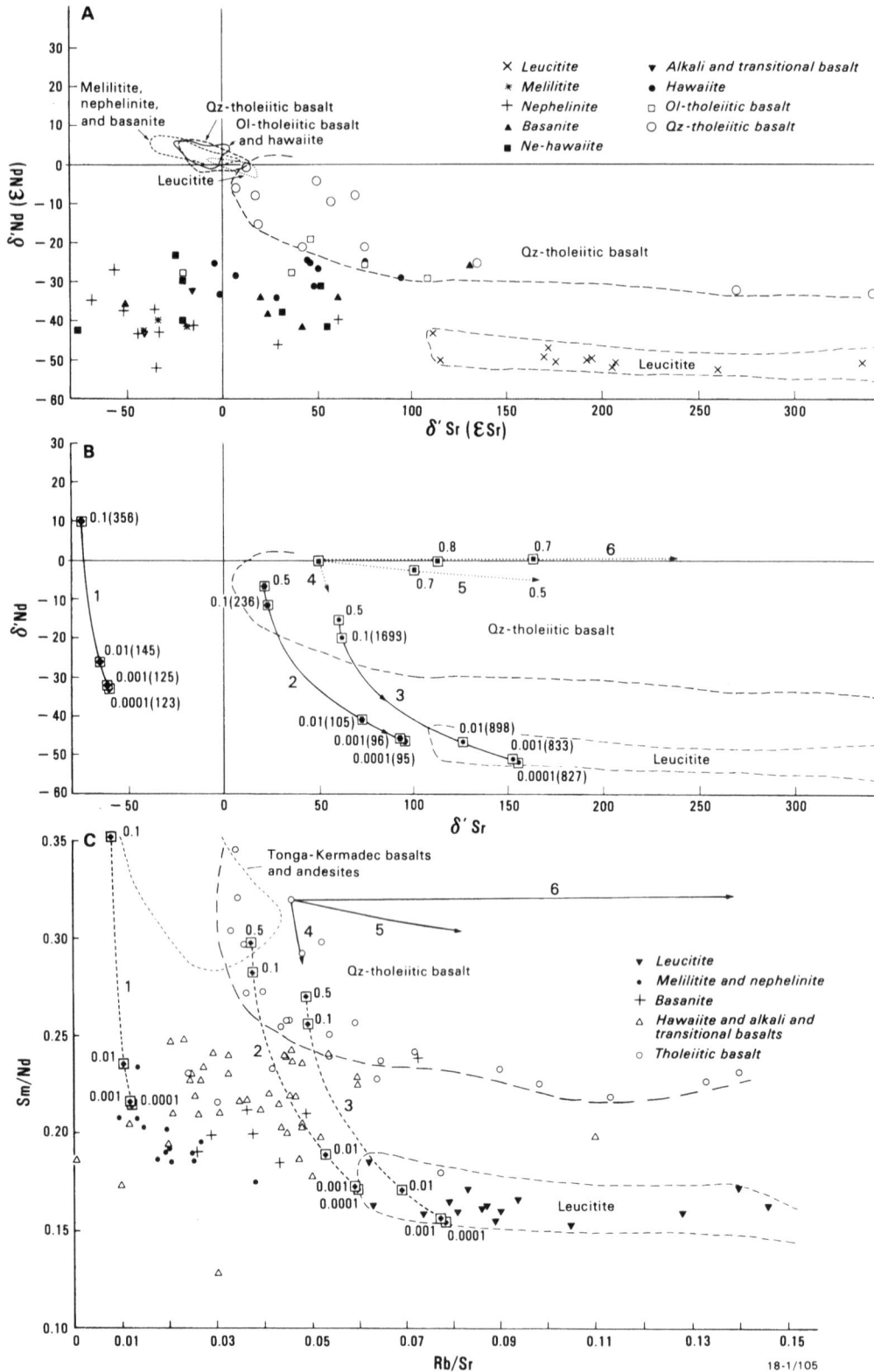

Figure 5.5.3 (A) δ'Nd (symbols) and ε_{Nd} (fields) versus δ'Sr and ε_{Sr} relationships for east-Australian basalts. δ' values are the fractional deviation of the measured parent-daughter ratios (^{147}Sm/^{144}Nd and ^{87}Rb/^{86}Sr) in the rock relative to Bulk Earth values (O'Nions et al., 1979a). (B) Melting curves for a garnet-peridotite source based on the melt-extraction equation of McKenzie (1985). Curve 1 is for a N-MORB source, curve 2 for primitive mantle, and curve 3 for inferred Dupal-type sub-continental lithosphere from Brazil (Hawkesworth et al., 1986). Numbers along curves refer to extracted-melt fractions. Figures in brackets are T_{DM} model ages calculated at each melt fraction. Modal proportions of the minerals olivine, orthopyroxene, clinopyroxene, and garnet, and their proportions entering the melt, are 0.55:0.20:0.15:0.10 and 0.10:0.10:0.40:0.40, respectively. Calculated fractionation curves 4 to 6 from a qz-tholeiitic-basalt parent to indicated F values, represent fractionation of (4) cpx, (5) cpx + plag (0.5:0.5 proportions), and (6) plag. (C) Sm/Nd versus Rb/Sr relationships for east-Australian mafic volcanic rocks. Tonga-Kermadec data are from Ewart & Hawkesworth (1987). Melting curves 1–3 and fractionation curves 4–6 are for the same conditions used in B.

both) assimilation processes, and (3) more complex fractionation processes, including very small degrees of melting of a garnet-bearing source (Nelson et al., 1986).

Fractionation curves based on the equations of McKenzie (1985) have been calculated for partial melting (garnet-peridotite model) of three assumed source compositions. These include N-type-MORB source mantle (Tarney et al., 1980) and primitive mantle (Wood, 1979), the latter representing an 'enriched mantle' composition (for example, Chen & Frey, 1983). The third model source composition is the estimated lithospheric mantle source for the continental flood basalts (HPT series) of Paraná, Brazil (Hawkesworth et al., 1986). The calculated curves support the interpretation that small-volume melt extraction can produce the more extreme type of observed fractionation behaviour (Fig. 5.5.3B–C), as proposed by McKenzie (1985) and Galer & O'Nions (1986).

The following points also arise from Figure 5.5.3:

(1) The ranges of δ'Nd and especially δ'Sr values are consistent with mantle sources that are heterogeneous in regard to trace-element composition. The leucitites are broadly consistent with the Paraná-type model source, whereas this same source is clearly inappropriate for the melilitites, nephelinites, and basanites which require a more depleted source composition in terms of Rb-Sr and Nd-Sm, perhaps even approaching that of an asthenospheric N-MORB source. This depletion is reflected only partially by the Nd-isotopic data (ε_{Nd}-ε_{Sr} fields in Fig. 5.5.3A) and so must represent a geologically relatively young event.

(2) The δ'Nd-δ'Sr data are consistent with the interpretation that the more alkaline lavas represent relatively small-volume melt fractions, whereas the less alkaline ones, including tholeiitic basalts, are the product of increasing degrees of melting.

(3) The δ'Nd-δ'Sr data for the qz-tholeiitic basalts define an anomalous group of values. One possible explanation is that these represent relatively higher melt fractions, subsequently modified by plagioclase+ clinopyroxene dominated fractionation (see calculated fractionation paths in Figs. 5.5.3B–C).

A further approach towards elucidating basaltic source characteristics is through Sm-Nd model age patterns (T_{DM}; DePaolo, 1981) which are compiled on the basis of whole-rock compositions (Fig. 5.5.4A). Apparent differences in model ages are seen for the different mafic rock groups: the more alkaline rocks have younger and narrower ranges of model ages (except for the leucitites which have relatively coherent model ages of 500–630 Ma, with two slightly higher values). In contrast, the qz-tholeiitic basalts have a wide range of model ages and all but one are greater than 600 Ma.

One explanation for these apparent model-age patterns is based on strong trace-element fractionation during partial melting. T_{DM} model ages in Figure 5.5.3B are given at weight-fraction intervals against each of the calculated melting curves. The ages increase with increasing melt fraction because of Sm/Nd fractionation (although only small changes take place between 0.0001 and 0.001 melt fractions). Note also that the calculated model ages along the melting

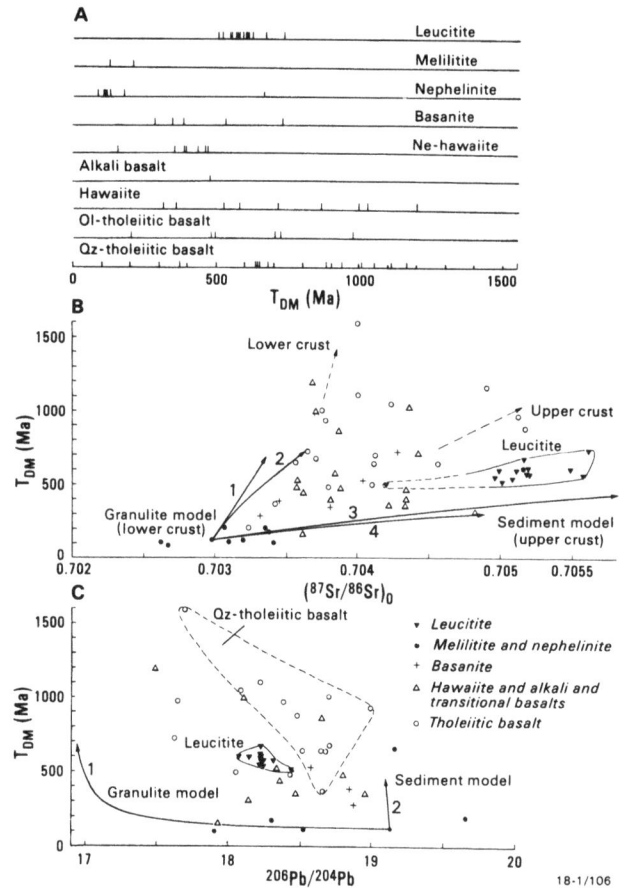

Figure 5.5.4 (A) Calculated T_{DM} model ages for the major, east-Australian basaltic types. (B) T_{DM} versus Sr-isotope relationships showing possible divergent trends, and AFC curves 1–4 which are based on an R-value of 0.8 and a nephelinite parent. Curves 1 (oliv + cpx) and 2 (oliv + plag) incorporate a granulite assimilate and extend to fractionation values (F) of 0.4. Curves 3 (oliv + cpx) and 4 (oliv + plag) extend to values of 0.5 and 0.8, respectively, and incorporate an upper-crustal (sediment) assimilate. Mineral proportions in each case are 0.5:0.5. (C) T_{DM} versus Pb-isotope relationships for mafic volcanic rocks, showing AFC curves 1 and 2 (R-value 0.8) for oliv+cpx fractionation (0.5:0.5 proportions, to F 0.4). Curve 1 incorporates the granulite contaminant, and curve 2 incorporates the upper-crustal contaminant (Table 5.5.1).

curve for the Paraná lithospheric source are similar to those for the New South Wales leucitites. A second possibility is that the T_{DM} model-age patterns reflect AFC or other mixing processes involving crustal components of different ages (which is feasible especially for the qz-tholeiitic basalts) or perhaps interaction of magmas with heterogeneous lithosphere.

Calculated T_{DM} model ages may be compared with the corresponding ^{87}Sr/^{86}Sr and ^{206}Pb/^{204}Pb compositions (Figs. 5.5.4B–C). Sr-isotopic compositions and T_{DM} model ages are rather broadly and positively correlated, and at least two different trends are identifiable within the tholeiitic-basalt and hawaiite groups. The more alkaline rocks, including leucitite, define a third and more coherent group. The hawaiite and tholeiitic-basalt data (Fig. 5.5.4B) may be explained by crustal assimilation, whereas the data for more alkaline rocks, although not excluding such end-members, seem more likely to define lithospheric-asthenospheric, time-integrated, mixing and enrichment trends resulting from melts injected into the lithosphere over an extended time span. No clear correlation is seen in the plot of T_{DM} versus $^{206}Pb/^{204}Pb$ (Fig. 5.5.4C), except for the qz-tholeiitic basalts which may define a trend of decreasing radiogenic Pb as T_{DM} increases. If correct, this may provide evidence for mixing with an old, unradiogenic Pb contaminant, most likely a lower-crustal component.

Some support for crustal assimilation is provided by the AFC curves shown in Figure 5.5.4B based on the model lower- and upper-crustal end-members (see above). Curves for R-values (mass-of-assimilate/mass-of-cumulate) of 0.8 are shown to illustrate gross trends, and are confirmation that increasing T_{DM} model ages can result from assimilation of either end-member. The greatest changes are produced by the lower-crustal model. The slopes of the curves obviously are dependent on the assimilate isotopic composition, and the models in the case of Pb also are extremely sensitive to Pb concentrations in the assimilate (Pb concentrations in most of the volcanic rocks are low, from 1 to 3 ppm).

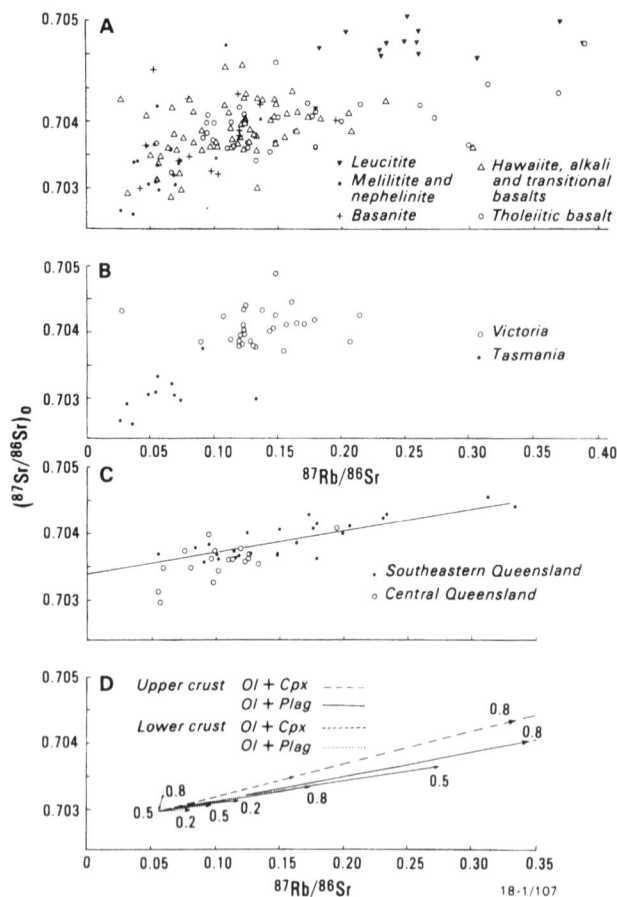

Figure 5.5.5 Initial $^{87}Sr/^{86}Sr$ versus $^{87}Rb/^{86}Sr$ relationships for east-Australian mafic rocks. Isochron in C is for mafic rocks from southeastern Queensland (mainly central volcanoes). (D) AFC curves calculated using a primitive (high Mg-ratio) nephelinite and two different crustal end-members, and all fractionating assemblages in 0.5:0.5 proportions. Curves are for three different R-values (0.2, 0.5, and 0.8) and each curve is extended to an F-value of 0.7, except for the two longest ones which extend beyond the graph limits.

A further line of evidence that needs to be considered concerns the covariance between initial $^{87}Sr/^{86}Sr$ and $^{87}Rb/^{86}Sr$ values (and the corresponding Sm/Nd and Nd-isotope system). O'Reilly & Griffin (1984) reported 'mantle isochrons' in basaltic provinces from New South Wales based on Rb/Sr-$^{87}Sr/^{86}Sr$ correlations, but attributed these to metasomatic processes affecting the mantle-source regions over extended geological time. A broad positive covariance does exist in which the leucitite and tholeiitic-basalt data tend to dominate the higher Rb/Sr region of the array (Fig. 5.5.5A). Moreover, there are marked regional differences, exemplified by the Victorian and Tasmanian data (Fig. 5.5.5B). The best correlation, however, is found in the southeast-Queensland provinces (Fig. 5.5.5C) which are dominated by central volcanoes. A remarkably good pseudoisochron can be fitted to this data set, giving a model age of 228 ± 19 Ma for a $(^{87}Sr/^{86}Sr)_0$ value of 0.70340 (this changes to 201 ± 12 Ma, $^{87}Sr/^{86}Sr$ 0.70346, where an outlying, radiogenic qz-tholeiitic-basalt point is also included). These apparent ages are close to the age of an extensive phase of felsic Triassic magmatism recognised throughout southern Queensland, and may be evidence that the initial Sr-isotopic ratio, calculated from the pseudoisochron, represents that of the source for the Triassic felsic magmas. However, as discussed below, a much older model age is obtained from a comparable Pb-Pb isochron, and the pseudoisochron therefore is unlikely to have a true age significance.

A more likely explanation for the covariance in Figures 5.5.5A–C is an assimilation, or AFC-type, process which, in the case of the southeast-Queensland central complexes, most likely involves a crustal end-member. Modelling of this process is illustrated in a general way (Fig. 5.5.5D) in terms of an AFC model using: (1) the greywacke (upper crustal) and granulite (lower crustal) contaminants, and (2) fractionation of olivine (Fo_{80}) + augite and olivine + plagioclase (An_{60}), to fractionation (F) values of 0.7. R-values are shown for 0.5 and 0.8. Three of the four model combinations result in strong covariance of Rb/Sr and Sr-isotopic compositions, and this is enhanced particularly by: (1) the presence of plagioclase as a fractionating mineral; (2) increasing values of R; and (3) the use of the more Rb-enriched greywacke as an assimilate. The main conclusion to be drawn is that appropriate AFC processes certainly can produce the type of trends observed in Figures 5.5.5C, although they are unlikely to provide an explanation, for example, for the leucitite data.

No clearly defined correlation is seen on a companion plot of $^{143}Nd/^{144}Nd$ versus $^{147}Sm/^{144}Nd$ (not shown) largely because of relative scatter of points representing hawaiite and tholeiitic-basalt data. However, a broad positive correlation exists for the more alkaline lava compositions.

5.5.4 Sr- and Nd-isotope and mg-ratio:SiO_2 relationships

The data presented above are indicative of mantle-metasomatic and crustal-mixing/assimilation processes having played a role in the genesis of at least some of

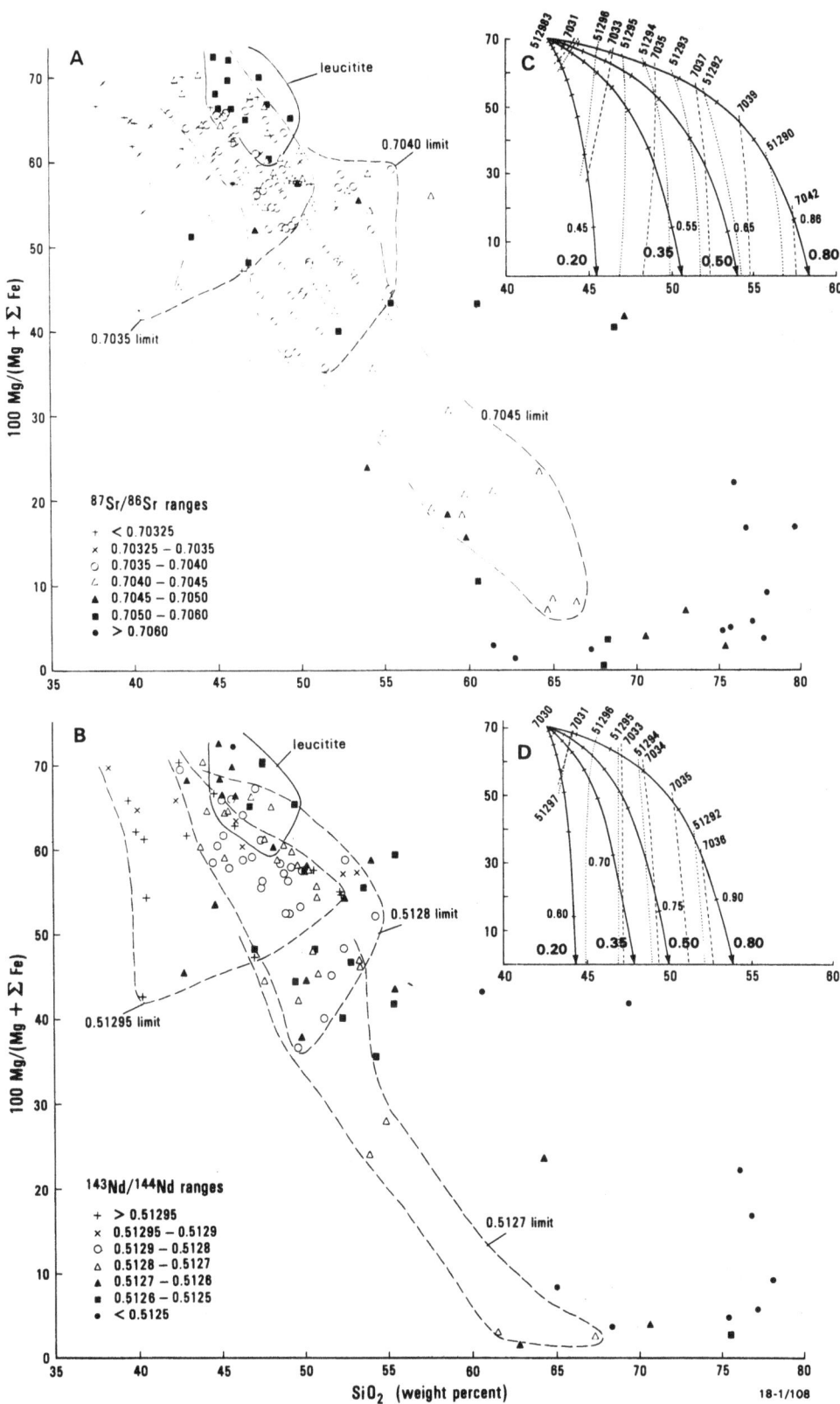

Figure 5.5.6 mg-ratio, SiO_2, and Sr- and Nd-isotope relationships for east-Australian rocks. Curves in C and D represent calculated values assuming AFC processes, using a primitive (high Mg-ratio) nephelinite and average Palaeozoic greywacke as end-members (Table 5.5.1). Four unbroken curves are for different R-values (heavy figures 0.2, 0.35, 0.5, and 0.8), and ticks along each curve represent F-values at 0.05 intervals (0.02 for the 0.80 R curve). Calculated Sr- and Nd-isotope values are shown as dashed and dot-dash curves, respectively. AFC fractionating assemblage in C is for oliv (Fo_{80}) + plag in 0.5:0.5 proportions, and in D is for oliv (Fo_{80}) + cpx in 0.5:0.5 proportions.

the diverse east-Australian Cainozoic rocks. An examination of the isotopic data therefore is warranted in relation to important major-element parameters such as SiO_2 and mg-ratio (Fig. 5.5.6). Two points may be stressed. First, the leucitites are quite unique compared to other east-Australian rock types. Second, the most alkaline lavas consistently have the least radiogenic Sr- and most radiogenic Nd-isotopic compositions. There is a general tendency, as SiO_2 increases and mg-ratio decreases, for $^{87}Sr/^{86}Sr$ values to become more radiogenic, and for $^{143}Nd/^{144}Nd$ values to decrease, reaching their more extreme limits in the rhyolites. These are illustrated by the dotted fields in Figure 5.5.6, representing the limits of selected Sr- and Nd-isotopic compositions. The isotopic compositional shifts seen in Figure 5.5.6 are consistent with AFC-type processes,

especially those involving crustal assimilation (as discussed above). However, this has to be considered in relation to the leucitites which are unlikely to have been produced by such a process (as discussed above). Considerable isotopic differences clearly exist in the more magnesian lavas, even excluding the leucitites, which is indicative of primary source heterogeneities, but even in these there is some correlation with silica concentration.

The modelling of AFC processes in Figures 5.5.6C–D is based on a set of simplified conditions using an isotopically 'primitive' nephelinite and the upper-crustal greywacke end-member. The most abrupt changes in isotopic composition at high R-values correlate with increasing SiO_2 and only secondarily with decreasing mg-ratio, precisely as observed in

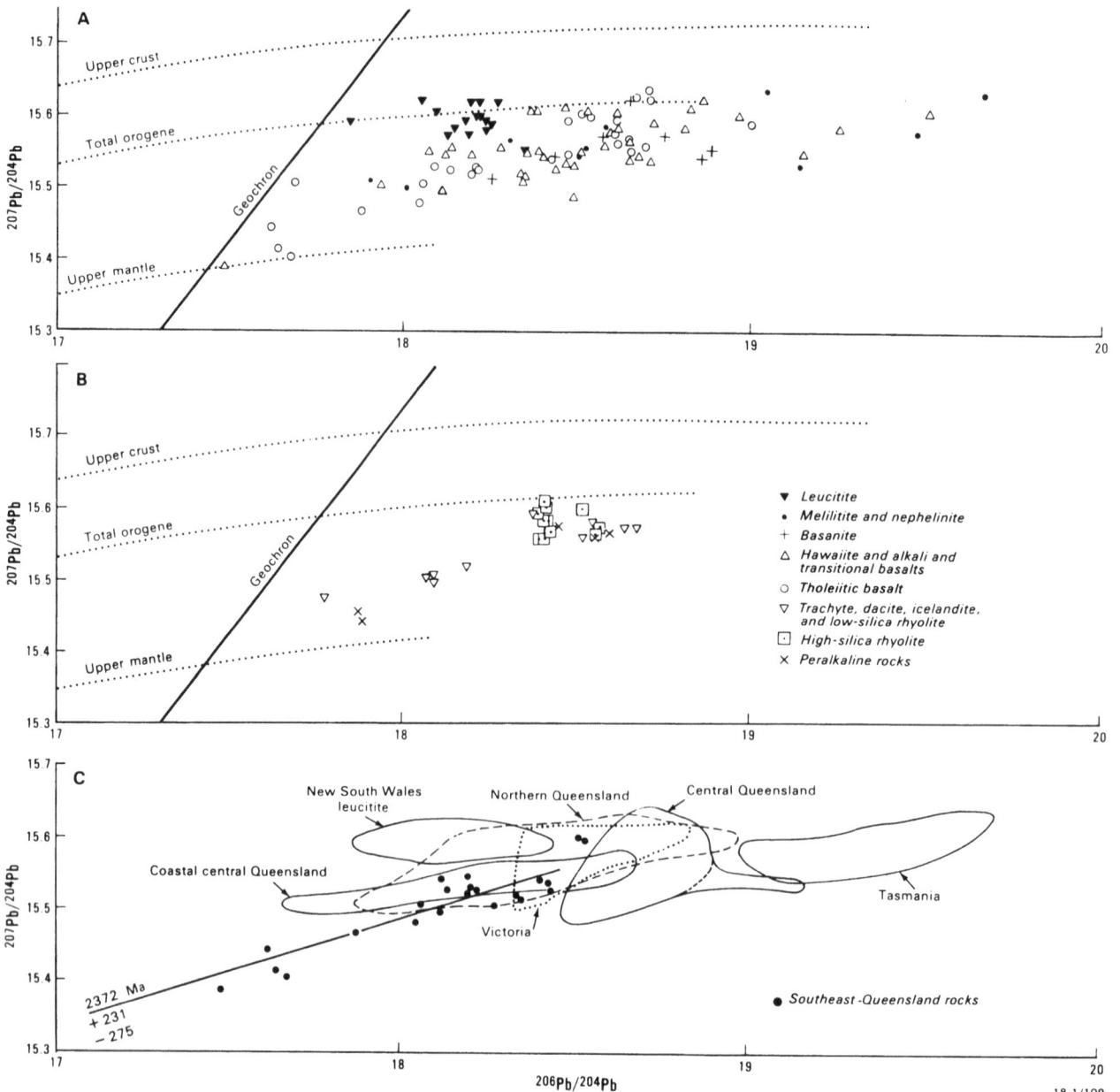

Figure 5.5.7 $^{207}Pb/^{204}Pb$ versus $^{206}Pb/^{204}Pb$ relationships for east-Australian mafic (A) and intermediate and felsic (B) volcanic rocks. Dotted lines in A and B represent model Pb-evolution curves (Doe & Zartman, 1979). (C) Generalised fields for basalts shown in relation to the southeast-Queensland isochron.

Figures 5.5.6A-B. In fact, small but significant isotopic changes can take place initially involving only small mg-ratio changes. The reverse takes place at R-values less than about 0.35.

The main differences between the olivine+plagioclase and olivine+clinopyroxene fractionating assemblages in Figure 5.5.6 are the control on the rate of change of mg-ratios and the limit on the relative degree of silica enrichment for the same R-values. AFC processes therefore offer a plausible explanation for isotopic differences observed in east-Australian mafic rocks, excepting the most alkaline types and noting that relatively high R-values (greater than 0.35) are required. Figures 5.5.6C-D are based on an upper-crustal contaminant, but the results may apply equally to a lower-crustal component, providing it is intermediate to felsic in composition and has similar ranges in Nd- and Sr-isotopic compositions. Note also that the data in Figure 5.5.6 seem broadly consistent with the concept of a sediment contribution to the source of the leucitite magmas (Weaver et al., 1986), although obviously not with AFC processes as such.

5.5.5 Pb isotopes

Mafic rocks

The Pb-isotope data presented here for mafic rocks are plotted separately from those for the evolved types, and the total lower- and upper-crustal, upper-mantle and total-orogene evolution curves of Doe & Zartman (1979) are included for reference purposes (Figs. 5.5.7-8). The overall differences in Pb-isotopic compositions of the mafic rocks are considerable: 11 percent for $^{206}Pb/^{204}Pb$ and 5.3 percent for $^{208}Pb/^{204}Pb$. The following points can be made:

(1) The leucitite data define a relatively restricted array that is displaced slightly from the main array towards higher $^{207}Pb/^{204}Pb$ values.

(2) All data points fall on the radiogenic side of the geochron, although several hawaiite and *ol*-tholeiitic basalt compositions (all from central volcanoes) plot close to it. The multiple evolution of the Pb-isotope characteristics of the different magma types is illustrated clearly.

(3) There is no particular isotopic signature that is characteristic of any of the individual basaltic types, except for the leucitites. There are, however, marked Pb-isotopic differences between regions (Fig. 5.5.7C). The most notable are the relatively radiogenic compositions of the Tasmanian mafic rocks which, as discussed above, are also somewhat distinctive in their Nd- and Sr-isotopic compositions (see also Fig. 5.5.9). Significantly, these Pb-, Sr-, and Nd- isotopic compositions overlap the MORB fields (Figs. 5.5.2,7,9), which perhaps is indicative of an asthenospheric mantle source. Equally significant, however, is the observation that the only other analysed rock that is similar isotopically to the Tasmanian rocks is the most primitive (highest mg-ratio) nephelinite analysed (from the Buckland-Mitchell region of central Queensland).

(4) The regional Pb-isotopic fields of the mafic rocks tend to be linear and indicative of mixing arrays. The best example is the southeast-Queensland region

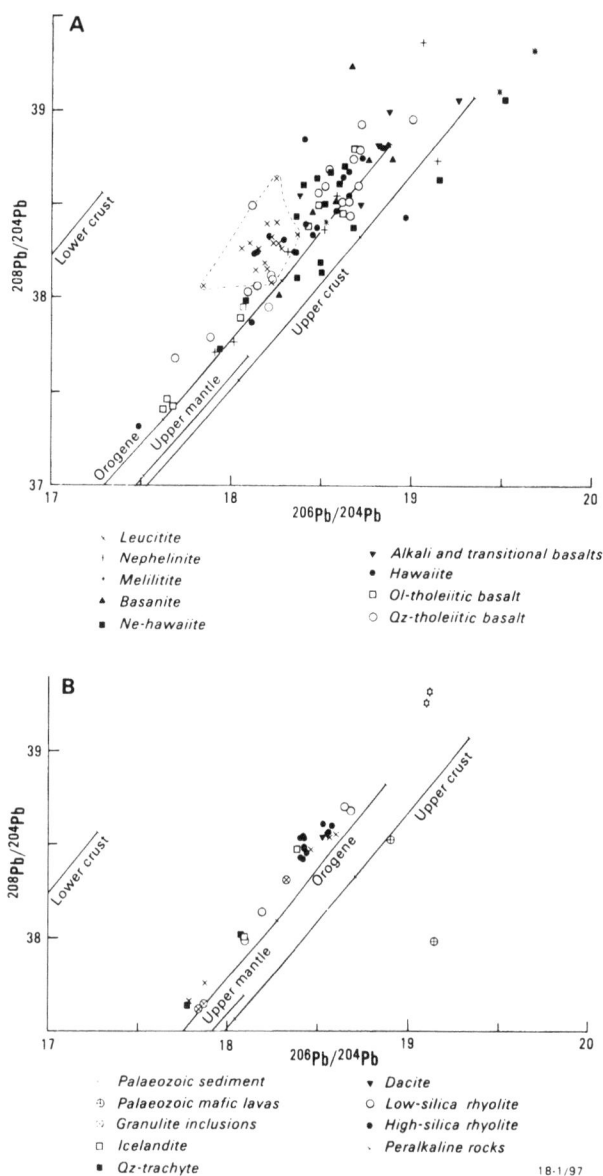

Figure 5.5.8 $^{208}Pb/^{204}Pb$ versus $^{206}Pb/^{204}Pb$ relationships for east-Australian mafic (A) and intermediate and felsic (B) volcanic rocks. Also shown are data for southeast-Queensland Palaeozoic greywackes and mafic lavas, and granulite xenoliths (Ewart, 1982a). The model Pb-evolution curves are those of Doe & Zartman (1979).

(Fig. 5.5.7C) where the least radiogenic compositions are found for any east-Australian Cainozoic province (Ewart, 1982a). An isochron fitted to these data has a model age of 2372 (+ 231 − 275) Ma (μ_1 7.91, an expected upper-mantle value). This age is much older than that obtained from the Rb-Sr data and both pseudoisochrons are unlikely to have any precise age significance, as discussed above. However, the possibility of a mixing array is indicated, incorporating a possibly Proterozoic, lower-crustal component, although this is not predicted from current tectonic models for the region. An isochron fitted to the linear data array for the rocks of central-Queensland coast (Fig. 5.5.7C) has a model age of 289 (+ 1071 − 288)

Ma (μ_1 8.07) which, if the array is also a mixing trend, possibly corresponds to a much younger end-member assimilate.

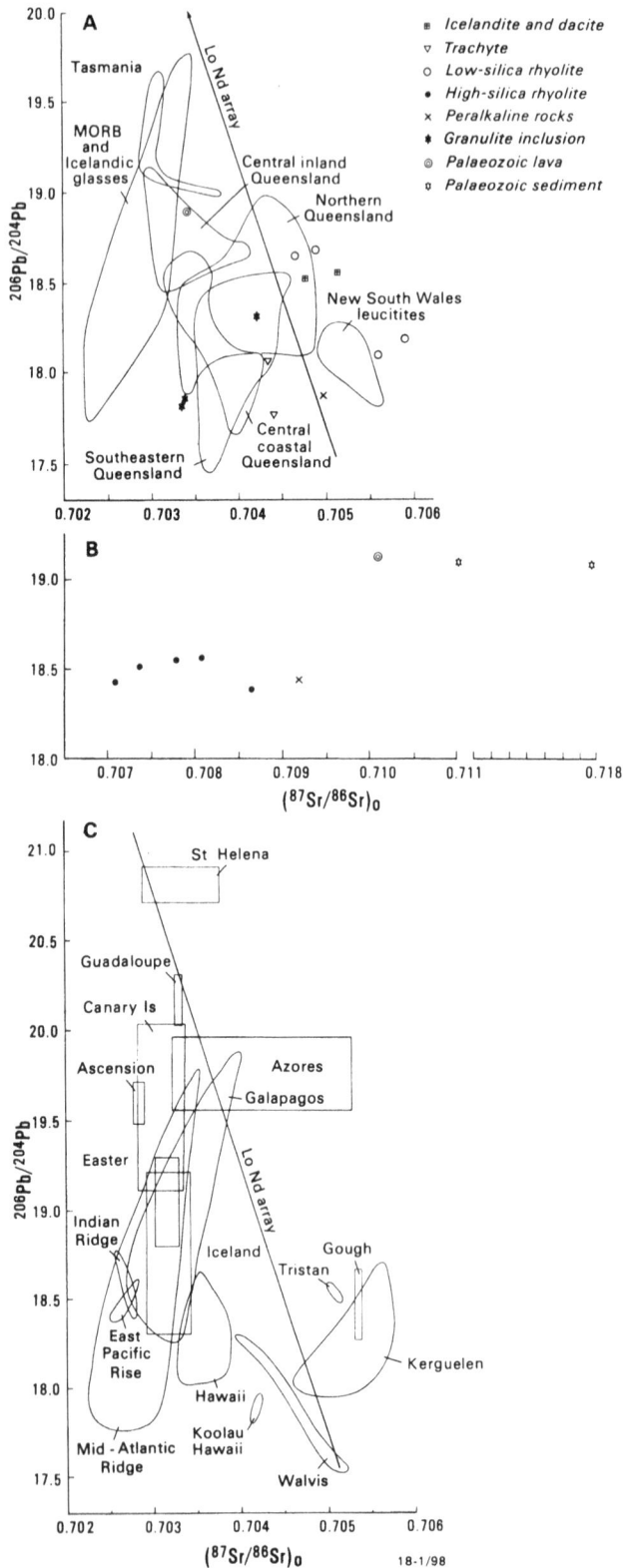

The east-Australian basaltic Pb-isotopic data generally follow the dual total-orogene/upper-mantle evolution curves (Doe & Zartman, 1979) within the $^{207}Pb/^{204}Pb$:$^{206}Pb/^{204}Pb$ plot, but tend to fall slightly above the orogene curve in the $^{208}Pb/^{204}Pb$:$^{206}Pb/^{204}Pb$ diagram, except for the Tasmanian data. The relatively higher $^{208}Pb/^{204}Pb$ data may be indicative of some U depletion, or Th enrichment, within the magma-source regions. In general, however, the basaltic magma sources are shown from the Pb-isotopic data to represent well mixed orogene/upper-mantle components. This is basically consistent with the extensive Palaeozoic-Mesozoic continental-margin magmatic history which has influenced the evolution of the east-Australian lithosphere. The leucitites, Tasmanian rocks, and rare other lavas are notable exceptions.

Felsic rocks

The trachytic, peralkaline, and rhyolitic rocks have Pb-isotopic compositions (Figs. 5.5.7B,8B) in the same range as those for the mafic lavas, although data are available only for southern- and central-Queensland volcanic centres (Ewart, 1982a). In detail, the trachytes, low-silica rhyolites, and most peralkaline lavas, although differing in their overall compositional ranges, have similar Pb-isotopic compositions to those of the mafic rocks from the same centres. This is expected if the evolved rocks have originated by predominantly fractional-crystallisation processes.

The high-silica rhyolites, however, are somewhat different in two respects: (1) they have closely similar Pb-isotopic compositions, despite their presence in at least three separate volcanic centres in southeastern Queensland; (2) the high-silica rhyolites have Pb-isotopic compositions distinctly different from, and more radiogenic than, those of the spatially associated mafic lavas (Ewart, 1982a). This is interpreted (Ewart, 1982a, 1985) to indicate a mode of formation for these rhyolites by localised crustal anatectic melting in response to basaltic-magma emplacement, followed by extensive crystal fractionation. The Pb- (and also Sr- and Nd-) isotopic compositions of these high-silica rhyolites are indicative of an upper-crustal type source.

Pb- and Sr-isotope relationships

The regional compositional fields of the mafic lavas compared in Figure 5.5.9A with individual data for evolved lavas and the Palaeozoic sediments, are compared further with MORB and ocean-island basalt data in Figure 5.5.9C. Three relationships are evident:

(1) The compositional fields of the east-Australian rocks do not overlap entirely with those of ocean-island basalts, but rather tend to have less radiogenic Pb and more radiogenic Sr. Moreover, the relatively low $^{206}Pb/^{204}Pb$ and $^{87}Sr/^{86}Sr$ characteristics of some of the southeast-Queensland rocks evidently are not matched by ocean-island basalt data.

(2) The Tasmanian rocks have apparently MORB-like isotopic characteristics. Only the Tasmanian qz-tholeiitic basalts deviate towards more radiogenic Sr compositions.

(3) The high-silica rhyolites are isotopically unique among the east-Australian volcanic rocks.

Figure 5.5.9 (A,B) $^{206}Pb/^{204}Pb$ versus initial-$^{87}Sr/^{86}Sr$ relationships for east-Australian volcanic rocks. Data for mafic rocks are shown as fields, and those for felsic rocks, Palaeozoic greywackes and mafic lavas, and granulite xenoliths are shown as individual points. B is the right-hand extension of A. Data sources for MORB and Icelandic glasses as in Figure 5.5.2. (C) Compositional fields of MORB and ocean-island basalts (after Staudigel et al., 1984). LoNd array after Hart et al. (1986).

The differences between the ocean-island basalt and east-Australian data are presumed to reflect important differences in the Phanerozoic lithospheric histories of the two types of environments, possibly in part reflecting the complex Phanerozoic history of subduction-related and extension-related volcanism.

5.5.6 Oxygen isotopes

Few oxygen-isotope analyses are published for east-Australian rocks. Data presented in Figure 5.5.10 are based on mineral and whole-rock analyses (Ewart, 1982a, 1985; H.P. Taylor et al., 1984). H.P. Taylor et al. (1984) showed that significant post-eruptive ^{18}O enrichment had taken place in New South Wales leucitites within both whole-rock and leucite samples, and therefore the data plotted in Figure 5.5.10 were recalculated to primary magmatic values. These range from 5.8 to 6.8, and therefore are essentially indistinguishable from the compositions of alkali basalts throughout the world. The data do not support the existence of an 'anomalous ^{18}O-rich' upper-mantle source from which these extreme potassic magmas were derived (H.P. Taylor et al., 1984). The remaining data plotted in Figure 5.5.10 are based on analyses of feldspar phenocrysts whose $\delta^{18}O$ compositions are equated with their magmatic values.

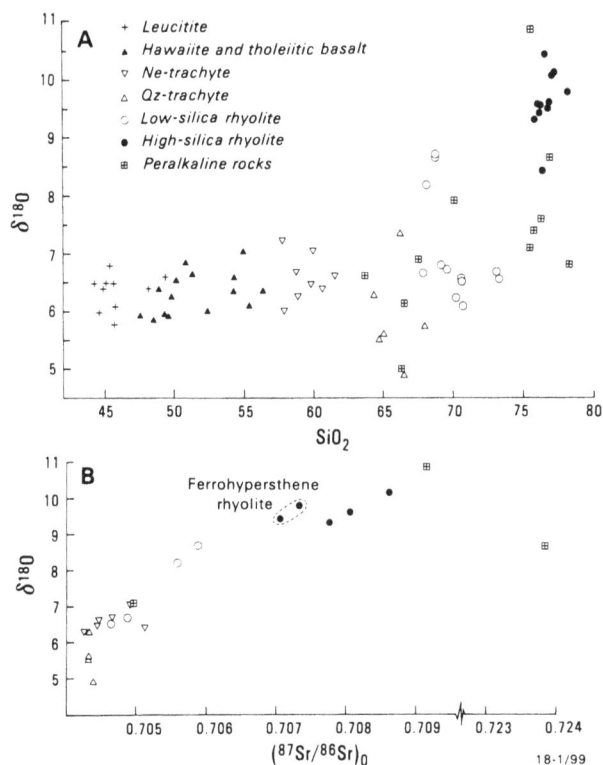

Several clear trends in $\delta^{18}O$ compositions can be seen for evolved rocks in Figure 5.5.10:

(1) The high-silica rhyolites are consistently ^{18}O-enriched. This correlates with their more radiogenic Pb- and Sr-isotopic compositions.

(2) Some trachytes have increasing ^{18}O-enrichment as SiO_2 increases and as their Sr-isotopic compositions become more radiogenic, whereas several other samples (a peralkaline trachyte from Nandewar volcano and trachytes from Fraser Island, southeastern Queensland) have anomalously depleted ^{18}O compositions. The 'normal' enrichment trend is interpreted by Ewart (1982a, 1985) in terms of fractional crystallisation and AFC processes. This is consistent with other isotopic and trace-element geochemical evidence discussed above. However, Ewart (1982a) noted that the AFC processes, as modelled, did not duplicate the strongly depleted total-Sr abundances in these more evolved felsic magmas. This was reconciled by suggesting a process involving at least two stages: (1) AFC processes, involving assimilates isotopically similar to the Palaeozoic greywacke sediment, followed by (2) isotopically closed-system fractional crystallisation of pockets of enclosed magma, perhaps isolated from further contamination effects by envelopes of already solidified magma.

The origin of the relatively ^{18}O-depleted trachytes is unexplained. One possibility is that they have equilibrated with ^{18}O-depleted lower-crustal granulites (see, for example, Wilson & Baksi, 1984).

(3) The peralkaline rocks have a range of compositions. Some have limited ^{18}O-enrichment as SiO_2 increases (consistent with fractional crystallisation and little or no accompanying assimilation). Other peralkaline rhyolites (the potassic types) have more enriched ^{18}O compositions that have been interpreted to represent extended ternary-minimum fractional crystallisation from high-silica rhyolite precursors (Ewart, 1982a).

5.5.7 Isotope and trace-element relationships

Patterns of trace-element and isotopic chemistry for east-Australian volcanic rocks are shown above to be indicative, with some notable exceptions, of fractional crystallisation and AFC processes. Possible correlations, or otherwise, are examined here between the behaviour of trace elements and isotopic compositional patterns beyond those specifically involved in daughter-parent relationships in the Nd-, Sr-, and Pb-isotopic systems.

Hf-Th-Ta data (used also in Fig. 5.4.2) are replotted and grouped according to Sr- and Nd-isotopic compositional ranges (Fig. 5.5.11). In general, those mafic lavas plotting in the 'within-plate alkaline field' (c) are characterised by the least radiogenic Sr- and most radiogenic Nd-isotopic compositions. These isotopic compositional trends become more accentuated, and have a wider range, as rock compositions plot closer to, and within, the 'within-plate tholeiitic field' (b). This is most pronounced for the Nd-isotopic data. An additional point is that even more strongly radiogenic Sr- and less radiogenic Nd-isotopic compositions are

Figure 5.5.10 (A) $\delta^{18}O$ versus SiO_2 relationships for east-Australian volcanic rocks. Leucitite (New South Wales) data after H.P. Taylor et al. (1984). Data for hawaiites and tholeiitic basalts (southeastern Queensland), trachyte, rhyolite, and peralkaline rocks (Victoria, New South Wales, and southern Queensland) after Ewart (1982a, 1985, unpublished data; based on separated-feldspar analyses). (B) $\delta^{18}O$ versus initial-$^{87}Sr/^{86}Sr$ relationships for feldspars separated from Victorian and New South Wales trachyte, ne-trachyte, and peralkaline rhyolite, together with data for trachyte, rhyolite, and peralkaline rhyolite from southern Queensland (after Ewart, 1985).

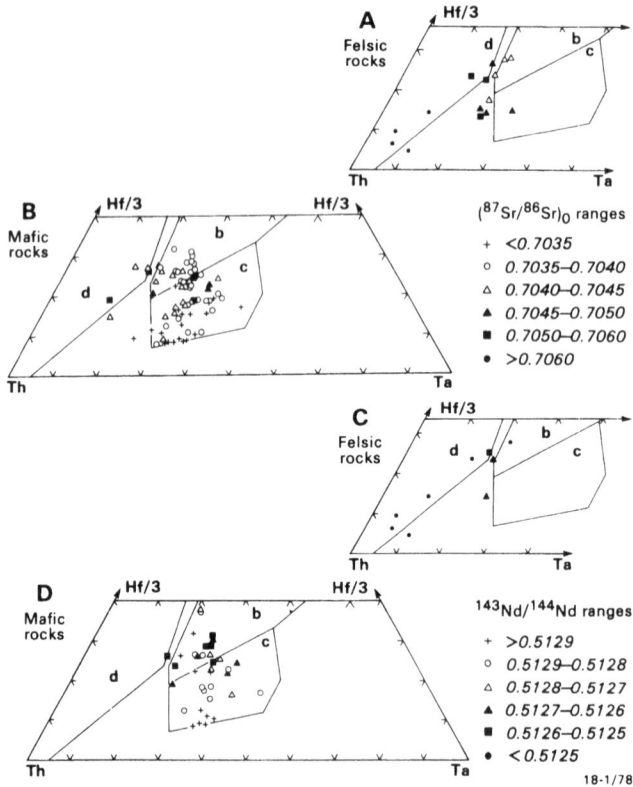

Figure 5.5.11 Hf:Th:Ta relationships for east-Australian volcanic rocks subdivided on the basis of Sr- and Nd-isotopic compositions (compare with Fig. 5.4.2). Field boundaries are those defined by D.A. Wood (1980). Field *b* refers to E-type MORB and tholeiitic within-plate rocks. Field *c* refers to alkaline within-plate rocks. Field *d* refers to calcalkaline destructive-plate-margin rocks.

represented by those data points extending from the 'within-plate field' towards the 'destructive plate-margin basalt field' (*d*). Data for the more felsic rocks are somewhat similar, as the more radiogenic Sr- and less radiogenic Nd-isotopic compositions characterise those rocks most displaced from the within-plate magmatic fields. The high-silica rhyolites represent the most extreme examples. The destructive-plate-margin trace-element signatures, and their corresponding isotopic shifts, are interpreted (see below) in terms of crustal-assimilation processes involving older arc (or active margin) and arc-derived crust (presumed to be dominantly Palaeozoic). This is considered to involve local anatexis of the sediments from these environments in the case of the high-silica rhyolites. Arc-modified lithosphere is a possible magma source for the tholeiitic basalts.

Maximum trace-element enrichment in the mafic rocks is found for Ba and Th (relative to MORB values), as seen in MORB-normalised diagrams (Fig. 5.4.1), and there is a relative shift from maximum Th enrichment in the more alkaline rocks to Ba in tholeiitic basalt and hawaiite (Fig. 5.4.1). The possibility that this trace-element-enrichment shift may relate to isotopic changes is explored in Figure 5.5.12. Some complex correlation patterns are evident. There is a broad positive covariance between $^{87}Sr/^{86}Sr$ and Ba/Th, whereas $^{143}Nd/^{144}Nd$ and Ba/Th have a broad negative correlation. The Pb-isotope data (Figs. 5.5.12C-D)

are correlated rather poorly where all the data are considered, but there is a well defined negative correlation where only the mafic rocks from the central volcanoes are considered.

These trends involving both isotopic and trace-element differences may represent mixing or AFC processes. The results of AFC modelling (Figs. 5.5.12E–F) are consistent with the general trends observed for Sr-isotope versus Ba/Th and Nd-isotope versus Ba/Th at the higher R-values, but clearly favour the lower-crustal end-member. This can be related to the behaviour of Th in crustal rocks where Th tends to be enriched in the upper crust (as in the greywackes, resulting in low Ba/Th values) or relatively depleted in the lower crust. 'Lower-crustal' assimilation therefore is inferred, if the trends in Figure 5.5.12 are indeed AFC-produced. This is supported further by the Pb-isotope data (Fig. 5.5.12C) as the trend requires a less radiogenic Pb component as contaminant, which is inconsistent with available upper-crustal Pb-isotope data. No specific model is presented for Pb because of its extreme sensitivity to the model data which, at this stage, are difficult to constrain.

A similar relationship is observed where Ba/Nb is plotted against Sr-, Nd-, and Pb-isotopic values (not shown here). Again, these general trends are reproduced by the AFC modelling given above, but Ba/Nb values do not discriminate so strongly between the two assimilate end-members as do Ba/Th values.

5.5.8 Summary

East Australian Cainozoic volcanic rocks represent a geochemically broad, diverse, complex, and heterogeneous group of magmas, and, predictably, these attributes extend to their Sr-, Nd-, and Pb- isotopic compositions. New South Wales leucitites are one group of rocks that have distinctive isotopic characteristics thus matching their overall geochemical uniqueness. Nelson et al. (1986) concluded that the mantle source of the leucitites was contaminated by a component in which $^{87}Sr/^{86}Sr$ was greater than (or equal to) 0.7057, ε_{Nd} was less than (or equal to) −4, and $^{207}Pb/^{204}Pb$ was about 15.6.

Other basaltic rocks have a complete continuity of isotopic compositions, and no particular isotopic compositional range is found to characterise a particular basaltic group. There are, however, some marked regional differences although, again, there is extensive overlap. The apparent overlap of the Tasmanian Sr-,

Figure 5.5.12 Sr-, Nd-, and Pb-isotope and Ba/Th relationships for east-Australian mafic volcanic rocks. D includes only mafic samples from central volcanoes. AFC curves 1–6 in E and F are for a nephelinite parent and upper-crustal (Palaeozoic sediments) and lower-crustal (granulite) end-members. Curves are for three different R-values (0.2, 0.5, and 0.8), and are extended to fractionation values (F) of 0.4 (except where extending beyond graph limits). Fractionating assemblages for each of the curves in E are (1,2) oliv + plag, (3,4) oliv + cpx, (5,6) oliv + plag, (7) oliv + cpx, and in F are (1,2) oliv + plag, (3) oliv + cpx, (4,5) oliv + plag, (6) oliv + cpx (all in 0.5:0.5 proportions).

KEY (A–C)

▼ Leucitite
● Melilitite and nephelinite
+ Basanite
△ Hawaiite, alkali and
   transitional basalts
○ Tholeiitic basalt

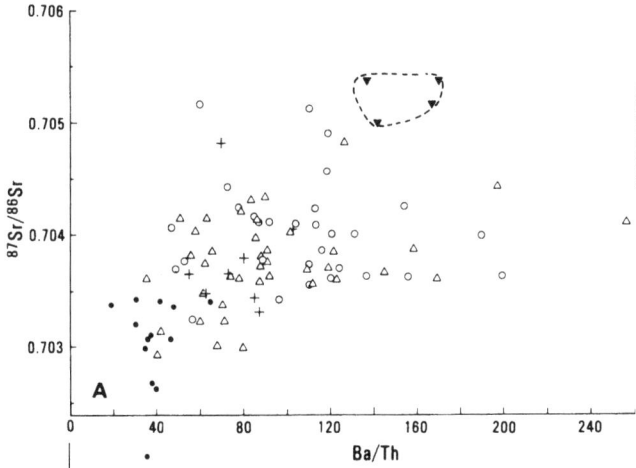

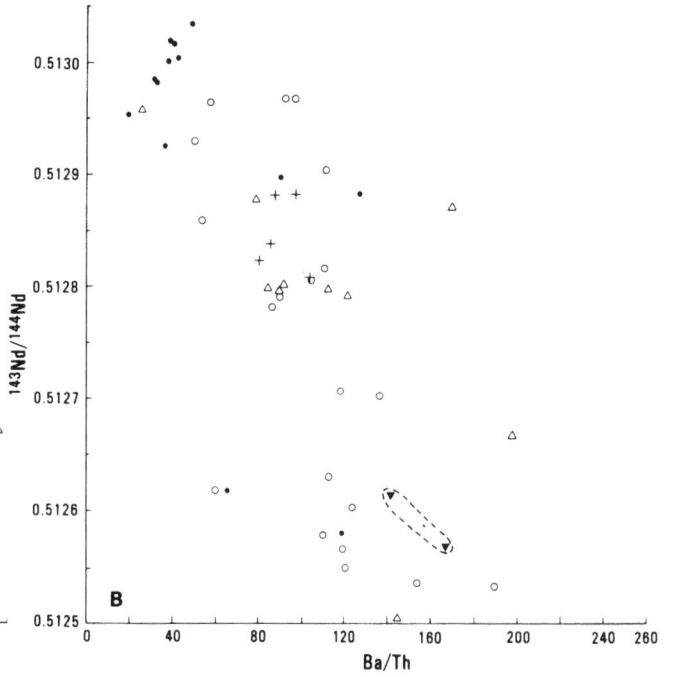

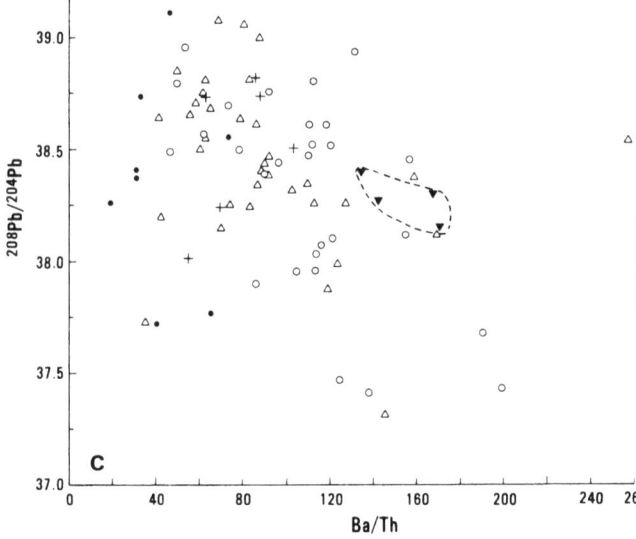

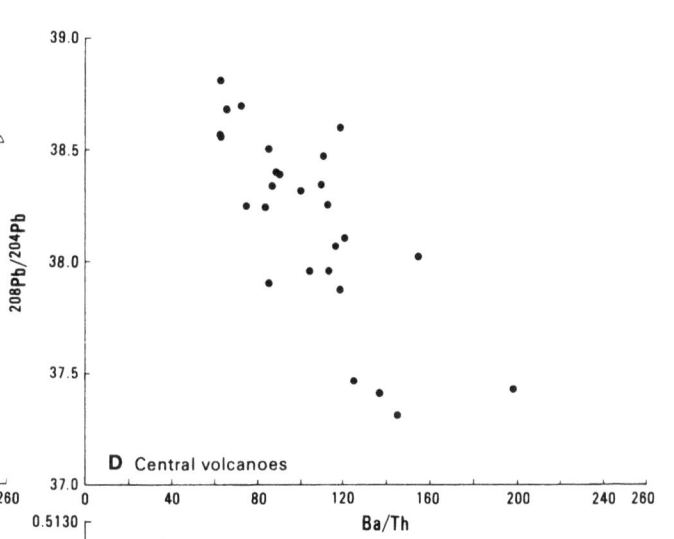

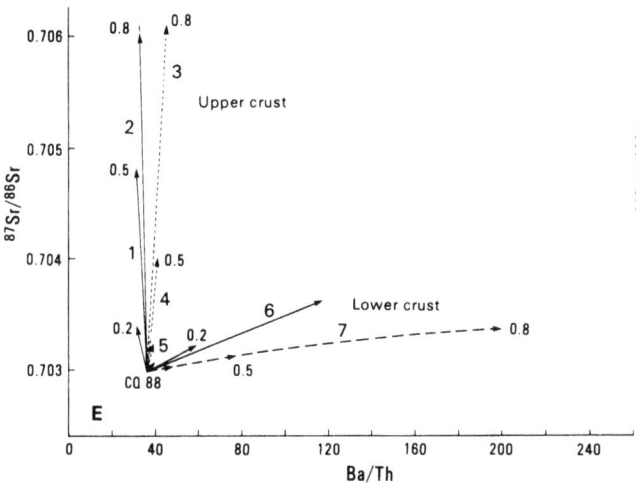

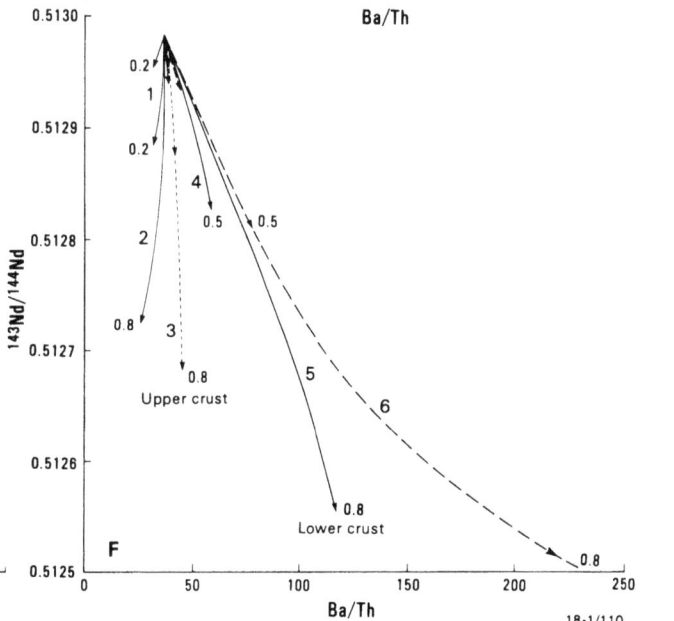

Nd-, and Pb-isotope data with MORB compositional fields (excepting the *qz*-tholeiitic basalts) is of particular interest. The only other analysed rock having these same features is a highly magnesian nephelinite from central Queensland. McDonough et al. (1986) proposed a model for the genesis of east-Australian mafic volcanic rocks involving the mixing of hotspot mantle plume-derived melts and lithospheric mantle-derived melts. The differences between tholeiitic and alkali basalts in this model are attributable to different degrees of mixing and exchange between the two mantle-melt components. This model obviously is attractive, although it is difficult to constrain the isotopic characteristics of the highly heterogeneous lithospheric mantle. However, the data synthesised in this account are evidence that the isotopic composition of the asthenospheric-mantle component is represented most closely by the analysed alkaline Tasmanian lavas and by the distinctive magnesian nephelinite from Queensland.

One complicating problem in the interpretation of all continental basaltic volcanism is the role of crustal assimilation. AFC processes, at least in principle, have the potential to explain much of the observed isotope and isotope/trace-element behaviour in the different mafic rocks. Indeed, the operation of assimilation during fractional crystallisation is not only possible, but perhaps inevitable. How the products of AFC processes can be distinguished from the melting products of heterogeneous subcrustal lithosphere and mantle metasomatism, is not altogether clear. Further complicating factors include possible crustal imbrication within the upper lithospheric mantle, and subduction-modified lithosphere containing an imprint of the subducted-sediment component. The isotopic ranges observed for east-Australian mafic rocks also need to be kept in perspective in terms of the isotopic differences now recognised within and between the Tasmantid Guyots (McCulloch, 1988). However, AFC processes are more clearly evident in the rocks from the central volcanoes, and lower-crustal sources seem to be implied by some of the individual sample data, in particular by some relatively unradiogenic Pb-isotope compositions.

The felsic rocks have relatively radiogenic Sr- and unradiogenic Nd-isotopic compositions, and are ^{18}O-enriched in different degrees, compared to associated mafic rocks. The high-silica rhyolites are a distinctive group, both isotopically and in other geochemical and mineralogical features, and are interpreted as local upper-crustal melts developed in response to mafic magma emplacement, but modified by subsequent crystal-liquid fractionation. Some of their trace-element characteristics correspond to sources that have a signature suggestive of a destructive-plate-margin type chemistry.

Trachyte and low-silica rhyolite have marked fractional crystallisation-controlled trace-element patterns. Their isotopic compositions correspond to different AFC processes, in some cases, however, accompanied by only minor assimilation input (Ewart, 1982a, 1985). This does not necessarily imply that their mafic parental magmas (inferred from the associated mafic rocks) may not have been modified by earlier AFC processes. The peralkaline rocks have a wide range of isotopic compositions, reflecting what is interpreted to be their different petrogeneses.

Further discussions on petrogenesis are given in Sections 7.5–7.

CHAPTER 6

Xenoliths and Megacrysts of Eastern Australia

6.1 Xenolith Types, Distribution, and Transport

6.1.1 Age and distribution

Xenoliths and megacrysts (discrete crystals or crystal fragments 0.5 cm or more across) are found in Jurassic to Holocene basaltic rocks in the Phanerozoic tectonic regime of eastern Australia (Fig. 6.1.1). The oldest occurrences are in the Jurassic diatremes of the Sydney Basin and the youngest in the approximately 4600 year-old flows at Mount Gambier (Section 3.7.4). This represents — except possibly for the basaltic provinces of eastern China — the largest and most widespread quasi-continuous sampling of a mantle section on the Earth's surface, both in space and time. Xenoliths are rare in basaltic rocks from the large, differentiated, central volcanoes that appear to represent hotspot trails (Wellman & McDougall, 1974a). They are most abundant in the basaltic lava-field provinces that do not have any distinctive pattern of age and geographic distribution (Section 1.7). There is within some lava-field provinces a bimodal age distribution of basaltic volcanism — for example, in the Sydney province (Section 3.6.3), Southern Highlands province (Section 3.6.4), Kandos province (Section 3.6.4), and east-central Queensland (Griffin et al., 1987). Mantle-derived xenoliths in hosts of different ages from these provinces are modally and geochemically similar despite significant age differences (up to 120 Ma for Kandos). However, younger, lower-crustal xenoliths from east-central Queensland have evidence of multiple thermal episodes, whereas older ones do not (see Section 6.3). Most xenoliths and megacrysts are considered to represent accidental (non-cognate) fragments, although some may be high-pressure precipitates from basaltic magmas, or their derivatives, in the same general magmatic episode (Section 6.2; Irving & Frey, 1984).

Mantle-derived xenolith suites include a relatively restricted range of rock types, but the proportions of types may differ widely between localities. Cr-diopside spinel lherzolite is the most abundant and ubiquitous xenolith type throughout eastern Australia, making up 80–100 percent of the population at most xenolith localities. Other mantle-derived rocks (mafic, ultramafic, granulitic, eclogitic) are distributed irregularly throughout the region. Lower-crustal xenoliths are relatively rare (about 40 localities so far recorded) but may

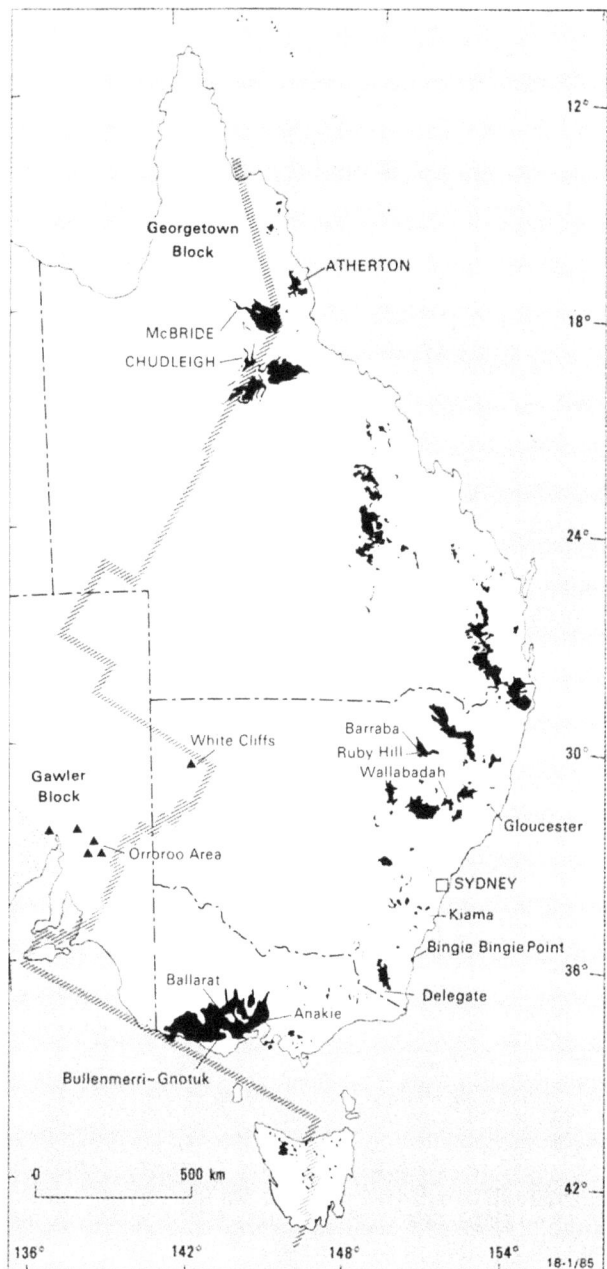

Figure 6.1.1 Some of the principal xenolith and megacryst localities in eastern Australia. Hachured strip represents approximate eastern limit of the cratonic areas of Australia. Volcanic areas shown solid.

dominate individual sites such as Anakie (Victoria; Wass & Hollis, 1983), Kellys Point (New South Wales; Halford, 1970) and, particularly, the McBride and Chudleigh provinces in northern Queensland (Kay & Kay, 1983; Rudnick et al., 1986; Rudnick & Taylor, 1987).

Xenoliths are found also in kimberlitic and lamproitic hosts in the western cratonic tectonic regime of Australia. Localities near White Cliffs (New South Wales; Edwards et al., 1979; Section 6.3.6) and near Orroroo in South Australia (Ferguson et al., 1979; Arculus et al., in press; Section 6.3.6) are on the eastern margin of the craton (Fig. 6.1.1). These localities have abundant xenoliths of lower-crustal origin, as well as upper-mantle eclogites and pyroxenites. Some of the Orroroo kimberlites contain xenocrysts of deformed olivine, Cr-diopside, garnet, and Cr-spinel (probably fragments of disaggregated mantle wall-rock) and microxenoliths of garnet-spinel lherzolite and xenocrysts of diamond (Scott Smith et al., 1984). Edwards et al. (1979) described xenoliths of garnet±spinel lherzolite from Kayrunnera (White Cliffs) diatremes.

The diamondiferous olivine lamproite of the Argyle AK1 pipe in northwestern Western Australia contains garnet lherzolite and harzburgite xenoliths, some of which are diamond-bearing (O'Neill et al., 1986). The primary mineral assemblages are inferred to consist of olivine ± orthopyroxene + garnet ± Cr-diopside ± picrochromite. All original garnet is replaced by pyroxene-spinel intergrowths, together with secondary calcite ± serpentine ± phlogopite ± K-feldspar ± illite. The xenoliths are altered in different degrees to talc, serpentine and, in some, montmorillonite.

Upper- and mid-crustal xenoliths are recognised rarely, though whether this is a human or a volcanic-sampling problem is not clear. Griffin & O'Reilly (1986a, 1987a) gave examples of how some crustal fragments may be overlooked because of extensive alteration. Quartz-rich, upper-crustal clasts and crystal debris can be identified in many flows and high-level intrusions. They may retain evidence of extensive fusion or reaction (or both) providing spectacular clinopyroxene reaction coronas. These have strongly-zoned, radial, clinopyroxene crystals which are aegirine-rich adjacent to the basalt and salitic adjacent to the quartz-rich xenolith. Incorporation and reaction of quartz-rich fragments in some flows from Kandos has resulted in the formation of pegmatitic schlieren and is reflected in the isotopic data (O'Reilly & Griffin, 1984).

The presence of high-level crustal fragments can be important in interpreting the stratigraphy of the upper crust buried beneath sedimentary basins (Emerson & Wass, 1980), as discussed in Section 6.3. Other scattered localities include examples of sampling from low-pressure intrusive complexes — for example, the pegmatoidal rock fragments at Bullenmerri (Victoria; Birch, 1984) and alkaline dolerite and gabbro in the Dubbo province (Ortez, 1976).

6.1.2 Occurrence

High-pressure xenoliths are found in all forms of volcanic deposits and high-level intrusions. Volcanic occurrences include isolated flows and flow sequences (for example, Southern Highlands, New South Wales; Wass, 1980) and cinder cones (for example, western Victoria; Irving, 1974a). They are found also in pyroclastic rocks, dykes, sills, and rarely in small differentiated intrusions. Xenoliths are found in the chilled margin of the Prospect intrusion, New South Wales, and in a 22 m-thick intrusion in the Scone area (Martin, 1984) in which reacted and partially dis-aggregated mantle-derived xenoliths are scattered throughout some differentiated layers.

Most xenolith occurrences represent discrete eruptive events of relatively small magma volumes, even in the case of some maars which represent large-scale explosive activity forming craters of the order of 1 km across (for example, Bullenmerri, Victoria; Griffin et al., 1984). Rare xenolith-bearing flows also have been produced in the extensive, differentiated central volcanoes.

Xenolith localities for the Newer Volcanics in western Victoria and South Australia, for many north-Queensland provinces (for example, Atherton, McBride, and Chudleigh), and for many east-central Queensland provinces (Griffin et al., 1987), are diatremes, maars, and cinder cones. Xenoliths are much easier to recognise and collect in the unconsolidated volcaniclastic debris of these localities than they are from the basaltic flows that dominate New South Wales localities. Xenoliths in flows normally can be seen only in quarries or by breaking large amounts of the tough, fine-grained, host rock. The apparently greater abundance of xenoliths at Victorian and Queensland localities there-fore may be a result of this difference in mode of occurrence.

Ease of collection may not be the only factor in giving a sampling bias. Human sampling may be selective, either purposely or inadvertently, as discussed by Griffin & O'Reilly (1987a) who gave, as one example, the Bullenmerri-Gnotuk locality in Victoria. This locality was particularly rich in garnet pyroxenites (Griffin et al., 1984) before being plundered in early 1985, but earlier investigators (Wass & Irving, 1976; Ellis, 1976) did not record any garnet-bearing assemblages. Jackson (1968) and Bloomer & Nixon (1973) both used a sound statistical approach in analysing xenolith assemblages to overcome sampling bias. Bloomer & Nixon (1973) clearly demonstrated that felsic granulites at Lesotho (southern Africa) were grossly under-represented in most collections because of their extensive alteration.

The volcanic sampling problem is well-known. Magmas incorporate high-pressure rocks erratically and at different levels by processes not fully understood, and adjacent localities commonly have different xenolith populations. One of many examples is in the Chudleigh province in northern Queensland where Airstrip Crater contains abundant large granulite xenoliths but no recorded garnet pyroxenite, whereas adjacent Batchelors Crater contains garnet pyroxenite and relatively rare, small granulite samples. A more striking example is in western Victoria where the Bullenmerri-Gnotuk locality (Griffin et al., 1984) has hydrous and anhydrous Cr-diopside suite lherzolites, amphibole and volatile-rich wehrlite-series xenoliths, and abundant garnet-bearing metapyroxenite-series xenoliths. In contrast, the Mount Leura locality only 1 km away is dominated by anhydrous lherzolite and there are no recorded metapyroxenite or amphibole-rich wehrlite-series xenoliths.

6.1.3 Nomenclature

The following nomenclature for mantle-derived xenoliths is based on a general consensus reached at a 1984 Penrose Conference (Nielson-Pike et al., 1985) and discussed in detail by O'Reilly & Griffin (1987).

(1) The *Cr-diopside suite* was defined originally by Wilshire & Shervais (1975) and subsequently designated 'Group I' type xenoliths by Frey & Prinz (1978). Distinguishing features are the presence of grass-green (in hand specimen) Cr-diopside and an olivine of composition Fo_{92-88}. Cr-rich pleonaste, where present, is also distinctive. This suite can be subdivided into spinel- and garnet±spinel-bearing assemblages that have similar compositional ranges, but different conditions of equilibration (see, for example, O'Neill, 1981). Rock types range from lherzolite to harzburgite, pyroxenite, and rare dunite. All Cr-diopside suite xenoliths have a wide range of metamorphic microstructures, from equigranular mosaic to highly sheared types (Section 6.2.2). Note that the nomenclature of Irving (1974a) is followed here. A further subdivision can be made on the basis of the presence or absence of volatile-bearing minerals.

(2) The *Fe-rich Cr-diopside suite* includes most of the rock types of the Cr-diopside suite defined above. However, differences are that the Cr-diopside is a darker, 'bottle-green' colour (Wilshire et al., 1985), that the suite is intermediate in composition between that of the Cr-diopside and Al-augite (see below) suites, and that the olivine — reflecting the bulk-rock composition — is more fayalitic (Mg-ratio less than Fo_{88}. This suite may represent Cr-diopside wall-rock that has undergone major-element metasomatism (dominantly a decrease in Mg-ratio) adjacent to veins of basalt (Irving, 1980; O'Reilly & Griffin, 1988; Griffin et al., 1988). Microstructures are metamorphic.

(3) An *Al-augite suite* also was defined originally by Wilshire & Shervais (1975), but then later designated 'Group II' by Frey & Prinz (1978). The suite is characterised by an Al-rich clinopyroxene that is black in hand-specimen and generally more Ti- and Fe-rich than those of the Cr-diopside suite. Spinel where present ranges from pale green to opaque. Al-augite suite xenoliths have a wide range in modal assemblages, microstructures, and geochemical characteristics. They are interpreted as having originated as precipitates from basaltic melts within the mantle (see, for example, Irving, 1980), and they include pyroxenites, wehrlites, harzburgites, and lherzolites that represent igneous cumulates, frozen liquids, and their recrystallised equivalents. The latter are metapyroxenites whose assemblages may include garnet, spinel, plagioclase, amphibole, biotite, ilmenite, rutile, and scapolite, as well as eclogites and granulites. Their origin probably involves several mechanisms of formation, ranging from precipitation of liquidus phases on conduit walls, differentiation of igneous bodies at high pressure, and

freezing of small basaltic veins. Such processes may have taken place repeatedly during magmatic episodes of different ages throughout the evolution of a given lithospheric volume (Wass & Hollis, 1983; Griffin et al., 1984; Griffin et al., 1987). Repetition of these processes accounts for the contact metamorphic effects observed in some xenoliths, as well as composite xenoliths in which there are contact relationships between different Al-augite suite rock types that have igneous and metamorphic microstructures, respectively.

Three distinct series are recognised within the Al-augite suite (O'Reilly & Griffin, 1987): (1) the *wehrlite* series, characterised by igneous microstructures and named because of the dominance of clinopyroxene+ olivine assemblages; (2) the *metapyroxenite* series, representing microstructural and mineralogical re-equilibration of wehrlite-series xenoliths; and (3) the *apatite/amphibole* series (described in detail by O'Reilly, 1987), representing precipitates from volatile-charged magma highly enriched in incompatible elements.

The nomenclature for lower-crustal xenoliths is more problematical and the following principles are based on pragmatism. The term 'granulite' has been interpreted as implying a lower-crustal origin, but this is not necessarily true. Granulites — that is, rocks containing modal plagioclase and equilibrated to granulite-facies conditions — have protoliths ranging from rhyolite through all types of basaltic compositions, and are stable up to pressures well within the upper mantle (Green & Ringwood, 1972; Griffin & O'Reilly, 1987b). Granulites have contact relationships with Cr-diopside suite rocks (see, for example, Wass & Hollis, 1983) and some have high equilibration pressures (O'Reilly & Griffin, 1985; Griffin & O'Reilly, 1986a; O'Reilly et al., 1988), so their origin is unequivocally in the upper mantle although the majority of granulites probably do come from the lower crust. Eclogites contain Na- and Al-rich clinopyroxene and have assemblages of garnet ± clinopyroxene ± orthopyroxene ± quartz ± kyanite. 'Mafic' refers to compositions derived from basaltic (in the broad sense) magmas and includes plagioclase-rich types such as anorthosite. 'Felsic' refers to quartzo-feldspathic rocks.

6.1.4 Host magma types

The xenolith- and megacryst-bearing host magmas are most commonly basanite and *ne*-hawaiite (Wass & Irving, 1976; Wass, 1980), but also include nephelinite, hawaiite, more rarely alkali basalt, mugearite, and trachyte, and one *ol*-tholeiitic basalt (Sutherland, 1974).

The host rocks forming pyroclastic breccias in diatremes such as at Jugiong, Delegate, and Bullenmerri are not 'kimberlitic' (see, for example: Stracke et al., 1979; Ferguson et al., 1979; Jaques et al., 1986), but most commonly are basanite or *ne*-hawaiite. South Australian kimberlites are micaceous and highly altered (Jaques et al., 1986). A breccia facies characterises the South Australian pipes, and a massive facies is restricted to dykes. A range of magma types is host to heavy minerals, megacrysts, and rare xenoliths in Western Australia. Kimberlite is found in the north and east Kimberley provinces, in a crystal-lithic, olivine-lamproite tuff at Argyle, in olivine lamproite at west Kimberley, and in 'kimberlitic' sills and tuffs in the Carnarvon Basin (Jaques et al., 1986).

6.1.5   Xenolith entrainment and transport

Introduction

Magmatic sampling of high-pressure rock types is unpredictably selective, as discussed above, and adjacent volcanoes, or even successive eruptions from one vent, may yield different xenolith assemblages (see, for example, Padovani & Carter, 1977). The mechanisms that result in this selectivity are not understood and significant factors may include the rheological characteristics and volatile contents both of the host magma and the country rock. Knowledge of these factors and their relative importance will reveal ultimately to what extent xenolith-based observations distort concepts of the constitution of the lower crust and upper mantle.

Brittle fracture in the mantle

An important observation on over 4000 high-pressure xenoliths in flows, dykes, sills, and cinder cones from Australia and Spitsbergen (O'Reilly, unpublished data; Skjelkvale et al., in press) is that they are typically angular to subangular. Rounding is restricted to xenoliths from breccias or to bombs associated with explosive eruptions such as at maars, diatremes, and some cinder cones. This rounding appears to have taken place during tumbling of xenoliths during turbulent ascent and eruption. The xenoliths at the time of entrainment therefore can be inferred to have been angular, and brittle fracture must have taken place. Furthermore, most spinel lherzolites subjected to mild hydraulic stress split along planar surfaces in up to three different orientations, so pre-existing planes of weakness presumably were present in the original upper-mantle wall rock.

Many metasomatised xenoliths contain veins (less than 0.1 mm to more than 10 cm wide) that have planar contacts, although many planar surfaces have no trace of fluid infiltration. Cross-cutting planar veins oriented in up to three directions are reported also from other localities world-wide (see, for example: Wilshire et al., 1980; Skjelkvale et al., in press). This reinforces the conclusion that planes of weakness exist in the upper mantle before entrainment. Basu (1980) described spinel lherzolites from San Quintin, Baja California, which have similar jointing. He concluded that these planar features formed because of high differential stress in the mantle, and that they predispose wall-rock thus affected to be entrained in ascending magma. One cause of such brittle failure could be intermediate-focus earthquakes such as those in Hawaii, eastern Australia, and China (MacGregor & Basu, 1974; Denham, 1985; Kirby et al., 1987). This possibility is consistent with shock features detected in olivine in mantle-derived spinel lherzolites from China (Kirby et al., 1987).

Volatile content of xenoliths

Evidence for volatiles and fluids within the mantle is provided by both the presence of volatile-bearing minerals and directly by fluid inclusions within mineral grains. This is discussed in detail in Section 6.2.6. Detailed studies on volatiles and fluids in mantle-derived xenoliths and megacrysts have been carried out on samples from eastern Australia and, particularly, from western Victoria by Andersen et al. (1984, 1987), Porcelli et al. (1986), O'Reilly (1987), and O'Reilly & Griffin (in press). Xenoliths from western Victoria contain up to 3 volume-percent cavities and fluid inclusions, all inferred to have been filled mainly with CO_2 fluid at high pressure. Andersen et al. (1984) surveyed samples from many xenolith-bearing localities world-wide and showed that xenoliths containing significant volumes of fluid cavities and fluid inclusions are confined to explosive-eruption vents such as Bullenmerri-Gnotuk (Victoria; Andersen et al., 1984), Anakie (Victoria; Wass & Hollis, 1983), Spitsbergen (Amundsen, 1986; Skjelkvale et al., in press), Craters 160 and 387, Arizona (Andersen et al., 1987); and Salt Lake Crater, Hawaii (O'Reilly, unpublished data). E. Roedder (personal communication, 1983) pointed out that such volumes of fluids represent high fluid pressure within the xenoliths relative to the magma, and that the direction of any volatile flow would be from the xenolith — and perhaps wall-rock — outwards into adjacent magma. The presence of large volumes of fluid in some deep-seated wall-rocks therefore may favour brecciation and entrainment of that horizon in the ascending magma caused by rapid outgassing.

Ascent velocity and transport

The concept of transport of dense mantle-derived xenoliths to the Earth's surface in less-dense basaltic (and kimberlitic) magmas provides some constraints on ascent velocities, including those at mantle depths. These constraints are based on physical, microstructural, and geochemical parameters.

Physical parameters depend on the rheological properties of the magma and the size and density of the xenoliths. Kushiro et al. (1976) carried out laboratory experiments on static settling rates in different magma compositions. These resulted in a minimum estimate of ascent times of 60 hours for xenoliths in basalts as the experiments simulated laminar-flow conditions. Spera (1980, 1984) used Stoke's law, rheological parameters from numerous sources, and the assumption of a Bingham behaviour for the magma, and showed, for example, that a minimum ascent rate of about 10 cm/s would allow transport of a 20 cm-diameter spherical xenolith of density 3.3 g/cc in a basaltic magma containing 15-percent phenocrysts.

Microstructural methods of assessing ascent rate include the application of experimental coarsening kinetics data to the size of olivine neoblasts inferred to have annealed during ascent in kimberlite (Mercier, 1979). These result in residence times in kimberlite hosts of 4–6 hours (equivalent to velocities of

10–20 m/s). Experimental determination of high-temperature annealing rates of fractures in olivine (Wanamaker et al., 1982) has been applied to secondary planes of CO_2 fluid inclusions that are assumed to have formed and healed during ascent. Magma residence times of 80 to 170 hours (ascent rates of about 0.5–1 m/s) are inferred.

There are numerous chemical criteria that can be used to estimate ascent rate — some only qualitative at this stage. Griffin et al. (1984) suggested that coarse, pyroxene-spinel coronas on garnet in xenoliths from western Victoria were formed because of relatively slow ascent rates (and decrease in pressure) during the initial stages of entrainment at high pressure, followed by a stage of rapid ascent. Wass & Pooley (1982) described altered lherzolite xenoliths up to 60 cm across from Wallabadah (New South Wales) which are inferred to have undergone concentric zonal alteration by carbonate-rich fluids during a similarly extended residence time in magma at mantle depths.

More quantitative methods rely on diffusion rates for given element/mineral systems. Hofmann & Magaritz (1977) obtained Ca-diffusion data and concluded that xenolith-melt patches retaining CaO heterogeneity required residence times of less than about three days (a minimum ascent rate from 60 km of 20 cm/s).

Ozawa (1983) developed a 'geospeedometric' method based on $Mg/(Mg+Fe^{2+})$ zoning in spinel in Cr-diopside lherzolite xenoliths to determine duration of heating. Selverstone (1982) previously determined that xenoliths of 10 cm-diameter acquire the temperature of the host magma in a minimum of three hours. These data allow total magma-residence time to be calculated. Bezant (1985) applied this geospeedometric technique to a range of lherzolites from eastern Australia, and found calculated heating durations ranging from 4 to 16 hours and 12 to 48 hours in different xenoliths. All calculated ascent rates are remarkably consistent considering the different precisions of the parameters used and the range of approaches.

In summary, explosive, xenolith-carrying eruptions resulting in pyroclastic deposits and commonly associated with diatreme formation, may rise at tens of metres per second whereas other xenolith-bearing alkaline magmas may rise at centimetres to metres per second. These rapid ascent rates reinforce the validity of recognising xenoliths as essentially unaltered fragments of deep-seated lithologies.

Volatile content of host magmas

High volatile contents in solution in magmas may act as propellants enhancing velocity of ascent. Spera (1984) carried out detailed fluid-dynamic modelling to test this hypothesis, and concluded that the propellant effect leads only to a small increase in magma-ascent velocity. Exsolution of volatiles also decreases magma viscosity and therefore works against suspension of xenoliths. However, magma volatiles are important in controlling the dynamics at the initiation of ascent and in determining the local magma pressure. Spera (1984) suggested that where the overpressure of a volatile-rich,

segregated magma exceeds a critical value, crack propagation ensues in overlying wall-rock, and magma ascends at a velocity of centimetres to metres per second. Note that these estimated velocities describe the ascent only up to below near-surface levels. Volatile saturation and a great increase in the molar volume of the volatile-phase at near-surface levels, will result in supersonic velocities followed by dramatic changes in rheology of the magma/volatile mix.

An additional significant effect of volatile content is that exsolution of CO_2 will result in heating of the magma, rather than the cooling that accompanies exsolution of H_2O. This is suggested by the pressure, volume, and temperature properties of CO_2 (Spera & Bergman, 1980), and it means that ascending CO_2-rich magma undergoing decompression does not freeze (Spera, 1984). These conclusions are a strong argument against the concept that kimberlitic and basaltic melts may rise slowly as kilometre-sized diapirs. Rather, once a small percentage of partial melt creates a sufficiently high magma pressure, it surges up at instantaneously high velocities. These results also bear on the observation that tholeiitic-basalt magmas are generally devoid of high-pressure xenoliths. Inferred lower volatile contents and greater percentages of partial melting for tholeiitic magmas may mean that the surrounding hot, weak, mantle wall-rock will not fail by crack propagation and that ascent initiation is probably by slow percolation. Final ascent velocities also will be slower and will allow settling of any entrained xenoliths.

Mantle-derived volatiles (especially CO_2) are considered to be the principal cause of formation of many maars, diatremes, and tuff-rings (Barnes & McCoy, 1979; Emerson & Wass, 1980). The rapid, near-surface decompression that drives the magma-volatile suspension to supersonic speeds will increase the conduit cross-section creating a typically funnel-shaped diatreme (Lorenz, 1975). Evidence for the important role of mantle-derived CO_2 includes: common cementation of breccia fragments by carbonate (see, for example, Emerson & Wass, 1980); common vesicle linings of Fe-Mg-rich carbonate; carbonation of some high-pressure xenoliths (Wass & Pooley, 1982); observations from recent Alaskan maars (Barnes & McCoy, 1979) that primary, outgassing CO_2 is a free fluid phase in the underlying mantle; and C-isotopic compositions that generally are consistent with a primordial origin.

There is also good evidence that a phreatic component has been critical in causing some of these explosive eruptions (see Section 2.5). Lorenz (1973) described groups of penecontemporaneous volcanic centres in the Eifel region formed by two modes of eruption. Volcanic centres at high levels are cinder or spatter cones, whereas those in valleys, on alluvial gravels, or in coastal regions, form diatremes, maars, tuff cones, or tuff rings. Further evidence for a phreatic volatile source is provided by associated vesiculated tuffs, accretionary lapilli, and well-bedded ash deposits (Lorenz, 1974). The criteria discussed above can be used to help discriminate between explosive eruptions caused by mantle-derived volatiles (especially CO_2) and by high-level phreatic phenomena.

6.2 Xenoliths and Megacrysts of Mantle Origin

6.2.1 Macroscopic features

Discrete xenoliths

Xenoliths of mantle origin have a wide range of size, shape, degree of disaggregation, and extent of interaction with enclosing magma. The typical size range at most localities is 5–10 cm, rare examples reach 60 cm (Wass & Pooley, 1982), whereas at some localities — for example Mount Elephant, Victoria — the xenoliths almost exclusively are small (less than 5 cm). Shapes range from sub-spherical or ovoid to more common angular forms that have rectangular to triangular cross-sections. Some xenoliths have evidence of partial fragmentation in the form of open fractures and therefore of disaggregation continuing to shallow levels.

There is abundant evidence for interaction between xenoliths and enclosing magmas or fluids (compare with, for example, Kuo & Kirkpatrick 1985). Some xenoliths carrying basaltic rinds or in massive flows, have scalloped surfaces indicative of magma infiltration or reaction. However, this rarely is seen in thin section to be extensive. Partial melting of clinopyroxene and amphibole — most commonly seen as sieved textures — probably reflects heating of the xenolith during transport (Selverstone, 1982). Progressive concentric hydration of some xenoliths and the development of coronas on some minerals (Nixon & Boyd, 1973; Griffin et al., 1984; Wass & Pooley, 1982) may be evidence for reaction with a fluid, either before or during entrainment and ascent in host magma.

Xenoliths of both the Cr-diopside and Al-augite suites in eastern Australia commonly have foliations on a millimetre to centimetre scale, some accompanied by mineralogical layering. Planar foliation in Cr-diopside suite xenoliths ranges from minor flattening of olivine granoblastic grains to mylonitisation with grain-sizes of 0.1 mm (Fig. 6.2.1A). Foliations are defined largely by tabular olivine crystals of 1–2 mm grain-size (Fig. 6.2.1B) and, more rarely, by discontinuous planar 'trains' of pyroxene or spinel grains or both (Fig. 6.2.2). More extensive mineralogical segregation may give rise to discontinuous pyroxenite layers in Cr-diopside suite lherzolite xenoliths (for example, at Lake Bullenmerri; Hollis, 1981; Griffin et al., 1984) and to rare lensoid concentrations of coarse (1–2 cm) olivine crystals (for example, at Mount Leura, western Victoria).

Pyroxene-rich xenoliths of the Al-augite suite commonly have mineralogical layering which, despite partial to complete recrystallisation, appears to reflect original igneous layering. The most common layered types amongst xenoliths that retain clear cumulate microstructures (the wehrlite series of Griffin et al., 1984, and O'Reilly & Griffin, 1987), are olivine-clinopyroxene-spinel orthocumulates. The metamorphosed analogues of the wehrlite series (the metapyroxenite series) may develop discontinuous concentrations of spinel or garnet, resulting from lamellar exsolution from original high-temperature pyroxenes (Griffin et al., 1984).

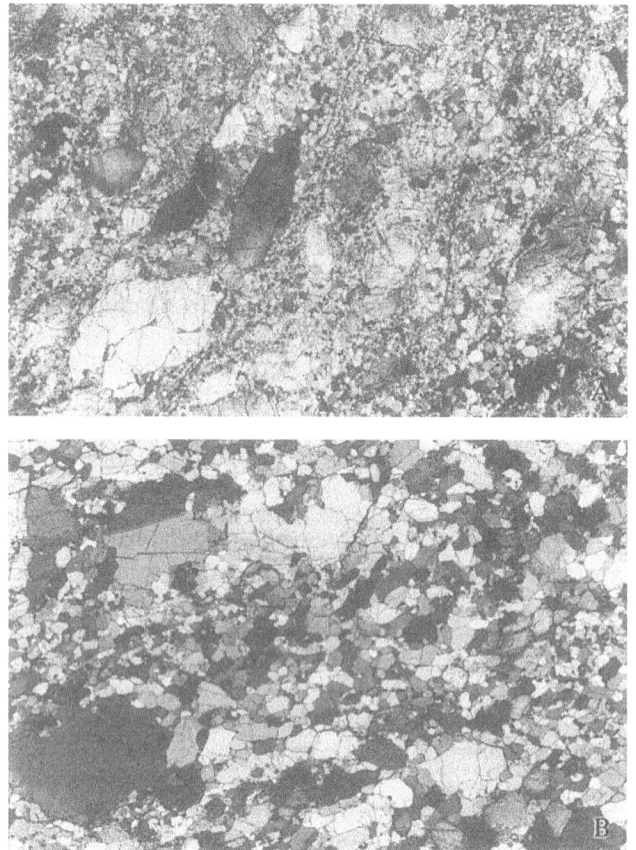

Figure 6.2.1 (A) Mylonitised Cr-diopside lherzolite xenolith from Lake Bullenmerri (Victoria). Note fine-grained sheared zones, relict strained porphyroclasts, and recrystallised mosaic domains. (B) Foliation in Cr-diopside lherzolite xenolith from Lake Bullenmerri (Victoria), defined by tabular olivine grains. Width photographed (with crossed polars) is 1.8 cm.

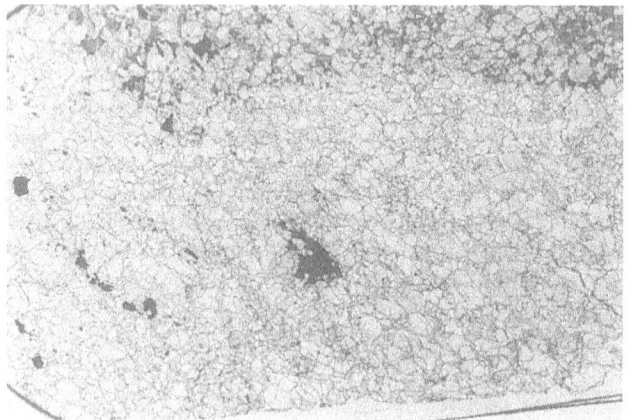

Figure 6.2.2 Foliation in Cr-diopside lherzolite xenolith from Lake Bullenmerri (Victoria) defined by preferred orientation of olivine and discontinuous planes of spinel grains. An Al-augite suite vein (amphibole + mica + clinopyroxene) cuts this foliation (parallel and near to upper edge). Width photographed (in plane polarised light) is 2.5 cm.

Both wehrlite- and metapyroxenite-series xenoliths have rare open cavities up to 1–2 cm in diameter, lined with pyroxene, amphibole, garnet, or olivine euhedra that appear to represent original fluid-filled volumes reflecting relatively large volumes of migrating fluid

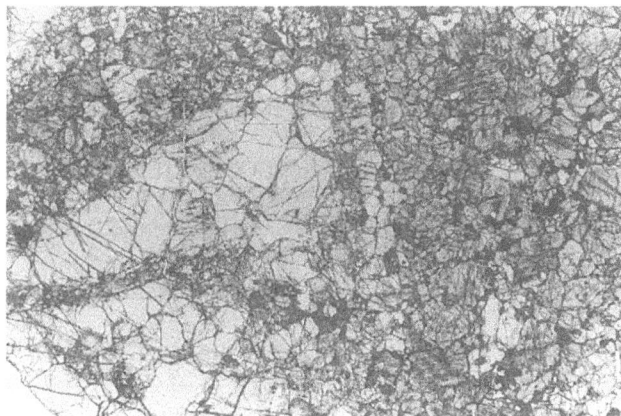

Figure 6.2.3 Cr-diopside lherzolite fragments are seen as coarse-grained areas on the lower left-hand side. This mantle wall-rock has been veined by Al-augite suite material (clinopyroxene + amphibole + Cr-poor spinel) seen here as the finer-grained area. This composite xenolith is from Lake Bullenmerri (Victoria). Width photographed (in plane polarised light) is 1.8 cm.

(Griffin et al., 1984; Wass & Hollis, 1983; Andersen et al., 1984). These may constitute up to 3 volume percent of the rock.

Composite xenoliths and veined xenoliths

Detailed descriptions of composite xenoliths that have contacts between different high-pressure rock types, have been published for examples throughout the east-Australian basaltic provinces (Irving, 1980; Hollis, 1981, 1985; Sutherland & Hollis, 1982; Wass & Hollis, 1983; Griffin et al., 1984). Relationships recognised include: (1) Cr-diopside suite host and parallel-sided veins or dykes of Al-augite suite material less than 1 to 10 cm in width; (2) extensive, irregular veining of Cr-diopside wall-rock by Al-augite material so that the former appears to be angular fragments within an Al-augite host (Fig. 6.2.3); (3) contact relationships (no discrete vein is evident) between Cr-diopside material and wehrlite or metapyroxenite series rocks (Griffin et al., 1984); (4) contacts between wehrlite- and metapyroxenite-series rock types. These veining and contact relationships are similar to those previously recorded for upper-mantle rocks by Wilshire & Trask (1971), Wilshire & Pike (1975), and Wilshire et al. (1980).

In addition, Cr-diopside suite xenoliths commonly have metasomatic veins 1–10 mm wide of amphibole, amphibole + Fe-rich clinopyroxene, Fe-rich clinopyroxene + apatite, or amphibole + phlogopite/biotite + apatite (O'Reilly & Griffin, 1987). Foliation in the Cr-diopside wall-rock material is commonly truncated by dykes, veins, and contact surfaces (Irving, 1980; O'Reilly & Griffin, 1987).

Megacrysts

Megacrysts are defined here as discrete crystals that have a dimension greater than 5 mm. They have been reported from more than 200 localities, mostly in eastern Australia (Wass & Irving 1976; Sutherland & Hollis, 1982; Hollis, 1985). Their distribution in terms

of relative abundances of the different types and their occurrences within individual volcanic provinces, was summarised recently by O'Reilly & Griffin (1987).

Megacryst assemblages are dominated by pyroxenes (Al-augite commonly much more abundant than Al-bronzite), followed by feldspar (mainly anorthoclase and rare plagioclase), spinel (Cr-Al spinel/Ti-magnetite), amphibole (pargasite-kaersutite), and mica (phlogopite, biotite). Rare megacryst types are Mg-ilmenite, apatite, zircon, garnet, and corundum. Megacryst-sized olivine crystals are present at many localities, but these are dominantly magnesian in composition and probably formed by disaggregation of Cr-diopside suite xenoliths. They tend not to be included in listings of megacryst assemblages, but this introduces a genetic connotation to the term 'megacryst'.

O'Reilly & Griffin (1987) confirmed that clino-pyroxene is the most abundant megacryst species, but that orthopyroxene abundance may be underestimated because it has been confused with clinopyroxene. There is little doubt that at some localities ortho-pyroxene megacrysts are at least as abundant as clinopyroxene, but this may be clear only from thin-section studies of basaltic hosts (for example, a basanite at Mount Elephant, western Victoria, contains roughly equal proportions of orthopyroxene and clinopyroxene).

A significant number of clinopyroxene megacrysts contain inclusions of sulphides (pyrrhotite, pentlandite, chalcopyrite, and rare pyrite; Andersen et al., 1987). These sulphide-bearing clinopyroxenes have a restricted compositional range (Fig. 6.2.4) and in general their compositions overlap with those in xenoliths.

The macroscopic characteristics of the major megacryst species of the east-Australian provinces are summarised in Table 6.2.1.

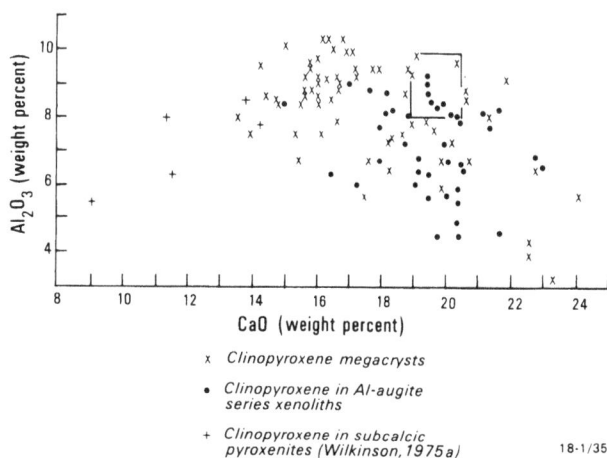

x   Clinopyroxene megacrysts

•   Clinopyroxene in Al-augite series xenoliths

+   Clinopyroxene in subcalcic pyroxenites (Wilkinson, 1975a)

18-1/35

Figure 6.2.4 Compositions of clinopyroxene megacrysts and clinopyroxenes from Al-augite suite xenoliths from eastern Australia. Enclosed area represents compositions of sulphide-bearing clinopyroxenes (Andersen et al., 1987).

6.2.2   Textures and microstructures

Introduction

The diverse microstructural characteristics of mantle-derived xenoliths in basaltic rocks, and their relationships to tectonic processes and magma genesis in the

Table 6.2.1 Macroscopic properties of megacryst species

	Size	*Shape*	*Colour, lustre, other distinctive features*
Clinopyroxene	1–5 cm typical	Subhedral prismatic	Black to light-brown reaction rim. Vitreous. Generally no cleavage, Conchoidal fracture.
Orthopyroxene	1–3 cm typical	Subhedral prismatic	Black to deep-brown green. Vitreous. Poor cleavage. Conchoidal fracture. Corroded.
Amphibole (kaersutite, pargasite)	1–5 cm typical. 7 cm maximum	Prismatic to subequant. Rounded.	Black-deep brown. Highly vitreous. Good prismatic cleavage.
Mica (Ti-phlogopite to Ti-biotite)	0.5–2 cm typical. 5 cm maximum rare.	Corroded plates	Bronze (phlogopite) to deep brown (biotite)
Olivine	1–3 cm	Subspherical to angular fragments	Grass-green to yellow-green
Anorthooclase/ plagioclase	2–5 cm typical. 8 cm maximum.	Subhedral-euhedral. Tabular to prismatic.	Water clear to white. Vitreous. Good cleavage.
Al-Cr spinel	To 2 cm	Corroded and rimmed. Octahedral.	Black-brown. Vitreous. Conchoidal fracture.
Ti-magnetite	To 1 cm	Equant	Black. Sub-metallic. Conchoidal fracture. Rounded.
Ilmenite	To 2 cm	Equant. Rhombohedra.	Black. Vitreous. Sub-metallic. Conchoidal corroded fracture.
Apatite	To 1 cm	Stumpy prismatic. Rounded.	Clear to turbid
Zircon	To 2 cm	Rounded and euhedral	Deep ruby red to white or clear
Garnet	To 1 cm	Octahedral/ dodocahedral. Rounded.	Deep red, pink, brown or orange
Corundum	To 5 cm	Rounded and euhedral	White-brown, yellow, green, black, or sapphire blue. May be strongly zoned.

mantle, have raised major interest in recent years (for example, Nicolas, 1986). Much of the early literature dealing with dominantly metamorphic microstructures was summarised by Harte (1977) who proposed a simple classification for olivine-rich xenoliths into four major microstructural types — coarse, porphyroclastic, mosaic-porphyroclastic, and granuloblastic. This classification will be used for xenoliths of the Cr-diopside suite, and for xenoliths of the Al-augite suite having dominantly metamorphic microstructures, but substituting the term 'granoblastic' for Harte's 'granuloblastic' (as suggested by R.H. Vernon, personal communication, 1987). Al-augite suite xenoliths having original or relict igneous microstructures (and similar vein phases in composite xenoliths) are interpreted in terms of classifications for igneous cumulates (for example, Cox et al., 1979).

Microstructural interpretation of xenoliths is fundamental to a full understanding of their petrogenetic significance. In particular, apparent but poorly understood relationships between deformation, recrystallisation, and the generation and migration of fluids and magmas within mantle rocks, are indications that trace-element and isotopic data for xenoliths should be interpreted against a background of information about microstructures and their origin.

Olivine-rich xenoliths of the Cr-diopside suite

Most Cr-diopside suite xenoliths from east-Australian localities have coarse equant to weakly porphyroclastic microstructures, and grain-sizes in the 2–10 mm range.

Coarse xenoliths constitute the entire assemblage at some localities, whereas others (for example, Mount Leura, Mount Shadwell, Lake Bullenmerri) have xenoliths covering a wide range of microstructures. These range from coarse equant and coarse tabular types, through porphyroclastic and mosaic-porphyroclastic types that have development (in different degrees) of planar fabrics and lenticular concentrations of Cr-spinel with or without Cr-diopside, to granoblastic types having 0.2–2.0 mm grain-size and pronounced planar fabrics. Features of the different microstructural types, and relationships between them, are summarised below, using mainly lherzolite examples from Victorian xenolith localities. Microstructures in most cases are dominated by olivine and orthopyroxene (as in the rarer harzburgites and dunites), but Cr-diopside may be sufficiently abundant in lherzolites (especially at the Anakies) to play a role similar to orthopyroxene. Microstructures of pyroxene-rich Cr-diopside suite xenoliths are described separately.

Harte (1977) recognised two types of *coarse* microstructures in olivine-rich xenoliths — *coarse equant* and *coarse tabular* — distinguished by the dominant olivine habit. Most east-Australian xenoliths having coarse microstructures contain mainly equant, granular to polygonal olivine. Some have significant tabular olivine, but well-developed, coarse, tabular microstructures are relatively rare. Even the best tabular examples have little or no mineral layering (the characteristic feature of 'laminated' microstructures, as defined by Harte). Nor do they have the Cr-diopside/ Cr-spinel lenticles or spinel 'trains' characteristic of 'disrupted' microstructures. However, these are common in porphyroclastic or especially mosaic-porphyroclastic xenoliths from some localities (for example, the Anakies).

The typical grain-size range for olivine (50–65 volume-percent) and orthopyroxene (20–35 percent) in coarse lherzolites is 2–5 mm. Olivines in tabular types have aspect ratios of about 3:1. Rare harzburgite-dunite xenoliths containing olivines up to 2 cm are found at some localities. Neoblast aggregates of equant polygonal olivines having grain-sizes of less than 1 mm are common in coarse xenoliths transitional to porphyroclastic types. Cr-diopsides are similar in size to orthopyroxenes where near their maximum abundance in coarse lherzolites (about 25 percent). Sub-prismatic 3×1 mm grains are typical, but rare grains reach 6×1.5 mm. Cr-diopside where less abundant commonly forms interstitial anhedral grains of less than 1 mm. Cr-spinel (1–5 percent) is also mainly an interstitial phase as lobate to subhedral 0.1–1.0 mm grains commonly closely associated with Cr-diopside. Sub-spherical embayed grains may reach 8 mm where spinel is unusually abundant.

The microstructures of coarse xenoliths, irrespective of grain-size, range from those dominated by irregular olivine grains, through grains having more regular but strongly curved boundaries, to polygonal grains characterised by straight to gently curved boundaries. Some localities have distinctive grain types within this spectrum, which may reflect differences in one or more of protolith microstructures (including whether these were of cumulate or metamorphic origin), different pressure-temperature conditions and degrees of recrystallisation, and different strain histories. Mercier & Nicolas (1975) and Harte (1977) noted that some xenoliths may represent mantle material that has undergone more than one cycle of recrystallisation, through coarse, porphyroclastic and granoblastic microstructures. The microstructural simplicity of most east-Australian coarse xenoliths is evidence that either relatively simple deformation histories or extensive annealing have been involved, probably at low strain rates and high temperatures.

Most coarse xenoliths have deformation and ex-solution features affecting olivine (broad kink bands) and pyroxene (common Ca-clinopyroxene exsolution lamellae). Most olivines other than small polygonal neoblasts have broad (0.5–1.0 mm), straight-sided (100) or (001) kink bands. Orthopyroxenes commonly have 5–10 micron-wide (100) Ca-clinopyroxene exsolution lamallae. Orthopyroxenes in the least-deformed coarse xenoliths rarely have evidence of strain, but as olivine neoblasts appear in the early stages of development of porphyroclastic microstructures, orthopyroxenes develop rare (001) kink bands which bend exsolution lamellae. Kink-bank spacing in both olivine and orthopyroxene decreases and undulose extinction within bands becomes more pronounced, with an increase in porphyroclastic character. In contrast, Cr-diopsides have appreciable strain features or exsolution lamellae (of orthopyroxene or Cr-spinel or both) only where they are unusually abundant and coarse. Grains affected by melting have dendritic networks of pale-brown glass blebs.

Examples of olivine-rich xenoliths with coarse equant and coarse tabular microstructures are given in Figures 6.2.5–6.

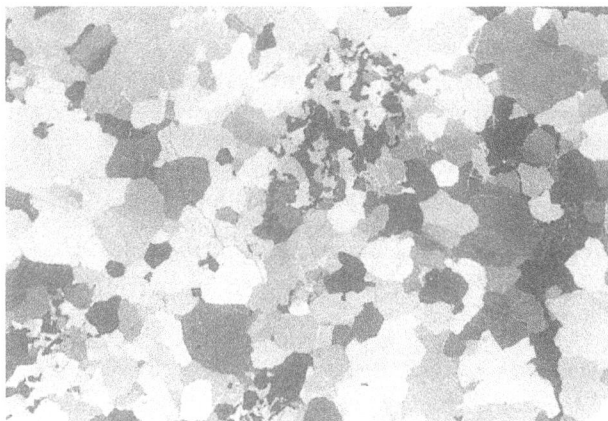

Figure 6.2.5 Coarse equant microstructure in a spinel dunite from Mount Shadwell (62817). Equant olivines in the 2–7 mm grain-size range have polygonal to less regular shapes, and common kink bands. A rounded spinel-pyroxene cluster is visible at top centre. Field of view (with crossed polars) is 35×22 mm. *All rocks illustrated in Figures 6.2.5–12 are from the Newer Volcanics province of western Victoria, and the five-digit number in parenthesis represents the Catalogue Number of the Department of Earth Sciences, Monash University.*

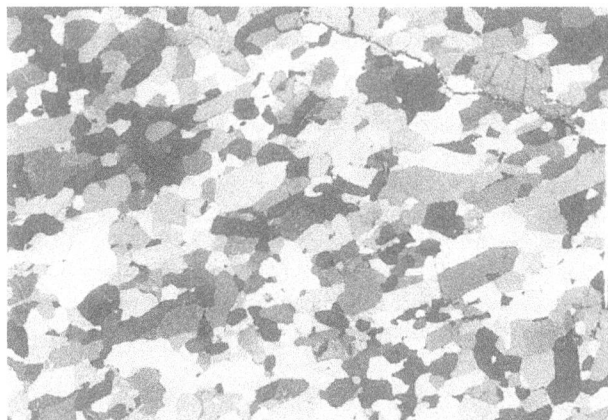

Figure 6.2.6 Coarse tabular microstructure in a spinel harzburgite from Mount Porndon (62818). Tabular olivines up to 7×1 mm defining the planar foliation have mainly straight grain boundaries and rare kink bands. Field of view (with crossed polars) is 35×22 mm.

Harte (1977) classified *porphyroclastic* microstructures in olivine-rich xenoliths as those in which at least 10 percent of olivine is present as coarse, commonly strained grains (porphyroclasts), set in a matrix of small, commonly strain-free recrystallised grains (neoblasts). Most east-Australian lherzolite examples also have porphyroclasts of orthopyroxene and less commonly Cr-diopside or pargasitic amphibole. About an order of magnitude range of grain-sizes of both porphyroclasts (2 to greater than 20 mm) and polygonal matrix grains (0.2–2 mm) is represented in strongly porphyroclastic xenoliths. East Australian examples rarely have sufficient tabular neoblastic olivine to produce pronounced planar fabrics.

The deformation, exsolution, and recrystallisation processes that lead ultimately to ultramafic 'mylonites' having fine-grained granoblastic microstructures, are illustrated clearly in porphyroclastic lherzolites that preserve most evidence of strain. Features include olivine porphyroclasts that have narrow wedge-shaped kink banks and marginal zones of subgrains, transitional to polygonal neoblasts that define the original grain outline. Orthopyroxene porphyroclasts and larger Cr-diopsides have broad kink bands, exsolution lamellae of the complementary pyroxene with or with Cr-spinel, patchy amphibole along cleavages, and minor neoblast development. Some pyroxenes are frayed and split along prismatic cleavages that may be filled with tiny olivine neoblasts.

An example of porphyroclastic microstructure is given in *Figure 6.2.7*.

Mosaic-porphyroclastic microstructures in olivine-rich xenoliths are defined as those that have porphyroclasts making up less than 10 percent of the total olivine. However, xenoliths of this group may contain higher proportions of other porphyroclast species, mainly orthopyroxene. Examples from Victorian localities such as the Anakies commonly have average grain-sizes (1–2 mm) little coarser than those of porphyroclast-free granoblastic xenoliths. Rather, they have like the latter, little strongly tabular olivine, although the arrangement of equant to weakly tabular olivine grains in 'brickwork' fashion may produce weak planar fabrics. Most lack mineral segregations, also like the granoblastic types, and Harte's terms 'laminated' and 'fluidal' denoting the presence of sub-parallel, semi-continuous layers or lenticles enriched in particular minerals, rarely can be applied. However, examples having linear or planar 'trains' of spinel grains (with or without enclosing Cr-diopside) parallel to planar fabrics defined by olivine, are common (Fig. 6.2.8). These trains are derived presumably from the deformation and recrystallisation of original coarse grains, and are characteristic of 'disrupted' microstructures.

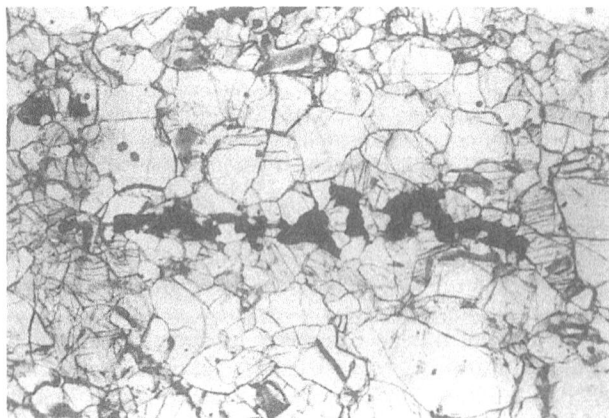

Figure 6.2.8 Mosaic-porphyroclastic disrupted microstructure in a spinel lherzolite from the Anakies (87139). Trains of dark Cr-spinel are enclosed by concentrations of enstatite and Cr-diopside. Field of view (in plane polarised light) is 11 × 7 mm.

The ubiquitous strain and exsolution features in coarse olivines and orthopyroxenes of strongly porphyroclastic xenoliths commonly are less pronounced in mosaic-porphyroclastic types, in which even the largest olivine and orthopyroxene porphyroclasts may have only a few broad kink-bands, and orthopyroxenes may lack exsolution lamellae. This presumably reflects high-temperature annealing. Small Cr-diopside porphyroclasts (which may represent original amphibole) are the grains most affected by partial melting. The dominant polygonal olivine neoblasts of granoblastic areas (the areas of 'mosaic' microstructure referred to in the term 'mosaic-porphyroclastic') are typically strain-free, but they also may have rare broad kink-banks.

Granoblastic microstructures in olivine-rich xenoliths are defined as having less than 5 percent of total olivine as porphyroclasts. Most minerals have narrow grain-size ranges (normally less than 2 mm) and dominantly polygonal shapes. The rare Cr-diopside suite xenoliths of this type at east-Australian localities are commonly just the least porphyroclastic members of a spectrum of mosaic-porphyroclastic to granoblastic types having widely different sizes and shapes of neoblast grains and degrees of mosaic metamorphic character.

All of the olivine, orthopyroxene, and Cr-diopside in the best-developed granoblastic microstructures form mainly polygonal 0.1-0.2 mm neoblast grains characterised by straight boundaries and 'triple-point' grain junctions. Rare olivine, orthopyroxene, and Cr-diopside porphyroclasts larger than 0.5 mm have few strain and exsolution features. Microstructures of this type represent the end-products of two main processes:

(1) Progressive recrystallisation and grain-size reduction affecting the major minerals in the order olivine, orthopyroxene, Cr-diopside, and Cr-spinel.

(2) Progressive exsolution of Ca-rich and Ca-poor pyroxene components and Cr-Al spinel from original, complex, high-temperature pyroxenes, followed by recrystallisation to discrete grains.

An example of granoblastic microstructures is given in Figure 6.2.9.

Pyroxene-rich xenoliths of the Cr-diopside suite

Xenoliths rich in magnesian pyroxenes that undoubtedly belong to the Cr-diopside suite are relatively rare at east-Australian localities. Mineralogical similarities to small-scale pyroxene-rich bands in lherzolite xenoliths in some cases are indicative of an origin as part of metamorphically differentiated mantle wall-rock assemblages. Others clearly represent recrystallised vein cumulate assemblages. Microstructural evidence for the origin of Cr-diopside suite pyroxenites has received little attention, unless they form part of composite xenoliths (Irving, 1980), compared to the more abundant xenoliths rich in more Fe-rich pyroxenes and belonging to the Al-augite suite (particularly the garnet-bearing metapyroxenites).

Wilkinson (1973) and Stolz (1984) described a wide range of clinopyroxenites, websterites, orthopyroxenites, and olivine-poor lherzolites from localities in New South Wales, some of which contain 'megacrystals' (up to 4 cm) of sub-calcic clinopyroxene or orthopyroxene, set in finer subpolygonal orthopyroxene and

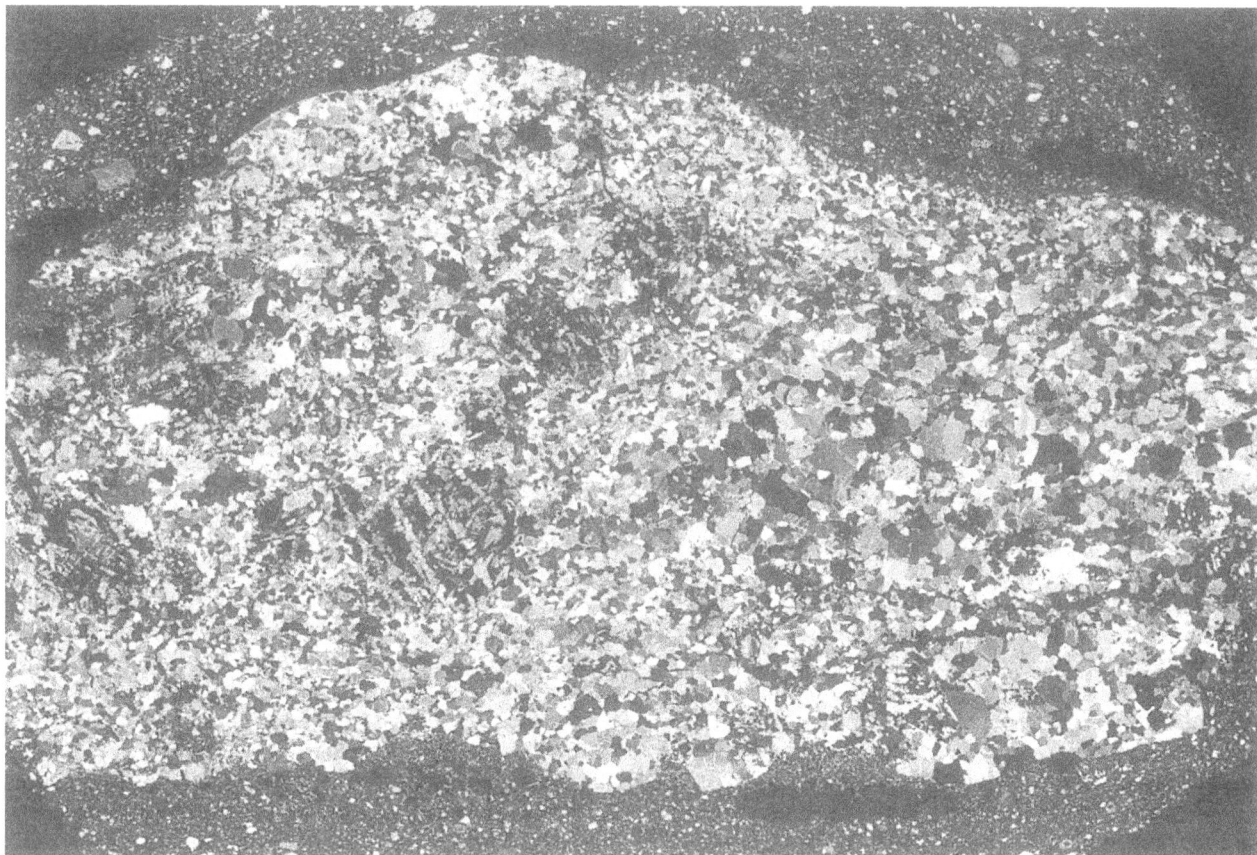

Figure 6.2.9 Granoblastic microstructure in a lherzolite with a basanite rind from Mount Shadwell (62750). Olivines and pyroxenes form a mosaic aggregate of polygonal neoblasts having an average grain-size of 0.2 mm. Coarser-grained areas of olivine correspond to lower Cr-diopside abundances. Just left of centre is a melt pool consisting of glass and sets of acicular skeletal diopsides (at an angle of about 120°) that may represent original amphibole. Field of view (with crossed polars) is 35 × 22 mm.

diopsidic clinopyroxene, and some containing Cr-Al spinel. Both pyroxenes are highly magnesian (Mg_{91-86}), and the diopsides contain significant chromium. The microstructures appear equivalent to those of porphyroclastic olivine-rich xenoliths, in which the matrix assemblages are produced by progressive exsolution and recrystallisation of high-temperature sub-calcic aluminous clinopyroxene similar to that of the porphyroclasts. Wilkinson & Binns (1977) interpreted xenoliths of this type as material derived from layered mantle wall-rocks, on the basis of mineral compositions closer to those of Cr-diopside suite olivine-rich xenoliths than to Al-augite suite pyroxenites. A similar interpretation was made by Griffin et al. (1984) for rare magnesian pyroxenites at Lake Bullenmerri, Victoria. However, Irving (1980) suggested an alternative origin for xenoliths whose bulk-rock compositions are similar to those of sub-calcic augite: as crystal cumulates formed within veins or dykes.

Al-augite suite

Al-augite suite xenoliths have a wider range of microstructures than do those of the Cr-diopside suite, reflecting both their greater mineralogical diversity and the range of magmatic, metamorphic, and metasomatic processes that may contribute to their formation. The wehrlite series is dominated by primary coarse poikilitic and other cumulate microstructures, interpreted to be of magmatic origin. The metapyroxenite series is characterised by partial to complete recrystallisation to metamorphic microstructures equivalent to those described for olivine-rich xenoliths. However, greater abundances of pyroxenes (which resist deformation) allow better preservation of transitional stages of microstructural development than in olivine-rich xenoliths.

Wehrlite-series xenoliths have widely different proportions of the dominant Ca-clinopyroxene and olivine. Spinel, orthopyroxene, and amphibole are common to abundant minerals, and phlogopite, ilmenite, apatite, feldspars, sulphides, and interstitial glass of different compositions, may also be present. The least-modified microstructures are analogous to those of igneous cumulates, containing common large oikocrysts of Ca-clinopyroxene, or more rarely amphibole (Fig. 6.2.10). Mineral layering is also common. The typical crystallisation sequence is olivine, orthopyroxene, clinopyroxene, spinel, amphibole, on the basis of inclusion relationships, and all may be cumulus minerals.

The best-documented occurrences of pyroxene-rich wehrlite-series xenoliths are in New South Wales and Victoria. Wilshire & Binns (1961), Irving (1974a), Wilkinson (1975b), Ellis (1976), and Griffin et al. (1984) described xenoliths ranging from dunite through wehrlite and olivine clinopyroxenite to clinopyroxenite that have features typical of igneous cumulates

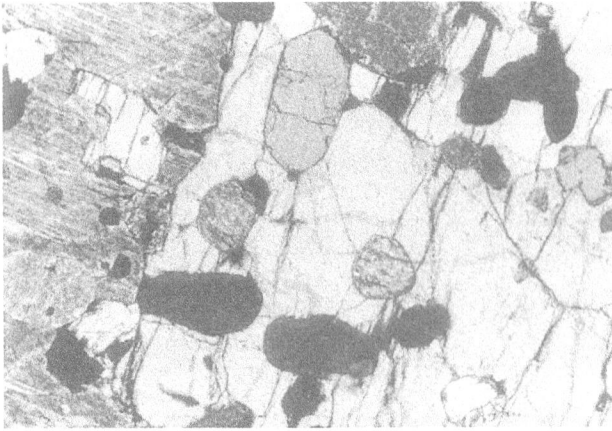

Figure 6.2.10 Cumulate microstructure in an olivine clino-pyroxenite from Lake Bullenmerri (62660). Interlocking exsolved Al-augite oikocrysts enclose small cumulus olivines. Field of view (with crossed polars) is 35 × 22 mm.

(especially Ca-clinopyroxene oikocrysts enclosing olivine) and superimposed development of mosaic metamorphic microstructures. Olivine, orthopyroxene, and Ca-clinopyroxene are the major cumulus species, and augite and some amphibole the main intercumulus species. The best examples containing intercumulus amphibole are found at Lake Bullenmerri (Griffin et al., 1984) where, in some cases, euhedral amphibole and clinopyroxene also line vugs up to 1 cm in diameter, indicative of the presence of a free fluid phase. Most examples have only minor intercumulus spinel, some containing phlogopite or biotite. Interstitial glass is believed to originate either as residual liquid from cumulate formation or more commonly by 'decompression' melting of amphibole or Ca-clinopyroxene (or both).

Typical wehrlite-series clinopyroxenites are ad-cumulates containing more than 90-percent augite, whereas websterites are mesocumulates containing cumulus augite (65–75 percent), orthopyroxene (15–25 percent), and intercumulus augite and minor pargasitic amphibole (5–10 percent).

Most investigators have emphasised the sub-calcic and aluminous nature of clinopyroxene in xenoliths having cumulate microstructures, and have noted that progressive recrystallisation proceeds through initial exsolution of orthopyroxene lamellae, and later development of discrete grains, to assemblages containing abundant orthopyroxene. Wehrlite-series clinopyroxenites and wehrlites having cumulate microstructures therefore may be compositionally equivalent to metapyroxenite-series, olivine-poor lherzolites and websterites.

Metapyroxenite-series xenoliths are formed by mineralogical and microstructural re-equilibration of pyroxene-rich wehrlite-series material in response to cooling from magmatic temperatures to the ambient geotherm (Griffin et al., 1984). Progressive exsolution of sub-calcic aluminous clinopyroxenes (more rarely orthopyroxenes) leads to the common neoblast assemblages clinopyroxene + orthopyroxene + spinel ± garnet ± plagioclase. Reaction coronas are present, commonly of garnet about spinel or symplectite spinel+pyroxene intergrowth about garnet.

Microstructures range from coarse equant and por-phyroclastic types incorporating relict clinopyroxenes up to 5 cm, and commonly having coarse exsolution lamellae of orthopyroxene ± garnet ± Al-spinel, to those dominated by granoblastic grains of the same minerals. Many metapyroxenite xenoliths have layering, probably of original cumulate origin (but possibly caused by metamorphic differentiation; for example, Hollis, 1981). Mineralogical and microstructural differences on the thin-section scale are common.

Metapyroxenites that have no aluminous phase, or which contain only minor spinel or plagioclase, commonly have microstructures similar to those of granoblastic olivine-rich xenoliths (Wilkinson, 1973, 1975a,b; Irving, 1974a; Ellis, 1976; Hollis, 1981; Griffin et al., 1984). Spinel (commonly a green Al-rich type) forms xenoblastic interstitial grains, even where relatively abundant (up to 10 percent), whereas rarer plagioclase forms interstitial patches of granoblastic grains (for example, Wilkinson, 1975b). Spinel may be present as millimetre-scale layers, probably representing coarse lamellae originally exsolved from clinopyroxene. Some have amphibole exsolution lamellae in clino-pyroxene or minor interstitial amphibole, or both.

Garnet and garnet-spinel metapyroxenite xenoliths at east-Australian localities have received considerable attention, and their mineralogy and microstructures have been studied as an aid to interpretation of complex reaction relationships between aluminous pyroxenes, spinel, and garnet, and associated changes in ambient pressure-temperature conditions. Increasing recrystal-lisation and grain-size reduction proceeds by progressive exsolution of original complex pyroxenes (to produce the complementary pyroxenes and garnet ± spinel), as in garnet-free metapyroxenites. This process is accompanied, or followed by, reactions between pyroxenes and spinel to produce further garnet.

Porphyroclastic xenoliths that contain coarse pyroxene oikocrysts having exsolution lamellae of pyroxenes + garnet ± spinel, are found at the New South Wales localities of Walcha (phlogopite-bearing garnet websterites and orthopyroxene 'megacrystals' up to 20 mm; Stolz, 1984) and Delegate (spinel-garnet websterites and clinopyroxenes up to 15 mm; Irving, 1974b). In contrast, only garnet clinopyroxenites (± hornblende, plagioclase, and scapolite) having grano-blastic, equant to weakly tabular microstructures and average grain-sizes of 1–2 mm are present at Gloucester, New South Wales (Wilkinson, 1974).

A broader range of types is represented within garnet metapyroxenites containing common amphibole at the Bullenmerri-Gnotuk complex. Here, differential recrystallisation of material with original mineral layering on thin-section to hand-specimen scales has produced considerable microstructural diversity, in-cluding in some cases interlayering of lamellar exsolved pyroxenes and granoblastic pyroxenes (Griffin et al., 1984). Microstructural complexities of Bullenmerri-Gnotuk metapyroxenites, and their contact relationships with Cr-diopside suite lherzolite material within composite xenoliths, were illustrated by Hollis (1981). Further distinctive features of some garnet meta-pyroxenites, including reaction coronas about garnets and spinels, common amphibole as exsolution lamellae in pyroxenes, and neoblasts within mosaic aggregates and in replacement/vein assemblages, were treated in detail by Griffin et al. (1984) who interpreted them as products of reactions taking place (with or without fluid activity and partial melting) after the main meta-pyroxenite-forming metamorphic event.

Evidence for relict cumulate features in garnet metapyroxenites having coarse microstructures comes from rare, coarse, poikilitic augites, and from T-shaped rather than triple-point grain-boundary junctions. More typical coarse equant microstructures, directly equivalent to those in olivine-rich xenoliths, have garnets up to 5 mm, set in matrices of finer, weakly granoblastic augite and orthopyroxene. Pyroxene-dominated microstructures equivalent to mosaic-porphyroclastic examples in olivine-rich xenoliths are the most common type amongst garnet- or garnet-spinel metapyroxenites.

In contrast, garnet metapyroxenites having well-developed granoblastic microstructures are rare, and even these have some garnet and lamellar, exsolved, pyroxene porphyroclasts. Some examples within the porphyroclastic to granoblastic microstructure range illustrate well the transition from lamellar-exsolved pyroxene porphyroclasts to granoblastic pyroxene-garnet aggregates through zones up to 5 mm wide. Subgrains and neoblasts within porphyroclast outlines are 0.5 mm or less and lobate, whereas outside they increase progressively to 0.5–1.5 mm and have polygonal shapes. Garnet lamellae change progressively outward to parallel lines of granules, then lose any alignment.

Most examples of all microstructural types in metapyroxenites have an almost total lack of preserved strain features, a situation quite different to that observed in olivine-rich xenoliths. Examples of microstructures in metapyroxenite xenoliths are given in Figures 6.2.11–*12*.

6.2.3 Composite xenoliths

Introduction

The most abundant, largest, and most extensively studied composite xenoliths are found in Victoria

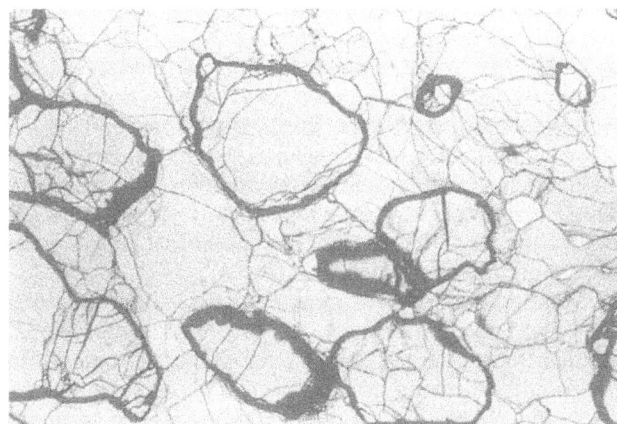

Figure 6.2.11 Coarse equant microstructure in garnet clino-pyroxenite from Lake Bullenmerri (62555). The rock consists of augite and pyropic-garnet grains (they have incipient kelyphitic rims) of 2 mm average grain-size. Field of view (in plane polarised light) is 11 × 7 mm.

(especially at Mount Shadwell, Lake Bullenmerri, and the Anakies) and Queensland (Lake Eacham and Sapphire Hill). Descriptions including microstructural information were given by Irving (1980) for examples from Mount Shadwell (Al-augite suite olivine clino-pyroxenite enclosing clasts of Cr-diopside spinel lherzolite), Lake Eacham (Al-augite suite spinel-amphibole-olivine clinopyroxenite in contact with Cr-diopside spinel lherzolite), and Sapphire Hill (spinel-websterite vein having a mineral assemblage intermediate to typical Al-augite and Cr-diopside suite examples, cutting Cr-diopside lherzolite). The pyroxene-rich phases of the three xenoliths are illustrative of the change from igneous cumulate to metamorphic microstructures. Irving (1980) interpreted the primary microstructures in terms of a model of flow-cumulate formation, including a treatment of chemical interaction between vein magmas and wall-rocks.

Griffin et al. (1984) and Wass & Hollis (1983) described composite xenoliths from Lake Bullenmerri and the Anakies (mainly associations of Al-augite suite wehrlite or metapyroxenite with Cr-diopside lherzolite), and concluded that in many xenoliths little chemical interaction had taken place between the phases.

Magmatic vein assemblages

Composite xenoliths from Victorian localities having veins consisting of pyroxene-rich cumulates represent strong evidence that, in many cases, much of the 'vein' assemblage is derived directly or indirectly from wall-rock material. As much as 40 volume percent of some vein wehrlites consists of strained olivine 'xenocrysts' from wall-rock lherzolite, and the typical sub-calcic aluminous clinopyroxene oikocrysts may be products of reactions involving wall-rock enstatite-bronzite, Cr-diopside, and Cr-spinel. The microstructural and chemical changes associated with these reactions in a few xenoliths are illustrated clearly in well-preserved transitional assemblages.

Olivine clinopyroxenite or wehrlite veins having cumulate microstructures only slightly modified by changes in ambient pressure-temperature conditions, are relatively common in Cr-diopside lherzolite xenoliths from Lake Bullenmerri and Mount Shadwell. A wehrlite vein in a Bullenmerri example contains abundant kink-banded, coarse olivine from the lherzolite host, in addition to 0.2–0.5 mm strain-free, cumulus olivines within a classic cumulate microstructure. Olivines are enclosed by twinned Al-augite oikocrysts up to 5 mm, with patchy orthopyroxene exsolution lamellae. The vein lacks obvious relict pyroxenes or Cr-spinel derived from the lherzolite host (in marked contrast to the situation for olivine) although this has dis-aggregated mechanically along the vein margins. The non-xenocrystic component of the vein is of olivine-clinopyroxenite composition. Interstitial pools of brown glass containing skeletal olivine and augite probably represent trapped liquid.

Olivine-clinopyroxenite veins in Mount Shadwell composite xenoliths described in part by Irving (1980) have cumulate textures similar to that of the Bullenmerri wehrlite, and modified by 'decompression melting'. All but the cores of the coarsest augite oikocrysts are riddled with blebs of brown glass, interpreted by Irving (1980) as representing liquid trapped during cumulate formation. However, augite cores that lack glass again have simple twins and exsolution lamellae that persist through even the densest zones of glass blebs. Melting therefore evidently postdated initial subsolidus adjustment after cumulate formation.

Reactions between basaltic liquid and Cr-diopside spinel lherzolite are best defined in another composite xenolith from Mount Shadwell. This consists of typical, coarse, equant Cr-diopside lherzolite wall-rock, containing small volumes of Cr-diopside orthopyroxenite (probably produced by an earlier vein-forming or metamorphic event) cut by sub-parallel veins of 1 mm grain-size spinel-olivine-plagioclase clinopyroxenite. Veins less than 5 mm in width have up to 40-percent coarse, kink-banded olivine derived from the host lherzolite, and pale-green augite and green-violet Al-spinel as the main 'magmatic' minerals. Olivine and plagioclase (An_{70-65}) are present interstitially together with brown glass, and form skeletal phenocrysts in 'pools' of quench-textured basalt where the veins widen.

Wall-rock enstatite grains in contact with spinel-clinopyroxenite vein material develop symplectite rims of olivine-augite intergrowth with interstitial glass. Cr-diopsides are replaced progressively by large pale-brown Ti-Al augites charged with glass blebs and small olivines, and brown Cr-spinels are converted to blue-green Al-spinel, some in symplectites with augite. Compositional change in both clinopyroxene and spinel is visible across single grains within the vein walls.

'Pools' within veins are zoned inward from 0.2–0.3 mm spinel-olivine-plagioclase clinopyroxenite, through coarser spinel wehrlite containing 20-percent glass and quench crystals, to quench-textured olivine/Ti-augite basalt having 50-percent plagioclase-rich glassy groundmass. The dominant clinopyroxenite zone has cumulus spinel and more interstitial olivine and plagioclase than that of the narrow veins.

Vein and vein-margin assemblages clearly were produced mainly by chemical interaction between wall-rock lherzolite and water-poor basaltic vein liquid, at pressures and temperatures at which plagioclase is stable. The details of reactions are illustrated by well-preserved intermediate assemblages. In contrast, the nature of the processes involved in most composite xenoliths is obscured by completion of reactions, metamorphic recrystallisation, or later hydrous vein activity (for example: Irving, 1980; Boettcher & O'Neil, 1980; Griffin, et al., 1984).

Metamorphosed or metasomatised vein assemblages

Microstructural descriptions of east-Australian composite xenoliths in which pyroxene-rich vein phases have been modified by metamorphism or hydrous metasomatism (or both) were given also by Irving (1980) and Hollis (1981).

Metamorphism of the clinopyroxene-rich vein phase is illustrated by a Sapphire Hill xenolith in which a spinel-websterite vein cuts Cr-diopside lherzolite. The vein has roughly equal amounts of deep-green augite and orthopyroxene (both containing oriented spinel exsolution), about 10-percent interstitial green Al-spinel, and minor pale-brown amphibole, ilmenite, and sulphide. The assemblage is finer-grained (about 0.5 mm), has more triple-point pyroxene grain junctions than typical cumulate veins, and lacks both olivine (of either 'xenocryst' or cumulus origin) and interstitial glassy material. These features are evidence for extensive exsolution and recrystallisation of primary complex pyroxenes and simultaneous reaction of olivine and glassy material. Metamorphism has produced strong similarities between the pyroxenes in the vein and the lherzolite host, and the overall assemblage is intermediate to typical Cr-diopside and Al-augite suite assemblages (Irving, 1980).

Metasomatic and metamorphic modification of a pyroxene-rich vein phase is illustrated by a sample from Lake Eacham, consisting of a spinel-amphibole-olivine clinopyroxenite in contact with Cr-diopside spinel lherzolite. The clinopyroxenite vein phase includes about 25 percent of olivine derived from the lherzolite host. The non-xenocrystic assemblage is about 4:1 pale-green augite and deep-green spinel, and has interstitial networks of brown Ti-pargasite. Microstructures are indicative of reactions between the primary phases of the vein and a hydrous fluid phase associated with amphibole formation. Olivine grains are yellow-brown (Fe-rich) and have either olivine overgrowths, thin orthopyroxene rims at contacts with augite, or porous zones of glass, olivine, green-brown spinel, and sulphide where enclosed by amphibole. The most obvious changes in the wall-rock lherzolite are also development of more Fe-rich olivine and Al-rich spinel and partial amphibole rims on spinels within about 1 cm of the vein.

Significance of composite xenoliths

The mineralogy and microstructures of the vein and wall-rock components of composite xenoliths provide the best available evidence for the nature of processes of partial melting and the movement of magmas within the mantle. However, definitive evidence is rarely available from east-Australian examples. Unambiguous evidence of reactions associated with emplacement of vein magma is preserved in only a few examples described by Irving (1980), and in this section, in which magma wall-rock interaction apparently took place at relatively low pressures (probably within the plagioclase stability field for pyroxenite assemblages). These xenoliths were transported to the surface before infiltration of vein magma and modification of wall-rock minerals had proceeded far.

Most examples have extensive microstructural and chemical re-equilibration of the vein assemblage. Evidence for an important role for wall-rock material in the origin of veins is obscured by total resorption (orthopyroxene) or re-equilibration (clinopyroxene, spinel), and only olivine survives almost unmodified. Microstructural and compositional interpretation of assemblages becomes a difficult proposition where 'primary' cumulate vein assemblages have undergone later overprinting caused by one or more of the activity of hydrous fluids (formation of amphibole or biotite, or both), subsolidus reaction, exsolution, and recrystallisation, and late-stage 'decompression' melting. Nevertheless, this interpretation is supported by the relatively common occurrence and splendid state of preservation of composite xenoliths in several east-Australian localities.

6.2.4 Primary mineral assemblages and conditions of equilibration.

Cr-diopside suite (including Fe-rich types)

Volatile-free rock types of the Cr-diopside suite are the dominant xenolith type in eastern Australia and consist of the four-phase assemblage olivine + Cr-diopside + orthopyroxene + spinel (Cr-rich pleonaste). Differences in modal proportions result in a wide range of rock types. The most common assemblage is within the range of 40–80 percent olivine, 5–20 percent diopside, 5–30 percent orthopyroxene, and 1–5 percent spinel — that is, lherzolite. Garnet+spinel lherzolites are found at five off-craton localities (Jugiong, Ferguson et al., 1977; Mount Shadwell and Mount Noorat, Irving, 1974a; Wass & Irving, 1976; Meredith Pipe, Day et al., 1979; and Bow Hill, Sutherland & Hollis, 1982). The South Australian kimberlite pipes (Scott Smith et al., 1984; O'Reilly and Griffin, unpublished data) and the Kayrunnera (White Cliffs) diatremes (Edwards et al., 1979) also have garnet+spinel lherzolite, but these are extremely altered. Garnet lherzolites (some diamond bearing) have been found in a restricted facies of the Argyle (AK1) diatreme in northwestern Australia (O'Neill et al., 1986).

Volatile-bearing spinel lherzolites may contain any combination of amphibole, mica, or apatite. These volatile-bearing minerals are found in several microstructural relationships, ranging from veins, to rims on pre-existing minerals (amphibole on spinel), to polygonal grains forming an integral part of the metamorphic fabric. These are discussed in more detail in Section 6.2.6.

Temperatures of formation are calculated readily for lherzolites by a range of methods if high-quality microprobe analyses are available. Calibration of calculated temperatures for spinel lherzolites for direct comparison with those derived for compositionally different rock types (for example, garnet pyroxenites) can be achieved using composite xenoliths (Griffin et al., 1984). There is no geobarometer that yet can be applied to spinel-lherzolite assemblages. However, a geobarometer based on Ca content of olivine (Finnerty, in press) is in the preliminary stages of development and is a potential solution. A pressure can be estimated at present by referring calculated temperatures to the east-Australian geotherm, which was determined empirically using data from garnet pyroxenites (O'Reilly & Griffin, 1985; Section 7.2.3). This method is valid for all east-Australian xenolith localities studied in enough detail to result in well-constrained geothermobarometric data (see summary in Section 7.2.3) provided the geothermometer used for the lherzolite temperature calculations is consistent with the pyroxenite temperatures.

Geothermometry has been carried out on over 180 spinel-lherzolite samples from eastern Australia using high-precision microprobe analyses and the olivine-orthopyroxene-spinel geothermometer of Sachtleben & Seck (1981; Bezant, 1985; O'Reilly and Griffin, unpublished data). Most temperature estimates for spinel lherzolites are between 850 and 1050°C, corresponding to a pressure range of 9–18 kb (27–50 km; see Fig. 7.2.1). This coincides well with the inferred depth interval from the crust/mantle boundary to the spinel- to garnet-lherzolite transition (Griffin & O'Reilly, 1987b). This transition may take place over a significant depth range (5–10 km) for different bulk-rock compositions (O'Neill, 1981). Pressure-temperature estimates on garnet lherzolites from Jugiong and Mount Shadwell plot on the higher extension of the east-Australian geotherm at 22–25 kb. The Bow Hill garnet peridotites (Sutherland et al., 1984a) appear to have equilibrated at relatively higher temperatures of up to 1280°C at 23.6 kb.

Temperature estimates using the Ellis & Green (1979) method for a garnet lherzolite from Calcutteroo are 1150 to 1200°C (at 20 and 30 kb, respectively). No pressure could be calculated as clinopyroxene and garnet are the only unaltered minerals. O'Neill et al. (1986) estimated equilibration pressure and temperatures for the Argyle garnet lherzolites (using reconstructed garnet compositions, as all garnet is replaced by symplectitic spinel+pyroxene aggregates). The estimated pressure-temperature range is about 46–58 kb (180–200 km) at 1100–1300°C.

Al-augite suite

The *wehrlite series* of the Al-augite xenolith suite is dominated by clinopyroxene+olivine assemblages, but there are large modal differences. Additional minerals in some rocks are orthopyroxene, amphibole, mica, feldspar, spinel, ilmenite, sulphides, and accessory apatite and zircon. Pressures can not be calculated from these assemblages. However, composite xenoliths of wehrlite-series rocks in contact with equilibrated garnet pyroxenite and spinel lherzolite of the Cr-diopside suite, have estimated depths of origin of about 28–58 km. The absence of composite xenoliths consisting of wehrlite-series rocks and garnet lherzolites of the Cr-diopside suite may simply reflect the rarity and small size of the garnet-lherzolite xenoliths in eastern Australia.

Temperatures of 950–1000°C have been calculated using two-pyroxene geothermometers (Wood & Banno, 1973; Wells, 1977) on wehrlite-series xenoliths from Bullenmerri-Gnotuk and Anakie, Victoria (Griffin et al., 1984; Wass & Hollis, 1983). However, the pyroxenes are compositionally heterogeneous and the temperatures are minimum values where compared with those calculated for the lherzolite and metapyroxenite in composite xenoliths. The wehrlites therefore may have crystallised at higher magmatic temperatures and partially re-equilibrated towards an ambient temperature similar to that represented by the constructed palaeogeotherm (Griffin et al., 1984). This interpretation is supported by the gradation in microstructures from unaltered igneous to thoroughly recrystallised types. Some of the wehrlite-series xenoliths contain co-equilibrated ilmenite and titano-magnetite, allowing calculation of oxygen fugacity (fO$_2$). These fO$_2$ values all plot near the quartz-fayalite-magnetite (QFM) buffer curve.

The *metapyroxenite-series* xenoliths of the Al-augite suite contain clinopyroxene ± garnet ± orthopyroxene ± amphibole ± ilmenite ± rutile ± mica ± plagioclase ± apatite ± sulphides. The 'granulites' that have equilibrated in the upper mantle (see Section 6.1.3.) can probably be included in this category. These are some of the Anakie xenoliths (Wass & Hollis, 1983) and some Ruby Hill, Gloucester, and Delegate xenoliths (O'Reilly & Griffin, 1985; Griffin & O'Reilly, 1986b; Stolz, 1984; O'Reilly et al., 1988). These mantle-derived granulites contain more plagioclase than do the metapyroxenites, and may contain anorthoclase or scapolite or both (Wass & Hollis, 1983; Lovering & White, 1969).

Both the metapyroxenites and the granulites have metamorphic microstructures dominated by polygonal grain shapes, and they commonly have evidence of deformation. Common layering may reflect development of foliation (Figs. 6.2.13–15) or original coarse grain-size. Metapyroxenites from Bullenmerri may have formed by exsolution and recrystallisation of coarse-grained (2–3 cm) clinopyroxene-rich cumulates (± orthopyroxene ± spinel). Relict clinopyroxene (Fig. 6.2.15) contains exsolution lamellae of orthopyroxene ± spinel ± garnet ± plagioclase. The presence of primary spinel is inferred for xenoliths in which modal spinel exceeds the amount that could have formed by exsolution. These spinels commonly are rimmed by garnet (*Fig. 6.2.14B*). Transition from relict lamellar to granoblastic microstructures may be observed on the scale of a thin section (Fig. 6.2.15–16). Feldspar-rich granulites may have recrystallised from frozen melts, cumulates, or tectonically-emplaced basaltic rocks (see Section 6.2.5).

Xenoliths of Bullenmerri and Gnotuk represent the most comprehensive assemblage of metapyroxenites in eastern Australia and their conditions of equilibration have been studied in detail by Griffin et al. (1984). They are ideal for geothermobarometric calculations because the minerals are chemically homogeneous within grains and from grain to grain, irrespective of

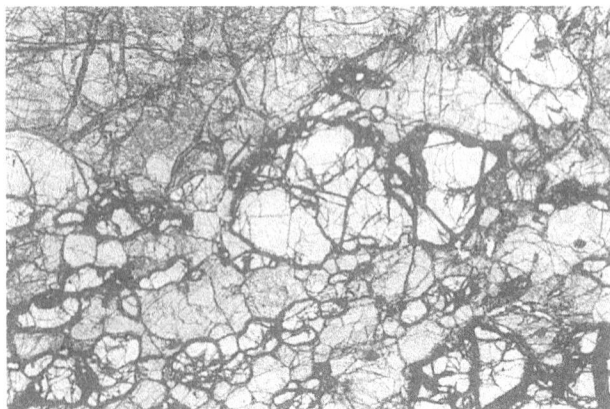

microstructure. This means that mineralogical re-equilibration took place more rapidly than did recrystallisation. In addition, the bulk-rock compositions are

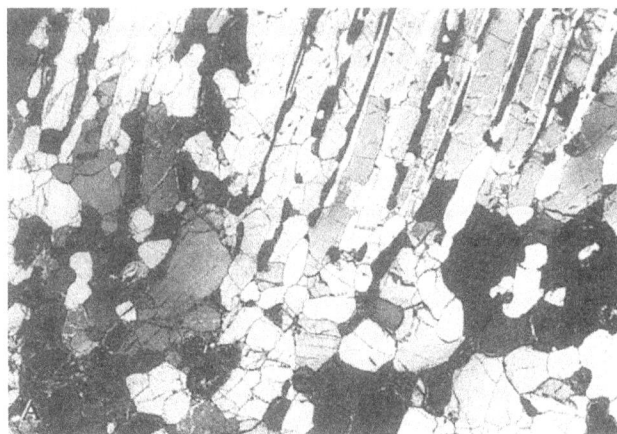

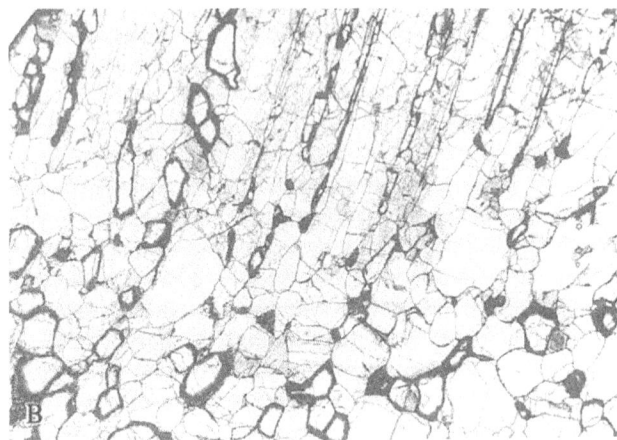

Figure 6.2.15   (A) Metapyroxenite showing transition from lamellar microstructure (large clinopyroxene with exsolution lamellae of garnet and orthopyroxene) to mosaic microstructure where subsolidus recrystallisation of these three minerals has taken place. All mineral grains in this example are unzoned and like minerals have identical compositions whether in the lamellar or mosaic domains. Width photographed (with crossed polars) is 2.5 cm. (B) Same sample in plane polarised light.

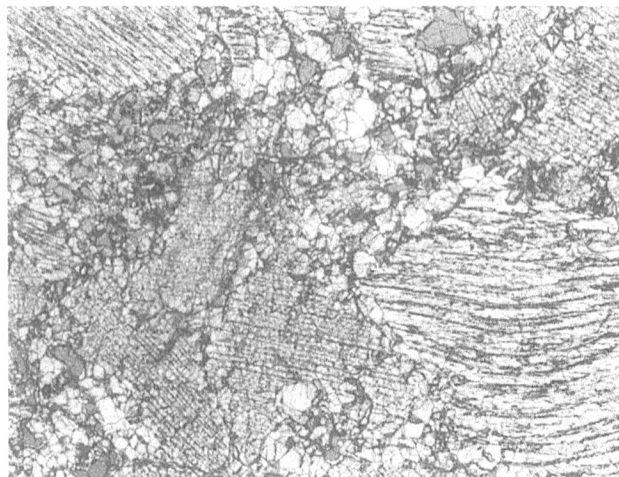

Figure 6.2.13   Metapyroxenite (garnet + clinopyroxene + spinel) showing pyroxene- and garnet-rich layers. Width photographed (in plane polarised light) is 2.5 cm.

Figure 6.2.16   Relict lamellar clinopyroxene showing some development of mosaic domains in metapyroxenite originating as a clinopyroxene-spinel cumulate. Width photographed (in plane polarised light) is 2.5 cm.

similar to those used in some experimental studies at pressures and temperatures appropriate for the Bullenmerri-Gnotuk xenoliths.

The geothermobarometric methods used require judicious screening if the resultant values are to be meaningful. This includes methods of accounting for possible Fe^{3+}, cross-calibration of different methods for consistency between different rock compositions, and use of high-quality microprobe data. These considerations are fully discussed by Griffin et al. (1984). Pressure-temperature values for the Bullenmerri-Gnotuk suite closely define an empirical geothermal gradient (see Fig. 7.2.1) which is interpreted as representing the ambient geotherm to which the metapyroxenites equilibrated before entrainment in the host magma. Pressure-temperature values for garnet metapyroxenite and mantle-derived garnet granulite from other localities (O'Reilly & Griffin, 1985; Stolz, 1984; Griffin et al., 1988) all plot along this empirical geotherm.

Fine-grained interstitial intergrowths and symplectitic coronas on garnet and spinel were formed by secondary reactions. The fine-grained areas appear to represent melt that has partially crystallised at high pressures. They commonly consist of vesicular glass containing euhedral phenocrysts and intergrowths of olivine + plagioclase ± spinel ± pyroxene. These mineral grains are homogeneous and have mineral compositions consistent with a high-pressure origin (Griffin et al., 1984).

The symplectitic coronas consist of coarse intergrowths of two-pyroxenes + spinel ± olivine or fine-grained pyroxene + plagioclase + spinel ± olivine. These represent garnet breakdown under high temperature or lower pressure, or both. Other symplectites have evidence for more complex reactions, such as garnet I reacting to garnet II + spinel + orthopyroxene + clinopyroxene. One complex garnet-spinel websterite has two apparently sequential reactions: (1) clinopyroxene + spinel to clinopyroxene + spinel + garnet, and (2) clinopyroxene + spinel + garnet to clinopyroxene + orthopyroxene + spinel + garnet. These interstitial patches and coronas have widely different degrees of development even within single xenoliths. They are considered to have formed rapidly compared with the time required to produce the mineralogical and microstructural equilibration of the exsolved metapyroxenites. A possible scenario for their formation during initial slow diapiric rise within the mantle before rapid ascent to the surface initiated at about 30–40 km was described by Griffin et al. (1984).

One Delegate garnet granulite contains co-equilibrated ilmenite and titanomagnetite and its calculated fO_2 is consistent with conditions defined by the FMQ buffer.

Amphibole/apatite-series xenoliths of the Al-augite suite contain amphibole, apatite, biotite, clinopyroxene, olivine, magnesian ilmenite, speneliferous titanomagnetite, spinel carbonate, rare rutile, and sulphides (including chalcopyrite, pyrrhotite, pyrite, and pentlandite) in different proportions. This suite is best represented at the Kiama locality described by Wass (1979), but small fragments are found throughout all provinces in eastern Australia. The Kiama locality is unique in both the abundance and range of amphibole/apatite-series rock types. Discrete mineral assemblages

commonly observed are: amphibole/clinopyroxene/ apatite ± ilmenite, spinel, biotite, olivine, carbonate; amphibole/biotite/apatite ± ilmenite, spinel, carbonate; biotite/apatite + ilmenite, spinel; amphibole/apatite ± carbonate, ilmenite, spinel; Fe-rich clinopyroxene/ apatite/spinel + biotite; spinel/apatite ± biotite, carbonate; and monomineralic rocks consisting of amphibole, biotite, apatite, and spinel. Sulphides are ubiquitous accessories. Layering is common in these xenoliths and is considered to reflect original cumulate processes. Microstructures range from igneous to granoblastic.

Some xenoliths appear to be original Cr-diopside spinel lherzolites that have been extensively metasomatised by the magmatic fluid from which the amphibole/apatite rocks crystallised. These altered lherzolites are inferred to be adjacent wall-rock, and therefore the amphibole/apatite series formed in the spinel-lherzolite zone of the upper mantle. There are no suitable assemblages to use for geothermobarometric calculations.

The amphibole/apatite xenoliths are interpreted as representing fragments of a high-pressure differentiation sequence crystallised in the mantle from a fractionated magma of kimberlitic or carbonatitic affinity. They may represent higher level, more evolved equivalents of the MARID suite (Dawson & Smith, 1977).

6.2.5 Geochemistry

Introduction

A large number of chemical and isotopic analyses are now available for east-Australian mantle-derived xenoliths. The great majority of these analyses are of Cr-diopside suite rocks, and most of the analysed samples are from a few cinder cones and maars in western Victoria. The next largest source of data is for samples from the cinder cones of northern Queensland. Few data are available from New South Wales, because the xenoliths are generally small and difficult to extract from the lavas in which they are found. The most important trends of major- and trace-element chemistry for the Cr-diopside suite and the Al-augite suite are described first in this section, and then are discussed and used to evaluate genetic relationships between the suites.

Geochemical data referred to here are derived from the following sources: Victoria — Frey & Green (1974), Archbald (1979), Irving (1980), Nickel & Green (1984), McDonough & McCulloch (1987a), O'Reilly & Griffin (1988), Griffin et al. (1988); New South Wales — Wilkinson & Binns (1977), Archbald (1979), O'Reilly (unpublished data); Queensland — Archbald (1979), Irving (1980), Griffin et al. (1987), O'Reilly and Griffin (unpublished data). The discussion is based largely on Victorian data, as these are the most complete and have a wider range of compositional variation than do the other data sets.

Cr-diopside suite

The great majority of Australian ultramafic xenoliths are anhydrous lherzolite and harzburgite. However, many — especially from western Victoria — have obvious modal metasomatism involving the growth of secondary amphibole and apatite, at the expense of pyroxene and spinel. Complex histories are implied from the microstructures, together with correspondingly complex effects on chemical composition. Many of the available analyses were carried out in an attempt to understand this complexity.

Frey & Green (1974) modelled the compositions of six Victorian lherzolites as mixtures of two components, A and B. Component A essentially describes major-element differences and was defined as a depleted lherzolite or harzburgite, residual after extraction of basaltic melt from a pyrolite-like composition. Component B was regarded as a percolating melt of basanite-like composition that is a source of incompatible elements (K, Ti, REE, and so on). Component B

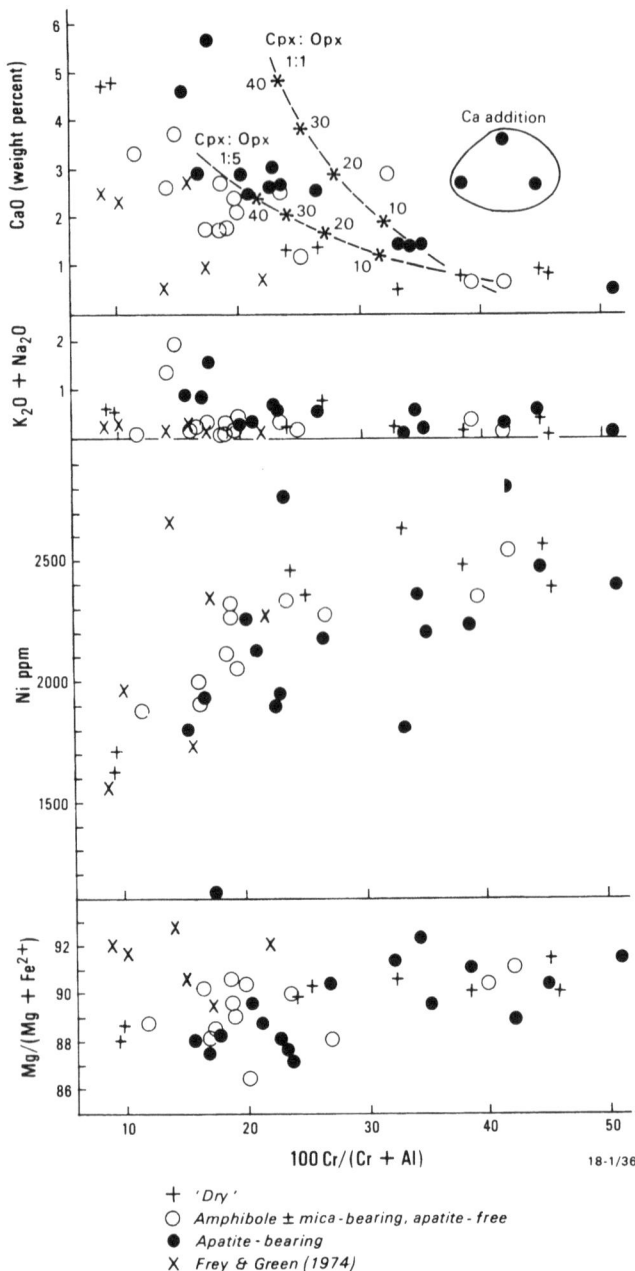

Figure 6.2.17 Selected compositional parameters for east-Australian lherzolites. Data are from O'Reilly & Griffin (1988). Numbers 10–40 represent calculated values for clinopyroxene (cpx) and orthopyroxene (opx).

was inferred to be present both in anhydrous rocks (cryptic metasomatism) and in those containing secondary, metasomatically introduced amphibole, mica, or apatite. This model is a useful framework for further research, but some of the conclusions have been modified by recent work (Griffin & O'Reilly, 1986a; O'Reilly & Griffin, 1988; Griffin et al., 1988).

Major and minor elements. Several chemical parameters are plotted in Figure 6.2.17 against $100Cr/(Cr+Al)$ values which are used as a measure of refractoriness (see also Table 6.2.2). Mg-ratios have a wide range (93–86) compared to most suites of mantle-derived lherzolites (92–89; Maaløe & Aoki, 1977). The lower Mg-ratios (less than 89) are found almost entirely among the modally metasomatised samples, and are interpreted as being caused by introduction of Fe during metasomatism (Irving, 1980; Wilshire et al., 1980, 1985; Griffin et al., 1984). The correlation between Ni and $100Cr/(Cr+Al)$ mainly reflects differences in the proportion of olivine to pyroxenes plus spinel. This is confirmed by the plot of CaO versus $100Cr/(Cr+Al)$. Ca-Cr-Al relationships in this suite can be explained by mixing of different proportions of clinopyroxene, orthopyroxene, and olivine. The interpretation is supported also by an excellent positive correlation of CaO and V (not illustrated). Exceptions to these generalisations are three apatite-rich, pyroxene-poor samples, to which probably Ca was added metasomatically. Na and K are generally low, even in the modally metasomatised samples.

The major-element data for the Victorian samples are consistent with the model of Frey & Green (1974) in which these rocks are considered to represent residua after partial melting (component A). The major-element chemistry of the residua is controlled by pyroxene: olivine:spinel proportions, and this signature persists even after strong modal metasomatism. The average composition of 36 lherzolite xenoliths from Bullenmerri and Gnotuk maars (Table 6.2.2) is similar to the worldwide average spinel-lherzolite composition of Maaløe & Aoki (1977). This Bullenmerri-Gnotuk suite was chosen to emphasise possible metasomatic effects: 30 of the samples contain secondary amphibole ± mica ± apatite. Even so, the average composition is not significantly different from the world-wide average. Slightly higher Al, Ca, and Na correspond to slightly lower Mg-ratio, which is consistent with world-wide trends, and may mean that the Victorian suite represents a slightly lower degree of original depletion than does the world average.

Two suites of samples from northern Queensland (O'Reilly and Griffin, unpublished data) and New South Wales (Archbald 1979; O'Reilly, unpublished data) also may be compared (Table 6.2.2). These suites have lower Mg-ratio and $100Cr/(Cr+Al)$ values than do the Victorian lherzolites or the world average. They also have higher Ca and Al, and lower Mg and Ni, on average. These differences reflect higher pyroxene/olivine proportions, and imply a lower degree of depletion in the basaltic component relative to the suite from western Victoria.

Trace elements. The most notable features of the available data (to 1986) on trace elements in Victorian lherzolites (Fig. 6.2.18) are the small dispersions for

Table 6.2.2 Average compositions of Cr-diopside spinel lherzolites

	Bullenmerri and Gnotuk[1]	Northern Queensland[2]	New South Wales[2]	Continental lherzolite[3]
Weight percent				
SiO_2	44.1 ± 1.4	44.6 ± 1.3	44.2 ± 1.1	44.1 ± 1.4
Al_2O_3	2.0 ± 1.5	2.7 ± 0.6	2.6 ± 0.8	2.0 ± 1.0
FeO	8.6 ± 0.9	8.7 ± 1.4	8.5 ± 0.7	8.3 ± 1.2
MnO	0.17 ± 0.04	0.13 ± 0.03	0.11 ± 0.02	0.12 ± 0.05
MgO	41.0 ± 3.9	39.7 ± 1.7	39.4 ± 3.2	42.3 ± 1.2
CaO	2.3 ± 1.3	2.5 ± 0.8	2.9 ± 1.0	2.1 ± 0.5
Na_2O	0.27 ± 0.25	0.19 ± 0.22	0.18 ± 0.11	0.18 ± 0.1
K_2O	0.05 ± 0.07	0.01 ± 0.03	0.06 ± 0.05	0.05 ± 0.07
Total	98.5	98.5	98.0	99.2
Ti (ppm)	650 ± 680	375 ± 170	630 ± 405	420 ± 60
Cr (ppm)	3060 ± 870	2890 ± 390	2870 ± 700	3010 ± 1000
Ni (ppm)	2210 ± 340	2020 ± 230	1975 ± 190	2120 ± 1000
Cr-ratio[4]	22.8	17.7	16.1	23.4
Mg-ratio	89.5	89.0	88.0	90.0
n[5]	36	44	12	301

[1] O'Reilly & Griffin (1988).   [2] O'Reilly and Griffin (unpublished data).   [3] Maaløe & Aoki (1977).   [4] Cr-ratio is 100Cr/(Cr+Al).   [5] Number of samples.

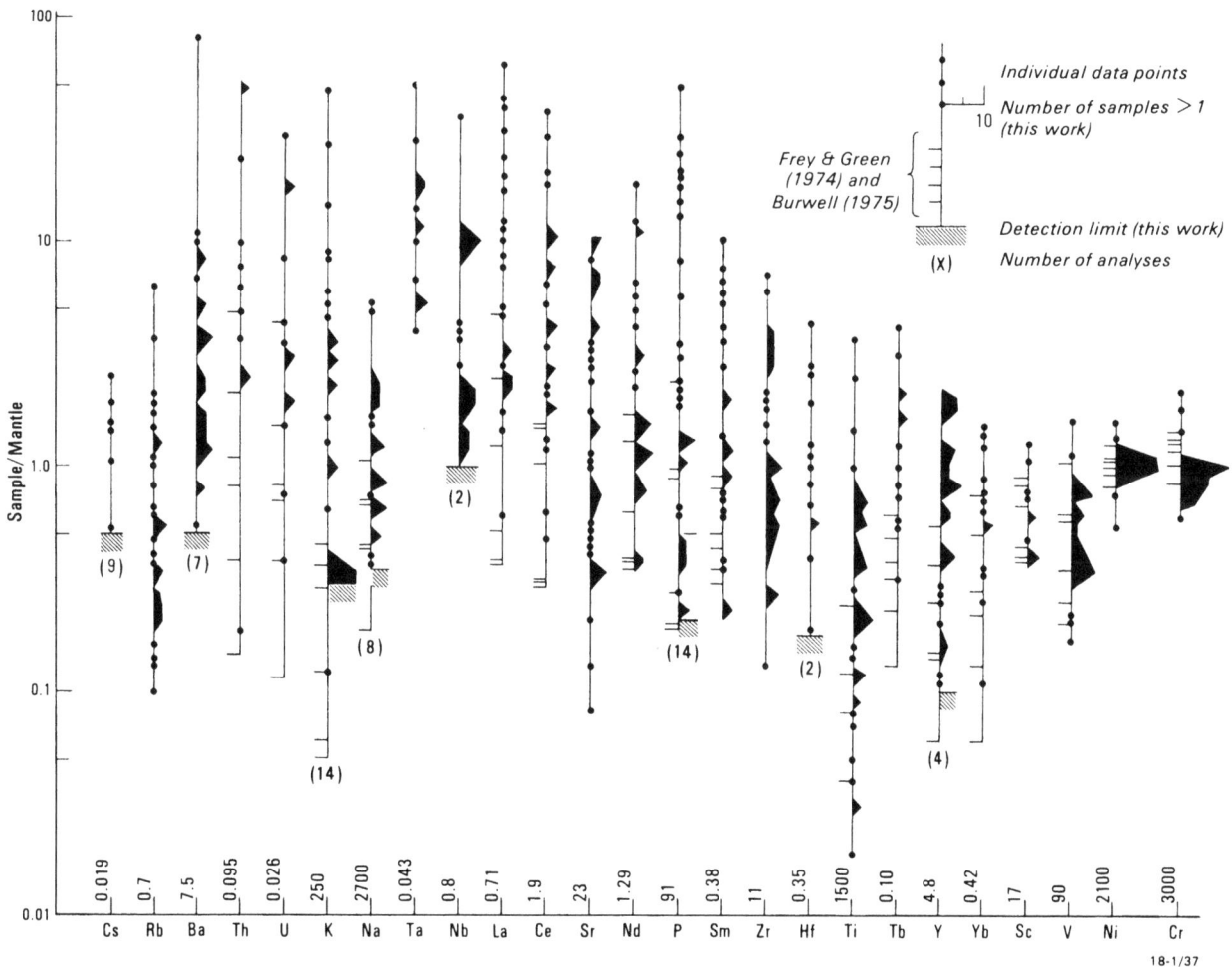

Figure 6.2.18 Trace-element data for xenoliths of Cr-diopside spinel lherzolite from western Victoria, normalised to values for primordial mantle. Normalisation values given on abscissa (from Wood, 1979, and Sun, 1982). Data sources are given in O'Reilly & Griffin (1988).

the refractory trace elements, the large dispersions for most of the incompatible elements, and the enrichments in some incompatible elements relative to a primordial-mantle composition. The overall pattern of enrichment and depletion is illustrated more clearly in Figure 6.2.19 where median values are plotted for each element (the median is used, rather than the average, because it is less strongly perturbed by the inclusion of a few samples that are greatly enriched or depleted in a given element). The most refractory elements, Ni and

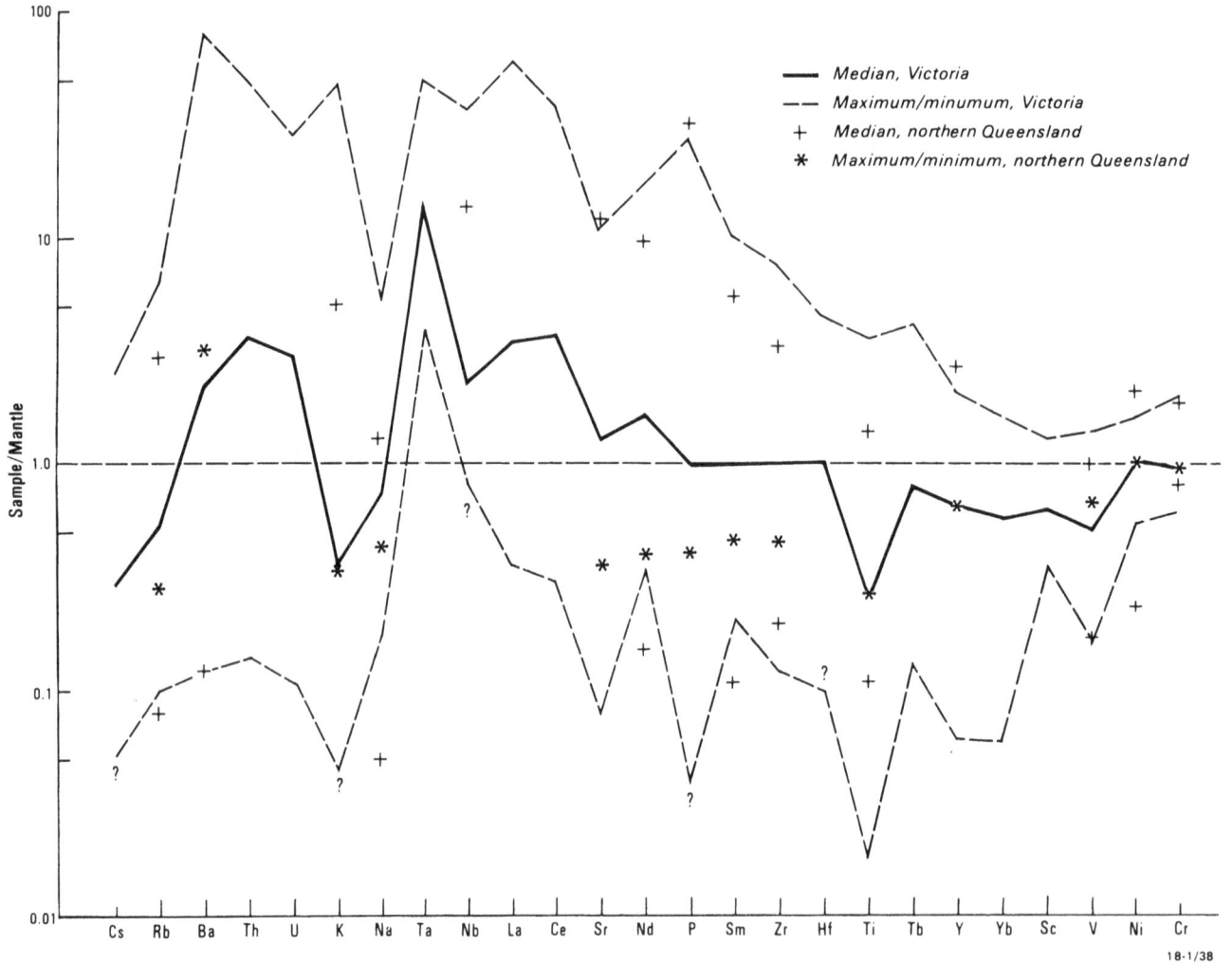

Figure 6.2.19 Median, maximum, and minimum values for data shown in Figure 6.2.18 for western Victoria, together with equivalent data for northern Queensland (O'Reilly and Griffin, unpublished data).

Cr, are close to 'primordial' values (Wood, 1979). This is not surprising because these values were originally derived from analyses of similar material. Depletion of Yb, Sc, and V relative to values for primordial mantle is consistent with the inferred residual nature of Component A and with inferred immobility of these elements in metasomatic processes.

The overall enrichment in the incompatible elements from Ba to Zr (Fig. 6.2.19) is striking relative to Yb, Sc, and V, and is strong support for the two-stage, depletion/enrichment model of Frey & Green (1974). However, note that overall incompatible-element enrichment is not paralleled by enrichment in Ti, K, Rb, or Cs. Component B has been described as 'very enriched in P, K, Ti, light REE, Th and U' (Frey & Green, 1974, page 1023) or 'KREEP' (Menzies & Murthy, 1980), but K, Ti, and Rb are decoupled from P, REE, Sr, and other incompatible elements during the enrichment process. Furthermore, Sr and the REE (exemplified by Nd) are correlated strongly, but these elements are not correlated clearly with other incompatible elements, or with Fe-enrichment (as shown by the Mg-ratios; Fig. 6.2.20).

This decoupling of the different groups of elements is caused by modal mineralogical differences, as demonstrated clearly by some mineral-element correla-

tions. High values of Sr (greater than 50 ppm) are uniquely related to the presence of trace amounts of apatite (with 1–2 percent Sr), and Sr is well correlated with P at levels above 50 ppm. Amphibole-bearing samples have K/Rb values greater than 800, whereas anhydrous and mica-bearing (± amphibole) samples have K/Rb values less than 600. High values of Ba, Rb, and Ta are confined to samples that have traces of mica. REE patterns are controlled by the presence of clinopyroxene, amphibole, or apatite. Anhydrous, clinopyroxene-rich rocks have light-REE-depleted patterns with high total-REE contents, whereas clinopyroxene-poor ones have low total and light-REE enriched patterns. Amphibole-rich lherzolites have smooth upwardly convex patterns, reflecting the tendency of amphibole to incorporate the middle REE (Nicholls & Harris, 1980; Wass et al., 1980; Irving, 1980; Irving & Frey, 1984). Apatite-rich samples have exceptionally high total-REE contents and linear light-REE enriched patterns.

Mineralogical control can be explained best by crystal/fluid partitioning in open systems during metasomatic events (Fig. 6.2.21). The different stages of this process can be related to distance from a hypothetical fluid source, represented in Figure 6.2.21 as a pyroxenite vein (see below). The primary sink for

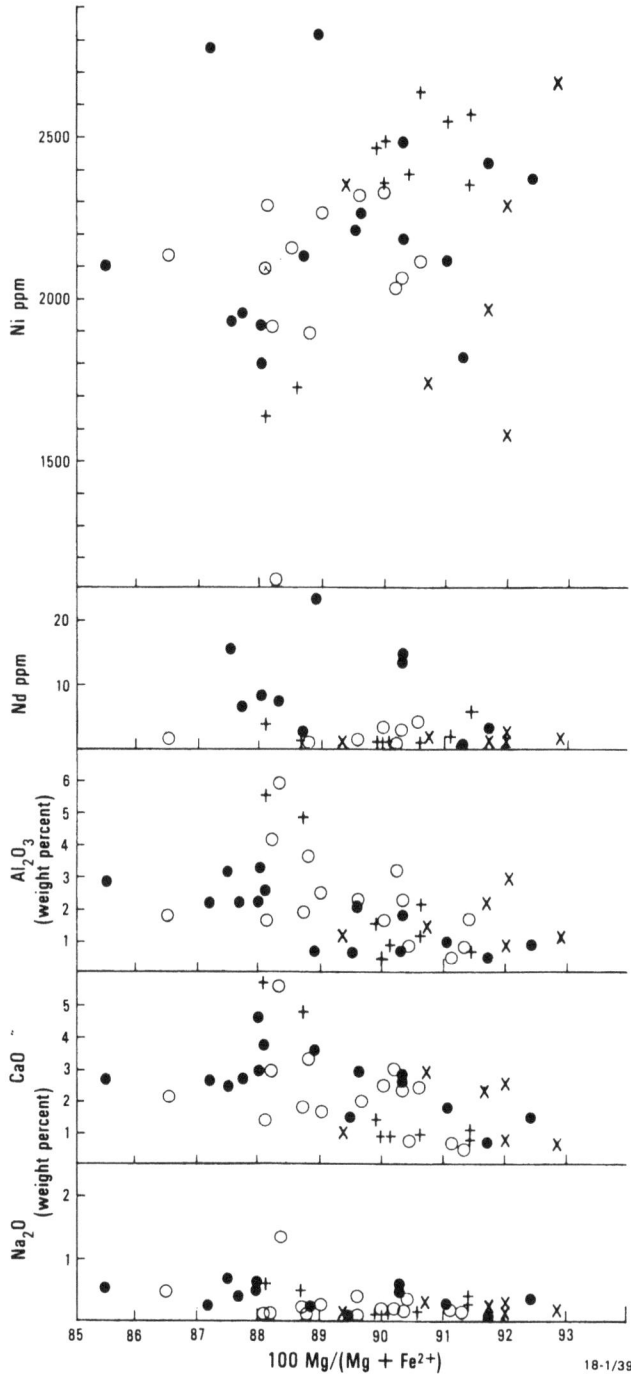

Figure 6.2.20 Selected oxide and trace-element abundances in lherzolites. Symbols and data sources as in Figure 6.2.17.

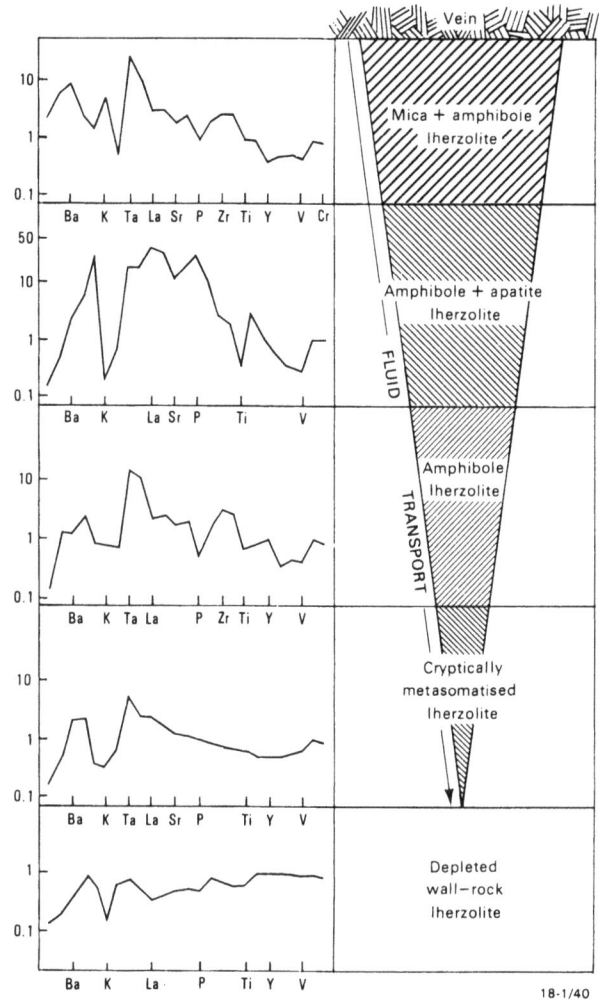

18-1/40

Figure 6.2.21 Schematic representation of the metasomatising action of fluid infiltration away from a vein into mantle wall-rock. Metasomatic types, their trace-element signatures, and inferred spatial relationships relative to the source of metasomatising fluids are shown. Median values (as in Figure 6.2.19) are characteristic of each metasomatic type, and are shown in the normalised element-abundance diagrams.

incompatible elements in a 'cryptic' stage of the metasomatism — where fluid activity is too low to initiate the growth of secondary phases — is clinopyroxene, and the trace-element signature of the rock will be controlled by the abundance of clinopyroxene and by clinopyroxene/fluid partitioning. The reaction clinopyroxene + spinel + fluid = 2 amphibole as fluid activity increases, will both increase the relative volume of the 'sink' and change the bulk crystal/fluid partition coefficients. The result will be a marked change in the trace-element signature of the rock. Similar changes will accompany the nucleation and growth of mica and apatite. Trace-element abundances

are therefore a consequence of the presence of specific minerals and will range widely for xenoliths containing different metasomatic minerals.

The modal control of incompatible-element enrichment during metasomatism also can explain some of the apparent anomalies seen in Figure 6.2.18. The number of mica-bearing samples is small, so these may represent a volumetrically insignificant part of the mantle volume, even beneath western Victoria. These assemblages may form as thin selvedges on pyroxenite veins or fluid conduits. If so, they would remove K, Rb, and Ti from the fluid before it could penetrate farther into rock. Many other incompatible elements would be unable to enter the mica structure, and would be carried farther from the veins before being incorporated in pyroxene, amphibole, or apatite.

Nucleation of mica at vein margins also may deplete the fluid in water, producing a $CO_2 \pm Cl$-rich nature of metasomatic apatites (O'Reilly & Griffin, 1987) and the OH-deficient nature of the metasomatic amphiboles (R. Oberti, personal communication, 1986; Boettcher & O'Neil, 1980).

North Queensland lherzolites (Fig. 6.2.19) have smaller dispersions for most elements (the number of analyses ranges from 12 for Rb, Sr, Sm, and Nd, to 44 for Ni, Cr, V, Ti, Zr, and K) and the maximum values for most elements are lower than in the Victoria suite (O'Reilly and Griffin, unpublished data). The two suites have similar median values for the refractory elements, but the Queensland suite is depleted relatively in Zr, P, Sr, light-REE, K, and Rb. Anomalous enrichments in Ba and Nb may be caused by the presence of trace amounts of Ba-Nb-Ti oxides (Haggerty, 1983).

Decoupling of major- and trace-element patterns in the Queensland and Victoria suites is a clear demonstration of the effects of mantle metasomatism. The Queensland suite in terms of major elements is less depleted (retains a higher basaltic component) than is the Victoria suite, but it has been less affected by later metasomatic activity than has the Victoria suite and therefore is more depleted in most incompatible elements. The differences are striking, but they are still a matter of degree. Modally metasomatised, incompatible-enriched samples are found in northern Queensland (Fig. 6.2.19) but are rare. Most Cr-diopside suite xenoliths in Victoria have some metasomatic effects, and truly primitive samples are relatively rare.

Al-augite suite

The Al-augite suite at most east-Australian localities can be divided petrographically into a 'wehrlite series', having igneous microstructures, and a 'metapyroxenite series' having equilibrated metamorphic microstructures. However, Wass & Hollis (1983) described a group of pyroxenites, granulites, and eclogites from Anakie (central Victoria) that spans this range of microstructures.

The wehrlite/metapyroxenite division in one well-documented xenolith population from Bullenmerri and Gnotuk maars in western Victoria, corresponds to a chemical dichotomy (Fig. 6.2.22). One transitional sample has wehrlite chemistry but contains garnet in a partially re-equilibrated microstructure. Both series have a wide range in composition, and the metapyroxenites can be divided into a low-Ca and a high-Ca group. Spinel pyroxenites are all in the former.

The wehrlite series is distinguished from the metapyroxenites primarily by higher Ti and Zr, and lower Al, comparing samples that have similar Mg-ratios. The Anakie xenoliths generally resemble the wehrlite series at Bullenmerri-Gnotuk, but extend to much lower Mg-ratios, probably corresponding to extensive differentiation within the mantle (Wass & Hollis, 1983).

Moderate levels of K_2O in the wehrlites and the Anakie samples reflect the presence of late magmatic biotite. Mica is absent from the Bullenmerri metapyroxenites, correlating with much lower K, Rb and Ba. High levels of Ti, Nb, and Zr in the wehrlites are related to the presence of cumulate ilmenite, as well as to late-magmatic amphibole and biotite. There is a negative correlation of Cr with Ca for the Bullenmerri metapyroxenites which relates to the presence of primary spinel. A good positive correlation between Ni

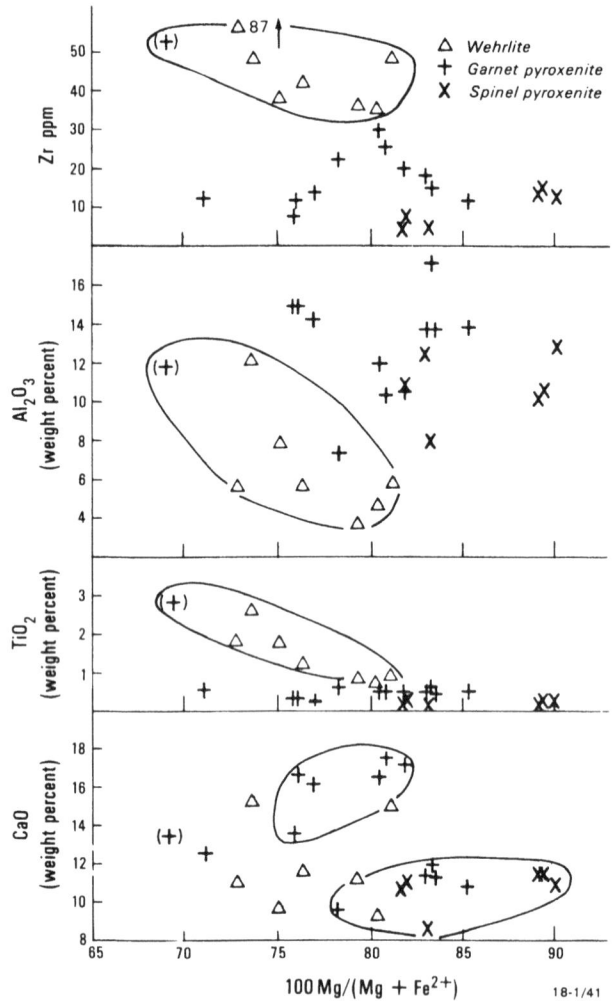

Figure 6.2.22 Selected chemical variations for Al-augite xenolith types. Data are from Griffin et al. (1988).

and Cr is seen for both series, and they differ significantly in REE patterns (Fig. 6.2.23). The wehrlites have smooth, upwardly convex patterns, whereas those for the metapyroxenites are upwardly concave. The low-Ca metapyroxenites have lower total-REE contents and less light-REE enrichment than do those of the high-Ca group.

Al-augite suite xenoliths are interpreted as cumulates from basaltic magmas (Irving, 1980; Griffin et al., 1984). Very few rocks have bulk compositions appropriate for those of quenched magmas. The wehrlites and the Anakie rocks are clinopyroxene+olivine cumulates containing significant proportions of trapped liquid crystallised as amphibole + mica + apatite. The high Ni and Cr and low Sr of the metapyroxenites may be evidence that they originated as clinopyroxene±spinel± olivine cumulates. This is consistent with their inferred petrographic evolution (Griffin et al., 1984).

Igneous microstructures, isotopic compositions (see below), and trace-element patterns (especially the high Ti, P, Ba, and K) of the wehrlites are similar to those of their host magmas, and the wehrlites therefore may have originated from magmas of similar composition. The same is true of the Anakie pyroxenite-granulite-eclogite suite, but their degree of microstructural evolution may correspond to that of a longer pre-eruption history. The older metapyroxenites may have been derived from less alkaline magmas. Calculated REE patterns for the parental magmas are strongly light-REE enriched, and resemble those of intraplate tholeiitic or alkali basalts. However, the metapyroxenites also have been modified by exchange with their wall rocks (Griffin et al., 1988), and this makes detailed discussion of their origin ambiguous.

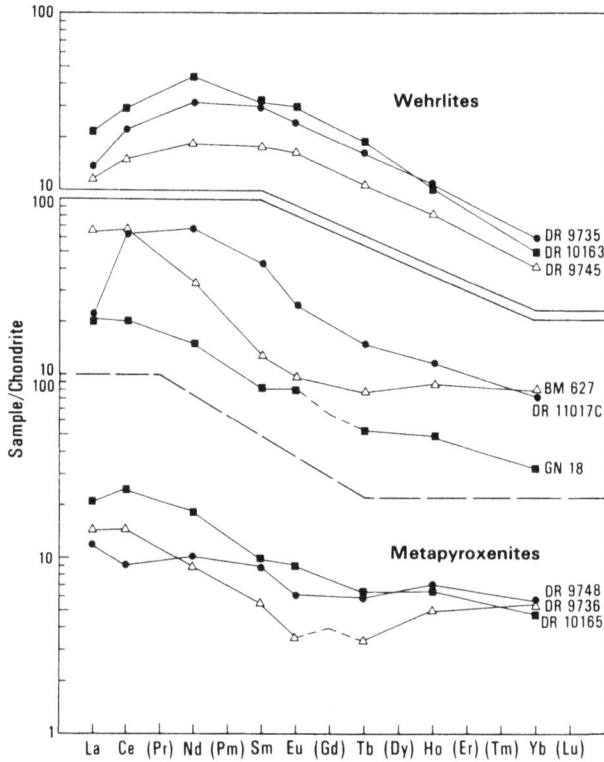

Figure 6.2.23 Chondrite-normalised REE patterns for wehrlite-and metapyroxenite-series xenoliths of the Al-augite suite.

Isotopic relationships

Sr-isotope ratios were measured in Victorian lherzolite xenoliths by Dasch & Green (1975) and Burwell (1975) and a recent restudy of Dasch and Green's samples has been carried out by McDonough &

McCulloch (1987a). A large data set, mainly of modally metasomatised samples, is presented by Griffin et al. (1988).

'Anhydrous' lherzolites have a relatively narrow range of $^{87}Sr/^{86}Sr$ and a median value of about 0.7043 (Fig. 6.2.24). This is distinctly higher than values (0.703–0.704) reported for anhydrous xenoliths from many other parts of the world. Modally metasomatised samples, containing amphibole ± mica ± apatite, or glass from breakdown of hydrous phases, have a greater range of $^{87}Rb/^{86}Sr$ and a median value of about 0.7049. Nearly all samples have $^{87}Rb/^{86}Sr$ values of less than 0.1 so that high values of $^{87}Sr/^{86}Sr$ are unsupported. The radiogenic Sr of the volatile-bearing lherzolites clearly reflects the metasomatic introduction of ^{87}Sr and the $^{87}Sr/^{86}Sr$ values of the anhydrous lherzolites presumably have the same cause.

$^{87}Sr/^{86}Sr$ values for wehrlites and for the chemically similar Anakie pyroxenites, have a narrow spread at the low end of the range for the Newer Volcanics (O'Reilly & Griffin, 1984, and unpublished data; McDonough et al., 1986). The metapyroxenites have a large range of unsupported $^{87}Sr/^{86}Sr$, extending up to typical 'crustal' values.

Nd-isotope data on lherzolites from Victoria were presented by McDonough & McCulloch (1987a) and Griffin et al. (1988) and here are combined with the Sr-isotope data (Fig. 6.2.25). The lherzolite isotopic data define a trend extending from 'depleted mantle' values through the Bulk Earth value to the 'enriched mantle' field. Most of these isotopic compositions reflect an enrichment in radiogenic Sr, and a depletion in radiogenic Nd, relative to other studied spinel-lherzolite suites (see, for example, Stosch & Lugmair,

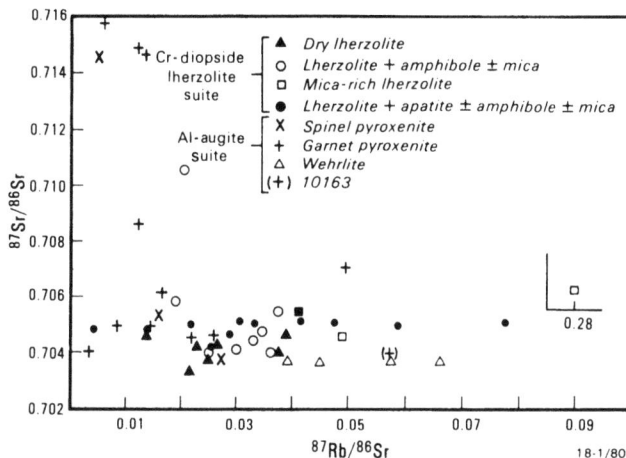

Figure 6.2.24 Sr-isotope relationships for Bullenmerri lherzolites and Al-augite suite samples.

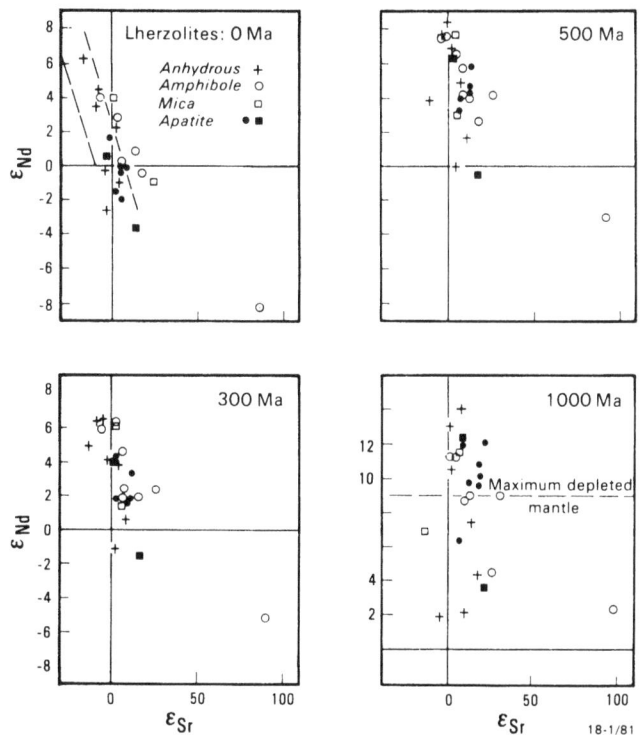

Figure 6.2.25 Nd- versus Sr-isotope relationships (ε units) for Bullenmerri lherzolites at different times (see text). Maximum depleted mantle is defined by a linear evolution with ε value of zero at 4.5 Ga and +12 at the present day.

1986). However, a similar trend is observed in some garnet-lherzolite suites from kimberlite pipes (Hawkesworth et al., 1983). Some of the most 'depleted' samples have marked light-REE enrichment, so this enrichment is a relatively recent event.

North Queensland xenoliths in general have depleted Sr-Nd isotopic compositions similar to those reported for spinel lherzolites from many other areas (O'Reilly and Griffin, unpublished data). These isotopic ratios are consistent with the relatively simple depletion history interpreted for the major- and trace-element data (Fig. 6.2.18). However, one volatile-free sample falls near the Bulk Earth composition, and one amphibole-mica lherzolite has even more 'enriched' Sr- and Nd-isotopic ratios than do any of the Victoria samples. The radiogenic ^{87}Sr in this sample is not supported by a high Rb/Sr value.

The metapyroxenites have a range in ε_{Nd} values that matches the spread in $^{87}Sr/^{86}Sr$. The most 'depleted' ones have Sr-Nd isotopic compositions similar to the anhydrous lherzolites, whereas the most 'enriched' ones resemble crustal material. The wehrlites and the Anakie pyroxenites have a relatively narrow range of ε_{Nd}.

Pb-isotope data for six Victorian lherzolites (Kleeman & Cooper, 1970) define linear trends in 207/204 versus 206/204 and 208/204 versus 206/204 plots. These can be interpreted either as mixing lines (lherzolite-basanite) or as differentiation of the lherzolites at about 2 Ga into subsystems that have similar Th/U but widely different Pb/U. The second alternative is favoured on the basis of trace-element data.

An unusual history or process is indicated by the isotopic compositions of the uppermost mantle beneath Victoria, compared to that sampled in other volcanic regions. The Sr- and Nd-isotopic data may be accounted for by the observed metasomatic activity having altered the isotopic composition of the mantle. These changes can be caused in two ways: (1) metasomatic changes in parent/daughter ratios, followed by an aging period, or (2) direct introduction of material that has a different isotopic signature.

The first mechanism was proposed as the explanation for enriched isotopic signatures in African garnet-lherzolite xenoliths (Hawkesworth et al., 1983). However, the Sr-isotopic composition of many radiogenic samples from eastern Australia cannot be accounted for in this way, because Rb/Sr values are so low. Griffin et al. (1988) argued that metasomatism involved *both* mechanisms, but if this is so the metasomatic episode becomes difficult to date directly. Burwell (1975) reported a 650 ± 125 Ma Rb-Sr errorchron for five of six lherzolites from Victoria. However, no isochron relationships are shown in either the Rb-Sr or Sm-Nd systems for later data and a much larger data set. Griffin et al. (1988) suggested several indirect methods of constraining the time or times of metasomatism. Nd model ages (T_{DM}) for the lherzolites cluster around 600 Ma and T_{CHUR} model ages average about 100 Ma (many are less than zero). The effect of metasomatism has been to lower Sm/Nd values, so these data imply maximum metasomatic ages of about 600 Ma. Lherzolite ε_{Nd} values become progressively

higher relative to Bulk Earth values where projected back in time, and by about 750 Ma ago most of them have ε_{Nd} values higher than those for the most extreme depleted-mantle models. This also is consistent with a metasomatic event much younger than 750 Ma ago.

O'Reilly & Griffin (1988) and Griffin et al. (1988) argued that the metapyroxenites represent the source of at least part of the metasomatic fluids that have affected the lherzolites. The pyroxenites in turn have been modified by reaction with their wall rocks, as observed in composite xenoliths. The Sr-Nd data for the lherzolites and pyroxenites, where projected back in time, should define a mixing hyperbola at the time of intrusion of the pyroxenites. This is defined most clearly at 300–500 Ma (Fig. 6.2.25).

The least-modified metapyroxenites in this model are those that have the most radiogenic Sr and the lowest $^{143}Nd/^{144}Nd$. These cumulates must have been derived from magmas having a large component of crustal material, or from a strongly enriched mantle source. Griffin et al. (1988) noted that 300–500 Ma coincides with the major period of crustal accretion in eastern Australia, and suggested that subducted crustal material, remelted at depth, may have been the source of the magmas that precipitated the pyroxenite. The incompatible-element enrichment seen in the Victorian lherzolites therefore may represent recycled crustal material.

6.2.6 Metasomatism

Introduction

The following points are evidence for mantle metasomatism in xenoliths from eastern Australia. They correspond to a range of metasomatic mechanisms from volatile fluxes to silicate melts and all combinations in between:

(1) The presence of volatile-bearing minerals (amphibole, mica, apatite, and carbonate) in Cr-diopside lherzolite suite xenoliths represents patent metasomatism (Dawson, 1980).

(2) Contact zones between Al-augite and Cr-diopside suite xenoliths may have geochemical gradients in both mineral and whole-rock compositions away from the contacts (Irving, 1980; Griffin et al., 1984). Geochemical exchange therefore seems to have taken place between the intrusive basaltic fluid (which crystallised as the Al-augite rock type) and the mantle wall-rock. This is a type of cryptic metasomatism (Dawson, 1980).

(3) Discrete Cr-diopside xenoliths may contain minerals that have high concentrations of incompatible elements, but no amphibole, mica, or apatite is present. These incompatible elements (for example, Sr and REE) are partitioned mainly into clinopyroxene and are detectable as anomalously high values and by distinctive isotopic signatures in the rocks. This is also cryptic metasomatism.

(4) Fluid inclusions trapped at mantle depths in both Al-augite and Cr-diopside series xenoliths represent direct samples of free fluids (likely metasomatic agents or their derivatives in the mantle; Andersen et al., 1984).

(5) Sulphide globules and rods included in minerals

of Cr-diopside and Al-augite suite xenoliths are important indications of the movement of sulphur and mobility of chalcophile elements in mantle processes (Andersen et al., 1987).

Complementary to the metasomatic effects recorded in xenoliths are the geochemical characteristics of primitive basaltic magma compositions in eastern Australia which reflect parent-mantle compositions. For example, high $^{87}Sr/^{86}Sr$ values in the rocks of particular basaltic provinces correlate positively with the presence of volatile-bearing minerals in entrained mantle-derived xenoliths (O'Reilly & Griffin, 1985). This aspect of magma compositions is discussed in Section 5.5.

An important point to note is that metasomatic effects recorded in Cr-diopside spinel lherzolites represent metasomatism at a higher level than the zone of magma generation. Intraplate tholeiitic and alkaline rocks in eastern Australia were generated in the garnet-lherzolite zone (Frey et al., 1978). Therefore the relevance of metasomatism in xenoliths from the spinel-lherzolite zone to the geochemical nature of the mantle which is the source of partial melts for the east-Australian basaltic rocks, relies on the assumption that metasomatism in the spinel-lherzolite zone is geochemically analogous to that affecting the underlying garnet-lherzolite zone.

Eastern Australia has a wide geographical distribution of mantle xenoliths in basaltic hosts, so this is an ideal basis for examining different scales of mantle heterogeneity. The characteristics of metasomatic events that took place at different times can be examined also because xenolith-transporting basaltic activity has recurred sporadically for about the last 200 Ma. This possibility is especially relevant where there have been at least two periods of volcanic activity distinctly separated in time within the same province — for example, the Kandos province (O'Reilly & Griffin, 1984) and western Victoria (O'Reilly & Griffin, 1988).

Both the spinel-lherzolite and garnet-lherzolite regions of the mantle beneath eastern Australia are inferred to be heterogeneous, based on the geochemical composition of spinel-lherzolite xenoliths and primary basaltic compositions, respectively. This heterogeneity in trace-element abundances and isotopic character appears to be on scales ranging from centimetres (observable in xenolith hand specimens) to about 1 km for small intraprovince differences in primary-basalt geochemistry, up to the order of 100 km for the larger differences between basaltic provinces (O'Reilly & Griffin, 1984).

Metasomatic processes are considered to be the cause of most of this geochemical heterogeneity. The timing of metasomatism relative to a particular volcanic episode is different throughout the region. It may be precursory (either recently or distantly), contemporaneous, or any combination of these. There is good isotopic evidence for the different times of basaltic activity identifiable in the southeast-Australian mantle (Section 6.2.5). In addition, the range of microstructures in Al-augite suite xenoliths from igneous to metamorphic in a given population (Griffin et al., 1984) is consistent with different times of formation.

Petrographic evidence

Two well-documented xenolith populations in eastern Australia have petrographic evidence for mantle metasomatism: the Bullenmerri-Gnotuk xenoliths (see, for example, O'Reilly & Griffin, 1988) and those from Kiama (for example, Wass & Rogers, 1980). Cr-diopside spinel lherzolite xenoliths at Bullenmerri-Gnotuk commonly have modal metasomatism indicated by the presence of one or more of amphibole, mica, and apatite. These volatile-bearing minerals are found in a range of different microstructural sites.

Amphibole may form polygonal grains that are an integral part of the granoblastic microstructure (some with foliations). It may form as subhedral to euhedral porphyroblasts either scattered throughout the rock or in segregations, or it may form as veins that commonly cut across pre-existing foliations. These veins also may contain one or more of clinopyroxene, mica, or apatite. They have evidence that their formation was associated with a free fluid phase because of abundant fluid inclusions in vein and adjacent wall-rock minerals. The veins also may anastomose forming invasive networks (Wass & Hollis, 1983; Fig. 6.2.3). The porphyroblastic amphibole grains generally enclose embayed spinel (Fig. 6.2.26) which may provide Al_2O_3 in a reaction that forms the amphibole.

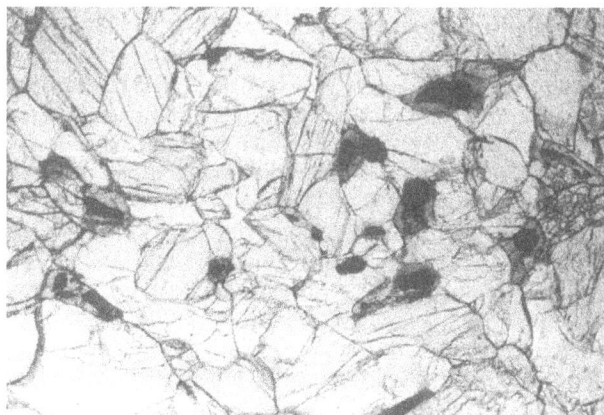

Figure 6.2.26 Amphibole around spinel grains in Cr-diopside lherzolite. Width photographed (in plane polarised light) is 3.5 mm.

Mica is found mostly as the same microstructural type. A strong foliation is defined commonly by preferred orientation of the mica grains in recrystallised rocks where the mica is dispersed throughout granoblastic domains.

Apatite is found as polygonal dispersed grains (commonly with amphibole ± mica) and in veins that have two distinct assemblages: apatite (up to 90 modal percent) + clinopyroxene, or apatite (commonly 1–10 percent) ± amphibole + mica ± clinopyroxene. Polygonal apatite grains may constitute up to 8 percent of the mode of the Cr-diopside lherzolite.

The amphiboles are dominantly pargasite. They range in colour in thin-section from pale cream, pale green, and pale brown through to dark brown and invariably have definite pleochroism. Micas (biotite and rare phlogopites) have pleochroism in straw yellow and brown. Apatites have anomalous optical properties because of high CO_2 substitution (O'Reilly & Griffin, 1987). They have second-order maximum interference colours and are identifiable only by common clouding which, in extreme cases, renders the grains almost opaque.

The Bullenmerri-Gnotuk area also has Al-augite suite xenoliths of both the wehrlite and metapyroxenite series. Both of these Al-augite types are found in contact with Cr-diopside rock types in composite xenoliths and both are considered to be sources for at least some metasomatic fluids that have infiltrated the upper-mantle wall-rock beneath western Victoria at different times.

Cr-diopside lherzolites containing volatile-bearing minerals are found also at many other east-Australian localities, including those in New South Wales and Queensland (see for example, Irving, 1980). However, they are more widespread and abundant in Victoria than in other regions. They are found also in intraplate basaltic provinces world-wide (O'Reilly, 1987) where most are associated with amphibole-rich rocks of the Al-augite suite.

A unique population of amphibole- and apatite-rich xenoliths at Kiama (Fig. 6.1.1) has evidence for the nature of one type of metasomatising fluid. These xenoliths are interpreted to represent crystallisation products of a fractionated magma body of kimberlitic-carbonatitic affinity that differentiated in spinel-lherzolite uppermost mantle (Wass, 1979). The direct link to its metasomatic effects is seen in xenoliths that have relict Cr-diopsides now highly altered to minerals equivalent to those in the discrete amphibole-apatite xenolith rock types. Minerals include amphibole, apatite, Al-augite (commonly Fe-rich), altered olivine, magnesian ilmenite, spineliferous titanomagnetite, Fe-rich dolomite, rutile, chalcopyrite, pyrrhotite, pyrite, and pentlandite. Analogous xenoliths and xenocrysts are distributed sparsely throughout basaltic rocks in eastern Australia and world-wide (O'Reilly, 1987), but the Kiama locality is unique because of the abundance, large size, and modal range of its amphibole/apatite xenoliths.

Geochemical evidence

Geochemical data for the large number of Cr-diopside suite xenoliths from western Victoria and northern Queensland represent a sound statistical basis for characterising the trace-element fingerprints and processes of mantle metasomatism (Section 6.2.5). The xenoliths include types that appear to be unmetasomatised as well as those having both cryptic and modal metasomatism. These data are additional to those used in the original study by Frey & Green (1974).

The main conclusions are that some mantle volumes beneath eastern Australia have been extensively modified by metasomatism whereas others are only slightly modified. However, the geochemical pattern of metasomatism and the processes affecting both severely and mildly altered upper mantle regions appear to be the same, differing only in degree. Multiple metasomatic events can be interpreted from the isotopic evidence from both the Cr-diopside and Al-augite suite xenoliths, and these events were significantly separated in time.

Fluid-inclusion evidence

Both Cr-diopside and Al-augite suite xenoliths from eastern Australia contain abundant evidence for the presence of significant volumes of a free fluid phase at high pressure. This is in the form of fluid inclusions 10–30 microns across and polygonal cavities up to 1.5 cm across (Andersen et al., 1984; O'Reilly & Griffin, in press). Common sulphide inclusions in clinopyroxene megacrysts and clinopyroxenes in Al-augite suite xenoliths (Andersen et al., 1987; MacRae, 1979) are indications of the presence of primary sulphide melts within the mantle.

The fluid inclusions now are filled dominantly by CO_2. The large cavities are inferred to have been filled with a compositionally identical fluid, but have leaked because of their large size. However, the presence of a Ca-Fe-Mg carbonate as inclusion linings as well as euhedral crystals of a Ca-Fe-rich amphibole containing significant Cl as secondary overgrowths on cavity walls, is evidence that the original fluid contained up to 10 per cent H_2O, minor Cl, and probably CO and F, before the post-entrapment reactions and precipitation.

Sulphide spheres and cylinders in clinopyroxene from upper-mantle rocks may contain, or adjoin, CO_2 fluid inclusions. They are interpreted as original immiscible droplets of sulphide melt commonly associated with silicate (basaltic) melt and a free CO_2-H_2O fluid (Andersen et al., 1987). They now consist of pyrrhotite and chalcopyrite, and of rare pyrite in some oxidised rock types. The sulphide inclusions have morphologies analogous to those typical of CO_2-rich fluid inclusions, such as decrepitation haloes.

A range of fluid species in small concentrations has been determined by mass-spectrometric analyses of both the condensable and non-condensable gas fractions obtained by crushing the samples (O'Reilly & Griffin, in press). These include N_2, H_2O, CO_2, Ar, SO_2, COS, and hydrocarbons. The hydrocarbons detected are aliphatic and not aromatic, so the possibility of contamination by surface or biological sources is eliminated.

Carbon-isotopic data for CO_2 from three mantle-derived xenoliths include $\delta^{13}C$ values in the range -2.9 to -8.7. These values are close to the $\delta^{13}C$ range of -4 to -8 considered characteristic of carbon-rich mantle-derived material such as diamonds and CO_2 (Pineau & Javoy, 1983). $\delta^{18}O$ values obtained for CO_2 in the same xenoliths range from $+36.1$ to $+51.5$. Pineau & Javoy (1983) suggested that $\delta^{18}O$ of inferred primary-mantle CO_2 trapped in mid-ocean-ridge basalts ranges from $+8.9$ to $+41.4$.

Porcelli et al. (1986) showed that mantle-derived xenoliths from Bullenmerri contain levels of primordial ^3He similar to those for mid-ocean-ridge basalts. $^3He/^4He$ ratios are 7–10 times atmospheric level and directly reflect pristine mantle values because of the young eruption ages. These ratios are within the range of those found for xenoliths from other continental areas (Porcelli et al., 1986). The presence of high levels of ^3He in the upper mantle is evidence for fluxing from a deep primordial reservoir that has not been fully degassed.

6.3 Xenoliths of Crustal Origin

6.3.1 Introduction

Most xenoliths of mid- and upper-crustal rocks are recognised easily. Metamorphic rocks are of amphibolite or lower grade, and most rock types — including sedimentary and felsic igneous rocks — have evidence for extreme reactions with the host magma, including extensive fusion. Recognition of lower-crustal rock types is not so clear-cut.

The lower crust is considered generally to represent a regime consistent with conditions in eclogite or granulite facies. The term 'granulite' is used commonly in a broad sense to include metamorphic plagioclase-bearing rocks that also contain pyroxene and garnet, or pyroxene alone. However, the converse that all eclogite or granulite-facies xenoliths represent lower-crustal rocks is not true. Granulites and eclogites may represent *mantle*-derived rock types (Section 6.1.3). Plagioclase is stable in a broad range of mafic to intermediate compositions over a large pressure-temperature field in the upper mantle (Green & Ringwood, 1972). Furthermore, geothermobarometric results on granulite, eclogite, and pyroxenite xenoliths commonly correspond to pressures greater than 15 kb and temperatures greater than 900°C, and therefore to mantle, rather than lower-crustal, conditions. An integrated approach using calculated temperatures for Cr-diopside lherzolites from the same lithospheric column, along with available geophysical information, must be used in these cases to resolve a crustal or mantle origin.

Granulite-facies rocks also may exist at higher crustal levels because of contact metamorphism or dehydration caused by fluxing of CO_2 (Newton et al., 1980). In addition, granulite and eclogite-facies rocks may have been tectonically emplaced to high crustal levels. Re-equilibration during uplift or subsequent metamorphism will produce mineral assemblages that are not related directly to lower-crustal conditions.

Griffin & O'Reilly (1986a) defined the minimum criteria necessary for the identification of xenoliths as lower-crustal fragments. These are: (1) the xenoliths must not resemble adjacent rock outcrops, as they do at Bournac (Dostal et al., 1980) and Hoggar (Leyreloup et al., 1982); (2) pressure and temperature estimates should be consistent with inferred lower-crustal conditions and depths (8–12 kb) for the given region.

The crust/mantle boundary is defined as the depth below which ultramafic rocks of the Cr-diopside suite become dominant (Griffin & O'Reilly, 1987a,b). The depth to this boundary can be constrained by geothermometric calculations on spinel-lherzolite xenoliths, if a realistic estimate of the geothermal gradient at the time of eruption is available.

The southeast-Australian empirical geotherm (Section 7.2.3; O'Reilly & Griffin, 1985; Griffin et al., 1987) is a useful reference for the pressure-temperature regime in this non-cratonic region throughout the last 200 Ma of basaltic activity. Geothermobarometric results are available for composite xenoliths that have contacts between Cr-diopside lherzolite and granulite or pyroxenite, and several xenolith assemblages containing granulites or eclogites (or both) appear to have been derived from the upper mantle rather than the lower crust. These include the following examples:

(1) Garnet pyroxenites (some containing plagioclase) from Bullenmerri and Gnotuk maars, Victoria (Griffin et al., 1984), and from the Walcha (Gloucester) locality, New South Wales (Stolz, 1984).

(2) Most of the garnet-granulite/eclogite suite from Anakie, Victoria (Wass & Hollis, 1983).

(3) Most of the garnet granulite and garnet pyroxenite samples from Delegate, New South Wales (Lovering & White, 1969; Irving, 1974c). There is microstructural evidence that many of the 'garnet granulites' and a sapphirine-bearing assemblage at this locality formed by exsolution of plagioclase and garnet (with or without sapphirine) from original tschermakitic pyroxenes (Griffin & O'Reilly, 1986b). Pressures and temperatures have been calculated, and re-equilibration is thought to have taken place under mantle conditions. However, calculated temperatures for granoblastic, two-pyroxene granulites from this locality correspond to those of lower-crustal conditions.

(4) The majority of cumulate rock types consisting of the assemblages clinopyroxene ± orthopyroxene ± olivine ± spinel ± amphibole, seen commonly in many xenolith populations from eastern Australia (see, for example, Irving, 1974a). These are found commonly in contact with spinel peridotite in composite xenoliths. Two-pyroxene temperatures for these xenoliths may reflect the igneous precipitation temperature but, where re-equilibration has taken place, the calculated temperatures are consistent with mantle depths derived by reference to the southeast-Australian geotherm. These igneous cumulates may precipitate at any depth between the site of magma generation and sub-surface levels. Indeed, Wilkinson (1975a) described an Al-spinel ultramafic xenolith suite that represents fractionates of a tholeiitic magma within the crust (Wilkinson & Taylor, 1980). However, fluid-dynamic (Herzberg et al., 1983) and magma heat-loss (Spera, 1984) calculations have been undertaken, and most intrusions are considered now to take place near the crust/mantle boundary and especially in the uppermost mantle.

(5) 'Eclogite' and 'granulite' (including kyanite- and quartz-bearing types from the eastern margin of the craton; Fig. 6.1.1) are found in kimberlites from South Australia and western New South Wales (Edwards et al., 1979; Arculus et al., in press), but whether they were entrained from crustal or mantle levels is not clear. The kyanite-bearing eclogites from White Cliffs (New South Wales) have calculated pressures of 18 to 21 kb (50–60 km) at 950–1000°C (O'Reilly et al., unpublished data). Therefore, either the eastern margin of the craton has a very thick crust, or the xenoliths were derived from mantle depths. Finlayson (1982) summarised available data corresponding to a refraction Moho at about 40 km for this cratonic region, which would mean that many of the eclogites were from the mantle. These examples are described here for convenience with other crustal xenoliths, but additional geophysical information is required in order to solve the current dilemma.

Table 6.3.1 Summary of critical mineral assemblages and microstructures in lower-crust xenoliths from eastern Australia

	Ol[1] +cpx +plag ±spin	Ol +2px +plag ±spin	2px +plag	2px +plag ±spin	2px +plag +gnt ±spin	Cpx +plag	Cpx +plag ±spin	Cpx +plag +gnt	Plag +gnt	Cpx +gnt ±opx ±spin	Age of host (Ma)[2]	References
Tasmania												
Pencil Point	–	I	G,I	–	–	G	–	–	–	–	Tertiary?	Sutherland, 1974
Corra Lin Bridge	G	–	–	–	–	–	–	–	–	–	Mid. Tert.	Sutherland, 1969a, 1971
Table Cape	–	–	G	G	–	–	–	–	–	–	Tertiary	Wass & Irving, 1976
Victoria												
Mount Franklin	–	–	G	G	–	–	M,G	–	–	–	Quaternary	Griffin & O'Reilly, 1986a
Mount Shadwell	–	–	G	–	–	–	–	–	–	–	Quaternary	Griffin & O'Reilly, 1986a
Anakie	–	G	–	G	–	–	–	–	G	–	Quaternary	Wass & Hollis, 1983; O'Reilly et al., 1988.
Hepburn Lagoon	M	–	–	–	–	G	G	–	–	–	Quaternary	J.D. Hollis, personal communication, 1985
Mount Bunninyong	–	–	–	–	–	G	G	–	–	–	Quaternary	J.D.Hollis, personal communication, 1985
Mount Leura	–	–	–	–	–	–	G	–	–	–	Quaternary	J.D.Hollis, personal communication, 1985
New South Wales												
Delegate	–	–	G	–	G	–	–	G	–	G	Ca. 194	Lovering & White, 1969; O'Reilly & Griffin, 1985.
Tumut-Eucumbene	–	–	–	G	G	–	–	G	–	–	?	Griffin & O'Reilly, 1986a
Jugiong	–	–	G	G	G	–	–	–	–	G	Tertiary?	Arculus et al., in press; Stracke et al., 1979.
Kellys Point	I,G,M	–	–	–	–	–	G	–	G	–	?	Wass & Irving, 1976
North Kiama	S	–	–	–	–	G	G	–	–	–	Ca. 200?	Griffin & O'Reilly, 1986a
South Bulli	–	–	–	–	–	G	G,S	–	–	–	Ca. 200?	Griffin & O'Reilly, 1986a
Stanwell Park	–	–	–	–	–	–	G,S	–	–	–	Ca. 200?	Griffin & O'Reilly, 1986a
Grabben Gullen	–	–	–	–	–	I	I	–	–	–	23–20	Griffin & O'Reilly, 1986a
Southern Highlands	G	–	–	G	–	–	G	–	–	–	45–35	Griffin & O'Reilly, 1986a; Wass, 1973.
Erskine Park	–	–	–	S	–	–	–	–	–	–	Jurassic	Griffin & O'Reilly, 1986a
Dundas	–	–	–	S	–	–	–	–	–	–	Jurassic	Griffin & O'Reilly, 1986a; Benson, 1911; Wilshire & Binns, 1961.
The Basin	S	–	G	–	–	I	–	–	–	–	Jurassic?	Osborne, 1920
Blue Mountains	–	–	–	–	–	I	–	–	–	–	18–14	Griffin & O'Reilly, 1986a
Glen Alice	–	–	G	–	–	G	–	–	G	–	180	Wass & Irving, 1976; Lovering & White, 1964.
Kandos	I,G	–	–	I,M	–	I	I	–	–	–	170/50	Griffin & O'Reilly, 1986a; O'Reilly & Griffin, 1984.
Dubbo	G	–	–	–	–	–	–	–	–	–	Eocene to Miocene	Wass & Irving, 1976

continued

Table 6.3.1 Summary of critical mineral assemblages and microstructures in lower-crust xenoliths from eastern Australia *(continued)*

	Ol[1] +cpx +plag ±spin	*Ol +2px +plag ±spin*	*2px +plag*	*2px +plag ±spin*	*2px +plag +gnt ±spin*	*Cpx +plag*	*Cpx +plag ±spin*	*Cpx +plag +gnt*	*Plag +gnt*	*Cpx +gnt +opx ±spin*	*Age of host (Ma)[2]*	*References*
New South Wales *(continued)*												
Wallabadah	G	–	–	G	–	–	G	–	–	–	50–30	Wass & Irving, 1976; O'Reilly et al., 1988.
Barrington	–	G	G	–	–	–	–	–	–	–	Ca. 50	Griffin & O'Reilly, 1986a
Lawlers Creek	–	–	G	–	–	–	–	–	–	–	Tertiary?	Wilkinson, 1975a
Boomi Creek	–	–	–	G	–	–	–	–	–	–	Tertiary?	Wilkinson, 1975b
Gloucester	–	–	G	–	G	–	–	G	–	–	160	Wilkinson, 1974
Comboyne	G	–	–	–	G	G	–	–	–	–	Ca. 16	Wass & Irving, 1976; Knutson & Green, 1975.
Ruby Hill	–	–	–	–	–	–	–	G	–	G	Jurassic?	Griffin & O'Reilly, 1986a; O'Reilly & Griffin, 1984; Lovering, 1964.
Kayrunnera	–	–	–	–	–	–	–	G(±ky) (±qtz)	–	G(±qtz)	Ca. 260	Edwards et al., 1979; O'Reilly et al., unpublished data; Stracke et al., 1979.
South Australia												
El Alamein	–	–	G	–	G	–	–	–	–	G	Ca. 170	Arculus et al., in press; McCulloch et al., 1982.
Calcutteroo	–	–	–	G	G	–	–	G(±ky) (±qtz)	–	G	Ca. 170	Arculus et al., in press; McCulloch et al., 1982; O'Reilly et al., unpublished data; Scott Smith et al., 1984.
Queensland												
Mount St. Martin	–	–	G	G	G	–	–	G	–	–	36–31	Ewart, 1982a; Sutherland & Hollis, 1982; Griffin et al., 1987.
Redcliffe Vale	–	–	–	–	–	–	–	G	–	–	28–25	Griffin et al., 1987
Mount Andrew	–	–	–	–	–	G	–	G	–	–	?	Griffin et al., 1987
Sheep Station Knob	–	–	–	–	G	G	–	G	–	–	26.7	Griffin et al., 1987
Mount Leura	–	–	–	–	–	G	–	G	–	–	19.8	Griffin et al., 1987
Bauhinia	–	–	G	–	G	–	–	–	–	–	27–26/25–21	Griffin et al., 1987
Boowinda Creek	–	–	G	–	G	–	–	–	–	–	27.3	Griffin et al., 1987
Brigooda	–	–	G	–	–	–	–	–	–	–	<0.01	Griffin et al., 1987
Hoy	G	–	G,M	–	–	–	–	–	–	–	55–14	Wass & Irving, 1976
Sapphire Hill	–	M,S	G,M	G,M	G,M	–	–	–	–	–	4–0.2	Griffin & O'Reilly, 1986a; Kay & Kay, 1983; Rudnick et al., 1986; Rudnick & Taylor, 1987.

continued

Table 6.3.1   Summary of critical mineral assemblages and microstructures in lower-crust xenoliths from eastern Australia *(continued)*

	Ol^1 +cpx +plag ±spin	Ol +2px +plag ±spin	2px +plag	2px +plag ±spin	2px +plag +gnt ±spin	Cpx +plag	Cpx +plag ±spin	Cpx +plag +gnt	Plag +gnt	Cpx +gnt ±opx ±spin	Age of host (Ma)[2]	References
Queensland *(continued)*												
Batchelor Crater	I,M	I,M	G,M	G,M	G,M	–	–	G,M	–	–	4–0.2	Griffin & O'Reilly, 1986a; Kay & Kay, 1983; Rudnick & Taylor, 1987; in press.
Airstrip Crater	–	G,M	G,M	S	–	G,S	–	–	–	–	4–0.2	Rudnick & Taylor, 1987; O'Reilly et al., 1988.
Lucies Crater	–	–	–	–	–	G	–	G	–	–	4–0.2	O'Reilly & Griffin, unpublished data; O'Reilly et al., 1988.
Hill 32	–	–	G,I,M	G,M	G,M	–	G,M	G,M	G,M (±qtz)	G,M (±qtz)	2.7–0.01	Griffin & O'Reilly, 1986a; Kay & Kay, 1983; Rudnick & Taylor, 1987, in press.
Mount Lang	–	–	G	–	–	–	–	G	–	–	4–0.2	O'Reilly & Griffin, unpublished data.
Atherton	–	–	G	–	–	–	–	–	–	–	3–0.001	Wass & Irving, 1976
Mount Quincan	–	–	G	–	–	–	–	–	–	–	<0.1?	O'Reilly & Griffin, unpublished data; O'Reilly et al., 1988.
Gillies Crater	–	–	G	–	–	–	–	–	–	–	<0.1?	O'Reilly & Griffin, unpublished data.

[1] Abbreviations: Ol, olivine; cpx, clinopyroxene; plag, plagioclase; spin, spinel; 2px, two pyroxenes; gnt, garnet; opx, orthopyroxene; ky, kyanite; qtz, quartz; G, granoblastic; I, igneous; M, recrystallised, relict igneous; S, symplectitic.

[2] Age sources: Wellman & McDougall (1974a), O'Reilly & Griffin (1984), references in Wass & Irving (1976), Griffin et al. (1987), Wass & Shaw (1984), Sutherland (1985), Bishop (1984), Stephenson & Griffin (1976a).

6.3.2 Occurrences of crustal xenoliths

Xenoliths derived from upper- and mid-crustal levels are generally less abundant than are mantle xenoliths. They are confined to relatively ubiquitous but sparse and small fragments of subjacent country rock that have evidence of advanced reaction. Few recorded localities contain abundant mid-crustal xenoliths. The Bullenmerri locality in western Victoria contains large (up to 40 cm) xenoliths of coarse-grained gabbroic and syenitic rocks. The Mogo Hill diatreme (Emerson & Wass, 1980) contains large angular xenoliths of low-grade metamorphic rocks. These are closely similar to exposed Silurian sequences west of the Sydney Basin and provide evidence that such rocks formed the basement for Permian deposition.

Emphasis is given in this section to xenoliths of lower-crustal origin. Over 50 localities of lower-crustal rocks have been documented from eastern Australia and from the eastern margin of the craton. These are summarised in Table 6.3.1, together with rock types and their microstructural types. Relatively few detailed studies of xenoliths from these localities have been reported, but those available include Delegate (Lovering & White, 1969; O'Reilly & Griffin, 1985; McDonough et al., 1988; O'Reilly et al., 1988, Arculus et al., in press), Kellys Point (Halford, 1970; Wass & Irving, 1976), Gloucester (Wilkinson, 1974; O'Reilly et al., 1988), Ruby Hill (O'Reilly & Griffin, 1985), Barraba (Wilkinson, 1975a; Wilkinson & Taylor, 1980; Griffin & O'Reilly, 1986a), the Sydney diatremes (summarised in Griffin & O'Reilly, 1987a), southern Queensland (Ewart, 1982a), central Queensland (Griffin et al., 1987), and northern Queensland (Jones, 1984; Kay & Kay, 1983; Rudnick et al., 1986; Griffin & O'Reilly, 1987a; Rudnick & Taylor, 1987). Cratonic localities include the South Australian El Alamein, Terowie, and Calcutteroo kimberlites, and the New South Wales Kayrunnera (White Cliffs) locality (Edwards et al., 1979; Ferguson et al., 1979). The following morphological, microstructural, and geochemical data are drawn from these xenolith assemblages.

6.3.3 Size, shape, and internal structure

Lower-crustal xenoliths are mostly angular and blocky or tabular. This appears to be a consequence of the common, well-defined, tectonic layering evident in these rock types. There is a size range from less than 1 cm to more than 50 cm and a higher relative incidence of larger crustal xenoliths than for mantle xenoliths. This is probably a consequence of the lower density — and hence greater buoyancy — of the lower-crustal rock types.

Differences in size and xenolith assemblage may be illustrated by crustal xenoliths from two of the scoria cones in the young volcanic areas of northern Queensland (Stephenson & Griffin, 1976a; Kay & Kay, 1983; Rudnick et al., 1986; Rudnick & Taylor, 1987). An exceptional suite of lower-crustal xenoliths is found at Batchelors, Airstrip, and Sapphire Hill craters in the Chudleigh volcanic province (Fig. 3.2.7). These xenoliths are generally fresh, blocky, and mafic (only one felsic xenolith out of 70 has been collected).

They range from 2 to 50 cm across, and most are greater than 10 cm. Compositional layering, defined by alternating plagioclase-rich and pyroxene-rich bands, is present in a few samples, whereas others contain augen of pyroxene set in a plagioclase-rich matrix.

Xenoliths from the McBride volcanic province (Fig. 3.2.4) are generally smaller and more altered than are those from the Chudleigh volcanic province. The McBride xenoliths range from 1 to 20 cm across, and there is considerable compositional diversity. The felsic xenoliths tend to be smaller and more friable than the mafic ones. Mineralogical layering is present in most of the felsic to intermediate xenoliths, whereas about half the mafic xenoliths are layered.

6.3.4 Mineral assemblages and conditions of equilibration

Most east-Australian localities are dominated by intermediate-pressure, two-pyroxene plus plagioclase assemblages (Table 6.3.1). Rarer garnet granulites of similar composition may represent higher-pressure and lower-temperature conditions, but their pressures of formation cannot be estimated because quartz is generally absent. Exceptions to this are some recently reported metasedimentary, quartz-bearing, garnet granulites from Hill 32, McBride province (Rudnick & Taylor, 1986), and quartz-bearing, two-pyroxene granulites from central Queensland (Griffin et al., 1987). These are described in more detail below. Assemblages containing olivine and plagioclase are interpreted to represent a high-temperature, relatively low-pressure environment (Herzberg, 1978). Many of these assemblages have evidence of partial reaction during cooling producing the lower-temperature assemblage of two pyroxenes $\pm$ plagioclase $\pm$ spinel. Xenoliths from Wallabadah, Erskine Park, and Dundas (New South Wales) have undergone complete re-equilibration to this assemblage, forming spectacular coarse-grained symplectites (*Fig. 6.3.1*).

Garnet clinopyroxenites are associated with garnet granulites of inferred crustal origin at Ruby Hill (O'Reilly & Griffin, 1985) and Sapphire Hill (Kay & Kay, 1983). Calculated temperatures for clinopyroxene-garnet pairs are between 875 and 950°C, after correction for Fe^{3+} in the clinopyroxene. Depths of origin for these rocks are near the base of the crust where reference is made to the southeast-Australian geotherm (Section 7.2.3).

Central Queensland (Redcliffe Vale, Mount Andrew, Mount St Martin, Sheep Station Knob, Mount Leura, Bauhinia, Boowinda Creek, and Brigooda) garnet granulites and abundant two-pyroxene granulites are inferred to be fragments of the lower crust (Griffin et al., 1987). The two-pyroxene granulite assemblages contain (in addition to pyroxenes) plagioclase $\pm$ ilmenite $\pm$ magnetite $\pm$ quartz. Many of these xenoliths have abundant prismatic apatite inclusions in all the constituent minerals. Modal pyroxene to plagioclase proportions range from 1:2 to 2:1. Garnet granulites contain two pyroxenes + plagioclase + garnet, or clinopyroxene + plagioclase + garnet + quartz $\pm$ accessory biotite, ilmenite, magnetite, and apatite. Several xenoliths have relict corona structures in which essentially continuous rings of garnet grains surround clinopyroxene grains, probably corresponding to an evolution from olivine + plagioclase or Al-clinopyroxene + plagioclase assemblages (Griffin & Heier, 1973).

Garnet in these granulite xenoliths is pseudomorphed commonly by a fine-grained almost opaque aggregate consisting of clinopyroxene ± orthopyroxene + plagioclase ± spinel ± olivine. Griffin et al. (1987) interpreted the aggregate as the product of contact metamorphism of original garnet at lower-crustal depths. In addition, some clinopyroxene has a spongey rim caused by incipient fusion, some have grain-boundary recrystallisation which also may correspond to a heating event, and some biotite has reacted to K-feldspar and sillimanite + magnetite. A contact-metamorphic interpretation is supported also by the exclusive distribution of these assemblages in localities where the volcanism associated with the xenolith entrainment post-dates the major central-volcano activity in the region. Calculated temperatures of equilibration for these central-Queensland crustal xenoliths range from about 700 to 900°C at pressures from about 5 to 9.5 kb (Griffin et al., 1987), and they plot closely along the east-Australian geotherm.

There is a range of xenolith mineral assemblages in the Chudleigh province although major-element rock compositions are similar. The mineral assemblages therefore have been interpreted to reflect different equilibration conditions (Kay & Kay, 1983; Rudnick & Taylor, 1988). Plagioclase and clinopyroxene are common to all xenoliths. Olivine, spinel, orthopyroxene, and garnet are present in different proportions, reflecting a continuum from lower pressure-temperature conditions (olivine-orthopyroxene-clinopyroxene-spinel-plagioclase — about 700°C, 5–6 kb) to higher ones (garnet-clinopyroxene-plagioclase — about 850-950°C, 12–14 kb; Kay & Kay, 1983; Rudnick et al., 1986; Rudnick & Taylor, 1988). Coronas of olivine rimmed by orthopyroxene which, in turn, is rimmed by symplectitic spinel plus pyroxene, and of spinel rimmed by garnet, are interpreted as the products of isobaric-cooling reactions. There is a divergence of opinion on the interpretation of symplectitic intergrowths of spinel and pyroxene in samples lacking olivine and garnet. Kay & Kay (1983) interpreted the intergrowths as the product of garnet breakdown, necessitating a pressure decrease for these samples, whereas Rudnick & Taylor (1988) suggested the symplectites are caused by olivine breakdown, or breakdown of a complex pyroxene, and can therefore be explained by isobaric cooling.

A limited number of accessory minerals is found in the Chudleigh xenoliths. Magnetite, ilmenite, and rutile are present in many of the samples, but amphibole is rare and, where present, appears to be secondary. Apatite was reported in most samples studied by Kay & Kay (1983), but was not observed in any of those examined by Rudnick & Taylor (1988). Corundum, zircon, and scapolite were observed in three separate samples (Kay & Kay, 1983; Rudnick et al., 1986). CO_2-rich fluid inclusions are present in different proportions in all samples.

The mineralogy of McBride xenoliths is wide-ranging and dependent on bulk composition. Felsic to intermediate xenoliths consist of quartz, feldspar (plagioclase or perthitic K-feldspar), and garnet. Rutile, orthopyroxene, clinopyroxene, ilmenite, apatite, and zircon are common accessory phases, and sillimanite and phlogopite are less common. All felsic to intermediate xenoliths contain negative crystal form CO_2-rich fluid inclusions. Temperatures of about 630 to 930°C and pressures of 8–10 kb were calculated for these xenoliths (Rudnick & Taylor, 1987).

Mafic xenoliths from the McBride province fall into two broad mineralogical categories: two-pyroxene granulite, and garnet-clinopyroxene granulite. Only a few mafic xenoliths contain two pyroxenes that coexist with garnet. Two-pyroxene granulites contain amphibole, phlogopite, ilmenite, apatite, and zircon as common accessory minerals. Two-pyroxene temperatures of 830 to 1000°C were obtained from adjacent mineral-rim analyses (Kay & Kay, 1983; Rudnick & Taylor, 1987) using the method of Wells (1977). Garnet-clinopyroxene granulite contains amphibole, phlogopite, ilmenite, and quartz as common accessory minerals and zircon, rutile, and scapolite are less common. Equilibration temperatures calculated from analyses of adjacent garnet-clinopyroxene rims range from 800 to 1000°C (Kay & Kay, 1983; Rudnick & Taylor, 1987). CO_2-rich fluid inclusions are common in both types of mafic granulites, as they are in the felsic xenoliths.

Two garnet-clinopyroxene granulite samples contain coronas of rutile rimmed progressively by ilmenite, phlogopite, and garnet. These coronas are interpreted to reflect reaction between garnet and rutile forming ilmenite and phlogopite caused by a decrease in temperature and pressure and influx of water and K_2O (Kay & Kay, 1983). Garnet in one intermediate granulite is rimmed by orthopyroxene, reflecting either a pressure decrease or a temperature increase (R.L. Rudnick, personal communication, 1986).

The lower crust of eastern Australia appears from all of these xenolith data to be mainly mafic, and to lie in the intermediate- to high-pressure granulite facies. In contrast, the xenolith assemblages from the eastern craton margin in South Australia and western New South Wales appear to represent a lower-temperature environment and to contain different xenolith types (although still dominantly mafic), including quartz-bearing (± kyanite) mafic eclogite and high-pressure granulite (Edwards et al., 1979; McCulloch et al., 1982; Arculus et al., in press; O'Reilly et al., unpublished data).

Three localities in the eastern craton have been examined in detail (O'Reilly et al., unpublished data): Calcutteroo and El Alamein in South Australia, and Kayrunnera in New South Wales. The xenoliths can be divided into four groups:

(1) *Garnet pyroxenite*. This group includes an Fe-rich 'eclogite' type. Assemblages are garnet + clinopyroxene + rutile ± quartz ± sphene ± biotite ± amphibole, together with secondary analcime or, very rarely, garnet + orthopyroxene + clinopyroxene + spinel. Garnets may be found as fish-net trains or as discrete aggregates and may have different degrees of kelyphitisation. Symplectite developed at clinopyroxene-orthopyroxene grain boundaries is up to 1 mm wide and composed of clinopyroxene + plagioclase ± K-feldspar

± amphibole ± analcime. Recrystallisation of the kelyphite rims around garnet produced aggregates of plagioclase + subcalcic pyroxene + spinel. Some original rutile is broken down to ilmenite, or ilmenite + sphene, and the ilmenite appears in a trellis pattern within the rutile. Clinopyroxene grains in the rare garnet websterites have exsolution lamellae of orthopyroxene and garnet, and fine exsolution lamellae of clinopyroxene are found in orthopyroxene grains.

(2) *Mafic granulite (kyanite-free)*. This group can be subdivided into: (1) two-pyroxene granulites consisting of clinopyroxene + orthopyroxene + plagioclase ± quartz ± K-feldspar ± ilmenite ± amphibole; (2) garnet+two-pyroxene granulites consisting of garnet + clinopyroxene + orthopyroxene + plagioclase ± quartz ± K-feldspar ± ilmenite ± amphibole; (3) garnet-clinopyroxene granulites consisting of garnet + clinopyroxene + plagioclase ± quartz ± amphibole ± rutile ± biotite ± scapolite ± sphene. Common features of the two-pyroxene and garnet+two-pyroxene granulites are garnet+clinopyroxene symplectites separating clinopyroxene, orthopyroxene, or clinopyroxene/plagioclase grains. Clinopyroxene grains may have exsolved garnet and orthopyroxene lamellae. Plagioclase is commonly antiperthitic and rare perthite is observed also. Some garnet+two-pyroxene granulites from El Alamein have corona structures developed between orthopyroxene and plagioclase grains. Successive layers consist of fine-grained granular clinopyroxene adjacent to the orthopyroxene, then clinopyroxene/garnet symplectite, followed by homogeneous garnet in contact with the plagioclase. The garnet-clinopyroxene granulites commonly have a fish-net microstructure defined by clinopyroxene, plagioclase, and quartz aggregates, separated by trains of garnet grains.

(3) *Kyanite+garnet+quartz granulite and pyroxenite*. Assemblages are kyanite + quartz + garnet + clinopyroxene ± plagioclase + rutile. These rocks are more aluminous and felsic than are the garnet pyroxenites and mafic granulites (Edwards et al., 1979), and their compositions are consistent with MORB or high-alumina basaltic protoliths.

(4) *Quartzo-feldspathic granulite*. This xenolith type is rare and consists of garnet + kyanite + K-feldspar + quartz ± biotite ± rutile ± zircon ± apatite. K-feldspar may be perthitic. Arculus et al. (in press) also recorded one garnet-kyanite gneiss which is geochemically similar to average upper-crustal compositions (Taylor & McLennan, 1985). However, it does not contain suitable minerals for determination of pressure and temperature, so its depth of origin cannot be estimated. Geothermobarometric results for the South Australian and Kayrunnera xenoliths define a lower geothermal gradient than that for eastern Australia. Kyanite granulites from White Cliffs cluster along the highest pressure equilibration region at temperatures of 950–1000°C and pressures of about 18–20 kb. Garnet+clinopyroxene granulites from Calcutteroo give temperatures of 850–1000°C at 13–17 kb, and those from White Cliffs spread from about 800–1050°C at 10–22 kb. Two-pyroxene granulites (plus a garnet websterite) from El Alamein and White Cliffs yielded temperatures of about 650–850°C at 7–11 kb.

6.3.5   Microstructures

Crustal xenoliths from non-cratonic eastern Australia have a wide range of microstructural types — from unmodified igneous microstructures (including gabbroic types and cumulates), through partially modified igneous microstructures, to thoroughly recrystallised metamorphic fabrics. Metamorphic microstructures include equigranular mosaic, coarsely foliated, and highly deformed and sheared types. Foliated and deformed microstuctures are the most common type, an observation relevant to lower-crustal granulites worldwide (see, for example: Griffin & O'Reilly, 1986a; Amundsen et al., 1987). This ubiquitous deformation of lower-crustal rock types possibly reflects the rheological contrast between the dominant granulites and the underlying ultramafic rock types of the uppermost mantle (Kirby, 1985; Passchier, 1986; De Rito et al., 1983).

Grain-sizes range widely, even in xenolith types that have the same mineral assemblage. This is illustrated well in the symplectitic two-pyroxene+spinel granulites. All examples of this xenolith type from Dundas and some from Wallabadah are coarsely foliated and have a grain-size of about 0.5 cm. However, some from Wallabadah are sheared or finely foliated and the grain-size is less than 1 mm (Fig. 6.3.2).

In contrast, anorthositic xenoliths in basaltic dykes at Bingie Bingie Point, New South Wales (Halford, 1970; locality formerly known as Kellys Point; Wass & Irving, 1976) have unmodified igneous grain shapes and relationships. Grain-sizes of these gabbroic xenoliths (consisting of different combinations of plagioclase, kaersutite, clinopyroxene, garnet, and Fe-rich olivine) are up to 5 cm across.

Granulites of inferred crustal origin from Gloucester, Ruby Hill, and Delegate all have relatively coarse granoblastic microstructures and most grain-sizes are 1 to 4 mm. Some of these xenoliths appear to have an isotropic fabric (Figs. 6.3.3-4) which could be an artefact of the small size of some xenoliths obscuring a relatively large-scale foliation. However, most are foliated on the hand-specimen scale (Fig. 6.3.5).

Crustal xenoliths from Victoria are rare and have not been described in the literature, but some information is available from J.D. Hollis (personal communication, 1986) and O'Reilly and Griffin (unpublished data). Most of these xenoliths (most abundant at Mounts Franklin and Bunninyong and at Hepburn Lagoon, near Daylesford, Victoria) are clinopyroxene+plagioclase assemblages that have coarsely-foliated granoblastic microstructures. They rarely have corona development or other evidence of reaction.

The microstructures of xenolith assemblages from New South Wales and Victoria contrast with those from central and southern Queensland which commonly have evidence of complex temperature and pressure histories (corona development, other reactions, and mineral zoning for the Queensland xenoliths were described in Section 6.3.4). The microstructures are more diverse because of these partial equilibrations.

Kay & Kay (1983) and Rudnick et al. (1986) described the Chudleigh province xenoliths as fine- to coarse-grained (0.3-5 mm) and noted deformation

Figure 6.3.2   A finely-foliated symplectitic two-pyroxene+ spinel granulite from Wallabadah (New South Wales). Width photographed (in plane polarised light) is 1.3 cm.

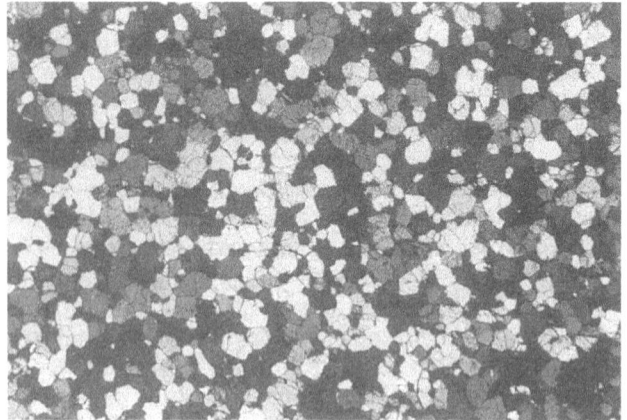

Figure 6.3.4   Fine-grained, isotropic, granoblastic microstructure in a granulite xenolith from Delegate (New South Wales). Width photographed (in crossed polars) is 2.4 cm.

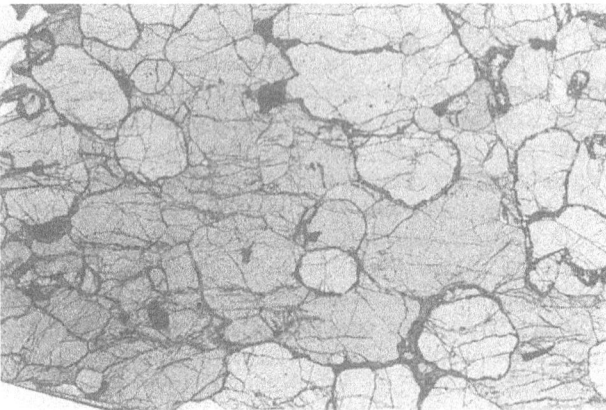

Figure 6.3.3   Granulite xenolith from Delegate (New South Wales) showing coarse-grained, isotropic, granoblastic microstructure. Width photographed (in plane polarised light) is 1.3 cm.

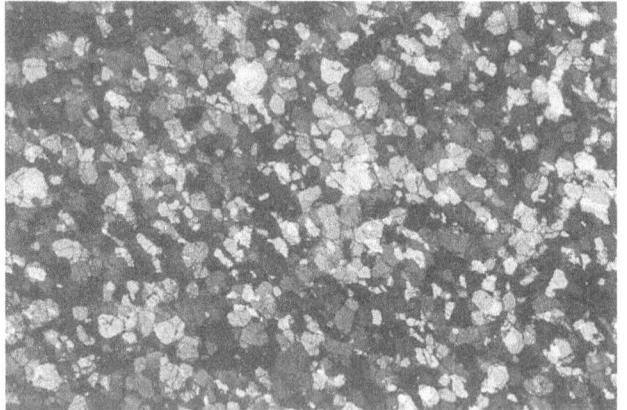

Figure 6.3.5   Fine-grained foliated microstructure in a granulite xenolith from Delegate (New South Wales). Width photographed (in crossed polars) is 2.4 cm.

features such as sub-grain development and bent plagioclase crystals superimposed on granoblastic microstructures. Many of these xenoliths also have well-defined gneissic foliations ranging from coarse to fine scale. Some xenoliths have incipient recrystallisation around relict plagioclase and orthopyroxene grains. Coronas and reaction rims are common.

McBride-province crustal xenoliths are fine- to medium-grained (0.1–3 mm) and generally have granoblastic microstructures. Evidence of deformation and relict igneous microstructures (for example, lath-shaped plagioclases) are found in only a few samples (Kay & Kay, 1983). In contrast to Chudleigh-province xenoliths, corona textures are uncommon in the McBride-province samples. Two types of coronas are: (1) rutile progressively rimmed by ilmenite, phlogopite, and garnet (Kay & Kay, 1983); (2) garnet surrounded by thin rims of orthopyroxene (R.L. Rudnick, personal communication, 1986).

A pronounced foliation commonly is evident in the central-Queensland crustal xenoliths (Griffin et al., 1987), defined by alignment of grain clusters or by layering of mafic and felsic domains on a scale of millimetres to centimetres. Microstructures within this foliation are granular mosaic. Most garnet granulites have extensive corona development and clusters

of clinopyroxene ringed by garnet. The garnet is extensively reacted (contact metamorphosed) to aggregates of clinopyroxene + orthopyroxene + plagioclase ± magnetite ± spinel ± olivine which may appear as a fine-grained black mass in thin section or as coarse intergrowths.

The South Australian and Kayrunnera mafic xenoliths have some microstructures similar to those of the eastern off-craton region, but kyanite granulites are distinctly different. The garnet pyroxenites are coarsely granoblastic and have an average grain-size of 3–5 mm but ranging up to about 8 mm. Foliation is not evident, possibly because of the coarse grain-size relative to xenolith size. The mafic granulites are finer-grained and generally have an average grain-size of about 1 mm. Foliation may be defined by grain elongation. Garnet-bearing mafic granulites commonly have a fishnet or corona microstructure (Fig. 6.3.6) and rings of garnet grains separating plagioclase and clinopyroxene domains. Quartz-bearing types generally have more equigranular grain shapes and little evidence of foliations or lineations. Fine-grained coronas are observed only in some of the El Alamein xenoliths. Some apparent rock contacts in xenoliths probably represent layering caused by differences in modal proportions of garnet and clinopyroxene.

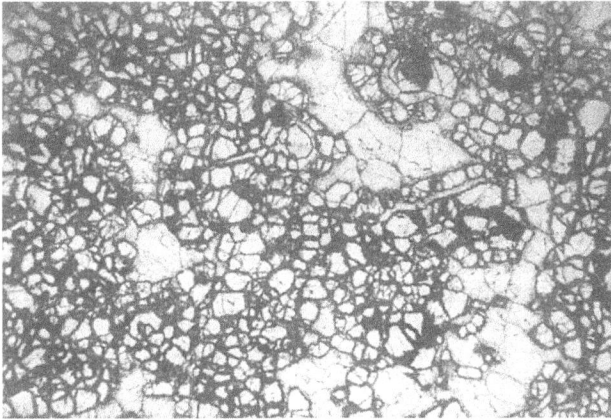

Figure 6.3.6 Fish-net or corona structure in garnet granulite from Calcutteroo (South Australia). Width photographed (in plane polarised light) is 1.3 cm.

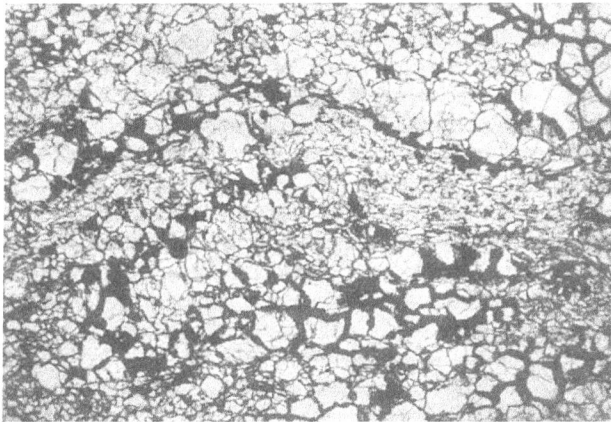

Figure 6.3.7 Kyanite parallel to microfolds in kyanite-garnet granulite from Kayrunnera (New South Wales). Width photographed (in plane polarised light) is 1.3 cm.

The kyanite-bearing and quartzo-feldspathic granulites are characterised by deformation structures. Strongly differentiated layering in the kyanite granulites on a 1–5 mm scale is defined by monomineralic lenses of clinopyroxene against layers of garnet + kyanite + clinopyroxene ± plagioclase ± quartz. Elongated rings of finer-grained garnets may form a fish-net microstructure within this foliation. Kyanite wraps round microfolds (Fig. 6.3.7). Subhedral to euhedral garnet porphyroblasts commonly have been rotated within the foliation as shown by different alignments of included kyanite needles.

6.3.6 Major- and trace-element geochemistry

Composition of the lower crust

The chemical composition of the lower crust is a critical factor relevant to many large-scale concepts. It is vital to geophysical interpretations, and it provides chemical constraints on the evolution of the crust-mantle system and mechanisms for crustal growth.

East Australian crustal xenoliths are overwhelmingly of mafic composition (Kay & Kay, 1983; Griffin & O'Reilly, 1986a; Rudnick et al., 1986; Rudnick &

Taylor, 1987). This observation can be extended to a world-wide generalisation (Griffin & O'Reilly, 1987a), and is a contradiction to the widely accepted 'andesite model' (Taylor & McLennan, 1985) in which the lower crust is intermediate in composition and has formed through accretion and differentiation of andesitic island-arc material.

There are four most likely possible origins for lower-crustal rocks: (1) burial or underthrusting of supra-crustals; (2) intrusion of mantle-derived melts and subsequent crystallisation of melts or formation of cumulates (or both); (3) tectonic accretion; (4) any of 1–3, together with a superimposed melting episode resulting in a residual lower-crustal lithology (restite). The following geochemical data form a basis for distinguishing between these possibilities.

Representative whole-rock analyses of about 80 granulites are available in the literature (Edwards et al., 1979; Rudnick et al., 1986; Griffin & O'Reilly, 1986a; Griffin et al., 1987; Rudnick & Taylor, 1987; O'Reilly et al., 1988; Arculus et al., in press) or as unpublished data (O'Reilly and Griffin). Isotopic data are available for granulite xenoliths from Delegate, South Australia, central Queensland, Chudleigh, Anakie (Victoria), Wallabadah, Ruby Hill, Gloucester, and diatremes in the Sydney Basin (Rudnick et al., 1986; McDonough et al., 1988; O'Reilly et al., 1988).

Many granulites (for example, Anakie and most Delegate samples) are interpreted to be of sub-crustal origin, as discussed in Section 6.3.1. However, there is no overall difference in major-element composition between granulite xenoliths of lower (crustal) or higher (mantle) pressure origin, so mantle-derived granulites are included in this discussion. Nevertheless, trace-element and isotopic compositions of granulites may be affected by geochemical interaction with wall-rocks (Rudnick et al., 1986; Griffin & O'Reilly, 1987a), and therefore crustal and mantle types must in this case be considered separately.

Most xenoliths have compositions consistent with derivation from a basaltic parent. Metasedimentary types are very rare, and known examples are confined to the McBride province (Rudnick & Taylor, 1987) and to South Australia (Arculus et al., in press). Note that both of these regions are underlain by Precambrian crust — the Georgetown Inlier and Gawler Block (on the eastern margin of the western craton), respectively (Fig. 6.1.1). The basaltic types range from nepheline-normative to hypersthene-normative (and rarely even to quartz-normative), overlapping the range of compositions typical for Cainozoic intraplate basaltic rocks in eastern Australia (Fig. 6.3.8). The lower K_2O values (less than 0.3 weight percent) are, however, anomalous for intraplate basalts. This may mean that many of the rocks formed as basaltic cumulates that had little trapped liquid, or that they lost K_2O during metamorphism.

Griffin & O'Reilly (1986a) noted that the presence of garnet in lower-crustal granulites appears to reflect bulk compositional differences, at least as much as different pressure-temperature conditions. Most of the nepheline-normative rocks are garnet-bearing, whereas the garnet-free pyroxene granulites are mainly hypersthene-normative (Fig. 6.3.8). There are exceptions to this, notably the Anakie xenoliths (Wass & Hollis, 1983) in which the appearance of garnet is independent of rock composition. This probably reflects the large depth range sampled relative to many other localities. The Anakie xenoliths have equilibrated at conditions ranging from granulite to eclogite facies and spanning from near the crust/mantle boundary to upper-mantle depths (about 12–20 kb). Another exception is the Boomi Creek suite (Wilkinson, 1975a) which consists of nepheline-normative, two-pyroxene+ spinel granulites. These appear to have crystallised at lower pressures (about 8 kb) than have garnet granulites of similar composition.

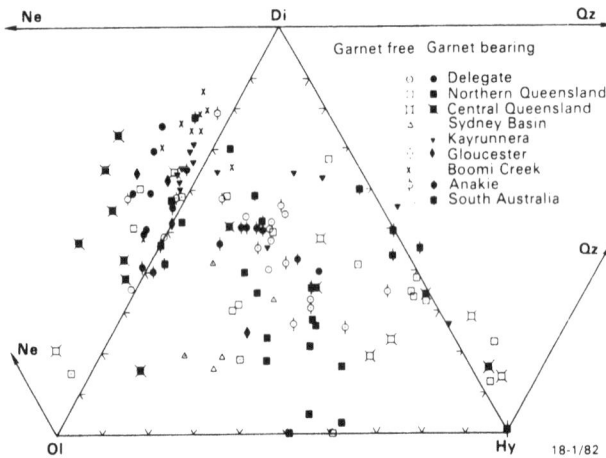

Figure 6.3.8 Granulite compositions from eastern Australia and east-cratonic terranes plotted in the system Ne-Ol-Di-Hy-Qz. All analyses were recalculated so that $Fe_2O_3/(Fe_2O_3+FeO)$ equals 0.2 before calculation of CIPW normative components.

The chemical data available are limited and the general spread of compositions is equivalent to that for intraplate basaltic rocks in eastern Australia, but some xenolith suites from individual localities plot in restricted fields (Fig. 6.3.8). Examples are Boomi Creek (Wilkinson, 1975a), Chudleigh province (Rudnick et al., 1986), and Gloucester (see below). Others, such as Anakie, spread across the whole range. This may correspond to derivation from the same or closely-related magmas for the former and to multiple parental sources (or protolith origins) for the latter.

Available detailed geochemical data for crustal-xenolith suites are summarised below. Trace-element and isotopic exchange with adjacent lithologies may have taken place, so these data for the Anakie, Delegate, and Bullenmerri-Gnotuk granulites, which are mantle-derived, are discussed in Section 6.2.

Gloucester, Sydney Basin, and Delegate

Garnet-bearing granulites from Delegate and Gloucester (New South Wales) form a tight cluster within the nepheline-normative field of Figure 6.3.8. Garnet-free granulites from these and from some Sydney Basin localities (Erskine Park diatreme and Wallabadah) are in the hypersthene-normative field. All compositions are consistent with derivation from basaltic liquids. Isotopic data for these are discussed below in Section 6.3.7.

Boomi Creek

Major- and trace-element data for the xenoliths of Boomi Creek (New South Wales) define trends of decreasing Cr, Ni, and Co, and increasing Ti, V, Cu, Ba, Zr, Hf, Nb, and Y, as Larsen Index increases (Wilkinson, 1975a; Wilkinson & Taylor, 1980). These trends are interpreted as reflecting progressive fractionation of clinopyroxene, spinel, and late-stage titanomagnetite from a common parent magma that may have been subalkaline (Wilkinson, 1975a). The REE patterns are light-REE enriched, and there are slight positive Eu anomalies possibly because of cumulate plagioclase.

Chudleigh province

Mafic lower-crustal xenoliths from the Chudleigh volcanic province (northern Queensland) have been subdivided into three groups based on mineralogy, major-element composition, and REE patterns: plagioclase-rich xenoliths, pyroxene-rich xenoliths, and xenoliths transitional between these two types (Rudnick et al., 1986). Plagioclase-rich xenoliths are characterised by high Al_2O_3 contents and by REE patterns that are light-REE-enriched and which have large positive Eu anomalies. Pyroxene-rich xenoliths have abundant modal pyroxene, low Al_2O_3 contents, and REE patterns that are light-REE depleted and which have no Eu anomaly. Transitional xenoliths have intermediate Al_2O_3 contents, and their REE patterns are slightly light-REE depleted and have small positive Eu anomalies. Incompatible elements (La, Ba, Th, U, and K_2O) in these xenoliths correlate negatively, and compatible trace elements (Ni and Cr) correlate positively, with Mg-ratios. Trace elements that are moderately incompatible to compatible in pyroxene (Zr, Hf, Y, V, and Ti) have a scattered distribution where plotted against Mg-ratio, but correlate negatively with Al_2O_3 content. These chemical differences are interpreted by Rudnick et al. (1986) to reflect genetic links between the three types of xenoliths. Strongly incompatible and strongly compatible trace-element concentrations are controlled primarily by the evolution of the coexisting melt, whereas trace elements that correlate with Al_2O_3 content are controlled by the proportion of pyroxene to plagioclase in the rock. The REE patterns, plus the wide range of incompatible trace-element contents and the low, fairly similar compatible trace-element abundances, may be indicative that the xenoliths are cumulates from a mafic melt rather than representing restites.

Sr- and Nd-isotopic compositions (Fig. 6.3.9) of Chudleigh xenoliths are wide-ranging ($^{87}Sr/^{86}Sr$ values from 0.70239 to 0.71467, and ε_{Nd} values from +9.6 to −6.1) and both $^{87}Sr/^{86}Sr$ and ε_{Nd} correlate with each other and with Mg-ratio. The isotopic ratios cannot be produced by simple igneous differentiation of an initially homogeneous magma, and require either an isotopically heterogeneous mantle source or progressive assimilation of crustal materials (Rudnick et al., 1986). The correlations between Sr- and Nd-isotopic compositions and Mg-ratio are not those expected to result from different degrees of partial melting in the mantle (either homogeneous or heterogeneous). The xenoliths therefore represent a series of cumulates from an evolving, mantle-derived magma that progressively assimilated crustal material as fractionation proceeded. The good correlations between radiogenic isotopes and Mg-ratio degrade as the isotopic ratios are back-calculated to earlier times, so the Chudleigh-province xenoliths are probably less than 100 Ma old and may be related to the Cainozoic basaltic magmatism (Rudnick et al., 1986).

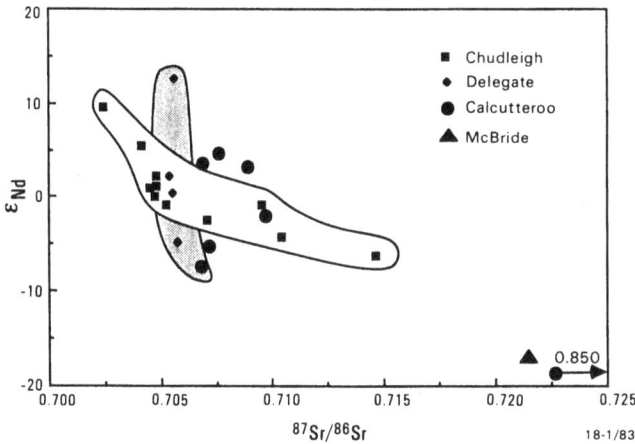

Figure 6.3.9 Nd- and Sr-isotope relationships for granulite xenoliths. Data for Delegate and Calcutteroo are plotted at age of eruption (about 170 Ma). Shaded field represents Delegate data. Unshaded field represents Chudleigh data.

McBride province

Chemical and isotopic data can be used to classify the xenoliths of McBride province (northern Queensland) into two broad categories: (1) intermediate to felsic, and (2) mafic. There is no evidence for a genetic relationship between the two groups.

The intermediate to felsic xenoliths range from metapelitic compositions (19–20 percent Al_2O_3 at 55–57 percent SiO_2) to dioritic and granodioritic compositions. REE patterns for these xenoliths are generally similar to average upper crust (post-Archaean Australian shale, PAAS; Taylor & McLennan, 1985), but the heavy REE are up to three times enriched relative to PAAS. They are similar in this respect to metasedimentary granulite xenoliths from Kilbourne Hole, New Mexico (Wandless & Padovani, 1985). Enrichment in Ni and Cr in these xenoliths is a primary feature (up to 118 ppm Ni at 57 percent SiO_2) and is suggestive of derivation from sedimentary rather than igneous protoliths.

The mafic McBride-province xenoliths range from silica undersaturated to oversaturated compositions, and trace-element characteristics have been used to subdivide the mafic xenoliths into four genetic groups: (1) possible melts; (2) cumulates; (3) residua; (4) those affected mainly by metamorphic differentiation.

Only one mafic xenolith has a major- and trace-element composition similar to that of a melt. This sample (85–108) is silica oversaturated and has enriched light-REE and a negative Eu anomaly (Fig. 6.3.10A). Xenoliths representing likely cumulates are characterised by REE patterns that are either light-REE depleted, having little or no Eu anomaly, or

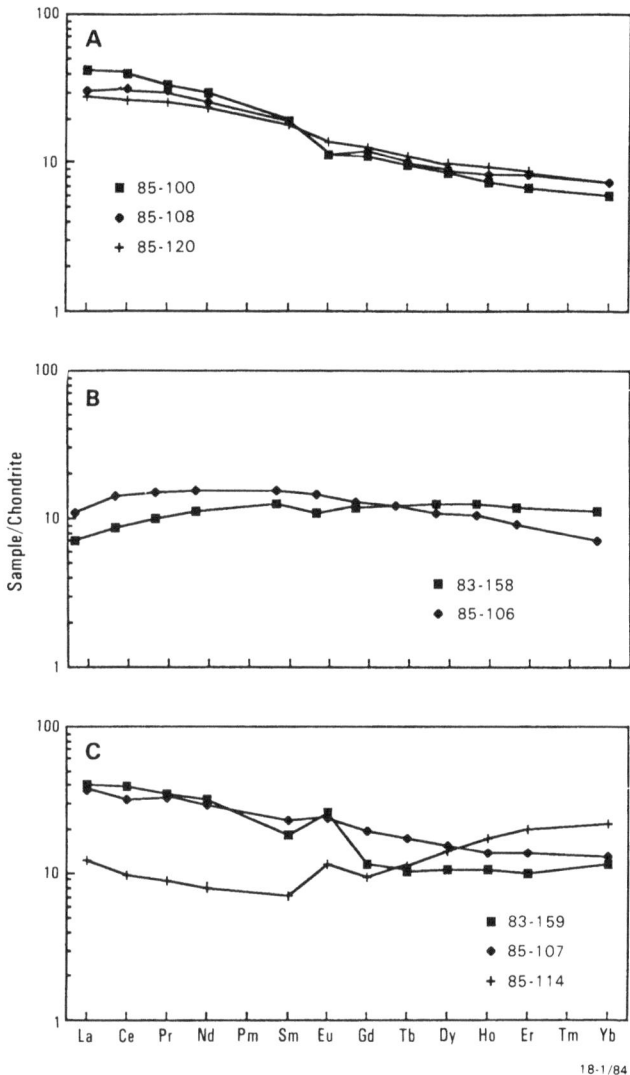

Figure 6.3.10 REE diagrams for north-Queensland lower-crustal xenoliths from Hill 32, McBride province. (A) Mafic melts. (B) Mafic cumulates. (C) Mafic restites or cumulates.

light-REE enriched and having a positive Eu anomaly (Fig. 6.3.10B).

Two mafic xenoliths are possible residua. Sample 85–114 is characterised by slightly enriched light-REE, a large positive Eu anomaly, and strongly enriched heavy-REE (Fig. 6.3.10C). Sample 85–107 is REE-enriched and has a positive Eu anomaly (Fig. 6.3.10C). This sample also has abundant zircon (0.3 normative percent) for a mafic rock (53-percent SiO_2). Zircon is not saturated in mafic melts until very high degrees (about 90 percent) of crystallisation (Dickinson & Hess, 1982) but may be a residual phase during formation of felsic melts (Watson & Harrison, 1983), so this rock is probably restite.

Finally, the composition of one mafic xenolith (85–100) from the McBride province appears to be controlled by the presence of thick, orthopyroxene-rich layers. The REE pattern of this sample (Fig. 6.3.10A) is nearly identical to that of 85–108 (the possible melt), but its major-element composition is distinctive: low Al_2O_3 (11 percent), moderate SiO_2 (52 percent), high MgO (14 percent), and low CaO (7 percent). This composition, in combination with very high Cr and Ni contents (1026 and 465 ppm, respectively) may be interpreted to reflect orthopyroxene enrichment through metamorphic processes. However, primary igneous layering would produce the same characteristics and would be consistent with the cumulate origin inferred for other xenoliths from McBride province.

The heterogeneous nature of the lower crust in the McBride region is illustrated by the chemistry of the McBride xenoliths. Intermediate to felsic metasediments, mafic melts, cumulates, and restites are found within a single xenolith suite. This compositional diversity may reflect a more protracted history of lower-crust formation and modification in this crustal block relative to the adjacent Chudleigh province.

East-central Queensland

Griffin et al. (1987) presented major- and trace-element data for granulite xenoliths from east-central Queensland. Most of these xenoliths are basaltic, ranging from basanitic through *ol*-tholeiitic to mildly quartz-normative compositions. Some have high Mg-ratios (greater than 70), low total-alkalies, and high Ca or Al (or both) indicative of a cumulate origin.

A cumulate origin for many of the granulites is supported also by trace-element data. Ni is generally low (about 50 ppm) despite high Mg-ratios. There is a rough correlation of Cr and V with Mg-ratio, corresponding to dominant clinopyroxene (rather than olivine) fractionation. Sr correlates roughly with Al_2O_3, but not CaO, which is also consistent with cumulate assemblages dominated by clinopyroxene and plagioclase.

Griffin et al. (1987) concluded that some xenoliths originated as troctolitic cumulates, some as norites, and one as an olivine+clinopyroxene-rich cumulate. One analysed xenolith closely resembles a basaltic liquid, containing high Ti, P, Zr, and Y, and having a moderate Mg-ratio (58).

These granulite xenoliths are broadly equivalent in composition to the Tertiary basalts erupted in the same area. Most of the nepheline-normative rocks are garnetiferous, whereas most of the hypersthene-normative ones are not. However, as discussed above, the appearance of garnet may be controlled in part by pressure as well as by bulk composition, as suggested by the presence of some silica-saturated (and over-saturated) garnet granulites.

Kayrunnera and South Australia

Most of the Kayrunnera (New South Wales) and South Australian xenoliths are broadly basaltic and may have originated as basaltic melts or cumulates. Edwards et al. (1979) showed that the Kayrunnera mafic xenoliths range from nepheline-normative to slightly hypersthene-normative (Fig. 6.3.8). The kyanite+quartz granulites (with or without K-feldspar) are hypersthene-normative and have affinities with high-alumina basalts and MORB.

McCulloch et al. (1982) and Arculus et al. (in press) presented major- and trace-element and isotopic data for seven diverse xenoliths from Calcutteroo. Arculus et al. (in press) referred to a larger unpublished data base and stated that the majority of samples are

basaltic, few are intermediate (SiO_2 about 55 percent), and rare xenoliths have SiO_2 contents of 60–68 percent. Trace-element compositions of the mafic xenoliths are wide-ranging, and there are no correlations with major elements. Several samples contain unusual enrichments of Ba, Sr, and Nb, which are of undetermined origin. One sample has a positive Eu anomaly, corresponding to preferential enrichment of plagioclase, either through cumulate or restite processes. Different xenolith types probably represent different origins and are not genetically related.

Nd- and Sr-isochron ages near 2.5 Ga (large uncertainties) were reported by McCulloch et al. (1982). However, these ages are heavily influenced by a single felsic xenolith which has a Nd model age younger than the Sr-isochron age (T_{CHUR} 1975; T_{DM} 2171). A younger age of 1300 Ma for the mafic xenoliths was proposed by Rudnick et al. (1986) based on back-calculation of Nd- and Sr-isotopic ratios to a time at which they fall along a mixing curve on a ε_{Nd} versus $^{87}Sr/^{86}Sr$ diagram. However, these xenoliths may not constitute a cogenetic suite and may have been formed by different processes at different times.

6.3.7 Isotopic characteristics

Sr- and Nd-isotopic data (O'Reilly et al., 1988) for granulite xenoliths from off-craton eastern Australia and from different geographic provinces and age groups, define distinct trends which imply a range of origins (Fig. 6.3.11).

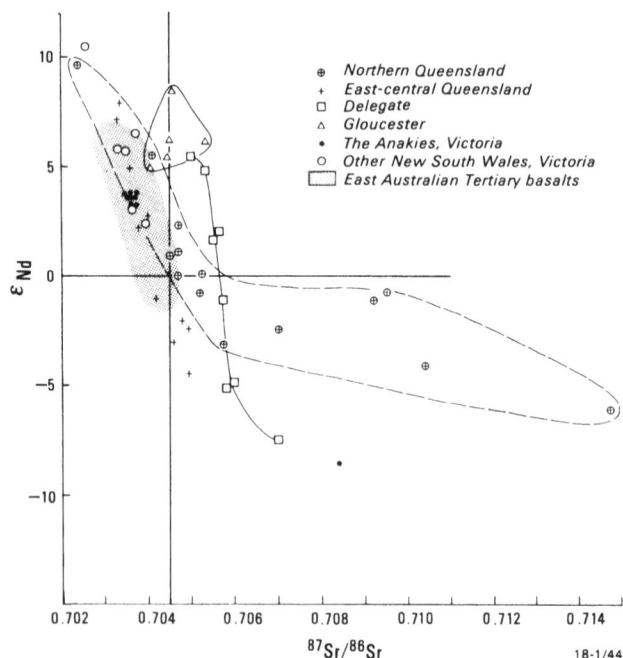

Figure 6.3.11 Nd- and Sr-isotope compositions for granulite xenoliths from eastern Australia. Data sources were given by O'Reilly et al. (1988).

Delegate diatreme and an alnöite near to the Gloucester diatreme in New South Wales were dated at 150–200 Ma (Lovering & White, 1969; Sutherland, 1985; Griffin and O'Reilly, unpublished data). The Gloucester xenoliths cluster at high ε_{Nd} values, but have higher $^{87}Sr/^{86}Sr$ than do most MORB or intraplate basalts. The Delegate xenoliths define a remarkable linear 'tail' extending from the Gloucester group to low ε_{Nd}, and all have high $^{87}Sr/^{86}Sr$ values (0.705–0.707). This tail includes both two-pyroxene granulites and garnet±sapphirine granulites of apparent mantle origin (Griffin & O'Reilly, 1986b). The pattern remains unchanged relative to Bulk Earth values where the data are calculated back to 300 Ma, but the Gloucester field becomes even smaller.

Granulite xenoliths from Tertiary and older centres in New South Wales (Wallabadah, Erskine Park, north Kiama) and the Anakies in Victoria fall on a 'mantle array' similar to that defined by east-Australian xenolith-bearing basalts (O'Reilly and Griffin, unpublished data). Granulites from Tertiary basalts in east-central Queensland (Griffin et al., 1987) also follow this array, but extend it to low ε_{Nd} values. Otherwise similar granulite xenoliths from northern Queensland (Kay & Kay, 1983; Rudnick et al., 1986) define a field extending to high $^{87}Sr/^{86}Sr$ and low ε_{Nd} (Figs. 6.3.9,11).

The isotopic compositions of the Delegate and Gloucester xenoliths resemble those of some island-arc basalts and seawater-altered MORB, plus an additional crustal component. Several of the samples yield pressure-temperature estimates that place them well within the mantle (O'Reilly & Griffin, 1985; Griffin & O'Reilly, 1987a). O'Reilly et al. (1988) therefore suggested that the granulite xenoliths from Gloucester and Delegate contain a component of tectonically underplated mafic material related to subduction processes active during the Palaeozoic continental accretion of eastern Australia. The other xenoliths from southeastern Australia are isotopically similar to the Tertiary basalts in the same region. This is consistent with the formation of the granulites by recrystallisation of cumulates from magmas intruded at or near the crust/mantle boundary. A similar origin may be proposed for the Queensland xenoliths, but these apparently have interacted with pre-existing crustal material. Those from east-central Queensland may have been in contact with a mafic lower crust, characterised by light-REE enrichment and low Rb/Sr. A model for the development of this type of crust was given by Ewart (1982a). The north-Queensland granulites apparently have undergone isotopic exchange with a pre-existing old crust containing a higher proportion of felsic and metasedimentary rocks (Rudnick et al., 1986). This suggestion is consistent with the inferred presence of Precambrian inliers in the Palaeozoic terrane of northern Queensland.

The isotopic data therefore are illustrative of the importance of mafic magmas in the generation of the lower continental crust. These may be emplaced both tectonically, during continental accretion, and as intrusions near the crust/mantle boundary.

6.3.8   Physical properties

Information on the physical properties of crustal xenoliths is necessary for the interpretation of geophysical data (see Section 7.2). Measurements and calculations of seismic velocities and magnetic charac-

teristics have been undertaken on east-Australian xenoliths.

Jackson & Arculus (1984) carried out laboratory measurements at 4 kb and 25°C on crustal xenoliths from the Calcutteroo kimberlite. The rock types range from garnet-clinopyroxene granulite to garnet pyroxenite and one kyanite+garnet+quartz+K-feldspar granulite. Compressional-wave velocities (Vp) range from 7.2–8.0 km/s, and shear-wave velocities (Vs) from 4.1–4.5 km/s. Corrections for the higher temperatures and pressures inferred for the xenolith-source regions by Jackson & Arculus (1984) yield Vp values of about 6.9–7.7 km/s and Vs values of 3.9–4.3 km/s. The specific-gravity range for the Calcutteroo xenoliths is 3.24-3.40 g/cc for the more mafic xenoliths and 2.94 g/cc for the kyanite-bearing granulite.

Rudnick & Jackson (1987) recorded preliminary results for laboratory determinations of Vp for granulites from the McBride and Chudleigh (northern Queensland) provinces. Vp ranges from 6.9 to 7.6 km/s at 10 kb and 25°C, and specific gravity ranges 2.8–3.0 g/cc.

O'Reilly & Griffin (1985) and Griffin et al. (1987) calculated seismic velocities for granulite xenoliths from southeastern Australia and central Queensland, respectively. This was done for xenoliths of known chemical composition and specific gravity using algorithms given by Manghnani et al. (1974) to give Vp at room temperature and pressure. Corrections for the effects of pressure and temperature were carried out using the dV/dT and dV/dP values of Anderson & Sammis (1970). The average Vp at 8 kb and 800°C of nine mafic granulites from southeastern Australia (O'Reilly & Griffin, 1985) is 6.95 km/s, and 12 mafic granulites from central Queensland gave Vp average of 6.66 km/s. The higher values for the southeast-Australian xenoliths reflect the inclusion of some dense samples from Delegate and Gloucester.

Magnetic characteristics of a wide range of lower-crustal granulite xenoliths have been determined (P.J. Wasilewski, personal communication, 1986). These granulites are weakly magnetic and have a Curie temperature consistent with that for magnetite.

6.3.9   Summary and conclusions

Xenolith data from eastern Australia correspond to a lower crust of dominantly mafic composition, and include rock compositions ranging from anorthosite/troctolite through gabbros to pyroxenites. These compositions are not unique. Griffin & O'Reilly (1987a) carried out a world-wide survey on available data for suites of granulite- or eclogite-facies xenoliths inferred to have been derived from the lower continental crust (and also deep levels beneath some oceanic intraplate basaltic centres), and showed that these xenolith suites have a similar range of mafic compositions whether they are derived from orogenic belts (Precambrian to Holocene) or from cratonic regions. Felsic to intermediate meta-igneous rocks or meta-sediments are rare and, where present, are subordinate in volume to the mafic granulites. There are only two known examples of possible supracrustal xenolith rock types in eastern Australia: McBride province (Rudnick & Taylor, in press) and Calcutteroo (Arculus et al., in press). Both of these localities are inferred to be underlain by cratonic crust. Most of the mafic xenoliths are consistent with a basaltic cumulate origin or, rarely, with being frozen melts. R.L. Rudnick (personal communication, 1986) identified rare probable restite from the McBride province.

The microstructures of the mafic xenoliths range from those of unaltered cumulates to thoroughly recrystallised types some of which have a high degree of deformation. Lower-crustal xenoliths from convergent plate margins typically have igneous microstructures that may retain evidence for partial readjustment and reaction. Well-equilibrated granoblastic microstructures and evidence of strong deformation are rare. The most common rock types in older continental and cratonic regions (for example: Colorado Plateau, Arculus & Smith, 1979; Lesotho, Griffin et al., 1979; South Australian and Kayrunnera kimberlite localities, this Section), are eclogite and granulite that have well-equilibrated microstructures and are commonly deformed. Younger fold belts (such as in eastern Australia), rifts, and other regions of continental volcanism, commonly have a mixture of equilibrated granulite-facies rocks and those having unequilibrated and partially-equilibrated igneous microstructures. In addition, contact relationships in xenoliths between recrystallised or foliated types and those having igneous microstructures, correspond to time differences and hence multiple, episodic, intrusion events (see, for example, Wass & Hollis, 1983).

Any mechanism of crust formation in eastern Australia must be consistent with the dominant mafic composition (including felsic types as well as differentiates) and the range of microstructures. Intrusion by basaltic melts around the crust/mantle boundary during multiple thermal events at different times in the evolution of eastern Australia satisfies these criteria. Tectonic emplacement of oceanic crust and island-arc material also may be a contributory factor as indicated by the isotopic character of some of the older xenolith suites (for example, Delegate and Gloucester). This is consistent with tectonic models for the evolution of eastern Australia (see, for example: Falvey & Mutter, 1981; Powell, 1983). However, note that the rock types are basaltic and not andesitic. This is in line with the concept of Heier (1973), Wasserburg & DePaolo (1979), Arculus (1981), and Karig & Kay (1981), that much of the material added to the continental crust since the Archaean is more mafic than 'andesitic'. The rare supracrustal rock types also may derive from tectonic underplating or from processes unique to cratonic evolution.

CHAPTER 7

Towards a General Model

7.1 Introduction

The foregoing six chapters are a summary of information on the nature of the intraplate volcanism of eastern Australia and New Zealand in terms of its geology, petrology, geochemistry, and tectonic setting. Interpretative material has been minimised in order that discussion of several issues can be brought together in this final chapter which is concerned largely with relationships between magma genesis and the tectonophysics of eastern Australia.

Authors of the following sections develop models based on the results of their individual research interests. Their accounts together represent an attempt to develop a general model for the intraplate volcanism. There are differences of opinion, but no special attempt is made to reconcile these. Rather, the volume is left open-ended in order that further research may be stimulated. Concepts of the relationship between magma genesis and tectonic setting in New Zealand are less developed than they are for eastern Australia, but some brief remarks are made in Section 7.8 (see also Section 4.9).

The first section (7.2) represents a continuation of Chapter 6 in that the results of xenolith and megacryst studies are used to establish a petrological model for the east-Australian lithosphere. In addition, pressure-temperature estimates obtained from xenoliths are used to construct an empirical geotherm for eastern Australia. The xenolith-derived geotherm has a higher temperature at any pressure less than 60 kb compared to conventional continental geotherms. It does not represent necessarily the pressure-temperature conditions beneath the whole of eastern Australia during the past 200 Ma, but may be a record of conditions at similar stages in the thermal evolution of each basaltic province containing the xenolith suites that have been studied.

A new approach in interpreting the tectonophysics of eastern Australia is taken in Section 7.3 where the tectonic, geomorphological, and magmatic effects of detachment faulting are discussed. The uplift that formed the eastern highlands is considered to be a consequence of the lithospheric distension and detachment that preceded seafloor spreading in the Tasman and Coral seas, and the relationship between the uplift mechanism and the Cainozoic volcanism is discussed

in detail. Magmatic underplating is considered to be important in the detachment/uplift model, as it is Section 7.2.

How similar is the tectonic setting of the Cainozoic volcanism in eastern Australia to other continental volcanic provinces world-wide? This question is addressed in Section 7.4 where the results of an extensive literature search are summarised. Much of the Cainozoic volcanism in eastern Australia can be classified as rift-shoulder type that has taken place after the cessation of seafloor spreading. It is perhaps similar to the volcanism in such places as the Red Sea/Gulf of Aden region and possibly Marie Byrd Land, Antarctica. The long, age-progressive, hotspot trail of felsic volcanoes in eastern Australia may be unique, as there appear to be no other similar examples elsewhere in the world. Lines such as the Snake River/Yellowstone province in the western USA and the Cameroon volcanic province in western Africa, are not strictly analogous as there the volcanism is not clearly age-progressive along each line.

The results of experimental petrology are important in defining conditions of basalt generation in the upper mantle. These are summarised in Section 7.5 where a generalised model for petrogenesis is presented. The main features of this model are: (1) the presence of volatiles (in the C-H-O system) in the source region which sensitively control the degrees of undersaturation of melts formed by very small degrees of melting; (2) diapirism from sub-lithospheric convecting mantle which produces local and transient perturbed geotherms that accompany lithospheric thinning (estimates of such 'perturbed' geotherms are said to be recorded in the xenolith-derived geotherm); (3) different ascent and cooling paths account for the ranges of basalt types found in different volcanic provinces.

The importance of polybaric fractional crystallisation during magma evolution is stressed also in Section 7.6. This section represents an extension of Chapter 5 where a wide range of petrological and geochemical data was presented. Combined AFC (assimilation/fractional-crystallisation) processes are regarded as a fundamental means of generating most of the evolved magmas of eastern Australia from a wide range of parental basaltic melts (see also Chapter 5). The high-silica rhyolites found in several central volcanoes are considered to be the products of crustal melting. An introductory discussion is given on the distinction between asthenospheric and lithospheric mantle sources for the basaltic magmas (see also Section 7.7), and attention is given to the degrees of partial melting of source rocks on the basis of REE geochemistry. Crustal underplating once again is stressed as an important phenomenon in the Cainozoic evolution of eastern Australia, and a preferred model is reconciled with the one presented in Section 7.2.

Penultimate Section 7.7 is concerned with the geochemical recognition of the main source components that may contribute to the composition of east-Australian basalts. Mantle source types identified from selected trace-element ratios and Sr-, Nd-, and Pb-isotopic data for the volcanic rocks, include an asthenospheric plume component, a shallow astheno-spheric component, and lithospheric mantle. The crust is recognised as a fourth component. A model is proposed whereby a southward tracking plume spreads out at the base of the lithosphere and contributes not only to central-volcano activity but to nearby lava-field volcanism as well. Lava-field volcanism far from , or south of, the inferred plume is considered to be derived mainly from shallow convecting asthenosphere. The possibility that the central-volcano and leucitite-suite activity can be accounted for by three plumes rather than one also is considered in Section 7.7.

Several unanswered problems remain to be addressed in studies of the intraplate volcanism of eastern Australia and New Zealand, and attention is drawn to these in the final Section 7.8.

7.2 Nature of the East Australian Lithosphere

7.2.1 Introduction

Formulation of a geologically realistic model for the nature of the lower crust and upper mantle requires the combined assessment of petrological and geophysical data. Relevant geophysical data include seismic refraction and reflection profiles, and heat-flow, magnetics, gravity, and conductivity measurements (Section 1.4). The geophysical data currently available for eastern Australia include seismic-refraction traverses (see, for example: Finlayson et al., 1979, 1984; Finlayson, 1983), seismic-reflection surveys (Mathur, 1983a,b,c; Finlayson et al., 1984), magnetic characteristics from MAGSAT data (Johnson et al., 1986), and some surface heat-flow measurements (Cull & Denham, 1979; Sass & Lachenbruch, 1979; Cull, 1982).

Xenoliths provide samples of the deep-seated lithologies. These can be used to obtain directly the physical characteristics of rock types (density, seismic-velocity characteristics, magnetic properties, conductivity) by calculation (for example, Vp, Vs) or by laboratory determination at appropriate pressures and temperatures where necessary. Some xenolith mineral assemblages also can be used for geothermobarometric calculations that can yield geotherms representing the thermal state of the lithosphere at the time of entrainment.

Xenolith data where possible also should be interpreted with the use of larger-scale analogues, such as deep-seated terranes tectonically emplaced into the upper crust (Griffin & O'Reilly, 1987a). Examples of these are the Musgrave and Fraser ranges (Fountain & Salisbury, 1981) and the Ivrea-Verbano zone in the Alps of northwestern Italy (Mehnert, 1975; Rivalenti et al., 1984). However, these sequences invariably have undergone deformation and metamorphic re-equilibration during emplacement and no longer retain original high-pressure mineralogy or physical properties. On the other hand, they have mesoscopic lithological relationships that are not observable in the smaller sample provided by xenoliths. Xenoliths in eastern Australia are the only samples of high-pressure rock types as there are no outcrops of terranes that originally were deep-seated.

7.2.2 Mantle and crust/mantle-boundary definition

The mantle is defined geophysically as the region beneath the Moho. The Moho represents a seismic discontinuity where compressional-wave velocity (Vp) increases from about 6.5–7.0 km/s to about 8.0 km/s (see, for example, Bott, 1982). The depth to this seismically-defined Moho differs regionally from 30–40 km beneath cratons such as in Western Australia (Drummond, 1981) to more than 50 km beneath some tectonically active regions as in eastern Australia (Finlayson et al., 1979; Finlayson, 1983). The original definition of the Moho specified it as a sharp dis-continuity, but the Moho or crust/mantle boundary is now known in some areas to be a transition zone 5–15 km thick where there is a strong positive Vp gradient. The seismic Moho definition does not specify crust and mantle rock types, but geophysical models commonly include the simplistic one of an andesitic lower crust underlain by a 'peridotite' (assumed to be 100-percent olivine) upper mantle.

The mantle petrologically is considered to consist dominantly of ultramafic rocks. It represents the original outer shell formed during the Earth's accretion (Ringwood, 1979) from which the present-day crust evolved by continuing magmatism and degassing (O'Nions et al., 1979a; Hanson, 1980). The present-day upper mantle is the residuum of this process of crust/mantle evolution. Subduction processes may return less buoyant layers of crust to the mantle where they may form eclogite bodies that can become megaliths, or parts of the convecting system, and ultimately become involved in further partial melting (Ringwood, 1982; McKenzie & O'Nions, 1983). The conceptual distinction of the crust and mantle as petrological units therefore should remain clear, as it is important for the interpretation of large-scale processes such as the nature of Earth's accretion, crust/mantle evolution, and the nature of crustal growth.

Geophysicists are identifying different types of 'Mohos' as data-acquisition and processing techniques are refined — especially with the recent development of vertical seismic-reflection profiling using long recording times, as well as the long-wavelength magnetic information collected by satellite. These different Moho types include the traditional refraction Moho, the transitional Moho (see, for example: Prodehl, 1977; Finlayson et al., 1979), the petrologic Moho (Bott, 1982), and the magnetic Moho (Wasilewski et al., 1979; Mayhew & Johnson, 1987). Such usage could blur further the essential petrological distinction between crust and mantle. Correlation of petrological with geophysical information is therefore important — especially using the physical properties of actual deep-seated rock types to provide realistic constraints for the geophysical models, and to obtain a clearer understanding of the petrological implications of these models.

The dominant crustal rock types in eastern Australian are mafic granulites (Section 6.3.1). There are many mantle rock types, but the dominant wall-rock type is considered to be a lherzolite consisting of olivine (50–70 modal percent), clinopyroxene (5–30), orthopyroxene (10–40), and spinel (less than 5 percent). The transition to garnet lherzolite takes place at depths greater than about 55 km. Xenoliths of garnet lherzolite from eastern Australia are too small for modal analyses to be useful, but in general the garnet-forming reaction of clinopyroxene I + spinel to garnet + clinopyroxene II + orthopyroxene produces 10–20 percent garnet.

The uppermost mantle also contains layers of mafic rocks commonly equilibrated to granulite or eclogite mineral assemblages, ranging from plagioclase-pyroxene granulites to garnet pyroxenite and eclogite. The upper mantle is therefore much richer in pyroxene than are most of the 'mantle' rock types used for laboratory measurement of Vp. In addition, mafic rocks contribute significantly to the physical characteristics of the upper mantle beneath eastern Australia, and so must be included in model calculations.

7.2.3   Geotherms

O'Reilly & Griffin (1985) constructed an empirical geotherm for eastern Australia, based on geothermobarometric calculations for garnet-websterite assemblages in xenoliths from the Bullenmerri and Gnotuk area of western Victoria. A detailed discussion of geothermobarometric methods and criteria for preferring certain methods was given by Griffin et al. (1984), and a justification for using such pressure and temperature results to construct a geotherm was given by Griffin & O'Reilly (1987a).

The xenolith-derived geotherm for eastern Australia (Fig. 7.2.1) is well-constrained from about 25 to 55 km. This geotherm has a higher temperature at any pressure (less than 60 kb) than conventional continental or even oceanic geotherms (Fig. 7.2.2). It is only slightly to the lower-temperature side of geotherms for active rifts, such as the Rio Grande Rift (Padovani & Carter, 1977), and active near-rift environments such as the Spitsbergen Fracture Zone (Amundsen et al., 1987). The southeast-Australian geotherm has a strong curvature from about 10 to 30 kb corresponding to advective heat transfer near the crust/mantle boundary. This is consistent with the observed basaltic volcanism and inferred magmatic overplating and underplating at the crust/mantle boundary.

In contrast, the east-Australian geotherm calculated assuming a steady-state heat-flow (simple conductive heat loss), and using sparse surface heat-flow measurements as the sole constraint (Sass & Lachenbruch,

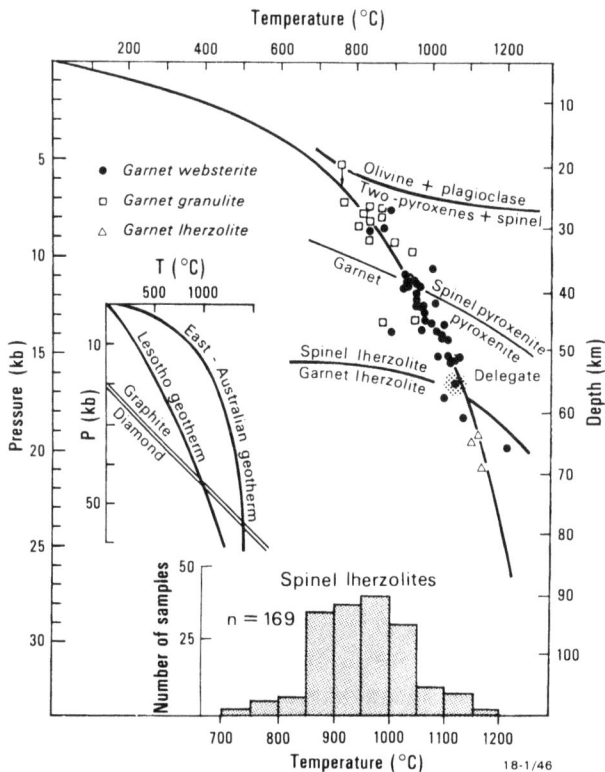

Figure 7.2.1   Empirical geotherm for southeastern Australia, based on pressure-temperature data for garnet-websterite xenoliths from Bullenmerri-Gnotuk (Griffin et al., 1984), as well as values available for xenoliths from other east-Australian localities. Other data sources are: Stolz (1984), Sutherland et al. (1984a), Griffin et al. (1987), and O'Reilly and Griffin (unpublished data). Delegate experimental data (stippled field) are from Irving (1974a). The spinel- to garnet-lherzolite transition is based on O'Neill (1981; Fo90 composition), and the spinel- to garnet-pyroxenite transition is from Herzberg (1978). The histogram represents the accumulated database for spinel-lherzolite temperatures (Bezant, 1985; O'Reilly et al., 1988; O'Reilly and Griffin, unpublished data).

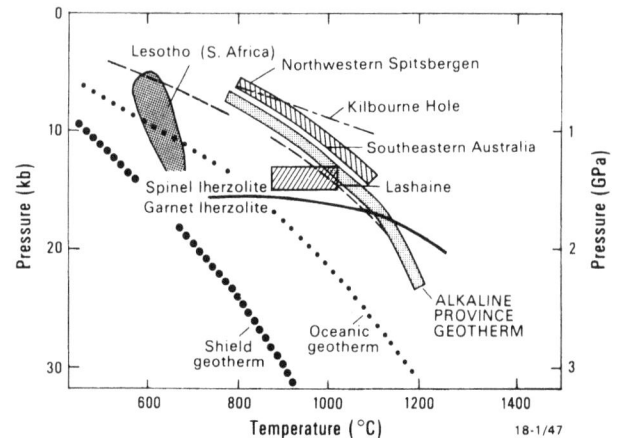

Figure 7.2.2   Pressure-temperature fields and geotherms derived by geothermobarometry on xenoliths from different tectonic environments, based on a diagram in Jones et al. (1983). Additions are the southeast-Australian geotherm (O'Reilly & Griffin, 1985) and the northwestern Spitsbergen geotherm (Amundsen et al., 1987). The spinel- to garnet-lherzolite (Fo90) transition of O'Neill (1981) is shown also.

1979), has a shallow curvature at low pressures. Pressure and temperature are constrained at depth for the xenolith-derived geotherms, whereas geotherms calculated from surface heat-flow measurements generally assume arbitrarily a steady-state conductive heat-flow and therefore are totally model-dependent below the surface. This may result in misleading extrapolation of heat-flow production to upper-mantle depths. The steady-state geotherm intersects the mantle solidus beneath eastern Australia at about 40 km, but this prediction is not substantiated by seismic data: no melt is detectable seismically at such depths. The high geothermal gradient inferred from the xenolith data is consistent with high heat-flow values of about 84 to a maximum of $120 \ mW/m^2$ for surface measurements in southeastern Australia (Cull, 1982).

The host rocks for the xenoliths were erupted at different times, ranging from about 200 Ma to 4000 years. However, the geotherm does not mean necessarily that these pressure-temperature conditions persisted beneath the whole of eastern Australia for the last 200 Ma. Nevertheless, there is the implication that since 200 Ma there has been episodic volcanism and that magmas from the different provinces sampled upper-mantle and lower-crustal xenoliths at similar stages in the thermal evolution of each province. The Bullenmerri-Gnotuk maar volcanic deposits that are host to the xenoliths used to construct the southeast-Australian geotherm, are a maximum of about 50 000 yr old, so the calculated geotherm should represent the present-day geotherm for that area. Furthermore, the thermal-decay constant is in the order of 10^7 yr, so young volcanic provinces (about 10 Ma) should have a concordance between empirical and present-day geothermal gradients.

The empirical geotherm was constructed using data from one lithospheric section beneath western Victoria. However, all other pressure-temperature data fulfilling the criteria specified by Griffin & O'Reilly (1987a), and calculated for appropriate xenolith assemblages from eastern Australia, fall along this geotherm. These include the garnet lherzolites from Bow Hill, Tasmania (Sutherland et al., 1984a), and Jugiong, New South Wales (Ferguson et al., 1977); garnet websterites and garnet granulites described by O'Reilly & Griffin (1985) from Anakie, Mount Leura, and Mount Shadwell, all in Victoria, and from Tumut-Eucumbene, Delegate, Jugiong, Southern Highlands, Ruby Hill, Gloucester, and Walcha (Stolz, 1984), all in New South Wales. In addition, Griffin et al. (1987) in a detailed study of a diverse xenolith assemblage including lower-crustal granulites and garnet websterites from a depth range of 15 to 55 km beneath central Queensland, obtained pressure-temperature values for equilibration of the xenoliths that coincide with those for the southeast-Australian geotherm.

A xenolith-derived geotherm for the eastern part of the Australian craton is about 150–200°C to the lower-temperature side of the east-Australian geotherm at any given pressure. The craton geotherm was constructed from garnet-granulite and garnet-websterite xenoliths from the South Australian and Kayrunnera kimberlite occurrences. The South Australian kimberlites are on the old continental suture (Cleary &

Simpson, 1971), which is seismically active, and the adjacent region to the east has a relatively high surface heat-flow (Cull & Denham, 1979; Cull, 1982). The kimberlitic intrusions in South Australia therefore probably predate the currently high heat-flow and may represent discrete, small-volume, magma pulses rather than a large-scale thermal mantle event such as those that produced the volcanic provinces in eastern Australia.

The xenolith-derived geotherm for eastern Australia defines a small field where diamond would be stable. This contrasts with the extrapolated steady-state geotherm prediction that the temperature is too high for diamond stability at all pressures (Ferguson et al., 1979).

7.2.4 Seismic properties of deep-seated lithologies

Two methods can be used to assess the seismic characteristics of samples of deep-seated rock types represented by xenoliths — calculation and laboratory measurement.

Calculations of Vp and Vs can be carried out using the algorithms of Anderson & Sammis (1970) which incorporate corrections for pressure and temperature. These refine Simmons' (1964) formulation of Vp as a function of mean atomic weight (M), CaO concentration, and density of a given rock type. The results of experimental measurements of Vp and Vs in granulite, eclogite, and peridotite (Birch, 1960; Birch, 1961; Manghnani et al., 1974; Christensen & Fountain, 1975; Christensen, 1982) are support for the general validity of this calculation.

Calculations have been undertaken for pyroxenite and lherzolite xenoliths from eastern Australia at 10 kb for 900, 1000, 1100 and 1200°C (O'Reilly & Griffin, 1985). The lherzolites have Vp values of 7.6–7.8 km/sec for temperatures consistent with the southeast-Australian, xenolith-derived geotherm. These are lower than expected because of two main factors: the presence of up to 60 percent pyroxene ± amphibole and, especially, the high geothermal gradient for eastern Australia. Wehrlite and granulite give even lower values, so any significant component of these rocks within the upper mantle beneath eastern Australia will result in a further decrease in average mantle Vp.

Jackson & Arculus (1984) measured Vp and Vs at 4 kb for a range of mainly mafic high-pressure xenoliths from the South Australian kimberlites. Vp values range from 6.9 to 7.7 km/s, and Jackson & Arculus (1984) suggested that these support interpretations of seismic data for southeastern Australia. However, these xenoliths are derived from the cratonic lithosphere of Australia and, as discussed in Section 6.3, many of them may be derived from greater depths than the seismic Moho in the region.

Ultrasonic velocity measurements were carried out in the laboratory of I. Jackson (Australian National University) by Bezant (1985) on four xenolith rock types representing uppermost-mantle lithologies beneath eastern Australia. Three of these are typical upper-mantle lherzolite in which microstructures range from a strong foliation to a weak preferred orientation of elongate grains. The other rock type is a garnet clinopyroxenite that has no apparent fabric. Vp values were obtained successively from 0 to 10 kb at 25°C. A rapid increase in Vp up to 4 kb in two samples, followed by a gradual increase, may be an indication that pore spaces were closed at this pressure. Vp values for the other two samples levelled off only at 8 kb, possibly because more friable xenoliths require higher pressure to close all spaces. Resulting maximum Vp values at 25°C were 8.37, 8.32, and 7.89 km/s for the three lherzolites, and 7.97 km/s for the garnet pyroxenite. These values would be reduced by 0.5–0.6 km/s for temperatures consistent with their depth of origin beneath eastern Australia.

Bezant (1985) found that Vp values calculated using density, M (mean atomic weight), and CaO concentration, are higher than those obtained by ultrasonic measurement. He applied an empirical correction to the calculated Vp determinations of O'Reilly & Griffin (1985) resulting in values of 7.4 to 7.6 for east-Australian, upper-mantle lherzolites under conditions consistent with the east-Australian, xenolith-derived geotherm.

Vp values obtained using single-crystal data and the mode of the rock correspond to within 2 percent for the four rocks (Bezant, 1985). Grids were constructed on this basis for the common mantle rock types so that Vp can be determined graphically from the mode in cases where the assemblage is dominantly three phase. The Vp-estimation grid for lherzolites containing less than 5-percent spinel is shown in Figure 7.2.3. Similar grids for garnet pyroxenite and garnet+clinopyroxene+olivine assemblages were given by Bezant (1985).

A further significant result from the work of Bezant is the characterisation of strong anisotropy in Vp (Fig. 7.2.4) in foliated xenoliths. Measured anisotropy

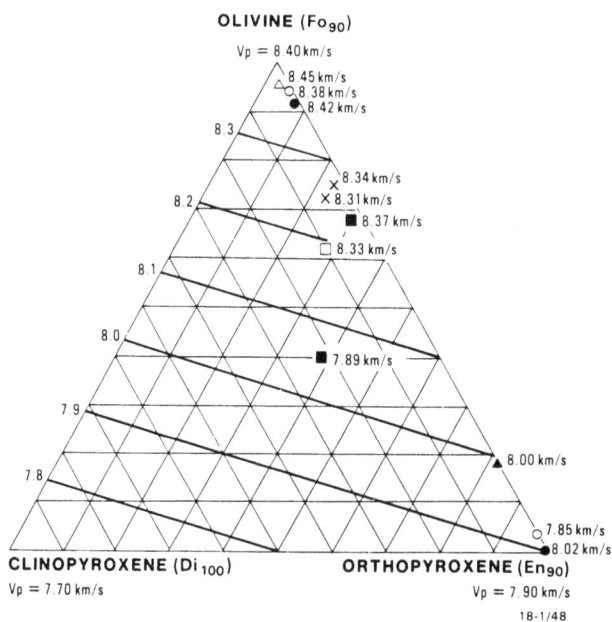

Figure 7.2.4 Measured sound-wave velocity of a spinel lherzolite (olivine 67.2 percent, orthopyroxene 30 percent, clinopyroxene 1.5 percent, spinel 1.3 percent). Levelling out of values towards higher pressures may be indicative of pore closure. The foliation caused by preferred orientation of tabular olivine grains results in anisotropy for the three measured orthogonal directions.

is up to 5.9 percent in this sample set. Any large-scale preferred orientation of minerals within upper-mantle domains therefore will result in dramatic differences within the same rock type for Vp measured in differently oriented seismic-refraction profiles.

7.2.5 Magnetic characteristics

Magnetic lows revealed in MAGSAT imagery for southeastern Australia (Mayhew & Johnson, 1987; Langel et al., 1982) may be evidence that the magnetic crust is thin. The southeast-Australian geotherm has the Curie point (about 550°C) at about 12 km depth in regions where this geotherm reflects the present-day thermal regime and this is consistent with the MAGSAT data.

Wasilewski & Mayhew (1982) and Wasilewski et al. (1979) demonstrated that Curie points for granulite xenoliths cluster around 560–570°C, and that eclogites and mantle-derived lherzolites are non-magnetic for MAGSAT purposes. The long-wavelength anomalies evident on MAGSAT maps therefore could be interpreted as differences in crustal thickness or regional heat-flow, or both (Mayhew, 1982; Mayhew & Johnson, 1987). The high geothermal gradient in southeastern Australia may be the most important constraint on the depth of magnetisation for this region.

7.2.6 Model for the crust/mantle boundary beneath eastern Australia.

Geothermobarometric results from xenolith assemblages allow construction of an empirical stratigraphy for the lower crust and upper mantle, and contact relationships in composite xenoliths provide further evidence.

The most abundant high-pressure xenolith rock type in eastern Australia (and world-wide) is mantle-derived spinel lherzolite. The mineral assemblage of this rock type readily yields temperature estimates by a range of

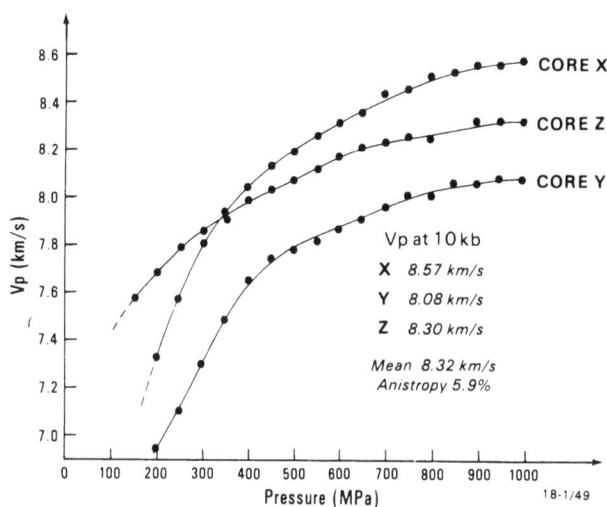

Figure 7.2.3 Triangular grid of wave velocities (Vp) in olivine+orthopyroxene+clinopyroxene assemblages calculated from measured single-crystal data by Bezant (1985) and corrected to 25°C and 1000 MPa. Plotted values for rocks measured in the laboratory (sources in Bezant, 1985) range by a maximum of 2 percent from the values predicted by the grid.

geothermometers but does not allow calculation of pressure. Griffin et al. (1984) used composite xenoliths that have spinel lherzolites in contact with pyroxenites to correlate geothermometers for both rock types. Pressure can be derived on this basis for a given temperature for spinel lherzolites by reference to the local xenolith-derived geotherm. Such pressure-temperature estimates for about 180 spinel-lherzolite xenoliths from eastern Australia have been obtained (O'Reilly & Griffin, 1985; O'Reilly et al., 1988), and this rock type appears to be present, probably as the dominant lithology, as a layer ranging from 15 to 30 km in thickness in different lithospheric sections. The crust/mantle boundary, defined by the dominant presence of spinel lherzolite, is found at depths ranging from 25 to 38 km in different domains of eastern Australia. Garnet±spinel lherzolites are present below about 55–60 km.

The integration of all available xenolith information (petrology, pressure-temperature estimates, physical properties) with geophysical data (seismic refraction, seismic reflection, magnetic data, heat-flow) has resulted in the model for the lower-crust and upper-mantle stratigraphy beneath eastern Australia. The crust/mantle boundary is found at about 30 km (in southeastern Australia) where ultramafic rocks (spinel lherzolites) first become dominant. The black horizontal lines in Figure 7.2.5B represent layers of mafic rocks that are frozen basaltic liquids or their cumulates. These may have re-equilibrated to granulite- or eclogite-facies conditions. They decrease in proportion both upward and downward from the crust/mantle boundary. The lower-crustal wall rocks are mafic to felsic granulites. The layers of mafic rocks are interpreted to represent repetitive basaltic underplating and overplating at the crust/mantle boundary by intrusive and tectonic processes.

Reversed seismic-refraction profiles for southeastern Australia (Finlayson et al., 1979) have a gradient in Vp between about 25 and 55 km where Vp is about 8 km/s (Fig. 7.2.6C). 55 km represents the seismic Moho, although spinel lherzolite is the dominant rock type from about 25 km according to the xenolith data. This highlights the discrepancy between the seismic Moho and the crust/mantle boundary.

Ferguson et al. (1979) and Finlayson (1982) interpreted the seismic data in terms of a thick (about 55 km) crust in eastern Australia. However, 55 km is the predicted depth at which the southeast-Australian geotherm crosses the spinel- to garnet-lherzolite boundary. The seismic Moho in this continental region of high geothermal gradient therefore represents the transition from spinel- to garnet-lherzolite rather than the crust/mantle boundary which in eastern Australia represents the change from granulite to lherzolite wall-rock. The gradual increase in velocity with depth around the crust/mantle boundary corresponds to a gradational mix of wall-rock containing mafic rock types (O'Reilly & Griffin, 1985).

Calculated and measured Vp values for xenoliths are consistent with this interpretation. A model seismic profile based on the xenolith data (Fig. 7.2.6D) is in agreement with the observed seismic-refraction profile. However, a similar calculation for the early crust/mantle model of Ferguson et al. (1979) and Finlayson (1982) produced a Vp/depth profile that is difficult to

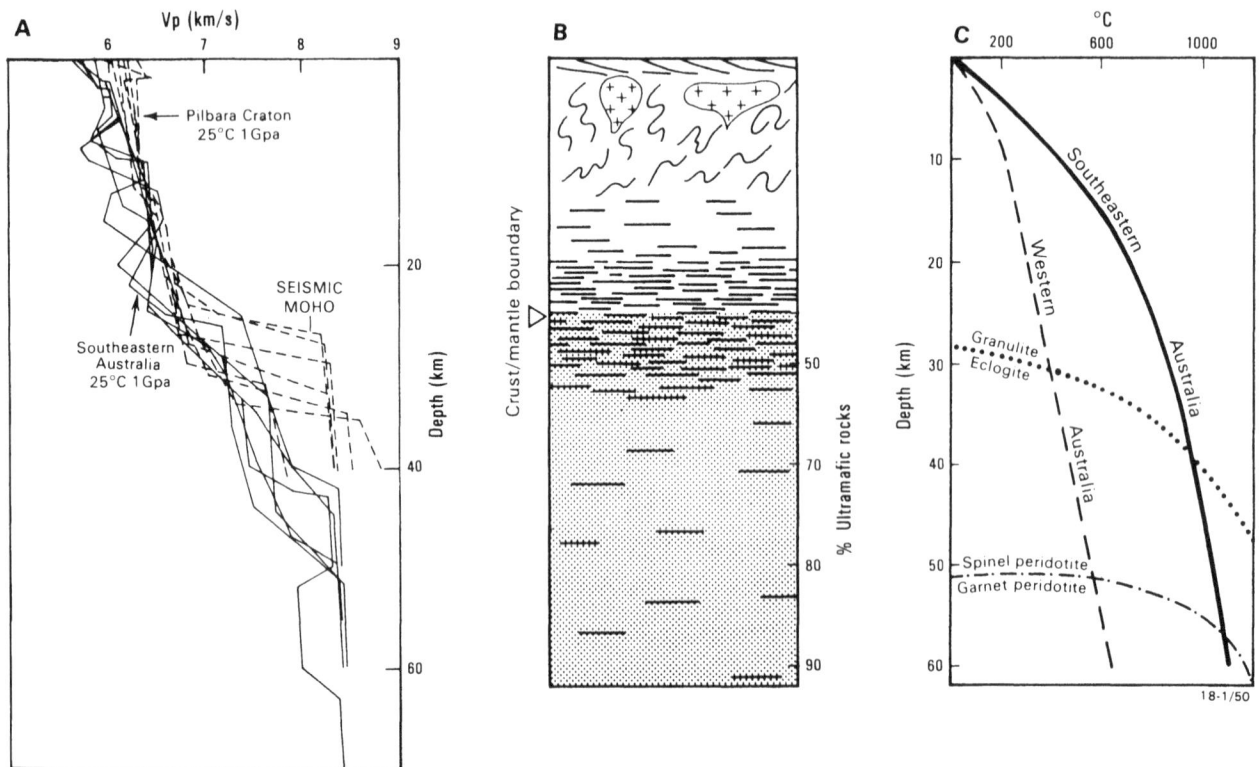

Figure 7.2.5 (A) Vp-depth models for Western Australia (Pilbara) and for southeastern Australia (after Drummond, 1982). (B) Model for the crust/mantle boundary stratigraphy in eastern Australia. (C) Geotherms for southeastern Australia (xenolith-derived; O'Reilly & Griffin, 1985) and Western Australia (extrapolated from surface heat-flow; Sass & Lachenbruch, 1979). The granulite-to-eclogite boundary is a conservatively high estimate (Griffin & O'Reilly, 1987b).

reconcile with the seismic data (Fig. 7.2.6A). The model also is not consistent with the abundance of spinel lherzolite in the southeast-Australian xenolith suites, or with the occurrence of spinel lherzolite as wall-rock below 25 to 38 km in different regions in eastern Australia, or with the relative rarity of mafic granulites in southeastern Australia. Finlayson's interpretation is based primarily on seismic parameters and the assumption that ultramafic rocks have Vp values greater than or equal to about 8 km/s. He does not consider the modes, compositions, and densities of the deep-seated rocks as revealed by the xenoliths, nor the high temperatures inferred from the empirical geotherm, heat-flow, and magnetic data.

Zones containing abundant horizontal reflectors (interpreted to represent the mafic lenses) are observed at depths of 15–35 km in many parts of the continental crust and are seen particularly well in seismic-reflection profiles from the Eromanga Basin (Mathur, 1983a,b,c; Section 1.4.7). These zones are described generally as 'lower crust' and the seismically transparent zone below as 'mantle'. The refraction Moho commonly coincides with the base of such layered zones (see, for example, McGeary & Warner, 1985), but this does not require that either the Moho or the base of the layered zone corresponds to the crust/mantle boundary. Uppermost mantle containing numerous subhorizontal lenses of mafic rocks (Fig. 7.2.6E) will appear as a layered zone on reflection profiles and will have a bulk density and Vp values intermediate between those of 'crust' and 'mantle'. The 'refraction Moho' and the 'reflection Moho' must coincide in such areas, even though part of the layered zone would be regarded as 'mantle' in the petrographic sense.

The detailed seismic-reflection data from the Eromanga and Bowen Basins (Finlayson et al., 1984; Mathur, 1983c) together with the lower-crust/upper-mantle profile constructed from xenoliths from central Queensland (Griffin et al., 1987) are a good illustration of this. There is a well-defined layered zone in the reflection profile from about 22 to 36 km. The seismic Moho defined from refraction data is at the base of this zone where Vp increases to about 8 km/s. However, the crust/mantle boundary is no deeper than about 28 km in this area — that is, within the zone of reflectors — on the basis of xenolith data.

7.2.7 Effect of geothermal profiles on seismic interpretation

The lithospheric model illustrated in Figure 7.2.5B for eastern Australia is consistent with the seismic evidence. In contrast, many cratonic areas such as Western Australia have a shallower, sharper Moho. These

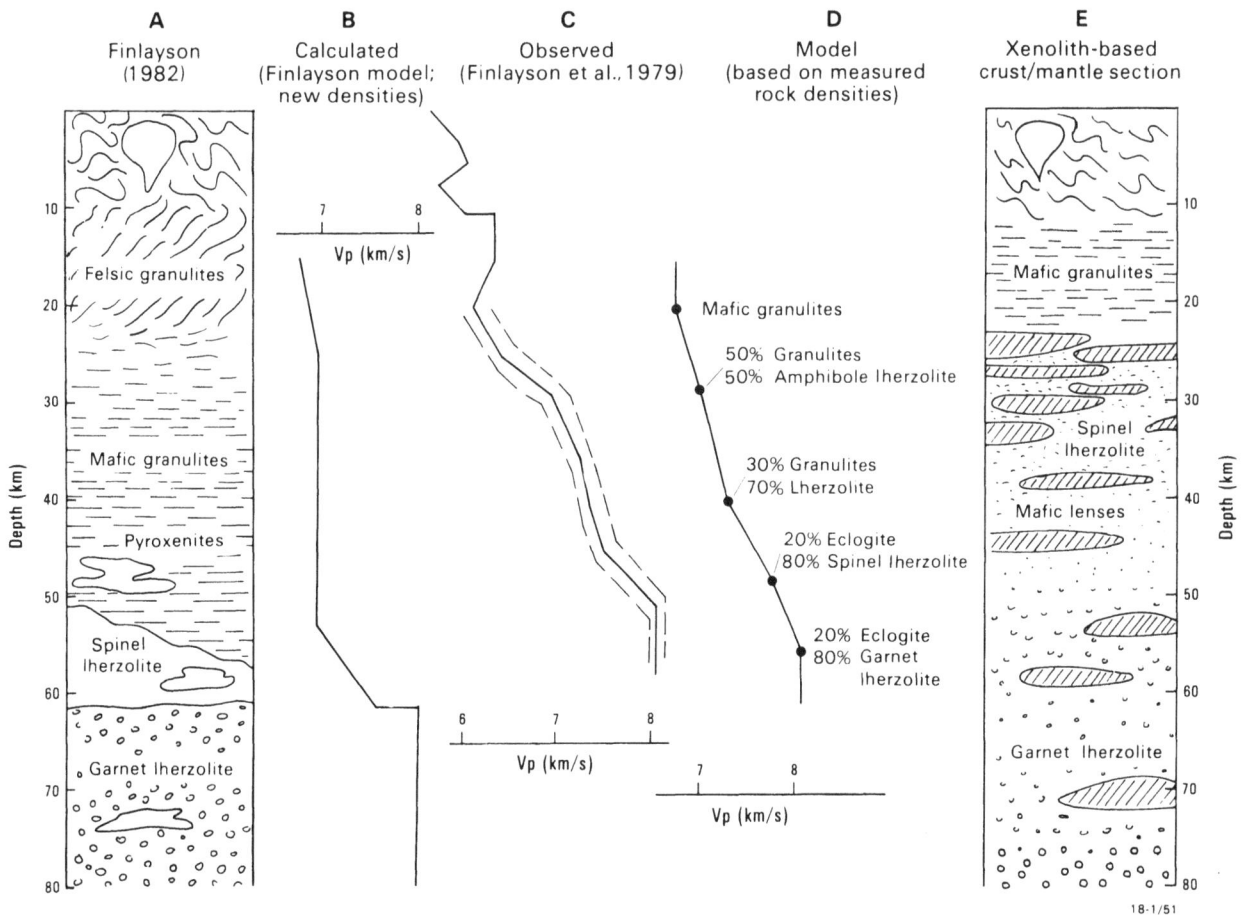

Figure 7.2.6 Interpretations of the seismic-refraction profile (C) of Finlayson et al. (1979), using data on the distribution, geochemistry, and seismic velocities of xenolith rock types and pressure-temperature estimates. Section A is an interpretation based on the early model of Ferguson et al. (1979). Note that the adjacent calculated seismic profile B is not compatible with the observed seismic profile C. Section E represents the interpretation of Griffin & O'Reilly (1985). This section is one likely mixture of rock types that is consistent with the observed seismic velocities as shown in the adjacent calculated seismic profile (D).

differences have been interpreted as being caused by contrasting types and thicknesses of crust. However, an important difference is the lower geothermal gradient of cratons. Different thermal profiles are critical for mafic rocks in determining whether or not the equilibrium mineral assemblage is in the eclogite or granulite facies, and therefore in determining the Vp of these deep-seated regions.

Formation of the crust/mantle transition by magmatic underplating and overplating inevitably will be accompanied by an elevated, strongly curved geotherm like that for southeastern Australia. This geotherm will decay where magmatic activity ceases towards a conductive geotherm — like the Western Australian one — with a time constant on the order of 10 Ma (Sass & Lachenbruch, 1979).

Mafic assemblages on successive cooling may convert from granulite to eclogite at shallower depths (Griffin & O'Reilly, 1987b; Fig. 7.2.7). Wood (1987) suggested on the basis of thermodynamic modelling that linear extrapolation of the experimental data of Green & Ringwood (1972) to low temperatures is probably not correct. His calculated 'plagioclase-out' curve for a *qz*-tholeiitic basalt composition is slightly concave towards the temperature axis and displaced to lower temperatures than is the corresponding curve in Figure 7.2.7A. Wood (1987) suggested on this basis that eclogite is not stable in the lower part of the continental crust. However, this conclusion applies only to the *qz*-tholeiitic basalt composition. Corresponding curves for more silica-undersaturated, alkali-basalt compositions, similar to those that dominate the mafic xenolith suites from eastern Australia, would plot at lower pressures and extend the stability range of eclogite well into the lower crust under the pressure-temperature conditions represented by cratonic geotherms. The transition to eclogite results in an increase in Vp of 0.5 to 1.0 km/s. The decrease in temperature causes the Vp of other rock types to rise, resulting in a maximum increase of

Vp near the crust/mantle boundary. The seismic Moho moves upwards and becomes much more pronounced as a result of the cooling (Fig. 7.2.7).

7.2.8   Implications for diamond occurrence in eastern Australia

The xenolith-derived geotherm for southeastern Australia provides more information relevant to the possible origin of diamond occurrences in eastern Australia (MacNevin, 1977; Section 3.10.3). This geotherm is more steeply curved at shallower depths than is the model-dependent geotherm extrapolated from surface heat-flow values (assuming steady-state conduction) so it lies within the diamond stability field at depths exceeding about 180 km. Diamonds may be stable at depths as shallow as 150–200 km (see, for example: Boyd et al., 1985; Haggerty, 1986), and this leaves a limited volume from which diamonds could be entrained. However, the top of the asthenosphere may be reached at depths of 150 km or less beneath eastern Australia according to geophysical data (Muirhead, 1984) and the probable intersection of the mantle solidus is at about 100 km according to the extrapolated geotherm.

Diamond morphologies from most individual occurrences in eastern Australia are distinctive (D. Robinson, personal communication, 1985) meaning the diamonds are unlikely to have been transported significant distances or to have undergone several erosion cycles. Possibly they are derived from cooler lithosphere that predates the general high geotherm that has persisted for at least 200 Ma, or they were entrained since then by deep melts rising through metastably preserved volumes of cool lithosphere.

Another possibility is that the occurrence of diamonds in eastern Australia is related to subduction processes. Helmstaedt & Doig (1975) advocated the role of subduction in diamond-forming processes, and Pearson

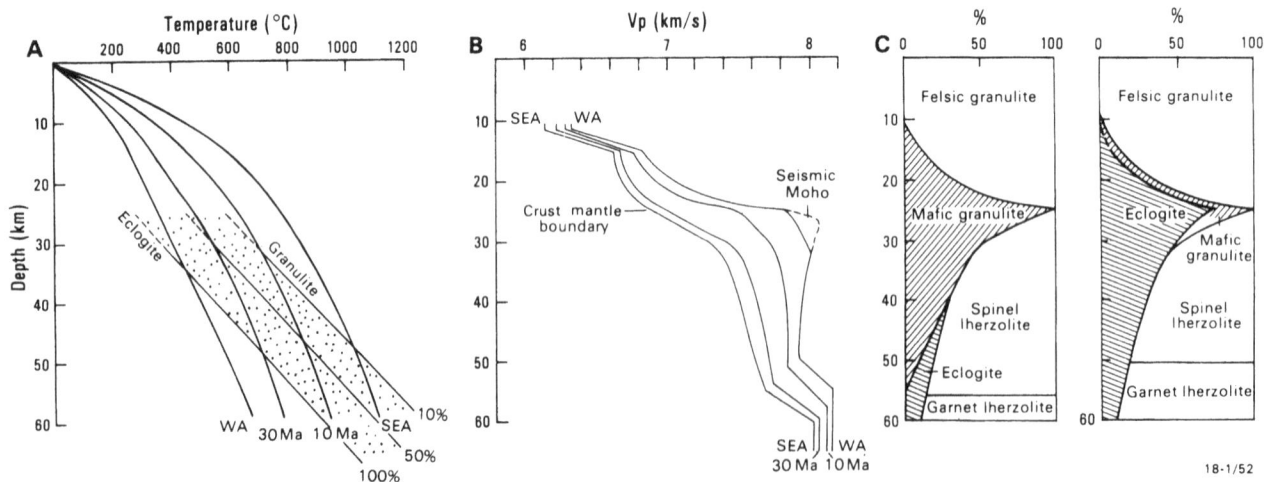

Figure 7.2.7   (A) Field of the transition from granulite to eclogite relative to geotherms for southeastern Australia (SEA) and western Australia (WA). The 10 Ma and 30 Ma curves represent decay of the SEA thermal anomaly, assuming a time constant of 10 Ma. The transitional granulite/eclogite field is based on the experimentally-determined 'plagioclase-out' line for a range of mafic to intermediate compositions (Green & Ringwood, 1972) extrapolated to lower temperatures using 15 bars/°C. (B) Vp-depth profiles for the crust/mantle section of Figure 7.2.5B. Vp is calculated (see text) using the rock-type distribution shown in C and the average densities and rock compositions for relevant xenolith rock types given by O'Reilly & Griffin (1985). The dashed part of the WA curve represents the effect of moving the 100-percent eclogite contour in A up by only 3 km. (C) Distribution of rock types in the model crust and mantle of Figure 7.2.5B for the two end-member thermal regimes of A.

et al. (1987) recorded graphite octahedra (interpreted as pseudomorphs of diamond) in tectonically-emplaced ophiolitic harzburgites from the Beni Bousera Massif, Morocco. There is considerable circumstantial evidence in support of an original subduction origin for many of the alluvial diamond deposits in eastern Australia: many are found in close proximity to exposed ophiolitic sequences (for example, in New England area of northern New South Wales); the mineral inclusions in east-Australian diamonds are compositionally unique and typical of minerals found in rhodingites rather than in high-temperature eclogites (Sobolev, 1984); and their unusual carbon isotopic compositions (δ^{13}C about +5; Sobolev, 1984) also are consistent with decarbonation processes during subduction. The Phanerozoic evolution of eastern Australia has involved many collision events (for example, Veevers, 1984; Fig. 7.2.8), and the appropriate tectonic setting has existed repetitively over a significant time span.

7.2.9 Implications for large-scale processes

An understanding of the lithologies and their stratigraphy in the lower crust and upper mantle provides a basis for interpreting large-scale lithospheric processes.

Recognition that the dominant lower-crustal rock types are mafic (including plagioclase-rich types) is strong evidence that magmatic and tectonic underplating have been significant crustal-growth mechanisms beneath eastern Australia (Fig. 7.2.8). The magmatic

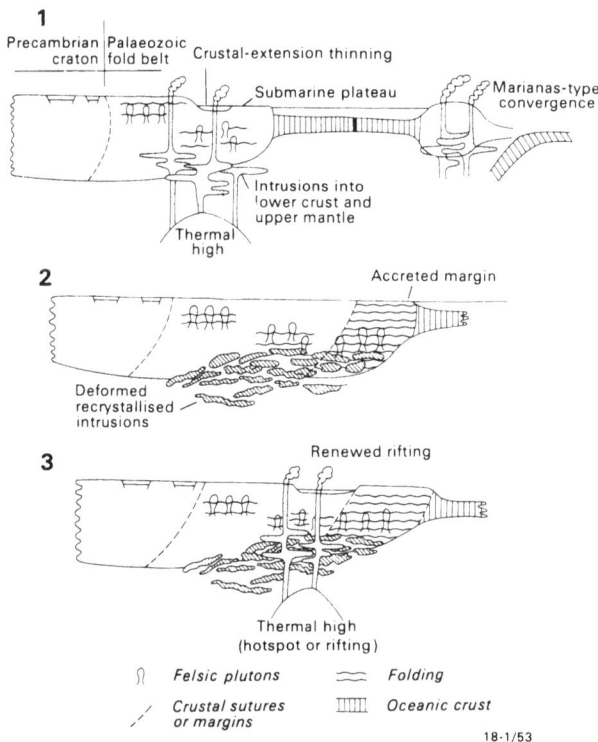

Figure 7.2.8 A possible scenario representing crustal development in much of eastern Australia, modified from Powell (1983). Parts 1, 2 and 3 represent successive stages, which may have been repeated. Two processes result in the accumulation of mafic rocks around the crust/mantle boundary: (1) intrusion of basaltic melts; (2) tectonic accretion of island arcs, submarine plateaux, and ocean-floor material. Both processes result in a mainly mafic lower crust.

component derives from repetitive intrusion and ponding of basaltic melts caused by density contrast between crustal and mantle rocks (Herzberg et al., 1983). The tectonic component is also basaltic in character and may include island-arc material (now considered to be mainly mafic; Arculus, 1981; Karig & Kay, 1981,), submarine plateaux, and ocean-floor basalts. Both a recycled and a primary origin for different xenolith suites are supported by isotope data (Griffin et al., 1987). Geochemical, petrological, and geophysical data therefore are all consistent with an underplating model.

These data are also evidence for a model for the evolution of the east-Australian margin by rifting associated with crustal inflation and thinning. Such rifting is an effective mechanism for initiating a high heat flux. Sass & Lachenbruch (1979) showed that a strain rate of 2 percent in a 60 km-thick layer can increase heat-flow by a factor of three over conductive-heat transport. The chronological coincidence between some of the basaltic activity in eastern Australia and the opening of the Tasman Sea, as well as the chronological and geographic coincidence with the tectonic uplift of the eastern continental margin (P. Wellman, 1979a,b; Jones & Veevers, 1982), are evidence for a strong link between rifting and magmatism. The thermal regime of eastern Australia probably represents a 'steady-state' off-ridge environment with superimposed hot-spot activity (Amundsen et al., 1987).

The nature of the lower crust, crust/mantle boundary, and upper mantle for eastern Australia is not unique. It is similar to that of many tectonically active regions world-wide. The common features are the mafic character of the lower crust, the general types of xenolith assemblages, the seismically transitional Moho, the layer of reflectors around the crust/mantle boundary, and high geothermal gradients.

7.3 Detachment Models for Uplift and Volcanism in the Eastern Highlands, and their Application to the Origin of Passive Margin Mountains

7.3.1 Introduction

This volume is concerned principally with intraplate volcanism, particularly with that along the eastern margin of Australia. However, as seen in many of the foregoing chapters, there is widespread opinion that the volcanism and the uplift that formed the eastern highlands are related to each other. Evidence for the spatial and, especially, the temporal relationships between volcanism and uplift are presented in Sections 1.3.1 and 1.7.2–3. Wellman (1987), among others, suggested that uplift resulted directly, or indirectly, from a mantle thermal event (or events) that produced the magmas erupted at the surface.

Ollier (1982a, 1985b) recognised that the eastern highlands of Australia are similar in character to several other coastal mountain ranges that border passive continental margins — for example, the Drakensburg of South Africa, the Eastern and Western Ghats in India, the Guyana Highlands of South America, and the spectacular Transantarctic Mountains. Ollier referred to these landforms as 'passive-margin mountains'. Passive-margin mountains are characterised by planated land surfaces that rise gradually from the interior of the continent to form plateaux, in places deeply dissected, and bordered on their seaward side by major erosional escarpments (Fig. 7.3.1).

The model presented in this section represents an attempt to link both uplift and igneous activity to the formation of the passive continental margin along eastern Australia. First, a brief review is given of the main mechanisms proposed previously for highlands uplift. Next, a theoretical model for uplift and igneous

activity is presented, based on continental extension by detachment faulting, which is compared with data from the east-Australian highlands.

7.3.2   Review of uplift mechanisms

Introduction

Mechanisms proposed for the origin of the eastern highlands of Australia — and of similar geomorphological features adjacent to other passive continental margins — fall into two broad categories: (1) the deeply eroded remnants of an ancient (older than 200 Ma) orogenic mountain chain; (2) the result of broad uplift of a planated surface sometime within the last 100 Ma. Evidence for the timing of uplift (reviewed in Sections 1.3.2 and 1.7.2–3) is not sufficiently unequivocal to be able to distinguish firmly between these two categories, but in general it is more supportive of the second alternative. A brief critique of the different mechanisms is provided here as an introduction to presentation of the uplift model.

Erosional remnants of older mountains

The proposition that the eastern highlands represent the remnants of an older orogen was made first by Craft (1933). It was taken up by Hills (1956) and most recently was modelled quantitatively by Stephenson & Lambeck (1985) and Lambeck & Stephenson (1986). The model is based on the premise that erosion rates are substantially slower than considered by some geomorphologists, and that erosion of an orogenic mountain range will result in isostatic rebound of the eroding surface. The average elevation of the mountain range decays with time, but only slowly, and more than half of the original elevation may remain after 200–300 Ma (see Fig. 3.3.3A).

The erosionally-induced isostatic rebound model has some important consequences. First, the area of highest present elevation is predicted to coincide generally with the area of maximum advection of material through the erosional surface. Second, there should be a correspondence between the position, height, and lateral extent of the eastern highlands and those of the parent orogenic mountain range. Third, average surface roughness is predicted to decrease with time, so there should be a relationship between relief and the age (and style) of the mountain-building event. The rebound model is consistent with the latest evidence that there must have been a substantial proportion of uplift prior to middle to early Tertiary times (Bishop, in press; Bishop & Young, 1980), and with the broad correspondence between the eastern highlands and the Tasman Fold Belt. However, it breaks down in detail on all three of the consequences listed above.

The surface of much of the eastern highlands, even in its highest parts, is a raised and little modified planated surface that is early Tertiary or older. Parts of the eastern highlands have undergone relatively little erosion since the middle Tertiary (Young, 1977, 1981, 1983). Fission-track data (Moore et al., 1986) are conclusive evidence that a major part of the eastern

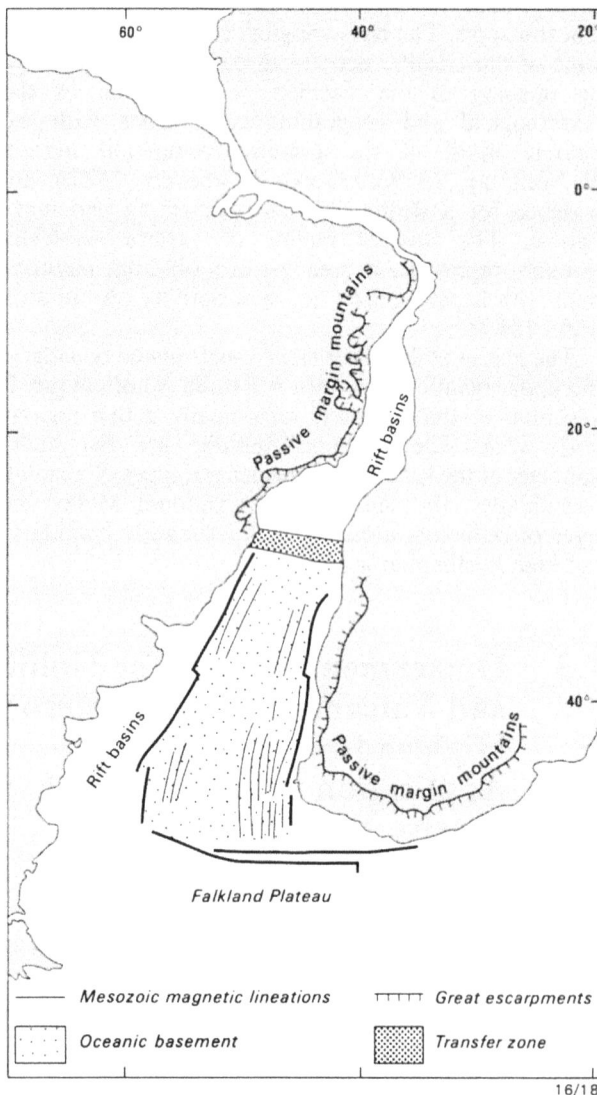

Figure 7.3.1   107 Ma reconstruction of the South Atlantic Ocean after Rabinowitz & LaBrecque (1979), showing the juxtaposition of the narrower upper-plate margins with passive-margin mountains against the broader lower-plate margins characterised by extensive rift basins. Note that the sense of asymmetry of the margin switches across a major transfer/transform fault near latitude 30°S. Heavy lines mark the inferred ocean-continent transition.

highlands has not undergone significant erosion since the Permian. The presence of this little-modified planated surface is inconsistent with the highest parts of the eastern highlands being continuously the locus of maximum material advection through the present erosional unconformity.

The Tasman Fold Belt has a long and complex history, including major orogenic events ranging from the Ordovician to Cretaceous (Section 1.3.1). Most importantly, the Lachlan Fold Belt in the south, which corresponds to the highest part of the eastern highlands, underwent its final orogenic development up to 200 Ma before those parts of southern and central-coastal Queensland that occupy the region of most subdued topography. There is no apparent correlation between the age of orogenesis and either present elevation or relief (Summerfield, 1986). In addition, the eastern highlands cut across major orogenic provinces and their bounding structures, in direct contrast to other mountain ranges that are related more obviously to compressional

orogenic processes — for example, the European Alps and the North American Cordillera. Even in the southeastern-highlands region, considered in detail by Stephenson & Lambeck (1985) and Lambeck & Stephenson (1986), the highlands clearly transect a region of (1) little or no Permo-Triassic activity, (2) the main Permo-Triassic fold belt, and (3) what is interpreted to have been the foreland molasse basin to the Permo-Triassic orogen, with little change of elevation, extent, or relief. There is little evidence to support the contention that the eastern highlands are essentially remnants of a Palaeozoic to early Mesozoic compressional orogen.

Consequence of Tertiary igneous activity

There is a clear spatial association between Cainozoic volcanism and the eastern highlands: most of the volcanic provinces are clustered near or along the crest of the antiformal divide (Fig. 1.1.6). This association

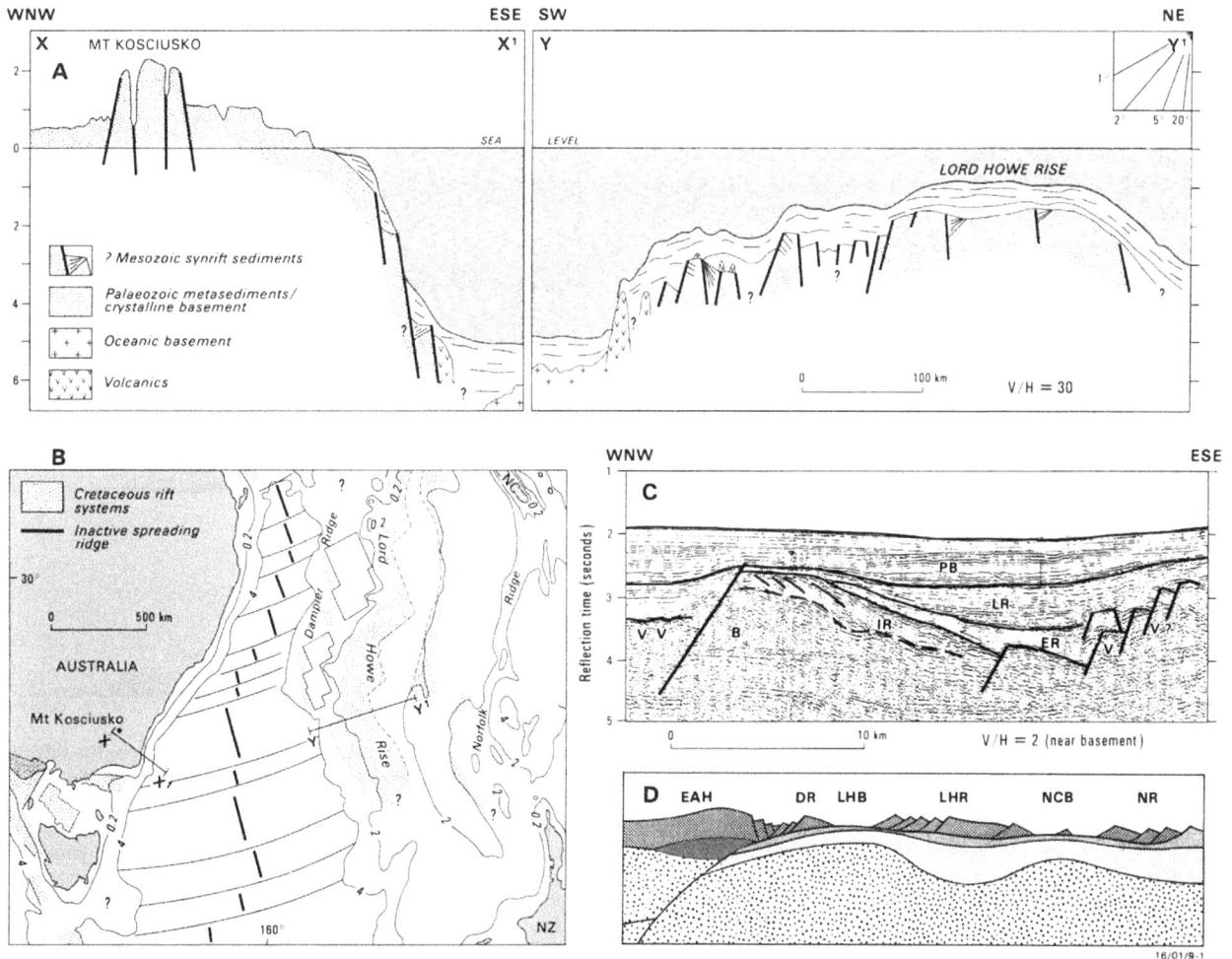

Figure 7.3.2 (A) Reconstructed schematic section across the Tasman Sea from Mount Kosciusko to the Lord Howe Rise, excluding about 800 km of the Tasman Basin oceanic crust. Location of sections X-X' and Y-Y' are shown in B. Note the passive-margin mountains of southeastern Australia, the absence of extensional (rift) structures beneath the east-Australian margin, the rift system beneath the western Lord Howe Rise, and the planated basement platform that underlies the eastern half of the rise. (B) The southeast-Australian and Tasman Sea region, showing major tectonic features. (C) A portion of the stacked multichannel seismic section across the rift system of the western Lord Howe Rise, showing the style of the half graben and tilt blocks within the rift. The corner of the tilt block, which is composed of pre-Mesozoic basement (B) and pre-rift or infra-rift sediment (IR), has been planated. The section also contains early-rift (ER), late-rift (LR), and post-breakup, sag-phase, sedimentary sequences, together with possible volcanic horizons (V). (D) Reconstruction prior to break-up: EAH — east-Australian Highlands; DR — Dampier Ridge; LHB — Lord Howe Basin; LHR — Lord Howe Rise; NCB — New Caledonia Basin; NR — Norfolk Ridge.

has led to proposals for a link between igneous activity and uplift (P. Wellman, 1979b, 1987; Jones & Veevers, 1982; Smith, 1982). Two specific mechanisms have been suggested: one by which the uplift results from thermal expansion at and near the source of melting (P. Wellman, 1979b; Smith, 1982); and the other whereby it results from the isostatic effect of underplating the crust with large volumes of mantle-derived magmas (P. Wellman, 1979b, 1987).

There are two main problems with purely thermal models. First, as pointed out by Wellman, the volume of volcanic material is small, and by itself cannot signify a thermal event large enough to produce the observed uplift. Second, thermal uplift decays exponentially with a time constant of a few tens of million years, so there should be a spatial relationship between elevation and age of the dominant volcanic rocks. The largest volume of very young volcanic rocks are at one of the lowest parts of the highlands (central and western Victoria), and the highest parts of the range contain mostly middle Tertiary volcanoes, so there is no obvious support for such a relationship.

P. Wellman (1979b) analysed gravity and regional elevation data in eastern Australia, in conjunction with evidence from seismic-refraction studies of crustal thickness and composition (from seismic P-wave velocity). He concluded that the eastern highlands are close to local isostatic equilibrium with the surrounding lowlands, and that their elevation reflects a thicker crust. This thicker crust is characterised by a lower crustal layer about 15 to 25 km thick, with P-wave seismic velocities consistent with largely mafic compositions. P. Wellman (1979b, 1987) suggested that some or all of the mafic lower crustal layer was underplated during the Cainozoic, reflecting a major series of igneous events of which only traces reached the surface as volcanism. He ascribed part of the eastern-highlands uplift (which he interpreted as having taken place through the Cainozoic) to this underplating and related thermal expansion.

The source of the Cainozoic volcanic rocks is predominantly within the upper mantle (Chapter 5), and much of it is considered to have been derived from a mantle plume or 'hotspot' over which the Indo-Australian plate has been migrating through much of the Cainozoic (Section 1.7). A mantle plume itself will result in uplift, but the scale problem raised by Lambeck & Stephenson (1986), and the absence of a clear relationship between local elevation and the time at which the plume (or plume swath) passed beneath any particular location, are arguments against it being the primary cause of eastern-highlands uplift.

Relationship between highlands and ancient rifts along the passive margin

The spatial relationship of the eastern highlands to the eastern passive margin of Australia has prompted searches for a genetic relationship between the rifting that preceded and accompanied seafloor spreading, and at least part of the uplift (P. Wellman, 1979b, 1987; Karner & Weissel, 1984; Weissel & Karner, in press; Lister et al., 1986, in press). Uplift is associated commonly with the flanks of both continental and

passive-margin rifts world-wide (Weissel & Karner, in press), and has been ascribed to (1) the thermal buoyancy that results from, or contributes to, lithospheric extension, (2) lateral conduction of heat to the rift flanks (Cochran, 1983), (3) small-scale convection (Buck, 1986), (4) heterogeneous stretching of the lithosphere (Royden & Keen, 1980), (5) dynamic support of rift-flank topography (Parmentier, 1987), (6) igneous underplating as the result of partial melting in the mantle (McKenzie, 1978, 1984, 1985; Lister et al., 1986, in press), (7) permanent support by lithospheric flexure (Weissel & Karner, in press).

Simple rifting models are grossly symmetrical, and similar uplift on both flanks of the extended region is predicted. This clearly is not so across the Tasman Sea, as there is about 3–4 km of present-day elevation difference between the eastern highlands and the Lord Howe Rise (Fig. 7.3.2A). A further problem with

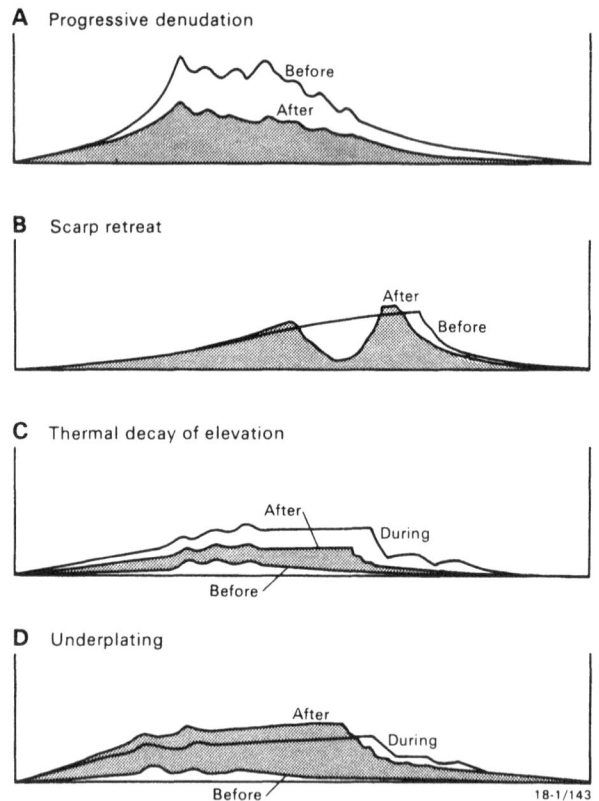

Figure 7.3.3 Different models may be used to predict different patterns of uplift and subsidence. The progressive-denudation model (A) predicts that the highest areas are eroded the fastest, so that topography progressively becomes more subdued. The retreating fault-scarp model (B) predicts that isolated patches of the peneplain will be raised above their original level. Topographic roughness will actually increase. The detachment model (C) predicts that the peneplain will be raised during continental extension, prior to break-up, and thereafter — as the fault scarp recedes inland — subsidence will take place so that elevation is decreased uniformly. The detachment model takes account of the uplift caused by intrusion of mantle-derived basaltic magmas into the crust during and after extension (D). This model predicts both thermal (transient) uplift which will begin to decay after break-up takes place and extension ceases, as well as permanent uplift related to underplating. Mantle diapirs triggered by the lithospheric extension process may continue to rise for 100 Ma or more subsequent to break-up, and melt generated where these rising columns intersect the solidus will cause step-wise uplift in the post-breakup period.

application of a simple rifting model to the eastern highlands is that the thermal peak would have been about 80 and 50 Ma ago, and most of the associated thermal uplift should have decayed by now (Fig. 7.3.3C). Wellman (1987) argued that the geomorphological evidence is in favour of stable or increasing uplift through the Cainozoic, although the evidence for increasing uplift is disputed (Bishop & Young, 1980; Bishop et al., 1986).

Asymmetric uplift has been suggested to be a general consequence of continental extension on major detachment systems, and a preliminary interpretation has been presented of the uplift of the eastern highlands, based on underplating beneath the upper plate or hanging wall of the detachment (Lister et al., 1986, in press). This model is presented in further detail in the following sections, and an account is given of how it can explain many of the enigmatic features of the eastern highlands, including the spatial and temporal relationship between uplift and volcanism.

7.3.3 Prediction of uplift history using detachment models for continental extension

Eastern Australia as an upper-plate passive margin

Traditional models for the continental rifting that precedes seafloor spreading and which leads to formation of passive continental margins, involves essentially vertical 'foundering' of continental blocks on steep normal faults. These faults dip and step downwards to the developing ocean floor, giving rise to a broadly symmetrical, steep-sided, graben character to the reconstructed, conjugate, margin pair. Detailed studies over the last decade in well-exposed rifted or extended regions, coupled with theoretical modelling of the thermal and subsidence behaviour of continental- and passive-margin rifts, have provided evidence that continental rifting involves large horizontal motions. Large horizontal motions require the operation of flat-lying or strongly rotational normal faults, at least in the upper to middle crust, and lead to quite different rift and passive-margin geometries than do the traditional models.

The new models for continental extension based on the existence of detachment faults and shallow-dipping shear zones, have considerable implications for the uplift-subsidence histories expected on passive margins (Davis & Hardy, 1981; Wernicke, 1985; Reynolds, 1985; Lister et al., 1986, in press; Davis & Lister, 1988; Lister & Davis, in press). In particular, a prediction of the new models is that extension is a fundamentally asymmetric process (Fig. 7.3.4) resulting in the formation of conjugate passive margins that are quite different in character (Lister et al., 1986, in press). The upper-plate margin occupies the hanging wall of the master detachments, and the lower-plate margin consists of the detachment footwall, commonly including highly faulted and extended segments of the upper plate (Fig. 7.3.4).

The model for an upper-plate passive margin involves the lower plate (that is, the lower crust and

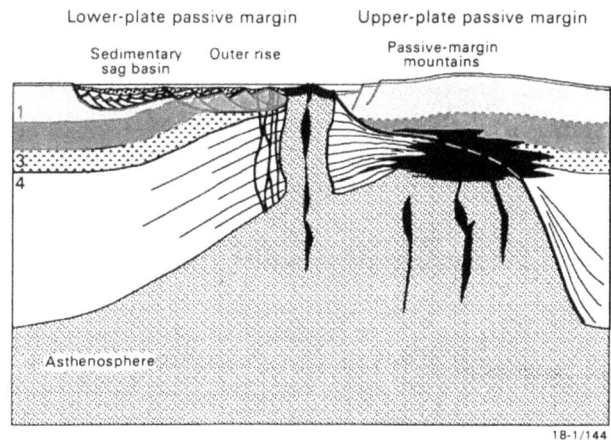

Figure 7.3.4 Gross architecture of upper-plate and lower-plate passive margins resulting from continental extension by detachment faulting. The opposing passive margins have marked but complementary asymmetry. The lower-plate margin (left) has a complex structure, including tilt-block remnants from the upper plate above bowed-up detachment faults. Multiple detachment has led to two generations of tilt blocks. The upper-plate margin (right) is relatively unstructured, and is raised to form passive-margin mountains as the result of thermal buoyancy related to rise of the asthenosphere, as well as of igneous underplating caused by intrusion of the mantle-derived melts. 1, 2, and 3 represent upper, middle, and lower crust, respectively, and 4 represents mantle lithosphere.

underlying lithosphere) being pulled out from beneath the margin during extension without significant (or any) extension of the upper crust taking place. Consequently, the continent bordering the passive margin should be raised as relatively cool, more dense lithosphere is replaced by hotter, less dense asthenosphere. The rocks at the surface will be subjected to a thermal pulse, but there need be little structure associated with these events. Melting of the upper mantle may accompany this process.

The lower-plate margin, on the other hand, will be highly structured, with large tilt blocks underlying a thermal sag-phase sedimentary cover, as observed in the western part of the Lord Howe Rise. Deep-seated metamorphic rocks will be overlain on the detachment by rocks of comparatively much lower metamorphic grade. They will tend to subside rather than be elevated during extension, because crustal thinning dominates on lower-plate margins. This is predicted for areas where tectonic denudation of the middle to lower continental crust has taken place.

The simple lithospheric-wedge or 'Wernicke' model shown in Figure 7.3.4 can be modified by allowing the lower plate to stretch uniformly below the detachment, and by allowing the detachment zone to terminate within the crust, in a zone of distributed deformation (Lister et al., 1986, in press). The most complex of such models involves the detachment (either a fault or a decoupling ductile shear zone) stepping downwards from one crustal level to another. This means that a highly structured lower-plate margin may be bordered by a now subsided marginal plateau (for example, the Queensland Plateau), and that such a marginal plateau may be separated from passive-margin mountains in the adjacent continental hinterland by a narrow rift basin. Such a model is illustrated in Figure 7.3.5. The

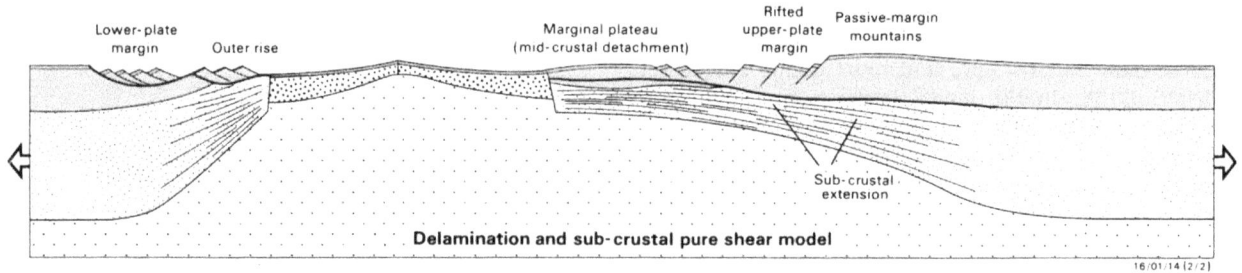

Figure 7.3.5 A model combining stepped detachment (fault plus a decoupling shear zone) and stretching of the crust and mantle beneath the detachment. The detachment system terminates at the crust/mantle boundary in a zone of subcrustal extension. Specific spatial relationships between the major architectural elements of passive margins are predicted from this model. There is a strong asymmetry of the uplift/subsidence history that results from the large lateral offset of the deep thinning from the upper-crustal extension. The lower-plate margin is highly structured and has at least two generations of tilt blocks. The marginal plateau will rise above sea-level during extension, but subside thereafter. This will be bounded by rift basins adjacent to passive-margin mountains. The uplifted area is relatively broad.

geometry of a detachment and the distribution of lithospheric deformation obviously have important consequences for the architecture of a margin, because in the case of a lithospheric wedge the zone of uplift is relatively narrow (Fig. 7.3.4), whereas in the case of a stepped detachment, with distributed pure shear, the zone of uplift is considerably wider (Fig. 7.3.5). Specific predictions from the detachment model concern the location of maximum igneous activity. Maximum partial melting should take place where uplift of the asthenosphere is greatest under the upper-plate margin.

Etheridge et al. (in press) interpreted several, major, passive-margin segments and conjugate margin pairs in terms of the detachment model. They presented detailed interpretations of the Australian southern margin (and its Antarctic conjugate) and the conjugate Atlantic margins of northwest Africa and the United States, as well as a schematic model of the extensional structure of the margins of the Tasman Sea. The Tasman Sea model — reproduced here in Figure 7.3.2 — involves a complex, westward-dipping, crustal detachment system beneath the Norfolk Ridge, New Caledonia Basin, and Lord Howe Rise. The detachment system is proposed to have extended into the mantle beneath eastern Australia which therefore occupies an upper-plate margin.

The amount of uplift or subsidence above an upper-plate margin can be estimated from idealised models simply by adding four terms:

(1) the negative buoyancy caused by any decrease in crustal thickness during detachment — for these calculations the crust is assumed to have an average density of 2.8×10^3 kg/m^3 and the mantle a density of 3.3×10^3 kg/m^3 at zero pressure;

(2) the positive buoyancy induced by any overall rise in the temperature-depth profile — for example, that caused by uplift related to movement on an inclined detachment system, or a steepening of the geothermal gradient caused by uniform attenuation of the lithosphere;

(3) the positive buoyancy induced by igneous underplating of the crust by basaltic magmas — these magmas are assumed to crystallise within, or at the base of, the crust as granulites having a density of 3.0×10^3 kg/m^3;

(4) effects related to flexure (for example, Weissel & Karner, in press) — such effects have not been taken

into account in the following analysis (see below).

The uplift history can be calculated in several different ways using the detachment model.

Thermal uplift predicted using the lithospheric-wedge model

The simplest model, termed the 'lithospheric-wedge' model, is that presented by Wernicke (1985). The amount of thermal uplift using the lithospheric-wedge model is not sufficient to account for the uplift of passive-margin mountains. The maximum uplift for a wedge-shaped separation geometry, assuming zero flexural rigidity, is expected in the upper plate, inboard of and above where the detachment system cuts through the crust/mantle boundary (Voorhoeve & Houseman, 1988). Movement on the detachment results in substitution of lithospheric mantle by relatively hot asthenosphere below that point. The maximum temperature increase in this case is constrained by the difference between the initial temperature at the base of the crust, and the temperature of the adiabatically cooled rising asthenosphere at this depth. The initial crustal thickness (t_c) is assumed to be 35–40 km, the thickness of the thermal lithosphere (t_l) to be 125 km, and the mantle adiabat (T) is described by the equation:

$$T = T_a + 0.5°C/km \cdot d \qquad (1)$$

where T_a is the temperature of the adiabat extrapolated to the surface. The ground surface is assumed to be at 0°C, and the maximum rise of temperature at the base of the crust caused by relative movement on the detachment is:

$$\delta T = T_a(1.0 - t_c/t_l) \qquad (2)$$

The average rise in mantle temperature over a vertical distance $t_l - t_c$ is half of this amount, so the thermally induced buoyancy is $0.5 (t_l - t_c).\alpha.\rho_m.\delta T$, where α, the coefficient of thermal expansion, is taken as $3.28 \times 10^{-5}/°C$. The density of the uplifted asthenosphere can be approximated by $\rho_m(1.0 - \alpha T_a)$. The maximum amount of thermal uplift that can be caused by detachment is therefore:

$$d = \left(\alpha T_a / 2(1.0 - \alpha T_a)\right) \cdot \left((t_l - t_c)^2 / t_l\right) \qquad (3)$$

where ρ_c is the density of the crust. A rise of the asthenosphere by 90 km (raising the asthenosphere to the base of the crust) induces an average temperature rise of about 480°C over a distance of 90 km in the mantle, for a 1350°C adiabat. Maximum uplift of the upper plate by 1.49 km is predicted by applying this formula (Voorhoeve & Houseman, 1988). Note, however, that such an uplift implies large displacements on the detachment. Raising a mantle adiabat of 1400°C leads to an increase in this estimate by only 55 m. The above amount of uplift is too small to explain passive-margin mountains which are 1–2 km high, 80–50 Ma after extension ceases, because the thermal anomalies caused by detachment would have decayed substantially in that time. The maximum uplift that can be explained using this model (80–50 Ma after break-up) is about 700–800 m, assuming a time constant of 60 Ma to estimate the effect of relaxation of thermal anomalies.

Thermal uplift by subcrustal stretching of the mantle

More complex calculations can be undertaken by considering the interaction of detachments with 'pure shear' of the crust or mantle (or both) above and below the detachment. The amount of thermal uplift for relatively small extensions can be increased if pure shear of the lithosphere under a detachment is considered, stretching the entire thermal lithosphere below the detachment zone (Fig. 7.3.5). The amount of uplift in such a model depends on the depth of the detachment in the crust, and the thickness of mantle that is subject to uniform attenuation. Maximum uplift is obtained if detachment takes place at the crust/mantle interface, and if only subcrustal stretching of the lithosphere takes place, because thermal-buoyancy terms then are not offset by the isostatic response to thinning of the less dense continental crust. This may be termed 'subcrustal extension', adopting the terminology used by Keen & Beaumont (in press). However, large amounts of uplift can be explained only if the thermal lithosphere is considered to be very thick, and if the geotherm is initially cool.

A geotherm for a thermal lithosphere 100 km thick, with a mantle adiabat of 1300°C, is shown in Figure 7.3.6A. The mantle under the continent is progressively stretched, giving the result shown in Figure 7.3.6B. The predictions made as the result of applying such a model are as follows: (1) the crust during the rift phase is raised by about 800m after a subcrustal stretch of

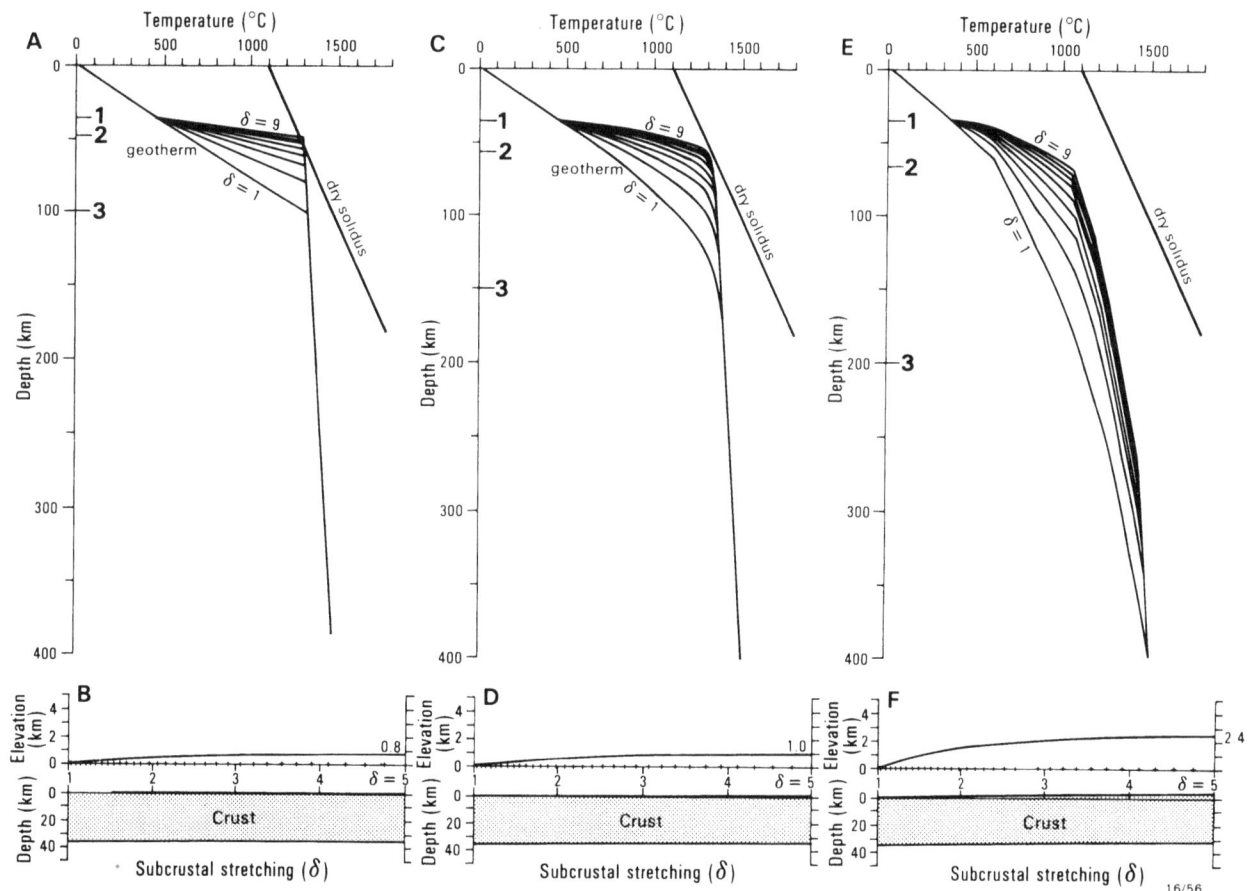

Figure 7.3.6 Calculated thermal and uplift effects of subcrustal stretching on mantle having a potential temperature of 1300°C. The solidus used is described by the equation $T(°C) = 1100°C + 3.8(°C/km)*d(km)$. 1 represents the base of the crust. The base of the lithosphere before and after heating is represented by 2 and 3, respectively. (A) Thermal lithosphere is initially 100 km thick. Subcrustal stretches (δ 1–9) transform the geotherm as shown. (B) Results of pressure-corrected buoyancy calculations for conditions in A. The solid line marks the total uplift during the extension phase for subcrustal stretches of 1–5. Total uplift reaches 0.8 km. Partial melting of the mantle will begin after a subcrustal stretch of 5 has been reached, but not before. (C–D) Same conditions as in A–B, except that the thermal lithosphere is initially 150 km thick. Total uplift reaches 1.0 km. (E–F) Same conditions as in A–B, except that the initial geotherm is that of an ancient shield having a 200 km-thick mantle root. Subcrustal stretching of such lithosphere markedly increases the thermal contribution to uplift — to 2.4 km.

400 percent (this uplift will decay as the thermal anomalies associated with extension are dissipated); (2) the asthenosphere at the end of the extension phase has been raised to the point at which partial melting has begun, and 0.3 km of igneous 'underplate' has been accreted to the crust. The amount of partial melting would increase markedly if deformation continued. However, a subcrustal stretch of 400 percent is already large.

The effects of stretching an initially thicker lithosphere are illustrated in Figures 7.3.6C–D. The thermal lithosphere is assumed to be 150 km thick for this calculation, again with a mantle adiabat of 1300°C in the underlying asthenosphere. The effects of subcrustal stretching on the geotherm are shown in Figure 7.3.6C. The thermal component of uplift will be greater, but greater subcrustal stretching must be undergone before partial melting begins. No melt is generated for the range of δ values considered here. The amount of thermal uplift has increased by 200 m to 1 km after a subcrustal stretch of δ equal to 5 has been applied.

The only way in which uplift of passive-margin mountains in excess of 1–2 km can be explained as the result of thermal uplift caused by subcrustal stretching of the lithosphere, seems to be to assume very thick mantle roots under the continents at a time prior to break-up. The geotherm used for this calculation is that of Minster & Archambeau (taken from Jordan, 1975) who suppose the geotherm to be non-adiabatic down to 400 km. The existence of thick mantle roots beneath the continental lithosphere therefore is assumed. The mantle root in Figures 7.3.6E–F is taken to be 200 km thick. Subcrustal stretching in the range δ 1–5 has the effect shown on the initial geotherm. The amount of thermal uplift is increased markedly (to a maximum value of 2.4 km). However, partial melting of the mantle will not take place unless very much larger subcrustal stretches are applied.

The amount of uplift caused by subcrustal extension indicated by the above calculations not surprisingly depends strongly on the assumed geotherm and on the thickness of the continental lithosphere immediately before extension begins. Jordan (1975, 1981), Sipkin & Jordan (1975, 1976), and Lerner-Lam & Jordan (1985) pointed to differences between seismic velocities beneath continental landmasses and those beneath the oceans. The differences can be explained only by substantial differences in mantle composition or thickness (or both) under the two regions. Jordan (1978) suggested that continents have lithospheric roots extending down to 200–300 km and which translate coherently during plate motions. The root zones are stabilised against convective disruption by the depletion of a basalt-like component and a consequent increase in effective viscosity. Lerner-Lam & Jordan (1985) concluded that significant continent-ocean shear-velocity heterogeneity persists to great depths, certainly greater than 220 km, and possibly approaching 400 km.

An explanation for the physical processes that might lead to these mantle roots may be found in periods of major magmatic activity that produced accretion of underplated layers to the continental crust during the Precambrian and the Palaeozoic (Drummond &

Collins, 1986). Multiple rifting events during the Precambrian were associated with large amounts of igneous activity, and presumably large volumes of mantle-derived basaltic rocks intruded and underplated the continental crust during these rifting events. These major igneous underplating events during the Proterozoic and early Palaeozoic may have led slowly toward a depleted and more refractory mantle, and the increase in viscosity which accompanied these changes may have been sufficient to terminate convection in the mantle root. Continental mantle roots cannot be explained as having formed subsequent to break-up of the supercontinents, because insufficient igneous activity has taken place to produce the observed vertical differentiation of the mantle in the relatively short time that has elapsed since break-up. Deep mantle roots of continental lithosphere must be remnant from the Palaeozoic or Precambrian.

Continental extension in the Mesozoic may have stretched mantle down to 200 km or greater depth, markedly increasing the component of thermally induced uplift, as indicated by the previous calculation, if mantle roots had survived the orogenic events throughout the history of the Tasman Fold Belt. The amount of uplift by now would have decayed to somewhat more than 1 km, but it is still sufficient to explain the present elevation of the eastern highlands. However, the highlands then must be assumed to have been considerably more elevated at the time of break-up, and they have been subsiding ever since. Neither of these two consequences is supported by the available data. The apparently stable (or increasing) uplift of the highlands can be used to argue against such models.

Highlands uplift as the result of igneous underplating during extension

Partial melting of rising asthenosphere leads to the production of basaltic magmas that segregate and migrate upwards. The driving force for upward migration diminishes when magmas reach the density drop defined by the crust/mantle boundary. The level of intrusion depends on magma density (Herzberg et al., 1983) and on the orientations and magnitudes of the principal stresses. These magmas may intrude the upper mantle and lower crust, and crystallise to form rocks with lower density than their parent. Buoyancy results because the weighted mean density of the mafic igneous rocks and their source residuum is less than the density of the original parent mantle source.

The thickness of the underplated layer can be estimated as follows (Cawthorn, 1975; Ahren & Turcotte, 1979; McKenzie, 1984). Extension is assumed to be rapid, so that adiabatic partial melting takes place. A constant melt fraction per degree elevation above the solidus is assumed. Klein & Langmuir (1987) argued that the thickness, t_m, of melt generated by uplift can be approximated by the formula:

$$t_m = 0.006(P_o - P_f)^2 \cdot 3.3 \text{ km} \qquad (4)$$

where the adiabatic melt coefficient is taken as 1.21 ± 0.4 percent for each kilobar the mantle is raised above the solidus. P_o is the pressure in kilobars for initiation of melting, and P_f is the pressure in kilobars for cessation of melting.

A similar function can be derived as follows. First, assume a dry solidus. This means that the volume of melt produced will be underestimated because hydrous melts may make up in excess of 2–5 percent by weight of the rising asthenosphere (I. A. Nicholls, personal communication, 1988). Second, assume a linear geotherm and a linear solidus in order to avoid calculations more complex than the accuracy of the input parameters would warrant. The adiabat is taken as:

$$T = T_a + c_p \cdot d \qquad (5)$$

where d is the depth in kilometers, c_p is the adiabatic compressibility, and T_a is the adiabat temperature at zero pressure. c_p is taken to be $0.5°C/km$, but note that estimates for this parameter differ by as much as ± 20 percent.

The dry solidus is taken as:

$$T = T_s + s_p \cdot d \qquad (6)$$

where d is the depth in kilometers, s_p is the pressure dependence of the solidus, and T_s is the solidus at zero pressure, here assumed to be $1100–1150°C$. s_p is $3.4–3.8°C/km$, using a value estimated from Takahashi (1986). Again, there is a wide range in estimates for this parameter.

Next, an arbitrary depth is taken at which melting ceases, because, for whatever reason, continued uplift of the asthenosphere is prohibited. Melting takes place in the region represented by the triangle shown in Figure 7.3.7, bounded by the solidus, the adiabat, and an arbitrary upper boundary where melting ceases because uplift of the mantle past that point is prohibited. Melting takes place at an arbitrary point in the rising

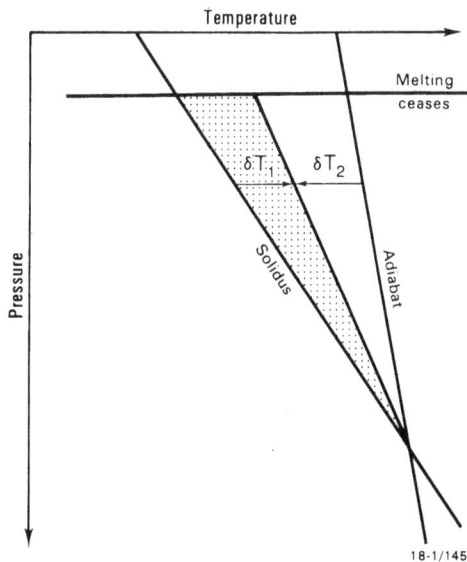

Figure 7.3.7 Melting will take place where material has risen along the adiabat, past the solidus. The energy available for melting is proportional to the shaded area shown. See text for further details.

column, and the rock cools by δT_1, as the latent heat of fusion is extracted. This temperature drop is

$$\delta T_1 = -k_1 \delta \chi \qquad (7)$$

where χ is the degree of partial melting, and T is the temperature. The solidus must rise at the same time as the composition of the restite changes. This temperature increase is

$$\delta T_2 = k_2 \delta \chi \qquad (8)$$

This means that the energy available for melting is reduced by the proportion (assuming linear behaviour):

$$\frac{k_1}{k_1 + k_2}$$

The area of the triangle bounded by the solidus, the adiabat, and the upper cap is then:

$$0.5(T_a - T_s)(P_o - P_f)^2/P_o \qquad (9)$$

The pressure at onset of melting can be calculated as follows:

$$P_o = (T_a - T_s)/(a_p - c_p) \qquad (10)$$

by solving the simultaneous equations (1) and (2).

The available energy for melting therefore is proportional to:

$$0.5(a_p - c_p)(P_o - P_f)^2 \qquad (11)$$

The amount of melt must be:

$$d_{km} = M \cdot \frac{k_1}{k_1 + k_2} \cdot \frac{(a_p - c_p)}{2} \cdot (P_o - P_f)^2 \cdot Z \qquad (12)$$

where d_{km} is the thickness of melt in kilometers, M is the percentage melt per $°C$ of adiabatic temperature drops, and Z is the number of kilometers per kilobar pressure rise. The value of

$$M \cdot \frac{k_1}{k_1 + k_2} \cdot \frac{(a_p - c_p)}{2}$$

is taken to be 0.006, assuming 1.2 percent per kilobar of adiabatic pressure release (Ahren & Turcotte, 1979) in order to be consistent with the calculations of Klein & Langmuir (1987).

Application of formula (4) leads to the prediction that the thickness of the underplated layer produced ranges from 3.3 km for the $1350°C$ adiabat, to 11.9 km for the $1450°C$ adiabat. Melting takes place late in the history of the extension event for the $1350°C$ adiabat, only after the asthenosphere has been raised to a depth of 69 km, but much more melting takes place for the $1450°C$ adiabat because melting begins after the asthenosphere has been raised to only 103 km.

There are many factors that can complicate the above calculations. First, the magnitude of the melt fraction per kilobar decompression may be in error by as much as 130 percent. Second, the value of s_p could be taken as high as 4°C/km. Third, there are different estimates of c_p, ranging by as much as 130 percent or more. In addition, small-scale convection cells (termed secondary convection by Buck, 1986) may be triggered by the extension, and therefore considerably increase the amount of melt that accumulates at this stage of the rift process. The results of these calculations therefore are illustrative of no more than the relative ease with which 5–15 km thickness of underplated gabbro can accumulate beneath an upper-plate passive margin during detachment faulting.

A 5–15 km-thick underplated layer will lead to 'permanent' uplift of the continental hinterland by about 0.5–1.4 km because accretion of mafic granulite to the base of the crust (as the result of underplating) effectively substitutes rock whose density is about 3.0×10^{-3} kg/m³ for mantle rocks that have a density of

about 3.3×10^{-3} kg/m³. There will be additional transient uplift in addition to this 'permanent' uplift because migration of the magmas upwards itself results in transfer of heat that increases thermally induced buoyancy.

Calculations of the amount of basaltic magma produced during stretching of the lithosphere (rather than simple asthenospheric uplift) have been made also, and large thicknesses of underplated material are found difficult to explain, unless large extensions are proposed. This is because most partial melting takes place as the result of asthenospheric uplift, and the greatest quantity of melt is generated by uplift to shallow levels. Lithospheric stretching does not cause sufficient uplift of the asthenosphere unless extensions are large (greater than 100 percent). Substantial igneous underplating will take place only late in the history of extension, and then only if (1) very large stretches are applied, or (2) the mantle adiabat (or potential temperature) is initially very high — for example, in the range 1450–1550°C. Subcrustal

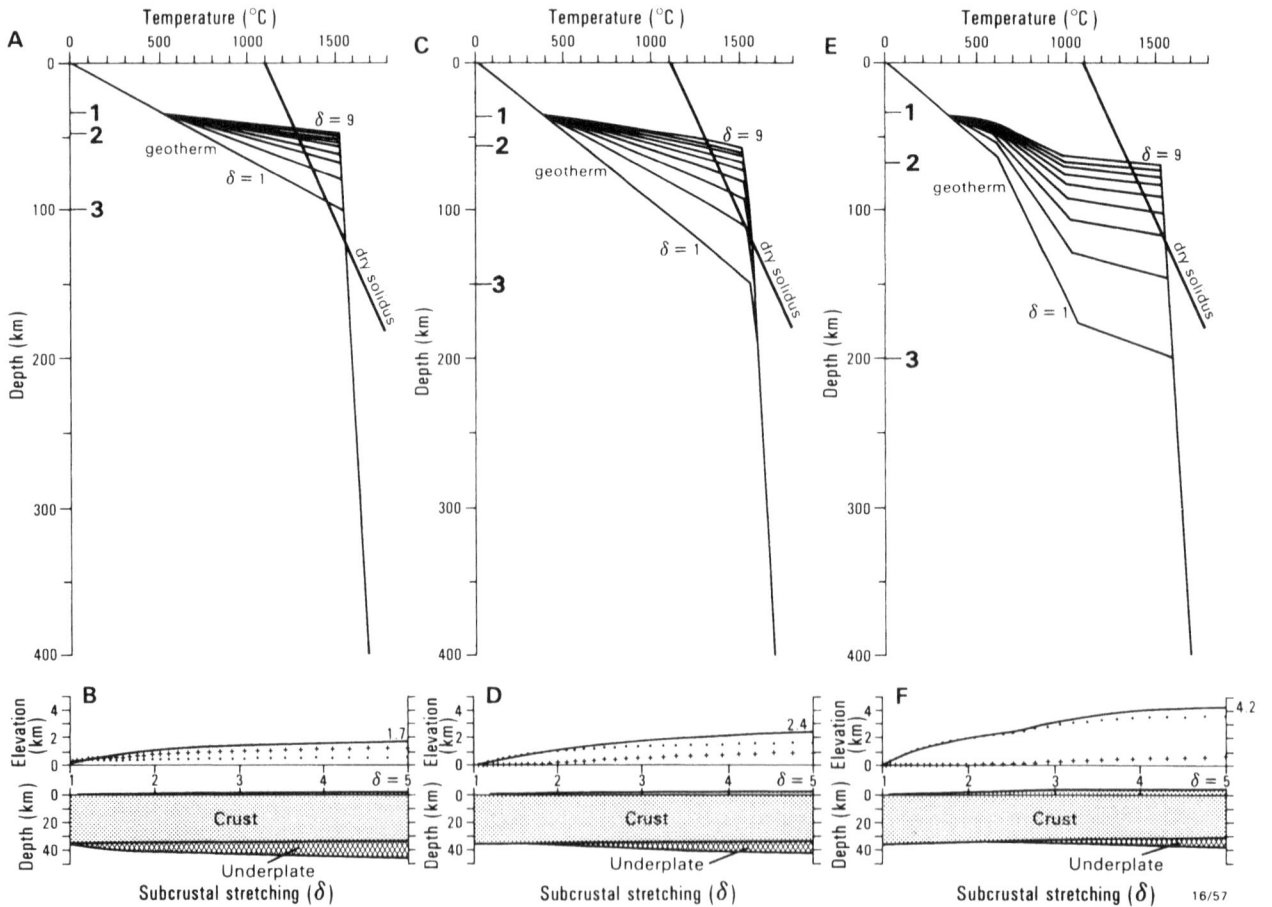

Figure 7.3.8 Calculated thermal and uplift effects of subcrustal stretching and igneous underplating on mantle having a potential temperature of 1500°C. 1 represents the base of the crust. The base of the lithosphere before and after heating is represented by 2 and 3, respectively. The solidus used is described by the equation T(°C) = 1100°C + 3.8(°C/km)*d(km). (A) Thermal lithosphere is initially 100 km thick. Subcrustal stretches (δ 1–9) transform the geotherm as shown. (B) Results of pressure-corrected buoyancy calculations for conditions in A. Dots represent the thermal contribution to uplift. Crosses mark the contribution by igneous underplating. Solid line represents the total uplift during the extensional phase for subcrustal stretches of 1–5. Total uplift reaches 1.7 km. (C–D) Same conditions and symbols as in A-B, except that the thermal lithosphere is initially 150 km thick. Total uplift reaches 2.4 km, but the 'permanent' contribution to uplift from underplating is reduced. (E–F) Same conditions and symbols as in A-B, except that the initial geotherm is that of an ancient shield having a 200 km-thick mantle root under which a transient temperature step has been introduced by the arrival of a convective upwelling, or a plume. Subcrustal stretching of such lithosphere markedly increases the thermal contribution to uplift, but again reduces the thickness of the underplated layer. Total uplift now reaches 4.2 km for a subcrustal stretch of 5.

stretching using a mantle adiabat of only 1300°C will not produce substantial melting, as discussed above. The situation can be reversed with a much higher mantle adiabat, as shown in Figures 7.3.8A–B. The effect of subcrustal extension on the geotherm associated with a 1500°C adiabat is shown in Figure 7.3.8A for a 100 km-thick thermal lithosphere, and in Figure 7.3.8C for a 150 km-thick thermal lithosphere. There is again a marked increase in the thermal component of uplift for the thicker lithosphere, but the amount of mantle that is raised through the solidus is correspondingly reduced. This effect is illustrated by the results of calculations shown in Figures 7.3.8B,D. The amount of igneous underplate accreted to the crust in Figure 7.3.8B for a 100 km-thick thermal lithosphere is 11.7 km for a subcrustal stretch of 5. The total uplift is 1.73 km, of which only 540 m is transient thermal uplift. The amount of igneous underplating shown in Figure 7.3.8D for a 150 km-thick thermal lithosphere has decreased to 8.41 km for a subcrustal stretch of 5, whereas the total uplift is 2.41 km, of which 1.55 km is transient thermal uplift.

The combination of the thermal effect of stretching a mantle root with the effects of igneous underplating also can be calculated. However, there is only one way in which both can exist: (1) an initial, thick mantle root and (2) substantial igneous underplating. This combination is to invoke a high adiabat temperature (or potential temperature) in the underlying asthenosphere. The hypothesis requires a transitory temperature step at the base of the thermal lithosphere, if the geotherm in the mantle root is initially low. One way to explain this would be to propose the arrival of a mantle plume or 'hotspot' upwelling. The calculation is illustrated in Figures 7.3.8E–F. The effect of the temperature step at the base of the thermal lithosphere is to increase uplift markedly in the initial stages. However, again, because the lithosphere is initially very thick, melting does not begin until late in the thermal history, and then the amount of melt is only limited. Underplating only 6.2 km thick is predicted, but the thermal contribution to uplift is now markedly larger. Total uplift exceeds 4.2 km in such a model.

The lithospheric-wedge model is somewhat more efficacious in producing melt. A detachment with a constant 30° dip through a 100 km-thick thermal lithosphere produces melt volumes as follows (assuming extension leads to relative movement of 170 km, bringing the asthenosphere to the base of the crust):

(1) the thickness of the melt column will be 14 km, and melting will begin at 121 km, for a mantle adiabat of 1500°C, the solidus at 1100°C at zero pressure, and a solidus slope of 3.8°C per kilometer.

(2) the thickness of the melt column will be 1.4 km, and melting will begin at 60 km, for a mantle adiabat of 1300°C, the solidus at 1100°C at zero pressure, and the solidus slope of 3.8°C per kilometer.

All these values are sensitive to the assumed parameters. If the solidus slope is reduced to 3.4°C/km then for the 1500°C adiabat the thickness of the underplate increases radically to 20 km, and melting begins at 140 km. The thickness of the underplate increases to 2.4 km for the 1300°C adiabat, and melting begins at 6.9 km. On the other hand, for a solidus slope of 3.8°C/km, but with a solidus starting at 1150°C, melt volumes are markedly reduced. The thickness of underplate for the 1500°C adiabat reduces to 9.7 km, with melting beginning at 106 km. The thickness of underplate decreases to 0.3 km, and melting starts only at 45 km depth, for the 1300°C adiabat. The modelling studies for partial melting illustrated in Figures 7.3.6 and 7.3.8 are equally sensitive to the values of these parameters.

Anomalously hot asthenosphere under eastern Australia during break-up

The cause of highlands uplift cannot be thermal uplift caused by subcrustal extension of the underlying lithosphere in the time preceding Tasman Sea opening, unless the eastern highlands once were considerably higher (by as much as even 3–4 km). This hypothesis would require subsidence to be continuing even now. This predicted uplift-subsidence history is shown in Figure 7.3.9A. However, the limited data available seem to be indicative of stable or increasing elevation, not the reverse. The remaining option is to consider uplift as the result of igneous underplating (Fig. 7.3.9B). However, too little melt can be produced to explain the observed uplift if melt production is considered to be the result of uplift of a normal mantle adiabat (Klein & Langmuir, 1987; McKenzie & Bickle, 1988). Substantial underplating can be explained only if the mantle adiabat is initially as high as about 1500°C.

The same conclusion has been reached independently by White et al. (1987b) considering the evolution of passive margins. Igneous underplating of both the upper- and the lower-plate margin is inherently capable of producing additional uplift of the correct amount, but only if anomalously hot mantle (in relation to that required to produce normal mid-ocean-ridge basalt) is raised. The remaining hypothesis requires much higher mantle adiabats in the initial period of extension than those supposed to exist in the asthenosphere at the present day beneath the mid-ocean ridges. Both crustal thicknesses and observed geochemistry at present-day mid-ocean ridges may be explained by partial melting of passively uplifted mantle with a potential temperature of about 1300°C (Klein & Langmuir, 1987). Anomalous crustal thicknesses such as those observed in Iceland at the present day, or inferred under passive margins in the past, require mantle adiabats in excess of 1500°C.

Why would anomalously hot mantle exist beneath the edge of Gondwanaland at the time of break-up? — because of the insulating effect of the super-continent, or because of the arrival of a mantle plume? A mantle plume could result because of (1) unrelated events in the deep mantle, (2) subduction and 'roll-back' of the Pacific slab immediately prior to extension which caused hot asthenosphere to migrate into the backarc region; or (3) lithospheric extension triggering either convective upwelling or diapiric rise of hotter, less dense mantle through the asthenosphere. Stretching results in the uplift of mantle with high potential temperatures, so that enhanced melting takes place. The locus of sea-floor spreading moves with the migrating lithosphere, and the zones of raised mantle become progressively cooler so that oceanic crust does

A Passive-margin mountains
(with sub-crustal extension)

Rift phase   *Sag phase*

Elevation

Time

B Passive-margin mountains
(with sub-crustal extension and igneous underplating)

Rift phase   *Sag phase*

Elevation

Time

Igneous underplating

C

Elevation

Gradual uplift
of the peneplain

Stepwise uplift
due to periods
of underplating

150   100   50   0
Time (Ma)

Continental
extension
commences   Inception
of breakup

Active uplift   Thermal component decays

Additional uplift caused by igneous underplating
18·1/147

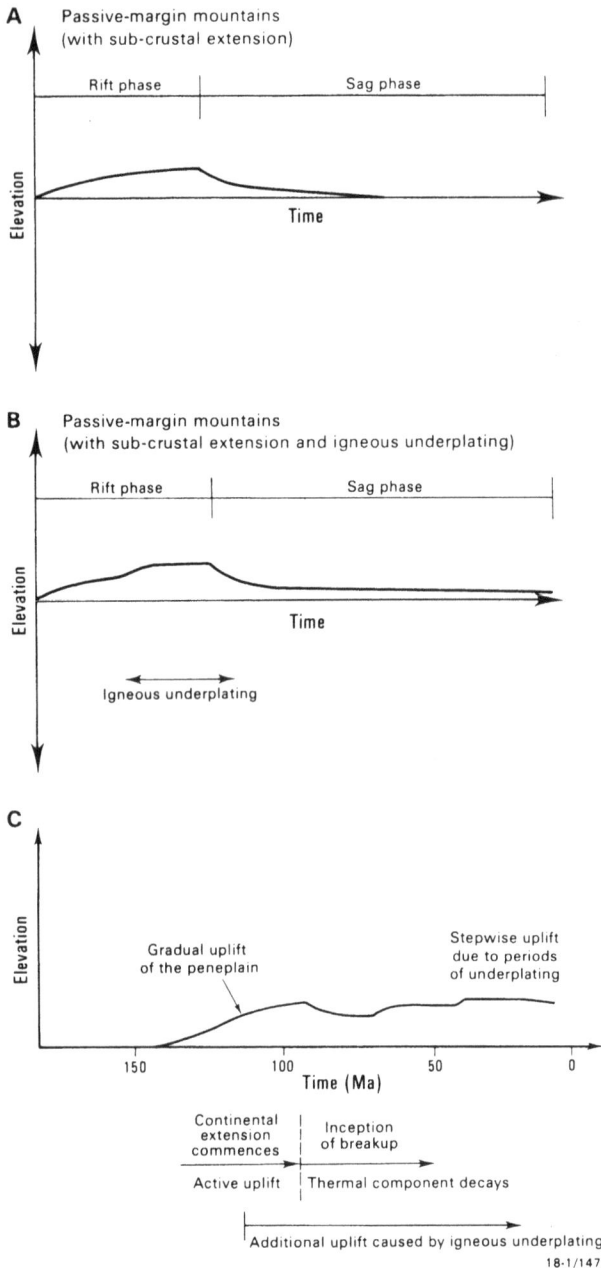

Figure 7.3.9 Schematic illustration of the uplift/subsidence history for an upper-plate margin. The land surface during the extension phase is elevated progressively (A), but once break-up has taken place the thermal anomalies caused by attenuating the lithosphere are relaxed slowly and subsidence takes place. However, underplating during extension, towards the end of the extension history, causes permanent uplift (B). A more complex history involving step-wise uplift during the subsidence phase will ensue (C) if mantle-derived melts continue to arrive for several million years.

not form today with the same thicknesses as does the initial igneous underplate.

Buck (1986) and Mutter et al. (1988) examined this same problem, and concluded that the conditions of 'incipient' break-up were favourable to enhanced small-scale convection. This is because the topography of the thermal lithosphere was considerable, because of localised stretching. The thicker, cooler lithosphere adjoining the stretched zone acts as cool 'wall rock' driving downward flow of the adjacent lithosphere. This

sets up small-scale convection cells for a limited period, until the thermal perturbation is reduced, or continued extension increases the width of the 'cell', making conditions for rapid circulation of the convecting asthenosphere less favourable (Buck, 1986; Mutter et al., 1988). The effect of reduced mantle viscosity would only accelerate the convection cell if the rising asthenosphere had reached the solidus. The result would be large volumes of magma produced during a short time interval, while the relatively rapid uplift of partially melting asthenosphere continued apace.

7.3.4 Detachment geometry and uplift location

Differences in detachment geometry influence the amount of offset of thermal uplift and underplating zones from the eventual continental margin, the shape of the uplift, and the relationship between uplift geometry and passive-margin structure.

The properties of a detachment system that have the greatest influence on upper-plate uplift are down-dip shape and extent. Two extremes of detachment geometry are illustrated in cross-section in Figures 7.3.4–5. The model shown in Figure 7.3.4 is not unlike the model proposed by Wernicke (1985) except that the effects of igneous underplating have been added, and the whole lithosphere fractures as the result of the switch from continental extension to seafloor spreading. Extension is accomplished almost entirely by movement on the detachment, and there is relatively little penetrative extensional strain. This example is modelled numerically above as the lithospheric-wedge model. Figure 7.3.5 represents the case where the detachment has several flats and ramps, and penetrative ('ductile') stretching of the crust and mantle takes place below the detachment shear zone. In particular, the detachment shear zone terminates as a long flat section along the crust/mantle boundary where it separates essentially unstretched crust from homogeneously extended mantle lithosphere.

The lithospheric-wedge model, in simple terms, is akin to renting of the mantle lithosphere, giving rise to localised, intense, thermal perturbation and increased likelihood of extensive mantle melting and consequent underplating, as outlined above. Such a geometry is likely to lead to uplift that has a relatively short wavelength and large amplitude. Passive-margin structure also is strongly influenced by detachment geometry, and the simple lithospheric-wedge model is predicted to result in narrow, relatively unstructured margins (Lister et al., 1986, in press). Steeper detachment dips within the mantle will exaggerate these effects, and reduce the offset of the uplift from the continental margin.

In contrast, the detachment plus pure-shear model results in a broader thermal anomaly and at times less mantle melting (and therefore less potential underplating). This will result in a broader, lower-amplitude uplift that generally is offset farther from the continental margin and which may decay completely as does the thermal anomaly. A stepped detachment such as that shown in Figure 7.3.5 having a mid-crustal flat also is predicted to give rise to a broader, upper-plate margin adjacent to a marginal plateau and possibly an upper-plate rift basin (Lister et al., in press).

The east-Australian margin has some of these relationships (see Fig. 1.3.4). The Great Divide south of about 30°S is on average about 130 km from the coast, and only 150–200 km from the edge of the continental shelf. The continental margin is very narrow, as is the wavelength of the uplift. The uplifted region between 20° and 30°S is broader, its elevation is generally only about a half that of the southeastern highlands, and the divide is between 200 and 300 km from the coast. This region of subdued uplift corresponds closely with a much broader continental margin consisting of several plateaux and rift basins. The uplift north of 20°S again narrows and approaches the coast, corresponding to a narrower continental margin. The coastal escarpment is only well developed where the uplift and the margin are both narrow.

The applicability of the detachment model to the differences in uplift and margin geometry along the east-Australian margin cannot be tested until more geophysical data are available to refine the structure and lithologies beneath both the highlands and the continental margin. However, there is clearly a correspondence between uplift geometry and margin structure, and this supports the view that the uplift is essentially an outcome of the continental extension that preceded the seafloor spreading in the Tasman Sea and Coral Sea system.

7.3.5 Detachment models and intraplate volcanism

Introduction

A question not yet considered is whether the Cainozoic intraplate volcanism of eastern Australia might be related to continental extension. A genetic relationship between at least some of the volcanism and the uplift of the eastern highlands is inferred from two observations. First, the volcanism is concentrated strongly along the Great Divide defining the axis of uplift. Second, there is relatively little volcanism that predates either the uplift or the continental extension. However, note that the volcanism predominantly postdated continental extension by a few tens of million years, whereas the maximum lithospheric thermal anomaly in the model presented here coincides with the switch from continental extension to seafloor spreading.

Wellman & McDougall (1974a) demonstrated that the Cainozoic volcanic rocks fall into two main groups in terms of petrology, eruptive style, and time-space distribution (Section 1.1.3). The central volcanoes and submarine guyots form linear arrays that decrease in age from north to south and appear to represent mantle hotspot traces. In contrast, the lava-field provinces have much less systematic time-space distributions, and appear to reflect near-random melting beneath the highlands, especially between about 50 and 20 Ma ago. The discussion that follows is concerned mainly with the lava-field volcanoes, although it bears also on the problem of the origin of the mantle hotspots.

There are two main requirements for intraplate basaltic volcanism. First, there must be a thermal anomaly in the mantle of sufficient magnitude to induce a gravitationally unstable and sufficiently mobile proportion of partial melt. Second, the stress field throughout the upper part of the lithosphere must be such that steeply dipping extensile fractures are permitted (Shaw, 1980), once the basaltic magma rises to a level within the thermal lithosphere such that rates of diapiric or compactional melt migration (McKenzie, 1984) are too slow.

The detachment model can be used to account for extensive partial melting of the mantle beneath an upper-plate margin, especially where rapid uplift of abnormally hot asthenosphere results, so there is not likely to be a problem of available melt volume. Eastern Australia was the site of an active convergent margin immediately prior to the beginning of the late Cretaceous extension (for example, Veevers, 1986). The underlying asthenosphere therefore was likely to be hotter than that beneath a stable craton, and therefore capable of producing large melt volumes (see Section 7.3.3). The volume of melt produced by raising the asthenosphere increases as a quadratic function of the temperature of the mantle, as well as in relation to the actual volume of material moved through the solidus. Hence, greatest melting takes place where the mantle is raised the most — that is, beneath an upper-plate margin.

Melts produced during continental extension were unable to pass through the crust, and were underplated, giving rise to an important component of eastern-highlands uplift. This is consistent with the paucity of late Cretaceous volcanism in the highlands region, but raises the question of why the basaltic magma induced by lithospheric extension did not breach the crust when the tectonic conditions (that is, active extension) apparently were ideal for the formation of vertical fractures. Further, if the volcanism is related largely to the period of continental extension, why was there a delay of several tens of millions of years between extension and much of the volcanism of both types?

An attempt is made here to answer both questions. The answer to the first question may be found in conventional fracture mechanics. A more difficult problem is to explain why intraplate volcanism should take place so long after lithospheric extension ceased.

Conditions under which melt penetrates to the surface

There is a paucity of late Cretaceous and earliest Tertiary volcanism in the highlands region at a time when extensive mantle melting and crustal underplating is considered to have taken place. The question then arises as to why this melt should underplate the crust, and not rise through it. Extension of the crust produces ideal conditions for the formation of vertical fractures, so surface evidence for magmatism might be expected.

The relation between the orientation of the principal stresses and the passage of magmas through the crust was dealt with thoroughly by Shaw (1980) and summarised in Section 1.6. The basic principle is that steeply dipping, magma-filled fractures can form only if (1) the magma fluid pressure is approximately equal to the lithostatic load, (2) the minimum principle compressive stress is horizontal. An additional, key, but commonly ignored factor is that open (tensile) fractures can form only where the differential stress magnitude is less than about six times the tensile strength of the wall rock (that is, less than about 50–100 MPa).

The widely held view that basaltic underplating commonly accompanies lithospheric extension provides an apparent paradox. There is a tendency to want to make a simple connection between an extensional tectonic environment, 'tensile' stresses, and the formation of vertical fractures. Drawing a direct relationship between tectonic setting (whether the orogeny was compressive or extensional) and the supposed ability of the magma to penetrate steeply dipping fractures in the crust, is dangerous. Such an approach ignores the problems of (1) maintaining magma fluid pressure at the tip of the advancing fracture (although buoyancy assists upward propagation of a vertical fracture at deeper levels), (2) the heterogeneity of the lithospheric stress field, and (3) situations where differential stresses exceed the maximum value to allow an open fracture.

The differential stress during active continental extension especially in the middle crust and upper mantle (Sibson, 1982; Lister et al., in press) is likely to be relatively high. The differential stress, within the stronger layers or stress guides, may be sufficiently great that the mode of brittle failure is restricted to the compressional shear-failure field, and in this case magma-filled fractures cannot propagate. These stronger layers therefore provide effective barriers to upward magma transport during active tectonism.

There are two cases that allow magma-filled fractures to propagate through stress guides: (1) magma must arrive at sufficiently high rates so that high deviatoric stress levels are relaxed by ongoing fracturing; (2) decrease in the rate of extension, so that deviatoric stress levels begin to fall. Each magma-filled fracture that terminates its upward movement at a stress guide leads to fracturing, and consequent stress release. Stress levels will build up at a rate determined by the rate of extension. The magnitude of deviatoric stress in the stress guides will begin to fall if stress drops related to the rate of magma arrival begin to dominate, and eventually magma-filled fractures will be able to penetrate the stress guide and continue migrating towards the surface.

Volcanism during extension might peak when cessation of active tectonism leads gradually to an overall reduction in the magnitude of differential stress, and extensional fractures begin to propagate to the surface. Such an explanation accounts for the apparent deficiency of volcanism during active continental extension beneath the eastern highlands. It is also consistent with the temporal coincidence between some of the oldest lava-field volcanism and the earliest periods of seafloor spreading.

Magma arrival as a trigger for seafloor spreading

The arguments presented in the preceding section are inconclusive. Why magmas might not penetrate through to the surface, and therefore why underplating will take place in preference to extrusion, can be explained, but not whether the required magmatic activity was actually taking place. A theoretical reason has been proposed why magmatic activity should start late in the extension cycle, but this does not account for the intraplate

volcanism continuing for several tens of millions of years after the start of seafloor spreading. Perhaps this is the expression of quite a general phenomenon, and there is a fundamental link between lithospheric extension and subsequent pulses of magmatic activity.

Continental break-up generally is a distinct event, quite separate from the period of continental extension that precedes it. This is logical, because continental extension requires penetrative deformation of parts of the crust and mantle, whereas break-up is essentially a brittle process requiring fracture of the lithosphere. Magma-filled fractures may tear either relatively ductile, convecting asthenosphere or stiffer lithosphere, such as oceanic lithosphere. The time constants of the two phenomena (ductile deformation and fracture migration) are dramatically different. Break-up will take place subject solely to the availability of sufficiently large volumes of melt, so that lithospheric fractures may form.

The arrival of magma in large volumes must be the process that triggers the switch from continental extension to progressive dilation of dyke swarms and continental break-up, followed by seafloor spreading. Magma production increases dramatically once the stretching and therefore rising geotherms encounter the solidus. Break-up, therefore, takes place late in the extension process, subject to the availability of sufficiently large volumes of melt. Moreover, magmas at some extensional margins like the eastern highlands may start to arrive 10–30 Ma after the extension has ceased, so continental break-up may significantly post-date the period of continental extension which precedes it, because no whole-lithosphere fractures can form until the arrival of these magmas. The question remains as to why magmas continued to arrive so long after the extension event has initiated.

Lithospheric extension as a trigger for the rise of deep-seated mantle diapirs

Why then do magmas begin to arrive beneath extensional terranes so long after the period of extension has ceased? Can lithospheric extension lead to the formation of local diapirs or 'hotspots' under the upper-plate margin, so that magmatic activity continues long after continental extension has ceased? Consider that the temporal and spatial relationships of volcanism in the eastern highlands reflect the time constant of mantle processes triggered by the rapid extension of the lithosphere beneath the upper plate. The rapid uplift of the base of the thermal lithosphere caused by stretching induces uplift of the asthenosphere at greater depths, if there is anomalously hot asthenosphere at depth (Fig. 7.3.10). This rapid uplift may initiate chains of mantle diapirs or elongate plumes rising along relatively hot adiabats until partial melting takes place. These chains of solid-state diapirs must rise from depths of 400-200 km (Fig. 7.3.11). Rising at rates of 5–10 km/Ma they will reach the mantle solidus anywhere from 20 to 100 Ma after the continental extension event that initiated their rise, and extensive partial melting then will begin. Impingement of the diapirs on the base of the stretched, but now thermally relaxing, lithosphere is likely to induce further partial melting at these higher

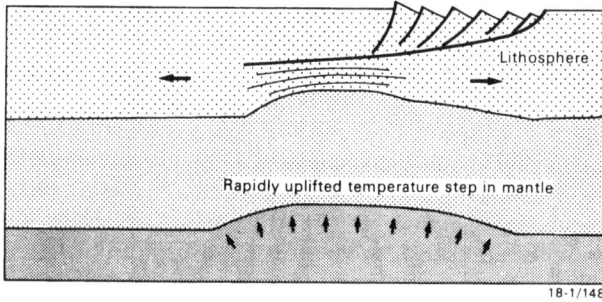

Figure 7.3.10 Rapid lithospheric extension could trigger chains of rising mantle diapirs if a temperature step in the mantle was bulged upwards. This could take place as shown if the displacement caused by lithospheric extension is rapid, whereas relatively slow extension would result in a more diffuse pattern of flow to accommodate the event.

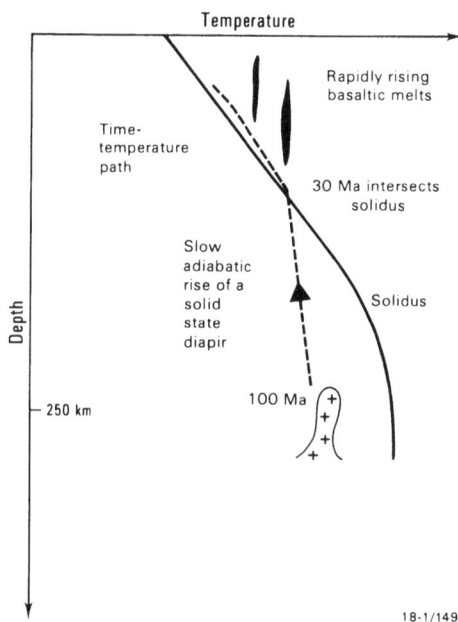

Figure 7.3.11 A rising mantle diapir might take 30–100 Ma to rise along its adiabat until it encounters the solidus, and basaltic melts form, segregate, and intrude the crust.

levels. What proportion of this melt reaches the surface depends on the lithospheric stress field, the thermal conditions of the melt and wall rocks, its density, and the speed of upwards migration of the magma.

There are, according to these arguments, three separate origins of basaltic melt that might result from detachment-related continental extension, all of which are predicted to be located preferentially beneath the eastern highlands and to post-date the continental extension phase by up to several tens of millions of years. The first is during rapid stretching of the continental lithosphere. These melts are interpreted to have been largely prevented from reaching the surface during extension by the high differential stress magnitudes accompanying rapid deformation. They were underplated, intruding the lower to middle crust during extension, and causing substantial uplift of the eastern highlands. These melts could have made their way to the surface in the waning stages of extension, or after extension ceased, and then continued to erupt until the thermal anomaly resulting from extension decayed to

the level where no further melt segregation was taking place. The source of these melts is within the lithosphere at the same location as the source of the underplated melts. Related volcanism therefore will migrate with the plate and be restricted to the eastern highlands and there will be no systematic time-space relationship along the axis of the highlands.

The second origin of melts is from asthenospheric mantle plumes and diapirs (hotspots) that are a secondary consequence of the lithospheric thinning. This source is predicted to provide melts some time after cessation of extension, and therefore will appear to migrate relative to the lithospheric plate, giving rise to volcanic traces beginning in the middle Tertiary.

The third origin is within the lithosphere, resulting from secondary melting induced by the arrival of the mantle diapirs mentioned above. The related volcanoes will have the same range of ages as the hotspot-related ones, but they will have a much less well-defined plate-migration pattern because they originate within the lithosphere. Opportunity will exist, in all three cases, and indeed be likely, for mixing of melts derived from different depths, and therefore they could be difficult to distinguish geochemically.

The hypothesis that all hotspots have their origin in the deep mantle, independent of the extensional process, requires remarkable coincidence. If hotspots originate randomly in the deep mantle, why should the hotspot activity be confined to the terrane immediately adjacent to a margin so recently subject to a major continental extension event? This would require phenomena in the deep mantle to 'recognise' events taking place in the lithosphere hundreds of kilometers above. An alternative hypothesis therefore is proposed: that age-progressive volcanic lines can be caused by melts segregating from the culminations of chains of mantle diapirs or convectional upwellings, whose rise was triggered by rapid uplift during a much earlier lithospheric-extension

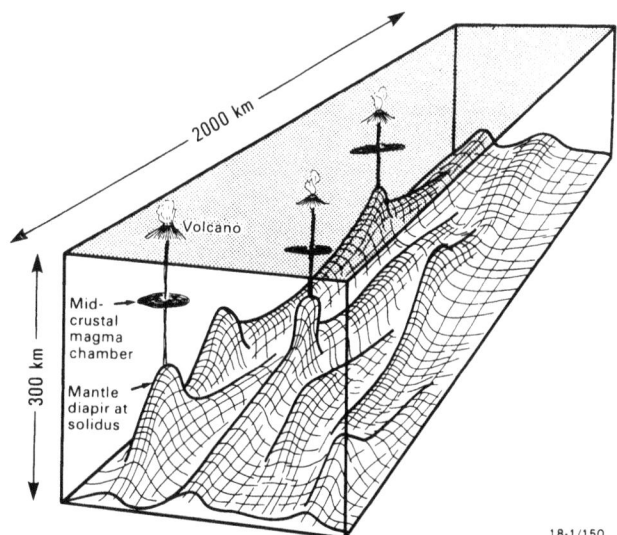

Figure 7.3.12 Chains of rising diapirs, or convective upwellings, could explain the concept of a plume swath which may have formed beneath the upper-plate margin of a detachment terrane. The effect is similar to that of deep-seated hotspots as the lithosphere drifts over the chain of rising diapirs. Three upwellings are shown, generating three hotspot traces on the lithosphere surface.

event. Convective upwellings, or diapiric rises in the asthenosphere, are shown in Figure 7.3.12. Some of these have brought asthenosphere above the solidus, so that melt segregation has begun, and there is consequent volcanic activity at the surface generating separate hotspot trails.

These hotspots would be confined almost exclusively to zones in which significant thinning of the lithosphere takes place during stretching — that is, in this model, to upper-plate margins or to marginal plateaux — if the plate was not moving relative to the asthenosphere. Melts generated during the extension process lead to tearing of the mantle root while the thermal lithosphere extends. The rising of magma bodies produced during subsequent intraplate volcanism eventually leads to additional underplating of the continental crust and, therefore, to additional uplift. The volume of melts extruded will be a poor indicator of the amount of melt intruded into the crust at depth during such events. The uplift-subsidence curve for the ancient planated surface would assume a more complex form than previously indicated, as illustrated in Figure 7.3.9C.

The hotspots would not be confined necessarily to upper-plate situations if the plate was moving. The Indo-Australian plate has moved northwards in the case of the eastern highlands, so the approximate spatial coincidence is preserved, as the ascending magmas will continue to underplate the upper plate. However, the magnitude of uplift might be expected to be correspondingly reduced, because some of the magmas will erupt to contribute only to seamounts in the Southern Ocean. Significantly, therefore, the Transantarctic Mountains defining the southern continuation of the eastern highlands have been elevated to considerably higher altitudes.

The difficulties with this hypothesis are (1) the paucity of seamounts in the southern ocean, and (2) the question as to what geodynamic events were taking place to the north — for example, during the Coral Sea opening — to trigger mantle rise and the production of the observed hot-spot traces. Answers to these questions are provided in Section 1.7. First, particular hot-spot traces are recognised: one starting in the South Tasman Rise, and other traces (the Tasmantid Guyots, the central-volcanoes trace (or traces), and the Lord Howe seamount chain) which started in the Tasman Sea and Coral Sea basins. Second, the past positions of the hotspots relative to the continent are calculated. The hotspots may well have originated in the asthenosphere beneath the opening Tasman Sea and Coral Sea basins (Fig. 1.7.3). The mechanism proposed here may need considerable revision but, given the available data, a novel interpretation can be supported — namely, that lithospheric extension can trigger the rise of mantle diapirs, or convective upwelling in the underlying asthenosphere, and that these mantle rises generate protracted histories of volcanism, visible in the overlying (migrating) lithosphere as hotspot trails.

Absence of intraplate volcanism on southern margin of Australia

The slow rise of solid-state diapirs could account for why intraplate volcanism begins so long after an extension event, as well as the observed spatial association. But then why is there no intraplate volcanism adjacent to Australia's southern margin? Two factors are recognised: (1) the southern margin may be an example of a lower-plate margin (Lister et al., in press; Etheridge et al., in press); and (2) the underlying asthenosphere may not have had as high a potential temperature, so that while there is considerable asymmetry in subsidence history between the southern margin of Australia, and the corresponding margin of Antarctica, both margins have subsided.

7.3.6 Evidence for an underplated layer beneath southeastern Australia

Uplift of the eastern highlands is attributed in the detachment model to a combination of the thermal anomaly generated during continental extension and the underplating of mantle-derived igneous rocks derived by partial melting of the upper mantle during extension. The thermal anomaly decays with a time constant of 80 to 50 Ma, whereas the uplift caused by underplating is permanent, unless transformation of the underplated mafic granulite to eclogite takes place after cooling. However, the underplating model has consequences for crustal thickness and depth-composition relationships, and for the timing and history of uplift, as discussed below.

An underplated layer may be recognised by means of geophysical observations and studies of xenoliths in the Cainozoic volcanic rocks. Finlayson et al. (1979) carried out seismic-refraction experiments in the southern part of the eastern highlands and described a layer of relatively high P-wave velocity (Vp) between about 25 and 50 km depth. There is a gradual increase in Vp from about 6 km/s at about 20 km to about 7.5 km/s at 45–50 km, followed by a jump in Vp to more than 8.0 km/s between 45 and 55 km. The depth region between 25 and about 50 km has been interpreted conventionally as lower crust, intruded and underplated by mafic material whose proportion increases with depth. The seismic Moho and the crust/mantle boundary are interpreted as corresponding to the base of this zone, where Vp increases to more than 8 km/s. Seismic-refraction data from the southeastern highlands therefore are consistent with underplating of an aggregated thickness of 15 to 20 km of mafic material. The ages of this material are poorly known, although some of it is known to be relatively young (less than 100 Ma) and some to be Palaeozoic or even Proterozoic (Rudnick et al., 1986 ; McDonough et al., in press), on the basis of isotopic data for mafic xenoliths from the northern and central parts of the eastern highlands.

More recent seismic-refraction, wide-angle reflection, and vertical-incidence reflection data are available from the central eastern highlands, and the crustal and uppermost mantle structure is known to be somewhat different there. A depth interval of increasing Vp from about 6 to about 7.2 km/s, followed by a relatively rapid increase to over 7.7 km/s is revealed by the refraction data, but the thickness of the transitional velocity layer and the depth to the seismic Moho are both considerably less than in the southern highlands. The transitional velocity layer corresponds to a band of prominent, discontinuous reflectors in the vertical-incidence profile, and the 'refraction' Moho is at the same depth as a distinctive reflector that separates the reflective zone from a transparent interval that extends to at least 70 km. What then is the significance of the difference between these crustal structural models, and how should they be interpreted in terms of underplating, uplift, and volcanism?

First, the interpretation of the transitional (high) velocity layer that commonly corresponds to a well-layered zone in reflection profiles will be considered. The most widely accepted interpretation is that this zone corresponds to felsic-to-intermediate lower crust intruded by increasing proportions of mafic material as depth increases. This layer near the base may be less reflective, and its high velocity could correspond to massive, mafic, unmodified mantle. This interpretation has been disputed by Griffin & O'Reilly (1987b) who proposed that the crust/mantle boundary corresponds to the upper part of the reflective zone, and that the reflection and refraction Moho represents the base of the uppermost mantle layer intruded by mafic sills. Both interpretations, however, include an aggregate thickness of mafic underplating of 10–20 km. The refraction data from the central eastern highlands were obtained in the New England region, nearly along the crest of the eastern-highlands divide, and the uplift along most of the shot line is similar in width and elevation to that in the southern highlands. P. Wellman (1979b) demonstrated that the elevated regions are close to isostatic equilibrium, and inferred that the highest regions were underlain by the thickest crust, which seems at first sight contradictory to the seismic evidence. However, the two data sets can be reconciled isostatically because the central region has a significantly lower, sub-Moho, P-wave velocity, and therefore density, than does the southern region. The refraction data cannot be used to determine velocity differences below the seismic Moho, but simple conversions of known velocity to density enable the two regions to be matched isostatically at a depth of about 70 km.

Large volumes of mafic intrusion into the lower crust and upper mantle (that is, effectively equivalent to underplating) are inferred for the model given in Section 7.2, which differs from the 'conventional' model only in that the crust/mantle boundary is much shallower (20 to 25 km). The higher Vp part of the reflective transitional layer is considered to be mantle (dominated by spinel lherzolite) whose Vp is reduced by its relatively high temperature and pyroxene content and by the intruded mafic material. The model in Section 7.2 is underpinned by petrological interpretation of the ambient pressure-temperature conditions of the spinel lherzolites that make up more than 75 percent of the xenolith suites in east-Australian Mesozoic to Cainozoic basalts.

Griffin & O'Reilly (1987a,b) also drew support for their thermal model from MAGSAT data. They interpreted the generally low magnetisation in eastern Australia as reflecting a shallow Curie point, and therefore a high regional geothermal gradient. There are three main anomalous areas within this area of generally low magnetisation, two of which coincide with areas of young volcanism (western Victoria and east-central Queensland), and which have been used by Griffin and O'Reilly to support their model. However, these areas have anomalously high magnetisation, which would infer a deeper Curie point temperature and therefore a lower geothermal gradient in a uniform crust, or different rock properties.

In particular, there are two critical assumptions. First, pressure cannot be determined from the spinel lherzolites, so temperatures calculated from the assemblages can be used only to infer depth of origin by fitting them to a geotherm modelled largely from the rarer garnet-bearing assemblages that yield both temperature and pressure estimates. This and other aspects of the petrological interpretations were criticised by McDonough et al. (in press) who considered that the petrological evidence is consistent also with the conventional model of a thick, underplated crust. The second assumption is that the thermal regime inferred from the xenolith studies is typical of the regional, steady-state, thermal structure, rather than being restricted in space and time to the immediate vicinity of the thermal perturbation associated with the volcano hosting the xenoliths. These two assumptions form the basis for the interpretation that the lithosphere beneath eastern Australian has been sufficiently hot for up to 200 Ma to produce mantle melts continuously. In contrast, a prediction of the detachment model is a regional thermal peak between 80 and 50 Ma that has been decaying ever since, except where locally perturbed by the hotspots (mantle plumes) that trigger relatively local volcanism.

Overall, neither the geophysical nor the xenolith data allow the construction of lithological and structural models of the lower crust and uppermost mantle beneath eastern Australia with the precision required to distinguish between the different models, except to support the view that the lower crust or the uppermost mantle (or both) contain substantial amounts of mafic material that probably was underplated. There is no compelling evidence that underplating took place in one event, nor that it was formed during Cretaceous to early Tertiary lithospheric extension along the eastern seaboard, as required by the detachment model, or that underplating is genetically related to the Cainozoic volcanism.

7.4 Comparison with other Intraplate Volcanic Areas

7.4.1 Introduction

The purpose of this section is to compare the east-Australian volcanic belt with other intraplate volcanic provinces world-wide in both continental and, to a lesser extent, oceanic settings. The account is based on an extensive literature search, but only those provinces considered to be of most relevance are considered. Furthermore, emphasis is given to provinces that have a similar age, and therefore degree of erosion, to the east-Australian volcanic belt. A summary of the volcanological, geochemical, and tectonic features of the areas to be compared, including the source references, is presented in Table 7.4.1. Comparative

Table 7.4.1 Comparison of intraplate volcanic provinces world-wide

Province	Age (Ma)	Tectonic environment	Type of activity[1]	Main chemical characteristics and xenoliths	References
East African Rift	22–0	Continental Rift	Central complexes, fissure lavas.	Strongly alkaline to transitional. Bimodal. Ab, Tb, Ne, Pc, Tr, pRh. Felsic > mafic. Pyroclastic deposits common. Lherz, harz, webst, gabbro, eclog.	Baker et al. (1977); Baker (1987); Lippard (1973); Price et al. (1985); Davies & McDonald (1987); Dautria & Girod (1987).
Deccan	70–63, mostly 69–65	Rifted continental margin, hotspot.	Central vents, fissure lavas.	Dominantly tholeiitic basalt. Ot, Qt, Pc, Rh, Ab, Ne, Cb. Mafic > felsic. Pyroclastic deposits minor.	Alexander (1981); Wellman & McElhinny (1970); Cox & Hawkesworth (1985); Beane et al. (1986); Devey & Lightfoot (1986); Lightfoot & Hawkesworth (1986); Duncan & Pyle (1988); Courtillot et al. (1988); K.V. Subbarao, personal communication (1989).
Paraná	149–119	Rifted continental margin, possible hotspot.	Fissure lavas.	Dominantly tholeiitic basalt. Ot, Qt, Rd, Rh. Mafic > felsic.	Basaltic Volcanism Study Project (1981); Bellieni et al. (1986); Petrini et al. (1987).
North Atlantic	60–0	Rifted continental margin, hotspot.	Central volcanoes, fissure vents, dyke swarms, layered complexes.	Dominantly tholeiitic basalt. Ot, Qt, Pc, Ic, Ab, Ha, Mu, Be, Da, Tr, Rh. Mafic > felsic. Pyroclastic deposits present.	Holland & Brown (1972); Moorbath & Thompson (1980); Basaltic Volcanism Study Project (1981); Palacz (1985); Cox (1983); Thompson & Morrison (1988).
Karoo	204–120, mostly 190–180	Rifted continental margin, possible hotspot.	Fissure lavas, sills, dykes, central complexes.	Dominantly tholeiitic basalt. Ot, Tha, La, Ab, Pc, Cb, Rh, Ph. Bimodal. Mafic > felsic. Pyroclastic deposits present.	Basaltic Volcanism Study Project (1981); Hawkesworth et al. (1983); Eales et al. (1984).
Rhine Graben	48 – < 1	Continental rift	Volcanic fields, maars, cones.	Strongly alkaline. Ba, Lc, Ph, Cb. Mafic > felsic. Pyroclastic deposits common. Splherz, cpxite, harz, wehr.	Wörner et al. (1985, 1986); Freundt & Schmincke (1986); Nixon (1987); Dewey & Windley (1988).
Massif Central	< 10	Continental rift, extensional horst and graben.	Volcanic centres (stratovolcanoes), fissure lavas.	Alkaline, minor subalkaline. Ba, Ab, Ha, Mu, Be, Tr, Ph, Rh. Mafic > felsic. Pyroclastic deposits abundant. Splherz, cpxite, gtlherz, gran.	Wimmenauer (1974); Chauvel & John (1984); Downes (1984, 1987); Nixon (1987).
Asia	Cainozoic	Continental rifts.	Fissure lavas, central volcanoes, shield volcanoes, cinder cones.	Alkali basalt dominant. Ba, Ne, Ab, Tb, Thb, Ha, Mu, Tr, Rh, pRh. Mafic > felsic. Pyroclastic deposits present. Splherz, gtwebst, amph, eclog.	Whitford-Stark (1983, 1987); Peng et al. (1986); Cao & Zhu (1987).
Red Sea and Gulf of Aden	30–20 < 10	Continental rift, rift-margin shoulder, possible hotspot.	Stratovolcanoes, plateau lavas, fissure lavas.	Alkali basalt, transitional to tholeiitic. Ab, Ha, Mu, Be, Tr, Rh, pRh. Mafic > felsic and felsic > mafic; differs between centres. Pyroclastic deposits common. Splherz, harz, dunite, gabbro.	Gass & Mallick (1968); Cox et al. (1970); Nixon (1987); Coleman & McGuire (1988); McGuire (1988).
Ethiopian Rift	Ca. 26–<7	Continental rift, possible hotspot.	Strato-central volcanoes, fissure lavas, domes, dykes.	Alkali basalt, transitional, tholeiitic. Ab, Ot, Qt, Ha, Mu, Be, Tr, pRh. Mafic > felsic and felsic > mafic; differs between centres. Pyroclastics present. Harz, dunite, lherz, cpxite, webst, troct, gabbro.	Barberi et al. (1970, 1975); Barberi & Varet (1970); Bizward et al. (1980); Ottonello (1980); Brotzu et al. (1981); Betton & Civetta (1984); Nixon (1987).

continued

Table 7.4.1 Comparison of intraplate volcanic provinces world-wide (continued)

Province	Age (Ma)	Tectonic environment	Type of activity[1]	Main chemical characteristics and xenoliths	References
Marie Byrd Land	30–0	Continental rift, rift margin shoulder.	Fissure lavas, shield volcanoes, cinder cones.	Alkaline to strongly alkaline. Ba, Ab, Ne, Ha, Mu, Be, Tr, Rh, pRh. Mafic > felsic. Pyroclastic deposits abundant. Splherz, pxite, dunite.	LeMasurier & Rex (1983, in press).
Southwestern Africa	136–35, episodic	Continental rift, rift-margin shoulder, possible hotspots.	Pipes, plugs, dykes, lava flows, central volcano.	Dominantly strongly alkaline. Kb, Cb, Ml, Ne, Lp, Ga, Sy, Ph, Tr. Pyroclastic deposits minor. Lherz, gtlherz, splherz, gran.	Moore (1976); Duncan et al. (1978); McIver (1981); Marsh (1987); Nixon (1987).
Cameroon line	66–0	Rift-related oceanic-continental magmatic line.	Central volcanoes, lava plateaux, seamounts.	Alkaline to strongly alkaline. Ne, Ba, Ab, Tb, Ph, Rh. Mafic > felsic. Splherz, plaglherz, harz, pxite, amph.	Fallick et al (1985); Fitton & Dunlop (1985); Fitton (1987); Nixon (1987).
Snake River	<16	Plate-margin convergence, possible hotspot.	Fissure lavas, plains lavas, central complexes.	Dominance of ol-tholeiitic basalt and rhyolite. Bimodal. Ot, Qt, La, Rh. Mafic > felsic. Crustal xenoliths.	Basaltic Volcanism Study Project (1981); Menzies et al. (1984).
Yellowstone	<2	Hotspot, extensional regime.	Central caldera complexes, marginal vents.	Dominance of rhyolite lavas. Minor extracaldera basalt. Thb. Rh. Felsic > mafic. Pyroclastic deposits common.	Doe et al. (1982); Hildreth et al. (1984).
Erebus	15.4–0	Continental rift, hotspot.	Lava fields, central complexes, domes, cinder cones.	Alkaline. Ba, Ab, Ha, Mu, Be, Ts, Ph, pTr. Felsic > mafic. Pyroclastic deposits present. Lherz, pxite, harz, wehr, dunite, gran.	Kyle (1981); Kyle et al. (1982).
Hoggar	Late Cretaceous, Miocene-Quaternary	Possible hotspot, continental swell.	Volcano-plutonic massifs, stratovolcanoes, lava flows, plugs, maars, cones.	Ba, Ne, Ab, Tr, Ph, Nesy, Di, Ga, Rh. Mafic > felsic. Pyroclastic deposits present. Splherz, spharz, webst, cpxite, opxite, gabbro, spwebst, gtwebst.	Duncan (1981); Nixon (1987), Dautria & Girod (1987); Dautria et al. (1988); Lesquer et al. (1988).

Abbreviations. Rocks: Ab, Alkali basalt; Ba, basanite; Be, benmoreite; Cb, carbonatite; Da, dacite; Di, diorite; Ga, gabbro; Ha, hawaiite; Ic, icelandite; Kb, kimberlite; La, latite; Lc, leucitite; Lp, lamprophyre; Ml, melilitite; Mu, mugearite; Ne, nephelinite; Nesy, nepheline syenite; Ot, ol-tholeiitic basalt; Pc, picrite; Ph, phonolite; pRh, peralkaline rhyolite; pTr, peralkaline trachyte; Qt, qz-tholeiitic basalt; Rd, rhyodacite; Rh, rhyolite; Sy, syenite; Tb, transitional basalt; Tha, tholeiitic andesite; Thb, tholeiitic basalt; Tr, trachyte; Ts, tristanite. Xenoliths: amph, amphibolite; cpxite, clinopyroxenite; dunite, dunite; eclog, eclogite; gabbro, gabbro; galherz, garnet lherzolite; gtwebst, garnet websterite; gran, granulite; harz, harzburgite; lherz, lherzolite; opxite, orthopyroxenite; plaglherz, plagioclase lherzolite; pxite, pyroxenite; spharz, spinel harzburgite; splherz, spinel lherzolite; spwebst, spinel websterite; troct, troctolite; webst, websterite; wehr, wehrite.

[1] Terminology is that used in quoted references.

age and rock-type data for the east-Australian provinces are given in Figures 3.2.1, 3.3.1, 3.4.1, 3.5.1, 3.6.1, 3.7.1, and 3.8.1.

An important conclusion that results from this literature search is recognition of the apparent uniqueness of intraplate volcanism in eastern Australia. First, the 3000 km length of the proposed east-Australian hotspot trace, as defined by the time-space relationship of the central-type and leucitite provinces, is highly unusual. Unambiguous continental hotspot traces of even a fraction of this length are unknown. Another striking feature is that the volume of volcanic rocks in eastern Australia is small (see Section 1.1.3) relative to volumes in many intraplate provinces (for example, continental flood basalts such as Karoo and Deccan, and oceanic islands such as Hawaii and Iceland), yet intermittent volcanism has taken place along the 4400 km-length of eastern Australia from about 70 Ma to the Holocene. This volcanism represents an exceptional example of long-term, widespread volcanic activity.

7.4.2   Tectonic environment

Intraplate volcanism can be triggered by a range of mechanisms, and a three-fold tectonic classification of intraplate volcanic provinces was presented in Section 1.6, namely: (1) continental rifting initiated either by extensional processes which can lead ultimately to new ocean floor (for example, the East African, Ethiopian, and Red Sea rifts, and the North Atlantic), or by continental collision resulting in compressive regimes and propagating fractures (for example, the Rhine Graben, Massif Central, and Baikal); (2) active backarc environments and associated basin-and-range type provinces (for example, Columbia River and northeastern China); (3) oceanic or continental mantle plumes, with or without relative movement of continental/oceanic crust (for example, Hawaii, Iceland, and eastern Australia). Only the first and third of these categories are considered to be of relevance here — namely, rifting or newly rifted continents, and volcanism well away from the influence of active plate margins (mainly 'hotspot'-type volcanism).

Features of the east-Australian Cainozoic volcanism that need to be considered in any comparison include (1) the time of onset of volcanism relative to the initiation of continental rifting, (2) its distribution adjacent to the east-Australian continental margin, (3) its association with the east-Australian highlands, (4) time/space trends of the central-type volcanism indicative of hotspot (or hotline) activity, and (5) the geochemical characteristics of the volcanism.

Lava-field type volcanism in eastern Australia was initiated about 70 Ma ago, and took place mainly between 50 and 30 Ma. It is considered to have been linked to the continental extension regime accompanying the opening of the Tasman Sea starting about 80 Ma ago. The opening of the Coral Sea between about 60 and 50 Ma falls within the period of maximum opening of the Tasman Sea (80–50 Ma). The lava-field type volcanism is dominantly alkaline to strongly alkaline, although subalkaline rocks are important in some provinces, particularly the young (less than 5 Ma)

Newer Volcanics province of western Victoria. Evolved alkaline rocks are volumetrically insignificant. The mildly alkaline to tholeiitic central-type volcanism in which evolved rocks are volumetrically important, has apparently been active on land only since about 34 Ma. Most of the Cainozoic volcanism on the east-Australia continent therefore was initiated about 10 Ma after the beginning of the phase of maximum ocean-floor production in the Tasman Sea, and became most productive after this activity ceased (at about 50 Ma).

The distribution of the east-Australian volcanic provinces on the uplifted flank of a rifted continental margin is strongly suggestive of control relating to the tectonics of the continental margin, even though the volcanism has been most active since cessation of continental rifting and seafloor spreading. The lava-field type volcanism in eastern Australia closely follows the trace of the rifted continental margin. Superimposed on this are the central-type provinces considered to reflect the passage of the Indo-Australian plate over a stationary mantle plume or plumes (Wellman & McDougall, 1974a; Section 1.7).

7.4.3   Continental rifting

Intracontinental rifts

Continental rifts are recognised as one of the major tectonic environments for the production of large volumes of igneous rocks having diverse volcanological and geochemical characteristics. An important example is the strongly alkaline *East African Rift* which, in contrast to east-Australian continental rifting, has not proceeded to the point of continental separation. It is also different with regard to the timing of volcanism and rifting, in that large-scale magmatic activity has taken place both before and during the period of continental rifting. The pre-rift volcanism in the East African Rift area consists of strongly alkaline melanephelinite lavas, dykes and clastic rocks, mostly emanating from localised central vents, followed by flood phonolites, trachytes, and nephelinites before or shortly after rift formation. Volcanism after graben formation was dominantly mildly alkaline in composition forming thick porphyritic trachytic and pantelleritic expanses of flood lavas and low shield volcanoes on the rift floor. The least alkaline magmas tended to be erupted within the well-developed rift in the central and deepest part of the rift zone (Baker, 1987). The temporal trend from strongly alkaline to mildly alkaline compositions is similar to that observed commonly in eastern Australia as a whole, reflecting the initial dominance of lava-field provinces and the later dominance of central-type provinces, but it contrasts with many individual provinces, such as the Newer Volcanics, where mildly alkaline to tholeiitic lavas tended to be erupted before the more strongly alkaline types.

The *Massif Central* and *Rhine Graben* provinces are examples of volcanism associated with a tensional tectonic regime resulting from continental collision, and the development of horst and graben structures and incipient rifting. The Tertiary and Quaternary volcanism in the Massif Central and the Rhine Graben and

Bohemian provinces of central Europe (Wimmeneuer, 1974; Downes, 1984, 1987; Wörner et al., 1986) have many similarities to the east-Australian lava-field type provinces. These central European examples of intraplate volcanism are represented by individual, well-defined volcanic fields of relatively small volume and limited time span distributed discontinuously over a distance of about 1500 km. The volcanic fields are dominated by alkaline to strongly alkaline lavas such as alkali basalt, basanite, and nephelinite, and there is no well defined correlation between age of volcanism and location. Minor evolved rocks include phonolite and rhyolite. The Rhine Graben volcanism has a progression from Tertiary alkali basalt to Quaternary basanite, nephelinite, leucitite, and melilitite, and in this respect mirrors the trend to more strongly alkaline compositions with time in a number of other provinces, including the Newer Volcanics in Victoria.

Cainozoic intraplate volcanism in continental *Asia*, as in central Europe, is dominantly alkaline, and there is a tendency for the degree of alkalinity to increase away from plate boundaries (Whitford-Stark, 1983, 1987). The most intense period of volcanism in the west-Asian areas coincided with the major collision phase of the Indian and Eurasian plates and the resulting development of major strike-slip systems, plus the Baikal and Shanxi graben systems. The strike-slip systems are characterised by strongly alkaline rocks, including leucitites, basanites, and phonolites (Whitford-Stark, 1983, 1987), whereas the graben systems are similar to many continental rift systems in which lavas transitional from mildly alkaline to subalkaline compositions are concentrated in regions of maximum extension, and the more strongly alkaline lavas are confined mostly to peripheral areas of reduced tension (Tapponnier et al., 1982; Kiselev, 1987).

Intracontinental flood basalts

The dominantly tholeiitic, large-volume, flood basalts associated with continental break-up such as *Deccan, Paraná, North Atlantic*, and *Karoo* (Cox & Hawkesworth, 1985; Bellieni et al., 1986; Moorbath & Thompson, 1980; Eales et al., 1984; Basaltic Volcanism Project, 1981) contrast markedly with much of the volcanism associated with intracontinental rifting, as well as with the rifted continental-margin volcanism of eastern Australia. These flood basalts have many geochemical similarities, but there are some differences in the timing of their emplacement relative to the timing of continental rifting. The main volcanic activity in both the Karoo and Paraná provinces was near or just prior to the beginning of rifting associated with the opening of the Indian and South Atlantic oceans (Eales et al., 1984; Bellieni et al., 1986). In contrast, volcanism in the North Atlantic and Deccan provinces was largely coincident with periods of maximum rifting (Norton & Sclater, 1979; Morgan, 1981; Pandey & Negi, 1987; Thompson & Morrison, 1988).

Large-volume flood basalts typically have low Mg-values relative to oceanic tholeiitic basalts, are dominated by qz-tholeiitic compositions, are made up of largely uniform eruptive units, have an absence of mantle xenoliths, and commonly are considered to be high-temperature melts (Cox, 1980). Miidly alkaline to transitional basalts are important in the North Atlantic and Karoo provinces, and to a lesser degree in the Deccan province. Strongly alkaline magmatism associated with flood-basalt provinces is rare, but alkali basalts, nephelinites, and carbonatites are present in the Upper Traps to the west of the Deccan province and are aligned broadly along two major rift zones that are younger than the Trap tholeiitic basalts (Basaltic Volcanism Project, 1981).

Rift margins

More directly comparable with eastern Australia is the volcanism in the Red Sea and Gulf of Aden area, Marie Byrd Land, Antarctica, and to a lesser degree southwestern Africa. Volcanic activity in both areas has continued over a considerable length of time forming discrete volcanic fields extending laterally along raised rift shoulders. Comparisons also can be made with the Cameroon-line volcanism, the continental part of which is on the uplifted flank of the Benue Trough.

There has been active volcanism in the *Red Sea* and *Gulf of Aden* area from late Oligocene to the present and continental rifting is considered to have begun about the same time (Voggenreiter et al., 1988). Transitional, mildly and strongly alkaline basalts form more than twenty separate fields along the western margin of the Arabian plate from the Gulf of Aden to the Mediterranean Sea and have an estimated volume of 10^3–10^5 km^3 (Coleman & McGuire, 1988). This volcanism extends inland for about 400 km where it is represented by alkali basalts containing upper-mantle xenoliths.

The Red Sea margin has strong asymmetry, and the height of the raised eastern margin averages between 1000 and 3000 m above sea-level (Martinez & Cochran, 1988) forming a steep scarp similar in form, but not in scale, to that along much of eastern Australia. Bosworth (1987) and Voggenreiter et al. (1988) suggested that the asymmetrical geometry of the Gulf of Suez possibly reflects a major detachment fault dipping eastwards under the Sinai Peninsula. Volcanism both pre-dates and post-dates the uplift, which is dated in the Sinai by fission-track methods at about 26 Ma. Steckler (1985) argued that the main phase of uplift was at least 8–10 Ma after initiation of rifting, and that the heat generated by rifting alone is not adequate to explain the uplift associated with the Gulf of Suez rifting. He suggested a model similar to that proposed by Karner & Weissel (1984) for eastern Australia whereby rifting induces thermal and velocity perturbations resulting from secondary small-scale convection beneath the rift. Supportive of this concept are seismic-tomography interpretations indicative that the Red Sea thermal anomaly (low-velocity zone) extends to a depth of more than 350 km (Anderson & Dziewonski, 1984). The possibility of widespread magmatic underplating associated with the Red Sea and Gulf of Suez also must be considered, and McGuire (1988) noted that Saudi Arabian xenolith and geophysical data are consistent with magma upwelling beneath the rift axis and flowing laterally outwards beneath the Arabian plate.

The volcanism of the African-Arabian rift systems, and particularly that associated with the Ethiopian and Red Sea rifts, has many geochemical similarities to both the lava-field and central-type volcanism of eastern Australia. There is a continuum of compositions ranging from strongly alkaline to transitional and tholeiitic basalts, through to undersaturated trachytes and phonolites, as well as saturated peralkaline trachytes and rhyolites considered typical of the transitional basaltic association (Coombs, 1963). More strongly alkaline lavas, equating with the east-Australian lava-field provinces, are dominant mostly on the rift escarpment, whereas those on the rift floor trend towards transitional, tholeiitic, and more evolved compositions (Barberi & Varet, 1977; Baker, 1987) and equate more closely with the east-Australian central-type provinces. The more strongly tholeiitic basalts of Aden, Little Aden, Erta'Ale, and Alayta (Gass, 1970; Barberi et al., 1970, 1972) are thought to exist where rifting has proceeded to the point of break-up of the continental crust and the production of MORB-type lavas and oceanic crust (Barberi et al., 1970; Civetta et al., 1975). $^{87}Sr/^{86}Sr$ values for the Red Sea and Gulf of Aden plateau alkali basalts are in the range 0.703–0.704 and are indicative of little or no upper-crustal contamination during their ascent through the crust (Coleman & McGuire, 1988). These values are similar to those for primary and near-primary basalts in both the lava-field and central-type provinces of eastern Australia (Sections 5.5, 7.7.4).

The *Marie Byrd Land* volcanic province rests on an uplifted zone between the early to middle Jurassic Transantarctic Rift, a major intracontinental rift system (LeMasurier & Rex, 1983, in press; Schmidt & Rowley, 1986), and the continental margin of western Antarctica. It could be described as being on the shoulder of the Transantarctic Rift, but equates more closely with the east-Australian volcanism if it is considered as rift-shoulder volcanism associated with the west-Antarctica continental margin. Major rifting in the middle Mesozoic along the Transantarctic Rift was reactivated during the development of the Pacific-Antarctic ridge system about 80 Ma ago and the separation of the New Zealand/Campbell Plateau from western Antarctica. The axis of the present rift system is parallel to the line along which the New Zealand/Campbell Plateau block broke away from western Antarctica. LeMasurier & Rex (in press) considered that much of the present-day topographic relief in Marie Byrd Land formed 25–30 Ma ago during the period of middle to late Cainozoic volcanic activity. A crustal thickness beneath coastal Marie Byrd Land of about 32 km has been estimated from seismic data, and positive gravity anomalies are consistent with geophysical modelling indicative of magmatic upwelling (LeMasurier & Rex, in press).

The Marie Byrd Land volcanic province extends along the west-Antarctica coast for 750 km and in places is up to 400–450 km wide. It consists of areas of basaltic volcanism, eighteen shield volcanoes composed of intermediate and felsic rocks, and small basaltic cinder and tuff cones containing spinel-lherzolite, pyroxenite, and dunite xenoliths (LeMasurier & Rex, in press). The cinder and tuff cones commonly are present on the caldera rims or flanks of the shield volcanoes and in this respect are similar to the post-shield volcanic activity of Hawaii and the younger cinder cone and maar activity of northern Queensland and the Newer Volcanics in western Victoria. Strongly alkaline basaltic rocks including basanite and *ne-hawaiite* make up about 90 percent of the total volume of the province, and the remainder consists of mugearite, benmoreite, trachyte, and rhyolite, dominated by peralkaline trachyte (LeMasurier & Rex, in press). Basaltic magmatism appears to have preceded the eruption of more felsic types by about 10 Ma, and the volcanism began about 30–28 Ma ago. $^{87}Sr/^{86}Sr$ values for the basalts range between 0.70258 and 0.70322 (Futa & LeMasurier, 1983). These values are similar to those for alkali basalts in Tasmania, but are mostly lower than values for comparable rocks on mainland eastern Australia. Also contrasting with eastern Australia is the absence of any Sr-isotopic evidence for crustal contamination in the more evolved rocks in Marie Byrd Land (Section 5.5).

There is an apparent serial change in the age of volcanic activity in the east-west and north-south chains of felsic shield volcanoes. However, the disparate directions of these trends and the absence of Antarctic plate motion during the period of the Marie Byrd Land volcanism, are evidence that these trends are not hotspot controlled.

Another possible example of rift-margin volcanic activity is the Tertiary magmatism which forms discrete volcanic fields over a distance of about 1000 km along the upwarped western margin of *southern Africa*. This magmatism ranges in age from 136 to 35 Ma, and mostly postdates the opening of the proto-Atlantic at around 127 Ma (Moore, 1976). Magmatic products are mostly strongly alkaline and include kimberlite, carbonatite, melilitite, nephelinite, syenite, phonolite and trachyte, which form pipes, plugs, tholoids, dykes, lava flows and, at Saltpetrekop in South Africa, a predominantly trachytic central volcano (Duncan et al., 1978; McIver, 1981; Marsh, 1987). The tectonic environment for this widespread magmatism has been attributed to either epeirogenic upwarping of the continental margin (Moore, 1976), or a series of hotspots now located in the South Atlantic (for example, Bouvet, Vema, and Gough; Duncan et al., 1978; Le Roex, 1986).

Southern African magmatism is somewhat similar compositionally to the more strongly alkaline lava-field provinces of eastern Australia, particularly those of Tasmania and the Bundaberg-Boyne area of southern Queensland, although Tertiary kimberlites and carbonatites are unknown in eastern Australia. However, unlike eastern Australia, where volcanic activity has been effectively continuous for the last 70 Ma, the magmatic activity in southwestern Africa is confined to four separate episodes at about 136–125, 90–83, 60–58, and 38–35 Ma (Moore, 1976). Such discrete volcanic episodes, as well as the presence of rock types ranging from kimberlite and carbonatite through to trachyte, is suggestive of different triggering mechanisms and origins for the individual volcanic fields.

The *Cameroon line* volcanic province of western Africa also has a number of similarities to the lava-field

Cainozoic volcanism in eastern Australia. These include the overall long duration of volcanism, the apparent absence of any time-space trends over a considerable lateral distance, association with a structural and topographic divide (for example, the Biu Plateau), and dominantly alkaline character.

The Cameroon line province consists of a 1600 km-long volcanic lineament of dominantly alkaline rocks ranging from nephelinite to transitional basalt which have evolved to phonolite and peralkaline rhyolite, respectively, and which form extensive lava plateaux as well as large central volcanoes (Fitton, 1987). There has been more or less continuous volcanism from 60 Ma to the present, although extrusive rocks are known only from 30 Ma. There is no evidence of any consistent age progression with time, but Halliday et al. (1988) suggested that the Cameroon line represents the continental intersection of a trace of the hotspot now located beneath St Helena. They suggested that magmas supplying current volcanism along the Cameroon line were derived from a lithospheric mantle source that was underplated about 120 Ma ago by the passing of the St Helena plume. However, the relationship between the tectonics and the Cameroon-line volcanism is unclear.

Fitton (1980,1983) pointed out the similarity in size and shape of the Cameroon line province to the adjacent Benue Trough, and suggested that this might result from the displacement of the African lithosphere relative to the asthenosphere. Thus, the Cameroon-line volcanism may represent rift-shoulder volcanism associated with the rifting that formed the Benue Trough. Other suggestions relating to the formation of the Benue Trough include a failed arm of a triple junction, and the continuation on the continent of the equatorial oceanic fracture zone (Dautria & Girod, 1987).

7.4.4 Mantle plumes

Oceanic

The mildly alkaline to subalkaline provinces of eastern Australia are by definition volcanoes containing significant proportions of felsic rocks. These central-type volcanoes define a southward younging indicative of hotspot control (Wellman & McDougall, 1974a; Section 1.7.2). An association in other continental settings of mildly alkaline to subalkaline volcanism with hotspot volcanism (that is, point-generated rather than rift generated) is not well documented, although hotspots have been postulated for the Aden and Afar provinces (Gass, 1970; Schilling, 1973). However, these transitional basaltic types are characteristic of many of the small intraplate oceanic islands such as *Gough, Ascension, Bouvet, Canary, Réunion, Easter,* and *Galapagos* (but not for *Tristan da Cunha* which is strongly alkaline), and mantle-plume origins have been postulated for them. These small ocean islands contrast markedly both geochemically and volumetrically with the 'type' hotspot volcanism of the Hawaiian Islands. Unlike Hawaii, they are not dominated by subalkaline basalts, and in many instances evolved compositions are important and include peralkaline types. They

therefore have many similarities to the east-Australian central-type provinces.

Continental

The presence of picritic basalts in the dominantly tholeiitic intra-continental flood provinces of *Deccan, Paraná, North Atlantic,* and *Karoo* are indicative of high-temperature melts. Proposed pressures and temperatures for the derivation of North Atlantic (Baffin Bay) primary picrites of 25 kb and greater than 1425°C respectively (Francis, 1985) are support for the suggestion that at least some of these volcanic provinces have a mantle-plume origin (Schilling, 1973; Petrini et al., 1987; White, 1988). An association is inferred between continental rifting and the triggering of thermal perturbations of either lower- or upper-mantle origin, but whether deep mantle plumes actually initiate rifting, or the rifting in turn triggers shallow mantle convection, or both, is open to debate.

Morgan (1972, 1981) suggested that the Deccan province represents an earlier episode of volcanism resulting from a hotspot now located beneath Réunion in the Indian Ocean. This proposed hotspot, like many Atlantic hotspot volcanoes, was active first in a continental environment, but it became oceanic as the Indo-Australian plate drifted northwards. The mildly alkaline to transitional Réunion lavas have a range in compositions, from alkali basalt to hawaiite, mugearite, benmoreite, and trachyte (Upton & Wadsworth, 1972), and are similar to those of the central-type provinces of eastern Australia. However, the Réunion volcanism contrasts with the dominantly tholeiitic Deccan province where differences in chemical and isotopic compositions have been attributed to the effects of crustal contamination (Cox & Hawkesworth, 1985). There is, therefore, the anomaly of a mantle plume which in a continental environment produced only local examples of felsic magmas, whereas the same plume in an oceanic environment produced a wide range of intermediate and felsic magmas.

One example of a proposed continental hotspot trace is that of the *Snake River* and *Yellowstone* volcanic province of the western United States. This is believed to be related to the migration of the American plate over a stationary hotspot (Morgan, 1971; Menzies et al., 1984). The Snake River volcanic province lies on an arc-shaped topographic depression, extending from Yellowstone to Oregon. It contrasts with the volcanic provinces of eastern Australia which are mostly related to highlands areas. Volcanism began in the Snake River area about 16–15 Ma ago and has continued through to the present. The Yellowstone area has had intermittent volcanic activity over the last 2 Ma. The Snake River and Yellowstone lavas are strongly bimodal in composition and there is a dominance of *ol*-tholeiitic basalts and subalkaline rhyolites compared with intermediate types (Menzies et al., 1984; Doe et al., 1982).

The presence of volcanism in both the Snake River and Yellowstone areas is indicative of underlying thermal anomalies, but some features are not consistent with derivation from a single mantle plume. First, the age of oldest volcanic activity in any specific portion of the province tends to increase with distance from

Yellowstone, but a correlation with predicted continental movement is poor. Additionally, there is recent volcanism of similar age to that in Yellowstone along a zone up to 500 km from Yellowstone (Basaltic Volcanism Project, 1981).

Important differences between the east-Australian central-type provinces and the Snake River and Yellowstone volcanism include the strictly subalkaline character and the widespread outpourings of flood basalts in the Snake River area, and the large-scale caldera-forming rhyolitic volcanic activity in the Yellowstone area. There are, however, remarkable chemical (including trace-element chemistry) and mineralogical similarities between the Snake River-Yellowstone rhyolites and the high-silica rhyolites of southeastern Queensland (Ewart, 1979; Ewart et al., 1985).

The large-volume *Erebus* volcanic province within the Ross Sea section of the Transantarctic Rift consists of strongly alkaline lavas of basanite to phonolite compositions ranging in age from older than 15.4 Ma to the present. It is part of the alkaline McMurdo Volcanic Group which also includes the late Cainozoic volcanic rocks of Victoria Land (Hallett and Melbourne volcanic provinces), islands in the western Ross Sea, and the Balleny Islands. The currently active Balleny Islands constitute a hypothetical hotspot, the trace of which projects back to eastern Tasmania (Fig. 1.7.3). Duncan (1981) identified Mount Erebus as the site of a further hotspot, indicating that two currently active hotspots may exist in the McMurdo area.

The Ross Sea area is believed to have undergone rifting and downwarping associated with major crustal fractures and faults developed during earlier periods of rifting and reactivated more than 26 Ma ago (Kyle, 1981). The Erebus province is in an area of marked crustal thinning adjacent to the Transantarctic Mountains. The strongly alkaline compositions of the Erebus lavas contrasts with the dominantly subalkaline to mildly alkaline compositions of most oceanic and continental hotspot provinces and the central-type provinces in eastern Australia. Erebus province more closely resembles the large stratovolcanoes of the East African Rift system, such as Mount Kenya, than it does the mildly alkaline to transitional central volcanoes of rift-margin shoulders such as those adjacent to the Gulf of Aden (Cox et al., 1970).

The Miocene to Quaternary *Hoggar* volcanic province is built on a basement high well away from the continental margin in northern Africa (Duncan, 1981). The nepheline-normative character of the basaltic rocks at Hoggar, like those of Erebus, is more closely similar to the east-Australian lava-field than to the central-type provinces. However, unlike the majority of east-Australian lava-field provinces the Hoggar province contains lava flows and plugs of evolved rocks such as trachyte and alkali rhyolite (Dautria et al., 1988). The Quaternary volcanoes are exclusively basaltic maars, small Strombolian cones, and lava flows containing upper-mantle and lower-crustal xenoliths of lherzolite, pyroxenite, and granulite (Dautria & Girod, 1987) that are more closely similar to the young lava-field type activity of northern Queensland and the Newer Volcanics in Victoria. There is no isotopic or minor-element evidence in Hoggar rocks for crustal contamination, and the presence of alkali basalts containing high incompatible-element contents is indicative of derivation from an incompatible-element enriched upper mantle rather than a hot deep-mantle plume (Dautria et al., 1988).

Recent thermal, gravity, and xenolith data are evidence that the lithosphere beneath Hoggar at the present day does not have elevated heat-flow, and most likely was modified by upwelling asthenospheric mantle at least 60 Ma ago at the time of late Mesozoic to early Cainozoic alkaline volcanism (Lesquer et al., 1988). A comparison can be made with the volcanic history of northern Queensland where granulite xenoliths in Pliocene and younger volcanics are considered to represent basaltic underplating in the lower crust during an earlier Cainozoic (possibly about 35 Ma) volcanic episode (Rudnick et al., 1986).

7.4.5   Discussion

There appears to be no consistent pattern in the volcanic provinces considered in this section between the onset of volcanism and the time of continental rifting. However, these volcanic provinces mostly fall into two groups. Volcanism and rifting in the first group were largely coincident and the main phase of volcanism was of relatively short duration. For example, the major phase of eruption in the flood basalt provinces of Deccan, Karoo, and Paraná, has been confined to a few million years either just prior or during the period of maximum rifting.

Volcanism in the second group was initiated either at about the time of rifting, or there was a delay in volcanism relative to initial rifting. Either way, however, volcanism has continued for a considerable time after the main period of rifting or, in a few instances, both rifting and volcanism have been of long duration. Initiation of rifting and volcanism in each of the east-Australian and the Red Sea/Gulf of Aden areas was coincident. However, east-Australian volcanism continued well after cessation of rifting, whereas rifting continues today in the Red Sea/Gulf of Aden areas. Marie Byrd Land volcanism started about 50 Ma after the separation of the Campbell Plateau from west Antarctica, but was approximately coincident with reactivation of the Transantarctic Rift. Both the Rhine Graben and Massif Central volcanism have been coincident with strong African-European plate convergence. The Rhine Graben volcanism was initiated soon after reactivation of this convergence (at about 51 Ma; Dewey & Windley, 1988), whereas that of the Massif Central was delayed to about 10 Ma.

The common feature of the flood-basalt group is the general consensus that these provinces represent early manifestations of major mantle plumes underlying moving plates (Morgan, 1981). In every instance a hotspot trace can be projected to a currently active ocean island representing the same, albeit much smaller, mantle plume. The mantle plumes associated with the flood-basalt provinces are obviously of a different order of magnitude to those responsible for much ocean-island volcanic activity, as well as that responsible for the central-volcano track in eastern Australia. The east-Australian hotspot trace is a strong indication that 34 Ma ago a small mantle plume was over-ridden by the Australian continental margin (at Hillsborough) and was only able to 'burn through' because the continental crust had been previously deformed, fractured, thinned, and heated by tectonism and the earlier lava-field volcanic activity (see Section 7.8.6).

A common feature for the second group is less apparent, possibly meaning that the volcanism is a response to different interacting conditions, the sum of which differs between provinces. However, the major triggering mechanisms in all instances are differential plate stresses (resulting from plate extension and collision) leading to thinning and fracturing of the lithosphere, which in turn causes instability and partial melting of the underlying mantle (Basaltic Volcanism Study Project, 1981).

There are apparently no direct analogies in continental settings to the Cainozoic volcanism of eastern Australia where lava-field volcanism, that formed in response to lithospheric plate stresses and resultant extension tectonics, has an apparent hotspot track superimposed upon it. However, the examples of rift-shoulder volcanism discussed in this section have relatively small, discrete areas of dominantly alkali-basaltic volcanism that are closely associated with mostly younger volcanic shields containing a wide range of volcanic products and compositions. These are located commonly in areas of most pronounced crustal thinning, and in some instances (for example, Afar, Aden, Erebus) a possible hotspot origin has been suggested for them. Shield volcanoes of these provinces most nearly equate with the east-Australian central-type provinces. The main difference is that no unambiguous age-progressive tracks are recognised. These shield volcanoes may represent hotspot igneous activity, and the absence of age-progressive tracks most

likely will reflect the absence of any substantial movement of the continental lithosphere relative to the asthenosphere. This is consistent with the African-Arabian and Antarctic plates currently being nearly stationary.

7.5 Experimental Petrology

7.5.1 Primary intraplate basalts

Experimental petrologists seeking to understand the origins of basaltic magmas have been attracted to study of the Cainozoic basalts of eastern Australia because of the wide diversity of magma types that demonstrably are derived from the upper mantle — that is, they contain xenoliths and megacrysts of high-pressure, mantle origin. Compositions from the highly under-saturated end of the basaltic spectrum (olivine melilitite, olivine leucitite, basanite, and *ne*-mugearite) have been studied using east-Australian samples (Table 7.5.1). The chemical similarity between these continental intraplate basalts from eastern Australia and oceanic intraplate basalts from Hawaii is sufficiently close that experimental study of Hawaiian *ol*-tholeiitic basalt to alkali-basalt compositions are broadly applicable to this part of the east-Australian basaltic spectrum (Frey et al., 1978).

Mantle-derived magmas may be primary melts or relatively evolved melts that have been modified by crystal fractionation, or by crystal fractionation together with contamination, or by magma mixing, provided these more complex processes took place within the upper mantle. Experimental studies in which the liquidus phases of mantle-derived melts are determined as functions of temperature, pressure, and different volatile contents, constrain the paths of crystal frac-

Table 7.5.1 Chemical analyses of intraplate basalts studied at high pressure

	1 Olivine tholeiite[1]	2 Olivine basalt	3 Picrite	4 Alkali olivine basalt	5 Olivine basalt	6 Olivine-rich basanite	7 Olivine melilitite	8 Ne-mugearite
SiO2	46.95	47.05	45.51	45.39	46.55	44.60	38.26	48.88
TiO2	2.02	2.31	1.93	2.52	3.18	2.90	2.72	2.19
Al2O3	13.10	14.17	12.44	14.69	12.70	11.70	9.58	15.41
Fe2O3	1.02	0.42	0.92	1.87	2.98	3.00	2.11	3.45
FeO	10.07	10.64	8.67	12.42	9.72	9.40	10.95	6.90
MnO	0.15	0.16	0.15	0.18	0.17	0.15	0.24	0.16
MgO	14.55	12.73	18.79	10.37	10.63	13.90	16.40	7.20
CaO	10.16	9.87	9.67	9.14	8.66	7.70	13.19	6.52
Na2O	1.73	2.21	1.64	2.62	2.95	3.65	3.77	5.74
K2O	0.03	0.44	0.08	0.78	0.95	2.00	1.43	2.87
P2O5	0.21		0.20	0.02	0.60	1.00	1.35	0.68
Mg-ratio	72	68	79	60	66	69	73	65.1

1. Model parental liquid for Kilauea (Green & Ringwood, 1967).
2. Transitional composition between 1 and 4 (Green & Ringwood, 1967).
3. Olivine-enriched version of 2 (Green & Ringwood, 1967).
4. Model, Hawaiian, fractionated composition (Green & Ringwood, 1967).
5. Transitional composition between olivine basalt and hawaiite, Mount Eden, Auckland Island (Green & Hibberson, 1970).
6. Model parental composition, Mount Leura, Newer Volcanics (Green, 1973).
7. Laughing Jack Marsh, Tasmania (Brey & Green, 1975, 1976).
8. Mount Anakie, Newer Volcanics (Irving, 1971; Irving & Green, 1976).

[1] Most rock names are those defined by Green & Ringwood (1967).

tionation of melts and establish whether a particular melt composition is a possible *primary* magma — that is, a melt derived by partial melting of mantle peridotite and segregated from refractory peridotitic residual phases. Primary magmas must have some condition of pressure, temperature, and volatile fugacities at which they are saturated in olivine and orthopyroxene. Residual olivine must have Mg-ratios equal to 89 or more. This is concluded on the basis of extensive collection and analysis of mantle harzburgite and lherzolite xenoliths, although samples of lower Mg-ratio are found in xenolith suites. Primary magmas may in addition be saturated with clinopyroxene ± amphibole ± phlogopite, and garnet, or spinel, or plagioclase, but these are minor minerals in mantle lherzolite, and a general model of mantle magma genesis must recognise the dominance of olivine and orthopyroxene in mantle xenoliths and in high-temperature, high-pressure peridotites. Absence of conditions of olivine- and orthopyroxene-saturation for particular melt compositions does not exclude models of special local sources (for example, phlogopite clinopyroxenite, eclogite, wehrlite) possibly produced by prior mantle metasomatism or crystal accumulation, but does exclude such magmas as members of the 'mainstream' of mantle petrogenesis — that is, products of partial melting of plagioclase-, spinel-, or garnet-lherzolite under different physical conditions.

The loci of melts derived by partial melting of anhydrous 'Hawaiian pyrolite' (Jaques & Green, 1980; Falloon & Green, 1988; Falloon et al., in press) are projected in Figure 7.5.1 into the basalt tetrahedron. These data provide a phase-petrology framework for an understanding of the intraplate basalts of eastern Australia. 'Hawaiian pyrolite' refers to a model source peridotite based on Hawaiian olivine-tholeiite liquid and harzburgite residue (Green & Ringwood, 1963, 1967). It is an appropriate composition in minor- and major-element contents (particularly K_2O, Na_2O, TiO_2, and P_2O_5) for genesis of the incompatible-element enriched intraplate basalts of eastern Australia. Comparison of Hawaiian pyrolite liquids with liquids derived from model 'MORB pyrolite' and refractory 'Tinaquillo lherzolite' may be found in Jaques & Green (1980) and Falloon et al. (in press).

Major-, minor-, and trace-element geochemistry of Cainozoic basalts from Victoria and Tasmania can be modelled by partial melting of Hawaiian pyrolite with appropriate residues which range from garnet lherzolite at 4–5 percent melting at about 30–35 kb, to spinel lherzolite, to harzburgite at about 20-percent melting at about 15 kb (Frey et al., 1978). The loci of melting in Figure 7.5.1 apply to anhydrous conditions and are evidence that melts from alkali olivine basalts and alkali picrites to *ol*-tholeiitic basalts and tholeiitic picrites can be derived from the 'Hawaiian' or 'enriched' pyrolite source in the 10–30 kb range. These magmas are products of greater than 10-percent melting, and the volatile ($H_2O \pm CO_2$) contents of these magmas are inferred to be low — of the order of 0.5–1 percent H_2O in parental *ol*-tholeiitic basalts to 1–2 percent H_2O in parental alkali basalt. H_2O contents in the source pyrolite of 0.1 to 0.2 percent are implied. This H_2O is held in pargasitic amphibole at pressures of less than or

equal to 30 kb and in interstitial melt at pressures greater than or equal to 30 kb and temperatures greater than 1000°C.

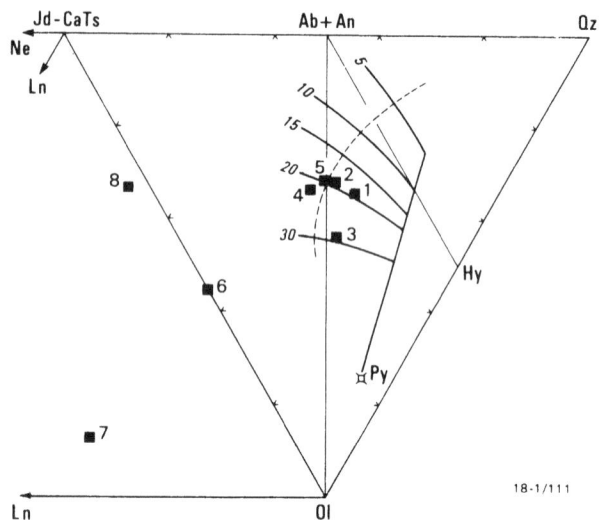

Figure 7.5.1 Compositions 1–8 (Table 7.5.1) projected onto the base of the basalt tetrahedron (Green, 1971). Abbreviations: Ne, nepheline; Jd-CaTs, jadeite plus calcium-tschermak's molecule; Ab+An, plagioclase; Qz, quartz; Ln, Larnite; Ol, olivine; Hy, hypersthene. Py represents the model 'Hawaiian pyrolite' investigated in experimental melting studies as the possible source for intraplate basalts 1–8. Lines 5 to 30 refer to differences in liquid compositions derived by anhydrous melting at 5, 10, 15, 20, and 30 kb. Melts produced by low-percentages of partial melting plot at the left-hand end of each curve and are in equilibrium with an olivine+orthopyroxene+clinopyroxene residual assemblage. Liquid compositions trend towards the right as temperature (and percentage-melting) increases. They pass through the clinopyroxene-disappearance boundary (dashed line) and along the olivine+orthopyroxene saturation surface, intersecting the olivine-control line that passes through the pyrolite-source composition. Compositions 1–5 can be accounted for by essentially anhydrous melting of 'Hawaiian pyrolite', but compositions 6–8 require models including amounts for H_2O with or without CO_2.

H_2O or H_2O+CO_2 contents assume increasing importance in the genesis of the extremely undersaturated basalts. Green (1973) showed that the Mount Leura (Victoria) olivine-rich basanite did not have a liquidus field for orthopyroxene at any pressure under anhydrous conditions, but was saturated with olivine + orthopyroxene + garnet + clinopyroxene at a pressure of about 27 kb and a temperature of about 1250°C, provided the basanite contained about 3–4 percent H_2O. Experimental study of a more strongly undersaturated olivine melilitite from Laughing Jack Marsh, Tasmania, was undertaken also, but neither anhydrous, nor H_2O-undersaturated, nor H_2O-saturated conditions produced a liquidus field for orthopyroxene. However, addition of CO_2 plus H_2O resulted in melt polymerisation and the appearance of orthopyroxene, together with olivine, garnet, and clinopyroxene as liquidus phases in the 27–30 kb and 1150–1200°C pressure-temperature range. The CO_2 and H_2O contents of the olivine melilitite were about 5 and 3–5 weight percent, respectively (Brey & Green, 1975, 1976).

An experimental study of olivine leucitite from central New South Wales also obtained liquidus fields for olivine, clinopyroxene, and phlogopite only, under both anhydrous and water-undersaturated conditions (A. Cundari, personal communication, 1985). This distinctive K-rich magma was proposed as the product of partial melting of phlogopite clinopyroxenite or phlogopite wehrlite. This model requires the prior formation of a distinctive mantle mineralogy by metasomatism of refractory peridotite or by local accumulation of phlogopite clinopyroxenite from a prior magmatic event. The effect of CO_2 on the liquidus phase relations of the olivine leucitite have not been explored yet, and this magma therefore could represent a possible primary melt from a K-enriched phlogopite lherzolite provided that both CO_2 and H_2O were available in the source region (see also the studies by Ryabchikov & Green, 1978, on an extremely potassic olivine leucitite from central Africa).

The experimental studies of both the model source composition ('Hawaiian pyrolite') and of the spectrum

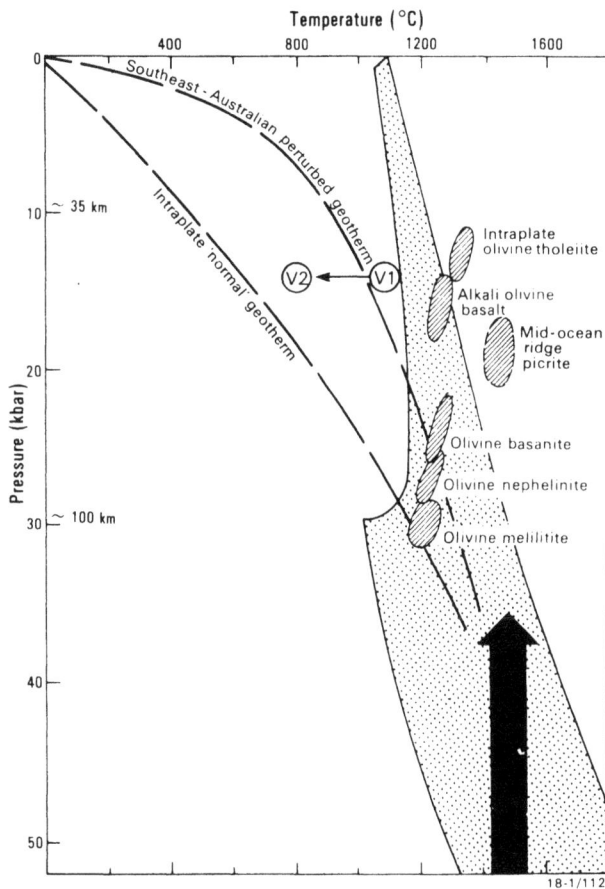

Figure 7.5.2 Petrogenetic model of magma segregation for primary intraplate basalts. The source rock is 'Hawaiian pyrolite' and the stippled area lies between the high-temperature anhydrous solidus and the lower-temperature dehydration solidus for amphibole-bearing pyrolite (pyrolite plus 0.3-percent H_2O). The solid arrow represents the adiabatic gradient within the asthenosphere. Mantle diapirism in regions of crustal tension leads to the perturbed geotherm (see O'Reilly & Griffin, 1985). Magma segregation from mantle diapirs is illustrated by the elliptical areas (compare with Fig. 7.5.1). V1 to V2 represents the cooling path for lherzolite xenoliths from Bullenmerri, western Victoria (Nickel & Green, 1984). This is assumed to represent a relaxation or cooling path from the perturbed geotherm in a region producing alkali- and tholeiitic-basalt volcanism through a waning phase of basanitic volcanism towards the stable intraplate geotherm.

of liquids from *ol*-tholeiitic basalt to olivine melilitite lead to a model for petrogenesis of the primary basalts of eastern Australia (Fig. 7.5.2). The anhydrous solidus and dehydration solidus (solidus for pargasite peridotite) in Figure 7.5.2 are as determined for Hawaiian pyrolite. The conditions of magma segregation for primary liquids are illustrated, although this figure does not allow presentation of differences in CO_2/H_2O which is a factor in producing olivine melilitite magmas (high CO_2/H_2O) as opposed to olivine nephelinite (low CO_2/H_2O) at about 30 kb and 1200–1250°C. Estimates of the perturbed geotherm in eastern Australia around the time of Cainozoic volcanism (derived from lower-crust and upper-mantle xenolith suites in Cainozoic volcanic provinces; O'Reilly & Griffin, 1985; see also Section 7.2.3) and of a possible cooling path deduced from mantle xenoliths from one locality, are shown also in Figure 7.5.2. These xenoliths contain exsolution textures and phase stabilities which are evidence for cooling from high to lower temperatures (Nickel & Green, 1984).

The petrogenetic model illustrated in Figure 7.5.2 is consistent with a regime of episodic and variable extension in which crustal uplift and lithosphere thinning is accompanied and caused by increased temperatures in the upper mantle. The accompanying volcanism derives from magmas segregated at depths of 50–90 km. The immediate cause of increased temperature in the lithosphere is mantle diapirism, the temperature and magmatic consequences of which are inferred to be constrained for the east-Australian region by a mantle adiabat passing through about 15 kb and 1300°C. Most magma generation takes place along transient geotherms between a 'normal' conductive geotherm for eastern Australia, approximately as illustrated, and the convective mantle adiabat penetrating to minimum depths of 40–50 km in the regions of most extensive volcanism (tholeiitic to transitional basalts).

The model for petrogenesis derived from the experimental studies is a generalised one. Results from detailed geochemical and isotopic studies of magmatic provinces and of mantle xenoliths are evidence that source compositions that are different from 'Hawaiian pyrolite' may be more appropriate for specific magma batches or suites. These localised source compositions may originate as refractory lherzolite or harzburgite compositions that have undergone one or more later enrichment events (Frey & Green, 1974). Key elements in the model are:

(1) the presence of volatiles in the C-H-O system (CO_2, H_2O, and CH_4 in the source region, thus lowering the solidus temperature and sensitively controlling the degree of undersaturation of liquids derived by very small degrees of melting;

(2) the role of diapirism from the convecting mantle beneath the lithosphere, producing local, transient, 'anomalous', or 'perturbed' geotherms caused by diapirism accompanying lithospheric thinning;

(3) the explanation of regional or time-changing magmatic provinces or suites that may range from tholeiitic or alkali basalt (high degrees of melting) to olivine melilitite or olivine leucitite (very low degrees of melting) as being caused by different ascent and cooling paths for individual diapirs. Individual diapirs also may

Table 7.5.2. Summary of high-pressure crystal fractionation of different intraplate primary magmas

Primary magma	Fractionating minerals[1]			Evolved magma
Mg-ratio greater than 68	*Stage 1* *Oliv[2] greater than Fo₈₅*	*Stage 2* *Oliv Fo₇₅ or greater*	*Stage 3* *Oliv less than Fo₇₅*	*Mg-ratio less than 68*
Tholeiitic picrite	Oliv			Olivine tholeiite
		(Oliv)[3] + opx		Olivine basalt
			(Oliv) + opx + cpx	Hawaiite
Alkali picrite	Oliv			Alkali olivine basalt
		(Oliv) + cpx + (opx)		Alkali olivine basalt to hawaiite
			Amph ± oliv ± cpx	Hawaiite-mugearite-benmoreite
Olivine-rich basanite	Oliv + (cpx)			Olivine basanite
		(Oliv) + (cpx) + amph ± phlog		*Ne*-hawaiite
			Amph ± phlog	*Ne*-mugearite *Ne*-benmoreite
			± anorthoclase	phonolite
Olivine nephelinite to olivine melilitite	Oliv + (cpx)			Olivine nephelinite
		Oliv + cpx + amph		Nephelinite
			Amph ± phlog	Ijolite
		Oliv + cpx ± melilite		Possibly phonolite and carbonatite

[1] Stages 1,2, and 3 are arbitrary divisions in a continuous crystallisation sequence from the liquidus of the primary magma through to evolved magmas having Mg-ratios of about 50 or less.

[2] Abbreviations for mineral names: oliv, olivine; opx, orthopyroxene; cpx, clinopyroxene; amph, amphibole; phlog, phlogopite.

[3] Brackets signify minor component.

differ subtly in their chemical composition, reflecting earlier processes of depletion or enrichment in the asthenosphere or lower lithosphere.

7.5.2 Fractionation of primary magmas within the upper mantle

Many of the Cainozoic volcanic centres of eastern Australia contain, in addition to xenoliths of spinel lherzolite or (rare) garnet lherzolite, large megacrysts of olivine, clinopyroxene, amphibole, orthopyroxene, and spinel and, less commonly, of anorthoclase, plagioclase, apatite, phlogopite, or ilmenite (see Section 6.2.1). The large size and single-crystal character of these megacrysts and their obvious reaction rims made up of minerals compatible with low-pressure crystallisation of the host magma, are possible evidence that some of these megacrysts may be high-pressure *phenocrysts* rather than xenocrysts (that is, accidentally included in the magma). This hypothesis of a genetic relationship between megacryst (high-pressure phenocryst) and host basalt can be tested experimentally and, if confirmed, provides evidence of magma evolution by crystal fractionation at high pressure within the lower crust or upper mantle.

An olivine basalt from Mount Eden, Auckland Island (Sub-Antarctic Islands; Section 4.4.2) contains spinel-lherzolite xenoliths together with large megacrysts of orthopyroxene, clinopyroxene, and minor olivine (Fo₈₇ or more; see also Section 4.5.3). Green & Hibberson

(1970) showed that the host olivine basalt (Mg-ratio 66.2) was saturated with olivine, orthopyroxene, and clinopyroxene at about 12.5 kb and 1320 ± 10°C, but while matching the natural megacrysts in Mg-ratio the liquidus pyroxenes had a much greater degree of mutual solid-solution ($Ca_{5.5}Mg_{79.7}Fe_{15.8}, Ca_{10.5}Mg_{15.5}Fe_{15}$) than the natural coexisting pyroxenes. Green & Hibberson (1970) showed that depression of the liquidus by 2–3 weight-percent H_2O extended the olivine-liquidus field from about 12.5 to about 15 kb, and in lowering the liquidus temperature to 1180 ± 20°C decreased the extent of pyroxene solid-solution to give compositions matching the observed megacrysts (that is, orthopyroxene $Ca_4Mg_{84}Fe_{12}$, clinopyroxene $Ca_{32}Mg_{59}Fe_9$). This experimental matching of liquidus phases with natural megacrysts was considered to demonstrate that the natural olivine basalt indeed was rapidly erupted from a magma chamber or conduit at pressures of 15 kb and temperatures of 1180 ± 20°C, and that the magma was fractionating at these depths from a parental, hypersthene-normative olivine basalt towards a nepheline-normative, olivine-rich, hawaiite composition. The study is important also in demonstrating the importance of small water contents (2–3 weight-percent H_2O) in the genesis of intraplate undersaturated basalts.

Amphibole in the water-bearing experimental study of the Auckland Island olivine basalt appeared at temperatures below the liquidus (at 13.5 kb, 1130°C) after crystallisation of olivine, orthopyroxene, and clinopyroxene. These anhydrous minerals reacted with liquid at lower temperatures to precipitate amphibole. There are many examples of xenolith-bearing basalts from eastern Australia that are more fractionated than the Auckland Island olivine basalt (hawaiite). Extreme examples are a 'mafic phonolite' from northeastern Otago, New Zealand (Price & Green, 1972) and a *ne*-benmoreite from Mount Mitchell, southeastern Queensland (Green et al., 1974). These evolved or fractionated mantle-derived magmas have Mg-ratios as low as 50 (Green et al., 1974, fig.1), instead of values of 75–70 which are appropriate for primary basalts derived from lherzolite with Mg-ratios of about 90.

Green et al, (1974), Irving (1971, 1974b), and Irving & Green (1976) used the compositions of natural amphibole megacrysts and of fractionated host magmas, together with the results of an experimental study (Irving, 1971) of a *ne*-mugearite from Mount Anakie, Victoria (Table 7.5.1; Fig. 7.5.1) to argue for a major role for kaersutitic hornblende in controlling a high-pressure (10–20 kb) fractionation trend for water-bearing intraplate basalts. In addition to the Auckland Island basalt discussed above, a more strongly under-saturated olivine basanite from Mount Leura, Victoria (Table 7.5.1; Fig. 7.5.1) was studied extensively at high pressure using different water and H_2O+CO_2 contents. In this composition also the role of olivine and clinopyroxene as near-liquidus phases in the 10–20 kb pressure range is supplanted by that of amphibole ± phlogopite at temperatures below 1100°C (Nickel & Green, 1984).

A generalised high-pressure fractionation trend for intraplate basalts is summarised in Table 7.5.2. The important role is seen for orthopyroxene and clino-pyroxene in primary magmas of tholeiitic or transitional affinity at intermediate pressures, in contrast to the dominant role for olivine, followed by plagioclase and clinopyroxene, at low pressures (less than about 5 kb). Small but very significant water contents (2–5 percent H_2O) in the more undersaturated primary magmas, particularly of picrite or olivine-rich basanite character, ensure that such magmas are water undersaturated at pressures greater than 5 kb, but typically begin to crystallise pargasitic to kaersutitic amphibole at temperatures of less than 1150°C in the 10–20 kb pressure range. The role of amphibole crystallisation is to increase K/Na and Na/Ca in the residual melts without rapid change of the degree of silica under-saturation — that is, alkali-basalt parents lead to hawaiitic or mugearitic compositions, whereas olivine-rich basanite parents yield *ne*-hawaiite to nepheline-rich phonolites.

7.6 Fractionation, Assimilation, and Source Melting: a Petrogenetic Overview

7.6.1 Introduction

The following characteristic features of the intraplate volcanic rocks of the east-Australian region have been established (Chapter 5): (1) the dominance of mafic magmas; (2) the highly diverse and complex nature of

magma compositions; (3) the overall general continuity of geochemistry throughout the range of magmas; (4) geochemical differences between central volcanoes and lava-field provinces, corresponding to important differences in magma evolution, and possibly magma sources, in these two environments; (5) the important role of fractional crystallisation in the development of the magmas, and which is particularly extreme in the trachytic, rhyolitic, and peralkaline rocks; (6) the relatively diverse Sr-, Nd-, and Pb-isotopic compositions.

A particularly relevant aspect of petrogenesis is the range of age, tectonic setting, and composition of the rocks through which the magmas have passed. These environments include known and inferred Proterozoic terrains (Section 1.3.1). Proterozoic rocks crop out in Tasmania, western New South Wales, and northern Queensland, and of note are the areas of inferred Proterozoic crust beneath much of Victoria and central and western New South Wales, possibly extending through to central Queensland (for example, Degeling et al., 1986). Two New South Wales central volcanoes — Canobolas and Warrumbungle — appear to be above inferred Proterozoic basement (Warrumbungle volcano may be on the margin of the basement), and rocks from both volcanoes have relatively radiogenic Sr-isotopic compositions compared to mafic lavas from other central volcanoes. Similar isotopic differences are not defined as clearly in the extensive lava fields of western Victoria and northern Queensland, and Tasmanian rocks retain no isotopic evidence for interaction with Proterozoic crust. The leucitites of central New South Wales may sit within an inferred Proterozoic rift zone.

Another aspect of potential significance is the possible involvement of subduction-modified lithosphere in the generation of the mafic lavas. Lithosphere of this type should be found widely beneath eastern Australia because of the long series of volcanic episodes, many interpreted as arc-related, from Cambrian to Triassic. These are confined evidently to the cratonic blocks and their margins, rather than to the intervening sedimentary basins (for example: Day et al., 1983; Degeling et al., 1986; Murray, 1986).

Major aims of this section therefore are to (1) attempt an identification of 'asthenospheric' (including 'plume') and 'lithospheric' source contributions to the mafic magmas (this topic is developed further in Section 7.7), (2) synthesise available data relevant to evaluation of the role of the crust in magma evolution.

7.6.2 Role and extent of fractional crystallisation

Mafic rocks

Compositional relationships on plots of mg-ratio:DI, Sc:mg-ratio, Ni-MgO, Cr-MgO, V-MgO, Ba-Rb, Rb-Sr, and REE abundance (Sections 5.2,4) are compatible with different degrees of fractionation within the alkaline lavas (*ne*-hawaiite to nephelinite) involving the assemblage Mg-olivine + augite ± orthopyroxene + Cr-spinel or magnetite (or both). Differences in mg-ratio:DI relationships for Tasmanian

and north-Queensland lava fields (both dominated by
alkaline lavas) are indicative that although lavas of
both areas have evidence of olivine+augite fractionation,
the Tasmanian lavas are more strongly controlled by
Mg-olivine, whereas the north-Queensland lavas have
more pronounced augite control. These effects are
likely to be pressure dependent, as deduced from
experimental data (Section 7.5). Augite fractionation is
favoured by slightly higher pressures (about 10–15 kb)
compared to olivine. Plagioclase is not identified
clearly by the trace-element data as a critical fractionating mineral in these alkaline mafic rocks. These
assemblages are consistent with fractionation stages 1
and 2 as outlined in Section 7.5.2 (Table 7.5.2),
although amphibole is included also as a major
additional mineral. However, amphibole megacrysts or
phenocrysts (or their pseudomorphed equivalents) are
very uncommon.

The more silica-saturated lavas, especially the
tholeiitic basalts, are controlled by the assemblage
olivine (Fo$_{80-70}$) + augite + plagioclase + magnetite +
ilmenite, and the olivine+augite assemblages are
inferred to become significantly Fe-enriched at more
advanced stages of fractionation in order to account for
the observed Sc:mg-ratio and V-MgO relationships.
The additional presence of orthopyroxene as a fractionation phase is not excluded on the basis of the
trace-element data, but is not supported by the results
of least-squares models using major-element and
observed mineralogical data (for example, Ewart et al.,
1980). However, orthopyroxene is listed in Section
7.5.2 as a likely fractionating mineral at moderate
pressures in primary magmas of tholeiitic-basalt
affinities.

Moderate degrees of fractional crystallisation result
in differences in absolute abundances but little change
in relative light-REE fractionation (Fig. 5.4.5). The
marked differences in REE-fractionation patterns
observed for the different basaltic types therefore are
either source-inherited (source heterogeneity, or different
degrees of partial melting, or both), or perhaps a result
of crustal contamination, although this is considered
unlikely for the more alkaline magmas (see below).

Intermediate and felsic rocks

Trace-element behaviour in the more evolved hawaiites,
tholeiitic basalts, mugearites, and benmoreites, through
to icelandites, can be modelled by the fractionation
assemblage olivine + augite + plagioclase + magnetite
+ ilmenite. Apatite becomes a critical phase as
fractionation proceeds, and increasingly Fe-enriched
olivine+clinopyroxene assemblages are required to
produce the observed data trends. This is entirely
consistent with observed mineralogical data. However,
note again the possible role at higher pressures of
amphibole in the genesis of at least some mugearites
and benmoreites (Section 7.5.2).

Trace-element patterns for the trachytic and rhyolitic
rocks and the equivalent peralkaline types become
increasingly fractionated, resulting in spectacular
depletion and enrichment patterns for the high-silica
and peralkaline rhyolites (Fig. 5.4.4). The importance
of sanidine and plagioclase (assuming crystal-liquid

fractionation processes) is demonstrated from the
modelling of these abundance patterns (see, for example,
Sections 5.4.3,5), but also other minerals — in
particular, Fe-rich olivine and clinopyroxene, magnetite,
and apatite (see also Ewart et al., 1985). Fractionation
in the high-silica rhyolites of the near-minimum
quartz+sanidine+plagioclase assemblage is found also
to be compatible with observed trace-element data.
The consistent depletion of P$_2$O$_5$ corresponds to apatite
fractionation during the development of the trachytic
and rhyolitic lavas (including the peralkaline types),
and zircon fractionation is inferred in some, but not all,
trachytes and low-silica rhyolites (zircon is not found in
the peralkaline lavas).

The high Nb concentrations found in peralkaline
lavas seem to preclude significant magnetite fractionation during their development. However, Nb concentrations for many of these peralkaline lavas are higher than
values predicted by simple, single-stage, closed-system,
fractional crystallisation, as noted in Section 5.4.6.
This problem was highlighted, in the case of samples
from Nandewar volcano, by Stolz (1985) who suggested
that light REE and U also have anomalous enrichments
in some samples that are not consistent with closed-
system fractionation. Stolz (1985) suggested the operation of late-stage volatile loss and volatile-enrichment
processes, specifically involving fluorine-rich vapour
phases, together with fluorine complexing (see below).

Amphiboles and micas are relatively rare in east-
Australian evolved lavas (Sections 5.3.4–5). However,
arfvedsonitic amphiboles are ubiquitous in the peralkaline lavas, and are fluor-bearing, having little or no
hydroxyl substitution. Similarly, micas in rhyolites are
commonly fluor-biotites. Fluorine, rather than H$_2$O,
therefore has tended to concentrate during the inferred
fractionation processes. The mafic lavas are characterised by the presence of fluor-apatite in all examples
so far analysed. This characteristic of halogen (and
possibly carbon gases) enrichment, and low H$_2$O
abundances, in magmas of continental-rift and related
environments (especially peralkaline magmas) was
discussed in detail by Bailey (1978). He related the
phenomenon ultimately to a complex sequence of deep-
mantle degassing processes. The strong dominance of
fluorine over H$_2$O inferred from the available data for
the east-Australian peralkaline magmas is certainly
indicative of essentially anhydrous parental (presumably
basaltic) magmas.

Mineralogical evidence pointing to strong liquid-
crystal fractionation processes for the evolved lavas is
provided by the progressive and finally extreme
Fe-enrichments noted especially in the pyroxenes
and olivines (Section 5.3.2), and by the systematically
changing compositional patterns observed for phenocryst
feldspars (except for the high-silica rhyolites). An
additional superimposed trend of late-stage pyroxene
Na enrichment is found in trachytes and low-silica
rhyolites, which reaches its most extreme development
in the peralkaline rocks.

Low- versus high-pressure fractionation

The trace-element fractionation patterns provide little
direct information on the relative depth at which the

inferred fractional-crystallisation processes took place. Information on this aspect is obtained from megacryst assemblages, experimental data, and from isotopic (and combined isotope/trace-element) signatures, possibly inherited from different parts of the lithosphere.

Megacrysts are common in the east-Australian lavas, as described in Section 6.2. They include olivine, aluminous and subcalcic augite, diopsidic augite, aluminous bronzite, plagioclase, anorthoclase, spinel (including magnetite), ilmenite, amphibole, mica, apatite, corundum, and zircon. They are interpreted conventionally either as accidental inclusions or high-pressure phenocrysts or cumulates (for example, O'Reilly & Griffin, 1987; see also Section 7.5.2). Ewart et al. (1980) calculated that megacryst assemblages in southeast-Queensland central volcanoes could have equilibrated with the erupted mafic lavas (hawaiites and tholeiitic basalts within a 'lower crustal' regime at pressures of 7.2–16.3 kb (average 12.4). These results were interpreted as the result of magma intrusion and fractionation at region of the crust/mantle boundary. Similar lower-crustal fractionation processes were proposed for northern Queensland (Rudnick et al., 1986) and Victoria and New South Wales (Duggan & Wilkinson, 1973; Irving, 1974b; Knutson & Green, 1975; Ellis, 1976). Similarly, a series of moderate- to high-pressure fractionation trends for intraplate basalts has been identified on the basis of megacryst occurrences and experimental data (Section 7.5.2).

Fractional crystallisation of mafic magmas clearly has taken place, at least in part, at moderate pressures (about 8–20 kb). However, fractionation crystallisation (specifically AFC) has taken place also within upper-crustal environments, as interpreted from the isotopic data (see below), and especially in the case of the more silica-saturated and oversaturated intermediate and felsic lavas, including some qz-tholeiitic basalts. Fractionation processes within, and throughout, the east-Australian provinces therefore are likely to represent a complex spectrum of polybaric processes (see also Section 4.6).

7.6.3   Continental crust and magma genesis

Assimilation/fractional-crystallisation (AFC)

The following are summary points that draw attention to some of the major aspects of the combined isotopic and chemical data presented in Chapter 5:

(1) The most alkaline lavas, except for the leucitites, tend to have the least radiogenic Sr-, and most highly radiogenic Nd-isotopic compositions. The hawaiites and tholeiitic basalts have wider isotopic ranges, especially in the central volcanoes.

(2) Intermediate and felsic rocks have trends towards radiogenic Sr- and unradiogenic Nd-isotopic compositions that are interpreted to represent upper-crustal characteristics.

(3) Linear trends are defined for many regions between $^{206}Pb/^{204}Pb$ and $^{207}Pb/^{204}Pb$ and between $^{87}Sr/^{86}Sr$ and $^{87}Rb/^{86}Sr$, and are especially well developed in southeastern Queensland.

(4) Correlations exist for the mafic rocks between isotopic compositions and certain trace- and major-element features. Examples include $^{87}Sr/^{86}Sr$ and Rb/Sr, $^{87}Sr/^{86}Sr$ and Ba/Th (and Ba/Nb), $^{143}Nd/^{144}Nd$ and Ba/Th (and Ba/Nb), and $^{206}Pb/^{204}Pb$ and Ba/Th (central volcanoes). Relatively systematic shifts in Nd- and Sr-isotopic compositions also correlate with whole-rock mg-ratios and SiO_2 contents.

(5) Nd-Sm model ages (T_{DM}) range widely for the different mafic lava types, and in hawaiites and tholeiitic basalts are correlated with Sr-isotopic compositions.

(6) Trachytic and rhyolitic rocks have correlations (mostly positive) between $\delta^{18}O$ and Sr-isotopic compositions.

The correlated isotopic, trace-, and major-element differences are, in principle, explicable in terms of AFC processes, based on simple numerical modelling using two end-members (felsic upper crust and lower intermediate granulite crust). In general, the tholeiitic basalts and hawaiites tend to retain evidence for the greatest modification by inferred AFC processes. Both crustal contaminants are thought to have been involved, on the basis of comparison of these trends with the two crustal models. The possibility of an early Proterozoic lower crust for southeast-Queensland provinces is supported by the Pb-Pb model age. This is not predicted on the basis of current tectonic reconstructions, but the possible existence of relatively old crust is supported further by Pb-isotope data from sulphide mineralisation in northeastern New South Wales (Gulson et al., 1985). The corresponding Rb-Sr model age of the same set of southeast-Queensland mafic lavas is much younger than the Pb age, and is indicative of decoupling of the two isotopic systems. This is attributable to the low total-Pb levels in these rocks compared to expected crustal-Pb abundances.

Another aspect of isotopic relationships is provided by a ΔNd versus $^{206}Pb/^{204}Pb$ plot (Fig. 7.6.1) in which ΔNd is a measure of the vertical deviation of a sample from the mantle plane which itself is based on combined Pb-, Sr-, and Nd-isotopic data (Zindler et al., 1982). The distinctive isotopic field of the central-volcano rocks is illustrated, as well as the contrasting isotopic compositions of the high-silica rhyolites (and one peralkaline lava) compared to other evolved lavas (see below). The strong positive deviations from the mantle plane are consistent with the involvement of upper-crustal rocks, whereas the negative deviation, particularly below the LoNd array, is indicative of lower-crustal involvement. The central volcanoes therefore have both lower- and upper-crustal geochemical signatures, superimposed on their source-mantle signatures, that evidently are inherited through AFC and melting of heterogeneous upper-lithospheric sources and which are detectable in the mafic and more evolved magmas alike.

Considerable emphasis is given in the above discussion to the apparently extensive role of AFC processes in producing the isotope-element correlations. However, how such processes can be distinguished from the melting products of heterogeneous subcrustal lithosphere (see Section 5.5.8) is not altogether clear, especially where the lithosphere has been overprinted geochemically by earlier subduction episodes (Section 5.4.2). In addition, a complex zone of interdigitation of lower crust and upper mantle may exist at the crust/mantle boundary, caused by both tectonic and magmatic processes over extended periods. These zones, where partially re-equilibrated, may provide heterogeneous lithospheric magma sources with distinctive 'crustal' signatures.

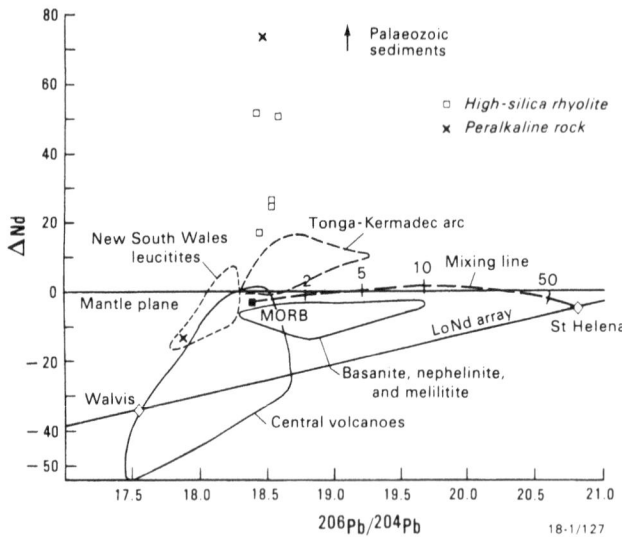

Figure 7.6.1 ΔNd versus ^{206}Pb/^{204}Pb relationships for east-Australian rocks. ΔNd is a measure of the vertical deviation from the mantle plane calculated using the method proposed by Hart et al. (1986). LoNd array is also from Hart et al. (1986), mean MORB composition is from Zindler et al. (1982), and Tonga-Kermadec field is from Ewart & Hawkesworth (1987). Field for the basanites, melilitites, and nephelinites is for basalts whose mg-ratios exceed 60. The curve between the MORB and St Helena end-members represents a simple mixing relationship. Numbers along mixing line refer to percentages of the St Helena component.

High-silica rhyolites as crustal melts

The following five points represent the distinctive features of the high-silica rhyolites of east-Australian intraplate provinces:

(1) The rhyolites are restricted mainly to four eruptive centres in southeastern Queensland.

(2) They have distinctive and relatively radiogenic Sr- and Pb-isotopic compositions (see, for example, Figs. 5.5.1,7,9) which are displaced towards strongly positive ΔNd values in Figure 7.6.1. Their Sr-, Pb-, and Nd-isotopic compositions are quite distinct from those of the associated mafic rocks, and there is no compositional overlap as found commonly within mafic lavas associated with low-silica rhyolites and trachytes.

(3) δ^{18}O values are consistently higher than those found for trachytes, low-silica rhyolites, and most peralkaline lavas (Fig. 5.5.10).

(4) Discontinuities in abundance patterns for such trace elements as Zr, Nb, light REE, Cs, and Ta, exist between the high-silica rhyolites and low-silica rhyolitic and trachytic rocks.

(5) Mineralogical discontinuities are recognised also — for example in phenocryst-feldspar trends, which are clear evidence that the high-silica rhyolites have not evolved from low-silica rhyolite or trachyte magmas.

The conclusion reached is that the high-silica rhyolites represent local, crustal, anatectic melts

developed in response to basaltic-magma intrusion. These magmas are visualised as developing as side-wall fractional melts along contact zones, leading to the development of lower-density, and thus upwardly convecting, boundary layers of migrating felsic liquids. This would result eventually in the accumulation of a rhyolitic roof zone that will remain discrete from the main underlying magma reservoir (see also, for example: McBirney, 1980; Turner & Gustafson, 1981; Sparks et al., 1984). Rice (1981) specifically advocated such a model for bimodal volcanic terrains such as Yellowstone, and suggested that some mixing and interlayering of the rhyolitic and mafic liquids may take place at their interface. The occurrence of dacites and icelandites of clearly hybrid character within the otherwise bimodal Lamington volcanic succession of Tweed volcano is therefore significant (Fig. 5.3.4; Ewart et al., 1977). Furthermore, the inferred rhyolitic liquid roof zone is presumed to convect actively (except for the outermost cooling zone) because of the net upward heat transfer from the underlying mafic-magma reservoir. This point is considered important in view of the observation that the high-silica rhyolites have exceedingly fractionated element-abundance patterns which are evidence that superimposed fractional-crystallisation processes have operated. Such fractionation is difficult to reconcile with a stagnant rhyolitic roof zone.

The isotopic data are evidence in support of an upper-crustal source for the high-silica rhyolites, although a distinct, basaltic, isotopic signature is recognisable too (Ewart, 1982a). Note also that the tops and bottoms of these intrusions have been estimated to be at 4 and 13 km depth, respectively, on the basis of geophysical modelling (Section 1.4.9), placing them wholly within the upper crust.

The relatively low Zr abundances in the high-silica rhyolites compared to those in the low-silica rhyolites and trachytes, are explicable in terms of zircon buffering during partial melting (Ewart, 1982a). Watson & Harrison (1983) showed from zircon-saturation experiments that Zr can be buffered in rhyolitic partial melts. The actual abundance levels in the melts are controlled by melt temperatures and the original zircon concentrations in the parental source rocks.

The peralkaline lavas are mostly isotopically and geochemically quite different from the high-silica rhyolites, but a single peralkaline rhyolite is known from Tweed volcano, in the Lamington basalt/high-silica-rhyolite succession, where the two groups of rhyolites are isotopically similar. This is interpreted as an example of the development of a peralkaline liquid through extended ternary-minimum fractionation of a high-silica rhyolite liquid (Ewart et al., 1977; Ewart, 1982a).

7.6.4 Basaltic magma sources

Lithosphere-asthenosphere interactions

Two sets of parameters are considered here: the possible nature of the mantle sources, and the degree (and depths) of partial melting. A distinction must be made between magmas of asthenospheric (including plume) and of lithospheric origin (or, perhaps, a mixture of both — see, for example, McDonough et al., 1985), and an evaluation made of whether the wide range of major- and trace-element chemistry is attributable to different degrees of mantle melting or to different source compositions.

The following points, taken together, are considered to be indicative of the involvement of lithosphere in the origin of east-Australian mafic magmas:

(1) The wide range of bulk compositions, coupled with a comparable range of trace-element and isotopic chemistry. These are more extreme than recognised typically in MORB or ocean-island environments (for example, Fig. 5.5.9). Recall also that the overall Pb-isotope data have trends consistent with well-mixed orogene/upper-mantle compositions based on the Doe & Zartman (1979) Pb-evolution model curves. This also is interpreted as consistent with lithosphere involvement.

(2) Correlations between different major-element, trace-element, and isotopic compositions. These can be modelled in principle by AFC processes which may take place in the complex region of the crust/mantle interface.

(3) The mafic magmas have evidence of modification by fractional crystallisation taking place at pressures equivalent to the crust/mantle boundary, except for the most alkaline lavas (Section 7.6.5). This does not provide direct evidence for a lithospheric origin for the basaltic melts, but the high probability of interaction and partial re-equilibration of the melts within the upper lithosphere is implied. This is particularly appropriate for the tholeiitic basalts for which lower pressures of melt segregation at lithospheric depths are inferred.

(4) Additional evidence is provided not only by the abundance of spinel-lherzolite xenoliths, but also by a comparison of temperature-fO_2 equilibration conditions between the east-Australian basaltic lavas and their Cr-diopside spinel-lherzolite xenoliths of lithospheric origin (Fig. 7.6.2). Equilibration essentially between QFM and WM buffers is supported by both sets of data, although equilibration temperatures tend to define

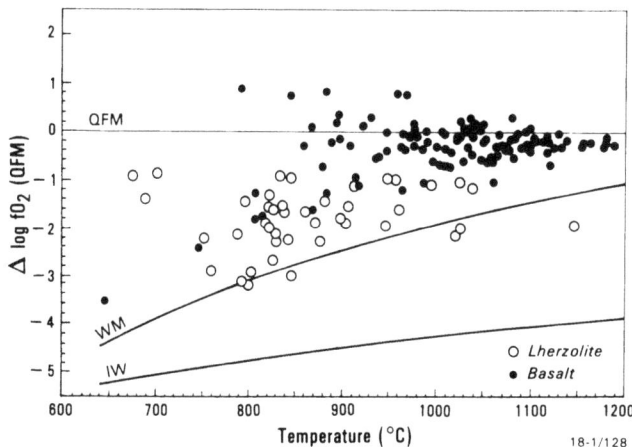

Figure 7.6.2 Comparison of fO_2-T°C data for east-Australian basalts using co-existing Fe-Ti-oxide compositions (Section 5.3.7), and for Cr-diopside spinel-lherzolite xenoliths (O'Neill & Wall, 1987) using the olivine-orthopyroxene-spinel geobarometer. QFM and WM buffers calculated for 15 kb.

different ranges. The xenolith data are based on the olivine-orthopyroxene-spinel geobarometer, whereas the basalt data are from coexisting Fe-Ti oxides. However, basanites, nephelinites, and melilitites lack coexisting Fe-Ti oxides and so are not represented in Figure 7.6.2 (Section 5.3.10). This is unfortunate as these rocks are the most likely to represent least-modified asthenospheric melts (see below).

(5) Both the spinel-lherzolite xenoliths (Section 6.2.5) and the basalts of the Victorian region have relatively radiogenic Sr-isotope compositions, and the basalts are dominated by hawaiites and tholeiitic basalts (Table 5.1.1). An unusual history or process has operated within the uppermost mantle beneath Victoria, based on the xenolith data (Section 7.2).

Direct observational evidence for asthenospheric sources is harder to identify. At least two discrete asthenospheric sources may have contributed — namely, a mantle plume component (or components), and a convecting asthenospheric component that has been emplaced locally by diapirism within (and through) the lithosphere. Identification of the isotopic characteristics of the plume component is discussed in detail in Section 7.7.5.

An integrated model of basalt generation is outlined in Section 7.5.1 in which progressively more silica-undersaturated lavas reflect increasing depths of magma segregation and plausibly represent the best samples of least-modified asthenospheric melts. Similarly, Ghiorso et al. (1983) used the olivine-orthopyroxene buffer exchange reaction to calculate isobaric curves for liquid silica activity as a function of temperature for a series of basaltic compositions. Similar calculations have been carried out on representative analyses of MgO-rich samples from all the major basaltic compositional groups from eastern Australia (using the procedures and program of Ghiorso, 1985). These data represent support for the point clearly made by Ghiorso et al. (1983) that equilibration of the progressively more silica-undersaturated magmas with a mantle olivine+ orthopyroxene assemblage requires progressively higher pressures. The following results are obtained on east-Australian samples, assuming the melts are anhydrous: *qz*-tholeiitic basalt, 9–14 kb; *ol*-tholeiitic basalt, 14–16 kb; hawaiite, alkali basalt, and *ne*-hawaiite, 16–21 kb; basanite, 20–23 kb; nephelinite, 29–31 kb; and melilitite, 40–50 kb. Brey (1978) showed on the basis of experimental results for melilitites, the presence of relatively high percentages of CO_2 and H_2O at source, and melting and segregation depths in the range 27–35 kb — that is, deeper lithospheric to asthenospheric depths.

A low-velocity layer may exist in the range 100–200 km (Section 1.4.7.), but how this zone relates to the lithosphere-asthenosphere boundary is unclear. However, it is certainly a region from which magma generation may take place. There are further geophysical arguments (Section 7.2.8) in support of the possible existence of the top of the asthenosphere at depths of 150 km (or less) beneath eastern Australia. The intersection of the mantle solidus with the extrapolated normal geotherm (Section 7.5.2) is estimated to be at about 100 km. Finally, note again that by far the most extensive development of melilitic and nephelinitic lavas is found in Tasmania (Table 5.1.1), indicative that this region has been a prime area of Cainozoic asthenospheric upwelling.

A significant point in this context is the relatively distinctive Tasmanian Pb-, Sr-, and Nd-isotopic compositions, which tend to group as the most radiogenic Pb and Nd, and least radiogenic Sr compositions analysed for east-Australian basalts (high-μ type), and to overlap MORB fields in conventional isotope plots (Figs. 5.5.1,7). The only other isotopically similar sample is the most magnesian nephelinite analysed, from central Queensland.

An alternative interpretation of the isotopic compositions of east-Australian basalts is gained from the Figure 7.6.1. The fields of the leucitites and the magnesian, strongly undersaturated alkaline lavas (basanites, nephelinites, and melilitites) are shown subparallel to the LoNd array. This pattern apparently is different from the ocean-island basalt, intra-suite differences shown by Hart et al. (1986), which run at steep angles to the LoNd array. The east-Australian alkaline field may represent a mixing array (a rather tentative interpretation, considering the large geographic spread of the volcanic centres) and, if so, one end-member quite feasibly is a radiogenic component along the LoNd array — that is, approaching the St Helena-type (high-μ) component.

The second component is more difficult to define but possibilities include: (1) the leucitite-source component (this, however, is unlikely because of trace- and major-element constraints); (2) a MORB component; (3) a pre-existing, mixed, MORB plus Walvis-type component; (4) sub-continental lithosphere having a sub-mantle-plane isotopic signature and a $^{206}Pb/^{204}Pb$ value less than about 18.3. The mixing line shown in Fig. 7.6.1 is illustrative of the potentially sensitive nature of mixing processes between these end-members. The crucial point, however, is that isotopically the Tasmanian lavas define the extreme end of the alkaline basaltic field (that is, they are most radiogenic in terms of $^{206}/Pb^{204}Pb$), which is interpreted as consistent with an asthenospheric isotopic signature. The involvement of such a source beneath Tasmania is possibly a consequence of the major phase of Jurassic dolerite magmatism found in Tasmania which may have resulted in extensive lithospheric thinning.

REE evidence for melt fractions

Frey et al. (1978) in their detailed study of a set of east-Australian alkali to tholeiitic basalts, concluded for their preferred model that the mantle source of the magmas had been enriched in strongly incompatible elements (Ba, Sr, Th, U, light REE) at 6–9 times chondritic levels, and less enriched in moderately incompatible elements (Ti, Zr, Hf, Y, heavy REE) at 2.5 times those of chondrites, prior to the partial-melting event. They suggested partial melting of the source ranging from 4–6 percent for olivine melilitite (see also Brey, 1978), 5–7 percent for olivine nephelinite and basanite, 11–15 percent for alkali basalt, and 20–25 percent for their ol-tholeiitic basalt composition, based on experimental and major-element data.

Relative light-REE and heavy-REE enrichment is illustrated in Figure 7.6.3. Normalised Yb abundances are fairly similar, but there is strong fractionation between the light- and heavy-REE which increases for the increasingly undersaturated lavas — that is, it is greatest in the melilitites, nephelinites, leucitites, and basanites. Conversely, the tholeiitic basalts — notably the quartz-normative ones — have the least REE fractionation. Similar styles of behaviour are found for

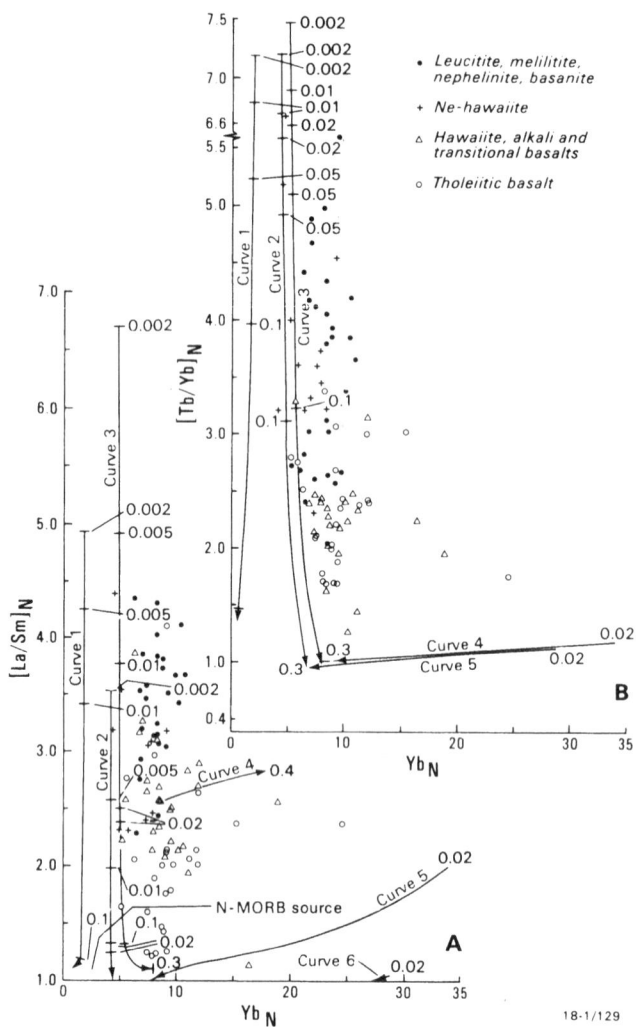

Figure 7.6.3 Relationships between mantle-normalised values for La/Sm, Tb/Yb, and Yb for east-Australian basalts (data from Frey et al., 1978; Stolz, 1985; and Ewart and Chappell, unpublished data). Partial-melting curves are as follows: curve 1 is based on the extraction equation of McKenzie (1985), curves 2–5 are based on the accumulated continuous melting model, and curve 6 is the trajectory of low-pressure fractionation from an ol-tholeiitic basalt (sample Q135) to an F-value of 0.4 (see text). Numbers along melting curves represent melt fractions. Different source compositions are used for curves 1–5: (1–2) primitive-mantle geochemistry and garnet peridotite; (3) N-MORB source geochemistry and garnet peridotite; (4) primitive-mantle geochemistry and spinel peridotite; (5) N-MORB source geochemistry and spinel peridotite. The modal proportions of peridotite minerals, and the proportions entering the melts, are as follows: (1) spinel peridotite — oliv + opx + cpx + spinel in 0.55:0.25:0.15:0.05 and 0.10:0.20:0.65:0.05 proportions, respectively; (2) garnet peridotite — oliv + opx + cpx + garnet in 0.55:0.20:0.15:0.10 and 0.10:0.10:0.40:0.40 proportions, respectively. Weight fraction of liquid remaining in residue is 0.02 (to F-value of 0.02) and 0.002 (to F-value of 0.005).

other incompatible elements (see, for example, Fig. 5.4.1).

The possible control of partial melting on relative REE fractionation is illustrated in Figure 7.6.3 by means of two sets of calculated curves, using garnet-peridotite and spinel-peridotite model mineralogy, and source compositions based on N-MORB and 'primitive mantle' (Wood, 1979; Tarney et al., 1980). The primitive mantle represents an 'enriched mantle' (for example, Chen & Frey, 1983). The calculated curves are based on accumulated continuous melting and on the melt-extraction equation of McKenzie (1985). The two sets of curves provide consistent results and represent support for the following points:

(1) The buffered Yb data are indicative of a garnet-bearing source, although the tholeiitic-basalt data possibly could be consistent with relatively high melt fractions from the spinel-peridotite model source.

(2) The wide range of light-REE enrichment, as exemplified by the [La/Sm]$_N$ values in Figure 7.6.3A, is consistent with a range of melt fractions. The most extreme fractionation requires extraction of melt fractions between about 0.005 and 0.02 (primitive-mantle source) or less than 0.002 for the N-MORB model source.

(3) The fractionation behaviour of the heavy REE can be interpreted similarly, although the most extreme fractionation requires slightly larger melt fractions (0.02 to 0.1 for both source models).

The most strongly incompatible-element ratios (La/Sm) therefore correspond to very small melt fractions, and the moderately incompatible-element ratios (Tb/Yb) to slightly higher degrees of melt extraction, whereas major-element chemical results (based on those of Frey et al., 1978) are indicative of even higher degrees of melt extraction (greater than 4 percent). Similar conclusions can be drawn from the δ'Nd-δ'Sr model melting curves (Fig. 5.5.3A). This apparent decoupling between the different trace and major elements is predictable from the McKenzie (1985) models of trace-element fractionation during partial melting, provided that melt fractions are sufficiently low.

A further requirement for these models is that the strongly undersaturated melts (for example, nephelinites) should have sufficiently low viscosities to be able to separate from their mantle sources in geologically realistic times. Moreover, these most strongly under-saturated melts are those inferred to have segregated from their mantle sources at greatest depths. There is, therefore, a correlation between strong incompatible-element fractionation, small melt fractions, and increasing depths of melting. This again is consistent with the models of isentropic mantle upwelling and hot rising jets (McKenzie, 1984, 1985; Galer & O'Nions, 1986) in which increasing degrees of melting at progressively shallower depths are predicted. It is also consistent with the petrogenetic model presented in Section 7.5.2.

7.6.5 Magmatic underplating

There seems to be general agreement that some degree of magmatic underplating has been taking place since before the Cainozoic in eastern Australia. Evidence comes from both geophysical and petrological studies, although there is some disagreement on the extent of underplating, both laterally and in terms of depth ranges, and its relationship to the lower crust and the Moho. There has been also a divergence of opinion regarding the relationships between uplift of the eastern highlands and underplating (Section 1.3.2).

Geophysical evidence stems in part from seismic data. Drummond & Collins (1986), for example, proposed that the lower crust increases with increasing crustal thickness and age, and attributed the increase in thickness and average velocity in Proterozoic and Palaeozoic areas to underplating of post-Archaean rocks through time. A lower crust having strong and coherent reflecting horizons is indicated by the seismic-reflection data (Section 1.4.7), and the overall properties of the lower crust are considered consistent with underplating and consequent crustal thickening. These types of studies follow from the earlier isostatic studies of P. Wellman (1979a) who deduced 20 km of underplated material to have accumulated at a constant rate beneath the southeastern highlands during the last 90 ± 30 Ma. This is not a unique model, but the general conclusions of Wellman nevertheless have important petrological implications.

A major line of petrological evidence comes from the detailed study of lower-crustal xenoliths and their lithological and geothermobarometric implications, as described in detail in Section 6.3. One of the major observations is that these xenoliths are overwhelmingly of mafic composition. Pressures appropriate to lower-crustal and upper-mantle depths are suggested, and no marked compositional differences are apparent. An important conclusion is that these xenoliths have compositions consistent with derivation from a basaltic parent (only very rare exceptions), and they are interpreted to represent recrystallised cumulates from magmas intruded at, or near, the crust/mantle boundary. Those from Queensland may have interacted with pre-existing crust (Section 6.3.7). These data overall are considered to be strong evidence for magmatic and tectonic underplating beneath eastern Australia, and this process therefore is believed to constitute an important crustal-growth mechanism (Section 7.2.9).

A second petrological approach (Ewart et al., 1980) was based on attempts to evaluate the temperatures and pressures at which megacrysts (or in some cases phenocrysts) could have been in equilibrium with their host lavas, using thermodynamic calculations. Minerals used were feldspar, olivine, pyroxene, and spinel. Calculated equilibration temperatures and pressures range from 995 to 1390°C (average 1190°C) and 7.2 to 16.3 kb (average 12.4 kb). These are interpreted as being indicative of intrusion and magma fractionation within the crust/mantle interface region, implying underplating and crustal thickening, and extensive cumulates remaining in the intrusive zones (Fig. 7.6.4). The calculations were applied specifically to the central volcanoes of southeastern Queensland, but the processes are applicable more widely in other volcanic provinces. Cox (1980) developed a closely similar lower-crustal fractionation model (from picritic parental magmas) for the much more voluminous Paraná and Karoo continental flood basalts of South America and southern Africa, respectively, which he suggested resulted in crustal underplating, involving sill-like intrusive bodies. The volume of concealed cumulates was considered to be at least as large as the volume of erupted lavas. However, the relative volume of concealed material in eastern Australia is believed to be substantially higher because of (1) a large, hidden, magma component, and (2) a long, pre-Cainozoic history of underplating (see also Section 7.2.6).

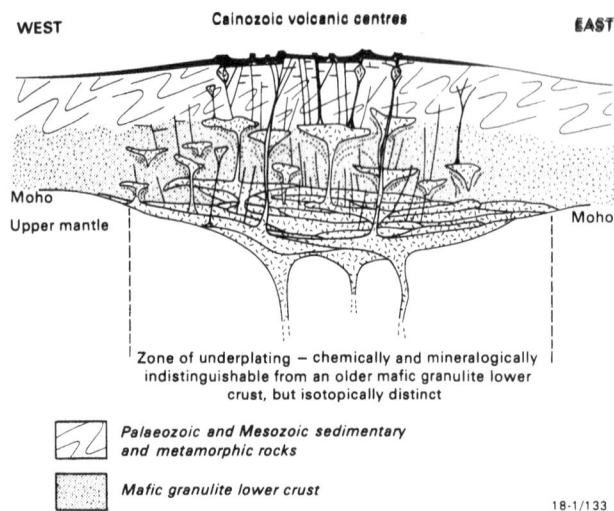

Figure 7.6.4 Schematic model for magmatic underplating for the central volcanoes of southeastern Queensland (after Ewart et al., 1980, fig. 13).

Figures 7.6.4 and 7.2.5–8 are similar schematic representations of proposed underplating structures. An important aspect of the model outlined in Section 7.2.6 is inclusion of a tectonic component, incorporating possible island-arc and ocean-floor materials. Some differences are apparent in detail, related mainly to crustal thickness and to the depths to the crust/mantle transition. The model in Figures 7.2.5–8 has an inferred crust/mantle boundary at about 25 km, and mafic lenses (representing basaltic underplating intrusions) ranging from depths of about 25 km to in excess of 60 km. The equilibrium pressures calculated by Ewart et al. (1980) correspond to depths of about 26–58 km, almost exactly the same to those inferred in Figures 7.2.5–8. These calculations also are indicative that the lower equilibration pressures correspond to the more silica-saturated (sub-alkaline) lavas. However, substantial uncertainties are involved in these calculations.

One of the implications of the underplating models is that basaltic magmas erupting from lower-crustal and subcrustal magma chambers will undergo substantial fractionation. Attention therefore is drawn to Table 5.1.1 where hawaiites are seen to be the dominant mafic rock type, except in Tasmania. Eruption from subcrustal magma chambers may account also for the occurrence of lherzolite xenoliths in some hawaiites.

Experimental anhydrous phase boundaries that are likely to control basaltic magma chemistry, both during initial magma generation (from 'fertile' and 'depleted' or 'MORB-source' mantle) and subsequent fractiona-

tion, are relevant to this discussion (Fig. 7.6.5). The compositional maxima from Figure 5.2.2 are superimposed on Figure 7.6.5, and the major data concentration bisects the line defining the locus of fertile-mantle, initial-melt compositions near the 15 kb locus. The tholeiitic basalts, on the same basis, could be interpreted as defining melts derived in part from depleted mantle (possibly at moderate to low pressures — less than 10 kb; see also Jaques & Green, 1980). These data therefore are broadly consistent with the underplating models, and with the presence of extensive mantle segregation and magma intrusion within the upper lithosphere, including the region of the crust/mantle boundary.

The underplating magmas are inferred to be mafic, and are believed to exist extensively at the base of the crust, but note that AFC processes, which are inferred to have modified magmas of the central volcanoes, require intermediate to felsic lower-crustal compositions. The implication is that these contaminants are part of pre-existing sedimentary-metamorphic crustal basements and are not the underplating intrusions.

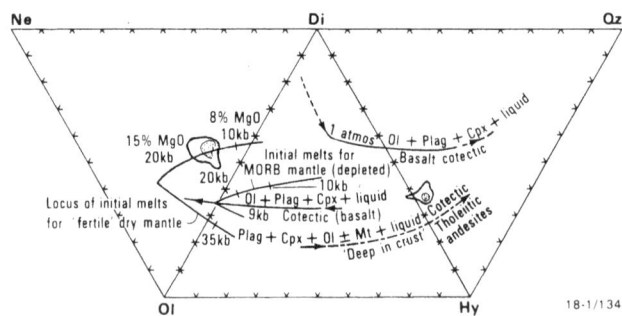

Figure 7.6.5 Anhydrous melting and fractionation phase boundaries projected in the system Ne-Ol-Di-Hy-Qz (after Thompson et al., 1983). Stippled areas represent compositional maxima for east-Australian mafic rocks taken from Figure 5.2.2. Abbreviations: Ol, olivine; Plag, plagioclase; Cpx, clinopyroxene; Mt, magnetite.

7.6.6 Tectonic aspects

The general picture that emerges from this discussion is of melting within the subcontinental lithosphere, and only localised eruption of relatively unmodified asthenospheric magmas, most notably in, but not exclusively confined to, Tasmania. This represents general support for the model proposed by McDonough et al. (1985) who envisaged mixtures of plume-derived melts and lithospheric-mantle derived melts (see also Section 7.7).

The correlations between greater incompatible-element enrichment, deeper magma-segregation depths, and inferred small melt fractions, are consistent with the models of isentropic mantle upwelling and hot rising jets (McKenzie, 1984, 1985; Galer & O'Nions, 1986). Increased melting at progressively shallow depths are predicted from these models. The hot rising asthenospheric jet model (McKenzie, 1984) appears to provide an appropriate mechanism for allowing melting to pass locally upwards into the subcontinental lithosphere, required by the hotspot-migration models. In contrast, the two-stage melting model of Galer & O'Nions (1986), although providing an explanation for the development of enriched, deep, small-volume melts, would seem to require a superimposed process of rapid diapiric uprise in order that these erupt within a continental environment. This may be expected to accompany an extensional stress field (Section 1.6.1) which would need to extend through lithosphere and crust.

One of the major features of intraplate volcanism in eastern Australia is the geochemical and petrological contrasts between the lava fields and central volcanoes. The evidence is strongly in favour of hotspot control for the central volcanoes (Section 1.7.2). These volcanoes also retain strong evidence of AFC processes, together with the development of strongly fractionated derivative magmas and, locally, the formation of crustal anatectic rhyolites. The central volcanoes evidently have lives of the order of 2–3 Ma, and at least some have prolonged periods of quiescence within their life span. They appear to develop from relatively large masses, or pulses, of basalt intruded and trapped within the crust and uppermost mantle of the lithosphere. Two of the larger complexes in southeastern Queensland are within extensive sedimentary basins between major cratonic blocks. A similar relationship seems to hold for the localisation of the other major central volcanoes (Section 1.6.1).

At the other extreme are the lava fields of, for example, northern Queensland, which are characterised by sporadic rapid eruption of voluminous mafic magmas. This situation again would seem to require crustal extensional stress fields, but coupled with relatively extensive and shallow lithospheric melting (the north-Queensland xenolith data support the interpretation of crustal interaction of these magmas; Section 6.3.7). The location of the north-Queensland lava fields is of interest as they lie close to what is apparently the intersection of the boundaries of major tectonic blocks (Fig. 1.3.1). Any readjustment between these blocks, perhaps caused by the northward movements of the Indo-Australian plate, presumably causes rifting between them and the opening of deep crustal fractures.

7.7 Four Component Dynamic Model for East Australian Basalts

7.7.1 Introduction

Intraplate alkali basalts from both continental and oceanic environments are known to share the same general chemical and isotopic characteristics, whereas the characteristics of intraplate tholeiitic basalts commonly are different. Continental tholeiitic basalts generally have higher $^{87}Sr/^{86}Sr$ and lower ε_{Nd} than do coexisting alkali basalts. They also may carry to some degree chemical and isotopic signatures that are suggestive of subduction-zone processes, and of assimilation/fractional-crystallisation (AFC). A major factor considered to be responsible for the contrasts between oceanic and continental intraplate volcanism is the lithosphere.

The oceanic lithosphere is relatively young (less than 200 Ma) and has developed through relatively simple processes. In contrast, the continental lithosphere since its early formation, has evolved discontinuously by means of a range of tectonic mechanisms, resulting in a diversity of chemical and isotopic characteristics. The extent of lithosphere involvement during intraplate magmatism depends on the regional tectonic history and thermal condition of the lithosphere below a given area. Small volumes of silica-undersaturated melt may be generated within the lithospheric mantle, under stable cratons by thermal perturbation and volatile flushing, whereas the convective asthenosphere may become the main magma source in areas where rifting and thinning of the lithosphere are well developed (for example, Perry et al., 1987). Tholeiitic basalt magmas having low concentrations of incompatible trace elements may be generated by magma separation at relatively shallow depths and ascend slower than alkali basalts. The continental lithosphere therefore may have a greater influence on the trace-element and isotopic composition of tholeiitic-basalt magmas than on alkali-basalt magmas (Section 7.6.4).

Detailed chemical and isotopic studies of the Cainozoic basalts of eastern Australia and New Zealand are summarised in Chapter 5 and Sections 4.5.6–8 and 7.6. This section is concerned with a dynamic model of the interaction between a mantle plume (or plumes), the asthenosphere, lithospheric mantle, and the crust. This four-component model may account for the origin and compositional diversity of the Cainozoic basalts of eastern Australia. The section represents an attempt to apply the model of mantle-plume upwelling and its interaction with shallow asthenosphere and lithospheric mantle through entrainment, thermal activation, and magma mixing. The model to some extent is necessarily speculative and some of the conclusions should be considered tentative. Acceptance of the model must await further tests.

The term 'shallow asthenosphere' used in this section is equated with the melting zone at the top of the asthenosphere corresponding to the seismic low-velocity zone (Fig. 7.7.1) which, beneath eastern Australia, is at depths of about 100 to 200 km (Section 1.4.7). Small-scale convection induced by rifting and extension of the lithosphere may draw material from greater depths into this melting zone (Fig. 7.7.2A).

Relevant ideas on the formation of oceanic and continental lithosphere and the possible consequences of lithosphere involvement in magma generation are reviewed. A discussion of the chemical and isotopic characteristics of central-volcano and lava-field type magmatism in eastern Australia then is presented in terms of this dynamic model. First, however, some major issues relevant to current models of mantle dynamic processes under different tectonic environments, as well as magma-genesis processes for east-Australian basalts, are identified:

(1) Contemporaneous tholeiitic and alkali basalts from the same area commonly have different chemical and isotopic characteristics. What are the implications with regard to either melting of a heterogeneous mantle or the interactions between melt components derived from the asthenosphere, lithospheric mantle, and crust?

(2) Basaltic volcanism resulting from hotspot activity involves interactions between mantle-plume and both

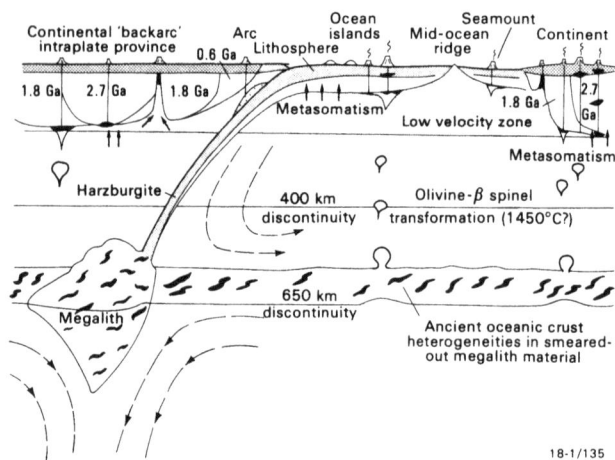

Figure 7.7.1 Generation of mantle heterogeneity through plate-tectonic processes, including lithospheric subduction, upward migration of silica-undersaturated melts from the asthenosphere, and growth of continental lithosphere through underplating of mantle-derived material, at different times and in different tectonic settings. Different components involved in magma generation in different tectonic environments are shown also. Differences of opinion on the issue of whole- or layered-mantle convection do not affect the discussion presented here. See Ringwood & Irifune (1988) for discussion on the 'megalith' concept.

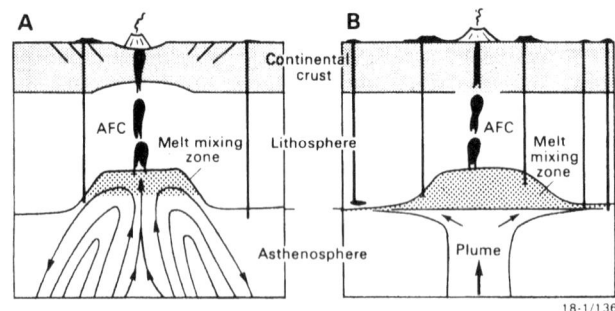

Figure 7.7.2 Two conceptual models for the interactions between different mantle components during the generation of alkali and tholeiitic basalts. (A) Convective asthenospheric upwelling beneath rifting continental lithosphere (adapted from Perry et al., 1987). (B) Deep-mantle plume spreading laterally at the base of continental lithosphere which is not necessarily being rifted (adapted from McDonough et al., 1985). Mantle upwelling, diapir, and plume generate basalts at higher temperature and larger degrees of melting than do basalts derived from mantle at shallow depths — for example, at the top of the asthenosphere.

the shallow asthenosphere and the lithospheric mantle. Can the chemical and isotopic characteristics of the mantle-plume, shallow-asthenosphere, and lithospheric components be identified?

(3) Are there systematic chemical and isotopic differences between the mantle sources for central-volcano and lava-field basalts? If so, how do they compare with basalts erupted in an oceanic environment, and do these differences support a hypothesis of a mantle-plume origin for hotspot-related central-volcano volcanism and a shallow-asthenospheric origin for the lava-field volcanism?

(4) A range of chemical and isotopic compositional domains may develop in the lithospheric mantle through

a range of plate-tectonic and other mantle processes. What are the diagnostic isotopic and trace-element criteria that characterise these different processes?

(5) To what extent do the compositions of the east-Australian and New Zealand volcanic rocks conform with the Dupal-anomaly concept of Hart (1984)?

(6) What is the significance of several Cainozoic occurrences of high-μ type basalts, characterised by high $^{206}Pb/^{204}Pb$ (about 20) and low $^{87}Sr/^{86}Sr$ (less than or equal to 0.7030) in both eastern Australia and New Zealand? Are they derived from metasomatised lithospheric mantle or from the asthenospheric mantle?

(7) To what extent can the geochemical and petrological data for east-Australian basalts be integrated with data from ultramafic mantle xenoliths, relevant geophysical observations, and tectonic reconstructions for the region? Can a consistent growth model be developed for the Australian continental lithosphere from such a data set? Is the Cainozoic volcanic activity in eastern Australia a result of regional hot-mantle upwelling (including mantle plumes) which may be the cause of the break-up of Gondwanaland, or is the volcanism a response to the surface tectonic activity as suggested in Section 7.3.4?

7.7.2 Heterogeneous mantle sources and lithosphere involvement

Mantle sources

A range of mantle domains having a diversity of chemical and isotopic characteristics may develop through geological time by different physical-chemical processes (Fig. 7.7.1). Major processes include partial melting, crust-mantle and intramantle differentiation, crust-mantle recycling, and intramantle mixing of melts or other components. All of these processes may be accompanied by continental-crust formation. Some mantle-melting events may be followed by homogenisation and differentiation of the mantle into MORB and ocean-island basalt reservoirs (for example, Hofmann et al., 1986). The subduction of altered oceanic crust (plus sediments) is another important mantle process for developing mantle heterogeneities. The subducted crust undergoes dehydration and partial melting, and these melts are injected into the overlying mantle wedge. The residual oceanic crust is resorbed along with other parts of the oceanic lithosphere, back into the convecting mantle (for example, Ringwood, 1982). Other possible processes include: upward migration into the lithospheric mantle of silica-undersatured melts from the asthenosphere; thermal erosion and delamination of the continental lithosphere and its mixing back into the convecting mantle; and mixing of mantle-plume material into the depleted upper mantle. The interaction of these different mantle processes can, in principle, produce the spectrum of geochemical characteristics observed in basaltic rocks.

Different degrees of partial melting of a 'plum-pudding' type shallow asthenospheric mantle in the oceanic environment may generate the chemical and isotopic heterogeneities observed in seamount basalts (Zindler et al., 1984). This concept requires that fertile 'plums' or veins are melted preferentially during generation of alkali basalts, whereas during more extensive melting both fertile 'plums' and more refractory matrix melt, generating tholeiitic basalts that have lower $^{87}Sr/^{86}Sr$ and higher ε_{Nd} values (that is, similar to MORB). This model is in contrast to mantle-plume related volcanism, such as in the Hawaiian Islands, which involves different relative contributions of the mantle plume, asthenospheric MORB source, and oceanic lithosphere (for example, Chen & Frey, 1985).

Shield-building tholeiitic basalts in Hawaii, considered to have been derived mainly from a mantle plume, have higher $^{87}Sr/^{86}Sr$ and lower ε_{Nd} values than do the overlying post-erosional alkali basalts. Alkali basalts of the post-erosional phase were generated 1–2 Ma later, after the island had migrated from the hotspot. Their isotopic and trace-element characteristics include a strong signature of a depleted MORB-type source, which is considered to be derived mainly from melting of oceanic-lithospheric mantle (for example, Chen & Frey, 1985).

Oceanic lithosphere

Oceanic lithosphere is formed continuously at mid-ocean ridges by sea-floor spreading processes (Fig. 7.7.1). The peridotite solidus moves deeper as the lithosphere migrates from the spreading center, because of continuous heat loss through conduction and interaction with sea water at the seafloor interface. The thickness of old oceanic lithosphere is about 100 km, and can be considered to be compositionally zoned (for example, Ringwood, 1982). The uppermost part of the oceanic lithospheric mantle, beneath the oceanic crust, is harzburgitic and represents the complementary residue to the oceanic crust. The lower part of the lithosphere consists of lherzolite accreted from the convecting asthenosphere by conductive cooling. A minor amount of melt may have been extracted from this peridotite. This lithosphere may be modified further by upward migration of silica-undersaturated melts derived from an underlying asthenosphere (or low-velocity zone; for example, Green, 1971), together with melt metasomatism and underplating of plume material caused by intraplate volcanism. Kay (1979) proposed that a mantle source enriched in incompatible elements could be generated in this way at the base of the oceanic lithosphere. Recycling of this oceanic lithosphere into the convecting mantle probably contributes to mantle heterogeneity.

Continental lithosphere

Formation of the continental lithosphere involves a more extensive range of physical processes compared to oceanic lithosphere. For example, stabilisation of a sizable Archaean continental crust would be accompanied by the formation of a thick, harzburgitic lithosphere (Nixon, 1987). The presence of Archaean diamonds in kimberlite pipes from the Kaapvaal craton, South Africa (Richardson et al., 1984) is evidence for the existence of Archaean lithosphere up to 200 km thick. Such old continental lithosphere also may contain eclogite bodies which represent trapped basaltic melts or subducted oceanic crust. Fertile or other refractory mantle components may be accreted either through cooling or from mantle diapirism. Localised zones of accreted mantle also may represent the refractory mantle wedge above a subduction zone. Multiple generations of crust-building episodes result in the modification of existing continental lithosphere and the formation of new lithosphere by intrusion and underplating. Thus, continental lithosphere may have a diversity of ages and may be chemically zoned both horizontally and vertically (for example, Brooks et al., 1976).

Events after the initial stages of continental lithospheric growth may leave a significant physical and chemical imprint on the lithosphere. The lithosphere may be thickened by continental collision or downward advection (for example, Jordan, 1981). Underplating of material derived from mantle plumes also may be important in growth of continental lithosphere (for example: Oxburgh & Parmentier, 1978; Ringwood, 1982; McDonough & McCulloch, 1987a,b). Further modification can take place by subduction-zone processes or by melt migration from the asthenosphere (or both).

Continental lithosphere in a rifting environment is continuously stretched and thermally thinned. Upwelling of asthenospheric mantle progressively replaces existing older lithosphere. Subsequent decay of such thermal anomalies allows for further growth of the continental lithosphere. The lowest parts of the continental lithosphere therefore may be relatively young and also chemically and isotopically similar to the underlying asthenosphere. Moreover, some continental lithosphere may be generated by underplating of mantle-plume material and therefore will be different from shallower lithospheric mantle.

Subduction-zone environments may induce modification of the continental lithospheric mantle by upward migration of fluids and melts. These fluids and melts could be derived from dehydration and partial melting of subducted, altered, oceanic crust with or without sediments, and the overlying mantle wedge. Subduction of sediments could be important under certain circumstances. Such a process could imprint distinctive crustal chemical characteristic on the lower lithosphere. Domains of refractory peridotite within the mantle wedge also may be involved in lithospheric growth during or after subduction.

The overall combined effects of these diverse processes inevitably lead to a chemically and isotopically heterogeneous continental lithosphere. Such a compositionally complex lithospheric mantle could have an age zonation, but this would not necessarily young downward, because of thermal and convective erosion processes taking place at the base of the lithosphere. Thermal and mechanical erosion of the continental lithosphere at its base may be an effective mechanism for buffering the thickness of the lithospheric mantle throughout geological time.

Hotspot, rift, and backarc environments

Plumes of deep-mantle origin penetrating either the continents or oceans are not expected to have systematic chemical and isotopic differences, whereas asthenosphere immediately underlying continental lithosphere may differ from that underlying oceanic lithosphere. This arises, for example, from a longer isolation time and may include material thermally transformed or eroded from the continental lithosphere.

Magma generation in hotspot and rift environments can involve mixing of melts derived from a mantle plume, the asthenosphere, and the lithosphere (for example: Chen & Frey, 1985; McDonough et al., 1985; Perry et al., 1987). The relative contribution of these components depends upon factors such as the thermal and mechanical condition of the lithosphere, the temperature of the mantle plume and shallow asthenosphere, depth of magma segregation, rate of magma generation, and the speed and volatile content of the ascending magma. Volatile-rich alkali basalts, in contrast to tholeiitic basalts, commonly carry mantle xenoliths, and erupt quickly through the crust to the surface. They are not completely free from contamination, but tend to be less contaminated by continental crust than tholeiitic basalts.

The continental lithosphere in well-developed rift environments is expected to be thinned (Fig. 7.7.2). Asthenospheric mantle, and the base of the lithospheric mantle, are perhaps the main or perhaps sole source for basaltic magmas. Mantle plume or convective upwelling in the asthenosphere raises the temperature at the asthenosphere/lithosphere boundary originally from about 1200° C (see Section 7.5.1 and Fig. 7.5.2) to higher temperatures and increases the degree of partial melting. Mixing of melts derived from the asthenospheric and lithospheric mantle sources might lead to decoupling of incompatible elements and Pb-, Sr-, and Nd-isotopic compositions from the major elements.

Primary ol-tholeiitic basalts from eastern Australia have high liquidus temperatures (about 1300°C) at about 50 km depth of magma generation (Section 7.5.1, Fig. 7.5.2) and low abundances of incompatible elements (for example, La about 10 ppm and Zr about 80 ppm). They are probably generated by relatively high degrees of partial melting, similar to those for MORB. Their major-element compositions probably are derived mostly from the mantle plume or deeper (perhaps greater than 300 km) and anomalously hot asthenospheric mantle having high potential temperatures (perhaps greater than 1350°C, compared to about 1280°C for N-MORB sources; McKenzie & Bickle, 1988). In contrast, their incompatible elements and Pb-, Sr-, and Nd-isotopic compositions may receive a major contribution from melts derived from both the lithospheric mantle and crust. These ol-tholeiitic basalts have low volatile contents and higher densities than do the associated alkali basalts. They tend to underplate at the crust/mantle boundary and intrude into the lower crust, resulting in magma differentiation accompanied by crustal assimilation, crustal melting, and the eruption of felsic magmas in the central volcanoes (for example, Ewart, 1982a).

Comment is required here on the above-mentioned concept of 'potential temperature' (Tp) in the upper mantle. McKenzie & Bickle (1988) defined Tp as the temperature a mantle source would have if it were brought to the surface adiabatically without melting — that is, Tp = T (at the source) − adiabatic gradient ×

depth. Thus, mantle at a depth of 400 km, a temperature of 1440°C, and an adiabatic gradient of 0.4°C/km, has a Tp of 1280°C. The temperature of the mass of the diapir must be further decreased if melting is involved, in order to account for the required latent heat of melting. This will translate for basalts to about a 30°C drop for 10-percent melting. The Tp of 1280°C estimated by McKenzie & Bickle (1988) for N-MORB sources may be too low, although it is probably not higher than 1350°C. Therefore, high Tp values estimated for the sources of ol-tholeiitic basalts from eastern Australia are equal to or higher than those expected in the ocean-ridge environment. If so, upwelling of anomalously hot mantle from great depth beneath eastern Australia could be inferred, as suggested by Miyashiro (1986). This may point to an 'endogenic' cause for break-up of Gondwanaland (Anderson, 1982).

The interaction of a mantle plume with shallow asthenosphere and the basal lithosphere is predicted to impose a strong influence on the chemical and isotopic characteristics of basalts. These effects may be manifested in different ways by contemporaneous regional volcanism in eastern Australia (Fig. 7.7.2). These interactions take place where upwelling plumes intersect the base of the continental lithosphere and spread out along the lithosphere/asthenosphere boundary. Channelling of plume material in preferred directions may be expected in this setting, producing a situation similar to that thought to exist in oceanic environments (for example, Schilling et al., 1985).

Possible lithospheric components involved in basaltic volcanism in continental backarc environments (Fig. 7.7.1) include subducted oceanic crust and sediments, a 'plum-pudding' type mantle wedge above the subduction zone, and mantle-plume and heterogeneous continental lithosphere of different ages and different chemical and isotopic compositions. The involvement of each component is likely to be different. For example, the effects of subduction-zone processes in a region where subduction has ceased can be detected in basaltic magmas a few million years later — for example, the shoshonites of Papua New Guinea (Johnson et al., 1978) and Fiji (Gill, 1984). Melting of either asthenosphere or lithospheric mantle previously contaminated by subducted sediment will produce primary basaltic melts that retain the strong chemical and isotopic characteristics of sediments — for example, enrichments in Pb, Ba, Rb, and K, but depletions in Sr, Nb, Ta, P, Ti, and Eu (up to 20-percent negative Eu anomaly). These strong sedimentary characteristics have been observed in continental basalts — for example, the Mesozoic Tasmanian dolerites (Hergt, 1987) and the 3.5 Ma-old ultrapotassic basalts in central Sierra Nevada (Van Kooten, 1981).

Upward migration of melts from the asthenosphere and later episodes of intraplate magmatism will continue to modify the lithosphere. Magmas migrating preferentially along major translithospheric faults eventually may overprint the effects of subduction-related processes within lithosphere near the fracture zones. This may result in a biased sampling of the lithospheric mantle through entrained xenoliths brought up by later alkali basalts.

A further example of the interplay of different factors controlling the geochemistry of intraplate basalts in backarc environments is provided by the studies of Cainozoic volcanism in the western United States. Differences in the chemical and isotopic composition of basalts from the Columbia River and Snake River regions, the Colorado Plateau, Basin and Range, and the Rio Grande Rift can be understood in terms of a model recognising (1) the effect of Mesozoic-Cainozoic subduction (including sediments) beneath the western north-American coast, (2) involvement of the mantle wedge affected by the subduction process, (3) occurrence of a deeper asthenosphere component of 'oceanic character' not affected by the subduction process, (4) a mantle plume for the basalts of Snake River and Yellowstone Park, (5) continental lithosphere of diverse ages and geochemical compositions (for example: Hart & Carlson, 1987; Perry et al., 1987).

Crustal contamination

Basalts erupted through continental rather than oceanic crust are more likely to be modified by crustal contamination, because of the lower density, lower melting temperature, and greater thickness of continental compared to oceanic crust. Magma chambers may form within the lithospheric mantle, the lower crust, or the upper crust, depending on crust and mantle density contrasts and the thermal and mechanical condition of the crust. Processes of crustal contamination may be of two different types (for example: Campbell, 1985; Dickin et al., 1987):

(1) *Bulk assimilation.* High-temperature, primitive tholeiitic basalts injected as dykes have the potential to assimilate a higher fraction of wall rock (Huppert & Sparks, 1985) than lower-temperature, more fractionated melts, and thereby may retain greater effects of crustal contamination.

(2) *Assimilation/fractional-crystallisation (AFC) processes* result from continued involvement of a crustal contaminant during differentiation processes in a magma chamber. More fractionated melts in this case preferentially retain the effects of crustal contamination.

These effects in many cases can be difficult to distinguish from mantle-source characteristics, except where detailed studies are undertaken. However, such studies are possible only where circumstances are favourable. A good example is the study of the early Tertiary basalts of northwestern Scotland (Thompson et al., 1984; Dickin et al., 1987) where chemical and isotopic trends correlate closely with the different types of basement rocks (Lewisian granulites and amphibolites).

7.7.3 East Australian continental lithosphere

There is a diversity of rock types in eastern Australia that record the numerous periods and tectonic settings of lithosphere development. Basement rocks of eastern Australia have ages ranging from early Proterozoic (for example, in central and southern Queensland) to probably early Palaeozoic in the New England Fold Belt (Section 1.3.1). Subduction-zone related volcanism

along most of eastern Australia (including Tasmania) was important during much of the Palaeozoic and early Mesozoic, prior to the opening of the Tasman and Coral seas. A major middle Mesozoic thermal event recognised throughout Gondwanaland was the eruption and emplacement of extensive flood basalts, dykes, sills, and layered complexes in South America, southern Africa, the Transantarctic Mountains, and Tasmania. Tholeiitic basalts of the same age and having chemical and isotopic compositions similar to those of the Tasmanian dolerites, are found in Kangaroo Island (South Australia) and western Victoria (Hergt, 1987). These occurrences provide evidence that this thermo-magmatic event in southeastern Australia was widespread. Other relatively contemporaneous Mesozoic intraplate magmatism in eastern Australia (for example, Jaques et al., 1985) also may be related to backarc tectonics.

The chemical and isotopic compositions of the Mesozoic Tasmanian dolerites include a strong crustal signature interpreted as being inherited from an early phase of sediment subduction (Hergt, 1987). These same chemical characteristics are shared by basalts in the Transantarctic Mountains and by some basalts from the southern Karoo province (for example: Hergt, 1987; Duncan et al., 1984; Hawkesworth et al., 1984). These basaltic lavas plausibly represent backarc volcanism correlated with Mesozoic subduction along the ancient western margin of Gondwanaland (for example, Cox, 1978). The continental lithosphere of southeastern Australia also may have been affected by this thermo-magmatic event. Cainozoic intraplate basalts of this area — in particular, the Older and Newer Volcanics of Victoria — therefore may carry some of the chemical and isotopic characteristics of the upper crust, if this modified lithosphere was involved in their generation. Similar situations could have existed in central and eastern Queensland and in eastern New South Wales, where late Palaeozoic subduction took place at the Pacific margin.

Evaluation of the chemical and isotopic characteristics of east-Australian lithosphere has been achieved with some success through the study of mafic-ultramafic xenoliths found in alkali basalts (Chapter 6, Section 7.2). However, as mentioned above, recurrent intraplate melt migration through major translithospheric faults may modify or erase the earlier 'memory' of subduction-related processes adjacent to such faults. This process may have resulted not only in biased sampling by younger alkali basalts, but also may have affected the geotherm estimates based on upper-mantle and crustal xenoliths.

The average thickness of the lithosphere of present-day southeastern Australia can be estimated by using the depth of the low-velocity zone as defined by seismic studies. The top of the low-velocity zone in this region is at a depth of about 90–100 km (Section 1.4.7). A temperature of about $1200°C$ can be assigned to it if the low-velocity zone represents the top of the convecting asthenosphere containing a small amount of melt (Section 7.5.1). Possibly also the top of the asthenosphere beneath individual lava-field and central volcanoes can be estimated using experimentally determined depths of separation and generation of primary basalts (Section

7.5.1). The depths of magma separation from the mantle for the basanites and nephelinites of the lava-field provinces are about 80–100 km, whereas *ol*-tholeiitic basalts have shallower depths of magma separation (about 50–70 km).

7.7.4   Key trace-element and isotopic characteristics

The elemental ratios of La/Nb, Ba/Nb, Ce/Pb, Nb/U, and Rb/Cs have been used in studies of oceanic basalts (1) as indicators of enrichment and depletion processes, caused by melt extraction and metasomatic enrichment in their mantle sources, and (2) to identify possible sediment-subduction effects (see Sun & McDonough, 1988, for a general review). Hofmann et al. (1986) pointed out that ocean-island basalts and MORB have constant values of Ce/Pb (25 ± 5), Nb/U (47 ± 10), and Rb/Cs (80 ± 20) which are almost independent of differences in isotopic compositions. These nearly constant ratios result from similar mineral/melt partition coefficients during partial melting and fractionation of both ocean-island basalt and MORB sources. Some fractionation is expected for other incompatible element pairs (for example, Ba/Nb and La/Nb), given small differences in their relative mineral/melt partition coefficients. In detail, however, the observed differences in La/Nb and Ba/Nb are considerably larger than would result from mantle melting processes. Additionally, a positive correlation between these ratios (Fig.

7.7.3) is opposite to what is expected (D_{Ba} is less than D_{Nb} which is less than D_{La}) from different degrees of partial melting. The positive correlation between Ba/Nb and La/Nb values and $^{87}Sr/^{86}Sr$ (Fig. 7.7.3) is evidence that these differences are related to long-term mantle processes. A crustal character (average crustal Ce/Pb about 4, Nb/U about 10, and Rb/Cs about 20) would be superimposed on the average MORB/ocean-island type mantle (ratios of 25, 47, 80, respectively), if mantle recycling of sediment is important locally. Note, however, that crustal contamination has a similar effect on the composition of basalts. These effects are caused by the high concentrations of U, Pb, Rb, and Cs and the relatively lower Nb concentration relative to U and the light REE in the crust. The result of crustal contamination or sediment addition to the mantle will be to decrease Ce/Pb, Nb/U, and Rb/Cs, and increase La/Nb and Ba/Nb values (Fig. 7.7.3).

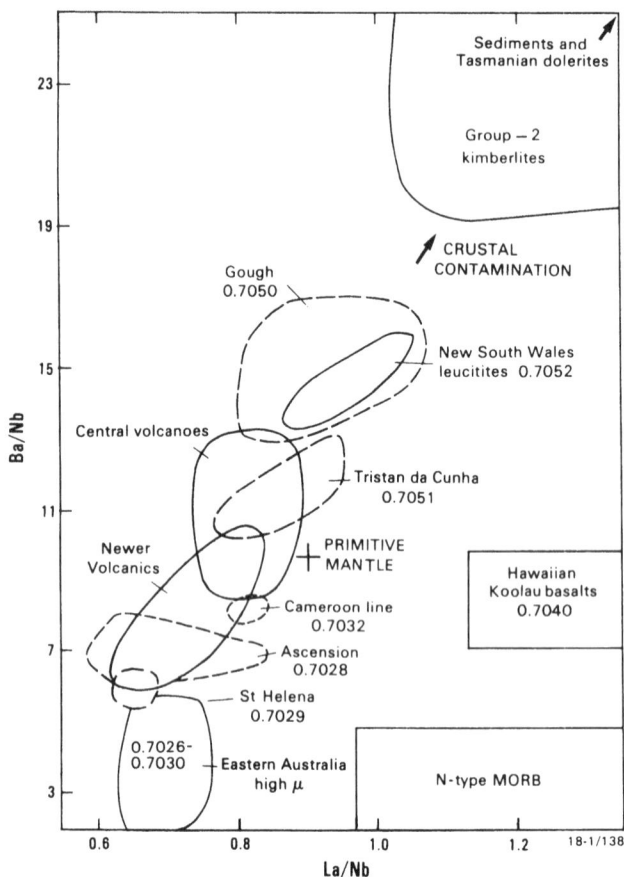

Figure 7.7.3   Positive correlations between Ba/Nb, La/Nb, and $^{87}Sr/^{86}Sr$ values for basalts from ocean islands and eastern Australia. Data sources are listed in Table 7.7.2.

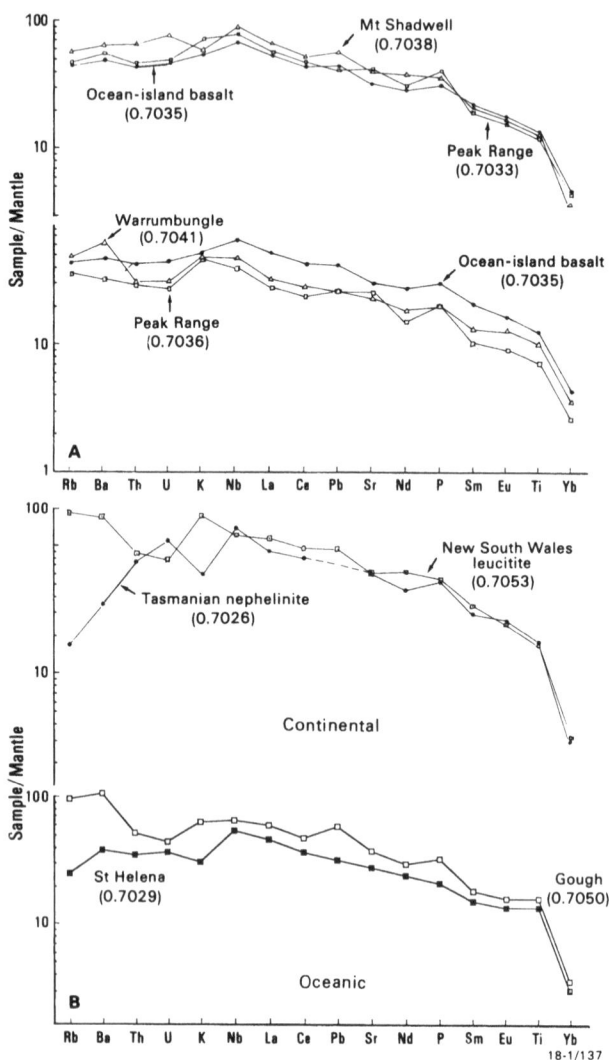

Figure 7.7.4   Representative mantle-normalised patterns for primitive basalts from ocean islands and eastern Australia. All data are normalised using the mantle values listed in Table 7.7.1. $^{87}Sr/^{86}Sr$ values are given in parenthesis. (A) Comparison of central-volcano and lava-field alkali and tholeiitic basalts with average ocean-island basalt. (B) Comparison of high-μ and Dupal-type alkali basalts from oceanic and continental environments. Data sources are listed in Table 7.7.1.

A critical examination of published and unpublished data for a few hundred, least-fractionated, east-Australian basalts has shown that they have chemical and isotopic compositions similar to those observed in ocean-island basalts (Fig. 7.7.4; see also Figs. 7.7.6–7). The ranges of Sr-, Pb-, and Nd-isotope compositions for these east-Australian basalts are $^{87}Sr/^{86}Sr$ 0.7026 to 0.7055, $^{206}Pb/^{204}Pb$ 17.8 to 19.7, and ε_{Nd} −4 to +8. Widespread isotopic heterogeneity in the upper-mantle beneath eastern Australia is indicated. Additionally, a good correlation exists between isotopic composition and elemental-abundance pattern for east-Australian and ocean-island basaltic lavas (Fig. 7.7.4).

The abundance patterns shown in Figure 7.7.4 were constructed using many high-quality chemical data. Good internal consistency of these patterns are found in each area represented. Consequently, each pattern is believed to represent a general feature. Only samples with high Mg-ratios (greater than 60) are used for this purpose in order that the effects of fractional crystallisation and crustal contamination are minimised. Some leucitite samples tend to be altered, causing an excessive scatter of mobile-element abundances, and therefore an 'idealised' mantle-normalised elemental pattern is shown based on relatively constant elemental ratios observed in the majority of samples: Nb/La 1.1, Ce/Nd 2.1, Ba/Nb 14, Ba/Rb 10, Nb/Th 11, Sr/Nd 15, and Th/U 5 (assumed). Chemical data are presented in Table 7.7.1.

Four samples are chosen as representative of primitive east-Australian tholeiitic and alkali basalts. There are close similarities between the patterns for these samples and those of ocean-island basalts (Fig. 7.7.4). A Warrumbungle volcano basalt has a $^{87}Sr/^{86}Sr$ value of 0.7041 and is representative of the majority of central-volcano basalts, although the Peak Range

basalt has a lower value of 0.7036 (Fig. 7.7.4A). The second basalt from Peak Range and the Mount Shadwell basalts are typical lava-field type alkali basalts, and both are primitive and ultramafic-xenolith bearing (Fig. 7.7.4A). A Tasmanian nephelinite (Fig. 7.7.4B) has greater depletions in Rb, Ba, and K and a lower $^{87}Sr/^{86}Sr$ value (0.70265) compared to the St Helena alkali basalt (0.70292). This nephelinite also has higher abundances of other incompatible elements such as Nb and light REE. One possible explanation for these differences is the presence of residual phlogopite in the source that held back K, Rb, and Ba, and which buffered the Ti content in the magma during generation of alkali basalt. This would cause depletion in these elements relative to Nb and La for such alkali basalts (for example, Sun & McDonough, 1988). Tholeiitic and transitional basalts (with presumably no phlogopite in the residue during magma generation) are characterised by higher K/Nb and Ba/Nb values than alkali basalts with similar $^{87}Sr/^{86}Sr$ values. The Warrumbungle sample has a more radiogenic $^{87}Sr/^{86}Sr$ value (0.7041), and higher relative Rb and Ba contents than the average ocean-island basalt with a $^{87}Sr/^{86}Sr$ value of 0.7035. These characteristics of the Warrumbungle sample are shared also by ocean-island basalts with higher $^{87}Sr/^{86}Sr$, such as a Gough Island basalt (Fig. 7.7.4B).

Positive correlations between $^{87}Sr/^{86}Sr$, La/Nb, and Ba/Nb values are observed for both ocean-island and east-Australian basalts having La abundances of greater than or equal to 20 ppm (Fig. 7.7.3). Nephelinites from Tasmania ($^{87}Sr/^{86}Sr$ 0.70265) also have low La/Nb (about 0.6) and the lowest Ba/Nb (about 2) so far observed in primitive alkali basalt. Lava-field and central-volcano basalts (in which La is greater than or equal to 20 ppm) generally have higher La/Nb values

Table 7.7.1 Trace-element and other analytical data for representative basalts from ocean islands and eastern Australia[1]

	Mantle normalising values	Ocean-island basalts			East Australian basalts					Leucitite
		St Helena 2882	Gough 111	Average OIB[2]	Peak Range CLM5	Warrumbungle WMB35	Tasmania 2854	Peak Range CLM100	Mount Shadwell 2679	(idealised)
Rb	0.65	16.1	62	29	23.5	31.5	12	30	36	122
Ba	7.23	275	760	350	230	445	250	395	460	1220
Th	0.092	3.01	4.52	4.0	2.7	2.8	6.6	4.1	5.7	7.9
U	0.022	0.86	1.05	1.02	0.6	(0.7)	2.3	(1.0)	1.63	1.6
K	230	6970	14110	12000	10380	10870	12870	15690	14940	37500
Nb	0.738	38.5	46	48	28	33	97	55	62	87
La	0.732	31.7	41.2	37	20	23	62.2	39	45.5	79
Ce	1.896	64.8	84.5	80	43	52	140	85	93	168
Pb	0.076	2.08	3.85	3.2	2	2		3	4	6.7
Sr	21.8	581	792	660	550	498	1250	885	830	1200
Nd	1.413	32.3	40.2	38.5	21	26	57.2	40	49.2	80
P	92	1720	2640	2700	1880	1800	4400	3250	3240	4400
Sm	0.456	6.68	8.08	10.0	4.5	5.9	11.9	8.0	9.1	13.6
Eu	0.172	2.25	2.66	3.00	1.54	2.1	3.9	2.6	2.99	3.75
Ti	1300	16800	19800	17200	9120	12720	20520	15420	16680	20400
Yb	0.498	1.55	1.81	2.16	1.3	1.8	1.47	2.1	1.74	1.45
MgO (wt%)		10.4	8.4	10	13.0	9.1	14.1	8.9	12.6	13.0
$^{87}Sr/^{86}Sr$		0.70292	0.7050	0.7035	0.70358	0.70409	0.70265	0.70331	0.70381	0.7052
ε_{Nd}		+4.0	(−1)	+6	+4.7	+2.7	+7.5	+5.5	+3.8	−1

[1] Data sources: Cundari (1973), Frey et al. (1978), McDonough et al. (1985), Knutson et al. (1986), Nelson et al. (1986), Ewart et al. (1988), and Sun & McDonough (1988).
[2] Ocean-island basalt.

Table 7.7.2 ^{87}Sr/^{86}Sr and some critical elemental ratios observed in basalts from ocean islands and eastern Australia[1]

	Average OIB[2]	Central volcano	Lava field	Leucitite suite
^{87}Sr/^{86}Sr	0.7035	0.7036–0.7043(?)	0.7026–0.7045	0.7052
Ba/Nb	7	8–13	4 ± 2 –13	15
La/Nb	0.8	Ca. 0.8 (up to 1.0)[3]	0.6–0.7 (up to 1.0)	1.1
Ce/Pb	25	Ca. 25 (down to 10)	Ca. 25 (down to 10)	Ca. 25 (?)
Nb/Th	12	11	15–11	11
Rb/Cs	80	Ca. 80 (down to 40)	Ca. 80 (down to 40)	

[1] Data sources: Cundari (1973), Frey et al. (1978), McDonough et al. (1985), Ewart et al. (1988), Sun & McDonough (1988), J. Knutson (personal communication, 1988).
[2] Ocean-island basalt.
[3] Values in parentheses are those for tholeiitic basalts characterised by low abundances of incompatible elements.

of 0.7–0.8. Leucitites from New South Wales have Ba/Nb and La/Nb values higher than the primitive-mantle ratios, but similar to the compositions of potassic Gough Island basalts (see also Ewart et al., 1988, fig. 7). However, some basalts with ^{87}Sr/^{86}Sr values in the range 0.7031–0.7036 and Ba/Nb values of about 6 from Barrington, Liverpool Range, and Oberon, appear to have lower La/Nb values of 0.5 to 0.6. Larger ranges of Nb/Th values for lava-field basalts (11 to 15; Table 7.7.2) correlate with a larger range of ^{87}Sr/^{86}Sr (0.7026 to 0.7045). Nevertheless, lava-field basalts within the ^{87}Sr/^{86}Sr range of 0.7035 to 0.7045 also have Nb/Th values of about 11.

Tholeiitic basalts from eastern Australia are characterised by low abundances of incompatible elements (La less than 15 ppm). They commonly have high La/Nb values of about 1.0, and their Ce/Pb values are as low as 10–15. An increase in the degree of partial melting can produce increases in La/Nb, because Nb is more incompatible than La (for example, Sun & McDonough, 1988). However, such a process will not cause a decrease in Ce/Pb (Hofmann et al., 1986). Combined changes in La/Nb and Ce/Pb values therefore may result from crustal contamination or the mixing of different mantle source regions. Crustal contamination in some instances can be demonstrated for the differentiates of tholeiitic basalts (for example, Ewart et al., 1988).

The interplay of source heterogeneity, magma-generation processes, and crustal contamination effects

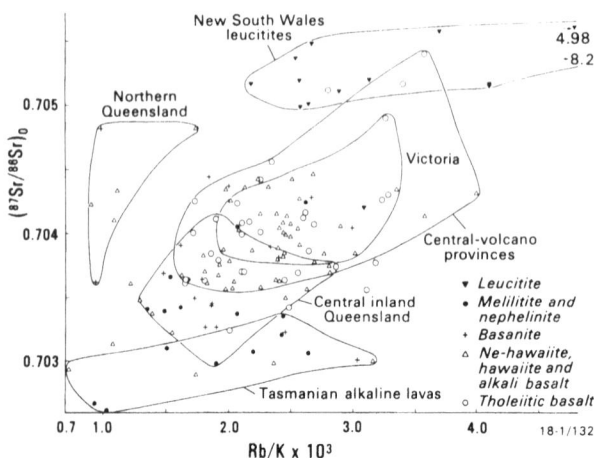

Figure 7.7.5 Initial- ^{87}Sr/^{86}Sr versus Rb/K relationships for east-Australian basalts.

can be illustrated further by a plot of ^{87}Sr/^{86}Sr initial ratios versus Rb/K values (Fig. 7.7.5). The Tasmanian alkaline basalts and the New South Wales leucitites have a range of Rb/K, and relatively constant ^{87}Sr/^{86}Sr values. This trend is interpreted as being controlled primarily by magmatic processes. However, possible alteration effects may have influenced the compositions of some leucitites. In contrast, the remaining basaltic data, considered in terms of broad regional groupings, tend to have correlations between increasing Rb/K and Sr-isotopic ratios. These features are interpreted as resulting from mixing of asthenospheric and lithospheric components (and in some cases including crustal contamination).

7.7.5 Isotope systematics and time-space relationships

Introduction

The southward migration of the locus of central-volcano magmatism during the past 35 Ma in eastern Australia is consistent with the hotspot model implying a 6.5 cm/yr northward migration of the Australian continent (Section 1.7). A deep-mantle plume origin for the hotspot is supported by seismological and P-wave residual data (Section 1.4.5,7). These data are indicative that there is at present deep melting of the mantle beneath the Tasman Sea northeast of Tasmania (Sections 1.4.5,7). The contemporaneous eruption of New South Wales leucitites and central volcanoes that are some 300 km to the east, may be a consequence of nearby hotspot activity and broad-scale structural controls (Section 7.5.4). Alternatively, multiple plumes may be responsible for central-volcano activity in eastern Australia. This model is discussed below (see also Section 7.3). The eruption of lava-field basalts has dominated east-Australian volcanic activity from 70 to 30 Ma. However, numerous, younger, lava-field provinces are contemporaneous with (or somewhat younger than) nearby central-volcano provinces. Some lava-field basalts (without associated felsic differentiates) may have been derived from the marginal parts of the plume and erupted alongside central volcanoes (Fig. 7.7.2B; see also Fig. 7.7.10).

Details of the dynamic model of mantle-plume, asthenosphere, and lithosphere interaction to produce the east-Australian intraplate volcanism may be explored by reference to well-documented studies in the oceanic environment using both geophysical and geochemical methods. Good examples of the interaction between the three components (plume, asthenosphere, and lithosphere) include studies of the Iceland and Reykjanes Ridge system (for example: Schilling, 1973; Sun et al., 1975), the St Helena and Tristan da Cunha plumes (off-ridge since 25 Ma ago), and the southern Mid-Atlantic Ridge (Schilling et al., 1985; Hanan et al., 1986). Additionally, intraplate hotspot activity in the Hawaiian Islands has been interpreted (for example, Chen & Frey, 1985) in terms of the interaction of a mantle plume (mainly for the shield-building tholeiitic basalts) with both oceanic lithosphere and shallow-asthenospheric mantle (mainly for the post-erosional alkali basalts). A similar model for intraplate alkaline and tholeiitic basalts from southeastern Australia has been proposed by McDonough et al. (1985).

The intersection of an upwelling mantle plume with the base of the lithosphere is visualised as resulting in flooding of the top of the adjacent shallow asthenosphere, accompanied by channelling in certain preferred directions (see Fig. 7.7.10B). The geochemical influence of the plume component on nearby lava-field basalts in eastern Australia will decrease as distance from the hotspot increases.

The following questions also must be addressed in the case of eastern Australia and in the context of these processes:

(1) How can the trace-element and isotopic compositions of four possible components — mantle plume, shallow asthenosphere, lithospheric mantle, and crust — be identified and distinguished?

(2) Is there geochemical evidence for dynamic mixing of plume-asthenosphere components, as seen for example by the changing isotopic composition of lava-field volcanism affected by passing plume activity?

(3) How do the inferred trace-element and isotopic characteristics of the east-Australian plume (or plumes) compare with the plume compositions inferred for the Tasmantid Guyots and other major oceanic islands, such as the Hawaiian Islands?

Component identification

The following is an attempt to identify the trace-element and isotope characteristics of mantle-plume, shallow-asthenosphere, lithospheric-mantle, and crustal components inferred to be involved in east-Australian basaltic volcanism. The four possible source com-

ponents for east-Australian basalts are listed in Table 7.7.3 and compared with equivalent components for the oceanic island of Hawaii.

The composition of the plume component is derived from the least-fractionated basalts in the central volcanoes. This screening process is checked by evaluating the possible effects of crustal contamination on Pb-, Sr-, and Nd-isotope compositions, and on Ce/Pb, La/Nb, and Ba/Nb values, as well as comparing central-volcano data with those of nearby lava-field basalts of similar age. These lava-field basalts have isotopic compositions similar to those of the central-volcano basalts, as discussed below. Plume compositions estimated this way may represent the mixing product of pure plume with some entrained shallow asthenosphere and lithospheric mantle.

The composition of the shallow-asthenosphere component is defined by lava-field alkali basalts older than the central volcanoes from a given area. Lava-field basalts of the same age, or younger, found far removed from the central volcanoes also have been used.

The composition of the lithospheric mantle component is difficult to define, except for the leucitites of New South Wales, as it is likely to differ in each region because of their different geological histories, as described above. The leucitites of New South Wales are characterised by having Sr-, Nd-, and Pb-isotopic compositions distinct from both the lava-field and central-volcano basalts of eastern Australia (see Figs. 7.7.8–9), but similar in chemical and isotope characteristics to some leucitites that are considered to be derived from long-term, metasomatised, continental lithosphere (for example, Nelson et al., 1986). A lithospheric origin for these leucitites is preferred, but a possible asthenospheric origin cannot be eliminated.

The gross effects of crustal contamination may be evaluated by documenting coupled changes in isotopic and trace-element characteristics (for example, La/Nb, Ce/Pb, and Rb/Cs). Crustal contamination is not expected to be a problem for alkali basalts, partly because their compositions are closer to those of expected primary magmas, and many are mantle-xenolith bearing.

The following conclusions are reached after application of these criteria. $^{87}Sr/^{86}Sr$ and ε_{Nd} values for the least-fractionated central-volcano basalts (and, in this model, the inferred plume component) have a limited range from 0.7036 to 0.7043 and from +5 to +1,

Table 7.7.3 Identification of source components for Cainozoic east-Australian basalts and comparison with Hawaiian basalts

Component	Eastern Australia — continental setting	Hawaii — oceanic setting
Mantle plume	Central volcanoes (Peak Range, Springsure, Warrumbungle)[1]	Shield-building tholeiitic basalts (Kilauea, Mauna Kea, Honomanu series of Haleakala)
Shallow asthenosphere	Old and/or distant, lava-field alkali basalts (Tasmania, New England, Liverpool Range, Blue Mountains, Monaro)	Major contribution to the post-erosional alkali basalts (Honolulu, Kauai, Hana)
Lithospheric mantle	Leucitites (El Capitan)	Major contribution to some caldera-stage alkali basalts (Kula series of the Haleakala volcano)
Crust	Continental	Oceanic

[1] Selected examples are given in parenthesis.

respectively. Examples are: Peak Range 0.7036, Springsure 0.7036–0.7039, Glass Houses 0.7041, Tweed 0.7036, Nandewar 0.7039, Warrumbungle 0.7041, Canobolas less than 0.7043, central Victoria less than or equal to 0.7043 (data from: Ewart, 1982a; Ewart et al., 1988; McDonough, 1987; H.-D. Hensel, personal communication, 1987). These compositions are similar to those measured for the Hawaiian shield-building tholeiitic basalts (0.7035–0.7042, ε_{Nd} +7 to +2; Tatsumoto et al., 1987) and the Tasmantid Guyots (0.7038–0.7047, ε_{Nd} +3 to −3; McCulloch, 1988). Pb-isotope compositions for the least-fractionated central-volcano basalts have the following ranges: $^{206}Pb/^{204}Pb$ 18.34–18.82; $^{207}Pb/^{204}Pb$ 15.52–15.61; $^{208}Pb/^{204}Pb$ 38.25–38.82 (Ewart, 1982a, Ewart et al., 1988). Hawaiian tholeiitic basalts have similar $^{206}Pb/^{204}Pb$ (17.8–18.7) but lower $^{207}Pb/^{204}Pb$ (15.42–15.50) and lower $^{208}Pb/^{204}Pb$ (37.7–38.2; Tatsumoto et al., 1987).

The shallow-asthenosphere component, as inferred from lava-field alkali basalts well removed from the influence of any plume component, have a wide range of $^{87}Sr/^{86}Sr$ values from 0.7026 (Tasmania and the New England region of New South Wales) to about 0.7044. The dominant ranges for $^{87}Sr/^{86}Sr$ and ε_{Nd} are 0.7030–0.7034 and +6 to +8, respectively. Examples are: Monto 0.7032, Chudleigh 0.7034, Rewan (central Queensland) 0.7030, Spring Mount 0.7026, Doughboy 07031, Barrington 0.7031–0.7035, Liverpool Range 0.7032–0.7034, Blue Mountains 0.7032–0.7035, Grabben Gullen 0.7031–0.7033, Kiama 0.7041, Moruya 0.7031, and Tasmania 0.7026–0.7033 (Menzies & Wass, 1983; O'Reilly & Griffin, 1984; McDonough et al., 1985; Rudnick et al., 1986; Ewart et al., 1988; H.-D. Hensel, personal communication, 1987). Higher $^{87}Sr/^{86}Sr$ values have been reported from lava-field basalts in southeastern Australia, notably Newer Volcanics (0.7037-0.7044), Kandos 0.7034–0.7037, Southern Highlands 0.7029–0.7044, Monaro 0.7036–0.7040, and Oberon 0.7034–0.7036 (McDonough et al, 1985; Menzies & Wass, 1983). The possible effects of underplating by earlier plume activity or Mesozoic subduction (related to the Tasmanian dolerite event) on the mantle sources of these basalts cannot be evaluated further without Pb-isotope, La/Nb, and Ce/Pb data for these basalts.

There appear to be some distinct differences in Pb-isotopic composition between lava-field and central-volcano basalts. Lava-field alkali-basalts from central Queensland with $^{87}Sr/^{86}Sr$ between 0.7030 and 0.7033 have $^{206}Pb/^{204}Pb$ values (18.50–19.10) similar to those measured for central volcanoes (18.34–18.82), but consistently lower $^{207}Pb/^{204}Pb$ (15.50–15.55 versus 15.52–15.61) and $^{208}Pb/^{204}Pb$ values (Fig. 7.7.6; see Ewart et al., 1988 for original data). The differences at a $^{206}Pb/^{204}Pb$ value of 18.7 are 15.54 versus 15.59, and 38.50 versus 38.75, respectively.

The isotopic composition of the inferred mantle-plume and shallow-asthenosphere (lava-field) components are distinct in Figure 7.7.6, but this is based on limited available data. No Pb-isotopic data are available yet from southeastern Australia for lava-field basalts that have higher $^{87}Sr/^{86}Sr$, except for the Newer Volcanics (see Fig. 7.7.8). However, they can be

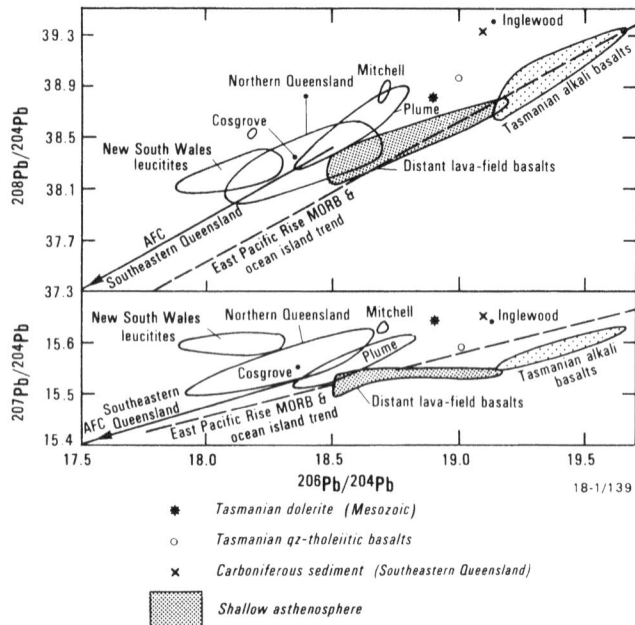

Figure 7.7.6 Pb-isotope fields for the mantle plume (inferred from the least-fractionated central-volcano basalts and nearby lava-field basalts of the same age), shallow asthenosphere (distant lava-field basalts), lithosphere (leucitites), and crustal contaminants for east-Australian volcanism. AFC represents trend of data for combined assimilation/fractional crystallisation processes. Data from Cooper & Green (1969), Nelson et al. (1986), Hergt (1987), and Ewart et al. (1988). East Pacific Rise MORB and ocean-island trend are practically the same as the Northern Hemisphere Reference Line of Hart (1984).

expected to be distinct from the high-μ samples and similar to the basalts of the central volcanoes — that is, having lower $^{206}Pb/^{204}Pb$, but higher $^{207}Pb/^{204}Pb$ and $^{208}Pb/^{204}Pb$ than the high-μ samples (Fig. 7.7.6). Note also that Ewart et al. (1988) showed that $^{206}Pb/^{204}Pb$ values for lava-field basalts with low $^{87}Sr/^{86}Sr$ (0.7030 or less) are relatively high (19.13–19.66), so there are some similarities to the high-μ ocean-island basalts from St Helena and Canary Islands (Sun, 1980). Basalts having these isotopic characteristics may be an important asthenospheric component beneath eastern Australia, especially considering the widespread occurrences of lava-field alkali basalts with low $^{87}Sr/^{86}Sr$ (0.7026–0.7031). Furthermore, occurrences of kaersutite xenocrysts with low $^{87}Sr/^{86}Sr$ (0.7025–0.7029) at Hill 32 in Queensland, and Wee Jasper and Spring Mount in New South Wales (Basu, 1978), are predicted to have relatively high $^{206}Pb/^{204}Pb$ values.

The nature of the lithosphere component in different areas can be evaluated potentially using Pb-isotope data (Fig. 7.7.6). The New South Wales leucitites apparently do not have a predominant mantle-plume component, although some contribution from the mantle plume is possible, if not likely. Nelson et al. (1986) and Ewart et al. (1988) suggested that they resulted from the melting of continental lithospheric mantle induced by the nearby passing mantle plume. The analcimite at Inglewood, southeastern Queensland, has Pb-isotope compositions similar to local Carboniferous sediments, and possibly represents melting of lithosphere previously modified by Palaeozoic subduction. Qz-tholeiitic basalts (MgO about 7 weight percent) from the Mitchell province, central Queensland, have higher $^{207}Pb/^{204}Pb$ and $^{208}Pb/^{204}Pb$ values than the least-fractionated central-volcano basalts (Fig. 7.7.6), and have low Ce/Pb values (about 10). *Ne*-hawaiites and hawaiites of northern Queensland (28–0 Ma) also have high $^{207}Pb/^{204}Pb$ and $^{208}Pb/^{204}Pb$. They are likely the products of AFC processes involving crustal contamination. However, the possibility of involvement of lithosphere previously modified by Palaeozoic subduction during generation of these Cainozoic basalts and differentiates cannot be eliminated. This interpretation appears to be consistent with the observation that spinel-lherzolite xenoliths collected from the same north-Queensland area have Nd- and Sr-isotope compositions similar to those of the hawaiites (W.L. Griffin, personal communication, 1988). They have higher $^{87}Sr/^{86}Sr$ values, possibly caused by subduction effects, than other Cainozoic basalts of eastern Australia with similar ε_{Nd} (Fig. 7.7.7). The same explanation is suggested also for a *qz*-tholeiitic basalt from Tasmania (MgO 8.8 weight percent) having relatively high $^{87}Sr/^{86}Sr$ (0.7038) and very low abundances of incompatible elements (La 7.7 ppm, Zr 92 ppm; Ewart et al., 1988). This sample has a Pb-isotope composition between those of the Cainozoic alkali basalts of Tasmania (low $^{87}Sr/^{86}Sr$, 0.7030) and the Mesozoic Tasmanian dolerites (0.710) which have a strong crustal signature in regard to their trace-element and isotope compositions (Hergt, 1987).

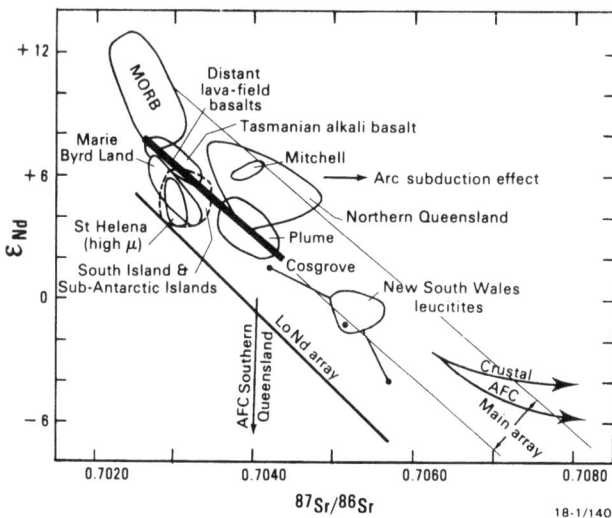

Figure 7.7.7 ε_{Nd} versus $^{87}Sr/^{86}Sr$ relationships for the inferred mantle plume, shallow-asthenosphere, and lithosphere components. Data from Menzies & Wass (1983), McDonough et al. (1985), Nelson et al. (1986), Ewart et al. (1988), and McDonough (1987). AFC represents assimilation/fractional-crystallisation processes.

Effect of the plume on lava-field provinces

The isotopic compositions of the lava-field basalts of similar or younger ages compared to nearby central-volcano basalts, provide evidence for the input of a mantle-plume component. Many of these lava-field basalts in Queensland have almost identical Pb-, Sr-, and Nd-isotopic compositions to those of the least-fractionated central-volcano basalts. Examples are Mount Fox, Mount Andrew, Exevale, Anakie, Silver

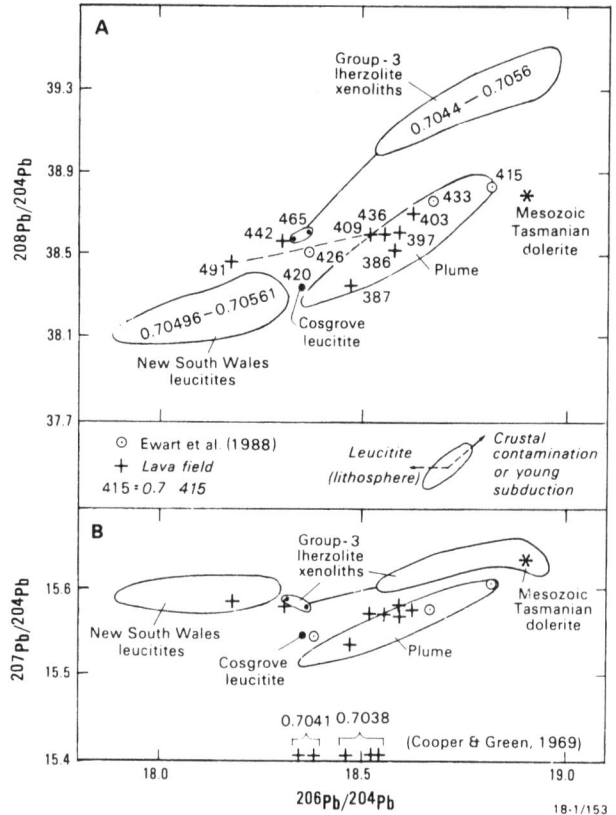

Figure 7.7.8 $^{208}Pb/^{204}Pb$ and $^{207}Pb/^{204}Pb$ versus $^{206}Pb/^{204}Pb$ relationships for basalts of the Newer Volcanics, western Victoria. Also shown are the fields for the mantle-plume component, leucitites of the New South Wales, and group 3 spinel-lherzolite xenoliths of Stolz & Davies (1988). $^{87}Sr/^{86}Sr$ values for the Newer Volcanics, leucitites, and lherzolite xenoliths are labelled (last three digits only in A). Only $^{206}Pb/^{204}Pb$ values for the Newer Volcanics reported by Cooper & Green (1969) are shown in B. Two trends of Pb-isotope values for tholeiitic basalts and their differentiates with $^{87}Sr/^{86}Sr$ values higher than 0.7043 are shown schematically in the legend between A and B. The samples analysed by Ewart et al. (1988) were identified by them as central-volcano type, although the samples are not from the Macedon-Trentham area.

Hill, Mount Schofield, Mount Boorambool, Stonecroft, and Bunya Mountains (Ewart et al., 1988).

The Newer Volcanics of western Victoria (mainly younger than 5 Ma) are of special interest because they crop out in the same general region as the Older Volcanics (70–15 Ma). The renewal of lava-field volcanism after a 10 Ma period of quiescence was dominated by tholeiitic basalts instead of alkali basalts (see Section 3.7.4), and may have been related to the nearby mantle-plume activity that produced the felsic complexes of the Macedon-Trentham area (about 6 Ma). On the other hand, the fact that the eruption of tholeiitic and transitional basalts peaked at about 2.5–2 Ma, considerably later than the Macedon-Trentham volcanism, is indicative that the Newer Volcanics activity was not related to a mantle plume (R.C. Price, personal communication, 1988). If so, high-temperature *ol*-tholeiitic basalts from the Newer Volcanics having low abundances of incompatible elements (La 10 to 15 ppm, Zr 100–120ppm) require an anomalously hot and deep asthenospheric source having a potential temperature of about 1350°C compared to about 1280°C for N-MORB source. In addition, this type of tholeiitic basalt is found also in young (less than 5 Ma) lava-fields in northern Queensland (Sections 3.2), in the Older Volcanics of Victoria (Section 3.7.2.), and in Tasmania (Section 3.8.3). Such high potential temperatures existed in the asthenospheric sources of lava-field basalts that have no obvious connection to plume activity. Upwelling of such hot mantle regions from great depths may be connected to the endogenic forces causing the break-up of Gondwanaland (Anderson, 1982). If this is correct, available isotopic data for the Newer Volcanics are evidence that the deep-mantle source region has similar isotopic compositions to the source for the central volcanoes.

Available Pb-, Sr-, and Nd-isotopic data for alkaline and least-fractionated tholeiitic basalts of the Newer Volcanics, western Victoria (Cooper & Green, 1969; McDonough et al., 1985; Ewart et al., 1988; Stolz & Davies, 1988; C.M. Gray, personal communication, 1987), as well as isotope data for the group–3 spinel lherzolites of Stolz & Davies (1988), are summarised in Figures 7.7.8 (Pb isotopes) and 7.7.9 (ε_{Nd} versus $^{87}Sr/^{86}Sr$). Some Newer Volcanics samples have a range of $^{87}Sr/^{86}Sr$ within the field inferred for the mantle-plume component (0.7038 to 0.7043). They also have the same range of Pb-isotope compositions ($^{206}Pb/^{204}Pb$ 18.38–18.82) and ε_{Nd} values (+4 to +2) as the inferred plume component, and therefore a genetic connection with the mantle plume is inferred.

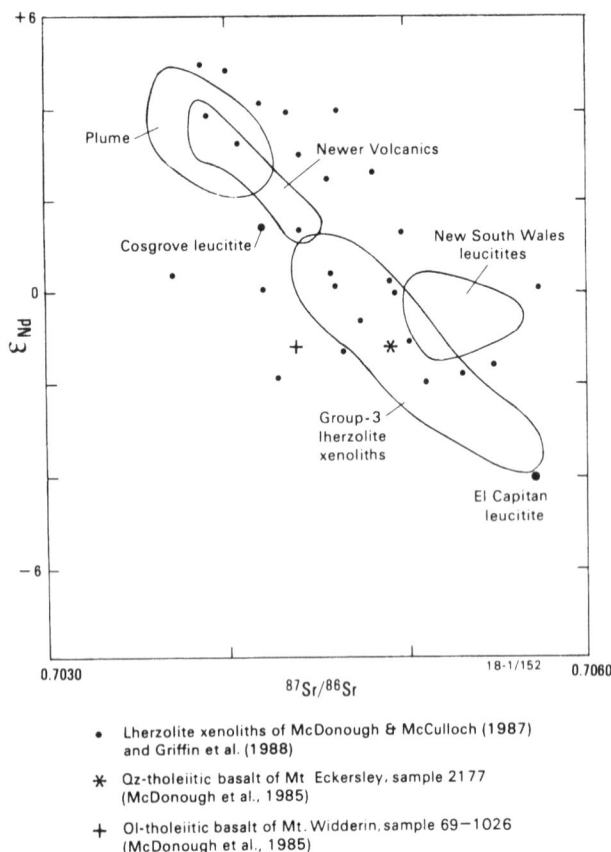

• Lherzolite xenoliths of McDonough & McCulloch (1987) and Griffin et al. (1988)

∗ Qz-tholeiitic basalt of Mt Eckersley, sample 2177 (McDonough et al., 1985)

+ Ol-tholeiitic basalt of Mt. Widderin, sample 69–1026 (McDonough et al., 1985)

Figure 7.7.9 ε_{Nd} versus $^{87}Sr/^{86}Sr$ relationships for the Newer Volcanics, inferred mantle plume, leucitites of the New South Wales, and group 3 spinel-lherzolite xenoliths, as well as for other ultramafic xenoliths from the Newer Volcanics.

McDonough et al. (1985) proposed a mantle-plume/lithospheric-mantle interaction model for generation of the Newer Volcanics.

Two trends are identifiable on Pb-isotope ratio plots for tholeiitic basalts and their differentiates from the Newer Volcanics field with $^{87}Sr/^{86}Sr$ ratios higher than 0.7043 (Fig. 7.7.8). The first trend points towards the field of the New South Wales leucitites — that is, a decrease of $^{206}Pb/^{204}Pb$ accompanied by a slight decrease of $^{208}Pb/^{204}Pb$, but nearly constant $^{207}Pb/^{204}Pb$ values. Pb-isotope data for clinopyroxene and amphibole separates from one group–3 lherzolite (BM18) also fall on this trend. Involvement of lithospheric mantle having isotopic compositions similar to the leucitites of New South Wales therefore is suggested for the generation of these tholeiitic basalts. The leucitite at Cosgrove, Victoria (northeast of Macedon), has Pb-, Sr-, and Nd-isotopic compositions intermediate between those for the New South Wales leucitites and the inferred plume component (Figs. 7.7.8–9). This again may represent a contribution of the plume component to the melting of the continental lithospheric mantle (Nelson et al., 1986).

The second trend follows an increase in $^{206}Pb/^{204}Pb$ and $^{208}Pb/^{204}Pb$ values towards a component with relatively high $^{207}Pb/^{206}Pb$ and $^{208}Pb/^{206}Pb$ represented by Pb-isotope compositions for the majority of the group–3 lherzolite xenoliths of Stolz & Davies (1988) and the Mesozoic Tasmanian dolerites. There is a trend in these Victorian tholeiitic basalts and differentiates of increasing La/Nb (0.7 to 1.0) and $^{87}Sr/^{86}Sr$ ratios (up to 0.7057) and decreasing Ce/Pb (25 to 10; D.A. Wallace, personal communication, 1988). Plains tholeiitic basalts have higher $^{87}Sr/^{86}Sr$ values than do nearby alkali basalts (for example, 0.7045 versus 0.7039). The involvement of continental lithosphere, perhaps affected by Palaeozoic-Mesozoic sediment subduction (the Tasmanian dolerite event), is suggested for the generation of these lavas if the changes are not caused by crustal contamination.

Concluding remarks

The general conclusion reached from the above geochemical and chronological survey is that east-Australian intraplate volcanism can be interpreted in terms of a model of mantle-plume, shallow-asthenosphere, and lithosphere (mantle and crust) interaction (Fig. 7.7.10). Primary and near-primary basalts have chemical and isotopic systematics similar to those recognised in ocean-island basalts, and differences, at least in theory, can be explained by the involvement of continental lithosphere (including crust). Tholeiitic basalts, especially highly fractionated ones, commonly have a crustal geochemical signature that normally is difficult to distinguish between the effect of crustal-assimilation/fractional-crystallisation and lithospheric mantle modified by subduction-zone processes. Central-volcano and some nearby lava-field basalts of the same (or slightly younger) ages are interpreted as derived mainly from the mantle plume. The apparent large deviations in the position of some central volcanoes up to 300 km from the main or inferred single hotspot track may be explained by the absence of deep, vertical,

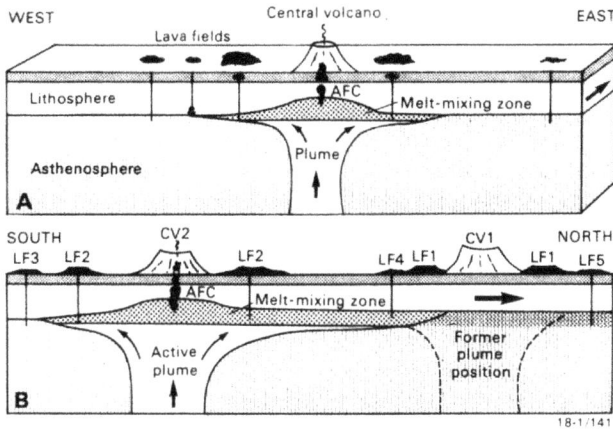

Figure 7.7.10 Model of interaction of mantle plume, shallow asthenosphere, and continental lithosphere during the generation of east-Australian basalts. Sections are (A) transverse to and (B) in the same plane as, the southward-tracking direction of the mantle plume. Plume material is channelled in favourable directions between the shallow asthenosphere and overlying lithosphere. Basalts of central volcanoes (CV) and nearby lava fields (LF) of the same age (CV1 and LF1, CV2 and LF2) are derived mainly from the mantle plume. Some lava-field basalts (LF3) near the front of the mantle plume are derived from sources that have not been affected yet by the passage of the plume, whereas others (LF4) that are more distant from, but in, the plume wake, are derived from the plume and erupt close to older plume-related lava fields (LF1). Later lava-field basalts (LF5) are erupted well after the passage of the plume and are derived from the shallow asthenosphere and overlying lower lithosphere. Both of these sources may have been strongly modified by plume material. AFC represents assimilation/fractional-crystallisation processes.

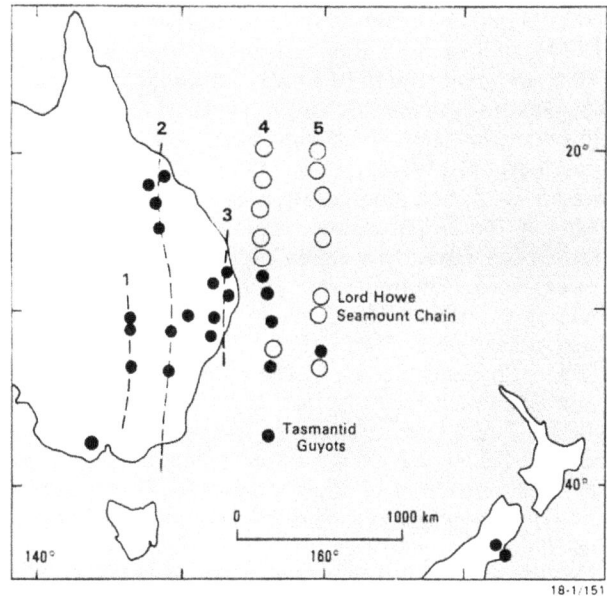

Figure 7.7.11 Traces of five proposed hotspots in eastern Australia and the Tasman Sea (adapted from Fig. 1.7.2; filled circles represent dated volcanoes). A prediction from the known age progressions of the central volcanoes in eastern Australia and the Tasmantid Guyots is that all four plumes are currently at about the same latitude (about 39°S).

translithosphere fractures that are able to tap magma directly from mantle depths. Plume material therefore may have been channelled to adjacent favourable fracture systems allowing central-volcano and lava-field eruptions to take place away from the main hotspot trace.

A possibly more reasonable explanation is that central volcanoes in eastern Australia were generated by more than one mantle plume (Wellman & McDougall, 1974a; Sutherland, 1983; Section 1.7.3). There may be three mantle plumes in eastern Australia (Fig 7.7.11), associated with the two plume-related hotspot traces in the Tasman Sea — the Tasmantid Guyots and the Lord Howe seamount chain. Each of these five hot-spot traces are separated by roughly equal distances of about 300 km (Fig. 7.7.11). Groupings of the central volcanoes in relation to the three continental plumes are, from west to east: (1) Macedon-Trentham, including possibly the leucitite-suite of New South Wales and Victoria; (2) Hillsborough, Nebo, Peak Range, Springsure, Nandewar, Warrumbungle, and Canobolas; (3) Glass Houses, Main Range, Tweed, Ebor, and Comboyne.

Incomplete hot-spot traces observed in eastern Australia in this model are attributed to the blanketing effect of thick continental lithosphere. Collectively, basalts of these central volcanoes and seamounts have similar isotope compositions but they also have temporal and secular isotopic differences similar to those observed in Hawaii. The plume component is interpreted to be the main magma source, but additions

from the lithospheric mantle for some basalts are recognised.

7.7.6   Leucitites and the Dupal anomaly

Leucitites in central New South Wales crop out along an inferred late Proterozoic rift zone (Ewart et al., 1988). They have a north-to-south age progression consistent with the hotspot model for the central volcanoes. A solely mantle-plume origin for these leucitites appears unlikely because their Pb-isotope systematics are quite different from those of the inferred plume component (Fig. 7.7.6). Instead, they are interpreted as melting products of continental lithosphere that was thermally reactivated by a mantle plume, as discussed above. Leucitites having similar trace-element and Sr- and Nd-isotope compositions are found at Nyiragongo and Bufumbira volcanoes, central Africa (Vollmer & Norry, 1983). These rocks also may have been derived mainly from continental lithosphere (metasomatised about 500 Ma ago), although with some contributions from an asthenospheric component. The African leucitites, however, have much higher Pb-isotope ratios than do their New South Wales counterparts ($^{206}Pb/^{204}Pb$ 19.2–19.4 versus 17.9–18.4; Nelson et al., 1986).

Pb-, Sr-, and Nd-isotopic compositions as well as the mantle-normalised trace-element abundance patterns (Fig. 7.7.4) and ratios such as Ba/Nb and La/Nb (Fig. 7.7.3) for the New South Wales leucitites, are closely similar to those of ocean-island basalts from the Dupal-anomaly region (for example, Kerguelen Islands, Tristan da Cunha, and Gough Island). These ocean-island basalts are characterised by high $^{87}Sr/^{86}Sr$ (0.7050 or greater), and high $^{207}Pb/^{204}Pb$ and $^{208}Pb/^{204}Pb$, relative to values for northern-hemisphere ocean-island basalts having similar $^{206}Pb/^{204}Pb$ (Hart, 1984). The term 'Dupal anomaly' was first introduced by Hart (1984), and initially was considered to be a circum-global anomaly related to a long-lasting (possibly since the Archaean) mantle convection system around 30°S. However, this anomaly is considered now to be neither circum-global (White et al., 1987a), nor limited to around 30°S, nor long-lasting (Sun et al., 1988). For example, the Dupal isotopic anomalies also appear to exist in the Japan Sea region and northeastern China (for example: Nakamura et al., 1985; Peng et al., 1986; Song et al., 1987; Zhi et al., in press). Basalts from these areas have $^{206}Pb/^{204}Pb$ values of about 17.5, $^{207}Pb/^{204}Pb$ about 15.45, $^{208}Pb/^{204}Pb$ about 37.8, and ε_{Nd} about 0, for rocks having $^{87}Sr/^{86}Sr$ values of about 0.7047. These late Tertiary and Quaternary alkali and tholeiitic basalts have low La/Nb (about 0.65) and low La/Ta (about 11), in contrast to those for Gough Island and Tristan da Cunha (about 0.9 and 15, respectively; Fig. 7.7.3).

Plume-related central-volcano and lava-field basalts, including basalts from the Newer Volcanics and from northern Queensland, also have distinct Dupal-type character, such as higher $^{207}Pb/^{204}Pb$ and $^{208}Pb/^{204}Pb$ relative to the North Hemisphere Reference Line (Fig. 7.7.6). An origin for the Dupal anomaly from the convecting mantle can be inferred for the plume-related central-volcano and lava-field basalts. These rocks have considerably higher $^{206}Pb/^{204}Pb$ (up to 18.8) than do New South Wales leucitites (17.9–18.2). The more radiogenic end-member (possibly contaminated by the continental lithosphere) has a Pb-isotope composition (18.8, 15.6, 38.8) similar to that of modern sediments. They therefore may have been derived from mantle sources that were modified by subduction of continental sediments around Gondwanaland.

Lithospheric mantle modified by sediment subduction also may have similar Dupal-type isotopic characteristics, such as observed in Mesozoic Tasmanian dolerites (Hergt, 1987) and group–3 spinel-lherzolite xenoliths from western Victoria (Stolz & Davies, 1988; see Fig. 7.7.8). Involvement of this lithosphere in magma generation will superimpose further Dupal-type isotopic characteristics on the basalts and their differentiates, such as observed for some Newer Volcanics and north-Queensland basalts. Note, however, that Stolz & Davies (1988) emphasized that their group–3 xenoliths have extreme ranges in Nb/Th (1–8), Nb/U (12–85), Nb/La (0.3 to 1), and Ce/Pb (63–258). The observed Ce/Pb values are much higher than in ocean-island basalts and in the Cainozoic basalts of western Victoria (10–30). These chemical characteristics are most likely the result of open-system metasomatism through fluid/melt infiltration (Stolz & Davies, 1988). They do not reflect the chemical characteristics of the mantle source where the fluid/melt originated. A similar conclusion was reached by O'Reilly & Griffin (1987) who suggested that the isotopic characteristics of metapyroxenites studied by them were created through Palaeozoic-sediment subduction. Whether or not this explanation also applies to the group–3 xenoliths of Stolz & Davies (1988) requires further evaluation.

The Tasmantid Guyots were built upon newly created oceanic lithosphere. Therefore contamination of the plume by direct interaction with the distant Australian continental lithosphere, or by mixing of the plume and recently eroded material from the lithosphere of eastern Australia is minimal. In this sense Pb-isotopic data for the Tasmantid Guyots should help in defining the Dupal character of this region. It will help also to evaluate the proposal that leucitites of New South Wales were derived from late Proterozoic lithospheric mantle rather than convecting asthenosphere.

7.7.7 Regional high-μ character

Some lava-field alkali basalts from central Queensland, New South Wales, and Tasmania, as well as the South Island and Sub-Antarctic Islands of New Zealand (McDonough et al., 1986; Barreiro & Cooper, 1987), and Marie Byrd Land (Futa & LeMasurier, 1983) and Ross Island in Antarctica (Sun & Hanson, 1975), are characterised by low $^{87}Sr/^{86}Sr$ (0.7026–0.7030, extending to 0.7033 for Ross Island) and high $^{206}Pb/^{204}Pb$ (19.2–20.6). The existence of these high-μ (high $^{238}U/^{204}Pb$) type isotopic characteristics is unusual and requires some comment.

High-μ samples generally fall on the low side of the 'main mantle array' on ε_{Nd} versus $^{87}Sr/^{86}Sr$ diagrams. This feature reflects a long-term high U/Pb and low Rb/Sr character for this type of mantle source. However, this is not shown clearly by the east-Australian samples (Fig. 7.7.7). Long-term isolation of a high-μ mantle is necessary for the generation of the low $^{87}Sr/^{86}Sr$ and high $^{206}Pb/^{204}Pb$ characteristics. One possible mechanism is through lithospheric metasomatism, involving addition of silica-undersaturated melt from the top of the convecting asthenospheric MORB source (Kay, 1979; Sun, 1980; Sun & McDonough, 1988; Menzies & Wass, 1983; Hart et al., 1986). Recycling of this enriched lithosphere component back into the convecting asthenosphere may provide high-μ sources.

Southeastern Australia, New Zealand, and Marie Byrd Land were closely juxtaposed prior to break-up of Gondwanaland (Section 1.2). Possibly the high-μ character of these basalts reflects a lithospheric mantle signature, and this character was developed by CO_2-rich, metasomatic melts upwelling from the asthenosphere about 200 Ma ago, as inferred from the widespread alkaline igneous activity that took place at that time. However, this process in Tasmania must have taken place after the Tasmanian dolerite event (about 185 Ma ago) to have been accessible during the Cainozoic volcanism. An alternative hypothesis would place the source of these high-μ reservoirs (at least for Tasmania and Ross Island) in the convecting mantle.

A suggestion of a lithospheric-mantle origin for high-μ type alkaline rocks was made by Halliday et al. (1988) for the Cameroon province (70 Ma to present) in western Africa. They suggested that this lithospheric source was created by underplating of material from the St Helena mantle plume during the initial break-up stage of the Atlantic Ocean. Furthermore, Taras & Hart (1987) proposed that there may be a systematic regional distribution of high-μ character volcanism in the Atlantic, including the New England seamounts, Hoggar (northwestern Africa), the Canary Islands, and the Azores. If this is correct, the region also should include Fernando de Noronha, Cape Verde, the Cameroon line, and Ascension and St Helena islands. Clearly, therefore, high-μ type volcanism, is not restricted to the Australasian-Antarctica region.

Pb-Pb isochrons (Sun & McDonough, 1988) for high-μ type ocean-island basalts range from 1.0 Ga (Ua Pou Island, Marquesas Archipelago) to about 1.8 Ga (St Helena). A model-Pb isochron age of about 2.2 Ga (see Fig. 7.7.6) obtained for five alkali basalts from Tasmania (Ewart et al., 1988), is indicative that the high-μ character, even if it was located in the lower lithosphere, was not developed in the Tasmanian lithosphere 0.2 Ga ago. In contrast, data for alkaline volcanic rocks (including carbonatites) from Westland, South Island (New Zealand), define low slopes (about 0.04) on a $^{207}Pb/^{204}Pb$ versus $^{206}Pb/^{204}Pb$ plot, perhaps corresponding to a relatively young enrichment event within the continental lithosphere (Barreiro & Cooper, 1987).

7.7.8 Some future tests

Studies of early Cainozoic lava-field basalts should be able to provide good evidence for lithospheric modification by Palaeozoic-Mesozoic subduction. Younger lava-field basalts can not be used for this purpose because there is the possibility that progressive modification of the base of the lithosphere by continued intraplate mantle underplating and igneous activity may have taken place. Detailed trace-element and isotopic studies of the Older Volcanics (70–15 Ma) in Victoria should be undertaken to test the concept that the lithosphere of southeastern Australia has been affected by the Mesozoic Tasmanian dolerite event or by Palaeozoic subduction (or by both). Comparison of results from the Older and Newer Volcanics also should provide a test for the alternative models of the proposed mantle-plume connection for the Newer Volcanics versus an origin from deep and hot asthenospheric source not related to the mantle plume activity. Similar comparative studies can be proposed for the relatively young lava-field basalts in northern Queensland.

Mesozoic tholeiitic basalts on Kangaroo Island, in South Australia, and from western Victoria, as well as metapyroxenite xenoliths in the Cainozoic lava-field basalts of Tasmania, should be studied and compared with results for the Tasmanian dolerites and with the metapyroxenite and spinel-lherzolite xenoliths from the Newer Volcanics (McDonough & McCulloch, 1987a; Griffin et al., 1988; Stolz & Davies, 1988) in order that the extent of chemical modification of the lithospheric mantle in southeastern Australia (as a result of the Palaeozoic-Mesozoic subduction events) can be defined more clearly. If the metapyroxenites have isotopic compositions similar to those of the Tasmanian dolerites, then the Tasmanian-dolerite characteristics should be considered as a possible major lithospheric compositional component, in addition to the lithospheric mantle sources for leucitites, in evaluations of the genesis of the Victorian Older and Newer Volcanics (Fig. 7.7.8). This metapyroxenite component also may have influenced other lava-field basalts of southeastern Australia. This kind of study can be applied also to

mantle xenoliths in young basalts (less than 5 Ma) from northern Queensland.

More effective tests are needed to identify the extent of involvement of subduction-modified lithospheric mantle or crustal contamination during generation of some tholeiitic basalts, hawaiites, and their differentiates, having crustal chemical characteristics. More Pb- and Nd-isotopic data are needed for basalts from the Newer Volcanics to define trends on isotope plots. Correlation of geographic position and geochemical data with isotopic composition may help to define better the number of components and processes involved in magma genesis.

More chemical and isotope data for high-μ type lava-field basalts (for example, Tasmania, and Spring Mount and Wee Jasper, New South Wales) are needed to define better the composition and occurrences of these basalt in eastern Australia and New Zealand. These data also should be used to further test the global implications of a regional high-μ anomaly.

The idea that Cainozoic volcanism in eastern Australia may have been a result of a regional upwelling of hot mantle from great depths (Miyashiro, 1986) deserves some serious consideration. More accurate estimation of pressure-temperature conditions for the generation of high-MgO (10 percent or more) primitive tholeiitic basalts containing low abundances of incompatible elements (for example, in the Newer Volcanics and north-Queensland provinces) would help in the evaluation of the need for high-temperature (potential temperatures of about 1350°C) mantle sources. Further evaluation of the proposed dynamic model of the interaction of mantle plume, asthenosphere, and lithosphere clearly is desirable.

7.8 Problems, Uncertainties, and Issues

7.8.1 Introduction

The preceding sections in this chapter contain discussions of several topics that continue to be points of debate or areas of uncertainty in studies of the intraplate volcanism of eastern Australia and New Zealand. Some of these issues and uncertainties are clearly stated in the foregoing sections or are self-evident. The purpose of this section is to list other issues that are less well defined in order that the general direction of future research can be identified a little more clearly. An opportunity is taken also to consider interpretations that are not dealt with in the foregoing sections.

Important petrogenetic issues are raised in the foregoing sections and need not be repeated here. However, two particular matters can be highlighted briefly. First, is the importance of continuing attempts to identify the geochemical characteristics and number of the main source components for the basalts. A list of possible tests for the four-component model is given in Section 7.7.8. Second, is to note that the importance of assimilation/fractional-crystallisation (AFC) processes is stressed in Chapter 5 and Section 7.6 in part as a result of the broad regional approach taken in these sections of the volume. Future attention should be given to more detailed studies of individual central volcanoes in order that other potentially important processes, such as liquid-state differentiation and volatile transfer, can be assessed more readily. Detailed sampling of carefully mapped sequences followed by comprehensive trace- and isotope-geochemical analyses are needed to clarify the relative importance of these alternative processes of magmatic differentiation.

This section is concerned more with geodynamic topics than petrological ones. It includes, particularly, discussion of the identification of hotspot versus hotline tracks, the importance of stress fields and geotherms, and a comparison of geodynamic events in eastern Australia and New Zealand.

7.8.2 Hotspot identification

An ongoing problem is the clear identification of central-type volcanoes and the age-progressive tracks that these volcanoes define. The distinction between central and lava-field volcanoes is based on the presence of a significant proportion of felsic rocks in central volcanoes, but this criterion is not unambiguous in all cases. For example, Wellman & McDougall (1974a) classified Dubbo province as central-volcano type, but in Section 1.7 Dubbo is recognised as lava-field type because of the uncertainty of the significance of the felsic rocks at Dubbo and even though the age of the province is consistent with the age of the relevant part of the single-hotspot track (Table 1.7.2; basalts at Dubbo also have low abundances of some incompatible elements, as do some of those from the central volcanoes). However, no clearly defined volcanic edifice of central type is recognisable at Dubbo, and the province does not contain the peralkaline rocks that are found in all of the central-volcano provinces. Felsic rocks are known also at Fraser Island (Section 3.4.2) and in the Central province (Section 3.5.3), but neither of these is classified as central-type because the volume of the felsic rocks is so small.

Another province classified as central type both in Section 1.7 and by Wellman & McDougall (1974a) is the Newer Volcanics because of the presence of the Macedon-Trentham felsic rocks (Section 3.7.3). The 6 Ma age of the felsic rocks is consistent with the age-progression of the southward-tracking hotspot. Furthermore, Newer Volcanics basalts have 'plume'-type chemical characteristics (Section 7.7). However, the relative volume of the Macedon-Trentham rocks is insignificant compared to that of the basalts of the Newer Volcanics. In addition, the Macedon-Trentham rocks are somewhat older (6 Ma) than the basalts (mainly 3–2 Ma), no large central-type edifice has been constructed, and the area is farther from the calculated single hot-spot track (Fig. 1.7.3) than is any other of the defined central volcanoes (the western end of the Newer Volcanics is more than 500 km west of the plume trace).

The Rockhampton province of central Queensland (Section 3.9.2) includes a significant proportion of trachytic rocks and is a central-type province in the sense used in this volume, rather than a lava-field type as listed in Table 1.7.2. However, its age is 70–67 Ma

and so it is twice as old as Hillsborough, the oldest known (34–31 Ma) of the Cainozoic central volcanoes. The relationship between the Mesozoic Rockhampton province and the Cainozoic hotspot track is therefore uncertain. However, an intriguing feature of Fig. 1.7.3 is the close proximity of Rockhampton to the 'back-tracked' part of the calculated hotspot track for the 80–60 Ma period. This closeness may be significant, allowing for the uncertainties involved in the calculation of the hotspot track for the 80–30 Ma period, perhaps lending some support to the interpretation that the Rockhampton province is part of the same hotspot track as the Cainozoic central volcanoes.

A point requiring emphasis in the recognition of central volcanoes and mantle hotspot tracks is that central volcanoes are defined solely on the basis of the presence of significant proportions of felsic rocks. These felsic rocks are believed to be the products of combined assimilation/fractional-crystallisation in crustal magma reservoirs and of partial melting in the crust. The ironic situation, therefore, is that crustally derived rocks are used as indicators for recognising the existence of basalt-generating, sub-lithospheric mantle hotspots (see below). The implication, of course, is that voluminous crustally derived melts are generated only in the presence of unusually large, mantle-derived, basalt fluxes.

The question of whether the most mafic basalts of central volcanoes are compositionally different from those of the lava-field provinces was addressed in Section 7.7. Central-volcano basalts low in incompatible-element abundances may be produced by large degrees of partial melting, therefore may be hotter than lava-field basalts, and so may carry a greater capacity for the melting of crustal rocks. However, no clear distinction could be made in places between the basalts of the two volcano types, especially where lava-field eruptions take place near central volcanoes of the same or similar age. In part, this is because of the masking effects of lithospheric melting and contamination. However, an attempt was made to distinguish 'plume' and 'shallow asthenosphere' mantle sources. No clear correlation can be made between these mantle-source types and the central and lava-field basalt types, because both central and some lava-field provinces have 'plume-type' chemical characteristics. The interpretation given in Section 7.7 is a model-dependent one in which the plume-type lava-field provinces are considered to derive from the wake of the plume after the main part of the hotspot has passed by. In addition, some basalts of the central volcanoes may have a signature corresponding to plume and shallow-asthenosphere mixing. This plume-wake interpretation needs to be tested, as proposed in Section 7.7.

7.8.3 Stress fields of active volcanism

A recurrent theme in this volume has been the interpretation that intraplate volcanism has taken place in a tensional environment permitting the ascent of basaltic magma to the surface (for example, Section 1.6.2). However, a tensional tectonic regime may not provide access for magma in all circumstances, as discussed in Section 7.3.4. Differential stress within strong layers or stress guides in the upper mantle or lower crust, may be sufficiently high that brittle failure is restricted to the compressional shear failure field in which magma-filled fractures cannot propagate. Furthermore, magmas may be able to erupt in slightly compressional settings, in which case the factor controlling eruption will be the interplay of lithostatic pressure and magma pressure.

Cloetingh & Wortel (1985, 1986) calculated the present-day intraplate stress field of the Indo-Australian plate by considering the effects of ridge push and slab pull along the margin of the plate (Section 1.7.3). The calculated stress field changes southwards from east-west tension in northern Queensland to lesser northeast-southwest tension in eastern New South Wales, to slight compression in Bass Strait. However, the calculated extensional effects are in direct conflict with the results of *in situ* stress measurements from boreholes, overcoring, hydrofracturing, and earthquake focal-mechanism solutions in eastern Australia (Section 1.4.7). All of these measurements are consistent with the interpretation that east-Australian upper crust is in compression, although there are local differences in direction. No stress measurements are available for the area of the young lava fields of northern Queensland where Cloetingh & Wortel (1985) predicted maximum tensional forces. The young lava fields of the Newer Volcanics are in an area of compression, as predicted by Cloetingh & Wortel (1985).

Measured stress, of course, applies only to the present-day tectonic regime and may not be relevant for earlier periods of volcanism, particularly the period 50–20 Ma ago when most of the lava-field volcanism took place. Indeed, there is the possibility of both lava-field and hotspot volcanism taking place mainly when conditions were tensional. Lava-field basaltic magmas under these conditions may have risen rapidly through the upper crust without fractionating to more evolved compositions. Central-volcano magmas, on the other hand, may have had greater difficulty in breaking through to the surface because of a difference in magma density. The low incompatible-element basalts of the central volcanoes may correspond to larger degrees of partial melting and to hotter plume sources compared to the lava-field sources, as discussed above. Magmas formed by larger degrees of melting will be less volatile-rich and denser, and therefore more likely to underplate, intrude, and fractionate, and to heat up the crust, resulting in the generation of anatectic melts and magma mixing. In contrast, the lower density of the volatile-rich lava-field magmas will aid their more rapid ascent to the surface.

Strongly compressional conditions may be a reason why comparatively little volcanism has taken place in eastern Australia during the Quaternary, particularly if the tectonically anomalous north-Queensland and Newer Volcanics areas are disregarded. It also may be the reason why there is no Quaternary volcanism in South Island, New Zealand, where the transpressional character of the Alpine Fault Zone appears to have 'turned off' the intraplate volcanism (Section 4.9.3). In contrast, stress conditions in the Northland-Auckland region appear to be mildly extensional (more 'relaxed'; Sections 1.6.1, 4.9.2) and Holocene volcanism has

taken place. The Northland-Auckland volcanism is primarily basaltic, and no large central-type volcanoes have been constructed, perhaps implying that largely tensional conditions have characterised this region for much of the Cainozoic.

7.8.4 Single and multiple hotspots and hotline models

Single-hotspot models

A single-hotspot model is preferred in Section 1.7 for the origin of the central volcanoes of eastern Australia, although the possibility of two or more hotspots could not be discounted. Furthermore, a single plume is visualised in Section 7.7 as having spread out laterally at the base of the continental lithosphere, effectively widening the volume of plume that could be used as a source for basaltic magmas (Fig. 7.7.10). This expanded plume is considered to underlie the entire width of the east-Australian belt. Its current position may be northeast of Tasmania, as discussed above.

A distinction between 'melting spots' and 'mantle source volumes' has been made in interpretations of the Hawaiian hotspot (for example, Dalrymple et al., 1973), and may be applicable to the east-Australian example. A 'melting spot' is a broad region in the mantle where basaltic magmas are generated, but 'mantle source volumes' are smaller specific areas of melting contained within the larger melting spot and which generate specific volcanoes (see also Duncan et al., 1986). Thus, a single, large melting spot may generate more than one volcano at any one time, possibly accounting for such time-equivalent volcano pairs as Nandewar and Ebor or Warrumbungle and Comboyne (Fig. 1.1.6). This concept may obviate the need for two or more independent hotspots beneath eastern Australia.

The principal difficulty with the single hotspot model is in accounting for the non-uniform distribution of the central volcanoes longitudinally — that is, the age of the volcanoes correlates closely with latitude (Figs. 1.7.1,3), but no straight north-south line of volcanoes is defined two-dimensionally (Fig. 1.1.6). Indeed, the overall trend of the central volcanoes is closer to the trend of the east-Australian coast than it is to a straight north-south line (Section 1.1.3). The implication is that structural features in the continental lithosphere above the plume control the positions of volcanoes, possibly by means of translithospheric fracture systems that permit east-west lateral migration of magmas in some cases. Control by lithospheric structure has been identified in the case of some oceanic hotspot traces — for example, in the Marquesas Islands of the central Pacific (McNutt et al., 1989).

Another point to note is that although the 600 km-wide single-hotspot model is proposed to account for the wide east-west extent of the central and leucitite-suite volcanoes at any one time period in eastern Australia, this 600 km width is not required to account for the narrow north-south extent of the same volcanoes. A north-south width of less than 100 km is all that is required, in which case the shape of the inferred single 'hotspot' is strongly elongate and, in effect, indistinguishable from a 'hotline' source (see below). Furthermore, the main structural trends of the fold belts in eastern Australia are north-south (Fig. 1.3.1) and therefore would be expected to favour the formation of north-south lines of volcanoes of the same age rather than east-west ones, if the diameter of the single hotspot was a uniform 600 km.

Translithospheric fracture systems in some areas may be reactivated older structures (Section 1.6.2), or they may be produced by the buoyant plume itself. However, the broad correlation between the lava-field volcanism and the hotspot track may be an indication that plume-derived magmas in general rise through those parts of the lithosphere that have been weakened previously by the tectonomagmatic processes that permitted lava-field volcanism. Thus, the plume diverts towards the coast at Buckland (Fig. 1.1.6) because its trend becomes too far towards the more stable and less weak continental interior, but it changes back into the continent where the lithosphere begins to thin at the continental shelf. Continental lithosphere may be weaker than oceanic lithosphere (see Vink et al., 1984, who propose that the properties of quartz and olivine, respectively, control the stress characteristics of the lower continental crust and oceanic upper mantle), and this may account for the way in which the east-Australian hotspot track is restricted to the weaker parts of the continent rather than migrating onto oceanic lithosphere, such as south of Comboyne volcano (13 in Fig. 1.1.6).

Multiple-hotspot models

Wellman & McDougall (1974a) suggested that the central volcanoes (excluding Macedon-Trentham and the leucitite-suite provinces) may have formed as a result of two hotspots rather than one (numbers 2 and 3 in Fig. 7.7.11). A third hotspot may be defined by the leucitite/Macedon-Trentham line (Section 7.7.5; number 1 in Fig. 7.7.11). This problem of the number of hotspots remains unresolved, but some general comments are worth making.

An appealing aspect of the three-hotspot model for eastern Australia is that the spacing between the three continental tracks and the Tasmantid Guyots and Lord Howe seamount chain is about the same (Fig. 7.7.11). Roughly equal spacings can be detected also in the distribution of inferred hotspots in Africa (see, for example, Duncan, 1981, fig. 1). Why such hotspots should be equally spaced in this way is not clear, particularly if each one represents an independent upwelling from the deep mantle. Furthermore, there is no obvious reason why the volcanism associated with the three proposed east-Australian hotspots should appear to be complementary. Thus, track 2 (Fig. 7.7.11) stops at Springsure and Buckland round about the time that track 3 starts to be recorded at Glass Houses, and by the time track 3 stops at Comboyne track 2 has started its activity again at Nandewar and Warrumbungle (see also Fig. 1.1.5).

Another difficulty with the three-hotspot model as outlined in Section 7.7 is the recognition of track 1 defined by the leucitite-suite centres and Macedon-Trentham. The leucititic magmas are considered in

Section 7.7 to have been derived from the continental lithosphere rather than a sub-lithospheric plume but, if so, track 1 cannot be a hotspot track in the conventional sense. Furthermore, Macedon-Trentham (not leucititic) is well to the west of the southern extrapolation of the leucitite track which ends at Cosgrove (Fig. 1.1.5).

A final difficulty is that only one of three proposed continental hotspots appears to be active at the present day — namely, track 2 which is located seismically and on the basis of P-wave residual data and electrical conductivity measurements (Sections 1.4.5,7–8). Proposed track 3 evidently stopped at the New South Wales coast about 16 Ma ago. Holocene volcanism on proposed track 1 took place in the western part of the Newer Volcanics, but this is not defined today by the same features as is the track–2 hotspot, and it is well west of the projected end of track 1.

Hotline models

An intriguing relationship to note is that points representing dated volcanism at any one time on each of four of the five hotspot tracks in eastern Australia and in the Tasman Sea (Fig. 7.7.11) fall in a linear zone that connects the four tracks orthogonally and which therefore trends at right angles to the direction of plate motion. The exception is the most easterly chain, containing Lord Howe Island which is about 10 Ma younger than the volcanoes at the same latitude on the hotspot tracks to the west. This latitude/age relationship forms the basis of the 'hotline' interpretation presented by Wellman (1983) who proposed that the Indo-Australian plate swept over stationary linear magma sources in the asthenosphere beneath both eastern Australia and the New Zealand/Campbell Plateau area, producing volcanism from different parts of the lines at different times.

Wellman (1983) noted that the line source was a sector of a great circle passing through the pole of rotation of the plate in the hotspot reference frame. The two areas of seismicity northeast of Tasmania and farther east out in the Tasman Sea (Fig. 1.4.5) also lie on a sector of a great circle. Wellman (1983) did not address the question of whether such a line source was rooted deep in the asthenosphere in the same way as the hotspots visualised by Morgan (1972), for example. However, such a deep rooting must be considered unlikely because the great-circle relationship between the hotline and the rotation pole then would be coincidental. Rather, this relationship is consistent with the interpretation that the orientation of the hotline source is controlled in some way by the passage of the lithosphere.

One speculative interpretation is that lithospheric movement causes generation of a roll-like cylinder in the shallow asthenosphere which points in the direction of the pole of rotation. The cylinder is long-lived, spalling diapirs that give rise to age-progressive volcanic chains at the surface. The equal spacing of the diapirs may be caused by some form of Rayleigh-Taylor type instability in the convective roll. A shallow asthenospheric source for the hotspot trails therefore is inferred from this interpretation rather than a deep one. The inferred Lord Howe hotspot track is difficult to reconcile with the interpretation, but dredging and dating of samples from the seamounts of Lord Howe Rise are required to clarify the age-latitude relationship of the chain and to determine whether the age of Lord Howe Island is consistent with the age of the seamounts. Convective rolls were visualised also for the Easter volcanic chain (or hotline) in the southeast Pacific (Bonatti et al., 1977), but these are parallel to the direction of plate motion rather than parts of great circles through a pole of rotation.

One problem with the shallow-asthenosphere convective-roll model is why only one hotline should be produced by the lithospheric motion. A series of such hotlines might be expected, in the manner visualised by Richter (1973) for 'transverse' convective rolls, all at right angles to the direction of plate motion. Furthermore, the model does not appear to be consistent with the high temperatures required for some ol-tholeiitic basalts (potential temperatures of about 1350°C).

The simplest interpretation for the east-Australian central volcanoes and leucitite-suite centres is that they define the passage of a single hotspot, but that the head of the hotspot has been deformed by the presence of the continental lithosphere moving over it. The plume is unable to readily break through the lithosphere so it spreads out at the lithosphere/asthenosphere boundary, enlarging its horizontal dimensions. The leading edge of the plume is bow-shaped (convex to the south) because of the southward migration of the plume, and a wake is drawn out behind it. The bow-shaped leading edge is the robust part of the plume and at times it is able to produce volcanism along different parts of its east-west length, simulating the effect of a hotline at right angles to the direction of plate movement.

New Zealand

The significance of the eastward younging of volcanism in the New Zealand region from South Island to the Antipodes Islands continues to remain unresolved (Section 1.7.4). Adams (1981) proposed that the younging corresponded to passage of the Pacific plate over a linear mantle source. Wellman (1983) identified volcanoes containing felsic rocks as 'central' type, adopting the same criteria used for eastern Australia, and demonstrated that their ages, except for Carnley volcano, correlate with azimuth from the rotation pole, as they do in eastern Australia. Wellman (1983) therefore favoured a 'hotline' source to account for these relationships. An alternative possibility raised in Section 1.7.4 is that dextral shear at the Pacific/Indo-Australian plate boundary is producing tensional forces across the Campbell Plateau, permitting the eruption of sublithospheric magmas. Such an interpretation may be applicable also to the origin of the intraplate volcanism of the Northland-Auckland area (for example, Section 1.6.1).

The main restriction on all these interpretations for the South Island and Campbell Plateau area is the small number of volcanoes that are available for dating. Furthermore, the deeper and greater parts of the island volcanoes are difficult to sample other than by drilling, so the ages of the oldest parts of the volcanoes are unknown. In addition, the pattern of eastward younging is less clear where new radiometric dates, obtained very recently from some of the subaerial volcanoes (and not available for consideration in this volume) are taken into consideration (Section 4.9.3). The eastward younging can be detected still in the Miocene to Holocene rocks but not in the early Tertiary ones. A larger geochronological database is required, including ages for the seamounts of the Campbell Plateau area that still remain undated.

Gravity-anomaly chain

Future considerations of hotspot and hotline traces in the east-Australian and New Zealand region will have to take into account the existence of a chain of gravity anomalies off the eastern coast of Tasmania (M.F. Coffin, personal communication, 1989). The anomaly chain is seen on the world gravity map produced by Haxby (1987), and it runs from about 38°S, 150.5°E, near Gascoyne seamount, southwestwards to about 44°S, 150°E on the East Tasman Plateau off southeastern Tasmania. The trend is roughly parallel to the coast of southeastern Australia, and the gravity-anomaly chain cuts across the oldest magnetic anomalies of the Tasman Basin. The chain requires further investigation to determine if it represents a hotspot track and, if so, why its trend is so different from the calculated tracks shown on Figure 1.7.3.

7.8.5 Geotherms, heat-flow, and potential temperature

The present-day geothermal regime of eastern Australia has continued to be a topic of active debate, especially since the establishment of the xenolith-derived empirical geotherm (Section 7.2.3) and the recognition of a considerable difference from the geotherm calculated assuming simple conductive loss (Sass & Lachenbruch, 1979). Determining palaeogeotherms and the present-day geotherm is of economic concern in relation to the origin of diamonds, as discussed in Sections 3.10.3 and 7.2.8.

The xenolith-derived geotherm does not represent necessarily the pressure-temperatures conditions beneath the whole of eastern Australia during the entire Cainozoic. However, the consistency of the data from different regions may be an indication that the empirical geotherm is a reflection of similar conditions at the time of the basaltic volcanism (Section 7.2.3). The xenolith-derived geotherm therefore may represent conditions that are 'perturbed' (Section 7.5.1) during basaltic volcanism towards temperatures that are higher than normal. The geotherm has a strong curvature at 10–30 kb corresponding to advective heat transfer near the crust/mantle boundary which is consistent also with magma emplacement below (underplating) and above the boundary. The xenolith-derived geotherm therefore may have regional significance if such underplating is widespread and not limited solely to the areas of the central volcanoes and lava fields.

Can the xenolith-derived geotherm be reconciled with present-day heat-flow in areas of Quaternary volcanism? The Bullenmerri-Gnotuk xenoliths that form the basis of the empirical geotherm are from volcanoes that are no older than about 50 000 years, so the geotherm may represent conditions in western Victoria where the heat-flow is higher than 'normal'. However, the significance of the heat-flow regime in the whole of eastern Australia remains a subject of debate. One viewpoint is that heat-flow values in general are close to 'world average', even though they are higher than those for the stable (presumably 'unperturbed') cratonic regions of central and western Australia.(Section 1.4.3). Another viewpoint is that the high geothermal gradient inferred from southeast-Australian xenoliths is consistent with 'high' heat-flow values in southeastern Australia (Section 7.2.3). Note that the estimate of about 1200°C as the liquidus temperature for olivine nephelinite and basanite (Section 7.5.1) based on combined data from experimental petrology and depth of the seismic low-velocity zone (about 100 km; Section 1.4.7), is not very different from that obtained from mantle xenoliths — 1200°C at 80 km. Furthermore, the apparent difference of about 20 km depth for 1200°C is consistent with the idea that the geotherm derived from mantle xenoliths represents a 'disturbed' geotherm instead of a 'steady-state' regional one.

More good-quality heat-flow measurements and additional theoretical modelling are required to reconcile the xenolith-derived geotherm with current heat-flow regimes, particularly in areas of Quaternary volcanism. In addition, more accurate mapping of the low-velocity zone (seismic tomography) is needed in order to measure the depths and inferred temperatures at which partial melting is taking place. This approach combined with a better defined relationship between decreases in V_p and V_s and degree of partial melting in mantle containing minor amounts of volatiles, may offer potentially more accurate temperature estimates for the base of the lithosphere in eastern Australia.

Another factor that requires investigation in relation to geotherms is the influence of cold lithospheric slabs subducted into the upper mantle. Such cold slabs may create pockets or layers that have the potential for diamond crystallisation at depths shallower than beneath cratons (P.H. Nixon, personal communication, 1988). Penecontemporaneous development of magma beneath the slab during a phase of tensional tectonics may provide a means of bringing diamond xenocrysts to the surface.

Finally, an important issue that needs to be properly addressed is the concept of potential temperature and the depths of mantle sources for the intraplate volcanism of both eastern Australia and New Zealand. Relatively high Tp values of about 1350° were proposed in Section 7.7.2 for mantle sources for the primitive basalts of the central volcanoes and some lava-field provinces — namely, the *ol*-tholeiitic basalts containing low abundances of incompatible elements. This Tp is higher than the estimated Tp of 1280°C for the mantle sources of normal MORB (McKenzie & Bickle, 1988). Even higher Tp values (greater than 1350°C) were considered in Section 7.3 where magma generation is considered in relation to detachment-faulting. Two relevant questions arise: (1) how can the high Tp values

be evaluated further, and (2) how did they originate?

Two mechanisms have been proposed to account for high Tp values: (1) that the break-up of eastern Gondwanaland, the opening of the Tasman Sea, and the intraplate volcanism are all related to a regional upwelling of hot mantle, or 'superswell', that is superimposed by mantle plumes (Section 7.7.2); (2) that the drawing up of hot mantle took place from great depth by rolling back and sinking a west-dipping, oceanic lithospheric slab to the east of Australia during the Mesozoic (for example, Section 7.3.3). Obviously, these two mechanisms are not mutually exclusive necessarily. However, whether the Tasman and Coral Sea system represent the backarc basins of such an arc-trench system is certainly open to question as evidence for the contemporaneous arc-trench system has yet to found.

7.8.6 Synthesis for eastern Australia

An attempt can be made to synthesise a simple geodynamic model for the Cainozoic volcanism of eastern Australia. The model is based on aspects of the several interpretations presented in the foregoing parts of the volume (see especially Section 1.7.3) and can be regarded as an alternative to the integrated detachment model presented in Section 7.3.

The basis of the model is that the Australian continent consisting of a major cratonic area in the centre and west and the relatively young Tasman Fold Belt down the eastern side, has been (and is continuing to be) deformed as a result of stresses in the Indo-Australian plate. The Tasman Fold Belt is underlain by hotter mantle than is the cratonic area, because it was adjacent to the mantle anomaly that produced late Mesozoic and early Tertiary seafloor spreading in the Tasman and Coral seas, and to an arc-trench system that was active at least as late as the early Mesozoic.

Deformation in the past is presumed to have included long periods, perhaps mainly 50–20 Ma ago, when extension was prevalent in the Tasman Fold Belt. This extension led to partial melting of the underlying hot mantle. Lava-field magmatism therefore took place along the fold belt, preferentially rising through trans-lithospheric fractures and other lines of weakness. Magma-residence times in the crust were short and therefore very few differentiated magmas were produced. The structure of the deep continental lithosphere is controlled by (1) reactivated fold-belt structures, (2) structures produced as a result of detachment faulting in the late Cretaceous, (3) structures produced by ongoing intraplate deformation. The surface distribution of the volcanic provinces therefore is controlled strongly by all of these influences which define the thermal and mechanical state of the continental lithosphere, in much the same way as visualised, for example, for the oceanic Marquesas Islands hotspot trace (McNutt et al., 1989).

The east-west orientation of the Newer and Older Volcanics lava fields in Victoria does not conform to the general north-south trend of the Tasman Fold Belt. A concept not considered in the previous chapters is that the Gippsland Basin in eastern Victoria (Fig. 1.1.5) represents the west-trending 'failed arm' of an originally three-branched rift system in the Tasman Sea (for example, Mutter & Jongsma, 1978). This failed arm may control lithospheric structure as far west as the Newer Volcanics province. Volcanism in Victoria therefore may arise from periods when intraplate extension took place along structures controlled by the failed arm, presumably as late as the early Quaternary, although apparently not at the present day as compression is indicated by *in situ* stress measurements.

The northeastern edge of the Australian continent began to override a hotspot about 35 Ma ago (perhaps 70 Ma ago if the Rockhampton centre is included). Well-defined continental hotspot traces are rare worldwide, presumably because subcontinental plumes are unable to readily penetrate the thick, cold, continental lithosphere and produce volcanism. This would have been the case for the east-Australian hotspot had it not under-ridden the hot Tasman Fold Belt which was actively producing lava-field volcanism and therefore providing opportunity for the hotspot volcanism to become well established.

The hotspot volcanism appeared on land at Hillsborough and for the next 18 Ma years or so it tracked southwards to what is now central New South Wales, following generally the zone of lava-field volcanism. The high-density and tholeiitic nature of the hotspot melts favoured the formation of crustal reservoirs, magma differentiation, and the production of anatectic melts. Lava-field volcanism began to die out about 20–15 Ma ago (except mainly in northern Queensland and western Victoria), and this also marks the time of waning activity at the major central volcanoes. The only hotspot volcanism since that time is the 12 Ma Canobolas activity and the minor Macedon-Trentham activity 6 Ma ago. The period 20–15 Ma therefore may correspond to a time when the current compressional tectonism of eastern Australia began to take over from a tensional regime, except apparently in northern Queensland and western Victoria where extensional tectonics may have persisted longer.

The east-Australian hotspot trace therefore represents the probably unique case of a hot fold belt at a continental margin that is (1) producing lava-field volcanism as a result of intraplate deformation, and (2) over-riding a hotspot that has created a substantial continental volcanic trace down the length of the fold belt.

7.8.7   Trans-Tasman relationships

The similarities between the intraplate volcanism of eastern Australia and that in the New Zealand region are quite striking. First, the age of the volcanism is the same — from late Mesozoic to Holocene. Second, the range of compositions of the erupted magmas is the same, from basalts of both alkaline and subalkaline types, through intermediate rocks such as mugearites and icelandites, to rhyolites of both peralkaline and non-peralkaline affinities. Leucitites and some other ultra-alkaline rock types found in eastern Australia have not been recorded from the New Zealand region, and no carbonatites such as are found in South Island

are known from the Cainozoic of eastern Australia. However, these rocks are unusual and volumetrically insignificant compared to the other rock compositions represented in the region.

Another similarity is the way in which the volcanoes of both regions can be classified as either central or lava-field type. Furthermore, the central volcanoes in each region have age progressions that may be indicative of linear thermal anomalies in the underlying mantle, as discussed above.

These close similarities are of course the reason why the intraplate volcanism of New Zealand has been included in this volume which has been concerned largely with eastern Australia. The similarities are such that a common tectonic cause is inferred for the two regions. However, a single tectonic cause affecting the entire region cannot be identified, although some broad generalisations can be made.

The opening of the Coral and Tasman seafloor-spreading complex 80–50 Ma ago appears to have been influenced the intraplate magmatism of eastern Australia, as inferred in a general sense throughout this volume. More specifically, the removal of lower crust and lithospheric upper mantle by detachment faulting has been proposed as a viable mechanism for introducing hot asthenosphere beneath the upper-plate margin of eastern Australia, so leading to the volcanism including, particularly, that of lava-field type (Section 7.3).

Tasman Sea detachment faulting is unlikely to have influenced intraplate volcanism in the Pacific-plate part of the New Zealand region because there the tectonics are quite distinct from events in the Tasman Sea. However, the early Tertiary volcanism on the Indo-Australian plate in South Island, such as the Arnott Basalt, took place close to the rifted margin on the Lord Howe Rise side (lower-plate side) of the Tasman Sea during the final stages of Tasman Sea opening (Section 4.9.3). This volcanism therefore is the counterpart to that of the same age on the opposite side of the Tasman Sea in eastern Australia. The extent to which other Cainozoic intraplate volcanism of Arnott Basalt type is represented on the Lord Howe Rise (other than the Lord Howe seamount chain) is unknown. Note, however, that a prediction of the detachment model (Section 7.4) is that volcanism is restricted to the upper-plate margin of the detachment and that none is expected on the lower plate. The Northland-Auckland volcanism in New Zealand is at the southern end of the Norfolk Ridge, east of the New Caledonia Basin, and therefore appears distinct from events directly connected with the Tasman Sea opening.

Similarities between the volcanism of eastern Australia and Marie Byrd Land, Antarctica, are so close that the Marie Byrd Land volcanism too can be recognised as a possible example of 'upper-plate' type (Section 7.4.3), although related to seafloor spreading between the Pacific and Antarctic plates rather than the Tasman Sea spreading. Volcanism on the Campbell Plateau therefore could be regarded as the counterpart of the Marie Byrd Land on the opposite side of the Antarctic Ridge. Again, however, there is a conflict with the detachment model for eastern Australia in which the volcanism should be restricted to one side (the upper-plate margin) of the detachment fault (a similar situation appears to exist in the North Atlantic; Table 7.4.1). Another apparent conflict is the absence of volcanism along the coastline of Antarctica opposite the 'lower plate' margin of southern Australia (but see Section 7.3.5).

The common phenomenon linking the separate areas of intraplate volcanism in the east-Australian, New Zealand, and Antarctic regions is the break-up of eastern Gondwanaland beginning in the late Mesozoic, and mantle melting related to a broad zones of tension developed down the Tasman Fold Belt in eastern Australia and on either side of the Indo-Australian plate boundary in the New Zealand region. A similar concept of tensional conditions induced by the passage of a hotspot was discussed in Sections 7.7.5–6 in relation to the leucitite-suite volcanoes of eastern Australia which are far removed from the track of the single hotspot.

7.8.8   Concluding remarks

Intraplate volcanism in the region evidently is related to the influences of upwellings of hot deep mantle connected with the break-up of Gondwanaland. Attempts to identify these separate influences are likely to be a continuing line of research for many years to come. However, a principal conclusion of this volume is that the term 'intraplate' should not carry the connotation that plate-boundaries have no part to play in understanding the generation of this type of volcanism, even though the volcanoes may be far removed from the nearest contemporaneous plate boundary.

A substantial database of information on the intraplate volcanism has been built up in recent years, and much of it has been summarised in this volume. Nevertheless, considerably more data are required on many aspects of the problem before a satisfactory general model can be adopted for the region. This volume therefore should be regarded as a starting point for ongoing research rather than an end-point.

References

Abbott, M.J. (1965). Petrology of the Nandewar volcano. Australian National University, unpublished PhD thesis.

Abbott, M.J. (1967). Aenigmatite from the groundmass of a peralkaline trachyte. American Mineralogist, 52, 1895–1901.

Abbott, M.J. (1969). Petrology of the Nandewar volcano, N.S.W., Australia. Contributions to Mineralogy and Petrology, 20, 115–134.

Abele, C., & R.W. Page (1974). Stratigraphic and isotopic ages of Tertiary basalts at Maude and Aireys Inlet, Victoria, Australia. Proceedings of the Royal Society of Victoria, 86, 143–150.

Abele, C., C.S. Gloe, J.B. Hocking, G. Holdgate, P.R. Kenley, C.R. Lawrence, D. Ripper, & W.F. Threlfall (1976). Tertiary. In: Douglas, J.G., & J.A. Ferguson (Editors), Geology of Victoria. Geological Society of Australia — Special Publication, 5.

Adams, C.J. (1975). Discovery of Precambrian rocks in New Zealand: age relations of the Greenland Group and Constant Gneiss, West Coast, South Island. Earth and Planetary Science Letters, 28, 98–104.

Adams, C.J. (1981). Migration of late Cenozoic volcanism in the South Island of New Zealand and the Campbell Plateau. Nature, 294, 153–155.

Adams, C.J. (1983). Age of the volcanoes and granite basement of the Auckland Islands, Southwest Pacific. New Zealand Journal of Geology and Geophysics, 26, 227–237.

Adams, C.J., & D.J. Cullen (1978). Potassium-argon ages of granites and metasediments from the Bounty Islands area, Southwest Pacific Ocean. Journal of the Royal Society of New Zealand, 8, 127–132.

Adams, C.J., P.A. Morris, & J.M. Beggs (1979). Age and correlation of volcanic rocks of Campbell Island and metamorphic basement of the Campbell Plateau, southwest Pacific. New Zealand Journal of Geology and Geophysics, 22, 679–691.

Adams, R.D. (1962). Thickness of the Earth's crust beneath the Campbell Plateau. New Zealand Journal of Geology and Geophysics, 5, 74–85.

Adams, R.D. (1968). The New Zealand Seismograph Network. Bulletin of the New Zealand Society for Earthquake Engineering, 1, 116–119.

Adrian, J. (1976). History of the Geological Survey of New South Wales. In: Johns, R.K. (Editor), History and Role of Government Geological Surveys in Australia. Adelaide, 31–45.

Ahren, J.L., & D.L. Turcotte (1979). Magma migration beneath an ocean ridge. Earth and Planetary Science Letters, 45, 115–122.

Alexander, P.O. (1981). Age and duration of Deccan volcanism: K–Ar evidence. In: Subbarao, K.V., & R.N. Sukheswala (Editors), Deccan Volcanism and Related Basalt Provinces in other parts of the World. Geological Society of India — Memoir, 3, 244–258.

Allen, J.M. (1974). Port Chalmers breccia and adjacent early flows of the Dunedin volcanic complex at Port Chalmers. New Zealand Journal of Geology and Geophysics, 17, 209–223.

Allen, M.E. (1972). The petrology and geochemistry of inclusion-bearing basalts and breccia from the Glen Alice area. Macquarie University, unpublished BA Honours thesis.

Alwar, M.A. (1960). Geology and structure of the middle Derwent Valley. Papers and Proceedings of the Royal Society of Tasmania, 94, 13–24.

Amoco Australia Petroleum Company (1986). Tilana No. 1. Well completion report.

Amundsen, H.E.F. (1986). Coexisting carbonatitic, ultramafic and mafic melts in the lithosphere: evidence from spinel lherzolite xenoliths, NW Spitsbergen. Fourth International Kimberlite Conference, Perth, 1986. Geological Society of Australia — Extended Abstracts, 16, 160–162.

Amundsen, H.E.F., W.L. Griffin, & S.Y. O'Reilly (1987). The lower crust and upper mantle beneath northwestern Spitsbergen: evidence from xenoliths and geophysics. Tectonophysics, 139, 169–185.

Andersen, T., S.Y. O'Reilly, & W.L. Griffin (1984). The trapped fluid phase in upper mantle xenoliths from Victoria, Australia. Contributions to Mineralogy and Petrology, 88, 72–85.

Andersen, T., W.L. Griffin, & S.Y. O'Reilly (1987). Primary sulphide melt inclusions in mantle-derived megacrysts and pyroxenites. Lithos, 20, 279–294.

Anderson, D.L. (1982). Hotspots, polar wander, Mesozoic convection and the geoid. Nature, 297, 391–393.

Anderson, D.L., & A.M. Dziewonski (1984). Seismic tomography. Scientific American, 251, 58–66.

Anderson, D.L., & C. Sammis (1970). Partial melting in the upper mantle. Physics of the Earth and Planetary Interiors, 3, 31–50.

Anderson, H.J. (1980). Geophysics of the Te Anau and Waiau Depressions, western Southland. New Zealand Department of Scientific and Industrial Research, Geophysics Division — Report, 180.

Andrews, E.C. (1910). Geographical unity of eastern Australia in late and post Tertiary time, with applications to biological problems. Journal and Proceedings of the Royal Society of New South Wales, 44, 420–480.

Angas, G.F. (1847). South Australia Illustrated. London.

Archbald, P.N. (1979). Abundance and dispersion of some compatible volatile and siderophile elements in the mantle. Australian National University, unpublished MSc thesis.

Archer, M., & M. Wade (1976). Results of the Ray E. Lemley Expeditions, Part I. The Allingham Formation and a new Pliocene vertebrate fauna from northern Queensland. Memoir of the Queensland Museum, 17, 379–397.

Arculus, R.J. (1974). Solid solution characteristics of spinels: Pleonaste-chromite-magnetite compositions in some island-arc basalts. Carnegie Institution of Washington Year Book, 73, 322–327.

Arculus, R.J. (1981). Island arc magmatism in relation to the evolution of the crust and mantle. Tectonophysics, 75, 113–133.

Arculus, R.J., & D. Smith (1979). Eclogite, pyroxenite and amphibolite inclusions in the Sullivan Buttes latite, Chino Valley, Yavapai County, Arizona. In: Boyd, F.R., & H.O.A. Meyer (Editors), The Mantle Sample: Inclusions in Kimberlites and other Volcanics. American Geophysical Union, Washington, D.C., 309–317.

Arculus, R.J., J. Ferguson, B.W. Chappell, D. Smith, M.T. McCulloch, I. Jackson, H.D. Hensel, S.R. Taylor, J. Knutson, & D.A. Gust (in press). Trace element and isotopic characteristics of eclogites and other xenoliths derived from the lower continental crust of southeastern Australia and southwestern Colorado Plateau. In: Smith, D.C. (Editor), Eclogites and Eclogite-Facies Rocks. Elsevier, Amsterdam, 335–386.

Ashcroft, J. (1986a). The Kerikeri Volcanics: a basalt-pantellerite association in Northland. In: Smith, I.E.M. (Editor), Late Cenozoic Volcanism in New Zealand. Royal Society of New Zealand — Bulletin, 23, 48–63.

Ashcroft, J. (1986b). The Kerikeri volcanics: a geochemical study. University of Auckland, unpublished MSc thesis.

Atkinson, A., T.J. Griffin, & P.J. Stephenson (1975). A major lava tube system from Undarra volcano, north Queensland. Bulletin Volcanologique, 39, 1–28.

Austin, P.M., R.C. Sprigg, & J.C. Braithwaite (1973a). Structure and petroleum potential of eastern Chatham Rise, New Zealand. American Association of Petroleum Geologists Bulletin, 57, 477–497.

Austin, P.M., R.C. Sprigg, & J.C. Braithwaite (1973b). Structural development of the eastern Chatham Rise and of the New Zealand region. In: Fraser, R. (Compiler), Oceanography of the South Pacific. New Zealand National Committee for UNESCO, Wellington, 201–215.

Aziz-ur-Rahman (1971). Palaeomagnetic secular variation for recent normal and reversed epochs, from the Newer Volcanics of Victoria, Australia. Geophysical Journal of the Royal Astronomical Society, 24, 255–269.

Aziz-ur-Rahman, & I. McDougall (1972). Potassium-argon ages on the Newer Volcanics of Victoria. Proceedings of the Royal Society of Victoria, 85, 61–70.

BMR (1976a). Gravity map of Australia, scale 1:5 000 000. Bureau of Mineral Resources, Australia.

BMR (1976b). Magnetic map of Australia, residuals of total intensity, scale 1:2 500 000. Bureau of Mineral Resources, Australia.

Bailey, D.K. (1978). Continental rifting and mantle degassing. In: Neumann, E.-R., & I.B. Ramberg (Editors), Petrology and Geochemistry of Continental Rifts. Reidel, Dordrecht, 1–13.

Baillie, P.W. (1984). A radiometric age for volcanic rocks at Musselroe Bay, northeastern Tasmania. Tasmanian Department of Mines — Report, 1984/46.

Baillie, P.W. (1985). A Palaeozoic suture in eastern Gondwanaland. Tectonics, 4, 653–660.

Baillie, P.W. (1986a). Radiometric ages for Circular Head, and the Green Hills basalt, north-western Tasmania. Tasmanian Department of Mines — Report, 1986/39.

Baillie, P.W. (1986b). A radiometric age for Moriaty Basalt, northwestern Tasmania. Tasmanian Department of Mines — Report, 1986/38.

Baillie, P.W. (1987). Geology and exploration history of the Tasmanian sector of the Gippsland Basin. Tasmanian Department of Mines — Report, 1987/23.

Baker, B.H. (1987). Outline of the petrology of the Kenya Rift alkaline province. In: Fitton, J.G., & B.G.J. Upton (Editors), Alkaline Igneous Rocks. Geological Society Special Publicication, 30, 293–311.

Baker, B.H., G.G. Goles, W.P. Leeman, & M.M. Lindstrom (1977). Geochemistry and petrogenesis of a basalt-benmoreite-trachyte suite from the southern part of the Gregory Rift, Kenya. Contributions to Mineralogy and Petrology, 64, 303–332.

Baker, C.K. (1983). Phytoclasts in metamorphic rocks. University of Newcastle, unpublished PhD thesis.

Baker, P.E., I.G. Gass, P.G. Harris, & R.W. Le Maitre (1964). The volcanological report of the Royal Society expedition to Tristan da Cunha, 1962. Philosophical Transactions of the Royal Society, 256, 439–578.

Ballance, P.F., & I.E.M. Smith (1982). Walks through Auckland's geological past: a guide to the geological formations of Rangitoto, Motutapu and Motiuhe Islands. Geological Society of New Zealand — Guidebook 5.

Ballance, P.F., & K.B. Sporli (1979). Northland allochthon. Journal of the Royal Society of New Zealand, 9, 259–275.

Ballance, P.F., J.R. Pettinga, & C. Webb (1982). A model of the Cenozoic evolution of northern New Zealand and adjacent areas of the Southwest Pacific. In: Packham, G.H. (Editor), The Evolution of the India-Pacific Plate Boundaries. Tectonophysics, 87, 37–48.

Banks, M.R. (1965). Geology and mineral deposits. In: Davies, J.L. (Editor), Atlas of Tasmania. Lands and Surveys Department, Hobart, 12–17.

Banks, M.R. (1972). General geology. In: Banks M.R. (Editor), The Lake Country of Tasmania. Royal Society of Tasmania, Hobart, 25–33.

Baragwanath, W. (1948). Diamonds in Victoria. Department of Mines, Victoria — Mining and Geological Journal, 3, 12–16.

Barberi, F., & J. Varet (1970). The Erta'Ale volcanic range (Afar, Ethiopia). Bulletin Volcanologique, 34, 848–917.

Barberi, F., & J. Varet (1977). Volcanism of Afar: small scale plate tectonics implications. Bulletin of the Geological Society of America, 88, 1251–1266.

Barberi, F., S. Borsi, G. Ferrara, G. Marinelli., & J. Varet (1970). Relations between tectonics and magmatology in the northern Danakil Depression (Ethiopia). Philosophical Transactions of the Royal Society of London, A267, 293–311.

Barberi, F., H. Tazieff, & J. Varet (1972). Volcanism in the Afar Depression: its tectonic and magmatic significance. Tectonophysics, 15, 19–29.

Barberi, F., R. Santacroce, G. Ferrara, M. Treuil, & J. Varet (1975). A transitional basalt-pantellerite sequence of fractional crystallisation, the Boina Centre (Afar Rift, Ethiopia). Journal of Petrology, 16, 22–56.

Barbetti, M., & M.J. Sheard (1981). Palaeomagnetic results from Mounts Gambier and Schank, South Australia. Journal of the Geological Society of Australia, 28, 385–394.

Barley, M.E. (1987). Origin and evolution of mid-Cretaceous, garnet-bearing intermediate and silicic volcanics from Canterbury, New Zealand. In: Weaver, S.D., & R.W. Johnson (Editors), Tectonic Controls on Magma Chemistry. Elsevier, Amsterdam, 247–267.

Barley, M.E., S.D. Weaver, & J.R. de Laeter (1988). Strontium isotope composition and geochronology of intermediate-silicic volcanics, Mt Somers and Banks Peninsula, New Zealand. New Zealand Journal of Geology and Geophysics, 31, 197–206.

Barnes, I., & G.A. McCoy (1979). Possible role of mantle-derived CO_2 in causing two 'phreatic' explosions in Alaska. Geology, 7, 434–435.

Barreiro, B.A., & A.F. Cooper (1987). A Sr, Nd and Pb isotope study of alkaline lamprophyres and related rocks from Westland and Otago, South Island, New Zealand. In: Morris, E.M., & J.D. Pasteris (Editors), Mantle

Metasomatism and Alkaline Magmatism. Geological Society of America — Special Paper, 215, 115–125.

Barron, L.M. (1987). Summary of petrology and chemistry of rocks from the sapphire project. New South Wales Geological Survey — Report, 1987/9 (GS1987/050).

Barton, C.E., & M.W. McElhinny (1980). Ages and ashes in lake floor sediment cores from Valley Lake, Mt Gambier, South Australia. Transactions of the Royal Society of South Australia, 104, 161–165.

Barton, C.E., & M.W. McElhinny (1981). A 10 000 yr geomagnetic secular variation record from three Australian maars. Geophysical Journal of the Royal Astronomical Society, 67, 465–485.

Barton, C.E., & H.A. Polach (1980). ¹⁴C ages and magnetic stratigraphy in three Australian maars. Radiocarbon, 22, 728–739.

Bartrum, J.A. (1925). The igneous rocks of North Auckland, New Zealand. Verhandelingen van het Geologisch — Mijnbonwkundig Genostschap voor Nederland en Kolenien's Gravenhage. Geologische Serie, 8, 1–16.

Bartrum, J.A. (1949). Geology of Auckland City and environs. In: Auckland Handbook, Seventh Pacific Science Congress, 5–11.

Basaltic Volcanism Study Project (1981). Basaltic Volcanism on the Terrestrial Planets. Pergamon, New York.

Bass, G. (1797). Historical Records of New South Wales, 3 (1796–99), December, 313.

Basu, A.R. (1978). Trace elements and Sr-isotopes in some mantle-derived hydrous minerals and their significance. Geochimica et Cosmochimica Acta, 42, 659–668.

Basu, A.R. (1980). Jointed blocks of peridotite xenoliths in basalts and mantle dynamics. Nature, 284, 612–613.

Baudin, N. (1802). The Journal of Nicholas Baudin, translated by C. Cornell. Libraries Board of South Australia, Adelaide, 1974.

Bean, J.M. (1974). The geology and petrology of the Mullaley area of New South Wales. Journal of the Geological Society of Australia, 21, 63–72.

Bean, J.M. (1975). Petrology and petrochemistry of igneous rocks in the Mullaley area of New South Wales. Journal and Proceedings of the Royal Society of New South Wales, 108, 131–146.

Beane, J.E., C.A Turner, P.R. Hooper, K.V. Subbarao, & J.N. Walsh (1986). Stratigraphy, composition and form of the Deccan Basalts, Western Ghats, India. Bulletin of Volcanology, 48, 61–83.

Beattie, R.D. (1978). A compilation and interpretation of aeromagnetic data from Bass Strait and Tasmania. Commonwealth Scientific and Industrial Research Organization, Australia, Mineral Research Laboratories, Division of Mineral Physics — Investigation Report, 124.

Beavis, F.C. (1962). The geology of the Kiewa area. Proceedings of the Royal Society of Victoria, 75, 349–510.

Beeson, M.H. (1976). Petrology, mineralogy, and geochemistry of the east Molokai volcanic series, Hawaii. United States Geological Survey — Professional Paper, 961, 1–53.

Beggs, J.M. (1978). Geology of the metamorphic basement and Late Cretaceous to Oligocene sedimentary sequence of Campbell Island, Southwest Pacific Ocean. Journal of the Royal Society of New Zealand, 8, 161–177.

Bell, J.M., & E.C. Clarke (1909). A geological reconnaissance of northern New Zealand. Transactions of the Royal Society of New Zealand, 42, 613–624.

Bellieni, G., P. Comin-Chiaramonti, L.S. Marques, A.J. Melfi, A.J.R. Nardy, C. Papatrechas, E.M. Piccirillo, A. Roisenberg, & D. Stolfa (1986). Petrogenetic aspects of acid and basaltic lavas from Paraná Plateau (Brazil):

geological, mineralogical and petrochemical relationships. Journal of Petrology, 27, 915–944.

Bembrick, C.S., C. Herbert, E. Scheibner, & J. Stuntz (1973). Structural subdivision of the New South Wales portion of the Sydney-Bowen Basin. Geological Survey of New South Wales — Quarterly Notes, 11, 1–13.

Bennett, D.J., & F.E.M. Lilley (1973). Electrical conductivity structure in the south-east Australian region. Geophysical Journal of the Royal Astronomical Society, 37, 191–206.

Bennett, F.D. (1974). On volcanic ash formation. American Journal of Science, 274, 1–120.

Benson, W.N. (1911). The volcanic necks of Hornsby and Dundas near Sydney. Journal and Proceedings of the Royal Society of New South Wales, 44, 495–555.

Benson, W.N. (1941a). Cainozoic petrographic provinces in New Zealand and their residual magmas. American Journal of Science, 239, 537–552.

Benson, W.N. (1941b). The basic igneous rocks of eastern Otago and their tectonic environment. Transactions of the Royal Society of New Zealand, 71, 208–222.

Benson, W.N. (1942a). The basic igneous rocks of eastern Otago and their tectonic environment. Transactions of the Royal Society of New Zealand, 72, 85–110.

Benson, W.N. (1942b). The basic igneous rocks of eastern Otago and their tectonic environment. Transactions of the Royal Society of New Zealand, 72, 160–185.

Benson, W.N. (1943). The basic igneous rocks of eastern Otago and their tectonic environment. Transactions of the Royal Society of New Zealand, 73, 116–138.

Benson, W.N. (1944). The basic igneous rocks of eastern Otago and their tectonic environment. Part IV. The mid-Tertiary basalts, tholeiites and dolerites of north-eastern Otago. Section B: Petrology, with special reference to the crystallization of pyroxene. Transactions of the Royal Society of New Zealand, 74, 71–123.

Benson, W.N. (1945). The basic igneous rocks of eastern Otago and their tectonic environment. Transactions of the Royal Society of New Zealand, 75, 288–318.

Benson, W.N. (1946). The basic igneous rocks of eastern Otago and their tectonic environment. Transactions of the Royal Society of New Zealand, 76, 1–36.

Benson, W.N. (1959). Dunedin volcanic complex. In: Fleming C.A. (Editor), Lexique Stratigraphique International, volume 6, Oceanie, Fascsimile 4, New Zealand, 91–93.

Benson, W.N. (1968). Dunedin District. New Zealand Geological Survey — Miscellaneous Serial, 1:50 000 Map. New Zealand Department of Scientific and Industrial Research.

Best, J.G. (1960). Some Cainozoic basaltic volcanoes in north Queensland. Bureau of Mineral Resources, Australia — Record, 78.

Betton, P.J., & L. Civetta (1984). Strontium and neodymium isotopic evidence for the heterogeneous nature and development of the mantle beneath Afar (Ethiopia). Earth and Planetary Science Letters, 71, 59–70.

Bezant, C. (1985). Geothermometry and seismic properties of upper mantle lherzolites from eastern Australia. Macquarie University, unpublished BSc Honours thesis.

Binns, R.A. (1966). Granite intrusions and regional metamorphic rocks of Permian age from the Wongwibinda district, north-eastern New South Wales. Journal and Proceedings of the Royal Society of New South Wales, 99, 5–36.

Binns, R.A. (1969). High-pressure megacrysts in basanitic lavas near Armidale, New South Wales. American Journal of Science, 267A, 33–49.

Binns, R.A., M.B. Duggan, & J.F.G. Wilkinson (1970). High pressure megacrysts in alkaline lavas from northeastern New South Wales. American Journal of Science, 269, 132–168.

Birch, F. (1960). The velocity of compressional waves in rocks to 10 kbar, 1. Journal of Geophysical Research, 65, 1083–1102.

Birch, F. (1961). The velocity of compressional waves in rocks to 10 kbar, 2. Journal of Geophysical Research, 66, 2199–2224.

Birch, W.D. (1976). Mineralogical note. The occurrence of a leucite-bearing lava at Cosgrove, Victoria. Journal of the Geological Society of Australia, 23, 435–437.

Birch, W.D. (1979). Mineralogy and geochemistry of the leucitite at Cosgrove, Victoria. Journal of the Geological Society of Australia, 25, 369–385.

Birch, W.D. (1980). Mineralogy of vesicles in an olivine leucitite at Cosgrove, Victoria, Australia. Mineralogical Magazine, 43, 597–603.

Birch, W.D. (1984). Late-stage crystallization trends in basic and alkaline volcanic rocks in Victoria. Geological Society of Australia — Abstracts, 12, 61.

Bishop, D.G., J.D. Bradshaw, & C.A. Landis (1985). Provisional terrane map of South Island, New Zealand. In: Howell, D.G. (Editor), Tectonostratigraphic Terranes of the Circum-Pacific Region. Circum-Pacific Council for Energy and Mineral Resources — Earth Science Series, 1, 515–521.

Bishop, P. (1984). Oligocene and Miocene volcanic rocks and quartzose sediments of the Southern Tablelands, New South Wales: definitions of stratigraphic units. Journal and Proceedings of Royal Society of New South Wales, 117, 113–117.

Bishop, P. (1985a). Early Miocene flow foot breccia from the upper Lachlan Valley, New South Wales: characteristics and significance. Australian Journal of Earth Sciences, 32, 107–113.

Bishop, P. (1985b). Southeast Australian late Mesozoic and Cenozoic denudation rates: a test for late Tertiary increases in continental denudation. Geology, 13, 479–482.

Bishop, P. (1986). Horizontal stability of the Australian continental drainage divide in south central New South Wales during the Cainozoic. Australian Journal of Earth Sciences, 33, 295–307.

Bishop, P. (1987). River profiles and highland evolution. In: Galloway, R.W. (Compiler), The Age of Landforms in Eastern Australia: Conference Summary and Field Trip Guide (7–12 September 1986). Commonwealth Scientific and Industrial Research Organization, Australia, Division of Water and Land Resources — Technical Memorandum, 87/2, 17–18.

Bishop, P. (in press). The eastern highlands of Australia: the evolution of an intra-plate highland belt. Progress in Physical Geography.

Bishop, P., & R.W. Young (1980). Discussion: on the Cainozoic uplift of the southeastern Australian highland. Journal of the Geological Society of Australia, 27, 117–119.

Bishop, P., R.W. Young, & I. McDougall (1985). Stream profile change and longterm landscape evolution: early Miocene and modern rivers of the east Australian highland crest, central New South Wales, Australia. Journal of Geology, 93, 455–474.

Bizward, H., F. Barberi, & J. Varet (1980). Mineralogy and petrology of Erta'Ale and Boina volcanic series, Afar Rift, Ethiopia. Journal of Petrology, 21, 401–436.

Black, L.P., T.H. Bell, M.J. Rubenach, & I.W. Withnall (1979). Geochronology of discrete structural-metamorphic events in a multiply deformed Precambrian terrain. Tectonophysics, 54, 103–137.

Black, P.M., & R.N. Brothers (1965). Olivine nodules in olivine nephelinite from Tokatoka, Northland. New Zealand Journal of Geology and Geophysics, 8, 62–80.

Blackburn, E., L. Wilson, & R.S.J. Sparks (1976). Mechanism and dynamics of strombolian activity. Journal of the Geological Society of London, 132, 429–440.

Blackburn, G., G.B. Allison, & F.W.J. Leaney (1982). Further evidence on the age of tuff at Mount Gambier, South Australia. Transactions of the Royal Society of South Australia, 106, 163–167.

Blackburn, G., G.B. Allison, & F.W.J. Leaney (1984). Errata. Transactions of the Royal Society of South Australia, 108, 130.

Blake, F. (1958). Geological atlas 1 mile series. Tasmanian Department of Mines — Sheet 76 (8412N).

Blake, M.C. Jr, & C.A. Landis (1973). The Dun Mountain ultramafic belt: Permian oceanic crust and upper mantle in New Zealand. United States Geological Survey — Journal of Research, 1, 529–534.

Blong, R.J. (1984). Volcanic Hazards — A Sourcebook on the Effects of Eruptions. Academic Press, Sydney.

Bloomer, A.G., & P.H. Nixon (1973). The geology of the Letseng-Laterae kimberlite pipes. In: Nixon, P.H. (Editor), Lesotho Kimberlites. Lesotho National Development Corporation, Maseru, 20–32.

Boettcher, A.L., & J.R. O'Neil (1980). Stable isotope, chemical and petrographic studies of high-pressure amphiboles and micas: evidence for metasomatism to mantle source regions of alkali basalts and kimberlites. American Journal of Science, 280A, 594–621.

Bolger, P. (1977). Explanatory notes on the Meredith and You Yangs 1:50 000 geological maps. Geological Survey of Victoria — Report, 1977/14.

Bonatti, E., C.G.A. Harrison, D.E. Fisher, J. Honnorez, J.-G. Schilling, J.J. Stipp, & M. Zentilli (1977). Easter volcanic chain (southeast Pacific): a mantle hot line. Journal of Geophysical Research, 82, 2457–2478.

Bonnichsen, B. (1982). Rhyolite lava flows in the Bruneau-Jarbridge eruptive center, south western Idaho. In: Bonnichsen, B., & R.M. Brechenridge (Editors), Cenozoic Geology of Idaho. Idaho Bureau of Mines — Bulletin, 26, 283–320.

Bosworth, W. (1987). Off-axis volcanism in the Gregory rift, east Africa: implications for models of continental rifting. Geology, 15, 397–400.

Bott, M.H.P. (1982). The Interior of the Earth: Its Structure, Constitution and Evolution. Arnold, London.

Bott, M.H.P., & J. Tuson (1973). Deep structure beneath the Tertiary volcanic regions of Skye, Mull and Ardnamurchan, north-west Scotland. Nature, 242, 114–116.

Boutakoff, N. (1963). The geology and geomorphology of the Portland area. Geological Survey of Victoria — Memoir, 22.

Bowen, F.E. (1974). The Parahaki volcanics and their associated clays. New Zealand Department of Science and Industrial Research — Bulletin, 215.

Bowen, K.G. (1975). Potassium-argon dates — determinations carried out by the Geological Survey of Victoria. Geological Survey of Victoria — Report, 1975/3.

Bowin, C., W. Warsi, & J. Milligan (1981). Free-air gravity anomaly map of the world. Geological Society of America — GSA MC–45.

Bowler, J.M. (1963). Tertiary stratigraphy and sedimentation in the Geelong-Maude area, Victoria. Proceedings of the Royal Society of Victoria, 76, 69–137.

Bowler, J.M., & T. Hamada (1971). Late Quaternary stratigraphy and radiocarbon chronology of water level fluctuation in Lake Keilambete, Victoria. Nature, 237, 330–332.

Boyd, F.R., J.J. Gurney, & S.H. Richardson (1985). Evidence for a 150–200 km thick lithosphere from diamond inclusion thermobarometry. Nature, 315, 387–389.

Bradley, G., E.K. Yoo, & P. West (1984). Geological mapping of Mesozoic to Tertiary diatremes and Illawarra Coal Measures east of Rylstone. Coal Geology Branch, New South Wales Department of Mineral Resources — File, 1984/203.

Bradshaw, J.D., C.J. Adams, & P.B. Andrews (1981). Carboniferous to Cretaceous on the Pacific margin of Gondwana: the Rangitata Phase of New Zealand. In: Cresswell, M.M., & P. Vella (Editors), Gondwana Five. Balkema, Rotterdam, 217–221.

Branagan, D.F. (1969). Cainozoic rocks outside the Murray Basin. Journal of the Geological Society of Australia, 16, 544–545.

Branagan, D.F. (1975). Samuel Stutchbury and Reverend W.B. Clarke. Not quite equal and opposite. In: Stanbury, P.J. (Editor), 100 Years of Australian Scientific Explorations. Holt, Rinehart & Winston, 89–98.

Branagan, D.F. (1983). The Sydney Basin and its vanished sequence. Journal of the Geological Society of Australia, 30, 75–84.

Branagan, D.F. (1986). Strzelecki's geological map of southeastern Australia: an electric synthesis. Historical Records of Australian Science, 6, 375–392.

Branch, C.D. (1966). Volcanic cauldrons, ring complexes, and associated granites of the Georgetown Inlier, Queensland. Bureau of Mineral Resources, Australia — Bulletin, 76.

Brey, G.P. (1978). Origin of olivine melilitites — chemical and experimental constraints. Journal of Volcanology and Geothermal Research, 3, 61–88.

Brey, G.P., & D.H. Green (1975). The role of CO_2 in the genesis of olivine melilitite. Contributions to Mineralogy and Petrology, 49, 93–103.

Brey, G.P., & D.H. Green (1976). Solubility of CO_2 in olivine melilitite at high pressures and role of CO_2 in the Earth's upper mantle. Contributions to Mineralogy and Petrology, 55, 217–230.

Briggs, R.M. (1986). Volcanic rocks of the Waikato region, western North Island, and some possible petrologic and tectonic constraints on their origin. Royal Society of New Zealand — Bulletin, 23, 76–91.

Briggs, R.M., & G.G. Goles (1984). Petrological and trace element geochemical features of the Okete volcanics, western North Island, New Zealand. Contributions to Mineralogy and Petrology, 86, 77–88.

Bristow, J.W., & E.P. Saggerson (1983). A review of Karoo vulcanicity in southern Africa. Bulletin of Volcanology, 46, 135–159.

Brodie, J.W. (1957). Marine geology of the Chatham Islands area. New Zealand Department of Scientific and Industrial Research — Bulletin 122, 28–30.

Brodie, J.W., & T. Hatherton (1958). The morphology of Kermadec and Hikurangi trenches. Deep-Sea Research, 5, 18–28.

Brooks, C., D.E. James, & S.R. Hart (1976). Ancient lithosphere: its role in young continental volcanism. Science, 193, 1086–1094.

Brothers, R.N. (1986). Upper Tertiary and Quaternary volcanism and subduction zone regression, North Island, New Zealand. Journal of the Royal Society of New Zealand, 16, 275–298.

Brothers, R.N., & M. Delaloye (1982). Obducted ophiolites of North Island, New Zealand: origin, age, emplacement and tectonic implications for Tertiary and Quaternary volcanicity. New Zealand Journal of Geology and Geophysics, 25, 257–274.

Brothers, R.N., & J. Golson (1959). Geological and archaeological interpretation of a section in Rangitoto ash on Motutapu Island, Auckland. New Zealand Journal of Geology and Geophysics, 2, 569–577.

Brotzu, P., M.T. Ganzerli-Valentini, L. Morbidelli, E.M. Piccirillo, R. Stella, & G. Traversa (1981). Basaltic volcanism in the northern sector of the main Ethiopian Rift. Journal of Volcanology and Geothermal Research, 10, 365–382.

Broughton, P.L. (1979). Economic geology of Australian gemstone deposits. Minerals Science Engineering, 11, 3–21.

Brown, A.V., M.P. McClenaghan, and others (1977). Geological Atlas 1:50 000 series, Sheet 32 (8415N), Ringaroonia. Tasmanian Department of Mines — Explanatory Report.

Brown, B.R. (1976). Bass Basin: some aspects of the petroleum geology. In: Leslie, R.B., H.J. Evans, & C.L. Knight (Editors), Economic Geology of Australia and Papua New Guinea, 3. Petroleum. Australasian Institute of Mining and Metallurgy — Monograph, 7, 67–82.

Brown, C.M. (1983). Discussion: a Cainozoic history of Australia's southeast highlands. Journal of the Geological Society of Australia, 30, 483–486.

Brown, D.A. (1955). The geology of Siberia Hill and Mount Dasher, north Otago. Transactions of the Royal Society of New Zealand, 83, 347–372.

Brown, E.H. (1964). The geology of the Mount Stoker area, eastern Otago. Part 2. New Zealand Journal of Geology and Geophysics, 7, 192–204.

Brown, F.H. (1970). Zoning in some volcanic nephelines. American Mineralogist, 55, 1670–1680.

Brown, M.C. (1983). Discussion: origin of coastal lowlands near Ulladulla, New South Wales. Journal of the Geological Society of Australia, 30, 247–248.

Brown, M.C., I. Clarke, K.G. McQueen, & G. Taylor (1988). Early Tertiary volcanism in the southern Monaro region, New South Wales. Ninth Australian Geological Convention, Geological Society of Australia — Abstracts, 21.

Browne, P.R.L., G.W. Coulter, M.A. Grant, G.W. Grindley, J.V. Lawless, G.L. Lyon, W.J.P. Macdonald, R. Robinson, D.J. Sheppard, & D.N.B. Skinner (1981). The Ngawha geothermal area. New Zealand Department of Scientific and Industrial Research — Geothermal Report, 7.

Browne, W.R. (1925). Notes on the petrology of the Prospect intrusion. Journal and Proceedings of the Royal Society of New South Wales, 58 (for 1924), 240–254.

Browne, W.R. (1933). Presidential address. An account of post Palaeozoic igneous activity in New South Wales. Journal and Proceedings of the Royal Society of New South Wales, 67, 9–95.

Browne, W.R. (1950, Editor). Geology of the Commonwealth of Australia by the late Sir T.W. Edgeworth David. Arnold, London.

Browne, W.R. (1969). Geomorphology. General notes. Journal of the Geological Society of Australia, 16, 559–569.

Bryan, W.B. (1969a). An olivine gabbro-microsyenite intrusion from the Carnarvon Range, Queensland, Australia. Carnegie Institution of Washington Year Book, 68, 247–250.

Bryan, W.B. (1969b). Volcanic rocks from the Bunya Mountains, Queensland, Australia. Carnegie Institute of Washington Year Book, 68, 244–247.

Bryan, W.B. (1983). Petrogenetic relations of basalts and related lavas, Bunya Mountains and Carnarvon Range, Queensland. Proceedings of the Royal Society of Queensland, 94, 69–84.

Bryan, W.B., & N.C. Stevens (1973). Holocrystalline pantellerite from Mt Ngun-Ngun, Glass House Mountains, Queensland, Australia. American Journal of Science, 273, 947–957.

Buck, R. (1986). Small-scale convection induced by passive rifting: the cause for uplift of rift shoulders. Earth and Planetary Science Letters, 77, 362–372.

Buddington, A.F., & D.H. Lindsley (1964). Iron-titanium oxide minerals and synthetic equivalents. Journal of Petrology, 5, 310–357.

Buller-Murphy, D. (1958). An attempt to eat the moon, and other stories recounted from the Aborigines. Georgian House, Melbourne.

Burns, K.L. (1965). Geological atlas 1 mile series, Sheet 29 (8115N), Devonport. Tasmanian Department of Mines — Explanatory Report.

Burr, T. (1846). Remarks on the Geology of South Australia. Murray, Adelaide.

Burton, S. (1973). The Milton monzonite. University of Sydney, unpublished BSc Honours thesis.

Burwell, A.D.M. (1975). Rb-Sr isotope geochemistry of lherzolites and their constituent minerals from Victoria, Australia. Earth and Planetary Science Letters, 28, 69–78.

Calhaem, I.M., A.J. Haines, & M.A. Lowry (1977). An intermediate-depth earthquake in the central region of the South Island used to determine a local crustal thickness. New Zealand Journal of Geology and Geophysics, 20, 353–361.

Camp, V.E., P.R. Hooper, D.A. Swanson, & T.L. Wright (1982). Columbia River basalt in Idaho: physical and chemical characteristics, flow distribution, and tectonic implications. In: Bonnichsen, B., & R.M. Breckenridge (Editors), Cenozoic Geology of Idaho. Idaho Bureau of Mines — Bulletin, 26, 55–75.

Campbell, I.H. (1985). The difference between oceanic and continental tholeiites: a fluid dynamic explanation. Contributions to Mineralogy and Petrology, 91, 37–43.

Cande, S.C., & J.C. Mutter (1982). Revised identification of the oldest sea floor spreading anomalies between Australia and Antarctica. Earth and Planetary Science Letters, 58, 151–160.

Cao, R.L., & S.H. Zhu (1987). Mantle xenoliths and alkali-rich host rocks in eastern China. In: Nixon, P.H. (Editor), Mantle Xenoliths. Wiley, Chichester, 167–185.

Carey, S.W. (1938). The morphology of New Guinea. Australian Geographer, 3, 3–51.

Carey, S.W. (1958). The isostrat, a new technique for the analysis of the structure of the Tasmanian dolerite. In: Dolerite — A Symposium. University of Tasmania, 130–164.

Carlson, J.R., J.A. Grant-Mackie, & K.A. Rodgers (1980). Stratigraphy and sedimentology of the Coalgate area, Canterbury, New Zealand. New Zealand Journal of Geology and Geophysics, 23, 179–192.

Carmichael, I.S.E. (1964). The petrology of Thingmuli, a Tertiary volcano in eastern Iceland. Journal of Petrology, 5, 435–460.

Carmichael, I.S.E. (1967). The iron-titanium oxides of salic volcanic rocks and their associated ferromagnesian silicates. Contributions to Mineralogy and Petrology, 14, 36–64.

Carmichael, I.S.E., F.J. Turner, & J. Verhoogen (1974). Igneous Petrology. McGraw-Hill, New York.

Carne, J.E. (1908). Geology and mineral resources of the Western Coalfield. Geological Survey of New South Wales — Memoir, 6.

Carter, R.M., M.D. Hicks, R.J. Norris, & I.M. Turnbull (1978). Sedimentation patterns in an ancient arc-trench-ocean basin complex: Carboniferous to Jurassic Rangitata orogen, New Zealand. In: Stanley, D.J., & G. Kelling (Editors), Sedimentation in Submarine Canyons, Fans and Trenches. Dowden, Hutchinson & Ross, Pennsylvania, 340–361.

Cas, R.A.F., & C.A. Landis (1987). A debris flow deposit with multiple plug flow channels and associated side accretion deposits. Sedimentology, 34, 901–910.

Cas, R.A.F., & J.V. Wright (1987). Volcanic Successions: Modern and Ancient. A Geological Approach to Processes, Products and Successions. Allen & Unwin, London.

Cas, R.A.F., R. Allen, S. Bull, & R. Sukyar (1984a). The Red Rock volcanic complex. In: Cas, R.A.F., E.B. Joyce, I.A. Nicholls, & R.C. Price (Editors), Volcanics Workshop 1984: Continental Volcanism, Western Victoria. Centre of Advanced Studies (Geoscience), Melbourne University.

Cas, R.A.F., R. Allen, S. Bull, I. Nicholls, & R. Sukyar (1984b). The Bridgewater volcanic complex. In: Cas, R.A.F., E.B. Joyce, I.A. Nicholls, & R.C. Price (Editors), Volcanics Workshop 1984: Continental Volcanism, Western Victoria. Centre of Advanced Studies (Geoscience), Melbourne University.

Cas, R.A.F., E. Fordyce, Y. Kawachi, & C.A. Landis (1986). The Eocene-Oligocene Oamaru volcano: an outstanding facies model for surtseyan volcanoes, possible subaqueous base surges, mass-flow deposits heralding sector collapse, and the importance of contemporaneous epiclastic deposits. International Volcanological Congress, New Zealand — Abstracts, 96.

Cassidy, J., C.A. Locke, & I.E. Smith (1986). Volcanic hazard in the Auckland region. In: Gregory, J.G., & W.A. Watters (Editors), Volcanic Hazards Assessment in New Zealand. New Zealand Geological Survey — Record, 10, 60–64.

Cawood, P.A. (1976). Cambro-Ordovician strata, northern New South Wales. Search, 7, 317–318.

Cawood, P.A. (1984). The development of the South West Pacific margin of Gondwana: correlations between the Rangitata and New England orogens. Tectonics, 3, 539–553.

Cawthorn, R.G. (1975). Degrees of melting in mantle diapirs and the origin of ultrabasic liquids. Earth and Planetary Science Letters, 27, 113–120.

Challis, G.A. (1961). Post-intrusion deformation of a dyke swarm, Awatere Valley, New Zealand. Geological Magazine, 98, 441–448.

Challis, G.A. (1963). Layered xenoliths in a dyke, Awatere Valley, New Zealand. Geological Magazine, 100, 11–16.

Challis, G.A. (1965). The origin of New Zealand ultramafic intrusions. Journal of Petrology, 6, 322–364.

Chalmers, R.O. (1967). Australian Rocks, Minerals and Gemstones. Angus & Robertson, Sydney.

Champion, D.C. (1984). Geology and geochemistry of the Mount Jukes intrusive complex. James Cook University of North Queensland, unpublished BSc Honours thesis.

Chauvel, C., & B.-M. Jahn (1984). Nd-Sr isotope and REE geochemistry of alkali basalts from Massif Central, France. Geochemica et Cosmochimica Acta, 48, 93–110.

Chayes, F. (1972). Silica saturation in Cenozoic basalt. Philosophical Transactions of the Royal Society of London, A271, 285–296.

Chayes, F. (1977). The oceanic basalt-trachyte relation in general and in the Canary Islands. American Mineralogist, 62, 666–671.

Chen, C.-Y., & F.A. Frey (1983). Origin of Hawaiian tholeiite and alkalic basalt. Nature, 302, 785–789.

Chen, C.-Y., & F.A. Frey (1985). Trace element and isotopic geochemistry of lavas from Haleakala volcano, East Maui, Hawaii: implications for the origin of Hawaiian basalts. Journal of Geophysical Research, 90, 8743–8768.

Chivas, A.R., I. Barnes, W.C. Evans, J.E. Lupton, & J.O. Stone (1987). Liquid carbon dioxide of magmatic origin and its role in volcanic eruptions. Nature, 326, 587–589.

Choubey, V.D. (1973). Long distance correlation of Deccan basalt flows, central India. Bulletin of the Geological Society of America, 84, 2785–2790.

Christensen, N.I. (1982). Seismic velocities. In: Carmichael, R.S. (Editor), Handbook of Physical Properties of Rocks, 2. CRC Press, Boca Raton, Florida, 1–228.

Christensen, N.I., & D.M. Fountain (1975). Constitution of the lower continental crust based on experimental studies of seismic velocities in granulites. Bulletin of the Geological Society of America, 86, 227–236.

Christiansen, R.L. (1982). Late Cenozoic volcanism in the Island Park area, eastern Idaho. In: Bonnichsen, B., & R.M. Breckenridge (Editors), Cenozoic Geology of Idaho. Idaho Bureau of Mines and Geology — Bulletin, 26, 345–368.

Christiansen, R.L., & P.W. Lipman (1972). Cenozoic volcanism and plate-tectonic evolution of the western United States. II. Late Cenozoic. Philosophical Transactions of the Royal Society of London, A271, 249–284.

Christoffel, D.A. (1971). Motion of the New Zealand Alpine Fault deduced from the pattern of sea-floor spreading. In: Collins B.W., & R. Fraser (Editors), Recent Crustal Movements. Bulletin of the Royal Society of New Zealand, 9, 25–30.

Civetta, L., M. De Fino, P. Gasparini, M.B. Ghiara, L. La Volpe, & L. Lirer (1975). Structural meaning of east-central Afar volcanism (Ethiopia, T.F.A.I). Journal of Petrology, 83, 363–373.

Clapperton, C.M. (1973a). Thrice threatened Heimaey. Geographical Magazine, 45, 495–500.

Clapperton, C.M. (1973b). Back home to work on a volcanic island. Geographical Magazine, 46, 83–90.

Clarke, W. (1984). The geology and hydrogeology of the Lancefield area. University of Melbourne, Geology, unpublished BSc Honours report.

Clarke, W.B. (1878). Remarks on the sedimentary formations of New South Wales. Fourth Edition. Government Printer, Sydney.

Cleary, J.R., & D.W. Simpson (1971). Seismotectonics of the Australian continent. Nature, 230, 239–241.

Cochran, J.R. (1983). Effects of finite rifting times on the development of sedimentary basins. Earth and Planetary Science Letters, 66, 289–302.

Cloetingh, S., & R. Wortel (1985). Regional stress field of the Indian Plate. Geophysical Research Letters, 12, 77–80.

Cloetingh, S., & R. Wortel (1986). Stress in the Indo-Australian plate. Tectonophysics, 132, 49–67.

Cohen, R.S., & R.K. O'Nions (1982a). Identification of recycled continental material in the mantle from Sr, Nd and Pb isotope investigations. Earth and Planetary Science Letters, 61, 73–84.

Cohen, R.S., & R.K. O'Nions (1982b). The lead, neodymium and strontium isotopic structure of ocean ridge basalts. Journal of Petrology, 23, 299–324.

Cohen, R.S., N.M. Evensen. P.J. Hamilton, & R.K. O'Nions (1980). U-Pb, Sm-Nd and Rb-Sr systematics of mid-ocean ridge basalt glasses. Nature, 283, 149–153.

Cole, J.W. (1978). Tectonic setting of Mayor Island volcano (note). New Zealand Journal of Geology and Geophysics, 21, 645–647.

Cole, J.W. (1984). Taupo-Rotorua depression — an ensialic marginal basin of North Island, New Zealand. In:

Kokelaar, B.P., & M.F. Howells (Editors), Marginal Basin Geology: Volcanic and Associated Sedimentary and Tectonic Processes in Modern and Ancient Marginal Basins. Geological Society of London — Special Publication, 16, 109–120.

Cole, J.W. (1986). Distribution and tectonic setting of Late Cenozoic volcanism in New Zealand. In: Smith, I.E.M. (Editor), Late Cenozoic Volcanism in New Zealand. Royal Society of New Zealand Bulletin, 23, 7–20.

Coleman, R.G., & A.V. McGuire (1988). Magma systems related to the Red Sea opening. Tectonophysics, 150, 77–100.

Colhoun, E.A. (1978). Recent Quaternary and geomorphological studies in Tasmania. Australian Quaternary Newsletter, 9, 5–13.

Collins, P.L.F., & E. Williams (1986). Metallogeny and tectonic development of the Tasman Fold Belt system in Tasmania. Ore Geology Reviews, 1, 153–201.

Compston, W., & B.W. Chappell (1979). Sr-isotope evolution of granitoid source rocks. In: McElhinny, M.W. (Editor), The Earth: Its Origin, Structure and Evolution. Academic Press, London, 377–426.

Condon, M.A. (1951). The geology of the lower Werribee River, Victoria. Proceedings of the Royal Society of Victoria, 63, 1–24.

Conolly, J.R. (1969). Western Tasman Sea floor. New Zealand Journal of Geology and Geophysics, 12, 310–343.

Coombs, D.S. (1963). Trends and affinities of basaltic magmas and pyroxenes as illustrated on the diopside-olivine-silica diagram. Mineralogical Society of America — Special Paper 1, 227–250.

Coombs, D.S. (1985). New Zealand terranes. Geological Society of Australia — Abstracts, 14, 45–48.

Coombs, D.S., & J.S. Dickey (1965). The early Tertiary petrographic province of north-east Otago: Waiareka and Deborah Volcanic Formations. New Zealand Department of Scientific and Industrial Research — Information Serial, 51, 38–54.

Coombs, D.S., & T. Hatherton (1959). Palaeomagnetic studies of Cenozoic volcanic rocks in New Zealand. Nature, 184, 883–884.

Coombs, D.S., & A. Reay (1986). Cenozoic alkaline and tholeiitic volcanism, eastern South Island. International Volcanological Congress 1986, New Zealand — Guide to Excursion C2, University of Otago.

Coombs, D.S., & J.F.G. Wilkinson (1969). Lineages and fractionation trends in undersaturated volcanic rocks from the east Otago province (New Zealand) and related rocks. Journal of Petrology, 10, 440–501.

Coombs, D.S., C.A. Landis, R.J. Norris, J.A. Sinton, D.J. Borns, & D. Craw (1976). The Dun Mountain ophiolite belt, New Zealand, its tectonic setting, construction and origin with special reference to the southern portion. American Journal of Science, 276, 561–603.

Coombs, D.S., R.A.F. Cas, Y. Kawachi, C.A. Landis, W.F. McDonough, & A. Reay (1986). Cenozoic volcanism in north, east, and central Otago. In: Smith, I.E.M. (Editor), Late Cenozoic Volcanism in New Zealand. Royal Society of New Zealand — Bulletin, 23, 278–312.

Cooper, A.F. (1979). Petrology of ocellar lamprophryes from western Otago, New Zealand. Journal of Petrology, 20, 139–163.

Cooper, A.F. (1986). A carbonatitic lamprophyre dike swarm from the Southern Alps, Otago and Westland. In: Smith, I.E.M. (Editor), Late Cenozoic Volcanism in New Zealand. Royal Society of New Zealand — Bulletin, 23, 313–336.

Cooper, A.F., B.A. Barreiro, D.L. Kimbrough, & J.M. Mattinson (1987). Lamprophyre dike intrusion and the age of the Alpine Fault, New Zealand. Geology, 15, 941–944.

Cooper, J.A., & D.H. Green (1969). Lead isotope measurements on lherzolite inclusions and host basanites from western Victoria, Australia. Earth and Planetary Science Letters, 6, 69–76.

Cooper, J.A., J.R. Richards, & A.W. Webb (1963). Some potassium-argon ages in New England, New South Wales. Journal of the Geological Society of Australia, 10, 313–316.

Cooper, R.A. (1974). Age of the Greenland and Waiuta groups, South Island, New Zealand. New Zealand Journal of Geology and Geophysics, 17, 955–962.

Cooper, R.A. (1979). Lower Palaeozoic rocks of New Zealand. Journal of the Royal Society of New Zealand, 9, 29–84.

Coote, J.A.R. (1987). Cenozoic volcanism in the Waiau area, North Canterbury. University of Canterbury, unpublished MSc thesis.

Cotton, C.A. (1944). Volcanoes as Landscape Forms. Whitcombe & Tombs, Christchurch.

Coulson, A. (1933). The Older Volcanic and Tertiary marine beds of Curlewis near Geelong. Proceedings of the Royal Society of Victoria, 45, 140–148.

Coulson, A. (1938). The basalts of the Geelong district. Proceedings of the Royal Society of Victoria, 50, 251–257.

Coulson, A. (1941). The volcanoes of the Portland district. Proceedings of the Royal Society of Victoria, 53, 394–402.

Coulson, A. (1954). The volcanic rocks of the Daylesford district. Proceedings of the Royal Society of Victoria, 65, 113–124.

Courtillot, V., G. Feraud, H. Maluski, D. Vandamme, M.G. Moreau, & J. Besse (1988). Deccan flood basalts and the Cretaceous/Tertiary boundary. Nature, 333, 843–846.

Coventry, R.J., P.J. Stephenson, & A.W. Webb (1985). Chronology of landscape evolution and soil development in the upper Flinders River area, Queensland, based on isotopic dating of Cainozoic basalts. Australian Journal of Earth Sciences, 32, 433–447.

Cox, K.G. (1978). Flood basalts, subduction and the break-up of Gondwanaland. Nature, 274, 47–49.

Cox, K.G. (1980). A model for flood basalt volcanism. Journal of Petrology, 21, 629–650.

Cox, K.G. (1983). The Karoo province of southern Africa: origin of trace element enrichment patterns. In: Hawkesworth, C.J., & M.J. Norry (Editors), Continental Basalts and Mantle Xenoliths. Shiva, Cheshire, 139–157.

Cox, K.G., & C.J. Hawkesworth (1985). Geochemical stratigraphy of the Deccan Traps at Mahabalashwar, Western Ghats, India, with implications for open system magmatic processes. Journal of Petrology, 26, 355–377.

Cox, K.G., I.G. Gass, & D.I.J. Mallick (1970). The peralkaline volcanic suite of Aden and Little Aden, South Arabia. Journal of Petrology, 11, 433–461.

Cox, K.G., J.D. Bell, & R.J. Pankhurst (1979). The Interpretation of Igneous Rocks. Allen & Unwin, London.

CRA Exploration Pty Ltd (1983). Exploration report E.L.'s 1943, 1944, 1945, Olinda-Mt Coricudgy-Mt Pomany area. New South Wales Geological Survey — File 65 1983/157.

Craft, F.A. (1933). The coastal tablelands and streams of New South Wales. Proceedings of the Linnaean Society of New South Wales, 58, 437–460.

Crawford, A.J., W.E. Cameron, & R.R. Keays (1984). The association boninite low-Ti andesite-tholeiite in the Heathcote Greenstone Belt, Victoria; ensimatic setting for the early Lachlan Fold Belt. Australian Journal of Earth Sciences, 31, 161–175.

Crawford, E., C. Herbert, G. Taylor, R. Helby, R. Morgan, & J. Ferguson, (1980). Diatremes of the Sydney Basin. In: Herbert, C., & R. Helby (Editors), A Guide to the Sydney Basin. Geological Survey of New South Wales — Bulletin, 26, 294–323.

Crohn, P.W. (1950). The geology, petrology and physiography of the Omeo district, northeastern Victoria. Proceedings of the Royal Society of Victoria, 62, 1–70.

Croker, P.J. (1984). The lava flows between Melbourne and Wallan: ages and distribution of the Newer Volcanics to the north of Melbourne. University of Melbourne, Department of Geology, unpublished BSc Honours report.

Cromer, W.C. (1972). The petrology and chemistry of the Circular Head teschenite. University of Tasmania, unpublished BSc Honours thesis.

Cromer, W.C. (1980). A late Eocene basalt date from northern Tasmania. Search, 11, 294–295.

Crook, K.A.W., & D.A. Feary (1982). Development of New Zealand according to the fore-arc model of crustal evolution. Tectonophysics, 87, 65–107.

Cross, W., J.P. Iddings, L.V. Pirsson, & H.S. Washington (1903). Quantitative Classification of Rocks. University of Chicago Press, Chicago.

Crough, S.T. (1978). Thermal origin of mid-plate hot-spot swells. Geophysical Journal of the Royal Astronomical Society, 55, 451–469.

Crough, S.T. (1981). The Darfur swell, Africa: gravity constraints on its isostatic compensation. Geophysical Research Letters, 8, 877–897.

Crough, S.T. (1983). Hotspot swells. Annual Review of Earth and Planetary Sciences, 11, 165–193.

Crowe, B.M., & R.V. Fisher (1973). Sedimentary structures in base surge deposits with special reference to cross-bedding, Ubehebe Craters, Death Valley, California. Bulletin of the Geological Society of America, 84, 663–682.

Cull, J.P. (1982). An appraisal of Australian heat-flow data. BMR Journal of Australian Geology and Geophysics, 7, 11–21.

Cull, J.P., & D. Conley (1983). Geothermal gradients and heat flow in Australian sedimentary basins. BMR Journal of Australian Geology and Geophysics, 8, 329–337.

Cull, J.P., & D. Denham (1979). Regional variations in Australian heat-flow. BMR Journal of Australian Geology and Geophysics, 4, 1–13.

Cull, F.A., & H.O.A. Meyer (1986). Oxidation of diamond at high temperature and 1 atm. total pressure with controlled oxygen fugacity. Fourth International Kimberlite Conference, Perth. Geological Society of Australia — Extended Abstracts, 16, 377–379.

Cullen, D.J. (1967). The Antipodes fracture zone, a major structure of the south-west Pacific. New Zealand Journal of Marine and Freshwater Research, 1, 16–25.

Cullen, D.J. (1969). Quaternary volcanism at the Antipodes Islands: its bearing on the structural interpretation of the southwest Pacific. Journal of Geophysical Research, 74, 4213–4220.

Cullen, D.J. (1970). A tectonic analysis of the southwest Pacific. New Zealand Journal of Geology and Geophysics, 13, 7–20.

Cundari, A. (1973). Petrology of the leucite-bearing lavas in New South Wales. Journal of the Geological Society of Australia, 20, 465–492.

Cundari, A., & C.D. Ollier (1970). Inverted relief due to lava flows along valleys. Australian Geographer, 11, 291–293.

Cundari, A., & G. Saliulo (1987). Clinopyroxenes from Somma-Vesuvius: implications of crystal chemistry and site configuration parameters for studies of magma genesis. Journal of Petrology, 28, 727–736.

Cundari, A., J.G.R. Renard, & A.J.W. Gleadow (1978). Uranium-potassium relationship and apatite fission-track ages for a differentiated leucitite suite from New South Wales (Australia). Chemical Geology, 22, 11–20.

Curran, J.M. (1891). A contribution to the microscopic structure of some Australian rocks. Journal and Proceedings of the Royal Society of New South Wales, 25, 179–233.

Curran, J.M. (1897). On the occurrence of precious stones in New South Wales and the deposits in which they are found. Journal and Proceedings of the Royal Society of New South Wales, 30, 228–237.

Curran, J.M. (1899). Geology of Sydney and the Blue Mountains. Angus & Robertson, Sydney.

Cuttler, L.G. (1972). Herberton Deep Lead. In: Blake, D., Regional and economic geology of the Herberton/Mount Garnet area-Herberton tinfield, north Queensland. Bureau of Mineral Resources, Australia — Bulletin, 124.

Daintree, R. (1872). Notes on the geology of Queensland. Quarterly Journal of the Geological Society of London, 28, 271.

Daintree, R. (1875). Notes on the microscopic structures of certain igneous rocks. New Zealand Institute Transactions, 7.

Dalrymple, G.B. (1979). Critical tables for conversion of K-Ar ages from old to new constants. Geology, 7, 558–560.

Dalrymple, G.B., E.A. Silver, & E.D. Jackson (1973). Origin of the Hawaiian Islands. American Scientist, 61, 294–303.

Dana, J.D. (1849). Geology. United States Exploring Expedition, Volume 10. Philadelphia.

Dasch, E.J., & D.H. Green (1975). Strontium isotope geochemistry of lherzolite inclusions and host basaltic rocks, Victoria, Australia. Americal Journal of Science, 275, 461–469.

Dasch, E.J., & D.J. Millar (1977). Age and strontium-isotope geochemistry of differentiated rocks from the Newer Volcanics, Mt. Macedon area, Victoria, Australia. Journal of the Geological Society of Australia, 24, 195–201.

Dautria, J.M., & M. Girod (1987). Cenozoic volcanism associated with swells and rifts. In: Nixon, P.H. (Editor), Mantle Xenoliths. Wiley, Chichester, 195–214.

Dautria, J.M., J. Dostal, C. Dupuy, & J.M. Liotard (1988). Geochemistry and petrogenesis of alkali basalts from Tahalra (Hoggar, Northwest Africa). Chemical Geology, 69, 17–35.

Davey, F.J. (1977). Marine seismic measurements in the New Zealand region. New Zealand Journal of Geology and Geophysics, 20, 719–777.

Davey, F.J., & A.G. Robinson (1978). Magnetic maps 1:1 million, Cook sheet. Total force magnetic anomalies. New Zealand Department of Scientific and Industrial Research, Geophysics Division — Oceanic Series.

Davey, F.J., & A.G. Robinson (1981). Magnetic maps 1:1 million, Bounty sheet. Total force magnetic anomalies. New Zealand Department of Scientific and Industrial Research, Geophysics Division — Oceanic Series.

Davey, F.J., & E.G.C. Smith (1983). The tectonic setting of the Fiordland region, south-west New Zealand. Geophysical Journal of the Royal Astronomical Society, 72, 23–38.

Davey, F.J. & A.B. Watts (1983, Compilers). Free Air Gravity Field of the New Zealand region, 1:3 million. Geological Society of America, Boulder, Colorado — Map and Chart Series, MC–48.

Davey, F.J., M. Hampton, J. Childs, M.A. Fisher, K. Lewis, & J.R. Pettinga (1986). Structure of a growing accretionary prism, Hikurangi margin, New Zealand. Geology, 14, 663–666.

David, T.W.E. (1887). Origin of the laterite on the New England district of New South Wales. Australasian Association for the Advancement Science — Report, 1, 233–241.

David, T.W.E. (1896). Note on the occurrence of diatomaceous earth in the Warrumbungle Mountains, New South Wales. Proceedings of the Linnaean Society of New South Wales, 21, 234–268.

David, T.W.E. (1932). Explanatory notes to accompany a new geological map of the Commonwealth of Australia. Australian Medical Publishing, Sydney.

Davies, G.R., & R. Macdonald (1987). Crustal influences in the petrogenesis of the Naivasha basalt-comendite complex: combined trace element and Sr-Nd-Pb isotope constraints. Journal of Petrology, 28, 1009–1031.

Davis, G.H., & J.J. Hardy (1981). The Eagle Pass detachment, southeastern Arizona: product of mid-Miocene listric (?) normal faulting the southern Basin and Range. Bulletin of the Geological Society of America, 92, 749–762.

Davis, G.H., & G.S. Lister (1988). Detachment faulting in continental extension: perspectives from the southwestern U.S. Cordillera. In: Clark, S.P. Jr, B.C. Burchfiel, & J. Suppe (Editors), Processes in Continental Lithospheric Deformation. Geological Society of America — Special Paper, 218, 133–159.

Davis, T.E., M.R. Johnston, P.C. Rankin, & R.J. Stull (1980). The Dun Mountain ophiolite belt in east Nelson, New Zealand. In: Panayiotou, A. (Editor), Ophiolites. Cyprus Geological Survey Department, Nicosia, 480–498.

Dawson, J. (1881). The Australian Aborigine. Melbourne.

Dawson, J.B. (1980). Kimberlites and their Xenoliths. Springer-Verlag, Berlin.

Dawson, J.B., & J.V. Smith (1977). The MARID (mica-amphibole-rutile-ilmenite-diopside) suite of xenoliths in kimberlite. Geochimica et Cosmochimica Acta, 41, 309–323.

Day, A.F. (1961). The geology of the Rylstone-upper Goulburn River district — with particular reference to the petrology and mineralogy of the alkaline intrusions. University of Sydney, unpublished PhD thesis.

Day, R.A. (1983). Petrology and geochemistry of the Older Volcanics of Victoria. Monash University, unpublished PhD thesis.

Day, R.A., I.A. Nicholls, & F.L. Hunt (1979). The Meredith ultramafic breccia pipe. CUMSEA Symposium Abstracts, Bureau of Mineral Resources, Australia — Record 1979/2.

Day, R.W., W.G. Whitaker, C.G. Murray, I.H. Wilson, & K.G. Grimes (1983). Queensland Geology. A companion volume to the 1:2 500 000 scale geological map (1975). Geological Survey of Queensland — Publication, 383, 1–194.

Dear, J.F., R.G. McKellar, R.M. Tucker (1971). Geology of the Monto 1:250 000 Sheet area. Geological Survey of Queensland — Report, 46.

Degeling, P.R., L.B. Gilligan, E. Scheibner, & D.W. Suppel (1986). Metallogeny and tectonic development of the Tasman Fold Belt system in New South Wales. Ore Geology Reviews, 1, 259–313.

De Keyser, F. (1964). 1:250 000 geological series, Innisfail, Queensland. Bureau of Mineral Resources, Australia — Explanatory Notes, SE/55–6.

De Keyser, F., & K.G. Lucas (1968). Geology of the Hodgkinson and Laura Basins, north Queensland. Bureau of Mineral Resources, Australia — Bulletin, 84.

Denham, D. (1985). The Tasman Sea earthquake of 25 November 1983, and stress in the Australian plate. Tectonophysics, 111, 329–338.

Denham, D. (in press). Australian seismicity — the puzzle of the not so stable continent. Seismological Research Letters.

Denham, D., L.G. Alexander, & G. Worotnicki (1979). Stresses in the Australian crust: evidence from earthquakes and in-situ stress measurements. BMR Journal of Australian Geology and Geophysics, 4, 289–295.

Denham, D., J. Weekes, & C. Krayshek (1981). Earthquake evidence for compressive stress in the southeastern Australian crust. Journal of the Geological Society of Australia, 28, 323–332.

Denison, R.E., & D.S. Coombs (1977). Radiometric ages for some rocks from Snares and Auckland Islands, Campbell Plateau. Earth and Planetary Science Letters, 34, 23–29.

DePaolo, D.J. (1981). Neodymium isotopes in the Colorado Front Range and crust-mantle evolution in the Proterozoic. Nature, 291, 193–196.

De Rito, R.F., A. Cozzarelli, & D.S. Hodge (1983). Mechanism of subsidence of ancient cratonic rift basins. Tectonophysics, 94, 141–168.

Devey, C.W., & P.C. Lightfood (1986). Volcanological and tectonic control of stratigraphy and structure in the western Deccan traps. Bulletin of Volcanology, 48, 195–207.

Dewey, J.F., & B.F. Windley (1988). Palaeocene-Oligocene tectonics of NW Europe. In: Morton, A.C., & L.M. Parson (Editors), Early Tertiary Volcanism and the Opening of the NE Atlantic. Geological Society — Special Publication, 39, 25–31.

Dibble, R.R., I.A. Nairn, & V.E. Neall (1985). Volcanic hazards of North Island, New Zealand — overview. Journal of Geodynamics, 3, 369–396.

Dickens, J.M., & E.J. Malone (1973). Geology of the Bowen Basin, Queensland. Bureau of Mineral Resources, Geology and Geophysics — Bulletin, 130.

Dickey, J.S. (1968a). Eclogitic and other inclusions in the Mineral Breccia Member of the Deborah Volcanic Formation at Kakanui, New Zealand. American Mineralogist, 53, 1304–1319.

Dickey, J.S. (1968b). Observations on the Deborah Volcanic Formation near Kakanui, New Zealand. New Zealand Journal of Geology and Gephysics, 11, 1159–1162.

Dickin, A.P., N.W. Jones, M.F. Thirwall, & R.N. Thompson. (1987) A Ce/Nd isotope study of crustal contamination processes affecting Paleocene magmas in Skye, Northwestern Scotland. Contributions to Mineralogy and Petrology, 96, 455–464.

Dickins, J.M., M.R. Johnston, D.L. Kimbrough, & C.A. Landis (1986). The stratigraphic and structural position and age of the Croisilles Melange, east Nelson, New Zealand. New Zealand Journal of Geology and Geophysics, 29, 291–301.

Dickinson, J.E. Jr., & P.C. Hess (1982). Zircon saturation in lunar basalts and granites. Earth and Planetary Science Letters, 57, 336–344.

Dieffenbach, E. (1943). Travels in New Zealand. London.

Dieseldorf, A. (1901). Beitrage zur kentniss der gesteine und fossilien der Chathaminseln, sowie einiger Gesteine und neuer Nephritfundorte Neuseelands. Marburg University, dissertation.

Di Paola, G.M. (1972). The Ethiopian Rift Valley (between 7°00′ and 8°40′ latitude North). Bulletin Volcanologique, 36, 517–560.

Dodson, J.R. (1974). Vegetation and climatic history near Lake Keilambete, western Victoria. Australian Journal of Botany, 22, 709–717.

Doe, B.R., & R.E. Zartman (1979). Plumbotectonics, the Phanerozoic. In: Barnes H.L. (Editor), Geochemistry of Hydrothermal Ore Deposits. Wiley, 22–70.

Doe, B.R., W.P. Leeman, R.L. Christiansen, & C.E. Hedge (1982). Lead and strontium isotopes and related trace elements as genetic tracers in the Upper Cenozoic rhyolite-basalt association of the Yellowstone Plateau volcanic field. Journal of Geophysical Research, 87, 4785–4806.

Dostal, J., C. Dupuy, & L.A. Leyreloup (1980). Geochemistry and petrology of meta-igneous granulite xenoliths in Neogene volcanic rocks in the Massif Central, France — implications for the lower crust. Earth and Planetary Science Letters, 50, 31–40.

Douglas, J.G. (1979). Explanatory notes on the Warragul 1:250 000 geological map. Geological Survey of Victoria — Report, 57.

Dow, D.B. (1954). The geology of the Waikaretu Valley and environs. University of Auckland, unpublished MSc thesis.

Downes, H. (1984). Sr and Nd isotope geochemistry of co-existing alkaline magma series, Cantal, Massif Central, France. Earth and Planetary Science Letters, 69, 321–334.

Downes, H. (1987). Tertiary and Quaternary volcanism in the Massif Central, France. In: Fitton, J.G., & B.G.J. Upton (Editors), Alkaline Igneous Rocks. Geological Society — Special Publication, 30, 517–530.

Drummond, B.J. (1981). Detailed seismic velocity/depth models of the upper lithosphere of the Pilbara craton, northwest Australia. BMR Journal of Australian Geology and Geophysics, 8, 35–51.

Drummond, B.J. (1982). Seismic constraints on the chemical composition of the crust of the Pilbara craton, northwest Australia. Revista Brasileira de Geociencas, 12, 113–120.

Drummond, B.J., & C.D.N. Collins (1986). Seismic evidence for underplating of the lower continental crust of Australia. Earth and Planetary Science Letters, 79, 361–372.

Drummond, B.J., K.J. Muirhead, C. Wright, P. Wellman (in press). A teleseismic travel time residual map of the Australian continent. BMR Journal of Australian Geology and Geophysics.

Duggan, M.B. (1968). Megacrysts and ultramafic xenoliths in the basalts of northeastern New South Wales. University of New England, unpublished BSc Honours thesis.

Duggan, M.B. (1974). The mineralogy and petrology of the southern portion of the Tweed Shield volcano, northeastern New South Wales. University of New England, unpublished PhD thesis.

Duggan, M.B. (1976). Primary allanite in vitrophyric rhyolites from the Tweed Shield volcano, northeastern New South Wales. Mineralogical Magazine, 40, 652–653.

Duggan, M.B. (1988). Zirconium-rich sodic pyroxenes in felsic volcanics from the Warrumbungle volcano, central New South Wales, Australia. Mineralogical Magazine, 52, 491–496.

Duggan, M.B., D.R. Mason (1978). Stratigraphy of the Lamington volcanics in far northeastern New South Wales. Journal of Geological Society of Australia, 25, 65–73.

Duggan, M.B., & A. Reay (1986). The Timaru Basalt. In: Smith, I.E.M. (Editor), Late Cenozoic Volcanism in New Zealand. Royal Society of New Zealand Bulletin, 23, 264–277.

Duggan, M.B., & J.F.G. Wilkinson (1973). Tholeiitic andesite of high-pressure origin from the Tweed Shield volcano, northeastern New South Wales. Contributions to Mineralogy and Petrology, 39, 267–276.

Duggan, N.T. (1972). Tertiary volcanic rocks of the Inverell area. University of New England, unpublished BSc Honours thesis.

Dulhunty, J.A. (1964). Our Permian heritage in central-eastern New South Wales. Journal and Proceedings of the Royal Society of New South Wales, 97, 145–156.

Dulhunty, J.A. (1973a). Mesozoic stratigraphy in central western New South Wales. Journal of Geological Society of Australia, 20, 319–328.

Dulhunty, J.A. (1973b). Potassium-argon dating and occurrence of Tertiary and Mesozoic basalts in the Binnaway district. Journal and Proceedings of the Royal Society of New South Wales, 105, 71–76.

Dulhunty, J.A. (1976). Potassium-argon ages of igneous rocks in the Wollar-Rylstone region, New South Wales. Journal and Proceedings of the Royal Society of New South Wales, 109, 35–39.

Dulhunty, J.A., I. McDougall (1966). Potassium-argon dating in the Coonabarabran-Gunnedah district, New South Wales. Australian Journal of Science, 28, 393–394.

Dulhunty, J.A., E.A.K. Middlemost, & R.W. Beck (1987). Potassium-argon ages, petrology and geochemistry of some Mesozoic igneous rocks in northeastern New South Wales. Journal and Proceedings of the Royal Society of New South Wales, 120, 71–90.

Duncan, A.R., A.J. Erlank, & J.S. Marsh (1984). Regional geochemistry of the Karoo igneous province. In: Erlank, A.J. (Editor), Petrogenesis of the Volcanic Rocks of the Karoo Province. Geological Society of South Africa — Special Publication, 13, 355–388.

Duncan, R.A. (1981). Hotspots in the southern oceans — an absolute frame of reference for motion of the Gondwana continents. Tectonophysics, 74, 29–42.

Duncan, R.A., & D.A. Clague (1985). Pacific plate motion recorded by linear volcanic chains. In: Nairn, A.E.M. (Editor), The Ocean Basins and Margins, Volume 7A, The Pacific Ocean. Plenum, New York.

Duncan, R.A., & R.B. Hargraves (1984). Plate tectonic evolution of the Caribbean region in the mantle reference frame. In: Bonini, W.E., R.B. Hargraves, & R. Shagam (Editors), The Caribbean-South American Plate Boundary and Regional Tectonics. Memoir of the Geological Society of America, 162, 81–93.

Duncan, R.A., & D.G. Pyle (1988). Rapid eruption of the Deccan flood basalts at the Cretaceous/Tertiary boundary. Nature, 333, 841–843.

Duncan, R.A., R.B. Hargraves, & G.P. Brey (1978). Age, palaeomagnetism and chemistry of melilite basalts in the Southern Cape, South Africa. Geological Magazine, 115, 317–396.

Duncan, R.A., M.T. McCulloch, H.G. Barsczus, & D.R. Nelson (1986). Plume versus lithospheric sources for melts at Ua Pou, Marquesas Islands. Nature, 322, 534–538.

Dunstan, B. (1901). Geology of the Dawson and Mackenzie Rivers — iii with special reference to the occurrence of anthracite coal. Queensland Government Mining Journal, 2, 212–216.

Dunstan, E. (1902). The sapphire fields of Anakie. Queensland Geological Survey — Report, 172.

Dupré, B., & C.J. Allègre (1980). Pb-Sr-Nd isotopic correlation and the chemistry of the North Atlantic mantle. Nature, 286, 17–22.

Du Toit, A.L. (1937). Our Wandering Continents. Oliver & Boyd, Edinburgh.

Eales, H.V., J.S. Marsh, & K.G. Cox (1984). The Karoo province: an introduction. In: Erlank, A.J. (Editor),

Petrogenesis of the Volcanic Rocks of the Karoo Province. Geological Society of South Africa — Special Publication, 13, 1–26.

Eaton, G.P. (1982). The Basin and Range province: origin and tectonic significance. Annual Review of Earth and Planetary Science, 10, 409–440.

Eaton, G.P. (1984). The Miocene Great Basin of western North America as an extending back-arc region. Tectonophysics 102, 275–295.

Edney, W.J. (1984a). The geology of the Tower Hill volcanic centre, western Victoria. Monash University, unpublished MSc thesis.

Edney, W.J. (1984b). Lake Purrumbete. In: Cas, R.A.F., E.B. Joyce, I.A. Nicholls, & R.C. Price (Editors), Volcanics Workshop 1984: Continental Volcanism, Western Victoria. Centre of Advanced Studies (Geoscience), Melbourne University.

Edney, W.J., & I.A. Nicholls (1984). Mount Leura volcanic centre. In: Cas, R.A.F., E.B. Joyce, I.A. Nicholls, & R.C. Price (Editors), Volcanics Workshop 1984: Continental Volcanism, Western Victoria. Centre of Advanced Studies (Geoscience), Melbourne University.

Edney, W.J., A.P. Kershaw, & J.A. Peterson (1985). Evidence of a Pleistocene age for Tower Hill, western Victoria. Search, 16, 302–303.

Edwards, A.B. (1938a). Tertiary volcanic rocks of central Victoria. Quarterly Journal of the Geological Society of London, 94, 243–320.

Edwards, A.B. (1938b). The age and physiographical relationships of some Cainozoic basalts in central and eastern Tasmania. Papers and Proceedings of the Royal Society of Tasmania, 175, 169–199.

Edwards, A.B. (1939). Petrology of the Tertiary Older Volcanic rocks of Victoria. Proceedings of the Royal Society of Victoria, 51, 73–98.

Edwards, A.B. (1941). The crinanite laccolith of Circular Head, Tasmania. Proceedings of the Royal Society of Victoria, 53, 403–415.

Edwards, A.B. (1942). Differentiation of the dolerites of Tasmania. Journal of Geology, 50, 451–480.

Edwards, A.B. (1945). The geology of Phillip Island. Proceedings of the Royal Society of Victoria, 57, 1–16.

Edwards, A.B. (1950). The petrology of the Cainozoic basaltic rocks of Tasmania. Proceedings of the Royal Society of Victoria, 62, 97–120.

Edwards, A.B., & W. Crawford (1940). The Cainozoic volcanic rocks of the Gisborne district, Victoria. Proceedings of the Royal Society of Victoria, 52, 281–311.

Edwards, A.B., J.F. Lovering, & J. Ferguson (1979). High pressure basic inclusions from the Kayrunnera kimberlitic diatreme in New South Wales, Australia. Contributions to Mineralogy and Petrology, 69, 185–192.

Eggins, S. (1984). Geology and geochemistry of the Barrington Tops batholith. University of New South Wales, unpublished BSc Honours thesis.

Eggins, S.M. (1988). Petrology and geochemistry of Tasmantid Seamounts. Geological Society of Australia — Abstracts, 21, 123.

Eiby, G.A. (1964). Earthquakes in Northland. New Zealand Engineering, 19, 125–129.

El Goresy, A., & H.S. Yoder Jr (1974). Natural and synthetic melilite compositions. Carnegie Institution of Washington Year Book, 73, 359–371.

Elliott, L.G. (1973). Geology of the Springwood area, Denison Trough east central Queensland. University of Queensland, unpublished BSc Honours thesis.

Ellis, D.J. (1971). The geology and petrology of the Mount Porndon volcanic complex. University of Melbourne, unpublished BSc Honours thesis.

Ellis, D.J. (1976). High pressure cognate inclusions in the Newer Volcanics of Victoria. Contributions to Mineralogy and Petrology, 58, 149–180.

Ellis, D.J., & D.H. Green (1979). An experimental study of the effect of Ca upon garnet-clinopyroxene Fe-Mg exchange equilibria. Contributions to Mineralogy and Petrology, 71, 13–22.

Ellis, P.L., & W.G. Whitaker (1976). Geology of the Bundaberg 1:250 000 Sheet Area. Geological Survey of Queensland — Report, 90.

Ellis, R.M., & D. Denham (1985). Structure of the crust and upper mantle beneath Australia from Rayleigh- and Love-wave observations. Physics of the Earth and Planetary Interiors, 38, 224–234.

Elphinstone, R. (1983). Sedimentology, palaeontology and stratigraphy of the Ulladulla area. University of Sydney, unpublished BSc Honours thesis.

Embleton, B.J.J. (1984). Magnetic properties of oriented rock samples from Sturge Island and Sabrina Island in the Balleny Group. In: Lewis, D. (Author), Voyage to the Ice. The Antarctic Expedition of Solo. Australian Broadcasting Commission, Sydney, 128–131.

Embleton, B.J.J., P.W. Schmidt, L.H. Hamilton, & G.H. Riley (1985). Dating volcanism in the Sydney Basin: Evidence from K-Ar ages and palaeomagnetism. In: Sutherland, F.L., B.J. Franklin, & A.E. Waltho (Editors), Volcanism in Eastern Australia, with Case Histories from New South Wales. Geological Society of Australia, New South Wales Division — Publication, 1, 59–72.

Emerson, D.W., & S.Y. Wass (1980). Diatreme characteristics — evidence from the Mogo Hill intrusion, Sydney Basin. Bulletin of the Australian Society of Exploration Geophysics, 11, 121–133.

Esso Australia Limited (1966). Bass-2 well completion report.

Esso Exploration and Production (N.Z.) Ltd. (1968). Geophysical report PPL712 (South Greymouth). Petroleum Report P.R. 400, lodged in the New Zealand Geological Survey Library, Lower Hutt.

Etheridge, M.A. (1967). The geology of the Tuross Lakes district. University of Sydney, unpublished BSc Honours thesis.

Etheridge, M.A., J.C. Branson, D.A. Falvey, K.L. Lockwood, P.G. Stuart-Smith, & A.S. Scherl (1985a). Basin-forming structures and the relevance to hydrocarbon exploration in Bass Basin, southeastern Australia. BMR Journal of Geology and Geophysics, 9, 197–206.

Etheridge, M.A., J.C. Branson, & P.G. Stuart-Smith (1985b). Extensional basin forming structures in Bass Strait and their importance for hydrocarbon exploration. APEA Journal, 25, 344–361.

Etheridge, M.A., P.A. Symonds, & G.S. Lister (in press). Application of the detachment model to reconstruction of conjugate passive margins. Tectonics.

Evans, A.L.I. (1970). Geomagnetic polarity reversals in a Late Tertiary lava sequence from the Akaroa volcano, New Zealand. Geophysical Journal of the Royal Astronomical Society, 21, 163–183.

Evans, B.W., & G. Moore (1968). Mineralogy as a function of depth in the prehistoric Makaopuhi tholeiitic lava lake, Hawaii. Contributions to Mineralogy and Petrology, 17, 85–115.

Evans, R. (1976). A study of the basic volcanic rocks of the Maleny-Mapleton area, southeast Queensland. University of Queensland, unpublished BSc honours thesis.

Evernden, J.F., & J.R. Richards (1962). Potassium-argon ages in eastern Australia. Journal of the Geological Society of Australia, 9, 1–50.

Ewart, A. (1979). A review of the mineralogy and chemistry of Tertiary-Recent dacitic, rhyolitic, and related salic volcanic rocks. In: Barker, F. (Editor), Trondhjemites, Dacites and Related Rocks. Elsevier, Amsterdam, 13–121.

Ewart, A. (1981). The mineralogy and chemistry of the anorogenic Tertiary silicic volcanics of S.E. Queensland and New South Wales, Australia. Journal of Geophysical Research, 86, 10242–10256.

Ewart, A. (1982a). Petrogenesis of the Tertiary anorogenic volcanic series of southern Queensland, Australia, in the light of trace element geochemistry and O, Sr, and Pb isotopes. Journal of Petrology, 23, 344–382.

Ewart, A. (1982b). The mineralogy and petrology of Tertiary-Recent orogenic volcanic rocks: with special reference to the andesitic-basaltic compositional range. In: Thorpe, R.S. (Editor), Andesites. Wiley, New York, 25–95.

Ewart, A. (1985). Aspects of the mineralogy and chemistry of the intermediate-silicic Cainozoic volcanic rocks of eastern Australia. Part 2: mineralogy and petrogenesis. Australian Journal of Earth Sciences, 32, 383–413.

Ewart, A., & A. Grenfell (1985). Cainozoic volcanic centres in southeastern Queensland, with special reference to the Main Range, Bunya Mountains, and the volcanic centres of the northern Brisbane coastal region. Papers of the Department of Geology, University of Queensland, 11, 1–57.

Ewart, A., & C.J. Hawkesworth (1987). The Pleistocene-Recent Tonga-Kermadec arc lavas: interpretation of new isotopic and rare earth data in terms of a depleted mantle source model. Journal of Petrology, 3, 495–530.

Ewart, A., S.R. Taylor, & A.C. Capp (1968). Geochemistry of the pantellerites of Mayor Island, New Zealand. Contributions to Mineralogy and Petrology, 17, 116–140.

Ewart, A., A. Mateen, & J.A. Ross (1976). Review of mineralogy and chemistry of Tertiary central volcanic complexes in southeast Queensland and northeast New South Wales. In: Johnson, R.W. (Editor), Volcanism in Australasia. Elsevier, New York, 21–39.

Ewart, A., V.M. Oversby, & A. Mateen (1977). Petrology and isotope geochemistry of Tertiary lavas from the northern flank of the Tweed volcano, southeastern Queensland. Journal of Petrology, 18, 73–113.

Ewart, A., K. Baxter, & J.A. Ross (1980). The petrology and petrogenesis of the Tertiary anorogenic mafic lavas of southern and central Queensland, Australia — possible implications for crustal thickening. Contributions to Mineralogy and Petrology, 75, 129–152.

Ewart, A., B.W. Chappell, & R.W. Le Maitre (1985). Aspects of the mineralogy and chemistry of the intermediate-silicic Cainozoic volcanic rocks of eastern Australia. Part 1: introduction and geochemistry. Australian Journal of Earth Sciences, 32, 359–382.

Ewart, A., N.C. Stevens, & J.A. Ross (1987). The Tweed and Focal Peak shield volcanoes, southeast Queensland and northeast New South Wales. Papers of the Department of Geology, University of Queensland, 11, 1–82.

Ewart, A., B.W. Chappell, & M.A. Menzies (1988). An overview of the geochemical and isotopic characteristics of the eastern Australian Cainozoic volcanic provinces. In: Menzies, M.A., & K.G. Cox (Editors), Oceanic and Continental Lithosphere: Similarities and Differences. Journal of Petrology — Special Publication, 225–274.

Exon, N.F. (1971a). Sheet SG/55–11, Mitchell, Queensland 1:250 000. Bureau of Mineral Resources, Australia — Explanatory Notes.

Exon, N.F. (1971b). Sheet SG/55–12, Roma, Queensland 1:250 000. Bureau of Mineral Resources, Australia — Explanatory Notes.

Exon, N.F., T. Langford-Smith, & I. McDougall (1970). The age and geomorphic correlations of deep weathering profiles, silcrete and basalt in the Roma-Amby region Queensland. Journal of the Geological Society of Australia, 17, 21–30.

Fallick, A.E., A.N. Halliday, A.P. Dickin, & J.G. Fitton (1985). Nd, Sr, Pb, and O isotopic evidence for the genesis of the Cameroon line volcanics. Transactions of American Geophysical Union, 1137.

Falloon, T.J. (1982). The geology of the Onawe-French Farm-Wainui area, Akaroa volcano, Banks Peninsula. University of Canterbury, unpublished BSc Honours thesis.

Falloon, T.J. & D.H. Green (1988). Anhydrous partial melting of peridotite from 8 to 35 kb and the petrogenesis of MORB. In: Menzies, M.A., & K.G. Cox (Editors), Oceanic and Continental Lithosphere: Similarities and Differences. Journal of Petrology — Special Publication, 379–414.

Falloon, T.J., D.H. Green, C.J. Hatton, K.L. Harris (in press). Anhydrous partial melting of fertile and depleted peridotites from 2 to 30 kb and applications to basalt petrogenesis. Journal of Petrology, 29.

Falvey, D.A., & J.C. Mutter (1981). Regional plate tectonics and the evolution of Australia's passive continental margins. BMR Journal of Australian Geology and Geophysics, 6, 1–29.

Farmer, N. (1985). Kingborough. Geological Atlas 1:50 000 Series, Sheet 88 (8311N). Tasmanian Department of Mines — Explanatory Report.

Farrar, E., & J.M. Dixon (1984). Overriding of the Indian-Antarctic ridge: origin of Emerald Basin and migration of Late Cenozoic volcanism in southern New Zealand and Campbell Plateau. Tectonophysics, 104, 243–256.

Faulks, I.G. (1962). The geology of a portion of the southern slopes of the Warrumbungle Mountains. University of New England, unpublished BSc Honours thesis.

Ferguson, A.K. (1977a). A note on a ramsayite-bearing pegmatoidal clot in a mela-nephelinite from the Older Volcanics near Bacchus Marsh, Victoria. Journal of the Geological Society of Australia, 24, 491–494.

Ferguson, A.K. (1977b). The natural occurrence of aegirine-neptunite solid solution. Contributions to Mineralogy and Petrology, 60, 247–253.

Ferguson, A.K. (1978a). Mineral chemistry and petrological aspects of some leucite-bearing lavas from Bufumbira, south-west Uganda, and related suites. University of Melbourne, unpublished PhD thesis.

Ferguson, A.K. (1978b). Ca enrichment in olivines from volcanic rocks. Lithos, 11, 189–194.

Ferguson, A.K. (1978c). A mineralogical investigation of some trachytic lavas and associated pegmatoids from Camel's Hump and Turritable Falls, central Victoria. Journal of the Geological Society of Australia, 25, 185–197.

Ferguson, A.K. (1978d). The crystallisation of pyroxenes and amphiboles in some alkaline rocks and the presence of a pyroxene compositional gap. Contributions to Mineralogy and Petrology, 67, 11–15.

Ferguson, J., D.J. Ellis, & R.N. England (1977). Unique spinel-garnet lherzolite inclusion in kimberlite from Australia. Geology, 5, 278–280.

Ferguson, J., R.J. Arculus, & J. Joyce (1979). Kimberlite and kimberlitic intrusives of southeastern Australia: a review. BMR Journal of Australian Geology and Geophysics, 4, 227–241.

Ferguson, J.A. (1969). Cainozoic erosion and sedimentation between Gympie and Brisbane. In: Campbell, K.S.W. (Editor), Stratigraphy and Palaeontology — Essays in Honour of Dorothy Hill. Australian National University Press, Canberra, 323–336.

Ferrar, H.T. (1925). The geology of the Whangarei-Bay of Islands subdivision. New Zealand Geological Survey — Bulletin, 27.

Field, B.D. & G.H. Browne (1986). Lithostratigraphy of Cretaceous and Tertiary rocks, southern Canterbury, New Zealand. New Zealand Geological Survey — Record, 14.

Finlayson, D.M. (1982). Geophysical differences in the lithosphere between Phanerozoic and Precambrain Australia. Tectonophysics, 84, 287–312.

Finlayson, D.M. (1983). The mid-crustal horizon under the Eromanga Basin, Eastern Australia. Tectonophysics, 100, 199–214.

Finlayson, D.M., & S.P. Mather (1984). Seismic refraction and reflection features of the lithosphere in northern and eastern Australia, and continental growth. Annales Geophysicae, 2, 711–722.

Finlayson, D.M., C. Prodehl, & C.D.N. Collins (1979). Explosion seismic profiles, and implications for crustal evolution in southeastern Australia. BMR Journal of Australian Geology and Geophysics, 4, 243–252.

Finlayson, D.M., C.D.N. Collins, & J. Lock (1984). P-wave velocity features of the lithosphere under the Eromanga Basin, eastern Australia, including a prominent mid-crustal (Conrad?) discontinuity. Tectonophysics, 101, 267–291.

Finnerty, A.A. (in press). Inflected mantle geotherms from xenoliths are real: evidence from olivine barometry. Proceedings of the Fourth International Kimberlite Conference, Perth. Geological Society of Australia — Special Publication.

Finnerty, A.A., & F.R. Boyd (1984). Evaluation of thermo-barometers for garnet peridotites. Geochimica et Cosmochimica Acta, 48, 15–27.

Fisher, R.V., & H.-U. Schmincke (1984). Pyroclastic Rocks. Springer-Verlag, Berlin.

Fisher, R.V., & A.C. Waters (1970). Base surge bedforms in maar volcanoes. American Journal of Science, 268, 157–180.

Fitton, J.G. (1980). The Benue trough and Cameroon line — a migrating rift system in West Africa. Earth and Planetary Science Letters, 51, 132–138.

Fitton, J.G. (1983). Active versus passive continental rifting: evidence from the West African rift system. Tectonophysics, 94, 473–481.

Fitton, J.G. (1987). The Cameroon line, West Africa: a comparison between oceanic and continental alkaline volcanism. In: Fitton, J.G., & B.G.J. Upton (Editors), Alkaline Igneous Rocks. Geological Society — Special Publication, 30, 273–291.

Fitton, J.G., & H.M. Dunlop (1985). The Cameroon line, West Africa, and its bearing on the origin of oceanic and continental alkali basalt. Earth and Planetary Science Letters, 72, 23–38.

Fleming, C.A. (1959). Geology of New Zealand by Ferdinand von Hochstetter, translated and edited by C.A. Fleming. Government Printer, Wellington.

Fleming, P.D., D.A. Steele, & A. Camacho (1985). A probable Cambrian migmatitic basement to the Wagga zone in North-East Victoria. In: Vandenberg, A.H.M. (Editor), Victorian Lithosphere Symposium, Melbourne — Abstracts, 14–15.

Flinders, M. (1799). Historical Records of New South Wales, Volume 3, 812.

Flint, J.C.E., C.G. Lancaster, R.E. Gould, & H.D. Hensel (1976). Some new stratigraphic data from the southern Clarence-Moreton Basin. Queensland Government Mining Journal, 70, 1–5.

REFERENCES

Flood, P.G. (1983). Tectonic setting and development of the Bowen Basin. In: Waterhouse, J.B. (Editor), 1983 Field Conference, Permian of the Biloela-Moura-Cracow area. Geological Society of Australia, Queensland Division, Brisbane, 7–21.

Ford, C.E., D.G. Russell, J.A. Craven, & M.R. Fisk (1983). Olivine-liquid equilibria: temperature, pressure and composition dependence of the crystal/liquid cation partition coefficients for Mg, Fe^{2+}, Ca and Mn. Journal of Petrology, 24, 256–265.

Forsyth, S.M. (1984). Geological Atlas 1:50 000 Series, Sheet 68 (83135), Oatlands. Tasmanian Department of Mines — Explanatory Report.

Fountain, D.M., & M.H. Salisbury (1981). Exposed cross-sections through the continental crust: implications for crustal structure, petrology and evolution. Earth and Planetary Science Letters, 56, 262–277.

Francis, D. (1985). The Baffin Bay lavas and the value of picrites as analogues of primary magmas. Contributions to Mineralogy and Petrology, 89, 144–154.

Francis, G., & G.T. Walker (1978). Silcretes of sub-aerial origin in southern New England. Search, 9, 321–323.

Frey, F.A., & D.H. Green (1974). The mineralogy, geochemistry and origin of lherzolite inclusions in Victorian basanites. Geochimica et Cosmochimica Acta, 38, 1023–1059.

Frey, F.A., & M. Prinz (1978). Ultramafic inclusions from San Carlos, Arizona: petrologic and geochemical data bearing on their petrogenesis. Earth and Planetary Science Letters, 129–176.

Frey, F.A., D.H. Green, & S.D. Roy (1978). Integrated models of basalt petrogenesis: a study of quartz tholeiites to olivine melilitites from south eastern Australia utilizing geochemical and experimental petrological data. Journal of Petrology, 19, 463–513.

Freundt, A., & H.-U. Schmincke (1986). Emplacement of small-volume pyroclastic flows at Laacher Sea (East-Eifel, Germany). Bulletin of Volcanology, 48, 39–59.

Futa, K., & W.E. LeMasurier (1983). Nd and Sr isotopic studies on Cenozoic mafic lavas from West Antarctica: another source for continental alkali basalts. Contributions to Mineralogy and Petrology, 83, 38–44.

Gaffney, E.S., G.C. McNamara (in press). A Meiolaniid turtle from the Pleistocene of northern Queensland. Memoir of the Queensland Museum. De Vis Symposium Special Volume.

Gage, M. (1957). The geology of Waitaki subdivision. New Zealand Geological Survey — Bulletin, 55.

Gair, H.A., & P.C. Rickwood (1965). The Timaru and Geraldine tholeiitic basalt sheets. New Zealand Department of Science and Industrial Research — Information Serial, 51, 34–38.

Galer, S.J.G., & R.K. O'Nions (1986). Magmagenesis and the mapping of chemical and isotopic variations in the mantle. Chemical Geology, 56, 45–61.

Galloway, R.W. (1967). Pre-basalt, sub-basalt, and post-basalt surfaces of the Hunter Valley, New South Wales. In: Jennings, J.J., & J.A. Mabbutt (Editors), Landform Studies in Australia and New Guinea. Cambridge University Press, 219–314.

Gamble, J.A. (1984). Petrology and geochemistry of differentiated teschenite intrusions from the Hunter Valley, New South Wales, Australia. Contributions to Mineralogy and Petrology, 88, 173–187.

Gamble, J.A., & C.J.D. Adams (1985). Volcanic geology of Carnley volcano, Auckland Islands. New Zealand Journal of Geology and Geophysics, 28, 43–54.

Gamble, J.A., P.A. Morris, & C.J. Adams (1986). The geology, petrology and geochemistry of Cenozoic volcanic rocks from the Campbell Plateau and Chatham Rise. In: Smith, I.E.M. (Editor), Late Cenozoic Volcanism in New Zealand. Royal Society of New Zealand Bulletin, 23, 344–365.

Garrick, R.A. (1968). A re-interpretation of the Wellington crustal refraction profile. New Zealand Journal of Geology and Geophysics, 11, 1280–1298.

Gass, I.G. (1970). The evolution of volcanism in the junction area of the Red Sea, Gulf of Aden and Ethiopia rifts. Philosophical Transactions of the Royal Society of London, A267, 369–381.

Gass, I.G., & D.I.J. Mallick (1968). Jebel Khariz: an Upper Miocene strato-volcano of comenditic affinity on the south Arabian coast. Bulletin Volcanologique, 32, 33–88.

Gee, R.D. (1971). Geological atlas 1 mile series. Sheet 22 (8016S) Table Cape. Tasmanian Department of Mines — Explanatory Report.

Gee, R.D. (1977). Geological atlas 1 mile series. Sheet 28 (8015N) Burnie. Tasmanian Department of Mines — Explanatory Report.

Gerard, V.B., & J.A. Lawrie (1955). Aeromagnetic surveys in New Zealand, 1947–1951. New Zealand Department of Scientific and Industrial Research — Geophysical Memoir, 3.

Ghiorso, M.S. (1984). Activity/composition relations in the ternary feldspars. Contributions to Mineralogy and Petrology, 87, 282–296.

Ghiorso, M.S. (1985). Chemical mass transfer in magmatic processes. 1. Thermodynamic relations and numerical algorithms. Contributions to Mineralogy and Petrology, 90, 107–120.

Ghiorso, M.S., & I.S.E. Carmichael (1981). A Fortran IV computer program for evaluating temperatures and oxygen fugacities from the compositions of coexisting iron-titanium oxides. Computers and Geosciences, 7, 123–129.

Ghiorso, M.S., I.S.E. Carmichael, M.L. Rivers, & R.O. Sack (1983). The Gibbs free energy of mixing of natural silicate liquids; an expanded regular solution approximation for the calculation of magmatic intensive variables. Contributions to Mineralogy and Petrology, 84, 107–145.

Gill, E.D. (1953). Geological evidence in Western Victoria relative to the antiquity of the Australian Aborigines. National Museum, Melbourne — Memoir, 18, 25–92.

Gill, E.D. (1978). Radiocarbon dating of the volcanoes of western Victoria, Australia. Victorian Naturalist, 95, 152–158.

Gill, E.D. (1979). The Tyrendarra lava flow, western Victoria, Australia. Victorian Naturalist, 96, 227–229.

Gill, E.D., & M.R. Banks (1956). Cainozoic history of the Manbray Swamp and other areas of north-western Tasmania. Victorian Museum — Record Quarterly, NS.6.

Gill, E.D., & L.K.M. Elmore (1973). Radiocarbon dating of the Mount Napier eruption, western Victoria, Australia. Victorian Naturalist, 90, 304–306.

Gill, J.B. (1981). Orogenic Andesites and Plate Tectonics. Springer-Verlag, Berlin.

Gill, J.B. (1984). Sr-Pb-Nd isotopic evidence that both MORB and OIB sources contribute to oceanic island arc magmas in Fiji. Earth and Planetary Science Letters, 68, 443–458.

Giret, A., B. Bonin, & J.M. Leger (1980). Amphibole compositional trends in oversaturated and undersaturated alkaline plutonic ring-complexes. Canadian Mineralogist, 18, 481–495.

Glaessner, M.F. (1950). Geotectonic position of New Guinea. Bulletin of American Association of Petroleum Geologists, 34, 856–881.

Gleadow, A.J.W., & C.D. Ollier (1987). The age of gabbro at The Crescent, New South Wales. Australian Journal of Earth Sciences, 34, 209–212.

Gloe, C.S. (1976). Brown Coal. In: Douglas, J.A., & J.A. Ferguson (Editors), Geology of Victoria. Geological Society of Australia — Special Publication, 5, 378–389.

Govey, A.L. (1974). The geology of an area between Nerriga and Sassafras, Southern Tablelands, N.S.W. University of New South Wales, unpublished BSc Honours thesis.

Grant, J. (1803). The Narrative of a Voyage of Discovery Performed in his Majesty's Vessel, the Lady Nelson — to New South Wales. Egerton, London.

Grapes, R.H. (1975). Petrology of the Blue Mountain complex, Marlborough, New Zealand. Journal of Petrology, 16, 371–378.

Gray, A.R.C. (1976). Hillsborough Basin. In: Leslie, K.B., H.J. Evans, & C.L. Knight (Editors), Economic Geology of Australia and Papua New Guinea. Australasian Institute of Mining and Metallurgy, 460–464.

Greeley, R. (1977). Basaltic plains volcanism. In: Greeley, R., & J.S. King (Editors), Volcanism of the Eastern Snake River plain, Idaho: A Comparative Planetary Geology Guidebook. NASA, Washington, D.C.

Greeley, R. (1982a). The Snake River plain, Idaho: representative of a new category of volcanism. Journal of Geophysical Research, 87, 2705–2712.

Greeley, R. (1982b). The style of basaltic volcanism in the eastern Snake River plain, Idaho. In: Bonnichsen, B. & R.M. Brekenridge (Editors), Cenozoic Geology of Idaho. Idaho School of Mines — Bulletin 26, 407–421.

Green, D.C. (1960). The geology of the South Arm-Sandford area. Paper and Proceedings of the Royal Society of Tasmania, 95, 17–35.

Green, D.C. (1974, Editor). University of Queensland, Department of Geology and Mineralogy, Isotope Geology Laboratory Report, 2, 1971–1974.

Green, D.C., & N.C. Stevens (1975). Age and stratigraphy of Tertiary volcanic and sedimentary rocks of the Ipswich district, southeast Queensland. Queensland Government Mining Journal, 76, 148–150.

Green, D.H. (1971). Composition of basaltic magmas as indications of conditions of origin: Application to oceanic volcanism. Philosophical Transaction of the Royal Society of London, A268, 707–725.

Green, D.H. (1973). Conditions of melting of basanite magma from garnet peridotite. Earth and Planetary Science Letters, 17, 456–465.

Green, D.H., & W.O. Hibberson (1970). Experimental duplication of conditions of precipitation of high presssure phenocrysts in a basaltic magma. Physics of Earth and Planetary Interiors, 3, 247–253.

Green, D.H., & R.C. Liebermann (1976). Phase equilibria and elastic properties of a pyrolite mantle for the oceanic mantle. Tectonophysics, 32, 61–92.

Green, D.H., and A.E. Ringwood (1963). Mineral assemblages in a model for the upper mantle. Journal of Geophysical Research, 70, 5259–5268.

Green, D.H., & A.E. Ringwood (1967). The genesis of basaltic magmas. Contributions to Mineralogy and Petrology, 15, 103–190.

Green, D.H., & A.E. Ringwood (1972). A comparison of recent experimental data on the gabbro-garnet granulite-eclogite transition. Journal of Geology, 80, 277–288.

Green, D.H., A.D. Edgar, P. Beasley, E. Kiss, & N.G. Ware (1974). Upper mantle source for some hawaiites, mugearites and benmoreites. Contributions to Mineralogy and Petrology, 48, 33–43.

Green, T.H., & N.J. Pearson (1986). Ti-rich accessory phase saturation in hydrous mafic-felsic compositions at high P, T. Chemical Geology, 54, 185–201.

Green, T.H., & N.J. Pearson (1987). An experimental study of Nb and Ta partitioning between Ti-rich minerals and silicate liquids at high pressure and temperature. Geochimica et Cosmochimica Acta, 51, 55–62.

Gregg, D.R. (1961). Volcanoes of Tongariro National Park. New Zealand Department of Scientific and Industrial Research — Information Series, 28.

Gregg, D.R. (1964). Sheet 18 Hurunui. Geological Map of New Zealand 1:250 000. New Zealand Department of Scientific and Industrial Research.

Gregory, A.C. (1879). On the geological features of the south-eastern districts of Queensland, Government Printer, Brisbane.

Gregory, J.W. (1903). The Geography of Victoria. Whitcombe and Tombs, Melbourne.

Grenfell, A.T. (1984). The stratigraphy geochronology and petrology of the volcanic rocks of the Main Range, southeastern Queensland. University of Queensland, unpublished PhD thesis.

Griffin, R.J. (1961). The Bugaldie (Chalk Mountain) diatomaceous earth deposit. New South Wales Department of Mines — Technical Report, 7, 19–36.

Griffin, T.J. (1977). The geology, mineralogy and geochemistry of the McBride basaltic province, northern Queensland. James Cook University of North Queensland, unpublished PhD thesis.

Griffin, T.J., & I. McDougall (1975). Geochronology of the Cainozoic McBride volcanic province, northern Queensland. Journal of the Geological Society of Australia, 22, 387–397.

Griffin, W.L., & K.S. Heier (1973). Petrological implications of some corona structures. Lithos, 6, 315–335.

Griffin, W.L. & S.Y. O'Reilly (1986a). The lower crust in eastern Australia: xenolith evidence. In: Dawson, J.B., D.A. Carswell, J. Hall, & K.H. Wedepohl (Editors), The Nature of the Lower Continental Crust. Geological Society — Special Publication, 24, 363–374.

Griffin, W.L., & S.Y. O'Reilly (1986b). Sapphirine in a mantle-derived xenolith from Delegate, Australia. Mineralogical Magazine, 50, 635–640.

Griffin, W.L., & S.Y. O'Reilly (1986c). Chemical and isotopic characteristics of multiple-metasomatised mantle xenoliths from western Victoria. Fourth International Kimberlite Conference, Geological Society of Australia — Extended Abstracts, 16, 247–249.

Griffin, W.L., & S.Y. O'Reilly (1987a). The composition of the lower crust and the nature of the continental Moho. In: Nixon, P.H. (Editor), Mantle Xenoliths. Wiley, London, 413–430

Griffin, W.L., & S.Y. O'Reilly (1987b). Is the continental Moho the crust/mantle boundary? Geology, 15, 241–244.

Griffin, W.L., D.A. Carswell, & P.H. Nixon (1979). Lower crustal granulites and eclogites from Lesotho, southern Africa. In: Boyd F.R, & H.O.A. Meyer (Editors), The Mantle Sample: Inclusions in Kimberlites and other Volcanics. American Geophysical Union, Washington, D.C., 59–86.

Griffin, W.L., S.Y. Wass, & J.D. Hollis (1984). Ultramafic xenoliths from Bullenmerri and Gnotuk maars, Victoria, Australia: petrology of a sub-continental crust-mantle transition. Journal of Petrology, 25, 53–87.

Griffin, W.L., F.L. Sutherland, & J.D. Hollis (1987). Geothermal profile and crust-mantle transition beneath east-central Queensland: vulcanology, xenolith petrology and seismic data. Journal of Volcanology and Geothermal Research, 31, 177–203.

Griffin, W.L., S.Y. O'Reilly, & A. Stabel (1988). Mantle metasomatism beneath western Victoria, Australia, II: Isotopic geochemistry of Cr-diopside lherzolites and Alaugite pyroxenites. Geochimica et Cosmochimica Acta, 52, 449–459.

Grimes, K.G. (1980). The Tertiary geology of north Queensland. In: Henderson, R.A., & P.J. Stephenson (Editors), The Geology and Geophysics of Northeastern Australia. Geological Society of Australia, Queensland Division, Brisbane, 329–347.

Grimes, K.G. (1982). Stratigraphic drilling report — GSQ Sandy Cape 1–3R, Queensland Government Mining Journal, 83, 224–233.

Grindley, G.W. (1961). Mesozoic orogenies in New Zealand. Proceedings of the Ninth Pacific Science Congress, 12, 71–75.

Grindley, G.W. (1980). Sheet S13 Cobb, Geological map of New Zealand. New Zealand Department of Scientific and Industrial Research.

Grindley, G.W., & F.J. Davey (1982). The reconstruction of New Zealand, Australia, and Antarctica. In: Craddock, C. (Editor), Antarctic Geoscience. University of Wisconsin Press, Madison, 15–29.

Grindley, G.W., C.J.D. Adams, J.T. Lumb, & W.A. Watters (1977). Paleomagnetism, K-Ar dating and tectonic interpretation of Upper Cretaceous and Cenozoic volcanic rocks of the Chatham Islands, New Zealand. New Zealand Journal of Geology and Geophysics, 20, 425–467.

Gulson, B.L., L.R. Bottomer, K.J. Mizon (1985). Precambrian source for granites and volcanics and source of gold-silver mineralization from the Drake area, New England, N.S.W. Commonwealth Scientific and Industrial Research Organization, Australia, Division of Mineralogy & Geochemistry — Research Review, 45–46.

Gunnarsson, A. (1973). Volcano ordeal by fire in Iceland's Westmann Islands. Iceland Review, Reykjavik.

Gunthorpe, R.J. (1970). Plutonic and metamorphic rocks of the Walcha-Nowendoc-Yarrowitch district, New South Wales. University of New England, unpublished PhD thesis.

Haast, H.F. von (1948). The Life and Times of Sir Julius von Haast. Wellington.

Haast, J. von (1879). Geology of the Provinces of Canterbury and Westland. Lyttelton Times.

Haggerty, S.E. (1976). Opaque mineral oxides in terrestrial igneous rocks. In: Rumble III, D. (Editor), Oxide Minerals. Mineralogical Society of America — Short Course Notes, 3, Chapter 8, Hg1–300.

Haggerty, S.E. (1983). The mineral chemistry of new titanates from the Jagersfontein kimberlite, South Africa: implications for metasomatism in the mantle. Geochimica et Cosmochim Acta, 47, 1833–1854.

Haggerty, S.E. (1986). Diamond genesis in a multiple constrained model. Nature, 320, 34–38.

Halford, G.E. (1970). Dykes and their inclusions from Kelly's Point, New South Wales. Australian National University, unpublished MSc thesis.

Halliday, A.N., A.P. Dickin, A.E. Fallick, & J.G. Fitton (1988). Mantle dynamics: a Nd, Sr, Pb and O isotopic study of the Cameroon line volcanic chain. Journal of Petrology, 29, 181–211.

Hamilton, D.L. (1961). Nephelines as crystallization temperature indicators. Journal of Geology, 69, 321–329.

Hamilton, P.J., N.M. Evensen, & R.K. O'Nions (1979). Sm-Nd systematics of Lewisian gneisses: implications for the origin of granulites. Nature, 277, 25–28.

Hanan, B.B., R.H. Kingsley, & J.-G. Schilling (1986). Pb isotopic evidence in the South Atlantic for migrating ridge-hotspot interactions. Nature, 322, 137–144.

Hanks, W. (1955). Newer Volcanic vents and lava fields between Wallan and Yuroke, Victoria. Proceedings of the Royal Society of Victoria, 67, 1–16.

Hanson, G.N. (1980). Geochemical evolution of the continental crust. In: Continental Tectonics, Studies in Geophysics. National Academy of Seismics, Washington, D.C., 151–155.

Harding, R.R. (1966). Catalogue of age determinations on Australian rocks, 1962–1965. Bureau of Mineral Resources, Australia — Report, 117.

Harland, W.B., A.V. Cox, P.G. Llewellyn, C.A.G. Pickton, A.G. Smith, & R. Walters (1982). A Geologic Time Scale. Cambridge University Press, Cambridge.

Harmon, R.S., A.N. Halliday, J.A.P. Clayburn, & W.E. Stephens (1984). Chemical and isotopic systematics of the Caledonian intrusions of Scotland and Northern England: a guide to magma source region and magma-crust interaction. Philosophical Transactions of the Royal Society of London, A310, 709–742.

Harrington, H.J. (1983). Correlation of the Permian and Triassic Gympie Terrane of Queensland with the Brook Street and Maitai Terranes of New Zealand. In: Permian Geology of Queensland. Geological Society of Australia, Queensland Division, Brisbane, 431–436.

Harrington, H.J., & R.J. Korsch (1985a). Late Permian to Cainozoic tectonics of the New England Orogen. Australian Journal of Earth Sciences, 32, 181–203.

Harrington, H.J., & R.J. Korsch (1985b). Tectonic model for the Devonian to middle Permian of the New England Orogen. Australian Journal of Earth Sciences, 32, 163–179.

Harrington, H.J., K.L. Burns, & B.R. Thompson (1973). Gambier-Beaconsfield and Gambier-Sorell fracture zones and the movement of plates in Australian-Antarctic-New Zealand region. Nature, 245, 109–112.

Hart, S.R. (1984). A large-scale isotope anomaly in the Southern Hemisphere mantle. Nature, 309, 753–757.

Hart, S.R. & C.J. Allègre (1980). Trace-element constraints on magma genesis. In: Hargraves, R.B. (Editor), Physics of Magmatic Processes. Princeton University Press, Princeton, 121–159.

Hart, S.R., & K.E. Davis (1978). Nickel partitioning between olivine and silicate melt. Earth and Planetary Science Letters, 40, 203–219.

Hart, S.R., D.C. Gerlach, & W.M. White (1986). A possible new Sr-Nd-Pb mantle array and consequences for mantle mixing. Geochimica et Cosmochimica Acta, 50, 1551–1557.

Hart, W.K., & R.W. Carlson (1987). Tectonic controls on magma genesis and evolution in the northwestern United States. Journal of Volcanology and Geothermal Research, 32, 119–135.

Harte, B. (1977). Rock nomenclature with particular relation to deformation and recrystallization textures in olivine-bearing xenoliths. Journal of Geology, 85, 279–288.

Harvey, M., & G.A. Joplin (1941). A note on some leucite bearing rocks from New South Wales with special reference to an ultrabasic occurrence at Murrumburrah. Journal and Proceedings of the Royal Society of New South Wales, 74, 419–442.

Hawkesworth, C.J., A.J. Erlank, J.S. Marsh, M.A. Menzies, & P. van Calsteren (1983). Evolution of the continental lithosphere: evidence from volcanics and xenoliths in southern Africa. In: Hawkesworth, C.J., & M.J. Norry (Editors), Continental Basalts and Mantle Xenoliths. Shiva, Cheshire, 111–138.

Hawkesworth, C.J., J.S. Marsh, A.R. Duncan, A.J. Erlank, & M.J. Norry (1984). The role of continental lithosphere in the generation of the Karoo volcanic rocks: evidence from combined Nd- and Sr-isotope studies. In: Erlank,

A.J. (Editor), Petrogenesis of the Volcanic Rocks of the Karoo Province. Geological Society of South Africa — Special Publication, 13, 341–354.

Hawkesworth, C.J., M.S.M. Mantovani, P.N. Taylor, & Z. Palacz (1986). Evidence from the Paraná of south Brazil for a continental contribution to Dupal basalts. Nature, 322, 356–358.

Hawkesworth, C.J., P. Van Calsteren, N.W. Rogers, & M.A. Menzies (1987). Isotope variations in recent volcanics: a trace element perspective. In: Menzies, M.A., & C.J. Hawkesworth (Editors), Mantle Metasomatism. Academic Press, London, 365–388.

Haxby, W.F. (1987). Gravity Field of the World's Oceans. National Geophysical Data Center, National Oceanic and Atmospheric Administration, Boulder, Colorado.

Haxby, W.F., G.D. Karner, J.L. La Brecque, & J.K. Weissel (1983). Digital images of combined oceanic and continental data sets and their use in tectonic studies. Transactions of the American Geophysical Union, 64, 905–1004.

Hector, J. (1865). On the geology of Otago, New Zealand. Quarterly Journal of the Geological Society, 21, 124–128.

Heier, K.S. (1973). A model for the composition of the deep continental crust. Fortschrift für Mineralogie, 50, 174–187.

Heiken, G.H., & K.H. Wohletz (1985). Volcanic Ash. University of California Press, Berkeley.

Heirtzler, J.R., G.O. Dickson, E.M. Herron, W.C. Pitman III, & X. Le Pichon. (1968). Marine magnetic anomalies, geomagnetic field reversals and motions of the ocean floor and continents. Journal of Geophysical Research, 73, 2119–2136.

Helby, R., & R. Morgan (1979). Palynomorphs in Mesozoic volcanoes. Geological Survey of New South Wales — Quarterly Notes, 35, 1–18.

Helmstaedt, H., & R. Doig (1975). Eclogite nodules from kimberlite pipes of the Colorado Plateau — samples of subducted Franciscan-type oceanic lithosphere. Physics and Chemistry of the Earth, 9, 95–111.

Heming, R.F. (1980a). Petrology and geochemistry of Quaternary basalts from Northland, New Zealand. Journal of Volcanology and Geothermal Research, 8, 23–44.

Heming, R.F. (1980b). Patterns of Quaternary basaltic volcanism in the northern North Island, New Zealand. New Zealand Journal of Geology and Geophysics, 23, 335–344.

Heming, R.F. (1980c). The Ngatutura diatreme. New Zealand Journal of Geology and Geophysics, 23, 569–573.

Heming, R.J., & P.R. Barnet (1986). The petrology and petrochemistry of the Auckland volcanic field. In: Smith, I.E.M. (Editor), Late Cenozoic Volcanism in New Zealand. Royal Society of New Zealand Bulletin, 23, 64–75.

Henderson, R.A. (1983). Early Ordovician faunas from the Mount Windsor subprovince, northeastern Queensland. Association of Australasian Palaeontologists — Memoir, 1, 145–175.

Hensel, H.D., M.T. McCulloch, & B.W. Chappell (1985). The New England Batholith: constraints on its derivation from Nd and Sr isotopic studies of granitoids and country rocks. Geochimica et Cosmochimica Acta, 49, 369–384.

Henstridge, D.A., & A.C. Hutton (1986). Geology and organic petrology of the Nagoorin oil shale deposit. Proceedings of the Third Australian Workshop on Oil Shale. Lucas Heights, Sydney, 40–44.

Henstridge, D.A., & D.D. Missen (1981). The geology of the Narrows Graben near Gladstone, Queensland, Australia. 51st ANZAAS Congress, Brisbane — Abstract.

Herbert, C. (1968). Diatomite deposits in the Warrumbungles. Geological Survey of New South Wales — Report, 1968/195.

Hergt, J.M. (1987). The origin and evolution of the Tasmanian dolerites. Australian National University, unpublished PhD thesis.

Herzberg, C.T. (1978). Pyroxene geothermometry and geobarometry: experimental and thermodynamic evaluation of some subsolidus phase relations involving clinopyroxenes in the system $CaO-MgO-Al_2O_3-SiO_2$. Geochimica et Cosmochimica Acta, 42, 945–957.

Herzberg, C.T., W.S. Fyfe, & M.J. Carr (1983). Density constraints on the formation of the continental Moho and crust. Contributions to Mineralogy and Petrology, 84, 1–5.

Hilde, T.W.C., S. Uyeda, & L. Kroenke (1976). Tectonic history of the western Pacific. In: Drake, C.L. (Editor), Geodynamics: Progress and Prospects. American Geophysical Union, Washington, 1–15.

Hildreth, W. (1979). The Bishop tuff: evidence for the origin of compositional zonation in silicic magma chambers. In: Chapin, C.E., & W.E. Elston (Editors), Ash-Flow Tuffs. Geological Society of America — Special Paper, 180, 43–75.

Hildreth, W., R.L. Christiansen, & J.R. O'Neil (1984). Catastrophic isotopic modification of rhyolitic magma at times of caldera subsidence, Yellowstone Plateau volcanic field. Journal of Geophysical Research, 89, 8339–8369.

Hills, E.S. (1940). The Physiography of Victoria. Whitcombe & Tombs, Melbourne.

Hills, E.S. (1956). A contribution to the morphotectonics of Australia. Journal of the Geological Society of Australia, 3, 1–15.

Hilmansyah, L. (1985). A volcanic hazard assessment of the Newer Volcanics in Victoria and South Australia. University of New South Wales, unpublished MSc thesis.

Hinz, K., J.B. Wilcox, M. Whiticar, H.R. Kudrass, N.F. Exon, & D.A. Feary (1986). The west Tasmanian margin: an underrated petroleum province? In: Glenie, R.C. (Editor), Second South-Eastern Australia Oil Exploration Symposium. Petroleum Exploration Society of Australia, Melbourne, 1985.

Hochstetter, F. von (1864). The Geology of New Zealand. Translated from German by C.A. Fleming, 1959. New Zealand Government Printer, Wellington.

Hocking, J.B. (1976). Gippsland Basin. In: Douglas, J.G., & J.A. Ferguson (Editors), Geology of Victoria. Geological Society of Australia — Special Publication, 5, 248–273.

Hockley, J.J. (1972). Alkaline rock lineages in the Warrumbungle shield volcano, eastern Australia. Nature, 236, 15–16.

Hockley, J.J. (1973). Differentiation trends in the Warrumbungle volcano, New South Wales, Australia. Geologischen Rundschau, 62, 179–187.

Hockley, J.J. (1974). The phonolite-trachyte spectrum in the Warrumbungle volcano, New South Wales, Australia. Journal and Proceedings of the Royal Society of New South Wales, 107, 87–89.

Hoffman, K.A. (1984). Late acquisition of 'primary' remanence in some fresh basalts: a cause of spurious palaeomagnetic results. Geophysical Research Letters, 11, 681–684.

Hoffman, K.A. (in press). Transitional field behaviour from southern hemisphere lavas: evidence for a two stage reversal process? Nature.

Hofmann, A.W., & M. Margaritz (1977). Diffusion of Ca, Sr, Ba and Co in a basalt melt: implications for the geochemistry of the mantle. Journal of Geophysical Research, 82, 5432–5438.

Hofmann, A.W., K.P. Jochum, M. Seufert, & W.M. White (1986). Nb and Pb in oceanic basalts: new constraints on mantle evolution. Earth and Planetary Science Letters, 79, 33–45.

Holland, J.G., & G.M. Brown (1972). Hebridean tholeiitic magmas: a geochemical study of the Ardnamurchan cone sheets. Contributions to Mineralogy and Petrology, 37, 139–160.

Hollis, J.D. (1981). Ultramafic and gabbroic nodules from the Bullenmerri and Gnotuk maars, Camperdown, Victoria. Proceedings of the Royal Society of Victoria, 92, 155–167.

Hollis, J.D. (1985). Volcanism and upper mantle-lower crust relationships: evidence from inclusions in alkali basaltic rocks. In: Sutherland, F.L., B.J. Franklin, & A.E. Waltho (Editors), Volcanism in Eastern Australia, with Case Histories from New South Wales. Geological Society of Australia, New South Wales Division, 1, 33–47.

Hollis, J.D., F.L. Sutherland, & R.E. Pogson (1983). High pressure minerals and the origin of the Tertiary breccia pipe, Ballogie Gem Mine, near Proston, Queensland. Australian Museum — Record, 35, 181–194.

Holmes, A. (1944). Principles of Physical Geology. Nelson, London.

Holmes, W.B.K., F.M. Holmes, & H.A. Martin (1983). Fossil eucalyptus remains from the Middle Miocene Chalk Mountain Formation, Warrumbungle Mountains, New South Wales. Proceedings of the Linnaean Society of New South Wales, 106, 299–310.

Hooper, P.R. (1982a). Structural model for the Columbia River basalt near Riggins, Idaho. In: Bonnichsen, B., & R.M. Breckenridge (Editors), Cenozoic Geology of Idaho. Idaho Bureau of Mines — Bulletin, 26, 129–136.

Hooper, P.R. (1982b). The Columbia River basalts. Science 215, 1463–1468.

Houghton, B.F., & H.-U. Schmincke (1986). A mixed deposit of simultaneous strombolian and phreatomagmatic volcanism: Rothenberg volcano, East Eiffel volcanic field. Journal of Volcanology and Geothermal Research, 30, 117–130.

Houghton, B.F., C.J.N. Wilson, & P. Van den Bogard (1986a). Pyroclastic fall deposits: eruptive mechanisms inferred from deposit characteristics. International Volcanological Congress, New Zealand — Abstract, 106.

Houghton, B.F., C.J.N. Wilson, I.E.M. Smith, & R.J. Parker (1986b). Crater Hill. International Volcanological Congress, New Zealand — Handbook, 68–83.

Houston, B.R. (1967). The post-Palaeozoic sediments and volcanics. In: Geology of the City of Brisbane. Geological Survey of Queensland — Publication, 324.

Howell, D.G. (1980). Mesozoic accretion of exotic terranes along the New Zealand segment of Gondwanaland. Geology, 8, 487–491.

Howitt, A.W. (1876). Notes on the microscopic examination of igneous rock specimens from south-eastern Gippsland. Geological Survey of Victoria — Progress Report, 1875, 175–177.

Hubble, T.C.T. (1983). The geology and petrology of the eastern part of the Warrumbungle volcano. University of Sydney, unpublished BSc Honours thesis.

Huppert, H.E., & R.S.J. Sparks (1984). Double-diffusive convection due to crystallization in magmas. Annual Reviews of Earth and Planetary Sciences, 12, 11–37.

Huppert, H.E. & R.S.J. Sparks (1985). Cooling and contamination of mafic and ultramafic lavas during ascent through continental crust. Earth and Planetary Science Letters, 74, 371–386.

Hutton, F.W. (1874). Report on the geology of the north eastern part of the South Island. Geological Survey New Zealand — Reports of Geological Explorations 1873-4, 27–58.

Idnurm, M. (1985a). Late Mesozoic and Cenozoic palaeomagnetism of Australia — I. A redetermined apparent polar wander path. Geophysical Journal of the Royal Astronomical Society, 83, 399–418.

Idnurm, M. (1985b). Late Mesozoic and Cenozoic palaeomagnetism of Australia — II. Implications for geomagnetism and true polar wander. Geophysical Journal of the Royal Astronomical Society, 83, 419–433.

Idnurm, M., & B.R. Senior (1978). Palaeomagnetic ages of Late Cretaceous and Tertiary weathered profiles in the Eromanga Basin, Queensland. Palaeogeography, Palaeoclimatology, Palaeoecology, 24, 263–277.

Irving, A.J. (1971). Geochemical and high pressure experimental studies of xenoliths, megacrysts and basalts from southeastern Australia. Australian National University, unpublished PhD thesis.

Irving, A.J. (1974a). Geochemical and high pressure experimental studies of garnet pyroxenite and pyroxene granulite xenoliths from the Delegate basaltic pipes, Australia. Journal of Petrology, 15, 1–40.

Irving, A.J. (1974b). Megacrysts from the Newer Basalts and other basaltic rocks of southeastern Australia. Bulletin of the Geological Society of America, 85, 1503–1514.

Irving, A.J. (1974c). Pyroxene-rich ultramafic xenoliths in the Newer Basalts of Victoria, Australia. Neues Jahrbuch für Mineralogie Abhandlungen, 120, 147–167.

Irving, A.J. (1978). A review of experimental studies of crystal/liquid trace element partitioning. Geochimica et Cosmochimica Acta, 42, 743–770.

Irving, A.J. (1980). Petrology and geochemistry of composite ultramafic xenoliths in alkalic basalts and implications for magmatic processes within the mantle. American Journal of Science, 280A, 389–426.

Irving, A.J., & F.A. Frey (1984). Trace element abundances in megacrysts and their host basalts: constraints on partition coefficients and megacryst genesis. Geochimica et Cosmochimica Acta, 48, 1201–1221.

Irving, A.J., & D.H. Green (1976). Geochemistry and petrogenesis of the Newer Basalts of Victoria and South Australia. Journal of the Geological Society of Australia, 23, 45–66.

Irving, A.J., & R.C. Price (1981). Geochemistry and evolution of lherzolite-bearing phonolitic lavas from Nigeria, Australia, East Germany and New Zealand. Geochimica et Cosmochimica Acta, 45, 1309–1320.

Isbell, R.F., P.J. Stephenson, G.G. Murtha, & G.P. Gillman (1976). Red basaltic soils in north Queensland. Commonwealth Scientific and Industrial Research Organization, Australia, Division of Soils — Technical Paper, 34.

Jack, R.L. (1886). Geological map of Queensland, on the scale of 32 miles to an inch. Queensland Department of Public Works and Mines.

Jack, R.L., & R. Etheridge (1892). The Geology and Palaeontology of Queensland and New Guinea. Government Printer, Brisbane.

Jackson, E.D. (1968). The character of the lower crust and upper mantle beneath the Hawaiian Islands. 23rd International Geological Congress, Prague — Proceedings, 1, 135–150.

Jackson, I., & R.A. Arculus (1984). Laboratory wave velocity measurements on lower crustal xenoliths from Calcutteroo, South Australia. Tectonophysics, 101, 185–197.

Jacobson, R., & T.R. Scott (1937). The geology of the Korkuperrimul Creek area, Bacchus Marsh. Proceedings of the Royal Society of Victoria, 50, 110–150.

Jakobsson, S. (1978). Environmental factors controlling the palagonitisation of the Surtsey tephra, Iceland. Bulletin of the Geological Society of Denmark, 27, 91–105.

James, E.A., & P.R. Evans (1971). The stratigraphy of the offshore Gippsland Basin. APEA Journal, 11, 71–74.

James, R.S., & D.L. Hamilton (1969a). Phase relations in the system NaAlSi$_3$O$_6$-KAlSi$_3$O$_6$-SiO$_2$-H$_2$O. Memoir of the Geological Society of America, 74, 1–153.

James, R.S., & D.L. Hamilton (1969b). Phase relations in the system NaAlSi$_3$O$_8$-CaAl$_2$Si$_2$O$_8$-SiO$_2$ at 1 Kilobar water vapour pressure. Contributions to Mineralogy and Petrology, 21, 111–141.

Jaques, A.L., & D.H. Green (1980). Anhydrous melting of peridotite at 0–15 kb pressure and the genesis of tholeiitic basalts. Contributions to Mineralogy and Petrology, 73, 287–310.

Jaques, A.L. & D.J. Perkin (1984). A mica, pyroxene, ilmenite megacryst-bearing lamprophyre from Mount Woolooma, northeastern New South Wales. BMR Journal of Australian Geology and Geophyics, 9, 33–40.

Jaques, A.L., J. Ferguson, & C.B Smith (1984). Kimberlites in Australia. In: Glover J.E., & P.G. Harris (Editors), Kimberlite Occurrence and Origin: a Basis for Conceptual Models in Exploration. University of Western Australia - Publication, 8, 227–274.

Jaques, A.L., R.A. Creaser, J. Ferguson, & C.B. Smith (1985). A review of the alkaline rocks of Australia. Transactions of the Royal Society of South Africa, 88, 311–334.

Jaques, A.K., J.D. Lewis, & C.B. Smith (1986). The kimberlites and lamproites of Western Australia. Geological Survey of Western Australia — Bulletin, 132.

Jennings, D.J., & F.L. Sutherland (1969). Geology of Cape Portland area with special reference to the Mesozoic (?) appinitic rocks. Tasmanian Department of Mines — Report, 13, 45–82.

Jennings, I.B. (1979). Geological atlas 1 mile series. Sheet 37 (8115S) Sheffield. Tasmanian Department of Mines — Explanatory Report.

Jennings, J.N. (1972). The age of Canberra landforms. Journal of the Geological Society of Australia, 19, 371–378.

Jensen, A.R., C.M. Gregory, & V.R. Forbes (1966). Geology of the Mackay 1:250 000 sheet area, Queensland. Bureau of Mineral Resources, Australia — Report, 104.

Jensen, H.I. (1903). The geology of the Glass House Mountains and district. Proceedings of the Linnaean Society of New South Wales, 28, 842–875.

Jensen, H.I. (1906a). Geology of the volcanic area of the East Morten and Wide Bay Districts, Queensland. Proceedings of the Linnaean Society of New South Wales, 31, 73–173.

Jensen, H.I. (1906b). Preliminary note on the geological history of the Warrumbungle Mountains. Proceedings of the Linnaean Society of New South Wales, 31, 228–235.

Jensen, H.I. (1907a). The geology of the Nandewar volcano. Proceedings of the Linnaean Society of New South Wales, 32, 842–914.

Jensen, H.I. (1907b). The geology of the Warrumbungle Mountains. Proceedings of the Linnaean Society of New South Wales, 32, 557–626.

Jensen, H.I. (1908). The alkaline petrographical province of eastern Australia. Proceedings of the Linnaean Society of New South Wales, 33, 589–602.

Jessop, A.M., M.A. Hobart, & J.G. Sclater (1976). The world heat flow data collection — 1975. Earth Physics Branch, Energy, Mines and Resources, Canada — Geothermal series, 5.

Jevons, H.S., H.I. Jensen, T.G. Taylor, & C.A. Süssmilch (1911). The geology and petrography of the Prospect intrusion. Journal and Proceedings of the Royal Society of New South Wales, 45, 445–553.

Jevons, H.S., H.I. Jensen, & C.A. Süssmilch (1912). The differentiation phenomenon of the Prospect intrusion. Journal and Proceedings of the Royal Society of New South Wales, 46, 111–138.

Johnson, B.D., M.A. Mayhew, S.Y. O'Reilly, W.L. Griffin, F. Arnott, & P.J. Wasilewski (1986). Magsat anomalies, crustal magnetization, heat flow and kimberlite occurrences in Australia. Fourth International Kimberlite Conference. Geological Society of Australia — Extended Abstracts, 16, 127–129.

Johnson, R.W., D.E. Mackenzie, & I.E.M. Smith (1978). Delayed partial melting of subduction-modified mantle in Papua New Guinea. Tectonophysics, 46, 197–216.

Johnston, M.R. (1981). Sheet 027AC Dun Mountain, Geological map of New Zealand 1:50 000. New Zealand Department of Scientific and Industrial Research.

Johnston, M.R., J.I. Raine, & W.A. Watters (1987). Drumduan Group of east Nelson, New Zealand: plant-bearing Jurassic arc rocks metamorphosed during terrane interaction. Journal of the Royal Society of New Zealand, 17, 275–301.

Jones, A.P., & A. Peckett (1980). Zirconium-bearing aegirines from Motzfeldt, South Greenland. Contributions to Mineralogy and Petrology, 75, 251–255.

Jones, A.P., J.V. Smith, J.B. Dawson, & E.C. Hansen (1983). Metamorphism, partial melting and K-metasomatism of garnet-scapolite-kyanite granulite xenoliths from Lashaine, Tanzania. Journal of Geology, 91, 143–165.

Jones, D. (1984). Difficulties associated with using indicator minerals for diamond exploration in north Queensland. Australasian Institute of Mining and Metallurgy — Annual Conference, Darwin, 127–139.

Jones, J.G., & I. McDougall (1973). Geological history of Norfolk and Philip Islands, southwest Pacific Ocean. Journal of the Geological Society of Australia, 20, 239–254.

Jones, J.G., & P.H.H. Nelson (1970). The flow of basalt lava from air into water — its structural expression and stratigraphic significance. Geological Magazine, 107, 13–21.

Jones, J.G., & J.J. Veevers (1982). A Cainozoic history of Australia's southeast highlands. Journal of the Geological Society of Australia, 29, 1–12.

Jones, J.G., & J.J. Veevers (1983). Mesozoic origins and antecedents of Australia's eastern highlands. Journal of the Geological Society of Australia, 30, 305–322.

Jones, P. (1979). The structural analysis and metamorphism of the Barnard and Barron River metamorphics in the Tully-Innisfail (N.Q.) area. James Cook University of North Queensland, unpublished BSc Honours thesis.

Joplin, G.A. (1971). A Petrography of Australian Igneous Rocks. Angus & Robertson, Sydney.

Jordon, R.H. (1981). Continents as a chemical boundary layer. Philosophical Transactions of the Royal Society of London, A301, 359–373.

Jordan, T.H. (1975). The continental tectosphere. Reviews of Geophysics and Space Physics, 13, 1–12.

Jordan, T.H. (1978). Composition and development of the continental tectosphere. Nature, 274, 544–548.

Joyce, E.B. (1974). Australia. In: Westgate, J.A., & C.M. Gold (Editors), World Bibliography and Index of Quaternary Tephrochronology. INQUA/UNESCO, 129–146.

Joyce, E.B. (1975). Quaternary volcanism and tectonics in southeastern Australia. In: Suggate, R.P., & M.M. Cresswell (Editors), Quaternary Studies. Royal Society of New Zealand — Bulletin, 13, 169–176.

Joyce, E.B. (1982). Australia. In: Vitaliano, D.B. (Editor), World Bibliography and Index of Quaternary Tephra, Supplement I. INQUA/UNESCO, 10–16.

Jukes, J.B. (1850). A Sketch of the Physical Structure of Australia. Boone, London.

Kamp, P.J.J. (1986a). Late Cretaceous-Cenozoic tectonic development of the southwest Pacific region. Tectonophysics, 121, 225–252.

Kamp, P.J.J. (1986b). The mid-Cenozoic Challenger Rift system of western New Zealand and its implications for the age of Alpine fault inception. Bulletin of the Geological Society of America, 97, 255–281.

Karig, D., & R.W. Kay (1981). Fate of sediments on the descending plate at convergent margins. Philosophical Transactions of the Royal Society of London, Series 1, 30, 233–251.

Karner, G.D., & J.K. Weissel (1984). Thermally induced uplift and lithospheric flexural readjustment of the eastern Australian highlands. Geological Society of Australia. Seventh Australian Geological Convention — Abstracts, 12, 293–294.

Katz, H.R. (1982). Plate margin transition from oceanic arc-trench to continental system: the Kermadec-New Zealand example. In: Packham, G.H. (Editor), The Evolution of the India-Pacific Plate Boundaries. Tectonophysics, 87, 49–64.

Kay, J.R. (1981). Thunder egg deposits of the Wycarabah district, central Queensland. Queensland Government Mining Journal, 82, 566–579.

Kay, R.W. (1979). Zone refining at the base of lithospheric plates: a model for a steady-state asthenosphere. Tectonophysics, 55, 1–9.

Kay, S.M., & R.W. Kay (1983). Thermal history of the deep crust inferred from granulite xenoliths, Queensland, Australia. American Journal of Science, 283A, 486–513.

Kear, D. (1960). Sheet 4 — Hamilton. Geological map of New Zealand 1:250 000. New Zealand Department of Scientific and Industrial Research.

Kear, D., & R.F. Hay (1961). Sheet 1, North Cape (1st Edition). Geological map of New Zealand, 1:250 000. New Zealand Department of Scientific and Industrial Research.

Keble, R.A. (1947). Notes on Australian Quaternary climates and migration. National Museum, Melbourne — Memoir, 15, 28–81.

Keen, C.E., & C. Beaumont (in press). Geodynamics of rifted continental margins. Geological Society of America, Decade of North American Geology — Special Publication.

Kemp, E.M. (1978). Tertiary climatic evolution and vegetation history in the southeast Indian Ocean region. Palaeogeography, Palaeoclimatology, Palaeoecology, 24, 169–208.

Kermode, L.O. (1986). The Auckland volcanic field. International Volcanological Congress, New Zealand. Handbook, 56–67.

Kershaw, A.P. (1970). A pollen diagram from Lake Euromoo, northeast Queensland. New Phytologist, 69, 785–805.

Kershaw, A.P. (in press). Long pollen records from the Atherton and Western Plains volcanic provinces: the basis for an Australian Quaternary stratigraphy. In: De Decker, P., & M.A.J. Williams (Editors), Cenozoic of the Australian Region: a Revaluation of the Evidence.

Kesson, S.E. (1968). The geology of an area west of Pambula, New South Wales. University of Sydney, unpublished BSc Honours thesis.

Kesson, S.E. (1972). Basic alkaline rocks. Australian National University, unpublished PhD thesis.

Kesson, S.E. (1973). The primary geochemistry of the Monaro alkaline volcanics, southeastern Australia — evidence for upper mantle heterogeneity. Contributions to Mineralogy and Petrology, 42, 93–108.

Kesson, S.E., & R.C. Price (1972). The major and trace element chemistry of kaersutite and its bearing on the petrogenesis of alkaline rocks. Contributions to Mineralogy and Petrology, 35, 119–124.

Kidd, P.R. (1975). The geology of Batemans Bay region. University of Sydney, unpublished BSc Honours thesis.

Kimbrough, D.L., & D.S. Coombs (1983). Uranium-lead ages from the Dun Mountain ophiolite belt, South Island, New Zealand. Geological Society of New Zealand — Miscellaneous Publication, 30A, 101.

King, R.L. (1985). Explanatory notes, Ballarat 1:250 000 geological map. Geological Survey of Victoria — Report, 75.

Kirby, S.H. (1985). Rock mechanic observations pertinent to the geology of the lithosphere and the localization of strain along shear zones. Tectonophysics, 119, 1–27.

Kirby, S.H., B.C. Hearn, H. Yongnian, & L. Chuanyong (1987). Geophysical implications of mantle xenoliths: evidence for fault zones in the deep lithosphere of eastern China. United States Geological Survey — Circular, 956, 63–65.

Kirkegaard, A.G. (1974). Structural elements of the northern part of the Tasman Geosyncline. In: Denmead, A.K., G.W. Tweedale, & A.F. Wilson (Editors), The Tasman Geosyncline — a Symposium. Geological Society of Australia, Queensland Division, Brisbane, 47–62.

Kirkegaard, A.G., R.D. Shaw, & C.G. Murray (1970). Geology of the Rockhampton and Port Clinton 1:250 000 Sheet Areas. Geological Survey of Queensland — Report, 38.

Kiselev, A.I. (1987). Volcanism of the Baikal rift zone. Tectonophysics, 143, 235–244.

Kitson, A.E. (1903). Volcanic necks at Andersons Inlet, South Gippsland. Proceedings of the Royal Society of Victoria, 16, 154–176.

Kitson, A.E. (1917). The Jumbunna and Powlett Plains district, South Gippsland. Geological Survey of Victoria — Bulletin, 40.

Kleeman, J.D., & J.A. Cooper (1970). Geochemical evidence for the origin of some ultramafic inclusions from Victorian basanites. Physics of the Earth and Planetary Interiors, 3, 302–308.

Klein, E.M., & C.H. Langmuir (1987). Global correlations of ocean ridge basalt chemistry with axial depth and crustal thickness. Journal of Geophysical Research, 92, 8089–8115.

Knutson, J. (1975). Petrology and geochemistry of igneous rocks in the Comboyne Plateau-Lorne Basin area, New South Wales. Macquarie University, unpublished PhD thesis.

Knutson, J., & T.H. Green (1975). Experimental duplication of a high-pressure megacryst/cumulate assemblage in a near-saturated hawaiite. Contributions to Mineralogy and Petrology, 52, 121–132.

Knutson, J., W.F. McDonough, M.B. Duggan, & B.W. Chappell (1986). Geochemical and isotopic characteristics of eastern Australian Cainozoic 'central' volcanoes. International Volcanological Congress, New Zealand — Abstracts, 174.

Kokelaar, B.P. (1983). The mechanism of Surtseyan volcanism. Journal of Geological Society of London, 140, 939–944.

Kokelaar, B.P. (1986). Magma-water interactions in subaqueous and emergent basaltic volcanism. Bulletin of Volcanology, 48, 245–289.

Korsch, R.J. (1984). Geological aspects of the Torlesse Complex, south coast of Wellington. Geological Society of New Zealand — Miscellaneous Publication, 31B, 67–90.

Korsch, R.J., & H.W. Wellman (1988). The geological evolution of New Zealand and the New Zealand region. In: Nairn, A.E.M., F.G. Stehli, & S. Uyeda (Editors), The Ocean Basins and Margins, Volume 7B, The Pacific Ocean. Plenum, New York, 411–482.

Kroenke, L.W., & P. Rodda (1984). Cenozoic tectonic development of the southwest Pacific. United Nations ESCAP, CCOP/SOPAC, Technical Bulletin, 6.

Kuntz, M.A., D.E. Champion, E.C. Spiker, R.H. Lefebvre, & L.A. McBroome (1982). The Great Rift and the evolution of the Craters of the Moon lava field, Idaho. In: Bonnichsen, B., & R.M. Breckenridge (Editors), Cenozoic Geology of Idaho. Idaho Bureau of Mines — Bulletin, 26, 423–437.

Kuo, L.-C., & R.J. Kirkpatrick (1985). Dissolution of mafic minerals and its implication for the ascent velocities of peridotite-bearing basaltic magmas. Journal of Geology, 93, 691–700.

Kushiro, I., H.S. Yoder, & B.O. Mysen (1976). Viscosities of basalt and andesite melts at high pressures. Journal of Geophysical Research, 81, 6351–6356.

Kyle, P.R. (1981). Mineralogy and geochemistry of basanite to phonolite sequence at Hut Point Peninsula, Antarctica, based on core from Dry Valley Drilling Project drillholes 1, 2 and 3. Journal of Petrology, 22, 451–500.

Kyle, P.R., R.R. Dibble, W.F. Giggenbach, & J. Keys (1982). Volcanic activity associated with the anorthoclase phonolite lava lake, Mount Erebus, Antarctica. In: Craddock, C. (Editor), Antarctic Geoscience. University of Wisconsin Press, Madison, 735–745.

Lambeck, K. & R. Stephenson (1986). The post-Palaeozoic uplift history of south-eastern Australia. Australian Journal of Earth Sciences, 33, 253–270.

Lambeck, K., H.W.S. McQueen, R.A. Stephenson, & D. Denham (1984). The state of stress within the Australian continent. Annales Geophysicae, 2, 723–742.

Lancaster, K. (1981). Thaumasite in Tasmania? Australian Gem and Treasure Hunter, November, 20–22.

Landis, C.A., & M.C. Blake Jr (1987). Tectonostratigraphic terranes of the Croisilles Harbour region, South Island, New Zealand. American Geophysical Union Geodynamics Series, 19, 179–198.

Landis, C.A., & D.S. Coombs (1967). Metamorphic belts and orogenesis in southern New Zealand. Tectonophysics, 4, 501–518.

Langel, R.A., J.D. Phillips, & R.J. Horner (1982). Initial scaler magnetic anomaly map from MAGSAT. Geophysical Research Letters, 9, 269–272.

Langford-Smith, T., G.H. Dury, & I. McDougall (1966). Dating the duricrust in southern Queensland. Australian Journal of Science, 29, 79–80.

Lapouille, A. (1982). Étude des bassins marginaux fossiles du Sud-Ouest Pacifique: bassin Nord-d'Entrecasteaux, bassin Nord-Loyauté bassin Sud-Fidjien. Travaux et Documents de l'Office de la Recherche Scientifique et Technique Outre-Mer, 147, 409–438.

Larsen, L.M. (1976). Clinopyroxenes and coexisting mafic minerals from the alkaline Ilimaussaq intrusion, south Greenland. Journal of Petrology, 17, 258–290.

Larsen, L.M. (1977). Aenigmatites from the Ilimaussaq intrusion, south Greenland: chemistry and petrological implications. Lithos, 10, 257–270.

Law, R.G. (1975). Radiocarbon dates for Rangitoto and Motutapu, a consideration of the dating accuracy. New Zealand Journal of Science 18, 441–451.

Lawrence, A.D. (1973). The geochemistry of the Featherbed volcanics, north Queensland. La Trobe University, unpublished BSc Honours thesis.

Leach, J.H.J. (1977). Geology of the eastern Corangamite region and the fossil fauna of Bald Hill, near Foxhow. University of Melbourne, unpublished BSc Honours thesis.

Leaman, D.E. (1975). Form, mechanism and control of dolerite intrusion near Hobart, Tasmania. Journal of the Geological Society of Australia, 22, 175–186.

Le Bas, M.J. (1977). Carbonatite-Nephelinite Volcanism, an African Case History. Wiley, London.

Leeman, W.P. (1982). Development of the Snake River plain-Yellowstone plateau province, Idaho and Wyoming: an overview and petrologic model. In: Bonnichsen, B., & R.M. Breckenridge (Editors), Cenozoic Geology of Idaho. Idaho Bureau of Mines — Bulletin, 26, 155–177.

Leeman, W.P., & D.J. Lindstrom (1978). Partitioning of Ni^{2+} between basaltic and synthetic melts and olivines; an experimental study. Geochimica et Cosmochimica Acta, 42, 801–816.

Leeman, W.P., M.A. Menzies, D.J. Matty, & G.F. Embree (1985). Strontium, neodymium and lead isotopic compositions of deep crustal xenoliths from the Snake River plain: evidence for Archean basement. Earth and Planetary Science Letters, 75, 354–368.

Leichhardt, L. (1847). Journal of an Overland Expedition in Australia. Boone, London.

Le Maitre, R.W. (1974). Partially fused granite blocks from Mt Elephant, Victoria, Australia. Journal of Petrology, 15, 403–412.

Le Maitre, R.W. (1984). A proposal by the IUGS subcommission on the systematics of igneous rocks for a chemical classification of volcanic rocks based on the total alkali silica (TAS) diagram. Australian Journal of Earth Sciences, 31, 243–255.

LeMasurier, W.E., & D.C. Rex (1983). Rates of uplift and the scale of ice level instabilities recorded by volcanic rocks in Marie Byrd Land, West Antarctica. In: Oliver, R.L., P.R. James, & J.B. Jago (Editors), Antarctic Earth Science. Australian Academy of Science, Canberra.

LeMasurier, W.E., & D.C. Rex (in press). Marie Byrd Land volcanic province and its relation to the Cainozoic West Antarctic rift system. In: Tingey, R.J. (Editor), Geology of Antarctica. Oxford University Press, Oxford.

Lensen, G.J. (1963). Sheet 16 Kaikoura. Geological map of New Zealand 1:250 000. New Zealand Department of Scientific and Industrial Research.

Lerner-Lam, A.L., & T.H. Jordan (1985). Comparison of continent-ocean structural differences derived from surface waveform inversion and body-wave analysis. Transactions of the American Geophysical Union, 66, 311.

Le Roex, A.P. (1986). Geochemical correlation between southern African kimberlites and South Atlantic hotspots. Nature, 324, 243–245.

Le Roex, A.P. (1987). Source regions of mid-ocean ridge basalts: evidence for enrichment processes. In: Menzies, M.A., & C.J. Hawkesworth (Editors), Mantle Metasomatism. Academic Press, London, 389–422.

Lesquer, A., A. Bourmatte, & J.M. Dautria (1988). Deep structure of the Hoggar domal uplift (central Sahara, south Algeria) from gravity, thermal and petrological data. Tectonophysics, 152, 71–87.

Lewis, K.B. (1980). Quaternary sedimentation on the Hikurangi oblique-subduction and transform margin, New Zealand. In: Ballance, P.F., & H.G. Reading (Editors), Sedimentation in Oblique-Slip Mobile Zones. International Association of Sedimentologists — Special Publication, 4, 171–189.

Leyreloup, A., J.L. Bodinier, C. Dupuy, & J. Dostal (1982). Petrology and geochemistry of granulite xenoliths from Central Hoggar (Algeria) — implications for the lower crust. Contributions to Mineralogy and Petrology, 79, 68–75.

Liggett, K.A., & D.R. Gregg (1965). Geology of Banks Peninsula. In: New Zealand Volcanology: South Island. New Zealand Department of Scientific and Industrial Research — Information Serial, 51, 9–23.

Lightfoot, P.C., & C.J. Hawkesworth (1986). Geochemistry of the South Deccan Trap basalts, India. International Volcanological Congress, New Zealand — Abstracts, 178.

Lilley, F.E.M. (1976). A magnetometer array study across southern Victoria and the Bass Strait area, Australia. Geophysical Journal of the Royal Astronomical Society, 46, 165–184.

Lilley, F.E.M., D.V. Woods, & M.N. Sloane (1981). Electrical conductivity profiles and implications for the absence or presence of partial melting beneath central and southeast Australia. Physics of the Earth and Planetary Interiors, 25, 419–428.

Lippard, S.J. (1973). The petrology of phonolites from the Kenya Rift. Lithos, 6, 217–234.

Lishmund, S.R. (1987). Regional distribution of sapphire, diamond, and volcaniclastic rocks. In: Tertiary Volcanics and Sapphires in the New England District. New South Wales Geological Survey — Report, GS1987/058, 9–12.

Lishmund, S.R., & G.M. Oakes (1983). Diamonds, sapphires and Cretaceous/Tertiary diatremes in New South Wales. New South Wales Geological Survey — Quarterly Notes, 53, 23–27.

Lister, G.S., & G.A. Davis (in press). The origin of metamorphic core complexes and detachment faults formed during Tertiary continental extension in the northern Colorado River region, U.S.A. Journal of Structural Geology.

Lister, G.S., M.A. Etheridge, & P.A. Symonds (1986). Detachment faulting and the evolution of passive margins. Geology, 14, 246–250.

Lister, G.S., M.A. Etheridge, & P.A. Symonds (in press). Detachment models for the formation of passive continental margins. Tectonics.

Lohe, E.M. (1980). The Neranleigh-Fernvale beds of southeastern Queensland: petrology, sedimentology, structure, metamorphism and tectonic evolution. University of Queensland, unpublished PhD thesis.

Lorenz, C.A. (1973). On the formation of maars. Bulletin Volcanologique, 37, 183–204.

Lorenz, C.A. (1986). On the growth of maars and diatremes and its relevance to the formation of tuff rings. Bulletin Volcanologique, 48, 265–274.

Lorenz, V. (1974). Vesiculated tuffs and associated features. Sedimentology, 21, 273–291.

Lorenz, V. (1975). Formation of phreatomagmatic maar-diatreme volcanoes and its relevance to kimberlite diatremes. Physics and Chemistry of the Earth, 9, 17–27.

Lovering, J.F. (1964). The eclogite-bearing basic igneous pipe at Ruby Hill near Bingara, N.S.W. Journal and Proceedings of the Royal Society of New South Wales, 21, 9–52.

Lovering, J.F. & J.R. Richards (1964). Potassium-argon age study of possible lower crust and upper mantle inclusions in deep-seated intrusions. Journal of Geophysical Research, 69, 1895–1901.

Lovering, J.F., & A.J.R. White (1964). The significance of primary scapolite in granulitic inclusions from deep-seated pipes. Journal of Petrology, 5, 195–218.

Lovering, J.F., & A.J.R. White (1969). Granulitic and eclogitic inclusions from basic pipes at Delegate, Australia. Contributions to Mineralogy and Petrology, 21, 9–52.

Luhr, J.F., I.S.E. Carmichael, & J.C. Varekamp (1984). The 1982 eruption of El Chichon volcano, Chiapas, Mexico: mineralogy and petrology of the anhydrite-bearing pumices. Journal of Volcanology and Geothermal Research, 23, 69–108.

Maaløe, S., & K.-I. Aoki (1977). The major element composition of the upper mantle estimated from the composition of lherzolites. Contributions to Mineralogy and Petrology, 63, 161–173.

Maaløe, S., & B. Hansen (1982). Olivine phenocrysts of Hawaiian olivine tholeiite and oceanite. Contributions to Mineralogy and Petrology, 81, 203–211.

Macdonald, G.A. (1960). Dissimilarity of continental and oceanic rock types. Journal of Petrology, 1, 172–177.

Macdonald, G.A. (1972). Volcanoes. Prentice-Hall, New Jersey.

Macdonald, G.A., & T. Katsura (1964). Chemical composition of Hawaiian lavas. Journal of Petrology, 5, 82–133.

Macdonald, R., & D.K. Bailey (1973). The chemistry of the peralkaline oversaturated obsidians. United States Geological Survey — Professional Paper, 440-N-1, 1–37.

MacGregor, I.D., & A.R. Basu (1974). Thermal structure of the lithosphere: a petrologic model. Science, 1007–1011.

Mackaness, G. (1956, Editor). The Discovery and Exploration of Moreton Bay and Brisbane River (1799–1823). Part 1. Australian Historical Monograph, Volume 43 (New Series).

Mackaness, G. (1965). Fourteen Journeys over the Blue Mountains of New South Wales, 1813–41. Horwitz-Grahame, Sydney, 143–165.

Mackenzie, D.E., & A.J.R. White (1970). Phonolite globules in basanite from Kiandra, Australia. Lithos, 3, 309–317.

MacKinnon, T.C. (1983). Origin of the Torlesse terrane and coeval rocks, South Island, New Zealand. Bulletin of the Geological Society of America, 94, 967–985.

MacNevin, A.A. (1972). Sapphires in the New England district, New South Wales. New South Wales Geological Survey — Record, 14, 19–35.

MacNevin, A.A. (1977). Diamonds in New South Wales. New South Wales Geological Survey — Mineral Resources, 42.

MacRae, N.D. (1979). Silicate glasses and sulphides in ultramafic xenoliths, Newer Basalts, Victoria, Australia. Contributions to Mineralogy and Petrology, 68, 275–280.

Macumber, P.G. (1978). Evolution of the Murray River during the Tertiary period. Evidence from northern Victoria. Proceedings of the Royal Society of Victoria, 90, 43–52.

Maillet, P., M. Monzier, M. Selo, & D. Storzer (1982). La zone d'Entrecasteaux (Sud-Quest Pacifique): nouvelle approche pétrologique et géochronologique. Travaux et Documents de l'Office de la Recherche Scientifique et Technique Outre-Mer, 147, 441–458.

Malahoff, A., R.H. Feden, & H.S. Fleming (1982). Magnetic anomalies and tectonic fabric of marginal basins north of New Zealand. Journal of Geophysical Research, 87, 4109–4125.

Malone, E.J., D.W.P. Corbett, & A.R. Jensen (1964). Geology of the Mount Coolon 1:250 000 Sheet area. Bureau of Mineral Resources, Australia — Report, 64.

Malone, E.J., F. Olgers, & A.G. Kirkegaard (1969). The Geology of the Duaringa and Saint Laurence 1:250 000 sheet areas, Queensland. Bureau of Mineral Resources, Australia — Report, 121.

Manghnani, M.H., R. Ramanantoandra, & S.P. Clark (1974). Compressional and shear wave velocities in granulite facies rocks and eclogites to 10 kbar. Journal of Geophysical Research, 79, 5427–5446.

Mansergh, G.D. (1965). A study of the Kerikeri volcanoes, north of Whangarei. University of Auckland, unpublished MSc thesis.

Marsh, B.D. (1978). On the cooling of ascending andesitic magma. Philosophical Transactions of the Royal Society of London, A288, 611–625.

Marsh, J.S. (1975). Aenigmatite stability in silica under-saturated rocks. Contributions to Mineralogy and Petrology, 50, 135–144.

Marsh, J.S. (1987). Evolution of a strongly differentiated suite of phonolites from the Klinghardt Mountains, Namibia. Lithos, 20, 41–58.

Marshall, B. (1969). Geological atlas 1 mile series. Sheet 31(8315N) Pipers River. Tasmanian Department of Mines — Explanatory Report.

Marshall, P. (1894). Tridymite-trachyte of Lyttelton. Transactions of the New Zealand Institute, 26, 368–387.

Marshall, P. (1906). The Geology of Dunedin (New Zealand). Quarterly Journal of the Geological Society of London, 62, 381–424.

Marshall, P. (1912). Geology of New Zealand. Government Printer, Wellington.

Marshall, P. (1914). The sequence of lavas at the North Head, Otago Harbour. Quarterly Journal of the Geological Society of London, 70, 382–408.

Martin, D.J. (1984). Microstructure, geochemistry and differentiation of a primary layered teschenite sill. Geological Magazine, 122, 335–350.

Martin, D.J. (1985). A small layered tholeiitic intrusion emplaced at shallow level, at Scone, New South Wales. In: Sutherland, F.L., B.J. Franklin, & A.E. Waltho (Editors), Volcanism in Eastern Australia, with Case Histories from New South Wales. Geological Society of Australia, New South Wales Division, 1, 107–140.

Martin, D.J. (1986). Geochemistry of a suite of teschenitic intrusions from the Sydney-Gunnedah Basin — evidence of crustal tension. Geological Society of Australia — Abstracts, 15, 132–133.

Martin, H.A., L.A. Worral, & J. Chalson (1987). The first occurrence of the Paleocene Lygistepollenites balmei zone in the Eastern Highlands Region, New South Wales, Australian Journal of Earth Sciences, 34, 359–365.

Martinez, F., & J.R. Cochran (1988). Structure and tectonics of the northern Red Sea: catching a continental margin between rifting and drifting. Tectonophysics, 150, 1–32.

Mason, B. (1966). Pyrope, augite and hornblende from Kakanui, New Zealand. New Zealand Journal of Geology and Geophysics, 9, 474–480.

Mason, B. (1968). Eclogitic xenoliths from volcanic breccia at Kakanui, New Zealand. Contributions to Mineralogy and Petrology, 19, 316–327.

Mason, D.R., & I. Kavalieris (1984). A preliminary note on the Barrington Tops granodiorite, New South Wales. University of Newcastle, Department of Geology — Research Report, 2.

Mathur, S.P. (1983a). Deep reflection experiments in north-eastern Australia, 1976–1978. Geophysics, 48, 1588–1597.

Mathur, S.P. (1983b). Deep reflection probes in eastern Australia reveal differences in the nature of the crust. First Break, July, 9–16.

Mathur, S.P. (1983c). Deep crustal reflection results from the central Eromanga Basin, Australia. Tectonophysics, 100, 163–173.

Matthews, W.H., & G.H. Curtis (1966). Date of the Pliocene-Pleistocene boundary in New Zealand. Nature, 212, 979–980.

Matthews, W.L. (1983). Geology and groundwater of the Longford Tertiary basin. Geological Survey of Tasmania — Bulletin, 59.

Mattinson, J.M., D.L. Kimbrough, & J.Y. Bradshaw (1986). Western Fiordland orthogneiss: early Cretaceous arc magmatism and granulite facies metamorphism, New Zealand. Contributions to Mineralogy and Petrology, 92, 383–392.

Mawson, D. (1950). Basaltic lavas of the Balleny Islands, ANARE report. Transactions of the Royal Society of South Australia, 73, 223–231.

Mayhew, M.A. (1982). Application of satellite magnetic anomaly data to Curie isotherm mapping. Journal of Geophysical Research, 87, 4846–4854.

Mayhew, M.A., & B.D. Johnson (1987). An equivalent layer magnetization model for Australia based on MAGSAT data. Earth and Planetary Science Letters, 83, 167–174.

McBirney, A.R. (1980). Mixing and unmixing of magmas. Journal of Volcanology and Geothermal Research, 7, 357–371.

McBirney, A.R., & T. Murase (1984). Rheological properties of magmas. Annual Review of Earth and Planetary Science, 12, 337–357.

McCallister, R.H., L.W. Finger, & Y. Ohashi (1976). Intracrystalline Fe^{2+}-Mg equilibria in three natural Ca-rich clinopyroxenes. American Mineralogist, 61, 671–676.

McClenaghan, M.P., N.J. Turner, P.W. Baillie, A.V. Brown, P.R. Williams, & W.R. Moore (1982). Geology of the Ringarooma-Boobyalla region. Geological Survey of Tasmania — Bulletin, 61.

McCulloch, M.T. (1988). Nd-Sr isotope geochemistry of the Tasmantid seamounts — evolution of a hotspot-trace. Ninth Australian Geological Convention, Brisbane — Abstracts, 21, 260–261.

McCulloch, M.T., & B.W. Chappell (1982). Nd isotopic characteristics of S- and I-type granites. Earth and Planetary Science Letters, 58, 51–64.

McCulloch, M.T., R.J. Arculus, B.W. Chappell, & J. Ferguson (1982). Isotopic and geochemical studies of nodules in kimberlite have implications for the lower continental crust. Nature, 300, 166–169.

McDonough, W.F. (1987). Chemical and isotopic systematics of basalts and peridotite xenoliths: implications for the composition and evolution of the Earth's mantle. Australian National University, unpublished PhD thesis.

McDonough, W.F., & M.T. McCulloch (1987a). The southeast Australian lithospheric mantle: isotopic and geochemical constraints on its growth and evolution. Earth and Planetary Science Letters, 86, 327–340.

McDonough, W.F., & M.T. McCulloch (1987b). Growth and evolution of the subcrustal lithosphere: oceanic versus continental. Terra Cognita, 7, 169.

McDonough, W.F., M.T. McCulloch, & S.-S. Sun (1985). Isotopic and geochemical systematics in Tertiary-Recent basalts from southeastern Australia and implications for the evolution of the sub-continental lithosphere. Geochimica et Cosmochimica Acta, 49, 2051–2067.

McDonough, W.F., M.T. McCulloch, R.A. Duncan, J.A. Gamble, P.A. Morris, & R.M. Briggs (1986). Geochemical and isotopic systematics of Cenozoic intraplate basalts from continental and oceanic regions in the south Pacific. International Volcanological Congress, New Zealand — Abstracts, 180.

McDonough, W.F., R.L. Rudnick, & M.T. McCulloch (in press). The chemical and isotopic composition of the lower eastern Australian lithosphere. In: Drummond, B. (Editor), The Eastern Australian Lithosphere. Geological Society of Australia — Special Publication.

McDougall, I. (1961). Determination of the age of a basic igneous intrusion by the potassium-argon method. Nature, 190, 1184–1186.

McDougall, I. (1963). Differentiation of the Tasmanian dolerites: Red Hill dolerite-granophyre association. Bulletin of the Geological Society of America, 73, 279–316.

McDougall, I., & Aziz-ur-Rahman (1972). Age of the Gauss-Matuyama boundary and of the Kaena and Mammoth events. Earth and Planetary Science Letters, 14, 367–380.

McDougall, I., & D.S. Coombs (1973). Potassium-argon for the Dunedin volcano and outlying volcanics. New Zealand Journal of Geology and Geophysics, 16, 179–188.

McDougall, I., & R.A. Duncan (1980). Linear volcanic chains — recording plate motions? Tectonophysics, 63, 275–295.

McDougall, I., & R.A. Duncan (1988). Age progressive volcanism in the Tasmantid seamounts. Earth and Planetary Science Letters, 89, 207–220.

McDougall, I., & E.D. Gill (1975). Potassium-argon ages from the Quaternary succession in the Warrnambool-Port Fairy area, Victoria, Australia. Proceedings of the Royal Society of Victoria, 87, 175–178.

McDougall, I., & P.J. Leggo (1965). Isotopic age determinations on granitic rocks from Tasmania. Journal of the Geological Society of Australia, 12, 295–332.

McDougall, I., & Z. Roksandic (1974). Total fusion $^{40}Ar/^{39}Ar$ ages using the HIFAR Reactor. Journal of the Geological Society of Australia, 21, 81–87.

McDougall, I., & G.C. Slessar (1972). Tertiary volcanism in the Cape Hillsborough area, North Queensland. Journal of the Geological Society of Australia, 18, 401–408.

McDougall, I., & G.J. Van der Lingen (1974). Age of the rhyolites of the Lord Howe Rise and evolution of the southwest Pacific Ocean. Earth and Planetary Science Letters, 21, 117–126.

McDougall, I., & P. Wellman (1976). Potassium-argon ages for some Australian Mesozoic igneous rocks. Journal of the Geological Society of Australia, 23, 1–9.

McDougall, I., & J.F.G. Wilkinson (1967). Potassium-argon dates on some Cainozoic volcanic rocks from northeastern New South Wales. Journal of the Geological Society of Australia, 14, 225–234.

McDougall, I., H.L. Allsop, & F.H. Chamalaun (1966). Isotopic dating of the Newer Volcanics of Victoria, Australia and geomagnetic polarity epochs. Journal of Geophysical Research, 71, 6107–6118.

McDougall, I., H.A. Pollach, & J.J. Stipp (1969). Excess radiogenic argon from young subaerial basalts from the Auckland volcanic field, New Zealand. Geochimica et Cosmochimica Acta, 33, 1485–1520.

McDougall, I., B.J.J. Embleton, & D.B. Stone (1981). Origin and evolution of Lord Howe Island, southwest Pacific Ocean. Journal of the Geological Society of Australia, 28, 155–176.

McElhinny, M.W. (1973). Palaeomagnetism and Plate Tectonics. Cambridge University Press, Cambridge.

McElhinny, M.W., B.J.J. Embleton, & P. Wellman (1974). A synthesis of Australian Cenozoic palaeomagnetic results. Geophysical Journal of the Royal Astronomical Society, 36, 141–151.

McElroy, C.T. (1962). The geology of the Clarence-Moreton Basin. Geological Survey of New South Wales — Memoir, 9.

McGeary, S., & M.R. Warner (1985). Seismic profiling of the continental lithosphere. Nature, 317, 795–797.

McGuire, A.V. (1988). The mantle beneath the Red Sea margin: xenoliths from western Saudi Arabia. Tectonophysics, 150, 101–119.

McIver, J.R. (1981). Aspects of ultrabasic and basic alkaline intrusive rocks from Bitterfontein, South Africa. Contributions to Mineralogy and Petrology, 78, 1–11.

McKenzie, D. (1984). The generation and compaction of partially molten rock. Journal of Petrology, 25, 713–765.

McKenzie, D. (1985). The extraction of magma from the crust and mantle. Earth and Planetary Science Letters, 74, 81–91.

McKenzie, D., & M.J. Bickle (1988). The volume and composition of melt generated by extension of the lithosphere. Journal of Petrology, 29, 625–679.

McKenzie, D., & R.K. O'Nions (1983). Mantle reservoirs and ocean island basalts. Nature, 301, 229–231.

McKenzie, D.A., R.J. Nott, & P.F. Bolger (1984). Radiometric age determinations. Geological Survey of Victoria — Report, 74.

McKenzie, D.P. (1978). Some remarks on the development of sedimentary basins. Earth and Planetary Science Letters, 40, 25–32.

McLennan, J.M., & S.D. Weaver (1984). Olivine-nephelinite at Mounseys Creek, Oxford, Canterbury. New Zealand Journal of Geology and Geophysics, 27, 389–390.

McNamara, G.C. (in press). The Wyandotte local fauna: a new, dated, Pleistocene vertebrate fauna from northern Queensland. Memoir of the Queensland Museum. De Vis Symposium Special Volume.

McNutt, M., K. Fischer, S. Kruse, & J. Natland (1989). The origin of the Marquesas fracture zone ridge and its implications for the nature of hot spots. Earth and Planetary Science Letters, 91, 381–393.

McQuillin, R., & J. Tuson (1963). Gravity measurements over the Rhum Tertairy plutonic complex. Nature, 199, 1276–1277.

McTaggart, N.R. (1962). The sequence of Tertiary volcanic and sedimentary rocks of the Mount Warning volcanic shield. Journal and Proceedings of the Royal Society of New South Wales, 95, 135–144.

Mehnert, K.R. (1975). The Ivrea zone: a model of the deep crust. Neues Jahrbuch für Mineralogie Abhandlungen, 125, 156–199.

Menzies, M.A., & V.R. Murthy (1980). Mantle metasomatism as a precursor to the genesis of alkaline magmas — isotopic evidence. American Journal of Science, 280A, 622–638.

Menzies, M.A., & S.Y. Wass (1983). CO_2 rich mantle below eastern Australia: REE, Sr and Nd isotopic study of Cenozoic alkali magmas and apatite-rich xenoliths, Southern Highlands province, New South Wales, Australia. Earth and Planetary Science Letters, 65, 287–302.

Menzies, M.A., W.P. Leeman, & C.J. Hawkesworth (1984). Geochemical and isotopic evidence for the origin of continental flood basalts with particular emphasis to the Snake River plain, Idaho, U.S.A. Philosophical Transactions of the Royal Society of London, A310, 643–660.

Mercier, J.-C.C. (1979). Peridotite xenoliths and the dynamics of kimberlite intrusion. In: Boyd, F.R. & H.O.A. Meyer (Editors), The Mantle Sample: Inclusions in Kimberlites and Other Volcanics. American Geophysical Union, Washington, D.C., 197–212.

Mercier, J.-C.C., & A. Nicolas (1975). Textures and fabrics of upper-mantle peridotites as illustrated by xenoliths from basalts. Journal of Petrology, 16, 454–487.

Merrill, R.B., & P.J. Wyllie (1975). Kaersutite and kaersutite-eclogite from Kakanui, New Zealand — water excess and water deficient melting to 30 kilobars. Bulletin of the Geological Society of America, 86, 555–570.

Middlemost, E.A.K. (1981). The Canobolas complex, New South Wales, an alkaline shield volcano. Journal of the Geological Society of Australia, 28, 33–49.

Middlemost, E.A.K. (1985). Miocene shield volcanoes of New South Wales. Geological Society of Australia, New South Wales Division — Publication, 1, 49–58.

Middleton, M.F. (1982). The subsidence and thermal history of the Bass Basin, southeastern Australia. Tectonophysics, 87, 383–397.

Middleton, M.F., & P.W. Schmidt (1982). Palaeothermometry of the Sydney Basin. Journal of Geophysical Research, 87, 5351–5359.

Mitchell, T.L. (1838). Three Expeditions in the Interior of Eastern Australia. Boone, London.

Miyashiro, A. (1986). Hot regions and the origin of marginal basins in the western Pacific. Tectonophysics, 122, 195–216.

Mohr, P.A. (1983). Ethiopian flood basalt province. Nature, 303, 577–584.

Mohr, P.A., & C.A. Wood (1976). Volcano spacings and lithospheric attenuation in the Eastern Rift of Africa. Earth and Planetary Science Letters, 33, 126–144.

Mollan, R.G. (1965). Tertiary volcanic rocks in the Peak Range, central Queensland. Bureau of Mineral Resources, Australia — Record, 1965/241, 1–97.

Mollan, R.G., J.M. Dickins, N.F. Exon, & A.G. Kirkegaard (1969). Geology of Springsure 1:250 000 Sheet area, Queensland. Bureau of Mineral Resources, Australia — Report, 123.

Mollan, R.G., V.R. Forbes, A.R. Jensen, N.F. Exon, & C.M. Gregory (1972). Geology of the Eddystone and Taroom 1:250 000 Sheet area. Bureau of Mineral Resources, Australia — Report, 142.

Molnar, P., T. Atwater, J. Mammerickx, & S.M. Smith (1975). Magnetic anomalies, bathymetry, and the tectonic evolution of the South Pacific since the Late Cretaceous. Geophysical Journal of the Royal Astronomical Society, 40, 383–420.

Moorbath, S., & R.N. Thompson (1980). Strontium isotope geochemistry and petrogenesis of the Early Tertiary lava pile of the Isle of Skye, Scotland, and other basic rocks of the British Tertiary igneous province. Journal of Petrology, 21, 295–321.

Moorbath, S., J.L. Powell, & P.N. Taylor (1975). Age and origin of the 'grey gneiss' complex of the southern Outer Hebrides. Journal of the Geological Society of London, 131, 213–222.

Moore, A.E. (1976). Controls of post-Gondwanaland volcanism in southern Africa. Earth and Planetary Science Letters, 31, 291–296.

Moore, J.G. (1967). Base surges in recent volcanic eruptions. Bulletin Volcanologique, 30, 337–363.

Moore, J.G., & D.L. Peck (1962). Accretionary lapilli in volcanic rocks of the western United States. Journal of Geology, 70, 182–193.

Moore, M.E., A.J. Gleadow, & J.F. Lovering (1986). Thermal evolution of rifted continental margins: new evidence from fission tracks in basement apatites from southeastern Australia. Earth and Planetary Science Letters, 78, 255–270.

Moore, W.R., P.W. Baillie, S.M. Forsythe, J.W. Hudspeth, R.G. Richardson, & N.J. Turner (1984). Boebyalla sub-basins, a Cretaceous on-shore extension of the southern edge of the Bass Basin. APEA Journal, 24, 110–117.

Morgan, W.J. (1971). Convection plumes in the lower mantle. Nature, 230, 42–43.

Morgan, W.J. (1972). Deep mantle convection plumes and plate motions. American Association of Petroleum Geologists — Bulletin, 56, 203–213.

Morgan, W.J. (1981). Hotspot traces and the opening of the Atlantic and Indian Oceans. In: Emiliani, C. (Editor), The Sea. 7. The Oceanic Lithosphere. Wiley, New York, 443–475.

Morgan, W.R. (1968). The geology and petrology of Cainozoic basaltic rocks in the Cooktown area, North Queensland. Journal of the Geological Society of Australia, 15, 65–78.

Morley, M.E., A.J.W. Gleadow, & J.F. Lovering (1981). Evolution of the Tasman Rift: apatite fission track dating evidence from the southeastern Australian continental margin. In: Cresswell, M.M., & P. Vella (Editors), Gondwana Five. Balkema, Rotterdam, 289–293.

Morris, J.C. (1987). The stratigraphy of the Amuri Limestone group, east Marlborough. University of Canterbury, unpublished PhD thesis.

Morris, P.A. (1979). Major and trace element geochemistry of volcanic rocks from the Chatham Islands. Victoria University of Wellington, Geology Department — Publication, 11.

Morris, P.A. (1982). Mapping and petrochemistry of the Chatham Island volcanics. Victoria University of Wellington, unpublished PhD thesis.

Morris, P.A. (1984). Petrology of the Campbell Island volcanics, southwest Pacific Ocean. Journal of Volcanological and Geothermal Research, 21, 119–148.

Morris, P.A. (1985). The geochemistry of Eocene-Oligocene volcanics on the Chatham Islands, New Zealand. New Zealand Journal of Geology and Geophysics, 28, 459–469.

Morris, P.A. (1986). Constraints on the origin of mafic alkaline volcanics and included xenoliths from Oberon, New South Wales, Australia. Contributions to Mineralogy and Petrology, 93, 207–214.

Morrison, M. (1904). Notes on some of the dykes and volcanic necks of the Sydney district. New South Wales Geological Survey — Record, 7, 241–281.

Muhlheim, M.M. (1974). Volcanic geology of the Kaikohe area, Northland, New Zealand. University of Auckland, unpublished MA thesis.

Muirhead, K.J. (1984). The base of the lithosphere under Australia. Geological Society of Australia — Abstracts, 12, 389.

Mulvaney, D.J. (1964). Prehistory of the Basalt Plains. Proceedings of the Royal Society of Victoria, 77, 427–432.

Murase, T., & A.R. McBirney (1984). Properties of some common igneous rocks and their melts at high temperatures. Bulletin of the Geological Society of America, 84, 3563–3592.

Murray, C.G. (1975a). Geology of Rockhampton, Queensland, 1:250 000 Sheet Series, SF/56–13. Bureau of Mineral Resources, Australia — Explanatory Notes.

Murray, C.G. (1975b). Geology of Port Clinton, Queensland, 1:250 000 Sheet Series, SF/56–9. Bureau of Mineral Resources, Australia — Explanatory Notes.

Murray, C.G. (1986). Metallogeny and tectonic development of the Tasman Fold Belt system in Queensland. Ore Geology Reviews, 1, 315–400.

Murray, C.G., & A.G. Kirkegaard (1978). The Thompson orogen of the Tasman orogenic zone. Tectonophysics, 48, 299–325.

Mutter, J.C. (1988). Convective partial melting: 1, A model for the formation of thick basaltic sequences during the initiation of spreading. Journal of Geophysical Research, 93, 1031–1048.

Mutter, J.C., & D. Jongsma (1978). The pattern of the pre-Tasman Sea rift system and the geometry of breakup. Bulletin of the Australian Society of Exploration Geophysicists, 9, 70–75.

Nakamura, E., H.I. Campbell, & S.-S. Sun (1985). The influence of subduction on the geochemistry of Japanese alkaline basalts. Nature, 361, 55–58.

Nakamura, K. (1977). Volcanoes as possible indicators of tectonic stress orientation — principle and proposal. Journal of Volcanology and Geothermal Research, 2, 1–16.

Nakamura, N. (1974). Determination of REE, Ba, Fe, Mg, Na and K in carbonaceous and ordinary chondrites. Geochimica et Cosmochimica Acta, 38, 757–775.

Nathan, S., H.J. Anderson, R.A. Cook, R.H. Herzer, R.H. Hoskins, J.I. Raine, & D. Smale (1986). Cretaceous and Cenozoic sedimentary basins of the West Coast region, South Westland, South Island, New Zealand. New Zealand Geological Survey — Basin Studies, 1.

Nelson, D.R., M.T. McCulloch, & S.-S. Sun (1986). The origins of ultrapotassic rocks as inferred from Sr, Nd, and Pb isotopes. Geochimica et Cosmochimica Acta, 50, 231–245.

New South Wales Department of Mineral Resources (1987). Extended abstracts from seminar on Tertiary volcanics and sapphires in the New England district. New South Wales Geological Survey — Report, GS1987/058.

Newbery, C. (1878). Report on the examination of black basalt and anamesite from Learmouth. Geological Survey of Victoria - Progress Report for 1877, 81–82.

Newhall, C.G., & W.G. Melson (1983). Explosive activity associated with the growth of volcanic domes. Journal of Volcanology and Geothermal Research, 17, 111–131.

Newton, R.C., J.V. Smith, & B.F. Windley (1980). Carbonic metamorphism, granulites and crustal growth. Nature, 288, 45–50.

Nicholls, I.A., & K.L. Harris (1980). Experimental rare earth element partition coefficients for garnet, clinopyroxene and amphibole coexisting with andesitic and basaltic liquids. Geochimica et Cosmochimica Acta, 44, 287–308.

Nicholls, J., & I.S.E. Carmichael (1969). Peralkaline acid liquids: a petrological study. Contributions to Mineralogy and Petrology, 20, 268–294.

Nickel, K.G., & D.H. Green (1984). The nature of the uppermost mantle beneath Victoria, Australia, as deduced from ultramafic xenoliths. In: Kornprobst, J. (Editor), II Kimberlites. Elsevier, Amsterdam, 161–178.

Nicolas, A. (1986). A melt extraction model based on structural studies in mantle peridotites. Journal of Petrology, 27, 999–1022.

Nielson-Pike, J.E., F.A. Frey, F.M. Richter, & B.O. Mysen (1985). Multistage mantle processes. Penrose Conference Report, Geology, 13, 742–744.

Nixon, P.H. (1987). Mantle Xenoliths. Wiley, Chichester.

Nixon, P.H., & F.R. Boyd (1973). Carbonated ultrabasic nodules from Sekemeng. In: Nixon, P.H. (Editor), Lesotho Kimberlites. Lesotho National Development Corporation, Maseru.

Noetling, F. (1911). The antiquity of man in Tasmania. Papers and Proceedings of the Royal Society of Tasmania (for 1910), 231–261.

Noon, A.T. (1972). Geology of the Mount Cooroy area, near Gympie, southeast Queensland. University of Queensland, unpublished BSc Honours thesis.

Noon, T.A. (1982). Stratigraphic drilling report — Geological Survey of Queensland, Monto, 5. Queensland Government Mining Journal, 83, 450–456.

Norris, R.J., & R.M. Carter (1980). Off-shore sedimentary basins at the southern end of the Alpine Fault, New Zealand. In: Ballance, P.F., & H.G. Reading (Editors), Sedimentation in Oblique-Slip Mobile Zones. International Association of Sedimentologists — Special Publication, 4, 237–265.

Norris, R.J., & R.M. Carter (1982). Fault-bounded blocks and their role in localising sedimentation and deformation adjacent to the Alpine Fault, southern New Zealand. In: Packham, G.H. (Editor), The Evolution of the India-Pacific Plate Boundaries. Tectonophysics, 87, 11–23.

Norris, R.J., & D. Craw (1987). Aspiring terrane: an oceanic assemblage from New Zealand and its implications for terrane accretion in the southwest Pacific. American Geophysical Union — Geodynamics Series, 19, 169–177.

Norton, I.O., & J.G. Sclater (1979). A model for the evolution of the Indian Ocean and the breakup of Gondwanaland. Journal of Geophysical Research, 84, 6803–6830.

Offenberg, A.C., D.M. Rose, & G.H. Packham (1968). Dubbo 1:250 000 Geological series sheet S155–4. Geological Survey of New South Wales, Sydney.

O'Hanlon, E.M. (1975). A petrographic and geochemical study of the Devonian and Tertiary volcanics of the Macedon district, central Victoria. La Trobe University, unpublished BSc Honours thesis.

Olgers, F., A.W. Webb, J.A.J. Smit, & B.A. Coxhead (1966). Geology of the Baralaba, Queensland 1:250 000 sheet area. Bureau of Mineral Resources, Australia — Report, 102.

Oliver, R.L., H.J. Finlay, & C.A. Fleming (1950). The geology of Campbell Island. New Zealand Department of Scientific and Industrial Research, Cape Expedition Series — Bulletin, 3.

Ollier, C.D. (1967a). Landforms of the Newer Volcanic province of Victoria. In: Jennings, J.J., & J.A. Mabbutt (Editors), Landform Studies from Australia and New Guinea. Australian National University Press, Canberra, 315–339.

Ollier, C.D. (1967b). Maars. Their characteristics, varieties and definition. Bulletin Volcanologique, 31, 45–73.

Ollier, C.D. (1969). Volcanoes. Australian National University Press, Canberra.

Ollier, C.D. (1981). A buried soil at Mount Eccles, western Victoria, and the date of eruption. Victorian Naturalist, 98, 195–199.

Ollier, C.D. (1982a). The Great Escarpment of eastern Australia: tectonic and geomorphic significance. Journal of the Geological Society of Australia, 29, 13–23.

Ollier, C.D. (1982b). Geomorphology and tectonics of the Dorrigo Plateau, New South Wales. Journal of the Geological Society of Australia, 29, 431–435.

Ollier, C.D. (1985a). Lava flows of Mount Rouse, western Victoria. Proceedings of the Royal Society of Victoria, 97, 167–174.

Ollier, C.D. (1985b). Morphotectonics of passive continental margins. Zeitschrift für Geomorphologie, Supplementband, 54, 1–10.

Ollier, C.D. (1988). Volcanoes. Blackwell, Oxford.

Ollier, C.D., & M.C. Brown (1965). Lava caves of Victoria. Bulletin Volcanologique, 28, 215–229.

Ollier, C.D., & E.B. Joyce (1964). Volcanic physiography of the western plains of Victoria. Proceedings of the Royal Society of Victoria, 77, 357–376.

Ollier, C.D., & E.B. Joyce (1973). Geomorphology of the Western District volcanic plains, lakes and coastline. In:

McAndrew, J., & M.A.H. Marsden (Editors), Regional Guide to Victorian Geology (second edition). School of Geology, University of Melbourne, 100–113.

Ollier, C.D., & E.B. Joyce (1976). Newer Volcanic landforms. In: Douglas J.G., & J.A. Ferguson (Editors), Geology of Victoria. Geological Society of Australia — Special Publication, 5, 342–348.

O'Neill, G.F.J. (1984). Geology of the Sunbury area — geochemistry of silcrete in the Sunbury area. University of Melbourne, Department of Geology, unpublished BSc Honours report.

O'Neill, H.St.C. (1981). The transition between spinel lherzolite and garnet lherzolite and its use as a geobarometer. Contributions to Mineralogy and Petrology, 77, 185–194.

O'Neill, H.St.C. (1987). Quartz-fayalite-iron and quartz-fayalite-magnetite equilibria and the free energy of formation of fayalite (Fe$_2$SiO$_4$) and magnetite (Fe$_3$O$_4$). American Mineralogist, 72, 67–75.

O'Neill, H.St.C., A.L. Jaques, C.B. Smith, & J. Moon (1986). Diamond-bearing peridotite xenoliths from the Argyle (AKI) Pipe. Fourth International Kimberlite Conference. Geological Society of Australia — Extended Abstracts, 16, 300–302.

O'Nions, R.K., S.R. Carter, N.M. Evensen, & P.J. Hamilton (1979a). Geochemical and cosmochemical applications of Nd isotope analysis. Annual Reviews of Earth and Planetary Sciences, 7, 11–38.

O'Nions, R.K., N.M. Evensen, S.R. Carter, & P.J. Hamilton (1979b). Isotope geochemical studies of North Atlantic ocean basalts and their implications for mantle evolution. In: Talwani, M., G.C. Harrison, & D.E. Hayes (Editors), Results of Deep Sea Drilling in the Atlantic: Ocean Crust. Maurice Ewing Series 2, American Geophysical Union, Washington, 342–351.

O'Reilly, S.Y. (1987). Volatile-rich mantle beneath eastern Australia. In: Nixon, P.H. (Editor), Mantle Xenoliths, Wiley, Chichester, 661–670.

O'Reilly, S.Y., & W.L. Griffin (1984). Sr isotopic heterogeneity in primitive basaltic rocks, southeastern Australia: correlation with mantle metasomatism. Contributions to Mineralogy and Petrology, 87, 220–230.

O'Reilly, S.Y., & W.L. Griffin (1985). A xenolith-derived geotherm for southeastern Australia and its geographical implications. Tectonophysics, 111, 41–63.

O'Reilly, S.Y., & W.L. Griffin (1987). Eastern Australia — 4000 kilometres of mantle samples. In: Nixon, P.H. (Editor), Mantle Xenoliths. Wiley, Chichester, 267–280.

O'Reilly, S.Y., & W.L. Griffin (1988). Mantle metasomatism beneath western Victoria, Australia, I: Metasomatic processes in Cr-diopside lherzolites. Geochimica et Cosmochimica Acta, 52, 433–447.

O'Reilly, S.Y., & W.L. Griffin (in press). The nature and role of fluids in the upper mantle: evidence in xenoliths from Victoria, Australia. In: Herbert, H.K. (Editor), Proceedings of Conference on Stable Isotopes and Fluid Processes in Mineralization. Geological Society of Australia — Special Publication, 13.

O'Reilly, S.Y., W.L. Griffin, & A. Stabel (1988). Evolution of Phanerozoic eastern Australia: isotopic evidence for magmatic and tectonic underplating. In: Menzies, M.A., & K.G. Cox (Editors), Oceanic and Continental Lithosphere: Similarities and Differences. Journal of Petrology — Special Publication, 89–108.

Ortez, N.A. (1976). Geochemistry and petrogenesis of the alkaline and associated igneous rocks of the Dubbo area, New South Wales. Macquarie University, unpublished BSc Honours thesis.

Osborne, R.A.L. (1979). Preliminary report: caves in Tertiary basalt, Coolah, New South Wales. Helictite, 17, 25–29.

Ottonello, G. (1980). Rare earth abundances and distribution in some spinel peridotite xenoliths from Assab (Ethiopia). Geochimica et Cosmochimica Acta, 44, 1885–1901.

Oversby, B.S., L.P. Black, & J.W. Sheraton (1980). Late Palaeozoic continental volcanism in northeastern Queensland. In: Henderson, R.A., & P.J. Stevenson (Editors), The Geology and Geophysics of Northeastern Australia. Geological Society of Australia, Queensland Division, 247–268.

Owen, H.B. (1954). Bauxite in Australia. Bureau of Mineral Resources, Australia — Bulletin, 24.

Oxburgh, E.R., & E.M. Parmentier (1978). Thermal processes in the formation of continental lithosphere. Philosophical Transactions of the Royal Society of London, A288, 415–429.

Oxley, J. (1820). Journals of Two Expeditions into the Interior of New South Wales in 1817–18. Murray, London.

Ozawa, K. (1983). Evaluation of olivine-spinel geothermometry as an indicator of thermal history for peridotites. Contributions to Mineralogy and Petrology, 82, 52–65.

Packham, G.H. (1969). The general features of the geological province of New South Wales. Journal of Geological Society of Australia, 16, 1–17.

Packham, G.H. (1973). A speculative Phanerozoic history of the south-west Pacific. In: Coleman, P.J. (Editor), The Western Pacific: Island Arcs, Marginal Seas, Geochemistry. University of Western Australia Press, Perth, 369–388.

Padovani, E.R., & J.L. Carter (1977). Aspects of deep crustal evolution beneath south central New Mexico. In: Heacock, J.G. (Editor), The Earths Crust, its Nature and Physical Properties. American Geophysical Union — Geophysical Monograph, 20, 19–55.

Page, R.W. (1968). Catalogue of radiometric age determinations carried out on Australian rocks in 1966. Bureau of Mineral Resources, Australia — Record, 1968/30, 24–25.

Pain, C.F. (1983). Geomorphology of the Barrington Tops area, New South Wales. Journal of the Geological Society of Australia, 30, 187–194.

Pain, C.F., & C.D. Ollier (1986). The Comboyne and Bulga Plateaus and the evolution of the Great Escarpment in New South Wales. Journal and Proceedings of the Royal Society of New South Wales, 119, 123–130.

Paine, A.G.L., D.E. Clarke, & C.M. Gregory (1970). Geology of the northern half of the Bowen 1:250 000 sheet area, Queensland (with additions to the geology of the southern half). Bureau of Mineral Resources, Australia — Report, 1970/50.

Palacz, Z.A. (1985). Sr-Nd-Pb isotopic evidence for crustal contamination in the Rhum intrusion. Earth and Planetary Science Letters, 74, 35–44.

Pandey, O.P. (1981). Terrestrial heat flow in New Zealand. Victoria University of Wellington, unpublished PhD thesis.

Pandey, O.P., & J.G. Negi (1987). A new theory of the origin and evolution of the Deccan Traps (India). Tectonophysics, 142, 329–335.

Park, J. (1905). On the marine Tertiaries of Otago and Canterbury, with special reference to the relations existing between the Pareora and Oamaru Series. Transactions of the New Zealand Institute, 37, 489–551.

Parmentier, E.M. (1986). Dynamic topography in rift zones; implications for lithospheric heating. Philosophical Transactions of the Royal Society of London, A321, 23–25.

Passchier, C.W. (1986). Mylonites in the continental crust and their role as seismic reflectors. Geologie en Mijnbouw, 65, 167–176.

Pearce, J.A. (1983). Role of the sub-continental lithosphere in magma genesis at active continental margins. In: Hawkesworth, C.J., & M.J. Norry (Editors), Continental Basalts and Mantle Xenoliths. Shiva, Cheshire, 230–249.

Pearce, J.A., T. Alabaster, A.W. Shelton, & M.P. Searle (1981). The Oman ophiolite as a Cretaceous arc-basin complex: evidence and implications. Philosophical Transactions of the Royal Society of London, A300, 299–317.

Pearson, D.G., G.R. Davies, & P.H. Nixon (1987). Diamond facies garnet pyroxenites of Beri Bousera, Morocco: recycled oceanic lithosphere. Oceanic and Continental Lithosphere, Similarities and Differences, Workshop, London — Abstracts, 618.

Pecover, S.R. (1987). Tertiary maar volcanism and the origin of sapphires in northeastern New South Wales. In: Tertiary Volcanics and Sapphires in the New England District. New South Wales Geological Survey — Report, GS1987/058, 13–21.

Peng, Z.C., R.E. Zartman, K. Futa, & D.G. Chen (1986). Pb-, Sr-, and Nd-isotopic systematics and chemical characteristics of Cenozoic basalts, eastern China. Chemical Geology, 59, 3–33.

Perry, F.V., W.S. Baldrige, & D.J. De Paolo (1987). Role of asthenosphere and lithosphere in the genesis of late Cenozoic basaltic rocks from the Rio Grande Rift and adjacent regions of the southwestern United States. Journal of Geophysical Research, 92, 9193–9213.

Peterson, D.W., & R.I. Tilling (1980). Transition of basaltic lava from pahoehoe to aa, Kilauea volcano, Hawaii: field observations and key factors. Journal of Volcanology and Geothermal Research, 7, 271–293.

Petrini, R., L. Civetta, E.M. Piccirillo, G. Bellieni, P. Comin-Chiaramonti, L.S. Marques, & A.J. Melfi (1987). Mantle heterogeneity and crustal contamination in the genesis of low-Ti continental flood basalts from the Paraná Plateau (Brazil): Sr-Nd isotope and geochemical evidence. Journal of Petrology, 28, 702–726.

Pettinga, J.R. (1982). Upper Cenozoic structural history, coastal southern Hawke's Bay, New Zealand. New Zealand Journal of Geology and Geophysics, 25, 149–191.

Philpotts, J.A., C.C. Schnetzler, & H.H. Thomas (1972). Petrogenetic implications of some new geochemical data on eclogite and ultrabasic inclusions. Geochimica et Cosmochimica Acta, 36, 1131–1166.

Pilger, R.H. Jr (1982). The origin of hotspot tracks: evidence from eastern Australia. Journal of Geophysical Research, 87, 1825–1834.

Pillans, B. (1986). A Late Quaternary uplift map for North Island, New Zealand. Royal Society of New Zealand — Bulletin, 24, 409–417.

Pineau, F., & M. Javoy (1983). Carbon isotopes and concentrations in mid-ocean ridge basalts. Earth and Planetary Science Letters, 62, 239–257.

Pinkerton, H., & R.S.J. Sparks (1976). The 1975 sub-terminal lavas, Mount Etna: a case history of the formation of a compound lava flow. Journal of Volcanology and Geothermal Research, 1, 167–182.

Pinkerton, H., & R.S.J. Sparks (1978). Field measurements of the rheology of lava. Nature, 276, 383–386.

Pittman, E.F. (1901). Diamonds in volcanic breccia. The Ruby Hill mine. New South Wales Department of Mines — Annual Report, 1900, 180–181, 206.

Polach, H.A., M.J. Head, & J.D. Gower (1978). Australian National University radiocarbon date list VI. Radiocarbon, 20, 360–385.

Polden, K. (1980). Gems of the basalt. Australian Lapidary Magazine, 16, 17–22.

Porcelli, D.R., R.K. O'Nions, & S.Y. O'Reilly (1986). Helium and strontium isotopes in ultramafic xenoliths. Chemical Geology, 54, 237–249.

Powell, C.McA. (1983). Tectonic relationship between the late Ordovician and late Silurian palaeogeographies of southeastern Australia. Journal of the Geological Society of Australia, 30, 353–373.

Powell, M., & R. Powell (1974). An olivine-clinopyroxene geothermometer. Contributions to Mineralogy and Petrology, 48, 249–263.

Prendergast, E.I. (1987). An early Palaeozoic subduction complex on the New South Wales south coast. International Conference on Deformation of Crustal Rocks. Geological Society of Australia — Abstracts, 19, 4–5.

Preusser, H. (1973). Der Vulkanausbruch auf Heimaey/Vestmannaeyjar und seine Asuswirkungan. Geographische Rundschau 25, 337–350.

Price, R.C., & B.W. Chappell (1975). Fractional crystallization and the petrology of Dunedin volcano. Contributions to Mineralogy and Petrology, 53, 157–182.

Price, R.C., & W. Compston (1973). The geochemistry of the Dunedin volcano: strontium isotope chemistry. Contributions to Mineralogy and Petrology, 45, 55–61.

Price, R.C., & D.S. Coombs (1975). Phonolitic lava domes and other features of the Dunedin volcano, east Otago. Journal of the Royal Society of New Zealand, 5, 133–152.

Price, R.C., & D.H. Green (1972). Lherzolite nodules in a 'mafic phonolite' from north-east Otago, New Zealand. Nature, 235, 133–134.

Price, R.C., & S.R. Taylor (1973). The geochemistry of the Dunedin volcano: east Otago, New Zealand: rare earth elements. Contributions to Mineralogy and Petrology, 40, 195–205.

Price, R.C., & S.R. Taylor (1980). Petrology and geochemistry of Banks Peninsula volcanoes, South Island, New Zealand. Contributions to Mineralogy and Petrology, 72, 1–18.

Price, R.C., & R.C. Wallace (1976). The significance of corona textured inclusions from a high-pressure fractionated alkalic lava, north Otago, New Zealand. Lithos, 9, 319–329.

Price, R.C., R.W. Johnson, C.M. Grey, & F.A. Frey (1985). Geochemistry of phonolites and trachytes from the summit region of Mt Kenya. Contributions to Mineralogy and Petrology, 89, 394–409.

Price, R.C., C.M. Gray, I.A. Nicholls, & R.A. Day (1988). Cenozoic volcanic rocks. In: Douglas, J.G., & J.A. Ferguson (Editors), Geology of Victoria. Geological Society of Australia, Victorian Division, 439–452.

Prodehl, C. (1977). The structure of the crust-mantle beneath North America and Europe as derived from explosion seismology. In: Heacock, J.G. (Editor), The Earth's Crust. American Geophysical Union, Washington, D.C., 349–369.

Protheroe, D.J. (1981). Possible structural controls in the Main Range volcanics near Crows Nest, southeast Queensland. Proceedings of the Royal Society of Queensland, 92, 49–55.

Purvis, A.C. (1965). The geology of an area north east of Central Tilba, New South Wales. University of Sydney, unpublished BSc Honours thesis.

Queensland Department of Mines Annual Report (1980).

Rabinowitz, P.D. (1979). The Mesozoic South Atlantic Ocean and evolution of its continental margins. Journal of Geophysical Research, 84, 5973–6002.

Rafferty, W.J. (1977). The volcanic geology and petrology of South Auckland. University of Auckland, unpublished MSc thesis.

Rafferty, W.J., & R.E. Heming (1979). Quaternary alkalic and sub-alkalic volcanism in south Auckland, New Zealand. Contributions to Mineralogy and Petrology, 71, 139–150.

Raggatt, H.G., & H.F. Whitworth (1930). The intrusive igneous rocks of the Muswellbrook-Singleton district, Part 1. Introduction. Journal and Proceedings of the Royal Society of New South Wales, 64, 78–82.

Raggatt, H.G., & H.F. Whitworth (1932). The intrusive igneous rocks of the Muswellbrook-Singleton district, Part 2. The Savoy sill. Journal and Proceedings of the Royal Society of New South Wales, 66, 194–233.

Raheim, A., & W. Compston (1977). Correlations between metamorphic events and Rb-Sr in metasediments and eclogite from western Tasmania. Lithos, 10, 271–290.

Raine, J.I. (1968). Geology of the Nerriga area with special reference to Tertiary stratigraphy and palynology. Australian National University, unpublished BSc Honours thesis.

Randall, M.J. (1971). Regional travel times of P and S seismic waves between Fiji and New Zealand. New Zealand Journal of Geology and Geophysics, 14, 133–152.

Reay, A. (1986). Andesites from Solander Island. In: Smith, I.E.M. (Editor), Late Cenozoic Volcanism in New Zealand. Royal Society of New Zealand — Bulletin, 23, 337–343.

Reay, A., & P.P. Sipiera (1987). Mantle xenoliths from the New Zealand region. In: Nixon, P.H. (Editor), Mantle Xenoliths. Wiley, Chichester, 347–358.

Reay, A., & C.P. Wood (1974). Ilmenites from Kakanui, New Zealand. Mineralogical Magazine, 39, 721–722.

Reay, M.B. (1980). Cretaceous and Tertiary stratigraphy of part of the middle Clarence Valley, Marlborough. University of Canterbury, unpublished MSc thesis.

Redicliffe, T.H. (1974). The geology of the Koorboora Creek area, northeast Queensland. University of Queensland, unpublished BSc Honours thesis.

Reed, A.H., & A.W. (1951, Editors). Captain Cook in New Zealand. Extracts from the Journals. Reed, Wellington.

Reed, A.W. (Editor) (1969). Captain Cook in Australia. Extracts from the Journals. Reed, Wellington.

Reilly, W.I. (1966). Sheet 21, Christchurch, Gravity map of New Zealand 1:250 000, Bouguer anomalies. New Zealand Department of Scientific and Industrial Research.

Reilly, W.I. (1972). Gravitational expression of the Dunedin volcano. New Zealand Journal of Geology and Geophysics, 15, 16–21.

Reilly, W.I., C.M. Whiteford, & A. Doone (1977). Gravity map of New Zealand, 1:1 million. North Island. Free air, Bouguer, isostatic vertical gradient anomalies. New Zealand Department of Scientific and Industrial Research, Geophysics Division.

Reynolds, S.J. (1985). Geology of the South Mountains, central Arizona. Arizona Bureau of Geology and Mineral Technology, Arizona Geological Survey — Bulletin, 195.

Rice, A. (1981). Convective fractionation: a mechanism to provide cryptic zoning (macrosegregation), layering, crescumulates, banded tuffs and explosive volcanism in igneous processes. Journal of Geophysical Research, 86, 405–417.

Richards, D.N.G. (1980). Palaeozoic granitoids of north-eastern Australia. In: Henderson, R.A., & P.J. Stephenson (Editors), The Geology and Geophysics of Northeastern Australia. Geological Society of Australia, Queensland Division, 229–246.

Richards, H.C. (1916). Volcanic rocks of south-east Queensland. Proceedings of the Royal Society of Queensland, 27, 105–204.

Richards, H.C. (1918). The volcanic rocks of Springsure, central Queensland. Proceedings of the Royal Society of Queensland, 30, 179–198.

Richards, H.C. (1926). Volcanic activity in Queensland. Australasian Association for the Advancement of Science — Report, 17, 275–299.

Richardson, R.M., S.C. Solomon, & N.H. Sleep (1979). Tectonic stress in the plates. Reviews of Geophysics and Space Physics, 17, 981–1019.

Richardson, S.H., J.J. Gurney, A.L. Erlank, & J.W. Harris (1984). Origin of diamonds in old enriched mantle. Nature, 310, 198–202.

Richter, F.M. (1973). Convection and the large-scale circulation of the mantle. Journal of Geophysical Research, 35, 8735–8745.

Rickwood, P.C., F. Colon, V.J. Dobos, J.V. Guy, C.G. Gwatkin, R. Poole, & I.E. Wainwright (1983). The origin of basalt in the Blue Mountains near Sydney, New South Wales, Australia In: Augustitis, S. (Editor), The Significance of Trace Elements in Solving Petrogenetic Problems and Controversies. Theophrastus, Athens, 124–135.

Ringwood, A.E. (1979). Origin of the Earth and Moon. Springer-Verlag, Berlin.

Ringwood, A.E. (1982). Phase transformations and differentiation in subducted lithosphere: implications for mantle dynamics, basalt petrogenesis and crustal evolution. Journal of Geology, 90, 611–643.

Ringwood, A.E., & T. Irifune (1988). Nature of the 650-km seismic discontinuity: implications for mantle dynamics and differentiation. Nature, 331, 131–136.

Rivalenti, G., A. Rossi, F. Siena, & S. Sinigoi (1984). The layered series of the Ivrea-Verbano igneous complex, Western Alps, Italy. Tschermaks Mineralogie und Petrologie Mitteilungen, 33, 77–99.

Roberts, P.S. (1984). Explanatory notes on Bacchus Marsh and Ballan 1:50 000 geological maps. Geological Survey of Victoria — Report, 76.

Robertson, A.D. (1979). Revision of the Cainozoic geology between the Kolan and Elliott Rivers. Queensland Government Mining Journal, 80, 350–363.

Robertson, A.D. (1982). Investigation of potential quarry rock reserves in Woongarra Shire. Geological Survey of Queensland — Record, 1982/19.

Robertson, A.D. (1985). Cainozoic volcanic rocks in the Bundaberg-Gin Gin-Pialba area, Queensland. Papers of the Department of Geology, University of Queensland, 11, 72–92.

Robertson, A.D., & C.G. Murray (1978). Olivine nephelinite from the Bundaberg area. Queensland Government Mining Journal, 79, 579–581.

Robertson, A.D., F.L. Sutherland, & J.D. Hollis (1985). Upper mantle xenoliths and megacrysts and the origin of the Brigooda basalt and breccia near Proston, Queensland. Papers of the Department of Geology, University of Queensland, 11, 58–71.

Robertson, C.S., K.L. Lockwood, E. Nicholas, & H. Soebarkah (1978). A review of petroleum exploration and prospects in the Gippsland Basin. Bureau of Mineral Resources, Australia — Record, 1978/110.

Robertson, D.J. (1976). A paleomagnetic study of volcanic rocks in the south Auckland area. University of Auckland, unpublished MSc thesis.

Robertson, D.J. (1983). Paleomagnetism and geochronology of volcanics in the northern North Island, New Zealand. University of Auckland, unpublished PhD thesis.

Robertson, W.A. (1966). Palaeomagnetism of some Cainozoic igneous rocks from south-east Queensland. Proceedings of the Royal Society of Queensland, 78, 87–100.

Robertson, W.A. (1979). Palaeomagnetic results from some Sydney Basin igneous rock deposits. Journal and Proceedings of the Royal Society of New South Wales, 112, 31–35.

Robinson, R. (1976). Relative teleseismic travel-time residuals, North Island, New Zealand, and their relation to upper mantle structure. Tectonophysics, 31, T41-T48.

Robinson, V.A. (1974). Geological history of the Bass Basin. APEA Journal, 14, 45–49.

Rock, N.M.S. (1981). How should igneous rocks be grouped? Geological Magazine, 118, 449–579.

Rodgers, K.A., & R.N. Brothers (1969). Olivine, pyroxene, feldspar and spinel in ultramafic nodules from Auckland, New Zealand. Mineralogical Magazine, 39, 375–390.

Rodgers, K.A., P.R. Spratt, & J. Grant-Mackie (1973). A reappraisal of the Ngatutura Volcanics and the Plio-Pleistocene boundary in southwest Auckland. New Zealand Journal of Geology and Geophysics, 16, 367–373.

Rodgers, K.A., R.N. Brothers, & E.J. Searle (1975). Ultramafic nodules and their host rocks from Auckland, New Zealand. Geological Magazine, 112, 163–174.

Roeder, P.L., & R.F. Emslie (1970). Olivine-liquid equilibrium. Contributions to Mineralogy and Petrology, 29, 275–289.

Roeder, P.L., I.H. Campbell, & H.E. Jamieson (1979). A re-evaluation of the olivine-spinel geothermometer. Contributions to Mineralogy and Petrology, 68, 325–334.

Ross, J.A. (1974). The Focal Peak shield volcano, southeast Queensland — evidence from its eastern flank. Royal Society of Queensland, 85, 111–117.

Ross, J.A. (1977). The Tertiary Focal Peak shield volcano, south-east Queensland. University of Queensland, unpublished PhD thesis.

Ross, J.A. (1985a). Tuff-lava from southeast Queensland, Australia. Current Science, 54, 272–274.

Ross, J.A. (1985b). Secondary silicification of rhyolites from the Focal Peak shield volcano, Australia. Current Science, 54, 1168–1171.

Royden, L., & C.E. Keen (1980). Rifting process and thermal evolution of the continental margin of eastern Canada determined from subsidence curves. Earth and Planetary Science Letters, 51, 343–361.

Rudnick, R.L., & I. Jackson (1987). Physical properties of a magmatically underplated lower continental crust. Terra Cognita, 7, 620.

Rudnick, R.L., & S.R. Taylor (1987). The composition and petrogenesis of the lower crust: a xenolith study. Journal of Geophysical Research, 92, 13981–14005.

Rudnick, R.L., & S.R. Taylor (in press). Petrology and geochemistry of lower crustal xenoliths from northern Queensland and inferences on lower crustal composition. In: Drummond, B. (Editor), The Eastern Australian Lithosphere. Geological Society of Australia — Special Publication.

Rudnick, R.L., W.F. McDonough, M.T. McCulloch, & S.R. Taylor (1986). Lower crustal xenoliths from Queensland, Australia: evidence for deep crustal assimilation and fractionation of continental basalts. Geochimica et Cosmochimica Acta, 50, 1099–1115.

Russell, R.E. (1965). A preliminary note on the lava succession near Spicer's Peak, southeast Queensland. Papers of the Department of Geology. University of Queensland, 5, 1–12.

Ryabchikov, I.D., & D.H. Green (1978). The role of carbon dioxide in the petrogenesis of highly potassic magmas. In: Problems on the Petrology of the Earth's Crust and Upper Mantle. Trudy Instituta Geologii Geofizik So An SSR, 403. Novosibirsk Nauka, 49–64.

Sachtleben, T., & H.A. Seck (1981). Chemical control of Al-solubility in orthopyroxene and its implications for

pyroxene geothermometry. Contributions to Mineralogy and Petrology, 78, 157–165.

Saito, K., A.R. Basu, & E.C. Alexander (1978). Planetary-type rare gases in upper mantle-derived amphibole. Earth and Planetary Science Letters, 39, 274–280.

Sass, J.H., & A.H. Lachenbruch (1979). Thermal regime of the Australian continental crust. In: McElhinny, M.W. (Editor), The Earth — it's Origin, Structure and Evolution. Academic Press, London, 301–352.

Saunders, A.D., J. Tarney, & S.D. Weaver (1980). Transverse geochemical variations across the Antarctic peninsula: implications for the genesis of calc-alkaline magmas. Earth and Planetary Science Letters, 46, 344–360.

Schairer, J.F. (1950). The alkali feldspar join in the system $NaAlSiO_4$-$KAlSiO_4$-SiO_2. Journal of Geology, 58, 512–517.

Scheibner, E. (1973). ERTS–1 geological investigations of New South Wales (Australia). Type III report for period January 1973-December 1973. Commonwealth Department of Supply, Australia — Report, GS1973/382.

Scheibner, E. (1976). Tectonic map of New South Wales, scale 1:1 000 000. Geological Survey of New South Wales — Explanatory Notes.

Scheibner, E. (1985). Suspect terranes in the Tasman Fold Belt System, eastern Australia. In: Howell, D.G. (Editor), Tectonostratigraphic Terranes of the Circum-Pacific Region. Circum Pacific Council of Energy and Mineral Resources, Houston, Texas, USA. Earth Science Series, 1, 493–514.

Schilling, J.-G. (1973). Iceland mantle plume: geochemical study of Reykjanes Ridge. Nature, 242, 565–571.

Schilling, J.-G., G. Thompson, R. Kingsley, & S. Humphries (1985). Hotspot-migrating ridge interaction in the south Atlantic. Nature, 313, 187–191.

Schmidt, D.L., & P.D. Rowley (1986). Continental rifting and transform faulting along the Jurassic Transantarctic Rift, Antarctica. Tectonics, 5, 279–291.

Schmidt, P.W., & B.J.J. Embleton (1981). Magnetic overprinting in southeastern Australia and the thermal history of its rifted margin. Journal of Geophysical Research, 86, 3998–4008.

Schmidt, P.W., & I. McDougall (1977). Paleomagnetic and potassium-argon dating studies of the Tasmanian dolerites. Journal of the Geological Society of Australia, 24, 321–328.

Schmincke, H.-U. (1967). Fused tuff and peperites in south-central Washington. Bulletin of the Geological Society of America, 78, 319–330.

Schmincke, H.-U. (1977). Phreatomagmatische phasen in quartaren Vulkanen der Osteifel. Geologishe Jahrbuch, (A)39, 3–45.

Schmincke, H.-U., R.V. Fisher, & A.C. Waters (1973). Antidune and chute and pool structures in the base surge deposits of the Laacher Sea area, Germany. Sedimentology, 20, 553–574.

Schofield, J.C. (1967). Sheet 3 Auckland (1st Edition). Geological map of New Zealand 1:250 000. New Zealand Department of Scientific and Industrial Research.

Scholz, C.H., J.M.W. Rynn, R.W. Weed, & C. Frohlich (1973). Detailed seismicity of the Alpine Fault zone and Fiordland region, New Zealand. Bulletin of the Geological Society of America, 84, 3297–3316.

Schön, R.W. (1985). Petrology of the Liverpool Range volcanics, eastern Australia. In: Sutherland, F.L., B.J. Franklin, & A.E. Waltho (Editors), Volcanism in Eastern Australia, with Case Histories from New South Wales. Geological Society of Australia, New South Wales Division — Abstracts, 73–85.

Schön, R.W. (1986). The petrology and geochemistry of extrusive rocks from the Liverpool Range, a volcanic centre in eastern New South Wales, Australia. University of Sydney, unpublished PhD thesis.

Scott Smith, B.H., R.V. Danchin, J.W. Harris, & K.J. Stracke (1984). Kimberlites near Orroroo, South Australia. In: Kornprobst, J. (Editor), Kimberlites I: Kimberlites and Related Rocks. Elsevier, Amsterdam, 121–142.

Searle, E.J. (1961). The petrology of the Auckland basalts. New Zealand Journal of Geology and Geophysics, 4, 165–204.

Searle, E.J. (1964). City of Volcanoes: a Geology of Auckland. Pauls Book Arcade, Auckland.

Searle, E.J., & R.D. Mayhill (1981). City of Volcanoes — a Geology of Auckland. Longman, Auckland.

Segerstrom, K. (1950). Erosion studies at Paricutin, State of Michoacan, Mexico. United States Geological Survey — Bulletin, 965A.

Self, S., & R.S.J. Sparks (1978). Characteristics of widespread pyroclastic deposits formed by the interaction of silicic magma and water. Bulletin Volcanologique, 41, 196–212.

Self, S., L. Wilson, & I. Nairn (1979). Vulcanian eruption mechanisms. Nature, 277, 440–443.

Selverstone, J. (1982). Fluid inclusions as petrogenetic indicators in granulite xenoliths, Pali-Aike volcanic field, Chile. Contributions to Mineralogy and Petrology, 79, 28–36.

Settle, M. (1978). Volcanic eruption clouds and the thermal output of explosive eruptions. Journal of Volcanology and Geothermal Research, 3, 309–324.

Sewell, R.J. (1985). The volcanic geology and geochemistry of central Banks Peninsula and relationships to Lyttelton and Akaroa volcanoes. University of Canterbury, unpublished PhD thesis.

Sewell, R.J. (1988). Miocene volcanic stratigraphy of central Banks Peninsula, Canterbury, New Zealand. New Zealand Journal of Geology and Geophysics, 31, 41–64.

Sewell, R.J., & I.L. Gibson (1988). Petrology andJ geochemistry of Tertiary volcanic rocks from inland Central and South Canterbury, South Island, New Zealand. New Zealand Journal of Geology and Geophysics, 31, 477–492.

Sewell, R.J., & S. Nathan (1987). Geochemistry of Late Cretaceous and Early Tertiary basalts from south Westland. New Zealand Geological Survey — Record, 18, 87–94.

Sharp, A. (1971, Editor). Duperrey's visit to New Zealand in 1824. Alexander Turnbull Library, Wellington.

Shaw, H.R. (1980). The fracture mechanisms of magma transport from the mantle to the surface. In: Hargraves, R.B. (Editor), Physics of Magmatic Processes. Princeton University Press, New Jersey, 201–254.

Shaw, H.R., D.L. Peck, & A.R. Okamura (1968). The viscosity of basaltic magma: an analysis of field measurements in Makaopuki lava lake, Hawaii. American Journal of Science, 266, 225–264.

Shaw, R.D. (1978). Sea floor spreading in the Tasman Sea: a Lord Howe Rise-eastern Australian reconstruction. Bulletin of the Australian Society of Exploration Geophysics, 9, 75–81.

Sheard, M.J. (1978). Geological history of the Mount Gambier volcanic complex, southeast South Australia. Transactions of the Royal Society of South Australia, 102, 125–139.

Sheard, M.J. (1983). Volcanoes. In: Natural History of the South East. Royal Society of South Australia, 7–14.

Sheard, M.J. (1986a). Some volcanological observations at Mt Schank, southeast South Australia. Geological Survey of South Australia — Quarterly Geological Notes, 100, 14–20.

Sheard, M.J. (1986b). Volcanic hazards in the southeast of South Australia. Geological Survey of South Australia — unpublished file report.

Shelley, D. (1987). Lyttelton 1 and Lyttelton 2, the two centres of Lyttelton volcano. New Zealand Journal of Geology and Geophysics, 30, 159–168.

Shelley, D. (1988). Radial dikes of Lyttelton volcano — their structure, form and petrography. New Zealand Journal of Geology and Geophysics, 31, 65–75.

Sheridan, M.F., & R.G. Updike (1975). Sugarloaf Mountain tephra — Pleistocene rhyolitic deposit of base-surge origin in northern Arizona. Bulletin of the Geological Society of America, 86, 571–581.

Shibaoka, M., & A.J.R. Bennett (1976). Effect of depth of burial and tectonic activity on coalification. Nature, 259, 385–386.

Sibson, R.H. (1982). Fault zone models, heat flow, and the depth distribution of earthquakes in the continental crust of the United States. Bulletin of the Seismological Society of America, 72, 151–163.

Simkin, T., & J.V. Smith (1970). Minor element distribution in olivine. Journal of Geology, 78, 304–325.

Simmons, G. (1964). Velocity of shear waves in rocks to 10 kilobars. Journal of Geophysical Research, 69, 1123–1130.

Singleton, O.P. (1973). Geology and petrology of the Macedon district. In: McAndrew, J., & M.A.H. Marsden (Editors), Regional Guide to Victorian Geology (second edition). School of Geology, University of Melbourne, 90–99.

Singleton, O.P., & E.B. Joyce (1969). Cainozoic volcanicity in Victoria. In: Brown, D.A. (Editor), Proceedings of Specialists' Meeting of GSA, Canberra. Geological Society of Australia — Special Publication, 2, 45–154.

Sipkin, S.A., & T.H. Jordan (1975). Lateral heterogeneity of the upper mantle determined from the travel times of ScS. Journal of Geophysical Research, 80, 1474–1484.

Sipkin, S.A., & T.H. Jordan (1976). Lateral heterogeneity of the upper mantle determined from the travel times of multiple ScS. Journal of Geophysical Research, 81, 6307–6320.

Sissons, B.A. (1979). The horizontal kinematics of the North Island of New Zealand. Victoria University of Wellington, unpublished PhD thesis.

Skeats, E.W. (1909). Volcanic rocks of Victoria. Australasian Association for the Advancement of Science — Report, 12, 173–235.

Skeats, E.W., & A.V.G. James (1937). Basaltic barriers and other surface features of the Newer Basalts of Western Victoria. Proceedings of the Royal Society of Victoria, 49, 245–278.

Skinner, D.N.B. (1986). Neogene volcanism of the Hauraki volcanic region. In: Smith, I.E.M. (Editor), Late Cenozoic Volcanism in New Zealand. Royal Society of New Zealand Bulletin, 23, 21–47.

Skjelkvale, B.L., H.E.F. Amundsen, S.Y. O'Reilly, W.L. Griffin, & T. Gjelsvik (in press). A primitive alkali basaltic stratovolcano and associated eruptive centres, NW Spitsbergen: volcanology, origin and tectonic significance. Journal of Volcanology and Geothermal Research.

Slade, M.J. (1964). A stratigraphic and palaeobotanical study of the Lower Tertiary sediments of the Armidale district, New South Wales. University of New England, unpublished MSc thesis.

Slater, R.A., & R.H. Goodwin (1973). Tasman Sea guyots. Marine Geology, 14, 81–99.

Slessar, G.C. (1970). The geology of Cape Hillsborough, Mackay region. James Cook University of North Queensland, unpublished BSc Honours thesis.

Sluiter, I.R.K. (in press). The age and depositional history of Cenozoic palaeodrainage systems in Central and Western Australia. In: De Deckker, P., & M.I.J. Williams (Editors), The Cainozoic of the Australian Region. Geological Society of Australia — Special Publication.

Smith, A.G. (1982). Late Cenozoic uplift of stable continents in a reference frame fixed to South America. Nature, 296, 400–404.

Smith, B.W., & J.R. Prescott (1987). Thermoluminescence dating of the eruption of Mount Schank, South Australia. Australia Journal of Earth Sciences, 34, 335–342.

Smith, G.C. (1986). Bass Basin geology and petroleum exploration. In: Glenie, R.C. (Editor), Second South-Eastern Australia Oil Exploration Symposium. Petroleum Exploration Society of Australia, Melbourne, 1985, 257–284.

Smith, I.E.M., B.W. Chappell, G.K. Ward, R.S. Freeman (1977). Peralkaline rhyolites associated with andesitic arcs of the south west Pacific. Earth and Planetary Science Letters, 37, 230–236.

Smith, I.E.M., R.A. Day, & J. Ashcroft (1986). Volcanic associations of Northland — International Volcanological Congress Tour Guide A4. In: Houghton, B.F., & S.D. Weaver (Editors), North Island Volcanism. New Zealand Geological Survey — Record, 12, 5–32.

Smith, W.D. (1971). Earthquakes at shallow and intermediate depths in Fiordland, New Zealand. Journal of Geophysical Research, 76, 4901–4907.

Sobolev, N.V. (1984). Crystalline inclusions in diamonds from New South Wales, Australia. In: Glover, J.E., & P.G. Harris (Editors), Kimberlite Occurrence and Origin: a Basis for Conceptual Models in Exploration. University of Western Australia, Department of Geology — Publication, 8, 213–226.

Solomon, P.J. (1964). The Mount Warning shield volcano. A general geological and geomorphological study of the dissected shield. Papers of the Department of Geology, University of Queensland, 5, 1–12.

Song, Y., F.A. Frey, & X. Zhi (1987). Geochemistry of basalts and peridotite xenoliths from Hannuoba region, eastern China: implication for subcontinental mantle heterogeneity. Transactions of the American Geophysical Union, 68, 1547.

Sparks, R.S.J. (1978). The dynamics of bubble formation and growth in magmas: a review and analysis. Journal of Volcanological and Geothermal Research, 3, 1–37.

Sparks, R.S.J. (1986). The dimensions and dynamics of volcanic eruption columns. Bulletin Volcanologique, 48, 3–16.

Sparks, R.S.J., & L. Wilson (1982). Explosive volcanic eruptions — V. Observations of plume dynamics during the 1979 Soufriere eruption, St Vincent. Geophysical Journal of the Royal Astronomical Society, 69, 551–570.

Sparks, R.S.J., H. Pinkerton, & G. Hulme (1976). Classification and formation of lava levees on Mount Etna, Sicily. Geology 4, 269–271.

Sparks, R.S.J., H.E. Huppert, & J.S. Turner (1984). The fluid dynamics of evolving magma chambers. Philosophical Transactions of the Royal Society of London, A310, 511–534.

Speight, R. (1908). On a soda amphibole trachyte from Cass's Peak, Banks Peninsula. Transactions of the New Zealand Institute, 40, 176–184.

Speight, R. (1924). The basic volcanic rocks of Banks Peninsula. Canterbury Museum Records, 2, 239–276.

Speight, R. (1944). The geology of Banks Peninsula — a revision. Transactions of the Royal Society of New Zealand, 74, 232–254.

Speight, R., & A.M. Finlayson (1909). The physiography and geology of Auckland, Bounty and Antipodes islands. In: Chilton, C. (Editor), The Sub-Antarctic Islands of New Zealand, 2, 705–744.

Spencer-Jones, D. (1970). Explanatory notes on the Geelong 1:63 360 geological map. Geological Survey of Victoria — Report, 1970/1.

Spera, F.J. (1980). Aspects of magma transport. In: Hargraves, E.B. (Editor), Physics of Magmatic Processes. Princeton University Press, New Jersey, 265–323.

Spera, F.J. (1984). Carbon dioxide in igneous petrogenesis III: role of volatiles in the ascent of alkaline magma with special reference to xenolith-bearing mafic lavas. Contributions to Mineralogy and Petrology, 88, 217–232.

Spera, F.J., & S.C. Bergman (1980). Carbon dioxide in igneous petrogenesis: I. Contributions to Mineralogy and Petrology, 74, 55–66.

Sporli, K.B. (1978). Mesozoic tectonics, North Island, New Zealand. Bulletin of the Geological Society of America, 89, 415–525.

Sporli, K.B. (1980). New Zealand and oblique-slip margins: tectonic development up to and during the Cainozoic. In: Ballance, P.F., & H.G. Reading (Editors), Sedimentation in Oblique-Slip Mobile Zones. International Association of Sedimentology — Special Publication, 4, 147–170.

Spratt, P.R., & K.A. Rodgers (1975). The Ngatutura volcanics, southwest Auckland. Journal of the Royal Society of New Zealand, 5, 163–178.

Spry, A.H. (1955). The Tertiary rocks of Lower Sandy Bay, Hobart. Papers and Proceedings of the Royal Society of Tasmania, 89, 153–168.

Spry, A.H. (1962). Igneous: In: Spry, A.H., & M.R. Banks (Editors), The Geology of Tasmania. Journal of the Geological Society of Australia, 9, 255–284.

Spry, A.H., & Banks, M.R. (1962, Editors). The Geology of Tasmania. Journal of the Geological Society of Australia, 9, 107–362.

Spry, A.H., & M. Solomon (1964). Columnar buchites at Apsley. Quarterly Journal of the Geological Society of London, 120, 519–545.

Staines, H.R.E. (1960). The Ipswich area. In: Geology of Queensland. Journal of the Geological Society of Australia, 7, 346–348.

Standard, J.C. (1961). Submarine geology of the Tasman Sea. Bulletin of the Geological Society of America, 72, 1777–1788.

Staudigel, H., A. Zindler, S.R. Hart, T. Leslie, C.-Y. Chen, & D. Clague (1984). The isotope systematics of a juvenile intraplate volcano: Pb, Nd, and Sr isotope ratios of basalts from Loihi Seamount, Hawaii. Earth and Planetary Science Letters, 69, 13–29.

Steckler, M.S. (1985). Uplift and extension at the Gulf of Suez: indications of induced mantle convection. Nature, 317, 135–139.

Steiger, R.H., & E. Jager (1977). Subcommission on geochronology: convention on the use of decay constants in geo- and cosmochronology. Earth and Planetary Science Letters, 36, 359–362.

Stellar Mining N.L. (1971). Reports on exploration of Exploration Licence 372. New South Wales Geological Survey — Unpublished File, GS1971/272.

Stephenson, P.J. (1956). The geology and petrology of the Mt Barney central complex. University of London, unpublished PhD thesis.

Stephenson, P.J. (1959). The Mount Barney central complex, south-eastern Queensland. Geological Magazine, 96, 125–136.

Stephenson, P.J. (1976). Sapphire and zircon in some basaltic rocks from Queensland, Australia. 25th International Geological Congress, Sydney, Australia — Abstracts, 2, 602–603.

Stephenson, P.J. (1985). Tertiary volcanic-plutonic rocks of the Cape Hillsborough-Mount Jukes area. In: Johnson, D.P., & A.W. Stevens (Editors), Guide to the Permian to Quaternary Geology of the Mackay-Collinsville-Townsville Region, Northeastern Queensland. Geological Society of Australia, Queensland Division — 1985 Field Conference, 46–61.

Stephenson, P.J. (1986). Landforms in north Queensland: aspects of their origin, age and evolution. In: Age of Landforms in Eastern Australia. Commonwealth Scientific and Industrial Research Organization, Australia (Canberra) — Abstract.

Stephenson, P.J., & R.J. Coventry (1986). Stream incision and inferred late Cainozoic tectonism in the Flinders River headwaters, north Queensland. Search, 17, 220–223.

Stephenson, P.J., & T.J. Griffin (1976a). Cainozoic volcanicity north Queensland. 25th International Geological Congress, Australia — Field Excursion Guidebook, 7A.

Stephenson, P.J., & T.J. Griffin (1976b). Some long basaltic lava flows in north Queensland. In: Johnson, R.W. (Editor), Volcanism in Australasia. Elsevier, Amsterdam, 41–51.

Stephenson, P.J., I.D.R. MacKinnon, & R.A. Young (1978a). Zoned basaltic plugs in the Mingela province, north Queensland. Third Australian Geological Convention, Townsville — Abstract, 35.

Stephenson, P.J., H. Polach, & D.H. Wyatt (1978b). The age of the Toomba basalt, north Queensland. Third Australian Geological Convention, Townsville — Abstract, 62.

Stephenson, P.J., T.J. Griffin, & F.L. Sutherland (1980). Cainozoic volcanism in northeastern Australia. In: Henderson, R.A., & P.J. Stephenson (Editors), The Geology and Geophysics of Northeastern Australia, Geological Society of Australia, Queensland Division, 349–374.

Stephenson, R., & K. Lambeck (1985). Erosion-isostatic rebound models for uplift: an application to southeastern Australia. Geophysical Journal of the Royal Astronomical Society, 82, 31–55.

Stern, T.A., & F.J. Davey (in press). A seismic investigation of the crustal and upper mantle structure within the central volcanic region of New Zealand. New Zealand Journal of Geology and Geophysics.

Stern, T.A., F.J. Davey, & E.G.C. Smith (1986). Crustal structure studies in New Zealand. In: Barazangi, M., & L. Brown (Editors), Reflection Seismology: a Global Perspective. American Geophysical Union — Geodynamics Series, 13, 121–132, Washington, D.C.

Stevens, N.C. (1954). A note on the geology of Panuara and Augullong, south of Orange. Proceedings of the Linnaean Society of New South Wales, 78, 262–268.

Stevens, N.C. (1959). Ring structures of the Mt Alford district, south-east Queensland. Journal of the Geological Society of Australia, 6, 37–50.

Stevens, N.C. (1960). Igneous rocks of the Kalbar district, south-east Queensland. University of Queensland Papers, Department of Geology, 5, 1–9.

Stevens, N.C. (1962). The petrology of the Mt Alford ring-complex, S.E. Queensland. Geological Magazine, 99, 501–515.

Stevens, N.C. (1965). Geological excursions in south-east Queensland. University of Queensland Press, Brisbane.

Stevens, N.C. (1969). The volcanism of southern Queensland. Geological Society of Australia — Special Publication, 2, 193–202.

Stevens, N.C. (1970). Miocene lava flows and eruptive centres near Brisbane, Australia. Bulletin Volcanologique, 34, 353–371.

Stewart, J.R. (1953). The geology of the Camden Haven district of New South Wales. University of Sydney, unpublished BSc Honours thesis.

Stewart, L.F. (1985). Geochemistry and petrogenesis of Newer Volcanics, Portland-Heywood district, Western Victoria. Monash University, unpublished BSc Honours thesis.

Stipp, J.J. (1968). The geochronology and petrogenesis of the Cenozoic volcanics of the North Island, New Zealand. Australian National University, unpublished PhD thesis.

Stipp, J.J., & I. McDougall (1968a). Potassium-argon ages from the Nandewar volcano, near Narrabri, New South Wales. Australian Journal of Science, 31, 84–85.

Stipp, J.J., & I. McDougall (1968b). Geochronology of the Banks Peninsula volcanoes, New Zealand. New Zealand Journal of Geology and Geophysics, 11, 1239–1260.

Stock, J., & P. Molnar (1982). Uncertainties in the relative positions of the Australia, Antarctica, Lord Howe, and Pacific Plates since the Late Cretaceous. Journal of Geophysical Research, 87, 4697–4717.

Stockdale Exploration Limited (1971). Final report on Exploration Licences 147 and 210, Gloucester area. New South Wales Geological Survey — Unpublished File, GS1971/541.

Stolz, A.J. (1984). Garnet websterites and associated ultramafic inclusions from a nepheline mugearite in the Walcha area, New South Wales, Australia. Mineralogical Magazine, 48, 167–179.

Stolz, A.J. (1985). The role of fractional crystallization in the evolution of the Nandewar volcano, north-eastern New South Wales, Australia. Journal of Petrology, 26, 1002–1026.

Stolz, A.J. (1986). Mineralogy of the Nandewar volcano, northeastern New South Wales. Mineralogical Magazine, 50, 241–255.

Stolz, A.J., & G.R. Davies (1988). Chemical and isotopic evidence from spinel lherzolite xenoliths for episodic metasomatism of the upper mantle beneath SE Australia. In: Menzies, M.A., & K.G. Cox (Editors), Oceanic and Continental Lithosphere: Similarities and Differences. Journal of Petrology — Special Publication, 303–330.

Stormer, J.C. (1973). Calcium zoning in olivine and its relationship to silica activity and pressure. Geochimica et Cosmochimica Acta, 37, 1815–1821.

Stormer, J.C. (1983). The effects of recalculation on estimates of temperature and oxygen fugacity from analyses of multicomponent iron-titanium oxides. American Mineralogist, 68, 586–594.

Stosch, H.-G., & G.W. Lugmair (1984). Evolution of the lower continental crust: granulite facies xenoliths from the Eifel, West Germany. Nature, 311, 368–370.

Stosch, H.-G., & G.W. Lugmair (1986). Trace element and Sr and Nd isotope geochemistry of peridotite xenoliths from the Eifel, West Germany, and their bearing on the evolution of the subcontinental lithosphere. Earth and Planetary Science Letters, 80, 281–298.

Stracke, K.J., J. Ferguson, & L.P. Black (1979). Structural setting of kimberlites in southeastern Australia. In: Boyd, F.R., & H.O.A. Meyer (Editors), Kimberlites, Diatremes and Diamonds: their Geology, Petrology and Geochemistry. American Geophysical Union, Washington, D.C., 71–91.

Strange, W.E., G.P. Woollard, & J.C. Rose (1965). An analysis of the gravity field over the Hawaiian Islands in terms of crustal structure. Pacific Science, 19, 381–389.

Strzelecki, P.E. (1840). Report by Count Streleski. Appendix C. In: Papers respecting New South Wales. Despatch

No. 2, Sir G. Gipps to Lord John Russell, 28 September, 1840.

Strzelecki, P.E. (1845). Physical Description of New South Wales and Van Diemen's Land. Longman, London.

Stutchbury, S. (1826). Journal of Pacific Pearl Fishery Expedition (Manuscript 1825–27). Alexander Turnbull Library, Wellington.

Subbarao, K.V., & R.N. Sukheswala (1981, Editors). Deccan Volcanism and Related Basalt Provinces in other parts of the World. Geological Society of India — Memoir, 3.

Suggate, R.P. (1958). The geology of the Clarence valley from Gore Stream to Bluff Hill. Transactions of the Royal Society of New Zealand, 85, 397–408.

Sukhyar, R. (1985). Volcanology and geochemistry of Newer Volcanics from the Portland area, Western Victoria. Monash University, unpublished MSc (preliminary) thesis.

Summerfield, M. (1986). Tectonic geomorphology: macroscale perspectives. Progress in Physical Geography, 10, 227–238.

Sun, S.-S. (1980) Lead isotopic study of young volcanic rocks from mid-ocean ridges, ocean islands and island arcs. Philosophical Transactions of the Royal Society of London, A297, 409–445.

Sun, S.-S. (1982). Chemical composition and origin of the earth's primitive mantle. Geochimica et Cosmochimica Acta, 46, 179–192.

Sun, S.-S., & G.N. Hanson (1975). Origin of Ross Island basanitoids and limitations upon the heterogeneity of mantle sources for alkali basalts and nephelinites. Contributions to Mineralogy and Petrology, 52, 77–106.

Sun, S.-S., & W.F. McDonough (1988). Chemical and isotopic systematics of oceanic basalts: implications for mantle composition and processes. In: Saunders, A.D., & M.J. Norry (Editors), Magmatism in the Ocean Basins. Journal of the Geological Society of London — Special Publication, 42, 313–345.

Sun, S.-S., M. Tatsumoto, & J.-G. Schilling (1975). Mantle plume mixing along the Reykjanes Ridge axis: lead isotopic evidence. Science, 190, 143–147.

Sun, S.-S., R.W. Nesbitt, & M.T. McCulloch (1988). Geochemistry and petrogenesis of siliceous, high magnesian basalts of the Archaean and Early Proterozoic. In: Crawford, A.J. (Editor), Boninites and Related Rocks. Allen & Unwin, London.

Süssmilch, C.A. (1905). On the occurrence of inclusions of basic plutonic rocks in a dyke near Kiama. Journal and Proceedings of the Royal Society of New South Wales, 39, 65–70.

Süssmilch, C.A. (1923). The history of volcanism in New South Wales. Journal and Proceedings of the Royal Society of New South Wales, 56, 12–53.

Süssmilch, C.A., & T.W.E. David (1919). Sequency glaciation and correlation of the Carboniferous rocks of the Hunter River district, New South Wales. Journal and Proceedings of the Royal Society of New South Wales, 53, 245–338.

Süssmilch, C.A., & H.I. Jensen (1909). The geology of the Canobolas mountains. Proceedings of the Linnaean Society of New South Wales, 34, 157–194.

Sutherland, F.L. (1966). Considerations on the emplacement of the Jurassic dolerites of Tasmania. Papers and Proceedings of the Royal Society of Tasmania, 100, 133–146.

Sutherland, F.L. (1969a). A review of the Tasmanian Cainozoic volcanic province. Geological Society of Australia — Special Publication, 2, 133–144.

Sutherland, F.L. (1969b). The mineralogy, petrochemistry and magmatic history of the Tamar lavas, northern Tasmania. Papers and Proceedings of the Royal Society of Tasmania, 103, 17–34.

Sutherland, F.L. (1971). The geology and petrology of the Tertiary volcanic rocks of the Tamar Trough, northern Tasmania. Queen Victoria Museum — Record, 36, 1–58.

Sutherland, F.L. (1972). Igneous rocks, Central Plateau. In: Banks, M.R. (Editor), The Lake Country of Tasmania. Royal Society of Tasmania, Hobart, 43–54.

Sutherland, F.L. (1973). The geological development of the southern shores and islands of Bass Strait. Papers and Proceedings of the Royal Society of Tasmania, 85, 133–144.

Sutherland, F.L. (1974). High-pressure inclusions in tholeiitic basalt and the range of lherzolite-bearing magmas in the Tasmanian volcanic province. Earth and Planetary Science Letters, 24, 317–324.

Sutherland, F.L. (1975). Magmatism associated with the dispersion of the Australian segment of Gondwanaland. In: Campbell, K.S.W. (Editor), Gondwana Geology. Australian National University Press, Canberra, 681–692.

Sutherland, F.L. (1976). Tacharanite from Tasmania, Australia. Mineralogical Magazine, 40, 887–890.

Sutherland, F.L. (1977a). Cainozoic basalts of the Mt Fox area, north Queensland. Australian Museum — Record, 30, 532–543.

Sutherland, F.L. (1977b). Zeolite minerals in the Jurassic dolerites of Tasmania: their use as possible indicators of burial depth. Journal of the Geological Society of Australia, 24, 171–178.

Sutherland, F.L. (1978). Mesozoic-Cainozoic volcanism of Australia. Tectonophysics, 48, 413–427.

Sutherland, F.L. (1980a). The geology and petrology of some Tertiary volcanic rocks in the Bowen-St Lawrence hinterland, northern Queensland. James Cook University of North Queensland, unpublished PhD thesis.

Sutherland, F.L. (1980b). Aquagene volcanism in the Tasmanian Tertiary, in relation to coastal seas and river systems. Papers and Proceedings of the Royal Society of Tasmania, 114, 177–199.

Sutherland, F.L. (1981). Migration in relation to possible tectonic and regional controls in eastern Australian volcanism. Journal of Volcanology and Geothermal Research, 9, 181–213.

Sutherland, F.L. (1983). Timing, trace and origin of basaltic migration in eastern Australia. Nature, 305, 123–126.

Sutherland, F.L. (1985). Regional controls in eastern Australian volcanism. In: Sutherland, F.L., B.J. Franklin, & A.E. Waltho (Editors), Volcanism in Eastern Australia, with Case Histories from New South Wales. Geological Society of Australia, New South Wales Division — Publication, 1, 13–31.

Sutherland, F.L. (1987). A major thermal cycle, contributing to late Palaeozoic-Mesozoic magmatism and mineralization, Pacific Rim, Australia. Proceedings of the Pacific Rim Congress, Queensland, 413–416.

Sutherland, F.L., & E.B. Corbett (1974). The extent of Upper Mesozoic igneous activity in relation to lamprophyric intrusions in Tasmania. Papers and Proceedings of the Royal Society of Tasmania, 107, 175–190.

Sutherland, F.L., & G.E.A. Hale (1970). Cainozoic volcanism in and around Great Lake, central Tasmania. Papers and Proceedings of the Royal Society of Tasmania, 104, 17–32.

Sutherland, F.L., & J.D. Hollis (1982). Mantle-lower crust petrology from inclusions in basaltic rocks in eastern Australia: an outline. Journal of Volcanology and Geothermal Research, 14, 1–29.

Sutherland, F.L., & R.C. Kershaw (1971). The Cainozoic geology of Flinders Island, Bass Strait. Papers and Proceedings of the Royal Society of Tasmania, 105, 151–176.

Sutherland, F.L. & P. Wellman (1986). Potassium-argon ages of Tertiary volcanic rocks, Tasmania. Papers and Proceedings of the Royal Society of Tasmania, 120, 77–86.

Sutherland, F.L., D.C. Green, & B.W. Wyatt (1973). Age of the Great Lake basalts, Tasmania, in relation to Australian Cainozoic volcanism. Journal of the Geological Society of Australia, 20, 85–94.

Sutherland, F.L., D. Stubbs, & D.C. Green (1977). K-Ar ages of Cainozoic volcanic suites, Bowen-St Lawrence hinterland, north Queensland (with some implications for petrologic models). Journal of the Geological Society of Australia, 24, 447–460.

Sutherland, F.L., J.D. Hollis, & L.M. Barron (1984a). Garnet lherzolite and other inclusions from a basalt flow, Bow Hill, Tasmania. In: Kornprobst, J. (Editor), Kimberlites II: The Mantle and Crust-Mantle Relationships. Elsevier, Amsterdam, 145–160.

Sutherland, F.L., J.D. Hollis, & L. Raynor (1984b). Origins of gem minerals in the basaltic regions of north-east New South Wales, particularly zircons, sapphires, and diamonds. Seventh Australian Geological Convention, Sydney — Excursion Notes.

Swanson, D.A., J.H. Wright, & R.T. Helz (1975). Linear vent systems and estimated rates of magma production and eruption of the Yakima basalt on the Columbia Plateau. American Journal of Science, 275, 877–905.

Takahashi, E. (1986). Melting of a dry peridotite KLB-1 up to 14 Gpa: implications on the origin of peridotitic upper mantle. Journal of Geophysical Research, 91, 9367–9382.

Takahashi, E., & I. Kushiro (1983). Melting of a dry peridotite at high pressures and basalt magma genesis. American Mineralogist, 68, 859–879.

Talandier, J., & E.A. Okal (1984). The volcanoseismic swarms of 1981–1983 in the Tahiti-Mehetia area, French Polynesia. Journal of Geophysical Research, 89, 11216–11234.

Tammemagi, H.Y., & F.E.M. Lilley (1971). Magnetotelluric studies across the Tasman Geosyncline, Australia. Geophysical Journal of the Royal Astronomical Society, 22, 505–516.

Tapponnier, P., G. Peltzer, A.Y. Le Dain, R. Armijo, & P. Cobbold (1982). Propagating extrusion tectonics in Asia: new insights from simple experiments with plasticine. Geology, 10, 611–616.

Taras, B.D., & S.R. Hart (1987). Geochemical evolution of the New England seamount chain: isotopic and trace-element constraints. Earth and Planetary Science Letters, 64, 35–54.

Tarney, J., D.A. Wood, A.D. Saunders, J.R. Cann, & J. Varet (1980). Nature of mantle heterogeneity in the North Atlantic: evidence from deep sea drilling. Philosophical Transactions of the Royal Society of London, A297, 179–202.

Tasmanian Department of Mines (1970). Catalogue of the minerals of Tasmania. Geological Survey of Tasmania — Record, 9.

Tasmanian Department of Mines (1976). Geological map of Tasmania, 1:500 000 scale.

Tate, R. (1893). Century of geological progress. Australasian Association for the Advancement of Science — Report, 5, 1–69.

Tate, R., & J. Dennant (1893). Correlation of the marine tertiaries of Australia. Transactions of the Royal Society of South Australia, 17, 203–226.

Tatsumoto, M., E. Hegner, & D.M. Unruh (1987). Origin of the West Maui volcanic rocks inferred from Pb, Sr and Nd isotopes and a multi component model for oceanic basalt. In: Volcanism in Hawaii. United States Geological Survey — Professional Paper, 1350, 723–744.

Tattam, C.M. (1976). Petrology of igneous rocks. In: Douglas, J.G., & J.A. Ferguson (Editors), The Geology of Victoria. Geological Society of Australia — Special Publication, 5, 349–374.

Taylor, G., G.R. Taylor, M. Bink, C. Foudoulis, I. Gordon, J. Hedstrom, J. Minello, & F. Whippy (1985). Pre-basaltic topography of the northern Monaro and its implications. Australian Journal of Earth Sciences, 32, 65–71.

Taylor, H.P. Jr, B. Turi, & A. Cundari (1984). $^{18}O/^{16}O$ and chemical relationships in K-rich volcanic rocks from Australia, East Africa, Antarctica, and San Venazo-Culpaello, Italy. Earth and Plantary Science Letters, 69, 263–276.

Taylor, P.N., N.W. Jones, & S. Moorbath (1984). Isotopic assessment of relative contributions from crust and mantle sources to the magma genesis of Precambrian granitoid rocks. Philosophical Transactions of the Royal Society of London, A310, 605–625.

Taylor, R. (1836). Richard Taylor's Diary. Transcribed copy. Alexander Turnbull Library, Wellington.

Taylor, S.R., & S.M. McLennan (1981). The composition and evolution of the continental crust: rare earth element evidence from sedimentary rocks. Philosophical Transactions of the Royal Society of London, A301, 381–399.

Taylor, S.R., & S.M. McLennan (1985). The Continental Crust — its Composition and Evolution. Blackwell Scientific Publications, Oxford.

Theile, B. (1983). Basement geology to the Lyttelton volcano. University of Canterbury, unpublished MSc thesis.

Thomas, D.E. (1967). Geology of the Melbourne district. Geological Survey of Victoria — Bulletin, 59.

Thompson, B.N. (1961). Sheet 2A, Whangarei (1st Edition). Geological map of New Zealand, 1:250 000. New Zealand Department of Scientific and Industrial Research.

Thompson, B.R. (1986). The Gippsland Basin — development and stratigraphy. In: Glenie, R.C. (Editor), Second South-Eastern Australia Oil Exploration Symposium. Petroleum Exploration Society of Australia, Melbourne, 1985.

Thompson, R.N., & W.S. MacKenzie (1967). Feldspar-liquid equilibria in peralkaline acid liquids: an experimental study. American Journal of Science, 265, 714–734.

Thompson, R.N., & M.A. Morrison (1988). Asthenospheric and lower-lithospheric mantle contributions to continental extensional magmatism: an example from the British Tertiary province. Chemical Geology, 68, 1–15.

Thompson, R.N., M.A. Morrison, A.P. Dickin, & G.L. Hendry (1983). Continental flood basalts... Arachnids rule OK? In: Hawkesworth, C.J., & M.J. Norry (Editors), Continental Basalts and Mantle Xenoliths. Shiva, Cheshire, 158–185.

Thompson, R.N., M.A. Morrison, G.L. Hendry, & S.J. Parry (1984). An assessment of relative roles of crust and mantle in rnagma genesis: an elemental approach. Philosophical Transactions of the Royal Society of London. A310, 549–590.

Thomson, A.A., & F.F. Evison (1962). Thickness of the Earth's crust in New Zealand. New Zealand Journal of Geology and Geophysics, 5, 29–45.

Thorarinsson, S. (1967). Surtsey: the New Island in the North Atlantic. Viking Press, New York.

Thorarinsson, S. (1968). On the rate of lava and tephra production and the upward migration of magmas in four Icelandic eruptions. Geologische Rundschau, 57, 705–718.

Thornton, C.P., & O.F. Tuttle (1960). Chemistry of igneous rocks I. Differentiation index. American Journal of Science, 258, 664–684.

Threlfall, W.F., B.R. Brown, & B.R. Griffith (1976). Gippsland Basin, offshore. In: Leslie, R.B., H.J. Evans, & C.L. Knight (Editors), Economic Geology of Australia and Papua New Guinea, 3. Petroleum. Australasian Institute of Mining and Metallurgy — Monograph, 7, 41–67.

Tilley, C.E. (1921). A tholeiitic basalt from eastern Kangaroo Island. Transactions of the Royal Society of South Australia, 45, 276–277.

Timms, B.V. (1976). Morphology of Lakes Barrine, Eacham and Euramoo, Atherton Tableland, north Queensland. Proceedings of the Royal Society of Queensland, 87, 81–84.

Trendall, A.F. (1965). Explanation of the geology of sheet 35 (Napak). Geological Survey of Uganda — Report, 12.

Truswell, E.M., I.R. Sluiter, & W.K. Harris (1985). Palynology of the Oligocene-Miocene sequence in the Oakvale–1 corehole, western Murray Basin, South Australia. BMR Journal of Australian Geology and Geophysics, 9, 267–295.

Tulloch, A.J. (1983). Granitoid rocks of New Zealand — a brief review. Memoir of the Geological Society of America, 159, 5–20.

Turner, F.J. (1942). Preferred orientation of olivine crystals in peridotites with special reference to New Zealand examples. Transactions and Proceedings of the Royal Society of New Zealand, 72, 280–300.

Turner, J.S., & L.B. Gustafson (1981). Fluid motions and compositional gradients produced by crystallisation or melting at vertical boundaries. Journal of Volcanology and Geothermal Research, 11, 93–125.

Tuttle, O.F., & N.L. Bowen (1958). Origin of granite in the light of exprimental studies in the system $NaAlSi_3O_8$-$KAlSi_3O_8$-SiO_2-H_2O. Memoir of the Geological Society of America, 74, 1–153.

Twidale, C.R. (1956). A physiographic reconnaissance of some volcanic provinces in north Queensland, Australia. Bulletin Volcanologique, 18, 3–23.

Ulrich, G.H. (1875). Geology of Victoria. A Descriptive Catalogue of the Specimens in the Industrial and Technological Museum (Victoria) Illustrating the Rock System of Victoria. Government Printer, Melbourne.

Uniacke, J. (1825). Narrative of Mr. Oxley's expedition to survey Port Curtis and Moreton Bay. In: Field, B. (Editor), Geographical Memoirs in New South Wales. Boone, London.

Upton, B.G.J., & W.J. Wadsworth (1972). Aspects of magmatic evolution on Réunion Island. Philosophical Transactions of the Royal Society of London, A271, 105–130.

Upton, B.G.J., P. Aspen, & N.A. Chapman (1983). The upper mantle and deep crust beneath the British Isles: evidence from inclusions in volcanic rocks. Journal of the Geological Society of London, 140, 105–121.

Utting, A.J. (1986). The petrology and geochemistry of the Ngatutura basalts. University of Waikato, unpublished MSc thesis.

Vallance, T.G. (1967). Palaeozoic low pressure regional metamorphism in southeastern Australia. Meddelelser fra Bansk Geologisk Forening, 17, 494–503.

VandenBerg, A.H.M. (1973). Geology of the Melbourne district (with contributions from M.A.H. Marsden, & J. McAndrew). In: McAndrew, J., & M.A.H. Marsden (Editors), Regional Guide to Victorian Geology (second edition). School of Geology, University of Melbourne, 14–30.

Van der Linden, W.J.M. (1967). Structural relationships in the Tasman Sea and south-west Pacific Ocean. New Zealand Journal of Geology and Geophysics, 10, 1280–1301.

Van der Lingen, G.J., & J.R. Pettinga (1980). The Makara Basin: a Miocene slope-basin along the New Zealand sector of the Australian-Pacific obliquely convergent plate boundary. International Association of Sedimentologists — Special Publication, 4, 191–215.

Van Kooten, G.K. (1981). Pb and Sr systematics of ultra potassic and basaltic rocks from the central Sierra Nevada, California. Contributions to Mineralogy and Petrology, 76, 378–385.

Veevers, J.J. (1984, Editor). Phanerozoic Earth History of Australia. Clarendon, Oxford.

Veevers, J.J. (1986). Break-up of Australia and Antarctica estimated as mid-Cretaceous (95 ± 5 Ma) from magnetic and seismic data at the continental margin. Earth and Planetary Science Letters, 77, 91–99.

Veevers, J.J., R.G. Mollan, F. Olgers, & A.G. Kirkegaard (1964a). The Geology of the Emerald 1:250 000 sheet area, Queensland. Bureau of Mineral Resources, Australia — Report, 68.

Veevers, J.J., M.A. Randal, & R.G. Mollan (1964b). The geology of the Clermont 1:250 000 sheet area, Queensland. Bureau of Mineral Resources, Australia — Report, 66.

Veitch, S.M. (1987). Evidence of major pre-basaltic relief at Cathcart. In: Galloway, R.W. (Compiler), The Age of Landforms in Eastern Australia: Conference Summary and Field Trip Guide. Commonwealth Scientific and Industrial Research Organization, Australia, Division of Water and Land Resources — Technical Memorandum, 87/2, 74–75.

Vink, G.E., W.J. Morgan, & W.-U. Zhao (1984). Preferential rifting of continents: a source of displaced terranes. Journal of Geophysical Research, 89, 10072–10076.

Voggenreiter, W., H. Hötzl, & J. Mechie (1988). Low-angle detachment origin for the Red Sea Rift system. Tectonophysics, 150, 51–75.

Vogt, P.R., & J.R. Conolly (1971). Tasmantid guyots, the age of the Tasman Basin, and motion between the Australian plate and the mantle. Bulletin of the Geological Society of America, 82, 2577–2584.

Voisey, A.H. (1942). The Tertiary land surface in southern New England. Journal and Proceedings of the Royal Society of New South Wales, 76, 82–85.

Vollmer, R., & M.J. Norry (1983). Possible origin of K-rich volcanic rocks from Virunga, East Africa, by metasomatism of continental crustal material: Pb, Nd and Sr isotopic evidence. Earth and Planetary Science Letters, 64, 374–386.

Voorhoeve, H., & G. Houseman (1988). The thermal evolution of lithosphere extending on a low-angle detachment zone. Basin Research, 1, 1–9.

Walcott, R.I. (1978). Present tectonics and late Cenozoic evolution of New Zealand. Geophysical Journal of the Royal Astronomical Society, 52, 137–164.

Walcott, R.I. (1984a). The major structural elements of New Zealand. In: Walcott, R.I. (Compiler), An Introduction to the Recent Crustal Movements of New Zealand. Royal Society of New Zealand — Miscellaneous Series, 7, 1–6.

Walcott, R.I. (1984b). Reconstruction of the New Zealand region for the Neogene. Palaeogeography, Palaeoclimatology, Palaeoecology, 46, 217–231.

Walcott, R.I. (1984c). The kinematics of the plate boundary zone through New Zealand: a comparison of short- and long-term deformations. Geophysical Journal of the Royal Astronomical Society, 79, 613–633.

Walker, G.P.L. (1973a). Lengths of lava flows. Philosophical Transactions of the Royal Society of London, 274, 107–118.

Walker, G.P.L. (1973b). Explosive volcanic eruptions — a new classification scheme. Geologische Rundschau, 63, 431–446.

Walker, G.P.L. (1980). The Taupo pumice: a product of the most powerful known (ultraplinian) eruption? Journal of Volcanology and Geothermal Research, 8, 69–94.

Walker, G.P.L. (1981a). Characteristics of two phreatoplinian ashes, and their water-flushed origin. Journal of Volcanology and Geothermal Research, 9, 395–407.

Walker, G.P.L. (1981b). Plinian eruptions and their products. Bulletin Volcanologique, 44, 223–240.

Walker, G.P.L. (1984). Characteristics of dune bedded pyroclastic surge bedsets. Journal of Volcanology and Geothermal Research, 20, 281–296.

Walker, G.P.L., S. Self, & L. Wilson (1984). Tarawera, 1886, New Zealand — a basaltic plinian fissure eruption. Journal of Volcanology and Geothermal Research, 21, 61–78.

Wallace, R.C. (1977). Anorthoclase-calcite rodding within a kaersutite xenocryst from the Kakanui Mineral Breccia, New Zealand. American Mineralogist, 62, 1038–1041.

Wanamaker, B.J., S.C. Bergman, & B. Evans (1982). Crack healing in silicates: observations on natural lherzolite modules. Transactions of the American Geophysical Union, 63, 437.

Wandless, G.A., & E.R. Padovani (1985). Trace element geochemistry of lower crustal xenoliths from Kilbourne Hole maar, New Mexico. Transactions of the American Geophysical Union, 66, 1110.

Wasilewski, P.J., & M.A. Mayhew (1982). Crustal xenolith magnetic properties and long wavelength anomaly source requirements. Geophysical Research Letters, 9, 329–332.

Wasilewski, P.J., H.H. Thomas, & M.A. Mayhew (1979). The Moho as a magnetic boundary. Geophysical Research Letters, 6, 541–544.

Wass, S.Y. (1973). Oxides of low pressure origin from alkali basaltic rocks, Southern Highlands, N.S.W., and their bearing on the petrogenesis of alkali basaltic magmas. Journal of the Geological Society of Australia, 20, 427–447.

Wass, S.Y. (1979). Fractional crystallization in the mantle of late-stage kimberlitic liquids — evidence in xenoliths from the Kiama area, New South Wales, Australia. In: Boyd, F.R., & H.O.A. Meyer (Editors), The Mantle Sample: Inclusions in Kimberlites and Other Volcanics. Proceedings of the Second International Kimberlite Conference, American Geophysical Union, 2, 366–373.

Wass, S.Y. (1980). Geochemistry and origin of xenolith-bearing and related alkali basaltic rocks from the Southern Highlands, New South Wales, Australia. American Journal of Science, 280A, 639–666.

Wass, S.Y., & J.D. Hollis (1983). Crustal growth in south-eastern Australia — evidence from lower crustal eclogitic and granulitic xenoliths. Journal of Metamorphic Petrology, 1, 25–45.

Wass, S.Y., & A.J. Irving (1976). XENMEG. A catalogue of occurrences of xenoliths and megacrysts in basic volcanic rocks of eastern Australia. The Australian Museum, Sydney.

Wass, S.Y., & G.D. Pooley (1982). Fluid activity in the mantle — evidence from large lherzolite zenoliths. Terra Cognita, 2, 229.

Wass, S.Y. & N.W. Rogers (1980). Mantle metasomatism — precursor to continental alkaline volcanism. Geochimica et Cosmochimica Acta, 44, 1811–1823.

Wass, S.Y., & S.E. Shaw (1984). Rb/Sr evidence for the nature of the mantle, thermal events and volcanic activity of the southeastern Australian continental margin. Journal of Volcanology and Geothermal Research, 21, 107–117.

Wass, S.Y., P. Henderson, & C.J. Elliot (1980). Chemical heterogeneity and metasomatism in the upper mantle — evidence from rare earth and other elements in apatite-rich xenoliths in basaltic rocks from eastern Australia. Philosophical Transactions of the Royal Society of London. A297, 333–346.

Wasserburg, I.G.J., & D.J. DePaolo (1979). Models of earth structure inferred from neodymium and strontium isotopic evidence. Proceedings of the National Academy of Science, 76, 3594–3598.

Watson, E.B. (1979). Zircon saturation in felsic liquids: experimental results and applications to trace element geochemistry. Contributions to Mineralogy and Petrology, 70, 407–419.

Watson, E.B., & T.M. Harrison (1983). Zircon saturation revisited: temperature and composition effects in a variety of crustal magma types. Earth and Planetary Science Letters, 64, 295–304.

Watson, J. (1985). Northern Scotland as an Atlantic-North Sea divide. Journal of the Geological Society of London, 142, 221–243.

Watters, W.A., & C.A. Fleming (1975). Petrography of rocks from the western chain of the Snares Islands. New Zealand Journal of Geology and Geophysics, 18, 491–498.

Watts, A.B., J.K. Weissel, & F.J. Davey (1977). Tectonic evolution of the South Fiji marginal basin. In: Talwani, M. & W.C. Pitman (Editors), Island Arcs, Deep Sea Trenches and Back-Arc Basins. Maurice Ewing Series, 1. American Geophysical Union, Washington, D.C., 419–427.

Weaver, B.L., & J. Tarney (1980a). Rare earth geochemistry of Lewisian granulite-facies gneises, northwest Scotland: implications for the petrogenesis of the Archaean lower continental crust. Earth and Planetary Science Letters, 51, 279–296.

Weaver, B.L., & J. Tarney (1980b). Continental crust composition and nature of the lower crust: constraints from mantle Nd-Sr isotope correlation. Nature, 286, 342–346.

Weaver, B.L., & J. Tarney (1981). Lewisian gneiss geochemistry and Archaean crustal development models. Earth and Planetary Science Letters, 55, 171–180.

Weaver, B.L., D.A. Wood, J. Tarney, & J.L. Joron (1986). Role of subducted sediment in the genesis of ocean-island basalts: geochemical evidence from South Atlantic Ocean basalts. Geology, 14, 275–278.

Weaver, S.D., & R.J. Sewell (1986). Cenozoic volcanology of Banks Peninsula. In: Houghton, B.F., & S.D. Weaver (Editors), South Island Igneous Rocks — International Volcanological Congress Tour Guides A3, C2, and C7. New Zealand Geological Survey — Record, 13, 39–63.

Weaver, S.D., R.J. Sewell, & C.J. Dorsey (1985). Extinct volcanoes: a guide to the geology of Banks Peninsula. Geological Society of New Zealand — Guidebook, 7.

Webb, A.W. (1969). K-Ar age determinations. In: Mollan, R.G., J.M. Dickins, N.F. Exon, & A.G. Kirkegaard, Geology of the Springsure 1:250 000 sheet area, Queensland. Bureau of Mineral Resources, Australia — Report, 123, 112–114.

Webb, A.W., & I. McDougall (1967). A comparison of mineral and whole rock potassium-argon ages of Tertiary volcanics from central Queensland, Australia. Earth and Planetary Science Letters, 3, 41–47.

Webb, A.W., & I. McDougall (1968). The geochronology of the igneous rocks of eastern Queensland. Journal of the Geological Society of Australia, 15, 313–346.

Webb, A.W., N.C. Stevens, & I. McDougall (1967). Isotopic age determinations on Tertiary volcanic rocks and intrusives of south-eastern Queensland. Proceedings of the Royal Society of Queensland, 79, 79–92.

Webb, P.K., & S.D. Weaver (1975). Trachyte shield volcanoes: a new volcanic form from south Turkana, Kenya. Bulletin Volcanologique, 39, 294–312.

Weissel, J.K., & D.E. Hayes (1977). Evolution of the Tasman Sea reappraised. Earth and Planetary Science Letters, 36, 77–84.

Weissel, J.K., & A.B. Watts (1979). Tectonic evolution of the Coral Sea Basin. Journal of Geophysical Research, 84, 4572–4582.

Weissel, J.K., D.E. Hayes, & E.M. Herron (1977). Plate tectonics synthesis: the displacements between Australia, New Zealand and Antarctica since the late Cretaceous. Marine Geology, 25, 231–277.

Wellman, H.W. (1979). An uplift map for the South Island of New Zealand, and a model for uplift of the Southern Alps. In: Walcott, R.I., & M.M. Cresswell (Editors), The Origin of the Southern Alps. Royal Society of New Zealand Bulletin, 18, 13–20.

Wellman, P. (1971). The age and palaeomagnetism of the Australian Cenozoic volcanic rocks. Australian National University, unpublished PhD thesis.

Wellman, P. (1973). Gravity fields associated with Cainozoic volcanic rocks. In: Symposium on Tertiary and Quaternary volcanism in Eastern Australia. Geological Society of Australia, Melbourne — Abstract, 51.

Wellman, P. (1974). Potassium-argon ages of the Cainozoic volcanic rocks of eastern Victoria, Australia. Journal of the Geological Society of Australia, 21, 359–376.

Wellman, P. (1975). Palaeomagnetism of two mid-Tertiary basaltic volcanoes in Queensland, Australia. Proceedings of the Royal Society of Queensland, 86, 147–153.

Wellman, P. (1976a). Gravity trends and the growth of Australia: a tentative correlation. Journal of the Geological Society of Australia, 23, 11–14.

Wellman, P. (1976b). Regional variation of gravity, and isostatic equilibrium of the Australian crust. BMR Journal of Australian Geology and Geophysics, 1, 297–302.

Wellman, P. (1978). Potassium-argon ages of Cainozoic volcanic rocks from the Bundaberg, Rockhampton and Clermont areas of eastern Queensland. Proceedings of the Royal Society of Queensland, 89, 59–64.

Wellman, P. (1979a). On the isostatic compensation of Australian topography. BMR Journal of Australian Geology and Geophysics, 4, 373–382.

Wellman, P. (1979b). On the Cainozoic uplift of the southeastern Australian highland. Journal of the Geological Society of Australia, 26, 1–9.

Wellman, P. (1983). Hotspot volcanism in Australia and New Zealand: Cainozoic and mid-Mesozoic. Tectonophysics, 96, 225–243.

Wellman, P. (1986). Intrusions beneath large alkaline intraplate volcanoes. Exploration Geophysics, 17, 30–35.

Wellman, P. (1987). Eastern highlands of Australia; their uplift and erosion. BMR Journal of Australian Geology and Geophysics, 10, 277–286.

Wellman, P., & H.M. McCracken (1979). Plate tectonics. Bureau of Mineral Resources, Canberra — Earth Science Atlas.

Wellman, P., & I. McDougall (1974a). Cainozoic igneous activity in eastern Australia. Tectonophysics, 23, 49–65.

Wellman, P., & I. McDougall (1974b). Potassium-argon ages of the Cainozoic volcanic rocks of New South Wales, Australia. Journal of the Geological Society of Australia, 21, 247–272.

Wellman, P., & M.W. McElhinny (1970). K-Ar age of the Deccan Traps, India. Nature, 227, 595–596.

Wellman, P., & A.S. Murray (1979). Free air gravity anomalies, Australia, 1:10 000 000. Bureau of Mineral Resources, Australia — Earth Science Atlas.

Wellman, P., M.W. McElhinny, & I. McDougall (1969). On the polar-wander path for Australia during the Cainozoic. Journal of the Royal Astronomical Society, 18, 371–395.

Wellman, P., A. Cundari, & I. McDougall (1970). Potassium-argon ages for leucite-bearing rocks from New South Wales, Australia. Journal and Proceedings of the Royal Society of New South Wales, 103, 103–107.

Wellman, P., A.S. Murray, & M.W. McMullan (1985). Australian long-wavelength magnetic anomalies. BMR Journal of Australian Geology and Geophysics, 9, 297–302.

Wells, P.R.A. (1977). Pyroxene thermometry in simple and complex systems. Contributions to Mineralogy and Petrology, 62, 129–139.

Wells, R.E., D.C. Engebretson, P.D. Snavely Jr, & R.S. Coe (1984). Cenozoic plate motions and the volcanotectonic evolution of western Oregon and Washington. Tectonics, 3, 275–294.

Wernicke, B. (1985). Uniform-sense normal simple shear of the continental lithosphere. Canadian Journal of Earth Sciences, 22, 108–125.

Westgarth, W. (1846). Observations on the geology and physical aspect of Port Phillip. Tasmanian Journal of Natural Science, 2, 402–409.

Westgarth, W. (1853). Victoria, late Australia Felix, or Port Phillip District of New South Wales. Edinburgh.

Whalen, J.B., K.L. Currie, & B.W. Chappell (1987). A-type granites: geochemical characteristics, discrimination and petrogenesis. Contributions to Mineralogy and Petrology, 95, 407–419.

Wheeler, B.F., & G.M. Kjellgren (1986). Amoco Australia Petroleum Company, Yolla No.1 Well completion report.

Whitaker, W.G., & W.F. Willmott (1969). The nomenclature of the igneous rocks of Torres Strait, Queensland. Queensland Government Mining Journal, 70, 530–536.

Whitaker, W.G., P.R. Murphy, & R.G. Rollason (1975). Geology of the Mundubbera 1:250 000 sheet area. Geological Survey of Queensland — Report, 84.

White, A.J.R., B.W. Chappell, & P. Jakeš (1972). Coexisting clinopyroxene, garnet and amphibole from an 'eclogite', Kakanui, New Zealand. Contributions to Mineralogy and Petrology, 34, 185–191.

White, D.A. (1965). The geology of the Georgetown/Clarke River area, north Queensland. Bureau of Mineral Resources, Australia — Bulletin, 71.

White, D.A., & I. Crespin (1959). Some diatomite deposits, north Queensland. Queensland Government Mining Journal, 60, 191–193.

White, G.P. (1983). Hydrothermal alteration and mineralization in a fossil geothermal system at Puhipuhi, Northland, New Zealand. University of Auckland, unpublished MSc thesis.

White, R.S. (1988). A hot-spot model for early Tertiary volcanism in the N Atlantic. In: Morgan, A.C., & L.M. Parson (Editors), Early Tertiary Volcanism and the Opening of the NE Atlantic. Geological Society Special Publication, 39, 3–13.

White, R.S., G.D. Spence, S.R. Fowler, D.P. McKenzie, G.K. Westbrook, & A.N. Bowen (1987). Magmatism at rifted continental margins. Nature, 330, 439–444.

White, W.M. (1988). Réunion to Deccan: history of a hotspot from ODP Leg 115 drilling. Transactions of the American Geophysical Union, 69, 479.

White, W.M., & A.W. Hofmann (1982). Sr and Nd isotope geochemistry of the oceanic basalts and mantle evolution. Nature, 296, 821–825.

White, W.M., A.W. Hofmann, & H. Puchett (1987). Isotope geochemistry of Pacific mid-ocean ridge basalt. Journal of Geophysical Research, 92, 4881–4893.

Whiteford, C.M. (1979). Gravity map of New Zealand, 1:1 million. South Island. Free air, Bouguer, isostatic vertical gradient anomalies. New Zealand Department of Scientific and Industrial Research, Geophysics Division.

Whitehead, P. (1986). The geology and geochemistry of the Mt Rouse and Mt Napier volcanic centres, Western Victoria. La Trobe University, unpublished BSc Honours thesis.

Whitford-Stark, J.L. (1983). Cenozoic volcanic and petrochemical provinces of mainland Asia. Journal of Volcanology and Geothermal Research, 19, 193–222.

Whitford-Stark, J.L. (1987). A survey of Cenozoic volcanism on mainland Asia. Geological Society of America — Special Paper, 213.

Wilkinson, C.S. (1891). Report on the mineral resources of the Mittagong, Bowral, and Berrima districts. New South Wales Department of Mines — Annual Report for 1890, 210–211.

Wilkinson, J.F.G. (1958). The petrology of a differentiated teschenite sill near Gunnedah, New South Wales. American Journal of Science, 256, 1–39.

Wilkinson, J.F.G. (1962). Mineralogical, chemical and petrogenetic aspects of analcite-basalt from the New England district of New South Wales. Journal of Petrology, 3, 192–214.

Wilkinson, J.F.G. (1963). Some natural analcime solid solutions. Mineralogical Magazine, 33, 498–505.

Wilkinson, J.F.G. (1965). Some feldspars, nephelines and analcimes from the Square Top intrusion, Nundle, New South Wales. Journal of Petrology, 6, 420–444.

Wilkinson, J.F.G. (1966). Residual glasses from some alkali basaltic lavas from New South Wales. Mineralogical Magazine, 35, 847–860.

Wilkinson, J.F.G. (1969). Mesozoic and Cainozoic igneous rocks. B. Northeastern New South Wales. In: Packham, G.H. (Editor), The Geology of New South Wales. Journal of the Geological Society of Australia, 16, 530–541.

Wilkinson, J.F.G. (1973). Pyroxenite xenoliths from an alkali trachybasalt in the Glen Innes area, northeastern New South Wales. Contributions to Mineralogy and Petrology, 42, 15–32.

Wilkinson, J.F.G. (1974). Garnet clinopyroxenite inclusions from diatremes in the Gloucester area, New South Wales. Contributions to Mineralogy and Petrology, 46, 275–299.

Wilkinson, J.F.G. (1975a). An Al-spinel ultramafic-mafic inclusion suite and high pressure megacrysts in an analcimite and their bearing on basaltic magma fractionation at elevated pressures. Contributions to Mineralogy and Petrology, 53, 71–104.

Wilkinson, J.F.G. (1975b). Ultramafic inclusions and high pressure megacrysts from a nephelinite sill, Nandewar Mountains, north-eastern New South Wales, and their bearing on the origin of certain ultramafic inclusions in alkaline volcanic rocks. Contributions to Mineralogy and Petrology, 51, 235–262.

Wilkinson, J.F.G. (1977a). Analcime phenocrysts in a vitrophyric analcimite — primary or secondary? Contributions to Mineralogy and Petrology, 64, 1–10.

Wilkinson, J.F.G. (1977b). Petrogenetic aspects of some alkali volcanic rocks. Journal and Proceedings of the Royal Society of New South Wales, 110, 117–138.

Wilkinson, J.F.G., & R.A. Binns (1969). Hawaiite of high pressure origin from northeastern New South Wales. Nature, 222, 553–555.

Wilkinson, J.F.G., & R.A. Binns (1977). Relatively iron-rich lherzolite xenoliths of the Cr-diopside suite: a guide to the primary nature of anorogenic tholeiitic andesite magmas. Contributions to Mineralogy and Petrology, 65, 199–212.

Wilkinson, J.F.G., & N.T. Duggan (1973). Some tholeiites from the Inverell area, New South Wales, and their bearing on low pressure tholeiite fractionation. Journal of Petrology, 14, 339–348.

Wilkinson, J.F.G., & S.R. Taylor (1980). Trace element fractionation trends of tholeiitic magma at moderate pressure: evidence from an Al-spinel ultramafic-mafic inclusion suite. Contributions to Mineralogy and Petrology, 75, 225–233.

Willan, T.L. (1925). Geological map of the Sydney district. New South Wales Department of Mines.

Williams, E. (1978). The Tasman Fold Belt system in Tasmania. Tectonophysics, 48, 159–205.

Williams, I.S., W. Compston, & B.W. Chappell (1983). Zircon and monazite U-Pb systems and the histories of I-type magmas, Berridale Batholith, Australia. Journal of Petrology, 24, 76–97.

Williams, J., & A.R. McBirney (1979). Volcanology. Freeman, Cooper & Company, San Francisco.

Williams, J.G., & I.E.M. Smith (1979). Geochemical evidence of paired arcs in the Permian volcanics of southern New Zealand. Contributions to Mineralogy and Petrology, 68, 285–291.

Williams, S.N. (1983). Plinian airfall deposits of basaltic composition. Geology, 11, 211–214.

Williamson, P.E., C.J. Pigram, J.B. Colwell, A.S. Scherl, K.L. Lockwood, & J.C. Branson (1987a). Review of stratigraphy, structure and hydrocarbon potential of Bass Basin, Australia. AAPG Bulletin, 71, 253–280.

Williamson, P.E., G.W. O'Brien, M.G. Swift, E.A. Felton, A.S. Scherl, J. Lock, N.F. Exon, & D.A. Falvey (1987b). Hydrocarbon potential of the offshore Otway Basin. APEA Journal, 27, 173–194.

Willmott, W.F. (1976). Coalstoun Lakes and Dundurrah lava tube. In: De Jersey, N.J., N.C. Stevens, & W.F. Willmott (Editors), Geological Elements of the National Estate in Queensland. Geological Society of Australia, Queensland Division, 7–10.

Willmott, W.F. (1983). Slope stability and its constraints on closer settlement on the Maleny-Mapleton plateau, southeast Queensland. Geological Survey of Queensland — Record, 1983/9.

Willmott, W.F., W.G. Whitaker, W.D. Palfreyman, & D.S. Trail (1973). Igneous and metamorphic rocks of the Cape York Peninsula and Torres Strait. Bureau of Mineral Resources, Australia — Bulletin, 135.

Willmott, W.F., M.L. O'Flynn, & D.L. Trezise (1986). 1:100 000 Geological map commentary, Rockhampton region, Queensland. Geological Survey of Queensland.

Wilshire, H.G. (1967). The Prospect alkaline diabase-picrite intrusion, New South Wales, Australia. Journal of Petrology, 8, 97–163.

Wilshire, H.G., & R.A. Binns (1961). Basic and ultrabasic xenoliths from volcanic rocks of New South Wales. Journal of Petrology, 2, 185–208.

Wilshire, H.G., & J.E.N. Pike (1975). Upper mantle diapirism: evidence from analogous features in alpine peridotite and ultramafic inclusions in basalt. Geology, 3, 467–470.

Wilshire, H.G., & J.W. Shervais (1975). Al-augite and Cr-diopside ultramafic xenoliths in basaltic rocks from the western United States. Physics and Chemistry of the Earth, 9, 257–272.

Wilshire, H.G., & N.J. Trask (1971). Structural and textural relationships of amphibole and phlogopite in peridotite inclusions, Dish Hill, California. American Mineralogist, 56, 240–255.

Wilshire, H.G., J.E.N. Pike, C.E. Meyer, & E.C. Schwarzmann (1980). Amphibole-rich veins in lherzolite xenoliths, Dish Hill and Deadman Lake, California. American Journal of Science, 280A, 576–593.

Wilshire, H.G., C.E. Meyer, J.K. Nakata, L.C. Calk, J.W. Shervais, J.E. Nielson, & E.C. Schwarzman (1985). Mafic and ultramafic xenoliths from volcanic rocks of the western United States. United States Geological Survey — Open-File Report 85-139.

Wilson, A.F., & A.K. Baksi (1984). Oxygen isotope fractionation and disequilibrium displayed by some granulite facies rocks from the Fraser Range, Western Australia. Geochimica et Cosmchimica Acta, 48, 423–432.

Wilson, L. (1972). Explosive volcanic eruptions II. The atmospheric trajectories of pyroclasts. Geophysical Journal of the Royal Astronomical Society, 30, 381–392.

Wilson, L. (1980). Relationships between pressure, volatile content and ejecta velocity in three types of volcanic explosion. Journal of Volcanology and Geothermal Research, 8, 297–313.

Wilson, L., & J.W. Head (1981). Ascent and emplacement of basaltic magma on the Earth and Moon. Geophysical Journal of the Royal Astronomical Society, 86, 2971–3001.

Wilson, L., R.S.J. Sparks, T.C. Huang, & N.D. Watkins (1978). The control of eruption column heights by eruption energetics and dynamics. Journal of Geophysical Research, 83, 1829–1836.

Wilson, L., R.S.J. Sparks, & G.P.L. Walker (1980). Explosive volcanic eruptions — IV. The control of magma properties and conduit geometry on eruption column behaviour. Geophysical Journal of the Royal Astronomical Society, 63, 117–148.

Wilton, C.P.N. (1830). An account of the Burning Mountain in Australasia. Edinburgh Journal of Science, 2, 270–273.

Wimmenauer, W. (1974). The alkaline province of Central Europe and France. In: Sørensen, H. (Editor), The Alkaline Rocks. Wiley, London.

Withnall, I.W., J.H.C. Bain, & M.J. Rubenach (1980). The Precambrian geology of northeastern Queensland. In: Henderson, R.A., & P.J. Stephenson (Editors), The Geology and Geophysics of Northeastern Australia. Geological Society of Australia, Queensland Division, 109–127.

Wohletz, K.H. (1986). Explosive magma-water interactions: thermodynamics, explosion mechanisms, and field studies. Bulletin Volcanologique, 48, 245–264.

Wohletz, K.H., & M.F. Sheridan (1983). Hydrovolcanic explosions II. Evolution of basaltic tuff rings and tuff cones. American Journal of Science, 283, 385–413.

Wones, D.R. (1972). Stability of biotite: a reply. American Mineralogist, 57, 316–317.

Wood, B.J. (1987). Thermodynamics of multicomponent systems containing several solid solutions. Reviews in Mineralogy, 17, 71–95.

Wood, B.J., & S. Banno (1973). Garnet-orthopyroxene and orthopyroxene-clinopyroxene relationships in simple and complex systems. Contributions to Mineralogy and Petrology, 42, 109–121.

Wood, C.A. (1980a). Morphometric evolution of cinder cones. Journal of Volcanology and Geothermal Research, 7, 387–413.

Wood, C.A. (1980b). Morphometric analysis of cinder cone degradation. Journal of Volcanology and Geothermal Research, 8, 137–160.

Wood, C.A. (in press). Maars. In: Fairbridge, R.W. (Editor), Encyclopedia of Volcanology.

Wood, D.A. (1979). A variably veined sub-oceanic upper mantle — genetic significance for mid-ocean ridge basalts from geochemical evidence. Geology, 7, 499–503.

Wood, D.A. (1980). The application of a Th-Hf-Ta diagram to problems of tectonomagmatic classification and to establishing the nature of crustal contamination of basaltic lavas of the British Tertiary volcanic province. Earth and Planetary Science Letters, 50, 11–30.

Woods, J.E. (1862). Geological Observations in South Australia. Longman, London.

Woodward, D.J. (1976). Sheet 25 Dunedin, Magnetic map of New Zealand, 1:250 000, Total force anomalies. New Zealand Department of Scientific and Industrial Research.

Woolley, D.R. (1978). Cainozoic sedimentation in the Murray drainage basin, New South Wales section. Proceedings of the Royal Society of Victoria, 90, 61–65.

Wopfner, H., & R.C.N. Thornton (1971). The occurrence of carbon dioxide in the Gambier Embayment. Geological Surveys of South Australia and Victoria — Special Bulletin, 377–384.

Wörner, G., J. Staudigel, & A. Zindler (1985). Isotopic constraints on open system evolution of the Laacher See magma chamber (Eifel, West Germany). Earth and Planetary Science Letters, 75, 37–49.

Wörner, G., A. Zindler, H. Staudigel, & H.-U. Schmincke (1986). Sr, Nd and Pb isotope geochemistry of Tertiary and Quaternary alkaline volcanics from West Germany. Earth and Planetary Science Letters, 79, 107–119.

Wright, J.B. (1966a). Olivine nodules in a phonolite of the east Otago alkaline province, New Zealand. Nature, 210, 519.

Wright, J.B. (1966b). Contributions to the volcanic succession and petrology of the Auckland Islands, New Zealand. I. West Coast Section through the Ross Volcano. Transactions of the Royal Society of New Zealand, 3, 215–229.

Wright, J.B. (1968). Contributions to the volcanic succession and petrology of the Auckland Islands, New Zealand. III. Minor intrusives of the Ross Volcano. Transactions of the Royal Society of New Zealand, 6, 1–11.

Wright, J.B. (1970). Contributions to the volcanic succession and petrology of the Auckland Islands, New Zealand. IV. Chemical analyses from the lower half of the Ross Volcano. Transactions of the Royal Society of New Zealand, 8, 109–115.

Wright, J.B. (1971). Contributions to the volcanic succession and petrology of the Auckland Islands, New Zealand. V. Chemical analyses from the upper parts of the Ross Volcano, including minor intrusions. Journal of the Royal Society of New Zealand, 1, 175–183.

Wright, J.V., A.L. Smith, & S. Self (1980). A working terminology of pyroclastic deposits. Journal of Volcanology and Geothermal Research, 8, 315–336.

Wyborn, D., & M. Owen (1986). Araluen, New South Wales, 1:100 000 Geological Map Commentary. Bureau of Mineral Resources, Australia.

Wyborn, L.A.I., & B.W. Chappell (1979). Geochemical evidence for the existence of a pre-Ordovician sedimentary layer in southeastern Australia. In: Denham, D. (Editor), Crust and Upper Mantle of Southeastern Australia. Bureau of Mineral Resources, Australia — Record, 1979/2.

Wyllie, P.J. (1988). Solidus curves, mantle plumes, and magma generation beneath Hawaii. Journal of Geophysical Research, 93, 4171–4181.

Yamagishi, H. (1985). Growth of pillow lobes — evidence from pillow lavas of Hokkaido, Japan, and North Island, New Zealand. Geology, 13, 499–502.

Yates, H. (1954). The basalts and granitic rocks of the Ballarat district. Proceedings of the Royal Society of Victoria, 66, 63–101.

Yim, W.W.-S., A.J. Gleadow, & J.C. van Moort (1985). Fission track dating of alluvial zircons and heavy mineral provenance in northeast Tasmania. Journal of the Geological Society of London, 142, 351–356.

Yimir, M. (1986). Geology of the Purerua Pensinsula, Northland. University of Auckland, unpublished MSc thesis.

Yoder, H.S. Jr (1964). Soda melilite. Carnegie Institution of Washington Year Book, 63, 86–89.

Young, R.W. (1977). Landscape development in the Shoalhaven River catchment of southeastern New South Wales. Zeitschrift für Geomorphologie, 21, 262–283.

Young, R.W. (1981). Denudation history of the south-central uplands of New South Wales. Australian Geographer, 15, 77–88.

Young, R.W. (1983). The tempo of geomorphological change: evidence from southeastern Australia. Journal of Geology, 91, 221–230.

Young, R.W., & P. Bishop (1980). Potassium-argon ages on Cainozoic volcanic rocks in the Crookwell-Goulburn area, New South Wales. Search, 11, 340–341.

Young, R.W., & I. McDougall (1982). Basalts and silcretes on the coast near Ulladulla, southern New South Wales. Journal of the Geological Society of Australia, 29, 425–430.

Young, R.W., & I. McDougall (1985). The age, extent and geomorphological significance of the Sassafras basalt, southeastern New South Wales. Australian Journal of Earth Sciences, 32, 323–331.

Zhi, X., Y. Song, F.A. Frey, J. Feng, & M. Zhai (in press). Geochemistry of Hannuoba basalts, Eastern China: constraints on the origin of continental alkali and tholeiitic basalt. Chemical Geology.

Zindler, A., & S. Hart (1986). Chemical geodynamics. Annual Review of Earth and Planetary Sciences, 14, 493–571.

Zindler, A., E. Jagoutz, & S. Goldstein (1982). Nd, Sr and Pb isotopic systematics in a three-component mantle: a new perspective. Nature, 298, 519–523.

Zindler, A., H. Staudigel, & R. Batiza (1984). Isotope and trace element geochemistry of young Pacific seamounts: implications for the scale of upper mantle heterogeneity. Earth and Planetary Science Letters, 70, 175–195.

Zoback, M.L., R.E. Anderson, & G.A. Thompson (1981). Cainozoic evolution of the state of stress and style of tectonism of the Basin and Range province of the western United States. Philosophical Transactions of the Royal Society of London, A300, 407–434.

Index

Printed in Australia
AUHW021933220921
352559AU00007B/29